Engineering Physics

Third Edition

Engineering Physics

Third Edition

SL Kakani
MSc (Physics)   PhD

Former Executive Director
Institute of Technology and Management
(Affiliated to Rajasthan Technical University, Kota)
Bhilwara 311001, Rajasthan, India

Shubhra Kakani
MSc (Physics)   PhD

MLV Government College
Bhilwara, Rajasthan, India

CBS Publishers & Distributors Pvt Ltd

New Delhi • Bengaluru • Chennai • Kochi • Kolkata • Mumbai
Bhopal • Bhubaneswar • Hyderabad • Jharkhand • Nagpur • Patna • Pune
• Uttarakhand • Dhaka (Bangladesh)

Engineering Physics
Third Edition

ISBN: 978-81-239-2841-8

Copyright © Authors and Publisher

Third Edition: 2016
  Reprint: 2020
Second Edition: 2010
First Edition: 2008

Published by Satish Kumar Jain and produced by Varun Jain for

CBS Publishers & Distributors Pvt Ltd

4819/XI Prahlad Street, 24 Ansari Road, Daryaganj, New Delhi 110 002, India.
Ph: 23289259, 23266861, 23266867   Website: www.cbspd.com
Fax: 011-23243014   e-mail: delhi@cbspd.com; cbspubs@airtelmail.in.
Corporate Office: 204 FIE, Industrial Area, Patparganj, Delhi 110 092

Ph: 4934 4934   Fax: 4934 4935   e-mail: publishing@cbspd.com; publicity@cbspd.com

Branches

- **Bengaluru:** Seema House 2975, 17th Cross, K.R. Road,
  Banasankari 2nd Stage, Bengaluru 560 070, Karnataka
  Ph: +91-80-26771678/79   Fax: +91-80-26771680   e-mail: bangalore@cbspd.com
- **Chennai:** 7, Subbaraya Street, Shenoy Nagar, Chennai 600 030, Tamil Nadu
  Ph: +91-44-26680620, 26681266   Fax: +91-44-42032115   e-mail: chennai@cbspd.com
- **Kochi:** 42/1325, 1326, Power House Road, Opposite KSEB Power House,
  Ernakulam 682 018, Kochi, Kerala
  Ph: +91-484-4059061-65   Fax: +91-484-4059065   e-mail: kochi@cbspd.com
- **Kolkata:** 6/B, Ground Floor, Rameswar Shaw Road, Kolkata-700 014, West Bengal
  Ph: +91-33-22891126, 22891127, 22891128   e-mail: kolkata@cbspd.com
- **Mumbai:** 83-C, Dr E Moses Road, Worli, Mumbai-400018, Maharashtra
  Ph: +91-22-24902340/41   Fax: +91-22-24902342   e-mail: mumbai@cbspd.com

Representatives

- **Bhopal**   0-8319310552   • **Bhubaneswar** 0-9911037372 • **Hyderabad** 0-9885175004 • **Jharkhand**   0-9811541605
- **Nagpur**   0-9421945513   • **Patna**   0-9334159340 • **Pune**   0-9623451994 • **Uttarakhand** 0-9716462459
- **Dhaka (Bangladesh)** 01912-003485

Printed at: Glorious Printers, Daryaganj, Delhi, India

Preface to the Third Edition

It gives us great pleasure to learn that the second edition of our book **Engineering Physics** received a great response from the students and the teachers. The feedback and also the revision made in the syllabi of several technical universities, resulted in updating of the text matter.

Two chapters, "Elements of Heat and Thermodynamics" and "Thermionic Emission and Vacuum Tubes", have been deleted, and two new chapters, "Bonding in Solids" and "Nanomaterials", have been added. Chapters on fibre optics, lasers, superconductivity and nuclear physics have been rewritten. Two appendices, "Composites, Shape Memory Alloys and Nonlinear Materials", and "Quantum Information and Quantum Computing" have been added. A good number of problems, multiple choice questions are included for self-evaluation by the students. We are sure that the additions and changes made in the third edition will receive appreciation from readers.

We are thankful to the publisher, CBS Publishers & Distributors Pvt Ltd, for bringing the third edition in a very short time.

SL Kakani
Shubhra Kakani

Preface to the First Edition

Physics has a key role to play in engineering and technology. Keeping this in view, various Universities and Boards of Technical Education have updated their syllabi of BE, BTech, Biotech, diploma and AMIE courses respectively. To keep pace with recent advancements in the field, they have included topics like optical fibres, lasers and holography, semiconductors and devices, superconductivity, etc.

The present text is an attempt to present the subject matter in a simple and systematic manner, covering latest syllabi of various universities and technical boards. Subject matter has been arranged such that a student should have no difficulty in understanding the topics. Each chapter is followed by good number of typical worked out and exercise problems together with review questions, problems, short answer questions and objective type questions. The book is complete with theory and solved and unsolved problems with all necessary details. We hope that it will definitely cater to the needs of students and teachers in every respect.

There cannot be much originality in a book of this kind. The authors take this opportunity to place on record their indebtedness to the large number of books that they have consulted in the preparation of this text.

The authors thank CBS Publishers & Distributors for their full cooperation to bring out the book in its present form. Suggestions for improvement will be gratefully received.

SL Kakani
Shubhra Kakani

Contents

1
Kinematics

1.1 INTRODUCTION

Physics is one of the fundamental natural sciences studying the laws of inanimate nature. Obviously, studying physics is an exciting and challenging adventure. The word *physics* comes from a Greek term meaning nature; and, therefore, physics should be a science dedicated to the study of all natural phenomena.

The nature or, say, universe consists of material bodies which are in permanent interaction and motion. All the observable natural phenomena are governed by certain laws. The study and the explanation of the laws governing the connections between various processes and phenomena are the fundamental aim of every branch of physics and in general of science. The analysis of the interaction of material bodies and of the laws of electromagnetic phenomena is the aim of physics.

The motion of the substance has various forms: mechanical, electromagnetic, thermal, etc. The mechanical form of motion of the matter is the simplest one. It consists of the movement of bodies or their parts relative to one another. We can see the movements of bodies or their parts relative to one another. We can see movements of bodies everywhere in our daily life. The laws of mechanical motion are studied in *mechanics*. The other branches of physics cannot be studied without mechanics since various displacements are observed almost in all physical phenomena.

Mechanics is generally subdivided into three parts: *kinematics, statics and dynamics*. In kinematics, the motion of the bodies is considered relative to the factors causing the motion or changing its characters. Statics deals with the laws of equilibrium of the bodies and dynamics studies the laws of motion and the causes producing the motion and changing it. On the other hand, if the laws of motion are known, it is possible to derive the laws of equilibrium as a special case of the former. Therefore, in physics, usually the laws of statics are not considered separately and are studied in connection with the general laws of dynamics.

Bodies are macroscopic systems consisting of a very large number of molecules or atoms, so that the sizes of these systems are many times larger than the intermolecular distances. We may note that a *material point* or *particle* is a body whose size and shape are of no consequence in the problem under consideration. For example, in studying the motion of the planets around the sun, they can be regarded as particles since the distances of the planets from the sun are many times greater than their sizes. *Classical* or *Newtonian* mechanics deals with the motion of bodies travelling at velocities that are very much less than that of light in vacuum. The investigation of the motion of bodies travelling at velocities commensurate with the velocity of light is taken up in *relativistic mechanics* which is based on the theory of relativity. Specific features of the motion of microparticles are dealt within *quantum* (wave) *mechanics*. Microparticles are particles whose rest mass is commensurate with, or smaller than, the rest mass of atoms.

Motion of a body or a particle occurs both in space and in time (space and time are in alienable forms of existence of matter). According to Newton, time is absolute and flows on without relation to the presence of any physical object or event. In this absolute space and time concept of Newton, motion of an object is characterized by change of its position in space as time evolves. Today, we know that in fact there is no object (in the universe) at rest in this 'absolute space', with respect to which one can define the position of another object. Obviously, at any instant of time, the position of an object is always defined by the *observer* with respect to himself. An observer, always, first creates a local space, in which positions of other objects are then described. This local space, to which is attached the observer, is called a *reference frame* or *frame of reference*.

We may note that the concept of rest and motion are relative, i.e. these are in relation to a frame of reference or observer. Absolute motion is a meaningless concept.

1.2 FRAME OF REFERENCE AND COORDINATE SYSTEMS

We have seen that both rest and motion are relative concepts, i.e. they depend on the condition of the object relative to the body that serves as reference. A tree and a house are at rest relative to the earth, but in motion relative to the sun.

To describe motion, therefore, the observer must define a *frame of reference* relative to which the motion is analysed. The best that can be done is to define a particular reference system at rest relative to a particular set of stars, called the *fixed stars*. All other frames of reference can be defined relative to this particular frame of reference.

A frame of reference is a real or conditionally *rigid body* with respect to which the motion of the body being studied is to be considered. Rigidly fixed in the frame of reference is some kind of coordinate system so that the position

of each point of a moving body can be uniquely determined by the three coordinates of the point. The following systems of coordinates are most frequently employed in mechanics:

1. Rectangular or cartesian coordinates: (x, y, z)
2. Spherical polar coordinates: (r, θ, ϕ)
3. Cylindrical coordinates: (ρ, ϕ, z).

Rectangular or Cartesian Coordinate System

In this coordinate system, the three dimensions are represented by the three axes X, Y and Z perpendicular to each other. The coordinates of any point in space are taken as distances from the origin along the three axes X, Y and Z and written as (x, y, z). This system is generally used where independent rectangular coordinate system is possible in space: (a) *the left-handed* and (b) *the right-handed* which cannot be made to coincide with each other by any means of translation or rotation. Cartesian coordinate system by convention is chosen to be right-handed (Fig. 1.1). Position vector r of a particle $P(x, y, z)$ relative to the origin is given by

$$r = x\hat{i} + y\hat{j} + z\hat{k} \qquad (1.1)$$

where $\hat{i}$, $\hat{j}$ and $\hat{k}$ denote constant unit vectors in the directions of X, Y and Z respectively. The position of a particle P relative to the origin O is given by the position vector r, characterized by its specific length and direction [Eq. (1.1)].

The vectorial increment dr in r can be written as

$$dr = dx\hat{i} + dy\hat{j} + dz\hat{k} \qquad (1.2)$$

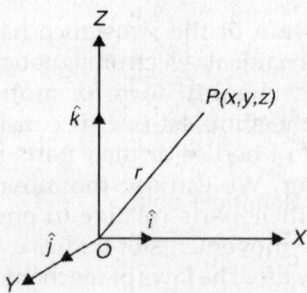

Fig. 1.1: Cartesian coordinate system $\hat{i}$, $\hat{j}$ and $\hat{k}$ are unit vectors in the directions of X, Y and Z axes respectively.

The instantaneous velocity and acceleration of P in cartesian coordinate system will be

$$V = \frac{d\boldsymbol{r}}{dt} = \frac{dx}{dt}\hat{i} + \frac{dy}{dt}\hat{j} + \frac{dz}{dt}\hat{k}$$

$$V = V_x\hat{i} + V_y\hat{j} + V_z\hat{k} \tag{1.3}$$

or the magnitude of V is given by

$$|\boldsymbol{V}| = \sqrt{V_x^2 + V_y^2 + V_z^2} \tag{1.4}$$

Acceleration is the rate of change of velocity. Thus differentiating Eq. (1.3) with respect to time, one obtains,

$$a = \frac{d\boldsymbol{V}}{dt} = \frac{dV_x}{dt}\hat{i} + \frac{dV_y}{dt}\hat{j} + \frac{dV_z}{dt}\hat{k}$$

$$= a_x\hat{i} + a_y\hat{j} + a_z\hat{k} \tag{1.5}$$

The magnitude of a is given by

$$|\boldsymbol{a}| = \sqrt{a_x^2 + a_y^2 + a_z^2} \tag{1.6}$$

Spherical Polar Coordinates

The coordinates of a point, say P, in this coordinate system are represented by radial vector r, the zenith, colatitude or polar angle θ, and azimuthal or longitudinal angle ϕ as shown in Fig. 1.2. These coordinates are related

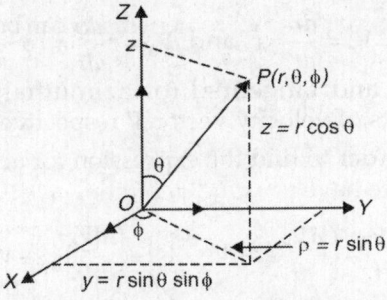

Fig. 1.2: Spherical polar coordinates and their relationship with rectangular or cartesian coordinates

to the rectangular or cartesian coordinates x, y, and z through

$$x = r\sin\theta\cos\phi, \ r = \sqrt{x^2 + y^2 + z^2} \tag{1.7a}$$

$$y = r\sin\theta\sin\phi, \tan\theta = \sqrt{\frac{x^2 + y^2}{z}} \tag{1.7b}$$

$$z = r\cos\theta, \quad \tan\phi = \frac{y}{z} \tag{1.7c}$$

The coordinate surfaces are:
i. Spheres concentric with the origin (r = constant)
ii. Cones with the apex at the origin and along the Z-axis (θ = constant)
iii. Planes through the Z-axis (ϕ = constant).

The unit vectors $\hat{e}_r$, $\hat{e}_\theta$, $\hat{e}_\phi$ extend respectively in the directions of r increasing, θ increasing and ϕ increasing (Fig. 1.3).

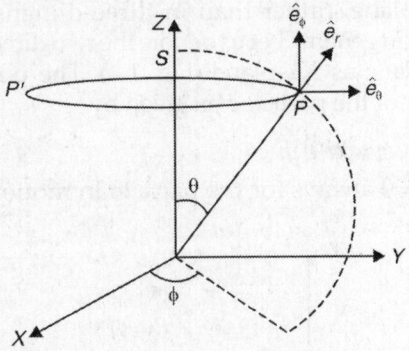

Fig. 1.3: The unit vectors

One can easily write the general differential displacement of particle P in spherical polar coordinates as

$$d\hat{r} = dr\hat{e}_r + rd\theta\hat{e}_\theta + r\sin\theta d\phi\hat{e}_\phi \tag{1.8}$$

The unit vectors $\hat{e}_r$, $\hat{e}_\theta$ and $\hat{e}_\phi$ can be expressed in terms of unit vectors $\hat{i}$, $\hat{j}$, $\hat{k}$ as follows:

$$\hat{e}_r = \sin\theta\cos\phi\hat{i} + \sin\theta\sin\phi\hat{j} + \cos\theta\hat{k} \tag{1.9a}$$

$$\hat{e}_\theta = \cos\theta\cos\phi\hat{i} + \cos\theta\sin\phi\hat{j} - \sin\theta\hat{k} \tag{1.9b}$$

$$\hat{e}_\phi = -\sin\phi\hat{i} + \cos\theta\hat{j} \tag{1.9c}$$

The unit vectors $(\hat{e}_r, \hat{e}_\theta, \hat{e}_\phi)$ in spherical coordinate system, unlike $(\hat{i}, \hat{j}, \hat{k})$ in cartesian coordinate system are not constant vectors but change in direction as coordinates θ and ϕ change. However, we can see that at each point they constitute an orthogonal right-handed coordinate system, i.e. we have

$$\hat{e}_r \cdot \hat{e}_\theta = \hat{e}_r \cdot \hat{e}_\phi = \hat{e}_\theta \cdot \hat{e}_\phi = 0$$

$$\hat{e}_r \times \hat{e}_\theta = \hat{e}_\phi, \hat{e}_\theta \times \hat{e}_\phi = \hat{e}_r \text{ and } \hat{e}_\phi \times \hat{e}_r = \hat{e}_\theta \quad (1.10)$$

The spherical polar coordinates are very convenient in those problems of physics where there is no preferred direction and the force in physical problem is spherically symmetrical, e.g. (i) coulomb force due to a point charge, and (ii) gravitational force due to a point mass. We may note that these are also examples of *central forces*.

Motion in Two Dimensions

Usually, we find a particle constrained to move in a plane, rather than in three-dimensional space in general. Let us denote the two dimensional plane as *XY*-plane (Fig. 1.4). The position vector of the particle *P* is given by

$$r = x\hat{i} + y\hat{j} \quad (1.11)$$

as $Z = 0$ always for the particle in motion.

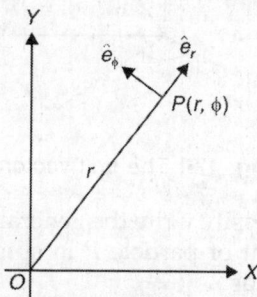

Fig. 1.4: Motion of a particle in two dimensional plane

Since $\theta = 90°$ for the particle *P* and hence the spherical polar coordinates (r, θ, ϕ) reduce to circular polar coordinates (r, ϕ). Now, the position of the particle *P* in terms of radial distance *r* from origin *O* and ϕ is given by

$$r = \sqrt{x^2 + y^2} \quad x = r\cos\phi \quad (1.11a)$$

$$\tan\phi = \frac{y}{x} \quad y = r\sin\phi \quad (1.11b)$$

The displacement vector dr in circular polar coordinates is obtained as (put $\theta = 90°$ and $d\theta = 0$ in Eq. (1.8))

$$dr = dr\,\hat{e}_r + rd\phi\,\hat{e}_\phi \quad (1.12)$$

where unit vectors $\hat{e}_r$ and $\hat{e}_\phi$ are given by

$$\hat{e}_r = \cos\phi\,\hat{i} + \sin\phi\,\hat{j} \quad (1.13a)$$

and $$\hat{e}_\phi = -\sin\phi\,\hat{i} + \cos\phi\,\hat{j} \quad (1.13b)$$

Velocity and Acceleration (in Circular Polar Coordinates System)

Let us find the velocity and acceleration of a point object moving in a plane in terms of circular polar coordinates (r, ϕ). Differentiating position vector $r = r\hat{e}_r$ with respect to time, we obtain

$$V = \frac{dr}{dt} = \frac{dr}{dt}\hat{e}_r + r\frac{d\hat{e}_r}{dt} \quad (1.14)$$

We may note that $\hat{e}_r$ is not a constant vector. One can obtain its time rate by using Eq. (1.13a). One obtains,

$$\frac{d\hat{e}_r}{dt} = -\sin\phi\frac{d\phi}{dt}\hat{i} + \cos\phi\frac{d\phi}{dt}\hat{j} = \frac{d\phi}{dt}\hat{e}_\phi \quad (1.15)$$

Similarly, one obtains

$$\frac{d\hat{e}_\phi}{dt} = -\frac{d\phi}{dt}\hat{e}_r \quad (1.16)$$

Using Eqs. (1.15) and (1.16), Eq. (1.14) takes the form

$$V = V_r\hat{e}_r + V_\phi\hat{e}_\phi \quad (1.17)$$

where $V_r = \dfrac{dr}{dt} = \dot{r}$ and $V_\phi = r\dfrac{d\phi}{dt} = r\dot{\phi}$ are the radial and tangential (or azimuthal) components of velocity vector *V* respectively.

In order to find the expression for acceleration, we have

$$a = \frac{dV}{dt} = \left(\frac{dV_r}{dt}\hat{e}_r + V_r\frac{d\hat{e}_r}{dt}\right) + \left(\frac{dV_\phi}{dt}\hat{e}_\phi + V_\phi\frac{d\hat{e}_\phi}{dt}\right)$$

$$= \left(\frac{dV_r}{dt} - V_\phi\frac{d\phi}{dt}\right)\hat{e}_r + \left(V_r\frac{d\phi_r}{dt} + \frac{dV_\phi}{dt}\right)\hat{e}_\phi$$

or $$a = a_r\hat{e}_r + a_\phi\hat{e}_\phi \quad (1.18)$$

where $$a_r = \ddot{r} - r\dot{\phi}^2 \quad (1.19a)$$

and $$a_\phi = 2\dot{r}\dot{\phi} + r\ddot{\phi} \quad (1.19b)$$

Obviously, a_r [Eq. (1.19a)] and a_ϕ [(Eq. (1.19b)] represent the *radial* and *tangential* components of acceleration vector a respectively.

Let us consider that the object is moving in a plane, in a circle of radius R (i.e. $r = R$, constant), one obtains

$$V_r = 0, \quad V_\phi = R\dot{\phi} \qquad (1.20a)$$

and

$$a_r = R\dot{\phi}^2, \quad a_\phi = R\ddot{\phi} \qquad (1.20b)$$

Here a_r denotes the centripetal acceleration (towards $\hat{e}_r$) and a_ϕ is tangential acceleration which is responsible for the increase in tangential speed V_ϕ.

Cylindrical Coordinates

In this system, the position of particle P relative to origin O is described by the two circular polar coordinates (ρ, ϕ) defined in the plane $Z = 0$, and the third coordinate being the z-coordinate itself. The cylindrical coordinates (ρ, ϕ, z) of P are related to cartesian coordinates (x, y, z) as

$$\rho = \sqrt{x^2 + y^2}, \quad \tan\phi = \frac{y}{x}, z = z \qquad (1.21)$$

Conversely,

$$x = \rho\cos\phi, y = \rho\sin\phi \text{ and } z = z \qquad (1.22)$$

The coordinate surfaces are:

  i. Cylinders coaxial with the Z-axis (ρ = constant)

 ii. Planes through the Z-axis (ϕ = constant)

iii. Planes perpendicular to the Z-axis (z = constant)

The unit vectors $\hat{e}_\rho$, $\hat{e}_\phi$ and $\hat{e}_z$ extend respectively outward along NP (in the direction of increasing ρ), perpendicular to the plane $ONPM$ (in the direction of increasing ϕ), and upward (in the direction of increasing z).

If r is the position vector of point $P(x, y, z)$, then we have

$$r = x\hat{i} + y\hat{j} + z\hat{k}$$

Substituting x, y and z from relation (1.22), one obtains

$$r = \rho\cos\phi\hat{i} + \rho\sin\phi\hat{j} + z\hat{k} \qquad (1.23)$$

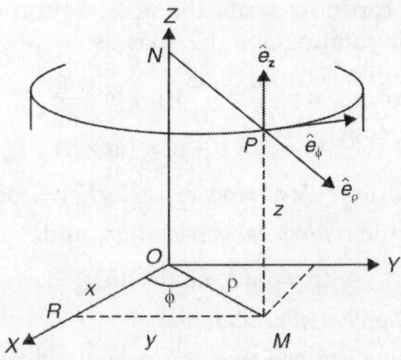

Fig. 1.5: Cylindrical coordinates

Thus, we obtain the general differential displacement of particle P in cylindrical coordinates as

$$dr = d\rho\,\hat{e}_\rho + \rho\,d\phi\,\hat{e}_\phi + dz\,\hat{e}_z$$

where $\hat{e}_\rho$ and $\hat{e}_\phi$ are related to $(\hat{i}, \hat{j}, \hat{k})$ as given by Eqs. (1.13a) and (1.13b) and $\hat{e}_z = \hat{k}$. The unit vectors $\hat{e}_\rho$, $\hat{e}_\phi$ and $\hat{e}_z$ constitute an orthogonal (right-handed) coordinate system, i.e.

$$\left.\begin{array}{l} \hat{e}_\rho \times \hat{e}_\phi = e_z; \ \hat{e}_\phi \times \hat{e}_\phi = \hat{e}_\rho; \\ \hat{e}_z \times \hat{e}_\rho = \hat{e}_\phi \end{array}\right\} \qquad (1.24a)$$

$$\hat{e}_\rho \cdot \hat{e}_\phi = \hat{e}_\rho \cdot \hat{e}_z = \hat{e}_\phi \cdot \hat{e}_z = 0 \qquad (1.24b)$$

Cylindrical coordinates are preferred to describe the system where there is an axis of symmetry. The (planar) rotation of a particle about an axis is good example of this system of coordinates.

If a particle moves in a plane, one can describe its motion in polar coordinates (ρ, ϕ), where ρ represents the radial position of the particle. Obviously, one can call the plane as $Z = 0$, or XY-plane. The velocity and acceleration of the particle are then given by Eqs. (1.14) to (1.19). Let us consider that particle moves in a circle of radius ρ, we have [Eq. (1.14)]

$$V = \rho\dot{\phi}\hat{\varepsilon}_\phi = \omega \times \rho$$

where $\rho = \rho\hat{e}_\rho$ and $\omega = \dot{\phi}\hat{e}_z$. Here ω represents angular velocity whose magnitude is $\dot{\phi}$ and direction is about the axis of rotation.

We can now write the acceleration of the particle rotating about Z-axis as

$$a = \frac{dv}{dt} = \frac{d\omega}{dt} \times \rho + \omega \times \frac{d\rho}{dt}$$

$$= \alpha \times \rho + \omega \times (\omega \times \rho)$$

where $\omega \times (\omega \times \rho) = \rho\dot{\phi}^2 \hat{e}_z \times (\hat{e}_z \times \hat{e}_\rho) = -\rho\dot{\phi}^2 \hat{e}_\rho$ is called the *centripetal acceleration*, and

$$\alpha \times \rho = \rho\ddot{\phi}(\hat{e}_z \times \hat{e}_\rho) = \rho\ddot{\phi}\hat{e}_\phi$$

is the *tangential acceleration*.

Now, suppose that the particle is rotating in the XY-plane. ρ will correspond to the position vector of the particle. Usually, in this case one denotes ρ by r. Thus, one can write

$$V = \omega \times r \tag{1.25a}$$

and

$$a = \omega \times (\omega \times r) + \alpha \times r \tag{1.25b}$$

We will make frequent use of these relations while studying the motion in rotating frame or rotating motion of an object.

1.3 GALILEO'S PRINCIPLE OF RELATIVITY AND GALILEAN TRANSFORMATIONS

Let us consider two reference frames S and S', where S' is moving relative to S with velocity v in X-direction (Fig. 1.6). The origins O and O' of two frames are considered at $t = t' = 0$. An event at P is specified by the coordinates (x, y, z, t) in S and the coordinates (x', y', z', t') in S'.

The position vectors of P at any instant t are related by the equation

$$r' - r = vt \tag{1.26}$$

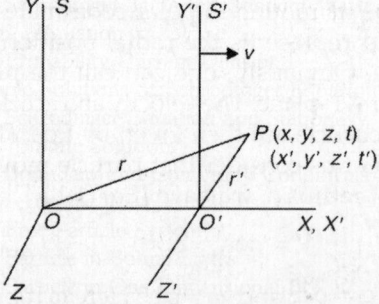

Fig. 1.6: Frames of reference in uniform relative translation motion

The vector Eq. (1.26) can be separated into its three components, keeping in mind that v is parallel to OX.

Therefore,

$$x' = x - vt, \ y' = y, \ z' = z \text{ and } t' = t \tag{1.27}$$

We have taken $t' = t$ in the three space equations in Eq. (1.27) to emphasize the assumption that the two inertial observers can measure time from the same "clocks" and therefore, use the same time, i.e. we implicitly assume that time measurements are independent of the observer's motion. The set of Eqs. (1.27) are called *Galilean transformations equations*. The first and last of Eqs. (1.27) are correct only for values of v that are small in comparison with the speed of light c (i.e. $v << c$). At values of v comparable with c, Galilean transformations must be replaced with the more general *Lorentz transformations*. Equations (1.27) are assumed to be accurate within the confines of Newtonian mechanics.

The *inverse Galilean transformations* can be written by changing primed into unprimed quantities, replacing v by $-v$ and *vice versa*. One obtains

$$x = x' + vt', \ y' = y, z = z' \text{ and } t = t' \tag{1.28}$$

or $\quad r' = r - vt \tag{1.28a}$

We may note that Galilean transformations are not the transformation equations between two coordinate systems within a single inertial frame.

Invariance of Galilean Transformations

The velocity V of particle P relative to O is defined by

$$V = \frac{dr}{dt}$$

and the velocity V' of A relative to O' is

$$V' = \frac{dr'}{dt}$$

We have not written $\dfrac{dr'}{dt'}$ because the assumption that $t = t'$ means that $\dfrac{dr'}{dt'}$ is the same as $\dfrac{dr'}{dt}$. Taking the derivative of Eq. (1.28a) relative to time t and noting that v is constant, we have

$$V' = V - v \qquad (1.29)$$

We may separate Eq. (1.29) into three velocity components:

$$V'_x = V_x - v, V'_y = V_y, V'_z = V_z \qquad (1.30)$$

Equations (1.29) and (1.30) give the rule of velocity addition in classical mechanics. We may note that Eq. (1.30), like any other vector equation, remains correct upon an arbitrary selection of the mutual directions of the coordinate axes of the frames S and S'. However, Eq. (1.30) is obeyed only when axes are chosen as shown in Fig. 1.6.

Equations (1.29) or (1.30) give the Galilean rule for comparing the velocity of a body or particle as measured by two observers in uniform relative translational motion. If P moves parallel to OX-axis, we have simply

$$V' = V - v \qquad (1.31)$$

the other components being zero.

In order to obtain the acceleration transformation, we differentiate Eq. (1.30) with respect to time such that

$$\frac{dV'_x}{dt} = \frac{d}{dt}(V_x - v) = \frac{dV_x}{dt}$$

Similarly $\quad \dfrac{dV'_y}{dt} = \dfrac{dV_y}{dt}$ and $\dfrac{dV'_z}{dt} = \dfrac{dV_z}{dt}$ (1.32)

Writing $\quad \dfrac{dV'_x}{dt} = a_{x'}, \dfrac{dV'_y}{dt} = a_{y'}, \dfrac{dV'_z}{dt} = a_{z'}$

and $\quad \dfrac{dV_x}{dt} = a_x, \dfrac{dV_y}{dt} = a_y, \dfrac{dV_z}{dt} = a_z$

one obtains

$$a'_x = a_x,\ a'_y = a_y, a'_z = a_z \qquad (1.33)$$

Writing Eq. (1.33) collectively, one obtains

$$a' = a \qquad (1.34)$$

Therefore, observers in uniform relative translational motion measure the same acceleration of a particle, i.e. the acceleration of a particle in all reference frames moving uniformly in a straight line relative to one another is the same. Obviously, if one of these frames is inertial (this signifies that in the absence of forces $a = 0$), then others will also be inertial (a' also equals to zero).

The fundamental equation of mechanics, i.e. $F = ma$ is characterized by containing only the acceleration of the kinematic quantities. It does not contain the velocity. As we have established above, however, the acceleration of a body in two arbitrary selected inertial reference frames S and S' is the same, i.e. acceleration remains invariant when passing from one inertial frame to another, i.e. in uniform relative translational motion. Hence it follows from Newton's second law that the force acting on a body in frames S and S' will also be same, i.e.

$$F = F' \text{ or } ma = ma' \qquad (1.35)$$

Consequently, *the equations of dynamics do not change upon transition from one inertial reference frame to another one, i.e. they are said to be invariant with respect to the transformation of the coordinates corresponding to the transition from one inertial reference frames to another.* From the view of classical mechanics, all inertial reference frames are absolutely equivalent, and none of them can be preferred to others.

The statement that all mechanical phenomena in different inertial reference frames proceed identically, owing to which no mechanical experiments allow us to determine whether the given reference frame is at rest or is moving uniformly in a straight line, is called *Galileo's relativity principle.* A further generalization of the relativity principle is treated in chapter on *special theory of relativity.*

Conservation of Momentum

If a force F is acting on a particle of mass m, then according to Newton's second law, we have

$$F = \frac{dp}{dt} = \frac{d}{dt}(mv) \qquad (1.36)$$

where $p = mv$ is the linear momentum of the particle. If the external force, F acting on the particle is zero, then

$$F = \frac{dp}{dt} = 0 \text{ or } p = mv = \text{constant} \quad (1.37)$$

Obviously, in the absence of an external force, the linear momentum of a particle remains constant.

Suppose that there are large number of particles in a system whose masses are m_1, m_2, m_3, ... The system can be a *rigid body* in which the particles are in fixed position with respect to each other, or it can be a collection of particles in which there may be all kinds of internal motion.

Let us suppose that the particles of the system are interacting with each other and are also acted by external forces. If $p_1 = m_1v_1$, $p_2 = m_2v_2$, ..., $p_n = m_nv_n$ are the momenta of particles of masses m_1, m_2,..., m_n respectively, then the total momentum (p) of the system of the particles is the vector sum of the momentum of individual particles, i.e.

$$p = p_1 + p_2 + ... + p_n$$
$$= m_1v_1 + m_2v_2 + ...+ m_nv_n \quad (1.38)$$

Differentiating Eq. (1.38) with respect to time t, we have

$$\frac{dp}{dt} = \frac{dp_1}{dt} + \frac{dp_2}{dt} + ... + \frac{dp_n}{dt}$$

$$= F_1 + F_2 + ... + F_n \quad (1.39)$$

where F_1, F_2,..., F_n are the forces acting on the particles m_1, m_2,..., m_n respectively.

We may note that these forces include external as well as internal forces. However, in accordance with Newton's third law, the internal forces exist in pairs of equal and opposite forces and they balance each other and hence they do not contribute any thing to the external force. This means that right hand side of Eq. (1.39) represents the result and force F_{ext} only due to the *external forces* acting on all the particles of the system. Thus, the sum of external forces is

$$F_{ext} = \frac{dp}{dt} + \frac{d}{dt}(p_1 + p_2 + ... + p_n)$$

If the resultant external force is zero, then

$$\frac{dp}{dt} = 0 \text{ or } p = \text{a constant}$$

i.e. $\quad p = p_1 + p_2 + ... + p_n = \text{a constant} \quad (1.40)$

Thus, if the *resultant external force acting on a system of particles is zero the total linear momentum of the system remains constant*. This simple but quite general result is called the *law of conservation of linear momentum* for a system of particles. We may note that momentum of individual particle may change but their sum, i.e. total momentum remains unaltered in the absence of external forces.

The law of conservation of momentum is a fundamental and exact law of nature and so far no violations of it have even been reported.

Law of Conservation of Angular Momentum

The angular momentum of a particle is defined as the moment of its linear momentum. Mathematically, the angular momentum J of a particle about a point is defined by

$$J = r \times p = m(r \times v) \quad (1.41)$$

where r is the vector distance of the particle from that point and $p = mv$ is the momentum in an inertial frame in which the point is stationary.

Differentiating Eq. (1.41) with respect to time t, one obtains

$$\frac{dJ}{dt} = \frac{d}{dt}(r \times p) = \frac{dr}{dt} \times p + r \times \frac{dp}{dt}$$

or $\quad \frac{dJ}{dt} = r \times \frac{dp}{dt} = r \times F$

$$\left[\because \frac{dr}{dt} \times p = v \times mv = 0 \right]$$

where $F = \frac{dp}{dt}$ is the force applied on the particle.

The vector product of r and F is called *torque* or *moment of force* about the reference point and is represented by τ. Thus,

$$\tau = \frac{dJ}{dt} = r \times F \quad (1.42)$$

Obviously, the torque is equal to the rate of change of angular momentum (J). Its unit is N/m.

Extending the above for a system of n particles in the form of a rigid body, we obtain

$$J = J_1 + J_2 + ... + J_n$$

or $\quad J = (r_1 \times m_1 v_1) + (r_2 \times m_2 v_2) + \ldots$

$$= \Sigma(r \times mv) = \Sigma(r \times p) \quad (1.43)$$

Clearly, *the rate of change of angular momentum of the system about any point is the sum of the torques about that point of all the external forces acting on the system.*

If $\quad \tau = 0$, then

$$\frac{dJ}{dt} = 0 \text{ or } J = \text{constant}$$

or $\quad J = J_1 + J_2 + \ldots = \text{constant}$

Thus, *the total angular momentum of a system of particles is constant, if the resultant external torque acting on the system is zero.* This is the principle of conservation of angular momentum.

1.4 NON-INERTIAL FRAMES

We have studied about the inertial frames of reference in which Newton's laws of linear motion are valid, i.e. the physical laws are covariant. Moreover, such reference systems are completely equivalent. However, if a frame of reference which is in an accelerated motion with respect to an inertial frame of reference, then the form of basic physical laws are completely different in two systems. *Such reference frames that experience relative accelerated motion are known as non-inertial frames of reference.* The observed acceleration of a body in a non-inertial frame, a' will differ from acceleration a with respect to frame S. Obviously, the formula $F = ma$ for the force acting on the body is for the non-inertial frame of reference. Let us use the symbol a_0 to denote the difference between the acceleration of body in an inertial (a) and a non-inertial frame (a'),

$$a - a' = a_0 \quad (1.44)$$

For a non-inertial frame in translational motion, a_0 is the same for all points of space (a_0 = constant) and is the acceleration of the non-inertial frame. For a rotating non-inertial frame, a_0 will be different at different points of space [$a_0 = a_0(r')$, where r' is the position vector relative to the non-inertial reference frame]. Obviously, all the accelerated and rotating frames are the non-inertial frames of reference.

1.5 REFERENCE FRAME WITH TRANSLATIONAL ACCELERATION AND FICTITIOUS OR PSEUDO OR INERTIAL FORCE

If no external force is acting on a particle, even then in the accelerated frame it will appear that a force is acting on it. Let us consider two non-inertial frames S and S' such that the frame S' is moving with acceleration a_0 with respect to S. Let the particle have an acceleration a with respect to frame S. Obviously to the observer in S', it will appear to have acceleration a' given by

$$a' = a - a_0 \quad (1.45)$$

If m be the mass of the particle (which is invariant under Galilean transformation), then the force on the particle in frame S' is

$$F' = ma' \quad (1.46)$$

Using Eq. (1.44), Eq. (1.45) takes the form

$$F' = m(a - a_0)$$

$$= ma - ma_0 = F - F_0 \quad (1.47)$$

where $F = ma$ is the force seen by an observer in frame S and $F_0 = ma_0$ is the force due to relative acceleration a_0 between the two frames. When $F = 0$, one obtains

$$F' = -F_0 = -ma_0 \quad (1.48)$$

Clearly, the particle seems to experience a force, $-F_0$ when viewed from frame S' even when there is no force on it in frame S. This means a force-free particle in frame S also appears to have an acceleration. Such an acceleration which a particle appears to possess on account of the acceleration of the observer itself and which does not originate because of any real forces acting on the particle is called *fictitious acceleration.* This accelerated motion in non-inertial frame may be due to linearly accelerated (uniform) motion or may be a uniform rotation of the frame of reference or both.

The force $F_0 = ma_0$ is called the *fictitious or pseudo force,* and is obviously given by the mass times the acceleration of the non-inertial frame with its sign changed. Hence, it follows that even when $F = 0$, the body will travel relative to the non-inertial reference frame with the acceleration $-a_0$, i.e. as if a force equal to $-ma_0$ acted on it.

What has been said above signifies that one can use Newton's equations in describing motion in non-inertial reference frames, if in addition to the forces due to the action of the bodies on one another, we take into account the so-called *forces of inertia* F_1, i.e. *fictitious forces* F_0. The latter should be assumed equal to the product of the mass of a body and the difference between its acceleration relative to the inertial and non-inertial frames taken with the opposite sign,

$$F_1 = -m(a - a') = -ma_0 \qquad (1.49)$$

The equation of Newton's second law for a non-inertial reference frame will accordingly be

$$ma' = F + F_1 \qquad (1.50)$$

The introduction of *inertial* or *pseudo* forces permit us to describe the motion of bodies in any (both inertial and non-inertial) reference frames using the same equations of motion.

One must understand distinctly that the forces of inertia may never be treated at par with such forces as gravitational, elastic and frictional ones, i.e. with forces produced by the action of a body on other bodies. Forces of inertia are due to the properties of the reference frame in which mechanical phenomena are being considered. In this sense they are called fictitious forces.

A feature of inertial forces is that they are proportional to the mass of a body. Owing to this property, inertial forces are similar to gravitational ones. The equivalence underlies *Einstein's general theory of relativity*.

To make the above discussion more clear, let S be an inertial frame and another frame S' moves with respect to S with a uniform acceleration a in $-\hat{k}$ direction (Fig. 1.7). Let the two frames coincide at $t = 0$ and S' starts from rest. A free falling box is an example of such a motion. Let r be the position of a point P with respect to S and r' with respect to S'.

Thus, we have

$$r' = r + \frac{1}{2}at^2\hat{k} \qquad (1.51)$$

Differentiating Eq. (1.51) with respect to t, one obtains

$$\frac{dr'}{dt} = \frac{dr}{dt} + at\hat{k} \qquad (1.52)$$

Differentiating Eq. (1.52), one obtains

$$\frac{d^2r'}{dt^2} = \frac{d^2r}{dt^2} + a\hat{k} \qquad (1.53)$$

If the mass of the particle situated at P be m, then

$$F = m\frac{d^2r'}{dt^2} = m\frac{d^2r}{dt^2} + ma\hat{k} \qquad (1.54)$$

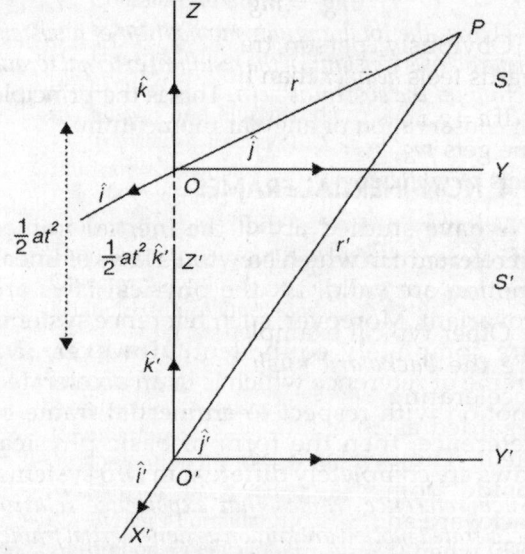

Fig. 1.7: Motion of S' frame w.r.t. S with a uniform acceleration

Thus, in addition to the impressed force $m\dfrac{d^2r}{dt^2}$, we have an *apparent* or *fictitious* or inertial force $ma\hat{k}$. Let $\hat{k}$ axis be vertical. Now, we consider two particular cases:

1. Downward Movement of Frame S'

Let frame S' be moving vertically downwards with an acceleration a. For example, we can consider a person of mass m travelling in a lift downwards. In each frame, i.e. in S and S' the weight of the person is inertial or impressed force. Hence, one can write

$$m\frac{d^2r'}{dt^2} = mg'(-\hat{k}) = -mg'\hat{k} \text{ and } m\frac{d^2r}{dt^2} = mg(-\hat{k}).$$

Making these substitutions in Eq. (1.54), one obtains

$$mg' = mg - ma \qquad (1.55)$$

Obviously, the person travelling in a lift downwards feels *lighter* than he is.

2. Upward Movement of Frame S′

Let the frame S' be moving vertically upwards with an acceleration a. Obviously, we can consider that a person of mass m is travelling in a lift upwards. In this case, we have

$$mg' = mg + ma \qquad (1.56)$$

Obviously, person travelling in a lift upwards feels *heavier* than he is.

If $a = g$, i.e. if the lift falls freely under gravity, one gets $mg' = 0$. In this situation, the person feels *weightless*.

We may note that the inertial force $ma\hat{k}$ in accelerated frame is also called a *frame dependent force*.

Other typical examples of fictitious forces are the *backward push* we feel inside the accelerating car with respect to the earth regarded as an inertial frame; that is a real push. The bob of a pendulum suspended inside from the ceiling of car is thrown backward and is brought to rest inside the car only when (horizontal component of) tension T in the string balances the inertial or fictitious force F_i, i.e. as seen in the accelerating car (here serving the purpose of frame S'), bob is at rest because net horizontal and vertical forces are zero. Thus, we have

$$T\sin\theta - F_i = 0$$

and $\qquad T\cos\theta - mg = 0$

where θ is the angle which the string makes with vertical at the equilibrium position of the bob. Thus, by observing θ, one can estimate the acceleration of car. We may note that this principle is used in the construction of *accelerometers*.

Now, we have

$$\tan\theta = \frac{F_i}{mg} = \frac{A}{g}$$

Obviously, one can determine the acceleration of a non-inertial frame by dynamics of objects observed from this frame.

Equation of Motion in a Rotating Reference Frame

Let us consider two reference frames $S(x, y, z)$ and $S'(x', y', z')$. S frame is stationary while S' is rotating in space about its Z'-axis. We assume that origins O and O' of two reference frames coincide and axis of rotation of S', i.e. Z'-axis also coincides with Z-axis of S-frame (Fig. 1.8).

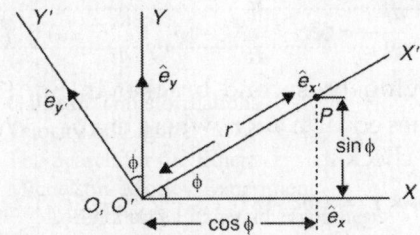

Fig. 1.8: Rotating frame of reference

Let us take a particle P whose position vectors in reference frames S and S' are r and r' at some instant t. We have

$$r = x\hat{e}_x + y\hat{e}_y + z\hat{e}_z$$

and $\qquad r' = x'\hat{e}_{x'} + y'\hat{e}_{y'} + z'\hat{e}_{z'} \qquad (1.57)$

Here $\hat{e}_x$, $\hat{e}_y$ and $\hat{e}_z$ are constant unit vectors in coordinate system $S(x, y, z,)$. $\hat{e}_{x'}$, $\hat{e}_{y'}$ and $\hat{e}_{z'}$ are unit vectors in coordinate system $S'(x', y', z')$ which rotate and hence change their directions with time. Let us assume that X', Y' axes are inclined at an angle ϕ with X, Y axes respectively at time t (Fig. 1.8).

We have

$$\left.\begin{array}{l} \hat{e}_{x'} = \cos\phi\,\hat{e}_x + \sin\phi\,\hat{e}_y \\ \hat{e}_{y'} = -\sin\phi\,\hat{e}_x + \cos\phi\,\hat{e}_y \\ \hat{e}_{z'} = \hat{e}_z \end{array}\right\} \qquad (1.58)$$

Since O and O' coincide, we have

$$r = r'$$

Hence $\qquad V = \dfrac{dr}{dt} = \dfrac{dr'}{dt}$

Now $\dfrac{dr'}{dt} = \left(\dfrac{dx'}{dt}\hat{e}_{x'} + \dfrac{dy'}{dt}\hat{e}_{y'} + \dfrac{dz'}{dt}\hat{e}_{z'} \right)$

$$+ \left(x'\dfrac{d\hat{e}_{x'}}{dt} + y'\dfrac{d\hat{e}_{y'}}{dt} + z'\dfrac{d\hat{e}_{z'}}{dt} \right) \quad (1.59)$$

We note that the first bracket in Eq. (1.59) can be interpreted as the velocity V' of the particle observed from frame S', i.e.

$$V' = \dfrac{dx'}{dt}\hat{e}_{x'} + \dfrac{dy'}{dt}\hat{e}_{y'} + \dfrac{dz'}{dt}\hat{e}_{z'} \quad (1.60)$$

The second bracket in Eq. (1.59) arises due to rotation of S'. From Eq. (1.58), one obtains

$$\dfrac{d\hat{e}_{x'}}{dt} = \dot{\phi}\hat{e}_{y'}, \quad \dfrac{d\hat{e}_{y'}}{dt} = \dot{\phi}\hat{e}_{x'}, \quad \dfrac{d\hat{e}_{z'}}{dt} = 0 \quad (1.61)$$

Obviously, second bracket in Eq. (1.59) becomes equal to $\omega \times r'$, where $\omega = \dot{\phi}\hat{e}_{z'}$. We can easily check it.

$$\omega \times r' = (\omega\hat{e}_{z'}) \times (e'\hat{e}_{x'} + y'\hat{e}_{y'} + z'\hat{e}_{y'})$$

$$= \omega x'\hat{e}_{y'} + \omega y'\hat{e}_{x'} \quad (\omega = \dot{\phi})$$

Thus, one obtains

$$V = V' + \omega \times r' \quad (1.62)$$

Differentiating Eq. (1.62) with respect to t, one obtains

$$a = \dfrac{dV}{dt} = \dfrac{dV'}{dt} + \dfrac{d\omega}{dt} \times r' + \omega \times \dfrac{dr'}{dt}$$

One can exactly evaluate $\dfrac{dV'}{dt}$ as $\dfrac{dr'}{dt}$ is found. Writing Eq. (1.60) as

$$V' = V_{x'}\hat{e}_{x'} + V_{y'}\hat{e}_{y'} + V_{z'}\hat{e}_{z'}$$

Now repeating the steps from Eqs. (1.58) to (1.62), one obtains

$$\dfrac{dV'}{dt} = a' + \omega \times V'$$

where $\quad a' = \dfrac{dV_{x'}}{dt}\hat{e}_{x'} + \dfrac{dV_{y'}}{dt}\hat{e}_{y'} + \dfrac{dV_{z'}}{dt}\hat{e}_{z'}$

Now, defining $\alpha = \dfrac{d\omega}{dt} = \ddot{\phi}\hat{e}_z$ and substituting the value of $\dfrac{dr'}{dt}$, one obtains

$$a = \ddot{\phi}\hat{e}_z + \omega \times (V' + \omega \times r')$$

$$= a' + 2\omega \times V' + \omega \times (\omega \times r') + \alpha \times r' \quad (1.63)$$

We note that V' and a' are the velocity and acceleration of the particle as measured by an observer standing in the rotating frame S'. Obviously, with respect to this observer in S' frame, S' frame is stationary or non-rotating.

Let us now consider that S' is rotating about Z (or Z') axis with constant angular velocity $\omega = \dot{\phi}$, then angular acceleration $\alpha = \ddot{\phi}\hat{e}_{z'} = 0$ ($\omega = \dfrac{d\phi}{dt}$, being constant), and hence the last term in Eq. (1.63) vanishes.

We note that Eqs. (1.62) and (1.63) represent the velocity and acceleration of a particle relative to frame S in terms of the variables observed in rotating frame S'.

Now, multiplying Eq. (1.63) by mass m of the particle and rearranging, one obtains the equation of motion as

$$ma' = ma - 2m(\omega \times V') - m\omega(\omega \times r') - m\alpha \times r' \quad (1.64)$$

1.6 CORIOLIS AND CENTRIFUGAL FORCES

Let us interpret Eq. (1.64) as the expression of Newton's second law in rotating frame S', $F' = ma'$. We note that the net force F' on a particle or object as seen from S' gets contribution from three sources, viz. (i) the ordinary force, $F = ma$ (usually referred to as real force in inertial frame), (ii) an inertial force $-2m(\omega \times V')$ called *Coriolis force*, and (iii) another inertial force $-m\omega(\omega \times r')$ called *centrifugal force*. We can see that both the inertial forces, i.e. Coriolis force and centrifugal force are in the direction perpendicular to the axis of rotation, $\omega = \omega\hat{e}_{z'}$; hence in XY plane.

The force represented by $-m\alpha \times r'$ is also an inertial force which arises if the frame S' is rotating with angular acceleration α.

The centrifugal force of inertia (F_{cf}) depends only on the variable r' (position) of the object and is always radially outwards,

$$F_{cf} = -m\omega(\omega \times r') = -m\omega^2 r'[\hat{e}_{z'} \times (e_{z'} \times e_{r'})]$$

$$= -m\omega^2 r'(\hat{e}_{z'} \times \hat{e}_{\phi'})$$

$$= -m\omega^2 r'\hat{e}_{r'} \quad (1.65)$$

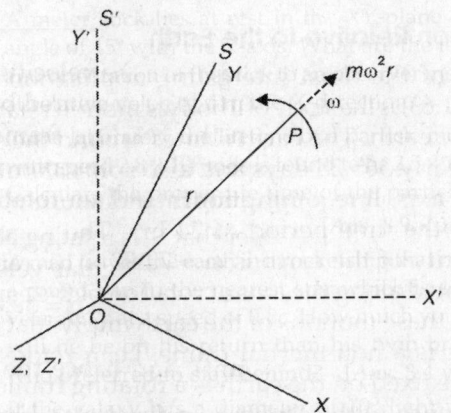

Fig. 1.9: Centrifugal force on particle *P*

Obviously, the magnitude of $|F_{cf}| = m\omega^2 r'$, which is same as the magnitude of ordinary centrifugal force acting on a particle moving in a circle of radius *r'* and perpendicular to ω. We may note that centrifugal force F_{cf} is an inertial force not seen in the stationary frame. This centrifugal force always acts on an object/point in a non-inertial rotating frame. This force is a frame dependent. This force does not exist in *S*. It is, therefore, called a *fictitious* or *pseudo* or *inertial force*. Figure 1.9 illustrates the centrifugal force on particle *P* in the *XY*-plane of an inertial frame *S*. Frame *S'* is a frame of reference fixed with respect to *P* and with its *Z*-axis coincident with that of *S*. The particle is describing uniform circular motion in *S*.

To make the concept of centrifugal force more clear, let us consider a passenger in a car which is closed on all sides so that the passenger cannot see anything outside the car. Let the car be travelling with uniform speed in a straight line. The passenger will not feel the movement of the car. Passenger will feel as if he were sitting in a chair in a room. Let us now suppose that the car takes a turn to the left. To keep the passenger going along the curved path the right side wall provides the *centripetal force*. Passenger is pushed against the right side of the car by the *inertial* or *fictitious* centrifugal force, F_{cf}. As a result, the passenger is at rest relative to the car.

Now, consider that the passenger is sitting in a moving car at a curved road. When tra-velling in a car or a bus, passenger is pushed away towards the outer wall of the vehicle when it moves along a curved road, the inner wall being that nearer to the centre of curvature of the road.

Coriolis Force

The inertial force, $-2m\omega \times V'$ is the Coriolis force and is perpendicular to both ω and *V'*. This is non-zero only when $V' \neq 0$ and the velocity of the point/object relative to the rotating reference frame *S'* must have a non-zero projection on a plane perpendicular to the axis of rotation. We may note that if the object has no velocity in the rotating frame, no Coriolis force acts on it. The effect of Coriolis force is to deflect any moving object from its instantaneous direction of motion, i.e. of *V'* in a rotating frame. We may note that Coriolis force is also proportional to the mass of the body to which it is applied and like other inertial forces arise due to non-inertial nature of the reference frame. In brief, we can say that *Coriolis force is velocity dependent force and arises due to rotation of earth.*

The action of the Coriolis inertial force can be demonstrated with the aid of a visual experiment (Fig. 1.10). Along the diameter of the rotating disk is placed a rubber pipe through which water is made to run as the disk rotates. The motion of the water is uniform relative to the disk and the water is acted upon by the Coriolis inertia force which bends the pipe in the direction opposite to the direction of the rotation of the velocity vector of the water particles. In this experiment the curvature of the pipe is increased.

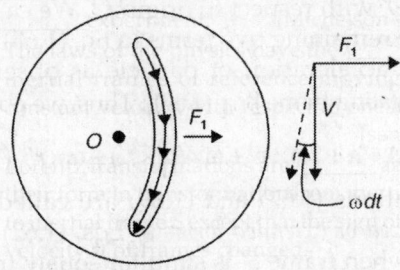

Fig. 1.10: The action of Coriolis inertial force

The same occurs when the velocity of the water flowing in the pipe is increased. One can observe both phenomena more easily when the disk is lit by a stroboscopic tube.

Generalized Equation of Motion in a Non-inertial Frame

Let us consider a non-inertial frame S' which may be simultaneously translating and rotating with respect to an inertial frame S. Let the origins of S and S' coincide at $t = 0$. At some later time t, let the origin O' of frame S' is at R from the origin O of frame S. At that instant, let the position vector of a point (particle) P be r and r' relative to O and O' respectively. Obviously,

$$r = R + r' \qquad (1.66)$$

Differentiating Eq. (1.66) with respect to time, one obtains

$$v = \frac{dr}{dt} = \frac{dR}{dt} + \frac{dr'}{dt}$$

where $\dfrac{dR}{dt} = V$, is the velocity of origin O' with respect to O; $\dfrac{dr'}{dt} = v' + \omega \times r'$ [Eq. (1.62)]; ω is the angular velocity of rotation of S' with respect to S. Thus, we have

$$v = V + v' + \omega \times r' \qquad (1.67)$$

Differentiating Eq. (1.67), one obtains

$$a = \frac{dv}{dt} = \frac{dV}{dt} + \frac{dv'}{dt} + \frac{d}{dt}(\omega \times r') \quad (1.68)$$

where, $\dfrac{dV}{dt} = \dfrac{d^2 R}{dt^2}$ is the acceleration A of origin O' with respect to origin O. We can see that the remaining two terms in Eq. (1.68) give the net acceleration of particle as observed from rotating frame [Eq. (1.63)]. Thus, we obtain

$$a = A + a + 2\omega \times v' + \omega \times (\omega \times r') + \alpha \times r' \quad (1.69)$$

Equations (1.67) and (1.69) are called the *generalized or extended forms of* Eqs. (1.62) and (1.63), when frame S' is simultaneously translating and rotating with respect to frame S.

Motion Relative to the Earth

One of the most interesting applications of Eq. (1.63) is the study of a body's motion relative to the earth. The earth rotates around the sun in nearly 365.25 days and also spins about its own axis, line joining north and south poles with the time period of 24 hrs. The angular velocity of the earth is $\omega = 7.292 \times 10^2$ rad/s oriented along the axis of rotation of the earth. Both these motions of the earth imply that the earth is a non-inertial frame, i.e. a reference frame fixed on the earth is a rotating frame of reference with respect to a fixed star frame. We may note that the rotation of the earth is about its axis (i.e. spin) does produce measurable effects on the motion of objects taking place on earth. Thus, a particle at rest or in motion on the earth is acted upon by the apparent or fictitious forces, i.e. the centrifugal and Coriolis forces. We now discuss these one by one.

Effect of Centrifugal Force: Correction to 'g'

Let us consider any particle of mass m on the surface of the earth (considered to be a uniform sphere of mass M_e and radius R_e). This particle is subjected to an acceleration due to gravity, g given by

$$g = G\frac{M_e}{R_e^2} \qquad (1.70)$$

Here g usually refers to the true value of acceleration due to gravity. We know from Newton's law of gravitation that the gravitational force of attraction between earth and particle is the cause of g: $F_g = mg$, where g is radially directed towards the centre of the earth.

However, when particle is observed from the rotating frame attached to the earth, the particle is also subjected to inertial forces, i.e. centrifugal and Coriolis forces. Let us consider that the particle is at rest at point at latitude λ on earth's surface. We may note that it is located at a radial distance $(R_e \cos \lambda)$ from origin O' of the reference frame with axes $O'X'$ and $O'Y'$ rotating about the axis $O'N$ (Fig. 1.11).

Obviously, there acts a centrifugal force on particle along the direction $O'B$ given by Eq. (1.65).

$$F_{cf} = m\omega^2(R_e\cos\lambda) \qquad (1.71)$$

where ω is the angular velocity of earth's rotation. Since the particle is at rest relative to earth, there is no Coriolis force acting on the particle.

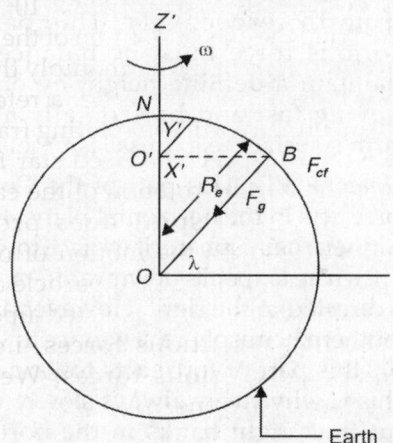

Fig. 1.11: Effect of centrifugal force

Obviously, for an observer on the earth, the particle is subjected to two forces: (i) $F_g' = mg$ (ii) F_{cf}. We may note that the resultant of these two forces would not be radially towards O, i.e. the centre of the earth. One can usually resolve F_{cf} into two components: (i) the radial component, $F_{r'} = F_{cf} \cos \lambda$ along OB and (ii) the component tangential to the earth's surface, i.e. $F_{r'} = F_{cf} \sin \lambda$. Thus, one finds

$$F_r' = mg' = F_g - F_{cen} \cos \lambda$$

$$= m(g - \omega^2 R_e \cos^2 \lambda) \qquad (1.72)$$

$$F_t' = F_{cf} \sin\lambda = m\omega^2 R_e \cos\lambda \sin\lambda \quad (1.73)$$

Here g' denotes the *effective acceleration* due to gravity, i.e. modified by centrifugal acceleration. We can now see that g' depends on λ as

$$g' = g - \omega^2 R_e \cos^2 \lambda \qquad (1.74)$$

The magnitude of the centrifugal acceleration is given by

$$\omega^2 R_e = \left(\frac{2\pi}{86,134}\right)^2 \times 6370 \times 10^5$$

$$= 3.4 \text{ cm/s}^2 = 0.034 \text{ m/s}^2$$

Here $R_e = 6370$ km is mean radius of the earth. We can see that $\omega^2 R_e \ll g$. The value of the centripetal acceleration is always very small when compared to the acceleration due to gravity, $g = 9.8$ m/s^2. Although a very small term, the centripetal acceleration accounts for the most of the observed variation in the value of the acceleration due to gravity with latitude.

The effect of the centrifugal acceleration on a freely falling body is to displace the body slightly from the radial direction towards the south in the northern hemisphere and towards the north in the southern hemisphere. Near the equator the centrifugal acceleration is about 0.3% of acceleration due to gravity, and therefore the variation of the centrifugal force during the free fall can be neglected in approximate calculations.

The deflection of the plumb-line at latitude λ from radial direction OB by an angle θ is obtained as

$$\tan \theta = \frac{F_s'}{F_r'} = \frac{\sin 2\lambda}{2(\gamma/\omega^2 R_e - \cos^2 \lambda)} \qquad (1.75)$$

We may note that the local plumb-line at position B defines the vertical direction there. Moreover, it is along the direction of plumb-line that the resultant force F' acts.

Horizontal and Vertical Components of Coriolis Force

Let us consider a point B on the surface of the earth at latitude λ (Fig. 1.12). The point B is located in the rotating frame S' with origin at O. The frame S' rotates with angular velocity ω about the line $O'N$. The Coriolis force acting on moving particle as seen by an observer at B on S' is as

$$F_{cor} = -2m\omega \times V' \qquad (1.76)$$

where V' is the velocity of the particle relative to frame S'.

One can resolve ω and V' into horizontal and vertical components to the surface of earth at B as

$$\omega = \omega_V + \omega_H$$

and $$V' = V_V' + V_H' \qquad (1.77)$$

From Fig. 1.12, we have $\omega_V = \omega \sin \lambda$ and $\omega_H = \omega \sin \lambda$.

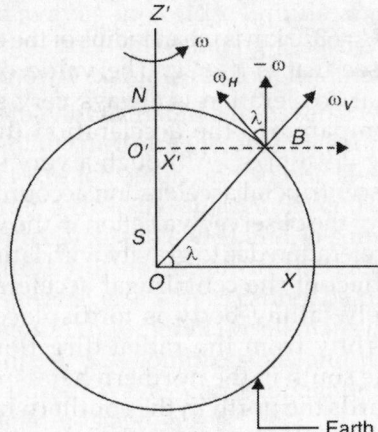

Fig. 1.12: Horizontal and vertical components of Coriolis force

ω_V points upward in the northern hemisphere, while in the southern hemisphere it points downward, to the earth's surface. We find that at any point on the globe, the component ω_H is always directed towards north along meridian (or great circle) passing through that point. Now, we can write Coriolis force as

$$F_{cor} = 2m\omega_V \times V'_H - 2m\omega'_H \times (V'_V + V'_H) \quad (1.78)$$

We may note that $\omega_V \times V'_V = 0$. However, $\omega_H \times V'_H$ need not vanish as horizontal component of V' is not always along ω_H, i.e. along north at B. Obviously, there are three contributions to net Coriolis force:

1. $-2m\omega_H \times V'_V$: For this part of Coriolis force, the vertical component of the velocity, i.e. V'_V is responsible. We note that ω_H and V'_V constitute the meridian plane $O'BN$ and $-2m\omega_H \times V'_V$ part of the Coriolis force is perpendicular to $O'BN$. Thus at B, the force acts in horizontal plane along east–west direction. However, if V'_V is downwards, the Coriolis force acts toward east whereas if V'_V is upwards, the force acts towards west. Clearly, a body falling vertically downwards from a height is deflected towards east by Coriolis force.

2. $-2m\omega_H \times V'_H$: The horizontal component of the velocity of the particle is responsible for two distinct parts of the Coriolis force, i.e. $-2m\omega_H \times V'_H$ and $-2m\omega_V \times V'_H$. The first part, i.e. $-2m\omega_H \times V'_H$ acts normal to horizontal plane, i.e. along the vertical direction as viewed by an observer at B. This part of the Coriolis force either pushes the particle towards earth or away from it, depending upon whether direction of V'_H is west or east of north respectively. This part of Coriolis force must be balanced to maintain a definite height by objects moving fast and horizontally above earth's surface, e.g. missiles or jets.

3. $-2m\omega_V \times V'_H$: This part of the Coriolis force acts in the horizontal plane. In the northern hemisphere, if we face towards V'_H, where ω_V points downward, $-\omega_V \times V'_H$ is directed to the right. However, in the southern hemisphere, if we face towards V'_H, this part of force acts towards left. This is why rivers always slowly wash out their right banks in the northern hemisphere and their left banks in the southern hemisphere and become steeper. This effect is independent of the direction of the flow. This has been actually verified that one of the river banks is always higher compared to the other. However, this is not the only source responsible for this erosion.

The Coriolis effect may be seen in two common phenomena:

1. *Whirling of wind in a hurricane.* If a low-pressure centre develops in the atmosphere, air will flow radically towards the centre. However, the Coriolis acceleration deviates the air molecules towards the right of their paths in the northern latitudes, resulting in the counter clockwise whirling motion. This pressure and temperature of the air also have a profound effect on its motion. This effect ultimately leads to the *cyclonic motion*.

2. Coriolis effect on the oscillations of a pendulum demonstrated with the help of Foucault's pendulum.

ILLUSTRATIVE EXAMPLES

Example 1

A bullet is fired horizontally in the north direction with a velocity of 500 m/s at 30°N latitude. Calculate the horizontal component of Coriolis acceleration and the consequent deflection of the bullet as it hits a target 250 m away. Also find the vertical displacement of the bullet due to gravity.

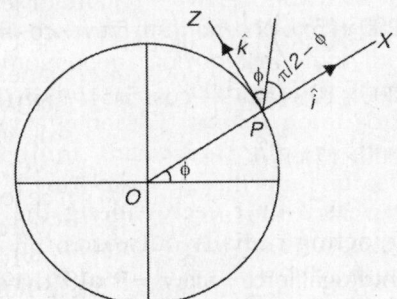

Fig. 1.13

Solution

Let us take X-axis vertically, Z-axis towards north and Y-axis along east. The velocity of the bullet

$$V = 500\,\hat{k} \text{ m/s}$$

and angular velocity

$$\omega = \omega(\hat{k}\cos 30° + \hat{i}\sin 30°)$$

Since the angular velocity ω of the earth is directed parallel to its axis and is inclined at 30° to the horizontal. Here

$$\omega = \frac{2\pi}{24 \times 60 \times 60} = 7.2 \times 10^{-5} \text{ rad/s}$$

$\because$ Coriolis acceleration $= 2\omega \times V$

$$= 2\omega(2\hat{k}\cos 30° + \hat{i}\sin 30°) \times 500\hat{k}$$

$$= -2 \times 7.2 \times 10^{-5} \times 10^2 \times \frac{1}{2}\hat{j}$$

$$= 0.036 \text{ m/s}^2 \text{ towards west.}$$

Time taken during the journey

$$= 250/500 = 0.5 \text{ s.}$$

$\therefore$ Deflection of the bullet due to Coriolis acceleration

$$= -\frac{1}{2}at^2 = \frac{1}{2} \times 0.036 \times \left(\frac{1}{2}\right)^2 = 4.5 \times 10^{-3} \text{ m}$$

Now, the vertical displacement of bullet due to gravity

$$= \frac{1}{2}gt^2 = \frac{1}{2} \times 9.8 \times \left(\frac{1}{2}\right)^2 = 1.23 \text{ m}$$

Now, Coriolis force $= -2m\omega \times V$

$$= 2 \times 0.1 \times 7.2 \times 10^{-5} \times 5 \times 10^2 \times \frac{1}{2}\hat{i}$$

$$= 3.6 \times 10^{-3}\,\hat{i} = 3.6 \times 10^{-3} \text{ N towards east}$$

Example 2

Prove that the observed acceleration due to gravity g_ϕ at the latitude ϕ is related to its real value g by the relation

$$g_\phi^2 = \left(g\cos\phi - \omega^2 R\cos\phi\right)^2 + \left(g\sin\phi\right)^2$$

Solution

If the particle is at rest at latitude ϕ, then it is not acted by Coriolis force.

Thus $a_i = a_r - \omega \times (\omega \times R)$

Now, we take Z-axis along the axis of the earth and X-axis perpendicular to it, then

$$a_r = a_i - \omega\,(\omega \times R)$$

or $\quad g_\phi = -g\left(\hat{i}\cos\phi + \hat{k}\sin\phi\right) - \omega\hat{k}$

$$\times\left[\omega\hat{k} \times R\left(\hat{i}\cos\phi + \hat{k}\sin\phi\right)\right]$$

$$= -g\left(\hat{i}\cos\phi + \hat{k}\sin\phi\right) + \omega^2 R\cos\phi\,\hat{i}$$

$$= -\hat{i}\left(g\cos\phi + \omega^2 R\cos\phi\right) - \hat{k}g\sin\phi$$

$\therefore g_\phi^2 = g_\phi \cdot g_\phi\left(g\cos\phi - \omega^2 R\cos\phi\right)^2 + \left(g\sin\phi\right)^2$

Example 3

A particle of mass 200 gm is stationary in an inertial reference system. Describe and interpret its motion in a frame rotating with angular speed 5π rad/s. The axis of rotation is

10 cm away from the particle. Also calculate the Coriolis and centrifugal contributions.

Solution

Let us choose the axes of inertial frame F and non-inertial frame F' as shown in Fig. 1.14. The coordinates of particle P in frame F are:

$$x = 10 \text{ cm}, y = z = 0.$$

Using transformation equations, the coordinates of P in frame F' are

$$X' = x\cos\omega t + y\sin\omega t = 10\cos(5\pi t)$$

$$Y' = -x\sin\omega t + y\cos\omega t = -10\sin(5\pi t)$$

$$Z' = z = 0$$

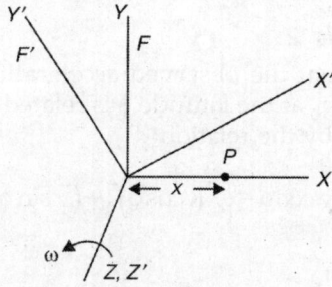

Fig 1.14

From these equations we note that the particle P will describe a circle of radius 10 cm with angular velocity 5π rad/s clockwise. The particle, thus, experiences a net centripetal force of magnitude

$$= m\omega^2 r'$$

$$= 200 \times (5\pi)^2 \times 10$$

$$= 5 \times 10^5 \text{ dynes} = 5 \text{ N (here } \pi^2 \approx 10)$$

We may now try to explain the origin of this force.

Net force on the particle P in frame F' is

$$F' = F - m\ddot{R} - m\omega \times (\omega \times r') - 2\omega \times V'$$

Here $\ddot{R} = 0$ and F, the force in the inertial frame = 0

$$\therefore \quad r' = 10\cos(2\pi t)\hat{i} - 10\sin(5\pi t)\hat{j}, \quad \omega = 5\pi\hat{k}$$

$$V' = \frac{dr'}{dt} = -50\pi\sin(5\pi t)\hat{i} - 50\pi\cos 5\pi t\,\hat{j}$$

$$= -50\pi\left\{\sin(5\pi t)\hat{i} + \cos(5\pi t)\hat{j}\right\}$$

$$= -50\pi\left\{\sin(5\pi t)\hat{i} + \cos(5\pi t)\hat{j}\right\}$$

$$= -\omega \times r'$$

Centrifugal contribution:

The centrifugal contribution towards the apparent force on the particle is

$$= -m\omega \times (\omega \times r')$$

$$= -200 \times \left(5\pi\hat{k}\right) \times \left\{-50\pi\left(\sin 5\pi t\hat{i} + \cos 5\pi t\hat{j}\right)\right\}$$

$$= -200 \times 10 \times (5\pi)^2 \left\{-\cos(5\pi t)\hat{i} + \sin(5\pi t)\hat{j}\right\}$$

$$= 2000 \times (5\pi)^2 \hat{j}\hat{r}'$$

where $\hat{r}'$ is a unit vector along the radius vector, acting radially outwards.

$\therefore$ Centrifugal force $= m\omega^2 r' = 5 \times 10^5$ dynes $= 5$N acting radially outwards.

Coriolis contribution:

$$= 2m\omega \times V'$$

$$= -2m\omega \times (\omega \times r') = 2m\omega^2 r$$

$$= -10 \times 10^5 \hat{r}' \text{ dynes} = -10\,\hat{r}'\,\text{N}$$

Obviously, Coriolis contribution is 10 N inwardly which is double the outward centrifugal force. Adding up the Coriolis and centrifugal contributions, we obtain the necessary centripetal force to maintain the observed circular motion.

Example 4

Calculate the direction and magnitude of the Coriolis force acting on a particle of mass 1.5 kg moving eastwards with a constant velocity of 100 m/s at 30°N. Also calculate horizontal deviation produced per second. Given, earth's angular velocity = 7.3×10^{-5} radian/s.

Solution

From Fig. 1.15, one can write

$$\omega = \omega\cos\lambda\hat{j} + \omega\sin\lambda\hat{k}$$

and $$V = V\hat{i}$$

∴ Coriolis acceleration

$$= -2\omega \times V$$

$$= -2(\omega\cos\lambda\hat{j} + \omega\sin\lambda\hat{k}) \times V\hat{i}$$

$$= -2\omega V(\sin\lambda\hat{j} - \cos\lambda\hat{k})$$

∴ Coriolis force

$$= -2m\omega \times V$$

$$= -2m\omega V(\sin\lambda\hat{j} - \cos\lambda\hat{k})$$

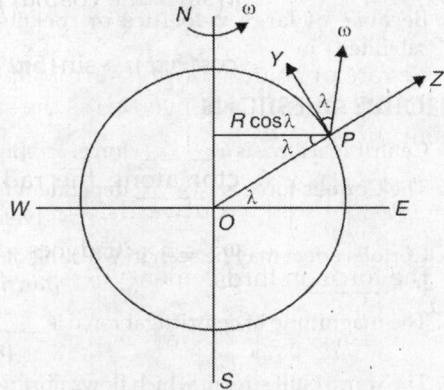

Fig. 1.15

Vertical component of Coriolis force

$$= 2m\omega V \cos\lambda\hat{k}$$

Magnitude of Coriolis force

$$= \sqrt{(2m\omega V \sin\lambda)^2 + (2m\omega V \cos\lambda)^2}$$

$$= 2m\omega V$$

$$= 2 \times 1.5 \times 7.3 \times 10^{-5} \times 100$$

$$= 2.19 \times 10^{-2}\,\text{N}$$

Direction of Coriolis force

$$= \tan^{-1}\left(\frac{\text{Horizontal component}}{\text{Vertical component}}\right)$$

$$= \tan^{-1}\left(\frac{\sin\lambda}{\cos\lambda}\right) = \tan^{-1}(\tan\lambda)$$

$$\lambda = 30°$$

Obviously, net Coriolis force is inclined at an angle of 30° with the vertical towards south.

The deviation produced per second in horizontal direction is

$$= \frac{1}{2}at^2 = \frac{1}{2} \times 2\omega V \sin(30°) \times (1)^2$$

$$= 7.3 \times 10^{-5} \times 100 \times \frac{1}{2}$$

$$= 3.65 \times 10^{-3}\,\text{m towards south}$$

REVIEW QUESTIONS

1. What are non-inertial frames and fictitious forces? Is the centrifugal force fictitious one?

2. Consider a non-inertial frame in translational motion and discuss how Newton's law can be extended to non-inertial frames. Hence explain the meaning of term 'inertial force'.

3. What is Coriolis acceleration? What is the effect of Coriolis force on a freely falling body?

4. Show that a frame rotating with a uniform velocity with respect to an inertial frame with their origins coinciding is not inertial.

5. What is the origin of fictitious forces in uniform rotational motion? Show that Coriolis force owes its existence to the motion of a particle with respect to a rotating frame of reference.

6. Derive an expression for all the fictitious forces acting on a particle moving with a velocity v' in a rotating frame which is itself rotating with a uniform angular velocity ω with respect to another inertial frame F. Identify the term representing Coriolis force.

7. Prove that the Coriolis acceleration with respect to the inertial frame of a particle moving with an instantaneous velocity v with respect to a frame rotating with angular velocity ω is $2\omega \times v$.

PROBLEMS

1. Consider planet Mars as a sphere of radius 1.7×10^3 km. It rotates about its axis in 24 hrs 37 min and acceleration due to gravity on its surface is 3.7 m/s². Find the maximum deflection of the plumb-line from the radial direction on the surface of Mars.

   [**Hint:** The deviation of the plumb-line from vertical is due to centrifugal force. At latitude, the angle of deviation θ from radial direction is given by

   $$\tan\theta = \frac{\sin 2\lambda}{2\left(g/\omega^2 R - \cos^2\lambda\right)}$$

   One obtains $(\tan\theta)_{max} = 0.0011$, $\theta_{max} = 0°\,5'$ (app.)

2. A body is thrown vertically upwards with a velocity V_0. Prove that it will fall back on a point displaced to the west by a distance equal to

    $(4/3)\omega\cos\lambda\sqrt{8h^3/g}$ where $h = V_0^2/2g$

3. Calculate fictitious and total force acting on a body of mass 5 kg relative to a frame moving vertically upwards on earth with an acceleration of 5 m/s². [*Ans.* 7.6 kg wt downwards]

4. A body of mass 5 kg is falling freely. Show that the fictitious force acting on it with reference to a frame moving with downward acceleration of 6 m/s² is 30 N acting upwards.

5. An object, fixed with respect to the surface of a planet identical in mass and radius to the earth, experiences zero gravitational acceleration at the equator. Show that the length of a day on that planet is 8.45 minutes.

    [**Hint:** We may note that in object's frame of reference the force of gravitational attraction due to the planet is just balanced by the centrifugal force,

    $$mR_e\,\omega^2 = \frac{GM_e m}{R_e^2} = mg$$

    or $\qquad \omega = \sqrt{g/R_e}$ and $T = 2\pi/\omega$

6. A pendulum is oscillating along N–S direction at a place in latitude 30° N. How long will it take to start oscillating along NE–SW direction?

    [*Ans.* 6 hrs]

SHORT ANSWER QUESTIONS

1. Explain the physical significance of inertial forces.

Ans. Inertial force makes it possible to take into account the acceleration of a body which is in uniform rectilinear motion relative to non-inertial frame, *i.e.* the acceleration resulting from the non-uniform motion of the system of reference.

2. What will be the effect of centrifugal acceleration on a freely falling body?

Ans. Centrifugal acceleration displace the body slightly from the radial direction toward the south in the northern hemisphere and toward the north in the southern hemisphere.

3. Why does the centrifugal acceleration decrease from the equator to the poles?

Answers

1. (d) 2. (d) 3. (a) 4. (a)

Ans. The centrifugal acceleration decreases from the equator to the poles because the radius of the circle described by the particle decreases as the latitude increases.

4. Why in the northern hemisphere the right-hand rails of railways are worn out faster then the left hand ones?

Ans. Due to Coriolis inertial force.

5. Although the effect of Coriolis force is negligible in the most cases, why it seriously affects the paths of rockets and satellites?

Ans. Because of large velocities of rockets and satellites.

OBJECTIVE QUESTIONS

1. Centrifugal force is a _____ force. [inertia]

2. The Coriolis force is _____ dependent force. [velocity]

3. Coriolis effect may be seen in whirling of wind in a _____ [hurricane]

4. The magnitude of centrifugal force is _____ $[m\omega^2 r]$

5. The warm Gulf stream which flows northwards is deflected towards the _____ due to Coriolis force effect. [east]

MULTIPLE CHOICE QUESTIONS

1. Centrifugal force is a
   (a) gravitational force   (b) viscous force
   (c) frictional force   (d) inertia force

2. Coriolis force is a
   (a) frictional force   (b) gravitational force
   (c) nuclear force   (d) inertia force

3. The Coriolis force is
   (a) velocity dependent force
   (b) gravitational attraction of the moon
   (c) gravitational attraction of sun
   (d) the force which arises due to stationary state of the earth

4. A pendulum is oscillating along the north–south direction at a place in latitude 30° N. Before the pendulum starts oscillating along NE–SW direction, the time elapse is
   (a) 6 hrs   (b) 4 hrs
   (c) 3 hrs   (d) 2 hrs

2

Special Theory of Relativity

2.1 INTRODUCTION

Probably no physical theory in 20th century has been the subject of more discussion amongst philosophers and scientists, and at the same time caught the imagination of the (intelligent) layman, than the Einstein's theory of relativity. This is essentially due to the fact that the concepts underlying the theory of relativity are not only radically new but also provide a framework which embraces practically all the branches of the physical sciences. We shall devote this chapter to the discussion of the special theory of relativity.

2.2 INERTIAL FRAME OF REFERENCE

Classical mechanics is based on the conception that the free space is empty and the position of a point in this free space can be given in terms of three real numbers (x, y, z), which are called the coordinates of a point. Since the position of a point or body can be stated only relative to some other points or bodies, the description of these observations requires a *coordinate system* or a *frame of reference. A* frame of reference is a technical term for the combination of a set of spatial coordinate axes and a time variable.

Galileo and other scientists realised that the form of the laws of nature depends on the choice of the frame of reference. Among all the possible frames of references, there exists a class called the *inertial frame of reference*, in which the laws of nature take a simple form.

Inertial frames of reference are those frames in which a particle or a body that is not acted upon by external forces, move with constant velocity with respect to each other, and if one of them is an inertial frame then the other frame will also be an inertial frame (Fig. 2.1). The laws of mechanics remain invariant in all inertial frames.

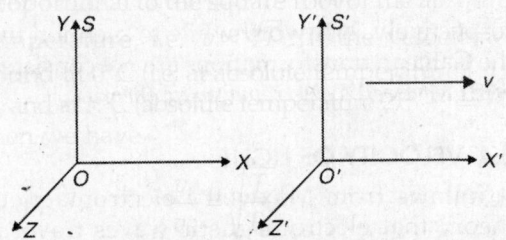

Fig. 2.1: Inertial frames of reference

No frame connected with any material object in the universe can serve the purpose of perfect inertial frame of reference for the earth and other planets as they revolve round the sun which itself is moving. Stars are also not fixed. Relation between the two coordinate systems is determined by the functional relation between their respective coordinates.

2.3 GALILEAN TRANSFORMATIONS

Consider two inertial frames of references S and S', such that their coordinate axes coincide at time $t = 0$, and S' moves with velocity v along the X-axis with respect to frame S.

Obviously, the relations between the coordinates in the two frames can be expressed by the following equations:

$$\left.\begin{array}{l} x' = x - vt \\ y' = y \\ z' = z \\ t' = t \end{array}\right\} \quad (2.1)$$

These equations are called Galilean transformation equations. In obtaining Galilean transformation equations, it is assumed that (i) it is possible to define a time t which is the same for all inertial frames of reference, and (ii) the distance between two points is independent of frames of reference.

It is easy to show from Galilean transformation that the velocities and acceleration in the two frames of reference are related by the equations

$$u' = u - v \quad (2.2)$$

where v is the relative velocity between two frames along the X-direction, and

$$a' = a \quad (2.3)$$

respectively. It is worthwhile to mention that the Galilean transformations are not consistent with *Maxwell's electromagnetic theory*.

2.4 VELOCITY OF LIGHT

It follows from Maxwell's electromagnetic theory that electromagnetic waves travel in vacuum with a speed equal to the ratio of electromagnetic unit to the electrostatic unit of charge. This ratio is essentially equal to the speed of light and hence light is considered as a form of electromagnetic radiation.

Now, the question arises, how does the velocity of light transform from one inertial frame to another? Galilean transformations reveals that the velocities are different in different frames of reference and are related by Eq. (2.2). However, Maxwell's equations for electromagnetic fields have no reference to the velocity of the inertial frame and hence imply that the speed of light is independent of the velocity of inertial frame. Michelson–Morley experiment also suggests that the speed of light is independent of the velocity of light.

2.5 THE SEARCH FOR THE ETHER

In spite of the great success of Maxwell's equations in describing the behaviour of electromagnetic phenomena in space and time, it was quite disturbing to find that they are not covariant under a Galilean transformations. Obviously, the Galilean transformations which permits a covariant description of mechanical forces, does not hold for electromagnetic forces.

Another serious difficulty arose from Maxwell's electromagnetic theory in connection with a medium which, was postulated to sustain wave character of light. We know that every wave motion has something that 'waves'. Sound waves have air and water waves. Surely it was argued that light waves must involve the waving of something even in free space. No one knew what it was, but it was given the name *luminiferous ether*.

Maxwell's equations and experimental observations, particularly on polarized light, show that light is transverse wave motion. This implies that the ether is a solid. Transverse waves involve shear force and can occur only in solids which can support shear. Obviously, the ether must be a rigid solid. The propagation velocity of mechanical waves in various materials depends on the elastic constants of the material. These are much greater for steel than for air. The very great velocity of light thus implies that they must have a very large shear modulus. It is rather hard on the imagination to suppose that all space is filled with this rigid solid and the material objects move through this solid with resistance, yet it was supposed to exist. It was then natural to assume that the ether was in a state of absolute rest, the so-called 'stationary ether'. Accordingly, the ether becomes an *absolute reference* through which the material bodies move without resistance.

If the ether is assumed to be at rest, then the interesting questions is; how fast are we moving through the ether? Since all speculations about the ether stem from its properties as a medium for carrying light, an optical experiment is indicated. It is not very difficult to know how sensitive the apparatus must be required in order to measure the ether drift. Assuming, for the sake of simplicity, that the sun has no ether drift, the velocity of the earth through the ether must be its orbital velocity. If the sun has an ether drift, then the drift of the earth will be given greater than its orbital velocity of some reasons.

Knowing that the radius of the earth's orbit is about 93 million miles, we find the orbital velocity to be about 18.5 miles/sec. (≈ 30 km/sec). By performing the experiment at the best season of the year, we know that we should be able to find an ether drift of at least 30 km/sec. The velocity of light is 3×10^8 m/sec. The orbital velocity of the earth is about 10^{-4} times the velocity of light. Obviously, we need a very sensitive experiment to detect the ether drift. A number of experiments were devised to measure the earth's motion through the ether, i.e. to detect an ether wind with respect to the earth. The most famous of them was the Michelson–Morley experiment which will be described below.

2.6 MICHELSON–MORLEY EXPERIMENT

Michelson, who became the first American to win a Nobel Prize in Physics in 1907, invented a new instrument of unprecedented sensitivity to determine the velocity of the earth through the ether, i.e. a medium for it to move in, then light should be dragged along by this ether as the earth moves along through space. In principle, the ether drift test consists simply of observing whether there is any shift of the interference fringes of light in Michelson interferometer when the entire instrument is turned through an angle of 90°.

The experimental arrangement is shown in Fig. 2.2. A monochromatic beam of light falls on a semi-silvered glass plate P placed at 45° to the beam and is partly reflected and partly transmitted. The reflected portion travels in a direction at right angles to that of the incident beam, falls normally at A on the plane mirror M_1 and is reflected back to P. The transmitted ray, travelling along the direction of the incident rays, falls normally at B on the plane

mirror M_2, and is reflected back to P. The two beams, thus turned to P, interfere on their final journey towards the telescope T so that an interference pattern can be observed and studied. The beam coming from S and entering into T after being reflected from M_1 traverses the plate P thrice while that reflected from M_2 traverses the plate P only once. Thus the optical paths of two interfering beams are not equal. They are made equal by introducing a compensating plate P' of the same thickness and material as P between P and M_2. Whole of the apparatus is floated on mercury, contained in a large vessel so that the interferometer may be rotated in a desired direction.

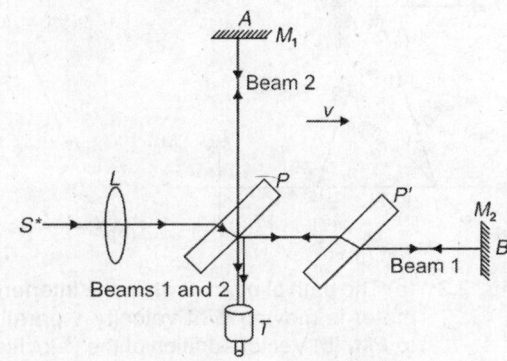

Fig. 2.2: Schematic representation of the Michelson–Morley experiment

If the whole apparatus were at rest in ether, the two rays would take the same time to return to P (distance PA and PB are kept equal). But in the actual experiment the whole apparatus is moving with the earth. Let us assume that the direction of motion of the earth coincides with the direction of the incident beam of light (in the short time of an experiment, this velocity can be considered as uniform and linear, since the earth makes one complete rotation round the sun in one year). On account of motion of the apparatus along with the earth in the direction PM_2 with velocity v relative to the ether, the paths of the two rays and the positions of their reflections from the mirrors will be as shown in Fig. 2.2. After reflection at P the ray proceeds towards the mirror M_1. In the meantime the ray reaches the mirror, it moves to the position M_1' so that the reflection occurs at this position of mirror M_1. If t_1 is time taken by the ray in traversing the path $PM_1'P_1'$, in the meantime plate P will move distance $PP' = vt_1$. From Fig. 2.3, we have

$$PM_1' = \sqrt{PN^2 + M_1'N^2} = \sqrt{\left(\frac{vt_1}{2}\right)^2 + L^2}$$

$\therefore$ Distance $PM_1'P_1' = 2PM_1' = 2\sqrt{L^2 + \dfrac{v^2t_1^2}{4}}$

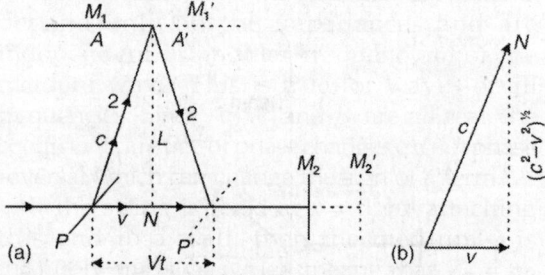

Fig. 2.3: (a) The path of beam 2 while the interferometer is moving wiht velocity v parallel to PM_1 (b) Vector addition of the velocities v and c.

This distance $PM_1'P_1'$ has been traversed by the beam in the ether or absolute or inertial frame, in which the velocity of light is c in all directions. Hence

$$ct_1 = 2\sqrt{L^2 + \frac{v^2t_1^2}{4}}$$

or $\qquad c^2t_1^2 = 4L^2 + v^2t_1^2$

or $\qquad t_1 = \dfrac{2L}{\sqrt{c^2 - v^2}} = \dfrac{2L}{c\sqrt{1 - v^2/c^2}}$

$$= 2L\left(1 - \frac{v^2}{c^2}\right)^{-1/2} = 2L\left(1 + \frac{v^2}{2c^2}\right) \quad (\because v \ll c)\,(2.4)$$

The beam transmitted through P and moving longitudinally towards M_2 has velocity $(c - v)$ relative to the mirror M_2

because the mirror is moving with the velocity of the earth (v). This beam after reflection at mirror M_2 will move path with a velocity $(c + v)$ relative to the plate P, because now the plate is moving opposite to the beam. If t_2 be the total time taken by this ray to return to the plate P, then

$$t_2 = \frac{L}{c - v} + \frac{L}{c + v} = \frac{2Lc}{c^2 - v^2}$$

$$= \frac{2L}{c}\left(1 - \frac{v^2}{c^2}\right)^{-1} = \frac{2L}{c}\left(1 + \frac{v^2}{c^2}\right) \quad (\because v \ll c)\ (2.5)$$

Thus, the difference between the time of travel of the longitudinal and transverse beams is

$$t_2 - t_1 = \frac{2L}{c}\left[\left(1 + \frac{v^2}{c^2}\right) - \left(1 + \frac{v^2}{2c^2}\right)\right]$$

or $\qquad \Delta t = \dfrac{L}{c} \times \dfrac{v^2}{c^2} = \dfrac{Lv^2}{c^3}$ \hfill (2.6)

This will introduce an optical path difference Δ between the two interfering waves, on account of velocity v of the laboratory frame, given by

$$\Delta = c\,(t_2 - t_1) = c\Delta t = L\frac{v^2}{c^2} \quad (2.7)$$

The interference pattern produced by the two components when they recombine will, thus, shift from the position it would have occupied if v were equal to zero. This is the principle of Michelson–Morley experiment. If Δ corresponds to the shifting of n fringes, then

$$\Delta = n\lambda \quad (2.8)$$

where λ is the wavelength of the light used. Equating Eqs. (2.7) and (2.8) for Δ, one obtains

$$n = \frac{c\Delta t}{\lambda} = \frac{Lv^2}{\lambda c^2} \quad (2.9)$$

The path difference Δ is made large by making the total path traversed by the component beams long by multiple reflecting. This way of elongating the paths is more convenient from the point of view of temperature control and mechanical rigidity than a direct increase in the length of the arms of the interferometer. To minimise the error due to vibrations, the whole apparatus is mounted on a block of stone resting on a special wooden ring which floats on mercury.

Now, if the whole apparatus is turned through 90° so that the other arm PM_1 becomes coincident with v. This causes the difference of path in the opposite direction and hence the displacement of fringes should be $2Lv^2/c^2$ corresponding to a path difference 2Δ. This procedure is adopted in experiments, since the different estimation of Δ from Eq. (2.7) involves the determination of the shift of the fringe system from its position under the condition $v = 0$ which cannot be realised. In actual experiment, distance L was taken nearly 11 m, the wavelength of light used was about 6000 Å. Taking $v = 3 \times 10^4$ m/s and $c = 3 \times 10^8$ m/s, one obtains,

$$2\Delta = \frac{2 \times 11 \times \left(3 \times 10^4\right)^2}{6 \times 10^{-7} \times \left(3 \times 10^8\right)^2}$$

$$= \frac{11}{3} \times \frac{1}{10} = 0.37 \text{ of a fringe} \simeq 0.4 \text{ fringe}$$

Thus, a shift of about 0.4 fringe is expected in the fringe pattern on turning the apparatus by 90°. Michelson–Morley experiment was sensitive enough to detect a fringe shift of the order of 0.01 fringe, but in spite of taking into consideration all experimental errors no fringe shift was observed. After Michelson and Morley, the experiment has been repeated with different wavelengths of light, using laser beam, at different places on the earth, at different altitudes, below the surface of the earth, etc. but the same negative result has been obtained every time.

The negative results of Michelson–Morley experiment has following two consequences:

1. The speed of light is the same in all directions.
2. The assumption of the *luminiferous ether* is not true as motion relative to it is undetectable.

Michelson tried to explain the negative result by assuming that the ether is dragged by the moving bodies, so that velocity of light is same in all directions. This idea was in contradiction to the original assumption of the ether as an all-pervasive, frictionless medium. Aberration experiment also contradicts this hypothesis.

Lorentz and Fitzgerald Hypothesis

Lorentz and Fitzgerald independently put forward another explanation to preserve the concept of ether frame. According to this hypothesis, along the direction of motion through the ether, the size of all material bodies contracts (Lorentz–Fitzgerald contraction). Thus, if we suppose that the distance in the direction of the motion of earth is shortened to l', such that

$$l' = l\sqrt{1 - \frac{v^2}{c^2}} \tag{2.10}$$

Thus $\quad t_1 = \dfrac{2L\left(1 - \dfrac{v^2}{c^2}\right)^{1/2}}{c\left(1 - \dfrac{v^2}{c^2}\right)} = \dfrac{2L}{c\sqrt{1 - \dfrac{v^2}{c^2}}} \tag{2.11}$

Thus, t_1 and t_2 will have same value, or

$$\Delta t = t_1 - t_2 = 0$$

No positive experimental confirmation could be provided for Lorentz–Fitzgerald hypothesis. The idea of such a contraction was later found to be correct, although the assumption of the presence of luminiferous ether was not found correct.

ILLUSTRATIVE EXAMPLES

Example 1

In actual Michelson–Morley experiment, the total distance from the partially silvered mirror to each of the two mirrors was 10 meters. The wavelength of the light used was 5000 Å. If the orbital velocity of the earth is taken as 30 km/s, calculate the expected total fringe shift when the apparatus is rotated through 90°.

Solution

Expected fringe shift

$$(\delta) = \frac{2Lv^2}{\lambda c^2}$$

$$L = 10 \text{ m}$$
$$\lambda = 5000 \text{ Å}$$
$$= 5000 \times 10^{-10} \text{ m}$$
$$c = 3 \times 10^8 \text{ m/s}$$
$$v = 30 \times 10^3 \text{ m/s}$$

$$\therefore \quad \delta = \frac{2 \times 10 \times (30 \times 10^3)^2}{5 \times 10^{-7} \times (3 \times 10^8)^2}$$

$$= \frac{2}{5} = 0.4 \text{ fringe.}$$

Example 2

What will be expected fringe shift in Michelson–Morley experiment, if the effective length of each part is 6 meters and wavelength of light used is 6000 Å? Velocity of earth is 3×10^4 m/s.

Solution

Expected fringe shift

$$\delta = \frac{2Lv^2}{c^2\lambda}$$

$$L = 6 \text{ m}$$
$$v = 3 \times 10^4 \text{ m/s}$$
$$c = 3 \times 10^8 \text{ m/s}$$
$$\lambda = 6000 \times 10^{-10} \text{ m}$$

$$\delta = \frac{2 \times 6 \times (3 \times 10^4)^2}{(3 \times 10^8)^2 \times 6 \times 10^{-7}}$$

$$= \frac{1}{5} \text{ of a fringe.}$$

Example 3

In Michelson–Morley experiment the length of the paths of the two beams is 11 meters each. The wavelength of the light used is 6000 Å. If the expected fringe shift is 0.4 fringe, calculate the velocity of earth relative to ether.

Solution

Expected fringe shift

$$\delta = \frac{2Lv^2}{c^2\lambda}$$

$$\delta = 0.4 \text{ fringe}$$
$$L = 11 \text{ m}$$
$$c = 3 \times 10^8 \text{ m/s}$$
$$\lambda = 6000 \times 10^{-10} \text{ m}$$

$$0.4 = \frac{2 \times 11 \times v^2}{(3 \times 10^8)^2 \times 6 \times 10^{-7}}$$

or

$$v^2 = \frac{0.4 \times (30 \times 10^8)^2 \times 6 \times 10^{-7}}{2 \times 11}$$

$$\therefore \quad v = \left[\frac{0.4 \times (30 \times 10^8)^2 \times 6 \times 10^{-7}}{2 \times 11} \right]^{1/2}$$

$$= 3.1 \times 10^5 \text{ m/s}$$

2.7 EINSTEIN'S SPECIAL THEORY OF RELATIVITY

The failure of such efforts as the Michelson–Morley and others to discover a preferred frame for Maxwell's electromagnetic equations suggested that the latter must confirm to a principle of relativity. However, the fact that the Galilean principle of relativity, which was known to be valid for classical mechanics, failed for Maxwell's equations was a considerable source of frustration to physicists at the turn of the century. After a critical examination of the concepts of space, time and simultaneity, Einstein discarded the Galilean principle of relativity and postulated instead a principle of relativity for all physical laws. His postulates may be stated as follows:

i. The laws of physics have the same form in all inertial frames of references moving with a constant velocity with respect to one another. Obviously, all physical phenomena (mechanical, electromagnetic, etc.) proceed in exactly the same way, under the same conditions, in all inertial frames of references or, in other words, it is impossible to ascertain, by means of any experiments whatsoever, conducted in a closed system of bodies, whether the system is at rest or is travelling at uniform velocity in a straight line with respect to some inertial frame of reference.

ii. The speed of light in vacuum is the same for all observers who are in uniform, rectilinear, relative motion and is independent of the motion of the source. Its free space value is the universal constant given by Maxwell's equations.

This postulate follows directly from the results of the Michelson-Morely experiment and many others.

Of these two postulates, the second represents an experimental fact, whereas the first is a generalization from a wide range of physical experience. The first postulate is in no way self-evident, is a hypothesis to be tested by experiment.

The above mentioned postulates are applicable to only inertial frames and due to this speciality the principle of relativity is applied to inertial frames is called *special relativity*. Einstein later generalized *special relativity* to include non-inertial frames also and the generalized theory is called *general relativity*.

2.8 LORENTZ TRANSFORMATIONS

Einstein re-examined the concepts of space and time on the basis of two postulates. An observer at rest at the origin of a coordinate system S can specify any event by giving the coordinates (x, y, z) in S of the place where the event occurs and the time t in S where the event occurs. The same event can be specified by another observer at rest at the origin of a coordinate system S' which is in uniform rectilinear motion relative to S, by giving the coordinates (x', y', z') and the time t' of the event in S'. The algebraic equations connecting two sets of coordinates and time are called the *transformation equations*.

Let us assume for our convenience that origin O and O' of the two inertial frames S

and S' coincide at $t = t' = 0$. Now, consider a light signal emitted in free space at the instant $t = 0$ from the coincident origins of the systems S and S'. When the light signal reaches the point P, let the position and times measured by the observers at O and O' be (x, y, z, t) and (x', y', z', t') respectively (Fig. 2.4). If the velocity of light signal in vacuum is c, then the time required by the light signal to traverse the distance OP in frame S is

$$t = \frac{OP}{c} = \frac{\sqrt{x^2 + y^2 + z^2}}{c}$$

or $\quad x^2 + y^2 + z^2 - c^2 t^2 = 0 \qquad (2.12)$

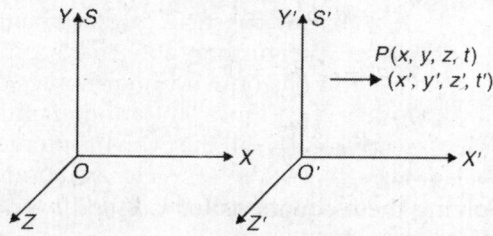

Fig. 2.4: S' is in uniform rectilinear motion to S

The time required by the same light signal in travelling the distance $O'P$ in S' frame, as seen by an observer at rest at the origin of S' is

$$t' = \frac{O'P}{c} = \frac{\sqrt{x'^2 + y'^2 + z'^2}}{c}$$

or $\quad x'^2 + y'^2 + z'^2 + c^2 t'^2 = 0 \qquad (2.13)$

The velocity of light (c) is same in both the inertial frames of reference.

The Galilean transformation connects measurement in the two frames according to Eq. (2.1). Using Eq. (2.1), Eq. (2.13) becomes

$$x^2 - 2xvt + vt^2 + y^2 + z^2 = c^2 t^2$$

and this is not in agreement with Eq. (2.13) in any way. This clearly indicates that the Galilean transformation fails. Therefore, we want new transformation equations which transform Eq. (2.13) to Eq. (2.12). The transformation must be linear in x and t, because we want to get a sphere which expands at a uniform rate. Let us try the relations:

$$x' = k(x - vt)$$
$$y' = y$$
$$z' = z \text{ and } t' = k'(t - bx) \quad (2.14)$$

Here k, k' and b are constants. Making these substitutions in Eq. (2.13), one obtains

$$k^2(x^2 - 2xvt + v^2t^2) + y^2 + z^2 = c^2k'^2 \times$$
$$(xt^2 + 2bxt + b^2x^2)$$

or $(k^2 - k'^2c^2b^2)x^2 - 2(k^2v + k'^2c^2b)xt + y^2z^2$

$$-\left(k'^2 - \frac{k^2v^2}{c^2}\right)c^2 t^2 = 0$$

This must be identical to Eq. (2.12). Equating the above with Eq. (2.12), one obtains

$$k^2 - k'^2b^2c^2 = 1$$

$$k^2v^2 - k'^2bc^2 = 0$$

$$k'^2 - \frac{k^2v^2}{c^2} = 1$$

Solving these equations for k, k' and b, one obtains

$$k = k' = \frac{1}{\sqrt{1 - \dfrac{v^2}{c^2}}}$$

and $$b = vc^2$$

The new transformation equations which are essentially based on the invariance of the velocity of light, is then

$$x' = k(x - vt) = \frac{(x - vt)}{\sqrt{1 - \dfrac{v^2}{v}}}$$
$$y' = y$$
$$z' = z$$
$$t' = k(t - bx) = \frac{\left(t - \dfrac{vx}{c^2}\right)}{\sqrt{1 - \dfrac{v^2}{c^2}}} \quad (2.15)$$

This set of transformation equations is known as the *Lorentz transformation*. These

equations are symmetrical and retain their form in transformation from S' to S except that the sign of v is changed. Thus,

$$x = \frac{x' + vt'}{\sqrt{1 - \dfrac{v^2}{c^2}}}$$
$$y = y', z = z'$$
$$t = \frac{\left(t' + \dfrac{vx}{c^2}\right)}{\sqrt{1 - \dfrac{v^2}{c^2}}} \quad (2.16)$$

These are *inverse transformations*.

The Lorentz transformations are linear and at low velocities $\left(\dfrac{v}{c} \ll 1\right)$, they become Galilean transformations.

From Lorentz transformation equations, one can easily confirm the statement that events which happen at the same place at different times, as viewed from one frame, may be seen from another frame to happen at a different place as well. Similarly, a difference in spatial position with respect to one frame may correspond to a difference in both space and time with respect to another frame. Thus a space difference can be converted partly into a time difference or *vice versa*, merely by changing the frame of reference that is being used. Thus Lorentz transformations represent a most profound conceptual change specially with regard to space and time. Writing,

$$\frac{v}{c} = b \text{ and } \left(1 - \frac{v^2}{c^2}\right)^{-1/2} = \gamma \quad (2.17)$$

Lorentz transformation equations can be expressed as

$$x' = \gamma(x - vt)$$
$$y' = y$$
$$z' = z$$
$$t' = \gamma\left(t - \frac{vx}{c^2}\right) \quad (2.18)$$

To understand the consequences of Lorentz transformations, we define an event, which is an occurrence taking place at a point (x, y, z) and at a time t. Thus, a point in the space–time complex, i.e. (x, y, z, t) is an event. If we consider two events in S defined by (x_1, y_1, z_1, t_1) and (x_2, y_2, z_2, t_2), then the two events are called simultaneous when $\Delta t = t_2 - t_1 = 0$, they are called *colocal* when $\Delta x = x_2 - x_1 = 0$, $\Delta y = y_2 - y_1 = 0$ and $\Delta z = z_2 - z_1 = 0$ and when the two events take place at the same position $(\Delta x = \Delta y = \Delta z = 0)$ and at the same time, i.e. $\Delta t = 0$, they are called *coincident*. The Lorentz transformations reveal that if two events are colocal in one inertial frame they may not be so in another frame (relativity of colocality) and if two events are simultaneous in one inertial frame they may not be observed at the same time in another inertial frame. This is called the *relativity of simultaneity*.

However, if two events are coincident in one inertial frame, they will be coincident in all inertial frames. One can easily check the above results with the help of following relations. These relations have been obtained by differentiating Eq. (2.15).

$$\left.\begin{aligned}\Delta x' &= \frac{\Delta x - v\Delta t}{\sqrt{1 - v^2/c^2}}\\[2mm]\Delta y' &= \Delta y\\[1mm]\Delta z' &= \Delta z\\[2mm]\Delta t' &= \frac{\Delta t - \dfrac{v\Delta x}{c^2}}{\sqrt{1 - \dfrac{v^2}{c^2}}}\end{aligned}\right\} \quad (2.19)$$

and

Obviously, if $\Delta x = \Delta y = \Delta z = 0$ but $\Delta t \neq 0$, then $\Delta x'$, $\Delta y'$ and $\Delta z'$ are not zero; if $\Delta t = 0$ but $\Delta x, \Delta y, \Delta z \neq 0$, then $\Delta t' = 0$; if $\Delta x = \Delta y = \Delta z = 0$ and also $\Delta t = 0$, then $\Delta x' = \Delta y' = \Delta z' = 0$ and $\Delta t' = 0$.

2.9 CONSEQUENCES OF LORENTZ TRANSFORMATION

Length Contraction

Consider a rod lying along the X-axis of a inertial frame S. Let the coordinates of the ends of the rod be x_1 and x_2 (Fig. 2.5). Then

the length L_0 of the rod in the inertial frame S in which it is at rest is given by

$$L_0 = x_2 - x_1 \qquad (2.20)$$

Let S' be another inertial frame with its X-axis along that of S and Y- and Z-axes parallel to the corresponding axes of S. Let S' be moving with a velocity v relative S to along X-axis. Let x_1' and x_2' be the coordinates of the ends of the rod as measured in S'. Then by Lorentz transformation equations, one obtains

$$x_1 = \frac{x_1' + vt'}{\sqrt{1 - \dfrac{v^2}{c^2}}}$$

and

$$x_2 = \frac{x_2' + vt'}{\sqrt{1 - \dfrac{v^2}{c^2}}} \qquad (2.21)$$

Subtracting one from the other, we obtain

$$x_2 - x_1 = L_0 = \frac{x_2' - x_1'}{\sqrt{1 - \dfrac{v^2}{c^2}}} = \frac{L}{\sqrt{1 - \dfrac{v^2}{c^2}}} \qquad (2.22)$$

where L is the length of the rod as measured in S'.

$$\therefore \qquad L = L_0 \sqrt{1 - \frac{v^2}{c^2}} \qquad (2.23)$$

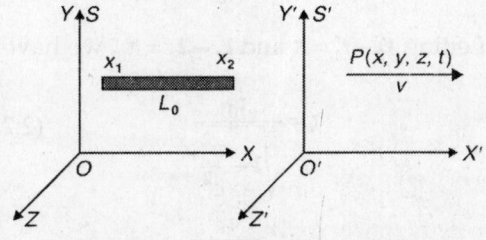

Fig. 2.5: Length contraction

Obviously, $L < L_0$. Thus, for an observer in motion with respect to an object, the object appears to be shorter in the direction of motion than when he is at rest. It is important to note that Lorentz–Fitzgerald contraction effect is mutually reciprocal for two different observers. The Lorentz contraction may be observed, but not 'seen' (since the eye and the instantaneous cameras record pictures formed photons that arrive together).

Time Dilation

Consider a clock at a point x' in an inertial frame S'. Let S' be moving with a velocity v with respect to another inertial frame S along the common X-axis. Let a clock be situated in frame S at x. Let clock in S frame give out signals at two instants of time t_1 and t_2 as measured by an observer in S. Let an observer in frame S' measure these instants of time t'_1 and t'_2 with his own clock. Then from Lorentz transformation equations, we obtain

$$\left.\begin{array}{c} t'_1 = \dfrac{t_1 - \dfrac{vx}{c^2}}{\sqrt{1 - \dfrac{v^2}{c^2}}} \\[2em] t'_2 = \dfrac{t_2 - \dfrac{vx}{c^2}}{\sqrt{1 - \dfrac{v^2}{c^2}}} \end{array}\right\} \quad (2.24)$$

Subtracting one from the other, we obtain

$$t'_2 - t'_1 = \frac{t_2 - t_1}{\sqrt{1 - \dfrac{v^2}{c^2}}} \quad (2.25)$$

Letting $t'_2 - t'_1 = \tau$ and $t_2 - t_1 = \tau_0$, we have

$$\tau = \frac{\tau_0}{\sqrt{1 - \dfrac{v^2}{c^2}}} \quad (2.26)$$

or
$$\tau_0 = \tau \sqrt{1 - \frac{v^2}{c^2}} \quad (2.27)$$

For the observer in frame S, the time interval between the two signals is $\tau_0 = t_2 - t_1$ and for the observer in S' the time interval between the same two signals is $\tau = t'_2 - t'_1$. Equation (2.26) reveals that τ is larger than τ_0. Obviously, for the moving observer, the time interval appears to be *elongated* or *dilated*. This phenomenon is called *time dilation*.

2.10 PROPER FRAME, PROPER LENGTH AND PROPER TIME

The inertial frame of reference in which the observed body is at rest is called the proper frame of reference. The length of a rod as measured in the inertial frame in which it is at rest is called the proper length. The relation between the *proper length* (L_0) and the apparent or non-proper length (L) is as follows:

$$L = L_0 \sqrt{1 - \frac{v^2}{c^2}} \quad (2.28)$$

The time interval recorded by a clock fixed with respect to the observed event is called the proper time interval. The relation between the proper time (τ_0) and apparent or non-proper time (τ) is as follows:

$$\tau = \frac{\tau_0}{\sqrt{1 - \dfrac{v^2}{c^2}}} \quad (2.29)$$

Obviously, proper time noted by a moving observer is always less than the corresponding interval of time in a stationary frame. A stationary observer finds that a moving clock runs slower than a stationary one. An event which repeats itself with a certain period in S' will appear to have a longer period when observed from S. It is important to note that this effect is mutually reciprocal between two observers.

The relation (2.29) suggests that any physical process occurring in S' appears, when viewed from S, to have slowed down compared with an identical process occurring in S. Thus, spectrum of radiation emitted by atoms moving relative to a spectroscope will appear shifted towards the longer wavelength or the red side of the corresponding spectrum emitted by atoms which are at rest relative to the spectroscope. This effect due to time dilation has to be considered while considering *Doppler shift in radiation*.

2.11 EXPERIMENTAL VERIFICATION OF TIME DILATION

μ-Meson Decay

Time dilation has been verified in experiments on a nuclear particle, called μ-meson. Fast moving μ-mesons are created in the cosmic rays at a height of about 10 kilometres from the surface of the earth and reach the earth in large numbers. These μ-mesons have a typical speed of 2.994×10^8 m/s, which is 0.998 of the speed of light c. A μ-meson is found to have an average lifetime of 2×10^{-6} s after which it decays into an electron. Obviously, a μ-meson in its lifetime can travel a distance of only 2.994×10^{-8} m/s $\times 2 \times 10^{-6}$ s $\simeq 600$ m or 0.6 km. Then the question arises, how do μ-mesons travel a distance of 10 km to reach the earth? Rossi and Hall in 1941 attributed this result to the time-dilation effect. The μ-meson has lifetime t_0 ($\approx 2 \times 10^{-6}$ s) in its own frame of reference. In the observer's frame of reference on the earth, however, the lifetime is lengthened owing to the relative motion, to the value t given by

$$t = \frac{t_0}{\sqrt{1 - \dfrac{v^2}{c^2}}} = \frac{2 \times 10^{-6}}{\sqrt{1 - (0.998)^2}}$$

$$= \frac{2 \times 10^{-6}}{0.063} = 3.17 \times 10^{-5} \text{ s}$$

In 3.17×10^{-5} s, a meson whose speed is $0.998\, c$ (= 2.994×10^8 m/s) can travel a distance

$$2.994 \times 10^8 \times 3.17 \times 10^{-5} \simeq 9500 \text{ m} \simeq 9.5 \text{ km}.$$

Hence, despite their brief lifetime it is possible for the μ-mesons to reach the ground from the large altitudes at which they are actually formed. More recently, the dilation caused by the thermal vibration of the nuclei in certain crystals has also been verified.

Example 4

What will be the apparent length of a meter stick measured by an observer at rest, when the stick is moving along its velocity equal to $\dfrac{\sqrt{3}}{2} c$.

Solution

We have
$$L = L_0 \sqrt{1 - \frac{v^2}{c^2}}$$

$$v = \frac{\sqrt{3}}{2} c, \; L_0 = 1 \text{ m}$$

∴
$$L = 1 \sqrt{1 - \frac{3}{4}} = \sqrt{\frac{1}{4}} = 0.5 \text{ m}$$

Example 5

A light pulse is emitted at the origin of a frame of reference S' at time $t' = 0$. Its distance x' from the origin after a time t' is given by $x'^2 = c^2 t'^2$. Use the Lorentz transformations to transform this equation to an equation in x and t and show that this is $x^2 = c^2 t^2$. Discuss the implication of this result.

Solution

$$x'^2 = c^2 t'^2$$

From Lorentz transformations, we have

$$x' = \frac{x - vt}{\sqrt{1 - \dfrac{v^2}{c^2}}} \text{ and } t' = \frac{t - \dfrac{xv}{c^2}}{\sqrt{1 - \dfrac{v^2}{c^2}}}$$

$$= \frac{(x - vt)^2}{1 - \dfrac{v^2}{c^2}} = \frac{c^2 \left\{ t - \dfrac{xv}{c^2} \right\}^2}{1 - \dfrac{v^2}{c^2}}$$

or
$$(x - vt)^2 = c^2 \left(t^2 + \frac{x^2 v^2}{c^4} + \frac{2xtv}{c^2} \right)$$

or
$$x^2 + v^2 t^2 - 2xvt = c^2 t^2 + \frac{x^2 v^2}{c^2} - 2xvt$$

or
$$x^2 - c^2 t^2 + v^2 t^2 - \frac{x^2 v^2}{c^2} = 0$$

or
$$(x^2 - c^2 t^2) - \frac{v^2}{c^2}(x^2 - c^2 t^2) = 0$$

or
$$(x^2 - c^2 t^2)\left(1 - \frac{v^2}{c^2}\right) = 0$$

Since $v \neq c$ and $\left(1 - \dfrac{v^2}{c^2}\right) \neq 0$

$\therefore$ $\qquad x^2 - c^2t^2 = 0$

$\therefore$ $\qquad x^2 = c^2t^2$

Obviously, the above result shows that the velocity of light is an absolute constant and independent of the frame of reference.

Example 6

Calculate the length and the orientation of a rod of length 5 m in a frame of reference which is moving with a velocity equal to $0.6c$, in a direction making in angle of 30° with the rod.

Solution

The proper length of the rod along the direction of motion of the frame is $L_{x0} = 5 \cos 30°$ m, and that perpendicular to this direction is $L_{y0} = 5 \sin 30°$ m.

The apparent length along the moving frame is given by

$$L_x = L_{x0} \sqrt{1 - \frac{v^2}{c^2}}$$

$$v = 0.6\,c$$

$$L_{x0} = 5\cos 30° = \frac{5\sqrt{3}}{2}$$

$$L_x = \frac{5\sqrt{3}}{2}\sqrt{1 - (0.6)^2}$$

$$= \frac{5\sqrt{3}}{2} \times 0.8 = 2\sqrt{3}\ \text{m}$$

The length perpendicular to the direction of motion remains the same, i.e.

$$L_y = L_{y_0} = 5\sin 30° = \frac{5}{2} = 2.5\ \text{m}.$$

$\therefore$ Length observed in the moving frame

$$L = \sqrt{\left(L_x^2 + L_y^2\right)} = \sqrt{\left(12 + \frac{25}{4}\right)} = 4.27\ \text{m}$$

If the rod appears to make an angle θ with the direction of motion, then

$$\tan\theta = \frac{L_y}{L_x} = \frac{\dfrac{5}{2}}{2\sqrt{3}} = \frac{5}{4\sqrt{3}} = 0.72$$

Example 7

Calculate the mean life of a burst of π-mesons travelling with $v = 0.3c$, if the proper mean lifetime is 2.5×10^{-7} s?

Solution

We have $\qquad \tau = \dfrac{\tau_0}{\sqrt{1 - \dfrac{v^2}{c^2}}}$

$$v = 0.73c$$

$$\tau_0 = 2.5 \times 10^{-7}\ \text{s}$$

$$= \frac{2.5 \times 10^{-8}}{\sqrt{1 - (0.73)^2}}$$

$$= \frac{2.5 \times 10^{-8}}{\sqrt{0.4671}} = 3.6 \times 10^{-8}\ \text{s}.$$

Example 8

Prove that $x^2 + y^2 + z^2 = c^2t^2$ is invariant under Lorentz transformations.

Solution

Using Lorentz transformations, we have

$$x = \frac{x' + vt'}{\sqrt{1 - \dfrac{v^2}{c^2}}}, y = y', z = z'$$

and $\qquad t = \dfrac{t' + \dfrac{x'v}{c^2}}{\sqrt{1 - \dfrac{v^2}{c^2}}}$

On making these substitutions in the given expression, $x^2 + y^2 + z^2 = c^2t^2$, we obtain

$$\frac{(x' + vt')^2}{1 - \dfrac{v^2}{c^2}} + y'^2 + z'^2 = \frac{c^2\left[x' + \dfrac{x'v}{c^2}\right]^2}{1 - \dfrac{v^2}{c^2}}$$

or $\dfrac{x'^2 + v^2 t'^2 + 2x't'v}{1 - \dfrac{v^2}{c^2}} + y'^2 + z'^2$

$= \dfrac{c^2 t'^2 + \dfrac{x'^2 v^2}{c^2} - 2x'vt'}{1 - \dfrac{v^2}{c^2}}$

or $y'^2 + z'^2 = \dfrac{1}{1 - \dfrac{v^2}{c^2}} \left[\left(c^2 t'^2 - x'^2 \right) \left(1 - \dfrac{v^2}{c^2} \right) \right]$

$= c^2 t'^2 - x'^2$

$\therefore \; x'^2 + y'^2 + z'^2 = c^2 t'^2$

Thus $x^2 + y^2 + z^2 - c^2 t^2 = x'^2 + y'^2 + z'^2 - c^2 t'^2$

Clearly $x^2 + y^2 + z^2 - c^2 t^2 = 0$ is invariant under Lorentz transformations.

Example 9

The length of a rocket is measured as 100 meters before launching. During the flight apparent length is found to be 96 meters when measured from the launching station. Calculate its speed.

Solution

Proper length of the rocket $L_0 = 100 \, \text{m}$
Apparent length $L = 96 \, \text{m}$

$\because \qquad L = L_0 \sqrt{1 - \dfrac{v^2}{c^2}}$

$\therefore \qquad 96 = 100 \sqrt{1 - \dfrac{v^2}{c^2}}$

or $\qquad 1 - \dfrac{v^2}{c^2} = \left(\dfrac{96}{100} \right)^2 = \left(\dfrac{24}{25} \right)^2 = \dfrac{576}{625}$

or $\qquad \dfrac{v}{c} = \sqrt{1 - \dfrac{576}{625}} = \sqrt{\dfrac{49}{625}} = \dfrac{7}{25}$

$\therefore \qquad v = \dfrac{7}{25} c = \dfrac{7}{25} \times 3 \times 10^8$

$= 8.4 \times 10^7 \, \text{m/s}.$

Example 10

μ-mesons with a proper life of 2.2 μs formed at the top of the earth's atmosphere have a velocity of 0.998 c. What is the vertical distance they travel in their mean lifetime before decaying as observed by a person on the surface of the earth?

Solution

Apparent lifetime as observed by the observer on the earth's surface

$$\tau = \dfrac{\tau_0}{\sqrt{1 - \dfrac{v^2}{c^2}}} = \dfrac{2.2 \, \mu s}{\sqrt{1 - (0.998)^2}}$$

$\therefore$ Vertical distance travelled= speed × time

$$= \dfrac{0.998 \times 3 \times 10^8 \times 2.2 \times 10^{-8}}{\sqrt{1 - (0.998)^2}}$$

$$= 164670 \, \text{m} = 164.67 \, \text{km}$$

Example 11

The average lifetime of a neutron as a free particle at rest is 15 minutes. It disintegrates spontaneously into an electron, proton and neutrino. What is the average minimum velocity with which a neutron must leave the sun in order to reach the earth before breaking up? Take the distance of earth from sun as 11×10^{10} m.

Solution

Let v be the velocity of the neutron. The average lifetime t of the moving neutron, as measured by an observer on earth would be

$$t = \dfrac{t_0}{\sqrt{1 - \dfrac{v^2}{c^2}}}$$

where t_0 is the proper lifetime of the neutron at rest. The time t is also the time for reaching the neutron from sun to earth before decay. Therefore,

$$t = \dfrac{d}{v} = \dfrac{11 \times 10^{10}}{v}$$

$$\therefore \quad \frac{11\times10^{10}}{v^2} = \frac{15\times60}{\sqrt{1-\dfrac{v^2}{c^2}}}$$

Squaring, we obtain

$$v^2 = \frac{121\times10^{20}\times9}{850\times10^4}$$

or $\qquad v = 1.13 \times 10^8 \text{ m/s}$

2.12 RELATIVISTIC VELOCITY TRANSFORMATION EQUATIONS

An observer in the stationary system would define velocity in his system as $U_x = \left(\dfrac{x_2 - x_1}{t_2 - t_1}\right)$, i.e. an interval of distance divided by an interval of time. Similarly, an observer in the moving inertial frame would define the velocity as $U'_x = \left(\dfrac{x'_2 - x'_1}{t'_2 - t'_1}\right)$. If an observer in S frame is to observe phenomenon in S' and wishes to express U' in terms of his position and time system, he would have to use the velocity transformation equations. Let us derive these relativistic velocity transformation equations. We have by definition

$$U_x = \frac{dx}{dt}, \quad U_y = \frac{dy}{dt}, \quad U_z = \frac{dz}{dt}$$

and $\quad U'_x = \dfrac{dx'}{dt'}, \quad U'_y = \dfrac{dy'}{dt'}, \quad U'_z = \dfrac{dz'}{dt'} \quad$ (2.30)

From the Lorentz transformation equations, we have $dy' = dy, dz' = dz$

$$dx' = \frac{dx - vdt}{\sqrt{1-\dfrac{v^2}{c^2}}} = \gamma(dt - vdt)$$

and $\quad dt' = \dfrac{dt - \dfrac{vdt}{c^2}}{\sqrt{1-\dfrac{v^2}{c^2}}} = \gamma\left(dt - \dfrac{vdx}{c^2}\right)$

$$\therefore \quad U'_x = \frac{dx'}{dt'} = \frac{\gamma(dx - vdt)}{\gamma\left(dt - \dfrac{vdx}{c^2}\right)}$$

or $\quad U'_x = \dfrac{\dfrac{dx}{dt} - v}{1 - \dfrac{v}{c^2}\dfrac{dx}{dt}} = \left(\dfrac{U_x - v}{1 - \dfrac{v}{c^2}U_x}\right) \quad$ (2.31)

Similarly, one obtains

$$U'_y = \frac{U_y}{\gamma\left(1 - \dfrac{vU_x}{c^2}\right)} \quad (2.32)$$

and $\quad U'_z = \dfrac{U_z}{\gamma\left(1 - \dfrac{vU_x}{c^2}\right)} \quad$ (2.33)

Equations (2.31) to (2.33) constitute the relativistic velocity transformation equations. It is important to note that the velocity components which are transverse to the relative motion of the reference frame, depend upon the relative motion but transverse distances do not.

From Eq. (2.31), it is clear that the denominator is less than 1. Obviously, the velocity as measured in the stationary system is less than $U'_x + v$, as we expect from classical theory. This reveals that if observed from rest, moving objects in a moving system are relatively slowed.

From Eq. (2.31) if $U_x = c$, we obtain $U'_x = c$. Obviously, the velocity of light represents maximum observable velocity for a physical particle.

By interchanging, the primed and unprimed quantities and writing $-v$ in place of v, inverse velocity transformation equations can be written as

$$U_x = \frac{(U'_x + v)}{\left(1 + \dfrac{vU_x}{c^2}\right)} \quad (2.34)$$

$$U_y = \frac{U'_y\sqrt{1 - \dfrac{v^2}{c^2}}}{\sqrt{1 + \dfrac{Uv'_x}{c^2}}} \quad (2.35)$$

and $\quad U_z = \dfrac{U_z'\sqrt{1 - \dfrac{v^2}{c^2}}}{\sqrt{1 + \dfrac{vU_x'}{c^2}}}$ $\qquad$ (2.36)

The relativistic velocity transformation equations take the form as $c \to \infty$, i.e.

$$U_x' = U_x, \; U_y' = U_y \text{ and } U_z' = U_z$$

Example 12

A spaceship moving away from the earth with velocity $0.5\,c$ fires a rocket whose velocity relative to spaceship is $0.8\,c$ (a) away from the earth (b) towards the earth. What will be the velocity of the rocket as observed from the earth in the two cases?

Solution

Let us regard the earth as S frame and spaceship as S' frame. The rocket is the object whose velocity in the S frame is to be determined.

We have $U = \dfrac{U' + v}{1 + \dfrac{U'v}{c^2}}$

$$U' = 0.8\,c$$
$$v = 0.5\,c$$

$\therefore \quad U = \dfrac{0.8\,c + 0.5\,c}{1 + 0.8 \times 0.5}$

In the second case, $U' = -0.8\,c$

$\therefore \quad U = \dfrac{U' + v}{1 + \dfrac{U'v}{c^2}} = \dfrac{-0.8\,c + 0.5\,c}{1 + \{(-0.8) \times (0.5)\}}$

$$\dfrac{-0.3\,c}{1 - 0.4} = -0.5\,c$$

Example 13

In the laboratory frame two particles are observed to travel in opposite directions with speed 2.8×10^8 m/s. Deduce the relative speed of the particles.

Solution

We have $\quad U_x' = \dfrac{U_x - v}{1 - \dfrac{U_x'v}{c^2}}$

$$U_x = 2.8 \times 10^8 \text{ m/s}$$
$$v = -2.8 \times 10^8 \text{ m/s}$$
$$c = 3 \times 10^8 \text{ m/s}$$

$$U_x' = \dfrac{(2.8 \times 10^8) - (-2.8 \times 10^8)}{1 - \dfrac{(2.8 \times 10^8)(-2.8 \times 10^8)}{(3 \times 10^8)^2}}$$

$$= 2.99 \times 10^8 \text{ m/s}$$

Obviously, the velocity of the first particle relative to the second is 2.99×10^8 m/s.

Example 14

A radioactive nucleus while moving with a velocity $0.2\,c$ in the lab-frame emits a β-particle. The β-particle moves with a speed $0.6\,c$ relative to the nucleus. What is the velocity and direction of the β-particle if it is emitted in a direction (a) parallel (b) perpendicular to the direction of motion of the nucleus in the lab-frame?

Solution

Let us consider that the frame of reference fixed on nucleus be S' and the lab-frame be S. The direction of motion of the nucleus be the X-direction.

(a) $\qquad U_x = \dfrac{U_x' + v}{1 + \dfrac{U'v}{c^2}}$

$$U_x' = 0.6c$$
$$v = 0.2c$$
$$U_y' = U_z' = 0$$

$\therefore \qquad U_x = \dfrac{0.6c + 0.2c}{1 + 0.12}$

$$= \dfrac{0.8c}{1.12} = 0.174c$$

β-particle is emitted in X-direction

$$U_y = \dfrac{U_y'\sqrt{1 - \dfrac{v^2}{c^2}}}{1 + \dfrac{U_x'v}{c^2}} = 0$$

Similarly, $\quad U_z = 0$.

Hence, in the laboratory-frame the β-particle will move in X-direction with a velocity 0.714c.

(b) When β-particle is emitted in Y-direction

$$U'_x = 0, \; U'_y = 0.6c, \; U'_z = 0; \; v = 0.2c$$

$$\therefore \quad U_x = \frac{U'_x + v}{1 + \dfrac{U'_x v}{c^2}} = \frac{0 + 0.2c}{1 + 0} = 0.2c$$

$$U_y = \frac{0.6c\sqrt{1-(0.2)^2}}{1+0} = 0.6c\sqrt{0.9c}$$

$$= 0.588c$$

$$U_z = 0$$

Resultant velocity $U = \sqrt{U_x^2 + U_y^2}$

$$= c\sqrt{0.04 + 0.36 \times 0.96}$$

$$= 0.621c$$

The direction of U in the XY-plane will be

$$\alpha = \tan^{-1}\frac{U_y}{U_x} = \tan^{-1}\frac{0.588c}{0.2c}$$

$$= \tan^{-1}(2.94) = 721°$$

2.13 RELATIVISTIC MOMENTUM

In classical mechanics, linear momentum, $p = mv$ is conserved in a system of particles not acted upon by outside forces. When an event, e.g. collision or an explosion occurs outside an isolated system, the vector sum of the momenta of the particles before the event is equal to their vector sum afterwards. Now, the question arises, whether this is valid in inertial frames in relative motion, and if not, what relativistic correction is required?

To begin with, let us consider an elastic collision (i.e. a collision in which KE is conserved) between two particles A and B as observed by observers in frames S and S' which are in uniform relative motion. The properties of A and B are identical when determined in reference frames in which they are at rest. Figure 2.6 shows that S' is moving in the +X direction with respect to S at velocity v.

Before the collision, the particle A had been at rest in frame S and particle B in frame S'.

Then, at the same instant, A was shown in the +Y direction at the speed v_A while B was thrown in the –Y' direction at the speed v'_B, where

$$v_A = v'_B \tag{2.37}$$

Obviously, the behaviour of A as seen from S is exactly the same as the behaviour of B as seen from S' frame.

Now, when A and B collide, A rebounds in the –Y direction at the speed v_A, while B rebounds in the +Y' direction at the speed v'_B. If the particles A and B are thrown from positions Y apart, an observer in S finds that the collision occurs at $y = \frac{1}{2}Y$ and one in S' finds that it occurs at $y' = y = \frac{1}{2}Y$. The round trip time T_0 for A as measured in frame S is given by

$$T_0 = \frac{Y}{v_A} \tag{2.38}$$

Obviously, T_0 is same for B in frame S', i.e.

$$T_0 = \frac{Y}{v'_B}$$

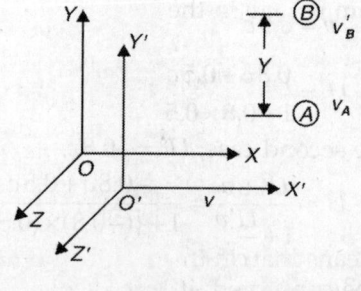

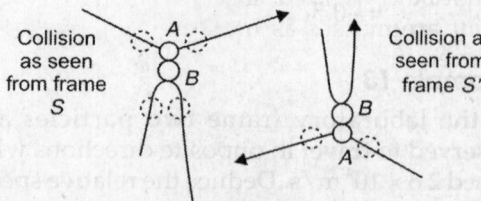

Collision as seen from frame S

Collision as seen from frame S'

Fig. 2.6: An elastic collision as observed in S and S' frames, where S' is moving with velocity **v** with respect to S. The balls A and B are initially Y apart, which is same in S and S' since S' moves only in X-direction.

The speed v_B in frame S is found to be

$$v_B = \frac{Y}{T} \qquad (2.39)$$

where T is the time required for B to make its round trip as measured in frame S. However, in frame S', B's trip requires the time T_0, where

$$T = \frac{T_0}{\sqrt{1 - \dfrac{v^2}{c^2}}} \qquad (2.40)$$

Although observers in both frames see the same event, they disagree about the length of time the particle thrown from the other reference frame requires to make the collision and return.

Now, replacing T in Eq. (2.39) with its equivalent in terms of T_0, one obtains

$$v_B = \frac{Y\sqrt{1 - \dfrac{v^2}{c^2}}}{T_0}$$

From Eq. (2.38),

$$v_A = \frac{Y}{T_0}$$

Now, using the classical definition of momentum $p = mv$ in the frame S,

$$p_A = m_A v_A = m_A \left(\frac{Y}{T_0}\right)$$

and

$$p_B = m_B v_B = m_B \sqrt{1 - \frac{v^2}{c^2}} \left(\frac{Y}{T_0}\right)$$

This means that, in this frame, momentum will not be conserved, if $m_A = m_B$, where m_A and m_B are masses as measured in frame S. However, if

$$m_B = \frac{m_A}{\sqrt{1 - \dfrac{v^2}{c^2}}} \qquad (2.41)$$

then momentum will be conserved.

In the collision shown in Fig. 2.6, both A and B are moving in both frames. Now, suppose that v_A and v'_B are small compared to v, the relative velocity of two frames. In this case,

an observer in S will see B approach A with the velocity v, make a glancing collision (since $v'_B \ll v$), and then continue. In the limit of $v'_B = 0$, if $m(v)$ is the mass of B in S, which is moving at the velocity v, then $m_B = m(v)$. Thus, Eq. (2.41) becomes

$$m(v) = \frac{m}{\sqrt{1 - \dfrac{v^2}{c^2}}} \qquad (2.42)$$

Now, if linear momentum is defined as

$$p = \frac{mv}{\sqrt{1 - \dfrac{v^2}{c^2}}} \qquad (2.43)$$

the conservation of momentum is valid in special theory of relativity. When $v \ll c$, Eq. (2.43) becomes just $p = mv$ the classical momentum, as required. Equation (2.43) is often written as

$$p = \gamma m v \qquad (2.44)$$

where

$$\gamma = \frac{1}{\sqrt{1 - \dfrac{v^2}{c^2}}} \qquad (2.45)$$

The principle of conservation of linear momentum in special theory of relativity is sometimes called as *Lorentz principle of conservation of linear momentum in special theory of relativity*.

2.14 RELATIVITY OF MASS

In classical physics, the mass of a particle is taken to be same at all velocities, i.e. independent of the velocity of the particle, i.e. when masses are involved we say that a constant force produces a constant acceleration. In relativistic mechanics, however, a constant force does not produce a linear increase in velocity. As the velocity increases, the factor $\gamma = \sqrt{1 - \dfrac{v^2}{c^2}}$ becomes significantly different from 1 and the same force produces a smaller increase in velocity, it is because the mass has increased. In fact, one of the consequences of the relativity is that mass increases with increase in velocity. We know that the principle of conservation of momentum is a fundamental law of physics and it must hold good in all inertial frames of references, under Lorentz transformations. We find that mass is not absolute but varies with velocity in such a manner that it makes the principle of conservation of momentum hold good in all inertial frames. An expression for the variation in mass with velocity can be derived as follows:

Consider a collision between two bodies of equal mass in an inertial frame S'. For simplicity, let us assume that the bodies are moving with velocities U' and $-U'$ along the X-axis and after collision they coalesce into one body. Let the mass of each body as observed in S' frame be m'. After collision the coalesced body will have mass $2m'$. From the principle of conservation of momentum, it will be at rest in S'.

Let S be another inertial frame such that S' is moving with a uniform relative velocity v in the X-direction with respect to S. The speeds of two bodies will not appear to be equal for an observer in frame S. If the velocities of the two bodies are observed as U_1 and U_2 in the frame S, then by the relativistic law of addition of velocities

$$U_1 = \frac{U'+v}{1+\dfrac{U'v}{c^2}} \text{ and } U_2 = \frac{-U'+v}{1-\dfrac{U'v}{c^2}} \quad (2.46)$$

If we now assume that mass is variable, the mass of the two bodies, in frame S will appear to be different. Let m_1 be the mass of the body moving with a velocity U_1 and m_2 be the mass of the body moving with a velocity U_2. After collision as the coalesced body is at rest in frame S', it will appear to be moving with a velocity v in frame S and its mass will be $(m_1 + m_2)$. Figure 2.7 depicts the situation as observed in the two frames of references. From the principle of conservation of momentum, we have

$$m_1 U_1 + m_2 U_2 = (m_1 + m_2)v \quad (2.47)$$

or $$m_1(U_1 - v) = m_2(v - U_2) \quad (2.48)$$

$\therefore$ $$\frac{m_1}{m_2} = \frac{v - U_2}{U_1 - v} \quad (2.49)$$

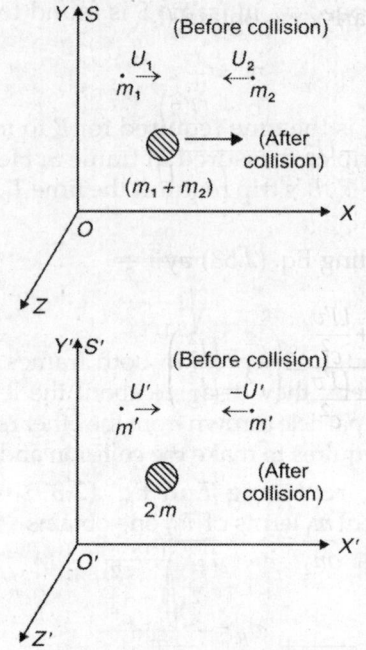

Fig. 2.7: Relativity of mass

Substituting the values of U_1 and U_2 from Eq. (2.46), we have

$$\frac{m_1}{m_2} = \frac{v - \dfrac{-U'+v}{1-\dfrac{U'v}{c^2}}}{\dfrac{U'+v}{1+\dfrac{U'v}{c^2}} - v} = \frac{1+\dfrac{U'v}{c^2}}{1-\dfrac{U'v}{c^2}} \quad (2.50)$$

Further from Eq. (2.46), we have

$$c^2 - U_1^2 = c^2 - \frac{U'+v}{1+\dfrac{U'v}{c^2}}$$

or $$c^2 - U_1^2 = c^2 - \frac{U'^2 + v^2 + 2U'v}{1 + \dfrac{U'^2 v^2}{c^4} + \dfrac{2U'v}{c^2}} \quad (2.51)$$

$$= \frac{\left(c^2 - U'^2\right)\left(1 - \dfrac{v^2}{c^2}\right)}{\left(1 + \dfrac{U'v}{c^2}\right)^2} \quad (2.52)$$

Similarly, we obtain (by replacing U' with $-U'$),

$$c^2 - U_2^2 = \frac{(c^2 - U'^2)\left(1 - \dfrac{v^2}{c^2}\right)}{\left(1 + \dfrac{U'v}{c^2}\right)^2} \qquad (2.53)$$

Dividing Eq. (2.53) by Eq. (2.52), we obtain

$$\frac{1 + \dfrac{U'v}{c^2}}{1 - \dfrac{U'v}{c^2}} = \left(\frac{c^2 - U_2^2}{c^2 - U_1^2}\right)^{1/2}$$

or

$$\frac{m_1}{m_2} = \frac{\left(1 - \dfrac{U_1^2}{c^2}\right)^{1/2}}{\left(1 - \dfrac{U_2^2}{c^2}\right)^{1/2}} = \frac{1 + \dfrac{U'v}{c^2}}{1 - \dfrac{U'v}{c^2}} \qquad (2.54)$$

or

$$m_1 \sqrt{1 - \frac{U_1^2}{c^2}} = m_2 \sqrt{1 - \frac{U_2^2}{c^2}} \qquad (2.55)$$

Generalising, we obtain

$$m \sqrt{1 - \frac{v^2}{c^2}} = \text{constant} \qquad (2.56)$$

For a particle at rest, i.e. $v = 0$, if its mass is m_0, then

$$m \sqrt{1 - \frac{v^2}{c^2}} = m_0 (1 - 0) = m_0$$

or

$$m = \frac{m_0}{\sqrt{1 - \dfrac{v^2}{c^2}}} \qquad (2.57)$$

Obviously, the mass of a body moving with speed v is always greater than its rest mass m_0. As v approaches c, the effective mass increases without limit (Fig. 2.8). No body with finite rest mass m_0 can quite achieve the speed of light, when $\dfrac{v}{c} \ll 1$, the difference between m and m_0 become insignificant.

As v approaches c, the observed mass approaches infinity (∞). The idea of infinite mass seems ridiculous. It actually means that

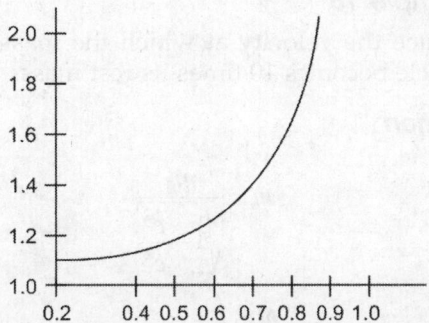

Fig. 2.8: Variation of mass with velocity

the velocity of the material body (whose rest mass m_0 is not zero) can never become equal to the velocity of light in vacuum, it has to be always less than it.

In 1909, Bucher determined the ratio e/m as a function of velocity for fast electrons emitted from radioactive nuclei and showed that the mass of the electron varies with velocity according to Eq. (2.57).

Example 15

With what velocity should a particle move so that the increase in its mass may be 25% of its rest mass?

Solution

We have

$$m = \frac{m_0}{\sqrt{1 - \dfrac{v^2}{c^2}}}$$

If m_0 is the rest mass of the particle, then its mass at the required velocity v is

$$m = m_0 + \frac{25}{100} m_0 = \frac{5}{4} m_0$$

$$\therefore \qquad \frac{5}{4} m_0 = \frac{m_0}{\sqrt{1 - \dfrac{v^2}{c^2}}}$$

or

$$1 - \frac{v^2}{c^2} = \frac{16}{25}$$

or

$$\frac{v^2}{c^2} = \frac{9}{25}$$

or

$$v = \frac{3}{5} c = 0.6 c$$

Example 16

Deduce the velocity at which the mass of a particle becomes 10 times its rest mass.

Solution

$$m = \frac{m_0}{\sqrt{1 - \dfrac{v^2}{c^2}}}$$

$$\frac{m}{m_0} = 10$$

We have $\dfrac{1}{\sqrt{1 - \dfrac{v^2}{c^2}}} = 10$

or $1 - \dfrac{v^2}{c^2} = \dfrac{1}{100}$

or $\dfrac{v^2}{c^2} = 1 - \dfrac{1}{100} = \dfrac{99}{100} = 0.99$

or $v = \sqrt{0.99}, c = 0.995 \times 3 \times 10^8$ m/s
$\qquad\qquad = 2.98 \times 10^8$ m/s

2.15 MASS–ENERGY EQUIVALENCE

The experimentally verified mass transformation equation has important consequences. In classical mechanics, we considered that kinetic energy could be increased only by increasing its velocity, but now dealing with relativistic mechanics we take mass variation into account. The relationship between mass and energy may be derived as follows:

If a force F, acting on a particle, produces a displacement dx, then the work done by the force = Fdx. The work done must be equal to the gain in the kinetic energy dE of the particle, i.e. $dE = Fdx$

But force, being rate of change of linear momentum of the particle, is given by

$$F = \frac{d}{dt}(mv) = m\frac{dv}{dt} + v\frac{dm}{dt}$$

(according to the theory of relativity)

$$dE = m\frac{dv}{dt}dx + \frac{dm}{dt}dx$$

$$= mvdv + v^2dm \quad \left(\because \frac{dx}{dt} = v\right) \qquad (2.58)$$

Now, from relativistic mass transformation relation

$$m = \frac{m_0}{\sqrt{1 - \dfrac{v^2}{c^2}}}$$

or $\qquad m^2 = \dfrac{m_0^2 c^2}{\left(c^2 - v^2\right)}$

or $\qquad m^2c^2 - m^2v^2 = m_0^2c^2$

Differentiating, we obtain

$$c^2 2mdm - (m^2 2vdv + v^2 2mdm) = 0$$

Dividing throughout by $2m$, one obtains

$$c^2dm - (mvdv + v^2dm) = 0$$

or $\qquad mvdv + vdm = c^2dm \qquad (2.59)$

Comparing Eqs. (2.58) and (2.59), one obtains

$$dE = c^2dm \qquad (2.60)$$

Equation (2.60) shows that in relativity a change in kinetic energy can be expressed in terms of mass as variable. This equation is valid for a change in any form whatever, although it was derived here for change in kinetic energy.

Since KE is zero, when $v = 0$, then it is also when $m = m_0$. Therefore, on integrating Eq. (2.60), one obtains

or $\qquad E = c^2 \displaystyle\int_{m_0}^{m} dm = c^2(m - m_0) \qquad (2.61)$

or $\qquad E = mc^2 - m_0c^2 \qquad (2.62)$

Equation (2.62) is the relativistic expression for kinetic energy. For $v \ll c$, it reduces to the classical expression. Putting $m = \dfrac{m_0}{\sqrt{1 - \dfrac{v^2}{c^2}}}$ in Eq. (2.62), one obtains

$$E = \frac{m_0c^2}{\sqrt{1 - \dfrac{v^2}{c^2}}} - m_0c^2$$

For $v \ll c$, we have

$$\left(1-\frac{v^2}{c^2}\right)^{-1/2}=\left(1-\frac{v^2}{2c^2}+\frac{3}{8}\frac{v^4}{c^4}+\cdots\right)$$

$$\therefore \quad E=m_0c^2\left(1-\frac{v^2}{2c^2}+\frac{3}{8}\frac{v^4}{c^4}+\cdots-1\right)$$

$$=\frac{1}{2}m_0v^2+\frac{3}{8}\frac{v^4}{c^2} \qquad (2.63)$$

The first term of Eq. (2.63) is the classical expression. Obviously, it is the only significant term at low velocities.

The equivalence between energy and mass proves to be universal. The best examples of conversion of mass into energy are the nuclear fission, nuclear fusion and nuclear reaction. Another most familiar example of mass energy equivalence is the phenomenon of pair production.

Example 17

What is the annual loss in the mass of the Sun, if the earth receives heat energy approximately $2\,\text{cal}/\text{cm}^2/\text{min}$. The earth–sun distance is about 150×10^6 km.

Solution

Rate of energy radiated

$$= 2 \times 4.2 \times 10^7 \text{ erg/cm}^2/\text{min}$$

Total energy radiated per minute

$$= 4\pi \times (150 \times 10^{11})^2 \times 2 \times 4.2 \times 10^7$$

Energy radiated per year

$$= 4\pi \times (150 \times 10^{11})^2 \times 2 \times 4.2 \times 10^7 \times 5.3 \times 10^5$$

Annual loss in mass is $\dfrac{E}{c^2}$

$$=\frac{4\pi\times(150\times10^{11})^2\times2\times4.2\times10^7\times5.3\times10^5}{9\times10^{20}}$$

$$= 1.4 \times 10^{14} \text{ tons per year}$$

Example 18

A nucleus of mass m emits gamma ray of frequency v_0. Show that the loss of internal energy by the nucleus is not hv_0 but it is

$$hv_0\left[1+\frac{hv_0}{2mc^2}\right].$$

Solution

The momentum of γ-ray photon is $p=\dfrac{hv_0}{c}$.

According to the law of conservation of momentum, the nucleus having mass m will recoil with the momentum $\dfrac{hv_0}{c}$ in the backward direction. Therefore, the loss of energy of the nucleus in the recoiling is

$$E=\frac{1}{2}mv^2=\frac{m^2v^2}{2m}=\frac{p^2}{2m}=\frac{(hv_0)^2}{2mc^2}$$

Energy of the emitted photon from the nucleus $= hv_0$

$\therefore$ Clearly, the total loss of internal energy of the nucleus

$$hv_0\frac{(hv_0)^2}{2m}=hv_0\left(1+\frac{hv_0}{2m}\right)$$

2.16 RELATION BETWEEN MOMENTUM AND ENERGY

The relativistic momentum of a particle moving with a velocity v is given by

$$p = mv \qquad (2.64)$$

where $m=\dfrac{m_0c^2}{\sqrt{1-\dfrac{v^2}{c^2}}}$, m_0 being the rest mass of the particle.

From relativity, we have

$$E = mc^2 \qquad (2.65)$$

From Eqs. (2.64) and (2.65), we have

$$E^2 - c^2p^2 = m^2c^4 - c^2m^2v^2$$

$$=\frac{m_0^2c^4}{1-\dfrac{v^2}{c^2}}-\frac{m_0^2}{1-\dfrac{v^2}{c^2}}c^2v^2$$

$$= m_0^2 c^4 \left[\frac{1 - \dfrac{v^2}{c^2}}{1 - \dfrac{v^2}{c^2}} \right]$$

$$= m_0^2 c^4$$

or $\qquad E^2 = c^2 p^2 + m_0^2 c^4 \qquad (2.66)$

2.17 PARTICLES WITH ZERO REST MASS

Photon and *graviton* are the familiar examples of particles with zero rest mass. A particle with zero rest mass always moves with the speed of light in vacuum.

According to Eq. (2.60), if $m_0 = 0$, we have

$$E = pc \qquad (2.67)$$

But relativistic momentum is given by

$$p = mv = v \frac{E}{c^2} \text{ (as } E = mc^2) \quad (2.68)$$

Hence, from Eqs. (2.67) and (2.68), we have

$$p = v \frac{E}{c^2} = v \frac{pc}{c^2}$$

or $\qquad v = c \qquad (2.69)$

i.e. a particle with zero rest mass always moves with the speed of light in vacuum.

The velocity of the particle observed in some other inertial frame S' is

$$U' = \frac{U - v}{1 - \dfrac{Uv}{c^2}}$$

where v is the velocity of the frame S' with respect to the frame S in which the velocity of the particle is v. Hence if $U = c$, we have

$$U' = \frac{c - v}{1 - \dfrac{cv}{c^2}} = c$$

Clearly, the particle will have the same speed c and zero rest mass for all observers in inertial frames.

2.18 SPEED LIMIT FOR MATERIAL PARTICLES

Equations (2.23) and (2.26) show that for $v = c$, a body's length contracts to zero and any time interval between events on it would dilate into eternity as measured by an observer in a frame relatively at rest. The mass increase relation (2.57) tells us that the mass of a body whose velocity equals c would become infinite and so would its energy. These consequences are clearly absurd and leads to the conclusion that the velocity of light in vacuum is the maximum possible velocity and that no material body can acquire this velocity. However, when light traverse in matter, its velocity reduces to $\dfrac{c}{\mu}$ where μ is the refractive index of the material, and is always greater than one. P A Cerenkov found that in transparent materials it is possible for particles like electrons to travel faster than light through them. This effect is called *Cerenkov effect*.

The possibility of a particle velocity exceeding the free space velocity of light is not considered as this makes particle mass imaginary. In 1962, Bilaniuk and Sudershan proposed that particles of imaginary mass might exist if it is assumed that they always travel at a speed faster than that of light. Feinberg in 1967 named these particles as *tachyons*. So far these particles have not been detected.

2.19 SPACE AND TIME IN RELATIVITY

Space and time are no longer separate concepts in relativity but are interwoven in such a way that makes the introduction of four dimensional space–time continuum very useful. In such a four dimensional space many of the laws of classical physics can be expressed in a particularly elegant form which greatly simplifies the shift from one inertial system to another.

Intervals

Four dimensional space is an imaginary concept of a space having four dimensions on whose axes the three coordinates x, y and z and the time t are plotted. Any event is represented by a point in four dimensional space is called *world point*. The motion of a certain particle in space and time is represented by a line in four dimensional space, then the quantity

$$S_{12} = \sqrt{c^2(t_2 - t_1)^2 - (x_2 - x_1)^2 - (y_2 - y_1)^2 - (z_2 - z_1)^2}$$

(2.70)

is called the *interval* between the two events.

The intervals between two infinitesimally close events is

$$ds = \sqrt{c^2 dt^2 - dx^2 - dy^2 - dz^2} \quad (2.71)$$

$$= \sqrt{c^2 dt^2 - dl^2}$$

where $dl^2 = dx^2 + dy^2 + dz^2$

If the variable $\tau = ict$ (where $i = \sqrt{-1}$) is plotted instead of t along the time axis, the value $-ds^2 = dx^2 + dy^2 + dz^2 + d\tau^2$ can be regarded as a square of an element in *four dimensional space*.

The interval between two events is the same in all inertial frames of reference (invariance of a four dimensional interval). The invariance of an interval is a mathematical expression of the invariance of the velocity of light.

If x_1, y_1, z_1, t_1 and x_2, y_2, z_2, t_2 are the world points of two events in certain frame of reference Σ, then under the condition that $S_{12}^2 > 0$ (realness of the interval), there exists a frame of reference Σ' in which both events take place at a single point in space ($x_1^1 = x_2^1, y_1^1 = y_2^1$ and $z_1^1 = z_2^1$). Real intervals are said to be the *time-like*. The time $t_{12} = t_2^1 - t_1^1$ that elapses between the two events in frame Σ', is equal to $t_{12}^1 = \dfrac{S_{12}}{c}$.

The condition

$$S_{12}^2 = c^2 t_{12}^2 - l_{12}^2 = \text{constant} > 0$$

where $t_{12} = t_2 - t_1$ and $l_{12}^2 = (x_2 - x_1)^2 + (y_2 - y_1)^2 + (z_2 - z_1)^2$, can be represented graphically as hyperbolas if $l = l_{12}$ and $ct = ct_{12}$ are plotted along the coordinate axes (Fig. 2.9), where l_{12} and t_{12} correspond to the two given events in an arbitrary inertial frame of reference. Points A and A_1 correspond to event 2 which occurs at the same point in space as event 1 (point 0), but either later (point A) or earlier (point A_1).

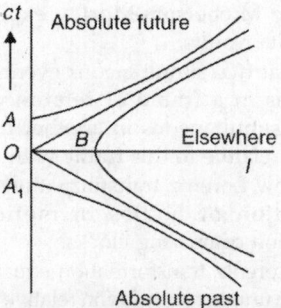

Fig. 2.9

For a *time-like* interval it is always possible to find an inertial frame in which the two events occur at the same place. However, it is impossible to find an inertial frame in which the two events occurs simultaneously since in such a frame S_{12}^2 would be negative.

There are events between which the interval, or separation is imaginary.

Thus,

$$S_{12}^2 = c^2 t_{12}^2 - l_{12}^2 < 0$$

Imaginary separations are said to be *space-like*. For any space-like interval, it is always possible to find some inertial frame in which the two events are *simultaneous*, but it is impossible to find a frame in which the events occur at the same place. Clearly the *space-like*, *time-like* character of an interval is independent of the inertial frame of reference.

The 'locus' of zero intervals in four dimensional space (x, y, z, ict) is called the *light cone* for the given *world point* 0.

REVIEW QUESTIONS

1. Describe Michelson–Morley experiment and show how the negative results obtained are interpreted.

2. Discuss the Michelson–Morley experiment to detect the motion of earth through ether. Give the difference in interpretation for the negative result as given by Einstein and Lorentz–Fitzgerald.

3. Show that the Lorentz–Fitzgerald contraction can account for the null result of the Michelson–Morley experiment.

4. State postulates of special theory of relativity and deduce them from the Lorentz transformations.

5. Describe Michelson–Morley experiment and discuss its results.

6. Show that two simultaneous events at different positions in a frame of reference are not in general simultaneous in another inertial frame moving relative to this frame with velocity v.

7. Show how Lorentz transformations account for contraction of bodies in motion and the retardation of moving clocks.

8. Derive Lorentz transformation equations for two frames in uniform translation relative to each other.

9. What is time dilation in special relativity? Explain it by Lorentz transformations. Describe briefly one experiment in support of time dilation.

10. Derive the formula giving the variation of mass with velocity.

11. Write a note on Einstein mass–energy relation.

12. How does the law of conservation of momentum lead to the relativistic relation of variation of mass with velocity?

13. What do you understand by space–time continuum? Show that the Lorentz transformation result from the rotation of the four-vector coordinate frame.

14. (a) Using principle of special theory of relativity derive mass–energy relation. Illustrate it with an example.

    (b) Calculate % contraction in length of a rod moving with velocity $0.8c$ in the direction inclined to $45°$ to its own length.

15. (a) State the postulates of the special theory of relativity and derive expression for velocity transformation.

    (b) A laboratory observation shows that a length of 2 metre is crossed by a beam of radioactive particle in time 1×10^{-8} second this process half of the particles disintegrate. Calculate the proper half-life of these particles and traversed length as seen by an observer and traversed length as seen by an observant fixed relative to the particle.

16. Write short note on mass–energy relation.

17. (a) Define an inertial frame of reference and derive the Lorentz transformations.

    (b) Calculate the velocity of a particle having kinetic energy three times the rest mass energy.

    (c) Prove that the particle having rest mass zero is moving always with velocity of light.

18. (a) Deduce Lorentz transformations.

    (b) Derive an expression for the variation of mass with velocity.

(c) Compute the mass m of an electron having kinetic energy 1.5 MeV. Given rest mass $m_0 = 9.11 \times 10^{-31}$ kg and velocity of light $c = 3 \times 10^8$ m/s.

19. Answer the following:

    (a) A stationary body explodes into two fragments of rest mass 1.5 kg each moving apart at speeds of $0.8c$. Find rest mass of the body.

    (b) Twins A and B are 20 years each. Twin A travels towards a star 30 lights years away at a speed $0.8c$. He then returns home. Twin B remains at the home. Why are the two observations different?

    (c) Find relativistic expression for the energy of a particle moving with momentum p. Is it true for a photon ?

20. (a) What is length contraction and time dilation in special theory of relativity? Explain.

    (b) In Michelson–Morley experiment the optical path of each beam was 11 m and wavelength of the light used was 5500 Å. Calculate the fringe shift. Velocity of earth is 30 km/s. Make the relevant diagram.

21. Derive mass–energy relation. Show that total energy E and momentum p are related, where m_0 is the rest mass and c is the speed of light.
$$E^2 = (pc)^2 + (m_0c^2)^2$$

PROBLEMS

1. Calculate the fringe shift in Michelson–Morley experiment. Given effective length of each path is 10 m, velocity of earth 3×10^4 m/s and wavelength of light used is 6000 Å.
   [*Ans.* 0.33 fringe]

2. In a Michelson–Morley experiment the optical path of each beam was 11 m and the wavelength of the light used was 5500 Å. Assuming $\dfrac{v}{c}$ to be 10^4. calculate the fringe shift that would have to be obtained if the ether existed.[*Ans.* 0.4 fringe]

3. In an experiment similar to the Michelson–Morley experiment the distance d of either mirror from the beam splitter is 25.9 m. The wavelength of the light used us 5890 Å and the apparatus is capable of detecting a time difference equal to one-hundredth of the period of this light. If no effect is observed when the apparatus is rotated by 90°, what is the maximum possible value of velocity of apparatus relative to the ether?
   [*Ans.* 3.25×10^3 m/s]

4. A meter stick lies at rest in the *XY*-plane at an angle of 45° with the *X*-axis. What are the length and orientation of the stick to an observer moving at relative velocity v in the *X*-direction?

5. In the laboratory the lifetime of a particle moving with speed 2.8×10^8 m/s is found to be 2.5×10^{-7} s. Calculate the proper life time of the particle.
[*Ans.* 8.9×10^{-8} s]

6. A man leaves the earth in a rocketship that makes a round trip to the nearest star which is 4 light year away at a speed of $0.8c$. How much younger will he be on his return than his twin brother who preferred to stay behind? [*Ans.* 3.4 years]

7. It the galaxy has a diameter of 10^5 light-years, what will be the diameter appear to be to cosmic ray particle travelling at the relative speed $\dfrac{v}{c} = 0.99$? How long will the trip take as measured by a clock riding with the particle?

8. Two particles are moving in opposite directions with speeds $0.9c$ as observed in laboratory frame. What is the velocity of one particle relative to the other? [*Ans.* $0.994c$]

9. L_0^3 is the rest volume of a cube. It is viewed from a reference frame moving with a uniform velocity v parallel to an edge of the cube, show that the apparent volume will be $L_0^3 = \sqrt{1 - \dfrac{v^2}{c^2}}$.

10. A cosmic ray μ-meson has a half-life 2.2×10^6 s in a reference frame in which the meson is at rest. What will be its half-life as measured by an observer on the earth, if it approaches the earth at a speed of $0.999c$? [*Ans.* 45.5×10^{-6} s]

11. A radioactive atom is moving with respect to the laboratory at a speed of $0.3c$ in the *X*-direction. If it emits an electron having a speed of $0.8c$ in the rest frame of the atom, find the velocity of the electron with respect to a laboratory observer when
(a) It is ejected in the *X*-direction
(b) It is ejected in the negative *X*-direction
(c) It is ejected in the *Y*-direction.

12. A beam of 10^4 π+-meson travels in a circular path of radius 20 m at a speed of $0.99c$. Proper mean lifetime of the π+-meson is 2.5×10^{-8} s. How many survive when the beam returns to the starting point after one revolution? [*Ans.* 925]

SHORT ANSWER QUESTIONS

1. How Lorentz transformation follows directly from the postulates of the special theory of relativity?

Ans. Any physical law which retains the same mathematical form under a Lorentz transformation automatically satisfies the principle of relativity.

2. What are the two important kinematical effects which derive from the special theory of relativity?

Ans. (i) The contraction of length parallel to the relative motion and (ii) Dilation of time.

3. What do you understand by Carenkov radiation?

Ans. Very energetic particles are capable of moving faster in a material medium (water, glass, etc.) than the velocity of light in that medium, though never faster than the velocity of light in free space. This phenomenon gives rise to the light waves known as Cerenkov radiation.

4. What do you understand by Tachyon?

Ans. Tachyon is a particle theoretically predicted and always travels with a speed greater than that of light. However, the existence of Tachyon has not been established so far.

5. What is the important dynamical effect of special theory of relativity?

Ans. An important dynamical effect of special theory of relativity is the increase of mass with increase in velocity, which leads to the statement of the equivalence of mass and energy, i.e. $E = mc^2$.

6. What is proper time interval?

Ans. The proper time interval is the time interval measured by a clock attached to the observed body. Proper time interval (dt') noted by a moving observer is always less than the corresponding interval of time in a stationary frame.

$$dt' = dt \sqrt{1 - \frac{v^2}{c^2}}.$$

OBJECTIVE QUESTIONS

1. The speed of light is same in all directions is the consequence of the negative results of _____experiment. [Michelson–Morely]

2. The laws of the physics have the _____ in all inertial frames of reference moving with a constant velocity with respect to one another. [same form]

3. Lorentz transformations are _____ and retain their form in transformation from inertial frame to inertial frame *S* except that the sign of relative velocity v of frames changed. [symmetrical]

4. Length contraction is one of the consequences of _____. [Lorentz transformations]

5. The phenomenon of time dilation reveals that for the moving observer the time interval appears to be _____. [elongated or dilated]

6. Momentum p can be expressed relativistically in terms of energy E as _____.

$$\left[p = \frac{\sqrt{E^2 - E_0^2}}{c} \text{ where } E_0^2 = m_0^2 c^2 \right]$$

7. Both the energy E and the momentum p of a particle have different values in different reference frames. However, the quantity $E^2 - p^2 c^2$ is_____and has the same value in different reference frames. [invariant]

8. In order to satisfy the postulates of special theory of relativity we have to replace the Galilean transformations by_____.

[Lorentz transformations]

9. The principle of constancy of speed of light postulates that the speed of light in vacuum, as measured by any inertial reference frame, is c regardless of the motion of the_____relative to that reference frame. [light sources]

MULTIPLE CHOICE QUESTIONS

1. An electron has a velocity $0.99c$. Its energy will be
   (a) 3.1 MeV   (b) 0.31 MeV
   (c) 31 MeV   (d) 0.031 MeV

2. A metre rod is moving parallel to its length. When its apparent mass becomes twice the rest mass, its apparent length will be
   (a) 0.5 m   (b) 1 m   (c) 1.5 m   (d) 0.1 m

3. A rod travels with a speed $v = 0.8c$ along its length. In accordance with special theory of relativity, the rod will shrink, i.e. l/l_0 will be (where symbols have their usual meaning)

   [**Hint:** $l/l_0 = \sqrt{1 - \frac{(0.8c)^2}{c^2}} = 0.6$]

   (a) 0.1   (b) 0.3   (c) 0.6   (d) 0.9

4. The total energy of a particle is exactly equal to its rest energy. The velocity of the particle in ms^{-1} is
   (a) 2.598×10^8 (b) 1.2×10^8
   (c) 3×10^8 (d) 0.8×10^8

5. The energy that can be obtained from complete annihilation of 1 gm of mass in joules is
   (a) 1×10^{14} (b) 3×10^{13}
   (c) 6×10^{13} (d) 9×10^{13}

Answers

1. (a)   2. (a)   3. (c)   4. (b)   5. (d)

3

Waves

3.1 INTRODUCTION

Basically, there are two ways of transporting energy from one place to another place: the first involves the actual transport of matter, e.g. a bullet fixed from a gun carries its kinetic energy with it which can be used at another place, and the second method involves a *wave process*. The wave carries the energy but in this process there is no transfer of matter. For example, when a drummer beats a drum its sound is heard at distant points. Obviously, the sound carries the energy as it can move the diaphragm of the ear. Light waves also carry energy, with the help of light waves one can transmit an electric signal (or a message) from one point to another. Clearly, these various processess of transport of energy from one place to another place are different, but they have a common feature which is usually termed as *wave process*. This method is much more useful and important.

We may note that in wave motion the key word is *disturbance* or *perturbation*. There are various ways of disturbing the physical state of a body and accordingly there are variety of wave motions, e.g. water waves in an organ pipe, waves in a stretched string, light waves, etc.

In a wave motion, the oscillations of the particles are similar but a *phase difference* exists between the oscillations of the different particles. In an elastic medium, the wave motion is generated by the tensile or the elastic forces between the particles of the medium.

Such waves propagating from one point to another in a material medium due to its elasticity are termed *elastic waves*. We may note that sound waves are also elastic waves.

A material medium is not necessary if some physical property of space can exist in vacuum. We know that electric and magnetic fields can exist in vacuum. Obviously, disturbance in this case could be a perturbation (or a change) in these fields. This distance travels in vacuum and needs no material medium for its propagation, e.g. light waves.

3.2 TYPES OF WAVES

In general, there are two types of wave motion:

1. Transverse Waves

When the particles of the medium vibrate at right angles to the direction of propagation of the wave, the wave is said to be a *transverse wave*. Waves along a stretched string, water waves, etc. are transverse waves.

Electromagnetic waves (which include light waves too) are transverse waves. However, in the case of electromagnetic waves the disturbance that wave travels is not a result of the vibration of material particles but oscillation of electric and magnetic fields which occur at right angles to the propagation of wave.

2. Longitudinal Waves

When the particles of the medium vibrate parallel to the direction of propagation of the wave, the wave is said to be *longitudinal wave*.

Sound waves in solids, liquids and gases, compressional waves along a string, etc. are well known examples of longitudinal waves.

We may note that in a transverse wave, the layers of the medium vibrate tangentially with respect to each other, and so one layer exerts a *shearing force* to the adjacent layer. However, such forces exist in solids and to some extent in liquids but not at all in the gases. This means, transverse waves cannot be produced in a gas. On the other hand, all material media are more or less compressible, and therefore, longitudinal waves can propagate through solids, liquids and gases.

In a wave propagation, the disturbance is a function of both *position* and *time*. If disturbance at a point is periodic function of time, the wave is said to have *time periodicity*. Now, if at a given instant of time, the disturbance has the same value at equidistant points on the path of the wave, the wave has a space periodicity as well. The waves having both time and space periodicity are called *periodic waves*. We may note that in actual practice, the magnitude of disturbance decays gradually resulting in a change of shape of disturbance. From the study point of view, *periodic waves of constant type*, i.e. which have no *attenuation*, i.e. waves which do not change shape and have a constant magnitude, form the basis of the study.

3.3 PERIODIC WAVES OF CONSTANT TYPE

Distance (space)–displacement curve of a periodic wave of period T is shown in Fig. 3.1. The space– displacement curve gives the

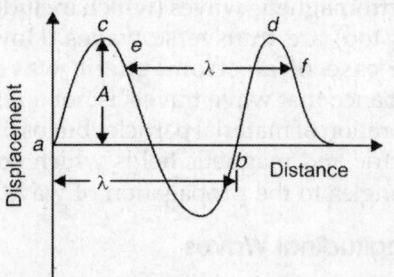

Fig. 3.1: Space–displacement curve for a periodic wave

shape or the form of the wave, and is known as *waveform* or the *wave profile*. We may note that for a wave of constant type, the waveform does not change.

The maximum displacement of particle on the path of the wave is called the *amplitude* and shown by A in the Fig. 3.1.

The time T needed for the generation of a complete wave, i.e. a full cycle of oscillation of a particle on the line of propagation, is called *period* of the wave. Period T can also be defined as the time needed for a vibrating particle on the line of propagation to undergo a complete cycle of oscillation.

The *frequency* (n or v) of a wave is defined as the number of complete waves produced in the medium per second. Obviously, the frequency of a wave is the number of complete oscillations per second of a vibrating particle on the line of propagation. We have

$$n = 1/T \text{ hertz (Hz)}$$

The *wavelength* denoted by λ *is the shortest distance between any two particles on the path of a wave having the same phase of vibration at any instant of time.* We may note that the period of a wave gives the time periodicity and the wavelength gives the space periodicity of the wave.

The period (T), frequency (n), wavelength (λ) and the velocity of propagation of a wave are related as follows:

$$\text{Velocity, } (v) = \frac{\text{Distance}}{\text{Time}} = \frac{\lambda}{T}$$

or $\qquad\qquad v = n\lambda \qquad\qquad$ (3.1)

This very important relation between wavelength, frequency and wave velocity holds for transverse as well as longitudinal waves.

The velocity of propagation of an unchanging waveform is referred to as *wave velocity* or *phase velocity*. Consider that a particle on the line of propagation at x_2 has at a latter instant of time t_2 the same phase of vibration as another particle at x_1 nearer to the source at an earlier instant of time t_2. Obviously, a given phase has propagated from

x_1 to x_2 in time $(t_2 - t_1)$. Thus, the phase velocity is

$$c = \frac{x_2 - x_1}{t_2 - t_1} \qquad (3.2)$$

When a wave is produced in a medium, the plane passing through the particles of medium in the same phase of vibration is referred to as the *wavefront*. The propagation of the wave through the medium can be looked upon as the movement of the wavefront with a velocity equal to the wave velocity. We may note that in a homogeneous isotropic medium, the wave velocity is the same in all directions. Therefore, for waves generated from a point source in the medium, the wavefronts are *concentric spherical* surfaces, i.e. with the source at the centre, i.e. wavefronts are *spherical wavefronts*. At a long distance from the source, a small position of a spherical wavefront can be considered to be a plane. Clearly, that small portion of the wavefront can be termed a *plane wavefront*. A wave with plane wavefronts is referred to as a plane wave. We may note that a plane wave moves in a fixed direction without spreading laterally.

3.4 GENERAL EQUATION OF WAVE MOTION

Let us consider that a wave train moves along the X-axis with velocity v. Now, we assume the source of wave or any point on X-axis as the origin $(x = 0)$. Thus, at any time t the displacement of a particle, situated at the origin, can be represented by the relation

$$y = f(t) \qquad (3.3)$$

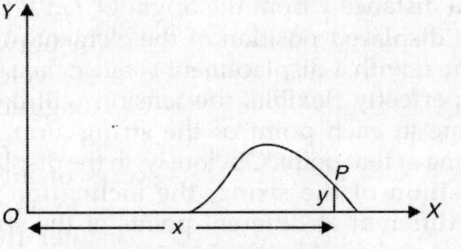

Fig. 3.2: A wave train

where $f(t)$ is any function of time. This wave will reach a point P, distant x from O, after x/v seconds. This means that at any instant t

the displacement y of the particle at P must be the same as the displacement at the origin x/v seconds earlier,

i.e.

$$y = f\left(t - \frac{x}{v}\right) \qquad (3.4)$$

Obviously, Eq. (3.4) is an equation for a wave, travelling along the positive direction of X-axis with constant velocity v. The function f determines the shape and size of the wave. Similarly, the equation of a wave moving along the negative direction of X-axis, can be represented as

$$y = f\left(t + \frac{x}{v}\right) \qquad (3.5)$$

Now, the *general equation of a wave*, moving along X-axis, may be expressed as

$$y = f\left(t - \frac{x}{v}\right) + g\left(t + \frac{x}{v}\right) \qquad (3.6)$$

This shows that $v dt - dx = 0$ or $v = \dfrac{dx}{dt}$, i.e. v is the phase velocity in the positive X-direction, f and g are any two functions. In Eqs. (3.4) and (3.5) y is the function of time and distance both. Now, partially differentiating Eq. (3.6), one obtains,

$$\frac{\partial y}{\partial t} = f'\left(t - \frac{x}{v}\right); \quad \frac{\partial y}{\partial x} = -\frac{1}{v} f'\left(t - \frac{x}{v}\right) \qquad (3.7)$$

$$\frac{\partial^2 y}{\partial t^2} = f''\left(t - \frac{x}{v}\right); \quad \frac{\partial^2 y}{\partial x^2} = \frac{1}{v^2} f''\left(t - \frac{x}{v}\right) \qquad (3.8)$$

where f' and f'' are the first and second differentials of f.

From Eq. (3.7), one obtains

$$\frac{\partial y}{\partial x} = -\frac{1}{v} \frac{\partial y}{\partial t}, \quad \frac{\partial y}{\partial t} = -v \frac{\partial y}{\partial x} \qquad (3.9)$$

$\dfrac{\partial y}{\partial t}$ represents the *particle velocity* at the time t and distance x from the origin and $\dfrac{\partial y}{\partial x}$ is the slope of the y–x curve at the same instant and at the same point. Hence, for those waves, travelling along +ve direction of X-axis, we have particle velocity = –wave velocity × slope of y–x curve.

From Eq. (3.8), one obtains

$$\frac{\partial^2 y}{\partial x^2} = \frac{1}{v^2}\frac{\partial^2 y}{\partial t^2} \quad \text{or} \quad \frac{\partial^2 y}{\partial t^2} = v^2\frac{\partial^2 y}{\partial x^2} \qquad (3.10)$$

This is called *one dimensional differential equation of wave motion.*

Equation (3.10) is very important equation. The striking feature of this equation is that the coefficient of derivative on the right hand represents the square of the wave velocity so that we need not to obtain the velocity of propagation of the wave.

Equation (3.10) holds only for plane waves of constant type. It is not applicable to waves of large amplitude and attenuated waves.

If we represent the disturbance by general symbol ψ, which is a function of $(t \pm x/v)$, i.e.

$$\psi = \psi\,(t \pm x/v) \qquad (3.11)$$

Obviously, Eq. (3.11) is a solution of the one dimensional wave equation

$$\frac{\partial^2 \psi}{\partial x^2} = \frac{1}{v^2}\frac{\partial^2 \psi}{\partial t^2} \qquad (3.12)$$

ψ is generally called the wave function of the wave and if the wave function is determined for a particular situation, one can easily know the form of the wave and its wavelength.

In three dimensions, the differential equation for wave takes the form

$$\frac{\partial^2 \psi}{\partial t^2} = v^2\nabla^2\psi \qquad (3.13)$$

where $\nabla^2 \approx \dfrac{\partial^2}{\partial x^2} + \dfrac{\partial^2}{\partial y^2} + \dfrac{\partial^2}{\partial z^2}$ is a *Laplacian operator.*

One of the solutions of Eq. (3.13) is a three dimensional plane harmonic wave, in which the displacement (or disturbance) at time t and at a point $r = \hat{i}x + \hat{j}y + \hat{k}z$ is given by

$$\psi = A\sin\,(\omega t - k\cdot r + \phi) \qquad (3.14)$$

or $\qquad \psi = A\sin\,(\omega t - k_1 x - k_2 y - k_3 z + \phi)$

where $k = \hat{i}k_1 + \hat{j}k_2 + \hat{k}k_3$ is called the propagation vector. The direction of vector k is the direction of wave motion and its magnitude, i.e. $|k| = 2\pi/\lambda$ is the propagation constant.

3.5 TRANSVERSE WAVES IN STRETCHED STRING

A string or wire in acoustics is a cord or wire whose length is large compared to its diameter, and which is perfectly uniform and flexible. The string is stretched between two fixed points with a large tension. This makes the effects of gravitational forces negligible.

Let us consider that a string be stretched between points A and B with a tension (Fig. 3.3). Now, the string is slightly displaced to one side at the centre and then released. The string starts to vibrate at right angles to its length as shown in Fig. 3.3. We take X-axis along the direction of the displacement of the particles at right angles to AB. Now, we consider a small

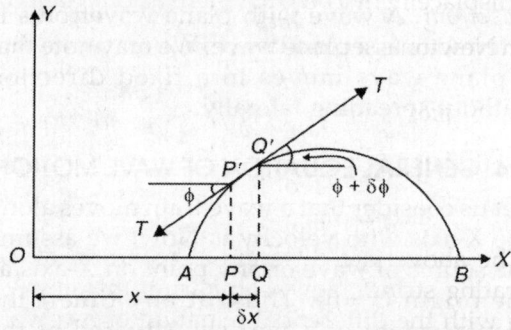

Fig. 3.3: Transverse wave in stretched string

element PQ of the string of length equal to δx at a distance x from the origin O. Let $P'Q'$ be the displaced position of the element at any time t, with a displacement y. Since the string is perfectly flexible, the tension will be the same at each point of the string along the string at that point. Obviously, in the displaced position of the string, the inclination will be different at different points of the string. Let ϕ and $\phi + \delta\phi$, (ϕ being very small) be the inclination with X-axis of the tangents at P' and Q' points respectively. Now, resolving the tensions acting at P' and Q' along X- and Y-axes, one obtains

$$\Sigma T_x = T\cos(\phi + \delta\phi) - T\cos\phi$$

$$\Sigma T_y = T\sin(\phi + \delta\phi) - T\sin\phi$$

Since ϕ and $\delta\phi$ are small, we have

$\cos(\phi + \delta\phi) \approx \cos\phi$, $\sin(\phi + \delta\phi) \approx \phi + \delta\phi$ and $\sin\phi = \phi$

Using these, we obtain

$$\Sigma T_x = 0 \text{ and } \Sigma T_y = T(\phi + \delta\phi) - T\phi = T\delta\phi$$

We have for small angles, $\phi = \tan\phi$.
Hence, the resultant force along Y-axis

$$= T\delta(\tan\phi) - T\delta\left(\frac{\partial y}{\partial x}\right) - T\frac{\partial}{\partial x}\left(\frac{\partial y}{\partial x}\right)\delta x$$

$$= T\frac{\partial^2 y}{\partial x^2}\delta x$$

Let m be the mass per unit length of the string, then the mass of small element δx will be $m\delta x$. Let the acceleration of this element at the displacement y be $\dfrac{\partial^2 y}{\partial t^2}$, then in accordance with Newton's second law of motion, we have

$$m\delta x \frac{\partial^2 y}{\partial t^2} = T\frac{\partial^2 y}{\partial x^2}dx$$

or $$\frac{\partial^2 y}{\partial t^2} = \frac{T}{m}\frac{\partial^2 y}{\partial x^2}$$

The above is a differential equation of a vibrating string. Now, comparing this equation with the differential equation of a wave motion $\dfrac{\partial^2 y}{\partial t^2} = v^2\dfrac{\partial^2 y}{\partial x^2}$, one obtains

$$v^2 = \frac{T}{m} \text{ or } v = \sqrt{\frac{T}{m}} \qquad (3.15)$$

This is the equation for the *velocity of transverse waves* in a stretched string. From Eq. (3.15), we note that velocity depends only on the applied tension and mass per unit length of the string.

We may note that the differential equation for the vibrating string can always be satisfied by the general equation

$$y = f\left(t - \frac{x}{v}\right) + g\left(t + \frac{x}{v}\right) \qquad (3.16)$$

The first part of Eq. (3.16) represents a disturbance (waves) travelling towards the positive direction of X-axis with speed v and the second part represents a disturbance, travelling towards the negative direction of X-axis with the same speed.

3.6 LONGITUDINAL WAVES IN GASES—PRESSURE VARIATION FOR PLANE WAVES

Let us consider that a plane progressive wave is moving through a gas. We may note that longitudinal waves can propagate through gases. The gaseous particles execute simple harmonic motion along the direction of propagation of the wave. The phases of these gaseous particles vary regularly. The distance between the gaseous particles so changes that at any instant particles are alternatively crowded and spreaded out. Obviously, pressure varies from particle to particle inside the gas.

Let us consider that a progressive wave be moving along the positive direction of X-axis. Imagine that cylindrical gas column of cross-sectional area α along X-axis. Now, consider in equilibrium position, two right planes A and B (very close to each other) of the gas column situated at positions x and $(x + \delta x)$ respectively from the origin as shown in Fig. 3.4. When the wave propagates, at some latter instant, the particles of plane A are displaced by an amount y to the position A' and those of the plane B are displaced by an amount $y + \delta y$ to the position B' as

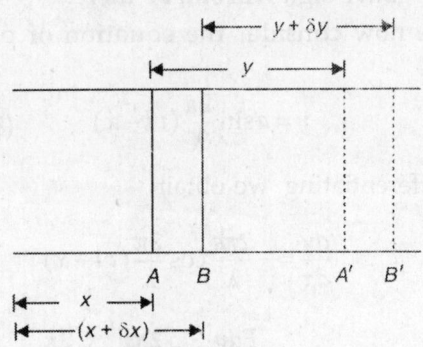

Fig. 3.4: Plane progressive wave through a gas

shown in Fig. 3.4. Since dy/dx is the rate of change of displacement with distance, we have

$$y + \delta y = y + \left(\frac{dy}{dx}\right)\delta x$$

Between the plane A and B, the initial volume of the gas column

$$= \alpha(AB) = \alpha\delta x$$

Its final volume $= \alpha(A'B') = \alpha(\delta x + \delta y)$

$$= \alpha\left[\delta x + \left(\frac{dy}{dx}\right)\delta x\right]$$

$\therefore$ Change in volume $= \alpha\left(\frac{dy}{dx}\right)\delta x$

Now, volume strain $= \dfrac{\alpha\left(\dfrac{dy}{dx}\right)\delta x}{\alpha\delta x} = \dfrac{dy}{dx}$ (3.17)

If E is the bulk modulus of the gas, then

$$E = \frac{\text{Volume stress}}{\text{Volume strain}} = \frac{\text{Decrease in pressure}}{\text{Volume strain}}$$

$$= -\frac{p}{dy/dx}$$

or $p = -E\dfrac{dy}{dx}$ (3.18)

In case of a longitudinal progressive wave travelling through a gas, Eq. (3.18) represents the *variation of pressure*. We may note that p will be positive if there is a decrease in volume, but negative sign will still be there.

We now consider the equation of plane wave,

$$y = a\sin\frac{2\pi}{\lambda}(vt - x)$$ (3.19)

Differentiating, we obtain

$$\frac{dy}{dx} = -\frac{2\pi a}{\lambda}\cos\frac{2\pi}{\lambda}(vt - x)$$

$\therefore$

$$p = -\frac{Edy}{dx} = E\frac{2\pi a}{\lambda}\cos\frac{2\pi}{\lambda}(vt - x)$$ or

$$= p_0\cos\frac{2\pi}{\lambda}(vt - x)$$ (3.20)

where v is the wave velocity and a is the amplitude of the wave. The quantity $p_0 = 2\pi\alpha E/\lambda$ is called the *pressure amplitude*. Since $dy/dt = y$ is the particle velocity and, therefore, we have

$$y = \frac{2\pi va}{\lambda}\cos\frac{2\pi}{\lambda}(vt - x)$$

$$p = E\frac{y}{v}$$ (3.21)

If the rate of change of pressure is dp/dx, then the pressure difference across the small element AB is

$$\delta p = \left(\frac{dp}{dx}\right)\delta x$$

$\therefore$ The force on this element AB is

$$= \delta p \times d = \left(\frac{dp}{dx}\right)\alpha\delta x$$

$$= \alpha\delta x\frac{d}{dx}\left(-E\cdot\frac{dy}{dx}\right) = -\alpha\delta x E\frac{d^2y}{dx^2}$$

Let ρ be the density of the gas, then the mass of the element $= \alpha\delta x\rho$.

Since the force on the element AB is in the direction $B'A'$ and hence the acceleration due to this force will be $-d^2y/dt^2$. In accordance with Newton's second law of motion,

$$-\alpha\delta x\rho\frac{d^2y}{dt^2} = -\alpha\delta x E\frac{d^2y}{dt^2}$$

or

$$\frac{d^2y}{dt^2} = \frac{E}{\rho}\frac{d^2y}{dx^2}$$ (3.22)

Equation (3.22) is a *differential equation* for the wave travelling in a gas. Comparing Eq. (3.22) with the standard differential equation of wave motion, one obtains

$$v^2 = \frac{E}{\rho}$$

or

$$v = \sqrt{\frac{E}{\rho}}$$ (3.23)

Obviously, the velocity of longitudinal waves in a gas depends upon the elasticity and the density of the medium. Formula (3.22) is known as *Newton's formula*. Newton assumed that during alternate compressions and rarefactions occurring when a sound wave travels through a gaseous medium, the temperature remains constant. Keeping this in view, Newton put the isothermal value of the elasticity E, which is equal to the pressure P of the gas. Thus,

$$v = \sqrt{P/\rho} \qquad (3.24)$$

Let us calculate the *velocity of sound in air at NTP* using Eq. (3.24). At $0°C$, the normal pressure $P = 0.76 \times 13,600 \times 9.8 \ N/m^2$ and density of air $= 1.293 \ kg/m^3$. Using these values, from Eq. (3.24), one obtains

$$v = \sqrt{\frac{0.76 \times 13,600 \times 9.8}{1.293}} = 280 \ m/s \ (app.)$$

However, the experimental value of velocity sound in air is 332 m/s, which does not agree with the theoretical value obtained above using Newton's formula [Eq. (3.24)]. The first satisfactory explanation of this large discrepancy between the calculated and observed values for the velocity of sound was provided by Laplace in 1817.

Laplace pointed out that when a sound wave propagates in a gas, the pressure changes, i.e. the compression and rarefactions occur so rapidly that there is no exchange of heat between the layers and the surroundings. Obviously, the process is *adiabatic* obeying the equation $PV^\gamma = constant$, where γ is the ratio of the specific heat of the gas at constant pressure to that at constant volume. For air $\gamma = C_P/C_V = 1.41$. The differentiation of the above equation yields

$$\gamma P V^{\gamma-1} dV + V^\gamma dP = 0$$

or
$$\gamma P dV = -V dP$$

or
$$\frac{dP}{dV} = -\frac{\gamma P}{V} \qquad (3.25)$$

The bulk modulus is

$$K = -\frac{p}{\delta V / V_0} = -V \frac{dP}{dV} \qquad (3.26)$$

Since $p = dP$, $\delta V = dV$ and $V_0 = V$. Now, substituting for $\dfrac{dP}{dV}$ from Eq. (3.25) into Eq. (3.26), one obtains $K = \gamma P$, so that the velocity of sound in the gas is

$$v = \sqrt{\frac{K}{\rho}} = \sqrt{\frac{\gamma P}{\rho}} = \sqrt{\frac{\gamma RT}{M}} \qquad (3.27)$$

For air at NTP, $\rho = 1.293 \ kg/m^3$, $P = 1.013 \times 10^5 \ Pa$ and $\gamma = 1.41$. Substituting these values in Eq. (3.27), one obtains $v = 332.3 \ m/s$, which is in good agreement with experimental value. We may note that the velocity of sound in a gas depends on the *density, temperature* and *humidity*. Sound travels faster in a lighter gas. The velocity of sound increases with increase of temperature and humidity of the gas. Moreover, since γ is greater for a monatomic gas, the velocity of sound in a monatomic gas is greater than that in a diatomic gas. From Eq. (3.27), it is clear that the velocity of sound does not depend on the pressure and it is proportional to the square root of the absolute temperature, i.e. $v \propto \sqrt{T}$. If the velocity of sound at $0°C$ (i.e. at absolute temperature T_0) is v_0 and at $t°C$ (absolute temperature $273 + t = T$), then we have

$$\frac{v_t}{v_0} = \sqrt{\frac{T}{T_0}} \qquad (3.28)$$

One can find a relationship between the velocity of sound and the root mean square (rms) speed of the gas molecules. According to kinetic theory of gases, the gas pressure $P = \dfrac{1}{3}\rho v_{rms}^2$, where ρ is the density of the gas and v_{rms} is the rms speed of the gas molecules. Thus, $v_{rms} = \sqrt{3P/\rho}$, and the speed of the sound is

$$v = \sqrt{\frac{\gamma P}{\rho}} = \left(\frac{\gamma}{3}\right)^{1/2} v_{rms} \qquad (3.29)$$

For air or a diatomic gas, $\gamma = 1.41$, so that $v = 0.69 v_{rms}$. Obviously, v and v_{rms} are of the *same order of magnitude*. One can easily understand this physically as follows. The compression and rarefaction are handed over from one layer of the gas to an adjacent layer through the collisions between the gas molecules moving with their thermal speeds. This is why the speed of propagation of compression and rarefaction, i.e. the speed of sound is thus expected to be of the same order as the rms speed v_{RMS} of the gas molecules.

3.7 SUPERPOSITION PRINCIPLE

This is a very important characteristic of all the waves. According to this principle, *when a number of waves are simultaneously propagated through a medium, the resultant physical disturbance, e.g. displacement, at any point is the sum of disturbance due to separate waves*, i.e.

$$\psi = \psi_1 + \psi_2 + \psi_3 + \psi_4 + ... \qquad (3.30)$$

where ψ represents the resultant disturbance at a point due to different waves, having ψ_1, $\psi_2, \psi_3, ...,$ etc. disturbances separately at the same point.

Let us consider a one dimensional case. Let $y_1, y_2, y_3, ...,$ etc. are the displacement at a point due to separate waves, then the resultant displacement due to all waves is

$$y = y_1 + y_2 + y_3 + y_4 + ... \qquad (3.31)$$

Now, if in general wave functions ψ_1, ψ_2, ψ_3, etc. are the solutions of wave equation

$$\frac{d^2\psi}{dt^2} = v^2 \frac{d^2\psi}{dt^2}$$

with certain restrictions, then any linear combination $(\alpha_1\psi_1 + \alpha_2\psi_2 + \alpha_3\psi_3 + ...)$ is also a solution. For example, when the sonometer wire is plucked, the all possible harmonics are simultaneously produced giving a complex wave. The transverse wave on a string will eventually get reflected at the fixed end and we have two waves travelling in opposite directions on the string. One uses the term *interference* to describe the physical effects of superposition of two or more waves travelling through the same region of space.

3.8 REFLECTION AND TRANSMISSION OF WAVES

When a wave travels in a medium I and meets the boundary of medium II, the wave is partly reflected and partly transmitted at the boundary. The reflected wave travels in medium I and the transmitted wave in medium II. A medium offers a characteristic *impedance* (Z) to the wave travelling in it. However, the characteristic impedance depends on the properties of the medium. Since the two media offer different impedances, it will be interesting for us to know how the waves will respond to an abrupt change of impedance at the boundary separating the two media, e.g. interface separating air from glass in the case of optical discontinuity of steel in water for the case of acoustic waves for under water transmission and detection devices. One can analyse this problem for all waves, e.g. transverse waves on a string, longitudinal (sound) waves in a medium, voltage and current waves on a transmission line and electromagnetic waves in a dielectric medium.

Reflection and Transmission of Transverse Waves on a String at the Discontinuity, i.e. at a Boundary between Two Strings, Reflection and Transmission Coefficients

At the junction of two media, whenever there is a change of impedance due to discontinuity, any type of wave, be it an acoustic wave on a string, voltage and current wave on a transmission line or an electromagnetic wave in any medium will suffer reflection at the boundary.

Let us suppose that a string consists of two sections of linear densities ρ_1 and ρ_2 which are joined smoothly at a point $x = 0$ as shown in Fig. 3.5. Let us assume that both parts of string are stretched with the same tension. The characteristic impedances of the strings are $Z_1 = \rho_1 c_1$ and $Z_2 = \rho_2 c_2$, where ρ_1 and ρ_2 are linear densities of the string and $c_1 \left(= \sqrt{T/\rho_1} \right)$ and $c_2 \left(= \sqrt{T/\rho_2} \right)$ are the wave velocities of in parts 1 and 2 of the string respectively.

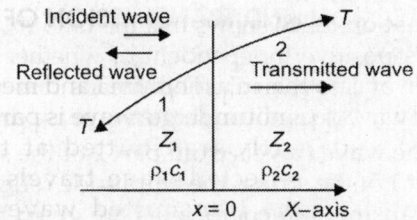

Fig. 3.5: Reflection and transmission of a wave at the boundary ($x = 0$), i.e. with a discontinuity in characteristic impedances Z_1 and Z_2. The string under a constant tension T along its entire length

Let us suppose that a wave (called incident wave) is travelling in the +X-direction on part 1 of the string. When this wave reaches the boundary (which is taken at $x = 0$) separating the two strings, it is partly reflected and partly transmitted at the boundary. The reflected wave travels on part 1 of the string in the negative X-direction and the transmitted wave travels on part 2 of the string in the +X-direction (Fig. 3.5). The incident, reflected and transmitted waves are given by the real parts of the following expressions:

$$\psi_i = A_1 \exp[i(k_1 x - \omega t)] \qquad (3.32)$$

$$\psi_r = B_1 \exp[i(k_1 x + \omega t)] \qquad (3.33)$$

$$\psi_t = A_2 \exp[i(k_2 x - \omega t)] \qquad (3.34)$$

where A_1 is the amplitude of the incident wave travelling in the positive X-direction with velocity c_1. B_1 is the amplitude of the reflected wave travelling in the negative X-direction with velocity c_1 and A_2 is the amplitude of the transmitted wave travelling in the positive X-direction with velocity c_2. $k_1 = 2\pi/\lambda_1 = 2\pi v/c_1$ is the wave number for the incident and reflected waves because reflection does not affect frequency of the wave as it involves only change in amplitude and direction of propagation. The wave number $k_2 = 3\pi/\lambda_2 = 2\pi v/c_2$ for transmitted wave is different from k_1 as only wavelength and velocity are changed and thereby no change in frequency occurs.

The point $x = 0$ at the boundary undergoes oscillations under the combined influence of the incident and reflected waves in part 1 of the string. It then acts as a source of transmitted waves travelling in part 2 of the string. Obviously, at $x = 0$, the waves have to satisfy the conditions of continuity called the *boundary conditions*. The boundary conditions to be satisfied at $x = 0$ are:

i. The displacement is the same immediately to the left and to the right of the boundary at $x = 0$, i.e. y is continuous across the junction $x = 0$ for all time. This is a geometrical requirement. Since y is continuous and hence the velocity $\dfrac{\partial y(x,t)}{\partial t}$ is also continuous.

ii. A dynamical condition that the restoring force or transverse component of tension $\left[-T\dfrac{\partial y}{\partial t}(x,t)\right]$ is continuous across the boundary at $x = 0$. This implies that the gradient $\dfrac{\partial y}{\partial t}$ must be continuous, i.e. same on both the sides. If this is not so, then finite transverse force will act on an infinitesimal small part and produce infinite acceleration which is not permitted.

In view of boundary condition (i), Eqs. (3.32), (3.33) and (3.34) become

$$\psi_i + \psi_r = \psi_t \qquad (3.35)$$

or $\quad A_1 e^{i(k_1 x - \omega t)} + B_1 e^{i(k_1 x + \omega t)} = A_2 e^{i(k_2 x - \omega t)}$

At $x = 0$, the above reduces to

$$A_1 + B_1 = A_2 \qquad (3.35a)$$

Now, applying the condition (ii), one obtains

$$T\frac{\partial}{\partial x}(\psi_i + \psi_r) = T\frac{\partial \psi_t}{\partial x}$$

At the boundary, i.e. at $x = 0$, this reduces to

$$k_1 T A_1 - k_1 T B_1 = k_2 T A_2$$

Putting $k_1 = \omega/c_1$ and $k_2 = \omega/c_2$ and using the relations $T_1/c_1 = \rho_1 c_1 = Z_1$ and $\dfrac{T_2}{c_2} = \rho_2 c_2 = Z_2$, one obtains

$$Z_1(A_1 - B_1) = Z_2 A_2 \qquad (3.36)$$

The ratio B_1/A_1, i.e. the ratio of reflected and incident wave amplitudes is called *reflection amplitude coefficient* (r_{12}) and that of the transmitted and incident amplitudes A_2/A_1 is known as *transmission amplitude coefficient* (t_{12}). We have,

reflection coefficient

$$(r_{12}) = \frac{B_1}{A_1} = \frac{Z_1 - Z_2}{Z_1 + Z_2} \qquad (3.37)$$

and transmission coefficient

$$(t_{12}) = \frac{A_2}{A_1} = \frac{2Z_1}{Z_1 + Z_2} \qquad (3.38)$$

The reflection and transmission coefficients depend only on the impedances and are independent of angular frequency ω of the incident wave. This is true for waves of all frequencies. Since A_1, A_2 and B_1 are all real, the coefficients are free of phase changes except phase reversal which can change the sign of a term.

If the string is fixed at $x = 0$, by attaching this end to a wall, then the medium 2 is infinitely massive which means that $Z_2 = \infty$. This means no transmitted wave, then from Eq. (3.37),

$$\text{reflection coefficient } (r_{12}) = \frac{B_1}{A_1} = -1$$

This shows that the incident wave is completely reflected with the phase reversal, we may note that this is a necessary condition for the existence of *stationary waves*. Now, if the incident waves are a group of waves, then at $Z_2 = \infty$, one finds that no component is transmitted and these will retain their shape and be accompanied by a change of phase of π radiations.

If $Z_2 = 0$, i.e. if the end of the second part of the string is free, then from Eqs. (3.37) and (3.38), one obtains

$$\frac{B_1}{A_1} = 1 \quad \text{and} \quad \frac{A_2}{A_1} = 2$$

This shows that the reflected wave has the same amplitude as the incident wave and there is no change of phase. This is why there is a flick at the free end of the string of a whip.

Equation (3.38) shows that the ratio A_2/A_1 is always positive independent of whether Z_2 is greater or less than Z_1. This means that transmitted wave does not undergo any phase change.

If the wave travels from part 2 of the string to part 1 of the string, the amplitude reflection and transmission coefficients are given by

$$r_{21} = \frac{Z_2 - Z_1}{Z_2 + Z_1} \qquad (3.38a)$$

$$t_{21} = \frac{2Z_2}{Z_2 + Z_1} \qquad (3.38b)$$

Reflection and Transmission of the Energy of the Waves at the Boundary

Waves transport energy and these are found to be very useful mechanism for the transport of energy in a medium. Now, we consider that when a wave meets a boundary between two media of different characteristics impedance, what happens?

When a wave travels along the string, each part of the string is thrown into harmonic oscillations with the passage of time. Considering each unit length of the string as a simple harmonic wave (SHM) of wave frequency ω and maximum amplitude A, we have the total energy carried per unit length along the string as

$$E = \frac{1}{2}\rho\omega^2 A^2$$

where ρ is the mass per unit length or the *linear density* of the string.

We now compute the rates at which energy is incident, reflected and transmitted at the boundary at $x = 0$. As the wave advances, each successive portion of the string takes up the oscillation and if c is the velocity of the wave, the rate at which energy is carried per unit length along the string, is energy times velocity and is given by $= \frac{1}{2}\rho\omega^2 A^2 c$.

Obviously, the rate of energy arriving at the boundary through the incident wave,

$$P_i = \frac{1}{2}\rho_1 c_1 \omega^2 A_1^2 = \frac{1}{2}Z_1\omega^2 A_1^2$$

Similarly, the rates of reflected and transmitted energies respectively are

$$P_r = \frac{1}{2} Z_1 \omega^2 B_1^2$$

and

$$P_t = \frac{1}{2} Z_2 \omega^2 A_2^2$$

The rate at which energy is leaving the boundary through the reflected and transmitted waves

$$= \frac{1}{2} Z_1 \omega^2 B_1^2 + \frac{1}{2} Z_2 \omega^2 A_2^2$$

or

$$P_r + P_t = \frac{1}{2} Z_1 \omega^2 B_1^2 + \frac{1}{2} Z_2 \omega^2 A_2^2$$

Now, substituting the values for

$$B_1 = A_1 \left(\frac{Z_1 - Z_2}{Z_1 + Z_2} \right) \quad \text{and} \quad A_2 = A_1 \left(\frac{2Z_1}{Z_1 + Z_2} \right)$$

from Eqs. (3.37) and (3.38), one obtains the total energy leaving the boundary takes the form

$$P_r + P_t = \frac{1}{2} Z_1 \omega^2 A_1^2 \left(\frac{Z_1 - Z_2}{Z_1 + Z_2} \right)^2$$

$$+ \frac{1}{2} Z_2 \omega^2 A_1^2 \left(\frac{2Z_1}{Z_1 + Z_2} \right)^2$$

$$= \frac{1}{2} Z_1 \omega^2 A_1^2 = P_i$$

This shows that the rate at which energy arrives at the boundary with the incident wave is equal to the rate at which energy leaves the boundary with the reflected and transmitted waves. Thus, energy is conserved at the boundary or junction of the two media. This means that all the energy arriving at the boundary with the incident wave leaves the boundary with the reflected and transmitted waves. The reflection and transmission energy coefficients are given by

$$\frac{\text{Reflected energy}}{\text{Incident energy}} = \frac{P_r}{P_i} = \frac{Z_1 B_1^2}{Z_1 A_1^2}$$

$$= \frac{B_1^2}{A_1^2} = \left(\frac{Z_1 - Z_2}{Z_1 + Z_2} \right)^2$$

$$\frac{\text{Transmitted energy}}{\text{Incident energy}} = \frac{P_t}{P_i} = \frac{Z_2 A_2^2}{Z_1 A_1^2}$$

$$= \frac{4 Z_1 Z_2}{(Z_1 + Z_2)^2}$$

We may note that,

reflection coefficient + transmitted coefficient = 1

If $Z_1 = Z_2$, we find that no energy is reflected and the impedances are said to be *matched*.

Reflection and Transmission of Longitudinal (Sound) Waves at a Boundary between Two Media

When a longitudinal wave like sound wave in a gas are incident at a boundary separating two media of different acoustic impedances $\rho_1 c_1$ and $\rho_2 c_2$, it is partly reflected and partly transmitted at the boundary. Boundary conditions to be satisfied at the interface are:

i. The displacement across the boundary is continuous

ii. The acoustic excess pressure across the boundary is continuous and hence the particle velocity is also continuous.

If it were not so, there will be finite force on an infinitesimal slice of the gas at the boundary, which is not permitted.

The acoustic impedances of the two media are $Z_1 = \rho_1 c_1$ and $Z_2 = \rho_2 c_2$, where c_1 and c_2 are speeds of sound in medium 1 and medium 2 respectively (Fig. 3.6). The incident, reflected and

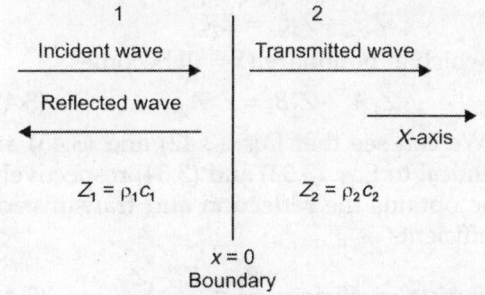

Fig. 3.6: Reflection and transmission of plane longitudinal sound waves at the plane boundary separating two media of acoustic impedances Z_1 and Z_2

transmitted waves are respectively given by

$$\xi_i(x,t) = A_1 e^{i(k_1 x - \omega t)} \qquad (3.39)$$

$$\xi_r(x,t) = B_1 e^{-i(k_1 x + \omega t)} \qquad (3.40)$$

$$\xi_t(x,t) = A_2 e^{i(k_2 x - \omega t)} \qquad (3.41)$$

where ξ is the symbol used for the longitudinal wave and represents the displacement for a particular wave which is designated by the suffix i, r and t for the incident, reflected and transmitted waves respectively.

The excess pressure $P = \Delta P = -E\dfrac{\partial \xi}{\partial x} = -\gamma P_0 \dfrac{\partial \xi}{\partial x}$

is continuous across the boundary, where $\gamma = C_P/C_V$ and P_0 is the equilibrium pressure. The boundary conditions give (at $x = 0$)

$$\xi_i + \xi_r = \xi_t \qquad (3.41a)$$

and $\qquad\qquad p_i + p_r = p_t$

or $\qquad \dfrac{\partial \xi_i}{\partial x} - \dfrac{\partial \xi_r}{\partial x} = -\dfrac{\partial \xi_t}{\partial x}$

At $x = 0$, Eq. (3.41a) becomes

$$A_1 + B_1 = A_2 \qquad (3.42)$$

Due to the wave,

the excess pressure $= B\, d\xi/dx = Z\dot{\xi}$,

Bulk modulus $(B) =$ impedance $(Z) \times$ particle velocity $(\dot{\xi})$. We may note that the specific acoustic impedance for an elastic medium $=$ (pressure due to wave)/(particle velocity) $= B\, d\xi/dx = Z\dot{\xi}$

The pressure boundary condition yields

$$Z_1\dot{\xi}_i - Z_1\dot{\xi}_r = Z_2\dot{\xi}_t$$

which at boundary $(x = 0)$ becomes

$$Z_1 A_1 - Z_1 B_1 = Z_2 A_2 \qquad (3.43)$$

We can see that Eqs. (3.42) and (3.43) are identical to Eqs. (3.33) and (3.34) respectively. One obtains the reflection and transmission coefficients as

reflection coefficient $\dfrac{B_1}{A_1} = \dfrac{Z_1 - Z_2}{Z_1 + Z_2} \qquad (3.44)$

transmission coefficient $\dfrac{A_2}{A_1} = \dfrac{2Z_1}{Z_1 + Z_2} \qquad (3.45)$

We may note that relations (3.44) and (3.45) are general for both kinds of waves, transverse as well as longitudinal. If $Z_1 > Z_2$, then $\dfrac{B_1}{A_1}$ is positive indicating that the incident and reflected displacements are in phase. But if $Z_1 > Z_2$, then $\dfrac{B_1}{A_1}$ is negative showing that the reflected wave undergoes a phase change of π with respect to the incident wave. A_2/A_1 remains positive and independent of whether Z_1 is less or more than Z_2 which means that the transmitted wave does not undergo any phase change.

At a rigid wall where Z_2 is infinity, $B_1 = -A_1$ shows that the wave is completely reflected.

We may note that the intensity of a wave is proportional to the product of characteristic impedance of medium and square of the amplitude in that medium. We have, $I = 2\pi^2 \nu^2 A^2 \rho c = 2\pi^2 \nu^2 A^2 Z$. Thus, intensity coefficient of reflection,

$$\frac{I_r}{I_i} = \frac{Z_1 B_1^2}{Z_1 A_1^2} = \frac{(Z_1 - Z_2)^2}{(Z_1 + Z_2)^2} \qquad (3.46)$$

and the intensity coefficient of transmission

$$\frac{I_t}{I_i} = \frac{Z_2 A_2^2}{Z_1 A_1^2} = \frac{4Z_1^2}{(Z_1 + Z_2)^2}\frac{Z_2}{Z_1} = \frac{4Z_1 Z_2}{(Z_1 + Z_2)^2} \qquad (3.46a)$$

We may note that $\dfrac{I_r}{I_i} + \dfrac{I_t}{I_i} = 1$ or $I_i = I_r + I_t$

$$(3.46b)$$

which means that the law of conservation of acoustic energy holds when a longitudinal (sound wave in air) meets an interface.

Experiments reveal that there is almost total reflection of sound energy at the air-water interface. However, in the case of steel-water interface, the reflection coefficient is about 0.85, i.e. 85%. This means only about 15% of sound energy is transmitted at a steel water interface. This severely limits the use of the transmission and detection devices in submarines using ultrasonic waves.

Standing Waves in Pipes

Prior to discussion of standing waves in air columns in pipes, we first determine the reflection and transmission coefficients for particle pressures and velocities at a boundary. At the boundary at $x = 0$, the particle velocities in incident, reflected and transmitted waves are

$$\left(\frac{\partial \xi_i}{\partial t}\right)_{x=0} = V_i = A_i \omega \cos \omega t$$

$$\left(\frac{\partial \xi_r}{\partial t}\right)_{x=0} = V_r = A_r \omega \cos \omega t$$

$$\left(\frac{\partial \xi_t}{\partial t}\right)_{x=0} = V_t = A_t \omega \cos \omega t$$

Velocity reflection coefficient,

$$\frac{V_r}{V_i} = \frac{A_r}{A_t} = \frac{Z_1 - Z_2}{Z_1 + Z_2} \qquad (3.47)$$

Velocity transmission coefficient,

$$\frac{V_t}{V_i} = \frac{A_t}{A_i} = \frac{2Z_1}{Z_1 + Z_2} \qquad (3.48)$$

Similarly, one can obtain, pressure reflection coefficient,

$$= \frac{P_r}{P_i} = \frac{-V_r}{V_i} = \frac{Z_2 - Z_1}{Z_1 + Z_2}$$

pressure transmission coefficient,

$$\frac{P_t}{P_i} = \frac{Z_2 V_t}{Z_1 V} \qquad (3.49)$$

$$\frac{P_t}{P_i} = \frac{2Z_1}{Z_1 + Z_2} \qquad (3.50)$$

Obviously, if $Z_1 > Z_2$ the incident and reflected particle velocities are in phase, but the incident and reflected acoustic pressures are out of phase by 180°. If $Z_1 > Z_2$, then the acoustic pressures are in phase, but the particle velocities are out of phase by 180°.

At a rigid wall where Z_2 is infinite, one obtains

$$\frac{V_r}{V_i} = 1, \quad \frac{P_r}{P_i} = 1$$

Standing Waves in a Closed Pipe

Let us consider a closed pipe of length L lying along the X-axis with its open end at $x = 0$ and closed end at $x = L$. There is a rigid boundary at the closed end, i.e. Z_2 is infinite, clearly, the particle velocity has a reflection coefficient -1 at the closed end which means that $A_i = -A_r = +A$ (say). Thus, the incident wave (travelling along the pipe in the $+X$-direction) and the reflected wave (travelling in the $-X$-direction) are expressed as

$$\xi_i (x, t) = A \sin[k(vt - x)]$$
$$\xi_r (x, t) = -A \sin[k(vt + x - 2L)]$$

These oppositely travelling waves on superposition gives standing waves in the pipe, which as in the case of a string are represented by

$$\xi(x, t) = 2A \sin[k(L - x)] \cos[k(vt - L)] \quad (3.51)$$

$$\therefore \frac{\partial \xi}{\partial x} = -2Ak \cos[k(L - x)] \cos[k(vt - L)] \quad (3.51a)$$

At time t, the excess pressure at x is given by

$$p = \Delta p = -\gamma P_0 \frac{\partial \xi}{\partial x}$$

$$= 2\gamma P_0 Ak \cos[k(L - x)] \cos[k(vt - L) \quad (3.52)$$

During the oscillation, the open end at $x = 0$ represents a condition of zero pressure variation. Obviously, at the open end, the excess pressure is zero, i.e. the pressure is equal to the equilibrium value P_0. Thus, at $x = 0$ we have $p = 0$ for all time. Now, putting $x = 0$ and $p = 0$ in Eq. (3.52), one obtains

$$0 = 2\gamma P_0 Ak \cos kL \cos[k(vt - L)]$$

We may note that this is true for all values of t, if $\cos kL = 0$

or $\quad kL = \dfrac{\pi}{2}, \dfrac{3\pi}{2}, \dfrac{5\pi}{2}, \dfrac{7\pi}{2}, \dots$

or $\quad k_m L = m\pi/2$

or $\quad k_m = m\pi/2L \qquad (3.53)$

or $\quad \lambda_m = \dfrac{4L}{m} \qquad \left(\because k_m = \dfrac{2\pi}{\lambda_m}\right)$

where $m = 1, 2, 3, 5, 7, \dots$

Thus, the frequency of vibration of air column inside the pipe is obtained as

$$\nu_m = \frac{v}{\lambda_m} = \frac{m}{4L}\sqrt{\frac{E}{\rho}} \qquad (3.54)$$

Since $E = \lambda P_0$ is the modulus of elasticity of the gas, one can obtain the particle displacement in the pipe by using Eq. (3.53) in Eq. (3.51) as

$$\xi(x,t) = 2A \sin\left[\frac{m\pi}{2L}(L-x)\right]\cos\left[\frac{m\pi}{2L}(vt-L)\right] \qquad (3.55)$$

We may note that the frequencies given by relation (3.54) are just the frequencies of the normal modes of a pipe closed at one end. Figure 3.7 shows the first three modes of the pipe. We may note that the particle displacements are longitudinal. We, further, note that a pipe closed at one end has only odd harmonics of frequencies 3, 5, 7, … times the fundamental frequency (ν_1). Clearly, all the even harmonics of frequencies $2\nu_1$, $4\nu_1$, $6\nu_1$, …, etc. are absent. We have the frequency of the fundamental mode ($m = 1$) as given by

$$\nu_1 = \frac{1}{4L}\sqrt{\frac{E}{\rho}} \qquad (3.56)$$

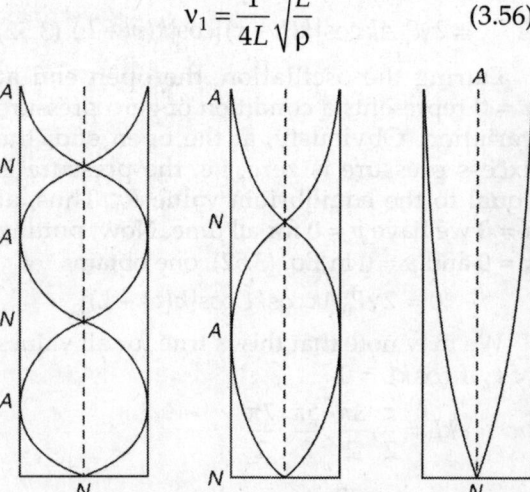

Fig. 3.7: First three normal modes of a pipe open at one end

Standing Waves in an Open Pipe

Let us consider the mechanism of reflection at open end of a pipe. We consider a pipe terminated by an opening into a large room. Obviously, the equilibrium pressure P_0 of the air in the pipe is equal to the pressure of the air in the room. Air can freely rush in or out at the open end of the pipe. When a wave of compression reaches the open end, the particles emerge out of the pipe spreading freely in all directions. Due to this the expansion takes place just outside the pipe in all directions tending to bring the pressure of the emerging air equal to the equilibrium pressure P_0. We may note that this happens at a certain distance from the open end, called as the *end correction*. However, the pressure at the opening is not exactly equal to P_0. This difference in pressure just inside the pipe and outside the pipe gives rise to a force. This force causes the expansion of air which spreads out sideways. The spreading of air in the outside region of the pipe produces rarefaction at the open end and therefore to annul this rarefacation air from behind the compression must rush forward. This again results in a rarefaction in the back region where the air has rushed forward and so on. In this way a wave of rarefaction starts from the open end of the pipe and travels backward, i.e. the wave of rarefaction travels in a direction opposite to the wave of compression. Obviously, a wave of compression on arrival at the open end of a pipe is reflected back into the pipe as a wave of rarefaction. However, the open end of the pipe represents a place of zero pressure variation (this happens at a certain point just outside the opening of the pipe), the excess pressure is zero at the opening of the pipe. Clearly, the pressure reflection coefficient is −1 at the open end. From Eqs. (3.47) and (3.49), it is clear that the velocity or amplitude reflection coefficient is +1. This means that there is no reversal of amplitude as a result of reflection at the open end. Clearly, a compression is reflected from an open end as a rarefaction and *vice versa* without any reversal of sign of the amplitude.

We shall now study the *standing waves* in a pipe of length L lying along the X-axis with its open end at $x = 0$ and $x = L$. Let $\omega(= kv)$ is allowed to pass through the pipe in the

+X-direction. One can represent the particle displacements due to this incident wave as

$$\xi_i\,(x,\,t) = A_i \sin[k\,(vt - x)]$$

at the end $x = L$, the reflected wave is given by

$$\xi_r(x,\,t) = A_r \sin[k(vt + x - 2L)]$$

We have the amplitude reflection coefficient A_r/A_i at the open end as +1. We have $A_r = A_i = A$. The resultant displacement is given by

$$\xi(x,\,t) = \xi_i + \xi_r = A\,[\sin\{k\,(vt - x)\}$$
$$+ \sin\{k\,(vt + x - 2L)\}]$$

$$= 2A\cos[k(L - x)]\,\sin[k(vt - L)]$$

Thus,

$$\frac{\partial \xi}{\partial x} = 2kA\sin[k\,(L - x)]\sin[k\,(vt - L)]$$

Now, the excess pressure is given by

$$p = -\gamma P_0\,\frac{\partial \xi}{\partial x}$$

$$= -2\gamma P_0 kA\sin[k(L - x)]\sin[k(vt - L)]$$

We may note that the excess pressure p at the other open end at $x = 0$ must also be zero. Now, putting $p = 0$ at $x = 0$, we obtain

$$-2\gamma P_0 k\,A\sin kL\,\sin[k\,(vt - L)] = 0$$

The above result is satisfied for all values of t if

$$\sin kL = 0 \text{ or } k_n L = n\pi \quad (n = 1, 2, 3, 4,...)$$

Thus, the frequency of vibration of air column in the pipe is given by

$$v_n = \frac{n}{2L}\sqrt{\frac{E}{\rho}} \tag{3.57}$$

The particle displacements are given by

$$\xi(x,t) = 2A\cos\left(\frac{n\pi x}{L}\right)\left(\frac{n\pi vt}{L}\right) \tag{3.58}$$

The first three modes of a pipe open at the ends are shown in Fig. 3.8. We may note that even and odd harmonics are present.

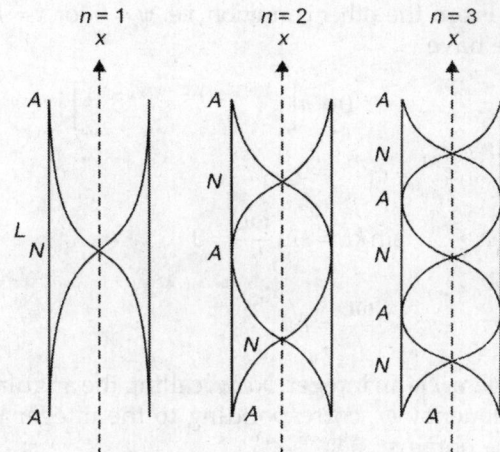

Fig. 3.8: First three modes of a pipe open at both ends

Stationary Waves on a String

Let us consider a string of length l rigidly fixed as both the ends. Obviously, these present an infinite impedance to the waves on it. We may note that a progressive wave is reflected completely from an infinite impedance with phase reversal. Let us assume that there is a monochromatic wave of frequency ω and amplitude a travelling in +X-direction. This wave gets reflected from the fixed end. Let us assume that b be the amplitude of the reflected wave traversing along the –X-direction.

We can represent the displacement of a point on the string as

$$\psi = a \exp[i(kx - \omega t)] + b \exp[-i(kx + \omega t)] \tag{3.59}$$

The boundary conditions to be satisfied at all times are

$$\psi = 0 \text{ for } x = 0 \text{ and } x = L$$

Now, the displacement ψ at $x = 0$ is zero. This gives

$$0 = ae^{-i\omega t} + be^{-i\omega t}$$

Thus, $a = -b$. This embodies the fact that the wave suffers a phase reversal from either end offering infinite impedance. We have

$$\psi = a\left[e^{i(kx - \omega t)} - e^{i(kx + \omega t)}\right] \tag{3.60}$$

From the other condition, i.e. $\psi = 0$ for $x = L$, we have

$$0 = a\left[e^{i(kL - \omega t)} - e^{i(kx + \omega t)} \right]$$

or $\quad e^{ikL} - e^{-ikL} = 0$

or $\quad \sin kL = \sin \dfrac{\omega L}{c} = 0$

or $\quad \dfrac{\omega L}{c} = n\pi$

where n is an integer. Now, calling the angular frequency ω_n corresponding to the integer n, one obtains

$$\dfrac{\omega_n^L}{c} = n\pi$$

or $\quad \nu_n = \dfrac{nc}{2L} = \dfrac{c}{\lambda}$

or $\quad L = \dfrac{n\lambda}{2}$

The corresponding functions which satisfy the boundary conditions are

$$\sin \dfrac{\omega_n x}{c} = \sin\left(\dfrac{n\pi x}{L} \right)$$

where ω_n are the normal frequencies or modes of vibration or the eigen frequencies of the string.

We may note that the set of functions of x, such as $\sin(\omega_n x / c)$ which satisfy the wave equation and the boundary conditions are called *characterstic functions* or more generally *eigen functions*.

Corresponding to $n = 1$, the lowest frequency is called the fundamental and those corresponding to $n > 1$, the harmonics. On the string, there are positions which will always be at rest and these are given by

$$\sin \dfrac{\omega_n x}{c} = \sin \dfrac{n\pi x}{L} = 0$$

or $\quad \dfrac{n\pi x}{L} = p\pi$ where $p = 0, 1, 2, 3, \ldots, n$

We may note that for the nth harmonic, there are, in addition to the fixed ends at $x = 0$ and $x = L$, $(n - 1)$ points which are equally placed in between the ends. These are called *nodes* or *nodal points*. Interestingly, the existence of complete nodal points presupposes the complete cancellation of the progressive waves by the reflected waves having equal and opposite amplitude. This means that the energy carried in one direction is exactly equal to the energy carried in the opposite direction with the consequence that the energy flux (the energy per unit area per second in the standing wave) is zero. The points of maximum displacement which lie in between the nodes, are named as *antinodes* (Fig. 3.9).

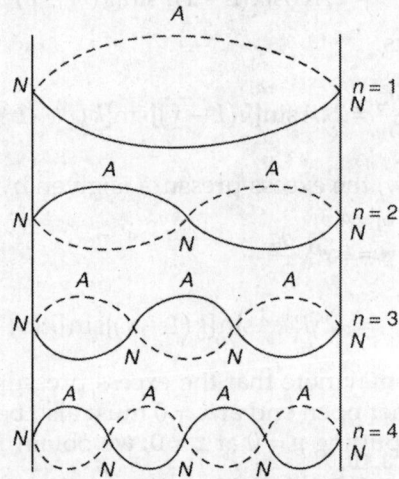

Fig. 3.9: Harmonics of standing waves on a string

Impedance Matching

Usually one faces a problem of connecting two wave-conducting media of different impedances so as to eliminate the reflections from the boundary back into the first medium with a view to ensure the maximum transfer of energy. This is called impedance matching. Impedance matching is of central importance for long distance cables for carrying electrical energy in order to eliminate wastage of power through energy reflection. It is found that the transfer of power from a generator to a load is maximum when there is a match of impedances. We match the loudspeaker to the output of the power amplifier with the help of coupling transformer of a suitable turn ratio. We may note that the coupling element which serves as an interface between two instruments of different impedances, is of great importance in electronic instrumentation and optical systems. Let us consider the problem of impedance matching in the case of two strings of different characteristic impedance. Let us represent the source string by subscript 1 and the load string by 3. The string connecting these two strings be represented by 2. Figure 3.10 shows the incident, reflected and transmitted waves in all three strings or media.

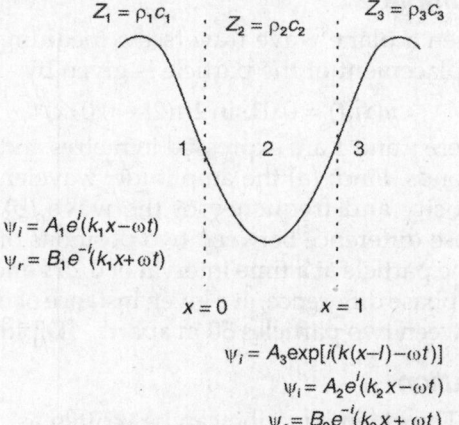

$$Z_1 = \rho_1 c_1 \qquad Z_2 = \rho_2 c_2 \qquad Z_3 = \rho_3 c_3$$

$$\psi_i = A_1 e^{i(k_1 x - \omega t)}$$
$$\psi_r = B_1 e^{-i(k_1 x + \omega t)}$$

$$x = 0 \qquad\qquad x = 1$$

$$\psi_i = A_3 \exp[i(k(x-l)-\omega t)]$$
$$\psi_i = A_2 e^{i(k_2 x - \omega t)}$$
$$\psi_r = B_2 e^{-i(k_2 x + \omega t)}$$

Fig. 3.10: A string of length l and impedance Z_2 connects two strings of impedances Z_1 and Z_3. Three strings are marked by 1, 2 and 3

The boundary conditions are:

i. At the interfaces $x = 0$ and $x = l$, the displacement ψ is continuous

ii. At the junctions, the transverse force $T \dfrac{\partial \psi}{\partial x}$ is also continuous.

With the help of these boundary conditions and simple mathematical calculation, one obtains the condition for impedance matching

$$Z_2 = \sqrt{Z_1 Z_3}$$

The perfect matching is achieved and the incident energy is transferred provided the following two conditions are properly met:

1. The impedance of the coupling element is the geometrical mean of impedances of the media to be matched

2. The coupling element has length $l = \lambda_2/4$ where λ_2 is the wavelength of the incident wave in this medium of coupling element.

The above principles find useful applications in the case of matching transmission lines and applying an ant-reflecting film of glasses and other optical materials so one can eliminate reflections as light passes from air to these media.

One can match two transmission lines of different impedances by inserting quarter wavelength stubs of the lines between them. We may note that the stub of the intermediate line acts like an impedance transformer or quarter - wave transformer.

There is also a important use in the application of dielectric coating of optical lenses for eliminating reflections.

3.9 BEL AND DECIBEL

Generally, intensities of sound waves are expressed in relative values in terms of some standard intensity. We way note that the response of the human ear is not proportional to the intensity, rather, the sensation is nearly proportional to the logarithm of the intensity.

One can determine the auditory sensation of loudness by the intensity of sound. Loudness increases or decreases as the intensity increases or decreases, but a precise quantitative relationship between them is quite difficult to obtain.

Weber–Fechner law gives a relationship between the intensity and the loudness. According to this law, the increase in loudness dL for an average listner is proportional to dI/I, where dI is a small increase in the intensity I. Thus,

$$dL = C_0 \frac{dI}{I} \tag{3.61}$$

where C_0 is a constant. On integrating Eq. (3.61), one obtains

$$L = C_0 \log I + C_1 = C \log_{10} I + C_1 \tag{3.62}$$

where C and C_1 are constants. So, if L_1 and L_2 correspond to the loudness for the intensities I_1 and I_2 respectively, the change in loudness corresponding to the change in intensity from I_1 to I_2 is

$$L_2 - L_1 = C \log_{10} \frac{I_2}{I_1} \qquad (3.63)$$

From Eq. (3.63), it is clear that a doubling of intensity does not give a doubling of loudness.

The quantity $\log_{10}(I_2/I_1)$ gives the intensity level in bel (B) of I_2 relative to the standard intensity I_1. Bel is a large unit and, therefore, the intensity level is usually expressed in decibel (dB). We have $1\,B = 10\,dB$, the intensity level in dB is $10 \log_{10}(I_2/I_1)$.

The threshold audibility, i.e. the lowest audibility for a note of frequency $1\,kHz$ corresponds to an intensity of $10^{-2}\,W/m^2$. We can take it as standard intensity. We may note that a sound having an intensity 10 times the threshold value has an intensity level of $1\,B$ or $10\,dB$.

ILLUSTRATIVE EXAMPLES

Example 1

Which of the following are the solutions to the one dimensional wave equation:

   i. $y = 2\sin x \cos vt$ ii. $y = 5 \sin 2x \cos vt$

iii. $y = x^2 - v^2 t^2$ iii. $y = 2x - 5t$ [BE]

Solution

The one dimensional wave equation is

$$\frac{\partial^2 y}{\partial t^2} = v^2 \frac{\partial^2 y}{\partial x^2}$$

i. $y = 2\sin x \cos vt$. Differentiating this equation twice partially with respect to t and x separately, i.e.

$$\frac{\partial^2 y}{\partial t^2} = -v^2 y \text{ and } \frac{\partial^2 y}{\partial x^2} = -y$$

$$\therefore \quad \frac{\partial^2 y}{\partial t^2} = -v^2 \frac{\partial^2 y}{\partial x^2}$$

Obviously, the given equation is the solution to the one dimensional wave equation

ii. $y = 5 \sin 2x \cos vt$

Now, $\dfrac{\partial^2 y}{\partial t^2} = -v^2 y$ and $\dfrac{\partial^2 y}{\partial x^2} = -4y$

$$\therefore \quad \frac{\partial^2 y}{\partial t^2} = \frac{v^2}{4} \frac{\partial^2 y}{\partial x^2}$$

Clearly, the given equation is not the solution of one dimensional wave equation. Similarly, one can show that (iii) is not the solution and (iv) is the solution of the wave equation.

Example 2

When a plane wave traverses a medium, the displacement of the particle is given by

$$y(x, t) = 0.01 \sin 2\pi (2t - 0.01x)$$

where y and x are expressed in metres and t in seconds. Find: (a) the amplitude, wavelength, velocity and frequency of the wave (b) the phase difference between two positions of the same particle at a time interval of $0.25\,s$ and (c) the phase difference, at a given instance of time, between two particles $50\,m$ apart.   [Diploma]

Solution

(a) The given equation can be written as

$$y(x,t) = 0.01 \sin \left\{ \frac{2\pi}{100}(200t - x) \right\}$$

Now, comparing with the wave equation

$$y(x,t) = A \sin \left\{ \frac{2\pi}{\lambda}(vt - x) \right\}$$

We obtain: amplitude $A = 0.01\,m$, wavelength $\lambda = 100\,m$, wave velocity $v = 200\,m/s$, and frequency

$$v = \frac{v}{\lambda} = \frac{200}{100} = 2\,Hz .$$

(b) Phase change in a time interval of Δt is

$$\Delta \phi = \frac{2\pi}{T} \cdot \Delta t = 2\pi v \Delta t$$

or phase difference $= 2\pi \times 2 \times 0.25 = \pi = 180°$. Clearly, the particle phase is reversed in a time $0.25\,s$ which is obvious since its period

$$T = \frac{1}{v} = 0.5\,s .$$

Phase difference corresponding to path difference Δx is

$$\Delta\phi = \frac{-2\pi}{\lambda}\Delta x = -\frac{2\pi}{100}\times 50 = -\pi = -180°$$

This means that particle located 50 m (which is half the wavelength) ahead of another particle lags in phase by 180°.

Example 3

The intensity level in a conversation is 70 dB above the threshold of 10^{-12} W/m². Determine the amplitude of vibration of the air particles in the sound wave. Given, velocity of sound = 350 m/s, density of air = 1.25 gm/litre, mean frequency = 500 Hz.

Solution

Let I is the intensity level of conversation. We have

$$dB = 10\log_{10}(I/I_0)$$

or

$$70 = 10\log_{10}(I/10^{-12})$$

whence

$$I = 10^7 \times 10^{-12} = 10^{-5} \text{ W/m}^2$$

Now,

$$I = \frac{1}{2}\rho_0 A^2 \omega^2 v$$

ρ_0 (air density) = 1.25 gm/litre = 1.25 kg/m³
v (velocity of sound) = 350 m/s
ω (angular frequency) = $2\pi f = 2\pi \times 500$
$$= 1000\pi \text{ rad/s}$$

$\therefore$ Amplitude of vibration of air particles

$$A = \sqrt{\frac{2I}{\rho_0\omega^2 v}}$$

$$= \left(\frac{2\times10^{-5}}{1.25\times(1000\pi)^2\times350}\right)^{1/2}$$

$$= 6.8 \times 10^{-8} \text{ m} = 0.068 \text{ μm}.$$

Example 4

A plane sound wave in air of density 1.29 kg/m³ falls on a water surface at normal incidence. The speed of sound in air and water are 334 m/s and 1480 m/s respectively. Determine (i) the ratio of the amplitude of sound wave that enters water to that of the incident wave (ii) fraction of the incident energy flux entering the water.

Solution

We have the characteristic impedances of air and water respectively as
$$Z_1 = \rho_1 v_1$$
$$Z_2 = \rho_2 v_2$$

$$\therefore \quad \frac{Z_1}{Z_2} = \frac{\rho_1 v_1}{\rho_2 v_2} = \frac{1.29\times334}{1000\times1480} = 0.000291$$

(i)
$$\frac{A_t}{A_i} = \frac{2Z_1}{Z_1+Z_2} = \frac{2Z_1/Z_2}{Z_1/Z_2+1}$$

$$= \frac{2\times0.000291}{0.000291+1} = 5.82\times10^{-4}$$

(ii)
$$\frac{I_t}{I_i} = \frac{4Z_1Z_2}{(Z_1+Z_2)^2} = \frac{4Z_1/Z_2}{(Z_1/Z_2+1)^2}$$

$$= \frac{4\times0.000291}{(0.000291+1)^2} = 1.16\times10^{-3}$$

Example 5

Sound waves are incident normally on water–steel interface. If the characteristic impedances of water and steel are 1.43×10^6 kg/m²/s and 3.90×10^7 kg/m²/s respectively, show that 0.86% of the incident energy is reflected at the interface.

Solution

We have $Z_1 = 1.43 \times 10^6$ kg/m²/s
and $Z_2 = 3.90 \times 10^7$ kg/m²/s
Now, the energy reflection coefficient

$$= \left(\frac{Z_1-Z_2}{Z_1+Z_2}\right)^2 = \left(\frac{1.43-3.90}{1.43+3.90}\right)^2$$
$$= 0.86\%$$

REVIEW QUESTIONS

1. Show that the general differential equation for a one dimensional wave equation is $\dfrac{d^2\psi}{dt^2} = v^2\dfrac{d^2\psi}{dx^2}$.

2. State two properties, a medium must possess in order to support a wave motion.

[**Hint:** Elasticity and density or inertia]

3. Solids can support both longitudinal and transverse mechanical waves, but only longitudinal mechanical waves can propagate in gases. Explain why ?

4. Can non-mechanical transverse waves propagate in gases ?

[**Hint:** Yes, electromagnetic waves travel in gases, although they are transverse in nature]

5. Can non-mechanical transverse waves propagate in gases ?

[**Hint:** Yes, although electromagnetic waves are transverse in nature and they travel in gases].

6. What are transverse and longitudinal waves ?

7. Starting with equation of simple harmonic disturbance, obtain the wave equation.

8. Obtain Newton's formula for the velocity of sound in a gas. [BE]

9. Which of the following are solutions to the one dimensional wave equation:

   (i) $y = 2x - 5t$ (ii) $y = (x - vt)^2$

  (iii) $y = 3\sin 2x \cos vt$ (iv) $y = x^3 - v^2 t^2$

   (v) $y = \exp[3(x - vt)]$

[*Ans.* (i), (ii) and (v)]

10. Derive an expression for the velocity of transverse waves in a stretched string. [BE]

PROBLEMS

1. The propagation constant of a wave is 280 per cm and its velocity is 400 m/s. Find its wave number, wavelength and frequency.

[*Ans.* 44.6 cm, 0.0224 cm, 17.84×10^5 cm/s]

2. Show that the superposition of two progressive waves of the same amplitude and frequency but travelling in opposite directions produces a stationary wave.

3. Calculate the intensity of a sound which is 12 dB lower than a sound of intensity 0.01 μW/m².

[*Ans.* 6.31×10^{-10} W/m²]   [BE]

4. A sound of intensity I_1 has the level p dB. Another of intensity I_2 has the level q dB. Show that the intensity level of $(I_1 + I_2)$ is

$$10 \log_{10} (10^{p/10} + 10^{q/10}) \text{ dB}.$$

5. Show that the phase difference between the two progressive waves:

$y_1 = a \sin(\omega t - cx)$ and $y_2 = a \cos(\omega t - cx)$ is $\pi/2$.

[**Hint:** $\sin(\omega t - cx) = \cos(\omega t - cx - \pi/2)$

$\therefore$ Phase difference $= \omega t - cx - (\omega t - cx - \pi/2) = \pi/2)$]

SHORT ANSWER QUESTIONS

1. Distinguish between transverse and longitudinal waves.

Ans. The particles of the medium in a transverse wave oscillate perpendicular to the direction of propagation of the wave. The particles of the medium in a longitudinal wave oscillate along the direction of propagation of the wave.

2. What is a dispersive medium?

Ans. A medium is said to be dispersive if the speed of a wave in that medium depends upon its frequency or wavelength. Obviously, the waves of different frequencies (or wavelength) travel with different speeds in a dispersive medium.

3. What are two properties which a medium must possess in order to support wave motion?

Ans. Elasticity and density (or inertia)

4. Why do electromagnetic waves travel in gases?

Ans. Because electromagnetic waves are non-mechanical transverse waves.

5. What type of waves are produced on the surface of a lake by dropping the stone?

Ans. Transverse.

6. What is a restoring force ?

Ans. The oscillating particle can remain at rest in its equilibrium position at which no net force acts upon it. When the particle is displaced from this position, periodic force acts upon it in such a direction as to return it to its equilibrium position. This force is called as the restoring force.

7. What does the mechanical energy of a oscillating particle consists of ?

Ans. Total mechanical energy (E) = kinetic energy (K) + potential energy (U).

OBJECTIVE QUESTIONS

1. Waves produced in air by a vibrating tuning fork are _____. [longitudinal]

2. Light waves travelling from the sun to the earth are _____ electromagnetic waves.

[transverse]

3. Solids can support both _____ and longitudinal waves. [transverse]

4. Newton's formula for velocity of sound in air is _____. $\left[v = \sqrt{E/\rho}\right]$

5. Laplace's relation for velocity of sound in gases is _____. $\left[v = \sqrt{\gamma p/\rho}\right]$

MULTIPLE CHOICE QUESTIONS

1. The displacement of a particle located at x at time t due to two waves are given by

   $y = a \sin(\omega t - kx)$ and $y_2 = a \sin(\omega t - kx + \phi)$.

   If the amplitude of the resultant wave formed by the superposition of these two waves is a, the phase constant ϕ is equal to

   (a) zero   (b) $\pi/2$

   (c) $2\pi/3$ (d) $3\pi/4$

2. A progressive wave in a medium is represented by the equation

   $$y = 0.1 \sin\left(10\pi t - \frac{5}{11}\pi x\right)$$

   where y and x are in cm and t in seconds. The maximum speed of a particle of the medium due to the waves in cm s^{-1} is

   (a) 1   (b) 10

   (c) π (d) 10π

3. Standing waves are produced by the superposition of two waves

   $$y_1 = 0.05 \sin(3\pi t - 2x)$$

   and $y_2 = 0.05 \sin(3\pi t + 2x)$

where x and y are expressed in metres and t is in seconds. What is the amplitude of a particle at $x = 0.5$ cm? Given $\cos 57.3° = 0.54$

(a) 2.7 cm   (b) 5.4 cm

(c) 8.1 cm   (d) 10.8 cm

4. A wave represented by the equation $y = a \cos(kx - ax)$ is superposed with another wave to form a stationary wave such that the point $x = 0$ is a node. The equation of the other wave is

   (a) $y' = a \sin(kx + ax)$

   (b) $y' = -a \cos(kx - ax)$

   (c) $y' = -a \cos(kx + ax)$

   (d) $y' = -a \sin(kx - ax)$

5. Which of the following functions represent a travelling wave ? Here a, b and c are constants.

   (a) $y = a \cos bx \sin ct$

   (b) $y = a \sin bx \cos ct$

   (c) $y = a \sin(bx + ct) - a \sin(bx - ct)$

   (d) $y = a \sin(bx + ct)$

Answers

    1. (c)   **2.** (c)   **3.** (b)   **4.** (c)   **5.** (d)

4 Nature of Light and Interference

4.1 NATURE OF LIGHT

From early recorded times, man had wondered about the mysteries of light and imagined its exact nature and relationship with human vision. Light is a form of energy (electromagnetic radiation) which evokes visual sensation of objects and on which our visual awareness of the universe and its contents relies. Each of these new discoveries is added to our growing knowledge on light. Based on these observations, laws have been formulated which predict the behaviour of light. Indeed, one of the basic question one needs to answer about light concerns its nature. Just what is light? When we scratch a match or switch on a light bulb in a dark room, not only we see the match or light bulb, but also the other objects. Moreover, these objects may come in different colours. Colour is a word we use to describe certain physiological and psychological sensation of vision. Just what is it that makes the difference we have between colours? How do these colours combine with one another?

The observed behaviour of light as it travels from one place to another, and its interaction with matter has led to the development of a lot of technological tools. Tools have been developed which allow us to see objects as tiny as single blood cells and bacteria, and also see bodies in galaxies long distances away. We have developed high-speed cameras used for taking photographs of fast-moving objects, and also have tools for taking photos during the night and under the depths of the sea. All these instruments use the intricate behaviour of light as it travels from one place to another, and as it interacts with material objects.

We can understand the nature of light when we relate light to vision. An object becomes visible when light coming from it meets the eye. A body is visible either because it is luminous, or because it is illuminated. Luminous objects produce their own light. Examples include lamps, the stars, and the sun. Most common objects we find around us are non-luminous, and they include such things as wood, concrete, and writing paper. These non-luminous objects are made visible by the reflection of light from their surfaces.

Air, glass, and water readily transmit light. They are said to be **transparent.** Other substances tend to diffuse the light, and scatter it in many directions. The result is that it is barely possible to make out the profiles of objects when viewed through such materials. We are not able to discern details of the object being viewed. These materials are said to be **translucent.** Frosted light bulbs, paper soaked in paraffin or vegetable oil are examples of translucent materials. Substances which do not transmit light at all are said to be **opaque.**

Energy can be transmitted from one point to another by two methods. In one of these, matter may move from one point to the other. This is the **corpuscular theory**. In the other method, a wave disturbance can travel between the two points in the medium. This is the **wave theory**. Two theories have been put forward to help explain the behaviour of light as it travels from one point to another, and as it interacts with matter. These are the **corpuscular theory**, and the **wave theory**. Both theories have been advanced to account for the nature, propagation, and interaction of light with matter.

The Corpuscular Theory

The corpuscular theory assumes light to be made up of particles, or **corpuscles**, originating from the light source, and moving in straight lines through space or through a medium. One of the most famous champions of this theory was Sir Isaac Newton himself. The small corpuscles passing through the eye are thought to stimulate the sensory nerves and produce the sense of vision. The corpuscles were later named **photons** (quantum of energy), the name they have even today.

The greatest success of the quantum theory lies in its account of the **photoelectric effect,** i.e. the production of electricity from certain metal surfaces when light of certain colour is incident on the surface. In this, it is thought that the energy of the photon is transferred to an electron in the metal surface, and the electron gets ejected. This leads to the electric current that is measured.

The Wave Theory

Christian Huygens (1629–1695) first conceived light as a wave motion. In his theory, he proposed light as a transverse wave motion. More than one hundred years later, this proposal was backed by the result of electromagnetic wave studies carried out by James Clarke Maxwell. In the wave theory, light is pictured as consisting of transverse waves of progressively changing electric and magnetic fields. The fields act at right angles to each other, and to the direction of propagation of the wave. In the propagation of light in a medium, Huygens proposed that each point on the wavefront acts as a secondary source of new disturbance that spreads out spherically. The new wavefront is the envelope of all the individual wavelets that originate from the old wavefront as shown in Fig. 4.1.

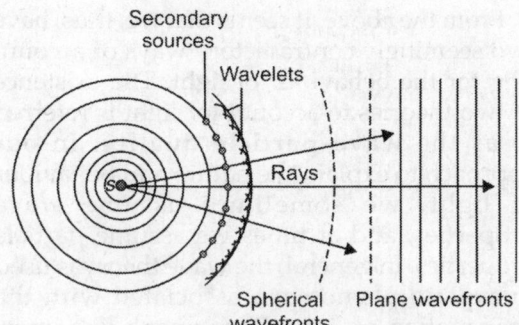

Fig. 4.1: Huygens' construction of secondary wavefronts. Spherical wavefronts from a point source become plane wavefronts at a great distance from the source

Many studies were carried out on the wave theory of light in the beginning of the nineteenth century as a result of discoveries made by Thomas Young of England, and Augustin Jean Fresnel of France. Young's theoretical work predicted the interference of light waves from two sources, just like surface waves in water. Fresnel carried out both theory and experiment much further, explaining diffraction by straight edges, obstacles and apertures, by a combination of Young's principle of interference, and Huygens' principle of constructing secondary wavefronts.

Wave–Particle Duality

There are two theories needed to explain all the properties and interactions of light. These are the wave theory, and the corpuscular (now quantum) theory. It is now known that light is a form of transverse waves in which the electric and magnetic vibrations are perpendicular to the direction of travel of the wave (Fig. 4.3). Thus, light waves have wavelength and frequency. This wavelength controls the colour of the light. Light also exhibits other phenomena, e.g. interference, diffraction, that are explained only if we assume that light is a wave motion. On the other hand, there are other observed phenomena that can only be satisfactorily explained if light is assumed to be made up of photons. The most important of these is the photoelectric effect.

From the above, it seems that we, thus, have two seemingly contradictory ways of accounting for the behaviour of light. The existence of two theories to account for light is referred to as the **wave–particle duality**. In our approach to explain the nature and behaviour of light, we sometimes assume wave properties, and at times we assume particle properties. In general, the wave theory is used to explain phenomena associated with the propagation of light. Phenomena like interference and diffraction are explained using the wave theory. The quantum theory comes into play when it is desired to explain phenomena dealing with the interaction of light with matter.

The Electromagnetic Spectrum

An orderly arrangement of radiations according to wavelength, frequency, or energy is known as a spectrum. The electromagnetic spectrum is such a spectrum. In it, the components of light are arranged in order of frequency and wavelength. The important parts of the electromagnetic spectrum are shown diagrammatically in Fig. 4.2

All the radiations constituting the electromagnetic spectrum travel at the same speed of 3.0×10^8 m/s. This speed is one of the fundamental constants of nature. For each frequency of the radiation, the energy E of the wave is given by the expression

$$E = h\nu$$

where ν is the frequency of the radiation, and h is a constant, known as Planck's constant with numerical value of 6.7×10^{-34} Js.

The most familiar frequency range is *visible light*, the narrow band of electromagnetic radiation to which the human eye is sensitive.

This extends from violet on the high frequency (or short wavelength) side to red on the low frequency (long wavelength) side. X-rays are highly energetic type of atomic radiation with high penetrating power. Because of their high penetration, these rays have applications in medical imaging, non-destructive testing of materials, and studying the crystal structure of minerals. X-rays are produced from specialised tubes and also from radioactive substances. Closely related to X-rays are gamma rays. These are highly penetrating radiations produced from certain radioactive sources. Gamma rays, in general, have more energy than X-rays.

From the longer wavelength side of the visible region, we start with the infrared radiation, known more commonly as heat radiation. Radio waves used for broadcast and other communication purposes have wavelengths ranging from about 1mm (for short radio waves) to 10^6 m. The most useful region for normal broadcast has wavelength between 10 m and about 400 m.

Becoming increasingly important in technology are microwaves. These are electromagnetic radiations having frequency in the range $10^9 - 10^{12}$ Hz, and a wavelength between 1.0 mm and 1.0 m. The popular application of these waves, nowadays, is in cooking. A microwave oven is a relatively small, box-like oven that raises the temperature of an electrically non-conducting material (the food and its container) by subjecting the material to a high frequency electromagnetic field. The heating occurs throughout the food, rather than from the outside first—as in most other cookers. This greatly reduces cooking time and energy consumption.

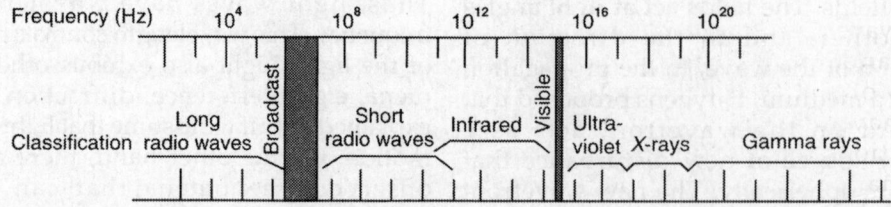

Fig. 4.2: The electromagnetic spectrum showing the component wavelengths and their frequencies

As stated above, the light wave is an electromagnetic wave consisting of periodically varying electric and magnetic fields oscillating at right angles to each other and to the direction of the propagation of waves. Figure 4.3 shows a portion of electromagnetic wave.

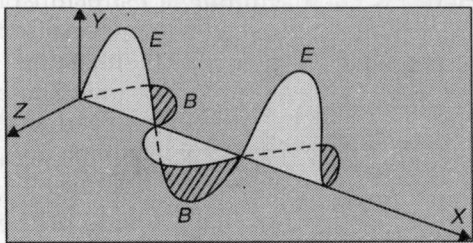

Fig. 4.3: An electromagnetic wave

The electric field in the wave is defined by the electric field strength E and the magnetic field by the magnetic induction vector B. Vectors E and B are mutually perpendicular to each other and also to the direction of propagation of wave. Both vectors E and B are of equal importance to the wave. However, a light wave is often represented by E wave, as many of the effects of light are mainly due to the electric field and can be easily explained with the help of vector E. The magnetic field is implied to be oscillating in a plane normal to the plane of the electric field oscillations and we have not shown it specifically in Fig. 4.3.

The light wave depicted as electromagnetic wave in Fig. 4.3 can be represented mathematically by the following expressions,

$$E = E_0 \sin(kx - \omega t)$$
$$B = B_0 \sin(kx - \omega t)$$

where k is wave vector, $\omega(= 2\pi\nu)$ is angular frequency and E_0 and B_0 are maximum amplitudes of E and B vectors respectively. We may note that Fig. 4.3 and the above equations represent an ideal plane, monochromatic and harmonic electromagnetic wave which travel with the speed of light c $(= 3 \times 10^8$ m/s$)$ in vacuum.

4.2 COHERENCE

Conventional light sources produce incoherent light. This means that the light that emerges from a conventional light source is mixture of waves at various frequencies that reinforce or cancel each other in a random fashion. Figure 4.4 depicts this situation. Obviously, the wavefront thus produced varies from point to point and changes from time to time. The wave from a laser is called almost *coherent* because it is an orderly wave of one frequency where the whole beam, is *spatially* in phase. There are, thus, two independent facets of this coherence: namely, *temporal coherence* and *spatial coherence*.

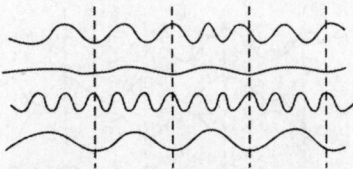

Fig. 4.4: Incoherent waves. There is no relationship between one wave and another

Temporal Coherence

This type of coherence refers to the correlation between the radiation field at a point and the radiation field at the same point at a later time, i.e. the relation between $E(x, y, z, t_1)$ and $E(x, y, z, t_2)$, where these represent the radiation field at the point (x, y, z) at times t_1 and t_2 respectively. If the phase difference between the two radiation fields is constant during the period normally covered by observation (~ few microseconds), the wave is said to have *temporal coherence*. If the phase difference between two radiation fields changes many times and in an irregular way during the shortest period of observation, the wave is said to be incoherent.

Time coherence comes about when each cycle of the wave takes exactly the same time to pass a point in space. This really means that the frequency of the wave is not varying and is given the name *monochromatic*.

Spatial Coherence

Two fields at two different points on a wavefront of a given electromagnetic wave are said to be space coherent if they preserve a constant phase difference over any time t, i.e. space coherency requires that the waves not only are of the same frequency, but also they are in phase in space. Figure 4.5 shows these conditions. In Fig. 4.5a, the waves are monochromatic (time coherent) but are not in sequence in space. This can occur when the source of the wave is physically broad rather than a point source. Figure 4.5b shows a spatially and time coherent wave. The whole wavefront is in step and each cycle takes the same length of time. The very nature of laser mechanism produces this coherent signal.

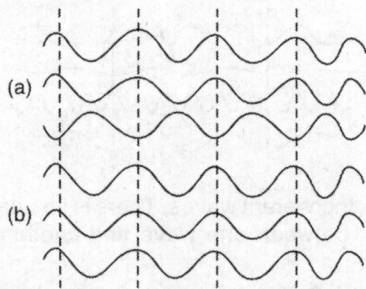

Fig. 4.5: (a) Time coherent space incoherent
(b) Time and space coherent

Spatial coherence is possible even when the two beams are individually time incoherent, as long as any phase change in one of the beams is accompanied by a simultaneous equal phase change in the other beam. With the ordinary light sources, this is only possible if the two beams have been derived from the same part of the source.

Time incoherence is a characteristic of single beam of light, whereas space incoherence concerns the relationship between two separate beams of light. Two beams of light produced in different parts of light source will have been emitted by different groups of atoms. Each beam will be time-incoherent and will suffer random phase changes, as a result of which the phase difference between the two beams will also suffer rapid and random changes. Two such light beams are said to be space incoherent.

Coherence Time and Coherence Length

Let us consider that the radiation field E, from a light source is an ideal sinusoidal function of time t, then at any given position it can be written as

$$E = a\cos(\omega_0 t + \phi) \qquad (4.1)$$

Here a is the magnitude of the field, ω_0 the angular frequency and ϕ the phase. Figure 4.6 depicts the field represented by Eq. (4.1)

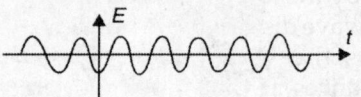

Fig. 4.6

However, no emitted light produces a perfect sinusoidal variation of field E with time t, and for any actual light source the magnitude a and phase ϕ will vary with time. One can represent a more realistic picture of the radiation from an emitting source of light as shown in Fig. 4.7.

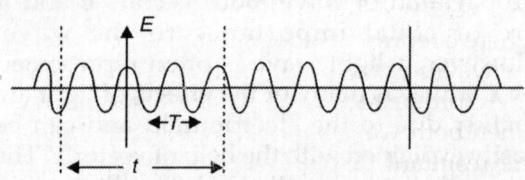

Fig. 4.7

Figure 4.7 represents the average time duration for which an ideal sinusoidal emission occurs. τ is called *coherence time* or *interval of coherence*.

Let T denotes the period of the oscillation, where $T = \dfrac{\omega_0}{2\pi}$. The spatial dimension, L, for which the light may be considered to be a perfect sinusoidal is given by

$$L \approx cT \approx \frac{\tau}{T}\lambda \qquad (4.2)$$

Here c is the velocity of light and L is called the *coherence length*. After time τ, there is no correlation between the phases of the waves.

In terms of purity Q of the spectral line ($Q \approx \lambda/\Delta\lambda$, where $\Delta\lambda$ is half width of the spectral line), we have

$$c_\tau \approx L \approx \lambda Q \qquad (4.3)$$

4.3 PRINCIPLE OF SUPERPOSITION

When two or more waves arrive at a point in space simultaneously, the disturbance at that point is the *vector sum* of the disturbance caused by those waves taken individually, i.e. the net wave disturbance at any given time and place is the vector sum of all the wave disturbances at that particular place at that particular time. This is called the principle of superposition.

Let Y_1, the displacement at the observation point due to one wave at any instant t, is given by

$$Y_1 = a_1 \cos \omega t = a_1 \cos 2\pi v t \qquad (4.4)$$

Let the second wave have the same frequency v and cause displacement in the same direction, represented by

$$Y_2 = a_2 \cos(\omega t + \delta) = a_2 \cos(2\pi v t + \delta) \quad (4.5)$$

where δ represents the *phase difference* between the waves at the point of consideration.

According to *Young's principle of superposition*, when both waves are present simultaneously the resultant displacement Y at that point is given by

$$Y = Y_1 + Y_2 \qquad (4.6)$$

It is worthwhile to mention that principle of superposition is of wider application in which Y_1 and Y_2 could have different directions. Obviously, then Eq. (4.6) is a vector equation. In discussing interference of light waves, we will limit ourselves to the situations where Y_1 and Y_2 are along the same line.

From Eqs. (4.4), (4.5) and (4.6), one obtains

$$Y = a_1 \cos 2\pi v t + a_2 \cos(2\pi v t + \delta)$$
$$= a_1 \cos 2\pi v t + a_2 \cos 2\pi v t \cos \delta - a_2 \sin 2\pi v t \sin \delta$$
$$= (a_1 + a_2 \cos \delta) \cos 2\pi v t - (a_2 \sin \delta) \sin 2\pi v t$$
$$= a \cos \phi \cos 2\pi v t - a \sin \phi \sin 2\pi v t$$
$$= a \cos(2\pi v t + \phi) \qquad (4.7)$$

where $\left. \begin{array}{l} a\cos\phi = a_1 + a_2\cos\delta \\ a\sin\phi = a_2\sin\delta \end{array} \right\} \qquad (4.8)$

Squaring and adding the terms of Eq. (4.8), one obtains

$$a^2 = a_1^2 + a_2^2 + 2a_1 a_2 \cos\delta \qquad (4.9)$$

and dividing Eq. (4.8), gives

$$\tan\phi = \frac{a_2 \sin\delta}{a_1 + a_2 \cos\delta} \qquad (4.10)$$

Constructive Superposition

When phase difference $\phi = 0, 2\pi, 4\pi, \ldots = 2n\pi$, where $n = 0, 1, 2, 3, \ldots$, we have from Eq. (4.9),

$$a^2 = a_1^2 + a_2^2 + 2a_1 a_2 = (a_1 + a_2)^2 \qquad (4.11)$$

This is called the case of *constructive superposition* or *additive interference*. In addition, if $a_1 = a_2$, then

$$a^2 = 4a_1^2 + 4a_2^2$$

i.e. intensity become 4 times.

Destructive Superposition

If $\phi = \pi, 3\pi, 5\pi, \ldots = (2n + 1)\pi$, where $n = 0, 1, 2, 3, \ldots$, one obtains

$$a^2 = a_1^2 + a_2^2 - 2a_1 a_2 = (a_1 - a_2)^2 \qquad (4.12)$$

This shows that the value of a is minimum. If $a_1 = a_2$, then $a = 0$. This is called the *destructive superposition*, $a = 0$ corresponds to zero intensity. Obviously, the change of intensity from maxima to minima is very sharp and can be observed clearly.

We, therefore, conclude that the phase difference between two vibrations must be (a) $\delta = 2n\pi$ for constructive superposition (b) $\delta = (2n + 1)\pi$ for destructive superposition.

We know that a phase difference of 2π corresponds to a path difference of λ, hence one can express path difference for constructive and destructive superposition as follows:

$$\Delta = \frac{\lambda}{2\pi} \cdot 2n\pi = 2n\lambda/2$$

[Constructive superposition]

$$\Delta = \frac{\lambda}{2\pi}(2n + 1) = (2n + 1)\lambda/2$$

[Destructive superposition] (4.13)

The above discussion demonstrates that first of all, superposition of waves results in a spatial redistribution of the oscillation energy. In other words, the intensity of waves is redistributed in space. Regions with zero intensity and those with intensity above the intensity of overlapping waves are produced. Second, this spatial redistribution of energy is found to be time-independent. This means that a stable pattern of fixed interference fringes is formed.

The phenomenon of wave interference reduces, therefore, to a spatial redistribution of intensities of waves which results in formation of a fixed pattern of interference fringes.

It is worthwhile to mention that interference of two plane waves constitutes one of the elementary cases of interference. It is also possible to analyse the cases of interfering plane and spherical waves. Obviously, the more complicated the wavefronts, more complex the interference fringe patterns are.

4.4 INTERFERENCE

A very important characteristic of wave motion is the phenomenon of interference. This occurs when two or more wave motions coincide in *space* and *time*, i.e. when two (or more) waves are in the same place at the same time, we say that they *interfere* with each other. The pattern of bright and dark fringes that results is called an *interference pattern*. By *interference*, we really mean that the waves are in the same place at the same time and apparently get in each other's way and follow the **principle of superposition**.

We have seen that as a result of superposition of two or more waves, the redistribution of energy takes place. This modification is called interference.

The interference in which there is an *increase of amplitude* and hence *intensity* is called *constructive interference* and when there is a *decrease* of *amplitude* and hence *intensity* is called the *destructive interference* (Fig. 4.8).

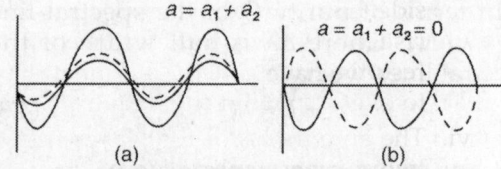

Fig. 4.8: (a) Constructive superposition (b) Destructive superposition

Interference of Light

In the case of light the observation of interference is not so easy as for other types of wave motion. One cannot observe interference pattern on a screen by placing two candles in front of it, i.e. by using two conventional sources of light we fail to observe any reinforcement or weakening of the light intensity in the illuminated space. In order to make the interferences in light persistent, *coherent sources* of light must be used. *Two waves are said to be coherent if their phase difference is independent of time.* This condition is met by two monochromatic waves of the same frequency.

The necessary conditions for producing persistent interference in light can be summarized as follows:

i. The two sources should be coherent, i.e. they should emit continuously the waves either in the same phase or with a constant phase difference.

ii. The two sources should emit continuously the waves of the same *wavelength*, *period* or *frequency*. Above two conditions are necessary for sustained interference.

iii. If the interfering waves are polarised, then their state of polarisation must be same.

iv. The two coherent sources should lie very close to each other so that the path difference between the waves reaching particular point on the screen is not very large. If this is not so, the dark and bright fringes will not be sharp.

v. The two sources must be very narrow. A broad source is equivalent to a large number of point sources lying side by side. Each pair of such sources will give rise to its own interference pattern resulting in a general illumination due to overlapping of the patterns.

vi. The amplitudes of the waves from the two sources should be equal. If it is not so, the intensity of minima will not be much different from that of maxima. This will result in general illumination.

vii. The background must be dark.

viii. The distance between the two coherent sources and the screen must be suitable.

Young's Double Slit Experiment

The classic experiment that demonstrates interference of light was first performed by Thomas Young in 1802. In the original experiment sunlight was used as the source, but any bright source such as a tungsten filament lamp or an arc would be satisfactory. Light is passed through a pinhole S so as to illuminate an aperture consisting of two pinholes or narrow slits S_1 and S_2 (Fig. 4.9). If a white screen is placed in the region beyond the slits, a pattern of bright and dark interference bands can be seen. The key to the experiment is the use of single pinhole S to illuminate the aperture. This provides necessary mutual coherence between the light from the two slits S_1 and S_2.

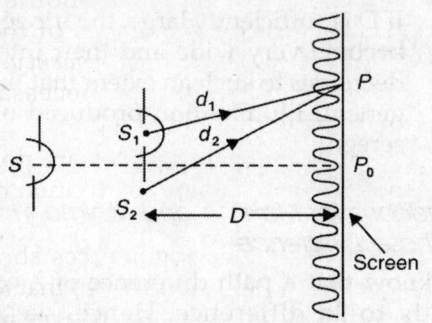

Fig. 4.9: Young's double slit experiment

The usual elementary analysis of the Young's experiment involves finding the difference in phase between the two waves arriving at a given point P over the distances d_1 and d_2 as shown in Fig. 4.9. Now, we shall

obtain expressions for path difference and phase difference.

Theory

Consider a source of light S emitting monochromatic waves of wavelength λ. Let S_1 and S_2 be two narrow slits equidistant from the source S separated by a distance $2d$ and acting as coherent sources. Let MN be the screen placed symmetrically to the slit perpendicular to the line SO at a distance D from S_2. Let P be a point distant x from O where the rays are from two slits S_1 and S_2, i.e. S_1P and S_2P. The intensity at P depends upon the path difference $S_2P - S_1P$. For the point O, the path $S_1O - S_2O = 0$, i.e. both waves reach at O in the same phase. The point O is, therefore, essentially bright. Whether point P is bright or dark depends upon the path difference $S_2P - S_1P$. From Fig. 4.10, we have

$$PQ = x - d \quad \text{and} \quad PR = x + d$$

$$S_1P^2 = D^2 + (x - d)^2 \tag{4.14}$$

and

$$S_2P^2 = D^2 + (x + d)^2 \tag{4.15}$$

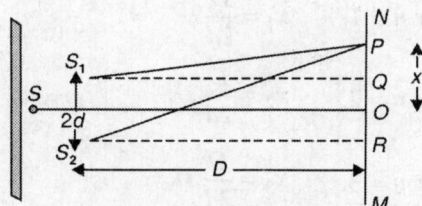

Fig. 4.10: Interference of two waves in the case of the double slit

Subtracting Eq. (4.14) from Eq. (4.15), one obtains

$$S_2P^2 - S_1P^2 = (x + d)^2 - (x - d)^2 = 4xd$$

or

$$S_2P - S_1P = \frac{4xd}{S_2P + S_1P} \tag{4.16}$$

Since the point P is very near to O and hence $D \gg x$ and $D \gg 2d$, therefore,

$$S_1P \approx S_2P = D$$

or

$$S_1P + S_2P = 2D \tag{4.16a}$$

From Eqs. (4.16) and (4.16a), one obtains

$$S_2P - S_1P = \frac{4xd}{2D} = \frac{2xd}{D} \quad (4.16b)$$

Now, if S_1 and S_2 be in the same phase, then P will have maximum intensity, i.e. $(S_1P - S_2P)$ is a whole number of wavelength, i.e. for maxima

$$\frac{2xd}{D} = n\lambda \quad (n \text{ is an integer}) \quad (4.16c)$$

Also P will have minimum intensity if $(S_2P - S_1P)$ is a whole number plus half wavelength, i.e. for minima

$$\frac{2xd}{D} = \left(n + \frac{1}{2}\right)\lambda = (2n+1)\frac{\lambda}{2}$$

$$(n \text{ is an integer}) \quad (4.16d)$$

Central Fringes

From Eq. (4.16c) when $n = 0$, $x = 0$. This gives the position of centre of the first bright fringe which coincides with O.

Bright Fringes

From Eq. (4.16c),

For $n = 1$, $\quad x_1 = \frac{D}{2d}\lambda$

For $n = 2$, $\quad x_2 = \frac{D}{2d}2\lambda$

For $n = 3$, $\quad x_3 = \frac{D}{2d}3\lambda$

For $n = m$, $\quad x_m = \frac{D}{2d}m\lambda \quad (4.17)$

If β is the distance between the two nearest bright fringes known as the fringe width, then

$$\beta = X_{n+1} - X_n = X_2 - X_1 = \frac{\lambda D}{2d} \quad (4.18)$$

As Eq. (4.18) is independent of n, therefore, the fringe width (β) for all bright fringes is the same.

For $n = 0$, let $x = x_0$, then $x_0 = \frac{D}{2d}\frac{\lambda}{2}$

For $n = 1$, $\quad X_1' = \frac{D}{2d}\frac{3\lambda}{2}$

For $n = 2$, $\quad X_2' = \frac{D}{2d}\frac{5\lambda}{2}$

For $n = n$, $\quad X_n' = \frac{D}{2d}(2n+1)\frac{\lambda}{2} \quad (4.19)$

$\therefore$ Fringe width

$$\beta = X_{n+1}' - X_n' = X_2' - X_1' = \frac{D}{2d}\lambda \quad (4.20)$$

This is again the same as that for a bright fringe. Thus, one finds that:

i. All fringes whether dark or bright are of equal width.

ii. The dark and bright fringes are alternately placed above and below the central bright fringe at O.

iii. If D and $2d$ are constants, then fringe width $\beta \propto \lambda$. Obviously, fringes produced by light of shorter wavelength will be *narrow* as compared to those produced by longer wavelength.

iv. If λ and D are kept constant, then $\beta \propto 1/2d$. Thus, if the distance between the two coherent sources $(2d)$ becomes sufficiently large, the fringes will become very narrow and general illumination will result.

v. If $2d$ and λ are kept constant, then $\beta \propto D$. If D is sufficiently large, the fringes will become very wide and their intensity decreases to such an extent that there is general illumination produced on the screen.

Conditions for Maxima and Minima in Terms of Phase Difference

We know that a path difference of λ corresponds to 2π difference. Hence, we write Eqs. (4.16c) and (4.16d) as below:

$$\left.\begin{array}{l} \Delta = \frac{2\pi}{\lambda}\frac{2xd}{D} = n\lambda \quad \text{[for maxima]} \\[2mm] \Delta = \frac{2\pi}{\lambda}\frac{2xd}{D} = \left(n + \frac{1}{2}\right)\lambda \quad \text{[for minimal]} \end{array}\right\} \quad (4.21)$$

Shape of the Interference Fringes

We have seen that the fringes formed at any point P on the screen will be bright or dark, if

$$S_2P - S_1P = n\lambda \qquad \text{(bright)}$$

$$S_2P - S_1P = \left(n + \frac{1}{2}\right)\lambda \qquad \text{(dark)}$$

For any value of n, the locus of maximum or minimum intensity point will be $S_2P - S_1P = \Delta$ (obviously Δ has to be less than $2d$).

The locus of such a point is a hyperbola. Since wavelength of the light is very small and hence the curves of intersection with the plane of the screen will be very nearly straight lines (Fig. 4.11).

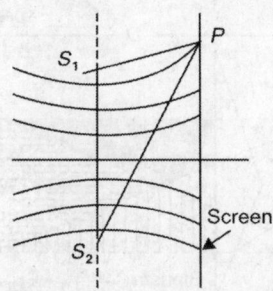

Fig. 4.11: Shape of the fringes

Intensity or Energy Distribution

Consider two waves from coherent sources S_1 and S_2 as

$$Y_1 = a\sin\omega t$$

and $\quad Y_2 = a\sin(\omega t + \delta)$

The resultant displacement is given by

$$Y = Y_1 + Y_2 = a\sin\omega t + a\sin(\omega t + \delta)$$

$$= 2a\sin\left(\omega t + \frac{\delta}{2}\right)\cos\frac{\delta}{2}$$

$$\left[\because \sin A + \sin B = 2\sin\frac{A+B}{2}\cos\frac{A-B}{2}\right]$$

or $\quad Y = 2a\cos\dfrac{\delta}{2}\sin\left(\omega t + \dfrac{\delta}{2}\right) \qquad (4.22)$

Thus, the amplitude of the resultant wave is

$$A = 2a\cos\frac{\delta}{2} \qquad (4.23)$$

Now the intensity at a point is proportional to the square of amplitude and if I_0 is the intensity due to the resultant wave and I due to the individual wave, then

$$I \propto 4a^2\cos\frac{\delta}{2}$$

and $\quad I_0 \propto a^2$

$$\therefore \quad \frac{I}{I_0} = 4\cos^2\frac{\delta}{2}$$

or $\quad I = 4I_0\cos^2\dfrac{\delta}{2} \qquad (4.24)$

For Maxima

When the path difference is 0, λ, 2λ,..., $n\lambda$, then

$$\delta = 0,\ 2\pi,\ 2(2\pi),\ ...,\ n(2\pi)$$

and $\quad \dfrac{\delta}{2} = 0,\ \pi,\ 2\pi,\ ...,\ n\pi$

Obviously, $\cos\dfrac{\delta}{2} = 1$

$\therefore \qquad\qquad I = 4I_0$, which is maximum.

Thus, the intensity is maximum when the path difference is an integral multiple of the wavelength (λ) and the fringes are bright.

For Minima

When the path difference is

$$\frac{\lambda}{2},\ \frac{3\lambda}{2},\ ...,\ (2n+1)\frac{\lambda}{2}$$

$\therefore$ The phase difference $\delta = \pi,\ 3\pi,\ ...,\ (2n+1)\pi$

and $\quad \dfrac{\delta}{2} = \dfrac{\pi}{2},\ \dfrac{3\pi}{2},\ ...,\ (2n+1)\dfrac{\pi}{2}$

Obviously, $\cos\dfrac{\delta}{2} = 0 \quad \therefore \quad I = 0$

Clearly, the intensity is minimum when the path difference is an odd multiple of $\lambda/2$, i.e. the fringes are dark.

Thus, the resultant intensity at a point varies from minimum value 0 to maximum value $4I_0$, i.e. 4 times the intensity due to the individual waves.

The intensity or energy distribution curve is as shown in Fig. 4.12 and the average value of intensity is $\dfrac{4I_0 + 0}{2} = 2I_0$, which is equal to the sum of intensity of two individual waves. This shows that there is no destruction of energy due to interference and thus the law of conservation of energy is not violated. The energy is merely transferred from places of minimum intensity to those of maximum intensity. As a result of this, the brightness at the maxima is greater than the brightness which would have been produced by the simple addition of two light beams without interference. Obviously, *interference is the phenomenon of redistribution of light energy*.

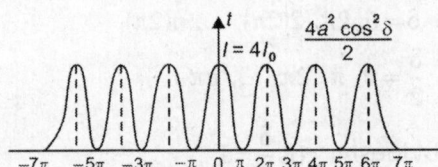

Fig. 4.12: Intensity or energy distribution

4.5 TYPES OF INTERFERENCE

The phenomenon of interference is usually divided into two classes, viz. interference due to division of wavefront and interference due to division of amplitude.

1. Interference Due to Division of Wavefront

The incident wavefront is divided into two parts, either by reflection or refraction. These two parts of the same wavefront travel unequal distances and reunite at small angle on the screen to produce interference fringes or bands. The devices based on the principle of division of wavefront to obtain interference are biprism, Fresnel's double mirror, Lloyd's single mirror, etc.

Fresnel's Biprism

It is a device for obtaining two coherent sources for producing sustained interference. It may be supposed to be made up of two prisms of very small refracting angles placed base to

base. In actual practice it is made from a single glass plate by grinding and polishing so that it becomes a single prism with one of its angles about 179° and other two about 30° each.

A horizontal section of the apparatus for production of interference fringes is shown in Fig. 4.13 in which S is a narrow vertical slit illuminated by monochromatic light. The light emerging from S is made to fall symmetrically on the biprism P having refracting edges parallel to S. The light emerging from the two virtual sources (images) S_1 and S_2 are superposed and the interference fringes are obtained in the overlapping region. These fringes can be seen through an eyepiece.

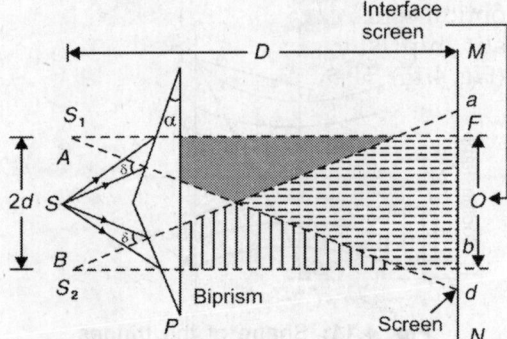

Fig. 4.13: Fresnel's biprism

Theory

The point O is equidistant from A and B and hence it has maximum intensity. On both sides of O, alternately *bright* and *dark* fringes are produced (Fig. 4.14). The spacing between two adjacent bright (or dark) fringes, i.e. fringe width is given by the relation (4.18) or (4.20), i.e.

$$\beta = \frac{D}{2d}\lambda$$

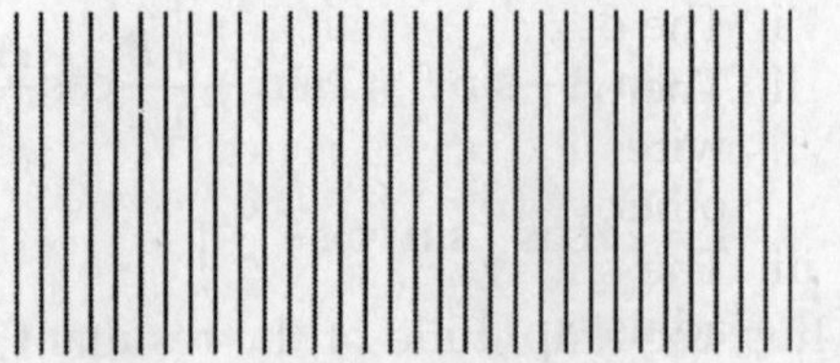

Fig. 4.14: Biprism fringes

where β is fringe width, D is distance between source and screen, $2d$ is distance between two coherent sources S_1 and S_2 and λ is wavelength of light used.

The above arrangement provides a very accurate method of finding the wavelength λ of light emitted by a monochromatic source. The relation is

$$\lambda = \frac{2d}{D}\beta \qquad (4.25)$$

From Eq. (4.25), it is clear that if $2d$, β and D are measured, the wavelength λ can be determined.

Experimental Arrangement for Determination of 2d, β and D

The experimental arrangement consists of an optical bench carrying three stands each for a slit, biprism and a micrometer eyepiece (Fig. 4.15). The biprism and slit can be rotated

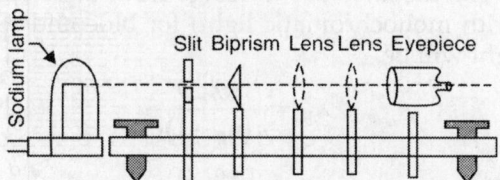

Fig. 4.15: Biprism experiment

in their own planes while the biprism and the eyepiece can also be moved at right angles to the length of the optical bench. The eyepiece can be moved with a micrometer screw and the distance moved can be accurately read on the attached circular and linear scales. The slit is illuminated with the monochromatic light whose wavelength is to be determined. The following adjustments are made prior to the observations:

i. The optical bench is levelled.
ii. The eyepiece is focussed on the cross wires and one of them is set vertical and other horizontal.
iii. The slit, eyepiece and the biprism are adjusted to the same height.
iv. The slit is made vertical by rotating it in its own plane and then it is made narrow.
v. To make the biprism at right angles to the length of the optical bench, the biprism

is moved at right angles to the optical bench until on looking through the biprism along the axis of the beam, the two equally bright images S_1 and S_2 appear in the field of view.
vi. To bring the overlapping region in the field of view the eyepiece is moved at right angles to the length of the optical bench.
vii. Now the lateral shift is removed, so that the line joining slit and the edge of the biprism becomes parallel to the length of the optical bench.

Now one can measure β, $2d$, D for determination of λ.

i. ***Determination of β***: For determination of fringe width (β), leaving the first few fringes the vertical crosswire of the eyepiece is set on a bright fringe and the position of the eyepiece is noted. Now by giving the eyepiece a lateral motion, the positions of the eyepiece corresponding to successive bright fringes are noted. From these observations, fringe width (β) is calculated.

ii. ***Determination of D***: The distance between S_1, S_2 and the focal plane of the eyepiece (D) is determined by noting the positions of the slit and eyepiece on the scale of the optical bench. By applying the bench error correction the corrected value of D is obtained.

iii. ***Determination of 2d***: For determination of $2d$, a convex lens whose focal length is less than $\frac{1}{4}$ th of the distance between the slit and the eyepiece is mounted between the biprism and the eyepiece on the optical bench. Now, by moving the convex lens along the length of the optical bench two such positions as in displacement method are obtained such that the real images S_1 and S_2 are observed in the eyepiece (Fig. 4.16). Let the separation between the two real images in the two cases be d_1 and d_2 respectively. In the first position of the

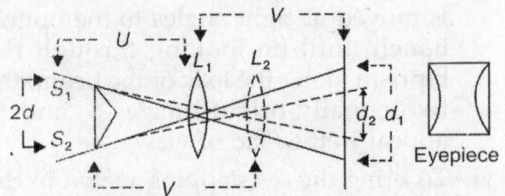

Fig. 4.16: Measurement of distance between virtual coherent sources

lens L_1, by applying magnification formula, we obtain

$$\text{magnification} = \frac{d_1}{2d} = \frac{V}{U} \qquad (4.26)$$

For the second position of lens L_2, V and U are interchanged. Again applying the magnification formula, we have

$$\text{magnification} = \frac{U}{V} = \frac{d_2}{2d} \qquad (4.27)$$

Multiplying Eqs. (4.26) and (4.27), we obtain

$$2d = \sqrt{d_1 d_2} \qquad (4.28)$$

For different positions of the eyepiece, several pairs of values of d_1 and d_2 are obtained and the mean value of $2d$ is found.

Location of Zero Order Fringes or White Light Fringes in Fresnel''s Biprism Experiment

With monochromatic light, all bright fringes are of the same colour. Therefore, it is impossible to locate the zero order fringe. But we know that with white light source, zero order fringe is white because the path difference for each wavelength of light reaching from the coherent sources at O (Fig. 4.13) is same. Therefore, by replacing monochromatic light by white light one can easily place the crosswire in the position of zero order fringe. Now replace the white light source by monochromatic light source, the crosswire in the eyepiece will be in the position of zero order band.

When monochromatic source of light is replaced by a white light source, a fringe system appears in the field of view with central white fringe surrounded by a few coloured fringes on both sides. The fringes of mixed colours are obtained due to overlapping of the maxima and minima of different colours at one place. The outer edges of the bands are violet whereas the inner edges are red. This is due to the fact that the fringe width for violet is minimum and maximum for red. Therefore, the first dark band of violet is obtained first and that of the red the last on either side of the central zero order band. The inner edge of the first minimum of violet receives sufficient light intensity from red because maximum of red falls in its vicinity and hence the edge is reddish. Similarly the first maximum of violet falls close to the inner edge of the minimum of red and hence the edge appears violet. For points at large distances from the centre, the maxima and minima due to large number of wavelengths overlap and uniform illumination is observed.

The distance of nth fringe from the centre (with monochromatic light) for blue and red light will be

$$X_b^n = \frac{n \lambda_b D}{a(\mu_b - 1)\alpha} \qquad (4.29)$$

$$X_r^n = \frac{n \lambda_r D}{a(\mu_r - 1)\alpha} \qquad (4.30)$$

Determination of the Thickness of a Thin Sheet of Transparent Material

Let a plate of thickness t and refractive index μ be introduced in the path, S_1P as shown in Fig. 4.17. The path S_1P is travelled by light partly in air and partly in plate. The path travelled in air $= S_1P - t$ and path travelled in plate $= t$. The optical path travelled by a ray is defined as:

Optical path = Distance travelled × Refractive index of the medium

$\therefore$ Optical path along $S_1P = (S_1P - t) \times 1 + t\mu$

$$= S_1P + t(\mu - 1)$$

and optical path along $S_2P = S_2P$

$\therefore$ The optical path difference $= S_2P - S_1P$

$$= S_2P - [S_1P + (\mu - 1)t]$$

Due to the introduction of a transparent plate of thickness, the optical path difference changes. The change of optical path difference causes fringe system to displace. White light fringe system is used to note this displacement of fringe system. Initially the crosswires are focussed on the central white without the plate G. Next the plate of thickness t is introduced in the path S_1P and the displacement is noted with the help of micrometer. It is worthwhile to mention that the thickness t of the plate should be small enough to displace fringes within the field of view.

In Fig. 4.17, O represents the position of zero order, i.e. central white fringe. Due to introduction of sheet G the central white fringe shifts from O to P. Therefore, the condition satisfied is

$$S_2P = (S_1P - t) + \mu t = S_1P + (\mu - 1)t \quad (4.31)$$

Lateral displacement $x = OP$ is measured.

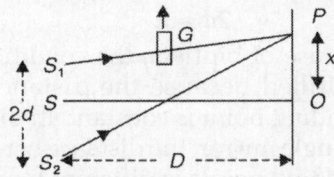

Fig. 4.17: Determination of the thickness of thin transparent film

Now we replace white light source by monochromatic light source. If β is the fringe width, then x/β represents the number of fringes of wavelength λ, which would have shifted due to introduction of plate in the path S_1P. The shift of one fringe corresponds to path difference λ and hence the path difference due to sheet, i.e.

$$S_2P - S_1P = \frac{x}{\beta}\lambda = (\mu - 1)t$$

But $$\beta = \frac{D}{2d}\lambda$$

$$\therefore \qquad (\mu - 1)t = x\frac{2d}{D} \qquad (4.32)$$

Knowing x, $2d$, D and μ, thickness t of the sheet can be calculated using Eq. (4.32).

From Eq. (4.32), $x = \dfrac{D}{2d}(\mu - 1)t$

If displacement $x = n \times$ fringe width, then $x = n\beta$

$$\therefore \qquad n\beta = \frac{D}{2d}(\mu - 1)t$$

or $$t = \frac{n\beta}{(\mu - 1)}\frac{2d}{D} = \frac{n\lambda}{\mu - 1} \qquad (4.33)$$

One can also calculate thickness of the plate knowing λ, μ and n.

The method is not suitable for determination of thickness of thick sheets due to (i) variation of μ with wavelength (ii) large lateral displacement.

Interference by Lloyd's Single Mirror

Lloyd's mirror is a plane, blackened glass plate about 30 cm in length and 6 to 8 cm in breadth (Fig. 4.18). Light incident on it is reflected from the front face. The light entering the plate is absorbed and hence there will be no reflection from the rear portion of the plate.

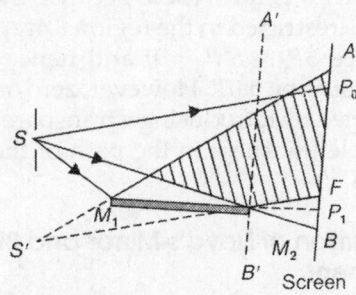

Fig. 4.18: Interference by Lloyd's single mirror

S is a slit source of light whose image is formed at S' by M_1M_2. The formation of image (S) and overlapping region is shown in Fig. 4.18. The light waves diverging from S and S' form two coherent sources. Interference takes place between the two waves diverging from S and S'. It is evident from Fig. 4.18 that the interference fringes are restricted to the region FA of the screen lying between the rays reflected from the two ends of the mirror.

We know that when light waves are reflected at the surface of a denser medium, while travelling from a rarer to a denser medium, they suffer a phase change of π. Obviously, the waves diverging from S and S' will have a permanent phase difference of π, in addition to the phase difference due to path difference. If the path difference between the interfering waves is $n\pi$ (n = 0, 1, 2, ...), then they meet in **opposite** phases (due to the permanent phase difference π) and cause **destructive** interference. If the path difference is $\left(n + \dfrac{1}{2}\right)\lambda$, then the phase difference, odd multiple of π due to this path difference, gets added to the permanent phase difference π, so that the resultant phase difference is even multiple of π. Obviously, the waves are then in phase, and they cause **constructive interference**. Clearly, the conditions for **darkness** and **brightness** on the screen are just the opposite of those in the case of biprism (if the conditions are expressed in terms of path difference).

Central Fringe

The central fringe which would be obtained at P_0 (Fig. 4.13) is not visible because the fringe system is restricted to the region FA. At P_0, the difference $SP_0 - S'P_0 = 0$ and hence central fringe would be dark. However, zero order can be obtained by introducing a transparent sheet of suitable thickness in the path of the actual source S.

Comparison of Lloyd's Mirror and Biprism Experiment

The fringes obtained from Lloyd's mirror differ from those obtained with biprism in the following respects:

i. In biprism experiment, full interference pattern is obtained while in Lloyld's mirror less than half the interference pattern is obtained.

ii. With white light source the zero order fringe is white in biprism while it is dark in Llyod's mirror.

iii. The zero order fringe in Llyold's mirror method is very sharp while in the biprism method it has a width equal to the width of either of the two coherent sources.

iv. In Llyod's mirror arrangement $2d$ is different from different pairs of coherent

sources and hence fringe width β is also different for different pairs of coherent sources. In the case of biprism experiment, the spacing between the two consecutive maxima is the same for all pairs of coherent sources.

Achromatic Fringes

Bright and dark fringes obtained by using white light source are called achromatic fringes. The fringe width is given by

$$\beta = \frac{D}{2d}\lambda \qquad (4.34)$$

This shows that fringe width is different for different wavelengths. However, β remains constant in heterogeneous light provided $\lambda/2d$ is kept constant. This means

$$2d \propto \lambda$$

In the case of biprism, the condition (4.34) is not fulfilled because the distance of the corresponding point is constant. In the case of Lloyd's single mirror, the distance between the corresponding points is different for different pairs and hence it can be employed to obtain achromatic fringes.

Fresnel's Double Mirror

One can obtain two close coherent sources by the method of reflection by using Fresnel's double mirror as shown by M_1NM_2 in Fig. 4.19. The parts M_1N and M_2N of double mirror are

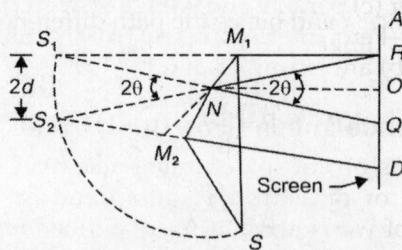

Fig. 4.19: Fresnel's double mirror

inclined with each other at very small angle. In actual practice, double mirror is made by grinding a single glass plate whose front surface is polished and silvered and back surface is blackened to avoid multiple reflection. S is a source, whose images are formed at S_1 and S_2 are the virtual coherent sources. Since the two mirrors M_1N and M_2N are inclined at small angle and hence the two virtual images S_1 and S_2 are quite close to each other. As is evident from Fig. 4.19, S, S_1 and S_2 lie on a circle with the pole of the mirror as centre. From Fig. 4.19, we have

$$2d = a \cdot 2\theta \qquad (4.35)$$

Here a is the distance of N from S_1, S_2 or S.

In actual experiment, the distance of S from N is kept quite small, so that $2d$ may be small. If b is the distance of N from screen, then

$$D = a + b \qquad (4.36)$$

where D is called the distance between the source and the focal plane of the eyepiece.

One can obtain the wavelength of the monochromatic light by the following relation:

$$\lambda = \frac{2a}{(a+b)} \theta \beta \qquad (4.37)$$

2. Interference Due to Division of Amplitude (Thin Films)

The amplitude of the incoming beam of light from the monochromatic source is divided into two parts either by partial reflection or refraction. These divided beams reunite after traversing different paths and interfere constructively or destructively. Due to use of broad sources of light, diffraction effects are minimised. The devices based on this principle are: (a) Newton's rings (b) Michelson's interferometer (c) parallel and wedge-shaped films (d) Fabry–Perot Interferometer. Truly speaking, (a) and (b) are examples of (c).

Interference in Thin Films

Thin films of transparent materials like soap babbles, or of a drop of oil spread on the surface of water show brilliant colours when exposed to an extended source of light. The phenomenon can be explained on the basis of interference. Interference takes place between rays reflected from the upper and internally from the lower surface of the film. It has been observed that interference in the case of thin films takes place due to (i) reflected light (ii) transmitted light.

Path Difference in Reflected Light

As shown in the Fig. 4.20, a ray of monochromatic light is incident from an extended source on a parallel sided film of thickness t and refractive index μ. At B, the ray divides

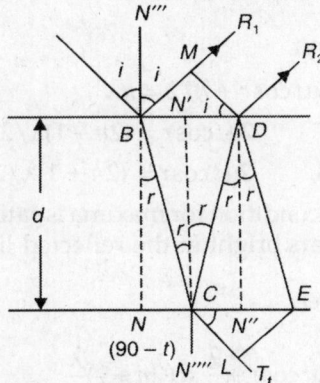

Fig. 4.20: Interference in thin films

into a reflected ray BR_1 and a refracted ray BC. At C, the ray is again partly reflected and partly refracted. Similar reflections and refractions occur at D, E, etc. Thus, one obtains a set of parallel reflected rays and parallel refracted rays. Let us calculate the effective path difference between the reflected rays BR_1 and DR_2. Let DM be $\perp^r$ to the reflected ray BR_1. Obviously, the difference between BR_1 and DR_2

= Path BCD in medium – Path BM in air

= $\mu(BC + CD) - BM$

= $2\mu BC - BM$ $\qquad (\because BC = CD)$

From Fig. 4.20, we have

$BC = BN \sec r / \cos r$

and $\quad BM = BD\cos(90° - i) = BD\sin i$

$\qquad = 2BN'\sin i = 2t\tan r \sin i$

$\qquad = 2t\dfrac{\sin r}{\cos r}\dfrac{\sin i}{\sin r}\sin r = 2t\mu \sec r \sin^2 r$

$\therefore$ The path difference

$\qquad = 2\mu\dfrac{t}{\cos r} - \dfrac{2\mu t\sin^2 r}{\cos r}$

$\qquad = 2\mu\dfrac{(1-\sin^2 r)}{\cos r} = 2\mu t\cos r$

The reflected ray BR_1 suffers a phase change of π on reflection at B and as we know that a phase change of π is equivalent to a path difference of $\lambda/2$, one obtains the effective path difference between BR_1 and DR_2

$$= 2\mu t \cos r + \lambda/2$$

For Maxima

$$2\mu t \cos r + \lambda/2 = n\lambda \qquad (4.38)$$

or $\qquad\qquad 2\mu t \cos r = (2n-1)\lambda/2$

In general, $\qquad 2\mu t \cos r = (2n \pm 1)\lambda/2 \qquad (4.39)$

When the condition for maxima is satisfied, the film appears bright in the reflected light.

For Minima

$$2\mu t \cos r + \frac{\lambda}{2} = (2n+1)\frac{\lambda}{2}$$

or $\qquad\qquad 2\mu t \cos r = n\lambda \qquad (4.40)$

(where $n = 0, 1, 2, 3, \ldots,$ etc.)

When the condition for minima is satisfied, the film appears dark in the reflected light.

It is worthwhile to mention that conditions (4.39) and (4.40) are in terms of *geometrical path difference* and not in terms of the total optical path difference. These conditions hold good for every successive pair of reflected waves.

When the *thickness of the film (t) is very small as compared to the wavelength (λ) of the light*, i.e.

$$t << \lambda$$

the total optical path difference is $\lambda/2$ only. Obviously, the rays will produce destructive interference and film will appear **dark** by reflected light.

Note: The interference pattern will not be perfect because the intensities of the rays will not be same and moreover their amplitudes are different. The amplitude depends on the amount of light reflected and transmitted through the film. It has been observed that for normal incidence, about 4% of the incident light **is** reflected and rest, i.e. 96% is BR_1 and DR_2. Obviously, the intensity never vanishes completely and perfect dark fringes will not be observed.

Path Difference in Transmitted Light

Path difference in the transmitted light

$$= \text{Path } CDE \text{ in film} - \text{Path } CL \text{ in air}$$
$$= \mu(BD + DE) - CL = 2\mu BC - CL$$

From Fig. 4.20, we have

$$CL = CE\cos(90° - i) = CE\sin i = BD\sin i$$

$\therefore$ Path difference between CT_1 and ET_2

$$= 2\mu BC - BD\sin i$$
$$= 2\mu BC - 2t\tan r \sin i \; (\because BD\sin i = 2t\tan r \sin i)$$
$$= \frac{2\mu t}{\cos r} - 2\mu t \sin r \tan r \; (\because \sin i = \mu \sin r)$$
$$= \frac{2\mu t}{\cos r} - 2\mu t \frac{\sin^2 r}{\cos r}$$
$$= 2\mu t \left(\frac{1 - \sin^2 r}{\cos r} \right) = 2\mu t \cos r$$

In the case of transmitted light, there will be no phase change due to reflection at C or D because in either case light is travelling from denser to rarer medium. Obviously, the conditions for maxima and minima in transmitted light are

$$2\mu t \cos r = n\lambda \qquad \text{(maxima)} \quad (4.41)$$
$$2\mu t \cos r = (2n+1)\lambda/2 \text{ (minima)} \quad (4.42)$$

A comparison of the conditions for maxima and minima in the reflected and transmitted light reveals that they are complementary.

Interference fringes in the case of transmitted light are less distinct because the difference in amplitude between CT_1 and ET_2 is very large. However, when the angle of incidence is nearly 45°, the fringes are more distinct.

Colours of Thin Films

When white light is incident on a thin film, the light which comes from any point from it will not include the colour whose wavelength satisfies the condition

$$2\mu t \cos r = n\lambda$$

in the reflected light. Obviously, the film will appear coloured and the colour will depend upon the thickness (*t*) and the angle of inclination. If t and r are constant, then the colour will be uniform. In the case of water or oil, different colours are seen because t and r vary.

Classification of Fringes Exhibited by Thin Films

The optical phase difference (δ) for a parallel thin film is given by

$$\delta = \frac{2\pi}{\lambda} = 2\mu t \cos r \qquad (4.43)$$

This shows that δ depends on λ, μt and r. On this basis, the fringe systems exhibited by thin films are classified into three classes, viz. **(i) Haidenger fringes or fringes of equal inclination (ii) Fringes of equal chromatic order (iii) Fizeau fringes or fringes of equal thickness**.

  i. *Haidenger Fringes or Fringes of Equal Inclination:* In this class, fringe of whatever shape it might be, corresponds to a particular value of r, while μt is constant. Constancy of μt automatically demands that light used must be almost monochromatic, i.e. λ be constant. Every fringe of the system corresponds to one value of r, i.e. characteristic of definite inclination. Familiar example of Haidenger fringes is circular fringes in Michelson interferometer with monochromatic light. These fringes are also called as **isocline fringes**.

  ii. *Fringes of Equal Chromatic Order:* When μt and r are kept practically constant, then each fringe corresponds to a particular wavelength λ. Obviously a coloured pattern of fringes is observed. These fringes are known as **fringes of equal chromatic order** (*FECO*).

  iii. *Localized Fringes or Fizeau Fringes or Fringes of Equal Thickness:* If a parallel or almost parallel pencil of rays of monochromatic light falls on a film whose thickness t varies at different places, then dark and bright interference fringes are seen in the reflected light at the top surface of the film. These are called **fringes of equal thickness**, or **isopachic fringes**, because each one passes through the points with the same values of t. Equal thickness fringes, localized on the surface of the film, can be observed on a screen as of a converging lens. Circular fringes as Newton's rings are the familiar example of this class. In white light, a system of differently coloured equal thickness fringes is observed.

Wedge-shaped Film

Let us suppose that the thin film is not parallel-sided but wedge-shaped, i.e. its surfaces make some angle with each other. One can obtain such film by placing two accurately plane pieces of glass in contact at one end, the far end being kept apart by a very thin object like a human hair or a cigarette paper. Such a film, when illuminated with a parallel beam of monochromatic light will produce evenly-spaced linear dark and bright fringes running parallel to the apex of the wedge. When such a film is illuminated with white light, coloured fringes are observed. If the light comes from an extended source, the interference pattern can be seen only at the portion of the wedge near the edge.

Consider a wedge-shaped film ABC of refractive index μ and whose surfaces are inclined at an angle θ (Fig. 4.21). The light from a sodium lamp is made to fall normally on the system by using a plane glass plate G and a travelling microscope is focussed on the fringes formed. Straight fringes parallel to the line of contact of the two glass plates will be observed through the microscope.

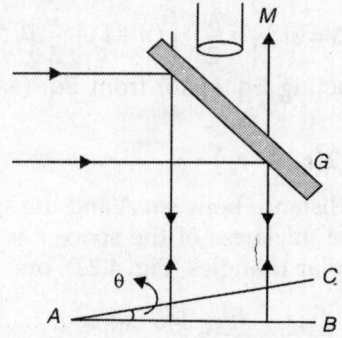

Fig. 4.21: Interference from thin wedge-shaped film

Rays of light reflected from the upper and lower faces of the air film between the plates suffer interference. The path difference between the interfering rays is given by

$$\Delta = 2\mu t \cos r + \lambda/2 \qquad (4.44)$$

Here t is the thickness of the air film at the point of the incidence of the incident ray. Since light is incident normally, we have $r = 0$ and $r = 1$. For air, $\mu = 1$. Using $r = 0$ and $r = 1$, Eq. (4.44) reduces to

$$\Delta = 2t + \lambda/2 \qquad (4.45)$$

The minimum thickness of the film is $t = 0$. The thickness varies as we move from A to B. At those points where t is such that

$$\Delta = 2n\lambda \qquad (n = 1, 2, 3, \ldots)$$

there will be brightness, and at points where t is such that

$$\Delta = \left(n + \frac{1}{2}\right)\lambda \qquad (n = 0, 1, 2, 3, \ldots)$$

there will be darkness. Since the thickness t remains constant at all points on a straight line parallel to the wire, one obtains alternately dark and bright straight fringes. These are called **fringes of equal thickness**. These fringes are formed at the upper surface of the air film and hence these are **localised**.

The nth dark fringe is formed where the thickness t_n of the air film satisfies the following condition:

$$\Delta = 2t_n + \frac{\lambda}{2} = \left(n + \frac{1}{2}\right)\lambda \qquad (4.46)$$

$(n + 1)$th dark fringe is formed when the thickness of the air film t_{n+1} satisfies the following condition:

$$\Delta = 2t_{n+1} + \frac{\lambda}{2} = \left((n+1) + \frac{1}{2}\right)\lambda \qquad (4.47)$$

Subtracting Eq. (4.46) from Eq. (4.47), one obtains

$$2(t_{n+1} - t_n) = \lambda \qquad (4.48)$$

If the distance between A and the space r is L and the thickness of the space r is D, then from similar triangles (Fig. 4.22), one obtains

$$\frac{t_{n+1}}{X_{n+1}} = \frac{t_n}{X_n} = \frac{D}{L} \qquad (4.49)$$

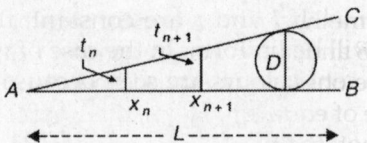

Fig. 4.22: Wedge-shaped film

where X_{n+1} and X_n are the distances of the $(n + 1)$th and nth dark fringes from A. We have

$$t_{n+1} = \frac{D}{L}X_{n+2} \quad \text{and} \quad t_n = \frac{D}{L}X_n \quad (4.50)$$

From Eqs. (4.49) and (4.50), one obtains

$$2\frac{D}{L}(X_{n+1} - X_n) = \lambda$$

or $\quad 2\dfrac{D}{L}\beta = \lambda\,[\beta = (X_{n+1} - X_n)$ is fringe width]

or $\quad \beta = \dfrac{L\lambda}{2D} \qquad (4.51)$

The total width of a known large number of fringes is measured with the travelling microscope and the fringe width (β) is calculated. The distance L between the end A and the space r is measured. Knowing the wavelength of light used, the thickness D of the space r is calculated.

The above method can be used to determine the thickness of the insulation on an enamelled copper wire. First, the experiment is performed using the wire width enamel as the space r. Then the enamel insulation is removed using sand paper and the experiment is repeated. Obviously, the difference between the diameters of the wire obtained in the two cases gives twice the thickness of the insulation.

Testing the Optical Flatness of Planes

If the two surfaces AB and AC (Fig. 4.22) are perfectly plane, the air film gradually varies in thickness from A to B. Equal thickness fringes will be observed if a beam of parallel monochromatic light is made to fall on the film because each fringe is the locus of the points at which the thickness of the film has a constant value. If the fringes are not of equal thickness it means the surfaces are not optically flat. The standard method is to take an optically plane surface *AB* and the surface to be tested *AC*. The fringes are observed through microscope and if they are of equal thickness, the surface *AC* is plane. If not, the plane *AC* is not optically flat. The surface *AC* is polished and the process is repeated. When the fringes observed are of equal thickness, it means that the plate *AC* is optically flat.

If a beam of parallel monochromatic light is made to fall on a film of absolutely uniform thickness, then *t* being the same at all points the film will appear to be uniformly tinted at all points and no interference fringes will be observed. This result provides the manufacturer of optical flats with a method of testing minute deviations from true optical flatness of the surface.

Necessity of an Extended Source

Let us consider that a thin film is viewed in reflected light from a point source *S* (Fig. 4.23). From Fig. 4.23a, it is evident that for each one angle of incidence, the interference rays are parallel to each other but for different angles of incidence the different pairs of interfering rays are obtained along widely different angles in accordance with the laws of reflection. Since the size of the pupil of the eye is limited, the interfering rays only forming a small portion

of the film can enter the eye. Obviously, it is clear that with a point source the entire film cannot be viewed by placing the eye in one position.

However, when the film is viewed with a broad source [Fig. 4.23b], the rays reflected from different parts of the film converge and hence by placing the eye in a suitable position, one can see the entire film simultaneously. It is worthwhile to mention that a narrow source is used to observe interference fringes in the case of a Fresnel's biprism, double mirrors and Llyod's single mirror.

Newton's Rings

When a convex lens of large radius of curvature is placed on a plane glass plate, an air film is formed between the upper surface of the plate and the lower surface of the lens. If monochromatic light be allowed to fall normally on this film, system of alternate bright and dark concentric rings with their centre dark is formed in the air film. As we move away from the centre, the spacing between the rings goes on decreasing. These are named as Newton's rings. These rings are formed as a result of the interference between the lens and the plate.

The experimental arrangement for observing Newton's rings is shown in Fig. 4.24. *L* is the lens kept in contact with the plane glass plate *E*. To get good fringes the radius of curvature of the face of the lens in contact with the glass plate must be fairly large, say about 1 m. *GG* is a glass plate fixed above the lens at 45° to the vertical. The light from a monochromatic source *S* (usually sodium lamp) is rendered

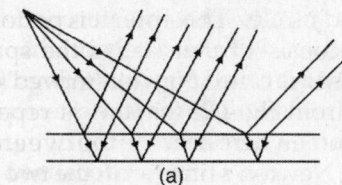

(a)

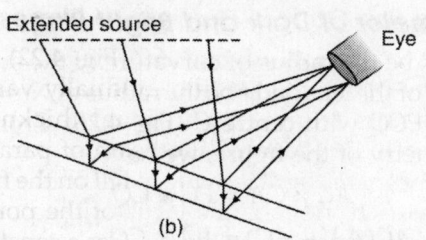

(b)

Fig. 4.23: A thin film illuminated by extended source of light

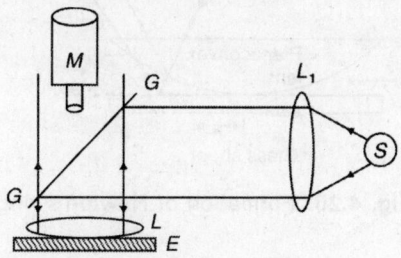

Fig. 4.24: Experimental arrangement for Newton's rings experiment

parallel by a condensing lens L_1 and made to fall on G in a horizontal direction. The light is practically reflected downwards towards the lens plate combination. After passing through the lens (L), part of it is reflected at the lower lens surface and part at the top surface of the glass plate (E). Then the light travels upwards through the glass plate G and is observed through a travelling microscope (M). There will be a path difference between the light reflected at the glass surface and the lens surface. As a result a maximum or minimum will be observed according as the path difference is an even multiple or an odd multiple of half of wavelength ($\lambda/2$). The locus of all points which have the same thickness is a circle. Therefore, circular fringes alternately bright and dark will be observed. The dark fringes are found to be more sharp than the bright fringes and therefore easier to measure.

Mathematical Analysis

In Fig. 4.25, AB is an incident ray and corresponding to it C and D are interfering reflected rays. The path difference between the reflected interfering rays is given by

$$\Delta = 2\mu t \cos(r+\theta) + \lambda/2 \quad (4.52)$$

where μ is refractive index of the film. For air, $\mu = 1$, t is the thickness of the air film at the point in question, r is the angle of refraction in the air film and θ is the angle of wedge. The experimental setup is adjusted so that $r = 0$ and θ is also very small.

$$\Delta = 2t + \lambda/2 \quad (4.53)$$

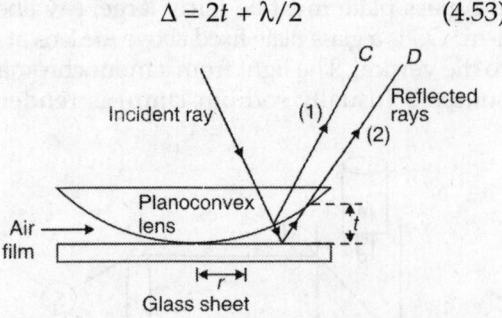

Fig. 4.25: Formation of Newton's rings

At the point of contact of the lens and the plate, $t = 0$, hence $\Delta = \lambda/2$. This is the condition for minimum intensity. Hence the central spot is dark.

At those points where the thickness of the air film satisfies the condition

$$\Delta = \left(n + \frac{1}{2}\right)\lambda \quad (4.54)$$

or $\quad 2t + \lambda/2 = \left(n + \frac{1}{2}\right)\lambda$

or $\quad 2t = n\lambda$ ($n = 1, 2, ...$) for dark rings $\quad$ (4.54a)

the rays suffer **destructive interference** and there will be **darkness**. At those points where the thickness satisfies the condition

$$\Delta = n\lambda \quad (4.55)$$

or $\quad 2t + \dfrac{\lambda}{2} = n\lambda \quad$ ($n = 1, 2, ...$) for bright rings

or $\quad 2t = \left(n - \dfrac{1}{2}\right)\lambda \quad (4.56)$

the rays suffer **constructive interference** and there will be **brightness**.

If we imagine concentric circles in the plane of the lower glass plate, with their centre at the point of contact between the plate and the lens, then the thickness of the air film remains constant for every circle. Obviously, the points of brightness and points of darkness lie along concentric circle. The interference fringes, therefore, consist of alternately bright and dark circle. These are called **Newton's rings**. One can easily see from Fig. 4.25 that the interfering rays intersect somewhere near the top of the air film. Therefore, Newton's rings are **localised** and can be seen through a microscope.

Diameter of Dark and Bright Rings

Let R be the radius of curvature of the surface PAQ of the lens and r be the radius of Newton's ring PCQ with centre C (Fig. 4.26). From the geometry of the figure, we have

$$AC \times CA' = PC \times CQ$$

or $\quad AC(AA' - AC) = PA \times CQ$

or $\quad t(2R - t) = r \times r = r^2$

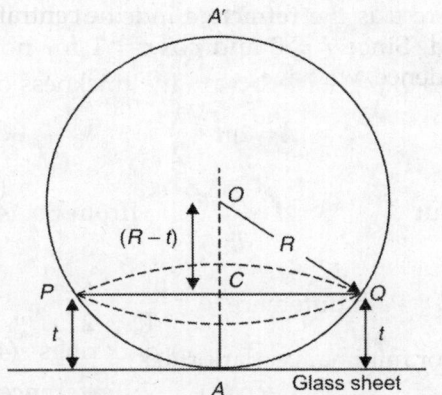

Fig. 4.26: Illustration to find diameter of rings

In practice $t >> R$, so that
$$2R - t \approx R$$
$$\therefore \qquad 2Rt = r^2$$

or
$$2t = \frac{r^2}{R} \qquad (4.57)$$

$\therefore$ Path difference,

$$\Delta = 2t + \frac{\lambda}{2} = \frac{r^2}{R} + \frac{\lambda}{2}$$

Condition for minima, i.e. dark rings

$$\frac{r^2}{R} + \frac{\lambda}{2} = (2n+1)\frac{\lambda}{2}$$

or
$$\frac{r^2}{R} = n\lambda \quad (n = 0, 1, 2, ...) \quad (4.58)$$

Equation (4.58) gives the radii r of the dark rings. The first dark ring is obtained with $n = 0$. Obviously, the radius of first dark ring is zero. Therefore, the central spot at the point of contact of the lens with the glass plate is dark. The first, second,..., nth dark rings are obtained with $n = 1$, $2,..., n$. Therefore, the radius of nth dark ring is given by

$$\frac{r_n^2}{R} = n\lambda \qquad (4.59)$$

or
$$r_n = \sqrt{Rn\lambda} \qquad (4.60)$$

If D_n is the diameter of nth dark ring ($D_n = 2r_n$), then

$$\frac{D_n^2}{4R} = n\lambda$$

or
$$D_n = \sqrt{4R\lambda}\sqrt{n}$$

or
$$D_n \propto \sqrt{n} \qquad (4.61)$$

Therefore, **diameters of dark rings are proportional to the square roots of the natural numbers**.

Condition for maxima, i.e. bright rings

$$\Delta = \frac{r^2}{R} + \frac{\lambda}{2} = n\lambda$$

or $\dfrac{r^2}{R} = \left(n - \dfrac{1}{2}\right)\lambda = (2n-1)\dfrac{\lambda}{2}$ $(n = 1, 2, 3, ...)$

$\therefore$ Radius of nth bright ring

$$r_n^2 = (2n-1)R\frac{\lambda}{2}$$

$$\therefore r_n = \sqrt{\frac{R\lambda}{2}}\sqrt{(2n-1)}$$

or diameter of nth bright ring ($D_n = 2r_n$)

$$\therefore D_n = \sqrt{2R\lambda}\sqrt{2n-1} \qquad (4.62)$$

or
$$D_n = \sqrt{2n-1} \qquad (4.63)$$

Therefore, **the diameters of the bright rings are proportional to the square roots of odd numbers**.

Determination of Wavelength

Equation (4.59) suggests a method of finding the wavelength of a given radiation if R is known or of finding R if the wavelength λ is known. If D_n be the diameter of nth dark ring, then from Eq. (4.59)

$$D_n^2 = 4Rn\lambda \qquad (4.64)$$

Similarly, the diameter of the $(n + m)$th ring is given by

$$D_{n+m}^2 = 4R(n+m)\lambda \qquad (4.65)$$

Subtracting Eq. (4.64) form Eq. (4.65), one obtains

$$\lambda = \frac{D_{n+m}^2 - D_n^2}{4Rm} \qquad (4.66)$$

After setting up the rings, the central spot must be observed to be dark. If, however, it is bright, the lens is tapped gently with a pencil till it becomes dark. The microscope is focussed on the rings. The diameters of the successive rings are measured. From Eq. (4.64), it is evident that the **fringe width decreases with the order of the fringe and the fringes (rings) get closer with increase in their order**. The wavelength λ is then calculated using Eq. (4.66). The radius of curvature R of the lower surface of the lens is found by Boy's method, or, if R is too large, by using a telescope.

If medium other than air forms the film, then Eq. (4.66) takes the form

$$\lambda = \frac{D_{n+m}'^2 - D_n'^2}{4Rm}\mu \qquad (4.67)$$

where μ is the refractive index of the medium and D_{n+m}' and D_n' are the diameters of $(n + m)$th and nth rings respectively.

Determination of Refractive Index of a Liquid using Newton''s Rings

Newton's rings method can be used for the determination of refractive index (μ) of a liquid available in small quantity. The transparent liquid whose refractive index is to be determined is placed between the lens L and the glass plate E (Fig. 4.24). If liquid is rarer than glass, a phase change of π will occur at the time of reflection from the lower face of the liquid film. If the liquid is denser than glass, then a phase change of π will occur due to reflection at the upper face of the film. Hence in either case, a path difference of $\lambda/2$ is introduced due to phase change. Therefore, path difference between the interfering waves in reflected light will be

$$\Delta = 2\mu t \cos r + \frac{\lambda}{2}$$

where μ is the refractive index of the liquid used. Since $r = 0$ and $\cos r = 1$ for normal incidence, we have

$$\Delta = 2\mu t + \frac{\lambda}{2}$$

But $\qquad 2t = \dfrac{r^2}{R} \qquad$ [from Eq. (4.57)]

$\therefore \quad$ Path difference $\Delta = \dfrac{\mu r^2}{R} + \dfrac{\lambda}{2}$

For minima, i.e. dark rings

$$\frac{\mu r^2}{R} + \frac{\lambda}{2} = (2n+1)\frac{\lambda}{2} \qquad (n = 0, 1, 2, \ldots)$$

or $\qquad \dfrac{\mu D^2}{4R} = n\lambda \qquad$ [where $D = 2r$]

or $\qquad D^2 = \dfrac{4Rn\lambda}{\mu}$

$\therefore \quad$ For nth dark ring,

$$D_n^2 = \frac{4Rn\lambda}{\mu} \qquad (4.68)$$

For $(n + p)$th dark ring

$$D_{n+p}^2 = \frac{4R(n+p)\lambda}{\mu} \qquad (4.69)$$

Subtracting Eq. (4.68) from Eq. (4.69), one obtains

$$D_{n+p}^2 - D_n^2 = \frac{4Rp\lambda}{4} \qquad (4.70)$$

or $\qquad \mu = \dfrac{4Rp\lambda}{D_{n+p}^2 - D_n^2} \qquad (4.71)$

Keeping the same value of p for air, one obtains

$$1 = \frac{4Rp\lambda}{D_{n+p}'^2 - D_n'^2} \quad (\mu = 1 \text{ for air}) \qquad (4.72)$$

From Eqs. (4.71) and (4.72), one obtains

$$\mu = \frac{D_{n+p}'^2 - D_n'^2}{D_{n+p}^2 - D_n^2} \qquad (4.73)$$

The expression shows that $D'_{n+p} > D_{n+p}$ and $D'_n > D_n$. Using Eq. (4.73), refractive index of the liquid can be determined.

Newton''s Rings Due to Transmitted Light

Arrangement for observing Newton's rings due to transmitted light is shown in Fig. 4.27. The condition for **maxima** and **bright** rings is

$$2\mu t \cos r = n\lambda$$

$$\therefore \qquad r = 0, \cos r = 1 \text{ and } \mu = 1 \text{ for air}$$

$$\therefore \qquad 2t = n\lambda \ (n = 0, 1, 2, 3, \ldots)$$

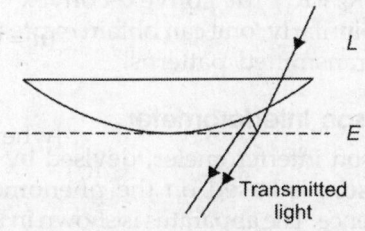

Fig. 4.27: Newton's rings due to transmitted light

When $n = 0$, for bright rings, $r = 0$. Obviously, in the case of Newton's rings due to transmitted light, the central ring is bright, i.e. just opposite to the rings due to reflected light.

For dark rings or **minima**, we have

$$2\mu t \cos r = (2n+1)\frac{\lambda}{2}$$

$$\therefore \qquad r = 0, \cos r = 1 \text{ and } \mu = 1 \text{ for air}$$

$$\therefore \qquad 2t = (2n+1)\frac{\lambda}{2}$$

Radius of **bright rings** is given by

$$r_n^2 = n\lambda R$$

For **dark rings,**

$$r_n^2 = \frac{(2n+1)\lambda R}{2}$$

Newton''s Rings with White Light

We have seen that with monochromatic light, Newton's rings are alternately dark and bright. The diameter of the ring depends upon the wavelength (λ) of the light used. When in place of monochromatic source of light, white light source is used, the diameter of the rings of the different colours will be different and coloured rings are observed. Only the first few rings are clear and after that due to overlapping of the rings of different colours of white light, one cannot see them.

Newton''s Rings with Both Curved Surfaces

In place of plane glass plate below the plane convex lens in Newton's rings experiment, one can employ curved spherical surfaces as shown in Fig. 4.28. In Fig. 4.28a, a concave surface is in contact with the convex surface and in Fig. 4.28b, two convex surfaces are

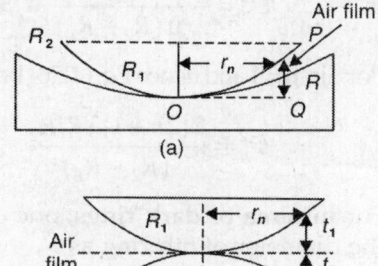

Fig. 4.28: Newton's rings with both curved surfaces

shown in contact with each other. In the former case, it is essential that the radius of curvature of the concave surface (say R_2), is greater than the radius of curvature of the convex surface (say R_1). In the second case, radii of curvatures of the two lenses in contact should be large to enclose a thin film in between.

i. In Fig. 4.28a, a convex surface of radius of curvature R_1 is in contact with a concave surface R_2 at point O ($R_2 > R_1$). Let us consider a Newton's ring of radius r_n, where the thickness of the air film is t. From Fig. 4.28a, it is evident that

$$t = PQ - QR = \left[\frac{r_n^2}{2R_1} - \frac{r_n^2}{2R_2}\right] \qquad (4.74)$$

For thin film and normal incidence, the path difference is given by

$$\Delta = 2\mu t = 2\mu \left[\frac{r_n^2}{2R_1} - \frac{r_n^2}{2R_2} \right] \qquad (4.75)$$

For **maxima** or **bright rings**, the path difference given by Eq. (4.75) must satisfy the condition

$$\mu r_n^2 \left[\frac{1}{R_1} - \frac{1}{R_2} \right] = (2n+1)\frac{\lambda}{2}$$

or $\quad \mu D_n^2 \left[\dfrac{R_2 - R_1}{2R_1 R_2} \right] = (2n+1)\dfrac{\lambda}{2}$

When $D_n = 2r_n$ is the diameter of the nth ring,

or $\qquad D_n^2 = \dfrac{2(2n+1)\lambda R_1 R_2}{\mu(R_2 - R_1)} \qquad (4.76)$

For air, $\mu = 1$ and hence Eq. (4.76) becomes

$$D_n^2 = \frac{2(2n+1)\lambda R_1 R_2}{(R_2 - R_1)} \qquad (4.77)$$

For **minima** or **dark rings**, one obtains the diameter of nth ring as

$$D_n^2 = \frac{4n\lambda R_1 R_2}{\mu(R_2 - R_1)} \qquad (4.78)$$

For air, $\quad D_n^2 = \dfrac{4n\lambda R_1 R_2}{(R_2 - R_1)} \qquad (4.79)$

ii. When two convex surfaces in contact with each other are used, then the thickness of the air film for convex-convex surfaces as shown in Fig. 4.28b is given by

$$t = t_1 + t_2 = \frac{r_n^2}{2R_1} + \frac{r_n^2}{2R_2} \qquad (4.80)$$

Proceeding exactly as in the previous case, one obtains the expression for the diameter of nth **bright** ring as

$$D_n^2 = \frac{2(2n+1)\lambda R_1 R_2}{\mu(R_1 + R_2)} \qquad (4.81)$$

For air, put $\mu = 1$ in Eq. (4.81).

The diameter of nth dark ring is given by

$$D_n^2 = \frac{4n\lambda R_1 R_2}{\mu(R_1 + R_2)} \quad \text{(for air } \mu = 1) \quad (4.82)$$

One can find the diameter of any order using Eqs. (4.78), (4.79), (4.81) and (4.82). Wavelength relation can be expressed as

$$D_{n+p}^2 - D_n^2 = \frac{4pR_1 R_2}{\mu(R_2 \pm R_1)} \qquad (4.83)$$

Here $(R_2 - R_1)$ is for concave–convex and $(R_2 + R_1)$ for convex–convex surface. Similarly, one can obtain expression for transmitted patterns.

Michelson Interferometer

Michelson interferometer, devised by Prof. A Michelson, is based on the phenomenon of interference. The apparatus is shown in Fig. 4.29. It consists of an optically flat glass plate G_1 semi-silvered at the back, two front silvered mirrors M_1 and M_2. Glass plate G_1 is set at 45° to the directions AM_1 and AM_2 which are perpendicular to each other. The mirror M_1 is mounted on a fine pitch screw (capable of reading upto 10^{-5} cm accurately) with whose help M_1 can be moved exactly parallel to itself.

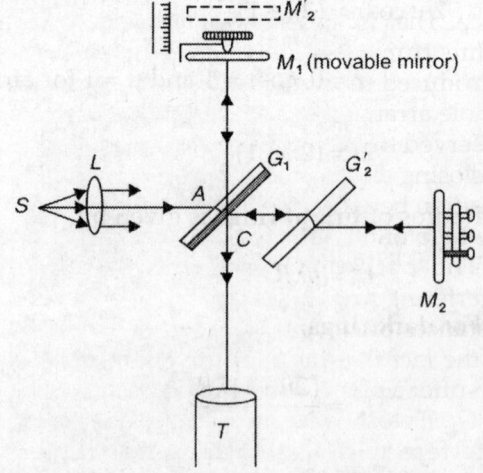

Fig. 4.29: Michelson interferometer

The mirror M_2 is provided with three levelling screws at its back. The plate G_2 is called the **compensating plate**. It is exactly similar to plate G_1 and kept parallel to G_1. The plane of G_2 can be slightly altered by means of a fine screw. The plate G_2 is essentially required to work with white light source.

The two rays AM_1 and AM_2 after reflection from M_1 and M_2 respectively produce interference when collected by a telescope T.

Working

Light from monochromatic source S after being rendered parallel by convex lens L, falls on G_1. The beam is divided into two parts, one being reflected along AM_1 and the other being transmitted along AM_2. The reflected part going along AM_1 falls on mirror M_1 normally, hence it is reflected back along its original path and then after passing through plate G_1 once again, proceeds along the path AT. The transmitted beam is also incident normally on mirror M_2, hence it is also reflected back along its original path. This suffers reflection at the lower surface of the plate G_1 and then proceeds along AT. Obviously, the telescope T receives two beams which are originally derived from the same single beam SA. Clearly, these two beams are in a position to produce interference fringes in the field of view of telescope T. The ray AM_1 passes through the plate G_1 thrice whereas the ray AM_2 passes through it only once. That is why the second plate G_2 of the same thickness and inclination as G_1 is introduced to compensate the two paths. The whole arrangement equivalent to interference observed from two reflecting faces M_1 and M'_2 enclosing an air film, where M'_2 is the virtual position between the faces can be varied with the help of micrometer screws with M_1 or M_2.

The path difference between the two interfering rays arises due to the difference in distances of M_1 and M_2 from G_1 and also due to the fact that the rays travelling from M_2 to G_1 suffer a phase change of π which is reflected at G_1. Two beams incident on the telescope interfere with each other and fringes are observed in the focal plane of the telescope.

Types of Fringes

Circular Fringes: Circular fringes are produced when the mirror M_1 and virtual mirror M'_2

which is the image of M_2 are parallel. If the distances of the mirrors M_1 and M_2 from G_1 differ by distance d, then the separation between M_1 and M'_2 will be $2d$ [Fig. 4.30a].

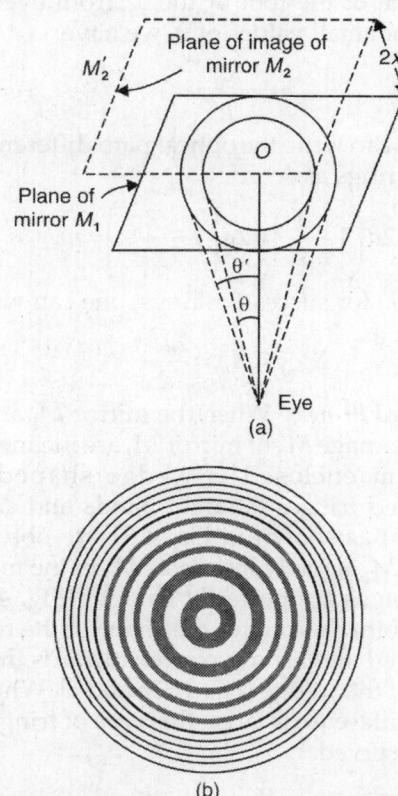

Fig. 4.30: Circular fringes in Michelson's interferometer

The path difference between the two beams will be $2d\cos\theta$. The path difference along a circle with centre O as the foot of the $\perp^r$ from eye is constant. The circumference of the circle subtends an angle θ on the eye. Obviously, the path differences for different values of θ (say θ, θ' ..., etc.) will be different, but same for one value of θ. Therefore, a ring which satisfies the condition for maxima, i.e.

$$2d\cos\theta = n\lambda$$

appears **bright** and one which satisfies the condition for **minima**, i.e.

$$2d\cos\theta = (2n + 1)\lambda/2$$

appears **black** with monochromatic light and a pattern of bright and dark rings resembling

Newton's rings in origin and are known as fringes of **equal inclination** or **Haidenger type of fringes**.

Let the radius of the nth ring be X'_n and distance of the foot of the $\perp^r$ from eye be L, then for small values of θ, we have

$$\theta = \frac{X_n}{L}$$

One can write the optical path difference for bright rings as

$$2d\left(1 - \frac{\theta'}{2}\right) = 2d\left(1 - \frac{X_n^2}{2L^2}\right) = n\lambda$$

Thus, for successive rings, one can write

$$X_{n-1}^2 = X_n^2 = \frac{\lambda L^2}{d}$$

Localized Fringes: When the mirror M_1 and the virtual image M'_2 of mirror M_2 are inclined, the air film enclosed is wedge shaped and localized fringes (straight bands and curved fringes) as shown in Fig. 4.31 are observed. When M_1 actually intersects M'_2 in the middle, the fringes are perfectly straight [Fig. 4.31b]. In the other positions, the shape of the fringes is curved and always convex towards the thin edge of the wedge [Fig. 4.31a and c]. When the path difference is large, this type of fringes are not observed.

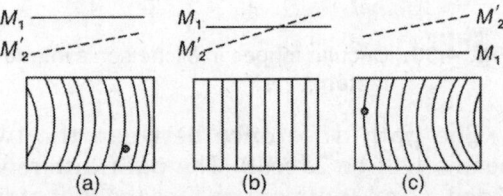

Fig. 4.31: Localized fringes by Michelson interferometer

White Light Straight Fringes: When white light source is used, the fringes are observed only when the path difference is small. Since the different colours overlap on each other hence only the first few coloured fringes are visible. The central fringe is dark or white depends upon whether the plate G_1 is unpolished or polished at the back. White light fringes are useful for the determination of zero path difference, especially in the case of measurement of the standard metre.

Setting Michelson Interferometer for Experiment

Before using Michelson interferometer, for any experiment, the following adjustments are made:

  i. First the interferometer is illuminated with a broad source of monochromatic light. Two mirrors are kept at different distances. A condensing lens is used.

 ii. A pin-hole is placed between the source S and plate G_1. Its four images are observed due to reflections from the silvered and unsilvered portions of the plate G_1 and mirrors M_1 and M_2.

iii. With the help of screws at the back of the one of the mirrors M_1 and M_2, the weaker images are made to coincide with the stronger ones, i.e. only two images of the pin-hole remain in the field of view. As soon as this adjustment is complete, the circular rings appear automatically.

If the centre of the fringe system is not in the field of view, then one can bring it in the centre of the field of view with the help of the special screws provided for the purpose at the top of the mirror.

Uses of Michelson" Interferometer

Michelson's interferometer has been used for a variety of purposes, some of the which are described below.

Determination of Wavelength of Monochromatic Light: First the fringes are obtained in the field of view with the help of a given monochromatic source of light and Michelson's interferometer. Telescope is focussed on one of the fringes. Now, the mirror M_1 is moved parallel to itself, then fringes sweep across the field of view of the telescope. For every displacement of $\lambda/2$ by mirror M_1, one bright fringe goes past the crosswire of the telescope. Number of fringes is counted. If m fringes move across, when the mirror traverses a known distance l, then

$$m\frac{\lambda}{2} = l \quad \text{or} \quad 2l = m\lambda$$

$$\therefore \quad \lambda = \frac{2l}{m} \tag{4.84}$$

Knowing l and m experimentally, λ can be found out. Conversely, if λ is known, then l can be found out.

Determination of Difference between Two Doublets or Resolution of Spectral Lines: Certain composite spectral lines, which ordinarily could not be easily separated, have been resolved by Michelson's interferometer by studying the change of visibility of fringes with increasing path difference. Suppose the source S of light emits composite light of wavelength λ_1 and λ_2. Both these wavelengths will give their own fringe systems and due to their superposition the fringes are not clearly visible. When mirror M_1 is moved, then these two sets of fringes get in step or out of step alternately. The two sets are said to be in step when bright fringe of one set falls on the bright fringe of the other. In that case the fringes become very distinct and well defined, i.e. maximum brightness is produced. The position of mirror M_1 is further moved and a stage is reached when the visibility decreases and becomes least when the two sets are out of step, i.e. the dark fringes of one system fall on the bright fringes of the other. On further moving the mirror, we again obtain a position of maximum brightness and so on.

If d be the displacement of the mirror M_1 for two consecutive positions of maximum distinctness or indistinctness, the path difference is $2d$. If the number of fringes of wavelength λ_1 lying in this path be n_1 then the number of fringes corresponding to fringes of wavelength λ_2 will be $(n + 1)$.

$$\therefore \quad 2d = n\lambda_1 = (n + 1)\lambda_2 = n\lambda_2 + \lambda_2$$

or $\quad n = \dfrac{2d}{\lambda_1}$

Also $n = \dfrac{2d - \lambda_2}{\lambda_2} = \dfrac{2d}{\lambda_2} - 1$

$\therefore \quad \dfrac{2d}{\lambda_1} = \dfrac{2d}{\lambda_2} - 1$

or $\quad 2d\left(\dfrac{1}{\lambda_2} - \dfrac{1}{\lambda_1}\right) = 1$

or $\quad \dfrac{\lambda_1 - \lambda_2}{\lambda_1\lambda_2} = \dfrac{1}{2d}$

or $\quad \lambda_1 - \lambda_2 = \dfrac{\lambda_1\lambda_2}{2d} \approx \dfrac{\lambda^2}{2d} \quad (\because \lambda_1\lambda_2 \approx \lambda^2)$

or $\quad \Delta\lambda \approx \dfrac{\lambda^2}{2d} \qquad (4.85)$

where λ is the geometric mean of two wavelengths. This method can also be used for testing whether a given source of light is monochromatic or not.

Determination of Refractive Index or Thickness of Thin Films: For the determination of refractive index of thin transparent plate or determination of thickness of thin transparent film or plate, white light fringes are obtained and the telescope is focussed on the central white fringe. The transparent plate or film is introduced in the path of one of the interfering beams, say AM_2. On doing so, the position of the central fringe shifts and mirror M_2 is moved through a distance x with the help of the micrometer screw attached to it till the central fringe is again brought back to the initial position. If t is the thickness of the film, then the path of beam is increased by $(\mu - 1)t$. Therefore, the path difference between the two beams becomes $2(\mu - 1)t$. Suppose due to this path difference, m fringes move across the field (in distance x), then

$$2(\mu - 1)t = m\lambda \qquad (4.86)$$

If t, m and λ are known, then μ can be calculated. If μ, m_2 and λ are known, then t can be calculated.

Calibration of Standard Metre: Michelson used a modified form of the interferometer for determining the number of wavelengths of a monochromatic light contained in the standard metre kept at the International Bureau of Weights and Measures at Sevres near Paris. For this experiment, he selected the red cadmium line of wavelength 6438 Å, because it is the simplest and the most homogeneous spectral line.

Fabry-Perot Interferometer

Principle

Sharp Haidinger fringes are produced here by means of multiple reflection from two partially silvered plates placed parallel to each other.

Construction

It essentially consists of two parallel glass plates A and B separated from each other (Fig. 4.32). Each plate is slightly wedge-shaped (Fig. 4.32a) to avoid inner reflecting rays to interfere.

The inner surfaces P_1, P_2 of the plates facing each other are optically plane and coated with a fine film of aluminium which reflects about 75% of the incident light. In the experimental arrangement both the plates are fitted on a straight carriage marked linearly and fitted with a micrometer scale. One of the plates B facing the observer is fixed and provided with three screws (Fig. 4.32b) with which the inner reflecting surface can be made exactly parallel

to the facing (inner) surface of plate A. The plate A can be moved along the carriage perpendicularly in its plane so that plate A is always parallel to path B and the thickness of air film between the two plates can be varied at will.

Light from a monochromatic extended source S (sodium lamp) is rendered parallel by collimating lens L_1 (Fig. 4.32c). The parallel light beam suffers multiple reflection in the film enclosed between the two plates A and B. The transmitted light $T_1 T_2$ interfere and results into circular fringes (Fig. 4.32c) of equal thickness called Haidinger fringes, which are formed in the focal plane of the telescope L and can be viewed directly.

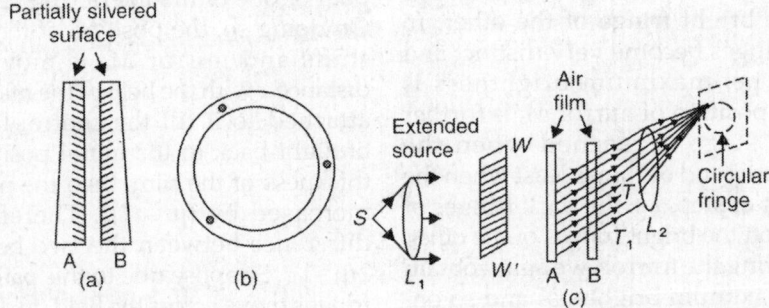

Fig. 4.32: Fabry–Perot interferometer showing formation of circular fringes

Theory of Intensity of Fringes

P_1, P_2 (Fig. 4.33) represent the plane of the two inner reflecting surfaces of the plates A and B. Light from the extended monochromatic source is incident over the plate P_1 at all angles. One such incident ray is shown in Fig. 4.33. Let the angle of incidence be θ. Here we have

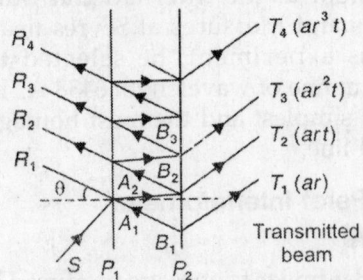

Fig. 4.33: Illustration for intensity distribution in Fabry–Perot interferometer

shown ray diagram for convenience. But actually it is the plane wavefront which is incident and undergoes both reflection and (transmission), thereby its amplitude goes on decreasing successively, which is not the case in Michelson interferometer.

Let SA_1 be the path of a plane wavefront of amplitude a incident at angle θ. As a result of the multiple reflections, consequent transmissions occur at the two boundary surfaces P_1 and P_2, and we get a set of parallel reflected waves (beam) $A_1R_1, A_2R_2, A_3R_3, \ldots$ and a set of parallel transmitted waves (beam) $B_1T_1, B_2T_2, B_3T_3, \ldots$, etc.

Suppose the reflection coefficient of each surface be r and similarly the transmission coefficient of each surface be t. Obviously, the fractions of amplitudes reflected and transmitted are $\sqrt{r}$ and $\sqrt{t}$ respectively The amplitude of wave along A_1B_1 is $(a\sqrt{t_1})$ and similarly amplitude of wave moving along B_1T_1 is $a\sqrt{t}\sqrt{t} = at$. The amplitude of wave B_1A_2 is $at\sqrt{r}$ and amplitude of wave A_2B_2 is atr. In a similar way, we can calculate the amplitudes of corresponding reflected and transmitted waves as shown in Fig. 4.33.

We shall consider the interference among the transmitted beams B_1T_1, B_2T_2, B_3T_3,..., etc. They are capable of interfering since they have been obtained from same incident beam SA_1, but their respective amplitudes are decreasing successively. Neglecting any phase change due to reflection from silvered surfaces, there is a constant phase difference between any two consecutive transmitted rays, due to path difference d between them, which is $2\mu d\cos\theta$.

Phase difference due to this path difference is

$$S = \frac{2\pi}{\lambda}[2\mu d\cos\theta] = \frac{4\pi}{\lambda}[d\cos\theta] \quad (4.87)$$

$$\text{(since for air } \mu = 1)$$

If the phase of first transmitted ray (wave) T_1 is assumed to be zero, then it represents the real part of $at\exp(j\omega t)$ and the successive transmitted rays (waves) are represented by the real part of $at\exp(j\omega t - \delta)$, $ar^2t\exp(j\omega t - 2\delta)$ respectively.

When all the transmitted rays (waves) are focussed, their resultant amplitude can be expressed as real part of

$$A = at[1 + r\exp(-j\delta) + r^2\exp(-2j\delta)$$
$$+ r^3\exp(-3j\delta) + ...] \quad (4.88)$$

R.H.S. of Eq. (4.88) represents a geometric series having ratio $r[\exp(j\delta)]$ between the consecutive terms. Sum of the terms is real part of amplitude

$$A = \left[\frac{at}{1 - r\exp(-j\delta)}\right] \quad (4.89)$$

The intensity of any wave is proportional to the square of its amplitude, i.e. A^2. So intensity

$$I \propto \frac{a^2t^2}{[1 - r\exp(-j\delta)][1 - r\exp(j\delta)]}$$

or $$I = \frac{ka^2t^2}{1 - r[\exp(-j\delta) + \exp(-j\delta)] + r^2}$$

where k is constant

or $$I = \frac{k^2a^2t^2}{1 - 2r\cos\delta + r^2} \quad (4.90)$$

The above expression can be further expressed as

$$I = \frac{ka^2r^2}{(1-r)^2}\left\{\frac{1}{1 + \frac{4r}{(1-r)^2}\sin^2\left(\frac{\delta}{2}\right)}\right\} \quad (4.91)$$

If we assume that there is no absorption of energy in the plate, then $1 - r = t$ and Eq. (4.91) can be expressed as

$$I = \frac{ka^2(1-r)^2}{(1-r)^2}\left\{\frac{1}{1 + \frac{4r}{(1-r)^2}\sin^2\left(\frac{\delta}{2}\right)}\right\} (4.92)$$

Equation (4.92) represents the expression for intensity in general. From this expression we can get expression for I_{max} and I_{min}.

Maximum intensity I_{max} is obtained when

$$\cos\delta = 1$$

or $$\delta = 2n\pi \qquad \text{where } n = 0, 1, 2, ...$$

So $$I_{max} = I_0 = \frac{ka^2}{(1-r)^2}(1-r)^2 = ka^2 \quad (4.93)$$

and minimum intensity is obtained when

$$\cos\delta = -1$$
$$\delta = (2n+1) \qquad \text{where } n = 0, 1, 2, ...$$

So $$I_{min} = \frac{ka^2(1-r)^2}{(1+r)^2} \quad (4.94)$$

Using Eq. (4.92), one obtains the general expression for intensity as

$$I = I_{max}\left[\frac{1}{1 + F\sin^2\frac{\delta}{2}}\right] \quad \text{where } F = \frac{4r}{(1-r)^2} \quad (4.95)$$

is called *coefficient of fineness* or *coefficient of sharpness.*

or $\qquad \dfrac{I}{I_{max}} = \dfrac{1}{1 + F\sin^2\left(\dfrac{\delta}{2}\right)}$ \hfill (4.96)

The conditions of maxima and minima in terms of path difference due to air film of thickness d are

   i. $\quad 2d\cos\theta = n\lambda$ \hfill (maxima)

   ii. $\quad 2d\cos\theta = (2n-1)\dfrac{\lambda}{2}$ \hfill (minima)

As d the separation between the two plates is constant for a given experimental set-up, so for particular values of n and λ, the angle θ must be constant. The locus of all points having same value of θ is circle (Fig. 4.32c).

Half Fringe Width and Sharpness of Fringes

Half fringe width is the width of fringe in terms of phase difference between points on either side of maxima where intensity falls to half its maximum value. So from Eq. (4.96), we have

$$\dfrac{I}{I_{max}} = \dfrac{1}{1 + F\sin^2\dfrac{\delta}{2}} = \dfrac{1}{2}$$

or $\quad 1 + F\sin^2\left(\dfrac{\delta}{2}\right) = 2$

or $\quad F\sin^2\left(\dfrac{\delta}{2}\right) = 1$

or $\quad \sin\left(\dfrac{\delta}{2}\right) = \sqrt{\dfrac{1}{F}} = \dfrac{1}{\sqrt{\{4r/(1-r)^2\}}} = \dfrac{1-r}{2\sqrt{r}}$

$\therefore \qquad\qquad \delta = 2\sin^{-1}\left[\dfrac{1-r}{2\sqrt{r}}\right]$ \hfill (4.97)

If δ is small, we have $\sin\left(\dfrac{\delta}{2}\right) = \dfrac{\delta}{2}$ (radians)

Equation (4.97) can be written as

$$\delta = \dfrac{1-r}{\sqrt{r}}$$ \hfill (4.98)

Sharpness of maxima depends upon half fringe width. Smaller the half fringe width, sharper is fringe.

From Eq. (4.97 or 4.98), it is clear that half fringe width decreases as the value of reflection coefficient r increases, and consequently maxima are sharper.

In general, here the maxima are sharper (Fig. 4.34) than those obtained in case of Michelson interferometer. Figure 4.34a represents the half width and Fig. 4.34b represents the variation of I/I_{max} with δ for different values of r.

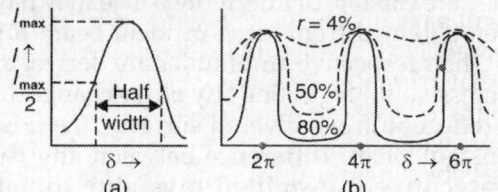

Fig. 4.34: Intensity curves for different values of reflection coefficient

Measurement of Wavelength of a Monochromatic Source of Light

The Fabry-Perot etalons are adjusted so that circular fringes are formed in the focal plane of telescope or eye [Fig. 4.32c]. Let n be the order of bright fringe at the centre of fringe system. At the centre the angle $\theta = 0°$. So path difference between two consecutive transmitted beams is

$$2t = n\lambda$$

When the movable plate A is moved parallel to B through a distance $\dfrac{\lambda}{2}$ on the carriage on which it is mounted, the path difference $2t$ changes by λ and hence we observe next order bright fringe at the centre of the system.

If the movable plate moves from position x_1 to x_2 on the Carriage Scale and N be the number of fringes appearing at the centre during this movement of plate, then

$$N\dfrac{\lambda}{2} = x_2 - x_1$$

$\therefore \qquad\qquad \lambda = \dfrac{2(x_2 - x_1)}{N}$ \hfill (4.99)

Hence, we can determine the unknown wavelength of light from Eq. (4.99).

Measurement of Difference in Wavelength (of Two Close Spectral Lines)

The difference in wavelength of two close spectral lines (sources) can be determined by the method of coincidence.

Let λ_1 and λ_2 be the wavelengths of two close spectral lines (D_1, D_2 in case of sodium light). At first the movable plate A is brought in contact with plate B. In this situation, the fringe system due to two sources (wavelength) λ_1 and λ_2 will partially coincide. By moving the plate A parallel to B, the path difference d increases, so that fringe system due to λ_1 and λ_2 will separate out and maximum discordance occurs when the rings system due to λ_2 are half way between those due to λ_1. Let for maximum discordance the separation between the plates be d_1. At the centre of system $\theta = 0°$ and $\cos \theta = 1$ so

$$2d_1 = n_1 \lambda_1 = \left(n_1 + \frac{1}{2} \right) \lambda_2 \text{ (here assume } \lambda_1 > \lambda_2\text{)}$$

$$(4.100)$$

or

$$n_1 \lambda_1 = n_1 \lambda_2 + \frac{\lambda_2}{2}$$

or

$$n_1 (\lambda_1 - \lambda_2) = \frac{\lambda_2}{2}$$

or

$$n_1 = \frac{\lambda_2}{2(\lambda_1 - \lambda_2)}$$

Substituting for n_1 in Eq. (4.100), we get

$$2d_1 = \frac{\lambda_2 (\lambda_1)}{2(\lambda_1 - \lambda_2)}$$

$$\therefore \qquad \lambda_1 - \lambda_2 = \frac{\lambda_1 \lambda_2}{4d_1} \qquad (4.101)$$

Since the two spectral lines are closely situated, so we have

$$\lambda_1 \lambda_2 = \lambda^2$$

$$\therefore \qquad \lambda_1 - \lambda_2 = \frac{\lambda^2}{4d_1} \qquad (4.101a)$$

If we further increase the separation between the two plates, then the rings system again coincides and then separate out and second maximum discordance appears.

Let the separation between plates be d_2.

$$2d_2 = n_2 \lambda_1 = \left(n_2 + \frac{3}{2} \right) \lambda_2 \qquad (4.102)$$

From Eqs. (4.101) and (4.102), one obtains

$$2(d_2 - d_1) = (n_2 - n_1)\lambda_1 = (n_2 - n_1)\lambda_2 + \lambda_2$$

or $(n_2 - n_1) (\lambda_1 - \lambda_2) \gtrless \lambda_2$

or

$$(n_2 - n_1) = \frac{\lambda_2}{\lambda_1 - \lambda_2} \qquad (4.103)$$

Substituting the value of $(n_2 - n_1)$ from Eq. (4.103) in Eq. (4.102), we get

$$2(d_2 - d_1) = \left[\frac{\lambda_2}{\lambda_1 - \lambda_2} \right] \lambda_1$$

or

$$\lambda_1 - \lambda_2 = \frac{\lambda_1 \lambda_2}{2(d_2 - d_1)}$$

$$\therefore \qquad \Delta \lambda = \lambda_1 - \lambda_2 = \frac{\lambda^2}{2(d_2 - d_1)} \qquad (4.104)$$

Here $\lambda^2 = \lambda_1 \lambda_2$ and λ is the geometric mean of two wavelengths.

Hence, by noting the separation d_1 and d_2 on the carriage scale the, difference in wavelength $\Delta \lambda$ can be calculated from Eq. (4.104).

Chromatic Resolving Power

As explained above, the fringes obtained in case of Fabry–Perot interferometer are sharper than those obtained in case of Michelson interferometer. On account of this characteristic, fine structure of a spectral line and its half width can be determined more accurately.

The chromatic resolving power ($\lambda / \Delta \lambda$) of an instrument depends on the fact, that how much the separation ($\Delta \lambda$) between two consecutive spectral lines, could be determined by knowing the mean wavelength 'λ' of the two lines ($\lambda_1 \lambda_2 = \lambda^2$).

According to Fabry–Perot interferometer

$$\frac{\lambda}{\Delta \lambda} = (-) \frac{m}{\Delta m_0}$$

and this change depends on the sharpness (intensity) of the fringe pattern which is given by

$$I = \frac{I_{max}}{1 + \left[\frac{4r}{(1-r)^2}\right]\sin^2\left(\frac{\delta}{2}\right)} \quad (4.105)$$

Here I_{max} represents the maximum intensity of the central maxima and I represents the intensity of any other maxima separated from the central maxima by δ.

Rayleigh's criterion for separation of two spectral lines is that the maxima of one should coincide with the minima of the other line as shown in Fig. 4.35.

In the above situation, the maximum intensity of the envelope in between the two maxima (Fig. 4.35) should be $\frac{8}{\pi^2}I_{max}$ ($\approx 0.8I_{max}$). It means contribution to the intensity of this part of the envelope due to separate spectral line be $\frac{4}{\pi^2}I_{max}$ ($\approx 0.4I_{max}$).

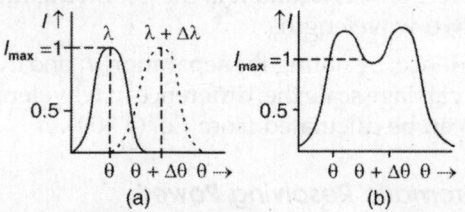

(a)   (b)

Fig. 4.35: Rayleigh's criterion for separation of two spectral lines

According to the above definition, Eq. (4.105) can be expressed as

$$\frac{I}{I_{max}} = \frac{4}{\pi^2} = 0.405 = \frac{1}{1 + \frac{4r}{(1-r)^2}\sin^2\left(\frac{\delta}{2}\right)}$$

or $\quad 0.405 + 0.405\left[\frac{4r}{(1-r)^2}\sin^2\left(\frac{\delta}{2}\right)\right] = 1$

or $\quad \sin^2\frac{\delta}{2} = \dfrac{0.595}{0.405\left[\dfrac{4r}{(1-r)^2}\right]} = \dfrac{1.469}{F}$

or $\quad$ where $F = \dfrac{4r}{(1-r)^2}$

or $\quad \delta = 2\sin^{-1}\left[\sqrt{\dfrac{1.469}{F}}\right] \quad (4.106)$

or $\quad 2\delta = 4\sin^{-1}\left[\sqrt{\dfrac{1.469}{F}}\right] \quad (4.106a)$

Hence the separation between two consecutive maxima which can be detected should be

$$\Delta m_0 = \frac{4\sin^{-1}\left[\sqrt{\dfrac{1.469}{F}}\right]}{2\pi}$$

Now by definition, resolving power

$$\frac{-\lambda}{\Delta\lambda} = (-)\frac{m_0}{\Delta m_0} = (-)\frac{m_0\pi}{2\sin^{-1}\left[\dfrac{1.21}{\sqrt{F}}\right]} \quad (4.107)$$

Substituting for F, we obtain

$$\text{R.P.} = \frac{\lambda}{\Delta\lambda} = -\frac{m_0\pi}{2\sin^{-1}\left[\dfrac{1.21(1-r)}{2\sqrt{r}}\right]} \quad (4.108)$$

Equation (4.108) shows that resolving power depends on (i) order of interference m_0 and (ii) coefficient of reflection of the surface.

ILLUSTRATIVE EXAMPLES
Example 1
The coherence length for sodium D_1 line is 2.5 cm. Determine: (a) the spectral width of the line, $\Delta\lambda$, (b) the purity factor, Q, (c) the coherence time, τ. Take wavelength of light = 6000 Å.

Solution

(a) $\quad \Delta\lambda \approx \dfrac{\lambda}{2L} \approx \dfrac{36\times10^{-10}}{5}\text{cm} \approx 7\text{ Å}$

(b) $\quad Q \approx \dfrac{\lambda}{\Delta\lambda} \approx \dfrac{6\times10^{-5}}{7\times10^{-8}} \approx 10^3$

(c) $\quad \tau \approx \dfrac{L}{C} = \dfrac{2.5}{3\times10^{10}}\text{sec} \approx 0.8\times10^{-10}\text{ s.}$

Example 2

In a Young's double slit experiment on interference, the distance between the slits is 1 mm; the distance between the slit and the screen is 1 m. The wavelength used is 5893 Å. Compare the intensity at a point distant 1 mm from the centre to that at the centre. Also, find the minimum distance from the centre of a point where the intensity is half of that at the centre.

Solution

i. Path difference

$$\Delta = \frac{x \cdot 2d}{D} = \frac{0.1 \times 0.1}{100} = 10^{-4} \text{ cm} = 10^{-5} \text{ m}$$

ii. Path difference

$$\delta = \Delta \frac{2x}{\lambda} = \frac{10^{-4} \times 2\pi}{5893 \times 10^{-8}} = 3.39\pi \text{ radians}$$

iii. $\dfrac{I}{I_0} = \cos^2 \dfrac{\delta}{2} = \cos^2 (1.697\pi)$

$$= \cos^2 (125.5°) = 0.3372$$

iv. When the intensity is half of the maximum, if δ is the phase difference, then we have

$$\cos^2 \frac{\delta}{2} = 0.5$$

or $\cos \dfrac{\delta}{2} = \sqrt{0.5} = 0.707$

$\therefore \qquad \dfrac{\delta}{2} = \dfrac{\pi°}{4}$ or $\delta = \dfrac{\pi°}{2}$

Obviously path difference $\Delta = \delta \dfrac{\lambda}{2\pi} = \dfrac{\lambda}{4}$

v. Distance of the point on the screen from the centre

$$x = \Delta \frac{D}{2d} = \frac{\lambda}{4} \frac{D}{2d} = \frac{5893 \times 10^{-8} \times 100}{4 \times 0.1}$$

$$= 1.47 \times 10 \text{ cm} = 1.47 \text{ m.}$$

Example 3

Two coherent sources whose intensity ratio is 100 : 1 produce interference fringes. Show that the ratio of maximum to minimum intensity in the fringe system is 121 : 81.

Solution

$$\frac{I_{max}}{I_{min}} = \frac{(a_1 + a_2)^2}{(a_1 - a_2)^2}$$

where a_1 and a_2 are amplitudes of two waves.

Given $\dfrac{I_1}{I_2} = \dfrac{100}{1}$, where I_1 and I_2 are intensities of two sources.

$\because \qquad I_1 \propto a_1^2$ and $I_2 \propto a_2^2$

$\therefore \qquad \dfrac{a_1}{a_2} = \dfrac{10}{1}$ or $a_1 = 10a_2$

$\therefore \qquad \dfrac{I_{max}}{I_{min}} = \dfrac{(10a_2 + a_2)^2}{(10a_2 - a_2)^2} = \dfrac{(11)^2}{(9)^2} = 121 : 81$

Example 4

The incident faces of a glass biprism of $\mu = 1.5$ makes angle of 2° with the flat face of the prism. The slit is at a distance of 10 cm from the biprism and is illuminated by light of wavelength 5890 Å. Calculate the band width of the fringes formed at a distance of 1 m from the biprism.

Solution

The distance $2d$ between the two virtual sources

$$2d = 2a(\mu - 1)\alpha$$

$$a = 10 \text{ cm}$$

$$\mu = 1.5$$

$$\alpha = 2° = \frac{\pi}{180} \times 2 \text{ radians}$$

$$= \pi / 90 \text{ radians}$$

$$D = 100 + 10 = 110 \text{ cm}$$

$\therefore \qquad 2d = 2a \times 10(1.5 - 1) \times \dfrac{\pi}{90} \text{ cm}$

$$= 0.349 \text{ cm}$$

Now, fringe width $\beta = \dfrac{D\lambda}{2d}$

$\therefore \qquad \beta = \dfrac{110 \times 5890 \times 10^{-8}}{0.349}$ cm $= 0.019$ cm .

Example 5

In Fresnel's biprism experiment, the fringe width of 0.185 cm is observed at a distance of 1 m from the slit. The image of the coherent sources is then produced at the same distance from the slit by placing a convex lens at 30 cm from the slit. Two images are found to be separated by 0.7 cm. Calculate the wavelength of the light used.

Solution

Let U and V be the distances of the virtual sources and their images from the convex lens. Let $2d$ and d_1 be the separation of sources and their images respectively, then we have

$\dfrac{U}{V} = \dfrac{2d}{d_1}$ or $2d = \dfrac{U}{V} d_1 = \dfrac{30}{70} \times 0.7 = 0.3$ cm

Fringe width $\beta = \dfrac{D\lambda}{2d}$

$\therefore \qquad \lambda = \beta \dfrac{2d}{D} = \dfrac{0.0185 \times 0.3}{100}$

$= 5550 \times 10^{-8}$ cm $= 5550$ Å.

Example 6

A thin slit illuminated by a monochromatic source of light is placed at a distance of 0.5 m from a biprism of $\mu = 1.5$. The distance between two consecutive bands formed on a screen placed at a distance of 1 m from the biprism is found to be 0.12 mm. If the wavelength of the light used be 5890 Å, find the magnitude of obtuse angle of biprism.

Solution

We have $\qquad \beta = \dfrac{D\lambda}{2a(\mu-1)x}$

$D = 50 + 100 = 150$ cm $= 1.5$ m

$\lambda = 5890 \times 10^{-8}$ cm

$a = 50$ cm

$\mu = 1.5$

$b = 0.012$ cm

$\therefore \qquad \alpha = \dfrac{D\lambda}{2a(\mu-1)\beta}$

$\therefore \qquad \alpha = \dfrac{150 \times 5890 \times 10^{-8}}{2 \times 50 \times (1.5-1) \times 0.012}$ radians

$= \dfrac{150 \times 5890 \times 10^{-8}}{2 \times 50 \times 0.5 \times 0.012} \times \dfrac{180}{\pi}$ degree

$= 0.84°.$

Example 7

A Llyold's single mirror has a length 5 cm. It is kept with plane at a distance of 1 mm from a slit illuminated by a monochromatic source of wavelength 5890 Å. The distance between the slit and the nearest edge of mirror is 15 cm. A screen is kept at a distance of 1.2 m from the slit. Find the bandwidth and width of the interference pattern.

Solution

i. $\qquad \beta = \dfrac{D}{2d}\lambda$

$2d = 2$ mm $= 0.2$ cm

$\lambda = 5890 \times 10^{-8}$ cm

$D = 120$ cm

$\therefore \qquad \beta = \dfrac{120}{0.2} \times 5890 \times 10^{-8}$ cm

$= 3.5 \times 10^{-2}$ cm

ii. In the Fig. 4.36, we have

$\Delta OSA \parallel\!\parallel \Delta CPA$

$\therefore \qquad \dfrac{PC}{SO} = \dfrac{AC}{OA}$

i.e. $\qquad PC = SO \times \dfrac{AC}{OA} = 0.1 \times \dfrac{110}{10} = 1.1$ cm

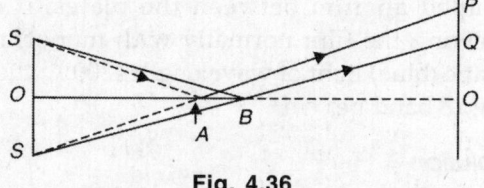

Fig. 4.36

Similarly from $\Delta OSB \; ||| \; \Delta CQB$

$$\therefore \quad \frac{OC}{SO} = \frac{BC}{BO}, \quad \text{i.e. } QC = SO \times \frac{BC}{BO}$$

$$= 0.1 \times \frac{105}{15} = 0.7 \text{ cm}$$

Obviously, the width of the interference pattern

$$QP = PC - QC = 1.1 - 0.7 = 0.4 \text{ cm}.$$

Example 8

The angle between the two planes of Fresnel's double mirror is 20′. The point of intersection of the mirrors is at a distance of 0.3 m from the slit illuminated by a monochromatic source of light of wavelength 5460 Å. The eyepiece is mounted at a distance of 120 cm from this point of intersection. Calculate the bandwidth and position of the 10th dark band from the centre.

Solution

i. $\qquad \beta = \frac{D}{2d} \lambda$

$$\theta = 20' = \frac{\pi}{540} \text{ radians}$$

$$a = 30 \text{ cm}$$

$$\lambda = 5460 \times 10^{-8} \text{ cm}$$

$$D = 30 + 120 = 150 \text{ cm}$$

$$\therefore \quad \beta = \frac{D\lambda}{2a\theta} = \frac{150 \times 5460 \times 10^{-8}}{2 \times 30 \times \pi/540}$$

$$= 0.0235 \text{ cm}.$$

ii. The distance of the 10th dark band from the centre

$$= 9.5\beta = 9.5 \times 0.0235 \text{ cm} = 0.223 \text{ cm}.$$

Example 9

While light is incident on a soap film at an angle $\sin^{-1}(4/5)$ and the reflected light on examination by a spectroscope shows dark bands. Two consecutive dark bands correspond to wavelengths 6.1×10^{-5} cm and 6×10^{-5} cm. If μ of the film be 4/3, calculate its thickness.

Solution

The path difference for dark bands in reflected light is given by

$$2\mu t \cos r = \mu\lambda \qquad (1)$$

Let the order of the bandwidth wavelength $\lambda_1 = 6.1 \times 10^{-7}$ m be n, then the order with wavelength $\lambda_2 = 6 \times 10^{-7}$ m will be $(n+1)$.

Thus, one can write (i) for the dark bands of two wavelengths as below:

$$2\mu t \cos r = n\lambda_1 \qquad (2)$$

and $\qquad 2\mu t \cos r = (n+1)\lambda_2 \qquad (3)$

From (2) and (3), one obtains

$$n\lambda_1 = (n+1)\lambda_2$$

or $\qquad n = \frac{\lambda_2}{\lambda_2 - \lambda_1} \qquad (4)$

From (2) and (4), we obtain

$$2\mu t \cos r = \frac{\lambda_1 \lambda_2}{\lambda_2 - \lambda_1}$$

or $\qquad t = \frac{\lambda_1 \lambda_2}{\lambda_2 - \lambda_1} \frac{1}{2\mu t \cos r} \qquad (5)$

From Snell's law, we have

$$\mu = \frac{\sin i}{\sin r} \quad \text{or} \quad \sin r = \frac{\sin i}{\mu}$$

$$\therefore \quad \cos^2 r = 1 - \sin^2 r = 1 - \frac{\sin^2 i}{\mu^2} = \frac{\mu^2 \sin^2 i}{\mu^2}$$

or $\qquad \mu \cos r = \sqrt{\mu^2 - \sin^2 i} \qquad (6)$

Substituting the value of $\mu \cos r$ in (5), one obtains

$$t = \frac{\lambda_1 \lambda_2}{\lambda_1 - \lambda_2} \frac{1}{2\sqrt{\mu^2 - \sin^2 i}} \qquad (7)$$

$\lambda_1 = 6.1 \times 10^{-7}$ m, $\lambda_2 = 6 \times 10^{-7}$ m, $\mu = 4/3$ and $\sin i = 4/5$

$$\therefore\ t = \frac{6.1 \times 10^{-7} \times 6 \times 10^{-7}}{\left(6.1 \times 10^{-7} - 6 \times 10^{-7}\right) 2 \sqrt{\left(\frac{4}{3}\right)^2 - \left(\frac{4}{5}\right)^2}}\ \text{m}$$

$$= \frac{6.1 \times 6 \times 10^{-7} \times 3 \times 5}{0.1 \times 2 \times 4 \times 4}\ \text{m} = 171.5 \times 10^{-7}\ \text{m}$$

Example 10

Intereference fringes are produced with monochromatic light falling on a wedge-shaped film of cellophane whose index of refraction is 1.4. The angle of the wedge is 40 seconds and the distance between the successive fringes is 1.25 mm. Calculate the wavelength of the light used.

Solution

The spacing between consecutive bands in a wedge viewed normally is given by

$$\Delta = \frac{\lambda}{2\mu\theta}$$

$$\Delta = 0.125\ \text{cm}$$

$$\mu = 1.4$$

$$\theta = 40'' = \left(\frac{40}{60 \times 60}\right)^{\circ}$$

$$= \frac{40}{3600} \times \frac{\pi}{180}\ \text{radian}$$

or $\quad \lambda = \Delta 2\mu\theta$

$$\therefore \quad \lambda = 0.125 \times 2 \times 1.4 \times \frac{40}{3600} \times \frac{\pi}{180}$$

$$= 0.125 \times 2 \times 1.4 \times \frac{1}{90} \times \frac{22}{7 \times 180} = 7 \times 10^{-5}\text{cm}$$

Example 11

Two pieces of plane glass are placed together with a piece of paper between the two at one edge. Find the angle, in seconds, of the wedge-shaped air film between the plates, if on viewing the film normally with monochromatic (blue) light of wavelength 4800 Å, there are 18 band per cm.

Solution

Let space occupied by nth band be $(x_2 = x_1)$, then the condition can be expressed as

$$2\mu\theta(x_2 = x_1) = n\lambda$$

$$\lambda = 4800\ \text{Å}$$

$$n = 18$$

$$\mu = 1\ \text{(air)}$$

$$x_2 - x_1 = 1\ \text{cm}$$

$$\theta = ?$$

$\therefore \quad 2 \times 1 \times \theta \times 1 = 18 \times 4800 \times 10^{-8}$

or $\qquad \theta = 9 \times 4800 \times 10^{-8}$ radian

$$\theta = 9 \times 4800 \times 10^{-8} \times \frac{180}{\pi}\ \text{degree}$$

$$= 9 \times 4.8 \times 10^{-5} \times \frac{180}{22} \times 7 \times 3600\ \text{second}$$

$$= 89.1\ \text{second}$$

Example 12

A vertical soap film of length 10 cm is viewed by reflected monochromatic sodium light of wavelength 5893 Å. Just before the film breaks there are (i) 12 black and 11 bright fringes, (ii) 12 black and 12 bright fringes. If the refractive index of a soap solution is 1.33, find the angle of wedge formed and thickness of the film at the base just before it breaks.

Solution

i. The bandwidth = 10/11, as the length of 10 cm is divided into 11 spacings. The first and the last fringe being black and in between fringes bright.

$$\therefore \qquad \beta = \frac{\lambda}{2\mu\theta}$$

$$\lambda = 5893 \times 10^{-8}\ \text{cm},\ \mu = 1.33$$

$$\beta = \frac{10}{11}$$

$$\therefore \qquad \beta = \frac{\lambda}{2\mu\theta}$$

$$\therefore \theta = \frac{5893 \times 10^{-8} \times 11}{2 \times 10 \times 1.33} = 2.44 \times 10^{-5} \text{ radian}$$

When the film breaks, obviously, the thickness of the top is zero.

$\therefore$ Thickness at the bottom

$$= x_0 \theta = 10 \times 2.44 \times 10^{-5} \text{ cm}$$

$$= 2.44 \times 10^{-4} \text{ cm}$$

ii. Bandwidth $\beta = \dfrac{10}{11.5}$ cm ($\because$ spacing $= 11.5$)

$\therefore$ Angle of wedge

$$= \frac{\lambda}{2\beta\mu} = \frac{5893 \times 10^{-8} \times 11.5}{2 \times 10 \times 1.33}$$

$$= 2.55 \times 10^{-5} \text{ radian}$$

$\therefore$ Thickness of the film at the base

$$x_0 \theta = 10 \times 2.55 \times 10^{-5} \text{ cm}$$

$$= 2.55 \times 10^{-4} \text{ cm}.$$

Example 13

Newton's rings are obtained by a source emitting $\lambda_1 = 6000$ Å and $\lambda_2 = 4500$ Å. If nth ring of λ_1 coincides with $(n + 1)$th ring of λ_2, then find the diameter of nth ring of λ_1 wavelength. The radius of curvature of the curved surface is 90 cm.

Solution

Let D be the diameter of nth ring for wavelengths λ_1 and λ_2, then we have

$$D^2 = 4n_1 R = 4(n + 1)\lambda_2 R \qquad (1)$$

(Here we have assumed that rings are formed in air for which $\mu = 1$)

$$\lambda_1 = 6000 \times 10^{-8} \text{ cm}, \lambda_2 = 4500 \times 10^{-8} \text{ cm}$$

$$\therefore \quad 4n \times 6 \times 10^{-5} R = 4(n + 1) \times 4.5 \times 10^{-5} R$$

or $\quad 6n = 4.5(n + 1) \quad \therefore \quad n = 3$

Now $\quad D = \sqrt{4n\lambda_1 R}$

$$= \sqrt{4 \times 3 \times 6 \times 10^{-5} \times 90} = 0.2545 \text{ cm}$$

Example 14

A thin plano-convex lens of focal length 1.8 m and of refractive index 1.6 is used to obtain Newton's rings. The wavelength of the light used is 5890 Å. Calculate the radius of the 10th dark ring by (i) reflection and (ii) transmission.

Solution

We have for the lens $\dfrac{1}{f} = (\mu - 1)\left(\dfrac{1}{R_1} - \dfrac{1}{R_2}\right)$

$$f = 180 \text{ cm}, \quad \mu = 1.6$$
$$R = ?$$

$$\therefore \qquad \frac{1}{180} = (1.6 - 1)\frac{1}{R}$$

or $\quad R = 0.6 \times 180 = 108 \text{ cm} = 1.08 \text{ m}$

$\therefore$ Radius of 10th dark ring by **reflection**,

$$D_n = \sqrt{4nR\lambda}$$

$$= \sqrt{4 \times 10 \times 108 \times 5890 \times 10^{-8}} = 0.504 \text{ cm}$$

Radius of 10th dark ring by **transmission**,

$$D'_n = \sqrt{\left(n - \frac{1}{2}\right)4R\lambda}$$

$$= \left(9.5 \times 4 \times 108 \times 5890 \times 10^{-5}\right)^{1/2} = 0.49 \text{ cm}$$

Example 15

A thin mica sheet of $\mu = 1.4$ is placed in one arm of Michelson's interferometer. If shift of 7 fringes occur for light of wavelength 5890 Å, calculate the thickness of mica sheet.

Solution

Let t be the thickness of the film. When it is put in the path of one of the interfering beams of Michelson's interferometer, an additional path difference of $2(\mu - 1)t$ is introduced. If N be the number of fringes shifted, we have

$$2(\mu - 1)t = N\lambda$$
$$N = 7$$
$$\lambda = 5890 \times 10^{-8} \text{ cm}$$
$$\mu = 1.4$$

$$\therefore \qquad t = \frac{N\lambda}{2(\mu - 1)}$$

$$\therefore \qquad t = \frac{7 \times 5890 \times 10^{-8}}{2(1.4 - 1)} = 7.74 \times 10^{-4} \text{ cm}$$

Example 16

Michelson's interferometer is used with sodium discharge tube which gives light of wavelength 5890 Å. It is observed that interference pattern disappears and reappears periodically as one moves the movable mirror. Calculate the change in path difference between the two successive reappearances of interference pattern.

Solution

If d be the distance moved by movable mirror between two successive positions of maximum distances or indistinctness, we have

$$\lambda_1 - \lambda_2 = \frac{\lambda_1 \lambda_2}{2d}$$

$$\lambda_1 = 5890 \times 10^{-8} \text{ cm}$$

$$\lambda_2 = 5890 \times 10^{-8} \text{ cm}$$

or $\qquad d = \dfrac{\lambda_1 \lambda_2}{2(\lambda_1 - \lambda_2)}$

or $\qquad d = \dfrac{5896 \times 5890 \times 10^{-16}}{2(5890 - 5890) \times 10^{-8}} = 0.028 \text{ cm}$

Example 17

The path difference between the mirrors of a Michelson's interferometer is 0.6 cm. When the light of wavelength 6000 Å is used, calculate the angular radius of the 10th bright fringe in the circular pattern of fringes obtained.

Solution

The condition to obtain bright fringe at the centre is

$$2d = n_1\lambda \qquad (1)$$

where d is the distance between the two mirrors and n_1 is an integer. Along a direction inclined at an angle θ, a bright fringe is obtained if

$$2d\cos\theta = n_2\lambda \qquad (2)$$

where θ is the angular radius of the circular bright fringe obtained and n_2 is another integer. Subtracting (2) from (1), one obtains

$$2d(1 - \cos\theta) = (n_1 - n_2)\lambda = n\lambda$$

$$d = 0.6 \text{ cm}$$

$$\lambda = 6000 \times 10^{-8} \text{ cm}$$

$$\therefore \qquad 1 - \cos\theta = \frac{n\lambda}{2d}$$

or $\qquad 1 - \cos\theta = \dfrac{10 \times 6 \times 10^{-5}}{1.2} = 5 \times 10^{-4}$

$$\therefore \qquad \cos\theta = 1 - 0.0005 = 0.9995$$

$$\therefore \qquad \theta = 1.812$$

Example 18

A Fabry–Perot interferometer is used to determine the difference between the wavelengths of two closely separated spectral lines, whose average wavelength is 6400 Å. It is found that at a given point near the centre of fringe, pattern of rings coincides and it again coincides when the separation is changed by 1.024 mm. Show that the **difference** between the wavelengths is 2 Å.

Solution

We have $\qquad \Delta\lambda = \lambda_2 - \lambda_1 = \dfrac{\lambda^2}{2(t_2 - t_1)}$

$$\lambda = 6400 \text{ Å}$$

$$= 6400 \times 10^{-10} \text{ m}$$

$$t_2 - t_1 = 1.024 \text{ mm}$$

$$= 1.024 \times 10^{-3} \text{ m}$$

$$= \frac{6400 \times 6400 \times 10^{-20}}{2 \times 1.024 \times 10^{-3}}$$

$$= 2 \text{ Å}$$

Example 19

In Young's double slit experiment, a source of light of wavelength 4200 Å is used to obtain interference fringes of width 0.64×10^{-2} m. What should be the wavelength of the light source to obtain fringes 0.46×10^{-2} m wide, if the distance between screen and slits is reduced to half the initial value?

Solution

In first case:

$$\lambda = 4200 \text{ Å} = 4200 \times 10\text{–}10 \text{ m}$$

$$\beta = 0.64 \times 10^{-2} \text{ m}$$

$$\therefore \quad 0.64 \times 10^{-2} = \frac{4200 \times 10^{-10} \times D}{2d}$$

$$\left[\because \beta = \frac{\lambda D}{2d} \right] \quad (1)$$

In second case:

$$\beta = 0.46 \times 10^{-2} \text{ m}, \ \lambda = ?, \ D = \frac{D}{2}$$

$$0.46 \times 10^{-2} = \frac{\lambda \times \left(\dfrac{D}{2} \right)}{2d} \fallingdotseq \frac{\lambda D}{2(2d)} \quad (2)$$

Dividing Eq. (1) by (2)

$$\frac{0.64 \times 10^{-2}}{0.46 \times 10^{-2}} = \frac{4200 \times 10^{-10} \times D \times 2d \times 2}{(2d)\,\lambda D}$$

$$\therefore \quad \lambda = \frac{4200 \times 10^{-10} \times 2 \times 0.46}{0.64} = 6037.5 \text{ Å}$$

Example 20

In a double slit interference arrangement, one of the slits is covered by a thin mica sheet whose refractive index is 1.58. The distance between the slits is 0.1 cm and the distance between the slits and the screen is 50 cm. Due to introduction of the mica sheet, the central fringe gets shifted by 0.2 cm. Determine the thickness of the mica sheet.

Solution

$x_0 = 0.2$ cm, $d = 0.1$ cm and $D = 50$ cm

Hence $\quad t = \dfrac{dx_0}{D(\mu - 1)}$

$$= \frac{0.1 \times 0.2}{50 \times 0.58} \text{ cm}$$

$$= 6.7 \times 10^{-4} \text{ cm} = 6.7 \text{ μm}$$

REVIEW QUESTIONS

1. Give the theory of interference fringes. What are the conditions to get sustained interferences pattern?
2. Illustrate the formation of interference fringes in Fresnel's biprism. Show, how would you see it to determine the wavelength of monochromatic light?
3. Calculate the displacement of fringes when a thin transparent plate is introduced in the path of one of the interfering beams of a biprism. Show how this method is used to find the thickness of mica sheet.
4. Describe Fresnel's biprism method of producing interference fringes and determining the wavelength of light.
5. Explain the formation of Newton's rings in reflected light. Prove that the diameters of the dark rings are proportional to the square root of the natural numbers.
6. How do the fringes obtained with Lloyd's mirror differ from those obtained with Fresnel's biprism? Explain the formation of fringes with white light in case of Lloyd's mirror.
7. (a) Explain the colours of thin films. Why should the film be thin in order to see the colours?
   (b) Explain what do you mean by:
   (i) Fringes of equal thickness
   (ii) Fringes of equal inclination (Haidinger fringes).
8. (a) Describe the formation of colours in thin films and show that with monochromatic light, the interference patterns of reflected and transmitted light are complementary.
   (b) With a suitable diagram explain why a broad source of white light is essential to observe colours of thin films. Prove that an excessively thin film appears black in reflected light.
9. Explain the formation of Newton's rings. What will happen if:
   (a) White light is used?
   (b) The lens is lifted slowly from the flat plate?
10. Explain the colours when a thin film of transparent material is observed in reflected white light using Fresnel's biprism. Why are colours not observed in case of thick films?
11. (a) Describe with necessary theory, the experiment to determine the wavelength of sodium light using Fresnel's biprism. Why is the angle kept so small?
    (b) How can you use the biprism to find the thickness of a thin sheet of mica?

12. Discuss the formation of Newton's rings by (i) transmitted light (ii) reflected light. Derive the expression for the radius of nth dark ring in the case of reflected light.

13. Briefly explain how the Michelson's interferometer is used to determine the difference in wavelength of the two spectral lines (D_1, D_2) of sodium.

14. Describe the construction with diagram and outline the theory of Michelson's interferometer. Discuss the nature of interference pattern produced.

15. Explain the phenomenon of interference for transmitted beam from an air film placed between two plain plates. What are the factors which determine the sharpness of these rings?

16. A plain parallel thin film is illuminated by a parallel beam of white light. Explain with necessary theory, the colour exhibited by the film in reflected light.

17. Describe, with a clear sketch, Michelson's interferometer and explain how it may be used to measure the wavelength of monochromatic light.

18. (a) Give the theory and the experimental arrangement of Lloyd's mirrors to measure the wavelength of monochromatic light.

    (b) Distinguish between the fringes produced by Lloyd's mirror and Fresnel's biprism.

19. (a) Describe coherent source. Discuss why two independent sources of light of same wavelength cannot show interference?

    (b) What is the role of compensating glass plate in Michelson's interferometer?

20. Describe an interference method for the measurement of radius of curvature of low power convex lens.

21. Describe and explain the formation of Newton's rings in reflected monochromatic light. How can these be used to determine the wavelength of light. Derive the formula used.

22. Discuss the formation and location of interference fringes in a thin wedged shaped film illumination by monochromatic light.

23. Describe the construction with diagram and outline the theory of Michelson's interferometer. How it may be used to measure the wavelength of monochromatic light?

24. Distinguish between temporal coherence and spatial coherence. Discuss why two independent sources of light of the same wavelength cannot produce interference fringes.

25. Explain the method of finding the wavelength of a monochromatic light with the help of a biprism. Give the theory of the method and adjustments.

26. Describe the construction and working of Michelson's interferometer. How would you use it to measure the wavelength of monochromatic light?

27. Give the theory of the measurement of thickness of a thin film transparent sheet by Fresnel's biprism.

28. Newton's rings are formed by reflected light of wavelength λ using a plano-convex lens of radius of curvature R. Derive expression for radius of nth bright ring. If Newton's ring lens and plate do not touch at the centre (due to, say, dust particles) such that the lens is Δ height above the plate at the centre, what is the radius of the ring in this case?

29. (a) How shall you measure thickness of thin silicon chip using biprism experiment? Derive the formula used.

    (b) White light is incident on a soap film at an angle $\sin^{-1}(4/5)$. In the reflected light two consecutive dark bands corresponding to wavelengths 6.1×10^{-5} cm and 6×10^{-5} cm are observed. If refractive index for the film be $\mu = 4/3$, what is thickness of the film?

    (c) The coherent time for Na light of wavelength 5890 Å is $\tau_c = 10^{10}$ s. What is the maximum thickness of film that could be measured using interference of Na light? [Raj. BE 2003]

30. (a) Draw labelled diagram of Michelson's interferometer, how is it used to find the wavelength of light? What is the order of the central fringe?

    (b) In a biprism experiment zero order and tenth order maxima are formed at 12.34 mm and 14.73 mm respectively on the screen, when its slit is illuminated with light of wavelength 6000 Å. Calculate the position of zero order and twentieth order fringes when illuminated with light of wavelength 5000 Å. Other arrangements remaining the same.

    [Raj. BE 2003]

31. (a) Prove that for the thin films, the interference patterns of reflected and transmitted light are complementary.

    (b) Explain why the excessively thin films seen by reflected light appear dark.

    (c) Light containing two wavelengths λ_1 and λ_2 falls normally on a plano-convex lens of radius of curvature R resting on a glass plate. If the nth dark ring due to λ_1 coincides with the $(n + 1)$th dark ring due to λ_2, prove that the radius of the nth dark ring of λ_1 is $= \sqrt{\left(\dfrac{\lambda_1 \lambda_2 R}{\lambda_1 - \lambda_2} \right)}$.

[Raj. BE 2002]

32. (a) A double slit of separation 1.5 mm is illuminated by white light (4000 – 8000 Å). On a screen 120 cm away coloured interference pattern is formed. If a pinhole is made on this screen at a distance of 3.00 mm from the central white fringe, what wavelength will be absent in the transmitted light?

(b) Explain with diagram the formation of Newton's rings. What will happen if white light is used?

(c) In what respects Haidinger's fringes in Michelson's interferometer are distinct in character from fringes in thin films (Newton's rings)? [Raj. BE 2001]

33. (a) With schematic diagram explain the working of Michelson's interferometer. How shall you use it to measure the wavelength separation between two closely spaced lines say D_1 and D_2 lines of sodium lamp?

(b) Evacuated tubes of 5 cm length are in the arms of Michelson's interferometer set to observe fringes of equal inclination using sodium light of wavelength 589.3 mm. If it is observed that ten finger pass a given point in the field of view as air is introduced into one of the tubes, determine pressure inside the tube. Given that refractive index of air increases as pressure increases as per relation $\mu = 1 + 3 \times 10^{-4} P$ where P is pressure in atoms. [Raj. BE 2001]

34. (a) Define spatial and temporal coherence.

(b) Explain the formation of the Newton's rings in reflected light. Prove that the diameter of dark rings are proportional to the square root of the natural numbers.

(c) In a biprism experiment the micrometer readings of zero order fringe and tenth order fringe are 2.37 mm and 3.55 mm respectively. Wavelength of light used is 5890 Å. Determine:

(i) The position of zeroth and 10th order fringes when λ changes to 7500 Å.

(ii) The position of zeroth order fringe when mica sheet of thickness 0.02 mm and $f = 1.5$ is introduced to cover half of the biprism. [Raj. BE 2000]

35. How will you use Fabry–Perot interferometer to determine the wavelength difference between the two spectral lines of sodium. Why the fringes formed in the case of Fabry–Perot interferometer sharper than those obtained in case of Michelson's interferometer.

PROBLEMS

1. A Fresnel biprism has an angle of 1° and index of refraction 1.5. Interference fringes are formed on a screen 8 m away from the biprism by the slit situated at the distance of 0.3 m away from the biprism. Find the fringe width of the fringes produced for the light of wavelength 6000 Å.

2. A soap film ($\mu = 4/3$ and $t = 2.26 \times 10^{-5}$ cm) is illuminated by white light incident at an angle of 45°. The light reflected by it is examined by spectroscope in which is found a dark band corresponding to wavelength of 6×10^{-5} cm. Calculate the order of interference of dark band.

3. Calculate the thickness of the thinnest film ($\mu = 1.4$) in which interference of violet component ($\lambda = 4000$ Å) of incident light can take place by reflection.

4. A biprism has the refracting angle of 1.2° for each part and refractive index 1.54. The distance between the slit and the biprism is 10 cm and that between the biprism and the eyepiece is 90 cm. If the wavelength of light is 5890 Å, calculate (i) the bandwidth (ii) the distance of the 4th dark band from centre and (iii) the distance between 10th bright band on one side and 10th dark band on the other side of the centre.

[*Ans.* (i) 0.026 cm (ii) 0.09 cm (iii) 0.5 cm]

5. A Lloyd's mirror has length of 5 cm. It is kept with its plane at a distance of 1 mm from a slit illuminated by a monochromatic source of wavelength 5890 Å. The distance between the slit and the nearest edge of the mirror is 10 cm. A screen is kept at a distance 1.2 m from the slit. Find the bandwidth obtained. Also calculate the width of the interference pattern obtained.

[*Ans.* 0.4 cm]

6. A glass plate of thickness 1.2 micron is placed in the path of interfering beams in a biprism experiment. The wavelength of the source is 6000 Å. The central band is found to shift by one bandwidth. Calculate the refractive index of the glass.

7. A man, whose eyes are 15 mm above the oil film on water surface, observes greenish colour at a

distance of 1 m from his feet. Calculate the probable thickness of the film. λ_{green} = 5000 Å, μ_{oil} = 1.4, μ_w = 1.5.   [**Hint:** $2\mu t = (2n + 1)\lambda/2$]

8. Two plane surfaces of glass in contact along one edge are separated at the opposite edge by a thin wire. 30 interference fringes are observed between the two edges when normally reflected sodium light is used. Find the diameter of the wire used. Wavelength of sodium light is 5893 Å.

[*Ans.* 8×10^{-6} m]

9. In a Newton's ring experiment, a parallel beam of light of wavelength 6560 Å and 5248 Å is incident normally on the arrangement. (i) It is found that the nth dark ring of the first coincides with $(n + 1)$th dark ring of the second, calculate n. (ii) If the radius of curvature of the convex surface is 0.8 m, calculate the diameter of this ring.

[*Ans.* (i) n = 4   (ii) 0.2892 cm]

10. In Newton's ring experiment, the diameter of the 5th ring is reduced to half of its value after introducing a liquid below the convex surface. Calculate the refractive index of the liquid.

11. A Newton's ring arrangement is used with a source emitting $\lambda_1 = 6 \times 10^{-7}$ m and $\lambda_2 = 4.5 \times 10^{-7}$ m and it is found that the nth dark ring of λ_1 coincides with $(n + 1)$th dark ring of λ_2. If radius of curvature of curved surface be 0.9 m, find the diameter of the nth dark ring for λ_1.

12. The radii of two consecutive bright rings obtained in Newton's ring experiment by reflection are 4.8 mm and 5.2 mm. If the wavelength used is 5460 Å, calculate the radius of curvature of the convex surface used.   [*Ans.* 183 m]

13. Newton's rings are obtained by reflection in the usual way. When a liquid is introduced in between the lens surface and plane glass plate, the radius of the 10th dark ring reduces by 0.3 mm. If the radius of curvature of the surface is 1 m and the wavelength used is 5600 Å, calculate the refractive index of the liquid.   [*Ans.* 1.3]

14. A plane beam of monochromatic light falls normally on a thin film of oil which covers a glass plate. The λ of source can be varied continuously. Complete destructive interference of reflected light is observed for $\lambda = 5 \times 10^{-7}$ m and 7×10^{-7} m and for no wavelength in between. If μ of oil and glass are 1.3 and 1.5 respectively, find the thickness of the film.

15. In a Michelson's interferometer, the path difference between the two mirrors is 4 mm and the wavelength of the light used is 6400 Å. Calculate the angular radius of the 5th bright fringe in the circular pattern of fringes obtained. [*Ans.* 1.69°]

16. In an experiment with Michelson's interferometer, the positions of the mirror for maximum distinctness give the reading on the scale 0.5828 mm and 0.8773 mm. If the mean wavelength is 5893 Å for sodium (light), calculate the difference between the wavelengths of the two components. [*Ans.* 5.9 Å]

17. In the experiment to measure the refractive index of air using a Michelsons's interferometer, a wavelength of 6000 Å is used. It is observed that the mirror has to be shifted by 100 fringes when the air inside the tube is completely removed. If the length of the tube is 20 cm, calculate the refractive index of air.   [*Ans.* 1.00015]

18. A thin film of glass of refractive index 1.5 is inserted in one arm of Michelson's interferometer. Find the thickness of the film if a shift of 20 fringes is obtained. The wavelength is 5460 Å.

[*Ans.* 1.09×10^{-5} m]

19. White light is incident at an angle of 30° to the normal on two parallel glass plates separated by an air film of 0.001 cm thickness and reflected light is examined by the spectroscope. Find the number of dark bands seen in the spectrum between the wavelength 4×10^{-5} cm and 7×10^{-5} cm.

20. In moving the mirror of the Michelson's interferometer through a distance of 0.141 mm, 500 fringes are counted. Calculate the wavelength of light.

21. Newton's rings are formed with reflected light of wavelength 5890 Å with a liquid between the plane and the curved surfaces. The diameter of 5th dark ring is 0.32 cm and radius of curvature of curved surface is 120 cm. Calculate the refractive index of the liquid.   [Jodhpur BE 97]

22. White light is incident on a soap film at an angle of $\sin^{-1}(4/5)$. In the reflected light two consecutive dark bands corresponding to wavelengths 6.1×10^{-5} cm and 6×10^{-5} cm are observed. If μ of the film is 4/3, calculate its thickness.

23. In a biprism experiment, the 5th order and 10th order maximum fall at micrometer readings 13.535 mm and 14.730 mm respectively, when wavelength of 6000 Å is used for producing interference pattern. If wavelength is changed to 5000 Å, calculate the positions of zero order and 20th order fringes, other arrangements remaining same.

24. Interference fringes are produced by monochromatic light falling normally on a wedge-shaped film of cellophane whose refractive index is 1.4. If the angle of the wedge is 20 seconds of arc and the distance between successive fringes is 0.25 cm, calculate the wavelength of light.

25. The convex surface of a plano-convex glass lens with curvature radius $R = 40$ cm comes into contact with a glass plate. A certain ring observed in reflected light has radius $r = 2.5$ mm. Watching the given ring, the lens was gradually raised from the plate by a distance $\Delta h = 5.0$ μm. What has the radius of that ring become equal to? [Raj. BE 97]

26. In biprism experiment, fringes are first observed with sodium light ($\lambda = 5893$ Å) and fringe width was measured as 347 mm. Then two thin transparent sheets A and B of thickness of 0.016 mm and 0.02 mm and refractive indices 1.65 and 1.45 respectively were introduced in two beams. Calculate the shift of fringe system. Is the shift towards A or B? [Raj. BE]

27. Newton's rings are observed between a spherical surface of radius of curvature 120 cm and a plane plate. The diameters of 5th and 16th bright rings are 0.314 cm and 584 cm. Calculate the diameters of 25th and 37th bright rings and also the wavelength of light used. [BE]

28. Biprism fringes are produced using a source of wavelength 5893 Å. The biprism is of refractive index 1.50 and refracting angles 1.04° and 1.23°. The distance of focal plane of eyepiece from the biprism is 56.1 cm and the distance from slit to biprism is 12.4 cm. Calculate (i) separation of the coherent slit images and (ii) the fringe width. [BE]

29. Michelson's interferometer is adjusted to form circular fringes using light of wavelength 5000 Å. When the difference of path lengths between the mirrors of interferometer is 2.5 mm, the fringe pattern having bright fringe at the centre is formed and we say it is the first bright fringe. (i) What is the angular radius of the 10th bright fringe? (ii) When mirror is moved slowly, 60 fringes cross the centre. How much path length is changed? [BE]

SHORT ANSWER QUESTIONS

1. Can two electric bulbs with point like filament of the same material each 15 W and lying close to each other produce interference?

Ans. As the two electric bulbs are two independent sources, i.e. they are not coherent sources. Obviously, they cannot produce intereference.

2. What happens to the energy of light waves in destructive interference?

Ans. The energy in the destructive interference is transferred to the region where intensity is further increased.

3. The phenomenon of interference can be observed only when the two sources have common characteristics. What are these characteristics

Ans. The two sources be monochromatic, vibrate with same amplitude and must bear constant phase difference.

4. The central part in Newton's rings seen in reflected light appears dark. Why?

Ans. In Newton's rings experimental arrangement, at the point of the contact between the lens and the glass plate the thickness of the air film is zero. Therefore, there is no path difference between the interfering rays due to difference in path lengths. But one of the rays suffers a phase change of π on reflection at the surface of the glass plate, i.e. denser medium. This is why the rays suffer destructive interference and the centre appears dark.

5. How is the interference pattern in Young's double slit experiment affected if one of the slits, say S, is covered with black opaque paper?

Ans. There is a bright central slit but no interference pattern is obtained.

OBJECTIVE QUESTIONS

1. The two sources of light are said to be _____, if they emit continuously the waves either in the same phase or with a constant phase difference. [coherent]

2. Colour of thin film is due to _____ of light. [interference]

3. In Young's double slit experiment, the fringes produced by light of _____ wavelength will be narrow as compared to those by _____ wavelength. [shorter, longer]

4. In the transmitted system, the film will appear bright if $2\mu t \cos r$ is an _____ multiple of $\lambda/2$ and it will appear _____ if it is an odd multiple of λ. [even, dark]

5. In the case of a transverse wave, an additional phase change of _____ or a path difference of _____ is introduced when reflection takes place at the surface of denser medium. [π, $\lambda/2$]

6. In Young's double slit experiment, the fringes become _____ if one of the slits is covered with cellophane paper. [indistinct]

MULTIPLE CHOICE QUESTIONS

1. Young's double slit experiment is first performed in air and then in a liquid. It is observed that the 10th bright fringe in liquid is replaced by 8th dark fringe in air. The refractive index of the liquid is

    (a) 3/2   (b) 4/3

    (c) 5/3   (d) 20/17

2. In Young's double slit experiment, if the slit widths are in the ratio 1 : 2, the ratio of the intensities at minima and maxima will be

    (a) 1 : 2   (b) 1 : 3

    (c) 1 : 4   (d) 1 : 9

    [**Hint:** $\dfrac{I_{max}}{I_{min}} = \left(\dfrac{A_1 + A_2}{A_1 + A_2}\right)^2$, $A_1 = 2A$ and $A_2 = A$

    $\therefore \quad \dfrac{I_{max}}{I_{min}} = \left(\dfrac{2A + A}{2A + A}\right)^2 = \dfrac{9}{1}$, $\therefore \dfrac{I_{min}}{I_{max}} = 1 : 9$]

3. The ratio of the intensities of the maxima and minima in an interference pattern is 49 : 9. What is the ratio of the intensities of the two coherent sources in the interference experiment ?

    (a) 7 : 3   (b) 49 : 9

    (c) 5 : 2   (d) 25 : 4

4. In a Young's double slit experiment, 12 fringes are observed to be formed in a certain region of the screen when light of wavelength 600 nm is used. If the light of wavelength 400 nm is used, the number of fringes observed in the same region of the screen will be

    (a) 12   (b) 18

    (c) 24   (d) 8

5. Interference pattern is obtaned with two coherent light sources of intensity ratio n. In the interference pattern, the ratio $\dfrac{I_{max} - I_{min}}{I_{max} + I_{min}}$ will be

    (a) $\dfrac{\sqrt{n}}{(n+1)}$ (b) $\dfrac{2\sqrt{n}}{(n+1)}$

    (c) $\dfrac{\sqrt{n}}{(\sqrt{n}+1)^2}$ (d) $\dfrac{2\sqrt{n}}{(\sqrt{n}+1)^2}$

Answers

   1. (b)   **2.** (d)   **3.** (d)   **4.** (b)   **5.** (b)

5

Diffraction

5.1 INTRODUCTION

Under the action of an electromagnetic wave, every molecule becomes a secondary radiator of electromagnetic waves. Due to the electric force, the electron cloud is displayed relative to the atomic nuclei and the molecule acquires a varying dipole moment having the same frequency as that of the incident wave. The behaviour of such a molecule differs in no way from the behaviour of the elementary dipole.

In a number of cases, the phenomenon of secondary radiation leads to the various phenomena of electromagnetic wave scattering By scattering, we mean any electromagnetic wave propagation phenomenon that is not included under reflection, refraction and rectilinear propagation.

The fact that the intensity increases sharply with radiation frequency explains why the effects of wave scattering by molecule are not detectable when the wavelengths are very long. The scattering intensity of visible light is quite sufficient to produce significant effects. Light wavelengths are hundreds and thousands of times greater than the the dimensions of ordinary molecule. Therefore, all the electrons of a molecule are made to vibrate in the same phase by the external field. For light waves, molecule behaves like an elementary dipole.

We have been discussing secondary radiation from a molecule, but often the secondary radiator is much bigger particle. The nature of wave scattering by particles is determined by the ratio of their dimensions to the wavelengths of the exciting electromagnetic wave. If the particle is small relative to the wavelength, the wave is scattered as by a single elementary dipole. If this is not the case, interference effects occur and the forward scattering predominates. When the scattering pattern has rather pronounced maxima and minima, we call it as *diffraction*. When the nature of interference pattern is not so evident, we call it as scattering. Diffraction is noticeable when a wave is distorted by an obstacle which has dimensions comparable to the wavelength of the wave. Mathematically, one can express the condition as $\lambda \le a$, where λ and a are wavelength of the wave and amplitude of obstacle respectively.

5.2 DIFFRACTION OF LIGHT WAVES

The diffraction of the light is the name given to the totality of phenomena which are due to the wave nature of light and are observed in its propagation in media with sharply defined inhomogeneities, e.g. a screen with a small opening or slit which allows only a small portion of the incident *wavefront* to pass, etc. In its narrower sense, diffraction refers to the bending of light around small opaque obstacles, i.e. its deviation from the laws of geometric optics. Diffraction is noticeable when a wave is distorted by an obstacle which has dimensions comparable to the wavelength of the wave. To illustrate the meaning of diffraction clearly, we consider the following example.

Consider a stream of particles falling on a screen which has a small opening, only those falling on the opening will be transmitted and allowed to continue their motion undisturbed (Fig. 5.1). The others will either be stopped or bounced back. Conversely, if an obstacle, i.e. object is placed in a stream of particles, it will block those particles falling on it, but the remaining particles will continue their motion undisturbed. But the waves behave in a different way, and that they extend around the obstacles interposed in their path (Fig. 5.2). This effect becomes more and more noticeable

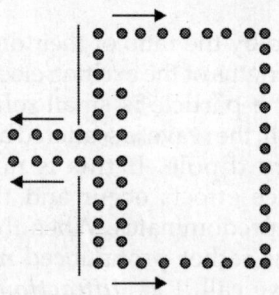

Fig. 5.1: Behaviour of a stream of particles impinging on a screen with a small

Fig. 5.2: Behaviour of a wave incident on a screen with a small opening

as the dimensions of the slits or the size of the obstacles approach the wavelength of the waves. Usually one cannot observe the diffraction of light with the naked eye, since most of the objects interposed in a beam of light are much larger than the wavelength of the light wave, whose magnitude is of the order of 5000 Å.

A strict mathematical solution of diffraction problems, based on the wave equation and having boundary conditions depending on the nature of the obstacle, proves to be exceptionally difficult. Hence, approximate methods are resorted to.

Young was the first physicist to use wave theory for explaining the diffraction of light but his argument was not convincing and was later rejected by Fresnel, who was first to give a rigorous mathematical treatment of the phenomenon. In this chapter, we shall discuss diffraction produced by certain apertures and screens of the simple geometry under two heads, namely **(i) Fresnel class of diffraction (ii) Fraunhofer class of diffraction**.

Before we actually proceed to the study of diffraction, we must try to know what is the difference between interference and diffraction. According to Feynman, no one has been able to define the difference between interference and diffraction satisfactorily. It is a question of usage, and there is no specific, important physical difference between interference and diffraction. The best way we can do, is to say that when there are only a very few interfering sources, say two, then the result is usually called interference, but if there are a number of interfering sources, the word diffraction is commonly used. The study of diffraction, in general, is the study of scattering from a more general point of view. The basic differences between interference and diffraction are as follows:

i. Interference is produced as a result of superposition of light waves coming from the same source whereas diffraction is due to the interaction of the light waves coming from different parts of the same wavefront.

ii. Interference fringes are of the same width while the diffraction fringes are not of the same width.

iii. The intensity of all bright fringes in interference pattern is the same while the intensity decreases successively in diffraction.

iv. The minimum intensity points are perfectly dark in interference pattern whereas this is not so in the case of diffraction.

v. In interference, one can observe easily a large number of fringes while in diffraction only a few fringes can be observed.

5.3 FRESNEL'S HALF PERIOD ZONES

According to Huygens' wave theory of light, each point in a light source sends out waves in all directions, thus setting ether particles in vibration (according to special theory of relativity, light can travel through vacuum and there is nothing like ether medium). The continuous locus of all ether particles vibrating in the same phase is called the **wavefront**. Each point on wavefront again becomes the source of disturbance and sends out secondary wavelets. Fresnel assumed that these wavelets are in a position to interfere and that the resultant intensity of light at any point is the result of interference of these wavelets. In order to calculate the resultant intensity at a point due to a wavefront, Fresnel assumed that:

i. Each element of wavefront sends secondary waves continuously.

ii. The effect of any element on the wavefront is maximum along a direction normal to it and decreases as the angle of inclination increases.

iii. The amplitude at any point is determined by combining the effects of the waves reaching there.

Fresnel divided the wavefront into a number of zones called Fresnel's half period zones.

In Fig. 5.3, *ABCD* is a plane wavefront. It is required to find out the effect of this wavefront at a point *P* situated at a distance *b* from it. Drop a perpendicular from point *P* on the wavefront. Let the foot of the wavefront be *O*. Then *O* is known as pole with respect to point *P*. With *P* as centre, draw a sphere of radius, $(b + \lambda/2)$. Intersection of this sphere and plane *ABCD* would be a circle M_1. Obviously,

$$PM_1 = b + \lambda/2 \text{ and } PO = b$$

$$\therefore \quad PM_1 - PO = \lambda/2$$

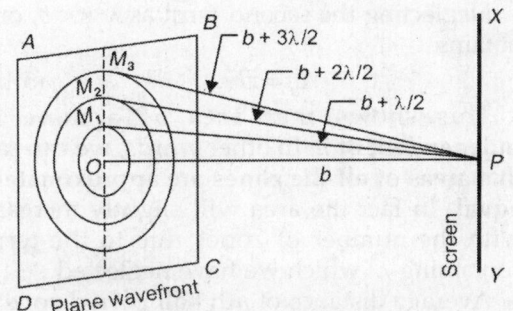

Fig. 5.3: Fresnel's half period zones

This means that wavelets originating from point *O* and that from the circumference of circle M_1 would differ in path by $\lambda/2$. In terms of time, it can be represented by $T/2$. Hence area enclosed by circle M_1 is called as Fresnel's half period zone. Similarly with *P* as centre and radii as $b + 2\lambda/2$, $b + 3\lambda/2,..$, etc. we draw other spheres which would cut the wavefront *ABCD* in different circles M_2, M_3, etc. All these zones are known as half period zones. They are so situated that path difference between waves arriving at point *P* (Fig. 5.3) from two consecutive zones is $\lambda/2$. Let us find out the area of *n*th half period zone. We have

$$A_n = \pi\left(r_n^2 - r_{n-1}^2\right)$$

$$= \pi\left[\left(PM_n^2 - b^2\right) - \left(PM_{n-1}^2 - b^2\right)\right]$$

$$= \pi\left[PM_n^2 - PM_{n-1}^2\right]$$

$$= \pi\left[\left(b + \frac{n\lambda}{2}\right)^2 - \left\{b + \frac{(n-1)\lambda}{2}\right\}^2\right]$$

$$= \pi\left[b^2 + bn\lambda + \frac{n^2\lambda^2}{4} - b^2\right.$$

$$\left. -b(n-1)\lambda - \frac{(n-1)^2\lambda^2}{4}\right]$$

$$= \pi\left[b\lambda + \frac{\lambda^2}{4}\left(n^2 - n^2 + 2n - 1\right)\right]$$

$$= \pi b\lambda + \pi(2n-1)\frac{\lambda^2}{4} \quad (5.1)$$

Neglecting the second term as $\lambda^2 \ll b$, one obtains

$$A_n = \pi b \lambda \qquad (5.1a)$$

This shows that **area of a zone is independent of n**. In other words, we can say that **areas of all the zones are approximately equal**. In fact the area will slightly increase with the number of zones due to the term containing λ^2 which we have neglected.

Average distance of nth half period zone

$$= \frac{(b + n\lambda/2) + \{b + (n-1)\lambda/2\}}{2}$$

$$= b + (2n - 1)\,\lambda/2 \qquad (5.2)$$

5.4 AMPLITUDE DUE TO A ZONE

The amplitude of the disturbance at P' due to any zone is:

i. Directly proportional to the area of the zone

ii. Inversely proportional to the average distance of the zone from P

iii. Directly proportional to the obliquity factor $(1 + \cos\theta)$, where θ is the angle between the normal to the zone and the line joining the zone to P.

∴ Amplitude due to nth zone is

$$R_n \propto \frac{\pi\left[b\lambda + \dfrac{\lambda^2}{4}(n-1)\right]}{b + (2n-1)\dfrac{\lambda}{4}}(1 + \cos\theta_n)$$

$$\propto n\lambda(1 + \cos\theta_n) \qquad (5.3)$$

As n (order of the zone) increases, θ_n and $\cos\theta_n$ decreases. Therefore, **the amplitude of the wave at P due to a zone decreases as n increases.**

5.5 RESULTANT AMPLITUDE DUE TO WAVEFRONT *ABCD*

Phase of the wavelet starting from point O is taken to be zero and that of wavelets from the circumference of M_1 is π. Hence, phase of the wave coming out of the first half period zone will be the average of these two phases, i.e.

$\dfrac{0 + \pi}{2} = \dfrac{\pi}{2}$. Similarly, the phase of the wavelet

starting from second half period zone will be the mean of π and 2π, i.e. $\dfrac{\pi + 2\pi}{2} = \dfrac{3\pi}{2}$.

Obviously, mean phases of the wavelets coming from successive outer zones are $\pi/2$, $3\pi/2, 5\pi/2, 7\pi/2$, etc. which means that waves arriving from two consecutive zones will be out of phase. If R_1, R_2, R_3, etc. are the amplitudes of waves at point P due to wavelets from first, second, third, etc. half period zones, then resultant amplitude due to their superposition is given by

$$R_p = R_1 - R_2 + R_3 - R_4 + \dots (-1)^{n-1}R_n \quad (5.4)$$

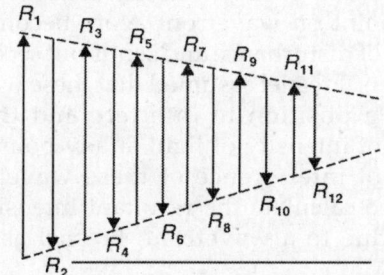

Fig. 5.4: Resultant amplitude of the wave

Now, the successive terms R_1, R_2, $R_3, \dots$ gradually decrease in magnitude. Thus R_2 is slightly smaller than R_1 but slightly greater than R_3. Hence to a close approximation, one can consider

$$R_2 = \frac{R_1 + R_3}{2}, \ R_4 = \frac{R_3 + R_5}{2}, \text{ etc.}$$

or $\ R_4 = \dfrac{R_1}{2} + \left(\dfrac{R_1}{2} - R_2 + \dfrac{R_3}{2}\right)$

$$+ \left(\frac{R_3}{2} - R_4 + \frac{R_5}{2}\right) + \dots + \frac{R_n}{2}$$

or $\ R_4 = \left(\dfrac{R_{n-1}}{2} - R_n\right)$ (accordingly as n is odd or even)

∵ $\ \dfrac{R_1}{2} - R_2 + \dfrac{R_3}{2} = 0$, $\dfrac{R_3}{2} - R_4 + \dfrac{R_5}{2} = 0$

∴ $\ R_p = \dfrac{R_1}{2} + \dfrac{R_n}{2} \qquad (n \to \text{odd})$

and $\ R_p = \dfrac{R_1}{2} + \dfrac{R_{n-1}}{2} - R_n \qquad (n \to \text{even})$

n is very large so that, on account of obliquity, one can write

$$R_{n-1} = R_n = 0 \qquad \therefore \ R_p = R_1/2 \qquad (5.5)$$

Hence, the resultant amplitude at point P is only half of the amplitude of the wavelet from first half period zone.

$$I_4 \propto (\text{amplitude})^2 \propto \left(\frac{R_1}{2}\right)^2 \propto \frac{R_1^2}{4}$$

Thus, the intensity at P is only one-fourth of that due to the first half period zone alone.

5.6 EXPLANATION OF APPROXIMATELY RECTILINEAR PROPAGATION OF LIGHT

Light does not travel perfectly in a straight line. The phenomenon of diffraction satisfies this statement. To understand the approximate rectilinear propagation of light, we shall take the help of Figs 5.5a and 5.5b. Both these diagrams represent the same thing from different angles of view. *EFGH* is a square slit and a plane wavefront is incident on it. We have to notice that the shadow cast by this slit be on the screen.

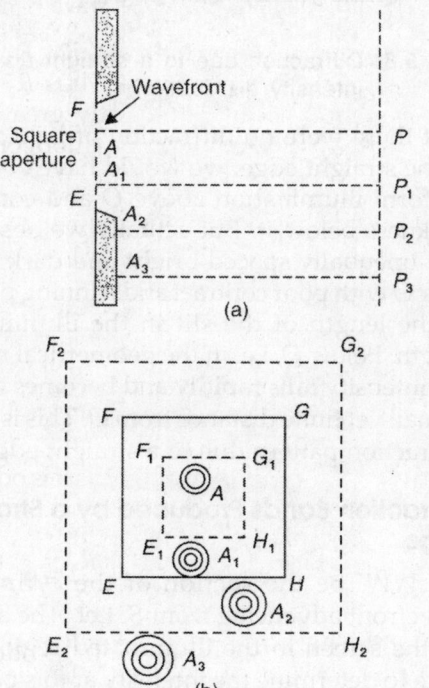

Fig. 5.5: Approximate rectilinear propagation of light

Let point A be the pole of wavefront with respect to a point P on the screen. Draw Fresnel's half period zones around A. Since the point A is well within the aperture *EFGH*, fairly large number of half period zones resultant at point P would be $R_1/2$, which is just half of the first half period zone will be transmitted. In other words, one can say that all the effective zones are transmitted. According to Fresnel's half period zone theory, resultant at P would be $R_{1/2}$, which is just the same even if the screen is removed. In the same way, we see that point A_3 is well outside the aperture *EFGH* and therefore all the effective half period zones are blocked. Therefore, point P_4 on the screen would be perfectly dark. Now we select points A_1 and A_2 on the wavefront which are poles with respect to points P_1 and P_2 on the screen. For these points we find that only some of the effective half period zones are transmitted and rest are blocked. Hence Fresnel's theory is not able to decide the intensity at these points.

One can now summarize the above findings as follows:

Rectilinear propagation of light holds good for all points on the screen whose poles lie within the dotted square $E_1F_1G_1H_1$ and outside the dotted square $E_2F_2G_2H_2$. The law fails when the poles lie within these squares, i.e. near the boundary of the obstacle. This proves that *the propagation of light is approximately rectilinear.*

5.7 ZONE PLATE

If an aperture is constructed so as to obstruct alternate Fresnel zones, say even numbered ones, then the remaining terms in Eq. (5.4) are of the same sign, i.e.

$$R_P = R_1 + R_3 + R_5 + \dots$$

Such an aperture is called a *zone plate*. It acts very much like a lens, because R_P and hence the intensity at P, is now much larger then if there were no aperture.

5.8 FRESNEL AND FRAUNHOFER DIFFRACTION

In the detailed treatment of diffraction it is customary to distinguish between two general cases. These are:

1. Fresnel Class of Diffraction

In this type of diffraction the source of light, screen or both are at finite distances from the obstacle or aperture. No lenses are used to make the rays parallel or convergent. The incident wavefront is not plane but is either spherical or cylindrical. As a result the phase of secondary wavelets is not the same at the point in the plane of the aperture or the obstacle causing diffraction. The resultant amplitude at any point on the screen is obtained by the mutual interference of secondary wavelets from different elements of unblocked portions of wavefront. Arrangement to see Fresnel's diffraction is shown in Fig. 5.6.

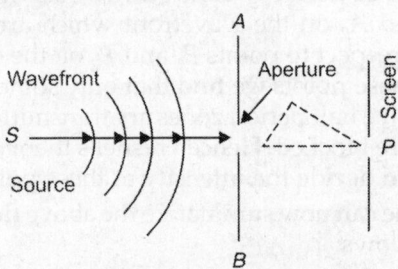

Fig. 5.6: Fresnel diffraction

2. Fraunhofer Class of Diffraction

In this type of diffraction the source of light and screen are effectively at infinite distances from the aperture or obstacle which causes diffraction. This may be achieved by using two convex lenses, one to make the light from the source parallel before it falls on the aperture and the other to focus the light after diffraction on the screen. This arrangement, in fact, removes the source and the screen to infinity. This class of diffraction is simpler to treat theoretically and easier to observe practically.

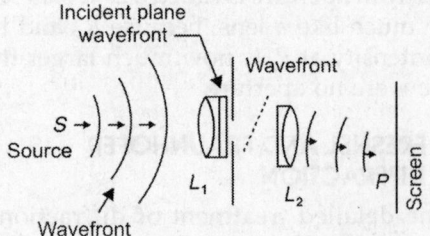

Fig. 5.7: Fraunhofer diffraction

Arrangement to see Fraunhofer diffraction is shown in Fig. 5.7. There is, however, no sharp line of distinction between Fresnel and Fraunhofer types of diffraction.

5.9 DIFFRACTION AT A STRAIGHT EDGE

Let S (Fig. 5.8) be a narrow slit perpendicular to the plane of the paper. Let this slit be illuminated with monochromatic light of wavelength λ. A is a straight edge of an opaque obstacle AC which is placed parallel to the slit (say the opaque obstacle is razor blade). RL is a screen placed parallel to the edge at finite distance.

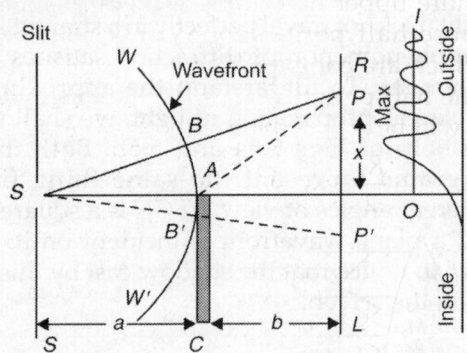

Fig. 5.8: Diffraction due to a straight edge and intensity distribution

If there were no diffraction of light waves at the straight edge, we would have obtained uniform illumination above O and complete darkness below it. But actually we observe a few unequally spaced bright and dark bands near O with poor contrast and running parallel to the length of the slit in the illumination length. Below O, i.e. in the geometrical region, the intensity falls rapidly and becomes zero at a small yet finite distance from O. This is called diffraction pattern due to a straight edge.

Diffraction Bands Produced by a Straight Edge

Let WW' be the section of the cylindrical wavefront advancing from S. Let P be a point on the screen in the illuminated region. We have to determine the intensity at this point P. Let us divide the WW' portion of the cylindrical wavefront into half **period zones**.

PS intersects *WW'* at *B*. *B* is called the pole of the wavefront with respect to *P*.

With *P* as centre and radii equal to $PB + \lambda/2$, $PB + 2\lambda/2$, $PB + 3\lambda/2, ..., PB + n\lambda/2$, we mark off points on the wavefront *WW'*. Now draw the lines through these points which are parallel to the slit.

Fig. 5.9: Slit

The wavefront is, thus, divided into half period strips (Fig. 5.10). From Fig. 5.10, it is evident that the half portions of the wavefront are *BW* and *BW'*. The point receives light from the entire upper half of the wavefront and from these half period zones of the lower half which are contained in *BW'*. If *OA* contains one half period strip, then the amplitude at *P* will be

$$\frac{R_1}{2} + R_1 \qquad \text{(first maxima)}$$

Fig. 5.10: Division of wavefront into half period zones with respect to the point *P*

It is to be noted that $R_1/2$ is the amplitude due to entire upper half wave from *BW* and R_1 is the amplitude due to the first half period zone of the lower half *BW'*.

As the point *P* moves away from *O*, the number of the half period zones increases in *BA*. If *BA* portion of the wavefront contains 2, 3, 4,..., etc. half period zones, then the respective amplitudes at *P* will be

$$\frac{R_1}{2} + R_1 - R_2 \qquad \text{(first minima)}$$

$$\frac{R_1}{2} + R_1 - R_2 + R_3 \qquad \text{(second maxima)}$$

$$\frac{R_1}{2} + R_1 - R_2 + R_3 - R_4 \quad \text{(second minima)}$$

and so on.

It is evident from the above discussion that the point *P* is a maximum or minimum according as *OA* contains an **odd** or **even** number of half period zones. Thus, on moving away from *O* in the illuminated region maxima and minima are obtained alternately. The amplitude, i.e. intensity of minima, are comparable to those of maxima and hence the bands have a poor contrast. At sufficient distance from *O* in the illuminated region, the intensity of maxima falls slowly and that of minima rises very slowly. This is why at a sufficient distance from *M* the diffraction merges into uniform illumination. In such a condition, the resultant amplitude at *P* is then

$$\frac{R_1}{2} + \frac{R_1}{2} = R_1$$

Conditions for Maxima and Minima

Let the distance between the slit and the straight edge and screen be *b* (Fig. 5.8). Let

Fig. 5.11: Diffraction at a straight edge

$OP = x$. *P* is a maximum or minimum according as *BA* contains an odd or even number of half period zones, i.e. according as the path difference $AP - BP$ is equal to an odd or even multiple of half wavelengths. Thus, we have to determine the path difference.

$$AP - BP \; (= \delta \text{ say})$$

$$\therefore \quad \delta = AP - BP = \left(b^2 + x^2\right)^{1/2} - (SP - SB)$$

$(SB \approx a, B \text{ being very near to } A)$

$$= \left(b^2 + x^2\right)^{1/2} \left[\left\{ (a+b)^2 + x^2 \right\}^{1/2} - a \right]$$

$$= b\left(1 + \frac{x^2}{2b^2} + \dots\right) - (a+b)\left[1 + \frac{x^2}{2(a+b)}\right] + a$$

$$= b + \frac{x^2}{2b^2} - a - b - \frac{x^2}{2(a+b)} + a$$

$$= \frac{x^2}{2}\left[\frac{1}{b} - \frac{1}{a+b}\right] = \frac{x^2}{2} \frac{a}{b(a+b)} \qquad (5.6)$$

Intensity at point P will be maximum, if

$$AP - BP = \delta = (2n+1)\lambda/2$$

$$\therefore \quad \frac{X_n^2}{2} \frac{a}{b(a+b)} = (2n+1)\frac{\lambda}{2}$$

$$\therefore \quad \frac{X_n^2}{2} = \frac{(2n+1)(a+b)b\lambda}{a}$$

or

$$X_n = \left[\frac{(2n+1)(a+b)b\lambda}{a}\right]^{1/2} \qquad (5.7)$$

where $n = 0, 1, 2, 3, \dots$ and X_n is the distance of the nth bright band from the point O. If P is minima, i.e. a point of minimum intensity, then

$$AP - BP = \delta(2n+1)\lambda/2$$

or

$$\frac{aX_n^2}{2b(a+b)} = \frac{2n\lambda}{2}$$

or

$$X_n = \left[\frac{2n(a+b)b\lambda}{a}\right]^{1/2} \qquad (5.8)$$

where $n = 0, 1, 2, 3, \dots$ and X_n is the distance of nth dark band from O. Obviously, $X_n \propto \sqrt{n}$ for dark bands and $X_n \propto \sqrt{2n+1}$ for bright bands.

Fringe Width

As $X_n \propto \sqrt{2n+1}$ for bright bands and $X_n \propto \sqrt{n}$ for dark bands, the fringe width goes on decreasing as we move away from O. For maxima, we have

$$X_n = \left[\frac{(2n+1)(a+b)b\lambda}{a}\right] = K\sqrt{2n+1} \qquad (5.9)$$

where $K = \left[\frac{(a+b)b\lambda}{a}\right]^{1/2}$

$$X_1 - X_0 = \left(\sqrt{3} - 1\right)K = 0.73K$$

$$X_2 - X_1 = \left(\sqrt{5} - \sqrt{3}\right)K = 0.05K$$

$$X_3 - X_2 = \left(\sqrt{7} - \sqrt{5}\right)K = 0.43K$$

Like this, one can also determine fringe width between minima. Obviously, the separation between successive maxima, i.e. $X_1 - X_0$ and $X_2 - X_1$ are decreasing. Similarly, one can also show that the separation between successive minima is also decreasing. Thus as one moves above O on the screen, the fringes come closer. The bands will disappear and uniform illumination occurs if P is at a large distance from O.

Intensity at a Point Inside the Geometrical Shadow

Let P' be a point below O (Fig. 5.8). Thus with reference to P', the pole of the wavefront is B', the lower half wavefront $B'W'$ and AB' being completely obstructed due to obstacle. Thus, the light at P' reaches only from AW portion of the wavefront. If $B'A$ cuts off only the first half period strip of the upper half wavefront, the resultant amplitude at P' will be $R_2/2$ (the resultant amplitude at P' will be $= R_2 - R_3 + R_4 ..$) As P' moves further into the geometrical shadow 2, 3, 4,..., etc. half period strips are successively cut off and the amplitude at P' takes on successive values $= \frac{R_3}{2}, \frac{R_4}{2}, \frac{R_5}{2}, \dots$ It is clear that the values of $R_1, R_2, R_3, \dots$ decrease rapidly at first, and slowly afterwards, the intensity in the geometrical shadow decreases rapidly at first and slowly afterwards. At a point for cut off, the intensity becomes zero. The variation of intensity on the screen due to a straight edge is shown in Fig. 5.8. From Fig. 5.8, it is evident that the intensity in the geometrical shadow falls off rapidly and outside it bright and dark fringes are observed in the illuminated portion of the screen.

However, with white light coloured bands are observed and the bands of shorter wavelengths are nearer to the point P'.

5.10 DIFFRACTION AT A CIRCULAR APERTURE

Let AB be a narrow circular aperture (Fig. 5.12). Let a parallel beam of monochromatic light be incident over it normally. On the screen XY,

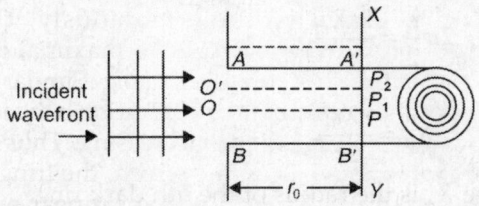

Fig. 5.12: Diffraction due to a circular aperture

kept on the other side of the aperture, the following diffraction pattern is obtained:

i. The centre P of the illuminated region $A'B'$ is bright or dark.

ii. In the illuminated region round P alternate bright or dark rings are observed.

iii. Above A' and below B', i.e. in the geometrical shadow, the intensity of light falls rapidly and then becomes zero (Fig. 5.13).

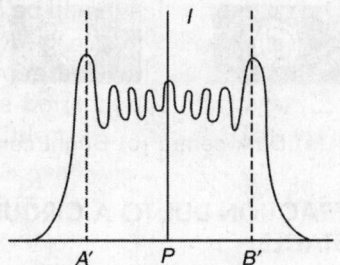

Fig. 5.13: Intensity distribution due to a circular aperture

Intensity at the Axial Point

Corresponding to the axial point P, there is pole O of the incident plane wavefront on aperture AB. Suppose the wavefront is divided into half period zones. The radius of the aperture AB is so small that only one half period zone can pass through it. Obviously, the amplitude at P will

be R_1 which is twice the half of whole wavefront amplitude $R_1/2$. Thus, the point P in this position will be the brightest point. If the size of the aperture is such that two half period zones can pass through it, then the amplitude at P will be almost zero ($R_1 - R_2 \approx 0$). Clearly in this situation, the point P will be a dark point. This explains, when the centre of diffraction pattern will be bright or dark. Obviously, this depends upon the number of half period zones exposed through the aperture (odd or even). If, keeping the size of the aperture fixed, the screen is moved towards the aperture, then the number of zones exposed through the aperture will be increased and the point P will be alternately bright or dark.

On taking the screen away from the aperture, the change in the intensity of the diffraction pattern is shown in Fig. 5.14.

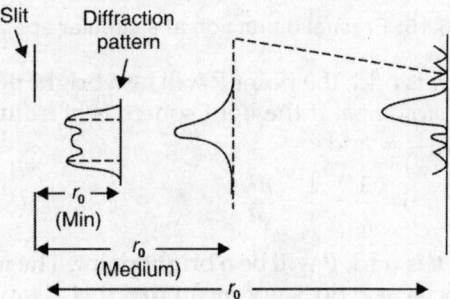

Fig. 5.14: The change in the intensity of the diffraction pattern when distance between aperture and screen is increased

Let r be the radius of the aperture and distance between O and P be r_0, then the area of each zone $= \pi r_0 \lambda$. The point P will be maxima or minima depending upon the following conditions:

Let δ be the path difference for the waves reaching at P along the paths SAP and SOP, then from Fig. 5.15, we have

$$\delta = SAP - SOP = SA + AP - SOP$$

$$= (a^2 + r^2)^{1/2} + (r_0^2 + r^2)^{1/2} - (a + r_0)$$

$$= a\left(1 + \frac{r^2}{2a}\right) + r_0\left(1 + \frac{r^2}{2r_0}\right) - (a + r_0)$$

$$= \frac{r^2}{2a} + \frac{r^2}{2r_0} = \frac{r^2}{2}\left(\frac{1}{a} + \frac{1}{r_0}\right) \qquad (5.10)$$

If the position of the screen is such that n full number of zones are exposed through the aperture, then

$$\delta = \frac{n\lambda}{2} \quad \text{or} \quad 2\delta = n\lambda$$

$$\therefore \quad \frac{1}{a} + \frac{1}{r_0} = \frac{n\lambda}{r^2} \qquad (5.11)$$

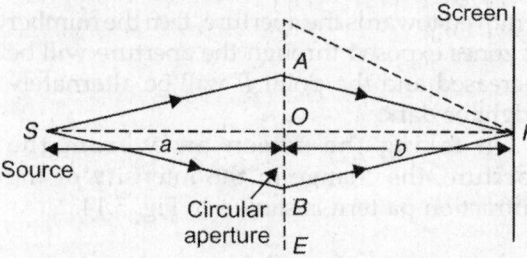

Fig. 5.15: Fresnel diffraction at a circular aperture

If n is odd, the point P will be a bright point and *vice versa*. If the light source is at infinity, then $a = \infty$ and

$$\frac{1}{r_0} = \frac{1}{f} = \frac{n\lambda}{r^2} \qquad (5.12)$$

If n is odd, P will be a bright point. The idea of focus at P does not mean that it is always a bright point.

Intensity at the Non-axial Points

Let us consider that due to diffraction at circular aperture AB, we want to find out the intensity at the point Q. Point Q is situated at a distance X_n from the point P (Fig. 5.16). Path difference between the secondary waves reaching at Q from A and B is given by

$$\delta = BQ - AQ$$

$$= \sqrt{r_0^2 + (X_n + r)^2} - \sqrt{r_0^2 + (X_n - r)^2}$$

$$= r_0\left[1 + \frac{(X_n + r)^2}{2r_0^2}\right] - r_0\left[1 + \frac{(X_n - r)^2}{2r_0^2}\right]$$

$$= \frac{(X_n + r)^2}{2r_0} - \frac{(X_n - r)^2}{2r_0} = \frac{4X_n r}{2r_0} = \frac{2X_n r}{r_0}$$

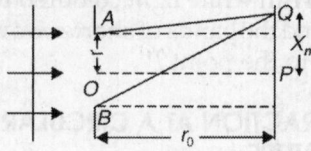

Fig. 5.16

The point P will be dark point if $\delta = 2n\lambda/2$ ($2n$ means the even number of zones)

$$\therefore \quad 2n\frac{\lambda}{2} = \frac{2rX_n}{r_0}$$

or $\qquad X_n = \frac{n\lambda r_0}{2r} \qquad (5.13)$

Here X_n is the radius of the nth dark ring.

If $\qquad \delta = (2n+1)\frac{\lambda}{2} \quad \text{or} \quad (2n+1)\frac{\lambda}{2} = \frac{2rX_n'}{r_0}$

$$\therefore \quad X_n' = \frac{(2n+1)r_0\lambda}{4r} \qquad (5.14)$$

where X_n' is the radius of the nth bright ring. Thus, it is evident that as we move away from the axis, the intensity falls in an alternate manner. Since the circular aperture is symmetrical around the axis, hence the maxima and minima are in the form of concentric rings round P (Fig. 5.17).

Fig. 5.17: (a) Dark centre (b) Bright centre

5.11 DIFFRACTION DUE TO A CIRCULAR OBSTACLE

If we take a circular obstacle in place of a circular aperture, then the diffraction pattern obtained is same as that at circular aperture with a difference that the centre is always bright in diffraction pattern due to circular aperture. This happens due to the fact that at the centre the contribution due to first unexposed half period zone is always positive. Diffraction arrangement due to a circular disc is shown in Fig. 5.18. Outside the geometrical shadow, brighter and broader rings are obtained.

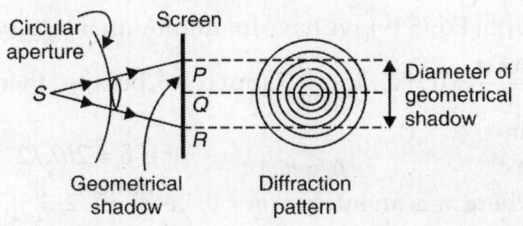

Fig. 5.18: Diffraction at a circular obstacle

5.12 COMPARISON OF FRESNEL DIFFRACTION PATTERNS AT A CIRCULAR HOLE AND CIRCULAR OBSTACLE

Fresnel Diffraction at a Circular Obstacle

When a circular obstacle is placed in the path of light from a point source, the diffraction pattern shows the following peculiarities:

i. A bright spot at the centre of geometrical shadow is formed.

ii. The smaller the size of the obstacle, the greater is the brightness of the spot.

iii. The central bright spot is surrounded by the faint rings in the shadow of the obstacle.

iv. Outside the shadow of the obstacle, brighter or broader rings are obtained.

Fresnel Diffraction at a Circular Aperture

When a narrow circular aperture is held in the path of light from a point source, the diffraction pattern shows the following characteristics:

i. The centre of illuminated ring is either bright or dark.

ii. In the illuminated region the centre is surrounded by dark and bright rings.

iii. The intensity in the geometrical shadow falls off rapidly.

5.13 FRAUNHOFER DIFFRACTION AT A SINGLE SLIT

Let AB be a narrow slit, whose width is e. Slit is placed perpendicular to the plane of the paper. Let a parallel beam of monochromatic light of wavelength λ be incident normally upon the slit from the source S. The diffracted light is focussed by a convex lens L_2 on a screen XY placed in the focal plane of the lens. The diffraction pattern thus obtained consists of a central bright band having alternate dark and weak bright bands of decreasing intensity on both sides of diffraction pattern.

We know that according to Huygen's wave theory, every point of the wavefront is a source of secondary wavelets. Thus, each point in AB sends out secondary wavelets in all the directions. The wavelets proceeding in the same direction as the incident wavefront are focussed at O. The diffracted wavefront through an angle θ is focussed at P_1. We are interested in determining the resultant intensity at P_1.

Let BL be perpendicular to AL. From Fig. 5.19a, it is clear that the optical paths from the plane AB to P are equal. The path difference between

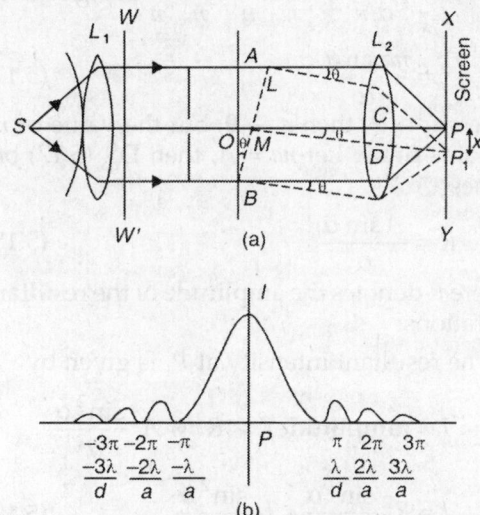

Fig. 5.19: (a) Diffraction at a single slit (b) Intensity distribution due to a single slit

the wavelets diffracted from A to B in the direction θ is

$$\delta = AL = AB\sin\theta = e\sin\theta$$

The corresponding phase difference

$$= \frac{2\pi}{\lambda} \times \text{path difference}$$

$$= \frac{2\pi}{\lambda} \times e\sin\theta$$

Let the slit width be divided into n equal parts. Since the screen is effectively at infinite distance from the source, therefore, the amplitude of vibration of all the wavelets reaching through the slit at P_1 is a. The path difference between the waves from any two consecutive parts is $\dfrac{1}{n}\dfrac{2\pi}{\lambda}e\sin\theta = d$ (say).

Hence, the resultant amplitude at P_1 is given by

$$R = \frac{a\sin(nd/2)}{\sin(d/2)} = \frac{a\sin(\pi e\sin\theta/\lambda)}{\sin(\pi e\sin\theta/n\lambda)} \quad (5.15)$$

Let $\quad \dfrac{\pi e\sin\theta}{\lambda} = \alpha \quad\quad\quad (5.16)$

then $\quad R = \dfrac{\pi e\sin\theta}{\sin(\alpha/n)}$

$$= \frac{a\sin\alpha}{\alpha/n}\left(\because \sin\frac{\alpha}{n}\approx\frac{\alpha}{n}, \frac{\alpha}{n}\text{ being small}\right)$$

$$= \frac{na\sin\alpha}{\alpha} \quad\quad\quad (5.17)$$

When $n \to \infty$, then $a \to 0$, but the value of na remains finite. Let $na = A$, then Eq. (5.17) becomes

$$R = \frac{A\sin\alpha}{\alpha} \quad\quad\quad (5.18)$$

where R denotes the amplitude of the resultant vibrations.

The resultant intensity at P_1 is given by

$$I \propto (\text{amplitude})^2 \propto R^2 \propto A^2\frac{\sin^2\alpha}{\alpha^2}$$

or $\quad I = R_0^2\dfrac{\sin^2\alpha}{\alpha^2} = I_0\dfrac{\sin^2\alpha}{\alpha^2} \quad (5.19)$

Here $I = R_0^2$ is the intensity of the principal maximum and the value α is given by Eq. (5.16).

Intensity distribution due to single slit is as shown in Fig. 5.19b.

Positions of Maxima and Minima

From Eq. (5.19), we have for minimum intensity

$\dfrac{\sin\alpha}{\alpha} = 0$, i.e. $\sin\alpha = 0$ (but $\alpha \neq 0$, because then $\sin\alpha/\alpha = 1$)

or $\quad\quad\quad \alpha = \pm m\pi$

where m is an integer ($m \neq 0$, i.e. $m = 1, 2, 3, \ldots$)

But $\quad\quad\quad \alpha = \dfrac{\pi e\sin\theta}{\lambda}$

$\therefore \quad \dfrac{\pi e\sin\theta}{\lambda} = \pm m\pi \quad\quad\quad (5.20)$

Equation (5.20) is a condition for minimum intensity. By putting $m = 1, 2, 3, \ldots$ in Eq. (5.20), we obtain positions of different minima. For maxima, we have

$$\frac{dI}{d\alpha} = 0$$

or $\quad\quad \dfrac{dI}{d\alpha}\left[I_0^2\left(\dfrac{\sin\alpha}{\alpha}\right)\right] = 0$

or $\quad I_0^2\dfrac{2\sin\alpha}{\alpha}\left[\dfrac{\alpha\cos\alpha - \sin\alpha}{\alpha^2}\right] = 0$

or $\quad\quad \dfrac{\alpha\cos\alpha - \sin\alpha}{\alpha^2} = 0$

or $\quad\quad\quad \alpha = \tan\alpha \quad\quad\quad (5.21)$

Equation (5.21) can be solved graphically by plotting the curves $y = \alpha$ and $y = \tan\alpha$ (Fig. 5.20).

The first of these equations represents a straight line through origin making an angle of $45°$, while second represents a discontinuous curve having a number of branches (Fig. 5.20).

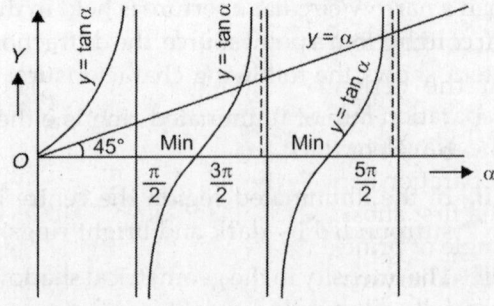

Fig. 5.20: Roots of equation $\alpha = \tan\alpha$

The points of intersection of these two equations give the values of α satisfying $y = \tan\alpha$. Approximate values of α are as follows:

$$\alpha = 0, \frac{3\pi}{2}, \frac{5\pi}{2}, \frac{7\pi}{2}, \ldots$$

More exactly, we have

$$\alpha = 0, 1.43\pi, 2.462\pi, 3.471\pi, \ldots$$

Substituting the approximate values of α in Eq. (5.19), one obtains the intensities for various maxima as below:

$$I = R_0^2 \left(\frac{\sin\theta}{\theta}\right)^2 = R_0^2 = I_0 \qquad (5.22)$$

Intensity of first subsidiary maxima is given by

$$I_1 = R_0^2 \left(\frac{\sin 3\pi/2}{3\pi/2}\right)^2 = \frac{R_0^2}{22} = \frac{I_0}{22} \quad (5.23)$$

Intensity of secondary subsidiary maxima

$$I_2 = R_0^2 \left(\frac{\sin 5\pi/2}{5\pi/2}\right)^2 = \frac{R_0^2}{61} = \frac{I_0}{61} \quad (5.24)$$

and so on.

It is obvious that most of the incident light is concentrated in the principal maxima which occurs in the direction given by $\alpha = 0$, i.e.

$$\frac{\pi e \sin\theta}{\lambda} = 0$$

$$\therefore \qquad \theta = 0$$

Thus, it is clear that the bright principal maxima is in the direction of the principal maxima. The intensity distribution pattern is as shown in Fig. 5.19b.

Effect of Width of Slit on Maxima

According to relation $\sin\theta = \pm\dfrac{m}{e}\lambda$, the width of the central maxima increases as the separation between the slits decreases. When the wavelength of the incident light and the separation between the slits become equal, then the first subsidiary maxima is situated at $\pi/2$ angle of principal maxima. In other words, no subsidiary maxima is formed and the complete space is occupied by central or principal

maxima. On putting $m = 1$ and $e = \lambda$, one obtains $\sin\theta = 1$ or $\theta = \pi/2$.

The diffraction pattern due to a narrow slit of width *e* differs much than the interference pattern due to a pair of narrow slits of separation *e*. In the diffraction pattern the principal maxima is the brightest, and angular width of λ/e. Thus, narrower the slit, broader is the maxima, i.e. the width of the principal maximum is inversely proportional to the width of the slit.

In the interference pattern, the maxima and minima are equally wide and equidistant. Moreover, all the maxima have the same intensity. The width of the fringes increases as the separation between the slits decreases.

5.14 FRAUNHOFER DIFFRACTION AT A CIRCULAR APERTURE

In Fig. 5.21 arrangement for Fraunhofer's diffraction at a circular aperture is shown. By replacing the list with a circular aperture, one observes the rings of maximum and minimum intensities around the central maximum.

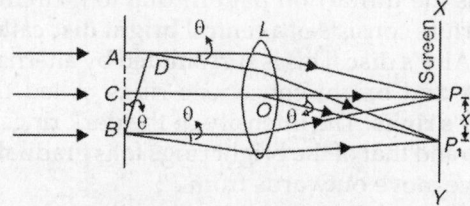

Fig. 5.21: Fraunhofer diffraction at a circular aperture

AB is a circular aperture of diameter *d*, having centre at *C*. *CP* is $\perp^r$ to the screen and screen is to the plane of the paper. A plane wavefront is incident on the circular aperture *AB*. The secondary waves travelling in the direction *CO* come to focus at *P* after passing through the lens *L*. Clearly *P* corresponds to the position of central maximum.

All the secondary waves emanating from points equidistant from *O* travel the same distance before reaching *P* and hence these waves reinforce one another. Now we consider the secondary waves travelling in the direction inclined at an angle θ with *CP*. All these waves meet at P_1 on the screen after passing through the lens *L*. Let $PP_1 = x$. The path difference between the secondary waves emanating from the extreme points *B* and *A* of the diameter is *AD*. From $\triangle ABD$, we have

$$AD = d \sin \theta \qquad (5.25)$$

If this path difference is equal to the integral multiple of λ, the point P_1 will be of minimum intensity, i.e.

$$d \sin \theta = n\lambda \text{ (minima)} \qquad (5.26)$$

$$n = 1, 2, 3, \ldots$$

and the point P_1 will be of maximum intensity if path difference *AD* is equal to odd multiple of λ/2, i.e.

$$d \sin \theta = (2n + 1)\lambda/2 \text{ (maxima)} \qquad (5.27)$$

The term $n = 0$ corresponds to central maximum at *P*.

Let us suppose that point P_1 is the point of minimum intensity. Then all the points at the same distance from *P* as P_1 and lying on the circle of radius *x* will be of minimum intensity. Thus the diffraction pattern due to a circular aperture consists of a central **bright disc** called the **Airy's disc**. This is surrounded by alternate dark and bright concentric rings called the **Airy's rings**. The intensity of the dark rings is zero and that of the bright rings falls gradually as we move outwards from *P*.

If the collecting lens is very near to the slit or the screen is at a large distance from the lens, then

$$\sin \theta \approx \theta \approx \frac{x}{f}$$

∴ For the first secondary minima,

$$d \sin \theta \approx \lambda$$

or $$\sin \theta \approx \theta \approx \frac{\lambda}{d}$$

∴ $$\frac{x}{f} = \frac{\lambda}{d}$$

or $$x = \frac{f\lambda}{d} \qquad (5.28)$$

Here *x* denotes the radius of the Airy's disc. But it is found that the radius of the first dark ring is slightly more than that given by Eq. (5.28).

The correction factor is 1.22. The corrected relation is given by

$$x = 1.22 \frac{f\lambda}{d} \qquad (5.29)$$

The intensity distribution in Fraunhofer diffraction at a circular aperture is as shown in Fig. 5.22.

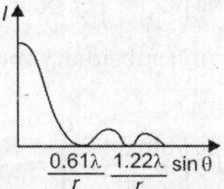

Fig. 5.22: Intensity distribution in diffraction pattern due to a circular aperture

5.15 FRAUNHOFER DIFFRACTION AT A DOUBLE SLIT

Let us have two parallel slits *AB* and *CD* of equal widths *a* and separated by opaque portion *b* (Fig. 5.23). Let a plane wavefront from a monochromatic source of light be incident on the surface of the slits (source is not shown in Fig. 5.23). Let the light diffracted from these slits be focussed on a screen by a convex lens *L*. *P* is a point on the screen such that $O'P$ is ⊥ᵣ to the screen. All the secondary waves travelling in a direction parallel to $O'P$ come to focus at *P*. Obviously, *P* corresponds to the position of central maxima.

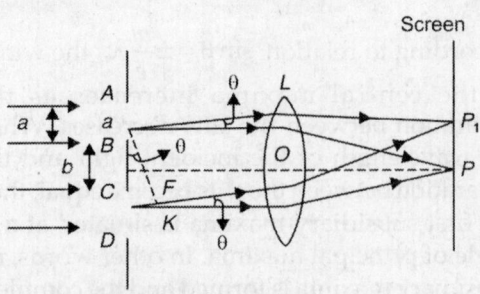

Fig. 5.23: Diffraction at a double slit

The diffraction pattern due to a double slit can be considered in two parts:

i. Interference phenomenon due to the secondary waves emanating from the corresponding points of the two slits.

ii. Diffraction pattern due to the secondary waves from the two slits individually.

According to Huygens' principle, every point in the slits AB and CD sends out secondary wavelets in all directions. From the theory of diffraction at a single slit, the resultant amplitude due to wavelets diffracted from each slit in θ direction is given by $A\sin\alpha/\alpha$, where $\alpha = \pi e\sin\theta/\lambda$. Obviously, one can consider the two slits AB and CD as equivalent to two coherent sources placed at their middle points. Each of these is sending a wavelet of amplitude at $A\sin\alpha/\alpha$ in the direction θ. The resultant amplitude P on the screen will be the result of the interference between two waves of same amplitude $A\sin\alpha/\alpha$. Let the phase difference between the two wavelets reaching at P be δ. The path difference from the two waves emanating from S and S_1 is $(a + b)\sin\theta$. Therefore,

phase difference $\delta = \dfrac{2\pi}{\lambda}\times$path difference

$$= \frac{2\pi}{\lambda}(a + b\sin\theta) \qquad (5.30)$$

The **resultant amplitude** (R) can be determined by making use of the vector **amplitude diagram** (Fig. 5.24). We have

$$R^2 = \left(\frac{A\sin\alpha}{\alpha}\right)^2 + \left(\frac{A\sin\alpha}{\alpha}\right)^2$$

$$+ 2\left(\frac{A\sin\alpha}{\alpha}\right)\left(\frac{A\sin\alpha}{\alpha}\right)\cos\delta$$

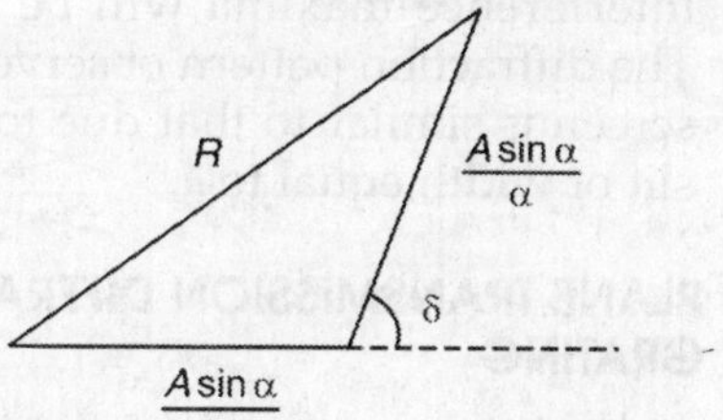

Fig. 5.24: Vector amplitude diagram

$$= \frac{2A^2\sin^2\alpha}{\alpha^2}(1 + \cos\delta)$$

$$= 4\frac{A^2\sin^2\alpha}{\alpha^2}\cos^2\frac{\delta}{2}$$

$$= 4A^2\frac{\sin^2\alpha}{\alpha^2}\cos^2\beta \qquad \left(\beta = \frac{\delta}{2}\right)$$

$\therefore$ Resultant amplitude at P

$$I \propto R^2 \propto \frac{4A^2}{1}\frac{\sin^2\alpha}{\alpha^2}\cos^2\beta$$

or $\quad I = 4R_0^2\dfrac{\sin^2\alpha}{\alpha^2}\cos^2\beta \qquad (5.31)$

From Eq. (5.31), it is obvious that the resultant intensity in the diffraction pattern depends on two factors:

i. $R_0^2\dfrac{\sin^2\alpha}{\alpha^2}$ which gives diffraction pattern due to each single slit

ii. $\cos^2\beta$ which gives the interference pattern due to light waves of the same amplitude from the two slits.

$\sin^2\alpha/\alpha^2$ gives a central maxima in the direction $\theta = 0$. On either side of it, alternately minima and subsidiary maxima of decreasing intensity are observed (Fig. 5.25).

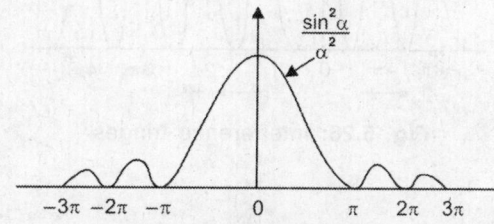

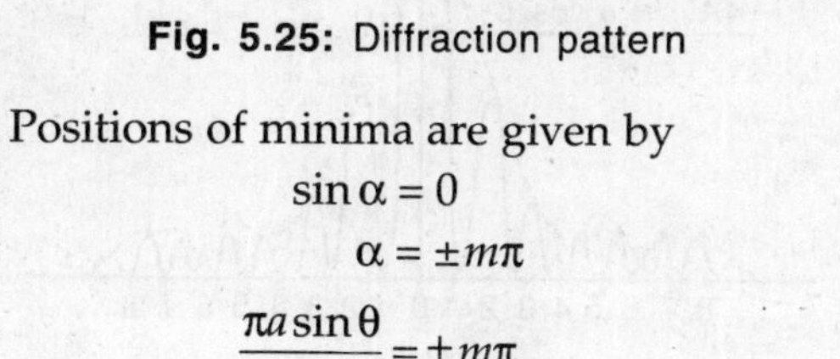

Fig. 5.25: Diffraction pattern

Positions of minima are given by
$$\sin\alpha = 0$$

or $\qquad \alpha = \pm m\pi$

or $\qquad \dfrac{\pi a\sin\theta}{\lambda} = \pm m\pi$

(where $m = 1, 2, 3, \dots$ but $m \neq 0$)

or $\qquad a\sin\theta = \pm m\pi \qquad (5.32)$

Thus, minima are obtained for
$$\alpha = \pm\pi, \pm 2\pi, \pm 3\pi, \dots$$

Positions of subsidiary maxima are given by
$$\alpha = \pm 3\pi/2, \pm 5\pi/2, \ldots$$

$\cos^2\beta$, the interference term, gives a set of equidistant dark and bright fringes as observed in Young's double slit experiment. The maxima, i.e. the bright fringes are obtained in the directions given by
$$\cos^2\beta = 1$$
or $$\beta = \pm nm$$
or $$\pi/\lambda(a + b)\sin\theta = \pm n\pi$$
or $$(a + b)\sin\theta = \pm n\lambda \ (n = 0, 1, 2, \ldots) \quad (5.33)$$

Corresponding to $n = 0, 1, 2, 3, \ldots$, the maxima are called as *first order, second order, third order,* ... maxima. Positions of minima are given by
$$\cos^2\beta = 0$$
or $$\beta = \pm(2n + 1)\pi/2, \text{ where } n = 0, 1, 2, \ldots$$
or $$\pi/\lambda(a + b)\sin\theta = \pm(2n + 1)\pi/2$$

For minimum value of θ, the width of minima and maxima are equal (Fig. 5.26). Resultant diffraction pattern is shown in Fig. 5.27 (for $b = 2a$).

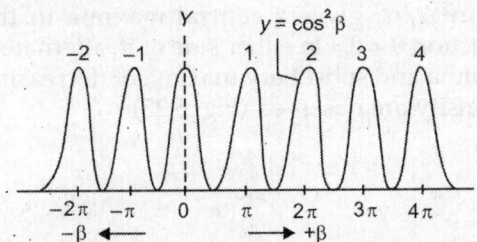

Fig. 5.26: Interference fringes

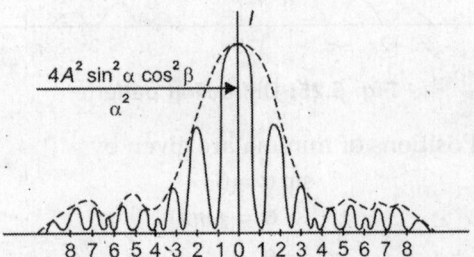

Fig. 5.27: Resultant diffraction pattern $b = 2a$

Effect of Increasing the Distance between Slits

If the slit width a is kept constant and the separation b between them is varied, then according to $(a + b)\sin\theta = \pm n\pi$, the separation between the interference maxima varies, while the size of the diffraction pattern due to single slit remains unchanged. For some values of b, interference maxima disappear.

$(a + b)\sin\theta = \pm n\pi$ (interference maxima condition)

$a\sin\theta = \pm m\pi$ (diffraction minima condition)

i. Let $\quad\quad a = b$
$\therefore \quad\quad 2a\sin\theta = \pm n\pi$
But $\quad a\sin\theta = \pm m\pi$
$\therefore \quad\quad n/m = 2 \quad \text{or} \quad n = 2m$

If $m = 1, 2, 3, \ldots$, then $n = 2, 4, 6$

Obviously, the orders 2, 4, 6, ..., etc. of the interference maxima will be missing in the diffraction pattern. There will be three interference maxima in the central diffraction maxima.

ii. Let $\quad\quad 2a = b$
$\therefore \quad\quad 3a\sin\theta = \pm n\pi$
Also $\quad a\sin\theta = \pm m\pi$
$\therefore \quad\quad a\sin\theta = \pm m\pi \quad \text{or} \quad n = 3m$

If $m = 1, 2, 3, \ldots$, then $n = 3, 6, 9, \ldots$

Obviously *third, sixth, ninth* orders of interference maxima will be absent in the diffraction pattern, i.e. 3rd, 6th, 9th,... orders of interference maxima will coincide with 1st, 2nd, 3rd,... orders of diffraction minima respectively. Clearly, the central diffraction maxima will have five interference maxima.

Thus, as b increases the number of interference maxima within the central diffraction maxima increases.

iii. If $a + b = a$ or $b = 0$, obviously the two slits join and all the orders of the interference maxima will be missing. The diffraction pattern observed on the screen is similar to that due to a single slit of width equal to a.

5.16 PLANE TRANSMISSION DIFFRACTION GRATING

A plane diffraction grating is an arrangement equivalent to a large number of parallel slits of equal widths and separated from one another by equal opaque spaces. When a wavefront from a light source is incident on a grating surface, light is transmitted through the slits and obstructed by the opaque portions. It is made by ruling a large number of finite, equidistant and parallel lines on an optically plane glass plate with a diamond point. If the spacing between the lines is of the order of wavelength of light, then an appreciable deviation of light is produced. Gratings used for the study of the visible region of the spectrum contain 10,000 lines per cm.

For practical purposes, repulses of the original grating are prepared. On the original grating surface a thin layer of collodion solution is poured and the solution is allowed to harden. The film of collodion is removed from the grating surface and then fixed between two glass plates. This is called as plane transmission grating. A large number of replicas are prepared in this way from a single original ruled surface.

Theory

XY is a grating surface and MN is a screen $\perp^r$ to the plane of the paper. All the slits are parallel to each other. $(a + b)$ is called the *grating element*. The points in the consecutive slits separated by the distance $(a + b)$ are called the **corresponding points**. The point P where all the secondary waves reinforce on another corresponds to the position of the central bright maximum.

Let us consider the secondary waves travelling in a direction inclined at an angle θ with the incident light. The direction of the collecting lens is also suitably rotated such that the axis of lens is parallel to the direction of the secondary waves. These secondary waves are focussed at P' on the screen. The intensity at P' depends on the path difference between the secondary waves originating from the corresponding point A and C of the two neighbouring slits. The path difference between the secondary waves emitting form A and C is $AC \sin \theta$.

But $\qquad\qquad AC = a + b$

$\therefore \quad$ Path difference $= (a + b) \sin \theta$

As we know that the point P' will be of maximum intensity if the above path difference is equal to integral multiple of wavelength λ. Here the secondary waves originating from the corresponding points of the neighbouring slits reinforce one another and θ gives the direction of maximum intensity, i.e.

$$(a + b) \sin \theta_n = n\lambda \qquad (5.34)$$

Here θ_n is the direction of the nth principal maxima. On putting $n = 1, 2, 3, \ldots$, one obtains the direction of principal maxima as θ_1, θ_2, $\theta_3, \ldots$, i.e.

$$(a + b) \sin \theta_1 = n\lambda$$
$$(a + b) \sin \theta_2 = 2\lambda \qquad \text{etc.}$$

If the incident light consists of more than one light wavelength (e.g. sodium light consists of 5890 Å and 5896 Å wavelengths), then the beam is dispersed and the angle of dispersion for each wavelength is different. Let λ and $\lambda + d\lambda$ are two wavelengths present in the incident light and the corresponding angle of scattering are θ and $\theta + d\theta$ respectively. For the first order principal maxima, we have

$$(a + b) \sin \theta = \lambda \qquad (5.35)$$

and $\quad (a + b) \sin (\theta + d\theta) = \lambda + d\lambda \qquad (5.36)$

Clearly, the number of principal maxima in any order corresponds to the number of wavelengths present. Corresponding to the different wavelengths the images of parallel slits will also be observed. When white light is used, a white central maxima is observed in the diffraction pattern. On both the sides of the central maxima a spectrum corresponding to the different wavelengths of light present in the incident beam will be observed for each other.

Secondary Maxima and Minima

θ_n, the angle of diffraction to nth principal maxima, can be determined by the following equation:

$$(a + b) \sin \theta_n = n\lambda$$

Let the angle of diffraction θ_n be increased by a small amount $d\theta$ such that the path difference between the waves (secondary) emanating from the points A and C increased by λ/N. Here N represents the total number of lines on the grating surface. The path difference between the secondary waves emanating from the extreme points of the grating surface will be $\lambda/N \times N = \lambda$. Let us assume that the whole of the wavefront is divided into two halves, the path difference between the corresponding points of the two halves will be $\lambda/2$. All the secondary waves will cancel each other's effect. Clearly, $(\theta_n + d\theta)$ represents the direction of the first secondary minimum after the nth primary maximum. If the path difference between the secondary waves is $2\lambda/N$, $3\lambda/N$, etc. for gradually increasing values of $d\theta$, then these angles will correspond to the directions of 2nd, 3rd, etc. secondary minima after the nth primary maxima corresponding to the value of $2\lambda/N \times N = 2\lambda$. Now considering the wavefront to be divided into 4 halves, the concept of second secondary minima can be understood. The number of secondary maxima in between any two primary minima is $N-2$ and the number of secondary minima is $N-1$. The intensity distribution is shown in Fig. 5.28.

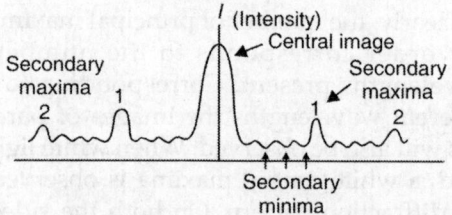

Fig. 5.28: Intensity distribution

For higher values of N, the secondary minima are weaker. In real grating the value of N is much larger and therefore these secondary maxima are not observed in the grating spectrum.

Width of Principal Maxima

We have $(a + b)\sin\theta_n = n\lambda$. From this relation, one can obtain the direction of the nth maxima.

Let $(\theta_n + d\theta)$ and $(\theta_n - d\theta)$ represent the direction of the first secondary minima on the two sides of the nth primary maxima. Now

$$(a + b)\sin(\theta_n \pm d\theta) = n\lambda + \frac{\lambda}{N} \qquad (5.37)$$

Dividing Eq. (5.37) by Eq. (5.34), one obtains

$$\frac{(a + b)\sin(\theta_n \pm d\theta)}{(a + b)\sin\theta_n} = \frac{n\lambda + \lambda/N}{n\lambda}$$

or $\quad \dfrac{\sin\theta_n \cos d\theta \pm \cos\theta_n \sin d\theta}{\sin\theta_n} = 1 \pm \dfrac{1}{Nn}$ $\quad$ (5.38)

or $\qquad 1 \pm \cot\theta_n d\theta = 1 \pm \dfrac{1}{Nn}$

(for small values of $d\theta$, $\cos d\theta \approx 1$ and $\sin d\theta \approx d\theta$)

or $\qquad d\theta = \dfrac{1}{Nn \cot\theta_n}$ $\qquad$ (5.39)

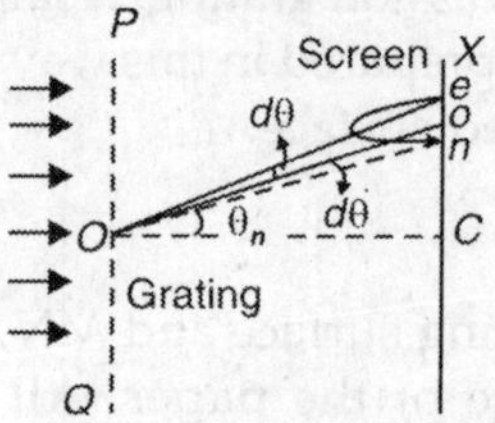

Fig. 5.29: Half angular width of principal maxima

$d\theta$ is the **half angular width** of the **principal maxima**. From Eq. (5.39), it is obvious that

(i) $d\theta \propto \dfrac{1}{n}$ $\qquad$ (ii) $d\theta \propto \dfrac{1}{N \cot\theta_n}$

Clearly, the value of $n \cot\theta_n$ is higher for higher orders and therefore half width of the principal maxima is less for higher orders.

Oblique Incidence

Let a parallel beam of light be incident obliquely on the grating surface as shown in Figs 5.30a and b.

The path difference between the secondary waves from A and C is equal to

$$FC + CE$$

From $\triangle AFC$ in Fig. 5.30b, we have

$$FC = (a + b)\sin i$$

and from $\triangle ACE$, we have

$$CE = (a + b)\sin\theta.$$

$\therefore$ Total path difference $= FC + CE$

$$= (a + b)(\sin\theta + \sin i) \quad (5.40)$$

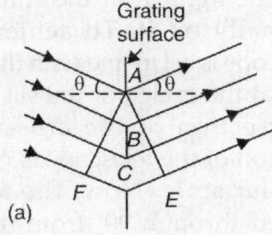

(a)

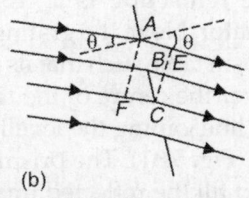

(b)

Fig. 5.30: Oblique incidence

The above path difference holds good if the beam is diffracted upwards. When the diffraction of the beam is downwards, the path difference is $(a + b)(\sin\theta - \sin i)$.

For the **nth primary maximum**, the condition can be expressed as

$$(a + b)[\sin\theta_n + \sin i] = n\lambda \quad (5.41)$$

or $(a + b)\left[2\sin\dfrac{\theta_n + i}{2}\cos\dfrac{\theta_n - i}{2}\right] = n\lambda$

or $\sin\dfrac{\theta_n + i}{2} = \dfrac{n\lambda}{2(a+b)\cos[(\theta_n - i)/2]} \quad (5.42)$

Clearly, the deviation of the diffracted beam is $(\theta_n + i)$.

For $(\theta_n + i)$ minimum, $\sin\dfrac{\theta_n + i}{2}$ should be minimum. It is possible when $\dfrac{\theta_n + i}{2}$ is maximum, i.e. $(\theta_n - i)/2 = 0$ or $\theta_n = i$.

Clearly, the deviation produced in the diffracted beam is minimum when the angle of incidence is equal to the angle of diffraction. If $D_m = \theta_n + i,$

where D_m is angle of minimum deviation, then $\theta_n = i$

$\therefore \qquad \theta_n = \dfrac{D_m}{2} \quad \text{or} \quad i = \dfrac{D_m}{2}$

$\therefore \quad (a+b)\left(\sin\dfrac{D_m}{2} + \sin\dfrac{D_m}{2}\right) = n\lambda$

or $\quad 2(a+b)\sin\dfrac{D_m}{2} = n\lambda \quad (5.43)$

This corresponds to the principal maxima of the nth order for a wavelength λ.

Condition for Absent Spectra

In the expression $(a + b)\sin\theta = \lambda$, if $(a + b) < \lambda$, then $\sin\theta > 1$. But $\sin\theta > 1$ is not allowed and hence for $(a + b) < \lambda$, the first order spectrum will remain absent. Similarly, for $(a + b) < 2\lambda$, $(a + b) < 3\lambda$, $(a + b) < 4\lambda$, …the 2nd order, 3rd order, the 4th order, etc. will remain absent. In general, $(a + b) < n\lambda$, the nth order spectrum will remain absent. The condition for absent spectrum can be obtained as below. For principal maxima of nth order, we have

$$(a + b)\sin\theta_n = n\lambda \quad (5.44)$$

If the value of a and θ_n is such that

$$a\sin\theta_n = n\lambda \quad (5.45)$$

then the effect of the wavefront from any particular slit will be nil. Considering each slit to be made of two halves, the path difference between the secondary waves from the corresponding points will be $\lambda/2$. These will cancel each other's effect.

If Eqs. (5.44) and (5.45) are simultaneously satisfied, then

$$\frac{(a + b)\sin\theta_n}{a\sin\theta_n} = \frac{n\lambda}{\lambda}$$

or $\qquad \dfrac{a + b}{a} = n \quad (5.46)$

$n = 1, 2, 3,…$ in Eq. (5.46) represent the absent principal maxima in diffraction pattern. If $\dfrac{a+b}{a} = 1$, i.e. $b = 0$, the first order spectrum will be absent and the resultant diffraction pattern will be same as single slit.

If $\dfrac{a+b}{a} = 2$, i.e. $a = b$, the second order spectrum will be absent.

5.17 OVERLAPPING OF SPECTRAL LINES

When the incident light on the grating surface consists of a large range of wavelengths, then the spectral lines of shorter wavelength of the higher order overlap on the spectral lines of the longer wavelength and of lower order. Let in the first order spectrum the wavelength of the spectrum line be λ_1, in the second order spectrum the wavelength of the spectrum line be λ_2, in the third order the wavelength of the spectrum line be λ_3, for all the value of angle of diffraction is same, i.e.

$$(a + b)\sin\theta = \lambda_1 = 2\lambda_2 = 3\lambda_3 \quad (5.47)$$

Obviously, in the visible region of spectrum there is no overlapping. This visible region of the spectrum extends from 4000 – 7200 Å.

5.18 DETERMINATION OF WAVELENGTH OF A SPECTRAL LINE USING PLANE TRANSMISSION GRATING

The principal maxima in the diffraction pattern of the grating can be obtained from the following condition,

$$(a + b)\sin\theta = n\lambda$$

Here n is the order of the spectrum. If the grating interval $(a + b)$ and the angle of diffraction $\theta \rightarrow$ is known, the wavelength λ can be determined.

Determination of Grating Element

The grating interval $(a + b)$ is determined by the number of lines marked per inch over the grating surface. For example, if the number of lines per inch on the grating is 10000, then the grating element is given by

$$a + b = \frac{2.54}{10000} \text{ cm}$$

$$= 2.54 \times 10^{-4} \text{ cm} = 2.54 \times 10^{-6} \text{ m}$$

Determination of Angle of Diffraction

The angle of diffraction θ is determined with the help of a spectrometer. The spectrometer slit is illuminated by the given monochromatic light. The following adjustments are made:

(a) The eyepiece of the telescope fitted in the instrument is focussed on the crossed wires.

(b) By Schuster's method using a prism, the collimator and the telescope are adjusted for the parallel beam.

(c) The grating is fixed on the prism table and then the adjustments are made so that the light from the collimator falls normally on it. To achieve this the telescope is set in line with the collimator so that the image of the slit falls on the intersection of the crosswires. The position of the telescope is noted on the circular scale. Now, the telescope is turned through 90° from this position and it is clamped. In this position the axis of the telescope is $\perp^r$ to that of the collimator. Now, the grating is placed on the prism table such that its ruled surface lies over the centre of the table and is $\perp^r$ to the line joining the levelling screws A and B (Fig. 5.31). The prism table is now rotated till the reflected image of the slit from the surface of the grating is obtained on the intersection of the cross wires (Fig. 5.32).

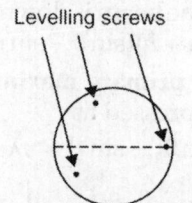

Fig. 5.31: Prism table

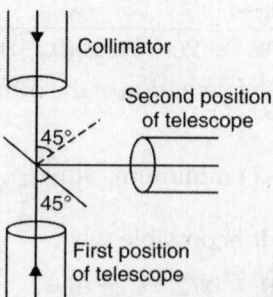

Fig. 5.32: Measurement of diffraction angle

The levelling screws A and B are now adjusted so that the image lies equally above and below the intersection of the cross wires. In this position of the prism table, the grating surface is 45° to the incident light rays. Now rotate the prism table through 45° in the proper direction so that the ruled surface of the grating is normal to the incident light and faces the telescope. Now clamp the prism table in this position.

(d) The ruling over the grating surface must be adjusted parallel to the spectrometer. To achieve it, the diffracted images of the slit are observed through the telescope. With the help of screw C adjustment is made until the centre of all the diffracted images lies at the same height in the field of view.

(e) The ruling over the grating surface must be adjusted parallel to the slit. For this purpose, rotate the slit in its own plane until the diffracted images are obtained as sharp as possible.

Now, for the determination of θ for the spectrum line whose wavelength is to be determined, rotate the telescope such that the spectrum line in the first order is visible on either side of the direct image. Adjust the telescope in such a manner that the spectrum lines falls on the intersection of the crosswires. At this position record the readings of both the verniers. Now, turn the telescope to the other side of the direct image and again record the reading of both the verniers for the same spectrum line. Now, determine the difference between the two readings of the same vernier. This gives 2θ. Thus, θ is determined. Knowing $2(a + b)$, θ and n, one can easily determine the wavelength λ by substituting these values in the relation $(a + b)\sin\theta = n\lambda$.

Maximum Number of Order Available with a Grating

We have

$$(a + b)\sin\theta = n\lambda$$

Maximum value of $\sin\theta_n$ is 1.

$$n_{max} = \frac{a+b}{\lambda}$$

If the grating element $(a + b) < 2\lambda$, then

$$n_{max} < \frac{2\lambda}{\lambda} < 2$$

This means only first order spectrum will be visible.

When $(a + b) < 3\lambda$, then

$$n_{max} < \frac{3\lambda}{\lambda} < 3$$

This means that only 1st and 2nd order spectrum will be visible.

5.19 DISPERSIVE POWER OF A GRATING

The dispersive power of a diffraction grating is defined as the rate of change of angle of diffraction with the incident wavelength of light and expressed as $d\theta/d\lambda$. The direction of the nth order principal maxima for a wavelength λ is given by the equation

$$(a + b)\sin\theta = n\lambda$$

Differentiating it with respect to λ, one obtains

$$(a + b)\cos\theta\, d\theta = n d\lambda$$

or

$$\frac{d\theta}{d\lambda} = \frac{n}{(a+b)\cos\theta} \quad (5.48)$$

This is the expression for the dispersive power of a diffraction grating. $1/(a + b)$ can be written as N'. Thus, Eq. (5.48) becomes

$$\frac{d\theta}{d\lambda} = \frac{nN'}{\cos\theta} \quad (5.49)$$

From Eq. (5.49), we note that:

i. $\dfrac{d\theta}{d\lambda} \propto n$, i.e. dispersive power of a grating is directly proportional to the order of the spectrum. This means that dispersive power is greater for higher order spectrum.

ii. $\dfrac{d\theta}{d\lambda} \propto N'$, i.e. dispersive power is directly proportional to the number of lines per cm on the grating surface, i.e. inversely proportional to the grating element.

This means that the dispersive power will be higher for a grating having more number of lines per cm.

iii. $\dfrac{d\theta}{d\lambda} \propto \dfrac{1}{\cos\theta}$, i.e. dispersive power is inversely proportional to the cosine of the angle of diffraction. This shows that when θ increases $\cos\theta$ decreases and dispersive power is higher. The angle of diffraction for the red light is greater than that for the violet in a given order spectrum, the dispersion in the red region is greater than that in the violet region.

In the above discussion, if we neglect the influence of $\cos\theta$, it is clear that the angular dispersion of any two spectral lines (in a particular order) is directly proportional to the difference in wavelength between the two spectral lines. Such spectrum is called **normal spectrum**.

5.20 PRISM AND GRATING SPECTRUM

In spectroscopy, in comparison to a prism, a diffraction grating is widely used due to following reasons:

i. With the help of grating a number of spectrum of different orders can be studied on both the sides of the central maxima whereas with a prism only one spectrum is obtained.

ii. The grating spectra are comparatively purer than that of a prism spectrum.

iii. In the case of grating spectra, knowing grating element $(a + b)$ and measuring the angle of diffraction θ, the wavelength (λ) of any spectral line can be determined very accurately. In the case of prism spectrum the angle of deviation are not directly related to the wavelength of spectral line. The angles of deviation are dependent on the refractive index (μ) of the prism material and which depends on the wavelength (λ) of the incident light.

iv. In a grating spectrum, the colours are in the order from violet to red. This is because the angle of diffraction is least for violet and maximum for red (Fig. 5.33). In the case of prism spectrum, the colours are in the order of red to violet. This is because the angle of deviation is least for red and greatest for violet (Fig. 5.34).

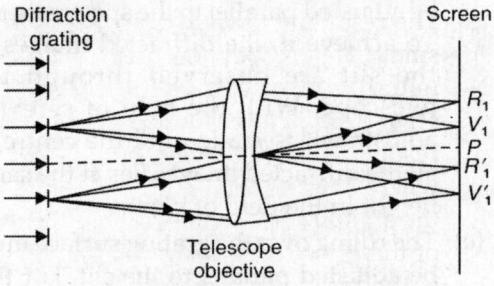

Fig. 5.33: Grating spectrum

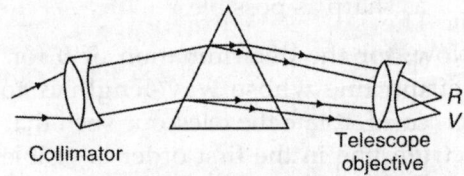

Fig. 5.34: Prism spectrum

v. The grating spectra are much fainter because most of the intensity of the incident light is associated with the undispersed central bright maxima and the rest of the energy is undistributed in the different order spectra on the two sides of central maxima. The prism spectrum is bright, since all the incident light is distributed only in a single spectrum.

vi. The dispersive power of a grating is given by $\dfrac{d\theta}{d\lambda} = \dfrac{n}{(a+b)}\cos\theta$. Since $\theta_V < \theta_R$ and therefore, $\cos\theta_V < \cos\theta_R$ and hence there is a greater dispersion in the red region than in the violet region. The dispersive power produced by the prism is given by $\dfrac{d\mu}{\mu-1}$ or $\dfrac{d\delta}{d\lambda} \propto \dfrac{1}{\lambda^3}$. Obviously, there is a great dispersion in the violet region (shorter wavelength) than in the red region. The spectrum obtained with a prism is said to be irrotational.

vii. The resolving power of grating is large. This is why the grating spectrum shows the fine structure of the spectral lines. The resolving power of a prism is very small. This is why the prism spectrum usually does not exhibit the fine structure of the spectral lines. The resolving power of the grating is Nn, where N is the total number of lines on the grating surface. The resolving power of the prism is equal to $t\, d\mu/d\lambda$. Here t is the base of prism and $d\mu/d\lambda$ is the variation of the refractive index with the wavelength.

viii. The spectrum lines in the grating spectrum are almost straight whereas the spectral lines in the prism spectrum are curved being convex towards the red end (Fig. 5.35).

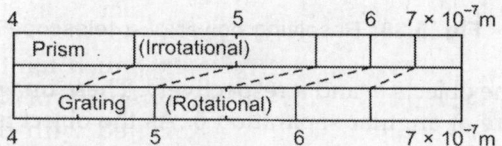

Fig. 5.35: Comparison of prism and grating spectra

5.21 RESOLVING POWER OF OPTICAL INSTRUMENTS

In any optical system, the image of an object is obtained by means of a beam of rays which are restricted by the entrance pupil, i.e. it is a result of the diffraction of light in the system. Hence the image cannot be absolutely stigmatic even in a system that is free from all possible kinds of aberration. Owing to diffraction, any point of a luminous object is seen as a central bright spot surrounded by alternate dark and bright interference rings.

This phenomenon limits the possibility of detecting fine details on the image of an object. For this purpose an optical instrument is used. *The ability of an optical instrument to produce distinctly separate spectral lines of light having two or more wavelengths very close to each other or to resolve the images of two nearby points is termed as its resolving power.*

If in any light radiations emitted from a light source, there are such radiations present bearing very small difference of wavelength, then the two spectral lines in a spectrum formed from a grating or prism will be very close to each other. In fact, these spectral lines are also the diffraction pattern. The ability of grating or prism to resolve these two nearby spectral lines is referred to as the resolving power of the grating or prism respectively.

5.22 RAYLEIGH'S CRITERION: LIMIT OF RESOULTION

Lord Rayleigh proposed the following criterion for deciding the resolving power of an optical instrument based upon the resultant intensity distribution curves of diffraction patterns obtained from two point sources. The criterion after his name is called as **Rayleigh's criterion**.

According to Rayleigh, *the images of two nearby self-luminous (incoherent) points can still be regarded as separate if the centre of the diffraction pattern corresponding to one point coincides with the first diffraction minimum for the other point.*

In Fig. 5.36a, the intensity curves of two patterns are shown. When the difference in the wavelengths of the two is large, the two images, i.e. the central maxima of the two images can be seen as separate ones (Fig. 5.36b). When the wavelength of the two spectral lines is so close, i.e. there is much overlapping of central maxima, then in such a situation the resultant curve of both the intensity curves exhibit only one maxima. In such a situation the optical instruments cannot resolve the spectral lines in two and both the spectral lines are viewed as only one spectral line.

If the difference of both the spectral lines is such that the principal maxima of one coincide with the first minima of the second (Figs 5.36b and c) then the eye will see the combined effect of two. The resultant curve shows a distinct 'dip' in the middle, indicating the presence of two different spectral lines (Fig. 5.36c). The lines are said to be just resolved.

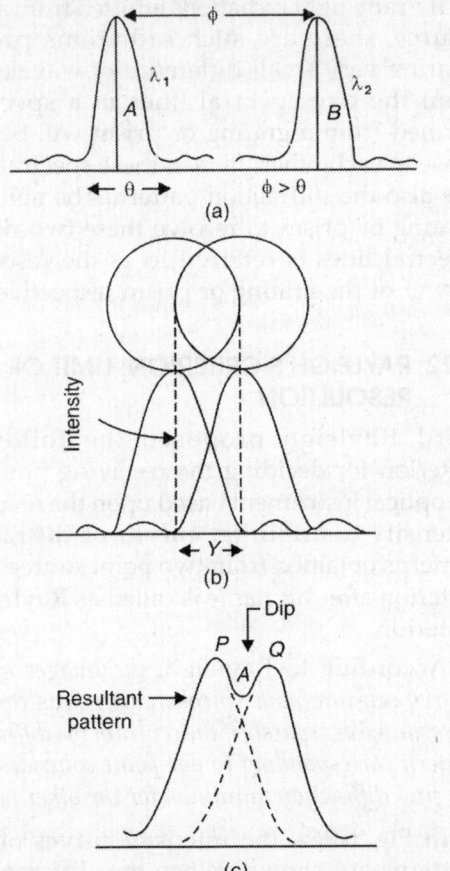

Fig. 5.36: (a) Distinctly resolved (b) Not resolved (c) Just resolved

In Fig. 5.37, the positions of Fig. 5.36 overlapping, limit of resolution and beyond resolution limit are shown.

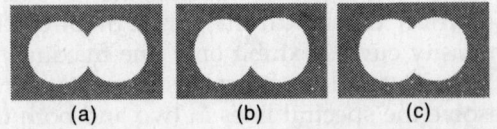

Fig. 5.37: Positions of (a) Overlapping (b) Resolution limit (c) Beyond resolution limit

5.23 RESOLVING POWER OF A TELESCOPE

Telescope is an instrument which is used to see the distant objects which subtend a small angle at its objective. Its resolving power is defined as:

The resolving limit of a telescope of the smallest angle subtended by two close and distant point objects at the objective of the telescope such that their images can just be seen as separate in the focal plane of the objective. If two objects subtend an angle less than this, their images cannot be separated by the telescopic objective. The smaller the value of this angle, the higher is the resolving power. The reciprocal of the smallest angle (θ_{min}) is expressed as the resolving power.

Two distant objects m and n subtend an angle θ at the telescopic objective AB (Fig. 5.38). P and F represent the centres of the images of

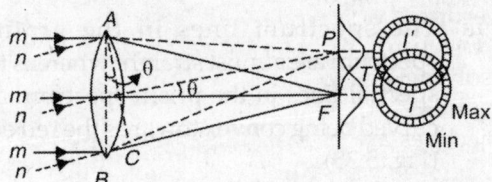

Fig. 5.38: Resolving power of a telescope

the objects m and n respectively. These images are at angular separation θ. As the object is a distant one, the wavefront entering the objective is practically plane. The boundary of the objective acts like a circular aperture and diffract light in various directions, resulting in a Fraunhofer diffraction pattern in the focal plane of the objective. The diffraction pattern consists of a central bright disc with centre at T surrounded by a number of alternate dark and bright bands. According to the theory of Fraunhofer diffraction at a circular aperture, the angular separation θ between the centre of the central bright disc and the first ring is given by

$$\sin \theta_{max} = 1.22 \frac{\lambda}{d} \qquad (5.50)$$

Here d is the diameter of the telescopic objective. Since $d \gg \lambda$, therefore $\sin\theta$ is small and hence we may take $\sin\theta \approx \theta$. Therefore, Eq. (5.50) reduces to

$$\theta_{min} = 1.22\frac{\lambda}{d} \text{ radian} \qquad (5.51)$$

Therefore, the resolving power of telescopic objective is given by

$$RP = \frac{1}{\theta_{min}} = \frac{d}{1.22\lambda} \qquad (5.52)$$

The resolving power of the entire instrument (R) depends as well on the resolving power of the receiver (R_r), i.e. eye, photographic emulsion, etc. It may be assumed, with some approximation, that

$$\frac{1}{R} = \frac{1}{R_0(\text{objective})} + \frac{1}{R_r(\text{receiver})} \qquad (5.53)$$

Equation (5.52) shows that larger is the aperture d of the telescopic objective, smaller is the limit of resolution, i.e. higher the resolving power. This is why for the higher resolving power, the telescopes with larger objectives are preferred.

5.24 RELATION BETWEEN MAGNIFYING POWER AND RESOLVING POWER OF A TELESCOPE

The magnifying power of an optical instrument represents the ability of the instrument to show a magnified view of the object of the eye.

The magnifying power of a telescope (M) is defined as the ratio of the angle subtended by the final image at the eye (β) to the angle subtended by the object in its optical position at the unaided eye (β), i.e.

$$M = \beta/\alpha \qquad (5.54)$$

The magnifying power of the telescope is defined as

$$M = \frac{D}{d} \qquad (5.55)$$

where D and d are diameter of the objective and diameter of the eye ring respectively.

When the diameter of the eye ring is equal to the diameter of the pupil of the eye, the magnifying power of the telescope is said to be normal. Therefore,

Normal magnifying power

$$= \frac{\text{Diameter of the objective}}{\text{Diameter of the pupil of the eye}}$$

Let us suppose that two object points subtending an angle α at the objective of the telescopes are just resolved by it, then

$$\alpha = \frac{1.22\lambda}{D} \qquad (5.56)$$

But we know that the final images are to be seen separated by the eye, they must subtend at the eye an angle at least α'. Here α' is the angle that can be just resolved by the eye and is given by

$$\alpha' = \frac{1.22\lambda}{d_e} \qquad (5.57)$$

Here d_e is the diameter of the pupil of the eye. Therefore, the smallest magnifying power which a telescope must possess is given by

$$M = \frac{\alpha'}{\alpha} = \frac{1.22\lambda}{d_e} \times \frac{D}{1.22\lambda} = \frac{D}{d_e} \qquad (5.58)$$

$$= \frac{\text{Diameter of the objective}}{\text{Diameter of the pupil of the eye}}$$

This is termed normal magnifying power. If the magnifying power of a telescope is less than this, the angle subtended by the final image will be less than α'. In such a situation, the images will not be resolved by the eye.

From Eq. (5.54), it is obvious that the images can be just resolved from a telescope when β is the resolving power of the eye and α is the resolving power of the telescope, then

$$M = \frac{\text{Resolving power of the eye}}{\text{Resolving power of the telescope}}$$

$\therefore$ Magnifying power of telescope × Resolving power of telescope = Resolving power of eye

The differences between the magnifying power and resolving power are given in the Table 5.1.

Table 5.1: Difference between magnifying power and resolving power

Magnifying power	Resolving power
i. This represent the ability of an instrument to magnify an image.	This represent the ability of an instrument to provide the details of the instrument.
ii. The distance between the centres of the maxima of the diffraction patterns due to two images is increased but it is not possible to obtain any information regarding the width of the maxima.	In addition to the increase in the distance between the centres of the maxima, the qualitative information is obtained regarding the width of the maxima.
iii. Magnifying power $= \dfrac{F}{f}$ $= \dfrac{\text{Focal distance of the objective}}{\text{Focal distance of the eye lens}}$	$\text{RP} = \dfrac{d}{1.22\lambda}$
iv. It is independent of the wavelength of the light source or sources.	It is low for smaller wavelengths, i.e. it depends on wavelength of the light source or sources.

5.25 RESOLVING POWER OF A MICROSCOPE

The resolving power of a microscope is characterized by δl or l, which is the shortest distance between two point objects whose images are just resolved by the objective of the microscope. Thus, *resolving power of a microscope is defined as its capacity to form separate images of two point objects lying close together*. Smaller the distance, higher the resolving power. In the case of self-luminous objects, the two points can be regarded as independent (incoherent) sources of light.

Let $S_1S_2S_3$ be a very small luminous object. The S_1 and S_2 of this luminous object will be said just resolved when first order minima of S_2 is coincident with zeroth order maxima of S_1 at I (Fig. 5.39a). The condition for minima due to S_1 at I is

$$S_2B + BI - (S_2A + AI) = 1.22\lambda$$

or $\qquad S_2B - S_2A = 1.22\lambda \quad (\because BI = AI)$

or $\quad (S_2B + S_1B) + (S_1A - S_2A) = 1.22\lambda$

$$(\because S_1B = S_1A)$$

To determine the value of $(S_2B - S_1B)$, the light emitted from S_1 and S_2 can be consid-

ered parallel. Draw a perpendicular from S_1 on S_2B (Fig. 5.39b).

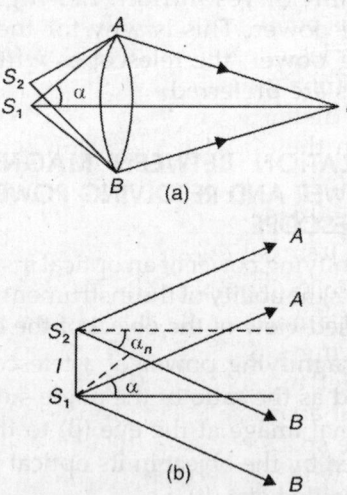

Fig. 5.39: Resolving power of a microscope (a) Zeroth order maxima of S_1 and first order minima of S_2 are coincident at I (b) In particular direction light emitted from S_1 and S_2 can be considered parallel

Therefore, $\quad S_2B - S_1B = l\sin\alpha \quad$ (Here $l = S_2S_1$)
Similarly, one obtains $S_1A - S_2A = l\sin\alpha$

$\therefore \qquad\qquad 2l\sin\alpha = 1.22\lambda$

or $$l = \frac{1.22\lambda}{2\sin\alpha} \quad (5.59)$$

If the object is situated in any medium whose refractive index is μ, then

$$l = \frac{1.22\lambda}{2\mu\sin\alpha} \quad (5.60)$$

$\mu \sin\alpha$ is called the **numerical aperture** of the objective of the telescope and generally written as NA. Therefore, Eq. (5.60) can be rewritten as

$$l = \frac{1.22\lambda}{2NA} \quad (5.61)$$

This measures the limit of resolution of the microscope. Its reciprocal is called as resolving power. The above expression of the resolving power of the microscope is based upon the supposition that the points S_1 and S_2 are self-luminous and their waves have no phase relationship.

Truly speaking, the objects examined in microscope are not self-luminous, but are illuminated. Hence, depending upon the condition of illumination, the light scattered by various points of the object is coherent to a greater or lesser extent. In this case, however, under optimal conditions of illumination, the shortest distance between two resolvable points on the object is (according to Abbe) given by

$$\delta l = \frac{0.5\lambda}{\mu\sin\alpha} = \frac{\lambda}{2\mu\sin\alpha} = \frac{\lambda}{2NA} \quad (5.62)$$

From Eq. (5.61), it is obvious that the resolving power of a microscope can be increased by (i) reducing wavelength (λ), i.e. applying ultraviolet microscopy (ii) increasing the numerical aperture (NA) of the objective by filling the space between the cover glass and objective with a liquid having a high absolute index of refraction (oil immersion objective). Usually for immersion liquids $\mu = 1.4$ to 1.6.

Nowadays, for high resolution electron microscope is used.

5.26 RESOLVING POWER OF A DIFFRACTION GRATING

The resolving power of a diffraction grating is defined as the ratio of the wavelength of

spectral line to the difference in wavelength between this line and a neighbouring line such that the two lines appear to be just resolved. The resolving power of grating is expressed as $\lambda/d\lambda$.

Let a parallel light beam of wavelength λ and $\lambda + d\lambda$ be incident normally over a grating surface (Fig. 5.40).

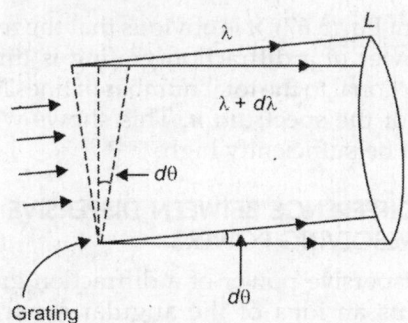

Fig. 5.40: Resolving power of a diffraction grating

If the diffraction angle of nth primary maxima corresponding to wavelength λ is θ_n, then

$$(a + b) \sin\theta_n = n\lambda \quad (5.63)$$

Corresponding to nth primary maxima, the angle of diffraction of first maxima is

$$N(a + b) = m\lambda \quad (5.64)$$

where m is an integer whose value can be any integer other than 0, N, $2N$,.., nN. Clearly corresponding to $m = 0$, N, $2N$,..., etc. one obtains zeroth, first, second, etc. principal maxima respectively. On the basis of the above discussion, we conclude that corresponding to the nth principal maxima, the first minima in the direction of increasing θ is obtained for $m = nN + 1$. If corresponding to nth maxima, the first minima is obtained in the direction $(\theta_n + d\theta_n)$, then from Eq. (5.63), we have

$$N(a + b)\sin(\theta_n + d\theta_n) = (nN + 1)\lambda$$

or $$(a + b)\sin(\theta_n + d\theta_n) = \frac{(nN + 1)\lambda}{N} \quad (5.65)$$

But according to the definition of RP of grating, the wavelengths λ and $\lambda + d\lambda$ will be just resolved if nth maxima of $(\lambda + d\lambda)$ is also in the direction of $(\theta_n + d\theta_n)$. In this situation, from Eq. (5.63), we have

$$(a + b)\sin(\theta_n + d\theta_n) = n(\lambda + d\lambda) \qquad (5.66)$$

Comparing Eqs. (5.65) and (5.66), one obtains

$$\frac{nN + 1}{N}\lambda = n(\lambda + d\lambda)$$

or

$$\frac{\lambda}{d\lambda} = Nn \qquad (5.67)$$

From Eq. (5.67), it is obvious that the resolving power of a diffraction grating is directly proportional to the total number of lines N and order of the spectrum n. This shows why N should be sufficiently high.

5.27 DIFFERENCE BETWEEN DISPERSIVE AND RESOLVING POWERS

The dispersive power of a diffraction grating gives us an idea of the angular separation between the lines of the spectrum produced by the grating. Dispersive power of a diffraction grating is measured by $d\theta/d\lambda$, where $d\theta$ is the measure of the angular separation between the two spectral lines whose wavelength difference is $d\lambda$. Dispersive power is expressed as

$$\frac{d\theta}{d\lambda} = \frac{n}{(a + b)\cos\theta} = \frac{nN'}{\cos\theta} \qquad (5.68)$$

From Eq. (5.68), it is obvious that higher the order n of the spectrum, and closer the rulings on the grating, the greater is the dispersive power of the grating.

The resolving power, on the other hand, expresses the degree of the closeness which the two spectral lines can have and yet be distinguished as separate. Resolving power is expressed by $\lambda/d\lambda$ and is equal to nN, i.e.

$$\frac{\lambda}{d\lambda} = nN \qquad (5.69)$$

Here $d\lambda$ is the smallest wavelength that can be just resolved at wavelength λ. N is the total number of lines on the grating. This shows that greater is the width of the ruled surface, higher is the resolving power. The higher resolving power results in sharp maxima.

The difference between dispersive power and resolving power is illustrated in Fig. 5.41,

in which the diffraction maxima of two wavelengths formed by two diffraction gratings of same grating element but different width of the ruled surface are shown. In both the cases, the angular separation is same but the resolving power in second case where the maxima is sharper, is greater than in first case.

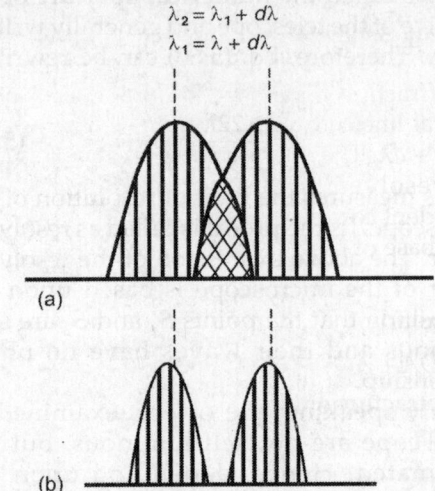

Fig. 5.41: (a) Intensity curves of two lines showing difference between dispersive and resolving powers (b) Width of maxima is very small and overlapping is not there

5.28 RESOLVING POWER OF A PRISM

The resolving power of the prism in a prism spectroscope, in which the prism is also the aperture stop, is defined as its capacity to form separate maxima of two wavelengths very close together. It is measured by $\lambda/d\lambda$, where $d\lambda$ is the difference in the two wavelengths of similar intensity which are just resolved by the prism and λ is the wavelength of either of them.

Let ABC (Fig. 5.42) represents a prism which is placed in the position of minimum deviation. Let a plane wavefront BD of composite light of wavelength λ and $\lambda + d\lambda$ be incident on the prism. Let CY and CQ' represent the emergent wavefront for λ and $\lambda + d\lambda$ respectively.

Let μ and $\mu - d\mu$ be refractive indices corresponding to wavelengths λ and $\lambda + d\lambda$ respectively. Let L_2 be objective of telescope. P_1 and P_2 are positions of central maxima in both

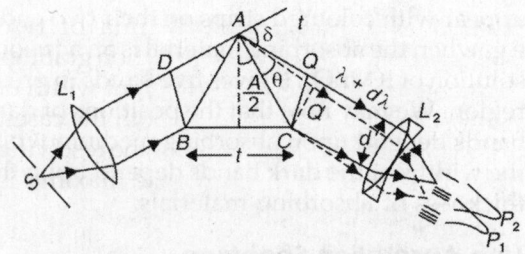

Fig. 5.42: Resolving power of a prism

the diffraction patterns of images. P_1 and P_2 are spectral lines corresponding to wavelengths λ and $\lambda + d\lambda$. The spectral lines will be observed just resolved when the position of P_2 is coincident corresponding to first maxima of P_1. If the base of the prism is t, then

$$DA + AQ = \mu BC = \mu t \qquad (5.70)$$

Similarly for $\lambda + d\lambda$, we have

$$DA + AQ' = (\mu - d\mu)BC = (\mu - d\mu)t \qquad (5.71)$$

Subtracting Eq. (5.70) from Eq. (5.71), one obtains

$$AQ - AQ' = QQ' = t\,d\mu \qquad (5.72)$$

According to the definition of resolving power, we have

$$QQ' = \lambda$$

or $\qquad t\,d\mu = \lambda$

$$\therefore \qquad \frac{\lambda}{d\lambda} = t\frac{d\mu}{d\lambda} \qquad (5.73)$$

Equation (5.73) is the expression for the resolving power of a prism. From Eq. (5.73), it is obvious that resolving power of a prism:

i. Varies directly as the length of the base (t) of the prism
ii. Varies directly as the rate of change of refractive index with the wavelength.

5.29 SPECTRUM

A spectrum may be defined as an orderly arrangement of something, e.g. momentum, energy, wavelength, mass, etc. and the corresponding spectra are referred to as momentum spectrum, energy spectrum, wavelength spectrum, mass spectrum, etc. We are mostly concerned with light spectrum. We know that when a beam of composite light say white light is allowed to pass through a prism

or grating, the light splits into its component colours. The regular arrangement of colours is known as spectrum. One can divide spectra into two following principal classes:

1. Emission Spectra

This kind of spectra are obtained when light coming directly from the source is examined with the help of a spectroscope.

2. Absorption Spectra

This type of spectra are obtained when the light from the source which gives continuous spectrum is passed through a substance, i.e. an absorbing material and the transmitted light is examined with a spectroscope. We may note that in absorption spectrum, the absorbing materials absorbs all those wavelengths which it emits in emission spectra. The dark lines in the absorption spectra corresponds to the lines which absorbing medium would have emitted on excitation. Now, one may ask, what is the utility of considering the absorption spectra? The answer is very simple, i.e. the substances which cannot be conveniently excited to give emission spectra are easily studied with the help of absorption spectra.

Both the emission and absorption spectra are further subdivided into three classes:
 i. Continuous spectra
 ii. Band spectra
 iii. Line spectra.

Division of Emission Spectra

Continuous Emission Spectrum

This spectrum contains all the wavelengths from one end to the other end of the spectrum. This type of spectrum appears to be unbroken and luminous. In visible region, this spectrum appears in the form of multicoloured strips in which no line of demarcation can be drawn between two colours. Intensity in this kind of spectrum depends upon temperature and the body itself. When a matter in bulk is heated, this type of spectrum is obtained. Spectrum of this type is also obtained when light from a hot solid, e.g. tungsten filament of an electric lamp, hot iron, charcoal, etc. is allowed to pass through the prism. Obviously, a continuous spectra cannot be the characteristic of a substance as produced by different substances, it contains all possible wavelengths.

Band Emission Spectrum

When the emission spectrum has coloured bands separated by dark spaces on the two sides of the spectrum, it is called as band emission spectrum. These bands are found to be sharply defined at one end and shade off gradually at the other end. The sharply defined edge of the spectrum is called the *head of the band* and the other edge as *tail*. When matter in molecular state is excited then this type of spectrum is produced. This is why this type of spectrum is also known as *molecular spectra*. We may note that for different types of molecules, band spectra are different.

Line Emission Spectrum

If the emission spectrum contains a series of fine sharp lines having dark spaces in between, the spectrum is called line emission spectrum. These lines are coloured in the visible spectrum. This type of spectrum is produced when the matter in the atomic state is excited. That is why this type of spectra is called atomic spectra. We may note that it is different for different types of spectra.

Division of Absorption Spectra

Continuous Absorption Spectrum

This is produced when the absorbing material absorbs a continuous range of wavelengths without any gap, e.g. when light from a filament of electric lamp is passed through a violet piece of glass. It absorbs all visible light except violet and the spectrum thus obtained will be continuous absorbing type. This type of spectrum is characteristic of the absorbing medium.

Band Absorption Spectrum

This type of spectrum is produced when light from a source of continous wavelengths passes through a medium in molecular state. At certain places in the spectrum, the dark bands appear with coloured strips on their two ends, e.g. when the absorbing material is an aqueous solution of $KMnO_4$, it gives five bands in green region. We may note that the positions of dark bands depend upon absorbing medium while the widths of the dark bands depend upon the thickness of absorbing materials.

Line Absorption Spectrum

This type of spectrum is produced when light is passed through the absorbing medium in atomic state. The characteristic dark lines appear in absorption at the places where the coloured lines would have been present when the absorbent is excited in atomic scale, e.g. when light is allowed to pass through sodium vapour in atomic state, two dark lines appear exactly at the places of D_1 and D_2 lines in the emission spectra. Solar spectrum is also an example of line absorption spectrum which consists of *Fraunhofer absorption lines* which correspond to vapours of different elements present in solar atmosphere.

ILLUSTRATIVE EXAMPLES

Example 1

In the diffraction pattern due to a straight edge, the separation between the first two maxima above the edge of the geometrical shadow was found to be 8×10^{-4} m when screen as well as the source were at a distance of 1 m from the edge. Calculate the wavelength of light emitted by the source.

Solution

The condition for maxima is

$$X_n = \sqrt{\frac{b(a+b)(2n+1)\lambda}{a}}$$

For $n = 1$, $X_1 = \sqrt{\frac{b(a+b)3\lambda}{a}}$

For $n = 2$, $X_2 = \sqrt{\frac{b(a+b)5\lambda}{a}}$

$\therefore$ $X_2 - X_1 = \sqrt{\frac{b(a+b)2\lambda}{a}}$ (1)

Here $X_2 - X_1 = 8 \times 10^{-4}$ m

$$a = 100 \text{ cm} \quad b = 100 \text{ cm}$$

Substituting these values in (1), we obtain

$$8 \times 10^{-4} = \sqrt{\frac{100 \times 200 \times 2\lambda}{100}}$$

or $\quad \lambda = \dfrac{64 \times 10^{-8}}{400} = 16 \times 10^{-10}$ cm $= 1600$ Å

Example 2

The first dark position obtained on a screen, when a straight edge is kept in the path of cylindrical waves is at a distance of 3.2 mm from the geometrical shadow. Find the position of nth dark band. If the distance between the linear source and the straight edge is 10 cm and that between the edge and the screen is 100 cm, calculate the wavelength of light used.

Solution

The nth bright position is given by

$$X_n = \sqrt{\frac{b(a+b)(2n-1)\lambda}{a}}$$

The nth dark position is given by

$$X'_n = \sqrt{\frac{2b(a+b)}{a}} = n\lambda$$

For $n = 1$, $\quad X'_1 = \left[\frac{2b(a+b)}{a}\lambda\right]^{1/2} = 0.32$ cm

$\therefore \quad \lambda = \dfrac{0.52 \times 0.32 \times a}{2b(a+b)} = \dfrac{0.32 \times 0.32 \times 10}{220 \times 110}$

$$= 4.655 \times 10^{-8} \text{ cm}$$

Distance of the nth dark band from the geometrical shadow

$$X'_n = X'_1 \sqrt{n} = \sqrt{n}\, 3.2 \text{ mm}$$

Distance of the nth bright band from the geometrical shadow

$$X_n = \sqrt{2n-1}\, X_1 \quad \text{and} \quad X_1 = \frac{X'_1}{\sqrt{2}}$$

$\therefore \quad X_n = \sqrt{\dfrac{2n-1}{2}} \times 3.2$ mm

Example 3

Light of wavelength 6000 Å passes through a narrow circular aperture of radius 9×10^{-4} m. At what distance along the axis from the aperture will the first intensity maximum be observed?

Solution

For the first maximum to appear only, the aperture should allow only one half period zone. Assuming that the incident wavefront is plane, the area of each of zone will be $nb\lambda$, where b is the distance from the circular aperture of the point at which the first maxima appear.

If r be the radius of the aperture, the area is πr^2. For the first maxima, we have

$$nb\lambda = \pi r^2$$

$$r = 9 \times 10^{-4} \text{ m}$$

$$\lambda = 6000 \times 10^{-10} \text{ m}$$

$\therefore \qquad b = \dfrac{r^2}{\lambda}$

or $\quad b = \dfrac{(9 \times 10^{-4})^2}{6000 \times 10^{-10}}$ m $= \dfrac{81}{60}$ m $= 1.35$ m

Example 4

The diameter of a narrow circular aperture is 0.09 cm. The light of wavelength 6000 Å is incident over it. Where the first maxima along the axis will be formed ?

Solution

Let the incident wavefront be plane. For the first maxima, we have

$$PA - PO = \lambda/2$$

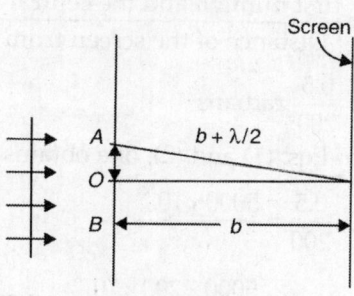

Fig. 5.43

$\therefore \qquad r^2 = (b + \lambda/2)^2 - b^2$

$$= b\lambda + \frac{\lambda^2}{4}$$

Neglecting $\lambda^2/4$ in comparison to λ, one obtains

$$r^2 = b\lambda$$

$\therefore \qquad b = \dfrac{r^2}{\lambda} = \dfrac{(0.09)^2}{6000 \times 10^{-8}}$

$$= 13.5 \text{ cm}$$

Example 5

Light of wavelength 5000 Å is incident normally on a slit. The first minimum of the diffraction pattern is observed to lie at a distance of 5 mm from the central maximum on a screen placed at a distance of 2 m from the slit. Calculate the width of the slit.

Solution

The direction of minima in a Fraunhofer diffraction due to a single slit can be obtained from the following relation

$e \sin \theta = n\lambda$ where $n = 1, 2, 3, \ldots$

The angular separation θ between the first minima and the central maxima is given by

$$e \sin \theta = \lambda$$

If θ be small and measured in radians, then $\sin \theta = \theta$

$\therefore \quad e\theta = \lambda$ or $\theta = \dfrac{\lambda}{e}$ radians

or $\quad \theta = \dfrac{5000 \times 10^{-8}}{e}$ radians $\qquad (1)$

Linear separation between the

But $\theta = \dfrac{\text{first minima and the central maxima}}{\text{Distance of the screen from the slit}}$

$$\theta = \frac{0.5}{200} \text{ radians} \qquad (2)$$

From Eqs. (1) and (2), one obtains

$$\frac{0.5}{200} = \frac{5000 \times 10^{-8}}{e}$$

or $\qquad e = \dfrac{5000 \times 200 \times 10^{-8}}{0.5} = 0.02 \text{ cm}.$

Example 6

The Fraunhofer diffraction pattern of a single slit is observed in the focal plane of a lens of focal length 1 m. The width of the slit is 0.4 mm. The incident light consists of wavelengths λ_1 and λ_2. The fourth minimum corresponding to λ_1 and fifth minimum to λ_2 occur at the same point, 5 cm from the central minimum. Compare λ_1 and λ_2.

Solution

In the Fraunhofer diffraction pattern due to a single slit, the minimum occurs in a direction θ, given by

$$a \sin \theta = n\lambda$$

where a is the width of the slit $= 0.4$ mm $= 0.04$ cm, λ is the wavelength composed of two wavelengths λ_1 and λ_2.

For the fourth dark band

$$\sin \theta_1 \approx \theta_2 = \frac{4\lambda_1}{0.04} \qquad (1)$$

Also $\qquad \theta_1 = \dfrac{x}{f_1} = \dfrac{0.5}{100}$ radian $\qquad (2)$

$\therefore \qquad \dfrac{4\lambda_1}{0.04} = \dfrac{0.5}{100}$

or $\qquad \lambda_1 = \dfrac{0.5 \times 0.04}{4 \times 100} = 5000 \times 10^{-8}$ cm

Similarly, for the fifth dark band

$$\frac{0.5}{100} = \frac{5\lambda_2}{0.04}$$

or $\qquad \lambda_2 = \dfrac{0.5 \times 0.04}{5 \times 100} = 4000 \times 10^{-8}$ cm .

Example 7

Calculate the angular separation between the first order minima on either side of the central maximum when the slit is 6×10^{-4} cm wide and light illuminating it has a wavelength of 6000 Å.

Solution

For the first order minimum, we have

$$a \sin \theta = \lambda$$

$$\therefore \quad \sin\theta = \frac{\lambda}{a} = \frac{6000 \times 10^{-8} \text{ cm}}{6 \times 10^{-4} \text{ cm}} = 0.6 \text{ radian}$$

$$\therefore \quad \theta = 36°52'$$

Therefore, the angular separation of the first order minima on either side of the central maximum is given by

$$2\theta = 73°44'$$

Example 8

Two parallel slits have widths 0.015 cm each and opaque portion between them is 0.03 cm. They are illuminated normally by light of wavelength 6×10^{-5} cm and the emergent light is focussed by a convergent lens of focal length 100 cm. Calculate the position of the first four interference maxima on one side in the focal plane of the lens.

Solution

Condition for interference maxima

$$(a + b)\sin\theta = n\lambda$$

$$a = 0.015 \text{ cm}$$

$$b = 0.03 \text{ cm}$$

$$\lambda = 6000 \times 10^{-8} \text{ cm}$$

$$\theta = \sin^{-1}\frac{n\lambda}{a+b}$$

From the given data, it is evident that $2a = b$ and hence 3rd, 6th and 9th orders of interference maxima are absent. Obviously, maxima lie at

$$\theta = \sin^{-1}\left(\frac{n \times 6 \times 10^{-5}}{0.015 + 0.03}\right) = \frac{4n}{3} \times 10^{-3} \text{ radian}$$

$$(n = 0, 1, 2, \ldots)$$

Clearly, apart from zero order maximum, i.e. $n = 0$, the fourth order maxima on either side occur at

$$\theta = \frac{4 \times 1}{3} \times 10^{-3},$$

$$\theta = \frac{4 \times 2}{3} \times 10^{-3},$$

$$\theta = \frac{4 \times 4}{3} \times 10^{-3}, \quad \theta = \frac{4 \times 5}{3} \times 10^{-3}$$

or $\quad \theta = \frac{4}{3} \times 10^{-3}, \quad \theta = \frac{8}{3} \times 10^{-3},$

$$\theta = \frac{16}{3} \times 10^{-3}, \quad \theta = \frac{20}{3} \times 10^{-3} \text{ radian}$$

One can obtain the position on the focal plane of the lens from

$$x = f\theta \mid f = 100 \text{ cm, focal length of the lens}$$

$$\therefore \quad x = 1.33 \text{ mm}, 2.66 \text{ m}, 5.33 \text{ cm}, 6.67 \text{ mm}.$$

Example 9

A transmission grating 4 cm long has 4000 lines per cm. Compute the resolving power of the grating for a wavelength of 5900 Å in the first order spectrum. Will the grating separate the two lines of wavelengths 5890 Å and 5896 Å, which constitute the sodium yellow doublet?

Solution

We have $\quad$ RP $= Nn$

$$\therefore \qquad \text{RP} = 16000 \times 1 = 16000$$

$$n = 1$$

$$N = 4000 \times 4 = 16000$$

$$\lambda = 5.9 \times 10^{-5} \text{ cm}$$

Smallest wavelength $d\lambda$ that can be resolved is given by

$$\frac{d}{d\lambda} = Nn = 16000$$

or $d\lambda = \dfrac{\lambda}{16000} = \dfrac{5890 \times 10^{-8}}{16000} = 368.1 \times 10^{-5} \text{ cm}$

The difference between the given two wavelengths

$$d\lambda' = 5896 \times 10^{-8} - 5890 \times 10^{-8}$$

$$= 0.006 \times 10^{-5} \text{ cm}$$

$d\lambda' << d\lambda$ and hence the given grating cannot separate the doublet of the sodium.

Example 10

Sodium light of mean wavelength 5893 Å is incident normally upon a plane transmission grating of 1000 lines per inch. Calculate the angular separation of sodium lines of wavelength 5896 Å in the spectrum of the first order as observed in a telescope of which the objective has a focal length of 24 cm and eye-piece of focal length 2 cm. [BE 2004]

Solution

We have $(a + b)\sin\theta = n\lambda$

Differentiating, one obtains

$$(a + b)\cos\theta\, d\theta = n\, d\lambda$$

$$\therefore \quad d\theta = \frac{n\, d\lambda}{(a+b)\cos\theta} = \frac{n\, d\lambda}{(a+b)\sqrt{1-\sin^2\theta}}$$

or $\quad d\theta = \dfrac{n\, d\lambda}{(a+b)\sqrt{\dfrac{1-n^2\lambda^2}{(a+b)^2}}} \quad \left[\because \sin\theta = \dfrac{n\lambda}{a+b}\right]$

$$d\lambda = 5896 - 5890 = 6 \text{ Å}$$

$$a + b = 1/6000$$

$$\lambda = \frac{\lambda_1 + \lambda_2}{2} = 5896 \text{ Å}$$

$$n = 2$$

or $\quad d\theta = \dfrac{2\times6\times10^{-8}}{\dfrac{1}{6000}\sqrt{\dfrac{1-4\times\left(5893\times10^{-8}\right)^2}{\left(1/6000\right)^2}}}$

$$= 1.1 \times 10^{-3} \text{ radian}.$$

Example 11

A plane transmission grating having 6800 lines per cm is used to obtain a spectrum of light from a sodium lamp in the second order. Find the angular separation between the two sodium lines whose wavelengths are 5890 Å and 5896 Å respectively.

Solution

We have

$$(a + b)\sin\theta = n\lambda$$

or $\qquad\qquad \sin\theta = Nn\lambda$

$$N = 6000$$

$$n = 2$$

$$\lambda_1 = 5890 \text{ Å}, \ \lambda_2 = 5896 \text{ Å}$$

Let $\quad \theta = \theta_1$ for $\lambda = \lambda_1$

and $\quad \theta = \theta_2$ for $\lambda = \lambda_2$

We have for λ_1, $\sin\theta_1 = Nn\lambda_1$

$$= 6000 \times 2 \times 5890 \times 10^{-8}$$

$$= 0.7068$$

$$\therefore \qquad \theta_1 = 44°59'$$

We have for λ_2, $\sin\theta_2 = Nn\lambda_2$

$$= 6000 \times 2 \times 5890 \times 10^{-8}$$

$$= 0.7075$$

$$\therefore \qquad \theta_2 = 44°2'$$

$\therefore$ Angular separation of two spectral lines

$$= \theta_2 - \theta_1 = 3'.$$

Example 12

Two stars are at a distance of 5 light years from earth. They are just resolved by a telescope having an objective of diameter 0.5 m. Calculate the distance between the two stars. Take $\lambda = 5600$ Å.

Solution

Angular separation between two stars is given by

$$\theta = \frac{1.22\lambda}{D} = \frac{1.22\times5.6\times10^{-5}}{50} \text{ radian}$$

$$= 1.366 \times 10^{-7} \text{ radian}$$

Distance between the two stars

$$= \theta \times 5 \text{ light years}$$

$$= 1.366 \times 10^{-7} \times 5 \times 365 \times 24 \times 60 \times 60$$

$$\times 3 \times 10^8 \text{ m}$$

$$= 6.46 \times 10^8 \text{ m}.$$

Example 13

The refractive index of a certain glass is given by $\mu_c = 1.6545$, $\mu_F = 1.6635$. The wavelengths of C and F lines in the solar spectrum are 6563 Å and 5270 Å respectively. Calculate the length of the base of a 60° prism made of glass which is just capable of resolving sodium lines of wavelengths 5890 Å and 5896 Å.

Solution

We have $\quad d\lambda = (5896 - 5890)$ Å

$$= 6 \times 10^{-8} \text{ cm}$$

Mean $\lambda = 5893$ Å

$$\frac{d\mu}{d\lambda} = \frac{\mu_F - \mu_C}{\lambda_C - \lambda_F}$$

$$= \frac{1.6635 - 1.6545}{(6563 - 5270) \times 10^{-8}}$$

$$= \frac{0.009}{1293 \times 10^{-8}}$$

Also we have $\dfrac{\lambda}{d\lambda} = t \dfrac{d\mu}{d\lambda}$

∴ The least wavelength difference $d\lambda$ that can be just resolved by a prism is

$$d\lambda = \lambda / t \frac{d\mu}{d\lambda}$$

or

$$t = \frac{\lambda}{d\lambda (d\mu/d\lambda)}$$

$$= \frac{5893 \times 10^{-8}}{6 \times 10^{-8} \times \dfrac{0.009}{1293 \times 10^{-8}}}$$

$$= 1.41 \text{ cm.}$$

Example 14

How many orders will be observed by a grating having 4000 lines/cm, if it is illuminated by a visible light of wavelength in the range of 4000 Å to 7500 Å ?

Solution

Given $(a + b) = \dfrac{1}{4000}$ cm

$$= \frac{1}{400000} \text{ m}$$

$\lambda_1 = 4000$ Å $= 4 \times 10^{-7}$ m

$\lambda_2 = 7500$ Å $= 7.5 \times 10^{-7}$ m

$\theta = 90°$ (for maximum order)

(a) $(a + b) \sin\theta = n_1 \lambda_1$

or

$$n_1 = \frac{(a + b) \sin 90°}{\lambda_1}$$

$$= \frac{1 \times 1}{400000 \times 4 \times 10^{-3}}$$

$$= 6.25$$

$$\cong 6$$

(b) $(a + b) \sin\theta = n_2 \lambda_2$

or

$$n_2 = \frac{(a + b) \sin 90°}{\lambda_2}$$

$$= \frac{1 \times 1}{400000 \times 7.5 \times 10^{-7}}$$

$$= 3.33$$

$$\cong 3$$

Hence, the order of spectrum varies from 3 to 6.

Example 15

Light of $\lambda = 500$ nm falls on a grating normally. Two adjacent principal maxima occur at $\sin\theta = 0.2$ and $\sin\theta = 0.3$ respectively. Calculate grating element.

Solution

The principal maxima of order n is given by

$$(a + b) \sin\theta = n\lambda$$

Two adjacent orders mean n and $(n + 1)$, they occured at $\sin\theta = 0.2$ and $\sin\theta = 0.3$ respectively.

Thus,

$$(a + b) 0.2 = n\lambda \tag{1}$$

$$(a + b) 0.3 = (n + 1)\lambda \tag{2}$$

Subtracting Eq. (1) from Eq. (2), we get

$$(a + b) 0.1 = \lambda = 500 \text{ nm}$$

$$(a + b) = 5 \times 10^{-6} \text{ m}$$

Example 16

What is the maximum number of lines of a grating which will resolve the third order spectrum of two lines having wavelengths 5890 Å and 5896 Å?

Solution

Given

$$n = 3$$

$$\lambda = 5890 \text{ Å}$$

$$d\lambda = 6 \text{ Å}$$

We know that the resoling power of grating is

$$\frac{\lambda}{d\lambda} = nN$$

$$N = \frac{1}{n}\frac{\lambda}{d\lambda}$$

$$= \frac{1}{3}\frac{5890\,\text{Å}}{6\,\text{Å}}$$

$$= 327.22$$

$$\cong 327$$

REVIEW QUESTIONS

1. Give an account of the phenomenon and the relevant theory of diffraction due to a straight edge.

2. A narrow circular aperture is held at one side of a final hole and a screen is placed at some distance on the opposite side. Explain the illumination observed on the screen when a beam of monochromatic light is made to pass normally.

3. Discuss qualitatively the diffraction pattern observed on the screen due to an opaque disc which is placed in the path of monochromatic light from a point source.

4. A small aperture of radius R is placed at a distance of a from a point source of light of wavelength λ. Diffraction pattern is observed on a screen placed at a distance of b from the aperture. Show that the axial point P on the screen will be bright or dark corresponding to n is odd or even respectively in the expression

$$\frac{1}{a}+\frac{1}{b}=\frac{n\lambda}{R^2}$$

5. Discuss the Fresnel type of diffraction of light due to a straight edge.

6. Discuss the phenomenon of diffraction at a straight edge and deduce the expression for the distance between nth and $(n-1)$th bright and dark bands.

7. A point source is viewed through a circular aperture. Draw a diagram showing the intensity distribution along the axis.

8. Explain why fringes are seen within fine shadows of a thin wire held vertically in front of narrow vertical slit illuminated by monochromatic light.

9. Explain the construction and significance of the half period zones on a cylindrical wavefront. Describe zone intensity distribution in the diffraction pattern due to a straight edge.

10. Give the theory of diffraction at a straight edge. Account for the intensity distribution pattern. Explain how you can find the wavelength of a monochromatic light using diffraction at a straight edge.   [Mysore, BE 98]

11. Explain Fresnel diffraction at a thin wire. How the experiment may be used to determine the diameter of the given wire? Will the diffraction pattern be affected if the diameter of the wire be increased? If yes, how?   [Kerela, BE]

12. Distinguish between Fresnel and Fraunhofer class of diffraction. Compare the diffraction pattern by a circular obstacle and a circular hole.

13. Discuss the phenomenon of Fraunhofer diffraction at a single slit and show that the relative intensities of successive maxima are nearly

$$1 : \frac{4}{9\pi^2} : \frac{4}{25\pi^2} : \frac{4}{49\pi^2}$$ [Roorkee, BE 97]

14. Discuss the Fraunhofer type of diffraction produced by a narrow slit illuminated by a monochromatic light. Draw diagram to indicate the distribution of intensity of light in the diffraction pattern.

15. Discuss Fraunhofer diffraction due to a single slit. Derive the expression for the intensity and show that the intensities of the first and secondary maxima are $\frac{1}{22}$ and $\frac{1}{61}$ of the intensity of central maximum respectively.

16. Derive a relation for intensity distribution in case of Fraunhofer diffraction pattern due to two parallel slits. How is this pattern affected by:

(a) the width of the slit
(b) separation of the two slits
(c) the wavelength of the light used ?

17. Find out an expression for the intensity at a point in the Fraunhofer's diffraction pattern due to a single slit. Draw the intensity distribution curve for it.

18. Give complete theory and experimental agreement of a plane transmission grating in its use to find the wavelength of light.

19. Define resolving power of a grating. Derive an expression for the resolving power of a plane transmission grating. How can it be increased?

20. (a) Derive an expression for the resolving power of a grating.

    (b) Distinguish between the grating and prism spectra.

21. Give the theory of plane transmission diffraction grating and show how would you use it to find the wavelength of light. How are spectral lines affected if the rulings are made closer? Show that only first order will be visible in grating element if double wavelength of light is used.

22. What are the absent spectra in case of diffraction grating? What is the condition which must be satisfied if the second order spectra are to be suppressed? If $b = 2a$, then which order will be absent ?

23. What do you mean by the resolving power of an optical instrument ? Derive an expression for the resolving power of a plane transmission grating.

24. Discuss Rayleigh's criterion of resolution. Derive an expression for the resolving power of a telescope. Distinguish between the prism and grating spectra.

25. Two gratings A and B have the same width of the ruled surface but A has greater number of lines. Giving reasons compare in two cases the (i) intensity (ii) width of the principal maxima (iii) dispersive power (iv) resolving power of gratings.

26. What particular spectra (produced by a grating) would be absent if the widths of the transparencies and opacities are the same?

27. Explain the action of diffraction grating and differentiate between the dispersive and resolving powers.

28. Explain Rayleigh's criterion of resolution. Derive an expression for the resolving power of a telescope. Find a relation between magnifying power and resolving power of a telescope.

29. Explain the formation of spectra by a plane diffraction grating.

    (a) What are the chief characteristics?

    (b) What do you understand by overlapping and absent spectra?

30. Explain the formation of spectra by a plane transmission grating. What are absent spectra in the case of diffraction grating? What particular spectra would be absent if the width of opacities be double that of transparencies in such a grating.

31. State Rayleigh's criterion for just resolution. Derive an expression for the resolving power of a plane transmission grating.

32. Give the theory of plane transmission diffraction grating and show how you would use it to find the wavelength of light. How are spectral lines affected if the rulings are made closer?

33. Explain the formation of multiple spectra by a plane transmission grating and mention the chief characteristics of such spectra.

    Find out an expression for intensity at a point in the Fraunhofer diffraction due to double slit. Draw the intensity distribution curve.

34. Explain, what is meant by resolving power of a diffraction grating. Deduce an expression for the same.

35. Derive an expression for the intensity of diffracted light in the Fraunhofer's diffraction due to a single slit.

36. (a) The intensity of light diffracted from a plane transmission grating is given by

    $$I = I_0 \left(\frac{\sin \alpha}{\alpha} \right)^2 \left(\frac{\sin N\beta}{\sin \beta} \right)^2$$

    where symbols have their usual meaning. Find the positions of maxima and minima and calculate:

      (i) Width of nth principal maximum

      (ii) Resolving power of grating.

    (b) A plane diffraction grating just resolves two spectral lines of wavelengths $\lambda_1 = 5140.34$ Å and $\lambda_2 = 5140.85$ Å in the first order. Will it resolve the spectral lines of wavelengths $\lambda_1 = 8037.2$ Å and $\lambda_2 = 8037.50$ Å in its second order?

37. (a) Derive an expression for angular width of principal maxima in the diffraction pattern of plane transmission grating.

    (b) KCl is a cubic crystal having density of 1.98 gm/cc and molecular weight 74.55. Find the distance between the adjacent atoms and the distance between two adjacent similar atoms.

    (c) Diffraction grating used at normal incidence gives a green line of wavelength 5400 Å in a certain order superimposed on the violet line of wavelength 4054 Å of the next higher order. The angle of diffraction is 30°. Calculate number of lines per cm in the grating.

38. (a) Distinguish between Fresnel and Fraunhofer class of diffraction.

    (b) What do you understand by 'resolution', explain. What is meant by resolving power of a grating? Deduce an expression for the same, and discuss its dependence on various constants of grating.

    (c) Diffraction pattern of a single slit of width 0.5 cm is formed by a lens of focal length 40 cm. Calculate the distance between the first dark and the next bright fringe from the axis. Wavelength = 4890 Å.

39. (a) Starting with the expression of intensity distribution for Fraunhofer diffraction in a plane diffraction grating, find the intensity and conditions for principal maxima and minima.

    (b) A set of 10 parallel equidistant slits of width 0.5 cm each and spacing 1.9 cm are used to study Fraunhofer diffraction of waves of $\lambda = 0.6$ cm falling normally on the plane of the slits. Deduce:

       i. Angular positions of first two maxima

       ii. Half widths of these maxima.

    (c) Derive Bragg's law.

    (d) Write short notes on Rayleigh's criterion of resolution and resolution power of a diffraction grating.

40. (a) With necessary theory explain the formation of spectrum by a plane transmission grating when composite light falls on it normally.

    (b) A sodium discharge lamp produces two intense wavelengths in the yellow region of visible spectrum at 589.0 nm and 589.6 nm. Can the transmission grating with 1200 elements resolve principal maxima in the first order?

41. (a) A set of 10 parallel equidistant slits of width 0.50 cm and opaque space 1.4 cm are used to study Fraunhofer diffraction of $\lambda = 0.60$ cm falling normally on the planes of the slits. Calculate:

       i. angular position and half width of first maxima.

       ii. the effect of covering up alternative slits on angular position and half width of first maxima.

    (b) Derive an expression for resolving power of a grating.

42. Derive Bragg's law. How will you verify it using Bragg's spectrometer?

43. Light composed of two spectral lines with wavelengths 6000 Å and 6000.5 Å falls normally on a diffraction grating 10 mm wide. At a certain diffraction angle θ these lines are close to being resolved (according to Rayleigh's criterion). Find θ.

PROBLEMS

1. Light of wavelength 6000 Å passes through a narrow circular aperture of radius 0.09 cm. At that distance along the axis will the first maximum intensity be observed? [*Ans.* 1.35 m]

2. In an experiment with straight edge diffraction, the slit to edge distance is 1 m, and edge to screen distance is 2 m. If the slit is illuminated by the light of wavelength 5000 Å, calculate the separation of first three bright fringes.

    [*Ans.* 0.07 cm, 0.055 cm]

3. A circular aperture of 0.5 mm radius is illuminated by plane waves of monochromatic light. The diffracted light is received on a screen, which gradually moved towards the aperture. The centre of the circular patch of light first becomes dark when the screen is 25 cm from the aperture. Find the wavelength of light used.

4. Find the diameter of the smallest circular obstacle which will cut off practically all the light from a point situated 33 cm behind the obstacle. Wavelength of the light used is 5500 Å. Assume that first ten zones are effective.

5. A monochromatic beam of light on passing through a slit 1.6 mm wide falls on a screen held close to the slit. The screen is then gradually moved away and the middle of the patch of light on it becomes dark when the screen is 50 cm from the slit. Calculate the wavelength of light. [*Ans.* 6400 Å]

6. Parallel light of wavelength 5400 Å passes through a narrow circular aperture of radius 0.9 mm. At what distance along the axis we get first maximum, and the second maximum intensity? Also where do we get the first minimum intensity position? [*Ans.* 1.5 m, 0.5 m, 0.75 m]

7. A plane transmission grating having 6000 lines per cm is used to obtain the spectrum of light from a sodium lamp in the second order. Calculate the angular separation between the two sodium lines whose wavelengths are 5890 Å and 5896 Å. [*Ans.* 3.5′]

8. Monochromatic light of wavelength 6560 Å falls normally on a grating. The spectral line is diffracted at an angle of 19° 9′ from the normal in the first order. Find the width of the element.

9. How many orders will be visible if the wavelength of incident light is 6000 Å and the number of lines on the grating be 15000 to an inch?

10. Light is incident on a diffraction· grating at normal incidence. It is found to give green line, $\lambda = 5400$ Å in a certain order superimposed on the violet line, $\lambda = 4050$ Å of the next higher order. If the angle of the diffraction is 30°, how many lines per cm are there in the grating?
   [*Ans.* 3086]

11. A parallel beam of monochromatic light is falling normally on a plane transmission grating having 1.25×10^5 lines per metre and a second order spectral line is observed at 30°. Calculate the wavelength of the spectral line.

12. Monochromatic light of wavelength 6560 Å falls normally on a grating 2 cm wide. The first order spectrum is produced at an angle 18°14′ from the normal. What is the total number of lines on the grating ? [*Ans.* 9540]

13. The limits of visible spectrum are approximately 400 mμ to 700 mμ (millimicron). Find the angular breath of the first order visible spectrum produced by a plane grating, having 15000 lines per inch, when light is incident normally on the grating. (Given 13°40′ = 0.237, sin 24° 30′ = 0.4150)

14. In a grating spectrum, which spectral line in fourth order will overlap with the third order line of 5461 Å? [*Ans.* 4095.7 Å]

15. The wavelength of the C-line of hydrogen is 6563 Å. What is the highest order of the C-line that can be observed with 6000 lines per cm? Will this line appear to be of the same colour in each order ? [*Ans.* Second, Yes]

16. Calculate the angular separation between two wavelengths with a difference of 1 Å, the mean being 5000 Å in a diffraction grating with 6000 lines per cm in the second order. Find the effective diameter of the telescope required for these lines. Calculate the minimum diameter of the telescope required for these lines.
   [*Ans.* 1.5×10^{-4} radian, 0.417 cm, 0.4 cm]

17. Calculate the number of orders which can be obtained for a wavelength of 5460 Å, when the number of lines per cm on the grating is 5000.
   [*Ans.* 3]

18. Calculate the minimum number of lines per cm in a 2 cm wide grating which will just resolve the sodium lines of wavelengths 5890 Å and 5896 Å in the second order spectrum. Examine whether these lines are also resolved in first order spectrum of the grating.

19. A grating has 100 lines ruled on it. In the region of wavelength $\lambda = 6000$ Å, find (i) the separation between two wavelengths that can be just resolved in the first order spectrum (ii) the resolving power in the second order.

20. A diffraction grating has 5000 lies per cm and the total ruled width is 5 cm. Calculate for a wavelength of 5000 Å in second order: (i) the dispersion (ii) the resolving power (iii) the smallest difference in wavelength is resolved. What happens if half the ruling width is · covered?

21. Light composed of two spectral lines with wavelengths 6000 Å and 6000.5 Å falls normally on a diffraction grating 10 mm wide. At a certain diffraction angle θ, these lines are close to being resolved (according to Rayleigh criterion). Find θ.

SHORT ANSWER QUESTIONS

1. Why diffraction of sound is more evident in daily life than light waves?

Ans. The wavelength of sound is much greater than that of light waves and is comparable with the size of the obstacle, which is essential condition for diffraction. This is why diffraction of sound is more evident in daily life than light waves.

2. A sharp razor blade is held vertically in a beam of white light diverging from a point source. Discuss the nature of fringes obtained on a screen placed behind the blade.

Ans. When white light is used, coloured diffraction bands are produced. Since $x \propto \sqrt{\lambda}$, the violet and blue bands are nearer the edge of the geometrical shadow than the red bands. Only a small number of bands are visible near the edge and then there is general illumination due to overlapping of bands of different colours.

3. Why one cannot get the diffraction from a wide slit illuminated by a monochromatic light?

Ans. As the number of half period zones or elements contained in a wide slit is very large and hence the resultant effect is general illumination.

4. A single-slit diffraction pattern is obtained using a beam of red light. What happens if the red light is replaced by blue light?

Ans. Diffraction fringes become narrower and crowded together.

5. Why optical diffraction grating cannot be employed to study the diffraction of X-rays?

Ans. Diffraction effect is possible if the width of the slit on the grating is of the order of wavelength of light. Since the width of the slits on the optical diffraction grating is quite large as compared to the wavelength of X-rays and hence optical diffraction grating cannot be employed to study the diffraction of X-rays.

OBJECTIVE QUESTIONS

1. The order of the absent spectra in a diffraction grating is given by _____. $\left[\dfrac{a+b}{a} = n\right]$

2. The dispersive power of grating is defined as the rate of change of _____ of light with wavelength. [angle of diffraction]

3. The resolving power of a telescope is given by the reciprocal of the _____. [angular limit]

4. Diffraction of sound waves is more evident in daily life than _____ waves. [light]

5. Radio waves diffract pronouncedly around buildings while light waves which are also electromagnetic waves do not because the wavelength of light waves is _____ compared to radio waves. [very small]

MULTIPLE CHOICE QUESTIONS

1. In a plane transmission grating, the angle of diffraction for the second order principal maxima for the wavelength 5×10^{-5} cm is 30°. The numbers of lines in one cm of the grating surface will be
   (a) 5000
   (b) 10000
   (c) 500
   (d) 100

2. A monochromatic beam of light of wavelength 5460 Å falls on the grating normally and gives a second order image at an angle 45°. The grating element is of the order of
   (a) 10^{-3} cm
   (b) 10^{-4} cm
   (c) 10^{-5} cm
   (d) 10^{-6} cm

3. Fresnel half period zone differs from each other by
   (a) $\pi/2$
   (b) $\pi/4$
   (c) π
   (d) 2π

4. A plane transmission grating is 5 cm wide and has 5000 lines per cm. The resolving power of the grating in the second order spectrum is
   (a) 50000
   (b) 25000
   (c) 10000
   (d) 5000

5. In a single slit diffraction pattern for a slit width (d) and wavelength (λ), the separation between the central maximum and the first minimum is
   (a) $\theta = \dfrac{\lambda}{2d}$
   (b) $\theta = \dfrac{\lambda}{d}$
   (c) $\theta = \dfrac{\pi}{2}$
   (d) $\theta = \dfrac{\lambda}{4d}$

6. A plane transmission grating having 5000 lines per cm is being used under normal incidence of light. The highest order spectrum that can be seen for the light of wavelength 4800 Å is
   (a) 4
   (b) 2
   (c) 3
   (d) 1

7. A grating resolves a given doublet in the first order and the other grating of same width resolves the same doublet in the same order. The ratio of number of lines on the two gratings is
   (a) 4 : 1
   (b) 1 : 4
   (c) 1 : 2
   (d) 2 : 1

Answers

1. (a)   2. (c)   3. (c)   4. (a)   5. (d)

6. (a)   7. (d)

6

Polarisation

6.1 INTRODUCTION

The phenomena of interference and diffraction prove that light is a form of wave motion, but they do not reveal the character of this wave motion, i.e. whether it is longitudinal or transverse. The new phenomenon called *polarisation* proved conclusively that light is a transverse wave motion, because *polarisability* is characteristic of transverse waves.

According to electromagnetic theory, light is an electromagnetic disturbance in which the electric and magnetic fields are in the transverse plane and the two fields vary continuously and rapidly with time. The propagation of an alternating electromagnetic field constitutes a pheno-menon called electromagnetic waves. Electromagnetic waves are transverse since the electric and magnetic intensity vectors E and B of the wave fields are mutually perpendicular and lie in a plane perpendicular to the velocity vector v of wave propagation. Vectors v, E and B form a right handed system (Fig. 6.1) where from the head of vector v the smallest angle of rotation from vector E to B is counter clockwise.

A ray is a line whose tangent at each point coincides with the direction of wave propagation at that point, i.e. with the direction of energy transfer.

The relation between E and B in an electromagnetic wave propagated in non-conducting medium is determined by Maxwell's equations.

$$\nabla \cdot D = \rho$$

$$\nabla \cdot B = 0$$

$$\nabla \times E = -\frac{\partial B}{\partial t}$$

$$\nabla \times B = J + \frac{\partial D}{\partial t} \quad (6.1)$$

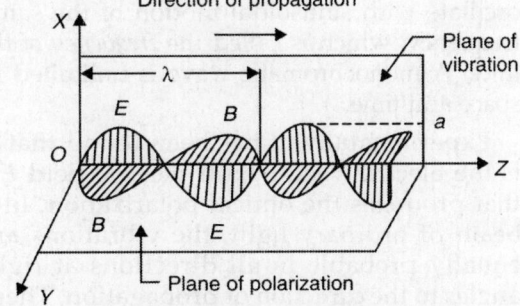

Fig. 6.1: Electromagnetic wave

Here ρ and j refer respectively to the charge density and current density. D and E, B and H and J and E are related with each other for an isotropic and homogeneous medium as

$$D = \varepsilon E, \, B = \mu H \text{ and } J = \sigma E$$

Solving Maxwell's equations for a homo-geneous isotropic non-conducting medium, one obtains

$$\nabla^2 E = \mu\varepsilon \frac{\partial^2 E}{\partial t^2}$$

and $$\nabla^2 B = \mu\varepsilon \frac{\partial^2 B}{\partial t^2} \quad (6.2)$$

Equation (6.2) represents a 3 dimensional vector wave equation where velocity is given by

$$v = \frac{1}{\sqrt{\mu\varepsilon}} \qquad (6.3)$$

If the medium is supposed to be vacuum, then $\varepsilon = \varepsilon_0$ and $\mu = \mu_0$ and we have

$$v_0 = \frac{1}{\sqrt{\mu_0\varepsilon_0}} = \frac{1}{\sqrt{4\pi \times 10^{-7} \times 8.8542 \times 10^{-12}}}$$

$$= 3 \times 10^8 \, \text{m/sec} = c \text{ (velocity of light)}$$

Obviously, *in free space, electromagnetic waves travel with the velocity of light.*

An electromagnetic wave is said to be plane if vectors E and B depend only on time and one cartesian coordinate, for instance x. In a plane wave, all rays are parallel to one another.

An electromagnetic wave is said to be *monochromatic* if the components of the vectors E and B of the wave's electromagnetic field oscillate with sinusoidal motion of the same frequency, which is called the *frequency of the wave.* A monochromatic wave is unlimited in space and time.

Experimentally, it has been found that it is the electric vector E (or electric field E), that produces the optical polarization. In a beam of ordinary light, the vibrations are equally probable in all directions at right angles to the direction of propagation. There is, thus, perfect symmetry around the direction of propagation (Fig. 6.2). Such light

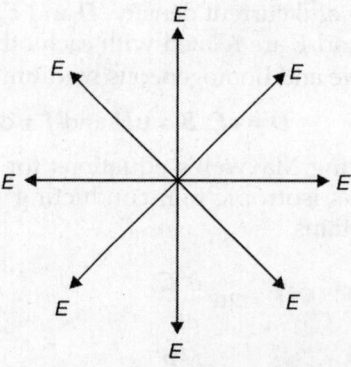

Fig. 6.2: Unpolarised light

waves are said to be **unpolarised** waves. If such perfect symmetry is lacking, i.e. asymmetry exists in any beam of light, the beam is said to be polarised. When we speak of vibrations in the light, we refer, then, to the variations of the electric vector E confined in planes transverse to the direction of propagation (such waves are said to be linearly polarised).

The plane in which the vibrations take place is called the *vibration plane* and the plane at right angles to the vibration plane is called the *plane of polarisation* (Fig. 6.3).

Figure 6.3 shows that light after passing through tourmaline crystal T has vibrations only in one plane. Such a light is called *plane polarised light*. We shall discuss about elliptically and circularly polarised light later.

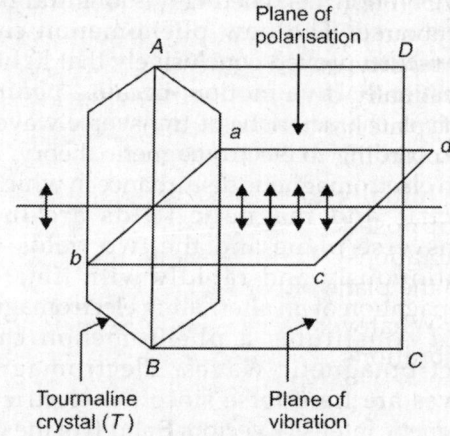

Fig. 6.3: Plane of polarisation and plane of vibration

ABCD is called the *plane of vibration*. This contains the crystallographic axis and the vibrations of transmitted light lie in this plane.

The plane *abcd* perpendicular to the plane of vibration in which no vibrations lie is called the *plane of polarisation*.

6.2 METHODS OF PRODUCING PLANE POLARISED LIGHT

There are following three important methods of producing plane polarised light:

1. Polarisation by Reflection

Malus in 1808 discovered that when natural light is incident on the glass surface in air, the reflected beam is partially polarised. The transmitted beams are also partially polarised. The degree of polarisation of the reflected beam depends on the angle of incidence and is maximum for a certain value called the *polarising angle* for the medium. The angle of maximum polarisation varies with wavelength. The angle of polarisation for glass of refractive index 1.51 is 57.5°. Complete polarisation is possible with monochromatic light and not with white light.

Let us consider two polished glass plates inclined equally to the horizon. A monochromatic beam of light *AB* incident at *B* is reflected along *BC* and after striking the second plate at *C* is reflected along *CD* at an angle of 57.5° (Fig. 6.4). If we now rotate the upper plate *C* about *BC* as an axis, then we observe that the intensity of the reflected beam along *CD* gradually decreases and becomes zero when the plate has been rotated through 90°. One can explain this as follows:

Initially the vibrations along *AB* can be supposed to consist of two mutually ⊥ʳ vibrations— in the plane of the paper and ⊥ʳ to the plane of paper.

Whatever be the angle of incidence, the vibrations ⊥ʳ to the plane of paper always

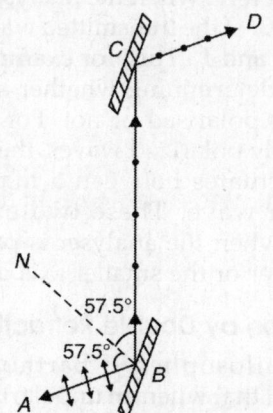

Fig. 6.4: Polarisation by reflection

remain ||ᶥ to the reflecting surface, i.e. their conditions of reflection are not affected by any change in the angle of incidence.

As the angle of incidence is changed, the vibrations in the plane of paper make different angles with the reflecting surface. When light is incident at the polarising angle on *B*, most of these vibrations are transmitted and are not reflected. Clearly, the reflected light is, therefore, completely polarised in the plane of incidence. It is worthwhile to note that the transmitted light is never completely plane polarised as some of the vibrations ⊥ʳ to the plane of incidence are also transmitted.

Brewster's Law

Sir David Brewster, a Scottish physicist, has shown that **when light is incident at the polarising angle, the reflected ray and the refracted ray are ⊥ʳ to each other.** This result is known as **Brewster's law**. If in this position, the angle of incidence is *i* and the angle of refraction is *r*, then by **Snell's law**,

$$\mu = \frac{\sin i}{\sin r} \qquad (6.4)$$

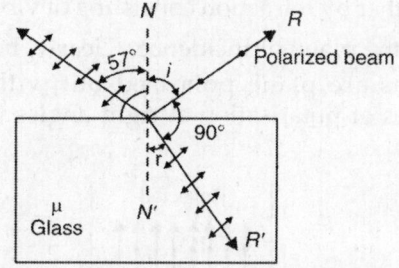

Fig. 6.5: Illustration of Brewster's law

But according to Brewster's law,

$$i + r = 90° \qquad (6.5)$$

or $\qquad r = (90° = -i)$

∴ $\qquad \sin r = \sin(90° = -i) = \cos i \qquad (6.6)$

From Eqs. (6.4) and (6.6), we have

$$\mu = \frac{\sin i}{\cos i} = \tan i \qquad (6.7)$$

2. Polarisation by Refraction—Pile of Plates

We have seen that with natural light incident at polarising angle on a glass plate, the reflected light is completely polarised in the plane of incidence, i.e. the reflected light consists of only $\perp^r$ vibrations. However, the intensity of the reflected beam is low because only 15% of $\perp^r$ vibrations are reflected and the remaining 85% are refracted. The reflected beam is not completely plane polarised but consists of 85% of the $\perp^r$ vibrations and all parallel vibrations.

However, by using a pile of glass plates, instead of a single glass plate, refracted beam by repeated reflections takes place.

The pile of plates consists of a number of thin glass plates placed one over the top of the other inside a wooden tube (Fig. 6.6). The plates are in contact with each other and are inclined at an angle of 32.5° (90 − 57.5°) to the axis of the tube. In this position, the light would be incident at an angle of 57.5° which is the polarising angle for glass. Obviously, this arrangement will give two beams of plane polarised light—one by reflection consisting of vibrations $\perp^r$ to the plane of incidence and the other by refraction consisting of vibrations $\|^l$ to the plane of incidence. Clearly, both the beams are plane polarised but with their planes of polarisation at right angles to each other.

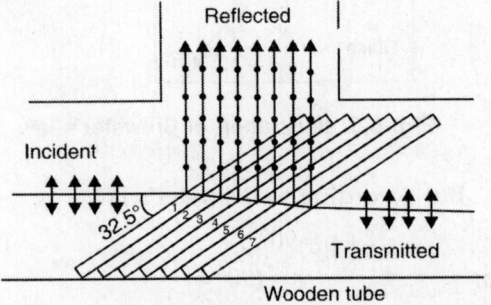

Fig. 6.6: Pile of plates

Degree of Polarisation

To obtain the degree of polarization or proportion of polarisation (p), let I_p and I_d denote the intensities of components with vibrations $\|^l$ and $\perp^r$ to the plane of incidence respectively in the transmitted light, then

$$p = \frac{I_p - I_d}{I_p + I_d} \tag{6.8}$$

Note that $p = 1$ for linearly polarised waves and that $p = 0$ for unpolarised waves. Provostaye and Desains have shown that Eq. (6.8) finally reduces to

$$p = \frac{n}{n + \left(2\mu/1 - \mu^2\right)^2} \tag{6.9}$$

where μ is the refractive index of the material of plates and n is the number of plates.

6.3 LAW OF MALUS

Malus found that the intensity of the transmitted beam varies with the angle between the plane of transmission and the plane of the polarizer. The *cosine law of Malus* can be stated as follows:

The intensity of light transmitted by the analyser is proportional to the square of the cosine of the angle between the planes of transmission of analyser and polarizer. We have the relation

$$I = a^2\cos\theta = I_0\cos^2\theta \tag{6.10}$$

where I_0 is the intensity of the incident wave and I is the intensity of the transmitted wave. When $\theta = 0$ or π, the intensity of transmitted light is maximum; when $\theta = \pi/2$ or $3\pi/2$, it is zero. Therefore, when the analyser is rotated, the intensity of the transmitted wave fluctuates between 0 and I_0. This, for example, affords a means of determining whether a wave, such as light, is polarised or not. For unpolarised or circularly polarized waves, the transmitted wave fluctuates between a maximum and minimum wave. These two extremes are obtained when the analyser is parallel either to the larger or the smaller axis of the ellipse.

Polarisation by Double Refraction

Dutch philosopher E Bartholin in 1669 discovered that when an unpolarized (natural) light is allowed to pass through a calcite (or quartz) crystal, it splits up into two reflected beams in place of the usual one as in glass. This phenomenon is known as double refraction. Many other crystalline substances like mica, selenite, sugar, topaz, aragonite, ice, etc. are found to exhibit the phenomenon of double refraction. Quartz and calcite are particularly important as they are used extensively in the manufacture of special instruments.

To demonstrate the phenomenon of double refraction, make an ink dot on a sheet of paper and view through a calcite crystal. We observe two images of the dot. Now rotate the crystal slowly as shown in Fig. 6.7a, we notice that one of the two images remain stationary while the second image rotates around the first. The image which remains stationary is called as the *ordinary image* and the *refracted ray* which produces this image is called as *ordinary ray* denoted by *O*-ray. *O*-ray obeys the ordinary laws of refraction. The second image which rotates round the first is called as the *extraordinary image* and the ray which produces this image is called *extraordinary ray* and denoted by *E*-ray. *E*-ray does not obeys the ordinary laws of refraction.

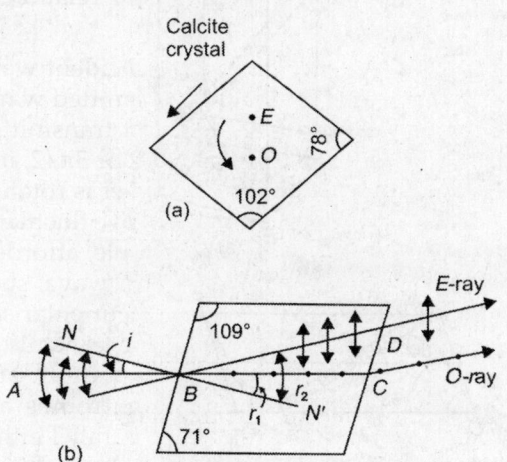

Fig. 6.7: Double refraction through calcite crystal

Let us consider a beam of light *AB* from a point source incident at an angle *i* on a calcite crystal (Fig. 6.7b). It splits into *O*-and *E*-rays inside the crystal. The vibrations in the ordinary ray are at right angles to the principal plane. Those in the extraordinary ray are at right angles to the principal plane. Obviously, the *O*-ray is plane polarized in the principal plane and the extraordinary ray is plane polarised at right angles to the principal plane.

The refractive indices for *O*-ray and *E*-ray are respectively

$$\mu_O = \frac{\sin i}{\sin r_1} \quad \text{and} \quad \mu_E = \frac{\sin i}{\sin r_2} \qquad (6.11)$$

where r_1 and r_2 are angles of refraction which *O*-ray and *E*-ray make inside the crystal respectively. For calcite crystal, $r_1 < r_2$ and hence $\mu_O > \mu_E$. Inside the calcite crystal, the velocity of light for the *O*-ray is less than for the *E*-ray.

6.4 CALCITE CRYSTAL

Calcite (Iceland spar) is crystalised calcium carbonate ($CaCO_3$). It exists in nature in several forms, but it cleaves very perfectly along three directions forming rhombohedra or rhombus. The six faces of the rhombohedra are parallelograms each having angles of 102° and 78° or more accurately 101°53′ and 78° 7′ (Fig. 6.8). At the two opposite corners *P* and *Q* of the rhombus, three angles of 102° meet, whereas at the other, angles of 72° meet. The corners where three obtuse angles meet are called as blunt corners.

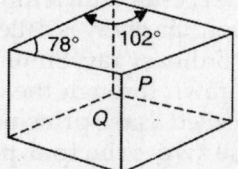

Fig. 6.8: Calcite crystal

6.5 OPTIC AXIS

*A line passing through any one of the blunt corners and equally inclined with the three faces, which meet at this corner, represents the direction of the crystallographic axis of the crystal (Fig. 6.9) and any direction parallel to it is called **optic axis**.* In Fig. 6.9, *xy* represents the optic axis. Any line parallel to *xy* is also an optic axis. It may be emphasized here that (i) *optic axis is a direction and not a particular ray* (ii) *if a ray of light falls along the optic axis or in a direction parallel to it, then it does not break up into O- and E-rays, i.e. there is no double refraction along optic axis.*

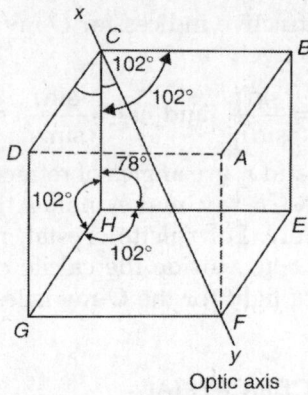

Fig. 6.9: Optic axis

6.6 PRINCIPAL SECTION

A plane which contains the optic axis and $\perp^r$ to the opposite faces of the crystal is called the *principal section* of the crystal. Since a crystal has six faces, therefore, for every point there are *three principal sections*. A principal section of calcite crystal is a parallelogram having angles of 71° and 109°.

6.7 PRINCIPAL PLANE

A plane in the crystal drawn through the optic axis and the ordinary ray is called as principal plane of the ordinary ray. Similarly a plane in the crystal drawn through the optic axis and the E-ray is called as the principal plane of the E-ray. These two principal planes do not coincide in general. In a special case, when the plane of incidence is a principal section, then the principal section of the crystal and the principal planes of the ordinary and extraordinary rays coincide.

6.8 NICOL PRISM

It is an optical device made from calcite and frequently used for producing and analysing plane polarised light. It is based on the phenomenon of double refraction and was invented by William Nicol in 1828. We have seen that when a beam of light is transmitted through calcite, it breaks up into two rays:

i. The O-ray which has its vibrations $\perp^r$ to the principal section of the crystal.

ii. The E-ray which has its vibrations $\parallel^l$ to the principal section. The O-ray is cut off by total internal reflection leaving the E-ray alone.

Construction of Nicol

Nicol prisms are constructed from a calcite rhombohedra, the edges of end faces of which are equal but one-third the length of the other longer edges (Fig. 6.10a). Let $A'BCDEFG'H$ represent such a crystal having A' and G' as its blunt corners where three obtuse angles of the three faces meet. $A'CG'E$ is one of the principal sections of the crystal with $A'CG' = 71°$.

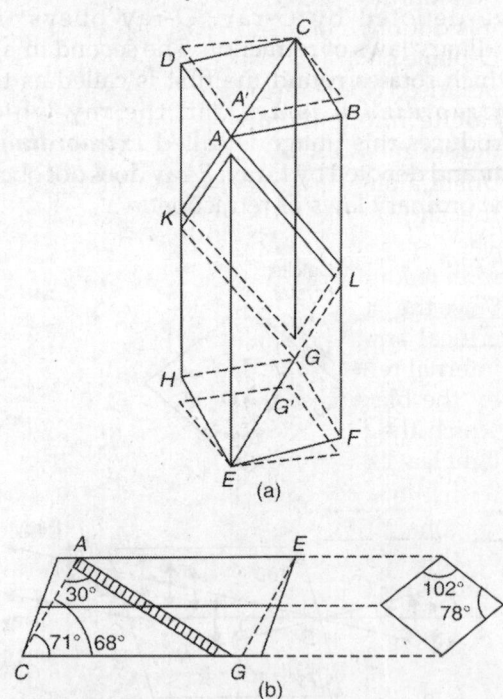

Fig. 6.10: (a) Calcite rhombohedron (b) Nicol prism

First the two short end faces $A'BCD$ and $EFG'H$ are grounded in such a way that the angle ACG becomes 68° instead of 71°. The rhombohedra is then cut along the plane $AKGL$ (Fig. 6.10a). The two surfaces are grounded and polished optically flat and then cemented together in their original position with Canada balsam, which is a transparent cement with a refractive index about midway between the refractive index of O-ray and that of E-ray (Fig. 6.10b). For sodium light of wavelength $\lambda = 5893$ Å, refractive index for Canada balsam is 1.55, $\mu_O = 1.658$, $\mu_E = 1.486$.

The sides of Nicol prism are blackened to absorb any light incident on the sides.

Action of Nicol Prism

Figure 6.11 represents the section $ACGE$ of the Nicol prism. The diagonal AG represents the Canada balsam layer. Canada balsam layer acts as a rarer medium for an O-ray and as a denser medium for the E-ray. When a beam of natural light enters one of the end faces in a direction parallel to the long side of the Nicol, it is doubly refracted and is decomposed into O-plane polarised beam and E-ordinary plane polarised beam. The E-ray, in passing from the calcite into Canada balsam, passes into a denser medium and emerges from the Nicol prism along $S'T'$. Its vibration will be $\parallel$ to the shorter diagonal of the end face EG. The O-ray on the other hand tries to enter the rarer medium, viz. Canada balsam from calcite. Since the angle of incidence is greater than the critical angle and hence it experiences total internal reflection at R. The O-ray is absorbed by the black internal surface of the tube in which the prism is mounted. Obviously, the light leaves the Nicol prism as plane polarised with the vibration plane $\parallel$ to the shorter diagonal of the end face. The salient features of the working of the Nicol prism are summarised as follows:

i. Refractive index (μ) for O-ray with respect to Canada balsam is given by

$$\mu = \frac{1.658}{1.550}$$

If θ be the critical angle for the O-ray, then

$$\sin\theta = \frac{1}{\mu} = \frac{1.550}{1.658}$$

or $\qquad \theta = \sin^{-1}\left(\frac{1.550}{1.650}\right) = 69°$

Hence, if the angle of incidence for the O-ray is greater than the critical angle, i.e. 69°, it is totally internally reflected and only the E-ray passes.

ii. If the angle of incidence is less than the critical angle for the O-ray, then it will also be transmitted through the Nicol prism.

Limitations of Nicol Prism as a Polarising Device

i. As a polarising device, Nicol prism is only effective with a slightly convergent or slightly divergent beam of light. Nicol fails to be effective polarizer with too convergent or too divergent incident beam.

ii. If the incident ray makes an angle much smaller than $\angle CQP$ with the face CA, the O-ray will strike the Canada balsam layer at an angle less than the critical angle (69°) and hence O-ray will also be transmitted and light emerging from Nicol will not be plane polarised. It is observed that the angle of incidence on the face CA is limited to $\angle PQP''$ which

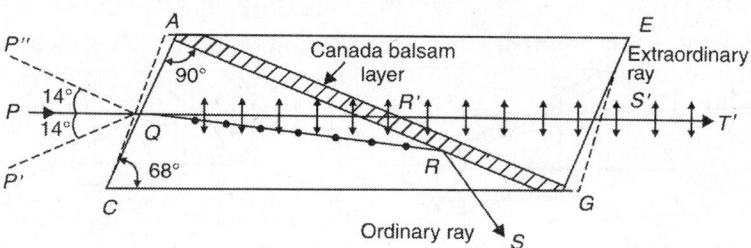

Fig. 6.11: Nicol prism

is nearly 14° beyond which the O-ray is transmitted through the Nicol.

iii. The E-ray also has a limit to the angle of incidence beyond which it is totally internally reflected by the Canada balsam surface. The refractive index for the E-ray, i.e. $\mu_E = 1.486$ when the E-ray is travelling at right angles to the direction of the optic axis. Along the optic axis, the E-ray travels with the same velocity as the O-ray and its refractive index is same as that of O-ray, i.e. $\mu_E = \mu_O = 1.658$. For intermediate angles, μ_E lies between the two limits 1.486 and 1.658. Obviously, for a particular angle μ_E may be greater than 1.55 and angle of incidence greater than the critical angle, then the E-ray will also be totally internally reflected at the Canada balsam surface. Clearly, no light will then emerge out of the Nicol.

iv. Nicol prism cannot be used for highly convergent or divergent beams.

Uses of Nicols

Two Parallel Nicols

When two Nicol prisms are lined up coaxially one behind the other, as shown in Fig. 6.12, one of them acts as a polariser and the other acts as an analyser. The first Nicol P_1 which produces the plane polarised light is called the *polariser* and the second Nicol P_2 which is used for examining the state of polarisation of transmitted light is called the *analyser*. The combination of two such Nicols is called a *polariscope*. When the two nicols are placed with their principal sections ∥ to each other as shown in Fig. 6.12a, then the E-ray transmitted by P_1 is freely transmitted by P_2.

If we rotate one of the two Nicols, the intensity of the transmitted beam decreases till it is reduced to zero.

However, if one of the two Nicols is further rotated so that it has turned through an angle of 180°, E-ray would be transmitted.

If I_0 be the intensity of transmitted beam when the principal sections of P_1 and P_2 are parallel and I be the intensity when the principal sections of P_1 and P_2 are inclined at an angle θ, then according to cosine law of Malus, we have

$$I = I_0 \cos^2 \theta \qquad (6.12)$$

One can utilise the above facts for detecting the plane polarised light. If the given light on examination through a rotating Nicol (say P_2) exhibits a variation in intensity with minimum intensity zero, the given light is clearly plane polarised.

6.9 ELLIPTICALLY AND CIRCULARLY POLARISED LIGHT

Let monochromatic light be passed through Nicol prism N_1. The outcoming beam is plane

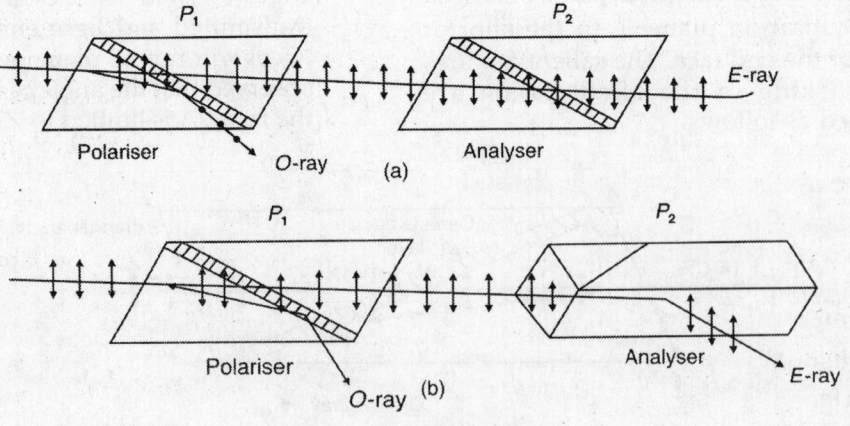

Fig. 6.12: Polariscope

polarised (Fig. 6.13). Now, this plane polarised light is allowed to fall normally on a uniaxial doubly refracting calcite or quartz plate *P*, whose faces have been cut $\parallel^l$ to the optic axis. On entering the doubly refracting calcite crystal, the plane polarised light is split up into two components *O* and *E*. Both these rays travel inside the plate along the same path but with different velocities. For calcite, $V_E > V_O$ and hence the *O*-ray falls behind the *E*-ray and a phase difference δ is introduced between them after transversing a thickness *d* of the calcite plate.

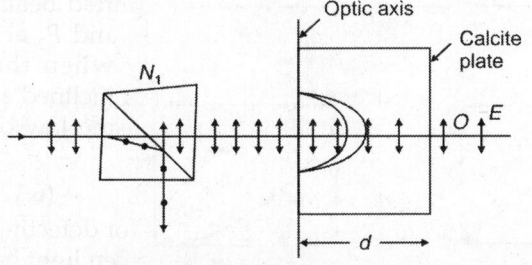

Fig. 6.13: Velocities of elliptically and circularly polarised light

Theory

Suppose the incident plane polarised light makes an angle q with the optic axis (Fig. 6.14). Let the amplitude of the incident plane polarised light is *A*. It can be resolved into two components—$A \cos\theta$ along the optic axis and $A \sin\theta \perp^r$ to the optic axis.

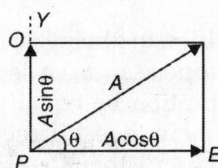

Fig. 6.14: Geometry of plane polarised light

We can represent the two vibrations by two simple harmonic motions at right angles to each other as

$$x = A\cos\theta \sin(\omega t + \delta) \text{ (E-wave vibrations) (6.13)}$$
$$y = A\sin\theta \sin\omega t \quad \text{(O-wave vibrations) (6.14)}$$

Put $A\cos\theta = a$ and $A\sin\theta = b$ (6.15)

Using Eq. (6.15), Eqs. (6.13) and (6.14) become

$$x = a\sin(\omega t + \delta) \tag{6.16}$$
$$y = b\sin\omega t \tag{6.17}$$

One can obtain the nature of resultant vibration by eliminating *t* from Eqs. (6.16) and (6.17), we have from Eq. (6.17)

$$\frac{y}{b} = \sin\omega t \quad \text{and} \quad \cos\omega t = \sqrt{1 - y^2/b^2} \tag{6.18}$$

Writing Eq. (6.16) as

$$\frac{x}{a} = \sin\omega t \cos\delta + \cos\omega t \sin\delta$$

or $$\frac{x}{a} = \frac{y}{b}\cos\delta + \sqrt{1 - y^2/b^2}\sin\delta$$

or $$\frac{x}{a} - \frac{y}{b}\cos\delta = \sqrt{1 - y^2/b^2}\sin\delta$$

Squaring and rearranging, we obtain

$$\frac{x^2}{a^2} + \frac{y^2}{b^2} - \frac{2xy}{ab}\cos\delta = \sin^2\delta \tag{6.19}$$

This equation represents an ellipse described in a rectangle of sides 2*a* and 2*b*. Obviously, the emergent light from the crystal is generally elliptically polarised. We now consider some particular cases of interest.

Case I. When $\delta = 0$, $\sin\delta = 0$, $\cos\delta = 1$, Eq. (6.19) gives

$$\frac{x^2}{a^2} + \frac{y^2}{b^2} - \frac{2xy}{ab} = 0$$

or $$\left(\frac{x}{a} - \frac{y}{b}\right)^2 = 0$$

or $$y = \frac{b}{a}x \tag{6.20}$$

Equation (6.20) is the equation of a straight line passing through the origin and having slope *b/a*. Obviously, the emergent light is *plane polarised* with vibrations in the same plane as in the incident light (Fig. 6.15a).

Case II. When $\delta = \pi/4$, Eq. (6.19) reduces to

$$\frac{x^2}{a^2} + \frac{y^2}{b^2} - \frac{\sqrt{2}xy}{ab} = \frac{1}{2}$$

This is an equation of oblique ellipse (Fig. 6.15b).

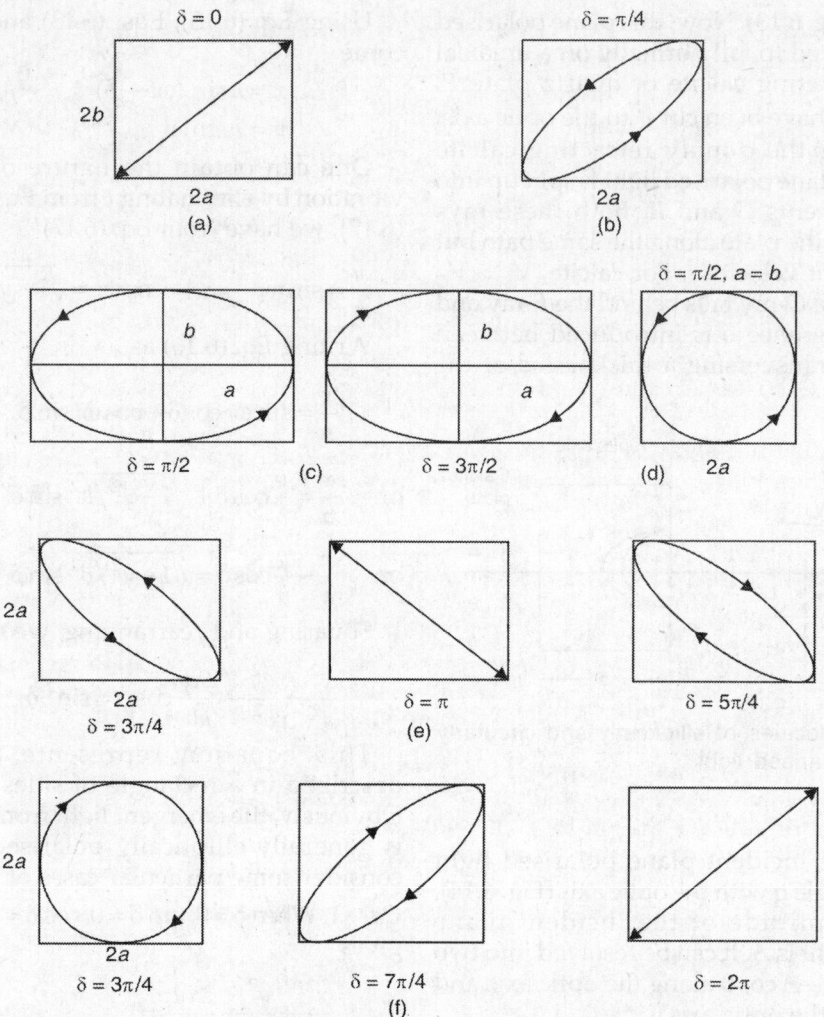

Fig. 6.15: Plane polarised light

Case **III.** i When the thickness (d) of the crystal plate is such that $\delta = \pi/2$, $3\pi/2$, then $\cos\delta = 0$, $\sin\delta = 1$ and Eq. (6.19) reduces to

$$\frac{x^2}{a^2} + \frac{y^2}{b^2} = 1$$

which represents an ellipse with the major and minor axes parallel to the two privileged directions of the crystal. Obviously, the emergent light is elliptically polarised ($a \neq b$) and the plane of the ellipse being normal to the direction of propagation (Fig. 6.15c).

ii. If an addition to δ being an odd multiple of $\pi/2$ suppose $a = b$, i.e. let the incident plane vibration be equally inclined to the two privileged directions of the crystal ($\theta = 45°$), then the emergent light can be expressed by the equation

$$x^2 + y^2 = a^2 \qquad (6.21)$$

which obviously represents a circle of radius a (Fig. 6.15d). Thus, the emergent light is circularly polarised. Such a crystal plate is called a *quarter wave plate*, which satisfies the general condition $\delta = (2K + 1)\pi/2$, where K is an integer.

Case **IV.** If δ = Kπ (K can have odd values), then cos δ = –1 and sin δ = 0. Equation (6.19) reduces to

$$\frac{x^2}{a^2} + \frac{y^2}{b^2} + \frac{2xy}{ab} = 0$$

or

$$\left(\frac{x}{a} + \frac{y}{b}\right)^2 = 0$$

or

$$y = -\left(\frac{b}{a}\right)x \qquad (6.22)$$

Equation (6.22) is again the equation of straight line passing through origin but with a slope (–b/a) in the negative direction of X-axis (Fig. 6.15e). Obviously, the emergent light is plane polarised with the vibration direction making an angle 2θ = 2tan⁻¹(b/a) with the incident light. The crystal plate behaves as a half wave plate.

Case **V.** For δ = π/4 or 7π/4, Eq. (6.19) represents the general equation of ellipse (Fig. 6.15f) with the major axis inclined in the positive and minor axis in the negative direction of the X-axis respectively.

Case **VI.** For all other values of δ, the nature of emergent vibrations will be as shown in Fig. 6.15.

Froms the above analysis, we conclude as follows:

i. Figure 6.15 represents the shape of the resultant transmitted vibrations for different values of δ.
ii. Elliptic vibrations represent the general case when the two components of vibrations have any value of δ other than a multiple of π except for the special case of circular vibrations. Obviously, the plane polarised and circularly polarised light are the special cases of elliptically polarised light.
iii. 0 < δ < 2π, the ellipse is described anti-clockwise and light is said to be *left handed elliptically polarised light*.
iv. π < δ < 2π, the ellipse is described in the clockwise direction and the light is said to be *right handed elliptically polarised*.
v. δ = 0 and δ = 2π, the plane of vibrations are parallel.

6.10 PHASE RETARDATION PLATES

Electromagnetic theory of light reveals that plane waves of light can pass through a crystal only if its vibration plane is ||ˡ to one of the two privileged directions which are mutually ⊥ʳ to each other. These directions are also known as *azimuthal directions*. The theory further shows that the velocities of propagation for these two possible vibrations are different. The direction along which the velocity is smaller is called the *slow axis* and the other is called the *fast axis*. Let the velocities along these axes be c' and c" and velocity in vacuum is c. Obviously, the two refractive indices corresponding to the vibrations are μ' = c/c' and μ" = c/c". If we cut the crystal with its faces ||ˡ to the optic axis, then the two privileged directions will be those of the *ordinary* (*O*) and *extraordinary* (*E*) vibrations.

Now consider that a thin crystal plate cut with optic axis is ||ˡ to the faces of the plate (Fig. 6.16). Let a plane polarized wave is incident normally on such a plate, the vibrations in the wave break at first face into

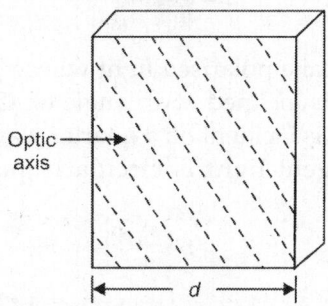

Fig. 6.16: A phase retardation plate

one vibration along the optic axis and the other vibrations ⊥ʳ to it. These components travel through the crystal plate in the same direction along the normal but with different velocities v_O and v_E. If the thickness of the crystal plate is d, the difference in time taken in traversing the plate by the two beams is

$$|\Delta t| = d\left(\frac{1}{v_O} - \frac{1}{v_E}\right) \qquad (6.23)$$

In terms of equivalent length in vacuum, the path difference or the relative retardation is

$$\delta \text{ or } \Delta x = c|\Delta t| = d|\mu_O - \mu_E| \qquad (6.24)$$

Obviously, a relative phase retardation $\frac{2\pi}{\lambda}\Delta x = \frac{2\pi}{\lambda}(\mu_O - \mu_E)d$ is created between two mutually $\perp^r$ vibrations. Due to this reason, such plates are called *phase retardation plates*. If we choose the thickness of the plate to make the retardation in path equal to $\lambda/4$ or $\lambda/2$, the plates are then called *quarter wave plate* or *halfwave plate* respectively.

Quarter Wave Plate

If the path difference in Eq. (6.24) satisfies the condition

$$d|\mu_O - \mu_E| = \left(n \pm \frac{1}{4}\right)\lambda \qquad (6.25a)$$

where n is an integer, the crystal plate is called quarter wave plate. Its thickness can be determined from

$$d = \frac{\left(n \pm \dfrac{1}{4}\right)\lambda}{|\mu_O - \mu_E|} \qquad (6.25b)$$

If the plane polarised light whose plane of vibration is inclined at an angle of 45° to the optic axis, is incident on a quarter wave plate, the emergent light is circularly polarised (Fig. 6.17).

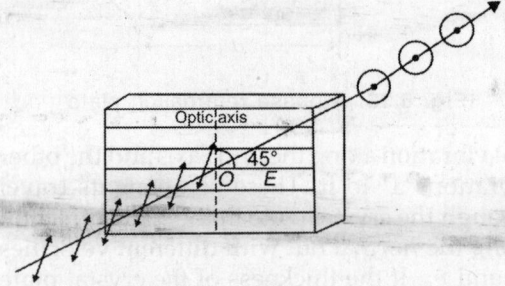

Fig. 6.17: Quarter wave plate

Half Wave Plate

If the path difference [Eq. (6.24)] satisfies the condition

$$d|\mu_O - \mu_E| = \left(n + \frac{1}{2}\right)\lambda \qquad (6.26)$$

the crystal plate is called half wave plate. Its thickness is given by

$$d = \frac{\left(n + \dfrac{1}{2}\right)\lambda}{|\mu_O - \mu_E|} \qquad (6.26a)$$

In practice, the quarter wave plate and half wave plate have to be supported between glass plates. When plane polarised light falls on a half wave plate with its plane of vibration making an angle θ, then the emergent light is also plane polarised with the only difference that its plane of vibration form that of the incident light (Fig. 6.18).

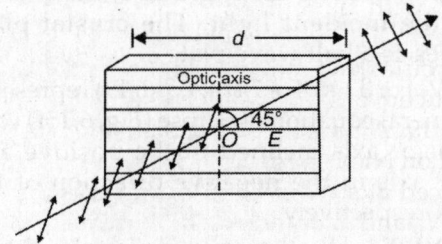

Fig. 6.18: Half wave plate

The QW and HW are usually made of thin sheets of split mica or of quartz cut $\|^l$ to optic axis. A major disadvantage of these retardation plates is that they are true for one particular wavelength because their thickness (d) depends upon wavelength (λ). Obviously, the same plate will not be equally effective for all the colours.

6.11 PRODUCTION OF PLANE, CIRCULARLY AND ELLIPTICALLY POLARISED LIGHT

Plane Polarised Light

It can be produced by the following methods:

   i. Polarisation by reflection from a reflecting surface (except metallic) at the polarising angle—Biot's polariscope
  ii. Polarisation by refraction—pile of plates
 iii. Polarisation by double refraction—Nicol prism
 iv. Polarisation by polaroid.

Usually, for producing plane polarised light, a Nicol prism is used. A beam of monochromatic light is passed through a Nicol prism. The beam split up into E-ray and O-ray is totally internally reflected back at the Canada balsam layer while the E-ray passes through the prism. The emergent light beam is plane polarised.

Circularly Polarised Light

One can produce circularly polarised light by the two waves of equal amplitudes vibrating $\perp^r$ to each other and having a path difference of $\lambda/4$ or phase difference of $\pi/2$. For this purpose plane polarised light is allowed to fall normally on a quarter wave plate such that vibrations in the incident light make an angle of $45°$ with the optic axis of the plate. The emergent light will be circularly polarised.

Circularly polarised light can also be produced by the arrangement shown in Fig. 6.19. A parallel beam of light is allowed to fall on Nicol prism N_1. The Nicol prism N_2 is placed at some distance from prism N_1 so that N_1 and N_2 are crossed. Obviously, the field of view will be dark. A quarter wave plate P mounted on a tube A is inserted in between N_1 and N_2 so that the field of view may be bright. The plane polarised light from N_1 falls normally on quarter wave plate P. The quarter wave plate P is rotated till the field of view is dark. This means that the optic axis of $\lambda/4$ plate is now parallel to the principal plane of polarising Nicol N_1. Keeping P fixed, A is rotated till the mark S on P coincide with zero mark on A. Afterwards by rotating P, the mark

S is made to coincide with $45°$ on A. In this position the vibrations of plane polarised light incident on the $\lambda/4$ plate make an angle $45°$ with the direction of the optic axis of the $\lambda/4$ plate. The plane polarised light entering the $\lambda/4$ plate is split up into O and E components of equal amplitudes and time period and as a result the emergent light from P is circularly polarised. If N_2 is rotated at this stage, the field of view shows no change in intensity. This is just similar to the case when ordinary light passes through the Nicol prism.

Elliptically Polarised Light

To produce elliptically polarised light, the two waves of unequal amplitudes vibrating at right angles to each other should have a path difference of $\lambda/4$ or a phase difference of $\pi/2$. One can use the arrangement of Fig. 6.19. A parallel beam of monochromatic light is allowed to fall on N_1 and N_1 and N_2 are crossed, i.e. field of view is dark. The plane polarised light from N_1 falls normally on $\lambda/4$ plate P. The field of view is illuminated and light emerging out of $\lambda/4$ plate P is elliptically polarised. In this case, one will have to be careful that the vibrations of the plane polarised light emerging out of $\lambda/4$ plate P should not make an angle of $45°$ with the optic axis of P. On rotating N_2, we will observe that the intensity of illumination of the field of view varies between a maximum and minimum. This just resembles the case when a mixture of plane polarised and unpolarised light is circularly polarised by rotating Nicol N_2.

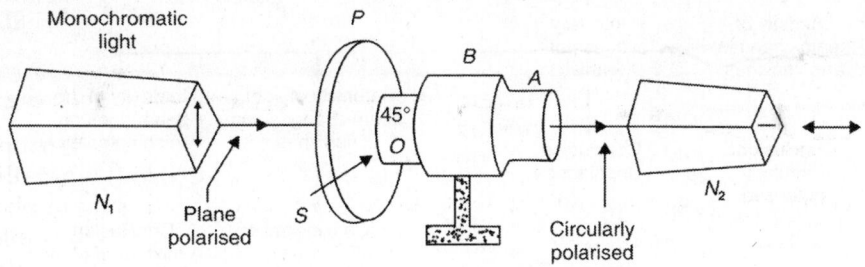

Fig. 6.19: Production of circularly polarised light

6.12 ANALYSIS OR DETECTION OF PLANE, CIRCULARLY AND ELLIPTICALLY POLARISED LIGHT

Suppose we are asked to find the state of polarisation of the supplied light, i.e. whether the light is linearly or elliptically or circularly polarised. We can adopt the following procedure for this purpose. Hold a Nicol prism in front of the light and rotate it about the incident light as the axis. The following three situations may arise:

1. On rotating the Nicol prism, the light intensity may vary between maximum value and zero. If it is so, then one can say that the given light is *plane pola-rised*.

2. If light intensity varies between maximum and minimum values but never becomes zero, the light under analysis is either *elliptically polarised* or *partially polarised,* i.e. it is mixture of plane polarised and unpolarised light. In the case of elliptically polarised light, maximum light would be transmitted when the principal section of the Nicol prism is parallel to the major axis and minimum, when it is $\|^l$ to the minor axis. In the case of partially polarised light, maximum intensity of transmitted light is obtained when the principal section of the Nicol prism is $\|^l$ to the plane of vibration of the plane polarised light and minimum when it is at right angles to it.

3. If on rotating the Nicol, there is no change in the intensity of light, then the light under examination is either completely *unpolarised* or *circularly polarised*.

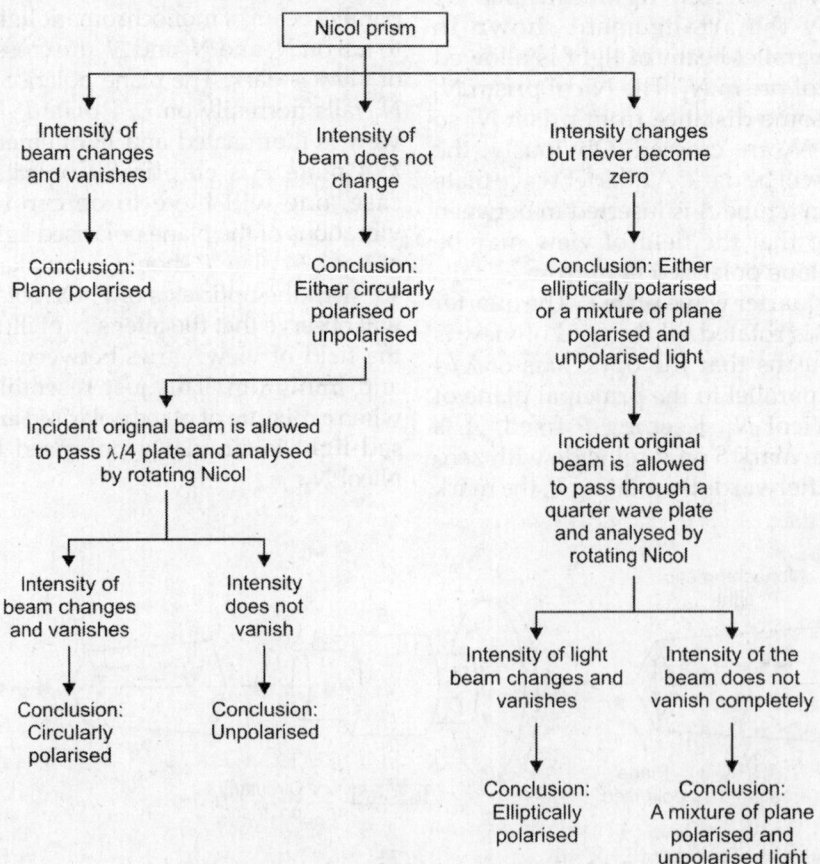

Fig. 6.20

To distinguish between the unpolarised and circularly polarised light, we introduce a λ/4 plate between the light and analysing Nicol. If the light under examination is circularly polarised, then it would emerge from the λ/4 plate as a plane polarised light which would be extinguished completely twice during one complete rotation of the Nicol prism.

If, on the other hand, the light under examination is *unpolarised*, then it would not become plane polarised on passing through λ/4 plate and would, therefore, not be extinguished for any position of the Nicol prism.

To distinguish between *elliptically polarised* and *partially polarised light*, we again introduce a λ/4 plate between the light and the Nicol prism. If the original light is *elliptically polarised*, then after passing it through λ/4 plate, it will become plane polarised and would be completely extinguished twice during each complete rotation of the Nicol prism. If the light under examination is *partially polarised*, then on passing it through the λ/4 plate, the plane polarised in the mixture would become elliptically polarised. Obviously, on rotating the Nicol prism, intensity of the transmitted light would again vary between maximum and minimum values and will never become zero in any orientation of the Nicol prism. The above results have been summarised in Fig. 6.20.

6.13 OPTICAL ACTIVITY

Arago in 1811 discovered that when a quartz plate is introduced between two crossed Nicol prisms, light is partially restored even if the crystal is cut with its faces normal to the optic axis. Arago further observed that by simply rotating the analyser Nicol through a suitable angle, the field of view could again made dark. Obviously, the quartz is found to rotate the plane of polarisation. This property is called *optical activity* and it is exhibited by many other crystals, e.g. sugar crystal, potassium phosphate crystal, etc.

The substances like quartz, sugar in solution, cinnabar, sulphate of strychnine, etc. which rotate the plane of vibration (and also plane of polarisation) are known as *optically active substances*. All the *optically active*

substances can be classified into two classes as mentioned below:

1. Dextrorotatory or Right-handed

The optically active substances which rotate the plane of polarisation in a clockwise direction when looking against the incoming light are called *dextrorotatory* or *right-handed substances*.

2. Laevorotatory or Left-handed

Those optically active substances which rotate the plane of polarisation in the anticlockwise direction when looking against the incoming light are called *laevorotatory* or *left-handed substance*. Some quartz and fruit sugars are examples of laevorotatory substances.

Quartz in crystalline form exhibits both laevo and dextro varieties, but when melted it becomes *inactive*. This shows that optical activity is associated with the crystalline structure of the substances. Many solutions like solution of tartaric acid, sugar, etc. in an optically inactive solvent are optically active. Obviously, the optical activity in these substances is due to their molecules.

Biot made a systematic study of the optical activity in crystals. The results of his study are summarised below:

i. For a given wavelength and material, the rotation is proportional to the thickness of the crystal, i.e. $\theta \propto l$.

ii. The rotation produced due to two crystals is the algebraic sum of those due to the individual crystals, i.e.

$$\theta = \theta_1 + \theta_2 + \theta_3 + \dots$$

where θ_1, θ_2, θ_3, etc. are rotations produced by individual specimens. The anticlockwise rotations are considered as positive while the clockwise rotations are negative. Obviously, the total rotation produced by two quartz plates—one dextrorotatory and the other laevorotatory and of equal thickness, is zero.

iii. The rotation produced by crystals depends on the temperature of the specimen.

iv. For a given crystal, the rotation is inversely proportional to the square of the wavelength of the incident light, i.e.

$$\theta \propto 1/\lambda^2$$

Obviously, the rotation for red (7000 Å) is least and greatest for violet (4000 Å). The above four conclusions are called *laws of rotation of plane of polarisation*.

6.14 SPECIFIC ROTATION

Certain liquids and solutions also exhibit the property of *optical activity*. For a given wavelength, the rotation produced by a solution at a given temperature is proportional to the length of the solution, i.e.

$$\theta \propto l \quad \text{and} \quad \theta \propto C$$

$$\therefore \qquad \theta \propto lC \quad \text{or} \quad \theta \propto SlC \qquad (6.27)$$

Here l is the length of the solution column in decimetres, C the concentration in gm/cc of the solution and S is a constant which depends upon the nature of the active substance and is known as the *specific rotation* of the solution. Thus,

$$S = \frac{\theta}{lC}$$

$$= \frac{\text{Rotation in degrees}}{\underset{\text{decimeters}}{\text{Length in}} \times \underset{\text{gm/cc}}{\text{Concentration in}}} \qquad (6.28)$$

$$= \frac{\theta}{l(m/V)} = \frac{\theta V}{lm} \qquad (6.29)$$

Here m (in gms) is the mass of the active substance present in V cc of solution. Thus, *specific rotation of a solution for a given wavelength of light at a temperature t is defined as the rotation produced by solution of unit length (diameter) containing a unit mass (1 gm) of the optically active substance per unit volume (1 cc) of the solution.*

From the above definition of specific rotation, it is clear that it is not a constant but it varies with the wavelength of light, the nature of the inactive solvent, the concentration of the solution and its temperature. Clearly, the specific rotation is greater for light of shorter wavelength and *vice versa*.

For quartz for which the rotation is very large, the specific rotation is defined as the rotation produced by 1 mm thick plate.

Molecular Rotation

The product of specific rotation and the molecular weight of the substance is called the *molecular rotation*.

6.15 POLARIMETERS

These are the instruments used for finding the optical rotation of different solutions. When these are used for finding the optical rotation of sugar, they are called *saccharimeters*. Polarimeter can be used to find the specific rotation of sugar solution or if specific rotation is known, they can be used to find its concentration. The following polarimeters are widely used.

Laurent's Half Shade Polarimeter

Figure 6.21 is a schematic representation of this polarimeter. N_1 and N_2 are two Nicol prisms. N_1 is a polariser and N_2 is an analyser. Behind Nicol prism N_1 there is a half wave plate of quartz Q which covers one half of the field of view while the other half G is a glass plate. G absorbs the same amount of light as the quartz plate Q. T is a glass tube containing the optically active substance and closed at the ends by cover slips and metal covers. This tube is mounted between the two Nicols N_1 and N_2. It is worthwhile to mention that there will be no air bubbles in the path of the light. The air bubbles (if any) will appear at the upper portion of the wide bore (T_1) of the tube T. S is a circular slit illuminated with mono-chromatic light.

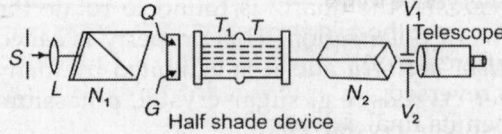

Fig. 6.21: Laurent's half shade polarimeter

The lens L renders the light from slit S parallel and directs it towards the polarising Nicol N_1. The light emerging from N_1 is plane polarised with its vibrations in the principal plane of the Nicol. This plane polarised light passes through Laurent's half shade device and then through the tube T containing the sugar solution (active substance). The emergent light, on passing through the analysing Nicol prism N_2, is viewed through a Galilean telescope to which is attached a scale V_1V_2.

Half Shade Device

Laurent's half shade device consists of a semicircular half wave plate of quartz Q cut with optic axis $||^l$ to its faces and at right angles to the vertical diameter. This is cemented together with another semicircular plate of glass G which is of such a thickness that its absorption of light is same as that of the quartz. Thus, a composite circular plate is formed (Fig. 6.22). Suppose a beam of plane polarised light with its vibration plane parallel to CD is incident on the half shade. The light passing through the glass portion is unaffected. The light passing through the quartz half plate Q is split into two vibrations, one vertical and the other horizontal. Vertical component called E-component is parallel to the optic axis YY' and the other O-component is $\perp^r$ to it, i.e. along XX'.

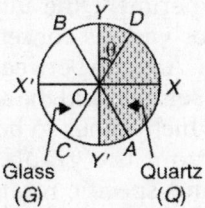

Glass   Quartz
(G)   (Q)

Fig. 6.22: Working of half shade device

Inside quartz, $V_O > V_E$ and hence on emergence the O-component will gain a phase of π over the E-component. Due to this phase difference of π, the direction of O-component is reversed, i.e. if OM is the initial direction, then its final direction will be ON (Fig. 6.23). Thus, if the component of initial incident vibrations along OD in quartz plate are along OL and OM, the emergent wave will be the resultant of the vibrations along OL and ON. Obviously, the resultant of OL and ON is OB

which makes the angle θ with Y-axis as incident vibrations along AD do but on the other side, i.e.

$$\angle DOL = \angle BOL = \theta$$

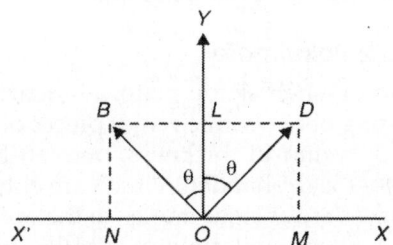

Fig. 6.23: Rotation of plane of polarisation by quartz plate

This shows that the plate *rotates the plane of polarisation by an angle* 2θ.

Thus, we notice that there are two plane polarised beams— one beam emerging from glass half plate with vibrations in the plane OD and other beam from quarter half with vibrations in the plane OB. If the analyser N_2 has its principal plane or section along YY', i.e. along the direction which bisects $\angle AOC$, the amplitudes of light incident on the Nicol N_2 from both the halves will be equal. Obviously, the field of view will be equally bright (Fig. 6.24a).

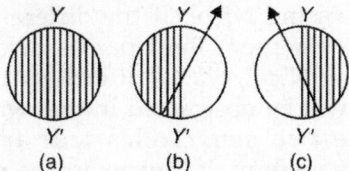

Y'   Y'   Y'
(a)   (b)   (c)

Fig. 6.24: Field of view

If N_2 is rotated to the right of YY', then the right half will be brighter as compared to the left half portion as shown in Fig. 6.24b). If N_2 is rotated to the left YY', then the left half is brighter as compared to the right half Fig. 6.24c). Obviously, the *half shade serves the purpose of dividing the field of view into two halves*. The half shade device is so accurate that if the analysing Nicol prism is rotated through even small angle from the position of equal brightness, a marked change in the intensity of the two halves of the field very accurately if they are simultaneously presented to the eye.

The disadvantage with half shade device is that it can be used only for special wavelength for which it has been designed.

Biquartz Polarimeter

Instead of a half shade plate, a biquartz plate consisting of two semicircular pieces of quartz (each 3.75 mm in thickness, one left-handed and other right-handed, which are cut so that their faces are at right-angles to the optic axis) is used. Such a polarimeter is called biquartz polarimeter. Two plates of biquartz are joined together to form a circular plate (Fig. 6.25).

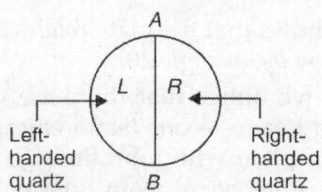

Fig. 6.25: Function of biquartz plate

Action of the Biquartz Plate

Each half portion of the biquartz circular plate produces equal rotation of the light incident normally upon them but in opposite directions. This device is placed just behind the polarising Nicol. If the different angles and the emergent light passes through the analysing Nicol, waves of different refractibilities will be obstructed in the two halves, and the two semicircles will appear of different colours. If, however, the principal planes of polarising and analysing Nicols are parallel, the light of the same colour will be extinguished in both the halves and other colours will be present in same proportion; the two parts will thus appear grey-violet coloured. This colour is named as the *tint of a passage* or *sensitive tint*. If the analyser is now rotated through a very small angle, one of the quartz piece becomes pink and other blue. This shows that a marked change is brought about in the appearance of the biquartz plate on slight rotation of the analyser.

Comparison with Half Shade Device

In comparison to half shade device, the biquartz device is much more sensitive because on slight rotation of the analyser a marked change is brought about in the appearance. For accurate determination of specific rotation, biquartz device is used.

6.16 DETERMINATION OF SPECIFIC ROTATION OF SUGAR

To determine the specific rotation of an optically active substance, say sugar solution, N_2 (analyser) is set in the position for equal brightness of the fields of view, first with water in the tube T. The readings of the verniers V_1 and V_2 are recorded. Now the water in the tube T is replaced by sugar solution. Due to this the vibrations from quartz half and glass half are rotated. Obviously, the field of view is not equally bright. The analyser is rotated in the clockwise direction and is brought to a position so that the entire field of view is equally bright. The new positions of the verniers V_1 and V_2 on the scale are again recorded. The angle (θ) through which the analyser has been rotated gives the angle through which the plane of vibration has been rotated by the sugar solution.

In actual experiment, the angle of rotation is determined for various concentrations of the sugar solution. A graph between concentration C of the sugar solution and the angle of rotation θ is plotted which comes to be a straight line (Fig. 6.26). From the graph, θ/C value is determined and specific rotation of sugar is calculated from the relation

$$S = \frac{100}{lC}$$

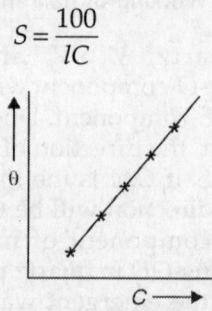

Fig. 6.26: Concentration (C) versus angle of rotation (θ) of sugar

ILLUSTRATIVE EXAMPLES

Example 1

Twenty parallel glass plates of refractive index 1.5 are used in a pile of plates. Calculate the degree of polarisation of the transmitted beam.

Solution

Given that $n = 20$, $\mu = 1.5$

$$P = \frac{n}{n - \left(2\mu/\mu^2 - 1\right)}$$

$$= \frac{20}{\left[20 + \dfrac{2\times 1.5}{(1.5)^2 - 1}\right]} = \frac{20}{20 + \left(\dfrac{3}{1.25}\right)^2}$$

$$= 0.776.$$

Example 2

The principal planes of two Nicol prisms form an angle of 30°. Prove that the intensity of the transmitted light diminished in the ratio of 3 : 2, if the angle between principal planes becomes 45°.

Solution

When the angle between the principal planes is 30°, intensity of the transmitted light is

$$I_1 = I_0 \cos^2 30° = \frac{3}{4} I_0 \ \left(\because \cos 30° = \sqrt{3}/2\right) \text{(1)}$$

When the angle between the principal planes is 45°, the intensity of the transmitted light is

$$I_2 = I_0 \cos^2 45° = \frac{I_0}{2} \tag{2}$$

From Eqs. (1) and (2), we have $\dfrac{I_1}{I_2} = \dfrac{3}{2}$.

Example 3

Two Nicols are first crossed and then one of them is rotated through 60°. Calculate the percentage of incident light transmitted.

Solution

Let us suppose that the intensity of incident unpolarized light beam is $2a^2$. When this beam enters the first Nicol, then it is broken up into O-and E-rays, each of amplitude a, i.e. intensity a^2. From the first Nicol, O-ray is lost by total internal reflection and only E-ray of amplitude a and intensity a^2 is transmitted.

When one of the two Nicols is rotated through 60° from the crossed positions, the angle between the principal planes of two Nicols will be

$$90° \pm 60° = 150° \text{ or } 30°$$

The intensity of finally emergent light from the second Nicol can be calculated from Malus law as

$$I = I_0 \cos^2 \theta = a^2 \cos^2 30° = \frac{3}{4} a^2$$

∴ Percentage of light transmitted from the system is

$$= \frac{a^2 \cos^2 30°}{2a^2} \times 100 = \frac{1}{2}\left(\frac{\sqrt{3}}{2}\right)^2 \times 100$$

$$= \frac{3}{8} \times 100 = 37.5\%$$

Example 4

The refractive index of a crystal is 1.5533 for the extraordinary ray and 1.5442 for the ordinary ray. Calculate the minimum thickness of the crystal plate which will behave like a quarter wave plate for radiation of wavelength 5000 Å.

Solution

We know that for a crystal plate to behave like a quarter wave plate the difference in the optical paths for the ordinary and extraordinary rays through the crystal must differ by an odd multiple of a quarter wave plate. Therefore, if the minimum thickness of this to happen is t, then we have

$$(1.5533 - 1.5442)t = \frac{\lambda}{\Delta} = \frac{5\times 10^{-5}}{4}$$

or $\quad t = \frac{5\times 10^{-5}}{4\times 0.0089}$ cm $= 0.0011$ cm

The next larger thickness for which the crystal plate behaves like a quarter wave plate will be

$$= 3t = 3 \times 0.0011 = 0.0033 \text{ cm.}$$

Example 5

Plane polarised light falls normally on a quarter wave plate. Explain what will be the nature of the emergent light if the plane of polarisation of the incident light makes the following angles with the principal plane of the quarter wave plate: 0°, 30°, 45°, 90°.

Solution

When a plane polarised light falls normally on a quarter wave plate and the optic axis is parallel to the crystal face, then the plane of polarisation of the incident polarised light is assumed to make 45° with the principal plane and the emergent light is used to be circularly polarised. The incident beam can be represented by

$$E_x = E_s = E_0 \sin 45° \sin \omega t = \frac{E_0}{2} \sin \omega t$$

and $\quad E_y = E_f = E_0 \sin 45° \sin \omega t = \frac{E_0}{2} \sin \omega t$

Here s and f represent slow and fast components. After the wave emerges out of the quarter wave plate, the slow component will be ahead in phase by $\pi/2$. Obviously, the emergent beam given by

$$E_x = \frac{E_0}{\sqrt{2}} \sin (\omega t + \pi/2) = \frac{E_0}{\sqrt{2}} \cos \omega t$$

and $\quad E_y = \frac{E_0}{\sqrt{2}} \sin \omega t$

represent a circularly polarised wave. Similarly, one can analyse for $\delta = 0°, 30°, 90°$.

Example 6

For calcite $\mu_O = 1.658$, $\mu_E = 1.486$. For sodium light ($\lambda = 5893$ Å), calculate the thickness of a quarter wave plate of calcite.

Solution

$$t = \frac{\lambda}{4(\mu_O - \mu_E)} = \frac{5893 \times 10^{-8}}{4(1.658 - 1.4860)}$$

$$= 8.58 \times 10^{-3} \text{ cm}$$

Example 7

A beam of linearly polarised light changes to a circularly polarised beam by passing it through a crystalline plate 0.03 mm thick. Assuming this thickness to be minimum, calculate the difference in refractive indices of two rays in the crystal for light of wavelength 6000 Å.

Solution

A linearly polarized light can become circularly polarised when the slice is quarter wave plate, i.e. we have

$$t = \frac{\lambda}{4(\mu_O - \mu_E)}$$

or $\quad \mu_O - \mu_E = \frac{\lambda}{4t}$

$$= \frac{6000 \times 10^{-8}}{4 \times 0.003} = 0.005 .$$

Example 8

The faces of quartz plate are parallel to the optic axis of the crystal. (a) What is the thinnest possible plate that would serve to put the ordinary and extraordinary rays of $\lambda = 5890$ Å a half wave apart on their exit? (b) What amplitudes of this thickness would give the same result? The indices of refraction of quartz are $\mu_E = 1.553$, $\mu_O = 1.544$.

Solution

(a) According to the question, the quartz plate behaves like a half wave plate, and hence its thickness is given by

$$t = \frac{\lambda}{2(\mu_E - \mu_O)}$$

$$= \frac{5890 \times 10^{-8}}{2(1.553 - 1.544)} = 3.27 \times 10^{-3} \text{ cm}$$

(b) The plates which would introduce a path difference of $\lambda/2$ or its odd multiples between the ordinary (O) and extraordinary (E) rays will produce the same effect. Obviously, the thickness of such plates is given by

$= (2n + 1)t \qquad$ where $n = 0, 1, 2, \ldots$

$= t, 3t, 5t, \ldots$

$= 3.27 \times 10^{-3}$ cm, 9.81×10^{-3} cm, $\ldots$

Example 9

The refractive index of calcite for ordinary ray is 1.6584, and the refractive index of Canada balsam is 1.550. A ray of light is incident parallel to the length side. Calculate the angle of incidence of the ordinary ray on the Canada balsam layer and show that it will suffer total internal reflection at that layer.

Solution

When the incident ray is parallel to the length side as shown in Fig. 6.27, then the angle of incidence is 22°.

Now,
$$\mu_O = \frac{\sin i}{\sin r}$$

$$\therefore \quad \sin r = \frac{\sin i}{\mu_O} = \frac{\sin 22}{1.6584} = 0.2259$$

$$\therefore \quad r = 13.05°.$$

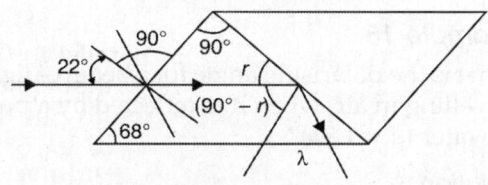

Fig. 6.27

Obviously, the angle of incidence on the Canada balsam layer = 90° − r = 90° − 13.05° = 76.95°.

Now, the critical angle of calcite with Canada balsam for ordinary ray,

$$C = \sin^{-1} \frac{1.550}{1.6584} = 69.17°$$

This shows that ordinary ray will suffer total internal reflection at the Canada balsam layer.

Example 10

Find the specific rotation of a given sample of sugar solution if the plane of polarisation is turned through 26.4°. The length of the tube containing 20% solution is 20 cm.

Solution

Given that θ = 26.4°, l = 2 decimeter
C = 20% = 0.2 gm/cc

$$S = \frac{\theta}{lC}$$

$$\therefore \quad S = \frac{26.4}{2 \times 0.2} = 66°$$

Example 11

A solution of camphor and alcohol in a tube 0.2 m long is found to rotate the plane of polarisation of light by 27°. What is the mass of the camphor in unit volume of the solution. The specific rotation of camphor is 54° decimeter/gm/cc.

Solution

Given that θ = 27°, S = 54°,
$$l = 0.2 \text{ m} = 2 \text{ decimeter}$$

$$S = \frac{\theta}{lC}$$

$$\therefore \quad C = \frac{\theta}{lS} = \frac{27}{54 \times 2} = \frac{1}{4} = 0.25 \text{ gm/cc} = 25\%$$

Example 12

If 20 cm length of a certain solution causes right handed rotation 38°, and 30 cm of another solution causes left handed rotation 24°, what optical rotation will be caused by 30 cm length of mixture of the above solution in the volume ratio 1:2? The solutions are not chemically reactive.

Solution

As given, 30 cm length of mixture solution will contain 10 cm of first and 20 cm of the second solution.

We know that dextro and laevo rotations are expressed by + and − sign respectively, we have:

Optical rotation produced by first solution

$$\theta_1 = +\frac{42°}{20} \times 10 = +21°$$

(i.e. right-handed rotation)

Optical rotation produced by second solution

$$\theta_2 = +\frac{27°}{30} \times 20 = -18°$$

(i.e. left-handed rotation)

∴ Total rotation produced is given by

$$\theta = \theta_1 + \theta_2 = +21° - 18° = +3°$$

Clearly, the mixture will cause right handed rotation of 3°.

Example 13

The optical rotation produced by a particular material is found to be 30° per mm at $\lambda = 5000$ Å and 50° per mm at $\lambda = 4000$ Å. In the visible region of the spectrum, the rotation of the plane of polarisation is given by $\theta = a + b\lambda^{-2}$. Find the values of the constants a and b for the given material. Calculate the possible thickness of the material which when interposed between two crossed Nicols will produce maximum transmission for $\lambda = 5500$ Å.

Solution

As per the problem, the rotation of plane of polarisation is given by

$$\theta = a + \frac{b}{\lambda^2} \qquad (1)$$

$$\lambda = 5000 \times 10^{-8}\ cm = 5 \times 10^{-4}\ mm$$

$$\theta = 30°/mm$$

$$\therefore \quad 30° = a + \frac{b}{\left(5 \times 10^{-4}\right)^2} \qquad (2)$$

When $\lambda = 4000 \times 10^{-8}$ cm = 4×10^{-4} mm, $\theta = 50°/mm$, then

$$50° = a + \frac{b}{\left(4 \times 10^{-4}\right)^2} \qquad (3)$$

From Eqs. (2) and (3), one obtains

$$a = \frac{50}{9} \quad \text{and} \quad b = \frac{8}{9} \times 10^{-5} \qquad (4)$$

Substituting the values of a and b in Eq. (1), one obtains the value of optical rotation for $\lambda = 5500$ Å = 5.5×10^{-4} mm as

$$\theta = \frac{50}{9} + \frac{8000 \times 10^{-8}}{9 \times \left(5.5 \times 10^{-4}\right)^2} = 24°/mm$$

Clearly, 1 mm thick plate produces a rotation of 24°. To obtain maximum transmission between two crossed Nicols, the plane of polarisation should be rotated through 90°. The desired thickness of the plate for the said purpose will be

$$t = \frac{90}{24}\ mm = 3.75\ mm.$$

Example 14

A glass plate is to be used as a polariser. Find the angle of polarisation and the angle of refraction, the refractive index of glass is 1.54.

Solution

Given $\qquad\qquad \mu = 1.54$

We know that $\tan i_p = \mu$

$$i_p = \tan^{-1} \mu$$
$$= \tan^{-1}(1.54) = 57°$$

If r be the angle of refraction, then

$$i_p + r = 90°$$

and $\qquad\qquad r = 90° - 57° = 33°$

Example 15

What is the polarising angle for a beam of light travelling in air, when it is reflected by a pool of water ($\mu = 1.33$)?

Solution

Using Brewster's law

$$\tan i_p = \mu_2$$
$$\tan i_p = 1.33$$
$$i_p = 53.1°$$

Example 16

Two polarising sheets have their directions so that the intensity of transmitted light is maximum. At what angle should either sheet be turned so that the intensity becomes one half of the initial.

Solution

Given $\qquad\qquad I = \dfrac{I_0}{2}$

From Malus' law, $I = I_0 \cos^2\theta$

$$\therefore \qquad\qquad \frac{I_0}{2} = I_0 \cos^2\theta$$

$$\therefore \qquad\qquad \cos\theta = \pm\frac{1}{\sqrt{2}} \quad \Rightarrow \quad \theta = 45° \text{ or } 135°$$

REVIEW QUESTIONS

1. What is double refraction? Describe how Huygens explained it. Describe the construction and action of a Nicol prism.

2. Describe the construction and working of a Nicol prism. Discuss how do you obtain a plane polarised beam with it.

3. Describe a Nicol prism. Show clearly how it is constructed. Explain its action. How can you use it as a polariser and analyser?

4. Give Huygens' construction of the wavefronts for the ordinary and extraordinary waves when a beam of light is reflected through a doubly refracted crystal. Optic axis as in the plane of incidence and is $\perp^r$ to the crystal surface.

5. Thin plates are cut from a negative doubly refracting crystal with various positions of the optic axis as described below. Explain your observations giving diagrams.

   i. Optic axis lying in plane of incidence but parallel to upper face.

   ii. Optic axis lying in plane of incidence but $\perp^r$ to upper face.

   iii. Optic axis lying in plane of incidence but inclined to upper face.

   iv. Optic axis $\perp^r$ to plane of incidence but parallel to upper face.

6. Explain Huygens' theory of double refraction in uniaxial crystals. Distinguish between negative and positive crystals.

7. Give the wave theory of the origin of the double refraction of light at a calcite crystal for normal and oblique incidence.

8. Define polarisation. Are sound waves polarised? What are elliptically and circularly polarised light? Plane polarised light is normally incident on a $\lambda/4$ plate. Sate conditions under which circularly or elliptically polarised light can be obtained.

9. Describe how (i) plane polarised (ii) elliptically polarised (iii) circularly polarised light can be produced and detected.

10. How can you produce elliptically and circularly polarised light? Explain fully how will you distinguish between (a) unpolarised light and circularly polarised light (b) partially polarised light and unpolarized light. State the polarisation in the emergent beam. What changes would occur if plate is $\lambda/8$ plate?

11. Light passes through a Nicol polarizer with its shorter diagonal vertical. The analyser is set for minimum intensity. How will you set a half wave plate between the two Nicols so that maximum intensity of light is seen through the analyser?

12. Unpolarised light falls on two polarising sheets so oriented that no light is transmitted. If third polarising sheet is placed between them, can light be transmitted? Explain how do you find (i) the polarising direction (ii) optic axis of a $\lambda/4$ plate.

13. What is quarter wave plate? Deduce its thickness for a given wavelength in terms of its refractive index.

14. How would you produce the following with the help of a Nicol prism and quarter wave plate: (i) plane polarised light (ii) circularly polarised light and (iii) elliptically polarised light.

15. Give two methods of producing elliptically polarised light. How would you distinguish it from partially polarised light?

16. Elliptically polarised light falls normally on a quarter wave plate. Explain the nature of emergent light if the major axis of the ellipse makes the following angles with the principal plane of the quarter wave plate: 0°, 30°, 90°.

17. Describe the construction and working of a phase retardation plate. What are quarter and half wave plates?

18. Define specific rotation. Describe the construction and working of Laurent's half shade polarimeter?

19. Define specific rotation. How will you determine it with the helf of a half shade polarimeter.

20. Give with full details the construction and working of Laurent's half shade polarimeter.

21. What is meant by rotatory polarisation? How can a right circularly polarized light be transformed to left circularly polarised light?

22. Discuss the phenomenon of rotation of the plane of polarisation of light by optically active materials. Give the necessary theory. Show that the rotation of plane of vibration

$$= \frac{\pi d}{\lambda}(n_A - n_C)$$

where n_A and n_C are the refractive indices of the crystal in the direction of the optic axis for anticlockwise circular polarised light and d is the thickness of the crystal lattice.

23. (a) What do you understand by a quarter wave plate and half wave plate?

 (b) Explain the functions of various parts of Laurent's half shade polarimeter.

24. Distinguish between circularly and elliptically polarized light. How would you produce and detect circularly polarized light?

25. Define specific rotation. Describe the construction and principle of a half shade polarimeter.

26. (a) Describe the construction and working of Nicol prism. How can you use it as a polariser and analyser?

 (b) What is quarter wave plate?

27. Describe how you will produce elliptically polarised light and distinguish it from partially polarised light.

28. A transparent plate is given. Using two Nicol prisms (or polaroids) how would you find whether the plate is a quarter wave plate, a half wave plate or simple glass plate? [Raj. BE 2000]

29. Write short notes on:

 (a) Specific rotation and its measurement.
    [Raj. BE 2000]

 (b) Laurent's half shade polarimeter.
    [Raj. BE 2001]

30. (a) What do you mean by optical rotation? Discuss how shall you measure optical rotation of sugar solution using a biquartz device.

 (b) Discuss the state of polarization of the following light waves:

 (i) $E = \hat{j} A \cos(kx - \omega t)$
    $+ \hat{k} 2A \cos(kx - \omega t + \pi/4)$

 (ii) $E = \hat{j} A \cos(kx - \omega t) + \hat{k} A \sin(kx - \omega t)$
    [Raj. BE 2003]

31. (a) How will you distinguish plane polarised, circularly polarised and elliptically polarised light?

 (b) A phase retardation plate of quartz has thickness 0.1436 mm. For what wavelengths in the visible region will it act as quarter wave plate?

 (c) What is bi-quartz device? [Raj. BE 2002]

32. (a) Describe the construction of a Nicol prism and explain its action in converting unpolarised light into plane polarised light.

 (b) Plane polarised light passes through a quartz plate with its optic axis parallel to the faces. Calculate the least thickness of the plate for which the emergent beam will be:

 i. Plane polarised
 ii. Circularly polarised

 Given: $\mu_E = 1.5533$, $\mu_O = 1.5442$ and $\lambda = 5000$ Å
    [Raj. BE 2002]

33. (a) Explain the working and construction of Nicol prism. Mention its limitation.

 (b) Show that linearly polarised light can be represented as a superposition of two circularly polarised lights of suitable amplitude and phases.

 (c) What is optical activity? Mention the laws of optical rotation. [Raj. BE 2001]

34. (a) How shall you differentiate the following:

 i. Three sources of light having the same physical appearance: partially polarized, elliptically polarized and mixture of unpolarized and circularly polarized light

 ii. Half wave plate and quarter wave light

 (b) How shall you measure specific rotation of glucose solution using a biquartz device?
    [Raj. BE 2001]

PROBLEMS

1. The refractive index of quartz is 1.5443 for the ordinary ray and 1.5334 for the extraordinary ray. Calculate the thickness of a quartz plate which will act as a quarter wave plate for light of wavelength 6000 Å. [*Ans.* 0.0014 Å]

2. The refractive index of calcite is 1.648 for the ordinary ray and 1.486 for the extraordinary ray. A plane parallel plate of calcite is cut of thickness 0.01 mm. For what wavelengths in the visible range will this plate behave as (i) a quarter wave plate (ii) a half wave plate?
    [*Ans.* (i) 7200, 5899, 4985 and 4320 Å,
    (ii) 6480 and 4629 Å]

3. How much a polariser and an analyser be oriented so that a beam of natural light is reduced to (i) 0.5 (ii) 0.25 (iii) 0.75 (iv) 0.125 of its initial value? [*Ans.* 45°, 60°, 30°, 69°]

4. The critical angle of glass is 40° and that of a liquid is 50°. Find the angle of polarisation when light is incident on glass from the liquid.
    [*Ans.* 50°]

5. Calculate the thickness of quarter wave plate from the following data $\mu_O = 1.6247$, $\mu_E = 1.5672$ and $\lambda = 5893$ Å. [*Ans.* 2.56 micron]

6. What is the maximum thickness that a crystal plate should have if it is to act as a quarter wave plate for radiation of wavelength 6000 Å. The refractive index of crystal is 1.553 for the ordinary ray and 1.554 for the extraordinary ray. [*Ans.* 0.0014 cm]

7. Two nicols are oriented with their principal planes making an angle of 30°. What percentage of incident unpolarized light will pass through the system? [*Ans.* 37.5%]

8. Calculate the thickness of a quarter wave plate for wavelength 5890 Å. The refractive indices of quartz (for $\lambda = 5890$ Å) for E-ray and O-ray are equal to 1.5533 and 1.5443 respectively. [*Ans.* 1.636×10^{-3} cm]

9. Find the thickness of a quarter wave plate when the wavelength of light is equal to 5880 Å and $\mu_O = 1.5818$, $\mu_E = 1.5508$. [*Ans.* 1.474×10^{-3} cm]

10. The thickness of a quarter wave plate for a certain wavelength of light is 2 micron. What will be the nature of light obtained when a plane polarised beam of light passes through it? What would happen if the thickness is 3 micron?

    [*Ans.* (i) $x^2 + y^2 = a^2$, circularly polarised (ii) if $x^2 + y^2 + \sqrt{2}xy = a^2/2$, elliptically polarised]

11. A cellophane sheet behaves as a half wave plate for light of wavelength 4000 Å. How would this sheet behave for light of wavelength of 8000 Å? It may be assumed that refractive indices do not change with wavelength.

    [*Ans.* $\lambda/4$ plate]

12. A plane polarised beam of light passes through a polarizer with its allowed direction inclined at 45° with the plane of vibration of the incident beam. A second polarizer is kept after this with its allowed direction at 45° with that of the first. Compare the intensity emergent from the second with that of incident on first. What will happen if the first polarizer is removed?

    [*Ans.* 0.25, zero]

13. The specific rotation of sucrose is 66.4. Calculate the rotation of a column of 20 cm of solution of concentration 12.5 gm of sucrose in 100 cc of water. [*Ans.* 16°36′]

14. On inserting a polarimeter tube 25 cm long containing a sugar solution of unknown strength, it is found that the plate of polarization its rotated through 10°. Find the strength of sugar rotation in per cc, given specific rotation of sugar rotation to be 60° per decimeter for unit concentration. [*Ans.* 0.066 gm/cc]

15. A solution of 0.3 gm of camphor in 10 cc of alcohol in a 20 cm long tube produces a rotation of 30°. Find the specific rotation of camphor.

    [*Ans.* 55]

16. A solution of an optically active solute produces a rotation of 20° of the plane of polarization in a path length of 10 cm when the concentration is 20 gm/litre. What is the concentration in a solution which produces a rotation of 30° in a path length of 5 cm? [*Ans.* 60 gm/litre]

17. A tube of sugar solution 20 cm long is placed between crossed Nicols and illuminated with light of wavelength 6000 Å. If the optical rotation produced is 13° and the specific rotation is 65°, determine the strength of the solution. [*Ans.* 0.1 gm/cc]

SHORT ANSWER QUESTIONS

1. Why light waves can be polarised while sound waves cannot be polarised?

Ans. Sound waves are longitudinal in nature whereas light waves are transverse in nature. Only transverse waves can be polarised.

2. How polarisation of light establishes that light waves are transverse in nature?

Ans. We know that longitudinal waves can have vibrations only along the direction of propagation of wave, i.e. one state of polarisation. Experiments reveal that light can have several different states of polarisation, which means that the vibrations are in plane perpendicular to the directions of the wave.

3. How does the polarisation of light afford a convincing evidence of transverse nature of light?

Ans. If light waves were not transverse in nature, then the vibrations would have passed through two tourmaline crystals with their optic axes perpendicular to each other, i.e. in crossed position.

4. What is the angle between the direction of propagation and plane of polarisation in the propagation of electromagnetic waves?

Ans. 0°

5. A calcite crystal is placed over a dot on a piece of paper and rotated. What will you observe through the calcite crystal?

Ans. One dot rotating about the other.

OBJECTIVE QUESTIONS

1. Nicol prism is an optical device made from the bifringent calcite crystal and is used both as a _____ and as an _____.

[polariser, analyser]

2. Along the optic axis, the velocities of the ordinary and extraordinary ray are the _____.

[same]

3. Polarising angle is the angle of incidence for which the reflected light is completely _____ with vibrations perpendicular to the plane of incidence.

[plane polarised]

4. If the natural light falls on a uniaxial crystal, the intensities of the ordinary and extraordinary wave as they enter the crystal are _____.

[same]

5. According to cosine law of Malus, the intensity of polarised light emerging from an analyser is proportional to the square of the _____ of the angle between the optic axis of the polariser and the _____.

[cosine, analyser]

MULTIPLE CHOICE QUESTIONS

1. Polarisation cannot occur in
   (a) light waves
   (b) radio waves
   (c) sound waves
   (d) X-rays

2. Polarisation of light proves the
   (a) longitudinal nature of light
   (b) quantum nature of light
   (c) corpuscular nature of light
   (d) transverse nature of light

3. The correct relation between the Brewster's angle i_p and the refractive index μ is
   (a) $\cos i_p = \mu$
   (b) $\sin i_p = \mu$
   (c) $\tan i_p = \mu$
   (d) $\cot i_p = \mu$

4. A beam of unpolarised light passes through a tourmaline crystal A, then it passes through a second tourmaline crystal B oriented so that its principal plane is parallel to that of A. The intensity of the emergent light is I_0. Now B is rotated through 45° about the ray. The emergent ray will have intensity
   (a) $2 I_0$
   (b) $I_0 \sqrt{2}$
   (c) $I_0/2$
   (d) $I_0/ \sqrt{2}$

5. Which of the following methods produce polarized light?
   (a) double refraction
   (b) refraction
   (c) selective absorption
   (d) all of the above

6. A calcite crystal is placed over a dot on a piece of paper and then rotated on viewing through calcite, we observe
   (a) a single dot
   (b) one dot rotating about the other
   (c) two rotating dots
   (d) two stationary dots

7. The phenomenon of rotation of plane polarised light is called
   (a) dichroism
   (b) refraction
   (c) double refraction
   (d) optical activity

Answers

1. (c)	2. (d)	3. (c)	4. (c)	5. (d)
6. (b)	7. (d)			

7

Laser and Holography

7.1 INTRODUCTION

We live in an age where laser light shows are common and lasers light up the sky. Within the past few years, lasers have come into widespread use in science, medicine, manufacturing, and in many other fields. The term laser stands for *Light Amplification by Stimulated Emission of Radiation*. Laser produces a highly intense, concentrated and parallel beam of monochromatic and coherent light. In the words of CH Townes, inventor of Maser and Laser, *what the laser does essentially is to give us electronic type control over light. It is a marriage of optics and electronics.*

Historically, the laser is the outgrowth of MASER, a device that uses microwaves instead of light waves. The word maser is an acronym for *Microwave (or Molecular) Amplification by Stimulated Emission of Radiation* and is also known as *optical maser*. The principle on which lasers operate undoubtedly place them in the family of masers, but in structure, properties and applications, lasers are quite different from masers. The major difference between the two devices is that of working frequency; the typical operating frequencies of masers are nearly 10^{10} Hz, while the frequencies in which lasers operate are of the order of 10^{15} Hz.

The basic idea behind laser and maser is the process of *stimulated emission* that was first postulated by Einstein in his derivation of Planck's radiation law in 1917. If his considerations has been applied to non-equilibrium situations, the operating principle of the laser and maser could have been proposed at that time. In fact, it took about 35 years for others to come up with the basic idea and make this idea work.

Before 1960, laser was just a twinkle in the eyes of scientists. Today it is brighter than the sun reality. The laser (it rhymes with razor) promises to be in measurement techniques, communications, military weaponry, and the conquest of space. It also figures in industrial processes, computers, medicine, chemistry and other fascinating possibilities. In brief, the applications of laser light range from straight forward (such as surveying and welding) to the sophisticated ones (such as optical communications and holography) and it may be that more interesting applications are still to come.

7.2 IMPORTANT PROPERTIES OF LASER LIGHT

The salient features of a laser light are *directionality, intensity, monochromaticity*, and *coherence*. These features are discussed below.

Laser Light is Highly Directional

A laser beam is highly directional, i.e. it spreads very little. A laser beam departs from strict parallelism only because of diffraction at the exit aperture of the laser. For example, a laser pulse used to measure the distance to the moon generates a spot on the moon's surface with a diameter of only a few metres. Light from an ordinary bulb can be made into an approximately parallel beam by using a lens, but the beam divergence is much greater than for laser light and a parallel beam of light from a torch on the earth shall spread in an area of few square kilometres. Each point on a light bulb's filament forms its own separate beam, and the angular divergence of the overall composite beam is set by the size of the filament. The directionality of a beam can be measured by the *full angle beam divergence* which is two times the angle subtended by the outer edge of the beam with the beam axis. The outer edge of the beam is considered to be the line where the beam intensity decreases to $1/\exp(I)$ times its value at the axis. Let us consider that an aperture of diameter d radiates a beam having a plane wavefront. Let the beam travels as a parallel beam through the distance of Rayleigh range, which is of the order of d^2/λ, where λ is the radiated wavelength. Obviously, the beam begins to spread with distance owing to diffraction as shown in Fig. 7.1. In general, the aperture diameter (d) control the angular spread $\Delta\theta$ radians of the far-field beam. The relationship between $\Delta\theta$ and d is given by

$$\Delta\theta = \frac{\lambda}{d} \tag{7.1}$$

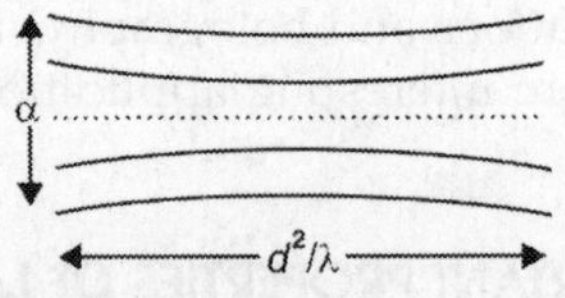

Fig. 7.1: Spreading of beam beyond Rayleigh range

Hence, shorter the wavelength and larger the aperture diameter, smaller the angular divergence of the beam. For a typical laser beam, the beam divergence is less than 10^{-5} radian, exhibiting that the beam spreading is less than 10^{-5} m for every meter.

Laser Light is Highly Intense, i.e. It can be Sharply Focussed

If two light beams transport the same amount of energy, the beam that can be focussed to the smaller spot will have the greater intensity at that spot. Lasers emit narrow beam of light, so that the energy is concentrated in a small region. The spatial and spectral confinement of energy in laser beam accounts for its high intensity. For laser light, the focussed spot can be so small that an intensity of 10^{17} W/cm^2 is readily obtained. An oxyacetylene flame, by contrast, has an intensity of only about 10^3 W/cm^2. Laser beams can generate power densities millions of times greater than those on the surface of the sun. It is found that when a laser beam is focussed by a lens, the tremendous intensity at the focal point can produce a radiation pressure of the order of 10^6 kg per sq. cm.

Laser Light is Highly Monochromatic

Light from an ordinary incandescent light bulb is spread over a continuous range of wavelength and is certainly not monochromatic. The radiation from a fluorescent neon sign is monochromatic, to about 1 part in about 10^6. However, the sharpness of definition of laser light can be many times greater, as much as 1 part in 10^{15}.

Laser Light is Highly Coherent

Individual long wave (wave trains) for laser light can be reversed hundred kilometres long. When two separated beams that have travelled such distances over separate paths their common origin and are able to form a pattern of interference fringes. The corresponding *coherence length* for wave trains emitted by light bulb is typically less than a metre. The wave from a laser is an orderly wave of one frequency when the whole beam is *spatially* in phase. There are two independent types of coherence namely (a) *temporal coherence* (b) *spatial coherence.*

Temporal Coherence

This refers to the correlation between the radiation field at a point and the radiation field at the same point at a later time, i.e. the relation between $E(x, y, z, t_1)$ and $E(x, y, z, t_2)$, where these represent the radiation field at the point (x, y, z) at times t_1 and t_2 respectively. If the phase difference between the two radiation fields is constant during the period normally covered by observations (~few microseconds), the wave is said to have *temporal coherence*. If the phase difference between two radiation fields changes many times and in an irregular way during the shortest period of observation (μs), the wave is said to be incoherent.

Time coherence comes about when each cycle of the wave takes exactly the same time to pass a point in space. This really means that the frequency of the wave is not varying and is given the name monochromatic.

Spatial Coherence

Two fields at different points on a wavefront of given electromagnetic wave are said to be space coherent if they preserve a constant phase difference over any time *t*, i.e. space coherence requires that the waves not only are of same frequency, but that they are in phase in space.

Spatial coherence is possible even when the two beams are individually time incoherent, as long as any phase change in one of the beam is accompanied by a simultaneous equal phase change in the other beam. With the ordinary light sources, this is only possible if the two beams have been derived from the same part of the source.

Coherence Time and Coherence Length

Let us consider the radiation field *E* from a light source which is an ideal sinusoidal function of time *t*, then at any given position one can write

$$E = a\cos(\omega_0 t + \phi) \qquad (7.2)$$

Here *a* is the magnitude of the field, ω_0 the angular frequency and ϕ the phase. Figure 7.2 depicts the field.

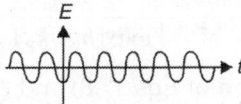

Fig. 7.2: Radiation field

However, no emitted light produces a perfect sinusoidal variation of field *E* with time *t*, and

for any actual light source the magnitude *a* and phase ϕ will vary with time. One can represents a more realistic picture of the radiation from an emitting source of light as shown in Fig. 7.3. In Fig. 7.3, τ represents the average time vibration for which an ideal sinusoidal emission occurs. τ is called *coherence time* or *interval of coherence*. Let *T* denotes the period of the oscillation, where $T = \dfrac{\omega_0}{2\pi}$. The spatial dimension *L*, for which the light may be considered to be a perfect sinusoidal is given by

$$L \approx cT \approx \frac{\tau}{T}\lambda \qquad (7.3)$$

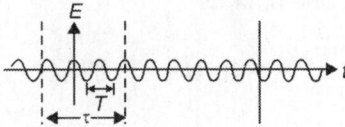

Fig. 7.3: Coherence time

Here *c* is the velocity of light and *L* is called the *coherence length*. After time τ, there is no correlation between the phases of the waves.

It terms of purity *Q* of the spectral line ($Q \approx \dfrac{\lambda}{\Delta\lambda}$, where $\Delta\lambda$ is half width of the spectral line), we have

$$c\tau \approx L \approx \lambda Q \qquad (7.4)$$

7.3 EINSTEIN'S THEORY OF ATOMIC TRANSITIONS AND EINSTEIN'S *A* AND *B* COEFFICIENTS

Transition between the atomic energy states is a statistical process and as such it is not possible to predict which particular atom will make a transition from one state to another at a particular instant. However, in an assembly of a very large number of atoms, it is possible to calculate the irradiative transition between two states, based on the laws of probability. Einstein in 1917, first calculated the probability of such transitions assuming the atomic system to be in equilibrium with electromagnetic radiation.

Let us consider an assembly of atoms at an absolute temperature T in which the atoms may be in different energy states. Let n_0 be the number of atoms per unit volume in the ground state ($E = 0$), then the number of atoms n per unit volume in an excited state of energy E is given by the Boltzmann distribution law

$$n = n_0 \exp[-E/k_B T] \qquad (7.5)$$

where k_B ($= 1.38 \times 10^{-23}$ J/K) is the Boltzmann constant. The maximum number of atoms in an assembly of large number of atoms at an absolute temperature T is in ground state. Let n_1 be the number of atoms per unit volume in a lower energy state E_1 and n_2 be the same in a higher energy state E_2 (Fig. 7.4), then from Eq. (7.5), one obtains

$$\frac{n_2}{n_1} = \exp[-(E_2 - E_1)/k_B T] \qquad (7.6)$$

If $E_2 < E_1$, then $n_2 < n_1$.

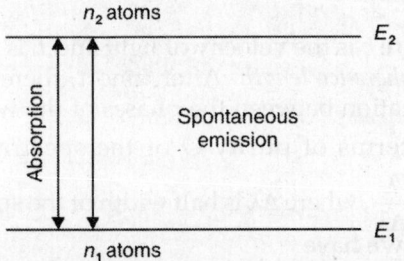

Fig. 7.4: Atomic transition

The atoms in the upper energy state E_2 makes *spontaneous transitions* to the lower energy state E_1 at a certain rate, determined by n_2 and the probability of spontaneous transition per unit time A_{21} from E_2 to E_1. Hence, the number of spontaneous transition per unit volume per unit time is $A_{21} n_2$. These transitions lead to the emission of electromagnetic radiation of energy given by

$$h\nu = E_2 - E_1 \qquad (7.7)$$

where h is Planck's constant and ν is frequency of the electromagnetic radiation. We must note that $1/A_{21}$ is a measure of the lifetime of the upper state against spontaneous decay to the lower state.

Now assume that the atoms are in equilibrium with electromagnetic radiation of frequency $\nu = (E_2 - E_1)/h$ having energy density u_ν. Due to absorption of energy from this radiation, some of the atoms in the lower energy state E_1 make upward transitions to the higher energy state E_2. The rate of such transitions is determined by the number n_1 and the energy density u_ν of the radiation and is given by $B_{12} n_1 u_\nu$ per unit volume per second. Obviously, B_{12} is the probability of the *radiation induced transitions* per unit from the lower state E_1 to the upper state E_2.

In addition to the above two types of transitions, there may also be induced transitions of atoms from E_2 to E_1 caused by interactions with the electromagnetic radiation. We call it as a *stimulated* or *induced emission* of radiation of frequency $\nu = (E_2 - E_1)/h$. The number of such induced or stimulated transitions per unit volume per unit time is $B_{21} n_2 u_\nu$, where B_{21} is the probability from E_2 to E_1 per unit time.

The coefficients A_{12}, B_{12} and B_{21} are known as Einstein's A and B coefficients. Under equilibrium, the time rate of transitions from E_2 to E_1 must be equal to that from E_1 to E_2. So one can write

$$A_{21} n_2 + B_{21} n_2 u_\nu = B_{12} n_1 u_\nu$$

or $\quad u_\nu = \dfrac{A_{21} n_2}{B_{12} n_1 - B_{21} n_2} = \dfrac{A_{21}}{B_{21}} \dfrac{1}{\left(\dfrac{B_{12}}{B_{21}} \right) \dfrac{n_1}{n_2} - 1}$

$$= \frac{A_{21}}{B_{21}} \frac{1}{\left(\dfrac{B_{12}}{B_{21}} \right) \exp\left(\dfrac{h\nu}{k_B T} \right) - 1} \qquad (7.8)$$

Equation (7.8) must agree with Planck's radiation formula

$$u_\nu = \frac{8\pi h \nu^3}{c^3} \frac{1}{\left[\exp(h\nu / k_B T) - 1 \right]} \qquad (7.9)$$

Comparison of Eqs. (7.8) and (7.9) yields

$$B_{12} = B_{21} \qquad (7.10)$$

and $\quad \dfrac{A_{21}}{B_{21}} = \dfrac{8\pi h \nu^3}{c^3} \qquad (7.11)$

Equations (7.10) and (7.11) are called Einstein's relations obtained by Einstein in 1917. The relation (7.10) shows that the induced emission and absorption probabilities per unit time are equal. The calculation does not allow us to obtain the values of A_{21}, B_{21} and B_{12}. They must be derived using quantum mechanical considerations. Using these relations, one obtains

$$\frac{A_{21}}{B_{21}u_v} = \exp\left(\frac{h v}{k_B T}\right) - 1 \qquad (7.12)$$

which is the ratio of the number of spontaneous to stimulated transitions. Therefore, if $h v \gg k_B T$, spontaneous emission is much more probable than induced emission, which can be completely neglected. This holds true in the case of electronic transitions in atoms and molecules and in the case of radiative transitions in nuclei. But if $h v \ll k_B T$, as it is in the microwave region of the spectrum, induced or stimulated emission may become important.

Induced emission is the result of the action of the incoming radiation on the atoms (or molecules) of the substance. Hence, the forced atomic oscillations bear a constant difference relative to the incoming radiation. This means that all atoms radiate in phase and, therefore, *induced emission is coherent*. On the other hand, spontaneous transitions occur at random with no correlation between the times at which atoms undergo transitions. Therefore, the phases of atomic radiations are distributed randomly. Obviously, spontaneous emission is incoherent.

7.4 PRINCIPLE OF LASER

The principle of laser exploits the phenomenon of stimulated transitions discussed in Section 7.3. We know that light is emitted from a source when the atoms in the source make transitions from an excited state (E_2) to a lower energy state (E_1) spontaneously. Normally, the mean life of excited atoms before spontaneous emission occurs is about 10^{-8} s. However, for some excited states, this mean life is perhaps as much as 10^5 times longer. We call such long-lived

states as *metastable*. They play an important role in laser operation. The energy of the emitted photon in the *spontaneous transitions* is

$$h v_{12} = E_2 - E_1 \qquad (7.13)$$

On other hand, if the atom is placed in an electromagnetic field that is alternating at frequency v, the atom can absorb an amount of energy $h v_{12}$ (photon) from that field and move to the higher energy state E_2. This process is called *absorption*.

Besides absorption and spontaneous emission, a third type of transition from an upper energy state E_1 may be induced by an incident photon of energy $h v_{12} = E_2 - E_1$ giving rise to the *stimulated emission* of radiation. This process is called stimulated because the event is triggered by the external photon. The emitting photon is in every way identical to the stimulating photon. Thus, the waves associated with photons have the same energy, polarization and direction of travel. All the above three types of transition are shown in Fig. 7.5.

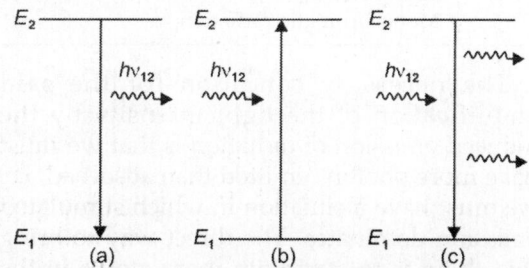

Fig. 7.5: (a) Spontaneous transition from an excited state E_2 to a lower state E_1 with an emission of photon $h v_{12}$ (b) Absorption (c) Induced or stimulated transition from an excited level E_2 to a lower state E_1. A light wave with photon energy $h v_{12}$ causes the atom to emit a photon of the same energy, increasing the energy of light wave

From Fig. 7.5c, it is obvious that we get two photons in this case. If these two photons are now incident on two other atoms in the excited state E_2, then it will result in the induced emission of two more photons so that there will be four coherent photons of the same energy result. The process goes on in a chain, yielding a large number of coherent photons of energy $h\nu_{12}$. This will increase the intensity of the coherent radiation enormously and amplification due to induced emission is achieved (Fig. 7.6).

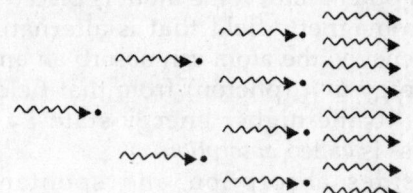

Fig. 7.6: Amplification due to induced emission

7.5 POPULATION INVERSION BY PUMPING

The physical mechanisms of populating and depopulating energy levels are many and diverse. Accordingly there exist various means of rendering the medium active, collectively known as the processes of pumping. Few of the population inversion processes are as follows:

Optical Pumping

In this process, the predominant population of the upper level is achieved by means of light energy delivered from approximately selected sources such as gaseous discharged flash tubes, or continuously burning tubes.

Table 7.1: Difference between stimulated and spontaneous emission

S.No.	Stimulated emission	Spontaneous emission
1.	Emission of a light photon takes place through an inducement, i.e. by an external photon.	Emission of a light photon takes place immediately without any inducement.
2.	It is not a random process.	It is a random process.
3.	The photons get multiplied through chain reaction.	The photons do not get multiplied through chain reaction.
4.	It is a controllable process.	It is an uncontrollable process.
5.	More intense.	Less intense.
6.	Monochromatic radiation.	Polychromatic radiation.

The necessary condition for the said amplification of the light intensity by the induced emission of radiation is that we must have more photons emitted than absorbed, i.e. we must have a situation in which stimulated emission dominates. The direct way to bring this about is to start with more atoms in the excited state than in the ground state. Normally the number of atoms in the excited upper energy state (E_2) is much lower than in the lower energy state (E_1), i.e. $n_2 << n_1$ due to thermal equilibrium. As a result the rate of downward induced transition $B_{12}n_2u_\nu$ is much less than that of the upward transitions $B_{12}n_2u_\nu$ between the same two energy levels due to absorption of photons. To increase the rate of induced emission from E_2 to E_1, it is necessary to achieve $n_2 >> n_1$. This is known as *population inversion*. The states with an inverse population are sometime referred to as states with a negative temperature. However, since such a population inversion is not consistent with thermal equilibrium, we must think clever ways to set up and maintain one.

Electrical Pumping

This is accomplished by means of a sufficiently intense electrical discharge in the medium and is particularly suited to gas media. The discharge converts the gas into a plasma where active centres collide inelastically with free electrons and cause the predominant populations of the upper pumping level. Inelastic collisions of active centres with other atoms and molecules purposefully introduced into the gas are also of importance for pumping as they provide resonance energy exchange.

Chemical Pumping

This raises active centres into the higher level by means of suitable exothermal chemical reactions in the active material.

Heat Pumping

In this scheme of pumping, the active material at first brought to a high temperature and then rapidly cooled down.

7.6 OPTICAL RESONATORS

The selections of some photon states and the suppression of other states can be realized by means of an *optical resonator*, a principal component of each laser. In its simplest version, the optical resonant cavity is a pair of mirrors set on an optic axis which defines the direction of the laser beam. The active material is placed in between these mirrors. Solid active materials are often in the shape of a cylinder whose axis is aligned with the axis of the optic resonator, the length of the cylinder being about ten times its diameters. At least one of the mirrors of the resonator cavity is made semitransparent to serve as an output element passing the light out of the resonator.

Figure 7.7 shows the schematic arrangement of a simple resonator. The photons produced spontaneously in the OO' direction sufficiently close to it will travel within the active material a relatively long way, which is elongated by multiple reflections from the resonator's mirrors. These photons interact with excited active centres to eventually initiate a powerful avalanche of stimulated photons constituting the laser beam. The photons emitted in other directions (and their stimulated avalanches) will traverse a relatively short path length in the materials and die out soon.

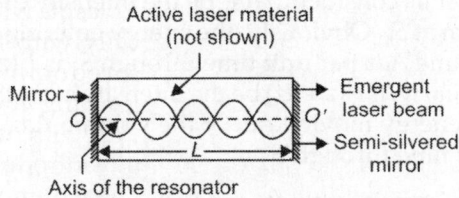

Fig. 7.7: The optical resonator determines the direction of laser action (along the resonator axis) and favours the process of stimulated emission in this direction

Thus, the optical resonant cavity provides the desired selectivity of photon states by primarily confining the possible direction of photon propagation. As a result, lasing action occurs in this direction.

Let L be the length of the resonator cavity. The change in one round trip between the mirrors is given by

$$\delta = \frac{2\pi}{\lambda} \cdot 2L + 2\gamma$$

where γ is the phase change at reflection of each mirror and λ is the wavelength of light. For resonance, i.e. reinforcement, δ must be an integral multiple of 2π. So, if m is an integer, one obtains

$$\delta = \frac{4\pi L}{\lambda} + 2\gamma + 2\pi m$$

or
$$\nu = \frac{c}{\lambda} = \frac{mc}{2L} - \frac{\gamma c}{2\pi L} \qquad (7.14)$$

where c is the speed of light. Obviously, the light having frequencies that satisfy the condition (7.14) will resonate and set up stationary waves in the cavity as shown in Fig. 7.7. Each mode corresponds to one integral value of m. Since one of the mirrors of the resonator is partially transmitting, i.e. half silvered, some of the radiations leaks out, thus providing the collimated laser output.

The optical resonator can provide selectivity in other properties of radiation. Of course, the selectivity in photon energy is secured by the choice of active centres with a system of levels is much more complicated than the laser scheme shown in Fig. 7.7. Actual active centres can have a few lasing transitions rather than one. To exclude undesired transitions, the resonant cavity may be provided with mirrors whose reflecting power is a function of frequency, so that the undesired transitions will be damped out, and the necessary selectivity will be ensured for photon states.

The resonant cavity plays, therefore, a key role as if guides the processes of stimulated emission induced by spontaneous photons in the active medium so that a laser radiation of high coherence properties results.

The Quality Factor (Q)

The most important characteristic of the optical resonator is its quality factor, Q. In the standing wave pattern, the reflectors give two nodes. The wavelength of light λ being much smaller, the optical resonator cavity length L, the number of half waves between the two mirrors is very large. One obtains from Eq. (7.14) the difference in frequencies between two consecutive modes as

$$\Delta v = v_{m+1} - v_m = \frac{c}{2L}$$

Obviously, there is a spread in the frequency of the laser beam, which is controlled by the spectral character of the emission line contributing to the gain. The sharpness of the cavity resonance is determined by Q, a dimensionless quantity. Q is used to show the ability of the cavity to store energy. The slower the energy of a radiation field falls off, i.e. smaller the losses in the resonator, consequently, higher the resonator's Q. The quality factor of a cavity is defined as

$$Q = 2\pi \frac{\text{Maximum energy stored per cycle in the mode}}{\text{Energy dissipated per cycle in the mode}}$$

In a passive resonant cavity, the light beam irradiance gradually declines with travel distance owing to various losses in the medium, namely, absorption, scattering, radiation through side surfaces, etc. This decreases Q and reduce the sharpness of resonance. If dv is the line width of the cavity mode, then one can easily show that

$$Q = v/dv$$

The smaller the line width, higher the Q of the cavity, and sharper the resonance. In the case of the He-Ne laser with a cavity of length 1 m, we have $\Delta v = c/2L = 1.5 \times 10^8$ Hz. Obviously, this is a small fraction of the mean frequency of the output radiation at 4.75×10^{14} Hz. This reveals the high degree of monochromaticity of the laser radiation.

7.7 AMPLIFICATION OF LIGHT AND THRESHOLD CONDITION

Let us consider a system of atoms through which a quasi-monochromatic light beam of frequency v propagates in the Z-direction. Let S_1 and S_2 represent the two planes of area A perpendicular to the Z-direction and located at

z and $z + dz$, respectively, as shown in Fig. 7.8. The volume of the active laser material between the planes S_1 and $S_2 = Adz$. If n_1 is the concentration of the atoms in the lower state, u_v is the energy density of the radiation and B_{12} is the Einstein's B coefficient, then the time rate of stimulated absorption $= B_{12}n_1u_v Adz$. Obviously, the energy absorbed per unit time in the volume Adz is, therefore, equal to $B_{12}n_1u_v hvAdz$. Similarly, one finds the energy gained per unit time in the volume Adz due to stimulated emission is $B_{12}n_2u_v hvAdz$, where n_2 is the concentration of atoms in the upper state. Here we have ignored the radiation due to spontaneous emission because such radiations are directed at random and are lost. Now, the net energy gain per unit time in volume Adz in the frequency range v and $v + dv$ is

$$(B_{21}n_2 - B_{12}n_1)u_v hvAdz = B_{12}(n_2 - n_1)u_v hvAdz$$
$$(\because B_{12} = B_{21})$$

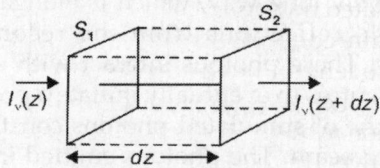

Fig. 7.8: Amplification of a light beam passing the planes S_1 and S_2 located at z and $z + dz$ respectively

Let us consider that I_v be the intensity of the beam at S_1. Obviously, the energy entering the volume Adz per unit time through S_1 is $I_v(z)A$. Similarly, if $I_v(z + dz)$ be the intensity at S_2, then the energy moving out of the volume Adz per unit time through S_2 is

$$I_v(z + dz)A = \left(I_v(z) + \frac{\delta I_v}{\delta z}dz\right)A$$

Obviously, net energy leaving the volume per unit time is $\dfrac{\delta I_v}{dz}Adz$. This must be equal to the energy gained by volume Adz due to stimulated transitions. Thus, one obtains

$$\frac{\delta I_v}{dz}Adz = B_{12}u_v hv(n_2 - n_1)Adz$$

$$\frac{\delta I_v}{dz} = B_{12}hv(n_2 - n_1)u_v \qquad (7.15)$$

Also, we have

$$I_v = v u_v = \frac{c}{n} u_v \qquad (7.16)$$

where $v = c/n$ is the velocity of light in the medium having refractive index n. Using Eq. (7.16), Eq. (7.15) reduces to

$$\frac{I}{I_v} \frac{dI_v}{dz} = \alpha_v \qquad (7.17)$$

where

$$\alpha_v = \frac{B_{12} h v n (n_2 - n_1)}{c} \qquad (7.18)$$

On integration, Eq. (7.17) gives

$$I_v(z) = I_v(0) \exp(\alpha_v z) \qquad (7.19)$$

Equation (7.19) represents the growth of the intensity of the light when it traverses a distance z through the active medium of laser. Obviously, the growth results from the population inversion created by pumping and stimulated transitions. The quantity α_v is called the gain constant having the dimension of reverse length. We can see that the gain constant α_v increases with the degree of population inversion $n_2 > n_1$ and hence α_v given by Eq. (7.18) is positive. This means that the intensity increases exponentially with distance z.

Let us consider that r_1 and r_2 be the reflection coefficients of two mirrors of the optical cavity. Obviously, in one complete round trip, the intensity of light decreases to $I_v(0) r_1 r_2$ where $I_v(0)$ is the initial intensity. Let us write $r_1 r_2 = \exp(-2\beta)$, where β is positive and can be considered to be a measure of the loss of intensity of light as it traverses from one mirror to the other. This means that on traversing a length L, the intensity drops to $I_v \exp(-\beta)$. We may assume that the reflection losses are the only losses present in the cavity. Thus, the intensity in a mirror to mirror passage becomes $I_v(0) \exp(\alpha_v L - \beta)$. Thus, for a substained laser action the amplification must compensate for energy losses. Obviously, for the lasing action to build up, we must have

$$\exp(\alpha_v L - \beta) > 1 \quad \text{or} \quad \alpha_v L > \beta$$

Consequently, one finds the *threshold condition* as

$$\alpha_v L = \beta \qquad (7.20)$$

From Eq. (7.20), we can easily see that if $\alpha_v L < \beta$, intensity dies out. At threshold, the population inversion is called the *critical inversion*. From Eqs. (7.18) and (7.20), one obtains the condition for critical inversion as

$$(n_2 - n_1)_{\text{crit}} = \frac{\beta c}{L} (\mu B_{12} h v)^{-1} \qquad (7.21)$$

7.8 SOLID-STATE LASERS

Ruby Laser

The first successful operation of the laser was achieved by TH Maiman of the Mughes Aircraft Company in July, 1960 using ruby crystals. Ruby is aluminium oxide in which few of the aluminium atoms have been replaced by chromium atoms ($Al_2O_3 : Cr^{3+}$), the chromium atoms being the ones that take part in a laser action. Maiman used 0.05% Cr^{3+} to get a pink colour, which is due to the fact that chromium absorbs ultraviolet, green and yellow, and transmits only red and blue. This absorption raises chromium atoms to excited states from which two steps are required to carry them back to the ground energy level. In the first step, chromium atoms give up some of their energy to lattice and fall in *metastable* state, characterized by a comparatively long lifetime. This transition is a non-radiative transition. In the second step, chromium atoms emit radiation of wavelength 6943 Å in the red region of the spectrum and fall to ground level. The metastable state has a lifetime of about 10^{-3} second compared to the extremely small lifetime of 10^{-7} second of the higher energy level. Thus, the necessary time delay between the useless relaxation transition and the useful stimulated transition for the laser action is readily obtained. The energy levels in the ruby laser are shown in Fig. 7.9.

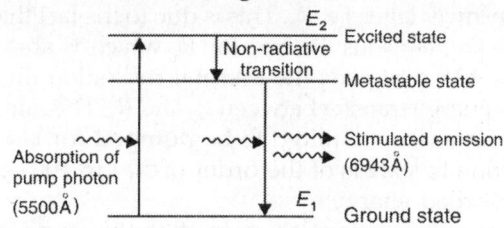

Fig. 7.9: Energy-band scheme for laser action in ruby [$Al_2O_3 : Cr^{3+}$]

Ruby optical laser apparatus is shown in Fig. 7.10. A ruby is machined into a rod typically 5 cm long and 0.5 cm in diameter with polished ends which are parallel and optically flat. This is placed near an electronic flash tube which provides the pumping energy. The ruby is cooled using liquid nitrogen. Up to a certain flash intensity, ordinary luminescent emission dominates, and a powerful beam of red light flashes out from the ends of ruby rod. This red light has the enormous power output of more than 10^4 watts over a beam of cross-section 10^{-4} square metre. The emitted band is within a wavelength interval of about 0.02 Å. The beam from the end face of the ruby rod has an angular spread less than one degree. Typical output of a ruby laser is 10^{-20} kilowatts in milli-second pulses.

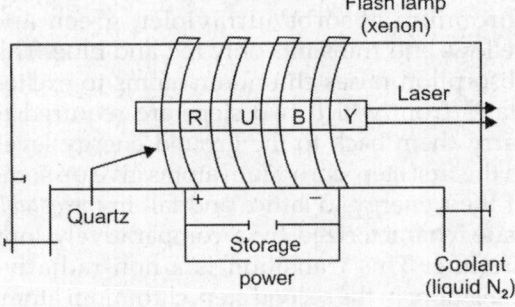

Fig. 7.10: Pulse ruby optical laser

The stimulated emission predominates over spontaneous emission when the population of metastable state exceeds considerably over the population predicted by Boltzmann distribution relation and in comparison with the ground energy state. The metastable state consists of closely spaced double level R_2 and R_1. The larger contribution to laser action, however, comes from the lower level, i.e. R_1. This is due to the fact that the spontaneous lifetime for R_1 which is about 4×10^{-2} s exceeds the thermal relaxation time for energy transfer between R_2 and R_1. The minimum energy required to be pumped for laser action to start is of the order of 50 J to 100 J of absorbed energy.

It is important to note that this type of oscillator action has losses which are determined by the flatness and parallelism of the end mirror.

The losses may be considerable, and may reduce the efficiency of the laser.

The maximum space coherence (direction-ality) obtainable is determined by the diffraction of the mirrors. The light beam spread angle $\delta\theta$, is given by

$$\delta\theta \cong \frac{\lambda}{D} \qquad (7.22)$$

where λ is the wavelength of the emitted radiation and D is diameter of the source, which is the mirror in this case.

Pumping Power

One can roughly estimate the pumping power necessary for the population inversion as follows:

Let us consider that P be the power of the pumping light of frequency ν_p incident on a unit area of the active material. Let σ_p be the absorption cross-section of the pump light. Thus, the pump power absorbed by the material = $\sigma_p P$. Obviously, this power is used in raising the atoms to the excited state. Let τ be the lifetime of the upper (excited state), the power consumed in the transition to the lower state is $h\nu_p/\tau$. For population inversion to occur, we have $\sigma_p P = h\nu_p/\tau$ or $P = h\nu_p/\tau\sigma_p$. In case of ruby laser, we have $h\nu_p = 3 \times 10^{-19}$ J, $\tau = 4$ ms and $\sigma_p = 10^{-23}$ square meter, one obtains $P = 7.5 \times 10^6$ W/s^2.

Apart from ruby, several other solid materials such as calcium fluoride-samarium, glass-niobium, etc. have been used as active materials for solid-state laser action. Basically, this class of lasers refers to the lasers based on ions in solid materials. These doped crystals gave pulsed output. The pulsed calcium fluoride laser has four levels and was first developed by PP Sorokin and MJ Stephenson during 1960–61. In general, most ordinary solid state lasers are of the four levels type as it is easier to establish population inversion in these systems. In a four-level laser system, excitation and radiation transfer proceed as in the case of a three level laser material, but there is now an additional energy level available above the ground level. Laser action, there-fore, begins as soon as there is a significant population of the level above this additional level, which in turn increase efficiency.

Next to ruby, Nd^{3+} is the most commonly used solid material for lasers. It is capable of producing infrared radiations over a large range, the most useful wavelength being 1.064 μm. Rare earth materials used as solid laser give coherent radiation in the range 0.6 – 2.6 μm. Another common solid laser is the YAG laser which has four levels instead of the three levels in ruby. The material consists of Ytterium Aluminium Garnet ($Y_3Al_5O_{12}$) doped with Nd^{3+} ions in place of Y. On pumping optically by flash lamps, the Nd^{3+} ions are excited to their characteristic bands. The excited ions decay non-radiatively into the metastable state $4F_{3/2}$ which serves as the upper laser level and has a lifetime of about 0.25 millisecond. The radiative (laser) transitions occur from this state predominantly to the $4I_{11/2}$ state below giving the 1.064 μm beam.

7.9 SEMICONDUCTOR (P-N JUNCTION) LASERS

Coherent laser radiation ranging from the infrared, through the ultraviolet regions of the spectrum has been obtained from semiconductors and semiconductor junction diodes. Laser radiation can be obtained by pumping a solid intrinsic semiconductor or a PN junction diode. The method of pumping varies with the type of semiconductor used. The most common type of semiconductor used is a III–V compound. Semiconductor lasers are different from other lasers in the following respects:

  i. Transition occurs between energy bands rather than between discrete energy levels, the emission being a result of electron transitions from the conduction to the valence bands.

 ii. The laser is very small in size, typically 50 μm × 250 μm × 50 μm.

iii. The characteristics of the laser beam are strongly influenced by the properties of the junction material.

 iv. In PN junction lasers, population inversion occurs in the very narrow region about the junction. The pumping of the PN junction laser is accomplished by the application of a forward bias to the diode.

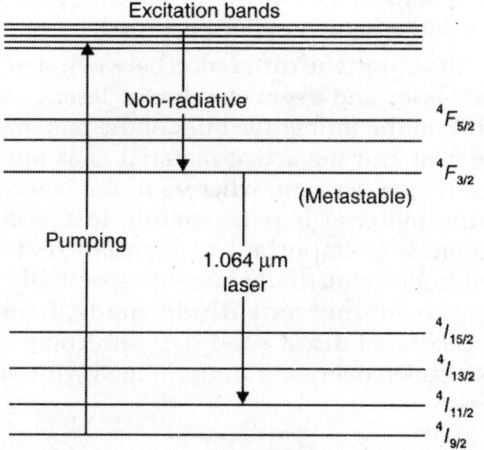

Fig. 7.11: Energy level diagram of Nd in YAG laser

It is a property of semiconductors that the downward transition of electrons recombining with holes may or may not result in the emission of optical energy. Based on this property, semiconductors can be classified as *indirect and direct*. This classification is related to the energy band structure: silicon and germanium are indirect types, whereas most of the III–V compounds are direct. In direct type, radiative transitions take place faster than non-radiative and impurity transitions. In indirect semiconductors, radiative recombinations occur slower than non-radiative recombinations, so that in essence photon emission is suppressed. In fact, some radiation is emitted by almost all semiconductor PN junctions, but in junctions using the indirect type of semiconductors, this radiation is very inefficient.

Figure 7.12a, shows the basic structure of a PN junction laser. The diode laser is made of gallium arsenide (GaAs) by the standard diffusion method. The operating wavelength for this laser is usually in the range 8400 – 8500 Å. The index of reflection of semi-conductors is usually so high that it is not necessary to polish the end faces to achieve reflection. Efficiencies as high as 70% have been observed in these lasers. A large number of materials are being used to get an operational range of wavelengths in the region from 0.33 μm to 30 μm. In these types of lasers due to band-to-band transitions involvement, the spectral purity is low and emission pattern is broad.

An important difference between a solid state laser and a semiconductor laser is that while in the former the bulk of the material is the host and the active material is as low as nearly one per cent, whereas in the latter the entire material is participating in the laser action. It is important to mention that the semiconductor diode laser is essentially an electroluminescent diode made from a degenerated direct band-gap semiconductor like GaAs, operating under a heavy forward bias.

Population Inversion in Semiconductor Laser

One can obtain the population inversion in a PN junction by heavily doping both the P and N regions and applying a forward bias. The heavy doping pushes the Fermi level E_F in the conduction band in the N-side and in the valence band in the P-side (Fig. 7.12b). The system is in thermodynamic equilibrium when there is no applied voltage and obviously there is no population inversion. To create population inversion, we will have to disturb the thermodynamic equilibrium, i.e. a forward bias has to be applied. For an applied forward voltage V, the band diagram of the PN junction is shown in Fig. 7.12c.

Let us assume that the charge carriers in the two bands are in thermal equilibrium with each other. This means that the population distributions in the conduction and in the valence bands are described by the quasi Fermi levels E_{FC} and E_{FV} respectively. One can express the occupation probability in the conduction band by

$$f_C(E_C) = \frac{1}{1 + \exp\left(E_C - E_{FC}\right)/k_B T} \quad (7.23)$$

and that in the valence band is expressed as

$$f_V(E_V) = \frac{1}{1 + \exp\left(E_V - E_{FV}\right)/k_B T} \quad (7.24)$$

For a light beam to be incident, one finds that the time rate of absorption of quanta R_a is proportional to (i) the probability per unit time of the transition from the valence band to the conduction band B_{VC} (ii) the density of incident radiation $u(\omega)$ (iii) the probability of occupancy

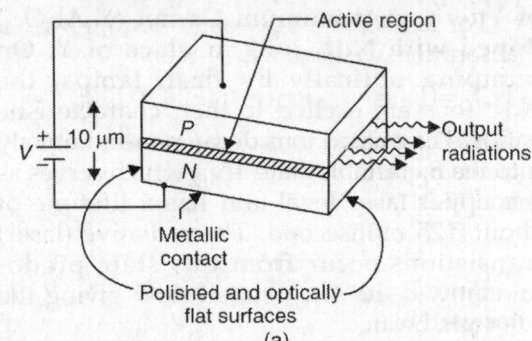

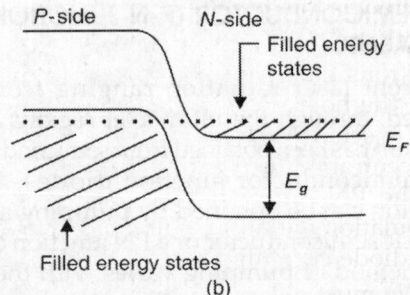

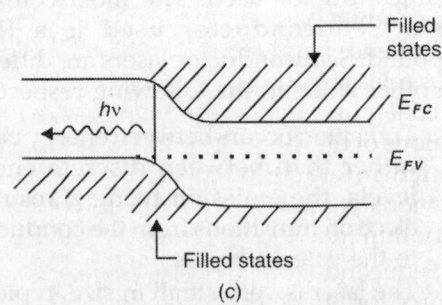

Fig. 7.12: (a) Semiconductor laser (b) Energy band diagram of a PN junction diode (unbiased heavily doped) used as a laser (c) Energy band diagram of an heavily doped PN junction diode (for an applied voltage *V*)

of the valence bands $f_V(E_V)$ (iv) the probability of vacancy of the conduction $[1 - f_V(E_V)]$. Thus, one can write

$$R_a = CB_{VC}u(\omega)f_C(E_V)[1 - f_C(E_C)] \quad (7.25)$$

Here C (constant of proportionality) includes the density of states in the two bands. Similarly, one can write the expression for the time rate of emission of the quanta by stimulated emission R_e as

$$R_e = CB_{CV}u(\omega)f_C(E_C)[1 - f_V(E_V)] \quad (7.26)$$

For laser action, the emission must exceed absorption, we have $R_e > R_a$, i.e.

$$f_C = (E_C)[1 - f_V(E_V)] > f_V(E_V)[1 - f_C(E_C)] \quad (7.27)$$

We can easily see that Einstein's B coefficients B_{CV} and B_{VC} are equal. Using Eqs. (7.23) and (7.24), Eq. (7.27) reduces to

$$E_{FC} - E_{FV} > E_C - E_V = h\nu \quad (7.28)$$

Since $E_{FC} - E_{FV} = eV$, where V is the applied forward bias, the inequality (7.28) reveals that the applied forward voltage must exceed $h\nu/e$, where ν is the frequency of the emitted light.

When PN junction is forward bias, the electrons flow from the N-side to the P-side, and the holes from the P-side to the N-side. The injected electrons and holes appear in high concentrations in a part of the depletion region at the PN junction. The consequent large population initiates the *lasing action* when the PN diode current exceeds a threshold value.

We must remember that the electrons from the conduction band in this region jump into the valence band to recombine with the holes, thus releasing the band energy gap in the form of coherent radiation. Moreover, the P and N regions of the diode absorbs the laser radiation which interacts with the lattice photons. Obviously, the laser action is restricted to the very thin junction region of the diode.

Semiconductor lasers are an important light source for fibre optic communications and are key components of such common appliances as compact disc players, supermarket scanners, laser printers, fax machines, and laser pointers. The lasers in these applications are so called double heterostructure lasers, essentially diodes consisting of an active semiconductor region sandwitched between doped semiconductor cladding layers, one N-type, the other P-type. The cladding regions supply electrons and holes to the active region when an appropriate bias voltage is applied. They also have a higher band gap and lower refractive index than the active layer, so that the injected electrons and holes as well as the photons generated by their annihilation are confined to the active region.

Double heterostructure lasers have demonstrated high performance and have been successfully commercialized for wavelengths ranging from blue to the near infrared (IR), up to about 1.6 μm. Unfortunately, few semiconductor materials emitting in the mid-IR (2–20 μm) are reliable, easily processed, and insensitive to temperature cycling—the repeated heating and cooling associated with laser operation.

7.10 GAS LASERS

Laser action can also be made to occur in some gases, such as CO_2 or a mixture of He and Ne. The credit for the discovery of first gas laser goes to A Javan, W Bennet and DR Herriot of Bell Labs. They used the electric discharge through a gas for ionisation and pumping action, i.e. resonant transmission of excitation energy in a gas discharge. Sometimes excitation or pumping may be provided externally by a separate electrode. The cavity is about 30 cm to 50 cm long having a tube diameter of about 5 mm and two concave reflecting mirrors at the ends. A steady potential of a few thousand volts is used for getting a discharge. The main advantage of such lasers resides in that they operate continuously, though some gas lasers are also capable of pulsed working. These lasers display exceptionally high monochromaticity, most pure spectrum and high stability of frequency. All these features make these lasers extremely useful in various branches of science and engineering. The tube used in these lasers is made up of glass or quartz. The ends of the tube terminate in flat glass windows inclined at Brewster's angle to eliminate reflections.

He–Ne gas lasers were the first to be developed and used. They operate in three distinct ranges, in the red at 6328 Å, in the near infrared around 1.15 μm and in the infrared at 3.39 μm. The partial energy levels diagram of He–Ne is shown in Fig. 7.13b, which explains the origin of these lines. With the help of a dispersive prism, these lines can be separated.

A gaseous mixture of He and Ne (about 10 : 1) is used as an active medium. When the helium atoms in the mixture are excited to the metastable state by means of a DC or radio frequency source, they collide with the ground state neon atoms and resonant energy transfer takes place. One of the excited states of the neon has nearly the same excitation energy as He metastable state. Hence, the Ne atoms are excited in the collision to a specific energy level as the helium atom returns to the ground level. This causes a population inversion in the energy states of neon and laser beam is obtained. He-Ne gas laser is shown in Fig. 7.13a.

Figure 7.13c shows the energy level diagram for He and Ne atoms. The energy states involved in the operation are labelled as E_0, E_1, E_2, and E_3. E_0 represents the common ground state for He and Ne atoms. E_1 represents the excited state of He atoms. He atoms being lighter get excited to energy state E_1 preferentially due to electron impact. The energy state E_1 is metastable and the atoms excited to this state cannot return to the ground state spontaneously. However, they can return to the ground state by transferring their energy to Ne atoms in ground state through collision with them. This energy transfer through collision, usually referred to as *resonant collision*, is possible as energy state E_1 of He (20.61 eV) and energy state E_3 of Ne (20.66 eV) are, by chance, very close to each other. The extra energy of 0.05 eV is provided by kinetic energy of He atoms. In this way, population of metastable energy state E_3 becomes much greater than the population of energy state E_2 of Ne, i.e. to say population inversion is established between E_3 and E_2. This population inversion is maintained as the energy state E_2 is depleted rapidly through the intermediate

stages not shown in the figure. Lasing transition occurs between E_3 and E_2 giving out red light at a wavelength of 632.8 nm. Thus, He atoms work as pumping medium and Ne atoms as lasing medium.

The energy level diagram shown in the Fig. 7.13b is over simplified. Actually, there are more energy states than depicted. Also, the energy states shown in the figure are not really single but a group of states close to each other

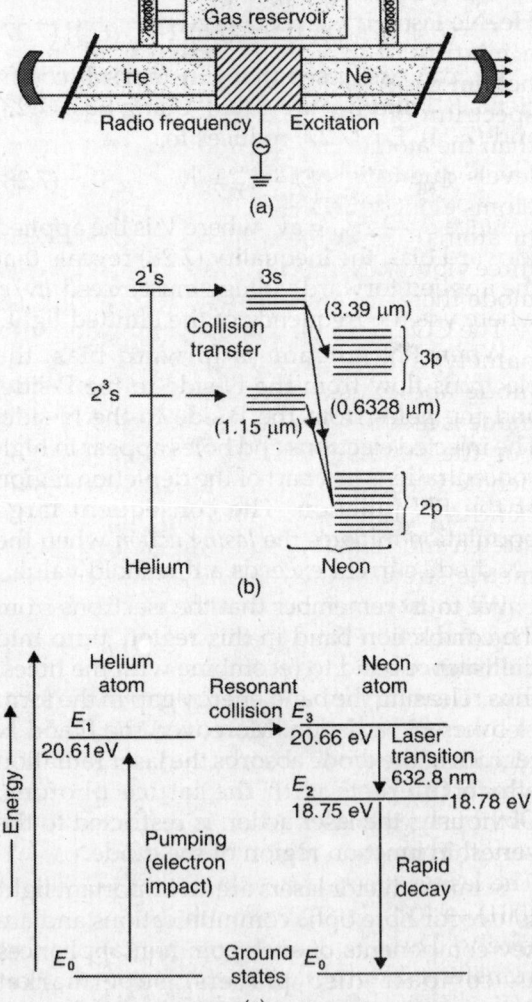

Fig. 7.13: (a) He–Ne gas laser (b) Partial energy levels of He–Ne gas laser (c) Energy level diagram for He and Ne atoms

having very small energy difference. Consequently, more than 130 lasing transitions are possible. However, the device can lase at any one desired frequency by appropriately designing the optical cavity.

Carbon Dioxide Laser

It is a molecular gas laser. The overall efficiency of most of the gas lasers is below 1%, but that of CO_2 laser is 20%. The unsual high efficiency of this laser was first recognized by CKN Patel in 1964.

The construction of CO_2 laser is similar to He–Ne laser. It is a discharge tube filled with a mixture of carbon dioxide, nitrogen and helium gases in the ratio 1 : 4 : 5. The energy spectrum of molecular gas is more complex than the atomic gas. The molecule has energy levels due to vibrations and rotations of the atoms along with the electronic energy levels of atoms. The carbon dioxide molecule has three vibrational modes (Fig. 7.14) and for each mode there are rotational modes.

The CO_2 molecule vibrates in three modes, namely, symmetric stretch mode, bending mode and asymmetric stretch mode. Each mode is associated with a set of energy levels. For the symmetric stretch mode they are denoted by (n00), where n is an integer; for bending mode (0n0) and for asymmetric stretch mode (00n). The first excited asymmetric stretch state (001) is alomost exactly equal to lowest vibrational level of nitrogen. The trick of energy transfer by resonant collision, used in He–Ne laser, is used here also. The nitrogen is excited electrically and it delivers the energy to CO_2 (001) state by way of collisions. If we attempt to populate (001) state of CO_2 directly by electron collisions, most of the energy goes to low-lying (010) and (020) states. Therefore, nitrogen is used as vehicle. The lasing occurs between (001) → (100) and (001) → (020), at 10.6 µm and 9.6 µm respectively. The vibrational-vibrational energy transfer pummping process results in high efficiency and enormous output power, increasing utility of this laser. Improvement in heat removal from the laser discharge is the main purpose of adding helium in the mixture.

It also helps in speeding up the transition from (100) to ground state (000) to maintain population inversion.

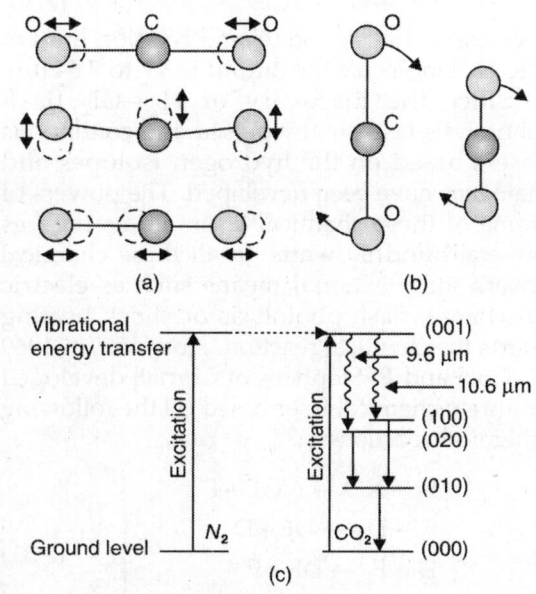

Fig. 7.14: (a) Vibrational modes of CO_2 molecule: symmetric stretch mode, bending mode, asymmetric stretch mode (b) Rotational modes of CO_2 molecule (c) Energy level diagram (not to the scale)

7.11 CHEMICAL LASERS

In chemical lasers, population inversion results from a chemical reaction. The high energy density that can be released in exothermic chemical reactions has led to the interest in converting this energy to coherent optical energy. If one could use the energy of such chemical reactions to obtain a population inversion directly, one could build compact and powerful lasers. Such lasers may also become a powerful tool in the study of kinetics of chemical reactions.

The first chemical laser was developed by JV Kasper and GC Pimentel in 1965. The active molecule was vibrationally excited hydrogen chloride (HCl*) molecule, formed by the following chemical reaction:

$$\left.\begin{array}{l} Cl + H_2 \rightarrow HCl^* + H \\ H + Cl_2 \rightarrow HCl^* + Cl \end{array}\right\} \qquad (7.29)$$

The initial chlorine molecule atoms were produced by flash photolysis according to the following reaction:

$$Cl_2 + h\nu \rightarrow 2Cl \qquad (7.30)$$

Corresponding to the several rotation lines of the HCl molecule the output is 3.7 to 3.84 μm.

Since the discovery of H_2—Cl_2 flash photolysis laser, many pulsed and continuous lasers based on the hydrogen isotopes and halogens have been developed. The powers of some of these chemical lasers go as high as several hundred watts. In all these chemical lasers some external means such as electric discharge, flash photolysis or shock heating starts the chemical reaction. However, in 1969 T. Cool and R. Stephens of Cornell developed a purely chemical laser based on the following chemical reactions:

$$\left.\begin{array}{l} F_2 + NO \rightarrow NOF + F \\ F + D_2 \rightarrow DF + D \\ D + F_2 \rightarrow DF + F \\ (DF)^* + CO_2 \rightarrow DF + CO_2 \end{array}\right\} \qquad (7.31)$$

In the first chemical reaction, F_2 is dissociated to give free fluorine atoms which on mixing with D_2 yield vibrationally excited DF. The excited DF transfers its energy to CO_2 and laser action results at the wavelength 10.5 microns. The power output from this laser is about 500–600 watts.

7.12 DYE LASERS

PP Sorokin and JR Lankard of IBM in 1966 obtained laser action in organic molecules by their efficient fluorescence. They also produced laser action in many dye solutions which were pumped with microsecond long flashes from high intensity lamps. As shown in the energy level digram (Fig. 7.15), the optical excitation and emission in organic dyes take place between the rotational vibrational levels of the ground singlet state S_0 and the excited singlet state S_1.

The active medium of an organic dye laser is an organic fluorescent material dissolved in a common solvent. These substances derive their colour from strong absorption in the visible region of spectrum. Sorokin obtained the laser action by using a solution containing a dye chloro-aluminium pthalocyanine at 765 Å. Dye laser can operate at room temperature and moreover they are tunable.

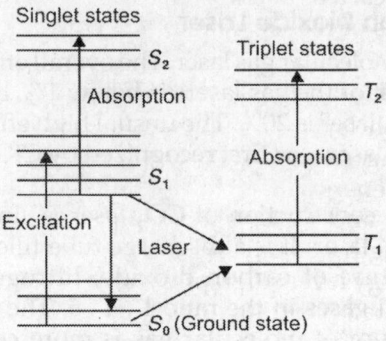

Fig. 7.15: Optical excitation and level emission between vibrational S_0 and S_1 in a dye laser

7.13 Nd: YAG LASER

This is a four-level solid state laser system. Yttrium Aluminium Garnet ($Y_3Al_5O_{12}$) commonly called as YAG doped with neodymium ions Nd^{3+} is the **active medium.** The active medium is taken in the form of a crystal and is drawn into a rod. The neodymium ions Nd^{3+} are the active centers.

The end faces of the Nd: YAG rod are ground polished and silvered to act as the **optical resonator** mirrors or the optical cavity can be formed by using two external reflecting mirrors M_1 and M_2.

A xenon flash lamp or a krypton flash lamp is used as optical pumping source.

The schematic diagram of a Nd: YAG laser is shown in Fig. 7.16.

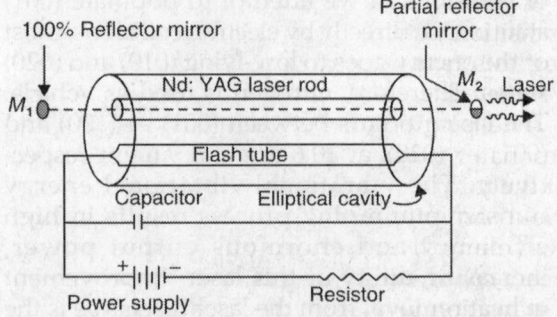

Fig. 7.16: Nd: YAG laser

Construction

A Nd: YAG rod and a krypton flash lamp are enclosed inside an ellipsoidal reflector. In order to make the entire flash radiation get focussed on to the laser rod, the Nd: YAG rod is placed at one focal axis and the flash lamp at the other focal axis of the ellipsoidal reflector.

Working

The flash lamp is switched on. The optical pumping excites the Nd^{3+} ions from the ground energy state E_0 to the higher energy level E_3 and E_4 by absorbing radiations of wavelength 0.80 μm and 0.73 μm respectively. The energy level diagram is shown in Fig. 7.17. The excited Nd^{3+} ions then make a transition from these energy levels. The transition from the energy level E_4 to E_2 is a non-radiative transition. The state E_2 is the metastable state. Upon continuous excitation, population inversion of Nd^{3+} ions is achieved at the metastable state E_2.

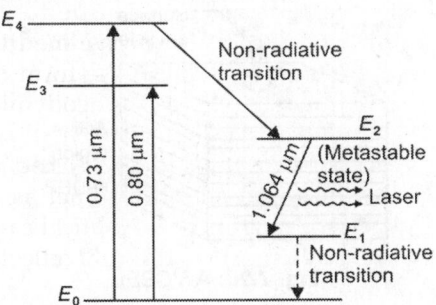

Fig. 7.17: Energy level diagram of Nd:YAG laser

Any of the spontaneously emitted photon will make the excited Nd^{3+} ions to undergo a transition between $E_2 \rightarrow E_1$ state. Thus, during this transition the stimulated photon is generated.

The photons travelling parallel to the resonator axis experience multiple reflections at the mirrors. As a result, the transition $E_2 \rightarrow E_1$ yields an intense and coherent laser beam of wavelength 1.064 μm. These lasers give beam continuously. The Nd^{3+} ions, then make a transition between $E_1 \rightarrow E_0$ which is a non-radiative transition.

Only a part of the energy emitted by the flash lamp is used to excite the Nd^{3+} ions, while the rest heats up the crystal. Thus, the system can be cooled by either air or water circulation.

Nd: YAG laser finds following applications:

i. These lasers are used for cutting, drilling, welding and surface hardening of the industrial products.

ii. These lasers are used for cataract surgery, to treat gastrointestinal bleeding and gall bladder surgery in medical field.

iii. These lasers are used as ranges finder and as target designations in military.

iv. These lasers are used in long hour communication.

v. These lasers are used in the study of inertial confinement fusion.

7.14 QUANTUM WELL LASER (QWL)

Quantum wells in which a smaller band gap semiconductor (such as GaAs) some tens of Å thick, sandwiched between the layers of a larger band-gap difference between the two materials have been produced due to recent development in crystal growth. There is an offset of the conduction and the valence band edges across the quantum well interfaces due to the band gap difference between the two materials as shown in Fig. 7.18.

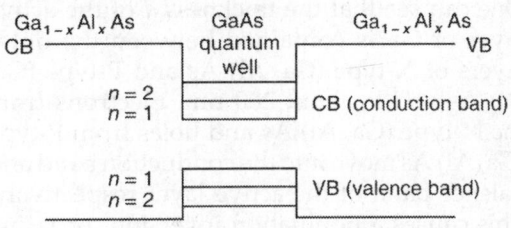

Fig. 7.18: Quantum well

When electrons are put into the quantum well, the confinement in the well forces electrons to have discrete energy levels designated by $n = 1, 2, ...$ In accordance with the quantum principle, obviously, the quantum size effects cause discrete energy levels in the direction perpendicular to the interfacial planes. A semiconductor laser based on the principle of quantum well is called as a *quantum well laser*.

The gain becomes larger and sharper for a quantum well laser than that for an ordinary semiconductor laser due to the formation of the discrete levels which reduces the spread of the electron. The threshold current of the quantum well laser is low and moreover, one can control the output wavelength by the well thickness. One finds that quantum well laser has an excellent temperature characteristic and a narrow spectral gain width.

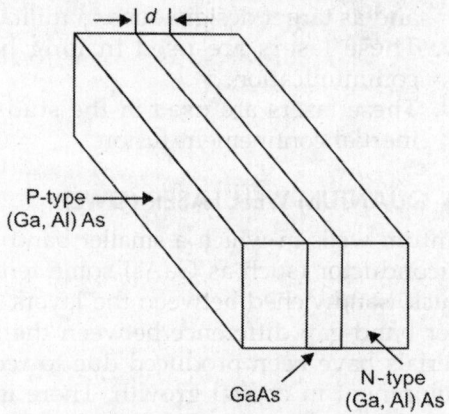

Fig. 7.19: Double hetrostructure laser

QWL is an improved version of the ordinary *double-hetro structure* laser shown in Fig. 7.19. One can see that the thickness d of the active layer of GaAs contained between the outer layers of N-type (Ga, Al) As and P-type (Ga, Al) As and is about 200 nm. Electrons from the N-type (Ga, Al) As and holes from P-type (Ga, Al) As move into the conduction band and valence band of the active layer respectively. This causes a population inversion for lasing action. The light is guided between the outer layers due to the larger refractive index of the active layer. This offers a narrow beam. We can easily see that this is an additional feature over the confinement of the carriers. We can also see that d is greater in the ordinary double hetro structure laser than in *QWL*. This means that the energy level quantization in the valence and conduction bands in the double hetro structure laser is of no significance.

7.15 VERTICAL-CAVITY SURFACE-EMITTING LASER (VCSEL)

VCSEL developed in 1990s, requires an active light emitting region contained between two mirrors. In this device, light is recirculated vertically in an optical cavity so that emission occurs normally from the chip surface. In VCSEL, the length of the active region is considerably shorter than in other laser devices.

The shorter active length in VCSEL makes it essential to use many more highly reflecting mirrors formed by a stack of alternating layers of quarter-wavelength-thick semiconductor material. These mirrors are called *distributed Bragg reflectors* (DBRs). Since the mirrors are made from semiconductor materials and hence the current injected into the active region flows directly through these mirrors. One can form a PN junction with current flowing vertically through the device by doping one DBR-P-type and other DBR-N-type as shown in Fig. 7.20.

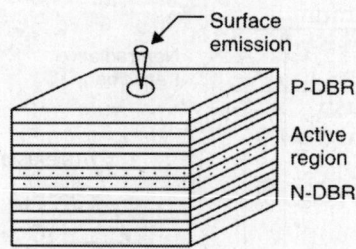

Fig. 7.20: A VCSEL

A VCSEL has compact size and the design flexibility of the output aperture. The small size of VCSEL ensures that the current required for lasing action is quite small. The low capacitance of the device makes the operating speed inherently high. One can tailor the emission pattern to match the input of an optical fibre more effectively than possible with an edge emitting laser. We can see that VCSEL's form an ideal choice for high-speed optical data transmission. For the wavelength window of 780–980 nm, (Al, In, Ga) As system is used as material for VCSEL whereas for the window of 1.3–1.55 μm, (In, Ga) (As, P) system is used.

7.16 QUANTUM DOT LASERS

Quantum dot lasers offer the potential to yield better properties as compared to quantum lasers. These emit light at wavelengths that are determined by the energy levels of dots. The research in nano-crystal quantum dots has opened the door for developing novel optical and opto-electronic devices, such as tunable lasers, optical amplifiers and light emitting diodes from assemblies of these invisibly small particles. Quantum dot lasers work like any other semiconductor lasers such as those found in home audio compact disk players. The goal of a quantum dot laser is to manipulate the material into a high energy state. The result is the net release of energy which emerges as a photon. European group of scientists have launched a project called BIG BAND to develop ultrawide band InP based quantum dot lasers amplifiers for telecom applications.

The primary goal of BIG BAND is to produce QD-based semiconductor optical amplifiers (SOAs) operating over the entire telecommunication window. SOAs are seen as potential replacement for erbium-doped fibre amplifiers in optical networks. Another goal is to push PNP device emission to longer wavelengths for gas sensing applications.

7.17 APPLICATIONS OF LASERS

Lasers, with their excellent characteristics, have found tremendous applications in every field of interest including research, medicine, communication, defence, entertainment, and industry. Some of the applications are as follows:

Medicine

The very first application of laser is in the medical field. Nowadays, lasers are playing very crucial role in the field of medicine. It has proved to be very useful in diagnosis as well as surgery. Lasers in association with fibre optic catheter has become a very fantastic tool in diagnosis as one can very easily see the interior parts of the body which otherwise would be very difficult. Focussed laser beam can be usd as a non-contact and hence sterile surgical tool. It can be used for cutting and destroying the diseased parts of the tissues as per necessity. It can also be used for welding the blood vessels being cut during surgery. Laser-assisted surgery is usually fast and less painful. As a result, the period of hospitalisation required is reduced. Following are some important applications of laser in medical field.

Ophthalmology: Retina, a light-sensitive part at back of the eyeball, may get detached due to some reasons. This may lead to blindness. In such cases, traditional open eye surgery was performed, which was very risky. The post-operation care period was long and critical. Nowadays, argon laser is very successfully used for treatment of eye. The green beam of argon laser is focussed on the desired part of the retina. The energy of the green beam is strongly absorbed by red blood cells in the retina causing thermal effects and thereby re-attaches the retina at appropriate position on the back of the eyeball.

Cataract: When natural lens in eye becomes cloudy, it becomes opaque. This is nothing but the formation of cataract which diminishes light entering the eye resulting in blindness. The cataract is needed to be replaced by artificial lens. Traditional open eye surgery is very risky. However, Nd: YAG (neodymium doped yttrium aluminium garnet) laser can be very successively used in cataract treatment.

Laser Photocoagulation: In case of diabetes patients, due to some reasons, abnormal growth of blood streamers takes place over the retina. Blood leaks through these streamers into vitrious chamber of eye. This results in gradual dimming of the vision. Laser photo-coagulation is used very effectively for destroying the mesh of blood streamers.

Keratomy: In order to correct the refractive power of the eye, shape of the cornea is needed to be changed. This can be efficiently achieved by drawing small incisions on the cornea, the outer transparent layer of the eye, with ArF (argon fluoride) excimer laser.

Angioplasty: This is one of the very important applications of laser. If, due to some reason, the cholesterol level in the blood increases, the inner diameter of the arteries decreases, preventing normal blood circulation. Laser-assisted angioplasty is found to be very effective in clearing the blocked arteries. A fibre optic catheter is inserted through the arteries at the blocked region and laser pulses are emloyed to burn out the unwanted extra growth and thus regulating the blood flow.

Lithotomy: Laser-assisted lithotomy is very popular these days. Laser pulses are employed through fibre optic catheter to shatter the kidney stone into small pieces, which may flow out with urine without pain. Gall stones are also destroyed with laser pulses in the similar way.

Oviduct Blockage: Sterility in women is mainly because of blockage of one or both of the oviducts, the opening of the fallopian tube is uterus. This blocks the entry of the egg in uterus resulting finally in sterility. Laser pulses can be employed to remove the blockage with very small risk compared to traditional methods.

Cancer Therapy: In cancer treatment, the first step is to locate the cancer-affected cells. For this, a dye called haematoporphyrin derivative (HpD) is injected into the body. The dye accumulates in the cancer-affected cells selectively. When illuminated with a laser of appropriate frequency, the cancerous cells having HpD absorb the radiation very strongly resulting in killing the cancerous cells. Tumors developed in brain and spinal cord can be treated with lasers with great ease compared with the traditional methods.

Dermatology: Lasers are used to remove the freckles, acne, warts, birth marks and tattoos, which appear due to the abnormal blood vessel network under the skin. When treated with argon laser, such areas are burned out and the blood vessels are closed, obstructing the excess blood supply.

Dentistry: Due to some reason, demineralisation of tooth enamels takes place. Laser treatment is found to reduce the rate of demineralisation of the enamels. Lasers are also employed to drill the cavities in the teeth particularly in root canal treatment.

Optical Communication

Rate of information transfer is a very crucial parameter in communication and is proportional to the bandwidth of the electromagnetic wave. As laser operates with a very large bandwidth, optical communication using laser is very attractive.

Open Space Communication: Laser beam being highly directional offers itself as a very good option for microwave communication.

Fibre Optic Communication: Traditionally used metallic cable is being replaced by fibre optic cable as it is advantageous. Few of them have very large bandwidth, electrical isolation, low transmission loss, high signal security, immunity to cross talk, small size and weight, ruggedness, flexibility, and low cost.

Defence

Applications of laser in defence mainly include ranging, guiding weapons to intended targets, and the laser beam itself acting as a weapon.

Ranging: Pulsed laser beam is used to determine the distance of the target that may be either stationary or in motion. This technique is similar to the one used in radar. If the target is stationary, time taken by the laser pulses to complete a round trip journey, from laser source to targen and back, is measured and is calibrated to read the distance of the target directly. If the object is moving, there is a change in frequency of the reflected laser beam. This change in frequency is known as *Doppler shift*. Knowing the Doppler shift, speed as well as direction of the moving object can be determined. With advanced techniques in range finding system, one can get information about shape, size, and nature of the target along with accurate ranging.

Guiding Weapons: In order to reduce the human element to minimum, in ground attacks, nowadays laser-guided weapons are used. The weapons are fitted with direction-sensing laser head and a direction controlling servo loop. Laser head illuminates the intended target, the radiations scattered from the target are received by the laser head again; from this information, the angle of line of sight of the target is determined. This angle is compared with the glide angle of the weapon, and an error signal, if any, is fed to servo loop, which corrects the direction of the weapon and directs it to the target. Carbon dioxide laser is used for this purpose as its infrared radiation (10.6 μm) is not absorbed by the fog, smoke, mist, etc. in the atmosphere.

Laser Weapons: Laser beam itself can be used as a weapon. It can be used to either disable or destroy the enemy weapons. Disabling the weapon is relatively easier as it usually involves damaging the direction-sensing and controlling unit fitted on the weapon. Electronic eye of the spy satellite can be damaged with laser beam, thereby making its normal functioning impossible. Both of these applications require relatively moderate power laser beams. Very-high-power (megawatt) laser beam can be used as a weapon to destroy the enemy weapons. However, it is difficult to keep track of the traget and focus the laser beam on it effectively. This is because of changes in atmospheric conditions produced due to passage of high-power laser beam.

Mechanical Industry

Material processing is of prime importance in mechanical industry. It mainly consists of cutting, drilling, and welding. These processes involve transfer of power from laser beam to a workpiece. Laser-assisted material processeing is very superior over traditional methods, it not only saves power but also avoids unnecessary thermal shocks, in turn leading to better control over quality of workpiece. This is mainly because laser beams can be focussed to a very fine spot and can deliver power in short-duration pulses. With laser, one can process all types of materials such as metal, non-metal, ceramic, and plastic. Laser cutting, drilling, and welding processes can be automated easily.

Drilling: Drilling is brought about by rapid evaporation of material. This is easily possible as the laser beam can be focussed on very fine spot and large power can be employed. Short-duration laser pulses avoid radial dissipation of heat and, in turn, assure better process control. Laser drilling is very effective in case of deformable materials such as plastic nozzles, nylon buttons, and rubber nipple for milk bottle of babies. Laser drilling is very advantageous over traditional methods in case of brittle materials such as ceramic and glasses as it avoids changes in shape, size, and position of the hole drilled. As laser beam can be directed in any direction, drilling is possible in any direction and at places difficult to reach. This is one more advantage over traditional drilling processes. Laser beam itself is used as a drill bit so the idea of physical drill bit is getting outdated.

Cutting: The process of cutting involves removal of material. Traditionally, cutting is carried out with the help of blades having thickness of the order of millimetre. This results in large curf losses. This can be minimised in orders of magnitude as fine focussing of laser beam is possible. After repeated use of the cutting blades, they become dull and needs replacement. This problem does not arise in cutting with laser beam. Cutting along a particular curvature is a very difficult task with conventional method. This is very easy with laser beam as a cutter. Laser cutting is very fine and precise, as it produces no mechanical distortion, no lateral dissipation of heat. Unnecessary heating of the workpiece is avoided during laser cutting. Thie ensures the quality of the job. Materials of various kinds such as metals, non-metals, alloys, composites, diamond, glasses, ceramics, wood, and cloth can be cut with required precision with laser cutter.

Welding: Laser welding is a non-contact process and hence introduces no impurity along the line of welding. With laser welding, mechanical distortion is at the minimal level compared with traditional methods. Thermal properties of the materials welded remains almost intact as the power employed is at the optimum level. Laser welding can be carried out with great case even along a curvature and at places difficult to reach.

Heat Treatment: Heat treatment is common in automobile industry. Heat treatment is a process in which the material is heated so as to strengthen it. Due to heat treatment, surface of the material gets converted into crystalline state, which is harder and hence more resistant to wear. Laser-assisted heat treatment is very useful as it can be selectively applied to the desired area only, which may be difficult to reach. With laser-assisted heat treatment, being a non-contact process, the whole workpiece remains stress free.

Electronic Industry

Lasers are very effectively and economically used for material processing in electronic industry. Some of the applications are as follows:

Scribing: Scribing of semiconductor and ceramic wafers is an important and a skillful step in electronic industry. This involves drawing fine lines on very thin and brittle semiconductor wafers, having thousands of integrated circuits on them, in order to break them along the line without damaging the other parts. This can be very effectively and economically done with lasers.

Soldering: In electronic industry, soldering of different materials is very common. Some materials such as platinum, and silver are difficult to solder. Laser beam can be very efficiently used in such cases. Laser offers a non-contact and fluxless soldering without damaging the nearby part of the sheet.

Trimming: Thin film resistor trimming is very effectively achieved using lasers compared to traditional methods. Resistance of the film inversely depends on its thickness. In order to have resistance of a particular value, a thick film is taken and is trimmed using pulsed laser while monitoring its thickness and hence the resistance. Use of pulsed laser minimises the thermal shocks to the films and hence assures the quality of the product.

Photolithography: Photolithography technique is used to create circuit patterns on semiconductor wafers. Fine laser beam plays a very important role in drawing complicated and dense circuit patterns with great accuracy, which is impossible with ordinary light sources.

Consumer Electronic Industry

Lasers are routinely used as bar code readers. Consumer items in super markets and books in the libraries are usually bar code labelled. Bar code consists of a series of dark and bright lines. When laser beam scans the bar code, the bright lines reflect the laser light while dark lines absorb it producing a particular pattern of light. This light modulation is registered by a photosensitive detector. It is further processed by the central computer, and the list of the items, their prices, and other relevant details are printed.

Lasers are also used to read and write the data on compact discs. Lasers find many applications in consumer electronic industry and the number is increasing day by day.

Nuclear Energy

Isotope Separation: Isotopes are the elements having same atomic number but different mass numbers. As the number of electrons is same, they have same chemical properties, and hence methods based on chemical properties cannot be used for separating them. Methods based on physical properties are used for separating them.

Natural uranium ore contains two main isotopes, U^{238} and U^{235}, approximately in the proportion 99.3% and 0.7% respectively. However, to operate the nuclear reactor, minimum percentage of U^{235} in the ore needs to be about 3%. The process of increasing the percentage of U^{235} in the natural ore is known as the *enrichment* of the ore. Traditionally used gascon diffusion technique for this purpose is costly and time consuming. Laser-assisted isotope separation technique is faster as well as cheaper. As isotopes have slightly different nuclear masses, their energy levels are also slightly different. Hence, the radiations they absorb also have slightly different frequency. If the natural ore containing a mixture of isotopes is irradiated with appropriate frequency, it is possible to excite one isotope without affecting the other. For this, highly monochromatic radiation source is essential. Only lasers can provide such small bandwidth radiation sources.

Isotope separation is a two-step process: Photoionisation and photodeflection. At first, the natural ore is irradiated with highly monochromatic radiation, and selective excitation of U^{235} is brought about without exciting U^{238}. By applying another high-eneryg laser pulse, selective ionisation of U^{235} is brought about. In the second step, selectively ionised U^{235} is separated from the mixture using electrostatic fields, and thus enrichment of the ore is achieved.

Thermonuclear Fusion: Fusion is a process in which two lighter elements (lighter nuclei) are compressed to such a great extent that their nuclei fuse to form a new heavier element (heavy nucleus). The mass of newly formed element is less than the sum of the masses of the fusing elements. The difference in the mass is converted into energy according to Einstein's mass–energy equivalence relation

$$E = mc^2$$

Controlled thermonuclear fusion reaction offers a viable option as a practical inexhaustible source of cheap and clean energy. However, extremely high temperature and pressure are required for this purpose. These conditions are very hard and can be achieved only with laser. In laser-assisted *inertial confinement* method, the fusion fuel pellet, comprising heavy isotopes of hydrogen, namely, deuterium and tritium, and of size of about 0.1 mm, placed in a reaction chamber, is suddenly bombarded with intense, high-energy laser pulses. As a result, the surface of the pellet is blast away with the rocket-like reaction forces, and the target pellet is compressed to required high densities (about 10^3 times the liquid density) and high temperatures (about 10^{10} K). This causes nuclei to fuse releasing large amount of energy.

In Shiva laser system, which was put in operation in 1978 at Lawrence Livermore Laboratories in USA, 20 neodymium laser beams were focussed on the pellet to start the fusion process. Nova laser, 10 times more powerful than Shiva laser, was put in operation at the same place on December 19, 1984.

Holography

Light scattered from an object carries information about the object in the form of amplitude and phase. In ordinary photographs, only amplitude of the light scattered from various parts of the object is recorded. The morphological information of the object which is contained in the phase of the light scattered is lost. Thus, the ordinary photograph does not contain the whole information of the object. It is a 2D record of the 3D object. Recording of the whole information of an object, i.e. recording amplitude and phase of the scattered light, is called *holography*. The Greek word *holos* means whole and *graphy* means recording. Dennis Gabor (1990 – 1979), an English physicist working in research laboratory of the British Thomson Company, UK, invented this entirely new technique of photography in 1947. He got Nobel Prize for the work on holography in 1971. Holography is fundamentally different from photography. Holography is possible only with highly coherent light. It is a 3D recording of the 3D object. Special optics is involved in holography. One more very interesting point about holography is that even if the holograph is broken into pieces, every piece carries the entire information though the quality of the image gets hampered.

The basic principle of recording a holograph is depicted in Figs 7.21a and b. To obtain a holograph, a laser beam is split into two using beam splitter. One is called *illuminating beam* and the other is called *reference beam*. The illuminating beam is used to illuminate the object to be holographed. The illuminating beam scattered from the object is called *object beam or carrier beam* as it carries information of the object. The holographic plate is exposed to the object beam and the reference beam simultaneously, and it records the interference pattern formed by them as they are mutually coherent. The holographic plate is developed and then reilluminated with the same interference beam. Most of the light from reference beam passes through the holograph directly; however, some of it is diffracted due to the interference pattern recorded on the

holographic plate. The diffracted beam is identical with an object beam, and for an observer, the diffracted beam appears to come from the object itself. A properly prepared and well reilluminated holograph appears as a window, and an observer gets an experience of viewing a real object through the window with different angles and elevations.

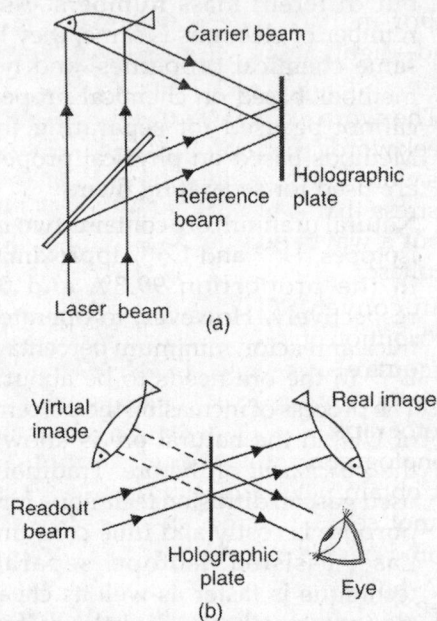

Fig. 7.21: The principle of holography (a) Construction of a hologram (b) Reconstruction of the 3D image

Fundamental Research

Some experiments of fundamental importance can be carried out more easily and with greater accuracy using laser sources. Few of them are Brownian motion, ether drift, absolute rotation of earth, counting of atoms, laser-assisted cooling, and trapping of atoms.

Lasers have revolutionised the field of optics. It is now possible to have laser-like sources capable of producing intense, highly directional, highly monochromatic, and coherent beam of atoms having same de Broglie wavelength. These sources of coherent matter waves are called *atom laser*. *Optical lasers* emit coherent electromagnetic waves (that is photons), whereas the atom lasers emit coherent matter waves. Atom laser promises to revolutionise the *atom optics*.

7.18 HOLOGRAPHY

Introduction

Holography is a way of recording and then reconstructing waves invented by Dennis Gabor in 1948. The waves may be of any kind—light, sound, X-ray, corpuscular waves, etc.

The word *holography* originates from the Greek word *holos* meaning the whole. By using this word, the inventor of holography wanted to stress that it records complete information about a wave—both about its amplitude and its phase.

In conventional photography, only the distribution of the amplitude (more exactly, of its square) is recorded in two-dimensional projection of an object onto the plane of the photograph. For this reason, when examining a photograph from various directions, we do not obtain new angles of approach, and we cannot see, for instance, what is happening behind objects in the foreground.

A hologram, on the contrary, regenerates not a two-dimensional image of an object, but the field of the wave which it scatters. By changing our point of observation within the confines of this wave field, we see the object from different angles, sensing its three dimensional and realistic nature.

It is worthwhile to mention that Gabor did not have a laser when he formulated the idea of holography. He performed his first experiment with mercury arc lamp as source of light. The technique of preparing holography became a practical proposition only after the advent of lasers. In 1963, E. Leith and J. Upatnieks prepared laser holograms for the first time. This opened the way to major advances in holography.

The principle of holography can best be explained in two steps:

1. Recording of the hologram
2. Reconstructing the image.

A schematic diagram showing how holograms are recorded is depicted in Fig. 7.22.

i. **The object whose hologram is to be illuminated with a laser beam:** The scattered light wave impinges on a photographic plate. The reference beam-part of the light from the same laser reflected from a mirror impinges on the same plate. The superposition of these two beams produces an interference pattern and this is recorded on that plate. The developed plate is called

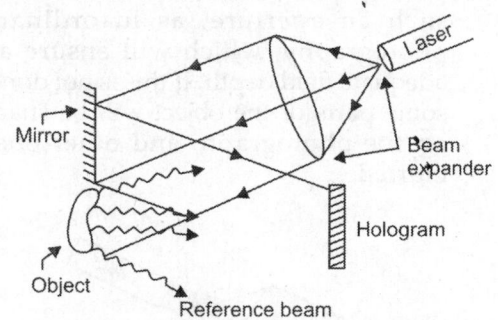

Fig. 7.22: Recording a hologram

as the *hologram*. Since the interference pattern of a hologram is generally very fine, a developed hologram does not differ in appearance from a uniformly exposed plate. A hologram often contains rings and strips. They are due to the diffraction of light on particles of dust getting onto the mirrors and lenses, and have nothing in common with the interference microstructure containing the record of the light wave scattered by the object. The hologram contains, however, enough information to provide a complete reconstruction of the object.

ii. **Reconstruction of the image from a hologram:** Remove the object and put the hologram in the place where it was when formed, then switch on the laser (Fig. 7.23). Now look through the hologram like through a window, you will see the object at its previous place, as if it were not removed, i.e. when light from a laser is directed at the hologram from the same direction as the reference beam, part of light is 'bent' so that it appears to come from the place once occupied by the object and the result is remarkably realistic 3 dimensional image. The visible object seems to real that we can detect parallax by changing the position of our head. When looking at the near and far parts of the object, we have to accommodate our eyes differently, and if we want to photograph it, we will have to choose such an aperture, as in ordinary photography, which will ensure an adequate field depth. If this is not done, some parts of the object will be sharp on the photograph, and other ones blurred.

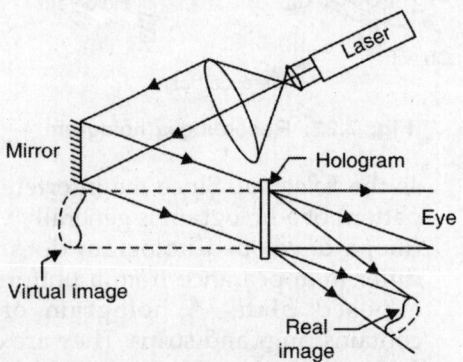

Fig. 7.23: Reconstruction of the image from a hologram

From Fig. 7.23, it is clear that two images are produced when the laser beam, which is now broken as the *readout wave*, interacts with the interference pattern on the holographic plate and two images are produced by the diffracted waves. One of them appears at the original position occupied by the object called the *virtual image* and the other is called the *real image* which can be photographed directly without using a lens. Since, the real image reverses foreground and background and hence the interest of the observer lies in the virtual image.

Theory

To understand the various characteristics of holograms it is essential to understand the theory on which recording of a hologram and reconstruction of the image is based.

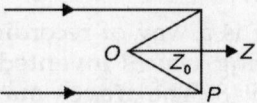

Fig. 7.24: Hologram of a point object O

Let us consider a small object O. Let it be illuminated by a source of light, but most of the light falls undisturbed on a photographic plate (Fig. 7.24). The light diffracted or scattered by the object also reaches the photographic plate P. The direct beam, so called the reference beam, interferes with the light reaching from the object. Let us consider a point on the photographic plate (say Q). We are interested in finding the intensity at this point Q. One can write the field arriving at Q as the sum of the two fields— the field due to the reference beam, E_r and and field scattered from the object, E_s, i.e.

$$E = E_r + E_s \tag{7.32}$$

It is worthwhile to mention that the scattered field E_s is not simple, both amplitude and phase of the field vary greatly with position. The field of such a wavefront can be represented by

$$E_s = \frac{A_0}{r_0} \exp\left[i(kr_0 - \omega t) \right] \tag{7.33}$$

Here $r_0 = PQ$ (Q is not shown in Fig. 7.24).

The reflected wavefronts are spherical and concentric around the point of origin. One can represent the field E_r by plane wave as follows:

$$E_r = A_r \exp[i(kZ_0 - \omega t)] \tag{7.34}$$

where Z_0 is the distance of the photographic plate from O.

The intensity at the point Q is given by

$$I = |E_r + E_s|^2$$

Substituting the values of E_s and E_r from Eqs. (7.33) and (7.34) respectively, one obtains

$$I = |A_r|^2 + \frac{|A_0|^2}{r_0^2} + \frac{A_0 A_r^*}{r_0} \exp\left[ik(r_0 - Z_0)\right]$$

$$+ \frac{A_0^* A_r}{r_0} \exp\left[ik(Z_0 - r_0)\right] \quad (7.35)$$

Choosing the constants k and ϕ suitably, one can combine last two terms in Eq. (7.35) and obtain

$$I = |A_r|^2 + \frac{|A_0|^2}{r_0^2} + k \frac{\cos\left(k(r_0 - Z_0) + \phi\right)}{r_0} \quad (7.36)$$

Equation (7.36) shows that the blackening of the photographic plate depends on the three terms, the last term being the most important as it gives the effect of the functions A_0 and ϕ. Obviously, the cosine term shows that the total intensity I as function of r_0 shows a series of maxima and minima. The interference of the plane wave E_0 with the spherical wave E_s thus produces a set of circular interference fringes on the photographic plate, which when developed forms the hologram.

If one assumes that the response is proportional to the intensity I, the power transmission of the plate represented by T^2 is given by

$$T^2 = 1 - I\alpha \quad (7.37)$$

where α is a constant. Equation (7.37) can be approximated as

$$T \cong 1 - \frac{1}{2} I\alpha \quad (7.38)$$

Reconstruction

Let us consider that this hologram is illuminated by a beam which is identical to the reference beam, the field of the transmitted wave will be given the expression

$$E_1 = T E_r \left(1 - \frac{\alpha}{2} I\right) A_r \exp\left[i(kZ_0 - \omega t)\right]$$

$$= \left[1 - \frac{\alpha}{2}|A_r|^2 - \frac{\alpha}{2} \frac{|A_0|^2}{r_0^2}\right] A_r \exp\left[i(kZ_0 - \omega r)\right]$$

$$- \frac{\alpha}{2} \frac{A_0 A_r^*}{r_0} \exp\left[ik(r_0 - Z_0)\right] A_r \exp\left[i(kZ_0 - \omega t)\right]$$

$$- \frac{\alpha}{2} \frac{A_0^* A_r}{r_0} \exp\left[ik(Z_0 - r_0)\right]$$

$$\times A_r \exp\left[i(kZ_0 - \omega t)\right] \quad (7.39)$$

$$= \left[1 - \frac{\alpha}{2}|A_r|^2 - \frac{\alpha}{2} \frac{|A_0|^2}{r_0^2}\right] A_r \exp\left[i(kZ_0 - \omega t)\right]$$

$$- \frac{\alpha A_0^* |A_r|^2}{2r_0} \exp\left[i(kr_0 - \omega t)\right]$$

$$- \frac{\alpha}{2} \frac{A_0^* |r|^2}{r_0} \exp\left[i(ikZ_0 - kr_0 - \omega t\right]$$

One can distinguish the following three terms on the RHS of Eq. (7.39):

i. The first term represents the attenuated incident wave.

ii. The second term represents the wave surface identical to the wave surface emitted by the object except for a constant factor. Thus, this wave surface when projected back toward the illuminating source, seems to emanate from an apparent object located from a place where the original object was located. We say that these waves produce a virtual image, and an observer who looks through the plate would see the object in three dimensions (Fig. 7.25). As stated earlier, one can take a photograph of the image by a camera.

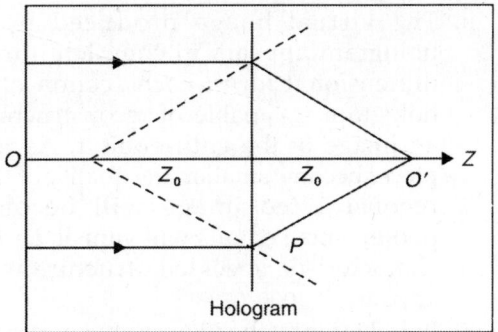

Fig. 7.25: Reconstruction of the image from a hologram

iii. The third term represents a wave surface which is also a replica of the regional wave, except that they have conjugate, or reversed curvature; originally diverging spherical waves from an object are converted into converging spherical waves. It converges at a point O' and produces a real image at this point. One can photograph this real image without a lens. Thus, a hologram produces a real image O' and a virtual image O.

Gas lasers have high coherence and continuous wave mode and hence they are often used for holography. Moreover, the emitting power of gas lasers is low and hence the time of exposure is large. Therefore, using gas lasers moving objects cannot be holographed. To prepare the holograph of moving objects and to record the development of a process in time, pulsed solid state lasers are often used. With the use of pulsed solid state lasers one can cut down the exposure time to about 10^{-3} sec.

Some Important Properties of a Hologram

i. The hologram acts somewhat like a simple periodic diffraction grating. It produces a beam direct and two first-order diffracted beams on either side of the direct beam. The zero order of the diffracted light is the direct beam, while the two first orders on either side comprise the virtual and the real images.

ii. The virtual image produced by a hologram appears in complete three dimensional form. Each section of a hologram is capable of reconstructing an image of the entire object. As the pieces become smaller, the quality of the reconstructed image will become poorer, small details will vanish, and a characteristic speckled structure will appear.

iii. In holography the viewer always sees a positive image whether a positive or a negative photographic transparency is used for the hologram. The reason for this is that a negative hologram merely produces a wave field that is shifted 180° in phase with respect to that of a positive hologram. Since the eye is insensitive to this phase difference, the view seen by the observer is identical in the two cases.

iv. A hologram is a reliable method of data storage. The information about a point object is recorded over the whole area of the hologram and each such point of a real object is recorded on the whole hologram and not in one of its points as is the case in photographic plate.

v. A hologram of a complex object can be considered as an interference (coherent) superposition of holograms from individual points or more complex portions of the object. The amplitudes of light waves are summated on account of the phase relations between them in such superposition.

We can also imagine a hologram that is an incoherent superposition of holograms of different objects or parts of the same object. In this case, the photolayer summates the illumination they create. If the number of such consecutive superpositions is not too great, the hologram will simultaneously reconstruct several consecutively recorded light waves without any appreciable distortions. This property of hologram is used for consecutive recording of the waves from several objects or several states of the same object on a single hologram.

vi. The curious property of the wavefront reconstruction process is that it does produce negative. The hologram itself would normally be regarded as a negative but the image it produces is a positive. A hologram copied by contact printing would be reversed in the sense that opaque areas would now become transparent and *vice versa*. The image reconstructed from the copy, however, would remain a positive and would be indistinguishable from the image produced by the original except for the small degradation in quality.

Applications of Holography

Holography has wide range of applications in science and technology. The main applications of holography are listed below:

  i. Three dimensional images: holographic, cinematography and television.
  ii. Holographic interferometry
  iii. Spatial filtration and character recognition
  iv. Spectroscopy
  v. In production processes and optical engineering.

The conventional holography needs coherent light. This is a drawback. Recent advances in techniques, e.g. *sandwich holography* offer a new method of storing confidential data. The *rainbow* (white light transmission) holography is a significant achievement because it uses while light.

ILLUSTRATIVE EXAMPLES

Example 1

A laser bean has a power of 59 mW. It has an aperture of 5×10^{-3} m and it emits light of wavelength 7200 Å. The beam is focussed with lens of focal length 0.1 m. Show that the area spread and the intensity of the image are $2.074 \times 10^{-10} \, m^2$ and $2.4 \times 10^8 \, W/m^2$ respectively.

Solution

$$d\theta = \frac{\lambda}{d}$$

$$\lambda = 7200 \times 10^{-10} \, m$$

$$d = 5 \times 10^{-3} \, m$$

$$d\theta = \frac{7200 \times 10^{-10}}{5 \times 10^{-3}} = 1.44 \times 10^{-4} \text{ radian}$$

Area spread $= (d\theta f)^2$

Here $\quad f = 0.1$ m

∴ Area spread $= (1.44 \times 10^{-4} \times 0.1)^2$

$$= 2.074 \times 10^{-4} \, m^2$$

$$\text{Intensity} = \frac{\text{Power}}{\text{Area}} = \frac{\text{Power}}{\text{Area spread}}$$

$$= \frac{50 \times 10^{-3} \, W}{2.074 \times 10^{-10} \, m^2} = 2.411 \times 10^8 \, W/m^2$$

Example 2

A laser beam has a wavelength of 8×10^{-3} m. The laser beam is sent to moon. The distance of the moon is 4×10^5 km from the earth's surface. Determine (i) the angular spread of the beam (ii) the areal spread when it reaches the moon's surface.

Solution

We have $\quad d\theta = \lambda/d$

$$\lambda = 8 \times 10^{-7} \, m$$

$$d = 5 \times 10^{-3} \, m$$

∴ Angular spread

$$d\theta = \frac{8 \times 10^{-7}}{5 \times 10^{-3}} = 1.6 \times 10^{-4} \text{ radian}$$

Areal spread $= (\Delta d\theta)^2 \mid \Delta = 4 \times 10^5$ km

$$= 4 \times 10^8 \, m$$

$$= (4 \times 10^8 \times 1.6 \times 10^{-4})^2 = 4.096 \times 10^9 \, m^2$$

Example 3

A laser beam of wavelength 6000 Å on earth is focussed by a lens (or mirror) of diameter 2 m on to a crater on the moon's surface. The distance of the moon from the earth is $4 \times 10^8 \, m$. Neglecting the effect of earth's atmosphere show that the angular and areal spreads of the spot on the moon will be 3×10^{-7} radian and $1.44 \times 10^4 \, m^2$ respectively.

Solution

Given that

$$\lambda = 6000 \times 10^{-10} \, m$$

$$d = 2 \, m$$

$$\Delta = 4 \times 10^8 \, m$$

Angular spread

$$d\theta = \frac{\lambda}{d} = \frac{6 \times 10^{-7}}{2} = 3 \times 10^{-7} \text{ radian}$$

Example 4

A laser beam has aperture $d = 1.8 \times 10^{-2}$ m and emits radiation of wavelength $\lambda = 5 \times 10^{-7}$ m. Calculate (i) the semi-angle of the cone of its beam (ii) the solid angle of this cone.

Solution

i. The semi-angle of the cone of the laser beam

$$\theta = \frac{\lambda}{\alpha} = \frac{5 \times 10^{-7} \text{ m}}{1.8 \times 10^{-2} \text{ m}} = 2.72 \times 10^{-5} \text{ radian}$$

ii. Solid angle of this cone

$$\phi = \pi \frac{\Delta^2 \theta^2}{\Delta^2} = \pi \theta^2 = \frac{22}{7} \times \left(2.72 \times 10^{-5}\right)^2$$

$$\approx 2.5 \times 10^{-9} \text{ starad}$$

Example 5

In Example 4, calculate the areal spread of the beam as it reaches the moon's surface. The distance of the moon from the earth is 4×10^8 m.

Solution

Areal spread $= \Delta^2 \phi = 4 \times 10^8$ m

Example 6

The coherence length for sodium light is 2.945×10^{-2} m. The wavelength of sodium light is 5890 Å. Determine (i) the number of oscillations corresponding to coherence length and (ii) the coherence time.

Solution

i. Given that
$$\lambda = 5890 \times 10^{-10} \text{ m}$$
$$c = 3 \times 10^{10} \text{ m/s}$$
$$L = 2.945 \times 10^{-2} \text{ m}$$

Number of oscillations in coherence length L,

$$n = \frac{L}{\lambda} = \frac{2.945 \times 10^{-2}}{5890 \times 10^{10}} = 5 \times 10^4$$

ii. Coherence time

$$= L/c = \frac{2.945 \times 10^{-2}}{3 \times 10^8} = 9.8 \times 10^{-11} \text{ s}$$

Example 7

The coherence length for sodium D_1 line is 2.5 cm. Determine (i) spectral width of the line $\Delta\lambda$ (ii) the purity factor, Q (iii) the coherence time τ. The wavelength of light is 6000 Å.

Solution

i. $\Delta\lambda = \dfrac{\lambda}{2L} = \dfrac{36 \times 10^{-10}}{5} \text{ cm} \approx 7 \text{ A}$

ii. $Q \approx \dfrac{\lambda}{\Delta\lambda} \approx \dfrac{6 \times 10^{-5}}{7 \times 10^{-8}} \approx 10^3$

iii. $\tau = \dfrac{L}{c} = \dfrac{2.5}{3 \times 10^8} \text{ sec} \cong 0.8 \times 10^{-10} \text{ s}$

Example 8

Find the ratio of population of the two energy states in a laser the transition between which is responsible for the emission of photons of wavelength 698.3×10^{-9} m. Assume the temperature to be 300 K.

Solution

Given, $T = 300$ K, $\lambda = 698.3 \times 10^{-9}$ m; $\dfrac{N_2}{N_1} = ?$

Let ΔE be the energy difference between the two energy states.

$$\therefore \quad \frac{N_2}{N_1} = e^{\left(\frac{-\Delta E}{k_B T}\right)}. \text{ But } \Delta E = \frac{hc}{\lambda}$$

Hence, $\dfrac{N_2}{N_1} = e^{-\left(\frac{6.625 \times 10^{-34} \times 3 \times 10^8}{698.3 \times 10^{-9} \times 1.38 \times 10^{-23} \times 300}\right)}$

$$= e^{-\left(\frac{1.9875 \times 10^{-25}}{2.89096 \times 10^{-27}}\right)} = e^{-68.748}$$

$$\therefore \quad \frac{N_2}{N_1} = 1.3892 \times 10^{-30}.$$

Example 9

A laser source emits light of wavelength 0.621 μm and has an output of 35 mW. Calculate how many photons are emitted per minute by this laser source.

Solution

Given, the power output, i.e. the total energy

$E = 3.5$ mW

$= 3.5 \times 10^{-3}$ J s^{-1} = $3.5 \times 10^{-3} \times 60$ J/minute

$\lambda = 0.621$ μm = 0.621×10^{-6} m

The frequency of the photon emitted is given by

$$\nu = \frac{c}{\lambda}$$

$$\therefore \quad \nu = \frac{3 \times 10^8}{0.621 \times 10^{-6}}$$

$$\nu = 4.8309 \times 10^{14} \text{ Hz.}$$

But we know, $E = h\nu$

$$\therefore \quad E = 6.625 \times 10^{-34} \times 4.8309 \times 10^{14}$$

$$E = 3.200 \times 10^{-19} \text{ J.}$$

This is the energy emitted by one photon.

∴ The number of photons emitted per minute is

$$n = \frac{\text{Total energy emitted per minute}}{\text{Energy of one photon}}$$

$$\therefore \quad n = \frac{3.5 \times 10^{-3} \times 60}{3.200 \times 10^{-19}} = \frac{0.21}{3.2 \times 10^{-19}}$$

$$n = 6.562 \times 10^{17} \text{ photons/minute.}$$

Example 10

Light from a 2.5 mW laser source of aperture diameter 1.8 cm and $\lambda = 500$ nm is focussed by a lens of focal length 20 cm. Compute:

(a) The area

(b) The intensity of the image.

Solution

Given

$$\lambda = 500 \text{ nm} = 5 \times 10^{-7} \text{ m}$$

$$2a = 1.8 \text{ cm}$$

$$= 0.018 \text{ m}$$

$$a = 0.009 \text{ m}$$

$$f = 20 \text{ cm} = 0.20 \text{ m}$$

(a) Area of the spot at focal plane $= \dfrac{\pi \lambda^2 f^2}{a^2}$

$$= \frac{\pi \times (5 \times 10^{-7})^2 \times (0.20)^2}{(0.009)^2}$$

$$= 3.88 \times 10^{-10} \text{ m}^2$$

(b) Intensity at the focus

$$I = \frac{P a^2}{\pi \lambda^2 f^2}$$

$$= \frac{2.5 \times 10^{-3} \text{ W}}{3.88 \times 10^{-10} \text{ m}^2}$$

$$= 6.44 \times 10^6 \text{ W/m}^2$$

Example 11

In a ruby laser, total number of Cr^{+++} ions are 2.8×10^{19}. If the laser emits radiation of wavelength 700 nm, what will be

(a) The energy of one emitted photon

(b) The total energy available per laser pulse?

Solution

Given

N = total number of photons = 2.8×10^{19}

$\lambda = 700$ nm = 7.0 nm = 7.0×10^{-7} m

(a) The energy of photon

$$E = \frac{12400}{\lambda(\text{Å})} = \frac{12400}{7000} = 1.77 \text{ eV}$$

$$= 1.77 \times 1.6 \times 10^{-19} \text{ J}$$

$$= 2.832 \times 10^{-19} \text{ J}$$

(b) Energy per pulse = Energy of one **photon**

$$\times \text{ Total number of photons}$$

$$= 2.832 \times 10^{-19} \times 2.8 \times 10^{19} \text{ J}$$

$$= 7.93 \text{ J}$$

Example 12

A He–Ne laser operating at 632.8 nm has **an** output power of 1 mW with 1.0 mm **beam** diameter. The beam comes out through **the** mirror which has 1% transmittance at **laser** wavelength. Find the ratio of stimulated emission to spontaneous emission $B_{21} u(\nu)/A_{21}$. The line width of laser line is 1.5×10^8 Hz.

Solution

The frequency of laser is

$$\nu = \frac{c}{\lambda} = \frac{3 \times 10^8}{632.8 \times 10^{-19}} = 4.74 \times 10^{14} \text{ Hz}$$

We know that

$$\frac{A_{21}}{B_{21}} = \frac{8 \pi h \nu^3}{c^3}$$

Therefore,

$$\frac{B_{21}}{A_{21}} = \frac{c^3}{8 \pi h \nu^3}$$

$$= \frac{(3 \times 10^8)^3}{8 \times 3.14 \times 6.63 \times 10^{-34} \times (4.74 \times 10^{14})^3}$$

$$= 1.52 \times 10^{13} \text{ m}^3/\text{Js}$$

$u(\nu)$ is obtained from the fact that intensity is the product of energy density and speed of light. Intensity is the energy per unit area per second. The energy within resonator is 99 mW.

$$\therefore u(\nu) = \frac{1}{c d \nu} = \frac{(99 \times 10^{-3})/(\pi \times 0.5 \times 10^{-3})^2}{3 \times 10^8 \times 1.5 \times 10^9 \times 1.5 \times 10^8}$$

$$= 2.80 \times 10^{-12} \text{ J/s/m}^3$$

The desired ratio

$$\frac{B_{21} u(\nu)}{A_{21}} = \left(\frac{B_{21}}{A_{21}} \right) u(\nu)$$

$$= 1.52 \times 10^{13} \times 2.80 \times 10^{-12} = 42.6$$

Example 13

Estimate the angular spread of a laser beam of wavelength 693 nm due to diffraction if the beam emerges through a 3 mm diameter mirror. How large would be the diameter of this beam when it strikes a satellite 300 km above the earth?

Solution

We know that diffraction effect limits the resolving power of an optical system. The minimum angle of resolution provided by a lens of diameter D is

$$\theta_{\min} = \frac{1.22 \lambda}{D},$$

Here $\theta_{\min} \approx \Delta \theta$

$$\lambda = 693 \times 10^{-9} \text{m}; \quad D = 3 \times 10^{-3} \text{ m}$$

Angular spread,

$$\Delta \theta = \frac{1.22(693 \times 10^{-9} \text{m})}{(3 \times 10^{-3}) \text{m}}$$

$$= 2.82 \times 10^{-4} \text{ rad}$$

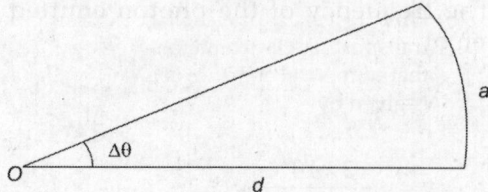

Diameter of the beam on the satellite,

$$a = \Delta \theta \cdot d$$

$$= (2.82 \times 10^{-4} \text{ rad}) \times (300 \times 10^3 \text{m})$$

$$= 84.6 \text{ m}$$

REVIEW QUESTIONS

1. Explain the terms absorption, spontaneous emission and stimulated emission of radiation. Obtain a relation between transition probabilities of spontaneous and stimulated emissions.

2. What is a laser ? Explain the salient features of a laser radiation. Mention some of its principal uses.

3. What are the important features of stimulated emission? Discuss the essential requirements for producing laser action?

4. What is the difference between temporal and spatial coherence? Explain coherence time and coherence length.

5. Give an elementary theory of the laser.

6. Describe the construction and working of a ruby laser with necessary diagrams.

7. Describe a He–Ne gas laser. How is population inversion achieved in this type of laser?

8. What are Einstein's A and B coefficients? Obtain a relationship between them.

9. Explain the role of an optical resonator in a laser. How population inversion is achieved in a laser?

10. Explain what do you understand by the modes of a cavity resonator. Explain quality factor.

11. Explain the principle and working of a semiconductor laser. Derive the condition for the predominance of the stimulated emission over absorption for this laser device.

12. Explain the principle of holography and discuss its salient features. Mention its uses.

13. Explain the basic principle of holography. What are salient features of holography. Mention some important uses of holography.

14. (a) Distinguish between spontaneous and induced emissions. How does induced emission dominates in He–Ne laser?

   (b) How does an optical fibre function in transporting electromagnetic energy? Show that numerical aperture of a step index fibre is given by

$$NA = \frac{\sqrt{2\Delta}}{n_0}$$

   where symbols have their usual meanings.

15. (a) Describe the construction and working of He–Ne laser. How is population inversion achieved in such a laser?

   (b) A laser beam has a wavelength of 700 mm and aperture of 5 mm. The laser beam is sent to the moon. The distance of the moon is 4×10^5 km from the earth. Calculate (i) the angular spread of the beam (ii) areal spread when it reaches the moon.
   [Raj. BE 2002]

16. (a) In He–Ne laser, what is the function of the He atoms? Explain the answer with the help of energy level diagram for He–Ne. Describe with a neat sketch the working of He–Ne laser.

   (b) Derive the relation between Einstein's coefficients and discuss the results.
   [Raj. BE 2002]

17. (a) Describe a ruby laser and explain its working, showing the effective energy levels.

   (b) Describe the method of recording the hologram and reconstruction of image from it.
   [Raj. BE 2001]

18. (a) What are essential requirements to produce a laser? How are these fulfilled in a He–Ne laser?

   (b) What is an optical fibre? What do you mean by numerical aperture of an optical fibre? Find an expression for the numerical aperture of a stop index optical fibre.

19. Explain the essential requirements for producing laser action. How are these requirements obtained in case of ruby laser? Draw a diagram to represent the component of ruby laser.
   [Raj. BE 2000]

20. (a) Explain how population inversion is obtained in ruby laser.

   (b) Explain the basic principles of holography.
   [Raj. BE 2000]

21. Describe the laser action in helium neon laser with energy diagram. Describe various applications of laser.
   [Raj. BE 1999]

22. Write short notes on the following:

   (a) Ruby laser
   (b) Gas laser
   (c) Semiconductor laser
   (d) Einstein's A and B coefficient
   (e) Optical resonator
   (f) Optical fibre
   (g) Holography
   [Raj. BE 2002]
   (h) Quantum well laser.

PROBLEMS

1. Determine the minimum difference between the two arms of a Michelson interferometer that seeks to measure two closely spaced wavelengths using He–Ne laser for which the coherence length is 600 km. [*Ans.* 300 km]

   [**Hint:** $2d = n_1 l_1 = n_2 \lambda_2$ or $2d\left(\dfrac{1}{\lambda_1} - \dfrac{1}{\lambda_2}\right) = n_2 - n_1 = m$,

   where n_1, n_2 and m are integers. Putting

   $\Delta\lambda = \lambda_1 - \lambda_2$, $m = \dfrac{2d\Delta\lambda}{\lambda^2} = \dfrac{2d\Delta v}{v\lambda} = \dfrac{2d}{L_c}$ where L_c is

   coherence length. Obviously, the minimum difference between the two atoms of the

   interferometer is $d = \dfrac{L_c}{2} = \dfrac{600}{2} = 300\,km$]

2. The output power of a gas laser is 1 mW and the emitted wavelength is 630 nm. Show that the number of photons emitted per second is about 3×10^{15} per second.

3. The coherence length of sodium D_2 line is 2.5 cm. If the velocity of light is 3×10^8 m/s, then show that the coherence time is 0.8×10^{-18} sec.

4. Calculate the spot size of a He–Ne laser beam at the centre of a confocal cavity if the length of the cavity is 1.24 m. The wavelength is 633 nm.
   [*Ans.* 0.5 mm]

5. Calculate the inversion density for He–Ne laser aperture at 633 nm if the temperature of the discharge is 150°C and the gain constant is 3 percent per metre. The lifetime of the upper state for spontaneous transition is 10^{-7} s.
   [*Ans.* $3.1 \times 10^{14}/m^2$]

6. The bandwidth of a He–Ne laser is 500 Hz. Obtain the coherence time and the longitudinal coherence length. [*Ans.* (i) 2 ms (ii) 600 km]

7. Calculate the gain constant β of a laser having the following parameters : wavelength = 650 nm, inversion density $(n_2 - n_1) = 5 \times 10^{22}/m^3$, lifetime for spontaneous emission 2×10^{-4} and line width $\Delta\lambda = 15$ Å. [*Ans.* 3.95]

8. Show that for a normal optical source with temperature about 10^3 K and wavelength 6000 Å, the emission is predominantly due to the spontaneous transitions.

9. A laser beam of wavelength 6000 Å, power 10 mW, and angular spread 5×10^{-5} radian is focussed by a lens of focal length 10 cm. Calculate the radius and power density of the image. What is lateral coherence width?
    [*Ans.* (i) radius = 2.5×10^{-4} cm (ii) power density = 5.1×10^4 W/cm^2 (iii) lateral coherence width = 1.2 cm].

10. For sodium line, the wavelength is 5890 Å and coherence time is 10^{-10}s. Show that the monochromaticity of the source is (5890 ± 0.0578) Å.

SHORT ANSWER QUESTIONS

1. Write salient features of laser radiation.
Ans. See text

2. What is the relation between the degree of non-monochromaticity (ζ) and longitudinal coherence length (L_c) or simply coherence length?
Ans. $\zeta = c/L_c \nu_0$

3. What is the relation between the coherence time (τ_c) and frequency spread of the wave $(\Delta\nu)$?
Ans. $\tau_c = 1/\Delta\nu$

4. In terms of quantum picture, when a light beam is said to be a perfectly coherent?
Ans. When all the constituent photons of a light beam have same energy, direction of momentum and identical polarization, the light beam is said to be perfectly coherent.

5. What is the relation between Einstein's A_{21} and B_{21} coefficients?
Ans. $A_{21}/B_{21} = 8\pi h\nu^3/c^3$

6. What do you understand by a metastable state?
Ans. Normally, the mean life of excited atoms before the spontaneous emission occurs is about 10^{-8} s. However, for some excited states, this mean life is perhaps as much as 10^5 times longer. We call such long-lived states metastable; they occupy an important role in laser action.

7. Mention any three methods of pumping for creating population inversion.
Ans. (i) Photon excitation (ii) Electron excitation (iii) Inelastic collision between atoms.

8. What is a hologram?
Ans. A hologram contains the encoded image of the object in the form of the interference pattern.

9. What is fundamental basis of laser operation?
Ans. Photons are bosons. Atomic decays accompanied by the emission of radiation are enhanced if there are already photons present that are identical to the emitted photon. This is the basis of laser operation.

OBJECTIVE QUESTIONS

1. A laser light is highly monochromatic and _____. [directional]

2. The two B coefficients of Einstein are _____. [equal]

3. The order of pumping power necessary to achieve the population inversion in a ruby laser is _____. [10^7 W/m^2]

4. Laser light is considered to be coherent because it consists of coordinated waves of exactly the _____ wavelength. [same]

5. A hologram is an encoded _____ of the object recorded on a photographic plate. [image]

6. A gyro laser detects slow rotational motions, down to _____ and less. [μrad/s]

MULTIPLE CHOICE QUESTIONS

1. The directionality of a laser beam is measured by
   (a) the size and aperture of laser source
   (b) the visibility of interference fringes
   (c) nature of lasing medium
   (d) the divergence angle of the beam with the distance from the source

2. The most important characteristic of a laser beam is
   (a) directionality   (b) polarization
   (c) coherence   (d) high intensity

3. The coherence of light is known by the
   (a) size and nature of the source
   (b) flickering of light beam
   (c) light intensity of the beam
   (d) visibility of the interference fringes it produces

4. A laser operates at a frequency of 3×10^{14} Hz and has an aperture of 10^{-2} m. The angular spread will be
   (a) 10^{-5} rad   (b) 10^{-3} rad
   (c) 10^{-2} rad   (d) 10^{-4} rad

5. The spontaneous emission is a
   (a) negative process
   (b) process that requires population inversion
   (c) non-equilibrium process that takes place when atoms are excited
   (d) natural and equilibrium process that takes place when atoms are excited

6. The function of helium atoms in the He–Ne laser is
   (a) to make neon atoms inactive
   (b) to provide energy to the neon atoms
   (c) to quench the neon atoms
   (d) none of the above

7. The wavelength produced by a He–Ne laser correspond to transition in
   (a) helium
   (b) neon
   (c) both helium and neon
   (d) neither helium nor neon

8. In He–Ne laser, the He and Ne are in the ratio
   (a) 1 : 10   (b) 10 : 1
   (c) 7 : 1   (d) 1 : 7

9. In a laser beam minimum angular divergence depends on
   (a) wavelength λ and diameter D of mirror

(b) only on wavelength λ
(c) only on diameter D of mirror
(d) alignment of mirrors only

10. For a laser beam $\lambda = 4400$ Å, and coherence time $= 4 \times 10^{-5}$ s, the coherence length will be
    (a) 1.2 km   (b) 12 km
    (c) 0.12 km   (d) 0.012 km

11. He–Ne gas laser discharge tube is fitted with $\lambda/100$ flat mirror for
    (a) getting coherent beam
    (b) pumping light into discharge tube
    (c) minimum losses and good look
    (d) getting polarised laser beam

12. Lasers used in holography are
    (a) He–Ne lasers
    (b) semiconductor lasers
    (c) solid state lasers
    (d) argon pulsed lasers

13. The useful material for semiconductor laser source is
    (a) gallium arsenide
    (b) germanium
    (c) silicon
    (d) none of the above

14. In He–Ne gas laser
    (a) the laser active centers are helium atoms
    (b) the laser active centers are neon atoms
    (c) the gases are maintained in 1 : 2 molar ratio
    (d) the inverted population density is achieved in helium atoms

Answers

1. (d)	2. (c)	3. (d)	4. (d)	5. (d)
6. (b)	7. (b)	8. (c)	9. (a)	10. (b)
11. (b)	12. (a)	13. (a)	14. (b)	

8
Fibre Optics

8.1 INTRODUCTION

Optical fibre is a wave guide through which light can be transmitted with very little leakage through the side walls. Optical fibres transport light signals from place to place as metallic conductors transport electrical signals. Optical fibres can guide light around bends and can carry light for long distances with very little attenuation (loss of light power). Telecommunications by modulating a beam of light (LED or Laser as a light source) and using optical waveguides has the inherent advantages of very large bandwidth, small dimensions, low losses bandwidth, and insensitivity to electrical interference.

An optical fibre is a glass or plastic conduits as thin as a human hair, designed to guide light waves along their length. Usually an optical fibre is a cylindrical wave guide system through which the optical wave can propagate. The principle by which light wave travel through the fibre is **total internal reflection** without loss of incident intensity.

An important structure used in optical system is the layered structure or the **wave guide structure**. As the name implies, these structures are used to confine the optical waves in a well defined region and guide their propagation. One can make the layered structures from non-crystalline materials or from crystalline materials. For example glass is used to produce optical fibres used in optical communication, whereas semiconductor wave guides are used in semiconductor laser.

An optical fibre is a very thin and flexible medium of cylindrical shape. It has three principal sections: (i) **the core** (ii) **the cladding** (iii) **the jacket**. The inner most section of the fibre is referred as the cladding. We must note that the optical properties of the cladding are different from those of the core. The outermost section of the fibre is known as the jacket. Jacket is made of plastic or polymer and other materials and this protects the fibre structure from moisture, abrasion, mechanical shocks, and other environmental hazards.

The core forms the actual working structure of the optical fibre. The radiation (say light) entering the core, suffers a number of familiar phenomenon of total internal reflection at the core-cladding interface. Obviously, the interface between the core and cladding acts as a mirror at which total internal reflection of the transmitted light takes place.

With the advent of optical fibres, an era of *photonics* has been ushered in. Apart from its use as communication channel, optical fibres find wide uses in other areas. Sensors for detecting electrical, mechanical, thermal energies are made using optical fibres. Copying machines, inexpensive and simple display systems, etc. also utilizes fibre optics. In medical diagnostics, fibroscopes are used in a variety of forms.

8.2 TOTAL INTERNAL REFLECTION

When a ray of light is incident at an angle i in denser medium (refractive index n_1), the

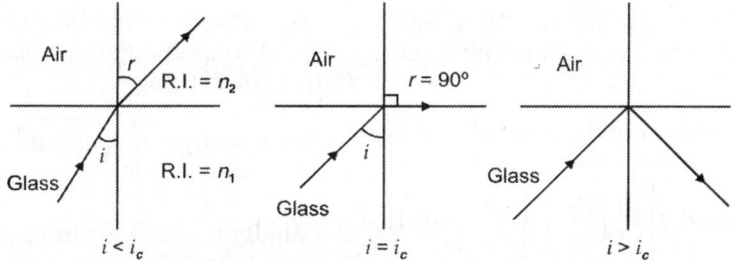

Fig. 8.1: Total internal reflection

refracted ray will bend away from the normal in rarer medium (refractive index n_2) and we have

Snell's law, $\quad \dfrac{\sin i}{\sin r} = \dfrac{n_2}{n_1}$

If the angle of incidence is i_c (say) when the angle of refraction $r = 90°$, then

$$\sin i_c = \frac{n_2}{n_1} \ (\sin r = \sin 90° = 1) \qquad (8.1)$$

The angle i_c is known as **critical angle**. The value of the critical angle (i_c) depends upon the refractive index of the denser medium (n_1). When the light in denser medium is incident at an angle $i > i_c$, then light will be reflected back into the denser medium. This is known as **total internal reflection** and this phenomenon is used in optical fibre communication. When light enters one end of the fibre, it undergoes successive total internal reflections from side walls and travels down the length of the fibre along the zigzag paths. A small fraction of light may escape through side walls but a major fraction emerges out from other end of the fibre.

8.3 THE OPTICAL FIBRE

Figure 8.2a shows a glass fibre consisting of a (cylindrical) central core cladded by a material of slightly lower refractive index. One can represent the corresponding refractive index distribution (in the transverse direction) as given by

$$n(r) = n_1 \qquad 0 < r < a \quad \text{Core}$$
$$\quad\ = n_2 \qquad r > a \qquad \text{Cladding}$$

and has been shown in Fig. 8.2b. Obviously, n_1 is the refractive index of the core of a fibre, and n_2 that of the cladding ($n_1 > n_2$). A ray of light OA, incident on one end face of the core, is refracted into the core along AB. If n_0 is the refractive index of air, we have from Snell's law

$$\frac{\sin i}{\sin \theta} = \frac{n_1}{n_0} \qquad (8.2)$$

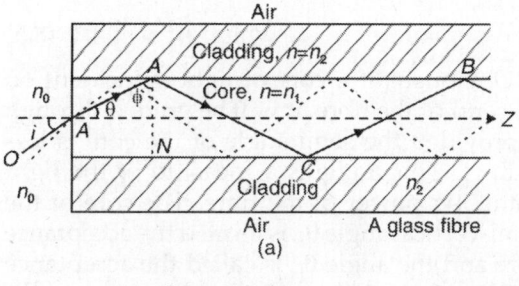

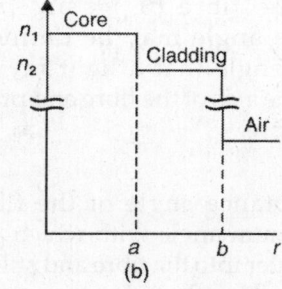

Fig. 8.2: (a) Refraction of a ray of light into core (b) Distribution of refractive index of a cladded optical fibre which consists of a cylindrical glass structure surrounded by a material of slightly lower refractive index

The ray AB is incident on the core–cladding interface at an angle $\phi = (90° - \theta)$. We can see that as i decreases, θ also decreases so that ϕ

increases. When this ray has to suffer total internal reflection at the corec–ladding interface,

$$\sin\phi(=\cos\theta) > n_2/n_1 \qquad (8.3)$$

Thus

$$\sin\theta < \left[1 - \left(\frac{n_2}{n_1}\right)^2\right]^{1/2} \qquad (8.4)$$

and one obtains

$$\sin i_m < \frac{n_1}{n_0}\left[1 - \left(\frac{n_2}{n_1}\right)^2\right]^{1/2} = \left[\left(n_1^2 - n_2^2\right)/n_0^2\right]^{1/2}$$

If $\left(n_1^2 - n_2^2\right) \geq n_0^2$, then for all values of i, total internal reflection will occur. Assuming $n_0 = 1$, one obtains that the maximum value of $\sin i$ for a ray to be guided is

$$\sin i_m = \left(n_1^2 - n_2^2\right)^{1/2} \quad \text{when } n_1^2 < n_2^2 + 1$$

$$= 1 \qquad \text{when } n_1^2 > n_2^2 + 1 \quad (8.5)$$

Obviously, if a core of light is incident on one end of the fibre, it will be guided through it provided the semi angle of the core is less than i_m. This angle is a measure of the light gathering power of the fibre. The core of the semi-vertical angle θ_m is termed the acceptance core and the angle θ_m is called the acceptance angle of the fibre $(\theta_m = \sin^{-1}[n_1^2 - n_2^2])$. Acceptance angle may be defined as the maximum angle that a light ray can have relative to the axis of the fibre and propagation down the fibre.

Thus,

(a) Acceptance angle of the fibre is the maximum angle with which a light ray can enter into the fibre and still be totally internally reflected.

(b) A cone of light incident at the entrance end of the fibre will be guided through the fibre, provided, the semi-vertical angle of the core is less than or equal to $\phi_{in\,(max)}$ (Fig. 8.2a)

Note: The angle $\phi_{in\,(max)}$ is unique only for a particular fibre. It differs from fibre to fibre and depends on the material and the core diameter.

The *numerical aperture* (NA), sometimes referred to as the *figure of merit* of an optical fibre, is defined as

$$NA = \sin\theta_m = \frac{1}{n_0}\sqrt{n_1^2 - n_2^2} = \sqrt{n_1^2 - n_2^2} \quad (8.6)$$

$$(\because n_0 = 1)$$

Usually $(n_1 - n_2)$ is not large and therefore, one can have

$$\left(\text{writing } n_1 - n_2 = n_1\left(1 - \frac{n_2}{n_1}\right) = n_1\sqrt{2\Delta}\right)$$

$$NA = n_1\sqrt{2\Delta} \qquad (8.7)$$

where

$$\Delta = \frac{\delta_n}{n_1} = \frac{n_1^2 - n_2^2}{2n_1^2}$$

$$= \frac{(n_1 - n_2)(n_1 + n_2)}{2n_1^2} = \frac{n_1 - n_2}{n_1} \quad (8.8)$$

The quantity Δ is usually referred to as the fractional difference of the refractive indices of the core and the cladding. Δ is also called *relative refractive index difference when $n_1 = n_2$*. For $n_1 = 1.48$ and $n_2 = 1.46$, one obtains NA ≈ 0.24. This gives $i_m \approx 14°$. Obviously, the acceptance angle i_m is less than 30°, so that NA < 0.5. We can easily see that if NA is too small, it is difficult to launch power into the fibre.

To prevent energy losses via absorption and scattering, the cladding should be at least a few wavelengths thick. One can see that a typical value for n_2/n_1 is 0.99. For this value, one finds the critical angle as 0.142 rads or 8.11°. Thus, rays travelling at an angle less than 8.11° relative to the reference axis will be totally internally reflected and guided by the fibre. Higher angle rays will enter the cladding and be lost due to high levels of scattering and absorption.

We may note that the dielectric cladding on glass core reduces scattering loss, protects core from absorbing external optical disturbances and provides mechanical strength to main core glass fibre. Sometimes there is buffer coating over cladding which adds further strength to main fibre and protects fibre from mechanical vibrations and impact.

8.4 CLASSIFICATION OF FIBRES

Usually the optical fibres are grouped under three categories, viz:

   i. Stepped index multimode fibre

   ii. Stepped index monomode fibre

   iii. Graded index multimode fibre.

Stepped Index Multimode Fibre

Stepped index is the simplest type of an optical fibre consisting of a thin cylindrical structure of transparent glassy material of uniform refractive index n_1 surrounded by a cladding of another material of uniform but slightly lower refractive index n_2 (Fig. 8.2). These fibres are referred to as stepped index fibres due to the step discontinuity of the index profile at the core-cladding interface. The light rays incident on the end face of the core at angles of incidence within the acceptance angle are guided along the core in different paths. It is observed that light rays of different colours of frequencies are also separated as they travel inside the core. This means multimode propagation is possible in the fibre as shown in Fig. 8.3.

From Fig. 8.3, we can see that the time taken by the different light rays propagating along paths of different lengths upto the destination are different. This means the variations of the successive pulses of light may overlap, thus distorting the information being carried. Such distortions are usually called as the transit time dispersion.

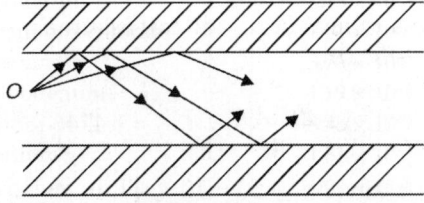

Fig. 8.3: Multimode propagation in a fibre

One can show that the time taken by a ray is function of the angle θ made by the ray with the Z-axis which leads to pulse dispersion and given by (Fig. 8.2)

$$\tau = \frac{n_1 L}{c \cos \theta} \tag{8.9}$$

where $L = AB$. If we assume that all rays lying between 0 and θ_c (critical angle) are present, then the time taken by the rays corresponding to $\theta = \theta_c \approx \cos^{-1}(n_2/n_1)$ would be given by

$$\tau_{min} = \frac{n_1 L}{c} \quad \text{and} \quad \tau_{max} = \frac{n_1^2 L}{n_2 c} \tag{8.10}$$

Table 8.1: Comparison between single mode fibre and multimode fibre

Single mode fibre	*Multimode fibre*
1. Light can propagate through the fibre in only one mode.	Light can propagate through the fibre with a large number of modes.
2. The fibre core diameter is very small (~ 10 µm) and also, the difference between the refractive indices of the core and cladding is very small.	The fibre core diameter is large and also, the difference between the refractive indices of the core and cladding is also large.
3. Since light propagates in single mode, no dispersion occurs (i.e. no degradation of light signal takes place during propagation through the fibre).	Dut to multimode propagation and material scattering, there is signal degradation.
4. Launching of light into the fibre and coupling process is not easy.	Launching of light into the fibre and coupling process are easy.
5. Used in long haul communication.	Used in LAN (Local Area Network).
6. Since the fabrication is difficult, the production cost is high.	Since the fabrication is easy, the production cost is low.

Thus, if all the input rays were excited simultaneously, the rays would occupy a time interval at the output end of duration

$$\Delta\tau = \tau_{max} - \tau_{min} = \frac{n_1 L}{c}\left(\frac{n_1}{n_2} - 1\right) \quad (8.11)$$

For a typical fibre, one can assume $n_2 = 1.46$ (pure silica),

$$\Delta = \frac{n_1 - n_2}{n_1} \implies n_1 \approx 1.4746$$

One obtains $\Delta\tau \cong 50 \times 10^{-9}$ s/km

$$\cong 50 \text{ ns/km}.$$

This shows that an impulse after traversing through the fibre of length 1 km broadens to a pulse dispersion. For this purpose, two alternative solutions exist, one involving the use of graded index fibres and the other involving single mode fibres.

Stepped Index Monomode Fibre

One can make core very thin, i.e. the core diameter is almost of the same order as the wavelength of the light to be propagated. This helps in minimizing the transit time dispersion problems. Figure 8.4 shows the propagation in a stepped index monomode optical fibre. Step index monomode fibre has low attenuation, small numerical aperature and wide bandwidths.

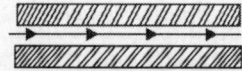

Fig. 8.4: Propagation in a stepped index monomode optical fibre.

One can show that when $V < 2.405$, only the LP_{01} mode can propagate, i.e. fibre becomes single mode fibre. Theoretically, LP_{01} mode will propagate no matter how small the V value, i.e. no matter how small the core radius a.

$$\because \qquad V = \frac{2\pi}{\lambda} a(\text{NA})$$

$$\therefore \qquad a = \frac{V\lambda}{2\pi(\text{NA})}$$

When $V < 2.405$, then

$$a < \frac{2.405\lambda}{2\pi(\text{NA})}$$

or

$$a < \frac{2.405\lambda}{2\pi\sqrt{n_1^2 - n_2^2}}$$

for single mode fibre.

We may note that, in general, in single mode fibre cores, tend to be only a few microns in diameter, but it has an advantage from several points of view if core can be made as large as possible. One can easily note that this may be achieved by reducing the NA value, i.e. by making core and cladding R.I. very close to each other. One of the major advantage of using single mode fibre is the inter model dispersion no longer exists, and so they are capable of carrying very large bandwidth signal for long distance communication.

Graded Index Multimode Optical Fibre

In comparison to stepped index monomode fibres, graded index multimode fibre is quite cost effective and offer a less expensive method of overcoming the problem of transit time dispersion. It is observed that in such fibres, the refractive index of the core falls off gradually from the axis towards the core-cladding interface. This means that the individual light rays are gradually refracted in the core itself. Figure 8.5 shows the wave propagation at different angles of incidence, travelling different distances from the central axis before suffering reflections to recross the central axis.

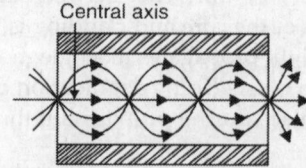

Fig. 8.5: Propagation in a graded index multimode optical fibre

We may note that the number of guided modes is dependent upon the physical parameters, i.e. the refractive index difference between core and cladding (i.e. $NA = \sqrt{n_1^2 - n_2^2}$), core radius of the fibre a and the wavelength (λ) of the transmitted light, are all included in the normalized frequency V' for the fibre, where $V = 2\pi/\lambda a$ (NA) = normalized frequency. One finds that the total number of guided modes for step index fibre is related to V by the following approximate expression

$$M_s = \frac{V^2}{2}$$

From Fig. 8.5, it is obvious that the light rays with large angles of incidence traverse greater paths than those with smaller angles of incidence. However, the velocity of propagation increases with a drop in the refractive index and, thus, all the rays recross the central axis at nearly the same instant of time. This helps in reducing considerably the transit time dispersion.

Fibre optic systems use optical fibres to transmit information, in the form of coded pulses or fragmented images (using bundles of fibres). Over moderate distances they are used in telecommunications, for which purpose they are becoming competitive with electric cables. In fibre communication, two

Table 8.2: Comparison between step index fibre and graded index fibre

Step index fibre	*Graded index fibre*
1. Refractive index of the core–cladding is uniform.	Refractive index of the core is not uniform. But the refractive index of the cladding is uniform.
2. Since there is an abrupt change in the refractive index at the core–cladding interface, the refractive index profile takes the shape of a step. Hence, called step index fibre.	In this fibre, the refractive index of the core is maximum at the centre and decreases gradually (parabolic manner) with distance towards the outer edge. Hence, called graded index fibre.
3. Pulse dispersion is more in single mode step index fibre.	Pulse dispersion is reduced by a factor of 200 in comparison to step index.
4. Attenuation is less for single mode step index fibre and more for multimode step index fibre.	Attenuation is less.
5. Number of modes of propagation for a multimode step index fibre is given by $$N_{step} = 4.9 \left(\frac{d \times NA}{\lambda}\right)^2$$ where d is diameter of the core, λ is optical wavelength and NA is numerical aperture.	Number of modes of propagation for a multimode graded index fibre is given by $$N_{graded} = \frac{4.9 \times \left(\frac{d \times NA}{\lambda}\right)^2}{2}, \text{ i.e. } N_{graded} = \frac{N_{step}}{2}.$$ Thus, the number of modes is half the number supported by a MMSI fibre.
6. The light rays propagate through the fibre in the form of meridional rays, i.e. the rays follow a zig-zag path when they travel through fibre and for every reflection it will cross the fibre axes.	The light rays propagate through the fibre in the form of skew rays, i.e. the rays follow a helical path around the fibre axis, and they would not cross the fibre axis at any time during their propagation.
7. SMSI fibre are expensive and difficult to manufacture. But MMSI are inexpensive and simple to manufacture.	Graded index fibres are easy to manufacture.
8. The bandwidth for single mode step index fibre is more than multimode step index fibre.	Multimode graded index fibre has a higher bandwidth.
9. Numerical aperture is very less for single mode step index fibre but is more for MMSI fibre.	Numerical aperture is high.

bands are extensively used— 800 nm to 900 nm and 1200 nm to 1400 nm. The losses in the fibres for the said two bands are very low. Lasers and LEDs (light emitting diodes) serve as light sources for fibres. Photodetectors are employed for processing the signal at the receiver end of the optical fibre link. Photodetectors convert the light signals into electrical waveforms. Optical fibres are also used in medical instruments (fibrescopes) to examine internal body cavities, such as stomach and bladder.

V-Number: An optical fibre is characterized by one more important parameter, called as *V*-number. This is more generally called as *normalized frequency* of the optical fibre and given by the relation

$$V = \frac{2\pi a}{\lambda} \sqrt{n_1^2 - n_2^2} \qquad (8.12)$$

where *a* is the radius of the core and λ is the free space wavelength. Using

$$NA = \sqrt{n_1^2 - n_2^2} = n_1 \sqrt{2\Delta}$$

one obtains

$$V = \frac{2\pi a}{\lambda}(NA) \qquad (8.13)$$

$$V = \frac{2\pi a}{\lambda} n_1 \sqrt{2\Delta} \qquad (8.14)$$

The maximum bar of modes N_m supported by a stepped index (SI) fibre is determined by

$$N_m \cong \frac{1}{2}V^2 \qquad (8.15)$$

Thus, for $V = 10$, $N_m = 50$.

For $V < 2.405$, the optical fibre can support only one mode and is classified as SMF. MMFs can have values $V > 2.405$ and can support many modes simultaneously.

The wavelength corresponding to the value of $V = 2.405$ is known as the *cut-off wavelength* (λ_c) of the optical fibre. We have,

$$\lambda_c = \frac{\lambda V}{2.405} \qquad (8.16)$$

From Eq. (8.14), one can deduce that single mode properties can be realized by decreasing the core diameter and/or decreasing Δ such that $V > 2.405$. We may note that either a larger core and/or a larger Δ will result in multimode properties.

In case of GRIN fibres, for large values of V, we have

$$N_m \cong \frac{V^2}{4} \qquad (8.17)$$

8.5 ADVANTAGES OF OPTICAL FIBRES

Fibre optic systems offer many advantages. Few of them are listed below:

i. Optical fibres are light in weight and moreover occupy less space than coaxial (bundles twisted pair) cables.

ii. Optical fibres are composed of dielectric materials and, therefore, they are totally immune to extraneous interfering electromegnetic signals.

iii. There is virtually no signal leakage from optical fibres and hence cross-talks between neighbours is not possible, whereas it is quite frequent in conventional metallic systems.

iv. The attenuation of signals in optical fibres is much smaller than in coaxial cables or twisted pair.

v. Optical fibres are immune to electromagnetic interference and do not pick up line currents. Obviously, these can be safely used in high voltage environments, e.g. in a power station, and can be laid alongside metallic power cables.

vi. The basic raw materials used in the fabrication of low loss fibres is silica, which is abundantly available in nature (although in an impure form), whereas copper constitutes the basic raw material in coaxial cables. Obviously, fibres are cost effective, i.e. more economical.

vii. Optical fibres can withstand environmental hazards better and also have longer life compared to copper cables.

vii. Optical fibres are non-conductive and non-inductive. This means there is no radiation and interference with other communication systems.

ix. The speed of transmission through optical fibres is very fast because the signals are carried by light.

x. The channel capacity of optical fibres is very large, compared to the audio band (5 – 10 Hz), the frequency of the carrier light wave is very high (~10^{14} Hz). Obviously, one can accommodate a large number of speech signal channels in the fibres whose bandwidth is larger than that of a wire transmission line.

8.6 DISADVANTAGES OF OPTICAL FIBRE SYSTEMS

i. Optical fibre systems are *virtually useless themselves*. To be practical, they must be connected to standard electronic facilities which often requires expensive interfaces.

ii. Optical fibres by themselves have a significantly lower tensile strength than coaxial cable. This can be improved by coating the fibre with standard Kevlar and a protective jacket of PVC.

iii. Occasionally it is necessary to provide electrical power to remote interface or regenerating equipment. This can not be accomplished with the optical cable so additional metallic cables must be included in the cable assembly.

iv. Optical fibre cable systems are relatively new and have not had sufficient time to prove their reliability.

v. Optical fibres require special tools to splice and repair cables and special test equipment to make routine measurements. Repairing fibre cables is also difficult and expensive and techniques working on optical fibre cables also require special skills and training.

In spite of all the attractive features that fibre optical communication systems possess, a fully developed fibre optical communication system started becoming available in the commercial market only by 1975. Ever since the technology in this area is progressing well. The attainment of very low pass fibre has also stimulated forceful efforts in the development of supporting technologies line, long-life semiconductor light sources, low loss optical fibre cable, fibre optic connectors, low pass splices, and highly sensitive fast response solid state photodetectors.

8.7 TYPES OF RAYS

The plane containing the central axis meridional rays remain the meridian plane which can be divided into two categories:

1. Bound rays that propagate along axis (Fig. 8.6a)

2. Unbound rays that are not bound along axis and are refracted out of the core and absorbed in cladding or buffer (Fig. 8.6b).

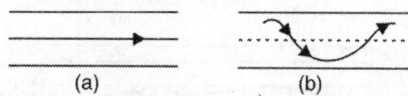

(a)   (b)

Fig. 8.6: Meridional rays propagation through optical fibre cable light

Skew Rays

These are the rays which enter the optical fibre obliquely and not confined to a plane and follow a helical path around the axis. These rays are soon scattered out of the core at bends and irregularities and thus do not contribute significantly in optical communication.

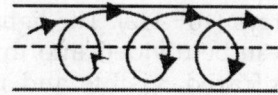

Fig. 8.7: Skewed ray propagation through optical fibre cable light

8.8 MODES OF PROPAGATION IN OPTICAL FIBRE

Light propagation through straight and curve step index optical fibre with cladding is shown in Fig. 8.8.

Here we have shown only one ray and, therefore, this is called as ray diagram. However, in reality an infinite number of such

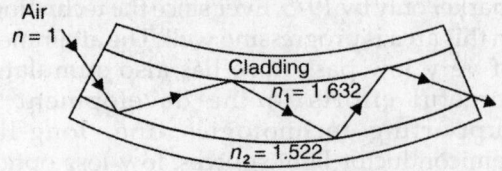

Fig. 8.8: Propagation of light through straight and curve step index optical fibre with cladding

rays (Figs. 8.9a and b) all slightly displaced from each other can be propagated at a time which are called different modes of propagation. At time of propagation they will follow different paths in the optical fibre (Fig. 8.10).

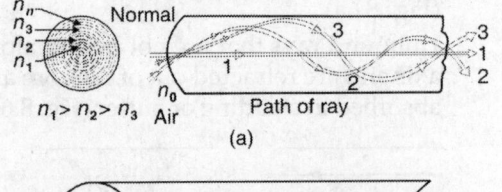

(a)

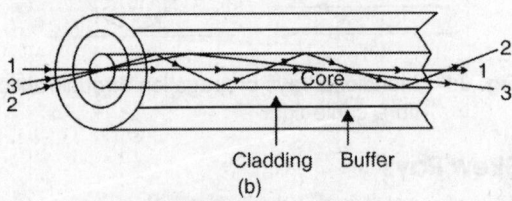

(b)

Fig. 8.9: Propagation of light in multimode graded index and step index fibre

We have read that light can be described as an electromagnetic wave. It can be seen that the frequency spectrum of electromagnetic waves extends from the subsonic (a few Hz) to cosmic rays (10^{22} Hz). The light frequency spectrum can be divided into three general bands— infrared, visible and ultraviolet. Electromagnetic wave consists of a periodically varying electric field E and magnetic field H which are perpendicular to each other. The transverse modes are shown in Fig. 8.10. These modes illustrates the case when the electric field E is perpendicular to the direction of propagation (z) and hence $E_z = 0$, but the corresponding component of magnetic field H is in the direction of propagation. These modes are, therefore, called transverse electric (TE_m) modes. Similarly, when a component of electric field E is in the direction of propagation but

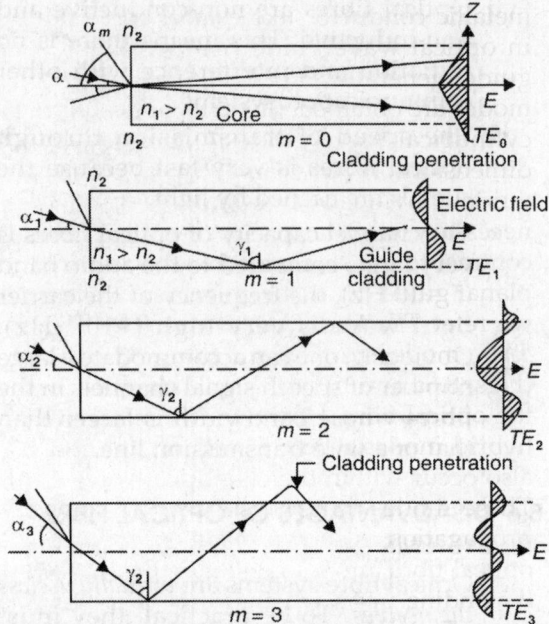

Fig. 8.10: Propagation of light in different modes through step index optical fibre, where α_m = acceptance angle, r_c = critical angle, and $\pi/2 > r_1 > r_2 > r_3 > ... > r_c$, whereas $\alpha_0 < \alpha_1 < \alpha_2 < \alpha_3 ... < \alpha_m$

$H_z = 0$, the modes so formed are called transverse magnetic (TM_m) modes. We have incorporated the mode number m into this nomenclature by referring to TE_m or TM_m modes, etc. We may note that each value of m is associated with a distinct wave pattern or mode as shown in Fig. 8.10. Within the wave guide they are following different paths. We also find that for each value of m, there is a value of γ (γ_m), where γ_m can take values in the range γ_c to $\pi/2$, which implies that the smaller the value of γ_m, i.e. γ_m closer to γ_c, higher the mode number and *vice versa*. We can see from Fig. 8.10 that the higher order modes, carry along with them a higher portion of mode energy within the cladding. This part of the wave which is carried in the cladding is generally referred to as the *evanescent wave*. We may note that (i) for free space communication, the total electromagnetic wave field lies in the transverse plane, where both $E_z = 0$, and $H_z = 0$. This is called *transverse electromagnetic wave* (TEM). However, TEM wave occur in metallic conductor like coaxial cable but not in optical wave guide (ii) for planar wave guide, generally $TE(E_z = 0)$ and TM ($H_z = 0$) modes are obtained. (iii) as in optical fibre the cylindrical wave guide is bounded in two dimensions, rather than one as in plane wave guide. Obviously, two integers l and m are necessary in order to specify the modes, in contrast to the single integer (m) required for planar guide. Thus, for cylindrical wave guide we refer TE_{lm} and TM_{lm} modes. The TE_{lm} and TM_{lm} modes correspond to meridional rays (Figs. 8.6a and b) which are travelling within the optical fibre. However, we may note that hybrid modes where E_z and H_z are non-zero also occur within the cylindrical wave guide. These modes which results from skew rays propagation as shown in Fig. 8.11, within the optical fibre are designated as HE_{lm} and EH_{lm} depending upon whether the components of H or E make larger contribution to the transverse field. These skew rays are known as *leaky rays* and corresponding modes are termed *leaky modes*, not bound modes because they can propagate only appreciable distance prior to being lost from the optical fibre.

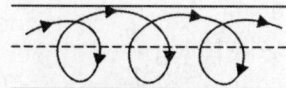

Fig. 8.11: Propagation of skewed rays through optical fibre cable light

In most practical wave guides, it is found that the refractive indices of core and cladding differ from each other by only a few percent and in that case the full set of modes (i.e. *EH*, *HE* and *TE*) can be easily approximated by a single set called linearly polarized (LP_{lm}) modes. An LP_{lm} mode in general has m field maxima along a radius vector and $2l$ field maxima round a circumference (Fig. 8.12).

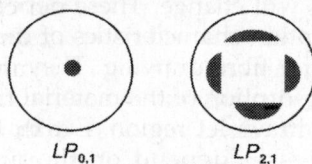

Fig. 8.12: Linearly polarised (LP_{lm}) modes

8.9 DISPERSION—INTERMODEL DISPERSION For Multimode Stepped Index Fibre

When the ray of light of different modes follow different paths, like some rays shown by AA' in Fig. 8.13, which propagate along the optical fibre axis and some of which propagate along the trajectories of the oblique rays, for example CC' or BB' (Fig. 8.13). Obviously, there will be difference in time taken for different modes to travel a distance l. We know that refractive index R.I. of a medium is simply a measure of velocity of propagation v of the light in medium. Thus, we can write

$$RI(n) = \frac{v_a}{v_m} \quad \text{or} \quad v_m = \frac{v_a}{n} = \frac{c}{n}$$

where c (= v_a) is the velocity of light in air and v_m is the velocity of light in medium.

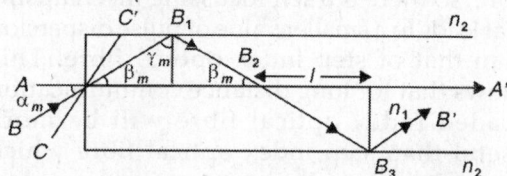

Fig. 8.13: Multipath time dispersion for multimode step index optical fibre

The axial ray AA' travels an axial distance l along the optical fibre in time whereas the oblique ray BB' travels the same axial distance l along the path B_2B_3 in the time

$$t_2 = \frac{n_1}{c} \frac{1}{\cos\beta_m} = \frac{n_1}{c} \frac{1}{\cos(90° - \gamma_c)}$$

$$\left(\because \sin\gamma_c = \frac{n_2}{n_1} \right)$$

$$= \frac{n_1}{c} \frac{1}{\sin\gamma_c} = \frac{n_1^2 l}{n_2 c}$$

The above results reveal that if two rays (AA' and BB') are launched together they will be separated on arrival at the other end by ΔT, where

$$\Delta T = t_2 - t_1 = \frac{n_1 l}{c} \left(\frac{n_1}{n_2} - 1 \right)$$

$$= \frac{n_1 l}{n_2} \cdot \frac{l}{c} \Delta n \quad \text{(where } \Delta n = n_1 - n_2\text{)}$$

Obviously, a pulse containing rays at all possible angles spreads out during propagation by an amount given by

$$\frac{\Delta T}{l} = \frac{n_1}{n_2} \cdot \frac{\Delta n}{c}$$

This is called as **multipath time dispersion** of the step index fibre and this causes pulse broadening.

For Graded Index Multimode Fibre

Different modes will follow different paths, sinusoidally. We may note the rays making larger angles (Fig. 8.14) traverse longer path, but they do so in lower refractive index area and hence they travel with higher speed of propagation. However, refractive index changes gradually in graded index optical fibre, so there is a self focussing mechanisms that leads to a smaller value of pulse dispersion than that of step index optical fibre. This shows that for long distance communication graded index optical fibre will be more useful than step index optical fibre which causes less time dispersion.

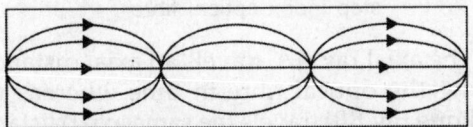

Fig. 8.14: Less multiple time dispersion for graded index optical fibre

8.10 LOSSES IN OPTICAL FIBRE CABLE

The losses in optical fibre, i.e. attenuation are mainly caused by absorption and scattering of rays. We may note that:

i. **Absorption** of rays is basically a material property and it occurs at all wavelengths when electronic transitions takes place within material and are followed by non-radiative relaxation processes. This causes an increase of thermal energy in the material.

ii. **Scattering** is also partly a material property but in optical fibres it is caused by the imperfection in their geometries. Scattering occurs when the mode of propagation of light is changed, such that some of the optical energy leaves the optical fibre.

iii. **Losses due to tight bending of fibres:** Tight bends of the optical fibre cause some of the light not to be internally reflected but to propagate into the cladding and be lost.

iv. **Losses due to imperfection in fibre:** The presence of imperfection in optical fibre material also cause power losses from the optical fibre cable.

v. **Losses due to connectors and splices:** Power losses are also caused for long distance communication cables due to connectors and splices which required to connect two optical fibres temporarily.

Attenuation in Optical Fibres

A part of the light energy being propagated through optical fibre is lost during its traversal from the source to destination. These losses are due to various reasons. The losses are measured through attenuation constant. Its unit is decibels per kilometer. Attenuation A is defined as

$$A = 10 \log \frac{p_i}{p_o} \qquad (8.18)$$

Attenuation constant α is attenuation per unit length. Hence

$$\alpha = A/l = \frac{10}{l} \log \frac{p_i}{p_o} \qquad (8.19)$$

where l is the length of the fibre.

The following are the loss mechanisms due to which the intensity of light decreases while propagation through fibre:

i. *Absorption Lasers:* Absorption losses could be intrinsic or extrinsic. Depending on the material used, intrinsic absorption losses will change. These depend on the absorption characteristics of the material used for fibre drawing. They depend on the absorption of the material in infrared and ultraviolet region. Extrinsic absorption losses depend on the purity and perfection of the material.

ii. **Scattering Losses:** The glass material used for the optical fibres will have many microscopic inhomogeneities. These are because of the defects and changes in the structure and result in light scattering.

(a) *Rayleigh Scattering:* This is because of small fluctuation in refractive index and changes in the density. These inhomogeneities are fundamental and intrinsic to the glass used for fibre preparation and cannot be completely avoided.

(b) *Mie Scattering:* This is due to the irregularities in the core–clad interface and refractive indices, diameter fluctuations, strains, bubbles, etc. These defects normally depend on the material, design and method and conditions of manufacture of the optical fibres. These defects can be reduced considerably by proper control of different processes during manufacture of the fibres.

(c) *Brillouin Scattering:* Modulation of light through thermal molecular vibrations within the fibre material. The incident photon (light packet) produces a phonon (quantized elastic wave having sound frequency) and a scattered photon. This type of scattering occurs only above certain minimum threshold frequency. Hence, it can be completely avoided by selecting proper light frequency below the threshold frequency.

(d) *Raman Scattering:* Raman scattering incident photon produces another photon (optical phonon). This can also be avoided as the Brillouin scattering.

(e) *Dispersion Losses:* Losses due to dispersion of the transmitted signal causes distortion of the signal. Dispersion losses could cause broadening of the pulses being transmitted resulting in overlap of the adjacent pulses and subsequent distortion. This limits the rate of bit transfer (bits transmitted per second).

(f) *Intermodal Dispersion or Mode Coupling*: This is chromatic dispersion and results in the overlap of signals of different wavelengths. Hence, this will limit the line width of the light source used.

Attenuation coefficient can be expressed as a function of some of the above loss mechanism as follows:

$$\alpha = A\lambda^{-4} + B + C(\lambda) \qquad (8.20)$$

A is Rayleigh scattering coefficient

B is wavelength independent losses like dispersions

C is wavelength dependent loss mechanisms like intermodal dispersion

8.11 OPTICAL FIBRE COMMUNICATION SYSTEM

Figure 8.15 shows a simplified block diagram of an optical fibre communications link. The three primary building blocks of the link are the transmitter, the receiver, and the fibre guide.

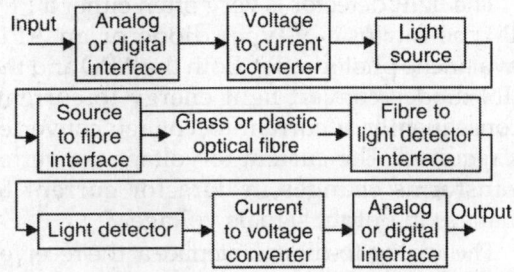

Fig. 8.15: Simplified fibre optic communication link

The transmitter consists of an analog or digital interface, a voltage to current converter, a light source, and a source to fibre light coupler. The fibre guide is either an ultra pure glass or plastic cable. The receiver includes a fibre to light detector coupling device, a photo detector, a current-to-voltage converter, an amplifier, and an analog or digital interface.

In an optical fibre transmitter, the light source can be modulated by a digital or an analog signal. For analog modulation, the input interface matches impedances and limits the input signal amplitude. For digital modulation, the original source may already be in digital form or, if in analog form, it must be converted to a digital pulse stream. For the latter case, an analog-to-digital converter must be included in the interface.

The voltage-to-current converter serves as an electrical interface between the input circuitry and the light source. The light source is either a light-emitting diode (LED) or an injection laser diode (ILD). The amount of light emitted by either an LED or an ILD is proportional to the amount of drive current. Thus, the voltage-to-current converter converts an input signal voltage to a current that is used to drive the light source.

The source-to-optical fibre coupler (such as a lens) is a mechanical interface. Its function is to couple the light emitted by the source into the optical fibre cable. The optical fibre consists of a glass or plastic fibre core, a cladding and a protective jacket. The fibre-to-light detector coupling device is also a mechanical coupler. Its function is to couple as much light as possible from the fibre cable into the light detector.

The light detector is very often either a PIN (P-type-intrinsic-N-type) diode or an APD (avalanche photodiode). Both the APD and the PIN diodes convert light energy to current. Consequently, a current-to-voltage converter is required. The current-to-voltage converter transforms changes in detector current to changes in output signals voltage.

The analog-to-digital interface at the receiver output is also an electrical interface. If analog modulation is used, the interface matches impedances and signal levels to the output circuitry. If digital modulation is used, the interface must include a digital-to-analog converter.

An optical communication system is quite similar in concept to electronic (microwave) communication system (10^9 Hz), only the frequency range is different (10^{15} Hz) and optical fibre instead of metallic fibre.

8.12 PHOTODETECTOR

The most common semiconductor photodetector is the photodiode. A photodetector detects the optical power incident upon it and converts the variation of this optical power into a correspondingly varying electric current. The

most commonly used system in optical fibre communication is *IM/SS*, where intensity of the optical carrier, modulated by biasing current in LED/ diode laser (IM) and that variation of optical power directly detected in photodetector (PD).

Photodiode Mechanism

The basic structure of extrinsic P-N junction silicon photo diode is shown in Fig. 8.16a. In normal operation a sufficiently large reverse bias voltage is applied across the diode, so that depletion region as well as barrier potential will be large. Now the light incident on the reverse biased P-N junction diode, cause a photon ($h\nu$) to be absorbed. If the energy of the photon is more than the band gap energy (E_g), i.e. $h\nu > E_g$, then the absorbed photon excite an electron from valence band (VB) to conduction band (CB) and leaving a hole in VB. The electron–hole pair created due to incident light that is absorbed in the depletion region of the photodiode is called photocarriers.

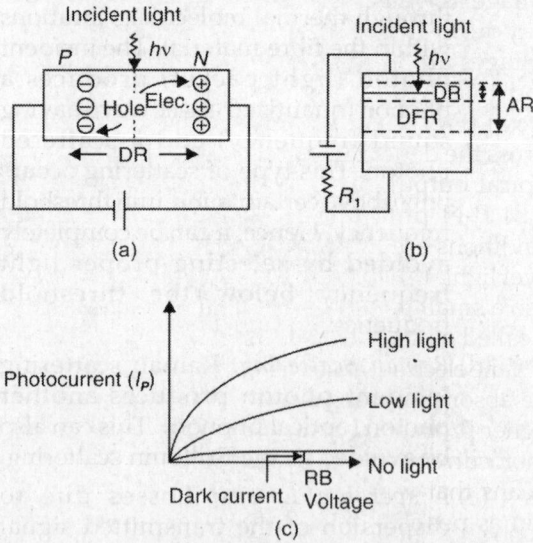

Fig. 8.16: Mechanism of P-N junction silicon photodiode

Under reverse bias voltage, electron and hole flow in the depletion region (DR) in opposite direction, some of the electron–hole pairs will recombine and hence lost. The distance they will traverse prior to their recombination is called diffusion length (L_n and L_p for electrons and holes respectively) and the time it takes for an electron and hole to recombine is known as carrier lifetime (T_n and T_p). The relation between the carrier lifetime and diffusion length can be expressed as

$$L_n = (D_n T_n)^{1/2}, L_p = (D_p T_p)^{1/2}$$

where D_n and D_p represents the electron and hole diffusion coefficients respectively, which are in cm²/sec.

We may note that photons can be absorbed in depletion region (DR) as well as in diffusion region (DFR) too, which is usually indicated as absorption region (AR) (Fig. 8.16b). Obviously, this region becomes void of free photocarriers. Now, within the deflection region those pairs of electron–hole remains which cannot recombine because of the existence of main electrostatic field due to the depletion region voltage. Any photon induced electron–hole pair therein will be separated and drawn in opposite directions by the combined effect of equivalent depletion region voltage as well as the reverse bias voltage. We may note that the polarities of these two voltages are in the same direction. This effect lead to flow of photocurrent in the external circuit (Fig. 8.16c), as photocarriers drift away across the depletion region. Figure 8.16c shows typical output characteristics for reverse bias (RB) P-N photodiode. Different operating conditions are indicated from no light to high level light. When there is no incident light, then also a small reverse current will be there. This is called as dark current.

If the electron–hole pairs are produced outside the absorption region (AR), then there will be higher probability available and in that case no photocurrent will be available. Obviously, this means that we have to make the top layer as thin as possible and effective depletion region as wide as possible in order to maximize the efficiency of photodiode as photodetector, which is termed *quantum efficiency*. One can increase the quantum efficiency of the photodiode if one goes from ordinary photodiode to PIN photodiode or avalanche photodiode (APD) or *reach through avalanche photo diode (RAPD)*.

Usually, there are two different structures for photodiodes possible as shown in Fig. 8.17. Figure 8.17a shows the photodiode with front illuminated operation at 0.8 – 0.9 µm range, whereas Fig. 8.17b shows the side illuminated photodiode structure operating at higher wavelength 1.09 µm range.

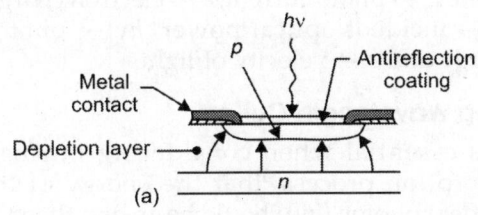

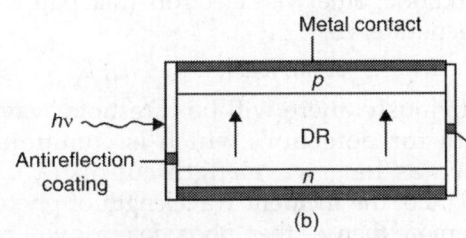

Fig. 8.17: Two different possible structures of photodiode

Quantum Efficiency (η) and Responsivity (R)

Quantum efficiency (η) and responsivity or response speed (R) for a photodetector are two important characteristics. These two parameters, i.e. η and R depend on the material band gap, operating wavelength, percentage of doping and thickness of P and N regions. In order to achieve high quantum efficiency, the depletion layer must be thick enough to permit a large fraction of the incident light to be absorbed. This is why PIN or APD photo-diode is better than ordinary photo-diode. One can define quantum efficiency and responsivity as follows:

Quantum efficiency,

$$\eta = \frac{\text{Rate of generation of electron – hole pair (electron/sec)}}{\text{Rate of photons incident (photons/sec)}}$$

or

$$\eta = \frac{I_p/e}{P_0/h\nu}$$

Responsivity,

$$R = I_p/P_0 = \eta e/h\nu = \frac{ne\lambda}{hc} \ \text{A/watt}$$

where $I_p \rightarrow$ photo current, $e \rightarrow$ electron charge, $P_0 \rightarrow$ incident optical power, $h\nu \rightarrow$ photon energy and $c \rightarrow$ velocity of light.

Long Wavelength Cut-off

It is essential when considering intrinsic absorption process, that the energy of the incident photon ($h\nu$) should be greater than the band gap energy (E_g) of the material used as photodiode, otherwise electron–hole pair will not generate, i.e.

$$h\nu \geq E_g \text{ or } hc/\lambda \geq \lambda \leq \lambda_c \leq hc/E_g$$

Obviously, there will be threshold wavelength for detection, which is commonly known as long wavelength cut-off (λ_c). If $\lambda > \lambda_c$, i.e. the incident wavelength of photon (λ) is more than λ_c, then photodetector will not work.

8.13 OPTICAL FIBRE CABLE CONSTRUCTION

A single optical fibre can transmit light energy very effectively (Figs. 8.8 and 8.9). However, in order to transmit an image of an object one requires assembly or bundles of fine glass fibres or plastic fibres because each fibre transmits rays from a very small region of the object (Figure 8.18). The quality of the image depends upon the diameter of the optical fibre. It can be as small as 10^{-6} m. A bundle of the optical fibre may consist of thousands of individual optical fibres.

Essentially, three varieties of optical fibres are available. All these three varieties are constructed of either glass, plastic or a combination of glass and plastic. The three varieties are:

1. Plastic core and cladding
2. Glass core with plastic cladding (often called PCS fibre, plastic-clad silica)
3. Glass core and glass cladding (often called SCS, silica clad silica)

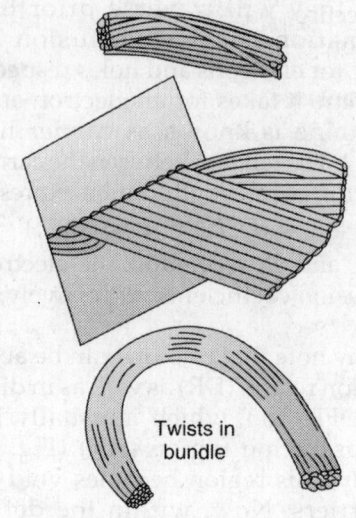

Fig. 8.18: Bundle of optical fibres

There is also a possibility of using fourth variety that uses a non-silicate substance, zinc chloride. Investigations have shown that fibres made of this substance will be as much as 1000 times as effective as glass—their silica based counterpart.

Plastic fibres have several advantages over glass fibres. First, plastic fibres are more flexible and consequently, more rugged than glass. They are easy to install, can better withstand stress, are less expensive, and weigh approximately 60% less than glass. The disadvantage of plastic fibres is their high attenuation characteristic. They do not propagate light as efficiently as glass. Consequently, plastic fibres are limited to relatively short runs, such as within a single building or a building complex.

Fibres with glass cores exhibit low attenuation characteristics, however, PCS fibres are slightly better than SCS fibres. Also, PCS fibres are less affected by radiation and, therefore, are more attractive to military applications. SCS fibres have the best propagation characteristics and they are easier to terminate than PCS fibres. Unfortunately, SCS cables are the least rugged, and they are more susceptible to increase in attenuation when exposed to radiation.

The selection of a fibre for a given application is a function of specific system requirements.

Cable Construction

There are many different cable designs available today. Figure 8.19 shows examples of several fibre optic cable configurations. Depending upon the configuration, the cable may include a **core**, a **cladding**, a **protective tube**, **buffers**, **strength numbers** and one or more **protective jackets**.

8.14 FIBRE OPTIC SENSORS

A fibre optic sensor is a transducer which can convert various input variables (physical quantity) into an electrical signal in a measurable form.

The increased applications of optical fibre sensors in comparison to traditional sensors is due to less cost and improved quality. Optical fibre sensors are used for measuring and sensing various parameters, e.g. temperature, pressure, electric field, current, humidity, acoustic vibrations, etc. The major advantages of using optical fibres in sensing applications are as follows:

i. They are small in size and light in weight.

ii. They have very good geometrical flexibility.

iii. These are free from risk of fire or sparks as they are made of silica.

iv. Their chemical and environmental ruggedness is more.

v. They are immune to electromagnetic interference, i.e. they are electrically positive. Moreover, they do not distort the surrounding electric and magnetic fields.

vi. Optical fibre sensors have large bandwidth and they are also highly sensitive.

Optical sensors are non contact and usually high accuracy devices and systems. In optical fibre sensors, the optical wave is the information carrier and sensor.

Whenever we use optical sensors for measuring a physical parameter, any one of the characteristics, e.g. intensity, amplitude, phase, polarisation, frequency and direction of propagation of wave gets modulated by the measured quantity.

However, the detected quantity is intensity and hence on demodulation it results in change of intensity, thus, relating to the measured quantity.

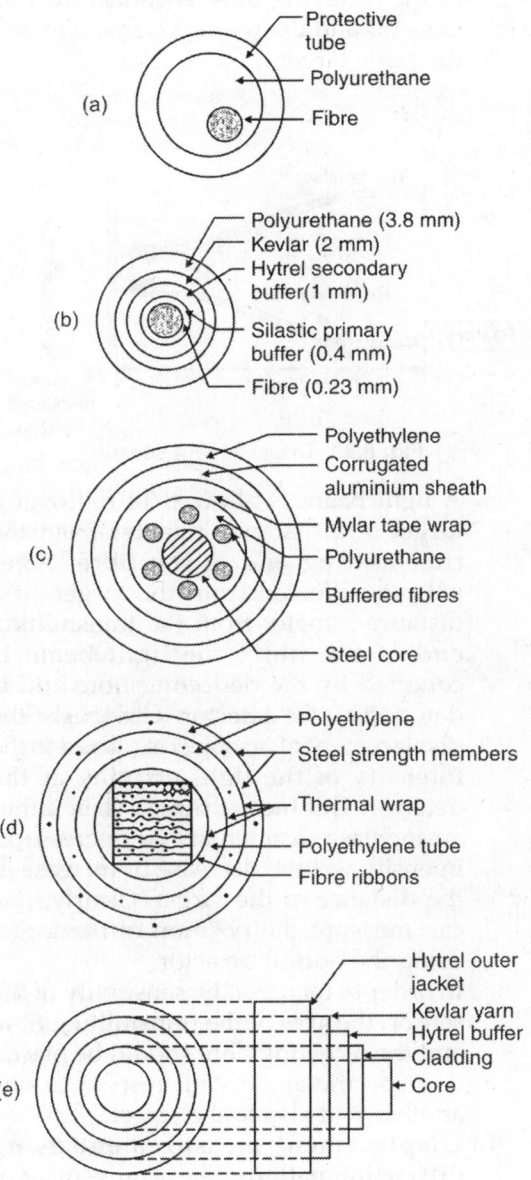

Fig. 8.19: Fibre optic cable configuration: (a) Loose tube construction (b) Constrained fibre (c) Multiple strands (d) Telephone cable (e) Plastic clad silica cable

Optical fibre sensors are of two types:

1. **Extrinsic sensors:** In this type of sensor, the interaction between the light and the measurand (the quantity under measurement) takes place outside the optical fibre. In this type of sensors, the fibre acts merely as a wave guide. This type of sensors have a sensor head, and the sensed optical signal is transferred to the point of measurement with low attenuation and increased mechanical stability for processing of signal.

   Extrinsic sensors are usually used for measurement of voltage, current, pressure, temperature, displacement, force, etc.

2. **Intrinsic sensors:** In these sensors, the measurand acts directly on the fibre itself and produces a change in the transmission characteristics. Liquid level sensors and faraday sensors or gyroscope are this type of sensors.

Table 8.3 gives a few of the measurand with their corresponding modulation effects in optical fibres.

Table 8.3: Modulation effects in optical fibres

Physical measurand	*Modulation effects in optical fibres*
1. Temperature	Thermoluminescence
2. Nuclear radiations	Radiation induced luminescence
3. Pressure	Piezoelectric effect
4. Mechanical force	Stress birefringence
5. Electric field	Electro-optic effect
6. Electric current	Electroluminescence

Interferometric sensors are widely used to measure displacement, temperature, pressure, electric field and magnetic field. These sensors are based on the principle of the interference pattern produced due to a phase shift on superposing a reference beam and a sensing beam.

(a) **Displacement sensor:** This sensor is based on the principle of intensity modulation of the transmitted light beam used to make a measurement of the displacement.

Figure 8.20 shows the arrangement of a displacement sensor. We can see that one end of the transmitting fibre in this type of sensor is coupled to a laser source and the other free end is made to face a target at a distance x meter. Similarly, one end of the receiving fibre is connected to a detector and its free end is made to face the same target.

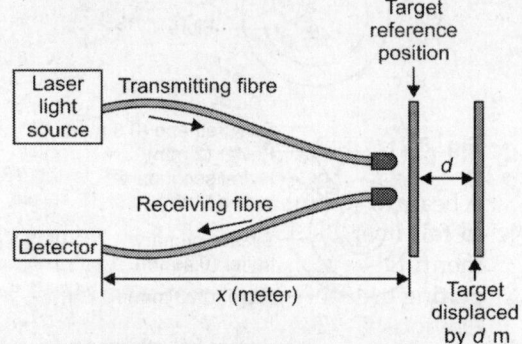

Fig. 8.20: Displacement sensor

A light beam is allowed to fall over a target at a distance x meter from the transmitting end of the fibre. After getting reflected from the target, at a distance x meter from the transmitting end of the fibre, the light beam is collected by the deflecting fibre and is detected by the detector. Obviously, the displacement of an object is related to the intensity of the light arriving at the detector. This means, there will be either an increase in intensity or decrease in intensity wrt the decrease or increase in the distance of the target. Clearly, one can measure the position of the target using the optical detector.

In order to increase the sensitivity of the sensor, the axes of the transmitting fibre and the receiving fibre should be placed at a specific angle with respect to one another and also to the target.

(b) **Displacement measurement using diffraction pattern:** The arrangement of an optical displacement sensor based on phase modulation is shown in Fig. 8.21. A pair of diffraction grating is used to sense the displacement. In order to form a new grating, two diffraction gratings are butted against each other.

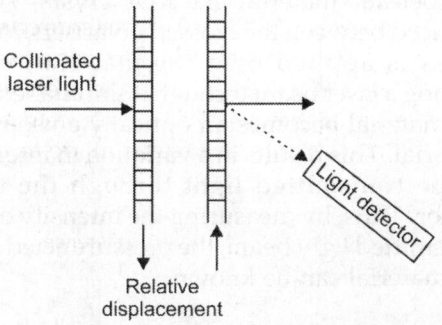

Collimated laser light

Relative displacement

Fig. 8.21: Displacement measurement

A beam of collimated laser light is allowed to fall over this butted grating arrangement. Detector receives the light after getting transmitted through the gratings. Obviously, when any one of the grating is transversely displaced, it causes the diffraction angle of the light beam to vary.

Clearly, the detector sensers this as an intensity variation. Thus, the displacement of the grating results in a high and low intensity beam on the digital detector, thereby giving a direct measurement of displacement.

Pressure sensor: Optical fibre sensors have been developed for measurement of temperature and pressure for downhole in oil wells.

These sensors are used in different fields, e.g. process control, aerospace, oceanography, meterology, hydrology, energy pressure exploration and laboratory instrumentation. These sensors are also used in the medical field to measure blood gas parameters and dosage levels.

The different types of pressure sensors are as follows:

i. **Microbending sensor:** These are based on measurement of the light losses caused by controlled microbending. A basic idea of this sensor is shown in Fig. 8.22. This type of sensor can be used for measuring pressure, temperature and displacement.

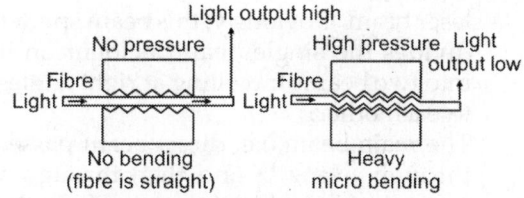

Light output high

No pressure

High pressure Light output low

Fibre Light

Fibre Light

No bending (fibre is straight)

Heavy micro bending

Fig. 8.22: Microbending sensor

A fibre is mounted in between a pair of plates containing parallel grooves. When pressure is applied over the plates, it causes microbends in the fibre, resulting in the leakage of light thereby reducing the power of the transmitted light. Clearly, the measurement of the output power gives the pressure acted over the fibre.

We may note that this type of microbending sensors require only an environment that cause the fibre to bend under certain conditions.

iii. **Pressure sensor based on phase modulation:** This type of sensor works on the principle of interference pattern produced due to phase shift, between the signals from a reference fibre and a test fibre in the measuring environment.

The basic principle of this type of sensor is illustrated in Fig. 8.23.

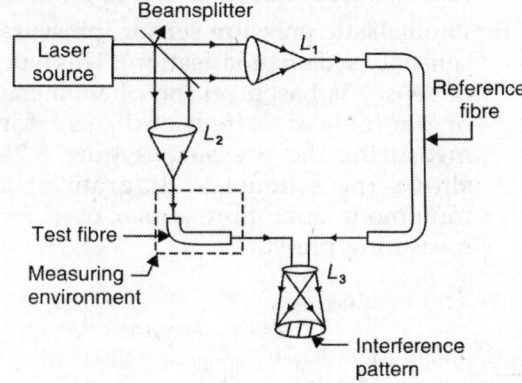

Beamsplitter

Laser source

L_1

Reference fibre

L_2

Test fibre

Measuring environment

L_3

Interference pattern

Fig. 8.23: Pressure sensor based on phase modulation

The light from a laser source is allowed to pass through a beam splitter. This beam splitter is a glass plate inclined at an angle of 45° to the direction of the laser beam. Obviously, this beam splitter divides the single beam incident on it into two beams travelling at right angles to each other.

The main beam, i.e. direct beam passes through a lens L_1 and then through a reference fibre which is isolated from the influence of any external environment. Finally, the beam from the other end of this reference fibre is allowed to pass through another lens L_3.

The splitted beam at right angle to the principal beam is allowed to pass through a lens L_3 and then through a test fibre which is kept in an varying environment to be measured. This means, an environmental effect like pressure or temperature when acts over this test fibre, a path difference is produced due to the change in fibre length, core diameter, and refractive index wrt the beam from the reference fibre. At last, this beam after getting transmitted through this test fibre is allowed to pass through the lens L_3.

An interference pattern is produced due to the phase difference between the light from the reference fibre and test fibre. Clearly, a very accurate measurement of pressure or temperature may be obtained from this interference pattern.

iii. **Photoelastic pressure sensor (pressure sensor based on polarisation):** This type of sensor is based on the phenomena of photoelastic effect and used for measuring the pressure. Figure 8.24 shows the schematic diagram of a multimode optic fibre sensor used for measuring pressure.

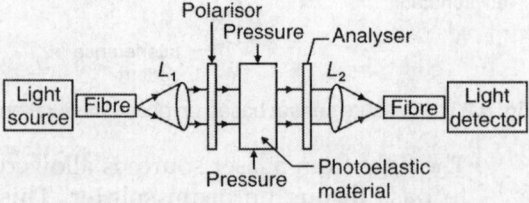

Fig. 8.24: Photoelastic multimode fibre optic pressure sensor

In this sensor device, the photoelastic effect is induced by a mechanical stress on a photoelastic material like KDP crystal which is placed between the crossed polarisers. When stress is applied over the material, upon passing a laser beam through it simultaneously the material becomes an optically anisotropic material. This results in a variation in intensity of the transmitted light through the fibre sensor. Thus, by measuring the intensity of the transmitted light beam, the pressure acted over the material can be known.

8.15 OPTICAL FIBRE ENDOSCOPE

This is a tabular optical instrument which is used to visualize the internal parts of the human body. One can also use it to study tissues and blood vessels far below the skin.

Figure 8.25 shows a schematic representation of a optical fibre endoscope.

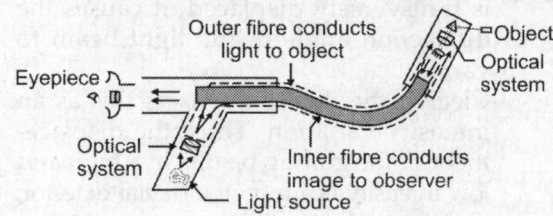

Fig. 8.25: Optical fibre endoscope

This type of fibre endoscope uses a bundle of fibres. The endoscope consists of two fibres: (i) an inner fibre (ii) an outer fibre. An optical light source is attached to the endoscope with suitable optical arrangements to transmit the light into the outer fibre. Suitable optical arrangements which are used to collect and examine (view) the object are arranged at the respective fibre ends. The outer fibre is used to carry the light and to illuminate the object under study. The inner fibre is used to collect the reflected light from the area under study, i.e. from the inner structure of the body.

After switching the light source, the suitable optical arrangements which are fixed at the respective ends of the inner and outer fibre collects the light, transmit and illuminate the area under study and also help to view the desired object. One can add a telescope system at the internal part of the telescope for a wider field of view and improved image quality.

The endoscopes are of following types:

Bronchoscope: This is used to check the presence of foreign bodies and or infection in trachea and larger air ways.

Gastroscope: This is used to check the presence of tumours, gastric ulcer and gastritis in the stomach.

Cytoscope: This is used to check the presence of tumours, inflammation and stones in the urinary bladder.

Nowadays, endoscopes are also used in the treatment of diseases and surgery.

8.16 APPLICATIONS OF OPTICAL FIBRES

Fibre optics technology has matured as a powerful alternative for information transport and sensing. Few important applications of optical fibres are:

Communication

Optical fibres are used to carry signals in optical communications due to their unique features, viz. large channel capacity, wide bandwidth, good electrical isolation, no cross talks, etc. Nowadays telephone, television, satellite and computer links of advanced communication systems rely heavily on this technology.

Scanning

Optical fibres can be used to measure the distribution of light intensity over an illuminated area. For this purpose, the input end of the optical fibre scans the area, and the light output from the optical fibres is fed to a photodetector. The variation of the photodetector output reveals the light intensity distribution.

Use in Medical Instruments

Optical fibres are used in medical instruments (fibres copes) to examine internal body cavities such as stomach and bladder. Optical fibres are also useful in the study of tissues and blood vessels far below the skin. For this purpose, a bundle of fibres is enclosed in a hypodermic needle. This flexible fibre bundles are used in medicine as endoscope. Endoscopes are employed to make observations at a place that can be approached along a curved path. Such types of fibre bundles are said to be passive, i.e. they merely convey light from one end to the other end. Optical fibres are also used for bloodless surgery laser opthalmoscope for retinal welding, etc.

Coupler

One can use the optical fibre for coupling two circuits without introducing a direct electrical link. An LED forms a load in the primary circuit, so that a light output is obtained. Light guide collects the light and transmit to a photodetector in the secondary circuit. The secondary circuit gives an electrical signal in this circuit corresponding to that in the first.

Display and Illumination

Optical fibres are employed to carry light to the display units. They are also used to illuminate dials when measuring instruments are used in dim light.

Transfer of Energy

Optical fibres are employed in the ultraviolet region for spectroscopic work and also used to transfer infrared energy from the source to the point of application of heat.

ILLUSTRATIVE EXAMPLES

Example 1

For a 3 μm diameter optical fibre with core and cladding indices of refraction of 1.545 and 1.510 respectively, determine the cut-off wavelength.

Solution

$$\lambda_c = \frac{2\pi a n_1 \sqrt{2\Delta}}{2.405} \ \mu m$$

where

λ_c : Cut-off wavelength (μm)
n_1 : Core index of refraction (unitless)
n_2 : Cladding index of refraction (unitless)
a : Core radius (μm)

$$\Delta = \frac{n_1 - n_2}{n_1}$$

$$\Delta = \frac{1.545 - 1.510}{1.545} = 0.023$$

$$a = \frac{3\mu m}{2} = 1.5 \ \mu m$$

$$\lambda_c = \frac{(2\pi)(1.5\mu m)(1.545)\sqrt{2(0.023)}}{2.405}$$

$$= 1.29 \ \mu m.$$

Example 2

For a multimode index fibre, $n_1 = 1.53$, $n_2 = 1.50$ and $\lambda = 1 \ \mu m$. If the core radius is 50 μm, then calculate the normalized frequency of the fibre (v) and the number of guided mode.

Solution

Normalized frequency, $v = \dfrac{2\pi a(NA)}{\lambda}$

$$= \frac{2 \times 3.14 \times 50 \times 10^{-6} \times \sqrt{(1.53)^2 - (1.50)^2}}{1 \times 60^{-6}}$$

$$= 94.72$$

Total number of guided mode

$$= M_s = \frac{v^2}{2} = \frac{(94.72)^2}{2}$$

$$= 4486.$$

Example 3

For a multimode stepped index fibre with a glass core ($n_1 = 1.5$) and a fused quartz cladding ($n_2 = 1.46$), determine the critical angle (θ_c), acceptance angle (θ_m) and numerical aperture (NA). The source to fibre medium is air.

Solution

Critical angle, $\quad \theta_c = \sin^{-1}\dfrac{n_2}{n_1} = \sin^{-1}\dfrac{1.46}{1.5}$

$$= 76.7°$$

$$\theta_m = \sin^{-1}\frac{\sqrt{n_1^2 - n_2^2}}{n_0}$$

and $\quad \theta_{m(max)} = \sin^{-1}\sqrt{n_1^2 - n_2^2}$

$$= \sin^{-1}\sqrt{(1.5)^2 - (1.46)^2}$$

$$= 20.2°$$

$$NA = \sin\theta_m = \sin 20.2 = 0.344.$$

The maximum diameter a single mode optical fibre can have is proportional to the wavelength of the light ray entering the cable and the numerical aperture of the fibre. Mathematically, the maximum radius of the core of a single mode fibre is

$$r_{max} = \frac{0.383 \text{Å}}{NA}$$

where r_{max} is maximum core radius in meters.

NA : numerical aperture (unitless)

λ : light ray wavelength (meters)

Example 4

The numerical aperture (NA) of an optical fibre is 0.5 and core refractive index 1.54. Determine (i) refractive index (RI) of cladding (ii) change in core–cladding refractive index per unit RI of the core.

Solution

i. $\quad NA = \sin\theta_m = \sqrt{n_1^2 - n_2^2}$

$\therefore \quad (0.5)^2 = (1.54)^2 - n_2^2 \quad$ or $\quad n_2 = 1.456$

ii. RI of the core

$$\frac{n_1 - n_2}{n_1} = \frac{1.54 - 1.456}{1.54} = 0.0542.$$

Example 5

A silica optical fibre has a core refractive index of 1.50 and a cladding refractive index 1.47. Compute:

(a) The critical angle at core–cladding interface

(b) The numerical aperture for the fibre

(c) The acceptance angle.

Solution

Given,

$$\mu_1 = 1.50$$
$$\mu_2 = 1.47$$

(a) The critical angle at the core–cladding interface is given by

$$\phi_c = \sin^{-1}\left(\frac{\mu_2}{\mu_1}\right)$$

$$= \sin^{-1}\left(\frac{1.47}{1.50}\right)$$

$$= \sin^{-1}(0.98)$$

$$\phi_c = 78.5°$$

(b) The numerical aperture (NA) is given by

$$NA = \sqrt{(\mu_1^2 - \mu_1^2)}$$

$$= \sqrt{(1.50)^2 - (1.47)^2}$$

$$= \sqrt{2.25 - 2.16}$$

$$= \sqrt{0.09}$$

$$NA = 0.30$$

(c) The acceptance angle

$$\theta_m = \sin^{-1}(NA)$$
$$= \sin^{-1}(0.30)$$
$$\theta_m = 17.4°$$

Example 6

For a step index fibre, the normalized frequency (V parameter) is 26.6 at a wavelength of 1300 nm. Determine the numerical aperture (NA) if the core radius is 25 µm.

Solution

Given,

$$V = 26.6$$
$$\lambda = 1300 \text{ nm} = 1.3 \text{ µm}$$
$$r = 25 \text{ µm}$$

The normalized frequency for a step index fibre is given by

$$V = \frac{2\pi r}{\lambda}(NA)$$

$$NA = \frac{V\lambda}{2\pi r}$$

$$= \frac{26.6 \times 1.3}{2 \times 3.14 \times 25}$$

$$= \frac{34.58}{157} = 0.2$$

Example 7

An optical fibre has an attenuation 3.5 dB/km. If 0.5 mW of optical power is initially launched into the fibre, what is the power level in µW after 4 km?

Solution

Given,

$$\alpha = 3.5 \text{ dB/km}$$
$$P_i = 0.5 \text{ mW}$$
$$L = 4 \text{ km}$$

We know that the attenuation of an optical fibre is given by

$$\alpha = \frac{10}{L} \log \frac{P_i}{P_o}$$

or

$$3.5 = \frac{10}{4} \log\left(\frac{0.5}{P_o}\right)$$

or

$$\frac{3.5 \times 4}{10} = \log \frac{0.5}{P_o}$$

or

$$1.4 = \log \frac{0.5}{P_o}$$

or

$$10^{1.4} = \frac{0.5}{P_o}$$

$$\Rightarrow \quad P_o = \log \frac{0.5}{10^{1.4}}$$

$$= \frac{0.5}{25.11}$$

$$= 19.9 \text{ mW}$$

Example 8

A glass clad fibre is made with core glass of refractive index 1.5 and the cladding is doped to give a fractional index difference of 0.0005. Calculate:

(a) The cladding index
(b) The critical internal reflection angle
(c) The external critical acceptance angle
(d) The numerical aperture

Solution

Given,

$$\mu_1 = 1.5$$

Fractional refractive index (FRI) difference, $\Delta = 0.0005$

(a) Let μ_2 be the refractive index of cladding, then

$$\Delta = \frac{\mu_1 - \mu_2}{\mu_1}$$

or $$0.0005 = \frac{1.5 - \mu_2}{1.5}$$

or $$\mu_2 = 1.5 - 1.5 \times 0.0005$$
$$= 1.49$$

(b) Let ϕ_c be the critical internal reflection angle, then

$$\sin \phi_c = \frac{\mu_2}{\mu_1}$$

or $$\phi_c = \sin^{-1} \left(\frac{\mu_2}{\mu_1} \right)$$

or $$\phi_c = \sin^{-1} \left(\frac{1.49}{1.5} \right)$$
$$= \sin^{-1} (0.99)$$

or $$\phi_c = 88.2°$$

(c) The external critical acceptance angle

$$\sin \theta_0 = NA = \frac{\sqrt{\mu_1^2 - \mu_2^2}}{\mu_0}$$

where $\mu_0 = 1$ (for air)

$$\theta_0 = \sin^{-1} \sqrt{\mu_1^2 - \mu_2^2}$$

$$= \sin^{-1} \sqrt{(1.5^2 - 1.49^2)}$$

$$= \sin^{-1} (0.047)$$

$$= 2.72°$$

(d) The numerical aperture (NA)

$$NA = \mu_1 \sqrt{2\Delta}$$

$$= \sqrt{(\mu_1^2 - \mu_2^2)}$$

$$= \sqrt{(1.5)^2 - (1.49)^2}$$

$$= 0.047$$

Example 9

An optical fibre core and its cladding have refractive indices of 1.545 and 1.495 respectively. Calculate the critical angle ϕ_c, acceptance angle $\phi_{in(max)}$ and numerical aperture.

Solution

Given,

$$n_1 = 1.545 \quad \phi_c = ?; NA = ?$$
$$n_2 = 1.495 \quad \phi_{in(max)} = ?$$

Critical angle $$\phi_c = \sin^{-1} \left(\frac{n_2}{n_1} \right)$$

$$\phi_c = \sin^{-1} \left(\frac{1.495}{1.545} \right)$$

$$\phi_c = \sin^{-1} (0.9676)$$

$$\therefore \qquad \phi_c = 75° \, 23'.$$

Acceptance angle $$\phi_{in(max)} = \sin^{-1} \sqrt{n_1^2 - n_2^2}$$

$$\therefore \quad \phi_{in(max)} = \sin^{-1} \sqrt{1.545^2 - 1.495^2}$$

$$= \sin^{-1} \sqrt{0.152}$$

$$= \sin^{-1} (0.3898)$$

$$\phi_{in(max)} = 22° \, 56'.$$

Numerical aperture $NA = \sin \phi_{in(max)}$

$$= \sin (22° \, 56')$$

$$\therefore \qquad NA = 0.3896.$$

Example 10

The refractive index of the core and cladding materials of an optical fibre are 1.54 and 1.5 respectively. Claculte the numerical aperture of the optical fibre.

Solution

Given $n_1 = 1.54$; $n_2 = 1.50$; NA = ?

Numerical aperture NA = $\sqrt{n_1^2 - n_2^2}$

$$= \sqrt{(1.54)^2 - (1.5)^2} = \sqrt{0.1216}$$

∴ NA = 0.3487.

Example 11

A silica optical fibre has a core of refractive index 1.55 and a cladding of fraction index of 1.47. Determine (a) the critical angle at the core–cladding interface (b) the numerical aperture for the fibre (c) the acceptance angle in air for the fibre.

Solution

Given, $n_1 = 1.55$; $n_2 = 1.47$; $\phi_c = ?$;

$$NA = ?; \phi_{in(max)} = ?$$

(a) The critical angle is

$$\phi_c = \sin^{-1}\left(\frac{n_2}{n_1}\right) = \sin^{-1}\left(\frac{1.47}{1.55}\right)$$

$$\phi_c = \sin^{-1}(0.94838)$$

∴ $\phi_c = 71° 30'$.

(b) The numerical aperture for the fibre is

$$NA = \sqrt{n_1^2 - n_2^2} = \sqrt{1.55^2 - 1.47^2}$$

$$= \sqrt{0.0891}$$

∴ NA = 0.2984.

(c) The acceptance angle in air for the fibre is

$$\phi_{in(max)} = \sin^{-1}(NA) = \sin^{-1}(0.2984)$$

∴ $\phi_{in(max)} = 17° 21'$.

REVIEW QUESTIONS

1. Define a optical fibre system. What is numerical aperature of an optical fibre. What is its significance?

2. Obtain the expression for NA is terms of RI of the core and cladding of an optical fibre.

3. Outline the primary building blocks of a fibre optic system.

4. Draw block diagram for optical communication through fibre optical cable. Mention the advantages of optical communication?

5. Explain the working of monomode and multimode optical fibres. For long distance communication which optical fibre is preferred and why?

6. Contrast glass and plastic fibre cables.

PROBLEMS

1. A multimode step index fibre has $n_1 = 1.53$, $n_2 = 1.50$ and $\lambda = 1$ μm. In order to make this fibre single mode, what will be core radius?

   [Hint: $a \le \dfrac{2.405\lambda}{2\pi(NA)}$, $NA = \sqrt{(1.53)^2 - (1.50)^2}$,

   $a \le \dfrac{2.405 \times 1 \times 10^{-6}}{2 \times 3.14 \times \sqrt{(1.53)^2 - (1.50)^2}}$, ∴ $a < 1.27$ μm]

2. Numerical aperature of a optical fibre is 0.5 and core refractive index is 1.48. Show that cladding refractive index is 1.393 and acceptance angle is 30°.

   [Hint: $NA = \sqrt{n_1^2 - n_2^2} = 0.5$, RI of core = $n_1 = 1.48$ that is $n_2 = 1.393$ that is RI of cladding = $n_2 = 1.393$. Acceptance angle $\theta = \sin^{-1}(NA) - 30°$]

3. A step index fibre in air has a NA of 0.16, a core refractive index of 1.45 and core diameter of 60 cm. Show that the normalized frequency for the fibre when light at a wavelength of 0.9 μm transmitted is 3.35 × 10⁵.

   [Hint: $v = \dfrac{\pi d}{\lambda_0}\sqrt{n_1^2 - n_2^2} = \dfrac{3.143 \times 0.6\,m}{9 \times 10^{-7}\,m} \times 0.16$

   $= 3.35 \times 10^5$]

4. An optical fibre has a NA of 0.20 and a cladding refractive index of 1.59. Show that the acceptance angle for the fibre in water which has a refractive index of 1.33 is 8.6°.

   [Hint: $NA = \dfrac{\sqrt{n_1^2 - n_2^2}}{n_0}$

When the fibre is in air, $n_0 = 1$, and

$$NA = \sqrt{n_1^2 - n_2^2} = 0.20$$

$$\therefore n_1 = \sqrt{(NA)^2 + n_2^2} = \sqrt{(0.20)^2 + (1.59)^2} = 1.6025$$

When the fibre is in water, $n_0 = 1.33$

$$\therefore NA = \frac{\sqrt{n_1^2 - n_2^2}}{n_0} = \frac{\sqrt{(1.6025)^2 - (1.59)^2}}{1.33} = 0.15$$

$$\therefore \theta_0(max) = \sin^{-1}(NA) = \sin^{-1}(0.15) = 8.6°]$$

5. Show that the core radius neccessary for single mode operation at 850 nm S.I. fibre with $n_1 = 1.480$ and $n_2 = 1.47$ is 1.89 µm. Also show that the NA and maximum acceptance angle of this fibre are 0.1717 and 9° 53′ 12″ respectively.

[**Hint:** (i) $v = \dfrac{\pi d}{\lambda_0} \sqrt{n_1^2 - n_2^2}$

$$2.405 = \frac{\pi d}{450 \times 10^{-9}\,m} \times 0.1717$$

or $\qquad d = 3.79$ mm $\quad \therefore r = 1.89$ mm

(ii) $NA = \sqrt{n_1^2 - n_2^2} = \sqrt{(1.48)^2 - (1.47)^2} = 0.1717$

(iii) $\sin\theta_0\,(max) = NA$

$$\therefore \theta_0(max) = \sin^{-1}(NA) = \left(n_1^2 - n_2^2\right)^{1/2}$$

$$= \sin^{-1}(0.1717) = 9° 53′ 12″$$

SHORT ANSWER QUESTIONS

1. To prevent energy losses via absorption and scattering, what should be the order of the cladding in an optical fibre?

Ans. At least a few wavelengths thick.

2. How numerical aperature (NA) of an optical fibre is defined?

Ans. $NA = \sqrt{n_1^2 - n_2^2}$ where n_1 is the refractive index of the core of a fibre and n_2 that of cladding $(n_1 > n_2)$. One can define the numerical aperature of a fibre as the physical area that light must enter in order to propagate through the fibre. Similarly, sources have a numerical aperature that describes the output pattern of the light. The numerical aperatures for light sources can be calculated in terms of angle involved with the output.

3. How will you define acceptance angle (θ_A) for an optical fibre?

Ans. $\theta_A = \sin^{-1}\sqrt{n_1^2 - n_2^2}$

4. What is major economic benefit offered by fibre optics in information technology?

Ans. Very high information transmission rate at a low cost per circuit-km.

5. What type of reflection is expected from an optical fibre?

Ans. Total internal reflection.

6. How is an optical fibre fabricated?

Ans. To fabricate fibre, one can use either different glasses with different refractive indices for core and cladding or use fused silica (quartz glass) whose refractive index is modified by doping. The type of glass used will also vary with application, but the purity (lack of any foreign materials as well as lack of imperfections such as cracks or bubbles) must also be kept very high.

7. What are the advantages of optical fibre communication over metallic cable communication?

Ans. Optical fibre communication systems:
   i. have extremely wide bandwidth
   ii. are immune to electrostatic interference
   iii. are light in weight and small in size
   iv. are more secured, higher safety and there is no cross talk interference.

8. What is kevlar?

Ans. This is a yarn type material whose tensile strength is high. This is used in fibre construction for providing additional strength to the cable.

9. Can a light beam propagate through a fibre cable with the angle of incidence at the entrance end greater than acceptance angle?

Ans. No light can propagate through the fibre cable by undergoing total internal reflection only when light enters at an angle less than the acceptance angle.

10. What is an optical fibre sensor?

Ans. This is a transducer which can convert various input variables, i.e. physical quantities into electrical signal in a measurable form.

11. Mention advantages of optical sensors in sensing applications.

Ans.   i. They are light in weight and small in size.
   ii. Their chemical and environmental ruggedness is high.
   iii. They are electrically positive.
   iv. They possess good geometrical flexibility.

OBJECTIVE QUESTIONS

1. Optical fibre is _____ carrier and immune to _____ interference.
   [non-electrical, electromagnetic]
2. Optical fibre does not exit any _____ that might interfere with signals. [noise]
3. Light has much _____ frequency than radio waves or modulated signals in copper wire.
   [higher]
4. Optical fibre does not rust or corrode and is much _____ and _____ than copper cable.
   [smaller, larger]
5. Scattering mechanism cause the transfer of _____ from one propagating mode into a different mode. [optical power]
6. Major implementation of optical fibre transmission involves some form of digital _____. [modulation]
7. Optical fibres are _____ pipes that guide light through them on the principle of total _____ . [light, internal reflection]
8. The refractive index of _____ region is always less than that of the core region.
   [cladding]
9. Optical fibres are divided into different groups based on their _____ profile. [modulation]

MULTIPLE CHOICE QUESTIONS

1. An optical fibre is based on the principle of
   (a) polarization of light
   (b) total internal reflection
   (c) interference of two beam
   (d) diffraction of light
2. The jacket of an optical enables
   (a) to prevent from mechanical abrasions
   (b) to prevent interaction with internal atomosphere
   (c) to prevent from moisture trapping
   (d) all of the above
3. In a graded index fibre, the refractive index
   (a) remains constant throughout the core
   (b) decreases parabolically throughout the core
   (c) increases linearly throughout the core
   (d) varies arbitrarily throughout the core
4. Pulse broadening can be minimized by using
   (a) single mode step index fibre
   (b) multimode step index fibre
   (c) multimode graded index fibre
   (d) none of the above

5. Attenuation in an optical fibre is measured by
   (a) $\text{loss} = -10 \log_{10} P_o$
   (b) $\text{loss} = 10 \log_{10} P_i$
   (c) $\text{loss} = -10 \log_{10} \dfrac{P_o}{P_i}$
   (d) $\text{loss} = -10 \log_{10} \dfrac{P_o}{P_i}$
6. Optical fibres transmit information signals in the form of
   (a) digital signals
   (b) analog signals
   (c) both analog and digital signals
   (d) binary and hexadecimal bits
7. Dispersion in optical fibres is due to
   (a) intermodal dispersion
   (b) chromatic dispersion
   (c) material dispersion
   (d) all of the above
8. The optical fibres are made of
   (a) metallic conductor
   (b) plastic doped with metallic impurities
   (c) dielectric material
   (d) magnetic oxide
9. The light can be guided along a curved stream of water was first shown by
   (a) Alexander Graham Bell
   (b) John Tyndall
   (c) TH Maiman
   (d) Albert Einstein
10. When V parameter is less than 2.405, then the fibre will support
    (a) one mode   (b) two modes
    (c) three modes   (d) infinite modes
11. In an optical fibre, the refractive index of the core is n_1, the refractive index of the clad is n_2, then
    (a) $n_1 < n_2$ (b) $n_1 > n_2$
    (c) $n_1 = n_2$ (d) $\dfrac{n_1 - n_2}{n_1} > 1$
12. Single mode step index fibres support propagation of
    (a) skew rays   (b) helical rays
    (c) meridian rays   (d) all the above rays
13. Loss of light intensity in optical fibre is due to
    (a) refraction   (b) absorption
    (c) scattering   (d) (b) or (c)

14. Optical fibre sensing applications mostly use
    (a) interference
    (b) polarization
    (c) refraction
    (d) reflection

15. Angle of acceptance is maximum for a fibre if
    (a) the critical angle is zero
    (b) the critical angle is maximum
    (c) the critical angle is minimum but not zero
    (d) the critical angle is negative

Answers

1. (b)	2. (d)	3. (b)	4. (c)	5. (c)
6. (a)	7. (d)	8. (c)	9. (b)	10. (a)
11. (b)	12. (c)	13. (d)	14. (a)	15. (c)

9 Acoustics

9.1 INTRODUCTION

The branch of physics that deals with the process of generation, reception and propagation of sound is termed acoustics. Truly speaking, this branch covers many fields and is closely related to various branches of engineering, e.g. (i) design of acoustical instruments (ii) electro acoustics, viz., the branch relating to the methods of sound production and recording (microphones, amplifiers, loudspeakers, etc.), (iii) architectural acoustics dealing with the design and construction of buildings, operas, music halls, recording rooms in radio and television broadcasting stations (in general architectural acoustics deals with the behaviour of sound waves in a closed space) and (iv) musical acoustics dealing with the design of musical instruments, etc.

The subject has developed to such an extent that it is classified as *acoustical engineering*.

9.2 ARCHITECTURAL ACOUSTICS

In a good auditorium particularly theatres, concert halls, classrooms, etc. sound produced by the speaker or by a source should be heard with sufficient loudness and clarity. It is found that some auditoriums are acoustically good and some bad, i.e. in some auditoriums the sounds produced are distinctly audible while in others the sounds lock in distinctness. In a good auditorium the following conditions should be satisfied:

i. The sound heard must everywhere be sufficiently loud and no echoes should be present.
ii. The 'quality' of sound, i.e. speech and music must remain unaltered, i.e. the relative intensity of several components of a complex sound must be maintained.
iii. The successive syllables spoken must be clear and distinct. Each syllable should die away sufficiently quickly to avoid overlapping with the next syllable.
iv. There should be no undesirable echoes. Reverbation should be quite proper.
v. There should be neither any concentration of sound nor any zone of silence in any part of the hall.
vi. There should be no undue noise and no resonance within the hall or building.
vii. There should be no echelon effect.

To achieve this, various factors need attention especially, the one known as *reverbation*.

9.3 REVERBATION

A common defect in halls, large rooms and auditoriums is the undue persistence or prolongation of the sound of the speaker or the singer. This arises due to the successive reflections of the sound from the walls, ceiling, floors, etc. of the hall or room. A short sound made in a hall reaches a listener directly as well as after successive reflections from the walls, ceiling and floor of the hall. The listener,

241

therefore, receives a series of sounds of diminishing intensity (since part of energy is lost at each reflection) and, instead of single sound, he/she hears a roll of sound. This prolongation or persistence of sound is termed *reverbation*. Reverbation can also be defined as the persistence of audible sound after the source has stopped to emit any sound.

Similarly, when a continuous source of sound starts sounding, some of the sound waves reach the listener directly, while the other reach successively later on after reflection and add to the direct waves still reaching there. During this time the energy is partly absorbed by walls and other materials in the room and partly escape through the open windows and ventilators, and after a few seconds a balance is reached between the energy emitted from the source and the energy lost. Then the intensity of sound attains a steady value. If now the source of sound is stopped, the sound does not stop instantaneously but the human ear continues to pick up the successive reflections until they fall below the minimum audibility. We may note that louder the original sound, the longer will this process take. Again this gradual decay of sound is known as *reverbation*.

Figure 9.1a shows the rise and fall of sound. When a succession of different notes are sounded the effect is as shown in Fig. 9.1b.

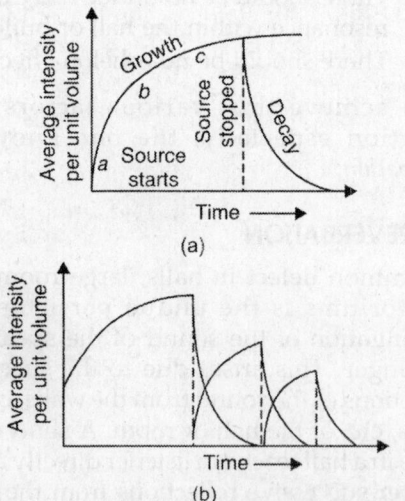

Fig. 9.1: Growth and decay of sound

The time gap between the initial direct note and the reflected note upon the minimum audibility level is called *reverbation time*. In other words, we can say that the time taken for the sound to fall below the minimum audibility, measured from the instant of its generation (in the case of short sounds) or the stopping of the source (in the case of continuous note) is termed *time of reverbation*.

In a hall or auditorium if the reverbation time is too large, then there is overlapping of successive sounds which causes confusion and results in loss of clarity in hearing. On the other hand, if the reverbation time is too small, then the loudness will be inadequate. Obviously, the reverbation time for a hall or auditorium should neither be too large nor too small. Obviously, the reverbation time must have a definite value which may be satisfactory for the speakers and audience. Thus, an adjustment of the appropriate time of reverbation time is essential requirement of good acoustics. The perfect time for reverbation is termed *optimum reverberation time*.

Prof W C Sabine made the extensive study of the problem of reverbation. He found that there is an optimum period of reverbation depending upon the size of the hall at which best results are obtained. This period is found greater for music than for speech. For example, in a hall of 400,000 cubic feet the maximum intelligibility of speech is obtained for a reverbation period between 1.0 s and 1.5 s, while the music is not appealing between 1.5 s and 2 s. Below this value the intensity of sound is weak and the music appears lifeless, while above this the syllables overlap and become indistinct.

Derivation for Reverbation Time

Sabine developed the reverbation time formula to express the rise and fall of sound in an auditorium. He observed that the reverbation time depends upon the size of the hall and the absorbing power of the objects in it and its walls. His formula for reverbation time is based on the assumption that the sound energy in the room is distributed uniformly all around. While deriving Sabine's formula, following steps have to be considered:

i. To calculate, in terms of energy density E, the rate at which the energy is incident upon the walls and other surfaces of the hall or auditorium and hence the rates at which it is being absorbed.

ii. To calculate the final steady value of energy density E in terms of the rate of emission of power P of the sound emitting source.

iii. To calculate the standard reverbation time T.

Now, we proceed to derive the relation.

Energy Density

Let us consider that ds is a small element of plane wall AB, then the rate of energy received from the source at ds is obtained as

$$E \cdot ds \cdot \frac{C}{4}$$

where C is the velocity of sound. Now, if a is the coefficient of absorption of the wall AB, then, one finds

Rate of energy absorbed by $ds = ECa\dfrac{ds}{4}$.

Obviously, the total absorption of all the surfaces of the wall where the sound is falling, obtained as

$$EC(\textstyle\sum a\, ds)/4 = ECA/4 \qquad (9.1)$$

Here $A = \sum a\, ds$, the total absorption of sound by all the surfaces of the wall where the sound falls.

Final Steady Value of E in Terms of P

If the rate of emission of energy from the source, i.e. the power output is P and V is the total volume of the room, then the total energy in the room at the instant when energy is E will be EV. Thus, we have

$$\text{Rate of growth of energy} = \frac{d}{dt}(EV) = \frac{VdE}{dt}$$

Now, the rate of growth of energy in space = rate of supply of energy by source – rate of absorption by all the surfaces,

i.e.
$$\frac{VdE}{dt} = P - ECA/4 \qquad (9.2)$$

Now, when steady state is reached,

$$\frac{dE}{dt} = 0$$

For steady energy E_m, we have
$$E_m = 4P/CA \qquad (9.3)$$

Therefore, Eq. (9.2) can be written as

$$\frac{dE}{dt} = \frac{P}{V} - \frac{CA}{4V}E$$

or
$$\frac{dE}{dt} + \alpha E = \frac{4P}{CA}\alpha \qquad (9.4)$$

where
$$\alpha = CA/4V \qquad (9.5)$$

Now, multiplying both sides of Eq. (9.4) by $\exp(\alpha t)$ and integrating with respect to t, one obtains

$$E\exp(\alpha t) = \frac{4P}{CA}\exp(\alpha t) + K \qquad (9.6)$$

Now, we study the growth and decay of sound energy density in a hall or auditorium.

Growth

We have the general distribution of sound energy in a hall or auditorium as

$$E\exp(\alpha t) = \frac{4P}{CA}\exp(\alpha t) + K$$

where A is the total absorption by the hall, P is the power transmitted, V is the volume of the hall, C is the velocity of sound, $\alpha = CA/4V$, t is time, K is the constant of integration and $A = \sum as$ is the summation of the product of different areas (s) and their respective absorption coefficients (a).

The growth of sound in a hall is controlled by all the above factors, one will have to know the value of K. We have at $t = 0$, $E = 0$. From Eq. (9.6), we have

$$K = -4P/CA$$

So
$$E = 4P/CA(1 - e^{-\alpha t})$$

At $t = \infty$,
$$E = 4P/CA = E_m \quad (\because e^{-\infty} = 0)$$

$\therefore$
$$E = E_m(1 - e^{-\alpha t}) \qquad (9.7)$$

Figure 9.2 shows the exponential growth of energy with time t in accordance with Eq. (9.7). We note that E ultimately attains the value E_m.

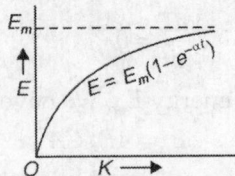

Fig. 9.2: Growth of sound with time in a hall

The sound gets reflected, absorbed and transmitted in the surroundings after it is transmitted from the speaker. When the absorption and transmission of sound is less, then most of the sound energy gets reflected back. In such a situation, the total energy in the hall will increase linearly and attain the ultimate energy in a very short time. However, due to the presence of absorbing materials in the hall sufficient absorption takes place and ultimately E attains the maximum value at $t = \infty$.

Decay

Suppose the source is cut off when $E = E_m$, then $P = 0$ at $t = 0$, and now the decay of sound energy will take place. Here $P = 0$ at $t = 0$ and $E = E_m$. Thus, from Eq. (9.6), one obtains $E_m = K$, i.e.

$$E = E_m \exp(-\alpha t) \qquad (9.8)$$

Equation (9.8) reveals clearly the decay of sound energy density with time after the source is cut off. Figure 9.3 shows the exponential decay of sound energy. Since $P = 0$, i.e. there is no transmission power, so the sound will decay exponentially with time from E_m to O.

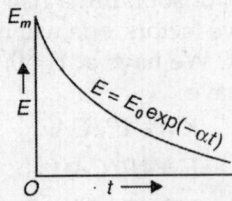

Fig. 9.3: Expontential decay of sound energy with time

Sabine's Reverbation Time Formula

Sabine developed the reverbation time formula to express the rise and fall of sound in an auditorium. The derivation is based upon the following main assumptions:

i. The average energy per unit volume is uniform. It is represented as σ.
ii. The energy is not lost in the auditorium. The energy is lost only due to the absorption of the material of the walls and ceiling. Energy is also lost due to escape through the window and ventilators. Both these factors are included in the *absorption* of energy.

Let us suppose that a source is producing sound continuously. This sound energy is propagated in all directions. Let σ represent the energy contained in a unit volume. Now, the energy contained in a *solid angle* $d\phi$ is

$$= \frac{\sigma d\phi}{4\pi}$$

Now, suppose that this energy be incident on a unit surface area of the wall at an angle θ. Let the velocity of sound be C, then the total energy falling per second on a unit surface area of the wall

$$= \left(\frac{\sigma d\phi}{4\pi} \right)(\cos\theta)C$$

Now, the total energy falling per second within a hemisphere

$$= \frac{\sigma C}{4\pi} \int \cos\theta \, d\phi$$

But $\phi = 2\pi(1 - \cos\theta)$ [*see* Fig. 9.4]

or $d\phi = 2\pi \sin\theta \, d\theta$

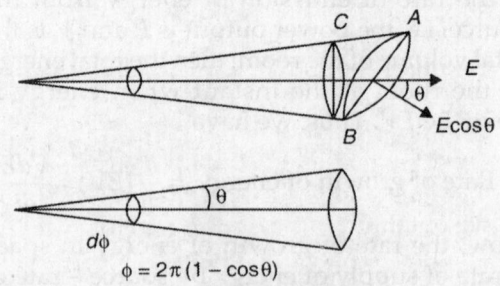

$\phi = 2\pi(1 - \cos\theta)$

Fig. 9.4: Solid angle

Now, substituting this value of $d\phi$, one obtains the total energy falling per second within a hemisphere

$$= \frac{\sigma C}{4\pi} \int_0^{\pi/2} 2\pi \sin\theta \cos\theta d\theta$$

$$= \frac{\sigma C}{2} \left[-\frac{\cos^2\theta}{2} \right]_0^{\pi/2} = \frac{\sigma C}{4}$$

Let us suppose that α is the absorption coefficient of the walls that refers to the fraction of the incident energy not reflected from the walls. The amount of energy absorbed per second per unit area $= \dfrac{\alpha \sigma C}{4}$. If A is the area of the walls and other absorbing materials including windows, ventilators, ceiling, etc. in a hall, the amount of energy absorbed per second $= \dfrac{A\alpha\sigma C}{4}$.

Let volume of the auditorium be V, then the total energy $= \sigma V$. The rate of increase of energy $= \dfrac{d}{dt}(V\sigma)$.

$$= V \frac{d\sigma}{dt}$$

Let us suppose that the source supplies energy at the rate of Q units per second. Thus, the rate of increase of energy

$$= Q - \frac{A\alpha\sigma C}{4} \qquad (9.9)$$

Equating Eq. (9.8) and Eq. (9.9), one obtains

$$V\frac{d\sigma}{dt} = Q - \frac{A\alpha\sigma C}{4} \qquad (9.10)$$

Taking $\dfrac{A\alpha C}{4} = K$,

$$\frac{K}{V} = \beta \quad \text{and} \quad \frac{Q}{K} = \frac{4Q}{A\alpha C}$$

we obtain from Eq. (9.10)

$$V\frac{V\sigma}{dt} = Q - K\sigma$$

or

$$\frac{d\sigma}{dt} = \frac{Q}{V} - \frac{K}{V}\sigma \qquad (9.11)$$

The general solution of Eq. (9.11) is

$$\sigma = B + be^{-\beta t} \qquad (9.12)$$

where

$$t = 0, \sigma = 0$$

Therefore, from Eq. (9.12), we have

$$0 = B + b$$

or

$$b = -B$$

$\therefore$

$$\sigma = B - Be^{-\beta t}$$

$$= B[1 - e^{-\beta t}]$$

Now, substituting the values of B and β, one obtains

$$\sigma = \frac{4Q}{A\alpha C} \left[1 - e^{-\frac{A\alpha C}{4V}t} \right] \qquad (9.13)$$

Equation (9.13) represents the rise of average sound energy per unit time from the time the source commences to produce sound. The maximum value of average energy per unit volume is obtained as

$$\sigma_{max} = \frac{4Q}{A\alpha C} \qquad (9.14)$$

Similarly, after the source ceases to emit sound, the decay of the average energy per unit volume is obtained as

$$\sigma = \frac{4Q}{A\alpha C} \exp\left[-\frac{A\alpha C}{4V}t \right] \qquad (9.15)$$

$$\sigma = \sigma_{max} \exp\left[-\frac{A\alpha C}{4V}t \right] \qquad (9.16)$$

The factor $\dfrac{A\alpha C}{4V}$ gives the *reverbation time* in the hall or auditorium. If σ_0 represents the minimum audible intensity after a time t_1, then one obtains from Eq. (9.16)

$$\sigma_0 = \sigma_{max} \exp\left[-\frac{A\alpha C}{4V}t_1 \right] \qquad (9.17)$$

where t_1 is the time interval between the cutting off the sound and the time at which intensity falls below the minimum audible level. From Eq. (9.17), we have

$$\sigma_{max} = \sigma_0 \exp\left[\frac{A\alpha C}{4V} t_1\right]$$

Taking logarithms, one obtains

$$\log_e\left(\frac{\sigma_{max}}{\sigma_0}\right) = \frac{A\alpha C}{4V} t_1 \qquad (9.18)$$

We may note that here α and σ_0 change with the frequency of sound.

Now, for calculating the reverbation time, a standard steady intensity is required. Sabine took the value $\frac{\sigma_{max}}{\sigma_0} = 10^6$. From Eq. (9.18), we have

$$\log_e\left(10^6\right) = \frac{A\alpha C}{4V} t_1$$

or $$2.303 \times 6 = \frac{A\alpha C}{4V} t_1$$

Now, taking the velocity of sound approximately at room temperature as 350 m/s, we have

$$2.303 \times 6 = \frac{A\alpha \times 350}{4V} t_1$$

or $$t_1 = \frac{2.303 \times 24V}{350 A\alpha}$$

$$= \frac{0.158V}{A\alpha} \qquad (9.19)$$

In general,

$$T = t_1 = \frac{0.158V}{\sum A\alpha} \qquad (9.20)$$

We may note that the quantities appearing in Eq. (9.20) are represented in MKS units.

Equation (9.20) is *Sabine's equation* for reverbation time. Equation (9.20) is in general good agreement with experimental value obtained by Sabine.

According to Eq. (9.19), the reverbation time T or t_1 is (i) directly proportional to the volume of the hall or auditorium (ii) inversely proportional to the ceiling, area of walls, etc.

and (iii) inversely proportional to the total absorption plus transmission through open surfaces. Experimentally it is found that the reverbation time of 1.03 s is most suitable for all rooms having approximately a volume of less than 350 cubic meters.

In order to decrease the reverbation time, the walls of the auditorium are usually covered with material having large absorption coefficients. In good cinema halls, the area of the surfaces of the walls is increased to decrease the reverbation time.

Limitations of Sabine''s Formula

i. Sabine's reverbation time formula is a special case of more general formula put forwarded by CF Eyring. We may note that Sabine's formula gives contradictory result in the case of a dead room. In the case of complete absorption, we have $a = 1$ and reverbation time T or t_1 should be 0, but according to Sabine's formula, we have $T = 0.161\ V/s$.

ii. Sabine's formula does not give correct result for absorption coefficient more than 0.2.

iii. In the derivation of Sabine's formula, it is assumed that there is uniform energy density and there is no loss of energy in the air.

However, this is not a practical reality.

Eyring's Formula

Eyring obtained a reverbation time formula on the assumption that sound images could be used to replace the walls of a room in calculating the rate of decay of sound energy after the sound source is cut off. Eyring's formula read as

$$T = \frac{KV}{-S\log_e\left(1 - \overline{a}\right)}$$

where K is a constant, S is the total area of the absorbing surface, and $\overline{a}$ is the average absorption coefficient, defined as

$$\overline{a} = \frac{\sum\left(a\delta S\right)}{S} = \frac{A}{S}$$

Eyring's formula is applicable to *live* as well as *dead* halls. For a perfectly dead hall (open space), $\bar{a} = 1$. Then the Eyring's formula yield

$$T = \frac{KV}{-S\log_e 0} = \frac{KV}{\infty} = 0$$

as expected.

For a *live* room or hall, the absorption coefficient $\bar{a}$ is very small. Then one can write $\log_e(1 - \bar{a}) = -\bar{a}$. Now, Eyring's formula takes the form

$$T = \frac{KV}{S\bar{a}}$$

which is same as Sabine's relation. Obviously, Sabine's relation is a special case of Eyring's formula.

9.4 ABSORPTION COEFFICIENT

The absorption coefficient (a) of a material is generally defined as the ratio of the sound energy absorbed by the surface to that of the total incident sound energy on the surface of the material, i.e.

$$a = \frac{\text{Sound energy absorbed by the surface of the material}}{\text{Total sound energy absorbed by the surface}}$$

To compare the relative efficiency of different absorbing materials, it is essential to first select some standard of absorption in terms of which one can assess all the substances. Sabine selected a unit area of open window, as standard of absorption, because all the sound falling on the open window passes out and none is reflected back. Obviously, open window is ideal, i.e. perfect absorber of sound.

Thus, one can define the *absorption coefficient* (*a*) *of a material as the reciprocal of its area which absorbs the same sound energy as absorbed by an unit area of an open window*. The unit of coefficient of absorption is Sabine or Open Window Unit (OWU). For example, suppose 8 sqm of certain carpet absorbs the same

amount of energy, as absorbed by 1 sqm of an open window. The coefficient of absorption of carpet is $1/8 = 0.125$. We may note that the absorption coefficient of a given material is different at different frequencies. Usually, it is higher at higher frequencies. The absorption coefficient of some common materials calculated with a source of frequency 512 Hz are given in Table 9.1.

Table 9.1: Absorption coefficient (*a*) of some common materials calculated with a source of frequency 512 Hz

Material	Absorption Coefficient (*a*)
Marble	0.01
Glass	0.027
Common plasters	0.03
Concrete	0.17
Cork	0.23
Asbestos	0.26
Carpet	0.30
Acoustic plaster	0.30
Acoustic felt	0.45
Fibre board	0.50
Heavy curtains	0.50
Hair or Felt	0.58
Fibre glass	0.75
Perforated cellulose fibre tiles	0.85

Measurement of Absorption Coefficient

There are different methods for the determination of absorption coefficient of materials as mentioned below.

Single Source Method

One can measure the absorption coefficient in terms of reverbation time. The reverbation time is first measured when the absorbing material is not inside the hall. Let its value of reverbation time be T_1. Thus,

$$1/T_1 = A/0.161V = \Sigma as/0.161V \qquad (9.21)$$

where V is the volume of the hall, Σs is the inside surface area and a is the average absorption coefficient of reverbation chamber. Now, when an absorber is present inside the hall, one can measure absorbtion coefficient in the following two ways:

1. *Absorber Suspended Inside the Hall*

Let us consider an absorbing material, e.g. a stage screen, which is suspended inside the room. The reverbation time T_2 is now measured. We have

$$1/T_2 = \frac{\sum as + 2a_2 S_2}{0.161V} \qquad (9.22)$$

where a_2 is the absorption coefficient of the material of area S_2. This material is suspended inside the hall.

Now, since the absorbing material is suspended inside the hall and hence absorption will take place from both the sides of the screen. Obviously, the absorption by the material in this case $= 2a_2 S_2$. Subtracting Eq. (9.21) from Eq. (9.22), one obtains

$$2a_2 S_2 = 0.161V \left(\frac{1}{T_2} - \frac{1}{T_1} \right)$$

or
$$a_2 = \frac{0.161V}{2S_2} \left(\frac{1}{T_2} - \frac{1}{T_1} \right) \qquad (9.23)$$

From Eq. (9.23), knowing the values of T_1, T_2, S_2 and V, the value of a_2 can be obtained.

2. *Absorber Spread on the Floor or on the Wall*

Let us consider that the absorber is spread on the floor of reverbation chamber, e.g. carpet or it may be fixed on the wall, e.g. window glass, curtain cloth. For such type of absorbing materials, absorption coefficient can be determined in the following manner. Let reverbation time with such a material be T_3, whose area is S_3 and absorption coefficient a_3, then we have

$$\frac{1}{T_3} = \frac{\sum as + a_3 S_3 - aS_3}{0.161V} \qquad (9.24)$$

Here, we have subtracted aS_3 because now the surface area S_3 of the floor or wall does not contribute to absorption which is included in $\sum as$, the total absorption by the chamber, and here it is $a_3 S_3$ and not twice of that, because the absorption takes place from one side only.

Now, subtracting Eq. (9.21) from Eq. (9.24), one obtains

$$a_3 = \frac{0.161V}{S_3} \left(\frac{1}{T_3} - \frac{1}{T_1} \right) + a \qquad (9.25)$$

Knowing other quantities, one can determine absorption coefficient a_3 of window glass or carpet.

Double Source Method

This method of measurement of absorption coefficient requires two sound sources of powers P_1 and P_2. Let the reverbation times of two sound sources be T_1 and T_2. Actual powers of two sources are not necessarily to be known but the knowledge of their ratio will be sufficient. The loudspeaker with measured audio-frequency current are used as the sources.

The steady energy densities maintained by two sound sources are

$$E_1 = \frac{4P_1}{AC} \quad \text{and} \quad E_2 = \frac{4P_2}{AC}$$

Let us consider that during the time of decay, they reach a value of bare inaudibility E_0 in times T_1 and T_2. We have

$$E_0 = \frac{4P_1}{AC} e^{-\alpha T_1} \quad \text{and} \quad E_0 = \frac{4P_2}{AC} e^{-\alpha T_2}$$

$$\therefore \quad \frac{P_2}{P_1} = e^{\alpha(T_1 - T_2)} \quad \text{where } \alpha = \frac{CA}{4V}$$

Thus $\quad \frac{CA}{4V}(T_1 - T_2) = \log_e \frac{P_2}{P_1}$

or $\quad A = \frac{4V \log_e (P_2/P_1)}{C(T_1 - T_2)} = \sum as$

where A is the total absorption by the areas of the hall, and is given by $\sum as$.

The average absorption coefficient of the hall is

$$A = \sum a = \frac{4V \log_e (P_2/P_1)}{CS(T_1 - T_2)} \qquad (9.26)$$

where S is the total internal surface area of the hall or auditorium.

Knowing the various quantities appearing on the RHS of Eq. (9.26), one can easily find out the average absorption coefficient A.

The experiment is carried out with empty chamber and then with various materials of different areas under test.

9.5 ACOUSTIC DESIGN

While constructing an acoustically good auditorium or hall a special consideration has to be made about reverbation time, absorption and reflections taking place inside. An acoustically good auditorium means an auditorium in which every syllable or musical note reaches an audible level of loudness, at every point of the auditorium, and then quickly dies away to make room for the next group of notes or syllable. It is found that an acoustically good hall or auditorium should satisfy the following conditions:

i. Adequate loudness
ii. Uniform distribution of sound, i.e. absence of echoes and focussing of sound
iii. The hall must be non-resonant
iv. Optimum reverbation
v. Exclusion of extraneous sound or noise
vi. Echelon effect.

Departure from these conditions makes the auditorium or hall defective. We shall now deal about these conditions.

Reverbation

When reverbation is large, there is overlapping of successive sounds, which results in loss of clarity in hearing in an auditorium. On the other side, if the reverbation is very small, the loudness is inadequate. This means that the time of reverbation for a hall or auditorium should neither be too large nor two small. Sabine's relation for standard time of reverbation is

$$T = 0.161V/A = 0.161V/\Sigma as$$

where A is the total absorption of the auditorium, V is the volume, Σa is the average absorption coefficient and s is the internal surface area of the hall.

On the basis of experimental observations it is found that reverbation time depends upon the size of the hall, loudness of sound, and kind of music for which hall is used. The best reverbation time for a frequency of 512 c/s, is controlled by the following factors:

i. To make the volume of reverbation time optimum, provision of windows and ventilators can be made and they can be opened or closed as per requirement
ii. Making use of heavy curtains with folds
iii. Walls of the auditorium can be decorated with pictures and maps
iv. Providing acoustic tiles and carpeting the floors in the auditorium
v. Making full capacity of audience within the hall
vi. The clothing worn by the persons in an auditorium also provides sound absorption. Obviously, the period of reverbation varies according to the number of persons present within the hall. To off set this variation, one can provide seats with cushions which themselves have a sufficiently large absorbent effect to make up when the seats are unoccupied.

Adequate Loudness

In an auditorium the sounds which reach an audience must be sufficiently loud. One can achieve this by placing large wooden boards near and behind the speaker. Then the sounds reflected from them reach the listener within 1/20 second of the direct sound and increase loudness without *creating confusion*. We may note that a hard plane wall behind the speaker and facing the audience in a hall also serves a very good reflecting surface. A *low* flat ceiling also plays a useful part in reflecting sound towards the audience. Large polished wooden reflecting surfaces immediately above the speakers in a hall are also quite helpful. If loud speakers are used, they should be fitted above the heads of the listeners and be pointed slightly downward.

Shape and Size of the Auditorium

There should be uniform distribution of the voice of the speaker in an entire hall or auditorium. For this purpose, the walls and ceiling of the hall should not have curved surfaces because they focus the sound in some parts, while in some other parts no sound reaches at all. Normally, reflection from the plane surfaces are quite helpful in increasing the loudness of sound and for distributing it uniformly, whereas the curved surfaces can be troublesome. It is found that for a uniform distribution of sound intensity, it is better to have parabolic reflector and the use of concave or spherical or cylindrical reflectors should be discouraged as such types of surfaces give rise to undesirable focussing effects (Fig. 9.5). A paraboloidal ceiling (a platform at the focus) is very useful in sending a uniform reflected beam of sound in the hall (Fig. 9.6). We may note that if the ceiling is not flat we may hear echoes.

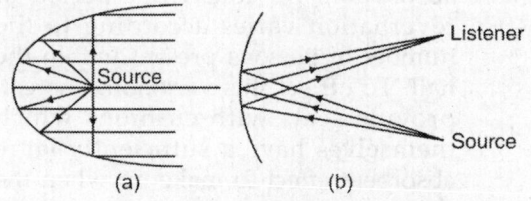

(a)   (b)

Fig. 9.5: (a) Parabolic reflection showing the uniform distribution of sound (b) Reflection at the concave surface showing the undesirable concentration of sound

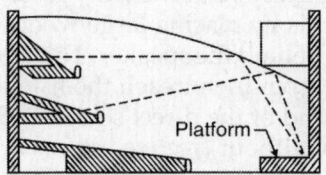

Fig. 9.6: Focussing of sound in a hall

It is found that if the centre of curvature of the ceiling of the hall is made twice the height of the room, the bad focussing of the sound can be removed. In the case of halls of larger dimensions, the echoes due to the reflections from the back of the speaker can be minimised by the use of suitable absorbents. We may note

that large size of hall results in the increase in reverbation time and therefore unnecessary large size should be avoided.

Absence of Echoes

It is found that the disturbing echoes are produced in a hall when the time interval between the sound received by the direct path and that by reflection from some walls or ceilings exceed about 1/10 sec. It has been estimated that an interval of 1/20 sec between direct and reflected sounds is the extreme limit of tolerance for speech and 1/15 sec for music. If the reflected sound in a hall is arriving earlier than that, it helps in arising the loudness, while those arriving later produce echoes, and cause confusion. Echoes may be avoided by covering distant walls and high ceilings with absorbent materials.

Absence of Echelon Effect

It is found that a set of railings or staircase or any regular spacing of reflected surfaces may produce a musical note due to regular succession of echoes of the original sound to observer (Fig. 9.7). This effect is called *echelon effect* and this makes original sound to appear confused or unintelligible. Obviously, such types of surfaces should be avoided or be properly covered with heavy stair carpet.

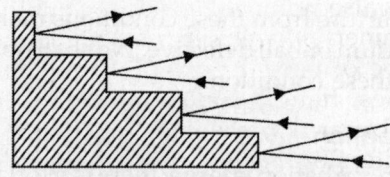

Fig. 9.7: Echelon effect

Freedom or Insulation from Noise

For good hearing, it is essential to take steps to guard against any noise reaching in the auditorium from outside or made by any instruments inside it. There are two types of noise: (1) air-borne noise (2) structure-borne noise.

Air-Borne Noise

The noises travelling through air come in from outside through open windows, doors, etc. One can reduce this type of noise using double doors and double or triple windows in separate windows. Noise entering the hall through the ventilating ducts can be minimised by packing the ducts with hairfelt baffles or bales of metal gauge. The noise travelling through the walls of the auditorium can be reduced by making the walls of layers of different materials, or by using double walls with air space between them. We may note that when sound passes from one medium to another medium, there is always a reflection of certain amount of energy that takes place at the boundary. One can also minimise the air-borne noise by using heavy glasses in doors, windows and ventilators in the auditorium and making perfect arrangement of the auditorium.

Structure-Borne Noise

Such type of noises are conveyed through the structure of the auditorium. They can be controlled by introducing discontinuity in the path of sound, e.g. noise from water pipe can be easily controlled by using rubber at the functions. Such type of noises can also be controlled by using double walls with an air space between them.

Inside Noise

Noise is also produced inside the auditorium by machinery, typewriters, etc. One can reduce this type of noise by proper lubrication of machines, hanging curtains of absorbent material near the machines, and placing the typewriters on rubber pads or compressed cork so that the variations are prevented from being transmitted through the floor and walls.

Freedom from Resonance

Sometimes powerful tones of right frequency produced in the auditorium set the windowpanes, sections of wooden portions and the walls lacking in rigidity, or volume of air contained in small rooms in resonant variations causing unpleasant results. Resonant frequency of a big hall or auditorium is normally well below the audible range because it is proportional to $1/\sqrt{\text{volume}}$. Such resonant

vibrations should be suitably damped. We may note that sometimes resonance can be useful. In order to improve the loudness of the hall, there are some halls famous for their acoustic properties which have a large area of resonant material in the form of wood panelling.

9.6 ACOUSTICAL MATERIALS

Usually, common building materials absorb sound only to a small extent and therefore, to meet the acoustical requirements, materials with better sound absorption property are to be incorporated in the halls. Such acoustical materials having more capacity to absorb the incident sound are called *absorbent* or *acoustical materials*.

In general, acoustical materials are soft and porous and work on the basic principle that the sound waves penetrate into the pores and the sound energy gets converted into other forms of energy. The absorbing capacity of the material depends on its density, thickness and the frequency of the incident sound wave. One can broadly classify the acoustical materials into the following classes:

   i. Porous materials

  ii. Cavity resonators

 iii. Resonant panels

 iv. Composite materials

ILLUSTRATIVE EXAMPLES

Example 1

A hall has dimensions $6 \times 4 \times 5$ cubic metres. Calculate: (i) mean free path of sound wave in the room (ii) the number of reflections made per second by the sound wave with the walls of the room. Given, velocity of sound in air = 350 m/s.

Solution

  i. One can define the mean free path of the sound waves as the average distance travelled by a sound wave through air between any two consecutive encounters with the walls of the room.

∴ Mean free path (L)

$$= \frac{4(\text{Volume of the room})}{\text{Total surface area}}$$

Volume of the room

$$= 6 \times 4 \times 5 = 120 \text{ m}^3$$

Total surface area

$$= 2[6 \times 4 + 4 \times 5 + 6 \times 5] = 148 \text{ m}^3$$

∴ $$L = \frac{4 \times 120}{148} = 3.2 \text{ m}$$

ii. Number of reflections made per second

$$N = \frac{\text{Velocity of sound}}{\text{Mean free path}}$$

$$= \frac{350}{3.243} = 107.9$$

Example 2

Define sound absorption coefficient of a material. A hall of volume 2000 m³ is found to have a reverbation time of 2 s. If the area of sound absorbing surface be 800 m², calculate the average absorption coefficient.

Solution

One can define the sound absorption coefficient of a surface as the ratio of the sound energy absorbed by the surface to that falling on the surface. Since all the energy falling on an open window passes through it, the absorption coefficient of open window is 1. Obviously, one can also define the absorption coefficient of any surface as the ratio of the sound energy absorbed by the surface to that absorbed by an open window.

Reverbation time, $T = \dfrac{0.161V}{A}$

where V is volume in m³ and A is the total absorption of sound. If S be the total area of the sound absorbing surface and a the average absorption coefficient, then $A = Sa$.

Given that $V = 2000$ m³, $T = 2$ s, $S = 800$ m²

∴ $$T = \frac{0.161V}{Sa}$$

∴ $$2 = \frac{0.161 \times 2000}{800 \times a}$$

or $$a = \frac{0.161 \times 2000}{800 \times 2} = 0.2$$

Example 3

The volume of a room is 600 m³. The wall area of the room is 220 m², the floor area is 120 m² and the ceiling area is 120 m². The average sound absorption coefficient for the (i) walls is 0.03 (ii) ceiling is 0.80 (iii) floor is 0.06. Find the average reverbation time.

Solution

The average sound absorption coefficient

$$a = \frac{\sum \alpha A}{\sum A}$$

$$= \frac{\alpha_1 A_1 + \alpha_2 A_2 + \alpha_3 A_3}{A_1 + A_2 + A_3}$$

Given $A_1 = 220$ m², $\alpha_1 = 0.03$

$A_2 = 120$ m², $\alpha_2 = 0.80$

$A_3 = 120$ m², $\alpha_3 = 0.06$

∴ $$a = \frac{0.3 \times 220 + 0.8 \times 120 + 0.06 \times 120}{220 + 120 + 120}$$

$$= \frac{109.8}{460} = 0.2389 \approx 0.24$$

Now, the total sound absorption of the room $= a\sum A = 0.24 \times 460 = 110.4$ metric Sabines

Reverbation time, $T = \dfrac{0.161V}{a\sum A} = \dfrac{0.161 \times 600}{110.4}$

$$= 0.86 \text{ s (approx.)}$$

Example 4

The reverbation time of a cubical hall of side 10 m is 0.7 s. If one of the walls is covered with felt, the reverbation time is reduced to 0.6 s. Show that the sound absorption coefficient of the felt is 0.7.

Solution

Reverbation time

$$T = \frac{0.161V}{A} = \frac{0.161V}{\sum as} = \frac{0.161V}{a\sum s}$$

where V is the volume of the hall in m^3 and $A = a\Sigma S$, a being the absorption coefficient and S the corresponding area. Here $T = 0.7$ s, $V = (10)^3$ m^3, A (for the four walls, ceiling and floor)

$$= a \times (10)^2 \times 6 = 600a$$

$$\therefore \quad 0.7 = \frac{0.161 \times (10)^3}{a \times (10)^2 \times 6}$$

or $\qquad a = 0.38$

Obviously, a is the absorption coefficient for the walls, ceiling and floor. If one of the walls is covered with felt whose absorption coefficient is a' (say), then we have

$$A = a \times (10)^2 \times 5 + a' \times (10)^2$$
$$= 0.38 \times (10)^2 \times 5 + a' \times (10)^2$$

Now, the reverbation time is reduced to 0.6 s. Again from Sabine's formula, we have

$$0.6 = \frac{0.161 \times (10)^3}{0.38 \times (10)^2 + 5 \times a' \times (10)^2}$$

$$= \frac{0.161 \times 10}{0.38 \times 5 + a'}$$

$$\therefore \qquad a' = 0.76.$$

Example 5

Find the reverbation time of an office which has a volume of 2000 m^3 and a total sound absorption of 90 metric Sabine. What is the additional sound absorption required for an optimum reverbation time of 1.5 s. [BE]

Solution

We have Sabine's formula

$$T = \frac{0.161V}{A}$$

where V is the volume of the room in m^3 and

A is the total absorption in metric Sabine. Now, substituting the given values in the above relation, one obtains

$$T = \frac{0.161 \times 2000}{90} = 3.6 \text{ s}$$

Now, the total absorption A for an optimum reverbation $T = 1.5$ s is given by

$$A = \frac{0.161V}{T} = \frac{0.161 \times 2000}{1.5}$$

$$= 216 \text{ metric Sabine}$$

∴ Additional sound absorption required

$$= 216 - 90 = 126 \text{ metric Sabine.}$$

Example 6

The volume of an auditorium is 12000 m^3. Its reverbation time is 1.5 seconds. If the average absorption coefficient of interior surfaces is 0.4 Sabine, find the area of interior surfaces.

Solution

Given, $\qquad V = 12000$ m^3, $T = 1.5$ s

$$\bar{a} = 0.4 \text{ Sabine}, S = ?$$

We know $\quad T = \dfrac{0.167V}{\bar{a}S}$

or $\qquad S = \dfrac{0.167V}{\bar{a}T}$

$$\therefore \qquad S = \frac{0.167 \times 12000}{0.4 \times 1.5} = \frac{2004}{0.6}$$

$$\therefore \qquad S = 3340 \text{ m}^2.$$

Example 7

The volume of a room is 1500 m^3. The wall area of the room is 260 m^2, the floor area is 140 m^2 and the ceiling area is 140 m^2. The average sound absorption coefficient for wall is 0.03, for ceiling is 0.8 and for the floor is 0.06. Calculate the average absorption coefficient and the reverbation time.

Solution

Given, $V = 1500$ m^3, $a_1 = 0.03$ Sabine

$$a_2 = 0.8 \text{ Sabine}, a_3 = 0.06 \text{ Sabine}$$

$S_1 = 260$ m^2, $S_2 = 140$ m^2, $S_3 = 140$ m^2.

$\bar{a} = ?, T = ?$

Average absorption coefficient is

$$\bar{a} = \frac{a_1 S_1 + a_2 S_2 + a_3 S_3}{S_1 + S_2 + S_3}$$

$$= \frac{0.03 \times 260 + 0.8 \times 140 + 0.06 \times 140}{260 + 140 + 140}$$

$$= \frac{7.8 + 112 + 8.4}{540}$$

$\bar{a} = 0.2374$ OWU

∴ The total sound absorption of the room is

$\bar{a} \, \Sigma S = 0.2374 \times 540 = 128.196$ OWU m^2.

Therefore, the reverbation time is

$$T = \frac{0.167V}{\bar{a} \, \Sigma S} = \frac{0.167 \times 1500}{128.196} = \frac{250.5}{128.196}$$

$T = 1.9540$ s.

Example 8

A hall has a volume of 12500 m^3 and reverbation time of 1.5 s. If 200 cushioned chairs are additionally placed in the hall, what will be the new reverbation time of the hall? The absorption of each chair is 1.0 OWU.

Solution

Given, $V = 12500$ m^3, $T_1 = 1.5$ s

$a_2 S_2 = 200$ Sabine m^2,

$\Sigma a_1 S_1 = ?, T_2 = ?$

Let $T_1 = \dfrac{0.167V}{\Sigma a_1 S_1}$

be the reverbation time before placing cushioned chairs.

∴ $\Sigma a_1 S_1 = \dfrac{0.167 \times 12500}{1.5} = \dfrac{2087.5}{1.5}$

$\Sigma a_1 S_1 = 1391.66$ Sabine m^2

The reverbation time after placing the cushioned chairs be,

$$T_2 = \frac{0.167V}{\Sigma aS + a_1 S_2} = \frac{0.167 \times 12500}{1391.66 + 200}$$

$$= \frac{2087.5}{1591.66}$$

$T_2 = 1.3115$ s

∴ The new reverbation time after placing the cushioned chairs is 1.3115 s.

Example 9

A hall has a volume of 2265 m^3. Its total absorption is equivalent to 94.85 m^2 of open window. What will be the effect on reverbation time if audience fills the hall and thereby increases the absoption by another 94.85 m^2?

Solution

Given, $V = 2265$ m^3, $\Sigma aS = 94.85$,

$\Sigma a_2 S_2 = 2 \times \Sigma aS = 189.7, T_1 = ?, T_2 = ?$

Let T_1 be the reverbation time in the hall without audience

∴ $T_1 = \dfrac{0.167V}{\Sigma a_1 S_1}$

∴ $T_1 = \dfrac{0.167 \times 2265}{94.85}$

$T_1 = 3.987$ s

Let T_2 be the reverbation time in the hall with audience

∴ $T_2 = \dfrac{0.167V}{\Sigma a_2 S_2}$

∴ $T_2 = \dfrac{0.167 \times 2265}{189.7} = \dfrac{378.255}{189.7}$

$T_2 = 1.993$ s

Thus, the reverbation reduces to half of its initial value when the audience fill the hall.

REVIEW QUESTIONS

1. What are the characteristics of an acoustically perfect hall or auditorium? What is the role of reverbation time? What steps would you take to improve the acoustics of a hall?

2. Discuss the various acoustic defects commonly found in larger halls. How are these defects removed? [BE]

3. Deduce Sabine's formula for the reverbation time of a hall or auditorium. Explain why does it fall in the case of dead room.

4. Investigate the growth of sound in hall, and then deduce Sabine's reverbation formula. Explain why does it falls in the case of dead room. Explain how the above result can be used to determine the sound absorption coefficient of a piece of felt.

5. Mention some effects of reverbation in daily life.

6. Give the theory of growth and decay of sound in a live room. Find the reverbation time. Why the theory fails in a dead room?

7. What are the acoustic requirements of a good auditorium? Discuss the theory of reverbation. What is the order of magnitude of optimum reverbation?

8. How absorption coefficient of a material is measured?

9. Enumerate the features that an auditorium should have for good acoustics. How can these features be incorporated in the design of an auditorium?

   What remedial measures would you suggest for the following defects in an existing auditorium?

   i. Uneven distribution of sound

   ii. Unintelligible audibility of speech.

PROBLEMS

1. The reverbation time for an empty hall is 1.5 s and 1s when a curtain cloth of 20 sqm is suspended at the centre of the hall. The dimensions of the hall are 10 m × 8 m × 6 m. Show that the absorption coefficient for the curtain cloth is about 0.6 Sabines. [BE]

2. A hall has dimensions 20 m × 15 m × 5 m. The reverbation time is 3.5 s. Determine the total absorption of its surface.
   [*Ans.* 67.7 metric Sabines]

3. The reverbation time for an empty hall of size 20 m × 15 m × 10 m is 3.5 s. Calculate (a) the average absorption coefficient of the hall (b) area of the wall that should be covered by curtains so as to reduce the reverbation time to 2.5 s. The absorption coefficient of curtain cloth is 0.5. [BE]
   [*Ans.* (a) 0.106 OWU (b) 140.0 m^2]

4. A hall has a volume of 7500 m^3. It is required to have reverbation time of 1.5 s. Determine total absorption of the hall. If the total absorption of the hall is increased, what will be its effect?
   [*Ans.* 7500 metric Sabine]

SHORT ANSWER QUESTIONS

1. For an acoustically perfect hall, is it essential that 'quality' of the sound must remain unaltered?

Ans. Yes.

2. How an acoustically perfect hall have marked effects on speech and music being delivered there in the hall?

Ans. Speech delivered is quite intelligible.

3. How echelon effect can be avoided?

Ans. By using a stair carpet.

4. Why an echo cannot be heard if the distance between the source and the obstacle is less then 17 m? [*V* = 340 m/s]

Ans. If a sound source be situated at a distance *d* from a reflecting obstacle, then the time interval *t* for the sound to go and return is given by

$$t = \frac{2d}{V} = \frac{2 \times 17\,\text{m}}{340\,\text{m/s}} = 1/10\,\text{s}$$

Obviously, if *d* is less than 17 m, then the time-interval *t* will be less than 1/10 s, and the echo cannot be heard.

5. What steps will you take to adjust the reverbation time to optimum value?

Ans. i. The windows and ventilators will be kept open.

   ii. Heavy curtains will be hung.

   iii. The floor will be covered with carpets or tapestries.

   iv. The walls and ceiling will be lined with absorbent material such as felt, celotex, fibre-board, glass wool, rock wool, mineral wool, etc.

   v. We will provide acoustic tiles.

   vi. The clothing worn by the persons in the hall also provides sound absorption. Therefore, the period of reverbation varies according to the number of persons present. To off set this variation, we will provide the seats with cushions which themselves have a sufficiently large absorbent effect to make up when the seats are unoccupied.

OBJECTIVE QUESTIONS

1. In an acoustically perfect hall, the sound heard must be sufficiently _____ in every part of the hall. [loud]

2. The undue persistence of sound of the speaker or the singer in an acoustic hall is called _____. [reverbation]

3. The Sabine relation for the period of reverbation in seconds is _____. $\left[T = \dfrac{0.161V}{A}\right]$

4. For adequate loudness in a hall, loudspeaker should be fitted _____ the heads of the listeners and pointed slightly _____. [above, downward]

5. The echoes in an acoustic hall may be avoided by covering distant walls and high ceilings with _____ materials. [absorbent]

6. Reverbation arises due to _____ reflection from the various surfaces in a hall. [multiple]

7. Reverbation time of a hall is _____ on its volume and nature of the _____ surface. [dependent, absorbing]

8. Sound waves are _____ waves and exhibit all _____ phenomena. [longitudinal, wave]

MULTIPLE CHOICE QUESTIONS

1. The unit of intensity of sound is
   (a) Wm^{-2} (b) Wm^{-1}
   (c) Wm^2 (d) Jm^{-2}

2. The minimum sound intensity which a human ear can sense is called
   (a) zero point energy
   (b) threshold energy
   (c) vibrational energy
   (d) rotational energy

3. A change in sound intensity level of 1 dB alters the intensity by
   (a) 74%   (b) 36%
   (c) 26%   (d) none of the above

4. The intensity of a source of sound is increased 20 times its value. The intensity level increase as
   (a) 1.301 dB   (b) 0.1301 dB
   (c) 13.01 dB   (d) 7.01 dB

   [**Hint:** $I_L = 10 \log_{10}\left(\dfrac{I}{I_0}\right) = 10 \log_{10}\left(\dfrac{20\, I_0}{I}\right)$
   $= 10 \log_{10} 20 = 10 \times 1.30102 = 13.01$ dB]

5. The ratio of intensity I and I_0 of a sound in a heavy traffic is $10^{-6}/10^{-12}$. The intensity level in dB is
   (a) 6 dB   (b) 30 dB
   (c) 60 dB   (d) 120 dB

   [**Hint:** $I_L = 10 \log_{10}\left(\dfrac{I}{I_0}\right) = 10 \log_{10}\left(\dfrac{10^{-6}}{10^{-12}}\right)$
   $= 10 \log_{10}(10^6) = 10 \times 6$ dB $= 60$ dB]

6. The time taken by the sound to fall below the minimum audibility level after the source stopped sounding is called
   (a) periodic time
   (b) reverbation time
   (c) decay time
   (d) growth time

Answers

1. (a)   2. (b)   3. (c)   4. (c)   5. (c)
6. (b)

10 Ultrasonics

10.1 INTRODUCTION

Ultrasonic waves are sound waves of frequency greater than that of audible sound, i.e. these waves are produced by an object vibrating at a frequency higher than the human ear can hear, i.e. greater than 20000 cycles/s = 20 kHz/s. On the basis of frequency range one have

Audio frequency range – 20 Hz to 20 kHz

Radio frequency range – 550 kHz to 22 MHz

TV frequency range – 47 MHz to 230 MHz

Above audio range – Ultrasonic (above 20 kHz)

Below audio range – Infrasonic (below 20 Hz)

With the help of modern techniques, one can produce ultrasonic waves of frequency upto 20 million Hz, which has a wavelength of 10^{-8} m (~X-ray wavelength).

Ultrasonic waves are highly energetic and have extremely short wavelength because of its high frequency and energy. Our ear is not sensitive to ultrasonic waves although some animals like dogs, bats, and some kinds of birds show response to these waves. These waves can be produced by the high frequency vibrations of a quartz crystal under an alternating fields (piezoelectric effect) or by the vibrations of a ferromagnetic rod under an alternating magnetic field (magnetostriction effect). Because of small wavelength and high energy, the ultrasonic waves find use especially in the field of medicine and in various industries. Due to this ultrasonic waves have great promises in future.

The longitudinal waves whose frequency lies below 20 Hz are called *infrasonic* waves. Human ear is not sensitive to these waves. These waves are produced by large vibrating bodies such as during an earthquake.

We may note that *supersonic* also means of very high frequency. However, we must differentiate ultrasonic from supersonic. Supersonic effects are produced by objects that travel through a medium at a faster speed than the waves they generate. Supersonic is essentially confined to aeroplanes, missiles, which fly through air at a speed of sound in air.

10.2 PRODUCTION OF ULTRASONIC WAVES

There are number of ways by which ultrasonic waves can be generated. Usually the method to be employed depends upon the power output necessary and the frequency range to be covered. Generation of mechanical type such as tuning forks of Galton's whistle can be used up to 10,000 cycles/s.

Similar to sonic range, ultrasonic range requires a source of energy. A device which transfers energy from one system to another generally with the change of the form of energy mainly from acoustical to mechanical or *vice versa* is known as *transducer*. We may note that a medium is also necessary for transmission of ultrasonic waves.

Galton's Whistle

Galton devised a miniature organ pipe in the form of a whistle to determine the limit of audibility. It consists of a closed end air column A whose length can be adjusted with the help of a moveable piston. With the help of a screw S_1, the piston P can be moved to the desired position. The open end of the pipe A is fitted with a lip L (Fig. 10.1). With the help of the screw S_2, the position of the pipe C can be adjusted. One can adjust the gap between the ends of A and C with the help of the screw S_2 as shown in Fig. 10.1.

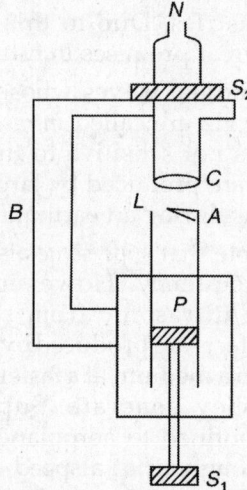

Fig. 10.1: Galton's whistle

An air blast is blown through the nozzle at the top and the blast of air coming out of C strikes against the lip L. Due to this the column of air in the pipe is set into vibration. Now, adjusting the length of air column in A, it is brought to the resonant position. The resonant frequency depends on the length and diameter of the pipe A. Now, if l is the length of the air column in A, x is the end correction, then the wavelength is obtained as

$$\lambda = 4(l + x) \qquad (10.1)$$

The frequency of sound

$$n = \frac{V}{\lambda} = \frac{V}{4(l+x)} \qquad (10.2)$$

One can produce frequencies of the order of 30,000 Hz with the help of whistle.

Magnetostriction Oscillator

Joule in 1847 observed that a rod of ferromagnetic substance such as nickel undergoes a change in length when magnetic field is applied along the length. The increase in length is proportional to the applied magnetic field. This increase in length is greater in the case of nickel in comparison to other ferromagnetic substances. We may note that the increase in length is independent of the direction of the magnetic field acting. If the magnetic field is alternating and applied parallel to the length of the nickel rod or tube, as magnetic field grows the length increases and the vibration sets up in the rod with a frequency double that of the frequency of the magnetic field. However, if a permanent steady polarizing field of suitable strength is applied from direct current supply, and a changing alternating current supply giving rise to alternating field is superimposed on it, the longitudinal vibration of the nickel rod will set up having frequency of vibration same as the frequency of the alternating field. However, resonance will occur when the frequency of alternation of the magnetic field be equal to the natural frequency of the bar which depends on the elastic constant and the dimension of the bar. The amplitude of vibration will be large. The rod vibrates with ultrasonic frequency and will emit ultrasonic waves.

The velocity of ultrasonic waves in a longitudinal bar is given by

$$c = \sqrt{Y/\rho} \qquad (10.3)$$

where, Y is Young's modulus and ρ is density of the material. If l denotes the length of the bar, the fundamental wavelength of the bar is $= 2l$ and hence the frequency $f = \dfrac{c}{2l}$. Thus, in general $f = \dfrac{sc}{2l} = \dfrac{s}{2l}\left(\sqrt{\dfrac{Y}{\rho}}\right)$ or $l = \dfrac{s}{2l}\sqrt{Y/\rho}$, where s is the order of harmonic. One can easily see from the above relation that at a frequency of 20000 c/s, the length of the nickel rod that will be resonant will be of the order of s'' and moreover, this length will be smaller and smaller with higher frequencies so that the length will become very small at a frequency greater than 60,000 c/s; hence the usual range of a magnetostriction ultrasonic oscillator will be from 5000 c/s to 60,000 c/s. We may note that magnetostriction has been used in laboratories to generate ultrasonic signals of frequency 2 M c/s but at higher frequencies the output is small that it cannot be utilized for any practical purposes.

An arrangement using magnetostriction for producing ultrasonic waves is shown in Fig. 10.2. A short rod XY of nickel is placed in a solenoid, fed by DC supply, which produces a steady polarizing magnetic field. L_1 and L_2 are two coils wound on the ends of the rod and are included in the grid- and anode-circuit respectively of a triode valve. The frequency of the oscillating anode-circuit is adjusted with the variable capacitor C. When the frequency equals the natural frequency of the rod, then the longitudinal oscillations of the rod are maintained and ultrasonics are produced in the surrounding medium.

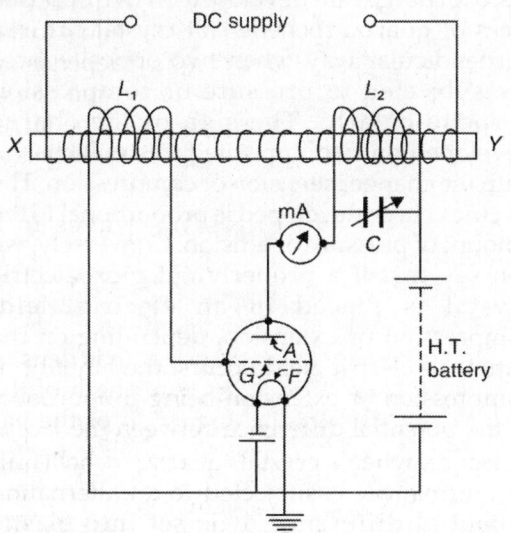

Fig. 10.2: Magnetostriction oscillator

The rod is already magnetised by the steady polarising magnetic field. The periodic variation in the anode-current passing in the coil L_2 causes a variation in magnetisation and a consequent variation in the length of the rod. Now, this variation in length causes a variation in the magnetic flux through the grid coil L_1. Obviously, by the converse magnetostriction effect, an induced emf is set up in coil L_1. This emf acts on the grid, and produces an amplified current variation in coil L_2. Hence, the oscillations of the rod are maintained. When the oscillations start, the milli ammeter (mA) will show an alteration. To make this alteration maximum, the capacitor C is adjusted, because at this stage the frequency of the rod and the vibrations of the rod will be most vigorous. A nickel rod of length 10 cm, gives out ultrasonic waves of frequency 25000 c/s. The oscillation frequency f of the nickel rod is given by

$$f = \frac{1}{2\pi\sqrt{LC}} \qquad (10.4)$$

When this frequency matches with the natural frequency of the rod $f = \frac{P}{2L}\sqrt{\frac{E}{D}}$, where L is the length of the rod, E is Young's modulus, D the density of the rod materials and P is the harmonic modes 1, 2, 3, etc. resonance will occur. With the help of this method, one can obtain frequencies of the ultrasonic waves upto 3×10^5 Hz. However, at higher frequencies the output is so small that it cannot be utilized for any practical purposes.

To secure maximum transfer of energy the *transducer* housing is fitted close to the medium. A common type of powerful magnetostriction transducer is shown in Fig. 10.3. A number of nickel tubing act as a driving element, each having a length equal to one quarter of the wavelength of the sound to be radiated. From Fig. 10.3, we can see that one end of these tubing is free and the other end is embedded in a circular sheet plate P whose resonant frequency is very nearly equal to that of the nickel tube. Nickel tubes are the driving elements. Each nickel tube is surrounded by its own driving coil C and all are driven in phase by the current from a very powerful oscillator. The permanent magnets

M mounted inside the water tight housing containing the tubes supply the optimum polarising field. The alternating force exerted on the plate by the reaction to the stress in the nickel tubes are transmitted by the plate.

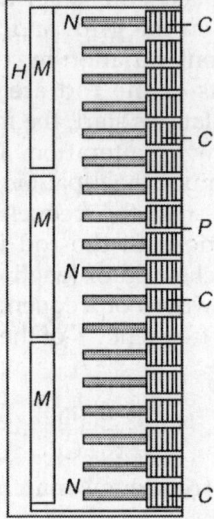

Fig. 10.3: Transducer

The power handling capacity of this type of transducer is higher. It has a sharper resonance curve which helps in reducing the disturbing effect of the background noise. However, it is difficult to produce optimum polarising field in this type of instrument.

Since the magnetostrictive effect is reversible and hence the same instrument can be used as a *receiver*. Sound waves impinging on the plate *P* set it into vibrations causing the Ni tubes to vibrate in imaginative field. The resulting alternating currents are induced in the surrounding coils which may be amplified for detection. We may note that the frequency range of magnetostriction is 5–60 kc/s beyond which rods of very small lengths are required but handling becomes difficult. Usually 25 kc/s are used for obtaining good results.

The main advantages of magnetostriction method are:

i. The method is simple and its cost of construction is low.

ii. At low ultrasonic frequency, large power output is possible without damage or risk to the oscillator circuit, even under temporary overload.

iii. In comparison to piezoelectric transducers, magnetostriction transducers have certain advantages. Magnetostriction transducers possess large power handling capacity and sharp resonance curve.

The main drawbacks of this method are:

i. One finds difficulty in imposing permanent polarising magnetisation of the Ni rod initially.

ii. One cannot use this method for very high frequency range, i.e. there is a limitation of frequency.

iii. Greater dependence of frequency on temperature and breadth of resonance curve causes changes in elastic constants of ferromagnetic substances with degree of magnetisation.

Piezoelectric Method

J and P Curie brothers in 1880 discovered that electric charges are developed on two opposite faces of quartz, rochelle salt crystal cut in a perpendicular way when two opposite faces are subjected to pressure or compression perpendicularly. The sign of the charge developed on two opposite faces will change with the change in tension or compression. The electric charge developed is proportional to the amount of pressure or tension. Conversely, we can say that if a properly cut piezoelectric crystal is placed in an electric field, compression or extension, depending on the nature of electric field occurs, the amount of compression or extension being proportional to the potential difference between the faces.

Hence, when a crystal (quartz, rochelle salt or tourmaline) is subjected to an alternating potential difference, it is set into elastic vibrations, i.e. it contracts and expands periodically and sets up mechanical vibrations in any acoustic medium in which it is placed. The frequency of the vibrations is within the ultrasonic range. Ordinarily the amplitude is very small. If the frequency of the electrical oscillations coincides with one of the natural frequency of the crystal, which are of the order of 250 to 10,000 kilocycles/s a large amplitude of vibrations results. This is adopted in the construction of piezoelectric transducer and the phenomenon is utilized to produce ultrasonic waves.

Quartz Crystal

Natural quartz crystal occurs in the shape of a prism of six sides with a pyramid attached to each end (Fig. 10.4a). The line joining the two opposite vertices of the pyramids is defined as the *optic axis* of the crystal. If the opposite corners of the crystal are joined together, the line is known as X-axis or also called as the *electrical axis* of the crystal. Figures 10.4b, c and d represent the sections of the crystal perpendicular to Z-axis. Line perpendicular to Z-axis and joining two opposite corners of hexagonal slice of the crystal as shown in Fig. 10.4b is the *electric axis* known as X-axis. A line represented by CD (Fig. 10.4c) joining the *opposite faces* of the hexagonal slice and perpendicular both to X-and Z-axes represents Y-axis or *mechanical axis*.

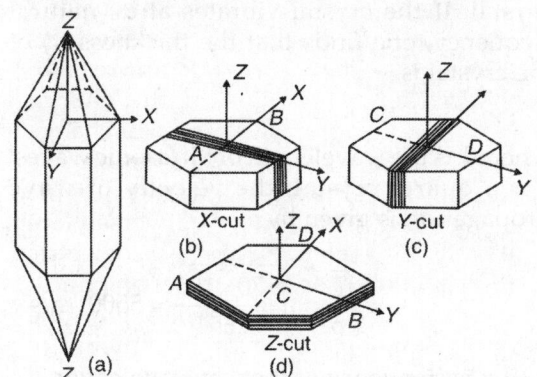

Fig. 10.4: Quartz crystal

One can obtain Y-cut crystal by cutting the hexagonal slice along X-axis, i.e. along the corners, such that the largest forces are perpendicular to Y-axis and Z-cut crystal is obtained by cutting the crystal perpendicular to Z-axis. X-cut crystal is obtained by cutting the hexagonal slice along Y-axis such that the

largest face are perpendicular to X-axis. These crystals are generally used for piezoelectric oscillators. Faces of the crystal are silvered to ensure proper electrical contact. The alternate electric field is applied by the metal plates pressing the faces, along X-axis. The crystal vibrates in the direction of Y-axis. If the frequency of the electrical oscillations coincides with one of the natural frequencies of the crystal, which are of the order of 250 to 10,000 kilocycles/s, a large amplitude of vibrations results.

Pressure along the direction of X-axis produces the charges on two surfaces normal to this axis but tension will produce opposite charges, opposite to what is produced when pressure is applied.

Tension along the direction of Y-axis charges two surfaces normal to X-axis as is produced when pressure is applied along X-axis and pressure along the direction of Y-axis charges oppositely two opposite surfaces normal to X-axis, as the charge is produced when tension is applied.

We may note that the pressure or tension along Z-axis produces no effect at all.

Electric field along X-axis only causes the crystal to expand along X-axis and to contract along Y-axis, reversal of the field produces the reversal of the effect. The deformation along Y-axis will produce longitudinal vibration and mechanical deformation along X-axis will produce surface vibration which is generally used as a source.

The necessary arrangement is shown in Fig. 10.5. A quartz slab Q is placed between two metal plates which are connected to the anode A and the grid G of a triode valve as shown in Fig. 10.5. Also connected to the anode is an inductance–capacitance (L–C) circuit and a high tension (H.T.) battery shunted by a bypass capacitor C_1. The bypass capacitor C_1 prevents high-frequency current from passing through the battery. The variable capacitor C in the anode-circuit is then adjusted to produce electrical oscillations of frequency equal to that of the modes of vibrations of the crystal.

A small alternating current is set up in the anode circuit of the triode valve. At some instant, the anode plate A is more negative and grid G is more positive. Since the anode plate A and the grid G are connected to the opposite plates of Q, these plates are oppositely charged. The charges reverse in sign when anode A becomes more positive and grid G more negative. Obviously, an alternating potential difference is applied on the opposite faces of the crystal. It sets the crystal into elastic vibrations of the frequency of AC supply. When the resonance occur between this frequency and the natural frequency of the plate, which depends upon dimension, the amplitude of vibrations will be large. We may note that such vibratory system is used as standard of frequency in electrical circuit.

The simplest form of the circuit is mostly the circuit as shown in Fig. 10.5 to maintain the oscillation of triode valve. Obviously, the crystal is thrown into resonant mechanical oscillations along MM' generating ultrasonic waves in the surrounding medium.

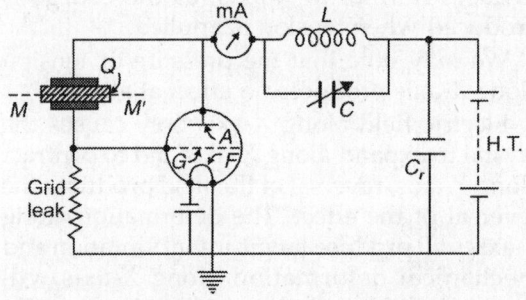

Fig. 10.5: Piezoelectric oscillator

It is found that the ultrasonic generator delivers maximum power when it is operated at the fundamental frequency of the crystal. In order to generate higher-frequency ultrasonic waves, the L–C circuit is made to oscillate at a frequency equal to one of the harmonics of the crystal. However, only odd harmonics of the crystal are used because piezoelectric effect can occur only when opposite charges appear on the electrodes.

The distribution of pressure and charge in the thickness of a crystal oscillating at three harmonics are shown in Fig. 10.6. It is found that the effect will occur in the fundamental and third harmonic, and not in the second harmonic.

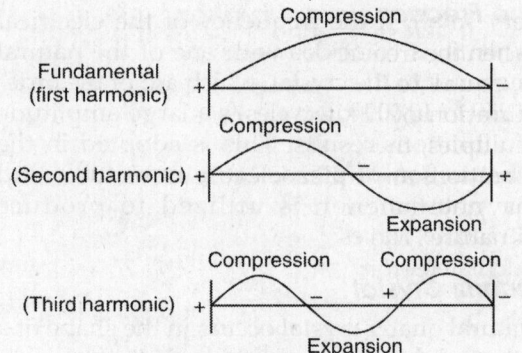

Fig. 10.6: Distribution of pressure and charge in the thickness of a crystal at three harmonics

Design of the Ultrasonic Crystal

One may cut the crystal along the X-axis or along the Y-axis. It is called the X-cut if it is cut along the X-axis and Y-cut if it is cut along the Y-axis. Usually the X-cut is used since it generates longitudinal or L-waves. For the production of shear waves, Y-cut crystals are also used. However, for the propagation of this type of waves through a liquid or a gas, the medium must have a shear elasticity. In case of an X-cut crystal, the longitudinal waves will occur along the thickness dimension of the crystal. If the crystal vibrates at its natural frequency, one finds that the thickness (t) of the crystal is

$$t = \lambda/2$$

where λ is the wavelength of ultrasonic waves. For a quartz crystal, the velocity of wave propagation is given by

$$c = \sqrt{\frac{E}{\rho}} = \sqrt{\frac{770 \times 10^9}{2.654}} = 5500 \text{ m/s}$$

If n is frequency of the wave, we have

$$n = \frac{c}{\lambda} = \frac{540 \times 10^3}{2t} = \frac{2700}{t} \text{ kc/s}$$

If t is expressed in millimetres, usually, the generated frequency is given by

$$n = \frac{2870}{t \, (\text{mm})} \text{ kc/s}$$

The Frequency of Vibration

When the frequency of the alternating voltage is equal to the natural frequency of the vibration of the crystal or its simple higher multiples, the crystal is thrown into resonant vibrations and the amplitude will be large. We may note that these vibrations are longitudinal in nature. The frequency of vibrations can be calculated from

$$n = \frac{p}{2l}\sqrt{\frac{E}{\rho}}$$

where $p = 1, 2, 3, ...,$ etc., E is the elasticity and ρ is the density of the crystal. The velocity of longitudinal wave propagation is

$$c = \sqrt{\frac{E}{\rho}}$$

$$= 5.5 \times 10^3 \text{ m/s for quartz}$$

For a crystal of length 0.05 m, the frequency for the first mode of vibration will be

$$n = \frac{1 \times 5.5 \times 10^3}{2 \times 0.05} = 5.5 \times 10^4 \text{ Hz}$$

The modes of frequency are simple integral multiples of 5.5×10^4 Hz. One can produce frequency upto 150,000 kHz by tourmaline crystal.

Piezoelectric oscillator has the following advantages:

i. Piezoelectric oscillator has to be used for higher frequency range
ii. Size of piezoelectric oscillator is small and they are economical
iii. When used in detector, these oscillators are very sensitive
iv. The waveform is good.

However, power handling capacity of piezoelectric oscillator is low. This is the disadvantage of these oscillators. We may also note that intense ultrasonic radiations has a disruptive effect on liquids by causing bubbles to be formed.

10.3 DETECTION OF ULTRASONICS

One cannot directly detect the ultrasonic waves, although some animals, specially the bat, can do so. Ultrasonic waves propagated through a medium can be detected indirectly by the following methods:

Piezoelectric Detector

The ultrasonic waves when fall on one pair of the faces of a quartz crystal, varying electric charges are produced on the other perpendicular faces. Though they are very small but can be amplified and detected with the help of some suitable means.

Kundt's Tube Method

One can detect ultrasonic waves with the help of Kundt's tube. When ultrasonic waves of relatively large wavelength, almost comparable to audible sound, were allowed to pass through Kundt's tube, then lycopodium powder sprinkled in the tube, collects in the form of heaps at nodal points and is blown off at antinodal points. The average distance between two adjacent heaps is equal to half the wavelength. One cannot use this method if the wavelength of ultrasonic waves is very small, i.e. less than a few millimetres. In the case of liquid medium, powdered coke is used instead of lycopodium powder to detect the position of nodes.

Acoustic Diffraction Method

According to Brillouin, when an ultrasonic wave passes through a liquid, the density of liquid varies from layer to layer due to periodic variation of pressure and the structure of the liquid thus agitated can be simulated to that of a light diffraction, i.e. if under this condition monochromatic light is passed through the liquid at right angles to the wave, then liquid behaves as diffraction grating. Moreover, this grating behaves in the same way as a ruled grating. Consequently, if a beam of monochromatic light be incident on such a simulated grating, there should be visible the diffraction pattern in the transmitted light. Such a grating is known as *acoustic grating*. The grating element is equal to the wavelength of the ultrasonic waves. Experimental arrangement is shown in Fig. 10.7. Light from a sodium lamp S is condensed by a lens L on the slit S_1 of a collimator C which renders it

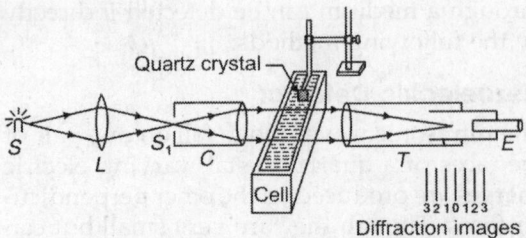

Fig. 10.7: Acoustic diffraction method

parallel. The parallel beam is allowed to pass through a glass cell filled with the experimental liquid. A quartz crystal fed by a radio-frequency oscillator and thus generating ultrasonic waves is suspended in the liquid such that the waves are propagated in the liquid at right angles to the beam of light. The waves get reflected back from the wall of the cell and a stationary wave-pattern is set up in the liquid. The light beam passing through this pattern, i.e. acoustic grating is diffracted. The diffracted beam of light is viewed through the telescope. It consists of a central maximum and principal maxima on either side. If θ_n be the angle of diffraction for the nth order maximum, then

$$d\sin\theta_n = n\lambda \qquad (10.5)$$

Here n = 1, 2, 3... etc., λ is the wavelength of sodium light and d is the distance between two adjacent nodals or antinodal planes. Knowing n, θ_n and λ, the value of d can be calculated using Eq. (10.5). Now, if λ_a is the wavelength of the ultrasonic waves through the medium, we have

$$d = \frac{\lambda_a}{2}$$

or $\qquad \lambda_a = 2d \qquad (10.6)$

If the resonant frequency of the piezoelectric crystal oscillator is N, the velocity of the ultrasonic waves can be determined from the following relation

$$V = N\lambda_a = 2Nd \qquad (10.7)$$

We may note that the theory of this method is not so simple as deduced above. This method is useful in measuring the velocities of ultrasonic waves through liquids and gases at various temperatures.

Thermal Detectors

For the detection of ultrasonic waves, thermal detectors method is most commonly used. A fine platinum wire is used in this method. This wire is moved through the medium. At the positions of nodes, due to alternate compression and rarefaction, adiabatic changes in temperature takes place and the resistance of the platinum wire changes with respect to time. One can detect this with the help of Callender and Garrifith's bridge arrangement. At the positions of antinodes, the temperature remains constant and the resistance of the platinum wire remains constant. The undisturbed balanced position of the bridge indicate this situation.

Sensitive Flame Method

Along the medium a narrow sensitive flame is moved. The flame is steady at the position of the antinodes. The flame flickers at the position of node because there is change in pressure. In this way, one can find the positions of nodes and antinodes in a medium. The average distance between two adjacent nodes is equal to half the wavelength. Knowing the value of the frequency of the ultrasonic wave, one can calculate the velocity of the ultrasonic wave through the medium.

Smoke Method

When one introduces light solid particles or liquid drops in the field of a sound wave, the particles take up the motion, the amplitude of which is given by Konig's following relation:

$$\frac{y_1}{y_2} = \sqrt{1+a^2} \quad \text{where} \quad a = \frac{2}{g}\frac{\rho_1}{\rho_0\lambda b^2}$$

where y_1 and y_2 are amplitudes of the particles when set into vibration and that of sound wave respectively, b and ρ are respectively the radius and density of detectors, ρ_0 the density of the gas and λ the wavelength of the sound wave. The method then demands in introducing small particles in the field to photograph them with long time of exposure and then measure the amplitude of the motion of the particles which appear as streaks in the negative. The above equation reveals that particles of very small radius must be introduced in the sound field. However, these limitations restrict the use of this method for detection of ultrasonic waves having wavelength smaller than 0.1 m. We may note that this method is not only useful for the detection but also for the measurement of the amplitude of sound wave.

10.4 PROPERTIES OF ULTRASONICS

i. Ultrasonic waves have a large energy content, i.e. they are highly energetic.

ii. The speed of propagation of ultrasonics depends upon their frequency, i.e. it increases with increase in frequency.

iii. Because of small wavelength, these waves show negligible diffraction and can be transmitted over long distances without any appreciable loss of energy. This is why these waves have been used in determining the depth of ocean by echo-sounding.

iv. When a plane stationary ultrasonic wave is set up in a liquid, a structure is developed in which the density of the liquid varies from layer to layer along the direction of propagation of the waves. This structure within the liquid can diffract light in the same way as the structure of a crystal diffract X-rays.

v. Intense ultrasonic radiations has a disruptive effect on liquid by causing bubbles to be formed.

10.5 APPLICATIONS OF ULTRASONICS

Ultrasonics is a very useful tool in the hands of physicists and chemists because from the nature of the absorption and dispersion of ultrasonics in liquids and gases much information can be obtained regarding the structure of molecules. Scientists have obtained very useful data regarding the superconducting state by propagation of ultrasonics through liquid helium. Besides these theoretical considerations, ultrasonics have found numerous applications in the following fields: (a) communication (b) industry (c) medical and biological fields (d) scientific research.

We shall consider some of these applications here.

Communication
Ultrasonic Signalling

As the wavelength of ultrasonics is very small, it is possible to produce a short beam of ultrasonics without diffraction. The high frequency sound waves may be readily formed into a beam and set in the desired direction. It is found that the beam of sound gets more and more narrow as the radius of the plate of the generator is increased in relation to the wavelength of radiated sound. A plate of 3.5 cm diameter is found to be sufficient to confine the sound with an angle of 5 degrees at 600 kc/s. We may note that such narrow beams can never be attained in the audible frequency range. This low power source can radiate very high intensity of ultrasonic sound waves, which can travel several kilometres in water prior being absorbed.

Detection of Submarines, Icebergs and Other Objects

Ultrasonics can be used to detect the reflecting objects like a submerged submarine and a hidden iceberg. For this purpose ultrasonic transmitters have been developed. A typical submarine ultrasonic transmitter uses piezo-electric resonator which generates powerful ultrasonic waves of frequency about 40 kHz. These submarine ultrasonic transmitter finds following uses:

Ultrasonic Signalling

It is used for signalling from ship-to-ship especially in submerged submarines and detecting a hidden iceberg.

Recently, ultrasonic microscope has been invented. It is used to detect concealed objects. The frequency is very high so that the wavelength is of the order of the wavelength of visible light.

Depth of Sea

One can determine the depth of the sea, position of a submerged submarine, position of a ship, etc. with the help of sub-marine ultrasonic transmitter. This uses the *eco principle*. The ultrasonic waves transmitted by the quartz crystal in ultrasonic transmitter are directed towards the bed of the sea and these waves are reflected back from the bed and the echo is detected by the crystal itself. We may note that the working principle is the same as that of a conventional pulsed radar. At short intervals, the pulses of ultrasonic signals are sent out at short intervals from a piezoelectric transmitter which is typically energised at a frequency of about 40 kc/s. As stated above, the returned echo is received in the ultrasonic receiver, which also uses piezo crystal followed by an amplifier and the time interval between the transmitted and received signals is recorded by time measuring instrument.

Let t be the time interval between the transmission of the ultrasonic wave and receipt of the echo, then the velocity of sound in sea water is

$$v = v_0 + 1.48 + 4.21T - 0.037T^2$$

where v is the velocity of sound at $t°C$ in sea water (m/s), v_0 is the velocity of sound at $0°C$ in water (m/s) = 1510 m/s, S is the salinity (gm/litre), T is the temperature of the sea water in $°C$. Knowing v, the velocity of sound waves through sea water, one can obtain the depth of the sea

$$h = \frac{vt}{2}$$

The method can also be used to detect the presence and depth of submarines, rocks, etc. from the surface of the water. The instrument directly calibrated to show the depth of sea is called a *fathometer* and *echometer*. This method of determining the depth of sea, the direction and distance of a submarine or iceberg is called *Sound Naviagation and Ranging* (SONAR).

Industrial Applications

Ultrasonic Flaw Detector

The strength of a component plays a significant role in most of the engineering materials. If there are defects in the material, the strength of the material gets reduced.

The defects in materials may be as large as cracks or as tiny as cavities produced during casting. The fine internal cracks and flaws in metals act as good sound reflectors of ultrasonics whose wavelength is small compared to the size of the crack or the flaw. Hence they can be detected and located by echo sounding techniques. A part of the surface of the metal is polished and a small ultrasonic generator is placed upon it which sends a beam into the specimen. The beam will be reflected at the far surface and also at the crack or flaw if any as shown in Fig. 10.8a. A receiver is also placed near the ultrasonic generator, which picks up the reflected echoes, which are amplified and displayed along the time-base of cathode-ray-oscilloscope (CRO). The time base of CRO shows a blip A in the beginning corresponding to the signal received directly, and a blip B at the end due to the signal reflected from the far surface. Any signal due to flaw appears between A and B as shown in Fig. 10.8b. The interval between the first and the last blip gives a length scale by which the exact location of the flaw can be known as *low power application*.

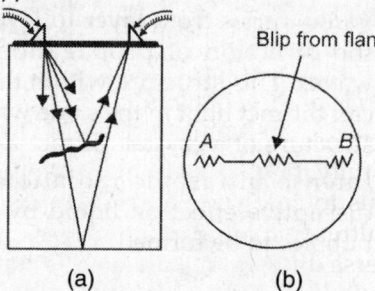

(a)　　　　　(b)

Fig. 10.8: Ultrasonic flaw detection

Non-homogeneties

One can use ultrasonic waves for the study of non-homogeneities in a medium such as metal or plastic. For visual presentation one can convert the ultrasonic waves coming from medium into corresponding light waves.

Thickness of the Gauge

One can use ultrasonic waves for measuring the thickness of boiler, atomic pile structure where one side is not accessible by the process of reflection of ultrasonic wave as low power application.

Cavitation

When one places an ultrasonic transducer in a liquid, it produces standing wave formation and it is observed that these result in the development and implosion of bubbles. When subjected to powerful ultrasonic wave, pressure develops at some points in a liquid, such that excessive stress breaks a part of the liquid and produces hollow bubbles. However, these bubbles collapse soon and at the instant of implosion, the pressure around the bubbles becomes very high (~ several hundreds of atmosphere). This extremely high pressure results in emulsification. Interestingly, the cavitation bubbles formed by ultrasonic vibration hamper the propagation of ultrasonic waves and produces noise. This action of ultrasonic waves has been successfully used in various industrial devices described as follows as *high power application of ultrasonic waves*:

Soldering and Metal Cutting

One can use the cavitation action of ultrasonic waves for soldering, metal cutting and drilling processes in metals.

Preparation of Uniform Alloy

The process of cavitation has been used successfully for dispersion of metals in molten materials to obtain uniform alloying. By the use of ultrasonic waves it has become possible to disperse different metals such as Na, K, Hg, Pb, Zn, Cu, etc. and also their fusible alloys in oil, water and alcohol.

Preparation of Very Stable Emulsion

The process of cavitation has been used for the preparation of very stable emulsion of two immiscible liquids such as oil and water.

Ultrasonic Cleaning Device

Once can clear contaminated articles by irradiating ultrasonically in sequences in suitable solvents. Nowadays, one uses this technique for washing textiles. The large variation of acoustic pressure actually breaks off the contaminated particles from the surface in the ultrasonic cleaning.

Scientific Applications

Photographic Emulsion

The homogeneity and stability of photographic emulsions are improved by the application of ultrasonic waves. Thorough dispersion of the dye in the emulsion under the influence of ultrasonic waves increases colour sensitivity of photographic emulsions.

Coagulation and Crystallisation

Using ultrasonics, one can bring the particles of a suspended liquid quite close to each other and thus coagulation may take place which helps in accelerating the rate of crystallisation.

Structure of Matter

Ultrasonics can be used for the study of structure of matter. This is a technique of ultrasonics displaying smaller and smaller in homogeneities in metal. It is reported that a limit of improvement is reached when the wavelength of the ultrasonic wave becomes comparable to the size of the crystal grain of the test materials for the frequency range of 1–20 Mc/s. Then due to the scattering of the ultrasonic waves at the boundaries of the crystal grain, multiple reflections and interferences are caused.

Medical and Biological Applications

Ultrasonography have a large number of applications in the field of medicine. Some important applications are:

Neuralgic Pain

Ultrasonic waves are useful for relieving neuralgic and rheumatic points. The affected portion of the body is exposed to ultrasonic waves. The waves produce a soothing massage action and relieves pain.

Detection of Abnormal Growth

Cerebral ventricles are explored by ultrasonic waves for locating extraordinary abnormal growth in the brain.

Arthritis

Ultrasonic waves are used to relieve pain due to arthritis. Here a small metal head, vibrating with a frequency of more than 10^6 Hz is moved over the skin of the patient. These vibrations after passing through the tissues, produce a deep massage action and the patient gets relieved of the pain.

Contracted Fingers

To restore the contracted fingers, ultrasonic waves are used. Ultrasonic waves are also used to loosen up the scar tissues in various parts of the human body.

Dental Cutting

Ultrasonic waves are used by dentists for the proper extraction of broken teeth. These waves have been found very useful by dentists because: (i) they make the cutting almost painless, (ii) they cut the hard material very easily, and (iii) they do not require any additional mechanical device for cutting purpose.

Bloodless Surgery

Ultrasonic waves are focussed on a sharp instrument and the tissues are destroyed without any loss of blood. Doctors have used such instruments for conducting bloodless brain operations.

Sterilization

Ultrasonic waves can destroy unicellular organisms. Under the action of ultrasonic waves bacteria perish. Ultrasonic waves are also used in the sterilization of water and milk.

Biological Effects

When some small animals like rats, frogs, fishes, etc. are exposed to high intensity ultrasonic waves, they become lame or killed. This is the only destructive application of ultrasonic waves.

Ultrasonic waves are finding more and more practical applications in various fields. Active research work is still in progress to study the effect of ultrasonic waves in various fields.

10.6 PZT (LEAD ZIRCONATE TITANATE BASE) CERAMICS

There are number of materials, e.g. barium titanate, lead metaniobate, lead titanate, lead zirconate, as well as materials on lead zirconate titanate base (PZT) resemble barium titanate. These are used as sintered ceramic materials because it is not possible to produce large single crystal from them. These materials exhibit piezoelectric property, some of them are ferroelectric also. We may note that in contrast to quartz, lithium sulphate and other neutral piezoelectric crystals, ceramics are giving their piezoelectric properties by polarization.

One can form ceramic piezoelectric crystals in different shapes as per requirement. The ground, raw material mixed with binders, is moulded as per requirement by pressing and sintering, above $1000\,°C$ and then get shaped accurately by grinding as per requirement. We may note that PZT ceramics are white to yellowish materials of lower hardness and resistance than quartz and one can get their dull surfaces to be silver plated by baking process.

PZT ceramic material is first heated to $350\,°C$ (curie temperature) which is higher than the curie temperature $120\,°C$ for $BaTiO_3$, and then the material is permitted to cool off, a direct voltage of few thousand volts/cm thickness is applied. Due to this voltage, the small elementary crystals in the ceramic material, which oriented randomly earlier, now align along one axis and get frozen. Thus, the PZT ceramic materials become polarized along one axis, provided the material is not reheated close to the curie temperature due to the alignment along one axis, the PZT ceramic materials thus remain piezoelectric in the frozen state.

One can use ceramic piezoelectric materials for obtaining transducers with curved surfaces by sintering and grinding, at every point, normal to the surface during polarization. We may note that for quartz crystal in transducer ground with a curve, is not possible. However, if the transducer ground is excessively concave or convex, then these faces no longer contribute to the radiation of longitudinal waves. This helps in chewing the considerable concentrations of the sound field along the axis of the cylinder by using PZT ceramic piezoelectric over natural crystal. Lead metaniobate and lithium sulphate are the best for non-destructive testing.

10.7 ULTRASONIC TRANSDUCER

These are actually the energy converters. These convert the energy from one transmission system where it is activated to second transmission system either in the same form or in another form. This can be used with electronic amplifier for development of intense acoustic fields in solutions and materials at frequency from few thousand Hz to several megahertz. Magnetostriction bar and the piezoelectric crystal are the common forms of ultrasonic transducers, both of which have been found useful for converting electric energy to mechanical acoustic energy.

ILLUSTRATIVE EXAMPLES

Example 1

Determine the frequency of the first and second modes of vibration for a quartz crystal of piezoelectric oscillator. The velocity of longitudinal waves in quartz crystal is 5.5×10^3 m/s. Given, thickness of quartz crystal is 0.05 m.

Solution

The distance between the two faces of the crystal (thickness t) in the lowest mode of vibration will be $\lambda/2$. Thus

$$\lambda = 2t = 2 \times 0.05 = 0.1 \text{ m}$$

∴ The lowest frequency

$$n_1 = \frac{v}{\lambda} = \frac{5.5 \times 10^3 \text{ m/s}}{0.1 \text{ m}} = 5.5 \times 10^4 \text{ Hz}$$

In the second mode, the frequency would be

$$n_2 = 2n_1 = 11 \times 10^4 \text{ Hz}.$$

Example 2

A quartz crystal of thickness 0.001 m is vibrating at resonance. Find the fundamental frequency. Given Y for quartz $= 7.9 \times 10^{10}$ N/m² and ρ for quartz $= 2650$ kg/m³.

Solution

For longitudinal vibration, we have

$$v = \sqrt{Y/\rho}$$

$$Y = 7.9 \times 10^{10} \text{ N/m}^2$$
$$\rho = 2650 \text{ kg/m}^3$$

∴
$$v = \sqrt{\frac{7.9 \times 10^{10}}{2650}} = 5460 \text{ m/s}$$

For the fundamental mode of variation, we have

Thickness $(t) = \lambda/2$

∴
$$\lambda = 2t = 2 \times 0.001 = 0.002 \text{ m}$$

Frequency $n = v/\lambda = \dfrac{5460}{0.002}$

$$= 2730,000 \text{ Hz}$$
$$= 2730 \text{ kHz.}$$

Example 3

An ultrasonic beam is used to determine the thickness of a steel plate. It was found that the difference in two adjacent harmonic frequencies is 50 kHz. The velocity of sound in steel is 5000 m/s. Show that the thickness of the steel plate is 0.05 m/s.

Solution

The velocity of ultrasonic wave

$$v = 2Nd$$

where N is the fundamental frequency and d is the thickness

∴
$$N = \frac{v}{2d}$$

We know that the harmonic frequencies are the multiples of the fundamental frequencies, we have

$$N_{(n)} - N_{(n-1)} = \frac{v}{2d}$$

$$v = 50,000 \text{ Hz}$$
$$= 5000 \text{ m/s}$$

∴
$$d = \frac{v}{2\left[N_{(n)} - N_{(n-1)} \right]}$$

$$= \frac{5000}{2 \times 50,000}$$

$$= 0.05 \text{ m.}$$

Example 4

Calculate the capacitance to produce ultrasonic waves of 10^6 Hz with an inductance of 1 henry.

Solution

We have

$$n = \frac{1}{2\pi}\sqrt{\frac{1}{LC}}$$

or

$$C = \frac{1}{4\pi^2 n^2 L}$$

$n = 10^6/s$

$L = 1$ henry

$$= \frac{1}{4 \times (3.14)^2 \times (10^6)^2 \times 1}$$

$$= 0.025 \times 10^{-12} \text{ farad}$$

$$= 0.025 \text{ } \mu F.$$

Example 5

A quartz crystal of thickness 0.001 m radiates ultrasonic waves of frequency 20 kHz and intensity 5×10^4 W/m² into water. Calculate (i) maximum acceleration (ii) maximum displacement. Velocity of sound in water = 1480 m/s and density of water = 10^3 kg/m³. Density of crystal = 2650 kg/m³.

Solution

We have

Sound intensity (I), $= \frac{P^2}{2\rho v}$ W/m², where P is the sound pressure, ρ is the density and v is the velocity of sound.

Now, sound pressure

$$P = \sqrt{2I\rho v}$$

$I = 5 \times 10^4$ W/m²

$\rho = 10^3$ kg/m³

$v = 1480$ m/s

$\therefore$

$$P = \sqrt{2 \times 5 \times 10^4 \times 10^3 \times 1480}$$

$$= 3.85 \times 10^5 \text{ N/m}^2$$

Force = Pressure × Area

= Mass × acceleration

Here $M \rightarrow$ Mass of the crystal

= Volume × density

$= A t D$

i. Maximum acceleration

$$a = \frac{P}{Dt} = \frac{3.85 \times 10^5}{1.45 \times 10^5} \text{ m/s}^2$$

ii. Maximum displacement

$$y = \frac{a}{\omega^2} = \frac{1.45 \times 10^5}{(2\pi f)^2}$$

where $f = 20$ kHz = 20,000 Hz

$$\therefore \quad y = \frac{1.45 \times 10^5}{(2\pi \times 20,000)^2} = 9.4 \times 10^{-6} \text{ m}$$

Example 6

Show that the natural frequency of 40 mm length of a pure iron rod is 49.75 kHz. Given the density of pure iron is 2.75×10^3 kg/m³ and $Y = 115 \times 10^9$ N/m².

Solution

$$n = \frac{1}{2L}\left(\frac{Y}{\rho}\right)^{1/2}$$

$$= \frac{1}{2 \times 40 \times 10^{-3} \text{ m}}\left(\frac{115 \times 10^9 \text{ N/m}^2}{7.25 \times 10^3 \text{ kg/m}^3}\right)^{1/2}$$

$$= 12.5 \times \left(15.86 \times 10^6\right)^{1/2} \frac{1}{\text{m}}\left(\frac{\text{N m}^3}{\text{m}^2 \text{ kg}}\right)^{1/2}$$

$$= 49750 \frac{1}{\text{m}}\left(\frac{\text{kg} - \text{m}}{\text{s}^2} \times \frac{\text{m}}{\text{kg}}\right) = 49.75 \text{ kHz} \cdot$$

Example 7

Find the frequency of the first and second modes of vibration for a quartz crystal of piezo-electric oscillator. The velocity of longitudinal waves in quartz crystal is 5.5×10^3 m s⁻¹. Thickness of quartz crystal is 0.05 m.

Solution

Given, $v = 5.5 \times 10^3 \text{ m s}^{-1}$; $t = 0.05$ m;

$$v_1 = ?; v_2 = ?$$

In the lowest mode of vibration, the distance between the two faces of the crystal of thickness t will be $\lambda/2$.

Therefore, $t = \dfrac{\lambda}{2}$

or $\lambda = 2t = 2 \times 0.05$

$\lambda = 0.1$ m

Therefore, the frequency in the first mode of vibration

$$v_1 = \frac{v}{\lambda} = \frac{5.5 \times 10^3}{0.1}$$

$$v_1 = 5.5 \times 10^4 \text{ Hz}$$

The frequency in the second mode of vibration is

$$v_2 = 2v_1 = 2 \times 5.5 \times 10^4$$

$\therefore \qquad v_2 = 110 \times 10^3 \text{ Hz}$

Example 8

An ultrasonic source of 0.09 MHz sends down a pulse towards the seabed which returns after 0.55 sec. The velocity of sound in water is 1800 m/s. Calculate the depth of the sea and wavelength of pulse.

Solution

Given, $f = 0.09 \text{ MHz} = 0.09 \times 10^6 \text{ Hz}$;

$$t = 0.55 \text{ sec}; v = 1800 \text{ m s}^{-1}$$

depth of the sea = ?; $\lambda_u = ?$

depth of the sea, $d = \dfrac{vt}{2} = \dfrac{1800 \times 0.55}{2}$

$\therefore \qquad d = 495$ m.

The wavelength of the ultrasonic pulse is

$$\lambda_u = \frac{v}{f} = \frac{1800}{0.09 \times 10^6}$$

$\therefore \qquad \lambda_u = 0.02$ m.

Example 9

Calculate the frequency to which piezoelectric oscillator circuit should be tunned so that a piezoelectric crystal of thickness 0.1 cm vibrates in its fundamental mode to generate ultrasonic waves. (Young's modulus and density of material of crystal are 80 GPa and 2654 kg m^{-3}).

Solution

Given, $E = 80 \text{ GPa} = 80 \times 10^9 \text{ Pa}$; $\rho = 2654 \text{ kg m}^{-3}$;

$$t = 0.1 \text{ cm} = 0.1 \times 10^{-2} \text{ m}$$

The frequency of vibration is given by

$$f = \frac{P}{2t}\sqrt{\frac{E}{\rho}} = \frac{1}{2 \times 0.1 \times 10^{-2}}\sqrt{\frac{80 \times 10^9}{2654}}$$

$$= \frac{5490.28}{2 \times 10^{-3}}$$

$$f = 2.7451 \times 10^6 \text{ Hz}$$

REVIEW QUESTIONS

1. Distinguish between audible, infrasonic and ultrasonic waves. Mention one source of each class.

2. What are ultrasonic waves? Describe a method for their production.

3. How are ultrasonics detected? Explain the magnetostriction method of producing ultrasonic energy and hence compare with piezoelectric method.

4. What is piezoelectric effect? Explain the generation of ultrasonics by piezoelectric generator. Show that only odd harmonics exist in this generator.

5. What is magnetostriction effect? Explain how is it used to design an ultrasonic generator. Give its working.

6. Describe a laboratory method to determine the velocity of ultrasonic waves in liquids.

7. Discuss the advantages and disadvantages of magnetostriction and piezoelectric methods in production of ultrasonic waves.

8. What are the various applications of the ultrasonic waves? Describe the non-destructive application in detail.

9. Bats, who are blind, fly about avoiding obstacles. Explain.

   [**Hint:** Bats have the natural power of emitting and detecting ultrasonic waves. When they fly, they emit continuously ultrasonic waves. These ultrasonic waves gets reflected from the obstacles which happen to come in the path of flying bats. The reflected waves are immediately detected by the bats who change their path avoiding the obstacle].

10. What are PZT ceramics? How do these help in the construction of ultrasonic transducers?

11. Draw circuit diagram for production of ultrasonic waves using piezoelectric effect. Explain its working.

12. Draw circuit diagram of magnetostriction oscillator. Explain its working.

13. Discuss briefly the scientific application of ultrasonics.

14. Write short notes on the following:
   i. Piezoelectric effect
   ii. Magnetostriction effect
   iii. Echo sounding
   iv. Cavitation
   v. PZT ceramics
   vi. Ultrasonic transducers

OBLEMS

1. A piezoelectric X-cut quartz crystal plate has a thickness of 1.6 mm. If the velocity of propagation of sound waves along the X-direction is 5760 m/s, calculate the fundamental frequency of the crystal. [*Ans.* 1.8 mega Hz]

2. A quartz crystal of thickness 0.005 m is vibrating at resonance. Calculate the fundamental frequency. Given Y for quartz is 7.9×10^{10} N/m^2 and ρ for quartz is 2650 kg/m^3.
[*Ans.* 5.46×10^5 Hz]

3. Ultrasonic pulse echo method is employed to detect possible defect in steel bar of thickness 40 cm. If the pulse arrival times are 30 and 80 microseconds, find the distance of the defect.
[*Ans.* 15 cm]

[**Hint:** Let x be the distance of the defect from the end of the bar at which the pulse enters the bar. The pulse arriving first is the one reflected from the defect and that arriving later on is the one reflected from the other end.
Obviously, the pulse takes 30 microseconds to cover a distance of $2x$, and 80 microseconds to cover a distance 80 cm.

Hence $30 = \dfrac{2x}{v}$ and $80 = \dfrac{80}{v}$

Solving, one obtains $x = 15$ cm.

4. Longitudinal standing waves are setup in a quartz plate with antinodes at opposite faces. The fundamental frequency of vibration is given by

$$N = \frac{2.87 \times 10^5}{t}$$

where N is in cycles/s and t is the thickness of the plate in cm. Compute (a) Young's modulus of the quartz plate (b) the thickness of the plate required for a frequency of 1200 kilocycles/s. The density of quartz is 2.66×10^3 kg/m^3.
[*Ans.* $Y = 8.76 \times 10^{10}$ N/m^2, $t = 0.24$ cm]

5. An ultrasonic source of 0.07 MHz sends down a pulse towards the sea bed, which returns after 0.61 s. The velocity of sound in sea water is 1700 m/s. Show that the depth of sea is 552.5 m and the wavelength of pulse is 0.0243 m.

[**Hint:** $2d = vt = 1700 \times 0.65$

$\therefore \qquad d = 552.5$ m

$V = n\lambda \therefore \lambda = V/n = 1700/0.07 \times 10^6$

$\qquad = 2.43 \times 10^{-2}$ m

$\qquad = 0.0243$ m]

SHORT ANSWER QUESTIONS

1. What are infrasonic waves?

Ans. Those longitudinal waves whose frequency lies below 20 Hz are called infrasonic waves. These waves are produced by large vibrating bodies such as during an earthquake and our ear is not sensitive to these waves.

2. What are hypersonic waves?

Ans. Longitudinal waves of frequency higher than 10^8 Hz are called hypersonic waves.

3. When does the ultrasonic generator delivers maximum power in piezoelectric method of production of ultrasonics?

Ans. The ultrasonic generator delivers maximum power when it is operated at the fundamental frequency of the crystal. In order to generate higher-frequency ultrasonic waves, the L–C circuit is made to oscillate at a frequency equal to one of the harmonics of the crystal.

4. What happens when ultrasonic waves pass through a substance?

Ans. Intense heat is produced.

5. Can a glass rod oscillating with ultrasonic frequency lose a hole through steel, diamond?

Ans. Yes.

OBJECTIVE QUESTIONS

1. Ultrasonic waves are used to detect tumours, breast cancer and also growth of _____ can be studied.   [foetus]

2. Weak ultrasonic waves are able to produce good _____ of two immissible liquids, so it is used for the manufacture of good _____.
                    [emulsion, paints]

3. The speed of propagation of ultrasonics increases with increasing _____.
                    [frequency]

4. Intense ultrasonic radiation has a disruptive effect on liquid by causing _____ to be formed.
                    [bubbles]

5. Bats, who are blind, fly about avoiding _____.
                    [obstacles]

6. Magnetostriction consists in changing of length of a ferromagnetic material subjected to a _____.   [magnetic field]

7. Sound waves having frequencies above 20 kHz are inaudible and are called _____ waves.
                    [ultrasonic]

8. The phenomenon of magnetostriction and piezoelectric effect are used to produce _____ waves.   [ultrasonic]

9. Certain crystals undergo periodic mechanical deformation when subjected to AC voltage. This is knows as _____.   [piezoelectric effect]

10. The materials that exhibit changes in their mechanical dimensions, create pressure variations in the surrounding air. These pressure variations of high frequency constitute _____.
                    [bubbles]

MULTIPLE CHOICE QUESTIONS

1. The frequency of ultrasonic sound wave is
   (a) greater than 5 kHz
   (b) greater than 10 kHz
   (c) greater than 20 kHz
   (d) less than 20 kHz

2. When a ferromagnetic material in the form of a rod is subjected to an alternating magnetic field, the rod undergoes alternate contractions and expansions at a frequency equal to the frequency of the applied magnetic field. This phenomenon is called
   (a) hysteresis
   (b) meissner effect
   (c) magnetostriction effect
   (d) piezoelectric effect

3. When pressure is applied to one pair of opposite faces of crystals like quartz, tourmaline, rochelle salt etc. cut with their faces perpendicular to its optic axis, equal and opposite charges appear across its other faces. This phenomenon is known as
   (a) meissner effect
   (b) superconductivity
   (c) magnetostriction effect
   (d) piezoelectric effect

4. If an alternating voltage is applied to one pair of opposite faces of the crystal, alternatively mechanical contractions and expansions are produced in the crystal and the cystal starts vibrating. This phenomenon is known as
   (a) piezoelectric effect
   (b) inverse piezoelectric effect
   (c) magnetostriction effect
   (d) meissner effect

5. Fathometer or echometer is a device which is directly calibrated to determine the
   (a) height of a satellite
   (b) depth of the sea
   (c) frequency of a tuning fork
   (d) flaws and cracks in metals

6. When ultrasonic waves propagate in a liquid medium, the alternating compressions and rarefactions vary the density of the medium. This change in density results in the variation of
   (a) density of the liquid
   (b) temperature of the liquid
   (c) pressure of the liquid
   (d) refractive index of the liquid

Answers

1. (c)   2. (c)   3. (d)   4. (b)   5. (b)
6. (d)

11 Magnetism and Magnetic Materials

11.1 MAGNETISM

This comprises physical phenomena involving magnetic fields and their effects upon materials. Magnetic fields may be set up on a macroscopic scale by electric currents or by magnets. On atomic scale, individual atoms cause magnetic fields when their electrons have a net magnetic moment as a result of their angular momentum. A magnetic moment arises whenever a charged particle has an angular momentum. It is the cooperative effect of the atomic magnetic moments which causes the macroscopic magnetic field of a permanent magnet. However, the underlying principles and mechanisms that explain the magnetic phenomenon are complex and subtle, and their understanding has eluded physicists until relatively recent times. Several of our modern technological devices rely on magnetism and magnetic materials; these include electric motors, electric power generators and transformers, components of sound and video reproduction systems, telephones, radios, televisions, computers, etc. Well known examples of magnetic materials which exhibit magnetic properties are iron, some steels and the naturally occurring mineral lodestone. The important facts about the magnetic materials are:

i. There are some materials which exhibit magnetic properties even without the application of any magnetic field and become more magnetic when a weak magnetic field is applied to them.

ii. There are many other materials which lose their initially strong magnetism when heated above a certain critical temperature and become comparatively weakly magnetised.

iii. There are some materials which show a magnetic response in a direction opposite to that of any externally applied field.

iv. Magnetic materials are all media capable of being magnetized in a magnetic field, i.e. of creating their own magnetic field. According to their magnetic properties, such materials are divided into three principal groups: *diamagnetic, paramagnetic* and *ferromagnetic materials*. From the applications point of view, all the magnetic materials can be placed under two groups:

1. *Soft* magnetic materials
2. *Hard* magnetic materials.

Iron, some steels, and the naturally occurring mineral lodestone are well-known examples of materials which exhibit magnetism. Ferro and ferrimagnetic materials are the most important magnetic materials from the point of view of practical applications. We will first consider the terms and definitions used in magnetism.

11.2 TERMINOLOGY

Magnetic Induction or Magnetic Flux Density (*B*)

In the presence of magnetic field in vacuum, the magnetic induction (*B*) is related to the field strength *H* (in units of Am^{-1}) as follows:

$$B = \mu_0 H \qquad (11.1)$$

where μ_0 is called the *permeability of free space*. *B* is expressed in units of tesla or Weber per square metre.

The units of permeability are

$$\mu_0 = \frac{B}{H} = \frac{Wb\,m^2}{Am^{-1}} \quad (\because Wb\,m^{-1} = NA^{-1}m^{-1})$$

$$= Hm^{-1}\ (H = Wb\,A^{-1}\ where\ H \to henry)$$

In SI units, the permeability of free space (μ_0) has a value of $4\pi \times 10^{-7}\ Hm^{-1}$.

Magnetic Field (*H*)

It is said to occupy a region when the magnetic effect of an electric current or of a magnet upon a small test magnet which is brought in the vicinity is detectable. Magnetic field strength is denoted by *H*. When a magnetic material is placed in a magnetic field *H*, it becomes magnetized, i.e. it itself becomes a magnet. Magnetic field strength (*H*) is expressed in units of Am^{-1}.

If the magnetic field is applied to a solid medium, the magnetic induction in the solid is given by a relationship

$$B = \mu H \qquad (11.2)$$

where μ is the permeability of the solid material through which the magnetic lines of force pass. In general, μ is not equal to μ_0. The ratio μ/μ_0 is the relative permeability of the medium and designated by μ_r.

Mathematically,

$$\mu_r = \frac{\mu}{\mu_0} \qquad (11.3)$$

Magnetization

This may be defined as the process of converting a non-magnetic bar into a magnetic bar. This term is almost analogous to the polarization in dielectric materials. The flux density

$$B = \mu H = \mu_0 \mu_r H$$
$$= \mu_0 \mu_r H + \mu_0 H - \mu_0 H$$
$$= \mu_0 H + \mu_0 H(\mu_r - 1) \qquad (11.4)$$
$$= \mu_0 H + \mu_0 M$$
$$= \mu_0(H + M) \qquad (11.5)$$

where $M = \mu_0(H + M)$ is called the magnetization of solid and expressed in ampere/metre. From the above relation, we find that if a magnetic field is applied to a material, the magnetic flux density is equal to the effect on vacuum and on the material. The magnetization (*M*) may thus be defined as the magnetic dipole moment per unit volume of the bar.

Magnetic Susceptibility (χ)

The magnitude of the magnetization, *M* is proportional to the applied field as follows:

$$M = \chi H \qquad (11.6)$$

where χ is called the magnetic susceptibility, which is *unitless*. The magnetic susceptibility and the relative permeability are related as follows:

$$\chi = \mu_r - 1 = M/H \quad or \quad \mu_r = 1 + \chi \quad (11.7)$$

We may note that *B*, *M* and *H* are vectors. Magnetic units and conversion factors for SI and CGS emu systems are given in Table 11.1. We may note that magnetic units may be a source of confusion because there are really two systems in common use. The ones used so far are SI [rationalized MKS (metre-kilogram-second)]; the others come from the CGS-emu (centimetre-gram-second-electromagnetic unit) system.

Intensity of Magnetization (*I*)

This is defined as the ratio of the magnetic moment (*M*) to the volume (*V*) of the magnetized material, i.e.

$$I = \frac{M}{V} \qquad (11.8)$$

Table 11.1: Magnetic units and conversion factors for the SI, CGS and EMU systems

| Quantity | Symbol | SI units | | CGS-emu unit | Conversion |
		Derived	Primary		
• Magnetic induction (Flux density)	B	tesla (Wb/m^2)	kg/sC	gauss	$1\dfrac{\text{Wb}}{\text{m}^3} = 10^4$ gauss
• Magnetic field strength	H	$\dfrac{\text{amp-turn}}{\text{m}}$	C/ms	oersted	$\dfrac{1 \text{ amp-turn}}{n}$ $= 4\text{p} \times 10^{-3}$ oersted
• Magnetization	M (SI) I (CGS-emu)	$\dfrac{\text{amp-turn}}{\text{m}}$	C/ms	Maxwell/cm^2	$\dfrac{1 \text{ amp-turn}}{\text{m}}$ $= 10^{-3}$ Maxwell/cm^2
• Permeability of vacuum	μ_0	henry/m	kgm/C^2	unitless	$4\pi \times 10^{-7}$ henry/m $= 1$ emu
• Relative permeability	μ_r (SI) μ'(CGS-emu)	unitless	unitless	unitless	$\mu_r = \mu'$
• Susceptibility	χ (SI) χ'(CGS-emu)	unitless	unitless	unitless	$\mu' = 4\pi\chi$

Note: Units of Weber (Wb) are volt-seconds. Units of Henry are Weber per ampere.

Now, if the given magnetic material is rectangular or cylindrical bar of length l and area of cross-section a, then $V = al$ and magnetic moment $M = ml$, where m is designated as pole strength, we have

$$I = \frac{ml}{al} = \frac{m}{a} \qquad (11.9)$$

Thus, the intensity of magnetization may also be defined as the pole strength per unit area at right angles to the direction of magnetization.

Intensity of Magnetization (I) and Susceptibility (χ)

For any magnetic material, the intensity of magnetization is found to be proportional to the measure of the capability of magnetizing field H, i.e.

$$I \propto M$$

or $\qquad I = \chi_m H \qquad (11.10)$

where χ_m is called the magnetic susceptibility of the medium and it is the measure of capability of the material to take up magnetization.

Magnetic Dipoles

Magnetism is *dipolar*, i.e. magnetism is characterized by having two opposite poles: north (N) and south (S). Magnetic dipoles are found to exist in magnetic materials, which in some respects are analogous to electric dipoles. The strength of a magnetic dipole is measured by the product of the pole strength and the distance between the poles. This is called *magnetic moment*. Magnetic dipoles are influenced by magnetic fields and within a magnetic field, the force of the field itself exerts a torque that tends to orient the dipoles with the magnetic field. One source of magnetism in an atom is the orbital motion of electrons. Each electron revolving around the nucleus in an atom constitutes a circulating electric charge or current and thus produces a small magnetic field. Moreover, each spinning electron on its axis can be conceived as a circulating charge and also produce small magnetic fields. Many times it is convenient to think of magnetic forces in terms of fields. Imaginary lines of force may be drawn to indicate the direction of the force at positions in the vicinity of the field source.

11.3 ORIGIN OF MAGNETIC MOMENTS

The macroscopic magnetic properties of a substance are a consequence of *magnetic moments* associated with individual electrons. Each electron in an atom has magnetic moments that originate from the following two sources:

1. Orbital magnetic moment of electrons
2. Spin magnetic moment of electrons.

We may note that permanent magnetic moments can also arise from *spin magnetic moment of the nucleus*. Of the three, spin dipole moments of electrons are important in most magnetic materials. Magnetic moments associated with (a) an orbiting electron (b) a spinning electron is shown in Fig. 11.1. Electron in an atom is continuously orbiting around the nucleus; being a moving charge, an electron may be considered to be a small current loop, generating a very small magnetic field, and having a magnetic moment along its axis of rotation (Fig. 11.1a). Moreover, each electron may also be thought of spinning around an axis; the other magnetic moment originates from this electron spin which is directed along the spin axis as shown in Fig. 11.1b. We may note that spin magnetic moments may be only in an "up" direction or in an antiparallel "down" direction. Obviously, each electron in an atom may be thought of as being a small magnet having permanent orbital and spin magnetic moments. The net magnetic moment due to electron spin in a sodium atom is one unit, called a *Bohr magneton* μ_B, which is of magnitude 9.27×10^{-24} A m². For each electron in an atom, the spin magnetic moment is $\pm \mu_B$ (+ sign for spin up and – sign for spin down).

The orbital magnetic moment contribution is equal to $m_l \mu_B$, m_l being the magnetic quantum number of the electron.

Magnetic properties of a material originate due to the imbalance of spin orientation in atoms. A number of atoms or molecules have paired and unpaired electrons. We may note that there is one electron in the outermost *s*-shell which is unpaired and align itself in an applied field, which gives rise to magnetism. This reveals that the magnetic moment of an atom in the solid state is only due to an incomplete inner shell. It is interesting to note that the electrons in the incomplete inner shells of transition elements have their spins aligned in the same direction, or one may say that the electrons arrange themselves among the energy levels to give the maximum possible total spin angular momentum. This is quite consistent with Pauli exclusion principle. Thus, the net magnetic moment, then for an atom is just the sum of the magnetic moments of each of the constituent electron, including both orbital and spin contributions, and taking into account moment cancellation. For an atom having completely filled electron shells or subshells, when all electrons are considered, there is total cancellation of both orbital and spin moments. This is why the materials composed of atoms having completely filled electron shells are not capable of being permanently magnetized. This type of materials include the inert gases (He, Ne, Ar, etc.) as well as some ionic materials.

One can obtain the atomic moments of 3rd transition elements in the solid state by adding the spin magnetic moments of the electrons in the unfilled shell.

11.4 CLASSIFICATION OF MAGNETIC MATERIALS

The three classes into which all the magnetic materials may be grouped according to their magnetic behaviour, although there is some overlap among groups are as follows:

1. Diamagnetic substances (Fig. 11.2a)
2. Paramagnetic substances (Fig. 11.2b)
3. Ferromagnetic substances (Fig.11.2c)

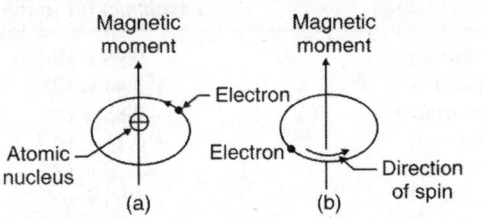

Fig. 11.1: Magnetic moment associated with (a) An orbiting electron (b) A spinning electron

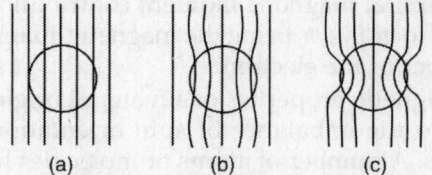

(a)   (b)   (c)

Fig. 11.2: (a) Diamagnetic solid: Lines of forces are slightly repulsed. Relative permeability is less than 1 (b) Paramagnetic solid: Lines of forces are attracted weakly. Relative permeability is slightly greater than 1 (c) Ferromagnetic solid: Lines of forces are attracted very strongly. Relative permeability is much greater than 1

In addition, *antiferromagnetism* and *ferrimagnetism* are considered to be the sub-classes of ferromagnetism.

Diamagnetism

Diamagnetism is a very weak form of magnetism exhibited by substances with a negative magnetic susceptibility ($\chi = \mu_0 M/B$), i.e. by substances which magnetize in a direction opposite to that of an applied magnetic field. A diamagnetic substance has a magnetic permeability less than 1, and is repelled when placed near a magnet. The examples are organic solids like naphthalene, benzene, etc; metals like silver, gold and copper; atoms with rare gas configurations like, Ar, He, Ne, etc. The magnetization of diamagnetic substances is associated with the currents induced on application of a magnetic field. According to Lenz's law, the flow of an induced current is in such a direction as to oppose the change of flux of inducing field; this accounts for the negative susceptibility. The diamagnetic susceptibility is invariably small, of the order of 10^{-5} cm³/mole. When placed between the poles of a strong electromagnet, diamagnetic materials are attracted toward regions where the field is weak.

The atomic magnetic dipole configuration for a diamagnetic material with and without an external field is shown in Fig. 11.3. The arrows in the figure represent atomic dipole moments. Figure 11.4 shows the schematic representation of the flux density B versus magnetic field strength H for magnetic materials.

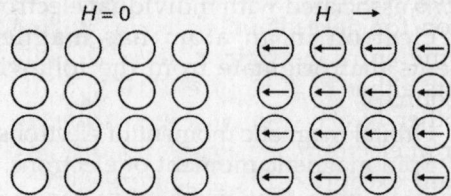

Fig. 11.3: Schematic illustration of the atomic dipole configuration for a diamagnetic material with and without a magnetic field. No dipoles exists in the absence of an external field, whereas in the presence of a field, dipoles are induced that are aligned opposite to the field direction

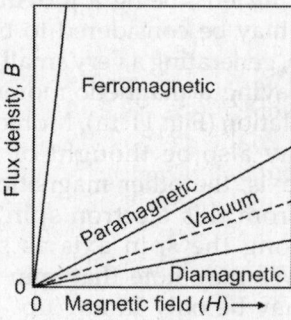

Fig. 11.4: Schematic representation of the flux density B versus the magnetic field strength H for diamagnetic, paramagnetic, and ferromagnetic materials.

Susceptibilities of several diamagnetic materials are given in Table 11.2.

Table 11.2: Magnetic susceptibilities for diamagnetic materials at room temperature

Material	Susceptibility, χ (volume) (SI units)
Copper	-0.96×10^{-5}
Gold	-3.44×10^{-5}
Mercury	-2.85×10^{-5}
Silicon	-0.41×10^{-5}
Silver	-2.38×10^{-5}
Zinc	-1.56×10^{-5}
Aluminium oxide	-1.81×10^{-5}
Sodium chloride	-1.41×10^{-5}

Diamagnetism is found in all materials, but because it is so weak, it can be observed only when other types of magnetism, i.e. para or ferromagnetism are totally absent. The condition for pure diamagnetism is that all electronic spins be paired and all orbital moments either be zero or effectively cancel one another. Nearly all molecules with an even number of electrons satisfy this condition; an important exception is O_2. The condition is also satisfied by most non-metallic solids, except compounds containing atoms with incomplete inner shell electron groups, such as the transition, rare-earth, and actinide elements.

The diamagnetic response of a substance is small; only a very small fraction of the applied magnetic field is shielded from the interior of the substance by the induced diamagnetic currents. There is one case, however, in which the inducing field is completely shielded (except for small surface effects). This is the perfect diamagnetism exhibited by superconductors and is known as Meissner effect. We may note that diamagnetism is of no practical importance.

Some important characteristics of diamagnetic substances are as follows:

i. When a diamagnetic substance is subjected to a strong magnetic field, the magnetism induced is in a direction opposite to that of the external magnetic field.

ii. Diamagnetic substances have a tendency to move from stronger to weaker part of the magnetic field, when the field is non-uniform.

iii. These substances are characterised by negative susceptibility.

iv. When a bar of diamagnetic substance is suspended between the pole pieces of a magnet, the bar stays at right angles.

v. A diamagnetic liquid in a U tube shows depression when placed between the pole pieces of a strong magnet.

vi. The diamagnetic susceptibility (χ) is independent of temperature.

Langevin Theory of Diamagnetism

We will now establish the fact that the effect of a magnetic field on the orbital motion of an electron is such as to produce diamagnetic susceptibility. Consider an electron rotating about the nucleus in a particular orbit, and let a magnetic field be applied perpendicular to the plane of paper as shown in Fig. 11.5a. Before this field is applied, we have

$$F_0 = m\omega_0^2 r \qquad (11.11)$$

where F_0 is the attractive coulomb force between the nucleus and the electron, which provides the necessary centripetal force. ω_0 is the angular velocity. The magnetic moment of electron is

$$\mu_0 = LA = \frac{e}{2}\omega_0 r^2 \qquad (11.12)$$

where r is the radius of the electron's orbit. This moment is perpendicular to the electron orbit, i.e. parallel to the field and sense of rotation shown in Figs 11.5a and b.

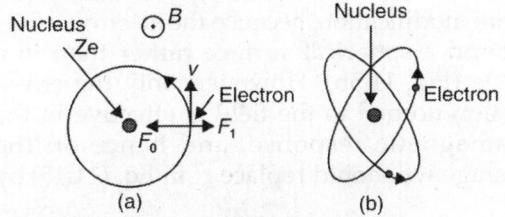

(a)   (b)

Fig. 11.5: Effect of magnetic field on the orbital motion of an electron

When the field is applied an additional force starts to act on the electron; the Lorentz force $-e(V \times B)$. For the geometry of Fig. 11.5a, the effect is to produce a radially outward force given by $eBr\omega$ and Eq. (11.11) takes the form

$$F_0 - eBr\omega = m\omega^2 r \qquad (11.13)$$

Thus, the angular frequency is now different from ω_0, and its value may be determined from this relation. The solution[1] of this quadratic equation in ω in the limit of the small field is given by

$$\omega = \omega_0 - \frac{eB}{2m} \qquad (11.14)$$

which shows that the rotation of the electron has been slowed down. This reduction in frequency produces a corresponding change in magnetic moment which according to Eq. (11.12) is

$$\Delta\mu_m = -\frac{e^2 r^2}{4m} \cdot B \qquad (11.15)$$

Since the moment parallel to the field has been reduced, the induced moment is opposite to the field, i.e. the responses of the electron is diamagnetic.

It can be readily appreciated that if we initially chose an electron which was rotating counter clockwise, the initial moment would be opposite to the Z-axis negatively. The effect of the field would then be to speed up the electron resulting in an even more negative moment, i.e. the induced moment would again be negative diamagnetic and given by Eq. (11.15). Thus, the diamagnetic response of an orbiting electron holds good in general and in fact may be shown to follow directly from the familiar Lenz's Law.

When applied to atom, Eq. (11.15) requires more modification, because the electron orbits around a spherical surface rather than in a circle (Fig. 11.5b). However, only the cross-section normal to the field is effective in the diamagnetic response, and hence on the average we should replace r^2 in Eq. (11.15) by $\frac{2}{3}r^2$ (the mean value of all possible orbital radii and for spherical charge distribution it is $\frac{2}{3}r^2$), the new r being the radius of sphere, which leads to

$$\Delta\mu_m = -\left(\frac{e^2 r^2}{6m}\right) B \qquad (11.16)$$

We can now readily evaluate magnetic susceptibility. Given that the atom has Z electrons and that there are N-atoms per unit volume, the susceptibility is

$$\chi = \frac{M}{H} = \frac{\mu_0 NZ\Delta\mu_m}{B}$$

or

$$\chi = -\frac{\mu_0 e^2}{6m}\left(NZ\bar{r}^2\right) \qquad (11.17)$$

where $\bar{r}^2$ is the average square radius of the electron. The averaging is done over all the occupied orbitals in the atom. This expression yields values which are of the same order ($\sim 10^{-5}$) of magnitude as those obtained by measurements.

Diamagnetic susceptibility is observed most clearly in those solids in which the atomic shells are completely filled. Examples of these are provided by rare-gas crystals, and also ionic crystals

In case of covalent crystals, which are also diamagnetic, Eq. (11.17) can be applied only to the core electrons. The electrons forming the bond have orbits which are far from circular, and hence the derivation leading to Eq. (11.17) does not apply here. The susceptibility of covalent crystal may be written as

$$\chi = \chi_i + \lambda \qquad (11.18)$$

where χ_i includes the effect of the core electrons (the ions) and λ the effect of the bonding electrons. One can determine the value of λ for a specific bond empirically in a given compound, and use this value in other compounds in which it occurs.

When some of the atomic shells in a solid are incompletely filled, the substance then has a paramagnetic contribution in addition to the diamagnetic contribution. The net suscepti-bility is the difference between the two contributions, but since the paramagnetic one is usually larger, it masks the diamagnetic contribution.

---

[1] $m\omega^2 r = F_0 - eBr\omega = m\omega_0^2 r - eBr\omega$ or $\omega^2 + \frac{eB}{m}\omega - \omega_0^2 = 0$

or $\qquad \omega = -\frac{eB}{2m} + \frac{1}{2}\sqrt{\frac{e^2 B^2}{m^2} + 4\omega_0^2} = \omega_0 + \frac{eB}{2m}$, neglecting $\frac{e^2 B^2}{m^2}$ and considering +ve sign.

Paramagnetism

A property exhibited by substances which when placed in a magnetic field, are magnetized parallel to the field to an extent proportional to the field (except at very low temperatures or in extremely large magnetic fields). Paramagnetic materials always have permeabilities greater than 1, but the values are in general not nearly so great as those of ferromagnetic materials.

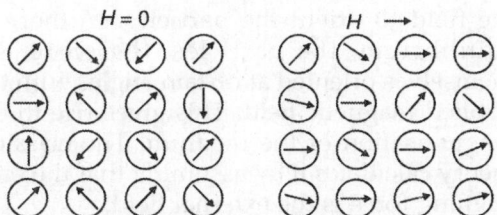

Fig. 11.6: Atomic dipole configuration with and without an external magnetic field for a paramagnetic material

For some solid materials, each atom possesses a permanent dipole moment by virtue of incomplete cancellation of electron spin and/or orbital magnetic moments. In the absence of an external magnetic field, the orientations of these atomic magnetic moments are random, such that a piece of material possesses no net macroscopic magnetization. These atomic dipoles are free to rotate, and *paramagnetism* results when they preferentially align, by rotation, with an external field as illustrated in Fig. 11.6. These magnetic dipoles are acted on individually with no mutual interaction between adjacent dipoles. In as much as the dipole align with the external field, they enhance it, giving rise to a relative permeability μ_r that is greater than 1, and to a relatively small and positive susceptibility. Susceptibilities for few paramagnetic substances are given in Table 11.3. Susceptibilities for paramagnetic substances range from about 10^{-5} to 10^{-2}. A schematic B–H curve for a paramagnetic substance is also shown in Fig. 11.4.

Table 11.3: Magnetic susceptibilities for paramagnetic materials at room temperature

Materials	*Susceptibility,* χ *(volume) (SI units)*
Aluminium	2.07×10^{-5}
Chromium	3.13×10^{-4}
Chromium chloride	1.51×10^{-3}
Manganese	3.70×10^{-3}
Molybdenum	1.19×10^{-4}
Sodium	8.48×10^{-6}
Titanium	1.81×10^{-4}
Zirconium	1.09×10^{-4}

The following types of substances are paramagnetic:

(a) All atoms and molecules which have an odd number of electrons. According to quantum mechanics, such a system cannot have a total spin equal to zero; therefore, each atom or molecule has a net magnetic moment which arises from the electron spin angular momentum. Examples are organic free radicals and gaseous nitric oxide.

(b) All free atoms and ions with unfilled inner electron shells and many of these ions when in solids or in solution. Examples are transition, rare-earth and actinide elements and many of their salts. This includes ferromagnetic and antiferromagnetic materials above their transition temperatures.

(c) Several miscellaneous compounds including molecular oxygen and organic biradicals.

(d) Metals, in this case, the paramagnetism arises from the magnetic moments associated with the spins of the conduction electrons and is called the *Pauli paramagnetism.*

Relatively few substances are paramagnetic. Aside from the Pauli paramagnetism found in metals, the most important paramagnetic effects are found in the compounds of the transition and rare-earth elements which have practically filled $3d$ and $4f$ electron shells respectively.

Most paramagnetic substances at room temperature have a static susceptibility which follows Langevin–Deby law [Eq. (11.19)]

$$\chi = \frac{Np^2\mu_B^2}{3kT + N\alpha} \tag{11.19}$$

where N is the number of magnetic dipoles per unit volume, p is the effective magneton dipoles per μ_B and μ_B is the Bohr magneton, k is Boltzmann's constant, T is absolute temperature and α is the temperature independent contribution of *Van–Vleck paramagnetism.*

Both paramagnetic and diamagnetic substances are considered to be non-magnetic because they exhibit magnetization only in the presence of an external field. Moreover, for both, the flux density B within them is almost the same as it would be in vacuum.

Salient features of paramagnetic substances are as follows:

  i. The relative permeability is slightly greater than unity and susceptibility is of the order of 10^{-6} and positive.

  ii. When a bar of a paramagnetic substance is suspended between the pole pieces of a magnet, it aligns parallel to the magnetic lines of force.

  iii. When a paramagnetic liquid in U-tube is placed between pole pieces of a strong magnet, liquid shows a size in U-tube.

  iv. The paramagnetic substance tends to move from weaker part to stronger part of a non-uniform magnetic field.

  v. A paramagnetic gas, when allowed to ascend between the pole pieces of a magnet, the gas spreads along the magnetic field.

  vi. The susceptibility of paramagnetic substances varies inversely to absolute temperature (T), i.e.

$$\chi \propto \frac{1}{T}$$

The law is known as *Curie law.*

Langevin''s Classical Theory of Paramagnetism

Classical theory of paramagnetism developed by Langevin, considers the paramagnetic solids in terms of a paramagnetic gas in which each particle is assumed to bear a permanent magnetic moment. When mass of such a gas is placed in a magnetic field, each particle tends to orient itself with its magnetic axis parallel to it. The thermal agitation to which each particle is always subjected at ordinary temperature, however resists this tendency of the field to orient the particles. In thermal equilibrium, the particles, therefore, set themselves oriented at certain angles with the applied magnetic field. This gives rise to the magnetisation of the medium. The classical theory calculates it by assuming that the only aligning force is the external field.

Let us consider a paramagnetic gas containing particles, each bearing a permanent magnetic moment μ_m. When the magnetic field in direction of B is applied, the potential energy of particles whose magnetic axis makes an angle θ with the field is $-\mu_m B\cos\theta$ (Fig. 11.7). The minimum dipole energy corresponds to $\theta = 0$. Therefore, all the dipoles tend to orient themselves in the direction of the external field, which is hampered by thermal motion.

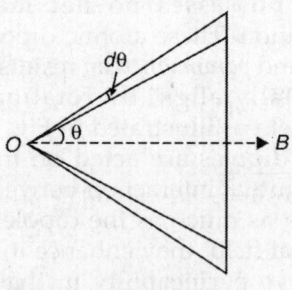

Fig. 11.7

The probability that a dipole is inclined at θ to the direction of the field B in thermal equilibrium is proportional to the Boltzmann factor

$$\exp\left(\mu_m B\cos\theta / k_B T\right)$$

This is in accordance with the classical Maxwell–Boltzmann distribution law. Next let us imagine that all the particles of the gas have been collected and placed with their centres at a point O, and their axes pointing in the same direction in which they actually pointed before they were collected (Fig. 11.7). Then the number dn of particles whose axes fall within the solid angle $d\omega$ between two hollow cones of semi angle θ and $\theta + d\theta$ is given by

$$dn = n_0 \exp(\mu_m B \cos\theta / k_B T) d\omega$$

where solid angle $d\omega = 2\pi \sin\theta\, d\theta$. Now each of the particles contributes a component of magnetic moment $\mu_m \cos\theta$ parallel to the field, while by symmetry, the component perpendicular to the field annul each other, hence the mean magnetic moment $<\mu_m>$ or $\bar{\mu}_m$ in the direction of the applied field is obtained by dividing the sum of the resolved components of the magnetic moments of all the dipoles in the direction of the field by the total number of dipoles, i.e.

$$\bar{\mu}_m = \frac{\int \mu_m \cos\theta \cdot dn}{\int dn}$$

$$= \frac{\mu_m \int_0^\pi \cos\theta \exp(\mu_m B \cos\theta / k_B T) \sin\theta\, d\theta}{\int_0^\pi \exp(\mu_m B \cos\theta / k_B T) \sin\theta\, d\theta}$$

Putting $\dfrac{\mu_m B}{k_B T} = a$

We obtain

$$\bar{\mu}_m = \frac{\mu_m \int_0^\pi \cos\theta \exp(a\cos\theta) \sin\theta\, d\theta}{\int_0^\pi \sin\theta \exp(a\cos\theta) d\theta}$$

$$= \mu_m \left(\coth a - \frac{1}{a} \right) = \mu_m L(a) \qquad (11.20)$$

where $L(a) = \left(\coth a - \dfrac{1}{a} \right)$ is called the *Langevin function*. Figure 11.8 shows the plot of Langevin function $L(a)$ with a.

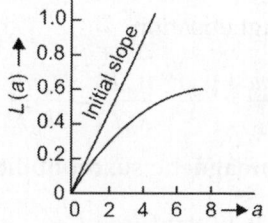

Fig. 11.8: Langevin function $L(a)$ versus a

Case 1: $\mu_B \gg k_B T$

At low temperatures of large applied field, we observe that the Langevin function tends to unity. Hence, magnetisation (magnetic moment per volume)

$$M = n\mu_m = M_s \qquad (11.21)$$

where n is the number of atoms per unit volume and M the saturation magnetisation. Although saturation effects have been observed at low temperatures in gadolinium sulphate $Gd_2(SO_4)_3 \cdot 8H_2O$, but the results do not fit the theory.

---

[1]Putting $\cos\theta = x$, the integral becomes

$$\frac{\int_{+1}^{-1} x e^{ax} dx}{\int_{+1}^{-1} e^{ax} dx} = \frac{\int_{-1}^{+1} x e^{ax} dx}{\int_{-1}^{+1} e^{ax} dx} = \frac{\left[\dfrac{x e^{ax}}{a} - \dfrac{e^{ax}}{a^2} \right]_{-1}^{+1}}{\left[\dfrac{e^{ax}}{a} \right]_{-1}^{+1}}$$

$$= \frac{\dfrac{e^a}{a} - \dfrac{e^a}{a^2} - \left(-\dfrac{e^{-a}}{a} - \dfrac{e^{-a}}{a^2} \right)}{\dfrac{e^a}{a} - \dfrac{e^{-a}}{a}} = \frac{\dfrac{1}{a}(e^a + e^{-a}) - \dfrac{1}{a^2}(e^a - a^{-a})}{\dfrac{1}{a}(e^a - e^{-a})}$$

Case 2: $\mu_B B \ll k_B T$

For $\dfrac{\mu_m B}{k_B T} \ll 1$, i.e. when T is large we can expand $\coth a$ in a power series

$$\coth a = a^{-1} + \frac{a}{3} - \frac{a^2}{45} + \ldots$$

For $a \ll 1$, we can limit ourselves with the first two terms of the expansion. Then

$$\bar{\mu}_m = \frac{\mu_m a}{3} \qquad (11.22)$$

and the magnetisation

$$M = \frac{\mu_m n a}{3} = \frac{\mu_m n}{3} \cdot \frac{\mu_m B}{k_B T} = \frac{\mu_m^2 n B}{3 k_B T} \qquad (11.23)$$

The paramagnetic susceptibility

$$\chi = \frac{M}{H} = \frac{\mu_0 \mu_m^2 n}{3 k_B T} = \frac{C}{T} \qquad (11.24)$$

where $C = \dfrac{\mu_m^2 \mu_0 n}{3 k_B}$ is called Curie constant. This is referred to as Curie law. If one substitutes $n = 10^{26}\,\text{m}^{-3}$ and $T = 100\,\text{K}$, one finds that $\chi \approx 10^{-5}$ in agreement with observation.

The susceptibility given by Eq. (11.24) is also referred to as the Langevin paramagnetic susceptibility. Note in particular that χ is inversely proportional to temperature. This is in marked contrast to the diamagnetic susceptibility, which is essentially temperature independent.

Shortcomings of the theory: At low temperatures and large applied field the theory predicts the saturation effects which have been observed at low temperatures in gadolinium sulphate but the results do not fit the theory.

Besides this the derivation of the Curie law is also open to a serious criticism which was pointed out by Min Van Leeuwan. It is not clear that if the permanent magnetic moment is to be associated with, and proportional to the angular momentum of moving electric charges, why should all the particles bear the same magnetic moment l/m. In fact, the magnetic moment of different particles should range from $-\infty$ to $+\infty$. But when the Maxwell Boltzmann statistics is applied to a system of such particles, the diamagnetic and paramagnetic contributions to the susceptibility cancel each other and susceptibility is thus zero. Recalling that paramagnetism does not exist alone, diamagnetism also exists at the same time which is usually much smaller. Hence, Langevin's theory must be treated as incomplete. A more satisfactory theory originates from quantum mechanics.

Ferromagnetism

Ferromagnetism is a property exhibited by certain metals, alloys and compounds of the transition (iron groups as bcc α ferrite and cobalt and nickel), rare earth metals such as gadolinium (Gd), and actinide elements below a certain temperature called the *Curie temperature* [Eq. (11.25)],

$$\chi = \frac{C}{T - \theta} \qquad (11.25)$$

We must note that the general behaviour of the susceptibility of ferromagnetic materials above the Curie temperature, T_C, follows the Curie–Weiss law [Eq. (11.25)]. The behaviour is followed in the region well above the ferromagnetic Curie temperature T_C. The paramagnetic Curie temperature θ is usually slightly greater than the temperature of transition T_C. Comparison of T_C and θ for three ferromagnetic metals is given in Table 11.4.

Table 11.4: Comparison of θ and T_C

Parameter	Fe	Co	Ni
θ (K)	1093	1428	650
T_C (K)	1043	1393	631

In a region just a fraction of degree above the critical point, or Curie temperature T_C, the susceptibility is found to approximate the following relation,

$$\chi = \frac{C'}{(T - T_C)^\gamma}$$

with γ generally very close to 1.33. We may note that the theory is extremely complicated and not entirely satisfactory, the atomic magnetic moments tend to line up in a common direction. All ferromagnetic materials exhibit paramagnetic behaviour above their ferromagnetic Curie point. The Curie temperature makes a transition between order and disorder of the alignment of the atomic magnetic moments. Some materials exhibit a special form of ferromagnetism below the Curie temperature called *ferrimagnetism*. The magnitude of the paramagnetic susceptibility is determined by the Curie constant C [Eq. (11.25)]. A typical value of C is 0.2 K/cm^3 for iron. Below the Curie point, the static susceptibility is not usually defined for a ferromagnetic substance, since the ferromagnet may have a finite magnetization in zero applied field. Magnetic susceptibilities as high as 10^6 are possible for ferromagnetic substances. Consequently, $H \ll M$, and from Eq. (11.26), we can write

$$B \cong \mu_0 M \qquad (11.26)$$

Some physical properties of ferromagnetic elements are summarized in Table 11.5.

Permanent magnetic moments in ferromagnetic materials result from atomic magnetic moments due to electron structure. There is also an orbital magnetic moment contribution that is small in comparison to the spin moment. Moreover, in a ferromagnetic material, coupling interactions cause net spin magnetic moments of adjacent atoms to align with one another, even in the absence of an external field (Fig. 11.9).

The characteristic property of a ferromagnet is that, below the Curie temperature, it can

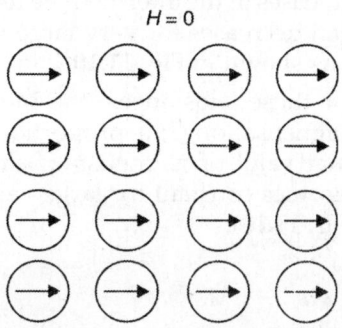

$H = 0$

Fig. 11.9: Mutual alignment of atomic dipoles for a ferromagnetic material, which exist even in the absence of external field ($H = 0$)

possess a spontaneous magnetization in the absence of an applied magnetic field. Upon application of a weak magnetic field, the magnetization increases rapidly to a high value called the saturation magnetization, which is a general function of temperature [Eq. (11.27)]

$$\frac{M}{M_{sat}} = \frac{T\alpha}{\theta} \qquad (11.27)$$

where M_{sat} is saturation value of magnetization, θ is the paramagnetic Curie temperature and $\alpha = \mu_0 \mu_B \gamma M / k_B T$, where γM is the Weiss field. Obviously, for a given temperature T, Eq. (11.27) is a plot of M/M_s versus α which represent a straight line with a slope equal to T/θ. For $T \geq \theta$, there is no spontaneous magnetization.

Some salient features of ferromagnetic substances are as follows:

i. These substances show almost all properties of paramagnetic substances.

ii. The susceptibility of these substances remain constant for smaller values of H,

Table 11.5: Some selected physical properties of ferromagnetic elements

Element	Electronic configuration	Crystal structure	Magnetization at 0 K (Amp/m)	Ferromagnetic Curie temp. T_C (K)	Melting temp. (K)
Fe	$3d^6 4s^2$	bcc	1.7×10^6	1043	1810
Co	$3d^7 4s^2$	hcp	1.0×10^6	1404	1750
Ni	$3d^8 4s^2$	fcc	0.48×10^6	632	1732
Gd	$4f^7 5d^1 6s^2$	hcp	5.66×10^6	290	586

increases in the intermediate region and then decreases for very large values of *H* as shown in Fig. 11.10.

iii. For these substances, the intensity of magnetisation (*I*) is proportional to the small value of magnetising field (*H*) and becomes constant for large values of *H* (Fig. 11.10).

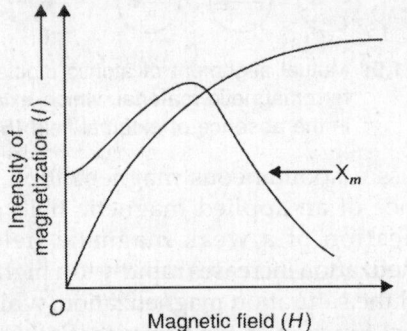

Fig. 11.10: Intensity of magnetization versus *H*

The task of a successful theory of ferromagnetism is to account for the spontaneous magnetism below the Curie point, the temperature dependence of the saturation magnetization, and the nature of the magnetization process, or magnetization curve. We now briefly present a discussion on the theories of ferromagnetism.

Weiss Theory

The Weiss molecular field theory of ferromagnetism represents the first realistic attempt to account for the properties of a ferromagnet. This theory rests on two hypotheses:

1. Below the Curie point, a ferromagnetic substance is composed of small, spontaneously magnetized regions called *domains*. The total magnetic moment of the material is the vector sum of the magnetic moments of the individual *domains*. It is now known that these assumed domains really exist and are usually 0.1 cm and 0.01 cm across.

2. Each domain is spontaneously magnetized because a strong molecular (magnetic) field tends to align the individual atomic magnetic moments within the domain.

The consequence of these assumptions is that, while each domain is spontaneously magnetized, the directions of magnetization of the domains do not coincide; therefore the overall magnetization of the sample may be much smaller than if it were composed of a single domain. Application of a relatively weak field (~100–1000 oersteds and often very much less) is sufficient to align the directions of the magnetization of the domains thereby achieving a large magnetization.

The second hypothesis of the Weiss theory leads to the existence of a Curie temperature below which a domain may be spontaneously magnetized in the absence of an applied magnetized field. According to the theory, the effective field acting on any atomic magnetic moment within the domain may be written as,

$$H = H_0 + \gamma M \tag{11.28}$$

where H_0 is an externally applied magnetic field and γM (M = magnetization) is the Weiss molecular field whose order of magnitude in Fe is 10^7 oersteds.

There is an overall general agreement between theory an experimental results of temperature dependence of spontaneous magnetization for Fe, Ni and Co. The low temperature behaviour, however, is better described by *magnon theory*. The theory satisfactorily describes the temperature dependence of susceptibility in the paramagnetic region provided the temperature is well above the Curie temperature. The Curie temperature determined from the theory of spontaneous magnetization differs by a few degrees from the experimentally determined value for the paramagnetic region.

Heisenberg Exchange Interaction Theory

The Heisenberg theory of ferromagnetism treats the origin of the Weiss molecular field on an atomic basis. It may be remarked that the ordinary dipole–dipole interactions among atomic magnetic moments are too much small to account for the Weiss field. The foundation of the Heisenberg theory is the Pauli exclusion principle. Due to Heisenberg exchange interaction, two neighbouring spins in a solid are coupled together with an exchange energy given by

$$E_{exch} = -2J\, S_i \cdot S_j \qquad (11.29)$$

where S_i and S_j are the spin angular momentum vectors of the two electrons i and j, and J is the so called exchange integral between the two electrons. The exchange integral decreases rapidly with distance between the electrons and depends in a complicated way upon the spatial distribution (wave function) of the electrons. It is extremely difficult to compute. The strength of J can be estimated from the Curie temperature T_C. If there are Z nearest neighbours to a central ith spin, the exchange energy for this spin is

$$E_{exch} = -2\sum_{j=1} J_{ij} S_i \cdot S_j \approx -2ZJS^2 \qquad (11.30)$$

where we have assumed that all the J_{ii} are equal to J and the spins are parallel. This exchange energy must be equal to kT_C since at T_C the magnetic order is destroyed. Thus,

$$J \approx \frac{k_B T_C}{2ZS^2} \qquad (11.31)$$

If the exchange integral is positive, the parallel arrangement is favoured, and if J is large enough, ferromagnetism should result. If J is large and negative, antiferromagnetism or ferrimagnetism supposedly arises. The order of magnitude of J given by $J \sim kT_C \sim 10^{-20}$ J.

There seems to be no question that the Heisenberg theory correctly accounts for the tendency of the electrons in the same ferromagnetic atom to exhibit parallel spins. However, whether or not, it leads to the correct explanation of the interatomic alignment of spins is still a subject of much controversy. In insulators it usually proves possible to express the coupling between atomic spins by Eq. (11.29), provided J is interpreted as an effective exchange integral. In metals the problem is much more complicated. There is little doubt, however that the basic Heisenberg idea is correct and

that the molecular field arises from the interplay between electrical forces and the effects of Pauli exclusion principle.

Crystalline Anistropy Energy

This accounts for the experimental fact that ferromagnets tend to magnetize along certain crystallographic axis, called directions of easy magnetization. For example, a single crystal of Fe, which is made up of a cubic array of iron atoms, tends to magnetize in the direction of the cube edges. It requires about 1.4×10^4 J/m^3 (at room temperature) to move the magnetization into a hard direction along a cubic body diagonal.

The Heisenberg exchange energy [Eq. (11.29)] is isotropic and cannot account for the observed anisotropy, which probably has its origin in a complicated interplay of spin-orbit coupling, crystalline electric fields, and overlap of orbital wave functions. Anisotropy energy depends on the strain of the crystal, giving rise to *magnetostriction*, i.e. changes in length of a substance when it is magnetized.

Ferromagnetic Domains

Small regions of spontaneous magnetization, formed at temperatures below the Curie point, are known as domains. Domains originate in order to lower the magnetic energy as illustrated in Fig. 11.11. In Fig. 11.11b, it is shown that two domains will reduce the extent of the external magnetic field, since the magnetic lines of force are shortened. On further subdivision (Fig. 11.11c), this field is still further reduced.

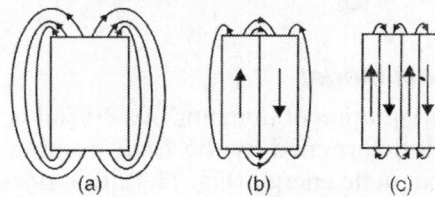

(a)   (b)   (c)

Fig. 11.11: Lowering of magnetic field energy by domains (a) Lines of force for a single domain (b) Shortening of lines of force by division into two domains (c) Reduction of field energy by further subdivision

Another way to describe the energy reduction is to note that the interior demagnetizing fields, coming from surface poles, are much smaller in the long, thin domain of Fig. 11.11c than in the fat domain of Fig. 11.11a.

The question arises as to how long this subdivision process continues. With each subdivision there is a decrease in field energy, but there is also an increase in Heisenberg exchange energy, since more and more magnetic moments are aligning antiparallel. Finally a state is reached in which further subdivision would cause a greater increase in exchange energy than it would cause decrease in field energy, and the ferromagnet would assume this state of minimum total energy.

Bloch Wall

Also because of exchange energy, the reversal of magnetization between domains does not occur abruptly but takes place gradually through a transition zone called the Bloch wall (Fig. 11.12). In iron, the wall is ~50 nanometers thick and has the total energy ~2×10^{-3} J/m^2.

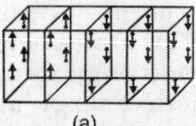 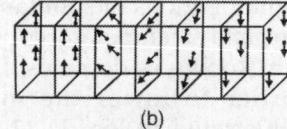

(a)   (b)

Fig. 11.12: Lowering of exchange energy by the transition zone known as Bloch wall. The reversal of magnetization between domains does not take place abruptly as is shown but by degrees as is illustrated

Domain Model

The orientation of domains in a crystal is primarily determined by the need to minimize the magnetic energy (Fig. 11.9). It is possible to eliminate all surface magnetic poles by forming flux closure domains (Fig. 11.13). Here the normal component of magnetization is continuous across all domain boundaries. The demagnetizating fields are zero every-

where, except for a trivial effect of surface poles in the Bloch walls. In a uniaxial crystal, that is, a crystal with a single easy direction, an arrangement as shown in Fig. 11.13b will be preferred since it has a lower density of magnetization normal to easy direction, or in the hard direction. Even in cubic crystals, in which all directions of magnetization in Fig. 11.13 may be easy, Fig. 11.13b will be preferred because of *magnetostriction*. In iron, for example, each domain increases in length along the direction of magnetization by a fraction ~2×10^{-5}. Thus, the domains of Fig. 11.13 can be fitted smoothly together only by straining whom elastically against this magnetostriction, and the required strain energy will be smaller in Fig. 11.13b than in Fig. 11.13a.

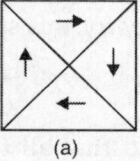 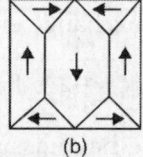

(a)   (b)

Fig. 11.13: Flux closure domains (single crystal). (a) Large domains at right angles. (b) Reduction in their size, causing reduction of anisotropy energy of uniaxial crystals or of strain energy of cubic crystals.

If an external magnetic field is applied to a ferromagnetic specimen, following two things can happen according to the domain model:

1. The domains parallel or nearly parallel to *H* can grow in size at the expense of the antiparallel domains shown in Fig. 11.14
2. The magnetic moment of the domains can rotate into the field direction.

   In either case, the sample acquires a magnetization, which increase as the

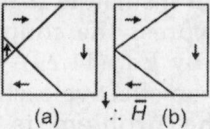

(a)   ∴ $\bar{H}$ (b)

Fig. 11.14: Growth of domains parallel to an applied magnetic field *H*

field increase, until all the domains are parallel to the allied field, at which point the material is said to be saturated and the magnetization is equal to M_S.

In polycrystals, the domain structure is more complicated (Fig. 11.15), depending upon such variables as grain orientation and grain boundaries. It is possible, however, for domains to cross grain boundaries.

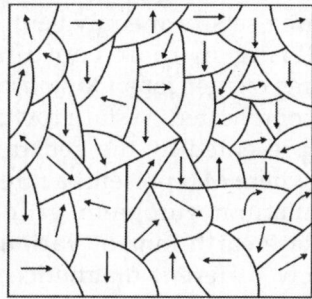

Fig. 11.15: Orientation of domains in a polycrystal

On minimizing the total contribution from (i) magnetic, or demagnetizing energy (ii) anisotropy (iii) magnetostriction (iv) elastic strain, (v) Bloch wall energy, it is found that, depending upon the composition and shape of the crystal, the theoretical domains thickness should vary from about 0.1 cm to 0.001 cm.

Direct experimental evidence of the existence of domains is furnished by magnetic powder patterns. It is also possible to observe domains by means of transmission or reflection of electrons in an electron microscope. The Kerr magneto-optic effect (rotation of plane of polarization of light reflected from a magnetic surface) has also been used to study domains. These two methods are particularly useful if the Bloch walls are extremely thick (because of small magneto crystalline anisotropy) and the resultant magnetic powder patterns very blurred.

Antiferromagnetism

This originates when the spin moments of the neighbouring atoms are ordered in an antiparallel arrangement (Fig. 11.16) or when the exchange integral is negative. This property is exhibited by some metals, alloys and salts of transition elements in which the atomic magnetic moments, at sufficiently low temperatures, form an ordered array which alternates or spirals so as to give no net total moment in zero applied field. The most direct way of detecting such arrangements is by means of neutron diffraction. A crystal exhibiting antiferromagnetism may be considered to be consisting of two independent sublattices A and B, one of which is spontaneously magnetized in the opposite direction. This type of magnetism was first observed in the crystals of MnO. Manganese oxide (MnO) is a ceramic material that is ionic in character having both Mn^{2+} and O^{2-} ions. No net magnetic moment is associated with the O^{2-} ions, since there is a total cancellation of both spin and orbital moments. However, the Mn^{2+} ions possess a net magnetic moment that is predominantly of spin origin. These Mn^{2+} ions are arrayed in the crystal structure of MnO crystal such that the moments of adjacent Mn^{2+} ions are antiparallel (Fig. 11.16). This is why that the opposing magnetic moments cancel one another, and as a consequence, the solid as a whole possess no net magnetic moment.

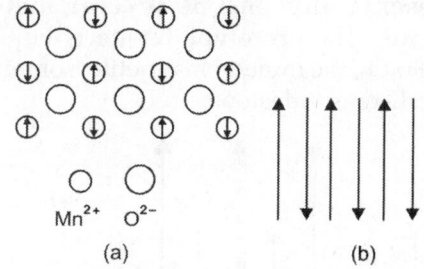

Fig. 11.16: The antiferromagnetism in manganese oxide

When a magnetic field is applied, a small magnetization appears in the direction of the field which increases further with temperature. Such a behaviour is typical of an antiferromagnetic material. The magnetization becomes the maximum at a critical temperature T_N, called the *Neel temperature*, which is analogous to the Curie temperature in the paramagnetic or ferromagnetic substances. This is the transition temperature below which the spontaneous magnetic ordering takes place. Above this temperature, the magnetization decreases continuously which is indicative of the paramagnetic state of the material. The variation of susceptibility with temperature is governed by

$$\chi = C/(T + \theta) \tag{11.32}$$

where $\theta = T_N$ is called the Neel temperature.

Ferrimagnetism

In some compounds, e.g. in some ceramics, etc. the constituent atoms may be antiferromagnetically coupled but with different magnetic moments. This would give rise to a net magnetic moment in each coupling and the some of the moments of all the coupling could result in magnetization which is comparable in order of magnitude to ferromagnets. This phenomenon is termed as ferrimagnetism. The distinction between the two lies in the source of the net magnetic moments (Fig. 11.17). This type of magnetism occurs in materials such as ferrites which are basically the oxides of various metal elements. These ionic materials may be represented by the chemical formula $M\,Fe_2O_4$, in which M represents any one of several metallic elements. The prototype ferrite is Fe_3O_4 or FeO. Fe_2O_3, the mineral magnetite, sometimes also called as lodestone.

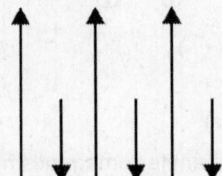

Fig. 11.17: Magnetic moment alignment in a ferrimagnet

Other than cubic ferrites, ceramic materials are also ferrimagnetic. These include the hexagonal ferrites and garnets. Hexagonal ferrites have a crystal structure similar to the inverse spinel, with hexagonal symmetry rather than cubic. We may write the chemical formula for these materials as $AB_{12}O_{19}$, in which A is divalent metal such as Ba, Pb or Sr, and B is a trivalent metal such as Al, Fe, chromium and gallium. $PbFe_{12}O_{19}$ and $BaFe_{12}O_{19}$ are two most common examples of hexagonal ferrites.

The most versatile of ferrimagnetic systems are the rare earth iron garnets. The garnet unit cell has three sets of inequivalent magnetic sites, differing in their coordination to neighbouring oxygen ions. Obviously, garnets have very complicated crystal structure, which may be represented by the general formula $M_3Fe_5O_{12}$, where M represents a rare earth ion such as samarium, europium, gadolinium or yttrium. Rare earth ions of various species have widely different magnetic moments, exchange interactions, and crystalline anisotropies. By a suitable choice of rare earth ions, it is possible to design ferrimagnetic systems with prescribed magnetisation and temperature behaviour. Yttrium iron garnet ($Y_3Fe_5O_{12}$), sometimes denoted by YIG, is the most common garnet material.

Ferromagnetic materials which find wide applications in electrical engineering have a disadvantage, that they have low electrical resistivity. The laminations used for electrical machines have a resistivity $\sim 14 \times 10^{-4}$ Ωm whereas the highest value obtainable in ferromagnetic alloys is less than 10^{-4} Ωm. Obviously, this advantage of the ferromagentic materials limit their application in high frequency alternating current applications, high eddy current losses and poor magnetic utilization of metals occur in sheets even at low frequencies. The saturation magnetization for ferrimagnetic materials are not as high as for ferromagnets. Moreover, ferrites being ceramic materials, are good electrical insulators. For some magnetic applications, such as high-frequency transformers, a low electrical conductivity is most desirable, because this prevents eddy currents in cores of coils.

11.5 MAGNETOSTRICTION

The change of length of a ferromagnetic material when it is magnetized is called magnetostriction. More generally, it is the phenomenon that the state of strain of a ferromagnetic sample depends on the direction and extent of magnetization. The phenomenon has an important application in magnetostriction transducers.

Magnetostriction results from the dependence of the crystalline anisotropy energy upon the state of strain of the crystalline lattice. If the crystal deforms (for example suffers a change in length), the anisotropy energy may be lowered more than the elastic energy is raised. Thus, a strained state will be favoured.

The total energy of a ferromagnetic substance depends upon the state of strain and the direction of magnetization through three contributions. The first two consist of the crystalline anisotropy energy of the strained lattice plus a correction which takes into account the dependence of the anisotropy energy on the state of the strain. The third contribution is that of the elastic energy, which is independent of magnetization direction and is minimum in the unstrained state. The state of strain of the crystal will be that which makes the sum of the three contributions to the energy a minimum. The result is that, when magnetized, the lattice is always distorted from the unstrained state, unless there is no anisotropy.

Since spontaneous magnetization occurs below the Curie temperature (T_C), there will always be a spontaneous lattice distortion which depends on magnetization direction in the ferromagnetic state. In Ni, the lattice spacing parallel to the magnetization is always smaller than the lattice spacing perpendicular to the magnetization.

Magnetostriction is an important property. This effect is exploited in *transducers* used for the reception and transmission of high frequency sound vibrations. Ni is often used for this application.

Magnetostriction or mechanical deformation of magnetic domains has also other important effects. When a ferromagnetic material is strained, we find that the domains tend to realign themselves into positions of lower energy. Because of this the permeability of the material is changed and it becomes easier or more difficult to magnetize. Realignment puts the domains in a better position to be polarized and permeability is increased in materials having positive magnetostriction, the opposite is true for materials with negative mechanical deformation.

We may note that due to thermal agitation magnetostriction is decreased, because at higher temperatures, the domains are hindered in their reorientation.

11.6 THE INFLUENCE OF TEMPERATURE ON MAGNETIC BEHAVIOUR OF MAGNETIC MATERIALS

Magnetic characteristics of materials can also be influenced by temperature. We have read that raising the temperature of a solid results in an increase in the magnitude of the thermal vibrations of atoms. Since the atomic magnetic moments are free to rotate and hence with rise in temperature, the increased thermal motion of the atoms tend to randomize the directions of any moments that may be aligned.

For ferromagnetic, antiferromagnetic, and ferrimagnetic materials, the atomic thermal motions counteract the coupling forces between the adjacent atomic dipole moments, thereby causing some dipole misalignment, regardless of whether an external magnetic field is present. This results in a decrease in the saturation magnetization for both ferro- and ferrimagnets. The saturation magnetization is maximum at 0 K, at which temperature the thermal vibrations are minimum. With rise in temperature, the saturation magnetization (M_s) diminishes gradually and then abruptly drops to zero at what is called the *Curie temperature* (T_C). Figure 11.18 shows the magnetization–temperature curves for Fe and Fe_3O_4. The mutual spin coupling forces are

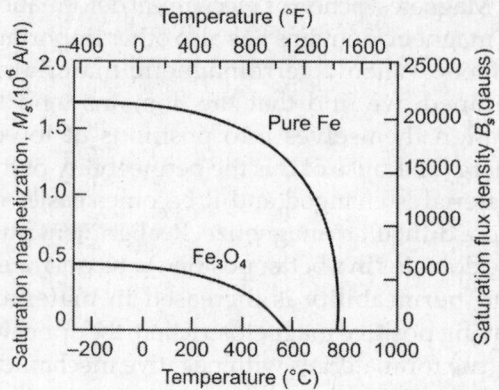

Fig. 11.18: Saturation magnetization (M_s) versus temperature curves for Fe and Fe_3O_4

completely destroyed at T_C such that above T_C both ferro- and ferrimagnetic substances are paramagnetic. The magnitude of Curie temperature varies from material to material, e.g. for Fe, Co, Ni and Fe_3O_4, the respective values of T_C are 768°C, 1120°C, 335°C and 585°C.

We may note that antiferromagnetism is also affected by temperature. At Neel temperature (T_N), this behaviour vanishes and above T_N, antiferromagnetic materials also become paramagnetic.

11.7 DOMAINS AND HYSTERESIS

At a temperature below T_C, any ferro- or ferrimagnetic material is composed of small-volume regions in which there is a mutual alignment in the same direction of all magnetic moments (Fig. 11.19). Such a region is called domain, and each domain is magnetized to its saturation magnetization. Adjacent domains are separated by *domain boundaries or walls*, across which the direction of magnetization gradually changes as shown in Fig. 11.20. As mentioned earlier, domains are normally microscopic in size, and for a poly-crystalline specimen, each grain may consist of more than a single domain. Obviously, in a macroscopic piece of specimen, there will be large number of domains, and all may have different magnetization orientations. Thus, for the entire solid, the magnitude of the M field is the vector sum of the magnetizations of all the

domains, each domain contribution being weighted by its volume fraction. We may note that for an unmagnetized specimen, the appropriately weighted vector sum of the magnetization of all the domains is zero.

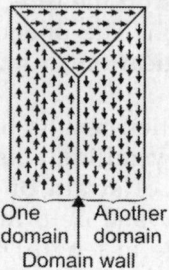

Fig. 11.19: Domains in a ferro- or ferrimagnetic material. The arrows represent atomic magnetic dipoles. Within each domain, all dipoles are aligned, whereas the direction of alignment varies from one domain to another domain

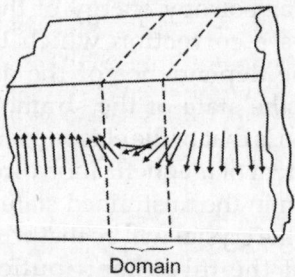

Fig. 11.20: The gradual change in magnetic dipole orientation across a domain wall or boundaries

When the ferromagnetic sample that is initially demagnetized is subjected to a continuously increasing magnetizing force H, the relation between H and flux density B is shown by the normal magnetization curve Oab (Fig. 11.21). The point a indicates that the magnetic condition as the increasing magnetic intensity has reached H_1. If H is increased to a maximum value H_2 and then decreased again to H_1, the decreasing flux density does not follow the path of increase, but decrease at a rate less than that at which it rose. The lag in the change of B behind the change of H is called *hysteresis*. If the value of H is further reduced from H_1 to zero, B is not reduced to zero but to a value B_r. The specimen has retained a permanent magnetism. This ordinate B_r is called the *retentivity* or *remanence*. The value of B may be reduced to zero at e by reversing the direction of H and increasing its value of H_C. This value of H_C is called the *coercive force* or *coercivity*.

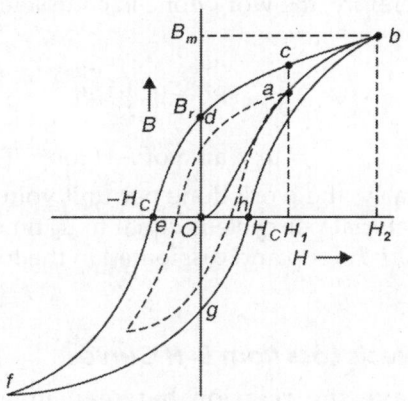

Fig.11.21: Magnetic flux density (B) versus the magnetic field strength (H) for a ferromagnetic material that is subjected to forward and reverse saturations (points b and f). Hysteresis loop *bdefghb*. The remanence B_r and the coercive force H_C are also shown

As H is increased in the negative direction, the magnetization proceeds along the curve of Fig. 11.21 until at f the values of B and H are the same as those at b, but opposite in direction. When reverse changes in H are made, the magnetization changes along the curve *fghb*. This entire loop *bdefghb* is called a *hysteresis loop*. If the hysteresis loop starts from another point on the normal magnetization curve, such as a, there will be a smaller hysteresis loop entirely within the larger, such as the dotted loop of the Fig. 11.21. These curves are sometimes called B–H curves and are used to describe magnetic materials. We may note that it is possible to reverse the direction of the magnetic field at any point along the curve and generate other hysteresis loops. To demagnetize the ferromagnetic or ferrimagnetic samples one has to repeatedly

cycle it in H field that alternates direction and decreases in magnitude. Magnetization curves are generally obtained from long, thin samples to avoid demagnetization. The demagnetization effects are extremely important in ferromagnetic resonance.

We may note that ferromagnetic materials in common use are polycrystalline, i.e. a piece of the material consists of tremendous number of single crystals of random orientation. In each single crystal, there are many domains. The magnetization of the whole body consists of magnetization of the various single crystals within the body.

In magnetizing the core, energy must be supplied. As H and B increase along *fghb* (Fig. 11.21), the energy gain is proportional to the area under that portion of the curve. Along the path *bcdef*, there is a loss in energy proportional to the area under *bcdef*. The net loss in energy per cycle per unit volume is given by Eq. (11.46) where

$$W = \oint dW = \oint HdB \qquad (11.33)$$

the integral is taken around the closed loop. But $\oint HdB$ is the area of the hysteresis loop, and the energy loss per unit volume per cycle is equal to the area of the hysteresis loop. This energy is converted into heat. If B is expressed in tesla and H in amp/m, the energy loss is in J/m^3 (cycle).

Steinmetz found an empirical relation between the energy loss per unit volume per cycle W and the maximum value B_m of the flux density during the cycle given by

$$W = 34/\eta(B_m)^n \qquad (11.34)$$

where η is called the *Steinmetz coefficient*. Steinmetz found a value of about 1.6 for the exponent n for many materials, but it varies from about 1.5 to 2.5 for others.

In alternating current machinery, masses of iron are in fields that are constantly reversing. Therefore, the iron is constantly being carried around hysteresis paths, and there is an energy loss per cycle that depends upon the hysteresis loop for particular iron that is used.

This hysteresis loss results in undesirable heating of the iron as well as waste of energy. Obviously, the hysteresis curves are quite important in determining the quality of magnetic material and selecting the material for a particular application. In accordance with their H_C, and W values, ferromagnetic materials are classified as *soft* and *hard*.

Energy Loss Due to Hysteresis

It is observed that a loss of energy is always involved in the process of magnetization of a ferromagnetic material through a hysteresis cycle. For aligning the molecular magnets in the direction of applied field, the work is done by the magnetizing field against the mutual attractive force acting between them. However, this energy spent in magnetizing the specimen is not completely recoverable on reversing the magnetizing field. Some of the molecular magnets remain aligned even when $H = 0$ due to the group forces. This is why a coercive force in the reverse direction is applied to demagnetize the ferromagnetic sample completely. Clearly, there is a loss of energy in taking a sample of ferromagnetic substance through a cycle of magnetization.

Hysteresis Loss from I–H Cycle

Consider a unit volume of the specimen with N elementary magnets. Let M be the magnetic moment of each elementary magnet. Further, let us suppose that the axis of molecular magnet be inclined at an angle θ with the magnetising field H.

The component of total magnetic moment per unit volume parallel to magnetic field is $\Sigma M \cos \theta$, and the component perpendicular to the magnetic field is $\Sigma M \sin \theta$. But from the definition of I, we have $\Sigma M \cos \theta = I$ and $\Sigma M \sin \theta = 0$, otherwise there will be a component of magnetic moment perpendicular to H, which carries no meaning.

Let us consider that I be increased by an amount dI, then the magnetic moment increases by

$$d\Sigma M \cos \theta = dI$$

or $\quad -\Sigma M \sin \theta \, d\theta = dI \quad\quad (11.35)$

Negative sign in Eq. (11.35) indicates that I decrease with increase in θ.

Now, the work done on all molecular magnets in unit volume in rotating them through an angle $-d\theta$ is

$$dW = \mu_0 \Sigma MH \sin \theta \, d\theta$$

$$= \mu_0 H dI \quad\quad (11.36)$$

Therefore, the work done in complete cycle is

$$W = \oint \mu_0 H dI = \mu_0 \oint H dI$$

$$= \mu_0 \times \text{area of } I\text{–}H \text{ loop} \quad (11.37)$$

Clearly, the work done per unit volume of the material per cycle is equal to μ_0 times the area of I–H loop and dissipated in the form of heat.

Hysteresis Loss from B–H Curve

We have the relation between magnetic induction B and H and I as

$$B = \mu_0 (H + I)$$

or $\quad\quad dB = \mu_0 (dH + dI)$

or $\quad\quad dI = \dfrac{dB}{\mu_0} - dH \quad\quad (11.38)$

Substituting the value of dI from Eq. (11.38) in Eq. (11.37), one obtains

$$W = \mu_0 \oint H \left(\frac{dB}{\mu_0} - dH \right)$$

$$= \oint H dB - \mu_0 \oint H \cdot dH \quad (11.39)$$

Since $\oint H \cdot dH = 0$, because the curve between H and H is a straight line and obviously, this will not enclose any area. Now, Eq. (11.39) gives

$$W = \oint H \cdot dB$$

$$= \text{area of } B\text{–}H \text{ loop} \quad\quad (11.40)$$

Clearly, the work done per unit volume for the material per cycle is equal to the area of B–H loop and dissipated in the form of heat.

Alternative Method

Consider a ring of ferromagnetic material having a coil of N turns wound uniformly over it. Let the circumferential length of the coil is l and its cross-sectional area is A. The permeability of the medium is μ. When a current i ampere is allowed to pass through the winding of the coil, it produces a flux of magnetic flux density B. Now, the total magnetic flux linked with the coil is

$$\phi = NBA \qquad (11.41)$$

The intensity of magnetization H in the material of the ring due to current flowing through its N turns is given by

$$H = \frac{Ni}{l} \text{ amp turns/m} \qquad (11.42)$$

With the increase in current i, the flux linked with the coil changes, and thus emf e is induced in the coil. The emf so induced is as follows:

$$e = -\frac{d\phi}{dt} = -\frac{d}{dt}(NBA) = -NA\frac{dB}{dt} \qquad (11.43)$$

The negative sign in Eq. (11.43) indicates that the induced emf in the ring opposes the increasing current. Now, the work done by the current i against the induced emf e in time dt is obtained as

$$dW' = eidt = \left(NA\frac{dB}{dt}\right)\left(\frac{Hl}{N}\right)dt$$

$$= HAl \, dB \qquad (11.44)$$

Therefore, the work done per unit volume of the ring of the material

$$dW = \frac{HAl}{Al}dB = HdB \qquad (11.45)$$

Now, by the B–H curve (Fig. 11.22), we have

$PQ = H$ and $PR = dB$

Area of the shaded portion $PQSR = HdB$

Since the amount of work done given by Eq. (11.45) is the loss of energy per unit volume in taking the material through the complete cycle of magnetization, we have the loss of energy per unit volume per cycle as

$$\oint H \cdot dB = \text{area of } B-H \text{ curve} \qquad (11.46)$$

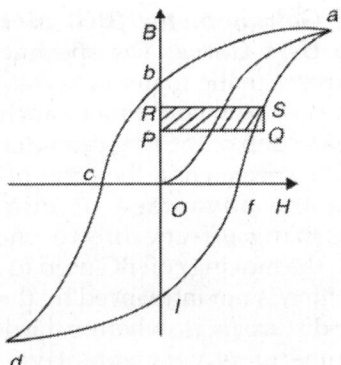

Fig. 11.22: *B–H* curve

Hysteresis Loss from I–H Curve

We have the magnetic flux density B as

$$B = \mu_0(I + H) \qquad (11.47)$$

Differentiating Eq. (11.47), one obtains

$$dB = \mu_0(dI + dH) \qquad (11.48)$$

Substituting the value of dB from Eq. (11.48) in Eq. (11.46), one obtains

$$\oint H \cdot dB = \mu_0 \oint HdI + \mu_0 \oint HdH \qquad (11.49)$$

Now, the term $\oint HdH = 0$ as the curve between H and H is a straight line. Hence, hysteresis loss is obtained as

$$\oint HdH = \mu_0 \oint HdI$$

$$= \mu_0 \times \text{area of } I-H \text{ curve} \qquad (11.50)$$

Methods of Plotting B–H or Hysteresis curve

For plotting the B–H curve, there are following three simple methods:

1. Magnetometer method
2. Ballistic galvanometer method
3. Bar and Yoke method

When the sample is in the form of a long rod or wire, the magnetometer method is used. The ballistic method is preferred for the sample in the form of a ring, whereas the Bar and Yoke method is used when the sample is a heavy rod. We will confine to the most simple and conventional method, i.e. Bar and Yoke method.

Ballistic Galvanometer (BG) Method for Drawing B–H Curve:

The specimen of the given ferromagnetic material is taken in the form of a ring (Rowland ring or anchor ring). The cross-section of the ring can be circular or square. The specimen in the form of a closed ring has the advantage of eliminating demagnetizing effects due to end poles. Secondly, the moving coil BG used to measure the deflection is not influenced by the outside magnetic disturbances, whereas the deflection magnetometer is very sensitive to such changes.

The experimental arrangement is shown in Fig. 11.23. A primary magnetizing coil P of N_1 turns is wound closely over the specimen ring. It is connected through a reversing key K and a two way key K_1 to a circuit which contains a battery, an ammeter, a rheostat R_1 and a resistance R'. One can remove the resistance K' from the circuit by pressing the key K'. A secondary coil S of N_2 turns is wound over a small part of the ring and is connected in wires through a key K_2 with a BG, variable resistance R and another secondary coil S' of total n_2 turns. Secondary coil S' is wound over a secondary solenoid P' of n_1 turns per meter. The two-way key K_1' connects either P or P' to the battery circuit. Across BG a tapping key is connected to arrest the motion of its coil whenever desired.

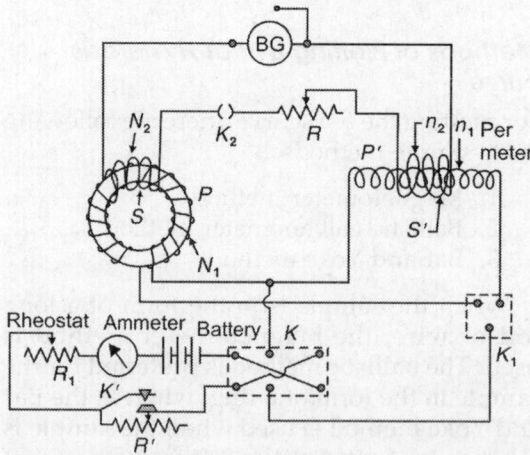

Fig. 11.23: Ballistic galvanometer method for drawing B–H curve

Theory: When the key K_1 is closed to the left, a current of i ampere passes through the primary coil of Rowland ring P to magnetize it. If r meter be the mean circumference of the ring having N_1 turns, the magnetizing field, H is given by

$$H = \frac{N_1 i}{2\pi r} \frac{\text{amp}}{\text{m}} \qquad (11.51)$$

The magnetizing field H is build-up from zero to H in certain time and obviously the magnetic flux density B also builds up from zero to B ($\because B = \mu_0 H$). Now, the flux linked with N_2 turns of secondary coil is equal to $N_2 BA$. We may note that the flux changes from zero to its maximum value during the time when B grows from zero to B. Now, the induced emf in the secondary coil is given by

$$e = -\frac{d\phi}{dt} = -\frac{d}{dt}(N_2 BA) = -N_2 A \frac{dB}{dt} \qquad (11.52)$$

Let R be the resistance of the galvanometer circuit. The current i through the galvanometer during the time interval dt is given by

$$i = \frac{|e|}{R} = \frac{N_2 A}{R} \frac{dB}{dt}$$

Since

$$i = \frac{dq}{dt} = \frac{N_2 A}{R} \frac{dB}{dt}$$

or

$$dq = \frac{N_2 A}{R} dB \qquad (11.53)$$

The total charge through the BG is then obtained as

$$q = \int dq = \int \frac{N_2 A}{R} dB = \frac{N_2 A}{R} B \qquad (11.54)$$

If θ be the first ballistic throw of the galvanometer, then the total charge through the BG is obtained as

$$q = K\theta \left(1 + \frac{\lambda}{2}\right) \qquad (11.55)$$

where K is the ballistic constant of the galvanometer and λ the logarithmic decrement. Now, equating Eqs. (11.54) and (11.55), one obtains

$$\frac{N_2 BA'}{R} = K\theta \left(1 + \frac{\lambda}{2}\right) \qquad (11.56)$$

One can determine the ballistic constant K by using a standard solenoid S. Let I be the current passed through the primary coil of the solenoid having n_1 turns per unit length, the magnetic flux density inside the solenoid is $\mu_0 n_1 I$. Thus, the change in flux through the secondary having n_2 as total number of turns is obtained as

$$n_2 a n_1 \mu_0 I \qquad (11.57)$$

where a is the area of cross-section of the primary of standard solenoid. This change in

the magnetic flux sends a charge $q' = \dfrac{n_2 a n_1 \mu_0 I}{R}$ through the BG. If θ' be the first ballistic throw in this case due to the flow of charge q', then we have

$$q' = \frac{\mu_0 n_2 n_1 a I}{R} = K\theta'\left(1 + \frac{\lambda}{2}\right) \qquad (11.58)$$

Dividing Eq. (11.56) by Eq. (11.58), one obtains

$$\frac{N_2 B A}{\mu_0 n_1 n_2 a I} = \frac{\theta}{\theta'}$$

or

$$B = \frac{\mu_0 n_1 n_2 a I}{N_2 A}\left(\frac{\theta}{\theta'}\right) \qquad (11.59)$$

Equation (11.59) gives the value of B induced in the specimen corresponding to the magnetic intensity H given by Eq. (11.51). Knowing B and H, one can draw B–H curve or hysteresis loop.

Uses of Hysteresis Curves

The knowledge of hysteresis curves of different ferromagnetic materials provide information about *retentivity, coercivity, susceptibility, permeability* and energy loss per cycle. On seeing the B–H loops of various ferromagnetic materials, one can select the material which gives minimum hysteresis loss when put to a cycle of magnetization. This type of ferromagnetic material will be most suitable for constructing the cores of the transformers and of armatures of dynamo and motors and telephone diaphragms. The idea of coercivity and retentivity of ferromagnetic materials helps one to select the proper material for the particular purpose.

For permanent magnets, the material should have the following characteristics:

i. High retentivity so that the magnet is strong

ii. High coercivity so that the magnetization is not wiped out by stray external fields, mechanical ill-treatment and temperature changes.

From these considerations *steel* is *more suitable for permanent magnets than soft iron.*

For *electromagnets,* the ferromagnetic material should have the following characteristics: (i) maximum flux density (ii) high initial permeability (iii) low hysteresis loss (iv) low coercivity (v) high retentivity. *Soft iron* is found to be an ideal material for electromagnets.

For *transformer cores, armature of dynamos and motors and chokes,* the ferromagnetic material should have the following characteristics. In these cases, the material goes through complete cycle of magnetization continuously.

i. The material must, therefore, have a low hysteresis loss to have less dissipation of energy and hence a small heating of the material, otherwise the insulation of the windings will break.

ii. A high permeability to obtain a large flux density at low field.

iii. High specific resistance to reduce eddy-current loses.

11.8 SOFT MAGNETIC MATERIALS

Soft magnetic materials are characterized by their low loss and high permeability. There are a variety of alloys used with various combinations of magnetic properties, mechanical properties, and cost. There are seven major groups of commercially important materials: iron and low carbon steels, iron-silicon alloys, iron-aluminimum-silicon alloys, nickel-iron alloys, iron-cobalt alloys, ferrites and amorphous alloys.

The behaviour of soft materials is controlled by the pinning of domain walls at heterogeneities such as grain boundaries and inclusions. In addition, eddy-current loss is minimized through alloying additions which increase the electrical resistivity. Initial permeability, important in electronic transformers and inductors, is improved by minimizing all sources of magnetic anisotropy. A high maximum permeability, necessary for motors and power transformers, is increased by the alignment of the anisotropy, for example through development of crystal texture of magnetically induced anisotropy. For this reason, the relative area within the hysteresis loop must be small; it is characteristically thin and narrow (Fig. 11.24).

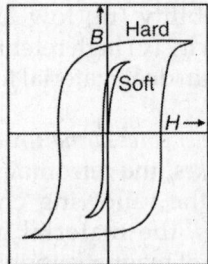

Fig. 11.24: Magnetization curves for soft and hard magnetic materials

The saturation field or magnetization is determined only by the composition of the material. For example in cubic ferrites, substitution of a divalent metal ion such as Ni^{2+} for Fe^{2+} in $FeO–Fe_2O_3$ will change the saturation magnetization. However, susceptibility and coercivity (H_C), which also influence the shape of the hysteresis curve, are sensitive to structural variables rather than to composition. We may note that a low value of coercivity corresponds to the easy movement of domain walls as the magnetic field changes magnitude and/or direction. Structural defects, e.g. particles of a non-magnetic phase or voids in the magnetic material tend to restrict the motion of domain walls, and thus increase the coercivity (H_C). Obviously, a soft magnetic material must be free from such structural defects.

Electrical resistivity is the another property which needs to be considered. In addition to hysteresis energy losses, energy losses may result from electrical currents that are induced in a magnetic material by a magnetic field that varies in magnitude and direction with time; these are called *eddy currents*. These energy losses in soft magnetic materials can be minimized by increasing the electrical resistivity. This is achieved in ferromagnetic materials by forming solid solution alloys—Fe-Si and Fe-Ni alloys, etc. The ceramic ferrites are commonly used for applications requiring soft magnetic materials because they are intrinsic insulators. However, their applicability is somewhat limited, in as much as they have relatively small susceptibilities.

The class of alloys used in largest volume is by far Fe and 1–3.5% Si-Fe for applications in motors and large transformers. In these applications the cost of the material is often the dominant factor, with losses and excitation power secondary but still important. A major improvement in these alloys occured around 1940 with a development of a {110}<001> crystal texture in 3.2% Si-Fe which greatly reduced losses and increased permeability. Since then the losses have decreased steadily, roughly logarithmically, through alloying, texture improvements, decreasing the thickness of strip and application of stressed insulating coatings.

Many special alloys find their use in special devices designed to exploit unusual properties. These include materials with high saturation induction B_s (the magnetic induction at very large values of H), usually Co-Fe alloys; and alloys with high μ, most often Ni-Fe alloys. These alloys are the mainstay of the telecommunication industry. Some of the devices require high initial permeability (μ at very low fields) and other may depend on high values of $μ_{max}$ (the maximum permeability).

In power equipment the major consideration is the power lost in the magnetic circuit under operating conditions. The total core loss P_C consists of the hysteresis loss P_h and eddy current loss P_e. Thus,

$$P_C = P_h + P_e = \text{const} \times f + \text{const} \times \frac{B^2 f^2}{\rho} \quad (11.60)$$

where f is the frequency and ρ is the electrical resistivity of the material.

The hysteresis loss may be kept low by using a material with a narrow hysteresis loop (one with low coercive force H_C the value of H required to reduce B to zero) and minimizing mechanical strain after the final stress-relief anneal. Eddy-current loss may be reduced by breaking up eddy-current paths, for example by using laminations rather than solid cores.

Audio-frequency devices require very thin laminations resin-bonded alloy powder cores. The very high electrical resistivity (1–10^6 Ωcm) qualifies many ferrites for very high frequency applications, for example in cores and transformers.

It is interesting to note that the hysteresis characteristics of soft magnetic materials may be enhanced for some applications by an appropriate heat treatment in the presence of a magnetic field. Using such a technique, one may produce a square hysteresis loop, which is desirable in some magnetic amplifier and pulse transformer applications. Moreover, soft magnetic materials are used in switching circuits.

11.9 HARD MAGNETIC MATERIALS

Hard magnetic materials are utilized in permanent magnets, which have a high resistance so that they strongly resist demagnetization once magnetized. In view of hysteresis behaviour, a *hard magnetic material* or permanent magnet has remanence, coercivity, and saturation flux density, as well as a low initial permeability, and high hysteresis energy losses. Figure 11.24 shows a comparison of the hysteresis characteristics of hard and soft magnetic materials.

The important characteristics relative to applications for hard magnetic materials are the coercivities H_C and what is termed the *energy product* $(BH)_{max}$. The energy product $(BH)_{max}$ corresponds to the area of the largest B–H rectangle that one can construct within the second quadrant of the hysteresis curve (Fig. 11.25). The units of $(BH)_{max}$ are kJ/m^3 (MGOe : 1MGOe = 10^6 gauss oersted = 7.96 kJ/m^3). We may note that the value of $(BH)_{max}$ is representative of the energy required to demagnetize a permanent magnet,

i.e. the larger $(BH)_{max}$ the harder is the magnetic material in terms of its magnetic characteristics. Although the overall quality of a permanent magnet depends on $(BH)_{max}$, but the design considerations, high H_C, high residual induction B_T (the magnetic induction when H is reduced to zero) and reversibility of permeability μ may also be controlling factors.

We may note that hysteresis behaviour of

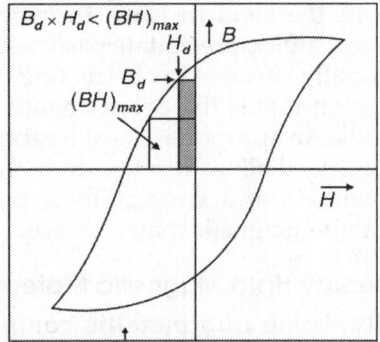

Fig. 11.25: Magnetization curve displaying hysteresis. Within the second quadrant, two B–H energy product rectangles are drawn, the area of rectangle labelled $(BH)_{max}$ is the largest possible, which is greater than the area defined by B_d–H_d

a material is related to the ease with which the magnetic domain boundaries move; by impending domain wall motion, the coercivity and susceptibility are enhanced, such that a large external field is required for demagnetization. Moreover, these characteristics are interrelated to the micro-structure of the material. To understand the relation between the resistance to demagnetization, i.e. the coercivity H_C, and the metallurgical micro-structure, it is essential to understand the mechanisms of magnetization reversal. The two major mechanisms are reversal against *shape anisotropy* and reversal through nucleation and growth of reverse magnetic domains against *crystal anisotropy*.

Conventional Hard Magnetic Materials

Hard magnetic materials can be studied under two main categories: *conventional* and *high-energy hard materials.* The conventional hard magnetic materials have $(BH)_{max}$ values that range between about 2 and 80 kJ/m^3 (0.25 and 10 MGOe). Ferromagnetic materials—magnet steels, cunife (Cu-Ni-Fe) alloys, alnico (Al-Ni-Co) alloys as well as hexagonal ferrites (BaO-6Fe$_2$O$_3$) are conventional hard magnetic materials.

The hard magnet steels are generally alloyed with tungsten and/or chromium. These two elements readily combine with carbon in the steel to form tungsten and chromium carbide precipitate particles, which are especially effective in obstructing domain wall motion under the proper heat-treating conditions. An appropriate heat treatment for the other metal alloys forms extremely small single domain and strong Fe-Co particles within a non magnetic matrix phase.

High-energy Hard Magnetic Materials

Recently, some intermetallic compounds having a variety of compositions have been developed. These permanent magnetic materials have $(BH)_{max}$ in excess of about 80 kJ/m^3. The two that have found commercial exploitation are SmCo$_5$ and Nd$_2$Fe$_{14}$B.

The magnetization and demagnetization behaviour of these hard magnetic materials is a function of domain wall mobility, which, in turn, is controlled by the final microstructure, i.e. the size, shape and the orientation of the crystallites and grains, as well as the nature and distribution of any second-phase particles that are present. We may note that microstructure will depend on how the material is processed. There are two different processing techniques available for the fabrication of Nd$_2$Fe$_{14}$B magnets: *powder metallurgy* (sintering) and *rapid solidification* (melt spinning). The powder metallurgical method is similar to that described for the SmCo$_5$ materials. In case of rapid solidification techniques, the alloy, in molten form, is quenched very rapidly such that either an amorphous or very fine grained and thin ribbon is produced. This ribbon material is then pulverized, compacted into the shape as per requirement, and subsequently heat treated. Rapid solidification is the more involved of the two fabrication processes; nevertheless, it is continuous, whereas powder metallurgy is batch process, which has its inherent disadvantages.

Hard materials are used in motors, loudspeakers, meters, holding devices and in a host of different devices in a variety of technological fields. Permanent magnets are far superior to electromagnets in that their magnetic fields are continuously maintained and without the necessity of having to expand electrical power. Moreover, no heat is generated during operation. This is why permanent magnets are preferred in motors. Motors with permanent magnets are much smaller than their electromagnet counterparts and are utilized extensively in fractional horsepower units. Few familiar motor applications include in cordless drills and screw drivers; in automobiles (starting, window winter, wiper, washer and fan motors); in clocks and in audio and video recorders. There are other common devices, e.g. speakers in audio systems, lightweight earphones, hearing aids, and computer peripherals, etc. that employ these hard magnetic materials.

11.10 MAGNETIC STORAGE

Recently, magnetic materials have become increasingly important in the field of *information storage*; in fact, magnetic recording has become virtually the universal technology for the storage of electronic information. The audio tapes, VCRs, computer hard disks, floppy disks, credit cards, etc. are few examples of magnetic storage. In computers, semiconductor elements serve as primary memory and magnetic disks are used for secondary memory because they are capable of storing larger quantities of information and also at a lower cost. The recording and television industries rely heavily on magnetic tapes for the storage and reproduction of audio and video sequences.

Computer bytes, sound, or visual images in the form of electrical signals are recorded on very small segments of the magnetic storage device—a tape or disk. Transference to and retrieval from the magnetic tape of disk is accomplished by means of an inductive read-write head, which consists basically of a wire coil wound around a magnetic material core into which a gap is cut. Using electrical signal, data are introduced (or written) within the coil, which generates a magnetic field across the gap. This magnetic field in turn magnetizes a very small area of the disk or tape within the proximity of the head. Upon removal of the magnetic field, magnetization remains, i.e. the signal has been stored. Figure 11.26 shows the essential features of this recording process.

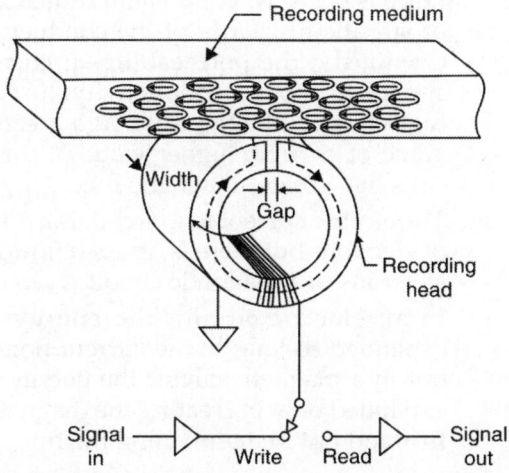

Fig. 11.26: Using magnetic storage medium, how information is stored and retrieved

We may note that the same head may be utilized to retrieve (or read) the stored information. A voltage is induced when there is a change in the magnetic field as the tape or disk passes by the head coil gap. One can amplify this and then converted back into its original form or character (Fig. 11.26).

More recently, hybrid heads that consist of an inductive-write and a magnetoresistive read head in a single unit have been introduced. In the magnetoresistive head, the electrical resistance of the magnetoresistive thin film element is changed as a result of magnetic field changes when the tape of disk passes by the read head. We may note that higher sensitivities and higher data transfer rates make magnetoresistive very attractive.

Magnetic media are of two principal types: *particulate* and *thin film*. First media consist of very small needle-like or acicular particles, normally of γ-Fe_2O_3 ferrite or CrO_2; these are applied and bonded to a polymeric film (for magnetic tapes) or to a metal or polymer disk. During manufacture, these particles are aligned with their long axis in a direction that parallels the direction of motion past the head. Each particle is a single domain that may be magnetized only with its magnetic moment lying along this axis. Corresponding to the saturation magnetization in one axial direction, and its opposite, two magnetic states are possible. These two states make possible the storage of information in digital from as 1's and 0's. In one system, 1 is represented by a reversal in the magnetic field direction from one small area of storage medium to another as the numerous acicular particles of each such region pass by the head. A lack of reversal between adjacent region is indicated by a 0.

The *thin-film storage technology* provides higher storage capacities at lower costs. It is used mainly on rigid disk drives and consists of multilayered structure. A magnetic thin film layer is the actual storage component. This thin film is normally either a Co-Pt-Cr or Co-Cr-Ta alloy, with a thickness between 10 nm and 50 nm (100 nm and 500 Å). A substrate layer below and upon which the thin film resides is either pure Cr or Cr alloy. The thin film is itself polycrystalline, having an average grain that is trypically between 10 nm and 30 nm (100 and 300 Å). Within the thin film, each grain is a single magnetic domain and it is highly desirable that grain shape and size be relatively uniform. For magnetic storage disk where these thin films are employed, the crystallographic direction of easy magnetization for each grain is aligned in the direction of disk motion (or the direction opposite). We may note that the mechanism of magnetic storage within each of these single domain grains is the same for the needle-shaped particles, i.e. the two magnetic states correspond to domain magnetization in one direction or its antiparallel equivalent.

We may note that the storage density of thin film is greater than for particulate media because the packing efficiency of thin-film domains is greater than for the acicular particles; particles will always be separated with void space in between. The known storage densities for particulate media are of the order of 1×10^8 bit/in^2 (1.5×10^5 bit/mm^2). For thin films, storage densities are approximately of an order of greater magnitude, i.e. ~5×10^{10} bit/in^2 (8×10^7 bit/mm^2).

The hysteresis loops for these magnetic storage media should be relatively large and square. Obviously, these characteristics ensure that storage will be permanent, and in addition, magnetization reversal will result over a narrow range of applied field strengths. For particular recording media, saturation flux density normally ranges from 0.4 tesla to 0.6 tesla. For thin film, B_s lies between 0.6 tesla and 1.2 tesla. H_C values are typically in the range $1.5 \times 10^5 - 2.5 \times 10^5$ A/m, i.e. 2000 – 3000°C.

11.11 MAGNETIC CIRCUIT

The magnetic flux always forms a closed path and this is called magnetic circuit. For example, in a bar magnet the lines of flux (induction) come out of north pole, passes through the air, enters the south pole, and complete their path through the magnet from south to north pole. Obviously, the path of a magnetic flux is magnetic circuit. Few examples of magnetic circuits as an illustration are shown in Fig. 11.27.

Comparison of Magnetic and Electric Circuits

i. Magnetic flux is analogous to electric current, magnetomotive force (mmf) is analogous to emf and reluctance is analogous to electrical resistance.

ii. The reluctance is $l/\mu A$ whereas resistance is $\rho l/A$, where ρ is the resistivity or specific resistance of the conductor. Obviously, the permeability μ corresponds to $1/\rho$, i.e. conductivity. Just as higher value of $1/e$ results in a greater value of current, higher value of μ results in a greater magnetic flux.

iii. The electric current in a circuit, is a flow of electrons but there is no such flow of electrons in a magnetic circuit.

iv. In an electric circuit, the energy is expanded so long as the current flows, but in a magnetic circuit the energy is expanded only in creating the magnetic flux and not in maintaining the flux.

11.12 MAGNETOMOTIVE FORCE (MMF), RELUCTANCE AND PERMEANCE

We have stated that an electric circuit has electromotive force (emf) whereas a magnetic circuit has a magnetomotive force (mmf).

One can define the mmf as the line integral of magnetizing force H around the closed circuit, i.e.

$$\oint_C H \cdot dl = \text{mmf} \qquad (11.61)$$

From Ampere's circuital law, we have

$$\oint H \cdot dl = NI \qquad (11.62)$$

where I is the current flowing through the windings which has N number of turns.

The magnetic flux (ϕ) is given by

$$\phi = BA \quad \text{or} \quad B = \frac{\phi}{A} \qquad (11.63)$$

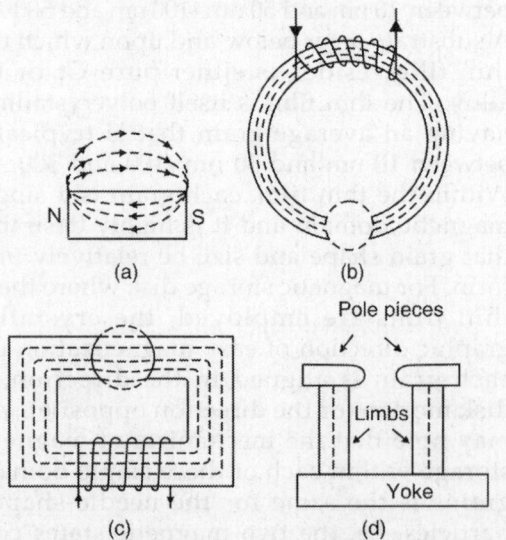

(a)

(b)

Pole pieces

Limbs

Yoke

(c)

(d)

Fig. 11.27: Magnetic circuits (a) Magnet (b) Iron ring (c) Core (d) Electromagnet

We also have $B = \mu H$ $\qquad$ (11.64)

$$\therefore \qquad \mu H = \frac{\phi}{A}$$

or $\qquad H = \frac{\phi}{\mu A}$ $\qquad$ (11.65)

Now, substituting the value of H from Eq. (11.65) in Eq. (11.61), one obtains

$$\text{mmf} = \oint \frac{\phi}{\mu A} dl = \oint \frac{\phi}{\mu A} dl$$

or $\qquad \phi = \dfrac{\text{mmf}}{\oint \dfrac{dl}{\mu A}}$ $\qquad$ (11.66)

We have stated that magnetic flux ϕ is equivalent to electric current of electric circuit and mmf of magnetic circuit is equivalent to electric circuit. Thus, comparing Eq. (11.66) with Ohm's law equation $i = \dfrac{\text{emf}}{R}$, one finds that the term $\oint \dfrac{dl}{\mu A}$ is similar to electrical resistance R in an electric circuit. Thus, the magnetic reluctance is given by

$$\text{mmf}(R) = \oint \frac{dl}{\mu A} \qquad (11.67)$$

In general, if we consider that A remains constant over the length l of the circuit, one can write Eq. (11.67) as

$$R = \frac{1}{\mu A} \oint dl = \frac{l}{\mu A} \qquad (11.68)$$

We may note that Eq. (11.68) is similar to the equation of resistance in electric circuit, i.e.

$$R = \frac{1}{\sigma A} \qquad (11.69)$$

where σ is called the conductivity of the material of the wire.

The reciprocal of reluctance (R) is called *permeance* (P). *Permeance* (P) is *the measure of easiness with which the magnetic flux is developed in the circuit.*

Units of mmf, reluctance (R) and permeance (P) are as follows:

Magnetomotive force (mmf): Ampere-turns

Magnetic flux: Weber

Reluctance (R): $\dfrac{\text{Ampere-turns}}{\text{Weber}}$

and Permeance (P): $\dfrac{\text{Weber}}{\text{Ampere-turns}}$

Reluctance in Series and Parallel

Series

When we join different reluctances in series in a magnetic circuit, one can obtain the resultant by a similar equation as we use for electrical resistances in series.

Let us consider that we have different magnetic materials having permeability $\mu_1, \mu_2, \mu_3, ...$, lengths $l_1, l_2, l_3, ...$ and area of cross-section $A_1, A_2, A_3, ...$ respectively. Let these are joined together together in series. Now, the equivalent magnetic reluctance of the circuit is obtained as

$$R = R_1 + R_2 + R_3 + ...$$

$$= \frac{l_1}{\mu_1 A_1} + \frac{l_2}{\mu_2 A_2} + \frac{l_3}{\mu_3 A_3} + ... \quad (11.70)$$

Parallel

Now, when different reluctances are joined in parallel, the equivalent reluctance is obtained as

$$\frac{1}{R} = \frac{1}{R_1} + \frac{1}{R_2} + \frac{1}{R_3} + ...$$

$$= \frac{\mu_1 A_1}{l_1} + \frac{\mu_2 A_2}{l_2} + \frac{\mu_3 A_3}{l_3} + ... \quad (11.71)$$

11.13 ROWLAND RING

Let us consider a coil of N turns wound uniformly on an iron ring of radius r as shown in Fig. 11.28. This is called Rowland ring. If we allow to pass i ampere current through it, then the reluctance (R) for the ring is given by

$$R = \oint \frac{dl}{\mu A} = \frac{1}{\mu A} \oint dl \qquad (11.72)$$

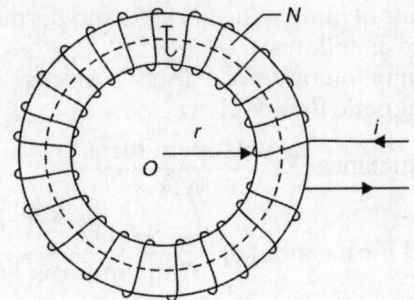

Fig. 11.28: Rowland ring

For the Rowland ring of radius r, we have

$$\oint dl = 2\pi r$$

$$\therefore \qquad R = \frac{2\pi r}{\mu A} \qquad (11.73)$$

Magnetomotive force (mmf) $= Ni = 2\pi rni$
$$(11.74)$$

where n is the number of turns per unit length of the ring. We have,

$$\text{Magnetic flux } (\phi) = \frac{\text{mmf}}{\text{Reluctance}}$$

$$= \frac{2\pi rni}{2\pi r / \mu A}$$

$$= \mu niA \qquad (11.75)$$

The flux density $(B) = \dfrac{\phi}{A} = \mu ni \qquad (11.76)$

We also have $B = \mu H$, therefore, the magnetic field,

$$H = ni \qquad (11.77)$$

11.14 VECTOR MODEL OF AN ATOM
Gyromagnetic Ratio

When we place a magnetic dipole in a magnetic field, its potential energy depends not only on the magnitude of its magnetic moment but also on its orientation with respect to the field. Let us suppose that a magnetic dipole of moment μ is placed in a magnetic field of flux density B.

Now, the torque of the dipole is

$$\tau = \mu B \sin\theta \qquad (11.78)$$

where θ is the angle between the dipole and the field. For $\theta = 90°$, the torque is maximum. When dipole is parallel to the field, the torque is zero. It only changes potential energy that are ever experimentally observed. Thus, we have chosen a reference configuration in which the potential energy is zero when $\theta = 90°$, i.e. when the dipole is at right angle to the magnetic field. We may note that the potential energy of the system in any other position of the dipole is given by the work done on the dipole in rotating it from the position $\theta = 90°$ to the position θ. Obviously, the potential energy is given by

$$PE = U = \int_{90°}^{\theta} \tau d\theta = B \int_{90°}^{\theta} \mu \sin\theta d\theta$$

$$= \mu B \cos\theta \qquad (11.79)$$

When $\theta = 0$, PE has its minimum value. This is obvious because the dipole while tending to align itself with the field direction moves towards the position of minimum potential energy.

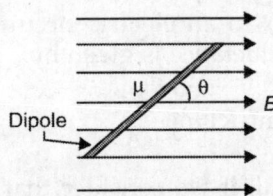

Fig. 11.29: Magnetic dipole in a magnetic field

We may note that the electron in a hydrogen atom going round its orbit is equivalent to *current loop* (Fig. 11.30). The current in the loop is usually referred to as *electric current in atoms*.

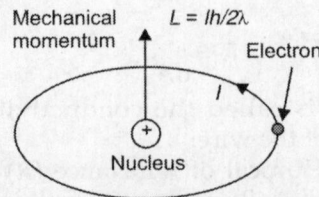

Fig. 11.30: Revolving electron in an orbit is equivalent to current loop

There is a magnetic moment associated with a current loop. This magnetic moment (μ) is equal to the product of current (I) and the area of the orbit (A), i.e.

$$\mu = IA \qquad (11.80)$$

where current $I = -e/T$, where T is the period of revolution of the electron round its orbit of area A. Now, we have the angular momentum L of the electron as

$$L = mvr = m\frac{d\theta}{dt}r^2$$

$\therefore$ The area of the orbit $(A) = \int_0^r \frac{1}{2}r^2\frac{d\theta}{dt}dt$

or $\int_0^T \frac{1}{2}\frac{L}{m}dt = \frac{1}{2}\frac{L}{m}T$

or $\mu = -\frac{e}{T}\frac{1}{2}\frac{L}{m}T = \left(-\frac{e}{2m}\right)L$ (11.81)

The quantity $-e/2m$ is called the *gyromagnetic ratio*. We may note that this involves only the charge and mass of the electron. We can see that *gyromagnetic ratio is equivalent to the ratio of the magnetic moment to the mechanical moment of the electron in the orbit*. The minus sign in Eq. (11.81) shows that the magnetic and mechanical moments are in opposite directions. We may note that the above result has been obtained by classical calculations but quantum mechanics also gives the same result.

Now, substituting for μ from Eq. (11.81) in Eq. (11.79) one obtains

$$U = -\left(\frac{e}{2m}\right)LB\cos\theta$$

$$= \left(\frac{e}{2m}\right)LB\cos\theta \qquad (11.82)$$

The Bohr Magneton

We know that a mechanical top precesses round the gravitational field. Similarly, the electron orbit precesses round the direction of the magnetic field in which it is placed. This precession of electron is called *Larmor precession*. Larmor precession is quantized. If $L = lh/2\pi$

is the mechanical momentum normal to the electron orbit (Fig. 11.30), then in accordance with the quantum condition the projection of L on the field direction can take only the integral values $l\hbar$, $(l-1)\hbar$, $(l-2)\hbar$, ... $2\hbar$, $\hbar$, 0, $-\hbar$, $-2\hbar$, ..., $-l\hbar$, where $\hbar = h/2\pi$. Quantization of the precession of the electron orbit is illustrated in Fig. 11.31. We denote the projection of l on the field direction by m_l. Then in one of the permitted orientation, we have

$$\cos\theta = \frac{m_l}{l} \qquad (11.83)$$

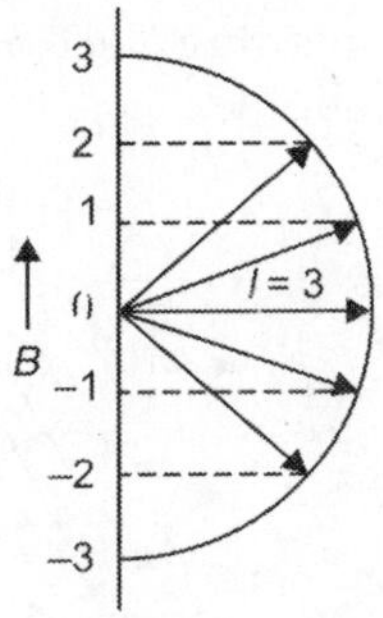

Fig. 11.31

Now, the potential energy (U) in this orientation is given by

$$U = \left(\frac{e}{2m}\right)LB\left(\frac{m_l}{l}\right)$$

Now, writing $L = l\hbar$, we obtain

$$U = \frac{e}{2m}\hbar B m_l = Bm_l B$$

where $B = \dfrac{e\hbar}{2m} = \dfrac{eh}{4\pi m}$ (11.84)

is called the *Bohr magneton*.

The Spinning Electron

Experimentally it was observed that in the spectra of the Balmer lines of hydrogen and that of alkali metals all the lines were doublets. Bohr's theory of atomic model as well as Sommerfeld theory could not explain the doublet nature of spectral lines. In order to explain this behaviour, Goudsmit and Uhlenbeck in 1925 postulated that an electron not only revolves round the nucleus but it also *spins* around its own axis. Obviously, electron possesses spin motion as well as orbital motion. The spin angular momentum of an electron is given by $p_s = s\dfrac{h}{2\pi}$ where $s = 1/2$. According to wave mechanics, an spin angular momentum p_s of the electron is given by

$$p_s = \sqrt{s(s+1)}\,\frac{h}{2\pi} \qquad (11.85)$$

As an electron carries an electric charge its spin produces a magnetic moment (Fig. 11.32). Obviously, the spinning electron thus behaves like a magnetic dipole.

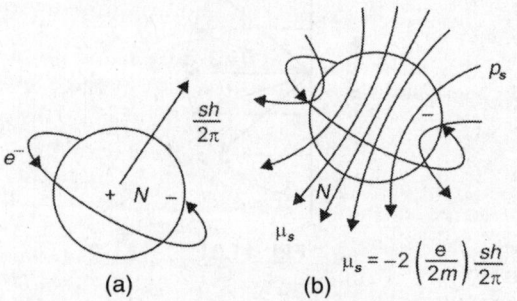

$$\mu_s = -2\left(\frac{e}{2m}\right)\frac{sh}{2\pi}$$

(a)   (b)

Fig. 11.32: Spinning electron

Magnetic Moment of an Orbital Electron (μ_s)

We have seen that there are two magnetic momenta associated with the electron, one due to its orbital motion and the other due to spin motion. As stated earlier, the orbital motion of an electron is just like the flow of current as an electron is negatively charged particle although the current is not confined in a wire. Due to the flow of electric current, a magnetic field is set up (Fig. 11.33).

Let us consider the simple case of an electron which is moving in a central orbit like an ellipse. The current i due to the motion of an electron round the nucleus is given by

$$i = \frac{e}{T} \quad \text{(MKS units)} \qquad (11.86)$$

where i is current in amperes, e is charge on an electron in coulomb and T is the periodic

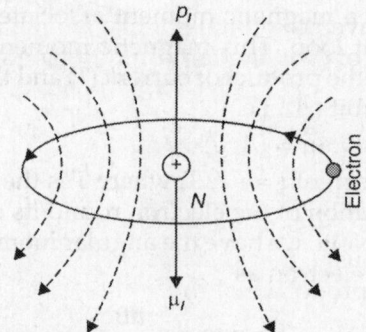

Fig. 11.33: Magnetic moment of an orbital electron

time in seconds. The current i gives rise to a magnetic moment μ_l given by

$$\mu_l = iA = \frac{eA}{T} \text{ amp}\,\text{m}^2 \qquad (11.87)$$

(in accordance with Ampere's theorem) where A is the area enclosed by the electron orbit.

If r and ϕ are the polar coordinates of electron in an elliptical path, then one can express area A as

$$A = \int_0^T \frac{1}{2} r^2 \left(\frac{d\phi}{dt}\right) dt \qquad (11.88)$$

where $\dfrac{p_l}{2m} = \dfrac{1}{2} r^2 \left(\dfrac{d\phi}{dt}\right)$ is the real velocity of motion. Now, angular momentum p_l is given by

$$p_l = mr^2 \left(\frac{d\phi}{dt}\right)$$

or

$$\frac{p_l}{2m} = \frac{1}{2} r^2 \left(\frac{d\phi}{dt}\right) \qquad (11.89)$$

Substituting the value of $\dfrac{1}{2} r^2 \left(\dfrac{d\phi}{dt}\right)$ from Eq. (11.89) in Eq. (11.80), one obtains

$$A = \int_0^T \frac{p_l}{2m} dt = \frac{p_l}{2m}[t]_0^T = \frac{p_l}{2m} T \qquad (11.90)$$

Now, substituting the value of A from Eq. (11.78) in Eq. (11.87), one obtains

$$\mu_l = \frac{e}{T}\frac{p_l}{2m} T = \frac{e}{2m} p_l \qquad (11.91)$$

We have the orbital angular momentum p_l in terms of orbital quantum number l as

$$p_l = l\frac{h}{2\pi}$$

$$\therefore \quad \mu_l = \frac{e}{2m}l \cdot \frac{h}{2\pi} = \frac{eh}{2\pi m}l \qquad (11.92)$$

Obviously, the *orbital magnetic moment* μ_l is directly proportional to l.

If $l = 0$, $\mu_l = 0$ and for $l = 1$, $\mu_l = \dfrac{eh}{2\pi m}$

The quantity $\dfrac{eh}{4\pi m}$ is known as *Bohr magneton* whose value is

$$9.27 \times 10^{-24}\ \mathrm{JT^{-1}} = 9.27 \times 10^{-24}\ \mathrm{amp\ m^2}.$$

The ratio $\dfrac{\mu_l}{p_l}$ is known as *gyromagnetic* ratio. Thus,

$$G = \frac{\mu_l}{p_l} = \frac{l(eh/4\pi m)}{l(h/2\pi)} = \frac{e}{2m} \qquad (11.93)$$

One can also express G as

$$G = g\frac{e}{2m} \qquad (11.94)$$

where g is called *Lande splitting factor*. $g = 1$ for orbital motion of electron.

Magnetic Moment due to Electron Spin (μ_s)

The spinning electron behaves like a spinning tinny magnet, i.e. every electron in an atom is subject to a magnetic field. As stated earlier, due to the spin of the electron it has spin magnetic moment. Figure 11.32b shows the magnetic field generated by the spinning electron and its spin magnetic moment μ_s. The spin magnetic moment of the electron is given by

$$\mu_s = 2\frac{e}{2m}p_s \qquad (11.95)$$

The spin angular momentum p_s is given by

$$p_s = s\frac{h}{2\pi} \qquad (11.96)$$

From Eqs. (11.95) and (11.96), one obtains

$$\mu_s = 2\frac{e}{2m}s\frac{h}{2\pi} = 2s\frac{eh}{4\pi m} = 2\left(\frac{e}{2m}\right)\frac{sh}{2\pi} \qquad (11.97)$$

But we know that $s = 1/2$

$$\therefore \quad \mu_s = \frac{eh}{4\pi m} = \text{one Bohr magneton.}$$

Obviously, *the magnetic moment due to electron spin is equal to one Bohr magneton.*

We may note that the doubling of energy levels in atoms having one electron is due to the interaction between the two fields, i.e. the magnetic field of electron orbit and magnetic field of electron spin.

Stern–Gerlach Experiment

A systematic arrangement of Stern–Gerlach experiment is shown in Fig. 11.34. O is an oven in which silver is vaporized. A beam of silver atoms issuing from the oven pass through a horizontal slit. They then pass through two more slits S_1 and S_2.

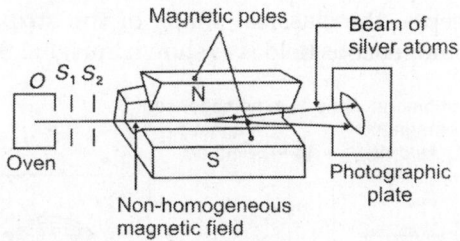

Fig. 11.34: Stern–Gerlach experimental arrangement

The collimated beam of silver atoms leaving the slit S_2 enters a strong nonhomogeneous magnetic field between the pole pieces of a magnet NS. The nonhomogeneous magnetic field is produced by having one of the pole pieces of the magnet flat with a cylindrical groove and other in the form of a knife edge, parallel to groove (Fig. 11.34). Obviously, the magnetic field is of much greater intensity near the knife edge than anywhere else in the gap, i.e. the intensity of the magnetic field increases as one moves from the centre towards the lower end (Fig. 11.35). After the passage of beam through the field, a photographic plate P records its configuration.

In order to avoid the deflection of silver atoms by gas molecules, the entire set-up is enclosed in a highly evacuated glass envelope.

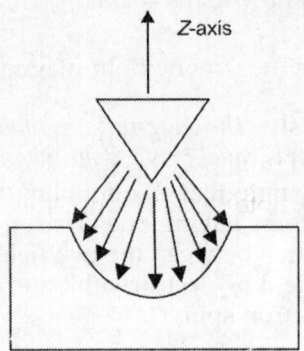

Fig. 11.35: Side view of magnet with magnetic lines of force

When the magnetic field is switched off, a trace of the form of a narrow strip (Fig. 11.36a) is obtained. When the field was switched on, the trace divided into two lines at the centre. (Fig. 11.36b). According to the classical physics concepts, the classical shape of the strip in inhomogeneous field is as shown in Fig. 11.36c.

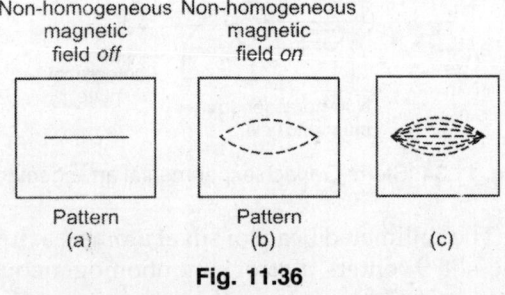

Fig. 11.36

The splitting of silver beam into two components in inhomogeneous magnetic field verifies the existence of electron spin and also the postulate of *space quantization.*

The atomic number of silver is $Z = 47$. Silver atom has a single optical electron which is normally in the $5 \, ^2s_{1/2}$ state. The atom itself is in the $^2s_{1/2}$ state. Obviously, the magnetic moment of the silver atom is entirely due to this single $5s$ electron. We have for an s state $l = 0$ and $s = 1/2$. Therefore $j = 1/2$. Using these values, one obtains Lande g factor = 2. Now,

the magnetic moment of the silver atom is $jg\beta = \dfrac{1}{2} \times 2\beta = \beta = 1$ Bohr magneton.

Moreover, the magnetic quantum number m_j can take up $(2j + 1) = (2 \times 1/2 + 1) = 2$ values, i.e. $+1/2$ or $-1/2$. The orientation θ of the vector j to the magnetic field direction is, therefore, given by

$$\cos\theta = \frac{m_j}{j} = +1 \quad \text{or} \quad -1$$

$$\therefore \qquad \theta = 0° \text{ or } 180°$$

Obviously, when a silver atom is placed in an inhomogeneous magnetic field the magnetic moment vector will be aligned either parallel to the magnetic field or antiparallel to the magnetic field.

Now, we consider a magnetic dipole of moment β placed in a nonhomogeneous, i.e. nonuniform magnetic field (Fig. 11.37).

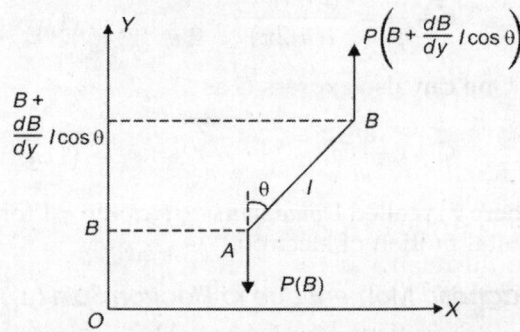

Fig. 11.37: Magnetic dipole in a nonhomogeneous field

If the dipole is aligned parallel to the magnetic field, it experiences a net force $\beta \dfrac{dB}{dy}$ in the direction of the field. If the dipole is aligned antiparallel to the magnetic field, it experiences an equal force but in opposite direction. According to classical physics, all orientations of the dipole with respect to the magnetic field are possible (we may note that in homogeneous magnetic field the magnet experiences only a turning moment, and no deflecting force. This is why in Stern–Gerlach experiment the field must be necessarily non-homogeneous). This means, when the field is applied only a broadening of the trace on the glass plate G should result. The splitting of the atoms into two distinct beams clearly shows that only two orientations are present. These correspond to the opposite orientations as permitted by space quantization. The space quantization is, thus, directly demonstrated.

The experiment also provides evidence for the existence of electron spin. The optical electron in the silver atom lies in an s level, for which the quantum number $l = 0$. Now, if there were no spin, then j would also be zero, i.e. $2j + 1 = 1$. In that case, there would be only one possible orientation of the silver atom in the magnetic field and only one trace would be obtained on the plate. We have seen that experiment, however, shows two traces. This means that we have to admit the existence of electron spin and a value $1/2$ is assigned to the quantum number. Thus, $j = l \pm s = 0 \pm 1/2 = 1/2$. Thus, $2j + 1 = 2 \times 1/2 + 1 = 2$. This is why we observe two traces.

ILLUSTRATIVE EXAMPLES

Example 1

A magnetic material has a magnetization of 3300 ampere/m and flux density of 0.0044 W/m². Calculate the magnetizing force and the relative permeability (μ_r) of the given material. [BE]

Solution

Given that $B = 0.0044$ W/m²

$$\mu_0 = 4\pi \times 10^{-7} \text{ henry/m}$$

$$M = 3300 \text{ amp/m}$$

We have $B = \mu_0(M + H)$

or $\dfrac{B}{\mu_0} = M + H$

or $H = \dfrac{B}{\mu_0} - M$

$\therefore \quad H = \dfrac{0.0044}{4\pi \times 10^{-7}} - 3300$

$$= 203 \text{ amp/m}$$

Now, $M = H(\mu_r - 1)$

or $\mu_r = \dfrac{M}{H} + 1 = \dfrac{3300}{203} + 1 = 17.26$

Example 2

The magnetic susceptibility of a medium is 2.14×10^{-11} N/m. Calculate its relative permeability.

Solution

Given that $\chi = 2.14 \times 10^{-11}$ H/m

$$\mu_r = 1 + \chi$$

$\therefore \quad = (1 + 2.14 \times 10^{-11})$

Example 3

A bar magnet has a coercivity of 5×10^3 amp/m. It is inserted inside a solenoid of length 10 cm and having number of turns 50, so that it is demagnetized. Show that current to be flown through the solenoid be 10 ampere. [BE]

Solution

We have the magnetic intensity inside a solenoid as given by

$$H = \frac{Ni}{l} \qquad (1)$$

where N is the total number of turns wound over a length l of the solenoid and i is the current flowing through it. Given that the bar magnet required $H = 5 \times 10^3$ amp/m in opposite direction to get it completely demagnetized. Thus, from Eq. (1), we have

$$H = \frac{50i}{0.10}$$

or $i = \dfrac{0.1}{50} H = \dfrac{0.1 \times 5 \times 10^3}{50} = 10 \text{ A}$

Example 4

An iron rod of volume 10^{-4} m³ and relative permeability 1000 is placed inside a long solenoid wound with 5 turns/cm. If a current of 0.5 A is passed through the solenoid, find the magnetic moment per unit volume. [BTech]

Solution

We have $\qquad B = \mu_0 H + \mu_0 I$

$\therefore \qquad I = \dfrac{B - \mu_0 H}{\mu_0}$

$\qquad\qquad = \dfrac{\mu H - \mu_0 H}{\mu_0}$

$\qquad\qquad = \dfrac{\mu_r \mu_0 H - \mu_0 H}{\mu_0} = (\mu_r - 1)H$

Now, for a solenoid of n turns per unit length and carrying a current i, the magnetic intensity is obtained as

$\qquad\qquad H = ni$

$\therefore \qquad\qquad I = (\mu_r - 1)ni$

Given that $\mu_r = 1000$

$\qquad\qquad n = 5$ turns/cm

$\qquad\qquad\quad = 200$ turns

$\qquad\qquad i = 0.5$ A

$\therefore \qquad\qquad I = (1000 - 1) \times 500 \times 0.5$

$\qquad\qquad\quad = 2.5 \times 10^5 \; \text{Am}^{-1}/\text{m}$

Since the magnetization I is the magnetic moment per unit volume, we have

Magnetic moment $= I \times$ volume

$\qquad\qquad = 2.5 \times 10^5 \dfrac{\text{A}}{\text{m}} \times 10^{-4} \text{m}^3$

$\qquad\qquad = 25 \; \text{Am}^2$

Example 5

An iron ring of mean circumferential length 0.3 m and cross-section 10^{-4} m² is wound uniformly with 300 turns of wire. When a current of 0.032 ampere flows in the windings the flux in the ring is 2×10^{-6} W. Find the flux density in the ring, the magnetic intensity and the permeability of iron.

Solution

Given that $\quad l = 0.3$ m

$\qquad\qquad A = 10^{-4}$ m²

$\qquad\qquad N = 300$

$\qquad\qquad i = 0.032$ A

$\qquad\qquad \phi_B = 2 \times 10^{-6}$ W

$\therefore \quad$ Flux density $B = \dfrac{\phi_B}{A} = \dfrac{2 \times 10^{-6}}{10^{-4}}$

$\qquad\qquad\qquad\qquad = 2 \times 10^{-2}$ W/m²

Magnetic intensity,

$\qquad H = \dfrac{Ni}{l} = \dfrac{300 \times 0.32}{0.3} = 32 \dfrac{\text{amp-turns}}{\text{m}}$

Permeability,

$\qquad \mu = \dfrac{B}{H} = \dfrac{2 \times 10^{-2}}{32} = 6.25 \times 10^{-4} \dfrac{\text{W}}{\text{A m}}$

Relative permeability,

$\qquad \mu_r = \dfrac{\mu}{\mu_0} = \dfrac{6.25 \times 10^{-4}}{4\pi \times 10^{-7}} = 500$

Example 6

A material core has 10 turns per cm of wire wound uniformly upon it which carries a current of 2 A. The flux density in the material is 1 W/m². Calculate H and I of the material. Also find the relative permeability of the core. Given $\mu_0 = 4 \times 10^{-7}$ W/Am.

Solution

$\qquad H = \dfrac{Ni}{l} = \dfrac{100 \text{ turns} \times 2 \text{ A}}{\text{m}}$

$\qquad\qquad = 2000$ A-turns/m

The flux density B is 1 W/m². We have

$\qquad\qquad B = \mu_0 H + \mu_0 I$

or $\qquad I = \dfrac{B - \mu_0 H}{\mu_0} = \dfrac{1 - (4\pi \times 10^{-7} \times 2000)}{4\pi \times 10^{-7}}$

$\qquad\qquad = 7.9 \times 10^5 \dfrac{\text{A-turns}}{\text{m}}$

Also $\quad \mu = \mu_r \mu_0 = \dfrac{B}{H}$

or $\qquad \mu_r = \dfrac{B}{\mu_0 H} = \dfrac{1.0}{4\pi \times 10^{-7} \times 2000} = 397$

Example 7

A iron rod of uniform area of cross-section $10^{-4} \, m^2$ and relative permeability 4000 is wound with 2000 turns of wire carrying a current of 2 ampere. The rod is bent into a close circular ring of diameter 0.2 m. Calculate the magnetic reluctance and flux passing through the rod.

Solution

Magnetic reluctance $R = \dfrac{l}{A\mu}$

Given that $d = 0.2$ m

$$l = \pi d = 0.628 \text{ m}$$
$$N = 200$$
$$i = 2 \text{ A}$$
$$\mu_r = 2000$$

$\therefore \qquad R = \dfrac{0.628}{10^{-4} \times 5.03 \times 10^{-3}}$

$$= 1.25 \times 10^5$$

Flux $\quad \phi = \dfrac{Ni}{R} = \dfrac{200 \times 2}{1.25 \times 10^5} = 3.2 \times 10^{-4} \text{ W}$

Example 8

A magnetic circuit is made of ferromagnetic material of $\mu = 7.3 \times 10^{-3}$ H/m. The average length of the circuit is 1 m and the area of cross-section is $10^{-2} \, m^2$. If the magnetic winding has 100 turns, show that the magnetizing current in order to produce a magnetic flux density of 0.2 W/m² is 0.27 amp. [BE]

Solution

We have magnetic flux,

$$\phi = \frac{\text{mmf}}{\text{Reluctance}} = \frac{Ni}{l/\mu A} = \frac{Ni\mu A}{l}$$

$\therefore \qquad \phi = BA$

$\therefore \qquad BA = \dfrac{Ni\mu A}{l}$

or $\qquad i = \dfrac{BAl}{N\mu A} = \dfrac{Bl}{N\mu}$

Given that $B = 0.20$ W/m²

$$A = 10^{-2} \, m^2$$
$$l = 1 \text{ m}$$
$$N = 100$$
$$\mu = 7.3 \times 10^{-3} \text{ H/m}$$

$\therefore \qquad i = \dfrac{0.2 \times 1}{100 \times 7.3 \times 10^{-3}} = 0.27 \text{ A}.$

REVIEW QUESTIONS

1. Define intensity of magnetization. Develop the relation $B = \mu_0(H + I)$. Explain the terms involved.

2. What do you understand by the magnetic permeability of a material medium. Establish the relation $\mu = \mu_0(1 + \chi_m)$.

3. Define the terms: (i) intensity of magnetization (ii) magnetic susceptibility (iii) magnetic permeability (iv) magnetic induction. Establish the relations connecting them.

4. Define magnetic susceptibility and magnetic permeability. How can you classify magnetic materials on the basis of these properties?

5. Explain diamagnetism, paramagnetism, ferromagnetism, antiferromagnetism and ferrimagnetism on the basis of concept of magnetic dipoles of the atoms. [BE]

6. Discuss Langevin's theory of paramagnetism deriving his equation for paramagnetic susceptibility. [BTech]

7. Discuss Langevin's theory of diamagnetism. Obtain an expression for the susceptibility of a diamagnetic material.

8. On the basis of Langevin's theory of paramagnetism show that susceptibility of a paramagnetic material is inversely proportional to the absolute temperature (Curie's law).

9. Define and explain the terms: retentivity, coercivity, hysteresis and hysteresis loop.

10. Explain what is hysteresis. Prove that the area of the B–H cycle denotes μ_0 times the energy dissipated per cc of the material during each magnetic cycle. Discuss the importance of hysteresis cycles. [BTech]

11. Explain how a magnetized piece of iron can be completely demagnetized.

12. Explain the ballistic galvanometer method of tracing the hysteresis curve for a sample of iron.

13. How would you use the hysteresis curve for selecting the material for construction of a permanent magnet?

14. Compare electric and magnetic circuits.

15. What are the factors responsible for hysteresis loss? Prove that the area of $B-H$ curve is equal to the hysteresis loss per unit volume of the specimen in one cycle.

16. How would you use the hysteresis curves to select material for the construction of (i) permanent magnets (ii) electromagnets, (iii) transformer cores.

17. Explain, what is meant by magnetic circuit? Define magnetomotive force and reluctance.

18. Explain how is the spin of the electron coupled with its orbital motion?

19. Obtain an expression for the magnetic moment of an electron due to its orbital motion.

20. Define the gyromagnetic ratio as applied to the orbital motion of an electron.

21. Explain what is meant by Bohr magneton. Obtain an expression for it.

22. Describe Stern–Gerlach experiment and show that it provides evidence for space quantization and also for electron spin.

PROBLEMS

1. A magnetic field of 1200 A/m produces a magnetic flux of 2.4×10^{-5} W in an iron bar of cross-sectional area 2×10^{-5} m². Calculate (i) permeability (ii) susceptibility of the bar.
   [BE] [*Ans.* (i) 10^{-3} N/A², (ii) 737]

2. The core of a generator armature is made of iron whose hysteresis loop under operating conditions has an area of 5×10^4 $B-H$ units. Find the rate of increase of temperature of the core if the number of rotations/s is 100, assuming that there is no loss of heat. Given density of iron = 7.7×10^3 kg/m³, and specific heat of iron = 0.11 kcal/kg K. [*Ans.* 1.41° C/s]

3. An iron rod of volume 10^{-3} m³ and relative permeability 1200 is placed inside a long solenoid wound with 5 turns/cm. If a current of 0.5 A is passed through the solenoid, show that the magnetic moment of the rod is 3×10^2 A m². [BE, BTech]

4. An iron ring of mean circumferential length 0.2 m and cross-sectional area 10^{-4} m² is wound uniformly with 200 turns of wire. When a current of 0.3 A flows in the windings, the flux in the ring is 2×10^{-6} W. Calculate: (i) flux density in the ring (ii) magnetic intensity and (iii) permeability of iron.
   [BE] [(i) $H = 30$ A-turn/m
   (ii) $\mu = 6.7 \times 10^{-4}$ W/A m (iii) $\mu_r = 534$]

SHORT ANSWER QUESTIONS

1. How diamagnetism results?
Ans. All materials are diamagnetic. This results from changes in electron orbital motion that are induced by an external field.

2. What are paramagnetic materials?
Ans. These are the materials having permanent atomic dipoles, which are acted on individually and are aligned in the direction of an external field.

3. Why dia and paramagnetic materials are considered to be non-magnetic?
Ans. The magnetization for these types of materials are relatively small and persist only when an applied field is present.

4. What are ferrimagnetic materials?
Ans. These are materials having an antiparallel arrangement of spins, but due to incomplete cancellation, there is a net magnetic moment that is comparable in order of magnitude to that of ferromagnetic materials.

5. How will you explain both hysteresis and permanent magnetization in terms of domain walls?
Ans. Hysteresis (the lag of B field behind the applied H field) and permanent magnetization (or remanence) result from the resistance to movement of domain walls.

6. A bar magnet and a current loop are magnetic dipoles. Write their magnetic dipole moments.
Ans. (i) $\mu_m = md$ (ii) $\mu_m = iA$

7. In SI units write the relation between B, H and M.
Ans. $B = \mu_0(H + M)$

8. What are diamagnetic materials?
Ans. These materials exhibit a weak, negative susceptibility and relative permeability less than 1. They are repelled by a magnetic field.

9. What are paramagnetic materials?
Ans. These materials possess a weak, positive susceptibility and relative permeability greater than 1. The paramagnetic susceptibility obeys Curie's law: $\chi_m^{\text{Para}} = C/T$.

10. When does ferromagnetic materials exhibit hysteresis?

Ans. These substances exhibit hysteresis when they are subjected to cyclic magnetization and demagnetization.

11. What is magnetostriction?

Ans. Magnetostriction is the change in the length of a ferromagnetic material in the direction of magnetization, when a magnetic field is applied.

12. What are soft ferromagnetic materials?

Ans. These are the materials characterized by narrow hysteresis loop, a high permeability and a low coercive force.

13. When magnetic materials are placed in alternating fields, how they suffer?

Ans. i. Hysteresis loss, which is the work dissipated in moving the domain boundaries during a magnetization cycle.

ii. Eddy current losses due to induced currents in the material.

14. What are hard magnetic materials?

Ans. These are characterized by large hysteresis loop, high coercive force and high saturation induction.

15. What are soft ferrites?

Ans. These are ceramic magnets which are electrically insulators. They are used for high frequency transformer cores.

OBJECTIVE QUESTIONS

1. Magnetic permeability is _____.
   [unitless]

2. Weber is equivalent to _____.
   [volt second]

3. The relation between magnetic susceptibility (χ) and relative permeability (μ_r) is _____.
   $[\mu_r = 1 + \chi]$

4. The susceptibility (χ) and Curie temperature (θ) for a magnetic material are related through the relation _____. $\left[\chi = \dfrac{C}{T - \theta}\right]$.

5. Relativity permeability μ_r = _____.
   $[\mu/\mu_0]$

6. According to Langevin's theory, the paramagnetic susceptibility is given by the relation _____. $\left[\chi_m^{\text{Para}} = \dfrac{N\mu_0\mu_m^2}{3kT}\right]$

7. In ferromagnetic materials, Weiss field is given by _____. $[H_i = H + \gamma M]$

8. One Bohr magneton = _____.
   $[9.27 \times 10^{-24}\,\text{A m}^2]$

9. The antiferromagnetic and ferrimagnetic materials become paramagnetic when they are _____ above Neel temperature.   [heated]

10. The ultimate source of magnetic properties of materials are _____ and _____ in an atom.
    [orbiting electrons, spinning electrons]

MULTIPLE CHOICE QUESTIONS

1. The unit of magnetic induction B is
   (a) Wb
   (b) N Wb
   (c) N/A m
   (d) W m^{-2}

2. The unit of permeability is
   (a) Am/Wb
   (b) Wb
   (c) Wb m^{-2}
   (d) Wb A^{-1} m^{-1}

3. The three vectors $\vec{B}$, $\vec{H}$ and $\vec{M}$ in a magnetic material are related as

   (a) $\vec{B} = \dfrac{\vec{M}}{\vec{H}}$
   (b) $\vec{B} = \mu_0(\vec{H} + \vec{M})$

   (c) $\vec{B} = \vec{H} + \vec{M}$
   (d) $\vec{B} = \mu_0\left(\dfrac{\vec{M}}{\vec{H}}\right)$

4. The electron revolves in an orbit of radius r meters inside the atom around the nucleus with a speed V ms^{-1}. The magnetic moment associated with it is

   (a) $\dfrac{eV}{r^2}$
   (b) $\dfrac{eVr}{2}$

   (c) $\dfrac{eV^2}{2}$
   (d) $\dfrac{eVr}{4}$

5. When a paramagnetic atom is placed in an externally applied magnetic field, the resulting magnetic moment of atom is

   (a) $g\,\mu_B\,(j + 1)$
   (b) $g\,\mu_B\,j$

   (c) $g\,\mu_B\,\sqrt{j(j+1)}$
   (d) $g\,\mu_B\,\sqrt{(j+1)}$

6. The materials that are useful as induction cores are

   (a) garnets

   (b) alnico magnets

   (c) barium-strontium ferrites

   (d) Ni-Zn ferrites

7. The material that is useful for magnetic recording is
   (a) $Y_3Fe_5O_{12}$ (b) γ–Fe_2O_3
   (c) nickel-zinc ferrite (d) cobalt ferrite

8. Ferrite possess
   (a) high permeability values compared to paramagnetism, high resistivity and low eddy currents
   (b) high hysteresis loss, low magnetocrystalline anisotropy and low permeability values
   (c) high resistivity, low permeability, low magnetocrystalline energy
   (d) anisotropic energy which is independent of the temperature and field.

9. The materials that are used in wind screen wiper motors, door catches, correction magnets, etc. are
   (a) hard ferrites   (b) steel
   (c) alnico   (d) Ni-Zn ferrite

10. The extrapolated line of $1/\chi$ *vs* T plot
    (a) passes through the origin
    (b) makes a positive intercept at a positive temperature θ_N on T-axis
    (c) makes an intercept on $1/\chi$ axis on the negative side
    (d) makes an intercept on the negative side of the T-axis

11. Magnetic induction B and magnetic field intensity H are related by
    (a) $B = \mu_0 H + \mu_0$ (b) $B = \mu_0 \mu_r H$
    (c) $B = \mu_0 H^2$ (d) $B = \mu_0 + H$

12. Which one of the following material does not have permanent magnetic dipoles?
    (a) paramagnetic
    (b) antiferromagnetic
    (c) ferromagnetic
    (d) diamagnetic

ANSWERS

1. (d)	**2.** (b)	**3.** (b)	**4.** (b)	**5.** (c)
6. (d)	**7.** (b)	**8.** (a)	**9.** (a)	**10.** (d)
11. (b)	**12.** (d)			

12

Electrostatics and Electromagnetism

ELECTROSTATICS

12.1 INTRODUCTION

Physics is based on the concept that all matter is made up of atoms and that these atoms in turn are made up of electrically charged particles whose behaviour is governed by electrodynamics.

There are two kinds of electric charges: *positive and negative*. Like charges repel each other while unlike charges attract each other. The algebraic sum of electric charges remains constant in a closed system. This is known as the *law of conservation of electric charges*.

The electric charge on any body consists of a whole number of elementary charges each of which equals 1.6×10^{-19} C charge. The smallest stable subatomic particle having a negative elementary charge is called the *electron*. The mass of an electron is 9.1×10^{-31} kg. The smallest stable subatomic particle having a positive charge is the proton. The mass of the proton is 1.67×10^{-27} kg. Electrons and protons are found in the atoms of any substance. A neutral body contains charges of opposite sign which are equal in absolute value. Electric charges are said to be point charges if the linear dimensions of the bodies on which the charges are concentrated are very much smaller than any other lengths pertinent to the problem under consideration. A body exhibiting positive electrification has a positive electric charge, and one with negative electrification has a negative electric charge. The net charge of a body is the algebraic sum of its positive and negative charges. A particle having a non-zero net charge is often called an *ion*. Since matter in bulk does not exhibit gross electrical forces, one may assume that it is composed of equal amounts of positive and negative charges. *The net or total charge does not change for any process occurring within an isolated system.* No exception has been found to this rule, i.e. *principle of conservation of charge*.

12.2 COULOMB'S LAW

The results of the electric interaction between two charged particles at rest in the observer's inertial frame of reference or, at most, moving with a very small velocity constitute what is called electrostatics. The electrostatic interaction for two charged particles is given by *Coulomb's law*.

Coulomb from experimental observations concluded that *the force of attraction or repulsion between two stationary point charges is* (i) *directly proportional to the product of the magnitude of the two charges and* (ii) *inversely proportional to the square of the distance between them, and its direction is along the line joining the two charges. This is Coulomb's law.*

Let '1' and '2' be two particles carrying charges q_1 and q_2 respectively and separated by a distance r_{12} (Fig. 12.1). According to Coulomb's law, the electric force exerted by the particle '1' on the particle '2' is

$$F_{12} \propto \frac{q_1 q_2}{r_{12}^2} \text{ i.e. } F_{12} = k\frac{q_1 q_2}{r_{12}^2} = -F_{21} \qquad (12.1)$$

where k is the constant of proportionality and F_{21} is the force exerted by the particle '2' on particle '1'. Expressing Eq. (12.1) in vector notation, one obtains

$$\boldsymbol{F}_{12} = k\frac{q_1 q_2}{|\boldsymbol{r}_{12}|^2} \hat{r}_{12} = k\frac{q_1 q_2}{|\boldsymbol{r}_{12}|^3} \boldsymbol{r}_{12} \qquad (12.2)$$

where $\hat{r}_{12}$ is the unit vector along $\boldsymbol{r}_{12}$.

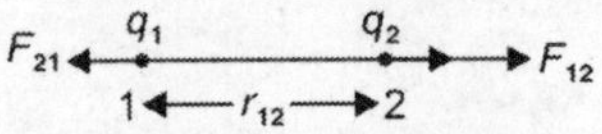

Fig. 12.1: Coulomb's law

Coulomb's law is very similar to the law of gravitational interaction. The sign of charges in Eq. (12.2) decides whether the force is *attractive* or *repulsive*. When both the charges are similar (i.e. positive or negative), F_{12} is positive and represents the force of repulsion; while if one is positive and other is negative, $\boldsymbol{F}_{12}$ is negative and is the force of attraction.

The numerical value of the constant k in the MKSC system is equal to $10^{-7} c^2 = 8.9874 \times 10^9$, where c is the velocity of light. For practical purposes, we may say that $k = 9 \times 10^9$. Then when the distance r_{12} is measured in metre and the force in newton, Eq. (12.1) becomes

$$F_{12} = 9 \times 10^9 \frac{q_1 q_2}{r_{12}^2} = -F_{21} \qquad (12.3)$$

Once we have decided on the value of k, the unit charge is fixed. This unit is called as coulomb and is designated by C. *The coulomb is that charge which when placed one metre from an equal charge in vacuum, repels it with a force of 8.987 $\times 10^9$ newtons.* Formula in Eq. (12.3) holds only for two charged particles in vacuum, i.e. for two charged particles in the absence of any other charge or matter. We must note that according to Eq. (12.2), the unit of k is Nm^2C^{-2} or $m^3 kg\, s^{-2}C^{-2}$. The constant k is then equal to

$$k = \frac{1}{4\pi\varepsilon_0}$$

where ε_0 is the vacuum permeability or *permittivity of free space*. It is also sometimes referred to as *capacitivity* in vacuum, or the dielectric constant of vacuum. Thus,

$$\varepsilon_0 = \frac{10^7}{4\pi c^2}$$

$$= 8.854 \times 10^{12}\, N^{-1}\, m^{-2} C^2 \text{ (or } m^{-3} kg^{-1} s^2)$$

Now, the Coulomb's law is written as

$$\boldsymbol{F}_{12} = \frac{1}{4\pi\varepsilon_0}\frac{q_1 q_2}{|\boldsymbol{r}_{12}|^3}\boldsymbol{r}_{12} \quad (N) \qquad (12.4)$$

The forces of electrostatic interaction are central and long range. One can obtain an idea about the relative magnitudes of electrostatic and gravitational forces by comparing the force of attraction between two electrons due to gravitation and the force of repulsion due to electrostatic repulsion, i.e.

$$\frac{\text{Electrostatic repulsion between two electrons}}{\text{Gravitational attraction between two electrons}}$$

$$= \frac{1}{4\pi\varepsilon_0}\frac{e^2}{r^2}\frac{r^2}{Gm^2} = \frac{e^2}{4\pi\varepsilon_0 Gm^2}$$

$$= 9 \times 10^9 \times \frac{\left(1.6 \times 10^{-19}\right)^2}{\left(6.67 \times 10^{-11}\right)\left(9.1 \times 10^{-31}\right)^2}$$

$$= 4.17 \times 10^{42}$$

Obviously, the electrical interaction is of the order of magnitude required to produce the binding between atoms to form molecules, or the binding between the electrons and protons to form atoms.

This reveals that the chemical processes (and in general the behaviour of matter in bulk) are due to electrical interactions between atoms and molecules.

Coulomb's law is based primarily on experiments. At this stage, one can raise a question whether the law is exactly that of inverse square, i.e. the force is proportional to $\frac{1}{r^n}$, is n exactly equal to 2?

Cavendish found that $n = 2 \pm 0.02$. Plimpton and Lawton in 1936 reported that n differs from 2 by not more than one part in 10^9. Lamb and Rutherford in 1947 found from their measurements of energy levels of the hydrogen atom that the exponent in Coulomb's law is correct top one part in 10^9 at distances of the order of 10^{-10} m. Evidence from nuclear experimental observation has shown that the electrostatic forces vary approximately according to inverse square law even at distances of the order of 10^{-15} m.

12.3 ELECTRIC FIELD

Any region where an electric charge experiences a force is called an *electric field*. The force is due to the presence of other charges in that region. For example, a charge q placed in a region where there are other charges q_1, q_2, q_3, etc. (Fig. 12.2) experiences a force $F = F_1 + F_2 + F_3 + \ldots$ and we say that it is in an electric field produced by the charges q_1, q_2, q_3.... We must note that charge q, of course, also exerts forces on q_1, q_2, q_3,..., but we are not concerned with them now. Since the force that each charge q_1, q_2, q_3, ... produce on the charge q is proportional to q, the resultant force F is also proportional to q. Obviously, the force on a

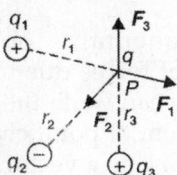

Fig. 12.2: Resultant electric field at P due to the presence of several charges in the region

particle placed in an electric field is proportional to the charge on the particle.

The intensity of the electric field at a point is equal to the force per unit charge placed at that point. Thus,

$$E = \frac{F}{q} \quad \text{or} \quad F = qE \qquad (12.5)$$

The unit of electric field intensity is newton/coulomb or $N\,C^{-1}$.

If q is positive, the force F acting on the charge has the same direction as the field E, but if q is negative, the force F has the direction opposite to E (Fig. 12.3).

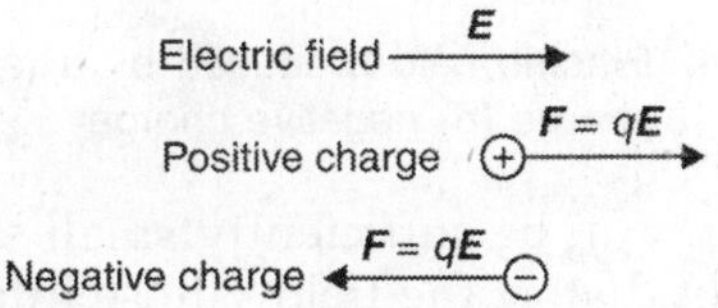

Electric field $\longrightarrow E$

Positive charge $\oplus \xrightarrow{F = qE}$

Negative charge $\xleftarrow{F = qE} \ominus$

Fig. 12.3: Direction of the force produced by an electric field on a positive and a negative charge

Obviously, if we apply an electric field to a region where positive and negative particles or ions are present, the field will tend to move the positively and negatively charged bodies in opposite directions, resulting in a charge separation, an effect sometimes referred to as *polarization*.

Let us write Eq. (12.3) in the form

$F = q_2 \left(\dfrac{q_1}{4\pi\varepsilon_0 r_{12}^2} \right)$. This gives the force produced by the charge q_1 on the charge q_2 placed at a distance r_{12} from q_1. Using Eq. (12.5), we may say that the electric field E at the point where q_2 is placed is such that $F = q_2 E$. Obviously, by comparing both expressions of F, we conclude that the electric field at a distance r_{12} from the charge q_1 is $E = \dfrac{q_1}{4\pi\varepsilon_0 r_{12}^2}$, or in the vector form,

$$E = \frac{q_1}{4\pi\varepsilon_0 r_{12}^2}\hat{a}_r \qquad (12.6)$$

where $\hat{a}_r$ is the unit vector in the radial direction away from the charge q, since F is along this direction. Equation (12.6) is valid for both positive and negative charges with the direction of E relative to $\hat{a}_r$ given by the sign of q_1. Obviously, E is directed away from a positive charge and towards a negative charge (Fig. 12.4).

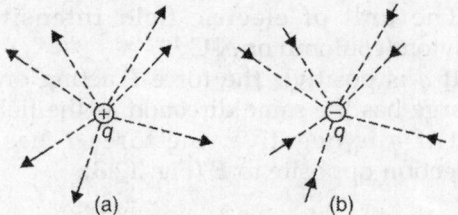

Fig. 12.4: Electric field produced by a (a) positive charge (b) negative charge

Let $q_1 = q_0$ be sufficiently small so that it does not distort the field whose intensity is to be measured. The resultant force F exerted on a test charge q by a field set up by a system of fixed charges $q_1, q_2, q_3, ..., q_n$ is equal to the vector sum of the forces F_i exerted on the test charge by each of the fields of charges q_i, i.e

$$F = \sum_{i=1}^{n} F_i \qquad (12.7)$$

This leads to the principle of superposition of electric fields as:

$$E = \sum_{i=1}^{n} E_i \qquad (12.8)$$

12.4 PRINCIPLE OF SUPERPOSITION

If there are more than two particles present with charges, say $q_1, q_2, q_3, ...$, then the total force on any one particle is the vector sum of forces it experiences due to all other particles present there separately. This is called the principle of superposition. Let there be three charges q_1, q_2 and q_3 as shown in Fig. 12.5. The force on q_3 is

$$F = F_{13} + F_{23}$$

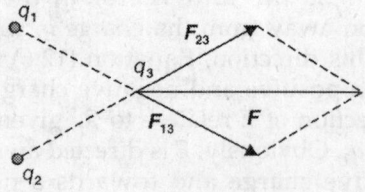

Fig. 12.5

$$= \frac{1}{4\pi\varepsilon_0} \frac{q_1 q_3}{|r_{13}|^3} r_{13} + \frac{1}{4\pi\varepsilon_0} \frac{q_2 q_3}{|r_{23}|^3} r_{23}$$

Generalising, one finds that the force acting on a charge q_j due to a number of other charges present in the region is

$$F_j = \frac{1}{4\pi\varepsilon_0} \sum_{i \neq j} \frac{q_i q_j}{|r_{13}^3|} r_{ij} \qquad (12.9)$$

Let r_i and r_j be the vectors representing the location of charges q_i and q_j respectively (Fig. 12.6). One can write the Eq. (12.9) in terms of these position vectors as

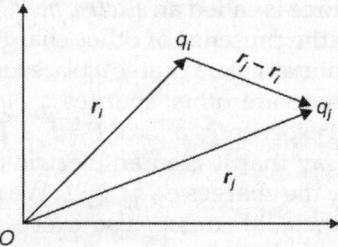

Fig. 12.6: Vector representation of location of charges

$$F_j = \frac{1}{4\pi\varepsilon_0} \sum_{i \neq j} \frac{q_i q_j}{|r_j - r_i|^3} (r_j - r_i) \quad (12.10)$$

The experimentally observed linear superposition of forces due to many charges means that one can write the electric field at X due to a system of point charges q_i located at X_i, $i = 1, 2, ..., n$ as a vector sum

$$E(X) = \frac{1}{4\pi\varepsilon_0} \sum_{i=1}^{n} q_i \frac{X - X_i}{|X - X_i|^3} \qquad (12.11)$$

If the charges are so small and so numerous that they can be described by a charge density $\rho(x')$ (if Δq is the charge in a small volume $\Delta x\, \Delta y\, \Delta z$ at the point X', then $\Delta q = \rho(X')\, \Delta x\,\Delta y\,\Delta z$), the sum in Eq. (12.11) is replaced by the integral

$$E(X) = \frac{1}{4\pi\varepsilon_0} \int \rho(X') \frac{X - X'}{|X - X'|^3} d^3 x' \qquad (12.12)$$

where $d^3x' = dx'dy'dz'$ is a three dimensional volume element at X'.

We have seen that the principle of superposition has facilitated considerably the mathematical handling of electromagnetic theory. However, this principle fails in nuclear interactions and this is one of the reasons why the nuclear theory is somewhat more complicated and troublesome than the theory of atomic interactions.

12.5 DIRAC DELTA FUNCTION

It is sometimes convenient to regard a point charge as a fictitious continuous charge distribution. One can achieve this with the help of Dirac delta function written as $\delta(x - a)$, which is a mathematically improper function having the following properties:

i. The delta function is an even function:
$\delta(-x) = \delta(x)$

ii. $x\delta(x) = 0$

iii. $x\delta(x - x_0) = x_0\delta(x - x_0)$

iv. $f(x)\,\delta(x - x_0) = f(x_0)\,\delta(x - x_0)$

v. $\delta(ax) = \dfrac{1}{a}\delta(x),\quad a > 0$

vi. $\delta(x^2 - a^2) = \dfrac{1}{2|a|}[\delta(x - a) + \delta(x + a)]$

vii. $\delta(x - b)\,\delta(a - x)\,dx = \delta(a - b)$

The Dirac delta function is defined by the conditions:

$$\delta(x) = \begin{cases} 0, & x \neq 0 \\ \infty, & x = 0 \end{cases}$$

such that

$$\int_{-\infty}^{\infty} \delta(x)\,dx = 1$$

We must remember that a delta function has the dimensions of an inverse volume whatever number of dimensions the space has.

A discrete set of point charges can be described with a charge density by means of delta functions. For example,

$$\rho(x) = \sum_{i=1}^{n} q_i\,(x - x_i) \tag{12.13}$$

represents a distribution of n point charges q_i, located at the points x_i. Using Eq. (12.13) in Eq. (12.12) and integrating using the properties of the delta function, yields the discrete sum [Eq. (12.11)].

12.6 LINES AND TUBES OF FORCE

Just as in the case of a gravitational field, an electric field may be represented by lines of force, which are lines that, at each point, are tangential to the direction of the electric field at that point, i.e. *lines of force* or *field lines* are parallel to the direction of the field everywhere. Since there is a single direction for the electric field at every point of the field and hence there is usually just one field line through any given point, i.e. two field lines can never intersect. The lines of force in Fig. 12.7a depict the electric field of a positive charge, and those in Fig. 12.7b show the electric field of a negative charge. We can see that they are straight lines passing through the charges.

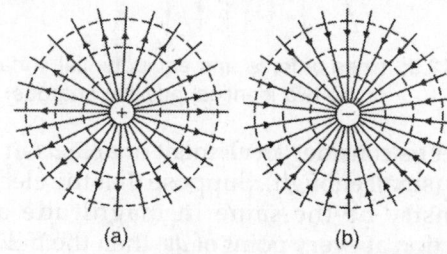

(a)   (b)

Fig. 12.7: Lines of force and equipotential surfaces of the electric field of a (a) Positive charge (b) Negative charge

Figures 12.8 and 12.9 show the lines of force near a pair of charges of equal magnitude, one positive and one negative and for two equal positive charges respectively.

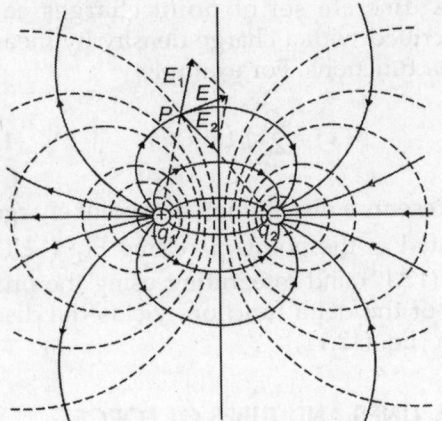

Fig. 12.8: Lines of force and equipotential surfaces of the electric field of two equal but opposite charges

In both figures, the lines of force of the resultant electric field produced by the two charges have also been represented.

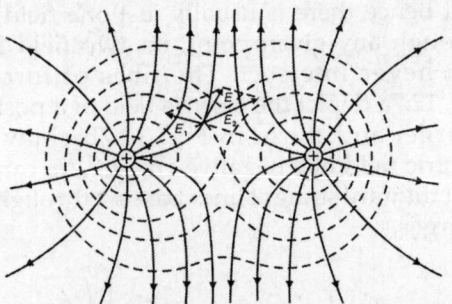

Fig. 12.9: Lines of force and equipotential surfaces of the two identical positive charges

Let us consider an element of area *ds* in the field (small enough). Suppose that the electric intensity be the same in magnitude and direction at every point of *ds*, then the field of lines or lines of force at all points on the boundary will be approximately all parallel.

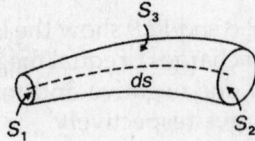

Fig. 12.10: Tube of flux

One obtains the tubular surface as shown in Fig. 12.10 and it is called a *tube of force*. We must note that the normal cross-section of the tube of force forms a part of an *equipotential surface*.

12.7 ELECTRIC FLUX

The concept of electric flux is of great usefulness in many physical problems. To understand the concept of electric flux, let us imagine that a fluid is flowing with a speed *v* through a small flat surface *ds* in a direction normal to the surface as shown in Fig. 12.11a. The rate of flow of the fluid, i.e. the volume of the fluid crossing the area per unit time is termed as the flux of fluid and it is equal to *vds*. Now, if the normal to the surface is not parallel to the direction of flow of the fluid and makes an angle with the surface as shown in Fig. 12.11b, then one finds the projected area in a plane perpendicular to *v* is $ds\cos\theta$. Thus, the flux *F* is

$$F = vds\cos\theta \qquad (12.14)$$

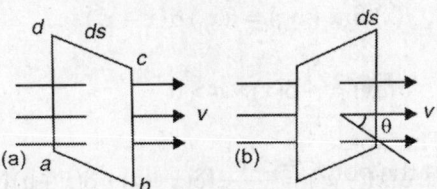

Fig. 12.11: Concept of electric flux

Let us represent the surface by vector $ds\,\hat{a}_n$ having magnitude *ds* and direction along the normal to the surface, $\hat{a}_n$ being the unit vector in this direction. One can express the flux as

$$F = v \cdot \hat{a}_n\,ds \qquad (12.15)$$

In an electrostatic field, there is nothing actually flowing, one can mathematically define a quantity analogous to the flux of a fluid. This is called as *electric flux* and is defined as

Electric flux, $\phi = E \cdot \hat{a}_n\,ds \qquad (12.16)$

The concepts of electric flux and tubes are sometimes useful in drawing the electric field lines about charges.

12.8 SOLID ANGLE

When every point on the boundary of a surface element, e.g. *CD* of area *ds* is joined to a given point *O*, then one finds that as a result of it, a cone is formed. Let us imagine a spherical surface with centre at *O* and of radius *r* (=distance of surface element from *O*), then the above cone will intersect this surface in the form of another surface *AB* of area *da*. We can easily see that the area *da* is the projection of the area *ds* perpendicular to *OP*, i.e. *da = ds* cos θ, where θ is the angle between the surfaces *AB* and *CD* (Fig. 12.12). Obviously, if we consider the surface element as vector whose direction is along the outward normal to the surface element and *O* is considered as origin, then the angle between *ds* and *r* will be θ. We have

$$da = \hat{r} \cdot ds = ds \cos \theta$$

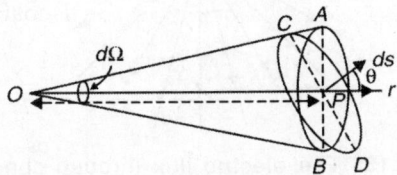

Fig. 12.12: Concept of a solid angle

The angle formed by this surface element *da* at the point *O* in three dimensions is termed *solid angle*. The solid angle formed by the area *ds* at a point *O* is obtained as the ratio of the normal area to the square of the distance, i.e.

$$d\Omega = \frac{da}{r^2} = \frac{ds \cos \theta}{r^2} \qquad (12.16a)$$

Solid angle has no dimension like an angle, but it is represented by the unit, *steradian*. If *r* = 1, then *dΩ = da*. Thus, *the solid angle formed by any surface at point O is equal in magnitude to the area intersected by the cone or pyramid formed by joining every point of that surface to the point O, on a spherical surface of unit radius and having centre at O.*

Let us now determine the solid angle formed by the element of the area *PQRS* situated on the spherical surface of radius *r*,

at the centre *O* (Fig. 12.13). Let us draw planes *TOM* and *TON* making angles φ and φ + *d*φ respectively from the plane *ZOX*, i.e. the angle formed by the side *SP* of area element *PQRS* on the Z-axis is *d*φ. From Fig. 12.13, it is evident that lines *OP* and *OQ* make angles θ and θ + *d*θ from the Z-axis and hence the angle formed at the centre *O* by the side *PQ* of the area element is *d*θ as shown in Fig. 12.13.

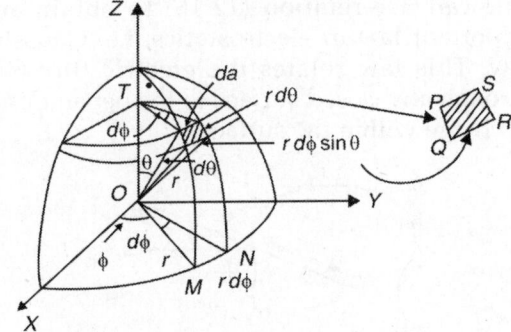

Fig. 12.13: Determination of solid angle

The radius of the spherical surface *PQRS* is *r*, and hence

$$PQ = rd\theta$$

From Δ*OPT*, we have

$$\frac{TP}{OP} = \sin \theta$$

∴ $$TP = OP \sin \theta = r \sin \theta \qquad (12.16b)$$

or $$SP = (r \sin \theta) d\phi$$

∴ Area of element *PQRS*

$$= PQ \times PS = (rd\theta) r \sin \theta \, d\phi$$

or $$da = r^2 \sin \theta \, d\theta \, d\phi \qquad (12.16c)$$

∴ Soild angle $$d\Omega = \frac{da}{r^2} = \frac{r^2 \sin \theta \, d\theta \, d\phi}{r^2}$$

$$= \sin \theta \, d\theta \, d\phi \qquad (12.16d)$$

Now, the solid angle formed at the centre by the entire spherical surface is equal to the sum of solid angles formed at the centre by all the area elements situated on the surface. Since the values of θ and φ vary from 0 to π and 0 to 2π respectively for the entire spherical surface and hence solid angle subtended by spherical surface at its centre is

$$\Omega = \int_0^{2\pi} \int_0^{\pi} \sin\theta \, d\theta \, d\phi = 2\pi \int_0^{\pi} \sin\theta \, d\theta$$

$$= 2\pi(-\cos\theta)_0^{\pi} = 4\pi \qquad (12.16e)$$

12.9 GAUSS'S LAW FOR THE ELECTRIC FIELD (INTEGRAL FORM)

One can use relation (12.16) to obtain an important law in electrostatics, i.e. Gauss's law. This law relates the electric flux (ϕ) through any closed surface to the net amount of charge within the surface.

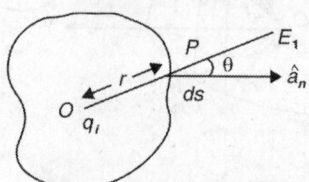

Fig. 12.14: A closed surface surrounding charge q_i

Let us now imagine a closed surface S surrounding a charge q_i as shown in Fig. 12.14. Let ds be an element of area around the point P on the surface and $\hat{a}_n$ an outward unit vector normal to it. Let θ be the angle between the electric field at P and unit vector $\hat{a}_n$. The electric flux through the element of area ds is

$$d\phi = E_i \cdot \hat{a}_n \, ds = E_i \cos\theta \, ds$$

$$= \frac{1}{4\pi\varepsilon_0} \frac{q_i}{r^2} \cos\theta \, ds$$

where r is the distance of the element ds from charge q_i.

Now $d\Omega = \dfrac{ds\cos\theta}{r^2}$ is the angle subtended by ds at q_i. Thus,

$$d\phi = E_i \cdot \hat{a}_n ds = \frac{1}{4\pi\varepsilon_0} q_i d\Omega \qquad (12.17)$$

Thus, the total flux through the entire surface S is

$$\phi = \int_S E_i \cdot \hat{a}_n ds = \frac{1}{4\pi\varepsilon_0} q_i \int d\Omega = \frac{q_i}{\varepsilon_0} \qquad (12.18)$$

Obviously, the electric flux through the surface S, then is proportional to the charge and independent of the distance r (in case of a spherical surface, r being the radius). Therefore, if we draw several concentric spherical surfaces S_1, S_2, S_3,... (Fig. 12.15) around the charge q_i, the electric flux through all of them is the same and equal to $\dfrac{q}{\varepsilon_0}$. The result is due to $\dfrac{1}{r^2}$ dependence of the field. This is also true in the case of gravitational field.

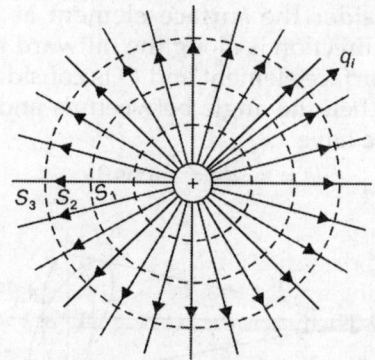

Fig. 12.15: The electric flux through concentric spheres surrounding the same charge is same.

If there is any arbitrary distribution of charges within the surface S, then by the principle of superposition

$$E = \sum E_i \qquad (12.19)$$

Equation (12.19) leads to the relation

$$\oint_s E_i \cdot \hat{a}_n ds = \frac{\Sigma q_i}{\varepsilon_0} = \frac{Q}{\varepsilon_0} \qquad (12.20)$$

where Q is the total charge ($= \Sigma q_i$) within the closed surface S. This result is valid for any closed surface, irrespective of the position of the charge within the surface.

In case of a continuous distribution of charge within the surface S, the relation (12.20) changes to

$$\oint_s E_i \cdot \hat{a}_n ds = \frac{1}{\varepsilon_0} \int \rho dv \qquad (12.21)$$

If no charges are present inside the closed surface S, or if the net charge is zero, the total electric flux through it is zero. The charges outside the closed surface do not contribute to the total flux. One can easily verify it. Let us suppose that charge q_i is outside the closed surface S (Fig. 12.16). The electric field vector E passes through two elements of area ds_1 and ds_2 cut out by a cone with its apex at charge q_i as shown in Fig. 12.16. $\hat{a}_{n_1}$ and $\hat{a}_{n_2}$ are the

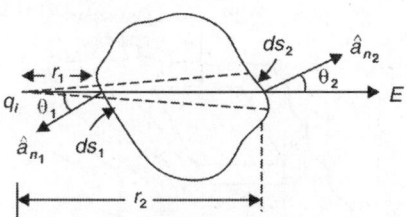

Fig. 12.16: Charge outside the closed surface

outward drawn unit vectors normal to ds_1 and ds_2 making angles θ_1 and θ_2 respectively with the field. The net outward flux through ds_1 and ds_2 area is

$$d\phi = \frac{1}{4\pi\varepsilon_0}\left[\frac{q_i}{r_1^2}\cos(180° + \theta_1)ds_1\right.$$

$$\left. + \frac{q_i}{r_2^2}\cos\theta_2 ds_2\right]$$

where r_1 and r_2 are the distances of the elements ds_1 and ds_2 from charge q_i. Thus,

$$d\phi = \frac{1}{4\pi\varepsilon_0}\left[-\frac{\cos\theta_1}{r_1^2}ds_1 + \frac{\cos\theta_2}{r_2^2}ds_2\right]$$

Now, the solid angle subtended by ds_1 and ds_2 at q_i are the same, i.e.

$$\frac{\cos\theta_1}{r_1^2}ds_1 = \frac{\cos\theta_2}{r_2^2}ds_2 = d\Omega$$

Hence $\qquad d\theta = 0$

One can see that this holds good for all the cones drawn from q_i through the surface, and hence, the total flux $\phi = 0$.

The results constitute what is known as Gauss's law for the electric field. One can state it as follows

$$\left.\begin{array}{l}\phi_S E \cdot \hat{a}_n ds = \dfrac{Q}{\varepsilon_0}\text{ for charge } Q \text{ inside } S\\[4mm] = 0 \text{ for charge } Q \text{ outside } S\end{array}\right\} \quad (12.22)$$

Relation (12.22) is the *integral form* of Gauss's law for the electric field.

Gauss's law is particularly useful when we wish to compute the electric field produced by charge distribution having certain geometrical symmetries.

12.10 GAUSS'S LAW FOR THE ELECTRIC FIELD (DIFFERENTIAL FORM)

One can express the Gauss's law in differential form also. We have the relation

$$\text{div } A = \lim_{dV \to 0} \frac{\oint A \cdot \hat{a}_n\, ds}{dV} \quad (12.23)$$

where S is the surface enclosing the volume element. Integrating over the finite volume, one obtains

$$\int_V \text{div} A\, dV = \oint A \cdot \hat{a}_n ds \quad (12.24)$$

We are familiar with Eq. (12.24), which is the well known *divergence theorem* usually used for transforming a volume integral into a surface integral and *vice versa*. Using Eq. (12.24), the Gauss's law can be expressed as

$$\oint E \cdot \hat{a}_n ds = \oint_V \text{div} E dV = \frac{1}{\varepsilon_0}\int_V \rho dV$$

i.e. $\qquad \displaystyle\int_V\left(\text{div } E - \frac{\rho}{\varepsilon_0}\right)dV = 0$

The above is true for any arbitrary volume V. Thus,

$$\text{div} E - \frac{\rho}{\varepsilon_0} = 0$$

or $\qquad\qquad \nabla \cdot E = \dfrac{\rho}{\varepsilon_0} \quad (12.25)$

Relation (12.25) is the differential form of Gauss's law for the electric field. In this form, the law shows that the sources of electric displacement are free electric charges (since $\rho \neq 0$), i.e. the electric flux lines either leave or enter a closed surface, therefore, there must be either a source or sink of flux lines. We know from convention that flux lines start from positive charges as sources and terminate in negative charges as sinks. This shows that positive and negative charges exist independently. Obviously, Gauss's law in differential form expresses a local relation between E and ρ. Thus, we may say that electric charges are the sources of the electric field, and that their distribution and magnitude determine the electric field at each point of space.

We have adopted Coulomb's law in electrostatics as the fundamental law. We have also seen that the Gauss's law in electrostatics is a consequence of the fact that the electric force between charged particles is inversely proportional to the square of the distance between them, i.e. Coulomb's inverse square law. One can easily see that any other law, say

$\dfrac{1}{r_n}$ with $n \neq 2$, would not give Gauss's law.

Obviously, we can take Gauss's law as also the fundamental law of electrostatics.

12.11 APPLICATIONS OF GAUSS'S LAW

This law provides a powerful method for evaluating the electric field intensity E and potential variation due to simple charge distribution whose *symmetry* is such that the field is constant and normal over the entire Gaussian surface. We shall discuss some useful applications of Gauss's law in this section. We must remember that *symmetry* is crucial to the application of Gauss's law. There are only three kinds of symmetry which are sufficient:

1. *Spherical symmetry*: One can make Gaussian surface a concentric sphere .

2. *Cylindrical symmetry*: One can make Gaussian surface a coaxial cylinder (Fig. 12.17).

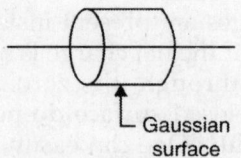

Fig. 12.17: Cylindrical symmetry

3. *Plane symmetry*: One can make use of Gaussian pill box, which straddles the surface (Fig. 12.18).

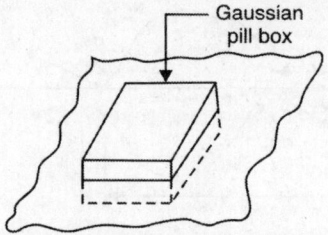

Fig. 12.18: Plane symmetry

Field Due to an Infinite Layer of Positive Charge with Uniform Surface Density (σ)

Let us consider that *ABCD* be a plane as shown in Fig. 12.19. Symmetry leads that the field lines are perpendicular to the plane. Now, we consider a right circular cylinder *PQR* with cross-section *ds*.

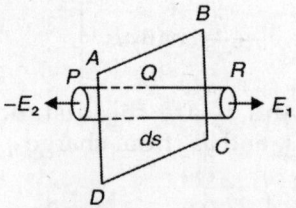

Fig. 12.19: Field due to an infinite layer of positive charge

One obtains from Gauss's law,

$$(E_1 - E_2)ds = \frac{\sigma ds}{\varepsilon_0}$$

$$\therefore \qquad E_1 - E_2 = \frac{\sigma}{\varepsilon_0} \qquad (12.26)$$

Obviously, the field changes across a charge layer and the change is equal to $\dfrac{\sigma}{\varepsilon_0}$.

Field Outside an Isolated Charged Sphere

Let us suppose that a sphere A is filled with a uniform charge distribution (Fig. 12.20). We are interested in finding the electric field at a point P outside the sphere.

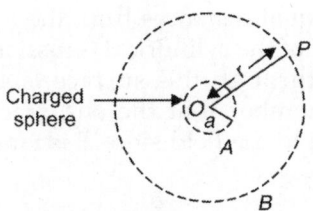

Fig. 12.20: Field outside an isolated charged sphere

Construct a spherical surface B concentric with the charged sphere A and passing through the point P. Let r be the radius of the imaginary sphere B and its surface area is $4\pi r^2$. By symmetry, we can say that the electric intensity E is same at every point of the surface. Obviously, the flux outwards through the surface is

$$\oint_s E \cdot \hat{a}_n ds = 4\pi r^2 E = \frac{Q}{\varepsilon_0} \quad \text{(Guass's law)}$$

where Q is the total charge within the sphere. Thus,

$$E = \frac{1}{4\pi\varepsilon_0} \frac{Q}{r^2} \qquad (12.27)$$

Equation (12.27) is the same as the field produced by the point charge Q at the centre of the sphere. If there is continuous charge distribution within the sphere, we have

$$Q = \frac{4}{3}\pi a^3 \rho$$

where ρ is the charge density and a the radius of the sphere. Thus,

$$E = \frac{1}{4\pi\varepsilon_0} \frac{4\pi a^3 \rho}{3r^2} = \frac{a^3 \rho}{3\varepsilon_0 r^2} \qquad (12.28)$$

Electric Field Due to a Spherical Shell of Charge

Following the arguments applied in the case of a field outside an isolated charged sphere, one can see that the field outside a thin spherical shell of charge is the same as if the total charge on the shell is concentrated at the centre. Now, we are interested in finding out the field at a point O inside the shell.

Let us imagine a cone with apex at O and extending on either side to cut surface elements ds_1 and ds_2 as shown in Fig. 12.21. Let r_1 and r_2 be the distances of ds_1 and ds_2 from O respectively. Let σ be the surface density of charge, the fields at O due to elements are $\dfrac{\sigma s_1}{r_1^2}$ and $\dfrac{\sigma ds_2}{r_2^2}$ and these will act in opposite directions. However,

$$\frac{ds_1}{r_1^2} = \frac{ds_2}{r_2^2} = d\Omega$$

($d\Omega$ is the solid angle subtended at O by ds_1 and ds_2).

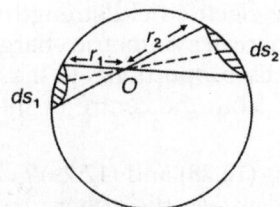

Fig. 12.21: Field due to a spherical shell of charge

Thus, the contribution to the electric field due to the two elements being equal and opposite cancel exactly. Extending this to cover the entire shell by balancing the field contributions of opposite differential areas, one can see that each pair of differential area gives a zero contribution. Thus, net field at O is zero.

Electric Field Strength at an Internal Point

Let P be an internal point at a distance r from the centre of the charge distribution O as shown in Fig. 12.22. Imagine a sphere of radius

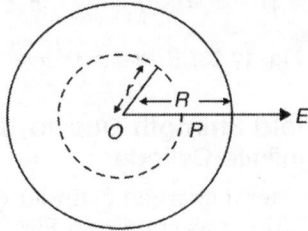

Fig. 12.22: Electric field at an internal point

r (= OP) concentric with the spherical charge. Let ρ be the volume charge density (i.e. charge per unit volume), i.e.

$$\rho = \frac{\text{Charge}}{\text{Volume}} = \frac{Q}{\frac{4}{3}R^3}$$

We must note that the charge in the shell of thickness $R - r$ does not contribute to the field at P since the point P lies inside the shell. If E is the field at P, the flux through the imaginary or Gaussian surface is

$$4\pi r^2 E = \frac{1}{\varepsilon_0}\frac{4}{3}\pi r^3 \rho \quad \text{(Guass's law)}$$

or
$$E = \frac{r\rho}{3\varepsilon_0} = \frac{1}{4\pi\varepsilon_0}\frac{Qr}{R^3}\frac{\sigma ds_2}{r^2} \quad (12.29)$$

Thus, the electric field strength at point P inside a spherical symmetric charge distribution is directly proportional to the distance of the point P from the centre of the spherical charge.

Equations (12.28) and (12.29) give the field outside and inside the sphere respectively. One can see that the two forms match, as they should, when $r = a$.

The variation of magnitude of electric field strength with distance from the centre of the spherically charge distribution is represented by the curve as shown in Fig. 12.23.

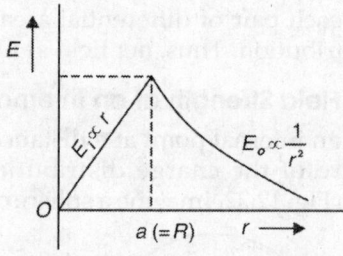

Fig. 12.23: E versus r curve

Electric Field Strength Due to Uniformly Charged Infinite Cylinder

Let us consider a charged cylinder of infinite length of radius r as shown in Fig. 12.24. We are interested in finding the field at a point P

at a distance d from the axis. For this, let us imagine a closed cylindrical surface coaxial with the cylinder and passing through P, its two end faces being perpendicular to the axis of the cylinder. One can easily see by symmetry that the field will be normal to the axis directed away from the axis and is the same at equal distances from the axis. If l is the height of the cylindrical Gaussian surface, the flux through this surface is $4\pi lE$. We must remember that the end faces do not contribute to the field since E is tangential to these faces.

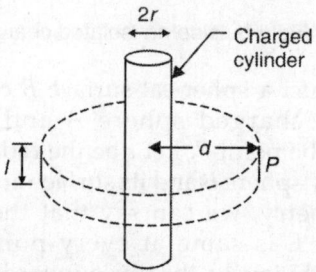

Fig. 12.24: Electric field due to uniformly charged infinite cylinder

Applying Gauss's law, we have

$$2\pi lE = \frac{\lambda l}{\varepsilon_0}$$

where λ is the charge per unit length. Thus,

$$E = \frac{\lambda}{2\pi\varepsilon_d} = \frac{1}{4\pi\varepsilon_0}\frac{2\lambda}{d} \quad (12.30)$$

Obviously, the electric field strength at a distance d from an infinite cylinder uniformly charged with linear charge density λ has

magnitude $\left(\dfrac{\lambda}{2\pi\varepsilon_0}\right)$ which is inversely

proportional to d. We can further see that the result is independent of the radius of charged cylinder and also holds for the rectilinear distribution of charges.

One could obtain the above result by direct integration. Let us consider a filament of finite length of constant line density of charge λ C/m as shown in Fig. 12.25. Now, the field at a point P due to a line element dz is

$$dE_d = \frac{\lambda dz \cos\theta}{4\pi\varepsilon_0 r^2}$$

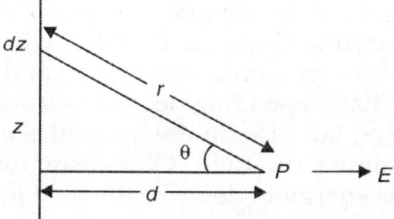

Fig. 12.25: Field at point due to a line element

Here we have assumed that the filament coincide with the Z-axis. The distance of the point P from the element is dz and the perpendicular distance from the Z-axis is d. One obtains the total field as

$$E_d = \frac{\lambda}{4\pi\varepsilon_0} \int_{-\infty}^{\infty} \frac{\cos\theta}{r^2} dz$$

We have $z = d\tan\theta$ and $r = d\sec\theta$

$$\therefore \quad E_d = \frac{\lambda}{4\pi\varepsilon_0} \int_{-\pi/2}^{\pi/2} \frac{\cos\theta}{d} d\theta = \frac{\lambda}{2\pi\varepsilon_0 d}$$

Electrostatic Field between Two Concentric Spheres with Equal and Opposite Charges

Figure 12.26 exhibits two concentric spheres 1 and 2 which have equal and opposite charges. Imagine a Gaussian surface drawn through P. Obviously, the field at P due to the outer sphere is zero and that due to the inner sphere is $E = \frac{1}{4\pi\varepsilon_0} \frac{q}{r^2}$.

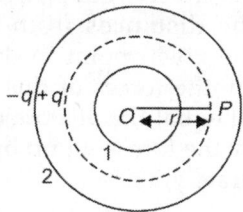

Fig. 12.26: Electrostatic field between two concentric spheres

Let us now consider the case of two nonconcentric charged spheres as shown in Fig. 12.27. The figure shows the redistribution of charge and field lines. Draw a cone with P as the apex. We can see that the two charge densities on ds_1 and ds_2 are not equal, $dE_2 > dE_1$, and hence the two different areas yield a net dE in the negative direction. One can find a net E_- due to the outer sphere in the negative X-direction by covering the entire sphere. The field due to the inner sphere at P is E_+. Thus, the effective field at P is $E = E_+ + E_-$.

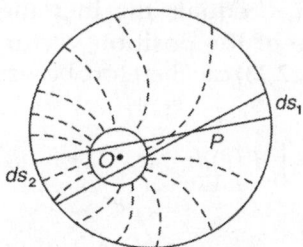

Fig. 12.27: Two nonconcentric charged spheres

12.12 ELECTROSTATIC POTENTIAL

A charged particle placed in an electric field has potential energy because of its interaction with the field. One can define the electric potential at a point as the potential energy per unit charge placed at that point. The electric potential is usually designated as V and the potential energy of a charge q as E_p, we have

$$V = \frac{E_p}{q} \quad \text{or} \quad E_p = qV \quad (12.31)$$

The electric potential is measured in joules/coulomb or JC^{-1}, a unit called volt and abbreviated as V. In terms of international units, $V = m^2 kg\, s^{-2} C^{-1}$.

Let us consider the field produced by a stationary point charge q. At any point in this field, the point charge q' experiences the force

$$F = \frac{1}{4\pi\varepsilon_0} \frac{q'q}{r^2} \hat{a}_r = F(r)\hat{a}_r \quad (12.32)$$

Here $F(r)$ is the magnitude of the force F and $\hat{a}_r$ is the unit vector of the position vector r

determining the position of the charge q' relative to the charge q.

The force [Eq. (12.32) is a central one. A central field of forces is conservative. Consequently, the work done by the forces of the field on the charge q' when it is moved from one point to another does not depend on the path. This work is

$$W_{12} = \int_1^2 F(r)\hat{a}_r \, dl \qquad (12.33)$$

where dl is the elementary displacement of the charge q'. Figure 12.28 shows that the scalar product $\hat{a}_r \, dl$ equals the increment of the magnitude of the position vector r, i.e. dr. Equation (12.33) can therefore be written in the form

$$W_{12} = \int_1^2 F(r) \, dr$$

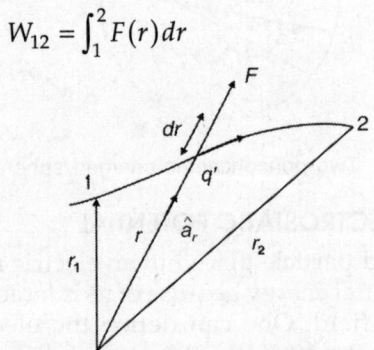

Fig. 12.28: Work done by the forces of the field is independent of the path

Using Eq. (12.32), one obtains

$$W_{12} = \frac{qq'}{4\pi\varepsilon_0} \int_1^2 \frac{dr}{r^2} = \frac{1}{4\pi\varepsilon_0}\left(\frac{qq'}{r_1} - \frac{qq'}{r_2}\right) \qquad (12.34)$$

The work of the forces of the conservative field can be represented as a decrement of the potential energy as

$$W_{12} = V_{p,1} - V_{p,2} \qquad (12.35)$$

A comparison of Eqs. (12.34) and (12.35) leads to the following expression for the potential energy of the charge q' in the field of charge q.

$$V_p = \frac{1}{4\pi\varepsilon_0}\frac{qq'}{r} + \text{const}$$

The value of the constant in the expression for the potential energy is usually chosen so

that when the charge moves away to infinity (i.e. when $r = \infty$), the potential energy vanishes. When this condition is observed, one obtains

$$V_p = \frac{1}{4\pi\varepsilon_0}\frac{qq'}{r} \qquad (12.36)$$

Let us use the charge q' as a test charge for studying the field. By Eq. (12.36), one obtains the potential energy which the test charge has depending not only on its magnitude q', but also on the quantities q and r determining the field. Obviously, one can use this energy to describe the field just like we used the force acting on the charge for this purpose.

Different test charges q'_t, q''_t, etc. will have different energies V'_p, V''_p, etc. at the same point of the field. But the ratio $\dfrac{V}{q_t}$ will be the same for all the charges [Eq. (12.36)]. The quantity

$$V \text{ or } \phi = \frac{V_p}{q_t} \qquad (12.37)$$

is called the *field potential* at a given point and is used together with the field strength E to describe electric fields.

It can be seen from Eq. (12.37) that the potential numerically equals the potential energy which a unit positive charge would have at the given point of the field. Substituting the value of potential energy in Eq. (12.37) its value from Eq. (12.36), one obtains the expression for the potential of a point charge as

$$V \text{ or } \phi = \frac{1}{4\pi\varepsilon_0}\frac{q}{r} \qquad (12.38)$$

Let us consider the field produced by a system of N point charges $q_1, q_2,.., q_N$. Let $r_1, r_2,.., r_N$ be the distances from each of the charges to the given point in the field. The work done by the forces of this field on the charge q' will equal the algebraic sum of the work done by the forces set up by each of the charges separately, i.e.

$$W_{12} = \sum_{i=1}^N W_i$$

By Eq. (12.34), each work W_i is given by

$$W_i = \frac{1}{4\pi\varepsilon_0}\left(\frac{q_i q'}{r_{i,1}} - \frac{q_i q'}{r_{i,2}}\right)$$

where $r_{i,1}$ is the distance from the charge q_i to the initial position of the charge q' and $r_{i,2}$ is the distance from q_i to the final position of the charge q'. Thus,

$$W_{12} = \frac{1}{4\pi\varepsilon_0}\sum_{i=1}^{N}\frac{q_i q'}{r_{i,1}} - \frac{1}{4\pi\varepsilon_0}\sum_{i=1}^{N}\frac{q_i q'}{r_{i,2}}$$

Comparing this equation with Eq. (12.35), one obtains the following expression for the potential energy of the charge q' in the field of a system of charges.

$$V_p = \frac{1}{4\pi\varepsilon_0}\sum_{i=1}^{N}\frac{q_i q'}{r_i}$$

from which it can be seen that

$$V \text{ or } \phi = \frac{1}{4\pi\varepsilon_0}\sum_{i=1}^{N}\frac{q_i}{r_i} \tag{12.39}$$

In general, for a set of point charges q_j, the potential at a point r_i is given by the algebraic sum of the individual potentials, i.e.

$$V(r_i) = \frac{1}{4\pi\varepsilon_0}\sum_{j}\frac{q_j}{r_{ij}} \tag{12.39a}$$

In case of a continuous distribution of charges,

$$V(r_i) = \frac{1}{4\pi\varepsilon_0}\int\frac{\rho d\tau_j}{r_{ij}} \tag{12.39b}$$

Comparing Eq. (12.39) with Eq. (12.38), we arrive at the conclusion that the *potential of the field produced by a system of charges equals the algebraic sum of the potentials produced by each of the charges separately.* Whereas the field strengths are added vectorially in the superposition of fields, the potentials are added algebraically. This is why it is usually much simpler to calculate the potentials than the electric field strengths.

Examination of Eq. (12.37) shows that the charge q at a point in the field with the

potential V or ϕ has potential energy

$$V_p = qV = q\phi \tag{12.40}$$

Hence, the work of the field forces on the charge q can be expressed through the potential difference as

$$W_{12} = V_{p,1} - V_{p,2} = q(V_1 - V_2) = q(\phi_1 - \phi_2) \tag{12.41}$$

Obviously, the work done on a charge by the forces of a field equals the product of the magnitude of the charge and the difference between the potentials at the initial and final points (i.e. the potential decrement).

If the charge q is removed from a point having the potential V or ϕ to infinity (where by convention the potential vanishes), then the work of the field forces will be

$$W_\infty = q\phi = qV \tag{12.42}$$

Obviously, the *potential numerically equals the work done by the forces of the field on a unit positive charge when the latter is removed from the given point to infinity*. Work of the same magnitude must be done against the electric field forces to move a unit positive charge from infinity to the given point of a field.

12.13. RELATION BETWEEN ELECTRIC FIELD STRENGTH AND POTENTIAL

An electric field can be described either with the aid of the vector quantity E, or with the aid of the scalar quantity V or ϕ. There must evidently be a definite relation between these quantities. If we bear in mind that E is proportional to the force acting on a charge and ϕ to the potential energy of the charge, it is easy to see that this relation must be similar to that between the potential energy and the force.

The force F is related to the potential energy by the expression

$$F = -\nabla V_p \tag{12.43}$$

For a charged particle in an electrostatic field, we have

$$F = qE$$

and

$$V_p = q\phi = qV$$

Introducing these values in Eq. (12.43), one obtains

$$qE = -\nabla(q\phi) = -q\nabla(\phi)$$

or $$E = -\nabla\phi = -\text{grad}\,\phi \qquad (12.44)$$

This is the relation between the field strength and potential. Since E has a finite value at any point in the field, the potential ϕ is a continuous function of the coordinates of points of the field. The rectangular components of the electric field E are given by

$$E_x = -\frac{\partial\phi}{\partial x}, \quad E_y = -\frac{\partial\phi}{\partial y} \text{ and } E_z = -\frac{\partial\phi}{\partial z} \quad (12.45)$$

Thus,

$$E = -\left[\hat{i}\frac{\partial}{\partial x} + \hat{j}\frac{\partial}{\partial y} + \hat{k}\frac{\partial}{\partial z}\right]\phi \qquad (12.45a)$$

In general, the component along the direction corresponding to a displacement ds is

$$E_s = -\frac{\partial\phi}{\partial s} \qquad (12.46)$$

Equation (12.45) or (12.46) is used to find the electric potential ϕ or V when the field E is known, and conversely.

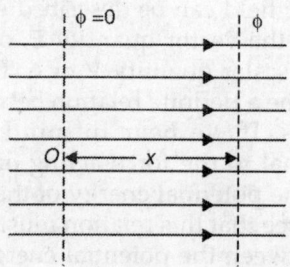

Fig. 12.29: Uniform electric field

Let us consider a simple case of a uniform electric field (Fig. 12.29).

The first of Eqs. (12.45) gives, for X-axis which is parallel to the field, $E = -\dfrac{d\phi}{dx}$. Since E is constant and we assume $\phi = 0$ at $x = 0$, by integration, one obtains

$$\int_0^\phi d\phi = -\int_0^x E\,dx = -E\int_0^x dx \text{ or } \phi = -Ex \quad (12.47)$$

Equation (12.47) is very useful relation represented graphically in Fig. 12.30.

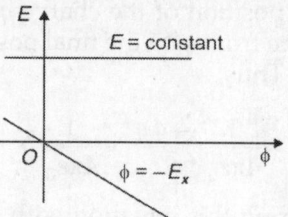

Fig. 12.30: Variation of E and ϕ for a uniform electric field

We may note that, because of the negative sign in Eq. (12.46) or Eq. (12.47), the electric field points in the direction in which the electric potential decreases. When we consider two points x_1 and x_2, Eq. (12.47) gives $\phi_1 = -Ex_1$ and $\phi_2 = -Ex_2$. Subtracting the two we have $\phi_2 - \phi_1 = -E(x_2 - x_1)$ or calling $d = x_2 - x_1$, we obtain

$$E = -\frac{\phi_2 - \phi_1}{d} = \frac{\phi_1 - \phi_2}{d} \qquad (12.48)$$

Although Eq. (48) is valid only for uniform electric fields, it can be used to estimate the electric field between two points separated by a distance d, if the potential difference $\phi_1 - \phi_2$ between them is known. If the potential difference $\phi_1 - \phi_2$ is positive, the field points in the direction from x_1 to x_2, and if it is negative it points in the opposite direction. Eq. (12.48) [or in fact also Eq. (12.45) or Eq. (12.46)] indicates that the electric field can also be expressed in volt/metre, a unit which is equivalent to newton/coulomb given before. One can easily verify it in the following way.

$$\frac{V}{m} = \frac{J}{Cm}$$

$$\frac{Nm}{Cm} = \frac{N}{C}$$

By common usages, the term volt/m, abbreviated Vm^{-1} is preferred to NC^{-1}.

We have seen that Eq. (12.44) allows us to find the field strength at every point from the known values of ϕ. One can also solve the reverse problem, i.e. find the potential difference between two arbitrary points of a field according to the given values of E. For this purpose, one can take advantage of the circumstance that the work done by the forces of a field on the charge q when it is moved from point 1 to point 2 be calculated as (Fig. 12.28)

$$W_{12} = \int_1^2 qEdl$$

At the same time in accordance with Eq. (12.41), this work can be written as

$$W_{12} = q(\phi_1 - \phi_2)$$

Equating these two expressions, one obtains

$$\phi_1 - \phi_2 = \int_1^2 E \cdot dl \qquad (12.49)$$

The integral can be taken along any line joining points 1 and 2 because the work of the field forces is independent of the path for circumvention along closed contour $\phi_1 = \phi_2$, and Eq. (12.49) becomes

$$\oint E \cdot dl = 0 \qquad (12.50)$$

(The circle on the integral sign indicates that integration is performed over a closed contour). In vector calculus, the line integral of a vector round a closed curve is called the circulation or the curl of the vector field. Eq. (12.50) indicates that the circulation of the electric field is zero. Its sources are given by Eq. (12.25). Other examples of such fields are the gravitational field, the magnetic field in region free from electric current and the velocity field of the flow of incompressible fluids free from viscosity. Such fields are termed *irrotational fields*. We must note that relation (12.50) is not applicable to the field of moving charges (i.e. a field changing with time), i.e. this field is not a potential one.

Now, we take curl of vectors in Eq. (12.44) and obtain

$$\nabla \times E = -\nabla \times \nabla\phi = 0 \qquad (12.50a)$$

Using Stokes' theorem, one can transform line integral in Eq. (12.50) into a surface integral

$$\oint E \cdot dl = \int_s \nabla \times E \cdot \hat{a}_n dS$$

$$\therefore \quad \int_s \nabla \times E \cdot \hat{a}_n dS = 0$$

and since S is arbitrary, $\nabla \times E = 0$. Thus, one obtains the following two important laws of electrostatics:

1. $\nabla \cdot E = \dfrac{\rho}{\varepsilon_0}$ \qquad (12.51)

2. $\nabla \times E = 0$ \qquad (12.52)

Equation (12.51) follows from Coulomb's inverse square law. Eq. (12.52) does not depend upon this law. All predictions of electrostatics follow from these laws and hence Eqs. (12.51) Eq. (12.52) can be considered fundamental laws of electrostatics.

12.14 EQUIPOTENTIAL SURFACES

The locus of the points in an electrostatic field having the same potential is called an equipotential surface. Its equation has the form

$$\phi(r) = c$$

where c is a constant. Different values of c generate a family of such surfaces.

The potential does not change in movement along an equipotential surface over the distance $dr (d\phi = 0)$. Hence, according to Eq. (12.46), the tangential component of the E to the surface equals zero. Let us consider a displacement dr on an equipotential surface.

$$\phi(r) - \phi(r + dr) = 0$$

i.e. $\quad \phi(r) - \left\{\phi(r) + \dfrac{\partial\phi}{\partial r} \cdot dr\right\} = 0$

$$\therefore \quad \dfrac{\partial\phi}{\partial r} \cdot dr = 0$$

i.e. $\qquad \nabla\phi \cdot dr = 0$

or $\qquad E \cdot dr = 0 \qquad (12.53)$

This shows that if dr is on the surface, E must be normal to the surface, i.e. the field lines are normal or orthogonal to the equipotential surface.

An equipotential surface can be drawn through any point in a field, consequently one can construct an infinitely great number of such surfaces. They are conventionally drawn so that the potential difference for two adjacent surfaces is the same everywhere. Thus, the density of the equipotential surfaces allows us to assess the magnitude of the field strength. Indeed, the denser are the equipotential surfaces, the more rapidly does the potential change when moving along a normal to the surface.

Figure 12.7 shows equipotential surfaces (more exactly their intersection with the plane of the drawing) for the field of a point charge. Figures 12.8 and 12.9 show the equipotential surfaces along with the lines of force for two equal but opposite positive charges and two identical charges respectively. In accordance with the nature of the dependence of E on r_i equipotential surfaces the denser, the nearer we approach a charge.

Equipotential surfaces for a homogeneous field are a collection of equispaced planes right angles to the direction of field.

12.15 ELECTROSTATIC ENERGY

The energy of electrostatic systems is solely the potential energy arising from the interaction between the charges.

Let us consider a charge q_1 situated at a certain point in space and charge q_2 which was initially at infinity, be brought up to a distance r_{12} from q_1. The potential of the charge q_1 at r_{12} is $\dfrac{q_1}{4\pi\varepsilon_0 r_{12}}$.

Thus, the work done in bringing a charge q_2 from infinity up to a distance r_{12} from charge q_1 is equal to is $\dfrac{q_1 q_2}{4\pi\varepsilon_0 r_{12}}$.

Now a third charge q_3 is added to the system. We will have to perform work against the field of q_1 and q_3. If r_{13}, r_{23} are the distances of q_3 from q_1 and q_2 respectively, then the

additional contribution to the potential energy is as follows:

$$\frac{q_1 q_3}{4\pi\varepsilon_0 r_{13}} + \frac{q_2 q_3}{4\pi\varepsilon_0 r_{23}}$$

Proceeding to build up the assembly in this way, one finds that the total energy of the assembly is

$$W = \frac{q_2}{4\pi\varepsilon_0}\left(\frac{q_1}{r_{12}}\right) + \frac{q_3}{4\pi\varepsilon_0}\left(\frac{q_1}{r_{13}} + \frac{q_2}{r_{23}}\right)$$

$$+ \frac{q_4}{4\pi\varepsilon_0}\left(\frac{q_1}{r_{14}} + \frac{q_2}{r_{24}} + \frac{q_3}{r_{34}}\right) + \cdots$$

$$= \frac{1}{4\pi\varepsilon_0}\sum_i q_i \sum_{j>i}\frac{q_i}{r_{ij}} \qquad (12.54)$$

The restriction $j > i$ ensures that the interaction between every pair is counted only once. One can write Eq. (12.54) in a different form, i.e.

$$W = \frac{1}{4\pi\varepsilon_0}\frac{1}{2}q_1\left(\frac{q_2}{r_{12}} + \frac{q_3}{r_{13}} + \cdots\right)$$

$$+ \frac{1}{4\pi\varepsilon_0}\frac{1}{2}q_2\left(\frac{q_1}{r_{21}} + \frac{q_3}{r_{23}} + \cdots\right)$$

$$+ \frac{1}{4\pi\varepsilon_0}\frac{1}{2}q_3\left(\frac{q_1}{r_{31}} + \frac{q_2}{r_{32}} + \frac{q_4}{r_{34}} + \cdots\right) + \cdots$$

$$= \frac{1}{4\pi\varepsilon_0}\frac{1}{2}\sum_{i=1} q_i \sum_{j\neq i}\frac{q_j}{r_{ij}}; \ i \neq j \qquad (12.55)$$

The factor $\dfrac{1}{2}$ appears in Eq. (12.55) because in this expression each pair is counted twice.

Since $\dfrac{1}{4\pi\varepsilon_0}\displaystyle\sum_{j\neq i}\frac{q_j}{r_{ij}}$ is the potential ϕ_i, produced by all the charges of the system except ith one at the point at which q_i is situated, we have

$$W = \frac{1}{2} \sum_i q_i \phi_i \qquad (12.56)$$

When there is a continuous distribution of charges and charges are not localized, we have

$$W = \frac{1}{2} \int \phi_\rho \, d\tau \qquad (12.57)$$

ELECTROMAGNETISM

12.16 INTRODUCTION

An important kind of interaction among the fundamental particles composing matter is called electromagnetic interaction. It is associated with a characteristic property of each particle called its *electric charge*. To describe the electromagnetic interaction, one will have to introduce the notion of *electromagnetic field*, characterized by two vectors, the electric field *E* and the magnetic field *B*, such that the force on an electric charge is given by

$$F = q(E + v \times B) \qquad (12.58)$$

The electric and magnetic fields *E* and *B* are, in turn, determined by the position of charges themselves and by their motions (or currents). The separation of the electromagnetic field into its electric and magnetic components depends on the relative motion of the observer and the charges producing the field. Also the fields *E* and *B* are directly correlated with each other by the Ampere-Maxwell and Faraday-Henry laws. All these relations are expressed by four laws, viz. Gauss's theorem, Bio-Savart law, Ampere's theorem and Faraday's law. Maxwell put these fundamental laws of electromagnetism into the form of four equations which have now become famous Maxwell's equations. The theory of the electromagnetic field is condensed in these four laws. These laws constitute the basic framework of the theory of electromagnetic interactions. The electric charge *q* and the current *I* are called the sources of the electromagnetic field since, given *q* and *I*, Maxwell's equations allow us to compute *E* and *B*.

According to Maxwell's theory, the rate of propagation of electric and magnetic interaction is equal to the velocity of light in the given medium. Maxwell's theory reveals the electromagnetic nature of light. The theoretically predicted velocity of these electromagnetic waves agrees with the experimentally determined velocity of light in vacuum. It is also found that the other types of waves, e.g. *X*-rays, ultraviolet rays and the infrared rays possess the same velocity and are essentially of the same nature as light waves. Maxwell's electromagnetic theory explains fairly satisfactorily most of the properties of these radiations. Apart from the mathematical deduction of equations, Maxwell proposed the idea of *displacement currents*. We know that the empty space is a non-conductor of electricity. Maxwell proposed that electric currents in empty space can exist in the form of *displacement currents*. However, the term displacement current is misleading and Maxwell's picture unnecessary, since there is no such current between the plates of the capacitor.

12.17 DISPLACEMENT CURRENT

Maxwell proposed the idea of displacement current while he was critically examining electromagnetic theory. He put forward this fact that the magnetic fields are generated not only due to circulating current but also due to time dependent charges in the electric field. This type of time varying field is equivalent to a current which Maxwell interpreted as *displacement current*.

To understand the concept of displacement current, let us consider two plates *P* and *Q* of a parallel plate capacitor separated by a dielectric of permittivity ε (Fig. 12.31). The plate *P* of the capacitor is charged positively with a surface charge density σ.

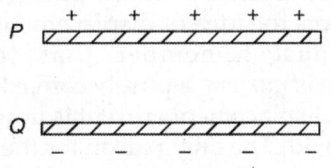

Fig. 12.31: Parallel plate capacitor

The plate P is charged negatively and its surface density is also σ. If the area of either plate is A, then the total charge on either plate will be $A\sigma$. Now join the two plates by a conducting wire. Obviously, electrons will flow from Q to P. In other words, this is equivalent to flow of a conventional current from P to Q. The magnitude of the resulting current in the connecting wire $= \dfrac{d}{dt}(A\sigma)$. Maxwell assumed that current is always circuital. Obviously, current equal to the conduction current in the connecting wire flows through the dielectric in the capacitor. Maxwell interpreted this as *displacement current*.

We have the electric intensity in dielectric as

$$E = \frac{\sigma}{\varepsilon} \tag{12.59}$$

or

$$\sigma = E\varepsilon$$

∴ Displacement current

$$= \frac{d}{dt}(A\sigma) = A\frac{d}{dt}(\varepsilon E)$$

But $\varepsilon E = D$ is electric displacement and hence displacement current $= A\dfrac{dD}{dt}$

Displacement current flowing per unit area, i.e. $\dfrac{dD}{dt}$ is called the *displacement current density*. Since E is a vector, and hence D (displacement) is also a vector. Obviously, one can write

$$\text{Displacement current} = \frac{dD}{dt} \tag{12.60}$$

From the above discussion, it is obvious that, the current in the circuit becomes continuous (even between the plates of the capacitor) if one accepts the idea of displacement current.

We must remember that the term *displacement current* is purely conventional. In essence, displacement current is time varying electric field. The only reason for the quantity given by Eq. (12.60) to be current is that the dimensions of this quantity coincides with that of current density. Of all the physical properties of a real current, a displacement current has only the ability of producing a magnetic field.

12.18 TOTAL CURRENT

Suppose that there is a volume distribution of charge and let the charge density at a point be ρ. Now, if the charges are in motion, the resulting current density will be ev, where v is the velocity of the charges at that point. In addition to the conduction current $J(= \rho v)$, if there is a displacement current $\left(\dfrac{\partial D}{\partial t}\right)$ at the concerned point, then the total current (I) is given by

$$I = \frac{\partial D}{\partial t} + J \tag{12.61}$$

The partial differentiation sign indicates that D may be a function of both space and time.

12.19 CONTINUITY EQUATION

Let us consider a closed surface S enclosing a volume V. If J is the current density at some point on the surface, then the current passing out of the surface is $\int_S J \cdot \hat{n} ds$, where $\hat{n}$ is the outward drawn unit normal at the point. We know that the charge is conserved, and hence the above current must be equal to the rate of decrease of charge within the volume V, i.e.

$-\dfrac{\partial}{\partial t}\int_V e dV$, where ρ is the charge density within the volume. For conservation of charge, we have

$$\int_S J \cdot \hat{n} ds = -\frac{\partial}{\partial t}\int \rho dV = -\int_V \frac{\partial}{\partial t}(\rho dV)$$

Making use of Gauss's theorem, one obtains

$$\int_S J \cdot \hat{n} ds = \int_V \nabla \cdot J dV = -\int_V \frac{\partial \rho}{\partial t} dV$$

$$\therefore \quad \nabla \cdot J = -\frac{\partial \rho}{\partial t}$$

or $\qquad \nabla \cdot \boldsymbol{J} + \dfrac{\partial \rho}{\partial t} = 0 \qquad\qquad$ (12.62)

Eq. (12.62) is known as the *equation of continuity* and it emphasizes the conservation of charge and energy.

In the steady state, $\dfrac{\partial \rho}{\partial t} = 0$

$\therefore \qquad\qquad \nabla \cdot \boldsymbol{J} = 0 \qquad\qquad$ (12.63)

This is valid in the region which does not contain a source or sink of current.

12.20 GAUSS'S THEOREM APPLIED TO ELECTROSTATICS

Let us consider a closed surface *S* enclosing a volume *V* and let the surface enclose an electric charge of volume density ρ. Let dV be an elementary part of the total volume *V* enclosed by *S*, then the total charge enclosed by surface $S = \int_V \rho dV$. Let $\boldsymbol{D}$ be the electric induction at some point in *S*, then the normal magnetic flux across an element ds of the surface *S* is $\boldsymbol{D} \cdot \hat{n} ds$, where $\hat{n}$ is the unit vector drawn outward normal to the surface at the point considered. Obviously, the total outward normal flux across the surface $S = \int_S \boldsymbol{D} \cdot \hat{n} ds$. According to Gauss's theorem, this must be equal to the total charge enclosed by surface *S*. Thus,

$$\int_S \boldsymbol{D} \cdot \hat{n} ds = \int_V \rho dV$$

But $\int_S \boldsymbol{D} \cdot \hat{n} ds = \int_V \nabla \cdot \boldsymbol{B} dV$ (Gauss's theorem)

$\therefore \quad \int_V \nabla \cdot \boldsymbol{D} dV = \int_V \rho dV$

Obviously, the integrals on both sides of the above equation are volume integrals taken over the same volume, and hence the two integrands must be equal, i.e.

$$\nabla \cdot \boldsymbol{D} = \rho \qquad\qquad (12.64)$$

In this form, the Gauss's theorem shows that the sources of electric displacement are free electric charges.

The application of Gauss's theorem in calculating $\boldsymbol{D}$ consists in choosing a closed surface of such shape that the electric flux ϕ can be readily calculated by simple means.

12.21 BIOT–SAVART LAW

This law establishes the magnitude and direction of vector of magnetic induction $d\boldsymbol{B}$ at any arbitrary point *P* of a magnetic field set up in vacuum by an element of a conductor of length dl carrying a current *I* (Fig. 12.32).

$$dB = \frac{\mu_0}{4\pi} Idl \frac{\sin\theta}{r^2} \qquad (12.65)$$

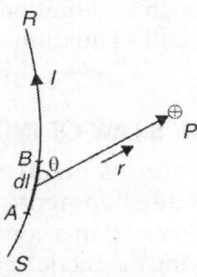

Fig. 12.32: Biot–Savart law

Let the position vector of *P* be $\boldsymbol{r} \cdot \boldsymbol{dl}$. It is the vector of the element of conductor numerically equal to dl and of same direction as the current. $dl\sin\theta$ can be written as

$$dl \times \hat{r} = \frac{dl \times r}{r}$$

$\therefore \qquad d\boldsymbol{B} = \dfrac{\mu_0}{4\pi} \dfrac{Idl \times r}{r^3}$

The total induction at *P* due to the entire conductor *RS* is given by

$$B = \frac{\mu_0}{4\pi} \int \frac{Idl \times r}{r^3} \qquad (12.66)$$

We must note that the direction of $\boldsymbol{B}$ is that of $dl \times r$. Taking the divergence on both sides of Eq. (12.66), one obtains

$$\nabla \cdot \boldsymbol{B} = \frac{\mu_0 I}{4\pi} \int \nabla \cdot \frac{dl \times r}{r^3}$$

But $\nabla \cdot \left(\dfrac{dl \times r}{r^3} \right) = \dfrac{r}{r^3} \cdot \nabla \times dl - dl \cdot \nabla \times \left(\dfrac{r}{r^3} \right)$

Since at the field point P, dl is independent of the coordinates of P and hence

$$\nabla \times dl = 0$$

$$\nabla \cdot dl \times \left(\frac{r}{r^3}\right) = -dl \cdot \nabla \times \left(\frac{r}{r^3}\right)$$

$$= -dl \cdot \nabla \times \nabla\left(\frac{1}{r}\right) = 0$$

Hence $\nabla \cdot B = 0$ (12.67)

This is the first law in magnetostatics corresponding to the relation $\nabla \times E = 0$ in electrostatics. The relation (12.67) shows that the magnetic field is *solenoidal* in contrast to the electric field which is irrotational. Eq. (12.67) is one of the Maxwell's equations for an electromagnetic field.

12.22 FARADAY'S LAW OF INDUCTION

The phenomenon of electromagnetic induction is the development of an induced electromotive force (ε) in a conducting circuit placed in a varying magnetic field. If the circuit is closed, a current called an *induced current* is produced in it. Faraday and Henry independently studied the phenomenon of electromagnetic induction and gave a law. According to this law of electromagnetic induction, the emf induced in a circuit is equal to the negative rate of increase of magnetic flux enclosed by the circuit. Negative sign takes account of Lenz's law. Let us consider a surface S which encloses a closed circuit C as its boundary (Fig. 12.33). Let E be the electric intensity at point P in the circuit C, then the work done in taking a unit charge through a small segment dl of the current at point P is $E \cdot dl$. Obviously, the total work done in taking

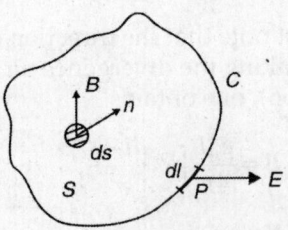

Fig. 12.33

a unit positive charge once round the closed circuit is $\oint E \cdot dl$. This is the induced emf in the circuit, i.e.

$$\varepsilon = \oint_C E \cdot dl \qquad (12.68)$$

Let ds be a small element of the surface. If the magnetic flux density at this point is B, then the normal flux across ds will be $B \cdot \hat{n}\, ds$, where $\hat{n}$ is unit normal outward to ds. Obvioulsy, the total magnetic flux enclosed by C is given by

$$\phi = \int_s B \cdot \hat{n}\, ds \qquad (12.69)$$

But Faraday's law is

$$\varepsilon = -\frac{\partial \phi}{\partial t}$$

$$\therefore \qquad \oint_C E \cdot dl = -\frac{\partial}{\partial t}\int_S B \cdot \hat{n}\, ds \qquad (12.70)$$

Time t and area S are independent variable quantities and hence can be interchanged, i.e.

$$\oint_C E \cdot dl = -\int_S \frac{\partial B}{\partial t} \cdot \hat{n}\, ds \qquad (12.70a)$$

But from Stoke's curl theorem,

$$\oint_C E \cdot dl = -\oint_S (\nabla \times E) \cdot \hat{n}\, ds$$

$$\therefore \quad \oint_S \nabla \times E \cdot \hat{n}\, ds = -\oint_S \frac{\partial B}{\partial t} \cdot \hat{n}\, ds$$

Obviously, on either side of the above equation we have a surface integral taken over the same surface. Thus, the two integrands must be equal.

$$\nabla \times E = -\frac{\partial B}{\partial t} \qquad (12.71)$$

Equation (12.71) is the differential form of Faraday's law of electromagnetic induction.

Equations (12.71) and (12.64) show that the electric field has a non-conservative part due to changing magnetic flux density as well as a conservative part due to electric charge density. Equation (12.71) or its equivalents express relation that must exist between the time rate of change of the magnetic field at a point and the electric field existing at the same point of space. It illustrates in a very obvious way the close inter-relationship between the electric and magnetic components of an electromagnetic field. Equation (12.71) is one of the Maxwell's equations for an electromagnetic field.

12.23 AMPERE'S LAW FOR MAGNETOMOTIVE FORCE

A magnetic field is one of the forms of an electromagnetic field. Its distinguishing feature is that it acts only on moving particles and bodies having an electric charge, as well as on magnetized bodies regardless of their state of motion. A magnetic field is produced by current carrying conductors, by moving electrically charged particles and bodies, magnetized bodies, or a variable electric field (by displacement current).

The force exerted by the magnetic field on the current carrying conductor is called *Ampere's force*. The relation between the magnetic field and the electric current is called *Ampere's law*. This law states that the line integral of the magnetic flux density taken round any closed path is equal to μ_0 times the total current crossing the area enclosed by the path.

Let us consider a closed circuit C as shown in Fig. 12.34. A surface S is shown which has C as its boundary.

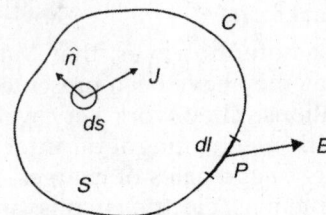

Fig. 12.34

Let B be the magnetic flux density at some point P in the circuit and dl be an elementary segment of the circuit at P. Let ds be an elementary part of the surface S. Let be is a current of density J at ds, then one finds that the normal current passing through ds will be $J \cdot \hat{n} ds$, where $\hat{n}$ is unit normal drawn outward to ds. Obvioulsy, the total current enclosed by the circuit C will be

$$I = \oint_S J \cdot \hat{n} ds \qquad (12.72)$$

We have seen that there exist a relationship between the magnetic field and the electric field in the same space [Eq. (12.70a)]. The close relationship that exists between the electric field and magnetic field suggests that an analogous relation should exist between the time rate of change of an electric field and a magnetic field at the same place. Eq. (12.70a) relates the circulation of the electric field to the time rate of change of the flux of the magnetic field. We might expect that a similar expression must relate the circulation of magnetic field to the time rate of change of flux of electric field. Thus,

$$I = \oint_C B \cdot dl = \mu_0 \oint_S J \cdot \hat{n} ds$$

We have from Stoke's theorem, the tangential line integral of any vector taken round a closed circuit is equal to the surface integral of the curl of that vector taken over any surface which has that circuit as its boundary. Hence,

$$\int_C B \cdot dl = \oint_S \hat{n} \cdot \nabla \times B ds$$

or $\quad \oint_S \hat{n} \cdot \nabla \times B ds = \mu_0 \oint_S J \cdot \hat{n} ds$

Since we have taken surface integrals on both the sides over the same surface and hence

$$\nabla \times B = \mu_0 I \qquad (12.73)$$

This is the second law of magnetostatics corresponding to $\nabla \times E = \rho / \varepsilon_0$ of electrostatics. The integral form of this law is obtained as

$$\oint B \cdot dl = \mu_0 I \qquad (12.74)$$

This is known as A*mpere's circuital law*. Since $B = \mu_0 H$, Eq. (12.73) can also be expressed as

$$\nabla \times H = I \qquad (12.75)$$

Using Eq. (12.61), one obtains

$$\nabla \times H = \frac{\partial D}{\partial t} + J \qquad (12.76)$$

$$\nabla \times B = \mu_0 \left(J + \varepsilon_0 \frac{\partial E}{\partial t} \right) \qquad (12.77)$$

Equation (12.77) expresses a relation between the electric current at a point in space and the electric and magnetic fields at the same point. In empty space, where there are no currents, $J = 0$, and Eq. (12.77) becomes

$$\nabla \times B = \mu_0 \varepsilon_0 \frac{\partial E}{\partial t} \qquad (12.78)$$

Physical Meaning of $\nabla \cdot E = 0$, $\nabla \times E = -\dfrac{\partial B}{\partial t}$, and $\nabla \times B = 0$

If the divergence and the curl of a vector field are known everywhere, that field can be calculated at any point. $\nabla \cdot E = 0$ and $\nabla \times E = -\dfrac{\partial B}{\partial t}$ enable us to do this for the electric field. Obvioulsy, they constitute two of the Maxwell's equations (out of four) describing the electromagnetic field. $\nabla \cdot B = 0$ is valid for both steady and time varying fields and states. This signifies that a conduction current as well as a changing flux produces a magnetic field.

12.24 MAXWELL'S EQUATIONS

The theory of electromagnetic field is condensed into four fundamental laws, i.e. (i) Gauss's law for electric field (ii) Gauss's law for magnetic field (iii) Faraday's law of electromagnetic induction (iv) Ampere's law of circuital current, and these are called Maxwell's equations. These equations may be written down, both in their differential and integral forms, as in Tables 12.1 and 12.2 respectively.

It has been found that above set of Maxwell's laws are in agreement with experimental results in all situations. Maxwell's equations are used in integral or differential form, depending on the problem to be solved. The solution of the equations may not be simple in all cases.

Gauss's law of electrostatics is a basic law in electrostatics. The differential form of Gauss's law is $\nabla \cdot D = \rho$. In this form, the law shows that the sources of electric displacement are free electric charges (since $\rho \neq 0$), i.e. the electric flux lines either leave or enter a closed surface. Obvioulsy, within the closed surface, there must be either a source or a sink of flux lines. We know from convention that a sink of flux lines start from positive charges as sources and terminate in negative charges as sinks. This shows that positive and negative charges exist independently.

The Gauss law for magnetic field in differential form, $\nabla \cdot B = 0$ means that the magnetic flux lines are continuous, i.e. they form loops. Obviously, there are no sources or sinks. This shows that magnetic poles always occur in pairs.

Maxwell's equations are compatible with the principle of relativity in the sense that they remain invariant under Lorentz transformations, that is, their form does not change when the coordinates x, y, z and time t are transformed according to the Lorentz transformation (*see* Chapter 1).

The synthesis of electromagnetic interactions as expressed by Maxwell's equations is one of the greatest achievements in physics, and that is what places these interactions in a unique position. They are the best understood of all interactions and the only ones, so far, that can be expressed in a closed, consistent, mathematical form.

We must note, however, that Maxwell's equations, as they have been presented, have their limitations. They work very well when dealing with electromagnetic interactions between large aggregates of charges, such as radiating antennas, electric circuits, and even beams of ionized atoms or molecules. But it has been found that the electromagnetic interactions between fundamental particles (especially at high energies) must be treated in a somewhat different way, and according to the laws of quantum mechanics, constituting a technique called *quantum electrodynamics*. This will not be considered in this book.

Table 12.1: Maxwell's equations in differential form

Law	Vacuum	Free space	Dielectric medium	Constant field (with time)
Gauss's law for electric field	$\nabla \cdot E = \dfrac{\rho}{\varepsilon_0}$	$\nabla \cdot E = 0$	$\nabla \cdot E = \rho/\varepsilon$	$\nabla \cdot E = \rho/\varepsilon$
Gauss's law for magnetic field	$\nabla \cdot B = 0$	$\nabla \cdot B = 0$	$\nabla \cdot B = 0$	$\nabla \cdot B = 0$
Faraday's law	$\nabla \times E = -\dfrac{\partial B}{\partial t}$	$\nabla \times E = -\dfrac{\partial B}{\partial t}$	$\nabla \times E = -\dfrac{\partial B}{\partial t}$	$\nabla \times E = 0$
Ampere's circuital law	$\nabla \times B = \mu_0\left(J + \varepsilon_0 \dfrac{\partial E}{\partial t}\right)$	$\nabla \times B = \mu_0 \varepsilon_0 \dfrac{\partial E}{\partial t}$	$\nabla \times B = \mu\left(J + \varepsilon \dfrac{\partial E}{\partial t}\right)$	$\nabla \times B = \mu J$

Table 12.2: Maxwell's equations in integral form

Law	Vacuum	Free space	Dielectric medium	Constant field (with time)
Gauss's law for electric field	$\int_S E \cdot ds = \int_V \dfrac{\rho}{\varepsilon_0} dV$	$\oint_S E \cdot ds = 0$	$\oint_S E \cdot ds = \oint_V \dfrac{\rho}{\varepsilon} dV$	$\oint_S E \cdot ds = \int_V \dfrac{\rho}{\varepsilon} dV$
Gauss's law for magnetic field	$\oint_S B \cdot ds = 0$	$\oint_S B \cdot ds = 0$	$\oint_S B \cdot ds = 0$	$\oint_S B \cdot ds = 0$
Faraday's law	$\int_C E \cdot dl = -\int_S \dfrac{\partial B}{\partial t} \cdot ds$	$\int_C E \cdot dl = -\int_S \dfrac{\partial B}{\partial t} \cdot ds$	$\int_C E \cdot dl = -\int_S \dfrac{\partial B}{\partial t} \cdot ds$	$\int_C E \cdot dl = 0$
Ampere's circuital law	$\int_C B \cdot dl = \mu\int_S\left(J + \varepsilon_0 \dfrac{\partial E}{\partial t}\right) \cdot ds$	$\int_C B \cdot dl = \mu_0\varepsilon_0\int_S \dfrac{\partial E}{\partial t} \cdot ds$	$\int_C B \cdot dl = \mu\int_S\left(J + \varepsilon_0 \dfrac{\partial E}{\partial t}\right) \cdot ds$	$\int_C B \cdot dl = \mu\int_S J \cdot ds$

Maxwell's equations are also consistent with the law of conservation of energy. For example, we consider Maxwell's fourth equation,

$$\nabla \times H = J + \frac{\partial D}{\partial t}$$

Let us take divergence of both sides of this equation, we obtain

$$\nabla \times (\nabla \times H) = \nabla \cdot \left(J + \frac{\partial D}{\partial t}\right) = \nabla \cdot J + \frac{\partial}{\partial t}(\nabla \cdot D)$$

but $\quad \nabla \cdot H = 0 \therefore \nabla \cdot \nabla \times H = \nabla \times \nabla \cdot H = 0$

Thus, one obtains

$$\nabla \cdot J + \frac{\partial}{\partial t}(\nabla \cdot D) = \nabla \cdot J + \frac{\partial \rho}{\partial t} = 0$$

which is the equation of continuity. Since equation of continuity is based on the law of conservation of charge and hence the Maxwell's equation under consideration is consistent with the laws of conservation of energy.

12.25 MAXWELL'S EQUATIONS UNDER STATIC CONDITIONS

One can obtain Maxwell's equations under static conditions by equating time rates to zero. Thus, one obtains

$$\left.\begin{aligned}\nabla \cdot D &= \rho \\ \nabla \times E &= 0\end{aligned}\right\} \text{(electrostatics)} \qquad (12.79)$$

$$\left. \begin{array}{l} \nabla \cdot B = 0 \\ \nabla \times H = J \end{array} \right\} \text{(magnetostatics)} \qquad (12.80)$$

Equations (12.79) involve only electric field whereas Eqs. (12.80) involve only magnetic field.

12.26 MORE ABOUT MAXWELL'S EQUATIONS

One can see the interdependence of Maxwell's equations and can judge that the equations involving curl are more fundamental than those involving divergence. Let us verify this statements.

Let us take divergence on both sides of Eq. (12.77), one obtains

$$\nabla \cdot \nabla \times H = \nabla \cdot J + \nabla \cdot \frac{\partial D}{\partial t} = \nabla \cdot J + \frac{\partial}{\partial t}(\nabla \cdot D)$$

But $\nabla \cdot \nabla \times H = 0$ and from equation of continuity, we have $\nabla \cdot J = -\dfrac{\partial \rho}{\partial t}$

$$\therefore \quad -\frac{\partial \rho}{\partial t} + \frac{\partial}{\partial t}(\nabla \cdot D) = 0$$

or $$\frac{\partial \rho}{\partial t} = \frac{\partial}{\partial t}(\nabla \cdot D)$$

$$\therefore \qquad \nabla \cdot D = \rho$$

Now we take divergence on both sides of Eq. (71), one obtains

$$\nabla \cdot \nabla \times E = -\nabla \cdot \frac{\partial B}{\partial t} = -\frac{\partial}{\partial t}(\nabla \cdot B)$$

But $\nabla \cdot \nabla \times E = 0$ and hence $\dfrac{\partial}{\partial t}(\nabla \cdot B) = 0$

or $\qquad \nabla \cdot B = \text{constant}$

The above must hold for every kind of magnetic field and must be true for a region where B = constant or zero. Obviously, constant must be zero. This means that

$$\nabla \cdot B = 0$$

This shows that Eq. (12.64) can be derived from Eq. (12.77) and Eq. (12.67) can be derived from Eq. (12.71).

Physical Meaning of $\nabla \times E = -\dfrac{\partial B}{\partial t}$, $\nabla \cdot E = 0$, $\nabla \cdot B = 0$ and $\nabla \times H = \partial D/\partial t$

We have seen that Maxwell's electromagnetic equations are a series of classical equations that govern the behaviour of electromagnetic waves in all practical situations. They connect vector quantities applying to any point in a varying electric or magnetic field.

From these equations, Maxwell demonstrated that each field vector obeys a wave equation. Maxwell further showed that where a varying electric field exist, it is accompanied by a varying magnetic field induced at right angles, and *vice versa*, and the two form an electromagnetic field that could propagate as a transverse wave. $\nabla \cdot E = \rho/\varepsilon_0$ and $\nabla \cdot B = 0$ are simply differential form of Gauss's law in electrostatics. $\nabla \cdot B = 0$ also signifies the non-existence of magnetic molecule. $\nabla \times E = -\dfrac{\partial B}{\partial t}$ represents the differential form of Faraday's law of electromagnetic induction. $\nabla \times B = 0$ or $\nabla \times H = J + \dfrac{\partial B}{\partial t}$ is the modified form of Ampere's circuital law. Obviously, these field equations represent generalisations of fundamental laws of electricity and magnetism.

$$\nabla \times E = -\frac{\partial B}{\partial t}$$

This describe the coupling between the electric and magnetic field vectors and their interaction with matter in space and time. This relation together with $\nabla \times H = \dfrac{\partial D}{\partial t}$ (where $D = \varepsilon_0 E$) in free space shows that time-variable E and H field cannot exist independently, e.g. if E is a function of time, then $D = \varepsilon_0 E$ will also be a function of time so that $\partial D/\partial t$ will be non zero. Consequently, $\nabla \times H$ is non-zero, and so a non zero H must exist. In a similar way, one can show that if H is a function of time, then there must be an E field present.

These relations also show that a static E field can exist in the absence of a magnetic field H, e.g. a capacitor with a static charge Q. Likewise, a conductor with a constant current I has a magnetic field H without an E field.

12.27 ELECTROMAGNETIC WAVES

Let us consider an isotropic and homogenous medium in which there are no charges and currents, i.e. $\rho = 0$ and $J = 0$. Maxwell's equations reduce to

$$\nabla \cdot D = 0 \quad \text{or} \quad \nabla \cdot E = 0 \tag{12.81}$$

$$\nabla \cdot B = 0 \quad \text{or} \quad \nabla \cdot H = 0 \tag{12.82}$$

$$\nabla \times E = -\frac{\partial B}{\partial t} = -\mu \frac{\partial H}{\partial t} \tag{12.83}$$

$$\nabla \times H = \frac{\partial D}{\partial t} = \varepsilon \frac{\partial E}{\partial t} \tag{12.84}$$

Take curl on both sides of Eq. (12.83),

$$\nabla \times \nabla \times E = -\nabla \times \mu \frac{\partial H}{\partial t} = -\mu \frac{\partial}{\partial t}(\nabla \times H)$$

But $\nabla \times \nabla \times E = \nabla(\nabla \cdot E) - \nabla^2 E$

We have from Eq. (12.81), $\nabla \cdot E = 0$

$$\therefore \quad \nabla^2 E = \mu \frac{\partial}{\partial t}(\nabla \times H) = \varepsilon\mu \frac{\partial}{\partial t}\left(\frac{\partial E}{\partial t}\right)$$

or $\quad \nabla^2 E = \varepsilon\mu \frac{\partial^2 E}{\partial t^2} \tag{12.85}$

Similarly starting with Eq. (12.84), one can obtain the equation

$$\nabla^2 H = \varepsilon\mu \frac{\partial^2 H}{\partial t^2} \tag{12.86}$$

By expanding the L.H.S. of Eq. (12.85) and Eq. (12.86), one can easily see that these are differential equations of wave motion, Obviously, the velocity of these waves is given by

$$c = \frac{1}{\sqrt{\varepsilon\mu}} \tag{12.87}$$

This reveals that the fields generated by moving charges can leave the source and travel through space in the form of waves. For free space,

$\varepsilon = \varepsilon_0 = 8.8542 \times 10^{-12}$ C^2/Nm^2 and

$\mu = \mu_0 = 4\pi \times 10^{-7}$ H/m $(= Ns^2/C^2)$

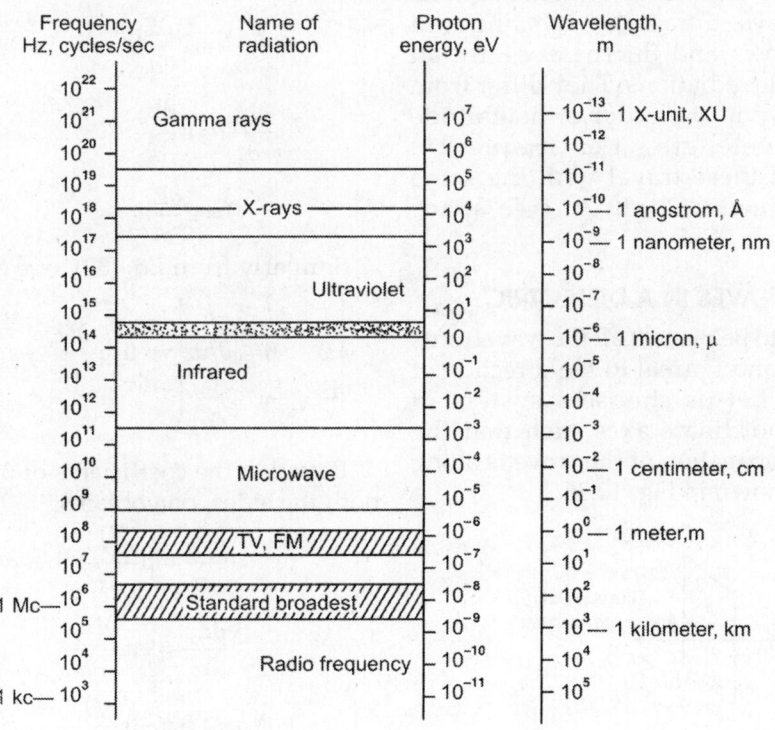

Fig. 12.35: Electromagnetic spectrum

Thus, $c = \dfrac{1}{\sqrt{8.8542 \times 10^{-12} \times 4\pi \times 10^{-7}}}$

$= 2.99794 \times 10^8 \, \text{m/s}$

represents the speed of propagation of electromagnetic waves in free space. Obviously, the theoretical value of the speed of electromagnetic waves in free space agrees very closely with the experimental value of the velocity of light in free space. This result is of immense importance in physics. This unites the subjects of optics and electromagnetism. It also gives the idea of the electromagnetic spectrum. Electromagnetic waves cover a wide range of frequencies of wavelengths, and may be classified according to their main source. The classification does not have very sharp boundaries, since different sources may produce waves in overlapping ranges of frequencies. Figure 12.35 relates the various sections of the electromagnetic spectrum in terms of energy, frequency and wavelength. Light is simply a form of electromagnetic radiation. X-rays, ultraviolet rays, infrared rays, radiowaves and microwaves are all electromagnetic radiations. They differ from each other only in the order of magnitude of their wavelength, i.e. frequency. Obviously, all these travel with the same speed, i.e. speed of light in free space (Fig. 12.35).

12.28 PLANE WAVES IN A DIELECTRIC

A wave is said to be plane if all the wavefronts in it are plane and normal to the direction of propagation. Let us choose a system of rectangular coordinate axes such that the direction of propagation of the wave is along the Z-axis as shown in Fig. 12.36.

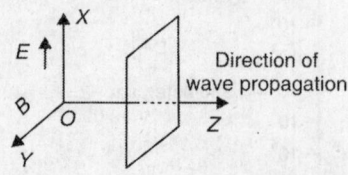

Fig. 12.36: Plane wave propagation

The components of E are E_x, E_y and E_z and similarly the components of H are H_x, H_y and H_z.

Let us consider that the wave is in a plane and direction of propagation is along the Z-axis. Obviously, the value of any component of either H or E must have the same value at all points in a plane normal to the Z-axis. This means all differential coefficients with respect to x and y must vanish for a given value of z.

In the light of this and expressing L.H.S. of $\nabla \times \boldsymbol{H} = \varepsilon \dfrac{\partial E}{\partial t}$ in the determinant form, one obtains

$$\begin{vmatrix} \hat{i} & \hat{j} & \hat{k} \\ 0 & 0 & \partial/\partial z \\ H_x & H_y & H_z \end{vmatrix} = \varepsilon\left(\frac{\partial E_x}{\partial t}\hat{i} + \frac{\partial E_y}{\partial t}\hat{j} + \frac{\partial E_z}{\partial t}\hat{k} \right)$$

(12.88)

where $\hat{i}, \hat{j}, \hat{k}$ are the unit vectors along X, Y and Z-axess respectively. Equating the coefficients of these unit vectors on both sides of Eq. (12.88), one obtains

$$-\frac{\partial H_y}{\partial z} = \varepsilon \frac{\partial E_x}{\partial t}$$

(12.89)

$$\frac{\partial H_x}{\partial z} = \varepsilon \frac{\partial E_y}{\partial t}$$

(12.90)

$$0 = \varepsilon \frac{\partial E_z}{\partial t}$$

(12.91)

Similarly from Eq. (83), one obtains

$$\begin{vmatrix} \hat{i} & \hat{j} & \hat{k} \\ 0 & 0 & \partial/\partial z \\ E_x & E_y & E_z \end{vmatrix} = -\mu\left(\hat{i}\frac{\partial H_x}{\partial x} + \hat{j}\frac{\partial H_y}{\partial y} + \hat{k}\frac{\partial H_z}{\partial z} \right)$$

(12.92)

Equating the coefficients of unit vectors on both the sides, one obtains

$$\frac{\partial E_y}{\partial z} = \mu \frac{\partial H_x}{\partial x}$$

(12.93)

$$\frac{\partial E_x}{\partial z} = -\mu \frac{\partial H_y}{\partial t}$$

(12.94)

$$0 = -\mu \frac{\partial H_z}{\partial t}$$

(12.95)

From Eq. (12.91) we note that E_z is either a constant or zero. Similarly, Eq. (12.95) reveals that H_z is either zero or a constant. However, the constant values of electric and magnetic fields are not relevant in the context of wave motion. Thus, $E_z = 0$ and $H_z = 0$, i.e. the electric and magnetic vectors have no components along the direction of propagation. This means that they are at right angles to the direction of propagation. Thus, the waves are transverse (Fig. 12.37).

The vectors E and H are at right angles to each other. Thus, if the direction of propagation is along the Z-axis and if E is assumed to point in the X-direction, then H will point in the Y-direction (Fig.12.37).

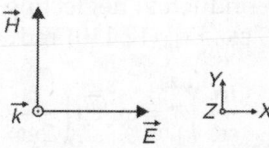

Fig. 12.37: If a plane wave is propagating in the Z-direction (which is coming out of paper) and if at any instant of time the electric vector is along the X-axis, then the magnetic vector will be along the Y-axis

12.29 RELATION BETWEEN E AND H

Let us express

$$E_x = f(z - ct) \qquad (12.96)$$

where c is the velocity of the wave. Differentiating Eq. (12.96) with respect to t, one obtains

$$\frac{\partial E_x}{\partial t} = -cf'(z - ct) \qquad (12.97)$$

Using Eq. (12.97), Eq. (12.89) becomes

$$\frac{-\partial H_y}{\partial z} = -c\varepsilon f'(z - ct) \qquad (12.98)$$

Integrating Eq. (12.98) with respect to z, one obtains

$$H_y = c\varepsilon f(z - ct) = c\varepsilon E_x \qquad (12.99)$$

We have $\qquad C = \dfrac{1}{\sqrt{\varepsilon\mu}}$

$$\therefore \qquad H_y = \frac{1}{\sqrt{\varepsilon\mu}}\varepsilon E_x = \sqrt{\frac{\varepsilon}{\mu}}E_x \qquad (12.100)$$

We can easily see that Eq. (12.94) also leads to the same result. This result shows that the electric and magnetic vectors are in phase with each other, i.e. E and H both attain the maximum and minimum values simultaneously. Fig. 12.38 represents a sinusoidal electromagnetic wave.

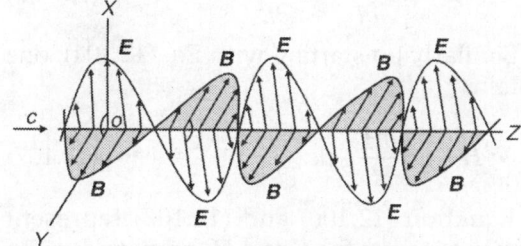

Fig. 12.38: Electric and magnetic fields in a harmonic plane electromagnetic wave

This theoretical prediction about the Maxwell's equations has been amply confirmed by experiment, and it results in several phenomena. Besides plane wave solutions to Maxwell's equations, there are also cylindrical and spherical electromagnetic waves. At a large distance from the source, a limited portion of a cylindrical or a spherical wave can practically be considered as plane, and in this case the electric and magnetic fields are also perpendicular to each other and to the direction of propagation (i.e. radial).

12.30 THE WAVE EQUATION IN A CONDUCTING MEDIUM

If there are no free charges in a medium, then $\rho = 0$. If the conductivity of the medium is σ, then for a conducting medium, we have $J = \sigma E$. Thus, Maxwell's equations become

$$\nabla \cdot E = 0 \qquad (12.101)$$

$$\nabla \cdot H = 0 \qquad (12.102)$$

$$\nabla \cdot E = -\mu\frac{\partial H}{\partial t} \qquad (12.103)$$

$$\nabla \times E = \sigma E + \varepsilon\frac{\partial E}{\Delta t} \qquad (12.104)$$

Taking the curl of Eq. (12.103), one obtains

$$\nabla \times \nabla \times E = -\mu\nabla \times \frac{\partial H}{\partial t}$$

$$\nabla(\nabla \cdot E) - \nabla^2 E = -\mu\sigma\frac{\partial E}{\partial t} - \mu\varepsilon\frac{\partial^2 E}{\partial t^2} \qquad (12.105)$$

Using Eq. (101), we get

$$\nabla^2 E = \mu\sigma\frac{\partial E}{\partial t} - \mu\varepsilon\frac{\partial^2 E}{\partial t^2} = 0 \qquad (12.106)$$

Similarly $\mu\sigma$ starting with Eq. (12.104), one obtains

$$\nabla^2 H - \mu\sigma\frac{\partial H}{\partial t} - \mu\varepsilon\frac{\partial^2 H}{\partial t^2} \qquad (12.107)$$

Equation (12.106) and (12.107) represent wave equations for E and H for a conducting medium. Considering a one dimensional case, say along the X-axis, Eq. (12.106) can be expressed in the form

$$\frac{\partial^2 E}{\partial x^2} - \mu\sigma\frac{\partial E}{\partial t} - \mu\varepsilon\frac{\partial^2 E}{\partial t^2} = 0 \qquad (12.108)$$

The solution of Eq. (12.108) is of the form
$$E = E_0\exp[i(\omega t - kx)] \qquad (12.109)$$

Substituting for E from Eqs. (12.109) and (12.108) and moving the common factor, one obtains

$$-k^2 + \varepsilon\mu\omega^2 - 2\omega\mu\sigma = 0$$

or $\qquad k^2 = \mu\varepsilon\omega^2\left[1 - \dfrac{i\sigma}{\varepsilon\omega}\right]$

$$k = \omega\sqrt{\varepsilon\mu}\left[1 - \frac{i\sigma}{\varepsilon\omega}\right]^{1/2} \qquad (12.110)$$

If $\sigma/\varepsilon\omega << 1$ (say $\gtrsim 0.01$), the medium can be classified as a dielectric and if $\dfrac{\sigma}{\varepsilon\omega} >> 1$ (say $\gtrsim 100$), the medium can be classified as a conductor. For

$$0.01 \lesssim \frac{\sigma}{\varepsilon\omega} \lesssim 100$$

the medium is said to be a quasi-conductor. Thus, depending on the frequency, a particular material can behave as a dielectric or conductor. For example, for fresh water $\varepsilon/\varepsilon_0 \approx 80$ and $\sigma \approx 10^{-3}$ mhos/m (both ε and σ can be assumed to be constants at low frequencies), one finds

$$\frac{\sigma}{\varepsilon} = \frac{10^{-3}}{80\times 8.85\times 10^{-12}} \approx 1.4\times 10^6 \text{ sec}^{-1}$$

For $\omega = 2\pi \times 10 \text{ sec}^{-1}$, $\dfrac{\sigma}{\omega\varepsilon} \approx 2\times 10^4$ and for $\omega \approx 2\pi \times 10^{10} \text{ sec}^{-1}$, $\dfrac{\sigma}{\omega\varepsilon} \approx 2\times 10^{-5}$. Thus, fresh water behaves as a good conductor for $v \lesssim 10^3 \text{ sec}^{-1}$ and as a dielectric for $v \lesssim 10^7 \text{ sec}^{-1}$. For copper, one may assume $\varepsilon \approx \varepsilon_0$ and $\sigma \approx 5.8 \times 10^7$ mhos/m and for $\omega \approx 2\pi \times 10^{10} \text{ sec}^{-1}$,

one obtains $\dfrac{\sigma}{\omega\varepsilon} \approx \dfrac{5.8\times 10^7}{10^{10}\times 8.9\times 10^{-12}} \approx 10^8$. Obviously, for such frequencies, copper behaves as an excellent conductor.

For good conductor, neglecting 1 in comparison to $i\sigma/\varepsilon\omega$, Eq. (12.110) reduces to

$$k = \omega\sqrt{\varepsilon\omega}\left(-\frac{i\sigma}{\varepsilon\omega}\right)^{1/2} = \omega\sqrt{\varepsilon\mu}\left(\frac{\sigma}{2\varepsilon\omega}\right)^{1/2}(-2i)^{1/2}$$

$$\because \quad (1-i)^2 = -2i \quad \therefore \ (-2i)^{1/2} = 1-i$$

or $k = \left(\dfrac{\mu\omega\sigma}{2}\right)^{1/2}(1-i) = \left(\dfrac{\mu\omega\sigma}{2}\right)^{1/2} - i\left(\dfrac{\mu\omega\sigma}{2}\right)^{1/2}$

$$\qquad\qquad\qquad\qquad\qquad (12.111)$$

Obviously, k is imaginary, writing
$$k = \alpha - i\beta, \text{ one obtains}$$

$$\alpha = \beta = \left(\frac{\mu\omega\sigma}{2}\right)^{1/2}$$

Equation (12.109) becomes
$$E = E_0\exp[i(\omega t - \alpha x + i\beta x)]$$

$$= E_0\exp(-\beta x)\exp[i(\omega t - \alpha x)] \qquad (12.112)$$

Equation (12.112) represents an attenuated wave. The attenuation is due to a Joule-loss. Obviously, the amplitude of the wave decreases with distance x exponentially. The constant β is called the *absorption coefficient*. From Eq. (12.112), it can be easily seen that the field decreases by a factor e in traversing a distance

$$\delta = \frac{1}{\beta} = \left(\frac{2}{\mu\omega\sigma}\right)^{1/2} \qquad (12.113)$$

which is known as the *penetration depth* or *skin depth*.

The relation (12.113) indicates that the skin depth goes to zero as the conductivity approaches infinity and is small for good conductors at high frequency currents. Thus, for copper $\sigma = 58 \times 10^6$ mhos/m. The skin depth at various frequencies is given in table 12.3.

Table 12.3: Skin depth at various frequencies

Frequency	Skin depth (δ)
60 Hz	8.5×10^{-3} m
1 MHz	6.6×10^{-5} m
30 GHz	3.8×10^{-7} m

A poor conductor can be made a good conductor with a thin coating of silver or copper. The rapid attenuation of waves indicates that in high frequency circuits current flows only on the surface of the conductor. The relatively high δ value in the case of sea water (10^{-1}m at 30 kHz) explains why the radio communication with submarines becomes difficult at the depth of several metres.

12.31 THE POYNTING VECTOR

Let us consider Eqs. (12.71) and (12.76), i.e

$$\nabla \times E = -\frac{\partial B}{\partial t}$$

$$\nabla \times H = J + \frac{\partial D}{\partial t}$$

Now $\quad \nabla \cdot (E \times H) = H \cdot \nabla \times E - E \cdot \nabla \times H$

$$= -H \cdot \frac{\partial B}{\partial t} - J \cdot E - E \cdot \frac{\partial D}{\partial t} \qquad (12.114)$$

For a linear material, we have

$$H \cdot \frac{\partial B}{\partial t} + E \cdot \frac{\partial D}{\partial t} = \mu H \cdot \frac{\partial H}{\partial t} + \varepsilon E \cdot \frac{\partial E}{\partial t}$$

$$= \frac{1}{2}\mu \frac{\partial}{\partial t}(H \cdot H) + \frac{1}{2}\varepsilon \frac{\partial}{\partial t}(E \cdot E)$$

$$= \frac{1}{2}\frac{\partial}{\partial t}(B \cdot H + D \cdot E)$$

Thus, Equation (12.114) can be rewritten in the form

$$\nabla \cdot S + \frac{\partial u}{\partial t} = -J \cdot E \qquad (12.115)$$

where $S = E \times H \qquad (12.116)$

is known as *poynting vector*. The vector $S = E \times H$ has the dimensions of $\dfrac{\text{energy}}{\text{area} \times \text{time}}$, and hence, it represents the energy that flows out of the boundary per unit time.

Also $\quad u = \dfrac{1}{2}(B \cdot H + D \cdot E) \qquad (12.117)$

Equation (115) resembles the equation of continuity and for a physical interpretation one can note that if a charge q (moving with velocity v) is acted on by an electromagnetic field, then the work done by the field in moving it through a distance ds would be $F \cdot ds$. Thus, the work done per unit time would be

$$F \cdot \frac{d}{dt} = F \cdot V = [qE + qV \times B] \cdot V$$

$$= qE \cdot V \qquad (12.118)$$

If there are N charged particles per unit volume, each carrying a charge q, then the work done per unit volume would be

$$NqV \cdot E = J \cdot E \qquad (12.119)$$

where J is the current density. The energy appears in the form of kinetic (or heat) energy of the charged particles. Thus, the term $J \cdot E$ represents the familiar Joules loss and, therefore, the quantity $J \cdot E$ appearing on R.H.S. of Eq. (12.115) would represent the rate at which energy is produced per unit volume per unit time. Obviously, one may interpret Eq. (12.115) as an equation of continuity for energy with u representing energy per unit volume. The quantities ½$(D \cdot E)$ and ½$(B \cdot H)$ represent the electrical and magnetic energies per unit volume respectively.

The direction of $E \times B$ for a plane electromagnetic wave is perpendicular to the wavefront and is therefore pointing in the direction of propagation of the wave (Fig. 12.39). The magnitude of poynting vector

$$|E \times B| = EB = \frac{1}{2}E^2$$

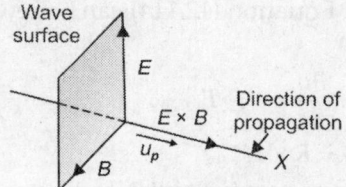

Fig. 12.39: Definition of the direction of energy flow in an electromagnetic wave

Thus, the vector $c\,E \times B$ has the magnitude E^2. Then, the flux $c^2 \varepsilon_0 E \times B$ (poynting vector) across a surface S is given by

$$\int_S c^2 \varepsilon_0 (E \times B) \cdot U_N ds = \frac{d(\text{energy})}{dt} \quad (12.120)$$

$\dfrac{d(\text{energy})}{dt}$ is the energy crossing the area S per unit time, and for that reason it has been designated so·

Since electromagnetic radiations propagate with velocity c, one may use the relation between energy and momentum $p = \dfrac{v(\text{energy})}{c^2}, (v = c)$ to obtain the momentum p per unit volume associated with an electromagnetic wave. Thus,

$$p = \frac{\text{energy}}{c} = \frac{\varepsilon_0 E^2}{c} = \varepsilon_0 |E \times B| \quad (12.121)$$

Since momentum must have the direction of propagation, one may express the above equation in vector form as

$$P = \frac{E}{c} u = \varepsilon_0 E \times B \quad (12.122)$$

where u is the unit vector in the direction of propagation.

If an electromagnetic wave has momentum, it also has angular momentum. The angular momentum per unit volume is

$$L = r \times p = \varepsilon_0 r \times (E \times B) \quad (12.123)$$

This is called the orbital angular momentum of radiation. In addition, electromagnetic radiation possesses an intrinsic angular momentum, or spin. Concluding, one can say that an electromagnetic wave carries momentum and angular momentum as well as energy.

ILLUSTRATIVE EXAMPLES

Example 1

A point charge of 2×10^{-6} C is placed at the centre of a hypothetical cube with the side 0.1 m each. Determine the electric flux through each face.

Solution

A cube has six equal faces. The charges is situated at a centre of the cube. Obviously, equal amount of flux will be passing through each face.

Electric flux from charge q is equal to

$$q/\varepsilon_0 = \frac{2 \times 10^{-6}}{\varepsilon_0}$$

Therefore, flux from each face

$$= \frac{2 \times 10^{-6}}{6 \times 8.85 \times 10^{-12}}$$

$$= 37.7 \times 10^3 \ \text{Nm}^2/\text{C}.$$

Example 2

Find the magnitude and direction of electric field to keep an electron in equilibrium in air.

Solution

$$Ee = mg$$

or

$$E = \frac{mg}{e} = \frac{3.1 \times 10^{-31} \times 9.8}{1.6 \times 10^{-19}}$$

$$= 55.74 \times 10^{-12} \ \text{N/C}$$

Example 3

A rectangular surface 0.2 m × 0.3 m is placed in an uniform electrostatic field 2000 N/C, such that normal to the surface makes an angle of 30° with direction of electrostatic field. Find the flux density through this surface.

Solution

$$d\phi = E \cdot ds = Eds \cos\theta$$

$$= 0.06 \times 2000 \times \frac{\sqrt{3}}{2}$$

$$= 60\sqrt{3} = 103.92 \ \frac{\text{N m}^2}{\text{C}}$$

Example 4

A large metal plate has a uniform surface charge density $\sigma = -2 \times 10^{-6}$ C/m². From what distance should the electron with energy 100 eV be fired so that it just fails to strike the plate?

Solution

Suppose that the electron be thrown towards the plate from a distance d. Electric field due to plate

$$E = \sigma/\varepsilon_0$$

Force experienced by electron due to electric field is

$$F = Ee = \frac{\sigma e}{\varepsilon_0}$$

Work done on electron in bringing it up to the plate is

$$F = \frac{e\sigma d}{\varepsilon_0}$$

If the work done is greater than electron energy, the electron will not be able to reach the plate. Thus,

$$\frac{e\sigma d}{\varepsilon_0} = 100 \text{ eV} = 100 \times 1.6 \times 10^{19} \text{ J}$$

or $\quad d = \dfrac{100 \times 1.6 \times 10^{-19} \times 2 \times 8.852 \times 10^{-12}}{1.6 \times 10^{-19} \times 2 \times 10^{-6}}$

$$= 4.42 \times 10^{-4} \text{ m} = 0.442 \text{ mm}$$

Clearly, if the electron is being thrown from a distance greater than 4.43×10^{-4} m, it will not be able to reach the plate.

Example 5

Two small spheres are positively charged. The amount of charge on each sphere is different. The combined charge on both the spheres is 5×10^{-5} C. When the spheres are 2 m apart, each sphere is repelled from the other by a force of 1 N. Find how the charges are distributed between the spheres.

Solution

Let the charges on both the spheres be q_1 and q_2 C.

$$\therefore \quad q_1 + q_2 = 5 \times 10^{-5} \text{ C} \qquad (1)$$

$$Fe = 1 = \frac{q_1 q_2}{4\pi\varepsilon_0 (2)^2} = \frac{q_1 q_2 \times 9 \times 10^9}{4}$$

or $\qquad q_1 q_2 = \dfrac{4}{9} \times 10^{-9} \qquad\qquad (2)$

From Eqs. (1) and (2), we have

$$q_1 - q_2 = \left[(q_1 + q_2)^2 - 4q_1 q_2 \right]^{1/2}$$

$$= \left[25 \times 10^{-10} - \frac{16}{9} \times 10^{-9} \right]^{1/2}$$

$$= \left(0.72 \times 10^{-9} \right)^{1/2} = 2.68 \times 10^{-5} \qquad (3)$$

Solving Eqs. (1) and (3), one obtains

$q_1 = 38.4 \times 10^{-6}$ C and $q_2 = 11.6 \times 10^{-6}$ C.

Example 6

An infinite line charge has a charge density 2 μC/m. Calculate the electric field at a point which is at a normal distance of 10 cm from the line charge.

Solution

We have $\qquad E = \dfrac{\lambda}{2\pi\varepsilon_0 r} \hat{r}$

$$= \frac{2 \times 2 \times 10^{-6} \times 9 \times 10^9}{(0.1)^2} \hat{r}$$

$= 36 \times 10^4$ N/C along the perpendicular direction of line charge.

Example 7

Two large metal plates of area 1 m² face each other. They are 5 cm apart and carry equal and opposite charges on their inner surfaces. If E between the plates is 55 N/C, what is the charge on the plates?

Solution

We have $\qquad E = \dfrac{\sigma}{\varepsilon_0} = \dfrac{q}{A\varepsilon_0}$

$\therefore \qquad q = EA\varepsilon_0 = 55 \times (1)^2 \times 8.55 \times 10^{-12}$

$$= 0.49 \times 10^{-9} \text{ C}$$

Example 8

A spherical charge distribution is given by

$$r = r_0(1 - r/a) \quad \text{when } r \le a$$

$$r = 0 \qquad \text{when } r > a$$

where a is the radius of the sphere. Find the electric field intensity at points outside and inside. Find the value of r for which E is maximum.

Solution

Electric field at external point

$$E = \frac{\rho_0}{\varepsilon_0} \frac{a^3}{12r^2}$$

Electric field at internal point

$$E = \frac{\rho_0}{\varepsilon_0}\left(\frac{r}{3} - \frac{r^2}{4a} \right)$$

E is maximum inside the charge distribution at a distance given by

$$r = 2a/3.$$

Example 9

A thin electric shell of metal has a radius of 0.25 m. Calculate the electric intensity at a point (i) inside the shell (ii) just outside the shell (iii) 3 m from the centre of the shell. Charge on the shell is 0.2 μC.

Solution

i. The electric intensity at a point inside a charged shell is zero.

ii. For external points, the shell behaves as if its entire charge is concentrated at the centre. Thus, if R be the radius of the shell, the intensity at a point just outside the shell is

$$E = \frac{1}{4\pi\varepsilon_0} \frac{q}{R^2}$$

Here q = 0.2 micro coulomb

$$= 0.2 \times 10^{-6}\,\text{C}, R = 0.25 \text{ m}$$

$$\therefore \quad E = 9 \times 10^9 \times \frac{0.2 \times 10^{-6}}{(0.25)^2} = 2.88 \times 10^4 \,\text{N/C}$$

iii. The intensity at an external point at a distance r from the centre of the shell is given by

$$E = \frac{1}{4\pi\varepsilon_0} \frac{q}{r^2}$$

$$r = 3 \text{ m}$$

$$= 9 \times 10^9 \times \frac{0.2 \times 10^{-6}}{(3)^2} = 200 \,\text{N/C}$$

Example 10

The distance between the parallel plates of a capacitor is 0.6 mm and the dielectric constant of the material placed between the plates is 15. Calculate the displacement current density if an alternating emf of 10 volts and frequency 1.5×10^9 Hz is applied between the plates of capacitor.

Solution

We have displacement current

$$J_d = \frac{\partial D}{\partial t}$$

Magnitude of displacement current density for alternating field

$$J_d = \omega D = \omega \varepsilon_0 \varepsilon_r E$$

$$\therefore \quad \text{Electric field } E = \frac{V}{d}$$

Here V is potential difference and d is the distance between the plates of the capacitor.

$$\therefore \quad J_d = \omega \varepsilon_0 \varepsilon_r \frac{V}{d}$$

$$= \frac{2 \times 3.14 \times 1.5 \times 10^9 \times 8.85 \times 10^{-2} \times 15 \times 100}{6 \times 10^{-4}}$$

$$= 2.08 \times 10^5 \,\text{A/m}^2$$

Example 11

Show that the following field vectors in free space ($\mu = \mu_0$, $\varepsilon = \varepsilon_0$, $\sigma = 0$, $\rho = 0$, $J = 0$) satisfy all the Maxwell's equations.

$$E = E_0\cos(\omega t - \beta z)\,\hat{i} \text{ and } H = \frac{E_0}{\eta}\cos(\omega t - \beta z)\hat{j}.$$

Solution

From Faraday's law, we have

$$\nabla \times E = -\mu_0 \frac{\partial H}{\partial t}$$

or

$$\begin{vmatrix} \hat{i} & \hat{j} & \hat{k} \\ \frac{\partial}{\partial x} & \frac{\partial}{\partial y} & \frac{\partial}{\partial z} \\ E_x & E_y & E_z \end{vmatrix} = -\mu_0 \frac{\partial H}{\partial t}$$

Then with $E_y = E_z = 0$, and $H_x = H_z = 0$, we have

$$\left(\frac{\partial E_{0z}}{\partial y} - \frac{\partial E_{0y}}{\partial z} \right) \hat{i} + \left(\frac{\partial E_{0x}}{\partial z} - \frac{\partial E_{0z}}{\partial x} \right) \hat{j}$$

$$+ \left(\frac{\partial E_{0y}}{\partial x} - \frac{\partial E_{0x}}{\partial y} \right) = -\mu_0 \frac{E_0}{\eta} \frac{\partial}{\partial t} [\cos(\omega t - \beta z)] \hat{k}$$

or

$$\frac{\partial E_x}{\partial z} a_y = \mu_0 \frac{E_0}{\eta} \omega \sin(\omega t - \beta z) \hat{j}$$

or

$$\beta E_0 \sin(\omega t - \beta z) \hat{j} = \omega \mu_0 \frac{E_0}{\eta} \sin(\omega t - \beta z) \hat{j}$$

This yields $\beta = \omega \mu_0 / \eta$ (1)

From Ampere's law, we have

$$\nabla \times H = J_0 + \varepsilon_0 \frac{\partial E}{\partial t} \qquad (2)$$

Expanding the curl, we obtain

$$\nabla \times H = \frac{\partial H_z}{\partial z} \hat{i} = \frac{\beta E_0}{\eta} \sin(\omega t - \beta z) \hat{i} \qquad (3)$$

Using Eq. (3) and (2) becomes

$$-\frac{\beta E_0}{\eta} \sin(\omega t - \beta z) \hat{i} = -\varepsilon_0 \omega E_0 \sin(\omega t - \beta z) \hat{i}$$

This gives $\beta = \omega \varepsilon_0 \eta$ (4)

Combining Eqs. (1) and (4), one obtains

$$\frac{\omega \mu_0}{\eta} = \omega \varepsilon_0 \eta \quad \text{or} \quad \eta^2 = \frac{\mu_0}{\varepsilon_0}$$

$$\therefore \quad \eta = \pm \sqrt{\frac{\mu_0}{\varepsilon_0}} \qquad (5)$$

Using (5), one obtains from (1) or (4)

$$\beta = \pm \omega \sqrt{\mu_0 \varepsilon_0} \qquad (6)$$

Let us now check Gauss's law. We have

$$\nabla \cdot D = \varepsilon_0 \nabla \cdot E = 0$$

or

$$\frac{\partial E_x}{\partial x} + \frac{\partial E_y}{\partial y} + \frac{\partial E_z}{\partial z} = 0$$

which we observe to be true since the only component of E, E_x is independent of x.

Now, $\nabla \cdot B = \mu_0 \nabla \cdot H = 0$

or

$$\frac{\partial H_x}{\partial x} + \frac{\partial H_y}{\partial y} + \frac{\partial H_z}{\partial z} = 0$$

This is also seen to be true since the only component of H, H_y is independent of y. Obviously, the given fields are valid ones, but only if η and β satisfy constraints in Eqs. (5) and (6).

Example 12

Compare the conduction and displacement current densities in copper ($\varepsilon = \varepsilon_0$, $\mu = \mu_0$) and $\sigma = 5.8 \times 10^7$ S/m at a frequency of 1 MHz.

Solution

Let us assume the sinusoidal variation of the electric field in the material, we have

$$E = E_0 \sin \omega t \text{ V/m}$$

Here $\omega = 2\pi v$ and $v = 10^6$ c/s, the conduction current density is

$$J_c = \sigma E = \sigma E_0 \sin \omega t \text{ A/m}^2$$

The displacement current density is

$$J_d = \frac{\partial D}{\partial t} = \varepsilon \frac{\partial E}{\partial t} = \omega \varepsilon E_0 \cos \omega t \text{ A/m}^2$$

The ratio of the magnitudes of these currents is

$$\frac{|J_c|}{|J_d|} = \frac{\sigma}{\omega \varepsilon}$$

For copper at 1 MHz, we find

$$\frac{\sigma}{\omega \varepsilon} = \frac{5.8 \times 10^7}{2\pi \times 10^6 \times \left(\frac{1}{3\pi} \right) \times 10^{-9}} = 1.04 \times 10^{12}$$

This is very large value and, therefore, even if the frequency is raised to an extremely high value (~100 GHz or 10^{11} Hz), the conduction current dominates the displacement current by an enormous amount. Obviously, for copper and other conductors it is reasonable to neglect displacement current.

Example 13

For a medium, conductivity $\sigma = 5$ mho/m and dielectric constant = 1. If an electric field $E = 250 \sin(10^{10}t)$ is applied, then find conducting current and displacement current in the medium. At what frequency both the currents will be equal?

Solution

We have the conducting current density

$$J_c = \sigma E$$

$$\sigma = 5 \text{ mho/m}$$

$$E = 250 \sin(10^{10}t)$$

$$\therefore \qquad J_c = 5 \times 250 \sin(10^{10}t)$$

$$= 1250 \sin(10^{10}t) \text{ A/m}^2$$

Displacement current density

$$J_d = \varepsilon \frac{\partial E}{\partial t} = \varepsilon_0 \varepsilon_r \frac{\partial E}{\partial t}$$

$$\therefore \qquad J_d = \varepsilon_0 \varepsilon_r \frac{\partial}{\partial t} 250 \sin\left(10^{10} t\right)$$

$$= \varepsilon_0 \varepsilon_r 250 \times 10^{10} \cos(10^{10}t)$$

$$= 8.85 \times 10^{-12} \times 1 \times 250 \times 10^{10}\cos(10^{10}t)$$

$$= 22.1 \cos(10^{10}t) \text{ amp/m}^2$$

For $|J_c| = |J_d|$, we have

$$\sigma E = \varepsilon_0 \varepsilon_r \omega E \therefore \omega = \frac{\sigma}{\varepsilon_0 \varepsilon_r} = \frac{5}{8.85 \times 10^{-12} \times 1}$$

$$= 5.65 \times 10^{11} \text{ radian/s}$$

REVIEW QUESTIONS

1. State Coulomb's law in electrostatics. What do you mean by the flux of an electric field?
2. State and prove Gauss's law in electrostatics. Apply it to calculate the electric field intensity due to a uniformly charged sphere (non conducting) at points (i) outside the sphere (ii) at the surface of the sphere and (iii) inside the sphere.
3. Using Gauss theorem discuss the electric field created by a line charge distribution.
4. Use Gauss's law to find the field intensity at a point near an infinite sheet of charge. How is the result modified if the charge lies only on one side of the sheet?
5. Deduce the Gauss's law in differential form, i.e. $\nabla \cdot E = \rho/\varepsilon_0$
6. Define electric field and intensity.
7. Derive Coulomb's law from Gauss's law. Would Gauss's law hold if the exponent in Coulomb's law were not exactly two?
8. Establish on the basis of Gauss's law the fact that when an excess charge is given to an insulated conductor, the charge resides entirely on the outer surface of the conductor.
9. Apply Gauss's theorem to show that the electric field near a charged conductor is twice that due to a non conducting thin sheet of charge, with the same surface charge density as the conductor.
10. State Faraday's law of electromagnetic induction. Use it to deduce, $\nabla \times E = -\dfrac{\partial B}{\partial t}$
11. How was the concept of displacement current helpful in removing discrepancy in Ampere's law? Prove that $\nabla \times H = J + D$.
12. What is the physical significance of equations (i) $\nabla \cdot B = 0$ (ii) $\nabla \cdot D = \rho$?
13. Using Maxwell's equations, prove that (i) $\nabla \cdot B = 0$ (ii) $\nabla \cdot D = \rho$.
14. Deduce the equations: (i) $\nabla \times E = -\dfrac{\partial B}{\partial t}$

    (ii) $\nabla \times H = J + \dfrac{\partial D}{\partial t}$.
15. Prove the continuity equation $\nabla J + \dfrac{\partial \rho}{\partial t} = 0$.
16. What is displacement current ? Show that the conduction current in a lead wire is identical with the displacement in the gap of the condenser.
17. Show that Maxwell's equation $\nabla \times H = J$ is inconsistent with the law of conservation of energy. How this has been modified in the light of the principle of conservation of energy?

18. Making use of Maxwell's equations, obtain the differential equation for an electromagnetic wave.

19. How electromagnetic waves are propagated? Show that these waves are transverse and travel with the velocity of light in vacuum.

20. Do the magnitudes of electric intensity (E) and magnetic intensity (H) vectors in the electromagnetic wave become maximum and minimum simultaneously? Explain it clearly.

21. From Maxwell's curl equations, derive the wave equation in H for a plane wave travelling in the positive X-direction in a medium with constants $\mu = \mu_0$, $\varepsilon = \varepsilon_0$ and $\sigma = 0$. The electric field is in the Y-direction.

22. Show that the energy stored in a magnetic field per unit volume is $\frac{1}{2}B^2/\mu_0$ and the corresponding expression for the energy density in an electric field is $\frac{1}{2}\varepsilon_0 E^2$.

23. Write Maxwell's equations in differential as well as integral form.

24. What is an electromagnetic spectrum? Write the names of its constituent members. Why does not classification have very sharp boundaries?

25. Why do the different parts of the electromagnetic spectrum behave in a different way when propagating through matter?

26. Why are visible light waves called electromagnetic waves? Justify your answer.

27. Show that when an electromagnetic wave enters a metal, it gets attenuated. Obtain an expression for the skin depth.

28. What is poynting vector? Obtain an expression for it.

29. An electromagnetic wave carries momentum and angular momentum as well as energy. Explain and justify your answer.

30. Show that the energy of an electromagnetic wave is equally shared between the electric and magnetic fields.

PROBLEMS

1. Point charges $+4q$ and $-q$ are separated by a distance a. Show that the only position where a third charge $+q$ could be in equilibrium is along the line joining the first two and at a distance a from $-q$.

2. An infinite line charge has a charge density 2 $\mu C/m$. Calculate the electric field at a point which is at a normal distance of 10 cm from the line charge. [*Ans.* 36×10^4 N/C]

3. The flux entering a closed surface is 2×10^3 N-m^2/C and flux leaving the closed surface is 8×10^3 Nm^2/C. Calculate the charge enclosed by the surface. [*Ans.* 5.31×10^{-8} C]

4. A thin spherical shell of metal has a radius of 0.25 m and carries a charge of 0.2 μC. Calculate the electric field at a point (i) 3 m from the centre of the shell (ii) just outside the shell (iii) inside the shell. Given $\varepsilon_0 = 8.85 \times 10^{-12}$ C^2/N^{-1} m^2.

[*Ans.* (i) 200 N/C (ii) 2.88×10^4 N/C (iii) 0]

5. If the electric field near the earth's surface be 300 V/m directed downward, what is the surface density of charge on earth's surface?

[*Ans.* 2.65×10^{-9} C/m^2]

6. An insulated soap bubble of radius 10 cm is given a charge of 4.8×10^{-8} C. Find the increase in radius due to this charge, neglecting the surface tension effect. Given atmospheric pressure = 10^5 N/m^2.

[*Ans.* 2.74×10^{-9}m]

7. The electric field of a plane electromagnetic wave in vacuum, using MKSC units, is represented

by $E_x = 0$, $E_y = 0.5\cos\left[2\pi \times 10^8\left(t - \dfrac{x}{c}\right)\right]$, $E_z = 0$.

(a) Determine the wavelength, the state of polarization, and the direction of propagation.

(b) Compute the magnetic field of the wave.

(c) Calculate the energy flux per unit area.

[*Ans.* (a) 3 m, linearly polarized in the XY-plane, propagates in the $+X$-direction.

(b) $B_x = B_y = 0$, $B_z = \dfrac{1}{6} \times 10^{-8}\cos\left[2\pi \times 10^{-8}\left(t - \dfrac{x}{c}\right)\right]$.

(c) 3.316×10^{-4} W/m^{-2}.]

8. Electromagnetic radiation from the sun falls on the earth's surface at the rate of 1.4×10^3 Wm^{-2}. Assuming that this radiation can be considered as plane waves, estimate the magnitude of the electric and magnetic field amplitudes in the wave. [*Ans.* 1.15×10^3 N/C, 3.84×10^{-6} T]

9. The conductivity of a medium is 2 mho/m and its dielectric constant is 2. An electric field $E = 220 \sin 10^6 t$ is applied on it. Calculate the conduction current and displacement current densities. [*Ans.* $440 \sin(10^6 t)$ A/m^2,

$5.53 \times 10^{-2}\cos(10^6 t)$ A/m^2]

10. If distilled water has constants $\mu_r = 1$, $\varepsilon_r = 81$, and power factor = 0.05 at 1 GHz, calculate (a) the $1/e$ depth of penetration (b) the 1 percent depth of penetration at this frequency.
[*Ans.* (a) $\delta(1/e) = 212$ mm (b) $\delta(1\%) = 977$ mm]

SHORT ANSWER QUESTIONS

1. Can an electric field be represented by lines of force.

Ans. Yes

2. How many electrons are there in one coulomb charge?

Ans. $\dfrac{1}{1.602 \times 10^{-19}} = 6.25 \times 10^{18}$

3. Does Coulomb's law of electric force obeys Newton's third law of motion?

Ans. Yes

4. Write the dimensional formula for ε_0.

Ans. $M^{-1}L^{-3}T^4A^2$

5. Can a point in space have a non-zero potential even when the electric field in the space is zero?

Ans. Yes, electric intensity at a given point is given by $E = -\dfrac{d\phi}{dx}$. If potential ϕ is constant in space, the potential gradient in any direction is zero, hence the intensity is zero. Obviously, potential can exist where there is no electric field

6. What is absolute index of refraction?

Ans. $\eta = \dfrac{c}{v} = \dfrac{1}{\sqrt{\varepsilon_0 \mu_0}} \sqrt{\varepsilon \mu} = \sqrt{\dfrac{\varepsilon \mu}{\varepsilon_0 \mu_0}} = \sqrt{\varepsilon_r \mu_r}$.

7. Are gamma rays electromagnetic radiations?

Ans. Yes, these waves are of nuclear origin.

8. Are there electromagnetic waves in cosmic radiations?

Ans. Yes, there are electromagnetic waves of even shorter wavelengths or longer frequencies.

9. Why does not we see the portion other than visible one of the electromagnetic spectrum?

Ans. The retina of the eye is sensitive only to colours in the visible region, i.e. wavelength lying between 3900 Å and 7800 Å. This region corresponds to visible part of the spectrum.

10. On what factors the velocity of electromagnetic wave depend?

Ans. (i) Permittivity (ii) Permeability, $c = \dfrac{1}{\sqrt{\varepsilon \mu}}$.

11. Do electromagnetic waves have same velocity in all transparent media?

Ans. No, as the refractive index is different for different media, $\eta = \dfrac{c}{v}$, v is different for different media.

12. An electromagnetic wave carries momentum. What it does signifies?

Ans. An electromagnetic interaction between two electric charges means an exchange of energy and momentum between the charges.

OBJECTIVE QUESTIONS

1. The work done in moving a positive charge on an equipotential surface is _____.   [zero]

2. The number of lines of force that radiate outward from one coulomb of positive charge is _____.   [$1/\varepsilon_0 = 1.13 \times 10^{11}$]

3. Two thin infinite parallel plates have uniform charge densities $+\sigma$ and $-\sigma$. The electric field in the space between them is _____.   [σ/ε_0]

4. When placed in a uniform field, a dipole experiences only _____.   [torque]

5. The work done in carrying a charge q once round a circle of radius r with a charge Q at the centre is _____.   [zero]

MULTIPLE CHOICE QUESTIONS

1. The ratio $\sqrt{\mu/\varepsilon}$ for a medium is called
(a) characteristic impedance
(b) reactance
(c) refractive index
(d) frequency

2. Poynting vector signifies
(a) current density vector producing electromagnetic field
(b) power density vector producing electrostatic field
(c) power density vector producing electromagnetic field
(d) current density vector producing electrostatic field

3. The characteristic equation of plane wave in E independent of two dimensions y and z is
(a) $\dfrac{\partial^2 E}{\partial x} = 0$ (b) $\dfrac{\partial E}{\partial t} = 0$
(c) $\dfrac{\partial E}{\partial x} = 0$ (d) $\dfrac{\partial^2 E}{\partial x^2} = 0$

4. Poynting theorem relating the electric field intensity E, magnetic field intensity H and the rate of energy flow per unit area at a point P is given by

   (a) $P = \dfrac{1}{E \times H}$
   (b) $P = E \times H$

   (c) $P = E/H$
   (d) $P = H/E$

5. For a electromagnetic wave propagated in a good dielectric having $\sigma/\omega\varepsilon \gg 1$, the attenuation factor α and phase shift factor β are given by

   (a) $\alpha = \dfrac{\sigma}{2}\sqrt{\dfrac{\mu}{\varepsilon}}$, $\beta = \sqrt{\omega\mu\varepsilon}$

   (b) $\alpha = \sqrt{\dfrac{\mu}{\varepsilon}}$, $\beta = \dfrac{1}{2}\sqrt{\omega\mu\varepsilon}$

   (c) $\alpha = \sqrt{\dfrac{\mu}{\varepsilon}}$, $\beta = \sqrt{\dfrac{\omega\mu}{\varepsilon}}$

   (d) $\alpha = \dfrac{\sigma}{2}\sqrt{\dfrac{\mu}{\varepsilon}}$, $\beta = \sqrt{\mu\varepsilon}$

6. The value of $\sqrt{\dfrac{\mu}{\varepsilon}}$ for a free space is about

   (a) 380 Ω
   (b) 38 Ω
   (c) 3.8 Ω
   (d) 3800 Ω

7. Poynting vector $P = E \times H$ has the dimensions
   (a) Watt m
   (b) Watt m^2
   (c) Watt/m
   (d) Watt/m^2

8. In a perfect dielectric, wave propagation occurs
   (a) with small attenuation
   (b) with large attenuation
   (c) with zero attenuation
   (d) none of the above

9. An electromagnetic wave incident on a perfect conductor is
   (a) fully transmitted
   (b) entirely reflected
   (c) partially transmitted
   (d) none of the above

10. The intrinsic impedance of free space is given by

    (a) $z_0 = \mu_0/\varepsilon_0$
    (b) $z_0 = \sqrt{\dfrac{\mu_0}{\varepsilon_0}}$

    (c) $z_0 = \mu_0\varepsilon_0$
    (d) $z = \sqrt{\mu_0\varepsilon_0}$

11. In a uniform plane wave, E and H are related by

    (a) $\dfrac{E}{H} = \sqrt{\dfrac{\mu}{\varepsilon}}$
    (b) $\dfrac{E}{H} = \sqrt{\dfrac{\varepsilon}{\mu}}$

    (c) $\dfrac{E}{H} = \dfrac{\mu}{\varepsilon}$
    (d) $\dfrac{E}{H} = \dfrac{\varepsilon}{\mu}$

Answers

1. (c)	2. (c)	3. (a)	4. (b)	5. (a)
6. (a)	7. (d)	8. (c)	9. (b)	10. (b)
11. (a)				

13

Scalar and Vector Fields

13.1 FIELDS

A physical quantity can be expressed as a continuous function of the position of a region of space. If the cartesian coordinates of a point P are (x, y, z) and the physical quantity at this is A, then if

$$A = A(x, y, z) \qquad (13.1)$$

this function is called a point function and the region in which it specifies the physical quantity is known as a *field*.

Fields are of two kinds: (1) *scalar field* (2) *vector field*, depending upon the nature of the physical quantity concerned.

13.2 SCALAR FIELDS

When a scalar physical quantity is expressed from point to point in a region of space by a continuous point function

$$\phi = \phi(x, y, z)$$

which gives the value of the quantity at each point, then the region is a *scalar field* and the function ϕ is a scalar point function. Distribution of density, temperature, electric potential, gravitational potential in space, etc. are few familiar examples of scalar fields.

One can represent scalar quantities graphically by drawing such surfaces at which the value of the quantity remains a constant. We must note that one point in space corresponds to only one value of the quantity, i.e. scalar fields are single-valued function at each point, i.e. two level surfaces cannot intersect (otherwise ϕ should have two values

at the point of intersection which contradicts our definition) (Fig. 13.1). These surfaces are called *equiscalar surfaces*. If we put a point charge at any place, then the electric potential around the charge will depend on the position of the point. Electric potential is a scalar quantity and hence the field around the charge is known as *scalar potential field*. If we join all such points at which the potential is constant, then such a surface is called a *equipotential surface*. Figure 13.1 shows S_1 and S_2 as equipotential surfaces, with constant potential ϕ_1 and ϕ_2 respectively.

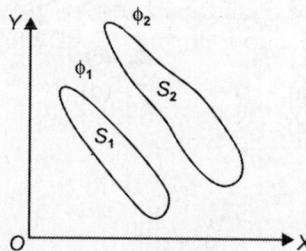

Fig. 13.1: Graphical representation of scalar fields

13.3 VECTOR FIELDS

When a vector physical quantity is expressed from point to point in a region of space by a continuous vector function $A(x, y, z)$, then the region is said to be a *vector field* and the function A is called a *vector point function*. At any given point in the field the function A is represented by a vector of definite magnitude, and distribution of gravitational or magnetic field intensity in space are few examples of vector fields.

Vector fields are graphically represented by lines and these lines are called *field lines* or *flux lines*. These lines are drawn in the field in such a way that the tangent at any point of the line represent the direction of the vector field at that point. The magnitude of A at any point on a flux line is given by the number of flux lines crossing unit area perpendicular to their direction drawn at that point. Two flux lines cannot intersect (otherwise the directions of the vector would become indefinite at the point of intersection).

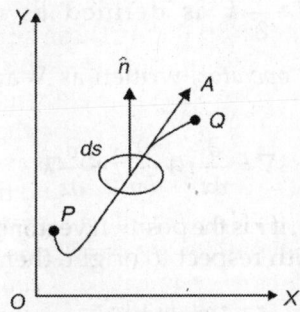

Fig. 13.2: Graphical representation of vector fields

Let us consider that the number of field lines, or flux going out of an area element ds perpendicular to field lines at any point is $d\phi$ (Fig. 13.2).

Obviously, the magnitude of the intensity of the vector field will be

$$|A| = \frac{d\phi}{ds} \qquad (13.2)$$

$$\therefore \qquad A = \left(\frac{d\phi}{ds}\right)\hat{n} \qquad (13.3)$$

where $\hat{n}$ is the unit vector perpendicular to the area ds. Thus,

$$\hat{n}\,ds = ds$$

Now,

$$A \cdot \hat{n} = \frac{d\phi}{ds} \cdot \hat{n} \cdot \hat{n} = \frac{d\phi}{ds}$$

or $\qquad d\phi = (A.\hat{n})ds = A \cdot ds \qquad (13.4)$

We must remember that point function $\phi(x, y, z)$ or $A(x, y, z)$ at every point of the field is single-valued.

13.4 PARTIAL DERIVATIVE

Let us consider that a physical quantity f is a positional function of the coordinates x, y and z of the point of observation, where x, y and z are independent variables, then

$$f = f(x, y, z) \qquad (13.5)$$

Keeping y and z constant, let us change the value of x by an infinitesimal amount Δx and the value of f changes by an amount Δf, then for constant y and z the rate of change of f with respect to x, $\left(\dfrac{\Delta f}{\Delta x}\right)$ is known as partial derivative of f with respect to x. Usually, it is expressed as $\left(\dfrac{\partial f}{\partial x}\right)_{y,z}$ or simply $\dfrac{\partial f}{\partial x}$. Obviously, Δx is considered infinitely small, i.e. $\Delta x \to 0$ and hence

$$\frac{\partial f}{\partial x} = \underset{\Delta x \to 0}{\mathrm{Lt}} \frac{f(x + \Delta x, y, z) - f(x, y, z)}{\Delta x} \qquad (13.6)$$

Similarly, if one keeps x and z constant and the value of y is changed by an infinitesimal amount Δy, then

$$\frac{\partial f}{\partial y} = \underset{\Delta y \to 0}{\mathrm{Lt}} \frac{f(x, y + \Delta y, z) - f(x, y, z)}{\Delta y} \qquad (13.7)$$

and if one keeps x and y constant and the value of z is changed by an infinitesimal amount Δz, then

$$\frac{\partial f}{\partial z} = \underset{\Delta z \to 0}{\mathrm{Lt}} \frac{f(x, y, z + \Delta z) - f(x, y, z)}{\Delta z} \qquad (13.8)$$

Now, consider that all the three independent variables x, y, z are simultaneously changed by infinitesimal amounts, say $\delta x, \delta y$ and δz respectively and the value of f is changed by an amount δf. When $\delta x \to 0$, $\delta y \to 0$ and $\delta z \to 0$, i.e. all the three vanish simultaneously, $\delta f \to 0$, i.e. δf will also vanish. Obviously, in this state, the total change in the value of f, i.e. δf is known as total derivative. Thus,

$$\delta f = \left(\frac{\partial f}{\partial x}\right)\delta x + \left(\frac{\partial f}{\partial y}\right)\delta y + \left(\frac{\partial f}{\partial z}\right)\delta z$$

One can write the total derivative for the limiting change as

$$df = \left(\frac{\partial f}{\partial x}\right)dx + \left(\frac{\partial f}{\partial y}\right)dy + \left(\frac{\partial f}{\partial z}\right)dz \quad (13.9)$$

The partial derivatives $\dfrac{\partial f}{\partial x}$, $\dfrac{\partial f}{\partial y}$ and $\dfrac{\partial f}{\partial z}$ are called *first partial derivatives* or partial derivatives of the first order. By differentiating these derivatives once more, one can obtain the *second partial derivatives* (or partial derivatives of the second order), e.g. $\dfrac{\partial^2 f}{\partial x^2}$, $\dfrac{\partial^2 f}{\partial y \partial x}$, $\dfrac{\partial^2 f}{\partial x \partial y}$, $\dfrac{\partial^2 f}{\partial y^2}$, $\dfrac{\partial^2 f}{\partial z^2}$, etc. Obviously, $\dfrac{\partial^2 f}{\partial x^2}$ is the partial derivative of $\dfrac{\partial f}{\partial x}$ with respect to x, $\dfrac{\partial^2 f}{\partial x \partial y}$ is the partial derivative of $\dfrac{\partial f}{\partial x}$ with respect to y, etc. If all the derivatives concerned are continuous, then $\dfrac{\partial^2 f}{\partial x \partial y} = \dfrac{\partial^2 f}{\partial y \partial x}$, i.e. the order of differentiation is immaterial. By differentiating the second partial derivatives again with respect to x, y and z respectively, one obtains the third partial derivatives and so on.

If A is a vector depending on more than one scalar variables such as cartesian coordinates x, y, z of a point in space, then $\dfrac{\partial A}{\partial x}$, $\dfrac{\partial A}{\partial y}$ and $\dfrac{\partial A}{\partial z}$ are partial derivatives of A with respect to x (when y and z remain constant), y (when x and z remain constant) and z (when x and y remain constant) respectively. If now x, y and z change simultaneously by small increments dx, dy and dz, then the total change or total derivative of A will be

$$dA = \frac{\partial A}{\partial x}dx + \frac{\partial A}{\partial y}dy + \frac{\partial A}{\partial z}dz$$

$$= \left(\frac{\partial}{\partial x}dx + \frac{\partial}{\partial y}dy + \frac{\partial}{\partial z}dz\right)A$$

$$= \left(\frac{\partial}{\partial x}\hat{i} + \frac{\partial}{\partial y}\hat{j} + \frac{\partial}{\partial z}\hat{k}\right)\cdot\left(dx\hat{i} + dy\hat{j} + dz\hat{k}\right)A$$

$$(13.10)$$

where $\hat{i}$, $\hat{j}$ and $\hat{k}$ are unit vectors in the directions of X, Y and Z axes respectively. $\dfrac{\partial}{\partial x}\hat{i} + \dfrac{\partial}{\partial y}\hat{j} + \dfrac{\partial}{\partial z}\hat{k}$ is defined as the *vector differential operator*, written as ∇ and read as *del*. Thus,

$$\nabla = \frac{\partial}{\partial x}\hat{i} + \frac{\partial}{\partial y}\hat{j} + \frac{\partial}{\partial z}\hat{k} \quad (13.11)$$

Further, if r is the position vector of the point (x, y, z) with respect to origin, then

$$r = \hat{i}x + \hat{j}y + \hat{k}z \quad (13.12)$$

so that

$$dr = \hat{i}dx + \hat{j}dy + \hat{k}dz \quad (13.13)$$

Equation (13.10) can be expressed as

$$dA = (\nabla\cdot dr)A \quad (13.14)$$

Here ∇ is a vector quantity possessing all the properties of an ordinary vector. Since it is an *operator*, its magnitude has no physical significance.

The differential operator $\nabla\cdot\nabla = \nabla^2$ is defined as

$$\nabla^2 = \frac{\partial^2}{\partial x^2} + \frac{\partial^2}{\partial y^2} + \frac{\partial^2}{\partial z^2} \quad (13.15)$$

∇^2 is a scalar and is known as *Laplacian operator*.

13.5 GRADIENT OF A SCALAR FUNCTION

A scalar field can be mapped out by a series of level surfaces. Let us consider that in a scalar field two surfaces S_1 and S_2 represent surfaces with constant scalar quantities ϕ and $\phi + \delta\phi$.

respectively (Fig. 13.3). r is the position vector of any point P on S_1 relative to the origin O.

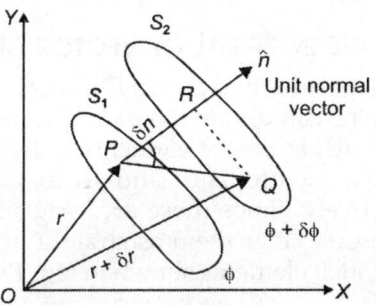

Fig. 13.3: Mapping of a scalar field by a series of level surfaces

The position vector of another point Q on S_2 is $r + \delta r$. Obviously, the displacement of Q with respect to P is $PQ = \delta r$. The rate of change of scalar function in the direction of PQ is $\dfrac{\delta \phi}{\delta r}$. Now, draw a perpendicular PR from P on surface S_2.

The minimum distance between the surfaces S_1 and S_2, i.e. $PR = \delta n$ will be in the direction of their normal. Therefore, the rate of change of scalar function ϕ in the direction of $PR = \dfrac{\delta \phi}{\delta n}$. If the distance δr is infinitesimal, then one can consider ΔPQR, a right-angled triangle. If Q is the angle between δr and δn, then

$$PR = PQ \cos \theta \quad \text{or} \quad \delta n = \delta r \cos \theta$$

$$\therefore \qquad PQ = \frac{\delta \phi}{\delta n} \cos \theta \qquad (13.16)$$

Obviously, Eq. (13.16) represents the rate of change of ϕ in the direction of PQ. Since the maximum value of $\cos \theta$ is 1 and hence the maximum rate of change of

$$\phi = \left(\frac{\delta \phi}{\delta r} \right)_{max} = \left(\frac{\delta \phi}{\delta n} \right)$$

$$\therefore \qquad \underset{\delta r \to 0}{\text{Lt}} \left(\frac{\delta \phi}{\delta r} \right)_{max} = \frac{d\phi}{dn}$$

Obviously, the rate of change of a scalar function is maximum in a direction perpendicular to the equiscalar surface.

Thus, *in a scalar field, the maximum rate of change of the scalar function is known as the gradient of the scalar field.*

Gradient is a vector quantity and its direction is perpendicular to the equiscalar surface, i.e. the direction in which the rate of change is maximum. If $\hat{n}$ is the unit vector normal to the equiscalar surface, then

$$\text{grad}\phi = \left(\frac{d\phi}{dn} \right) \hat{n} = \left(\frac{d\phi}{dr} \right)_{max} \hat{n} \qquad (13.17)$$

Obviously, the gradient of a scalar field ϕ depends only on the way the values of ϕ are distributed in space. It is entirely independent of the position and direction of the coordinate axes. Thus, grad ϕ is an invariant vector field with respect to transformation of coordinates.

To make the above statement more clear, let us consider an example. Suppose ϕ is the potential in an electric field. The intensity of the field at any point is in the direction of the greatest rate of fall of potential, i.e. normal to the equipotential surface and is equal to the rate, i.e.

$$E = -\frac{\partial \phi}{\partial n} \hat{n} = -\text{grad}\phi \qquad (13.18)$$

The negative sign is used because the direction of the field is opposite to the direction of the increase of potential.

The Gradient of a Scalar Field in Rectangular Cartesian Coordinates

Suppose the scalar field ϕ is a function of (x, y, z), then with the use of partial differentiation, one can write

$$d\phi = \frac{\partial \phi}{\partial x} dx + \frac{\partial \phi}{\partial y} dy + \frac{\partial \phi}{\partial z} dz \qquad (13.19)$$

Now, the gradient of ϕ is a vector given by

$$\nabla \phi = \frac{\partial \phi}{\partial n} \hat{n}$$

where $\hat{n}$ is a unit vector. Let us take its scalar product with an element of radius vector dr, one obtains

$$(\text{grad}\phi) \cdot dr = \frac{\partial \phi}{\partial n} \hat{n} \cdot dr$$

$$= \frac{\partial \phi}{\partial n} dr \cos\theta \quad (\because \hat{n} \cdot dr = dr \cos\theta)$$

$$= \frac{\partial \phi}{\partial n} \cdot dn$$

$$\left(\because \frac{\partial \phi}{\partial n} \text{ is the rate of change of } \phi \text{ in the}\right.$$

direction $\hat{n}$)

$$= d\phi$$

$$= \left(\frac{\partial \phi}{\partial x}\right) dx + \left(\frac{\partial \phi}{\partial y}\right) dy + \left(\frac{\partial \phi}{\partial z}\right) dz$$

$$= \left(\frac{\partial \phi}{\partial x} \hat{i} + \frac{\partial \phi}{\partial y} \hat{j} + \frac{\partial \phi}{\partial z} \hat{k}\right) \cdot \left(dx\hat{i} + dy\hat{j} + dz\hat{k}\right)$$

$$= \left(\frac{\partial \phi}{\partial x} \hat{i} + \frac{\partial \phi}{\partial y} \hat{j} + \frac{\partial \phi}{\partial z} \hat{k}\right) \cdot dr$$

$$\therefore \nabla\phi = \frac{\partial \phi}{\partial x} \hat{i} + \frac{\partial \phi}{\partial y} \hat{j} + \frac{\partial \phi}{\partial z} \hat{k} = \text{grad}\phi \quad (13.20)$$

Obviously, $\nabla\phi$ is such a vector which represents the maximum rate of change of scalar function ϕ.

It is worthwhile to note that the gradient of a scalar field is always a vector field, however, its converse is not always true. All vector fields cannot be expressed as gradient of scalar fields.

A vector field which can be expressed as gradient of a scalar field is called a *lamellar* or *non-curl field*.

A pure electrostatic field is lamellar vector field, because it can be expressed as (negative) gradient of a scalar potential field. If V is the electric potential, then one can write the electric field as

$$E = -\nabla V \quad (13.21)$$

A scalar field $\phi(x, y, z)$ evaluated at a particular point is independent of the coordinates of that point. For example, the temperature at a point is not dependent on whether coordinates (x, y, z) or (x', y', z') are used. This means a scalar field is invariant with respect to the transformation of coordinates. Its gradient

which is a vector field, must therefore be also invariant with respect to the transformation.

13.6 LINE INTEGRAL OF VECTOR FIELD

Let us consider two points P and Q in a vector field. One can divide the path between P and Q into displacement elements of lengths dr_1, dr_2, dr_3, ..., etc. (not shown in Fig. 13.4) respectively. Since these are very small and one can consider them as straight lines, dr is one of such elements shown in Fig. 13.4. Let A

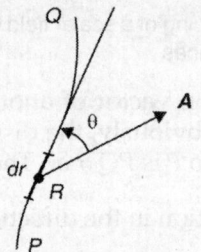

Fig. 13.4: Line integral of vector field

be the vector at R in a direction making an angle θ with dr. If A varies in magnitude and direction from point to point along the curve PQ, then the integral

$$\int_B^A A \cdot dr = \int_P^Q (A\cos\theta) dr$$

$$= \int_P^Q \left(A_x dx + A_y dy + A_z dz\right) \quad (13.22)$$

is defined as the line integral of vector A along the curve AB. Few familiar examples of line integral are:

i. If A denotes a force acting on a particle which moves along the curve PQ, then the integral $\int_P^Q A \cdot dr$ represents the work done by the force in moving from P to Q.

ii. If A represents the electric field intensity at any point, then the line integral represents the potential difference between P and Q.

iii. If A represents the velocity at any point in a fluid, then the line integral of A round a closed path is known as circulation of the fluid.

If the value of the line integral depends only upon the coordinates of the two points in the vector field and not upon the actual path taken between them, then the vector field A (x, y, z) is called a *conservative* or *non-curl* or *lamellar field*.

13.7 SURFACE INTEGRAL

Let S be a surface drawn in a vector field, and ds an infinite small element of the surface (Fig. 13.5). Let $\hat{n}$ be unit positive vector normal to ds. The surface element of area ds is represented by a vector ds whose magnitude is $|ds|$ and direction is that of $\hat{n}$. Thus,

$$ds = \hat{n}\,ds$$

ds is called the **area vector**.

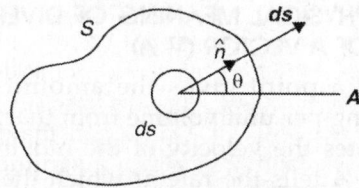

Fig. 13.5: Surface integral

Now, if A be a vector at the middle of the element ds in a direction making an angle θ with $\hat{n}$, then the integral

$$\iint A \cdot ds \quad \text{or} \quad \int_S A \cdot \hat{n}\,ds = \int_S (A\cos\theta)\,ds \quad (13.23)$$

is defined as the surface integral or flux of A across the surface S.

If A denotes the velocity of a moving fluid in which a fixed surface S is drawn, then the surface integral $\int_S A \cdot ds$ gives the amount of fluid flowing per unit time normally through the surface S. In case of a closed surface, a positive surface integral means that the fluid is flowing outwards, if negative then inwards.

Other examples of surface integrals are $\int_S A \times ds$ and $\int_S \phi\,ds$ where ϕ is a scalar function.

13.8 VOLUME INTEGRAL

Let $dV = dx\,dy\,dz$ denotes the element of volume V. The volume element dV is a scalar. If A be a vector function inside it, then the integral $\int_V A\,dV$ is called the volume integral $\iiint_V A\,dV$ (we will write it as $\int_V A\,dV$) of A over the volume V, i.e.

$$\int_V A\,dV = \int_V \left(A_x\hat{i} + A_y\hat{j} + A_z\hat{k}\right)dx\,dy\,dz \quad (13.24)$$

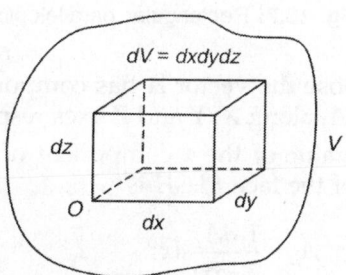

Fig. 13.6: Volume integral

In case of scalar function ϕ, the integral $\int_V \phi\,dV$ represents the volume integral over V.

13.9 DIVERGENCE OF A VECTOR FIELD

The divergence of a vector field at any point is defined as the amount of flux per unit volume diverging from that point. Since the divergence is the amount of flux, it is essentially a scalar. One can express the divergence of a vector field $A(x, y, z)$ as

$$\nabla \cdot A = \frac{\partial A_x}{\partial x} + \frac{\partial A_y}{\partial y} + \frac{\partial A_z}{\partial z} = \operatorname{div}A \quad (13.25)$$

Equation (13.25) is the expression for divergence in cartesian coordinates. Divergence of a vector field has important applications in hydrodynamics. Now, we derive the expression for divergence in cartesian coordinates.

Let us consider a small rectangular parallelopiped whose centre is C (x, y, z) and whose sides have lengths dx, dy and dz parallel to the coordinate axes X, Y and Z respectively as shown in Fig. 13.7.

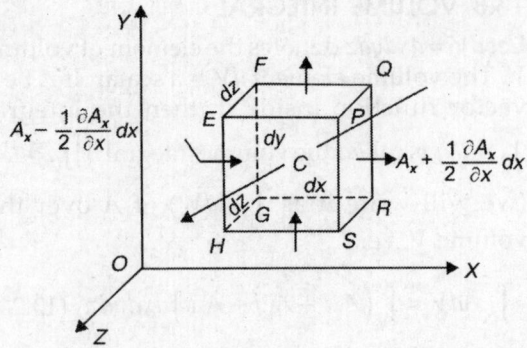

Fig. 13.7: Rectangular parallelopiped

Suppose the vector A has components A_x, A_y and A_z along X, Y and Z axes respectively.

The value of the x component of A at the centre of the face $EFGH$ is

$$A_x - \frac{1}{2}\frac{\partial A_x}{\partial x}dx \qquad (13.26)$$

and that at the centre of the face $PQRS$ is

$$A_x + \frac{1}{2}\frac{\partial A_x}{\partial x}dx \qquad (13.27)$$

Now, the volume of the fluid flowing per unit time through a face is given by the product of the area of the face and the normal component of the velocity upon it. We call it the *flux* through the face. One finds that the area of each of the faces $EFGH$ and $PQRS$ is $dydz$.

Flux entering the face $EFGH$

$$= \left(A_x - \frac{1}{2}\frac{\partial A_x}{\partial x}dx \right) dydz \quad (13.28)$$

Flux leaving the face $PQRS$

$$= \left(A_x + \frac{1}{2}\frac{\partial A_x}{\partial x}dx \right) dydz \quad (13.29)$$

Obviously, the excess of the flux leaving the parallelopiped over that entering it in the X-direction is

$$\left(A_x + \frac{1}{2}\frac{\partial A_x}{\partial x}dx \right)dydz - \left(A_x - \frac{1}{2}\frac{\partial A_x}{\partial x}dx \right)dydz$$

$$= \frac{\partial A_x}{\partial x}dxdydz \quad (13.30)$$

Similarly, one obtains that the net flux leaving the parallelopiped in the Y and Z directions are $\dfrac{\partial A_y}{\partial y}dxdydz$ and $\dfrac{\partial A_z}{\partial z}dxdydz$ respectively. Therefore, the total net flux diverging from (i.e. leaving) the parallelopiped is

$$\left(\frac{\partial A_x}{\partial x} + \frac{\partial A_y}{\partial y} + \frac{\partial A_z}{\partial z} \right)dxdydz \quad (13.31)$$

Since $dxdydz$ is the volume of the elementary parallelopiped and hence the amount of flux diverging per unit volume, which is defined as divergence of A is

$$\nabla \cdot A = \text{div}A = \frac{\partial A_x}{\partial x} + \frac{\partial A_y}{\partial y} + \frac{\partial A_z}{\partial z} \quad (13.32)$$

13.10 PHYSICAL MEANING OF DIVERGENCE OF A VECTOR ($\nabla \cdot A$)

$\nabla \cdot A$ at a point gives the amount of flux diverging per unit volume from that point. If A denotes the velocity of the moving fluid, then $\nabla \cdot A$ tells the rate at which the fluid is diverging from the point per unit volume.

If the divergence of a vector is positive at a point in a fluid, then either the fluid is expanding and its density at that point is falling with time or the point as a source of fluid. In case, if the divergence of a vector is negative, then the fluid is contracting and its density is rising or the point is a negative source, i.e. *sink*.

If the flux entering any element of field space is exactly balanced by that leaving it, the quantity $\nabla \cdot A = 0$. Obviously, there is no source or sink, nor its density is changing, i.e. the fluid is incompressible. If the flux entering and leaving an element are equal, the lines of flow of vector A should form either closed curves (e.g. in case of magnetic field due to a current) or extend to infinity. A vector which satisfies this condition is called *solenoidal*.

In the case of non-material fluxes (e.g. magnetic or electric flux), the existence of divergence means the presence of source or sink of flux at that point.

13.11 GAUSS'S DIVERGENCE THEOREM

This theorem states that the volume integral of the divergence of a vector field A taken over any volume V is equal to the surface integral of A taken over the closed surface surrounding the volume, i.e.

$$\iiint_V (\nabla \cdot A) dV = \iint_S A \cdot ds \qquad (13.33)$$

Proof: In cartesian coordinates,

$$\nabla \cdot A = \left(\hat{i} \frac{\partial}{\partial A} + \hat{j} \frac{\partial}{\partial y} + \hat{k} \frac{\partial}{\partial z} \right) \cdot \left(A_x \hat{i} + A_y \hat{j} + A_z \hat{k} \right)$$

$$= \frac{\partial A_x}{\partial x} + \frac{\partial A_y}{\partial y} + \frac{\partial A_z}{\partial z}$$

and $dv = dx\,dy\,dz$

Also

$$A \cdot ds = \left(A_x \hat{i} + A_y \hat{j} + A_z \hat{k} \right) \cdot \left(dS_x \hat{i} + dS_y \hat{j} + dS_z \hat{k} \right)$$

$$= A_x dS_x + A_y dS_y + A_z dS_z$$

$$= A_x dy dz + A_y dx dz + A_z dx dy$$

Using the above results, one can express Gauss's theorem in cartesian coordinates as

$$\iiint_V \left(\frac{\partial A_x}{\partial x} + \frac{\partial A_y}{\partial y} + \frac{\partial A_z}{\partial z} \right) dx\,dy\,dz$$

$$= \iint_S \left(A_x dS_x + A_y dS_y + A_z dS_z \right) \qquad (13.34)$$

Let us consider the first part of the LHS integral of Eq. (13.34). As shown in Fig. 13.8, integrate it with respect to x along a stripe of

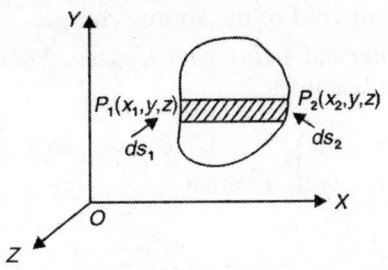

Fig. 13.8: A stripe of cross-section *dy dz*

cross-section $dydz$ from $P_1(x_1, y, z)$ to $P_2(x_2, y, z)$, i.e.

$$\iiint_V \frac{\partial A_x}{\partial x} dx\,dy\,dz = \iint_S |A_x|_{P_1}^{P_2} dy\,dz$$

$$= \iint_S \left[A_x(P_2) - A_x(P_1) \right] dy dz$$

$$= \iint_S \left[A_x(x_2, y, z) - A_x(x_1, y, z) \right] dy dz$$

Evidently, at P_1, $A_x\,dydz = -A_x\,dS_x$ and at P_2, $A_x\,dydz = A_x\,dS_x$

($\because$ at P_1 and P_2, the x-components of area have opposite directions)

$$\therefore \quad \iiint_V \frac{\partial A_x}{\partial x} dx\,dy\,dz = \iint_S A_x dS_x \qquad (13.35)$$

We should not confuse that we have written for

$$\iiint_V A_x(P_2) - A_x(P_1) dy dz = \iint_S (A_x dS_x + A_x dS_x)$$

only $\iint_S A_x dS_x$ because if we take all the stripes, then each integral $\iint_S A_x dS_x$ account for the half integral. Hence, $\iint_S A_x dS_x$ accounts for the total surface integral for x-component. Similarly, one obtains

$$\iiint_V \frac{\partial A_y}{\partial y} dx dy dz = \iint_S A_y dS_y \qquad (13.36)$$

and $$\iiint_V \frac{\partial A_z}{\partial z} dx\,dy\,dz = \iint_S A_z dS_z \qquad (13.37)$$

Adding Eq. (13.35) to Eq. (13.37), one obtains

$$\iiint_V \left(\frac{\partial A_x}{\partial x} + \frac{\partial A_y}{\partial y} + \frac{\partial A_z}{\partial z} \right) dx\,dy\,dz$$

$$= \iint_S (A_x dS_x + A_y dS_y + A_z dS_z)$$

or $$\iiint_V (\nabla \cdot A) dv = \iint_S A \cdot ds$$

or $\quad \iiint_V (\text{div} A) dv = \iint_S A \cdot ds \quad$ (13.38)

This proves the *Gauss's divergence theorem*.

Now we apply this theorem to an electric field E. Let us consider that there exist throughout the volume V enclosed by the surface S, a charge distribution of varying density ρ, then

$$\frac{1}{\varepsilon_0} \int_V \rho dv = \int_S E \cdot ds = \int_V (\nabla \cdot E) dv \quad (13.39)$$

The two volume integrals in Eq. (13.39) are equal, whatever be the volume over which the integration takes place. It, therefore, follows that integrands themselves must be equal, i.e.

$$\nabla \cdot E = \frac{\rho}{\varepsilon_0} \quad (13.40)$$

Equation (13.40) is sometimes referred to as the *differential form of Gauss's law in electrostatics*. According to this, the divergence of the electric field at a point equals ρ/ε_0 in the immediate neighbourhood of that point. If the point is situated in a material of dielectric constant K, then RHS of Eq. (13.40) must be divided by K.

The differential or point form Eq. (13.40) is valid only where the derivatives exist. The integral form Eq. (13.39) is valid everywhere.

13.12 THE LAPLACIAN OPERATOR

Let ϕ be a scalar potential function. If ϕ represents an electric field E, then

$$E = -\nabla \phi = -\text{grad}\phi \quad (13.41)$$

According to Gauss's law in electrostatics, the divergence of an electric field emerging from a closed surface is equal to $\frac{1}{\varepsilon}$ times the charge density enclosed in the surface , i.e.

$$\nabla \cdot E = \frac{\rho}{\varepsilon_0} \quad (13.42)$$

Here ε is the permittivity of the vacuum. Substituting Eq. (13.41) in Eq. (13.42), one obtains

$$-\nabla \cdot \nabla \phi = \frac{\rho}{\varepsilon_0} \quad (13.43)$$

But

$$\nabla \cdot \nabla \phi = \text{div}(\text{grad}\phi)$$

$$= \left(\hat{i} \frac{\partial}{\partial x} + \hat{j} \frac{\partial}{\partial y} + \hat{k} \frac{\partial}{\partial z} \right) \cdot \left(\hat{i} \frac{\partial}{\partial x} + \hat{j} \frac{\partial}{\partial y} + \hat{k} \frac{\partial}{\partial z} \right) \phi$$

$$= \left(\frac{\partial^2}{\partial x^2} + \frac{\partial^2}{\partial y^2} + \frac{\partial^2}{\partial z^2} \right) \phi$$

$$= \nabla^2 \phi \quad (13.44)$$

Thus, $\nabla^2 = \dfrac{\partial^2}{\partial x^2} + \dfrac{\partial^2}{\partial y^2} + \dfrac{\partial^2}{\partial z^2}$ is known as *Laplacian operator,* or simply the *Laplacian* $\nabla^2 = \nabla \cdot \nabla$ is usually pronounced as del squared.

13.13 POISSON'S AND LAPLACE'S EQUATIONS

Substituting Eq. (13.44) in Eq. (13.43), one obtains

$$\nabla^2 \phi = -\frac{\rho}{\varepsilon_0} \quad (13.45)$$

Equation (13.45) is known as *Poisson's equation*. One can calculate the electric potential at a point due to a charge distribution with the help of this equation. In a medium of dielectric constant K, Eq. (13.45) takes the form

$$\nabla^2 \phi = -\frac{\rho}{K \varepsilon_0} \quad (13.46)$$

In a charge free region, volume charge density $\rho = 0$, so the Eq. (13.45) or Eq. (13.46) becomes

$$\nabla^2 \phi = 0 \quad (13.47)$$

Equation (13.47) is known as *Laplace's equation*. By solving this equation, one can calculate the potential in an electric field, which is devoid of the source charge.

In *spherical polar coordinates*, Poison's equation becomes

$$\frac{1}{r^2} \frac{\partial}{\partial r} \left(r^2 \frac{\partial \phi}{\partial r} \right) + \frac{1}{r^2} \frac{1}{\sin\theta} \frac{\partial}{\partial \theta} \left(\sin\theta \frac{\partial \phi}{\partial \theta} \right)$$

$$+ \frac{1}{r^2 \sin^2\theta} \frac{\partial^2 \phi}{\partial \theta^2} = \frac{\rho}{K \varepsilon_0} \quad (13.48)$$

In *cylindrical polar coordinates*, Poisson's equation becomes

$$\frac{1}{r}\frac{\partial}{\partial r}\left(r\frac{\partial \phi}{\partial r}\right) + \frac{1}{r^2}\frac{\partial^2 \phi}{\partial \theta^2} + \frac{\partial^2 \phi}{\partial z^2} = \frac{\rho}{K\varepsilon_0} \quad (13.49)$$

13.14 THE CURL OF A VECTOR FUNCTION

We have seen that divergence of a vector field is scalar. Now, we define a vector called curl A, also written as $\nabla \times A$, associated with a vector field A.

If a vector field is derived as the gradient of a scalar field, then the line integral of the vector around any closed path in the field is zero. Such a field is called *lamellar* or *non-curl field*. There are, however, such vector fields for which the line integral round any closed path is non-zero and have a finite value. Such vector fields cannot be derived as gradient of any scalar field and they exhibit the property of curl.

Suppose a non-lamellar vector field is represented by several lines of flow as shown in Fig. (13.9). Let us consider a plane rectangular area in this field. When the plane *abcd* is perpendicular to the field (position 1), no line of flow lies along its boundary so that the line integral round it is zero. When the plane *abcd* is parallel to the field (position 2), lines of flow lie along the upper and lower edges of the plane. Since the value of the vector at the upper edge is different from that at the lower edge, the line integral round the boundary has a finite value. This value will be different for different orientations of the plane and shall

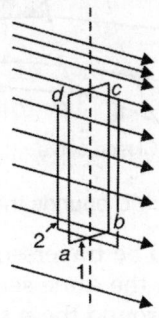

Fig. 13.9: Non-lamellar vector field

be maximum for a particular orientation. This maximum line integral computed per unit area along the boundary is called the curl of the vector field. Thus, one can define the curl of a vector field as follows:

The curl of a (non-lameller) vector field at any point is defined as the maximum line integral of the vector component per unit area along the boundary of an infinitesimal area at that point. Obviously, a curl of a vector field is a vector.

Curl in Cartesian Coordinates

We now calculate curl A, where A is function of cartesian coordinates (x, y, z). Let us consider a rectangular shaped infinitely small element of area $dxdy$ in the (X, Y) plane as shown in Fig. 13.10.

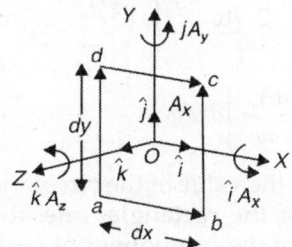

Fig. 13.10: Curl in cartesian coordinates

Let the magnitude of the vector A at a point O is $|A|$ and A_x, A_y and A_z be its components along X, Y and Z axes respectively. Let the sides dx, dy of the rectangle be parallel to the X and Y axes respectively, so the normal to the area is along the Z-axis. The arrow heads on the sides of the rectangle show the directions in which the component of A act.

Since the rectangle is infinitely small, the average value of A along any side of it may be taken as the value at the middle point of that side. Obviously, the average values along the four sides will be as follows:

Along *ab*; $\quad A_x - \dfrac{1}{2}\dfrac{\partial A_x}{\partial y}dy$

Along *bc*; $\quad A_y + \dfrac{1}{2}\dfrac{\partial A_y}{\partial x}dx$

Along dc; $\quad A_x + \dfrac{1}{2}\dfrac{\partial A_x}{\partial x}dy$

Along ad; $\quad A_y - \dfrac{1}{2}\dfrac{\partial A_y}{\partial x}dx$

Here we have considered only the first order of small quantities dx, dy in the Taylor's expansion.

Therefore, the closed path line integral around the boundary $abcd$ of the rectangle is given by

$$= \left[\left(A_x - \frac{1}{2}\frac{\partial A_x}{\partial y}dy\right)-\left(A_x + \frac{1}{2}\frac{\partial A_x}{\partial y}dy\right)\right]dx$$

$$+ \left[\left(A_y + \frac{1}{2}\frac{\partial A_y}{\partial x}dx\right)-\left(A_y - \frac{1}{2}\frac{\partial A_y}{\partial x}dx\right)\right]dy$$

$$= \left(\frac{\partial A_y}{\partial x}-\frac{\partial A_x}{\partial y}\right)dx\,dy$$

On dividing the value of the circulation by the area $dx\,dy$ of the rectangle, one obtains the magnitude of the component of curl A along Z-axis as

$$(\text{curl } A)_z = \left(\frac{\partial A_y}{\partial x}-\frac{\partial A_x}{\partial y}\right)\hat{k}$$

Similarly, one obtains the components of curl A along X and Y axes as

$$(\text{curl } A)_x = \left(\frac{\partial A_z}{\partial y}-\frac{\partial A_y}{\partial z}\right)\hat{i}$$

and $\quad (\text{curl } A)_y = \left(\frac{\partial A_x}{\partial z}-\frac{\partial A_z}{\partial x}\right)\hat{j}$

Adding the three components, one obtains

$$\text{curl } A = \left(\frac{\partial A_z}{\partial y}-\frac{\partial A_y}{\partial z}\right)\hat{i}+\left(\frac{\partial A_x}{\partial z}-\frac{\partial A_z}{\partial x}\right)\hat{j}$$

$$+\left(\frac{\partial A_y}{\partial x}-\frac{\partial A_x}{\partial y}\right)\hat{k} \qquad (13.50)$$

One can express $\nabla \times A$ in the form of a determinant as

$$\text{curl } A = \begin{vmatrix} \hat{i} & \hat{j} & \hat{k} \\ \dfrac{\partial}{\partial x} & \dfrac{\partial}{\partial y} & \dfrac{\partial}{\partial z} \\ A_x & A_y & A_z \end{vmatrix} \qquad (13.51)$$

We must note that Eq. (13.51) is independent of axes. The name curl indicates that a vector field at a point in space has circulation. The determinant in Eq. (13.51) is equal to the vector product of del operator (∇) and vector A, i.e.

$$\text{Curl } A = \nabla \times A \qquad (13.52)$$

13.15 STOKE'S CURL THEOREM

This states that *the flux of the curl of a vector field A over any surface of any shape is equal to the line integral of the vector field A over the boundary of that surface.* We must note that the sense of line integral is related with the positive normal to the surface.

Mathematically, one can express it as

$$\iint_S \text{curl } A \cdot ds = \oint A \cdot dl \qquad (13.53)$$

Here the line integration is being carried out for the boundary of the surface.

The path C bounds the surface S as shown in Fig. 13.11. Let the surface S be divided into large number of small areas ds_1, ds_2, ds_3, .., etc. Since

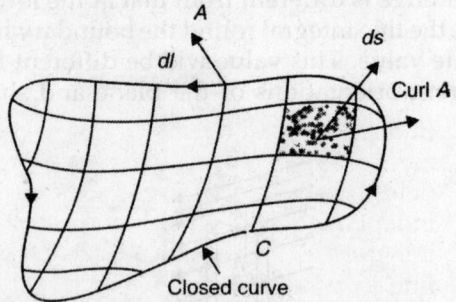

Fig. 13.11: Path C bounds the surface S

all the areas are to be traversed by the curves bounding them in the same sense, the sum of the line integrals around these areas is equal to the line integral around the closed path C, i.e.

$$\oint A \cdot dl = \Sigma_i \oint_C A \cdot dl$$

From the definition of curl A, at a point inside ds_i, we have

$$\text{Curl } A \cdot ds_i = \oint c_i \, A \cdot dl$$

Since the number of these areas is very large and hence in the limiting case

$$\oint A \cdot dl = \left[\underset{\partial s \to 0}{\text{Lt}} \, \Sigma_i \frac{\partial}{\partial s_i} \oint A \cdot dl \right] \partial s_i \qquad (13.54)$$

Since the curl is the line integral per unit area in the limiting case, when area tends to zero, we have

$$\oint A \cdot dl = \iint_S \text{curl} A \cdot ds \qquad (13.55)$$

Equation (13.55) represents the *Stoke's curl theorem. Obviously, the line integral of a vector field along the limiting boundary of the area enclosed by a closed path is equal to the surface integral of the curl of the vector field.* This theorem is used in converting line integral into surface integral and *vice versa.*

13.16 PHYSICAL MEANING OF CURL OF A VECTOR FIELD

The existence of the curl of a vector field is always due to its rotation or circulation. A vector field having non-zero curl has circulation and has curly or rotational field lines about the point of non-zero curl. In German literature, curl is written as *rot* which is derived from rotation. The following examples illustrates clearly the physical meaning of curl.

i. In an electrostatic field E, the curl E is zero everywhere. The line integral of the field, E between two points A and B is independent of the path. Therefore, line integral along A to B is negative of the line integral along B to A or in other words,

$$\int_A^B E \cdot dl = -\int_B^A E \cdot dl$$

Hence $\quad \int_A^B E \cdot dl + \int_B^A E \cdot dl = 0$

Obviously, the line integral of the field E around any closed path is zero. Also the circulation is zero around any closed path, then from the Stoke's theorem, the surface integral of curl E is over an element of any size and shape. This clearly reveals that curl E should also be zero **every**where, i.e.

$$\text{curl } E = 0 \qquad (13.56)$$

The above condition is for a field to be conservative. Since the line integral of a vector around the closed path is zero when the vector is derived by taking the gradient of a scalar function, therefore one conclude that electric field can be described as the gradient of some potential function.

ii. Let us consider an example of a vector field having nonzero curl. We can show that the curl of the magnetic induction vector is proportional to the current density vector at the given point, i.e.

$$\nabla \times B = \mu_0 J \qquad (13.57)$$

where μ_0 is the permeability of free space ($\mu_0 = 4\pi \times 10^{-7} \, \text{N A}^{-2}$). However, we must note **that** Eq. (13.57) holds good only for the magnetic field in a vacuum in the absence of time-varying electric fields.

A comparison of Eqs. (13.56) and (13.57) shows that electrostatic and magnetic fields are of an appreciably different nature. The curl of an electrostatic field is zero, i.e. an electrostatic field is conservative and can be characterized by the scalar potential ϕ. The curl of a magnetic field is non-zero, i.e. proportional to current density vector in vacuum. Accordingly, the circulation of B is proportional to the current enclosed by a loop. This is why we cannot ascribe to a magnetic field a scalar potential that would be related to B by an equation similar to Eq. (13.41). This potential would not be unique upon each circumvention of the loop and return to the initial point it would receive an increment equal to $\mu_0 I$. A field whose curl differs from zero is called a *vortex* or a *solenoidal* one.

Since $\nabla \cdot \boldsymbol{B}$ is zero everywhere and hence $\boldsymbol{B}$ can be represented as the curl of another vector $\boldsymbol{A}$. Thus,

$$\boldsymbol{B} = \nabla \times \boldsymbol{A} \qquad (13.58)$$

The divergence of a curl is always zero. $\boldsymbol{A}$ is called the *vector potential*.

13.17 GREEN'S THEOREM

Green's theorem is a corollary of the divergence theorem. The theorem states that if ϕ and Ψ are two scalar point functions such that these functions and their first derivatives are continuously differentiable, then

$$\iiint_V \left(\phi \nabla^2 \psi + \nabla\phi \cdot \nabla\psi\right) dV = \iint_S \left(\phi \nabla \psi\right) \cdot ds \quad (13.59)$$

and

$$\iiint_V \left(\phi \nabla^2 \psi - \psi \nabla^2 \phi\right) dV = \iint_S \left(\phi \nabla \psi - \psi \nabla \phi\right) \cdot ds$$
$$(13.60)$$

Equations (13.59) and (13.60) are respectively known as first and second form of Green's theorem. One can derive it as follows:

We start with Gauss's divergence theorem which states

$$\iiint_V \nabla \cdot \boldsymbol{A}\, dV = \iint_S \boldsymbol{A} \cdot ds \qquad (13.61)$$

Let us take $\boldsymbol{A} = \phi \nabla \psi$ or

$$\hat{i} A_x + \hat{j} A_y + \hat{k} A_z = \phi \left(\hat{i} \frac{\partial \psi}{\partial x} + \hat{j} \frac{\partial \psi}{\partial y} + \hat{k} \frac{\partial \psi}{\partial z} \right)$$

Obviously,

$$A_x = \phi \frac{\partial \psi}{\partial x}, \ A_y = \phi \frac{\partial \psi}{\partial y} \text{ and } A_z = \phi \frac{\partial \psi}{\partial z}$$

Now,

$$\nabla \cdot \boldsymbol{A} = \frac{\partial A_x}{\partial x} + \frac{\partial A_y}{\partial y} + \frac{\partial A_z}{\partial z}$$

$$= \frac{\partial}{\partial x}\left(\phi \frac{\partial \psi}{\partial x}\right) + \frac{\partial}{\partial y}\left(\phi \frac{\partial \psi}{\partial y}\right) + \frac{\partial}{\partial z}\left(\phi \frac{\partial \psi}{\partial z}\right)$$

$$= \left(\phi \frac{\partial^2 \psi}{\partial x^2} + \frac{\partial \phi}{\partial x}\frac{\partial \psi}{\partial x}\right) + \left(\phi \frac{\partial^2 \psi}{\partial y^2} + \frac{\partial \phi}{\partial y}\frac{\partial \psi}{\partial y}\right)$$

$$+ \left(\phi \frac{\partial^2 \psi}{\partial z^2} + \frac{\partial \phi}{\partial z}\frac{\partial \psi}{\partial z}\right)$$

$$= \phi \left(\frac{\partial^2 \psi}{\partial x^2} + \frac{\partial^2 \psi}{\partial y^2} + \frac{\partial^2 \psi}{\partial z^2} \right) + \frac{\partial \phi}{\partial x}\frac{\partial \psi}{\partial x}$$

$$+ \frac{\partial \phi}{\partial y}\frac{\partial \psi}{\partial y} + \frac{\partial \phi}{\partial z}\frac{\partial \psi}{\partial z}$$

$$= \phi \nabla^2 \psi + \nabla\phi \cdot \nabla\psi$$

Making use of these results in Eq. (13.61), one obtains

$$\iiint_V \left(\phi \nabla^2 \psi + \nabla\phi \cdot \nabla\psi\right) dV = \iint_S \left(\phi \nabla\psi\right) \cdot ds$$
$$(13.62)$$

This is Green's first equation.

Interchanging ϕ and ψ in Eq. (13.62), one obtains

$$\iiint_V \left(\psi \nabla^2 \phi + \nabla\psi \cdot \nabla\psi\right) dV = \iint_S \left(\psi \nabla\phi\right) \cdot ds$$
$$(13.63)$$

Subtracting Eq. (13.63) from Eq. (13.62), one obtains

$$\iiint_V \left(\psi \nabla^2 \psi - \psi \nabla^2 \phi\right) dV = \iint_S \left(\phi \nabla\psi - \psi \nabla\phi\right) \cdot ds$$
$$(13.64)$$

This is Green's second equation. This is also referred to as *symmetrical theorem*.

In the RHS of Eq. (13.64), we have

$$\nabla\psi \cdot ds = \frac{\partial \psi}{\partial n} ds$$

and

$$\nabla\phi \cdot ds = \frac{\partial \phi}{\partial n} ds$$

where $\dfrac{\partial \psi}{\partial n}$ and $\dfrac{\partial \phi}{\partial n}$ are the directional derivatives of ψ and ϕ respectively along the outwards normal to ds. In the light of these results, Eq. (13.64) can also be expressed as

$$\iiint_V \left(\phi \nabla^2 \psi - \psi \nabla^2 \phi\right) dV = \iint_S \left(\phi \frac{\partial \psi}{\partial n} - \psi \frac{\partial \phi}{\partial n} \right) \cdot ds$$
$$(13.65)$$

Equation (13.65) is an alternative form of Green's symmetrical theorem.

One can easily derive Gauss's divergence theorem from Green's theorem. We have Green's symmetrical theorem [Eq. (13.64)] if ϕ is constant, then we have

$$\phi \iiint_V \nabla^2 \psi \, dV = \phi \iint_S \nabla \psi \cdot ds$$

or $\quad \iiint_V (\nabla \cdot \nabla \psi) dV = \iint_S \nabla \psi \cdot ds$

Putting $\nabla \psi = A$, one obtains

$$\iiint_V (\nabla \cdot A) dV = \iint_S A \cdot ds \qquad (13.66)$$

which is *Gauss's divergence theorem*.

ILLUSTRATIVE EXAMPLES

Example 1

A field is expressed by the potential function $V = 3x^2z - xy^3 + z$. Calculate the potential at the point $(1, -2, +2)$.

Solution

We have
$$V = 3x^2z - xy^3 + z$$

Potential at the point $(1, -2, +2)$ is obtained as

$$V_{(1, -2, +2)} = 3(1)^2 \times 2 - (1)(-2)^3 + 2 = 16 \text{ volts.}$$

Example 2

A vector field A is represented by the function $A = \hat{i}(2x^2y - x^4) + \hat{j}(yz^2) - \hat{k}(xy^2)$, then obtain

the values of $\dfrac{\partial A}{\partial x}, \dfrac{\partial^2 A}{\partial x^2}$ and $\dfrac{\partial^2 A}{\partial y \partial z}$.

Solution

We have

$$A = \hat{i}(2x^2y - x^4) + \hat{j} yz^2 - \hat{k} xy^2$$

$$\therefore \quad \frac{\partial A}{\partial x} = \frac{\partial}{\partial x}\left[\hat{i}(2x^2y - x^4) + \hat{j} yz^2 - \hat{k} xy^2\right]$$

$$= \hat{i}(4xy - 4x^3) + \hat{j}0 - \hat{k}y^2$$

Now $\quad \dfrac{\partial^2 A}{\partial x^2} = \dfrac{\partial}{\partial x}\left[\hat{i}(4xy - 4x^3) - \hat{k}y^2\right]$

$$= \hat{j}(4y - 12x^2) - \hat{k}0$$

and $\quad \dfrac{\partial^2 A}{\partial y \partial z} = \dfrac{\partial}{\partial z}\left(\dfrac{\partial A}{\partial y}\right)$

$$= \frac{\partial}{\partial z}\left[\frac{\partial}{\partial y}\left\{\hat{i}(2x^2y - x^4) - \hat{j} yz^2 - \hat{k} xy^2\right\}\right]$$

$$= \frac{\partial}{\partial z}\left[\hat{i}(2x^2) + \hat{j} z^2 - \hat{k} 2xy\right] = 2\hat{j} z$$

Example 3

Find the value of div $(r^n \hat{r})$, where $\hat{r}$ is the unit vector.

Solution

We have $\qquad r = \hat{i} x + \hat{j} y + \hat{k} z$

$$\therefore \qquad |r| = \sqrt{x^2 + y^2 + z^2}$$

$$\therefore \quad \text{div}(r^n \hat{r}) = \nabla \cdot (r^n \hat{r})$$

$$= \left(\hat{i}\frac{\partial}{\partial x} + \hat{j}\frac{\partial}{\partial y} + \hat{k}\frac{\partial}{\partial z}\right) \cdot (r^{n-1} r \hat{r})$$

$$= \left(\hat{i}\frac{\partial}{\partial x} + \hat{j}\frac{\partial}{\partial y} + \hat{k}\frac{\partial}{\partial z}\right) \cdot (r^{n-1} r)$$

$$= \left(\hat{i}\frac{\partial}{\partial x} + \hat{j}\frac{\partial}{\partial y} + \hat{k}\frac{\partial}{\partial z}\right) \cdot r^{n-1}(\hat{i} x + \hat{j} y + \hat{k} z)$$

$$= \frac{\partial}{\partial x}(xr^{n-1}) + \frac{\partial}{\partial y}(yr^{n-1}) + \frac{\partial}{\partial z}(zr^{n-1})$$

$$= \left[r^{n-1} + x\frac{\partial}{\partial x}(x^2 + y^2 + z^2)^{n-1/2}\right]$$

$$+ \left[r^{n-1} + y\frac{\partial}{\partial y}(x^2 + y^2 + z^2)^{n-1/2}\right]$$

$$+ \left[r^{n-1} + z\frac{\partial}{\partial z}(x^2 + y^2 + z^2)^{n-1/2}\right]$$

$$= r^{n-1} + x\left(\frac{n-1}{2}\right)\left(x^2 + y^2 + z^2\right)^{n-3/2} 2x + \dots$$

$$= 3r^{n-1} + (n-1)\left[x^2 + y^2 + z^2\right]^{n-1/2}$$

$$= 3r^{n-1} + (n-1)r^{n-1} = (n+2)r^{n-1}$$

Example 4

Find: (a) $\nabla \cdot r$, where r is the position vector

(b) $\nabla \cdot \left[\hat{i}(x+y) + \hat{j}(x-y) + \hat{k}(4z)\right]$

Solution

(a) $\nabla \cdot r = \left(\hat{i}\dfrac{\partial}{\partial x} + \hat{j}\dfrac{\partial}{\partial y} + \hat{k}\dfrac{\partial}{\partial z}\right) \cdot \left(\hat{i}\,x + \hat{j}\,y + \hat{k}\,z\right)$

$$= \frac{\partial x}{\partial x} + \frac{\partial y}{\partial y} + \frac{\partial z}{\partial z} = 3$$

(b) $\nabla \cdot \left[\hat{i}(x+y) + \hat{j}(x-y) + \hat{k}(4z)\right]$

$$= \left(\hat{i}\frac{\partial}{\partial x} + \hat{j}\frac{\partial}{\partial y} + \hat{k}\frac{\partial}{\partial z}\right) \cdot \left[\hat{i}(x+y)\right.$$

$$\left. + \hat{j}(x-y) + \hat{k}(4z)\right]$$

$$= \frac{\partial}{\partial x}(x+y) + \frac{\partial}{\partial y}(x-y) + \frac{\partial}{\partial z}(4z)$$

$$= 1 - 1 + 4 = 4$$

Example 5

If r is the position vector of a point, calculate the gradient of $\dfrac{1}{r}$.

Solution

$$r = \hat{i}\,x + \hat{j}\,y + \hat{k}\,z$$

$$\therefore \quad |r| = \sqrt{x^2 + y^2 + z^2}$$

Now $\quad \nabla\left(\dfrac{1}{r}\right) = \nabla\left(x^2 + y^2 + z^2\right)^{-1/2}$

or $\quad \nabla\left(\dfrac{1}{r}\right) = \left(\hat{i}\dfrac{\partial}{\partial x} + \hat{j}\dfrac{\partial}{\partial y} + \hat{k}\dfrac{\partial}{\partial z}\right)\left(x^2 + y^2 + z^2\right)^{-1/2}$

$$= \hat{i}\frac{\partial}{\partial x}\left(x^2 + y^2 + z^2\right)^{-1/2} + \hat{j}\frac{\partial}{\partial y}\left(x^2 + y^2 + z^2\right)^{-1/2}$$

$$+ \hat{k}\frac{\partial}{\partial z}\left(x^2 + y^2 + z^2\right)^{-1/2}$$

$$\therefore \quad \frac{\partial}{\partial x}\left(x^2 + y^2 + z^2\right)^{-1/2}$$

$$= -\frac{1}{2}\left(x^2 + y^2 + z^2\right)^{-3/2}(2x) = \frac{-x}{r^3}$$

Similarly,

$$\frac{\partial}{\partial y}\left(x^2 + y^2 + z^2\right)^{-1/2} = \frac{-y}{r^3}$$

and $\quad \dfrac{\partial}{\partial z}\left(x^2 + y^2 + z^2\right)^{-1/2} = \dfrac{-z}{r^3}$

$$\therefore \quad \nabla\left(\frac{1}{r}\right) = -\frac{\hat{i}\,x + \hat{j}\,y + \hat{k}\,z}{r^3} = \frac{-r}{r^3}$$

Example 6

The electric field due to a point charge is expressed as $E = \dfrac{Q}{r^2}\hat{r}$. Show that the divergence of electric field due to that point charge is zero.

Solution

$$\nabla \cdot E = \nabla \cdot \left(\frac{Q}{r^2}\hat{r}\right) = \nabla \cdot \left(\frac{Q}{r^3}\hat{r}r\right) = \nabla \cdot \left(\frac{Q}{r^3}r\right)$$

$$= \left(\hat{i}\frac{\partial}{\partial x} + \hat{j}\frac{\partial}{\partial y} + \hat{k}\frac{\partial}{\partial z}\right) \cdot Q\left(\frac{\hat{i}x + \hat{j}y + \hat{k}z}{\left(x^2 + y^2 + z^2\right)^{3/2}}\right)$$

$$\left(\because r = \hat{i}\,x + \hat{j}\,y + \hat{k}\,z \therefore |r| = \sqrt{x^2 + y^2 + z^2}\right)$$

$$= Q\left[\frac{\partial}{\partial x}\left\{\frac{x}{\left(x^2 + y^2 + z^2\right)^{3/2}}\right\}\right]$$

$$+ Q\left[\frac{\partial}{\partial y}\left(\frac{y}{\left(x^2 + y^2 + z^2\right)^{3/2}}\right)\right]$$

$$+ Q\left[\frac{\partial}{\partial z}\left\{\frac{z}{\left(x^2 + y^2 + z^2\right)^{3/2}}\right\}\right]$$

$$= Q\left[\frac{1}{\left(x^2 + y^2 + z^2\right)^{3/2}} - \frac{3x^2}{\left(x^2 + y^2 + z^2\right)^{5/2}}\right.$$

$$+ \frac{1}{\left(x^2 + y^2 + z^2\right)^{3/2}} - \frac{3y^2}{\left(x^2 + y^2 + z^2\right)^{5/2}}$$

$$\left. + \frac{1}{\left(x^2 + y^2 + z^2\right)^{3/2}} - \frac{3z^2}{\left(x^2 + y^2 + z^2\right)^{5/2}}\right]$$

$$= 0$$

Example 7

Show that the potential function

$$V = q\left(x^2 + y^2 + z^2\right)^{1/2}$$

does not satisfy the Laplace's equation.

Solution

The Laplace's equation is $\nabla^2 V = 0$, i.e.

$$\frac{\partial^2 V}{\partial x^2} + \frac{\partial^2 V}{\partial y^2} + \frac{\partial^2 V}{\partial z^2} = 0 \tag{1}$$

Using $V = q\left(x^2 + y^2 + z^2\right)^{1/2}$, LHS of Eq. (1) becomes

$$= q\frac{\partial}{\partial x}\left[\frac{\partial}{\partial x}\left(x^2 + y^2 + z^2\right)^{1/2}\right]$$

$$+ q\frac{\partial}{\partial y}\left[\frac{\partial}{\partial y}\left(x^2 + y^2 + z^2\right)^{1/2}\right]$$

$$+ q\frac{\partial}{\partial z}\left[\frac{\partial}{\partial z}\left(x^2 + y^2 + z^2\right)^{1/2}\right]$$

$$= q\frac{\partial}{\partial x}\left[x\left(x^2 + y^2 + z^2\right)^{1/2}\right]$$

$$+ q\frac{\partial}{\partial y}\left[y\left(x^2 + y^2 + z^2\right)^{1/2}\right]$$

$$+ q\frac{\partial}{\partial z}\left[z\left(x^2 + y^2 + z^2\right)^{1/2}\right]$$

$$= q\left\{-\frac{1}{4}\left(x^2 + y^2 + z^2\right)^{-3/2}4x\right.$$

$$+ \left(x^2 + y^2 + z^2\right)^{-1/2}$$

$$- \frac{1}{4}\left(x^2 + y^2 + z^2\right)^{-3/2}4y + \left(x^2 + y^2 + z^2\right)^{-1/2}$$

$$\left. - \frac{1}{4}\left(x^2 + y^2 + z^2\right)^{-3/2}4z + \left(x^2 + y^2 + z^2\right)^{-1/2}\right\}$$

$$= q\left\{\left(x^2 + y^2 + z^2\right)^{-3/2}\left(x^2 + y^2 + z^2\right)\right.$$

$$\left. + 3\left(x^2 + y^2 + z^2\right)^{-1/2}\right\}$$

$$V = 2q\left(x^2 + y^2 + z^2\right)^{-1/2} \neq 0$$

Obviously, the given potential function does not satisfy the Laplace's equation.

Example 8

Show that curl $r = 0$, where r is position vector.

Solution

$$\nabla \times r = \left(\hat{i}\frac{\partial}{\partial x} + \hat{j}\frac{\partial}{\partial y} + \hat{k}\frac{\partial}{\partial z}\right) \times (\hat{i}x + \hat{j}y + \hat{k}z)$$

$$= \begin{vmatrix} \hat{i} & \hat{j} & \hat{k} \\ \dfrac{\partial}{\partial x} & \dfrac{\partial}{\partial y} & \dfrac{\partial}{\partial z} \\ x & y & z \end{vmatrix} = 0$$

Example 9

A potential function in an electric field is given by, $\phi = (2x^2 - y^2 + 3z^2)$ volts. Show that the intensity of electric field at point $(1, -1, 0)$ is $-\left(4\hat{i} - 2\hat{j}\right)$ V/m.

Solution

$$E = -\nabla\phi = -\left(\hat{i}\frac{\partial\phi}{\partial x} + \hat{j}\frac{\partial\phi}{\partial y} + \hat{k}\frac{\partial\phi}{\partial z}\right)$$

Now $\dfrac{\partial\phi}{\partial x} = 4x$, $\dfrac{\partial\phi}{\partial y} = -2y$ and $\dfrac{\partial\phi}{\partial z} = 6z$

$$\therefore \quad E = -\left(\hat{i}\,4x - \hat{j}\,2y + \hat{k}\,6z\right)$$

$\therefore$ E at the point $(1, -1, 0)$

$$= -(4\hat{i} + 2\hat{j}) = -(4\hat{i} + 2\hat{j}) \text{ V/m.}$$

Example 10

Show that curl grad $\phi = 0$, where ϕ is any scalar function.

Solution

$$\nabla \times \nabla\phi = \left(\hat{i}\frac{\partial}{\partial x} + \hat{j}\frac{\partial}{\partial y} + \hat{k}\frac{\partial}{\partial z}\right)$$

$$\times \left(\hat{i}\frac{\partial\phi}{\partial x} + \hat{j}\frac{\partial\phi}{\partial y} + \hat{k}\frac{\partial\phi}{\partial z}\right)$$

$$= \begin{vmatrix} \hat{i} & \hat{j} & \hat{k} \\ \dfrac{\partial}{\partial x} & \dfrac{\partial}{\partial y} & \dfrac{\partial}{\partial z} \\ \dfrac{\partial\phi}{\partial x} & \dfrac{\partial\phi}{\partial y} & \dfrac{\partial\phi}{\partial z} \end{vmatrix}$$

$$= \hat{i}\left(\frac{\partial^2\phi}{\partial y\partial z} - \frac{\partial^2\phi}{\partial z\partial y}\right) + \hat{j}\left(\frac{\partial^2\phi}{\partial z\partial x} - \frac{\partial^2\phi}{\partial x\partial z}\right)$$

$$+ \hat{k}\left(\frac{\partial^2\phi}{\partial x\partial y} - \frac{\partial^2\phi}{\partial y\partial x}\right)$$

$$= 0 \left(\because \frac{\partial^2\phi}{\partial y\partial z} = \frac{\partial^2\phi}{\partial z\partial y} \text{ and so on}\right)$$

Example 11

Show that div grad $(R^n) = n(n+1)r^{n-2}$, where R is position vector.

Solution

$$\text{div grad } (R^n) = \nabla \cdot \nabla (R^n)$$

Now $$\nabla\left(R^n\right) = \left(\hat{i}\frac{\partial}{\partial x} + \hat{j}\frac{\partial}{\partial y} + \hat{k}\frac{\partial}{\partial z}\right)R^n$$

$$= \hat{i}nR^{n-1}\frac{\partial R}{\partial x} + \hat{j}nR^{n-1}\frac{\partial R}{\partial y} + \hat{k}nR^{n-1}\frac{\partial R}{\partial z}$$

$$= nR^{n-1}\left(\hat{i}\frac{\partial R}{\partial x} + \hat{j}\frac{\partial R}{\partial y} + \hat{k}\frac{\partial R}{\partial z}\right)$$

$\because$ $$R = \hat{i}\,x + \hat{j}\,y + \hat{k}\,z$$

$\therefore$ $$R^2 = R \cdot R = x^2 + y^2 + z^2$$

$\therefore$ $$2R\frac{\partial R}{\partial x} = 2x$$

$\therefore$ $$\frac{\partial R}{\partial x} = \frac{x}{R}$$

Similarly, $$\frac{\partial R}{\partial y} = \frac{y}{R} \quad \text{and} \quad \frac{\partial R}{\partial z} = \frac{z}{R}$$

$\therefore$ $$\nabla R^n = nR^{n-1}\left(\hat{i}\frac{x}{R} + \hat{j}\frac{y}{R} + \hat{k}\frac{z}{R}\right)$$

$$\nabla \cdot \nabla R^n = \left(\hat{i}\frac{\partial}{\partial x} + \hat{j}\frac{\partial}{\partial y} + \hat{k}\frac{\partial}{\partial z}\right) \cdot nR^{n-2}$$

$$\left(\hat{i}\,x + \hat{j}\,y + \hat{k}\,z\right)$$

$$= \frac{\partial}{\partial x}\left(nR^{n-2}x\right) + \frac{\partial}{\partial y}\left(nR^{n-2}y\right) + \frac{\partial}{\partial z}\left(nR^{n-2}z\right)$$

$$= n\left\{R^{n-2} + x(n-2)R^{n-3}\frac{\partial R}{\partial x} + R^{n-2}\right.$$

$$\left. + y(n-2)R^{n-3}\frac{\partial R}{\partial y} + R^{n-2} + z(n-2)R^{n-3}\frac{\partial R}{\partial z}\right\}$$

$$= 3nR^{n-2} + n(n-2)R^{n-3}\left\{x\frac{\partial R}{\partial x} + y\frac{\partial R}{\partial y} + z\frac{\partial R}{\partial z}\right\}$$

$$= 3nR^{n-2} + n(n-2)R^{n-3}\left\{x\frac{x}{R} + y\frac{y}{R} + z\frac{z}{R}\right\}$$

$$= 3nR^{n-2} + n(n-2)R^{n-3}\left(\frac{x^2 + y^2 + z^2}{R}\right)$$

$$= 3nR^{n-2} + n(n-2)R^{n-3}R$$

$$= 3R^{n-2}(3 + n - 2) = n(n+1)R^{n-2}$$

REVIEW QUESTIONS

1. Describe scalar and vector fields. How can they be represented graphically?

2. What do you mean by the field of any physical quantity? What are scalar and vector fields? Give examples.

3. Define gradient of a scalar field. Show that the gradient of a scalar field is a vector. Give an example.

4. Prove that the gradient of a scalar field is normal to the surface, ϕ = constant.

5. Express the gradient of a scalar field $\phi(x, y, z)$ in terms of cartesian coordinates.

6. Show that when a vector field can be derived as the gradient as a scalar field the line integral of the vector taken round any closed path in the vector field is zero.

7. Explain the meaning of line integral of a vector field using at least one example.

8. Explain the physical meaning of the divergence of a vector field A. Express divergence A in orthogonal coordinates and show that div $A = \nabla \cdot A$.

9. Define curl of a vector field and give its physical significance. Derive expression for a vector field and show that curl $A = \nabla \times A$.

10. What is Laplacian operator? Express div and grad ϕ in cartesian coordinates, where ϕ is a scalar quantity.

11. Derive Laplace's and Poisson's equations starting from the differential form of Gauss' law.

12. State and prove Green's theorem.

PROBLEMS

1. For a position vector $R = \hat{i}x + \hat{j}y + \hat{k}z$, find the values of (i) grad $\dfrac{1}{R}$ (ii) div $\left(\dfrac{R}{R^3}\right)$ and

(iii) curl $\left(\dfrac{R}{R^3}\right)$.   [*Ans.* (i) $-\dfrac{R}{R^3}$ (ii) 0 (iii) 0]

2. A potential field is represented by the equation $\phi = 4yz^2 + 3xyz - z^2 + 2$. Calculate the potential at the point $(1, -1, -2)$.   [*Ans.* –36 V]

3. A vector field is represented by the equation, $A = \hat{i}x^2yz - \hat{j}2xz^3 + \hat{k}xz^2$. Calculate the values of the partial derivatives (i) $\dfrac{\partial A}{\partial x}$ (ii) $\dfrac{\partial A}{\partial z}$.

[*Ans.* (i) $\hat{i}2xyz - \hat{j}2z^3 + \hat{k}z^2$

(ii) $\hat{i}x^2z - \hat{j}6xz^2 + \hat{k}2xz$]

4. The temperature of a body at any point changes according to the relation, $T = 4x^2 + 3y^2 - 2z^2$. Show that the grad T at the point $(1, 2, 3)$ is equal to $-8\hat{i} - 12\hat{j} - 12\hat{k}$.

5. Show that the function $\phi = x^2 - y^2$ satisfies the Laplace's equation.

6. Show that the potential functions:

    i. $V = x^2 - y^2 + z$

    ii. $V = x^2 + y^2 - 2z$

    iii. $V = ax + by + cz$

    iv. $V = ax^2 - ay^2 + 4z$

satisfy Laplace' equation.

7. Calculate the curl of the following fields:

(a) $E = a\left\{3x^2yz\hat{i} + \left[x^3z + yz^2(yz)^{1/2}\right]\hat{j} \right.$

$\left. + \left[x^3y + y^2z(yz)^{1/2}\right]\hat{k}\right\}$

(b) $B = y\hat{i} - x\hat{j}$

(c) $B = a\left\{\dfrac{y\hat{i}}{\left(x^2 + y^2\right)^{3/2}} - \dfrac{x\hat{k}}{\left(x^2 + y^2\right)^{3/2}}\right\}$

[*Ans.* (a) **0** (b) $-2\hat{k}$ (c) $\dfrac{a\hat{k}}{\left(x^2 + y^2\right)^{3/2}}$]

8. Calculate the divergence at a point (x, y, z) for the fields in problem **7**.

[*Ans.* (a) $6axyz + \dfrac{3}{4}a(yz)^{1/2}$ (b) zero (c) zero]

9. If $A = xz^3\hat{i} - 2x^2yz\hat{j} - 2yz^4\hat{k}$, show that curl A at the point $(1, -1, 1)$ is $\left(3\hat{i} + 4\hat{k}\right)$.

10. Evaluate $\iint_S F \cdot ds$, where $F = \hat{i} 4xz - \hat{j} y^2 + \hat{k} yz$ and S is the surface of the cube bounded by $x = 0$, $x = 1$, $y = 0$, $y = 1$, $z = 0$ and $z = 1$. [*Ans.* 3]

11. Evaluate $\iint_S r \cdot \hat{n} ds$, where S is a closed surface and V is the volume enclosed by S. [*Ans.* 3V]

12. Show that $\iint_S \hat{n} ds = 0$ for any closed surface S.

13. Show that $\iint_S \text{curl} F \cdot dS = 0$ for any closed surface S.

14. A fluid of density $\rho(x, y, z, t)$ moves with velocity $v(x, y, z, t)$. If there are no sources or sinks, prove that

$$\nabla \cdot J + \frac{\partial \rho}{\partial t} = 0 \text{, where } J = eV$$

15. Show that the necessary and sufficient condition for $\oint_C A \cdot dr = 0$ to be true for every curve C is that curl $A = 0$.

16. Prove Stoke's theorem for the vector $A(x + y, 2x - z, y + z)$ taken over the triangle cut from the plane $3x + 2y + z = 6$ by the coordinate planes.

SHORT ANSWER QUESTIONS

1. How many kinds of fields are there?
   [*Ans.* Two: (i) scalar (ii) vector]

2. Give two examples of vector field.
   [*Ans.* (i) Distribution of velocity in a fluid (ii) distribution of gravitational, magnetic or electric field in space]

3. If $\phi(x, y, z)$ is a scalar field, is $\hat{i} \frac{\partial \phi}{\partial x} + 2\hat{j} \frac{\partial \phi}{\partial y} + 3\hat{k} \frac{\partial \phi}{\partial z}$ a vector field?
   [*Ans.* We know that a scalar field $\phi(x, y, z)$ evaluated at a particular point is independent of the coordinates of that point. Obviously, scalar field, is invariant with respect to the transformation of coordinates. Its gradient, which is a vector field, must also be invariant with respect to the transformation. Now, on applying coordinate transformation, the vector

   $\hat{i} \frac{\partial \phi}{\partial x} + 2\hat{j} \frac{\partial \phi}{\partial y} + 3\hat{k} \frac{\partial \phi}{\partial z}$ does not retain its form on

applying a coordinate transformation. Thus, it cannot be a vector field corresponding to given scalar field $\phi(x, y, z)$]

4. What are lamellar fields?
   [*Ans.* Certain vector fields can be expressed as gradient of a scalar field ϕ and such fields are known as lamellar fields]

5. What is name of the theorem which provides a method of reducing volume integrals to surface integrals?
   [*Ans.* Gauss's theorem of divergence]

OBJECTIVE QUESTIONS

1. In two dimensions, the divergence transforms as a _____. (scalar)

2. The curl of a vector function A is _____. (vector)

3. The divergence of a curl is always _____. (zero)

4. The curl of a gradient is always _____. (zero)

5. $\nabla^2 = \frac{\partial^2}{\partial x^2} + \frac{\partial^2}{\partial y^2} + \frac{\partial^2}{\partial x^2}$ is known as _____ operator. (Laplacian)

MULTIPLE CHOICE QUESTIONS

1. Which one of the following is an example of scalar field:
   (a) gravitational field
   (b) electric field
   (c) electric potential
   (d) magnetic field

2. The maximum rate of change of the scalar function is known as the
   (a) gradient of a scalar field
   (b) divergence of a scalar field
   (c) curl of a vector field
   (d) none of the above

3. A vector, A which satisfies the condition, $\nabla \cdot A = 0$, is called
   (a) scalar   (b) solenoidal
   (c) tensor   (d) none of the above

Answers

   1. (c)   2. (a)   3. (b)

14

Thermoelectricity

14.1 SEEBECK EFFECT

Seebeck in 1821, discovered that if two dissimilar metals or alloys are joined together to form a closed circuit, and their two junctions are maintained at different temperatures, an emf is developed and an electric current flows in the circuit. A pair of junctions flows in the circuit. A pair of junctions of this kind is called a *thermocouple* (Fig. 14.1) and the effect is known as Seebeck effect.

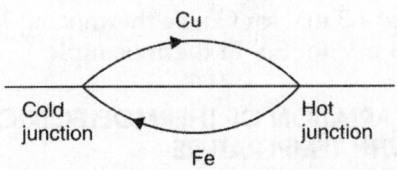

Fig. 14.1: Iron–copper thermocouple

The emf generated in a thermocouple is called *thermocouple* emf and the current which flows is termed *thermoelectric current*. The magnitude of thermo emf depends on the nature of the two metals and on the temperature difference of their junctions. The thermoelectric emf will exist and thermoelectric current will flow so long there is temperature difference between two junctions of a thermocouple.

The highest thermoelectric emfs are obtained with antimony and bismuth (about 1.2×10^{-3} volt per degree temperature difference) and the current flows from antimony to bismuth through the cold junctions.

Whenever the temperature of the two junctions of a thermocouple becomes equal, emf will be zero and current stops flowing and with the increase of temperature difference between the two junctions of a thermocouple will increase. Cold junction is maintained at $0°C$, then the emf developed would be a measure of the temperature of the hot junction of a thermocouple.

14.2 EXPLANATION OF SEEBECK EFFECT

One can explain the origin of thermoelectric effect with the help of free electron theory. A conductor or metal possesses a large number of free electrons and the concentration of free electrons per unit volume, and their average velocity varies from one metal to another, though all the metals are good conductors. This means that at junction of two dissimilar metals, there may be a tendency of electrons to migrate from one metal to the other metal across the junction of these metals, from higher concentration to lower concentration. However, such migration of electrons cannot continue for long as it quickly sets up an opposite electric field that prevents any further movement of electrons across the boundary. As a consequence, a fixed potential arises at the junction between two dissimilar metals, termed *contact potential*. It is found that the velocity of an electron in a metal depends to some extent on temperature, as a result the contact potential between any two given metals varies with temperature. When two junctions of dissimilar metals (say Fe and Cu) are kept at two different tem-peratures, the difference between the two contact potential V and V' will act to derive current round the closed circuit as shown in Fig. 14.2. The resultant emf (E) is given by

$$E = V - V' \qquad (14.1)$$

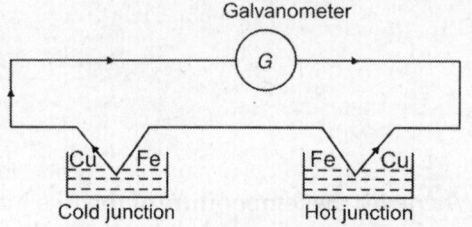

Galvanometer

Cu Fe | Fe Cu
Cold junction | Hot junction

Fig. 14.2: Iron–copper (Fe–Cu) thermocouple

When the temperature of two junctions of a thermocouple is same, the contact potential is also same, i.e. $V = V'$ across the two junctions. Obviously, as a result there will be no resultant emf between the two junctions of a thermocouple and hence there will be no flow of current in the circuit. This means, when one of junctions of thermocouple is heated up and the other junction is cooled, then the contact potential at the heated junction of the thermocouple becomes more than that of cold junction, and a *potential difference* is developed between the two junctions. Obviously, the contact potential difference is strongly affected by the temperature. This potential difference between the two junctions acts as the resultant emf and is responsible for the flow of current in the outer circuit (Fig. 14.2). Clearly, the thermo emf is the cause of thermoelectric current.

14.3 THERMOELECTRIC SERIES

Seebeck made a series of metals, known as *Seebeck series*, such that if a couple is constructed with any two metals in this series, the thermocurrent flows across the hot junction from the metal appearing earlier in the series to the one appearing later. Seebeck series is as follows:

Bi, Ni, Co, Pd, U, Cu, Mn, Tl, Hg, Pb, Sn, Cr, Mo, Rh, Ir, Au, Zn, W, Cd, Fe, As, Sb, Te.

Greater the separation of the two metals in the series, higher the thermo emf for a given temperature difference between the junctions. Obviously, the thermo emf for Ni–Fe couple is greater than for Cu–Fe couple. The direction of the current will be from a metal occuring earlier in this series to a metal occurring later in the series through the cold junction, for example in Cu–Fe thermocouple, the current flows from Cu to Fe through the cold junction and from Fe to Cu through the hot junction.

The metals to the left side of Pb are called thermoelectrically *positive* metals, while those to its right side are called thermoelectrically *negative* metals.

The magnitude of the thermo emf is of the order of a few micro volts per degree temperature difference between the two junctions. Magnitude of thermo emf depends upon how far the metals are separated in Seebeck series. As a general rule, more the separation, greater will be the magnitude of thermo emf. The magnitude of thermo emf for a difference of $100°$ C temperature is found to be about 1.3 mV for Cu–Fe thermocouple and about 8 mV for Sn–Bi thermocouple.

14.4 VARIATION OF THERMOELECTRIC EMF WITH TEMPERATURE

Experimentally, it has been observed that if the temperature of the cold junction of a thermocouple be kept at $0°C$ and the thermoelectric emf e be plotted against the temperature T of the hot junction, the graph is a parabolic curve as shown in Fig. 14.3. The thermo emf increases with the temperature of the hot junction and becomes maximum at a

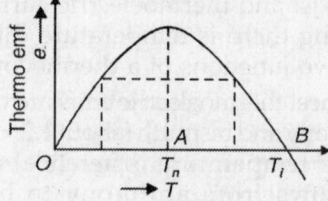

Fig. 14.3: Variation of thermo emf with temperature

particular temperature, T_n called the *neutral temperature* which is constant for the given pair of metals forming the thermocouple. If the temperature of hot junction be raised above the neutral temperature, the emf decreases and becomes zero at a particular temperature of the hot junction, called *temperature of inversion* (T_i). The temperature of inversion (T_i) is not fixed. It is much above the neutral temperature as the cold junction is below the neutral temperature. Beyond the temperature of inversion, the emf again increases but in the reverse direction.

The thermo emf varies with temperature according to following relation

$$e = at + \frac{1}{2}bt^2 \qquad (14.2)$$

where a and b are Seebeck constants for the thermocouple. Equation (14.2) is known as *Seebeck equation*.

Differentiation of Eq. (14.2) gives

$$\frac{de}{dT} = a + 2bT$$

At $T = T_n$, e is maximum, i.e. $\dfrac{de}{dT} = 0$

Thus $\qquad 0 = a + 2bT_n$

or $\qquad T_n = -\dfrac{a}{2b} \qquad (14.3)$

Further at $T = T_i$, $e = 0$. Thus, from Eq. (14.2), we have

$$0 = aT_i + bT_i^2$$

or $\qquad T_i = -\dfrac{a}{b} \qquad (14.4)$

From Eqs. (14.3) and (14.4), we have

$$T_i = 2T_n = -\frac{a}{b} \qquad (14.5)$$

Obviously, the inversion temperature T_i is as much above the neutral temperature as the temperature of the cold junction (0°C) is below it. Clearly, T_i is not constant for the given thermocouple but depends on the temperature of the cold junction. If the cold junction temperature changes, then inversion

temperature will also change. However, in that situation also T_n will be same for a particular thermocouple.

14.5 SEEBECK COEFFICIENTS

Generally the values of the Seebeck coefficients are expressed with respect to lead (Pb). The values can be positive or negative (Fig. 14.4)

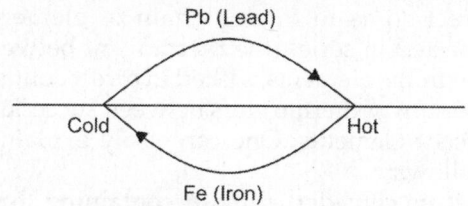

Fig. 14.4: Iron–lead thermocouple

One can express Seebeck coefficients for a thermocouple formed with metals X and Y as a_{x-y} and b_{x-y}. We can write

$$a_{x-y} = a_x - a_y \qquad (14.6)$$

$$a_{(x-Pb)-(y-Pb)} = a_{(x-Pb)} - a_{(y-Pb)} = (a_x - a_y) \ (14.7)$$

Similarly $b_{x-y} = b_x - b_y \qquad (14.8)$

We may note that Seebeck coefficient's value will be positive when current flows at hot junction from Pb to other metals used in thermocouple.

14.6 LAW OF SUCCESSIVE TEMPERATURES

For a given thermocouple, the thermo emf for any number of successive temperatures is the sum of thermo emfs for any number of thermo steps into which the given range of temperature may be divided. Let us suppose that $\theta_1, \theta_2, \theta_3, ..., \theta_n$ are successive temperatures between θ_1 and θ_n, then

$$E_{\theta_1}^{\theta_n} = E_{\theta_1}^{\theta_2} + E_{\theta_2}^{\theta_3} + E_{\theta_3}^{\theta_4} + ... + E_{\theta_{n-1}}^{\theta_n} \ (14.9)$$

There can be three possible fixed temperatures between 0°C and 100°C: (i) 0°C (ii) room temperature (say 30°C) (iii) 100°C, which will not change during experiment. It can be proved experimentally

$$E_0^{100} = E_0^{30} + E_{30}^{100} \qquad (14.10)$$

14.7. LAW OF INTERMEDIATE METALS OR SUCCESSIVE CONTACTS

According to this law, the insertion of any other metal into a thermocouple circuit does change the total emf provided that the added metal is entirely at the temperature of the part of the circuit where the metal is inserted. In general, one can say that if at a given temperature, a number of metals are in successive contact so as to form a chain of elements connected in series, the thermo emf between the extreme elements, placed in direct contact, is the sum of thermo emfs between successive adjacent elements. One can easily explain it as follows:

Let us consider a circuit containing three metals A, B and C as shown in Fig. 14.5. In case if all the junctions in the circuit are at the same temperature, then there will be no possible source of energy that could derive current in the circuit. This means the algebraic sum of the three contact potentials taken in the same direction must be zero, i.e.

$$V_1 + V_2 + V_3 = 0 \qquad (14.11)$$

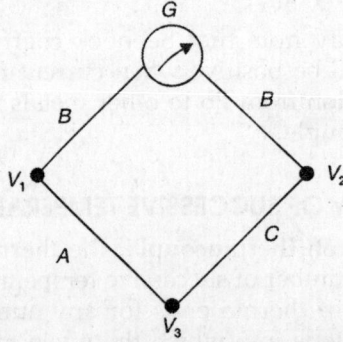

Fig. 14.5: Junction of three metals A, B and C illustrating law of intermediate metals

Now, if we change the temperature of the junction AB, the contact potential at that point also changes to new value V_1', but V_2 and V_3 remains unaltered. Now, the thermo emf E, acting in the circuit will become

$$E = V_1' + V_2 + V_3 \qquad (14.12)$$

Using Eq. (14.11), we have $V_2 + V_3 = -V_1$. Substituting this in Eq. (14.12), we obtain

$$E = V_1' - V_1 \qquad (14.13)$$

which is same as one will obtain, if metal C is eliminated from the circuit by bringing its junctions into contact with A and B. This is an experimental fact. Obviously, we can cite the result that breaking the result at some point or inserting a third metal, whose temperature is the same as that of the wire at the break does not alter the total emf in the given circuit.

14.8 THERMOELECTRIC POWER

When the cold junction of a thermocouple is maintained at a constant temperature, and the temperature of the hot junction is changed, the thermo emf (e) of the thermocouple also changes. The rate of change of thermo emf with temperature of the hot junction, de/dT, is called *thermoelectric power* (P) of the thermocouple at the particular temperature. Thermopower of a thermocouple is given by the slope of the tangent to the e–T curve at that temperature (Fig. 14.6).

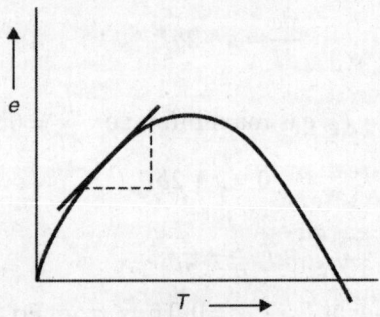

Fig. 14.6: e–T curve

The relation between the thermoelectric emf, e and the temperature of the hot junction, T when the cold junction is kept at 0°C, is given by

$$e = aT + bT^2 \qquad (14.14)$$

where a and b are constants for a given thermocouple. Differentiating Eq. (14.14), we have thermoelectric power (P) given by

$$P = \frac{de}{dT} = a + 2bT \qquad (14.15)$$

Obviously, the graph between the thermoelectric power (P) and the temperature is a straight line not passing through the origin. A graph plotted between the thermoelectric power $(P = \dfrac{de}{dT})$ and temperature T of the junction is usually a straight line, called the *thermoelectric power line* or *thermoelectric diagram*. Figure 14.7 shows thermoelectric diagrams for two different metals Cu and Fe. These diagrams clarify that the rate of change of thermo emf with temperature is positive for thermoelectrically positive metals like copper and it is negative for thermoelectrically negative metals like Fe.

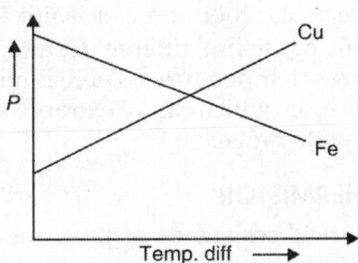

Fig. 14.7: Thermoelectric power lines or thermoelectric diagrams

14.9. PRACTICAL APPLICATIONS

There are many useful applications of thermoelectric effect, e.g. Boy's radio pyrometer, thermogalvanometer, thermopile, thermodynamic pyrometer. Some of the important applications are discussed below.

Thermoelectric Thermometer

This is the simplest arrangement to measure the temperature by using thermocouple wires. The different parts of a complete thermoelectric thermometer outfit are:

(a) A thermocouple consisting of two elements

(b) Protecting tubes and electrical insulation of wires

(c) Thermo emf measuring instrument, e.g. galvanometer or millivoltmeter and or potentiometer

(d) Suitable arrangement for controlling cold junction temperature.

Different Ranges of Temperatures

Thermoelectric thermometers can be used for various ranges of temperatures. However, their requirement of thermocouple, etc. are also different. For low temperature range up to 300°C, the thermocouple of base metals such as iron–constantan and copper–constantan are found satisfactory, as they develop a large emf per degree centigrade. However, for high temperature measurements, base metal thermocouples cannot be used, as these get oxidised and melt. Ni–Fe thermocouple may be used up to 600°C, but above 600°C platinum and alloys of platinum can be used. We may note that the thermoelectric thermometers are useful in the range –200–1600°C. This range may be extended upto 2100°C. Obviously, the choice of metals and alloys forming thermocouple depends upon the range of temperature to be measured. We may note that within the temperature range be large, and to avoid the reversal of emf, we should take care that the neutral temperature should be far remote from the temperature to be measured with the thermocouple.

Construction and Working of Thermoelectric Thermometer

Figure 14.8a shows the construction of thermoelectric thermometer. This thermometer consists of two elements in the form of a wire and is insulated with fire clay near the hot junction. To keep them in their positions, these wires are passed through mica discs. The entire arrangement is enclosed in a porcelain or quartz tube. The free ends of these wires are joined with two terminals C_1 and C_2. The terminals C_1 and C_2 are connected to extension leads, leading to cold junction as shown in Figure 14.8b. Figure 14.8c illustrates the circuit connection. The cold junction is placed at a convenient place and maintained at a constant low temperature (say 0°C), because one have to maintain cold temperature at points C_1 and C_2 (Fig. 14.8a), which just above the hot junction (say at temperature 2200°C) is not possible. A calibrated galvanometer (G) is connected in between these two extension wires, from which one can measure the temperature of the specimen using emf particular thermo (Fig. 14.8).

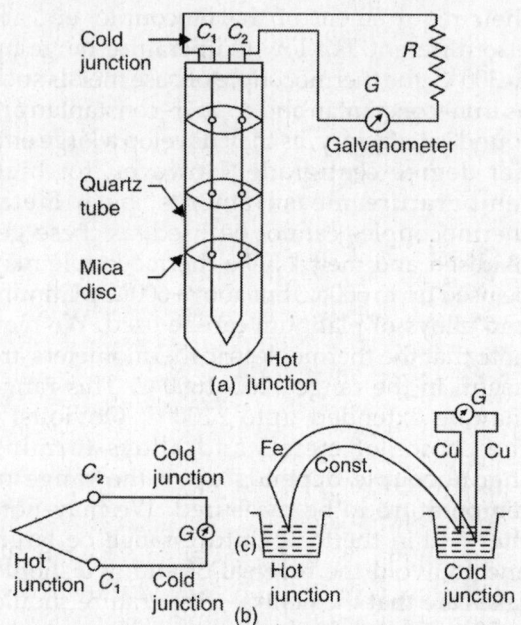

Fig. 14.8: Thermoelectric thermometer (a) Construction (b) Leads (c) Circuit connection

Working

The specimen whose temperature is to be measured is brought in contact with the welded hot junction of the thermoelectric thermometer. The deflection of the galvanometer G is noted when the junction attains the temperature of the specimen. Since the galvanometer G is calibrated for temperature and hence its reading directly gives the temperature. One can measure temperature between $-200°C$ and $1600°C$ by choosing suitable metallic wires.

The merits and demerits of a thermocouple thermometer are as follows:

Merits

i. The operating range is very wide, $-200 - 2100°C$.

ii. In a thermocouple thermometer, the surface area of the hot function is small and the thermal capacity of the hot junction is also small and hence it can be quickly heated up. Obviously, time lag is virtually absent.

iii. One can measure rapidly varying temperatures.

iv. This thermometer can measure temperature at a point and also cheap and easy to construct.

v. One can measure a very high temperature of the order of furnace with the help of this thermometer.

Demerits

i. Over the wide range of temperatures a particular thermometer is not accurate.

ii. One has to use different formula for different temperatures. There is no general relation, which can be extrapolated over a wide range.

14.10 THERMISTOR

The electrical resistance of most metals and alloys changes with temperature in accordance with the following relation:

$$R_t = R_0[1 + \alpha(T - T_0)] \qquad (14.16)$$

where R_t is the resistance at $T°C$, R_0 is the resistance at $0°C$ which is usually called as reference resistance at $0°C$ and α is the temperature coefficient of resistance per $0°C$.

For most of the metals, the temperature coefficient (α) is positive, and thus with the increase in temperature, the resistivity of most metals also increases. However, in the case of semiconductors, the resistivity generally decrease with increase of temperature. Due to this property, a semiconductor finds wider applications including thermometers, in the measurement of microwave frequency, power, as a thermal delay and in control devices functioned by changes in temperature.

In the case of semiconductors, e.g. Si, Ge resistance decreases 6 to 8% per degree C, whereas in the case of most metals the resistance increases about 0.4% per degree C. This reveals that metals have positive coefficient of resistance in smaller magnitude whereas semiconductor thermistor have negative coefficient of resistance of higher magnitude.

However, the temperature of the thermistor coefficient β can also be positive. Interestingly, the thermistor with +β are generally used in temperature measuring instruments. One finds that the resistance of thermistor is given by

$$R_t = R_0 \exp \beta \left(\frac{1}{T} - \frac{1}{T_0} \right) \qquad (14.17)$$

Depending upon the property of the material, temperature coefficient β can be positive or negative (Fig. 14.9). The active material of thermistor is usually a transition metal oxide or mixture of oxides, i.e. oxides of manganese, cobalt, nickel, copper, titanium, iron, etc. It is a type of semiconductor. Usually, thermistors are manufactured in various shapes and sizes, e.g. rods, washers, discs, beads and glass sealed probes. Miniature beads are well suited for applications where fast response and minimum thermal loading are required. The time for response in the case of glass encapsulation of the thermistor to sudden temperature variation increases significantly (from millisecond to one second).

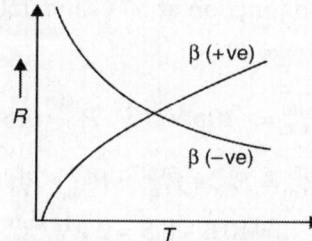

Fig. 14.9: Variation of resistance in a thermistor with temperature

Characteristic Curve

Figure 14.10 shows a typical characteristics curve, i.e. resistance versus temperature for a thermistor with –ve β (negative temperature coefficient). From Fig. 14.9, it is clear that resistance decreases nonlinearly with temperature coefficient β as

$$\beta = \frac{1}{x} \left(\frac{dx}{d\theta} \right) \qquad (14.18)$$

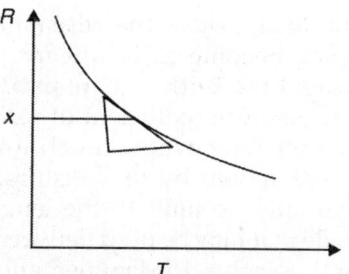

Fig. 14.10: A typical characteristic curve, i.e. resistance versus temperature curve for a thermistor

Linear approximation can be considered to be valid over a small range of temperatures. One can use the thermistor over the range of temperatures from 0 to 100°C. One can also measure higher temperature with the same success. When temperature changes, resistance of thermistor does not jump immediately to a new value and takes a small time to stabilize at the new resistance value. One can express this in terms of the time constant of the thermistor similar to the charging of a capacitor in an *R–C* electric circuit. Knowing the characteristic of a thermistor, one can easily relate the current measurement to the actual temperature and one can easily use it as temperature measuring device. We may note that a special type of thermistor has been developed for measuring very low temperatures in the vicinity of absolute zero.

Advantages of Thermistors

i. Using thermistor with high sensitivity is lower cost, precision temperature measurement and control.

ii. Nowadays, thermistors are available in small sizes and their response time is also faster.

iii. In case if thermistors are not exposed to high temperature, they have long-term stability. The stability of most thermistors is degraded above 300°C.

iv. Nowadays, there are wide range of high-resistance value thermistors available. This makes remote two-wire measurement possible without any need for temperature compensation of the leads, since the resistance of the leads become insignificant. We may note that both the resistivity and temperature coefficient of resistance of copper wire are so much lower than those shown by thermistors, there is virtually no limit to the length of the code that may be used between a sensor and associated instrumentation.

ILLUSTRATIVE EXAMPLES

Example 1

A thermocouple is made of iron and constantan. Determine the emf developed per 0°C difference of temperatures between the junctions, given that thermo emfs of Fe and constantan against platinum are +1600 and –3400 μV per 100°C difference of temperatures.

Solution

We have according to the law of intermediate metals,

$$e_{const}^{Fe} = e_{Pt}^{Fe} + e_{const}^{Pt}$$

$$= e_{Pt}^{Fe} - e_{Pt}^{const}$$

$$\therefore \quad e_{const}^{Fe} = 16 - (-34)$$

$$= 50 \ \mu V/°C$$

$$e_{Pt}^{Fe} = 1600 \ \mu V/100°C$$

$$= 16 \ \mu V/°C$$

$$e_{Pt}^{const} = -3400 \ \mu V/100°C$$

$$= -34 \ \mu V/0°C.$$

Example 2

The emf of a thermocouple, one junction of which is kept at 0°C is given by $e = at + bt^2$. Determine the neutral temperature.

Solution

$$e = at + bt^2$$

$$\therefore \quad e = a(T - 273) + b(T - 273)^2$$

[∵ $t°C = T - 273°C$, where T is in absolute degrees]

Differentiating,

$$\frac{de}{dT} = a + 2b(T - 273)$$

and

$$\frac{d^2e}{dT^2} = 2b$$

We have at the neutral temperature,

$$T = T_n \text{ and } \frac{de}{dT} = 0.$$

Thus

$$0 = a + 2b(T_n - 273)$$

or

$$T_n - 273 = -\frac{a}{2b}$$

or

$$T_n = \left(273 - \frac{a}{2b}\right)$$

Neutral temperature in 0°C

$$t_n = T_n - 273 = -(a/2b)°C.$$

Example 3

The thermoelectric powers of Fe and Cu with respect to lead are +10.5 and +8.5 μV respectively at 100°C. Find the emf of a copper–iron couple with junction at 50°C and 150°C.

Solution

Given $P_{Cu}^{Pb} = +10 \mu V/°C$, $P_{Fe}^{Pb} = +8.5 \mu V/°C$

Here $P_{Cu}^{Fe} = P_{Cu}^{Pb} + P_{Pb}^{Cu} = P_{Cu}^{Pb} - P_{Fe}^{Pb}$

$$= 10.5 - 8.5 = 2 \ \mu V/°C$$

Now $P = \dfrac{de}{d\theta}$

For $\theta = 150 - 50 = 100°C$

$$e = P_{Cu}^{Fe} \times \theta = 2 \times 100 = 200 \ \mu V$$

REVIEW QUESTIONS

1. What do you understand by Seebeck effect in thermoelectricity? Give some practical applications of this effect.

2. What are the laws of intermediate metals and intermediate temperatures in thermoelectricity?

3. What is meant by thermoelectric power? Describe an experiment you would perform to obtain thermoelectric power versus temperature graph.

4. Explain the terms: neutral temperature, inversion temperature, thermoelectric neutral temperature and temperature of inversion in the study of thermoelectric emf.

5. Explain thermoelectric power. Explain how it helps to determine the a and b coefficients for the thermocouple in the relation $e = at + bt^2$.

6. Write short notes on:

   i. Thermoelectric power

   ii. Thermoelectric thermometer

   iii. Thermistor

   iv. Seebeck effect.

PROBLEMS

1. It is observed that thermo emf for Fe–Cu thermocouple is zero when one of the junctions is kept at 20°C, and the other junction is kept at some higher temperature. If the neutral temperature is 285°C, show that the higher temperature and hence the temperature of inversion at cold junction temperature of – 20°C is 590°C.

   [**Hint:** T_n = 285°C, T_c = 20°C

   $\therefore$ $T_n = (T_n + T_c)/2$ or $T_i = 550°C$

   Now $T_c = -20°C$, $T_i = 2T_n - T_c = 570$

   $+ 20 = 590°C$]

2. The emf in microvolts (e) of a thermocouple, one junction of which is at 0°C is given by

   $$e = 1600\, t - 4t^2$$

   where t °C is the temperature of the hot junction. Find the neutral temperature. [*Ans.* 200°C]

3. The thermo emf generated in a thermocouple, whose cold junction is at 0°C and hot junction is at t°C is given by $e = at + bt^2$, where a and b are constants of the thermocouple. At two known temperatures 20°C and 240°C, the emfs generated by this thermocouple are respectively 791 µV and 9482 µV. Determine (i) constants a and b (ii) value of t at $e = 4160$ µV (iii) neutral temperature of the couple.

   [**Hint:** (i) $e = at + bt^2$, $e_1 = 791$ µV, $t_2 = 20°C$,

   $e_2 = 9482$ µV, $t_2 = 240°C$

$\therefore$ $791 = 20a + (20)^2 b$ (1)

and $9482 = 240\, a + (240)^2 b$ (2)

Solving Eqs. (1) and (2), one obtains $a = 42.8\,\mu V/°C$

and $b = -0.012\ \mu V/(°C)$

(ii) Using the values of a and b and $e = 4169$ µV in the equations $e = at + bt^2$, we obtain,

$$4160 = 42.8t - 0.012\, t^2$$

The positive root of this quadratic equation in t is obtained as $t = 100°C$

(iii) Neutral temperature t_n is the value of t when $de/dt = 0$, i.e. e attains a stationary value. Differentiating the equation $e = at + bt^2$ and setting $de/dt = 0$, one obtains $0 = a + 2bt$. When this happens $t = t_n$. Hence, $0 = a + 2bt_n$

$$t_n = -\frac{a}{2b} = -\frac{42.8}{2 \times -0.012} = 1783°\mathrm{C}$$

4. A cu-constantan thermocouple produces an emf of 40 µV per degree C in the linear range of temperatures. A galvanometer of resistance 10 Ω and capable of detecting currents of the order of 1 µV is used. Show that the smallest temperature that can be detected by this galvanometer is 0.25°C.

5. When the temperature difference between the hot and the cold junctions of a thermocouple is 100 K, an emf of 1 mV is generated. When the temperature of the cold junction is raised by 20 K, determine the percentage by which the emf is changed. Assume that the thermoelectric power remains constant over the whole temperature range.

   [**Hint:** $\dfrac{de}{dt}$ = constant and hence the variation of

e with the temperature difference t between the junctions is linear. Thus, $e = at$, $e = 1$ mV when $t = 100$ K = 100°C

$\therefore$ $a = (1/100)$ mV/°C

When the temperature of the cold junction increases by 20 K, t becomes 80°C. The corresponding e.m.f. is

$$e = \frac{1}{100} \times 80 = 0.8\,\mathrm{mV}$$

The percentage change in emf is

$$\left(\frac{1-e}{1}\right) \times 100 = 0.2 \times 100 = 20\%$$

SHORT ANSWER QUESTIONS

1. What are thermoelectrically positive metals?

Ans. The thermoelectric series for a selection of metals is Fe, Cd, Zn, Ag, Au, Sn, [Pb], Hg, Mn, Cu, Pt, Co, Ni, Bi.

The metals to the right of Pb are called thermoelectrically positive metals.

2. Mention the thermocouple in the thermoelectric series which produces the largest emf.

Ans. Sn-Bi thermocouple

3. How thermoelectric emf (E) varies with temperature?

Ans. $E = at + 1/2bt^2$

Obviously, when the cold junction is maintained at constant temperature then E will vary almost as a parabolic junction of the temperature difference between the two junctions as represented by the Seebeck equation.

4. What inference can you draw from *Thermoelectric diagrams*?

Ans. Thermoelectric diagrams for different metals clarify that the rate of change of thermo emf with temperature is positive for thermoelectrically positive metals like Cu and it is negative for thermoelectrically negative metals like Fe.

5. What will be the effect of inserting a third metal in between a thermocouple circuit?

Ans. Breaking the thermocouple circuit at some point or inserting a third metal, whose temperature is the same as that of the wire at the break, does not after the total emf in the circuit.

OBJECTIVE QUESTIONS

1. For Fe–Cu couple it is observed that thermo emf is zero when one of the junction is at 20°C, and the other is at some higher temperature. The neutral temperature is 285°C. The temperature of inversion at cold junction temperature of –20°C is _____.   [590°C]

2. The values of Seebeck coefficients a and b are generally given with respect to _____, and it can be positive or _____.   [lead, negative]

3. The relation between Seebeck coefficients a and b and neutral temperature T_n is _____.

$$[T_n = -a/b]$$

4. The relation between Seebeck coefficients a and b and temperature of inversion T_i is _____.

$$[T_i = -2a/b]$$

5. The relation between neutral temperature T_n, temperature of inversion (T_i) and cold junction temperature (T_c) is _____.

$$[T_n = (T_i + T_c)/2]$$

MULTIPLE CHOICE QUESTIONS

1. The emf generated in a thermocouple is called
   (a) ampere
   (b) thermocouple emf
   (c) internal resistance
   (d) temperature gradient

2. In the Seebeck series,
   Bi, Ni, Co, Pd, U, Cu, Mn, Tl, Mg, [Pb], Sn, Cr, Mo, Rh, Ir, Au, Zn, W, Cd, Fe, As, Sb, Te, the metals to the left side of Pb are called thermoelectrically
   (a) negative elements   (b) positive elements
   (c) neutral elements   (d) none of the above

3. Thermo emf varies with temperature as
   (a) $e = at$ (b) $e = at + \dfrac{1}{2} bt^2$
   (c) $e = at^2$ (d) $e = at^2 + bt$

4. Inversion temperature (T_i) of a given thermocouple depends on
   (a) temperature of hot junction
   (b) temperature of cold junction
   (c) neutral temperature
   (d) none of the above

5. Thermoelectric power is given by
   (a) $P = \left(\dfrac{de}{dt}\right) = a + 2bT$

   (b) $P = \left(\dfrac{de}{dt}\right) = bT^2$

   (c) $P = \left(\dfrac{de}{dt}\right) = (a + 2b)T$

   (d) $P = \left(\dfrac{de}{dt}\right) = aT/2$

Answers

1. (b)	**2.** (b)	**3.** (b)	**4.** (b)	**5.** (a)

15

Motion of Charged Particles in Electric and Magnetic Fields

15.1 INTRODUCTION

Electron, proton, etc. are charged particles and their motion is affected by electric, magnetic and gravitational fields. When a charged particle is in electric or magnetic field, it experiences a force, and hence the particle is accelerated. Provided that the forces acting on the charged particle are known, one can determine the trajectory precisely by Newton's laws of motion. This property of the charged particles has been used to make *particle accelerators* and *radiation detecting instruments*. Charge on an electron (e) has been determined by numerous experiments, e.g. Millikan's oil drop experiment, etc. However, the direct measurement of the mass m of an electron cannot be made, but one can measure e/m (electron charge/electron mass) ratio for an electron, from which one can estimate the mass of an electron. Recent experiments give a value of e as 1.602×10^{-19} C and mass of electron as 9.1×10^{-31} kg.

15.2 MOTION OF A CHARGED PARTICLE IN A PARALLEL ELECTRIC FIELD

Let us consider that an electron is situated between two parallel plates of large surface area contained in an evacuated envelope as shown in Fig. 15.1.

A potential difference V is applied between the two plates A and B and the direction of the field is as indicated in the Fig. 15.1. The distance between the plates A and B is d. Considering that d is small compared to the

dimensions of the plate and hence we can consider the field between the plates as uniform. Therefore, the field between the plates is

$$E = \frac{V}{d} \qquad (15.1)$$

Obviously, the lines of force are pointing along negative X-axis.

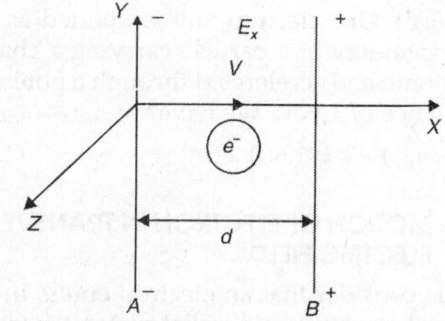

Fig. 15.1: Motion of charged particle (electron) in parallel electric field

We can see that there is no force along the Y and Z directions, and therefore electrons emitted from −ve plate A are accelerated along X direction towards the +ve plate B. If e is charge on the electron and m is mass of the electron, then the electron force acting on the electron is

$$eE = ma_x \quad \text{or} \quad a_x = \frac{eE}{m} \qquad (15.2)$$

or
$$a_x = \frac{V}{d}\frac{e}{m} \qquad (15.3)$$

where a_x is the electron's acceleration along X-axis and it is constant for a particular value of V and d.

Let the electron strike the positive plate B with a velocity v_x which is called as terminal velocity, then the kinetic energy T of the electron will be

$$T = \frac{1}{2}mv_x{}^2 \qquad (15.4)$$

which is equal to work done W.
We have
$$W = \text{force} \times \text{distance}$$
$$= eEd = (eV/d) \times d = eV$$

Thus, $\quad \dfrac{1}{2}mv_x^2 = eV \qquad (15.5)$

or the terminal velocity v_x

$$v_x = \sqrt{2eV/m} \qquad (15.6)$$

The unit of work or energy is called electron volt (eV). One electron volt is defined as the kinetic energy of a particle carrying a charge e coulomb and accelerated through a potential difference of 1 volt. We have

$$1 \text{ eV} = 1.602 \times 10^{-19} \text{ J} \qquad (15.7)$$

15.3 MOTION OF ELECTRON IN TRANSVERSE ELECTRIC FIELD

Let us consider that an electron enters in the region between two parallel plates separated by a distance d as shown in Fig. 15.2. Let the initial velocity of the electron be v_x in +ve X direction. A uniform electric field $\vec{E}_y$ is given

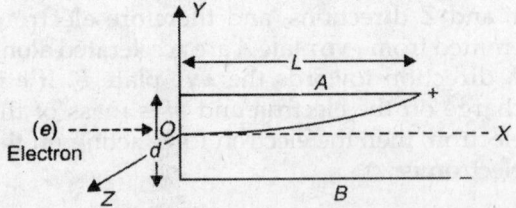

Fig. 15.2: Motion of a charged particle in a transverse electric field

between these two plates A and B. From Fig. 15.2, we can see that the length of the field (E_y) is L and the direction of the electric field is along –ve Y-axis. We may note that no other fields exist in this region.

Now, we want to study the motion of the charged particle under this transverse field E_y. We have the initial velocities of charged particle along X, Y and Z directions as

$$u_x = v_x$$
$$u_y = 0 \quad \text{when} \quad t = 0$$
and $\quad u_z = 0 \qquad (15.8)$

Since there is no force along X and Z directions, and hence velocity along Z direction remains zero and velocity along X direction v_x. The distance travelled along X direction at any time t is

$$x = v_x t, \quad \therefore \ t = x/v_x \qquad (15.9)$$

However, there is constant electric field along Y-axis, so there is constant acceleration along Y-axis. Thus, the velocity at any instant of time t is

$$v_y = u_y + a_y t$$
$$v_y = a_y t \qquad (\because \ u_y = 0)$$

or $\quad v_y = \dfrac{eE_y}{m}\dfrac{x}{v_x} = \dfrac{e}{m}\dfrac{V}{d}\dfrac{x}{v_x} \qquad (15.10)$

Now, the distance travelled along Y-axis by the time t within the applied electric field (E_y) is obtained as

$$y = u_y t + \frac{1}{2}a_y t^2$$

$$\therefore \quad y = \frac{1}{2}a_y t^2 \ (\because \ u_y = 0) \qquad (15.11)$$

The force acting on the electron of charge e along +Y-axis is given by

$$ma_y = eE_y$$

or $\quad a_y = eE_y/m = \dfrac{eV}{md} \qquad (15.12)$

From these equations we can see that in the region between the plates the electron is accelerated upwards with the velocity component v_y varying from point to point, whereas the velocity component v_x remains unchanged in the passage of the electron between the two plates A and B. One obtains the resultant velocity as

$$v = \sqrt{v_x^2 + v_y^2} \qquad (15.13)$$

One can determine the path of the particle in the electric field with respect to the point O by combining Eqs. (15.9) and (15.11). By eliminating t, this yields

$$y = \frac{1}{2} a_y \frac{x^2}{v_x^2} \qquad (15.14)$$

From Eq. (15.14), it is clear that particle moves in a parabolic path in the region between the charged plates A and B.

The angular deflection of the electron from original path produced by the transverse electric field at any instant of time t is obtained as

$$\tan\theta = \frac{v_y}{v_x} = \frac{eE_y}{m} \frac{x}{v_x^2} = \frac{eV}{dm} \frac{x}{v_x^2} \qquad (15.15)$$

Using Eqs. (15.12) and (15.14), one obtains the displacement of the electron due to transverse electric field, after travelling length $x = L$, i.e. at the end point of the applied electric field area as

$$y = \frac{1}{2} a_y t^2$$

or

$$y = \frac{1}{2}\left(\frac{eE_y}{m}\right)\left(\frac{L}{v_x}\right)^2 \qquad (15.16)$$

We may note that at the end of the applied electric field area, electron emerges out tangentially to the parabolic path at the point of emergence.

15.4 MOTION OF A CHARGED PARTICLE (ELECTRON) IN TRANSVERSE MAGNETIC FIELD

Let us study the motion of an electron with initial speed v entering a uniform transverse magnetic field acting downward along Z-axis, i.e. perpendicular to the plane of the paper as shown in Fig. 15.3. The magnetic field is perpendicular to initial velocity v, and thus to

electron's motion at every instant. This means, no work is done by the electron. This shows that kinetic energy of the electron will not increase and electron speed remains unchanged within the magnetic field. Now, the force acting on the electron due to transverse magnetic field is

$$F_m = Bev \qquad (15.17)$$

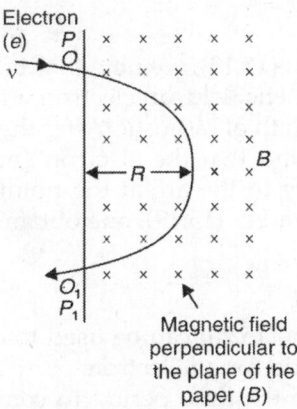

Fig. 15.3: Motion of a charged particle in a transverse magnetic field

Since B, e and v are constant in magnitude, therefore F_m is constant in magnitude and perpendicular to the direction of motion. Due to this force, electron will be continuously deflected in accordance with Fleming left hand rule and finally one finds the resultant motion in circular path within the transverse magnetic field and electron moves with constant speed within this. For the downward magnetic field, movement of the electron, i.e. charged particle will be clockwise, whereas it will be anticlockwise for upward magnetic field. We may compare it with a mass tied to a rope and twirled around with constant speed. We may note that this force which is actually the tension in the rope, remains constant in magnitude and is always directed towards the centre of the circle, and so is normal to the motion. We call this force as centripetal force.

Now, we compare the motion of the electron of mass m and charge e moving in a circular path of radius R with a constant speed v having an radial acceleration towards the centre in the tranverse magnetic field with the above cited example. We find that the magnetic force acting on the charged particle is

$$F_m = \frac{mv^2}{R} \qquad (15.18)$$

From Eqs. (15.17) and (15.18), one obtains

$$F_m = Bev = \frac{mv^2}{R} \quad \text{or} \quad R = \frac{mv}{eB} \qquad (15.19)$$

Equation (15.19), we note that in the transverse magnetic field, an electron will move in a circular path of radius R. After the magnetic field, we find that the electron emerges out tangentially to the arc at the point of emergence. From Eq. (15.19), one obtains

$$\frac{e}{m} = \frac{v}{RB} \qquad (15.20)$$

Equation (15.20) can be used to determine the e/m ratio for an electron.

Let T be the time period to complete one revolution, then

$$T = 2\pi R/v = \frac{2\pi R m}{BeR} = (2\pi/B)(m/e) \qquad (15.21)$$

Obviously, T is independent of v and R.
Now, we obtain the resonance frequency,

$$n = \frac{1}{T} = \frac{Be}{2\pi m} \qquad (15.22)$$

15.5 MOTION OF A CHARGED PARTICLE IN CROSSED ELECTRIC AND MAGNETIC FIELDS

Let us consider that an electron is moving with an initial speed v entering in a crossed electric and magnetic field. The initial direction of the speed of the electron is along X-axis. Electric field E is applied along Y-direction. Obviously, the direction of electric lines of force is along the Y-axis. As is clear from Fig. 15.4 that the magnetic field of flux density B is applied along Z direction, the direction of magnetic lines of force is along $+Z$-axis. In Fig. 15.4, the circle represents the area of uniform magnetic field. A screen is placed at a distance D (= OS) from the centre O of electric and magnetic fields.

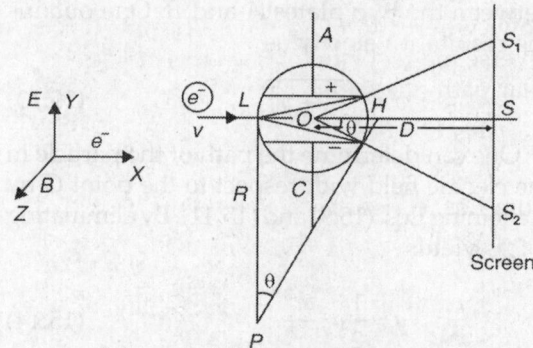

Fig. 15.4: Motion of a charged particle in crossed electric and magnetic fields

If the electron is made to travel within these crossed electric and magnetic fields along X-direction, then due to electric field alone it will deflect upwards along Y-axis as shown in Fig. 15.4. On the other hand, due to magnetic field alone it will deflect downward along Y-axis (Fig. 15.4). After crossing the magnetic or electric field, electron will move tangentially to the arc at point of emergence of field. Obviously, if S is striking position of electron on the screen when there is no field, then due to electric field alone, electron will deflect by an amount SS_1 upwards on the screen as shown in Fig. 15.4. Now, due to magnetic field alone, electron will deflect by an amount SS_2 downward just in the opposite direction on the screen (Fig. 15.4). Now, if force acting on the electron by electric field and magnetic field simultaneously are same, then clearly there will not be any deflection by the electron or electron beam. However, the force acting on the electron due to electric field is Ee, whereas the force acting due to magnetic field is Bev. Now, if these two forces are equal, then $SS_1 = SS_2$. In this situation, there will not be any deviation of the electron from its original path. In that condition, we have

$$Bev = Ee \qquad (15.23)$$
or $$v = E/B \qquad (15.24)$$
But $$E = V/d$$
where V is the voltage applied between two plates A and C separated by a distance d. Now, we have

$$v = V/dB \qquad (15.25)$$

Now, when the electron is moving in a transverse magnetic field, then it will travel a circular path of radius R. Thus, we have

$$mv^2/R = Bev$$

or $$v = BRe/m \qquad (15.26)$$

From Eqs. (15.23) and (15.26), one obtains

$$v = V/dB = BRe/m$$

or $$\frac{e}{m} = \frac{V}{dB^2R} \qquad (15.27)$$

where V and d are known. Knowing other parameters from the experiment, one can calculate e/m of an electron.

Cathode ray tube (CRO) is the best example where we use motion of charged particle in crossed electric and magnetic fields. From Fig. 15.4, we have

$$R = \frac{LH \times OS}{SS_2} \qquad (15.28)$$

15.6 THOMSON'S METHOD FOR MEASURING e/m RATIO FOR ELECTRON

A discharge is struck in a rarefied gas in a glass cathode ray tube (CRT) G between the cathode C and the anode A by means of source of a high voltage as shown in Fig. 15.5. C may be a small aluminium disc, while A is a brass cylinder pierced with a fine bore along its axis. The cathode rays strike the anode A and some of them emerge through the long, narrow hole in A as a fine, well collimated beam. The cathode ray beam then continues its forward journey in the field free space with a uniform velocity v determined by the potential difference V between cathode C and anode A, till it strikes normally the fluorescent screen S of zinc sulphide or barium platinum cyanide kept at the end of the tube.

P and Q are two parallel plates of length L, between which electric field is applied. These plates are separated by a distance d. On applying electric field along Y-axis perpendicular to the motion of the electron (along X-axis), the deflection of the spot occurs, which is marked by S_1. Now, by placing the CRT between two pole pieces of a magnet, a downward magnetic field can be applied simultaneously along Z-axis, perpendicular to the motion of the electron (along X-axis), as well as $\perp^r$ to the direction of electric field (Y-axis). The dotted circular area shown in Fig. 15.5 represents the region of uniform magnetic field. As stated earlier the direction of deflection will be downward. The crossed electric and magnetic fields are acting on the electron beam simultaneously. This will cancel each other's effect and the spot will come back to its original position S. Using results of previous section, one can calculate e/m ratio of electron.

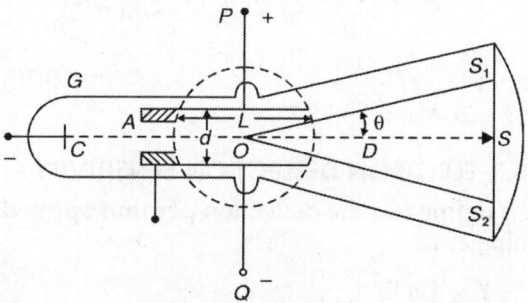

Fig. 15.5: Thomson's method for measuring e/m for electron

15.7 VERTICAL DEFLECTION DUE TO ELECTRIC AND MAGNETIC FIELDS

Consider that an electron is moving with initial velocity $v_x \perp^r$ to the electric field ($E_y = V/d$), then the vertical deflection on the screen, which is at a distance D from the centre O of the applied electric field, can be calculated as mentioned below.

Let accelerating anode voltage is V_A. The energy of the electron moving with initial velocity v_x from the accelerating anode is

$$\frac{1}{2}mv_x^2 = eV_A$$

$$v_x = \sqrt{\frac{2eV_A}{m}}$$

One obtains, the vertical deflection (Fig. 15.5)

$$y = SS_1 = OS\tan\theta = D\tan\theta$$

Now $\tan \theta = \dfrac{v_y}{v_x} = \dfrac{\dfrac{e}{m}\dfrac{V}{d}\dfrac{L}{v_x}}{v_x} = \dfrac{e}{m}\dfrac{V}{d}\dfrac{L}{v_x^2}$

Because $vy = \dfrac{e}{m}\dfrac{V}{d}\dfrac{L}{v_x}$ \hfill (15.29)

where L = length of the plates P, Q in CRT where we have applied voltage V.

d = distance between two plates in CRT where vertical voltage V is applied.

$D = OS$ = distance from O to the screen.

Vertical deflection,

$$Y = D \tan\theta = D\frac{e}{m}\frac{V}{d}\frac{L}{v_x^2} = D\frac{e}{m}\frac{V}{d}\frac{Lm}{2eV_A}$$

or $\quad Y = \dfrac{1}{2}D\dfrac{L}{d}\dfrac{V}{V_A}$ \hfill (15.30)

15.8 ELECTRON DEFLECTION SENSITIVITY

It is defined as the deflection per unit applied voltage, i.e.

$$\frac{Y}{V} = \frac{1}{2}\frac{DL}{dV_A}$$ \hfill (15.31)

15.9 VERTICAL DEFLECTION DUE TO PERPENDICULAR MAGNETIC FIELD OF MAGNETIC INDUCTION *B*

We have (from Fig. 15.5)

$$Y = SS_2 = \frac{OS \cdot LH}{R} = \frac{(OS \cdot LH \cdot B \cdot e)}{mv}$$

or $\quad y = D\dfrac{BeL}{mv}$

Now, $v = v_x = \sqrt{\dfrac{2eV_A}{m}}$

so $\quad Y = DBL\sqrt{\dfrac{e}{2mV_A}}$ \hfill (15.32)

where $D = OS$ is the distance from centre O of the magnetic field to the screen.

$L = LH$ = length of the applied magnetic field.

Magnetostatic Deflection Sensitivity

It is defined as the deflection per unit magnetic induction

$$\frac{Y}{B} = DL\sqrt{\frac{e}{2mV_A}} = \frac{Dl}{v}\left(\frac{e}{m}\right)$$ \hfill (15.33)

15.10 ASTON'S MASS SPECTROGRAPH

This is an apparatus of high accuracy designed by Aston, which enables the measurement of the mass of single atomic ion and is useful for investigation of isotopes. This method is an improvement over Thomson's method of mass spectrograph.

Principle

The positive rays emerging from perforated cathode are made into a fine pencil by using slits. They are then subjected to an electrostatic field in a direction perpendicular to the direction of rays with the help of electrically charged plates P_1 and P_2. The beam is not only deflected but also dispersed because the particles are having different velocities. The dispersed beam is then subjected to a magnetic field whose direction is perpendicular to the direction of electrostatic field. Thus, the magnetic field produces dispersion and deviation in an opposite direction but in the same plane. If a photographic plate is held in the direction of deflected beam, line images are obtained. Each line corresponds to a particular value of q/m. The number of lines correspond to the number of isotopes present in the element.

Theory

The different parts of the Aston's mass spectrograph are shown in Fig. 15.6. AO is the direction of positive rays entering the electrostatic field, S_1 and S_2 are slits which provides a fine pencil of positive rays. The electrostatic field is maintained by plates P_1 and P_2 and the direction of the field being from P_1 to P_2. The beam is deflected and dispersed downwards. Let θ and $d\theta$ be the angles of deviation and dispersions produced by electrostatic field. Using a diaphragm, D some of the rays are selected and are allowed to pass between the poles of an electromagnet, the magnetic field being perpendicular to the plane of the paper and inward. According to the Fleming left hand rule, the beam will be deflected upwards. This magnetic field annuals the dispersion produced by electric field and recombines the particle which are brought to focus in the form of sharp lines on a photographic plate *CD*. The lines are similar to those of spectral lines.

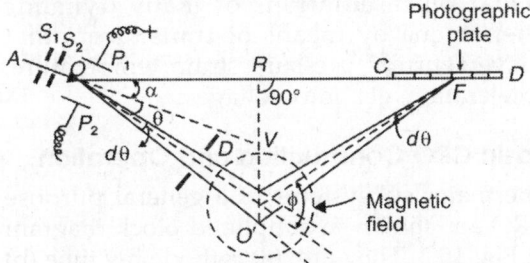

Fig. 15.6: Aston's mass spectrograph

Let q' = Charge on positive ray particle

m' = Mass of each particle

E = Electrostatic field

B = Magnetic field strength

v = Velocity of each particle

ϕ = Angle of deviation produced by magnetic field

$d\phi$ = Angle of dispersion produced by magnetic field.

Considering that the deflection in electrostatic field is small, the curve near the vertex may be considered as circular disc of radius r, we have

$$Eq' = \frac{m'v^2}{r} \quad \text{or} \quad \frac{1}{r} = \frac{Eq'}{m'v^2}$$

Hence, the deflection θ, which is proportional to $1/r$ is given by

$$\theta = C\frac{Eq'}{m'v^2} = C_1 \frac{q'}{m'v^2}$$

(where $C_1 = CE$ because E = constant)

$\therefore$ Dispersion $\dfrac{d\theta}{dv} = -2C_1 \dfrac{q'}{m'v^3} = -2\dfrac{\theta}{v}$ (15.34)

If r' is the radius of curvature in magnetic field, then

$$Bq'v = \frac{m'v^2}{r'} \quad \text{or} \quad \frac{1}{r'} = \frac{Bq'}{m'v}$$

$$\therefore \quad \phi = C'\frac{B'}{m'v} = C_2 \frac{q'}{m'v} \quad (\because B \text{ is constant})$$

Again dispersion $\dfrac{d\phi}{dv} = -C_2 \dfrac{q'}{m'v^2} = -\dfrac{\phi}{v}$ (15.35)

From Eqs. (15.34) and (15.35), one obtains

$$\frac{d\theta}{\theta} = 2\frac{d\phi}{\phi} \quad (15.36)$$

Obviously, for a given deflection, the dispersion due to the electric field is twice that due to magnetic field. The small changes $d\theta$ and $d\phi$ refer to the particles with identical mass and charge but possessing velocities differing by dv.

In the absence of magnetic field, dispersion produced in the beam for a distance $(a + b)$ is given by

$$= (a - b)d\theta \quad (15.37)$$

where a = distance OO' and b = distance $O'F$.

The magnetic field acts in a direction perpendicular to the electric field and produces the same dispersion in a distance b but in the opposite direction.

Dispersion produced by the magnetic field is $b\,d\phi$

As all the ions are focussed to the same position

$$(a + b)d\theta = b\,d\theta$$

and $\dfrac{d\theta}{d\phi} = \dfrac{b}{(a+b)}$ (15.38)

From Eq. (15.36), $\dfrac{d\theta}{d\phi} = \dfrac{2\theta}{\phi}$

$$\therefore \quad \frac{b}{(a+b)} = \frac{2\theta}{\phi}$$

or $\qquad b\phi = (a+b)2\theta$

$$b(\phi - 2\theta) = 2a\theta \quad (15.39)$$

This is the condition of focussing.

Let $O'R$ be the perpendicular to the line CD produced and $\angle ROV = \alpha$. Then from $\Delta ROO'$, we have

$$RO' = OO'\sin(\alpha + \theta) = a\sin(\alpha + \theta)$$

In $\Delta RO'F$, $RO' = O'F \sin RFO'$

$$= b\sin[180° - (\phi - \alpha - \theta)] = b\sin(\phi - \alpha - \theta)$$

$$\therefore \qquad a\sin(\alpha + \theta) = b\sin(\phi - \alpha - \theta)$$

For small angles, $a(\alpha + \theta) = b(\phi - \alpha - \theta)$ (15.40)

Comparing Eqs. (15.39) and (15.40), it is observed that two equations are same when $\alpha = \theta$. Thus the focussing condition is that the photographic plate must be placed at an angle θ with the direction of the incident positive ray beam.

Thus, we find that in Aston's apparatus:

i. All particles of the same q'/m' are brought to the same focus irrespective of the velocities

ii. Particles of different masses are brought from different foci.

15.11 CATHODE RAY OSCILLOSCOPE (CRO)

CRO is an extremely useful and versatile instrument used for measurement and analysis of electrical signal waveforms and other phenomena related to electronics. It is

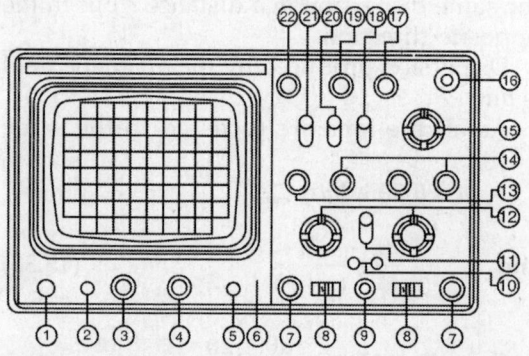

1. Power, 2. Power lamp, 3. Intensity, 4. Focus, 5. Trace rotation, 6. Calibrator, 7. Input (X), 7. Input (Y), 8. AG-GND-DC, 9. Ground terminal, 10. Step Att Bal, 11. Mode, 12. Volts/Div, 13. Variable, 14. V. Position, 15. Time/Div, 16. Ext. Trig, 15. Variable Pull × 5 Mag, 18. Mode Selector, 19. H. Position, 20. Source Selector, 21. Slope, 22. Level.

Fig. 15.7: Double beam CRO

an important instrument available in every physics laboratory, and electrical/electronics industry.

CRO is basically very fast X-Y plotter that displays an input electrical signals versus time or another signal. The styles of this plotter is a luminous spot generated due to electron beam on a fluorescent screen (Fig. 15.7), that moves over the display area in response to input voltages.

In addition to voltages, the CRO can present visual representations of many dynamic phenomena by means of transducers that convert current, pressure, strain, temperature, acceleration, etc. into voltages.

Basic CRO Construction and Operation

The major subsystems of a general purpose CRO are shown in simplified block diagram in Fig. 15.8. They are: (a) cathode ray tube (b) vertical amplifier (c) delay line (d) time base generator (e) horizontal amplifier (f) trigger circuit (g) power supply— high tension (H.T.) low tension (L.T.).

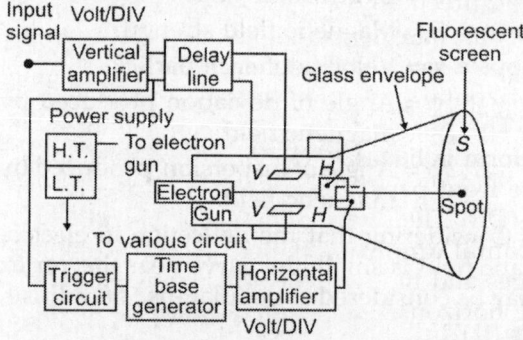

Fig. 15.8: Simplified block diagram of a CRO

Cathode Ray Tube

It is the main part of a CRO with the rest of the CRO consisting of circuitry to operate the tube.

It comprises of an electron gun (Fig. 15.9) which produces a sharply focussed beam of electrons, accelerated to a very high velocity. This highly focussed and accelerated beam of electrons travels through two pairs of deflecting plates, i.e. vertical deflecting plates

VV and horizontal deflecting plates *HH*, and strikes the fluorescent screen *S*, with sufficient energy to cause the screen to light up in the form of a small spot.

When voltage is applied across the horizontal deflection plates *HH*, the spot moves in a horizontal direction (*XX*) since the electron beam while passing through plates gets deflected. Similarly, when voltage is applied across the vertical deflecting plates *VV*, the spot moves in a vertical direction *YY*. These two movements are independent of each other, so that the spot on the screen *S* can be positioned anywhere on the screen by simultaneous application of appropriate vertical and horizontal voltage inputs.

The signal waveform to be viewed on VR tube screen is applied to the vertical amplifier input. The gain of this amplifier is set by a calibrated input attenuator usually marked Volts/Div (Nob 12_1, 12_2) on the front panel (Fig. 15.7). The output of the amplifiers is applied across the vertical deflection plates *VV* (Fig. 15.8) via so called delay line with sufficient power to drive the CR tube spot in vertical direction.

The time base generator or sweep generator develops a saw tooth waveform that is used as horizontal deflection voltages of the CR tube. The positive going part of saw tooth waveform is linear, and its rate of rise is set by the front panel control (Nob 15) marked Time/Div. The saw tooth wave is fed to the horizontal amplifier which produces two voltages and the two, in turn, are connected to the horizontal deflecting plates *H* and *H*. These voltages cause electron beam to be swept horizontally across the screen from left to right and right to left in time units and spot sweeping the screen appears as a horizontal line on it due to persistence of vision.

If the unknown signal connected to vertical input is of recurrent nature, a stable (CR tube) display on screen can be maintained by starting each horizontal sweep at the same point on the signal waveform. To achieve this, a sample of the input waveform is fed to the trigger circuit which produces a trigger pulse at same selected point on the input waveform.

This trigger pulse is used to start the time base generator which in turn starts horizontal sweep as explained above.

The purpose of delay line is to retard the arrival of input waveform at the vertical deflection plates until the trigger and time base circuits have had a chance to start the sweep of the screen. The delay line introduces a total delay of 0.25 ms approximately in the vertical deflection channel.

The power supply consists of two types: (1) high tension (voltage) of the order of 1 kV or more to operate the CR tube specially the electron gun (2) low tension (voltage) of the order of 230 V to supply the electronic circuitry like the amplifiers saw tooth generator, etc. of the oscilloscope.

Cathode Ray Tube Structure and Operation
The internal structure of a cathode ray tube is as shown in Fig. 15.9. The main components of the general purpose CR tube are:

(a) Electron gun assembly
(b) Deflection plate assembly
(c) Fluorescent screen
(d) Glass envelope and base.

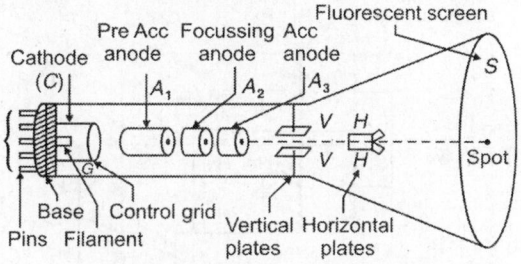

Fig. 15.9: Cathode ray tube

(a) *Electron Gun:* The name electron gun is derived from the analogy between the motion of electrons emitted from CRT gun structure and the travel path of a bullet fired from a conventional gun.

Electrons are emitted from an indirectly heated cathode *C*. The cathode is completely surrounded by a control grid *G* which consists of a nickel cylinder(Fig. 15.9) with a small centrally located hole, coaxially located with the axis of the tube axis. The electrons which manage to pass through the small hole of grid together make up the so called beam current. The magnitude of the beam current can be varied by a front panel control (Fig. 15.7) marked intensity which varied the negative voltage of the control grid (Fig. 15.10) with respect to cathode. An increase in control grid bias reduces the beam current and hence intensity of the spot on screen. The electrons coming out of the grid hole are accelerately high positive potential of the order of 1 kV applied to the two accelerating anodes. These two anodes are separated by a focussing anode (Fig. 15.10), which provides a method for focussing the electrons into a narrow and sharply defined beams since two accelerating anodes (A_1 and A_3) and focussing anode (A_2) are also in cylindrical form with small openings in the centre of the circular face of each cylinder. The various holes are coaxial with the axis of the tube. The holes in these cylinders (electrodes) allow the electron beam to travel past the vertical and horizontal deflection plates to the screen.

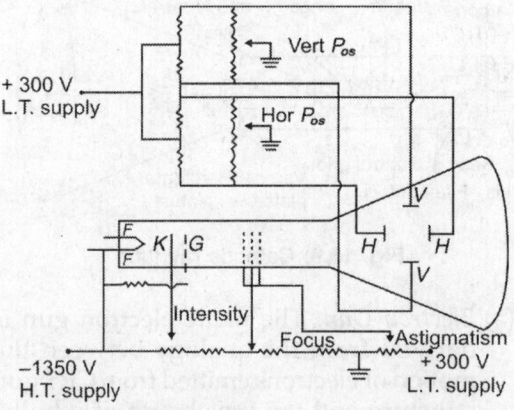

Fig. 15.10: Cathode ray tube connections

(b) *Screen:* The screen S of the glass tube is nearly circular. A specific material is pasted on the inner surface that produces the bright spot when electrons strike at it. The bright spot is due to phosphorescence produced in the material pasted. The phosphor (material) absorbs kinetic energy of striking electrons and reemits the energy at a lower frequency in visible range of spectrum due to atomic transition in the material pasted on it.

The length of time during which phosphorescence or after glow occurs is called *persistence of the phosphor* (material). The intensity of light emitted from cathode ray tube screen is called luminance which depends upon several factors.

The bombarding electrons striking the phosphor (material) in the inner surface of screen, release secondary electrons, thus keeping the screen in a state of equilibrium. The secondary emission (low velocity electrons) is collected by a conducting coating known as a quadag (Fig. 15.9) on the inside conical surface of tube which is electrically connected to second accelerating anode (A_3). In some tubes, this conducting point is used as anode finally.

(c) *Graticule:* The waveform display on the face of CR tube can be visually measured against a set of horizontal and vertical scale marks on a plastic plate called graticule (Fig. 15.7). These scale marks can be placed external so the face of CR tube in which case we speak of an external graticule. Generally external graticule is used. It consists of a clear and transparent plastic plate with inscribed divisions (mm or cm) on it. It is mounted on the outside of the face of tube and has the advantage of being replaced when it gets fade.

(d) *Deflection sensitivity:* The deflection sensitivity of a CR tube is defined as the reflection on the screen (in metres) per volt of deflection voltage.

$$S = \frac{D}{V} = \frac{L l_d}{2 d E_a} \text{ (metre/volt)}$$

where D = Deflection screen
V = Voltage applied across the vertical plates

L = Distance of screen from mid-point of vertical deflection plate

l_d = Length of the deflecting plates

d = Distance between the plates

E_a = Accelerating voltage.

15.12 SWEEP GENERATOR

Oscilloscopes are generally used to display a waveform that varies as a function of time. If the waveform is to be accurately reproduced, the electron beam must have a constant horizontal velocity. Since the beam velocity is a function of the deflecting voltage, the voltage must increase linearly with time. Such a voltage is called a *ramp voltage*. If the voltage decreases suddenly (quickly) to zero with waveform repeatedly reproduced as shown in Fig. 15.11, the pattern is called a *saw tooth wave*.

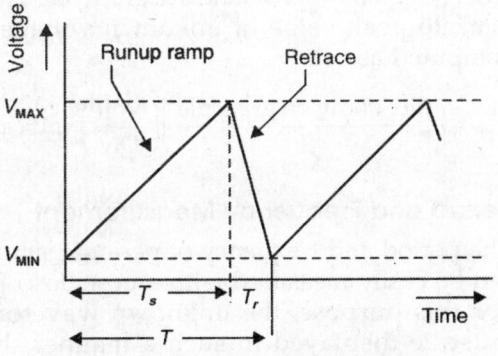

Fig. 15.11: Linear saw tooth wave

During sweep time (travel time) T_S, the beam moves from left to right across the CR tube screen and during retrace time T_r, the beam returns quickly to the left side of the screen. To prevent an undesirable retrace pattern from appearing on the screen during retrace, the control grid of tube is generally gated off which blanks out the beam during retrace time.

Since signal (wave shapes) of varying frequencies will be observed with the oscilloscope, the sweep rate must be adjustable, this

can be changed in steps by switching different capacitors in the circuit. For this the front panel control adjustment marked Time/Div (S.No 15 Fig. 15.7) is used.

The simultaneous application of deflection voltages to both sets of deflection plates, i.e. *VV* and *HH* causes the CR tube spot to trace an image on the screen. Production of the image is shown in Fig. 15.12, where a saw tooth or sweep voltage is applied to the horizontal plates (*HH*) and the signal which is a sine wave applied to the vertical plates (*VV*).

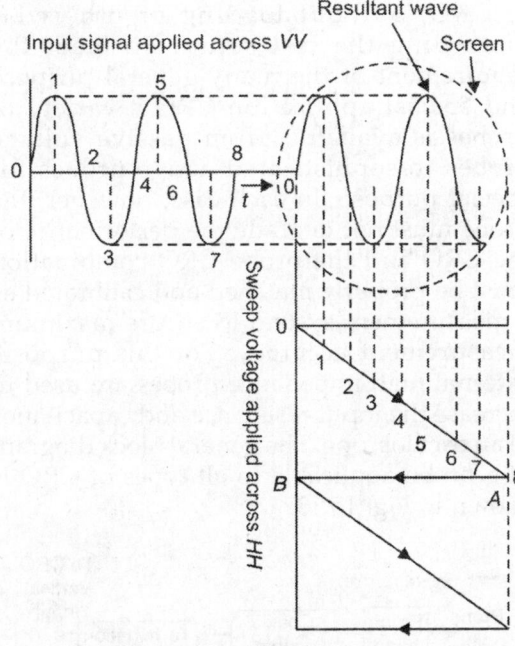

Fig. 15.12: Image production by CRO

Since the horizontal sweep voltage increases linearly with time, the CR tube spot moves across the screen with a constant velocity from left to right. At the end of the sweep, when the saw tooth voltage drops suddenly to minimum (zero), the CRT spot also returns quickly to its starting point on the LHS of screen and remains there until a new sweep is initiated. When simultaneously with horizontal sweep voltage, an input signal (sine wave in the case) applied to vertical deflection plates the electron beam will be under the influence of two fields, i.e. one along horizontal (XX) and another along vertical at (YY), thereby the spot moves up and down as per polarity also moving forward horizontally. The resultant display is with time. If the input signal is of recurrent nature, a stable display can be maintained by starting each sweep at the same point as shown in Fig. 15.12.

15.13 CRO PROBES

The CRO probe performs the important function of connecting the circuit under investigation to the vertical input terminal of CRO, without loading or otherwise disturbing the test setup. To meet the requirement of the many general purpose and special application CRO a variety of probes is available, from passive voltage probes to sophisticated active probes for special purpose. In each case, however, the probe must not degrade the performance of the CRO and the probe CRO combination must be properly matched and calibrated as a measurement system to ensure maximum measurement accuracy. For this purpose, external high impedance probes are used to increase the input resistance and capacitance of an oscilloscope. The general block diagram of a probe applicable to all types of CRO is shown in Fig. 15.13.

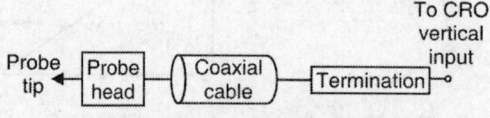

Fig. 15.13: CRO probe

The probe head contains the signal sensing circuiting. This circuitry may be passive, such as 10 mΩ resistor shunted by 7 pF capacitor or it may be active such as FET source follower and associated elements. A coaxial cable is used to complete probe head to termination circuitry, which also can be active or passive. The termination circuit provides the CRO with source impedance it requires and terminates the coax cable in its characteristic impedance.

For measuring high voltage (kV or more) we have special types of probes. Here the probe head is of high impact strength, thermoplastic material and of special design to protect from shock.

15.14 APPLICATIONS OF OSCILLOSCOPES

The range of applications of oscilloscopes varies from basic voltage measurements and waveform observations to highly specialized applications in all areas of science, engineering and technology. A few applications are mentioned below.

Voltage Measurement

The direct voltage measurement made with an oscilloscope is peak to peak value. The rms value of the voltage can easily be calculated from the peak to peak measurement as desired. To arrive at value from CR tube display, we set the vertical attenuator control expressed in Volt/Div till the display, if (wave) signal is persistent, are graticule. Then peak to peak value of unknown voltage is computed as below

$$V_{pp} = \left(\frac{\text{No. of div on graticule}}{1} \right)\left(\frac{\text{Volt}}{\text{Div}} \right) \text{ indicator}$$

Period and Frequency Measurement

The period and frequency of periodic signals can be easily measured with an oscilloscope. For this purpose, the unknown waveform must be displayed in such a manner that complete cycle is displayed by adjusting the attenuator control expressed in Time/Div on the screen. The time period is computed as

$$T = \left(\frac{\text{Time}}{\text{Div}} \right)\left(\frac{\text{No. of div}}{\text{Cycle}} \right) \text{ sec}$$

Accuracy is generally improved if single cycle displayed fills as much of horizontal distance across the screen as possible.

Knowing the time period, the frequency of the signal can be easily calculated, since

$$f = \frac{1}{T} \text{ Hz}$$

Determination of Frequency with Lissajous Figure

The oscilloscope can be used in the XY mode to determine the frequency of a signal. It is determined by applying the signal of unknown frequency to either X or Y input terminal and a signal of known frequency to other input terminal. In this way two periodic (e.g. sine) signals are superimposed perpendicularly to each other thereby to produce a resultant (periodic function) pattern called Lissajous pattern/figure. The particular Lissajous pattern observed on the screen depends on the ratio of two frequencies and the phase difference between the two signals superimposed as explained below:

i. If the frequencies of the applied signals are equal, i.e. if their ratio is 1 : 1 (a) a circular pattern is displayed on the screen provided signals are out of phase by 90° or 270° and their amplitudes are same. (b) symmetrical elliptical pattern is displayed on the screen provided the signals are out of phase by 90° and their amplitudes are unequal. An inclined ellipse will have different phase difference (c) a straight line inclined equally to X- and Y-axes is displayed when the phase difference is 0° or 180°.

ii. If the ratio of the frequencies of the two applied signals is 2 : 1 and phase difference is 90°, a figure of 8 (eight) is produced as shown in Fig. 15.14a, i.e. when the frequency of signal applied to horizontal input is twice that of the applied to vertical input. (b) when the frequency of signal applied to vertical input is twice that of horizontal input signal, the pattern is as shown in Fig. 15.14b.

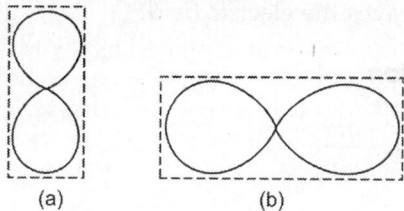

(a)   (b)

Fig. 15.14: Lissajous figure

Another way to know, which signal has higher frequency is to draw a tangent against the top edge of the pattern, and a tangent against the side edge of the pattern. The ratio of number of points of contacts of the tangents gives the ratio of the frequencies of signals. In the above case for Fig. 15.14a, the number of points of contact of top tangent is one and that of side tangent is two. So the ratio of the frequencies is $f_1 : f_2 :: 1 : 2$. But for Fig. 15.14b, the number of points of contact is reversed, i.e. for top tangent is two and for side tangents is one. So the ratio of frequencies $f_1 : f_2 :: 2 : 1$.

Interesting patterns result when the two signals have frequency ratio 2 : 1, but different phase difference as shown in Figs 15.15a, b, c and d.

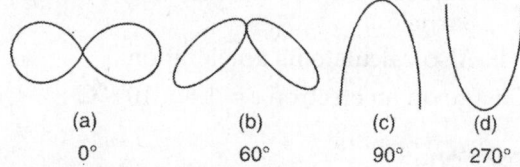

(a)   (b)   (c)   (d)

0°   60°   90°   270°

Fig. 15.15: Lissajous figures ($f_1 : f_2 :: 1 : 2$)

Computation of Phase Angles

Regardless of the relative amplitudes of the applied voltages, but having same frequency, the resultant elliptical pattern provides a simple method of finding the phase difference between the two signals. This method is illustrated in Fig. 15.16.

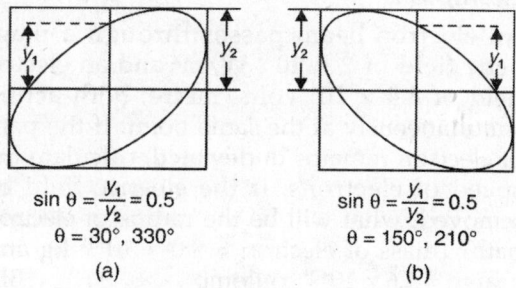

$$\sin \theta = \frac{y_1}{y_2} = 0.5 \qquad \sin \theta = \frac{y_1}{y_2} = 0.5$$

$$\theta = 30°, 330° \qquad \theta = 150°, 210°$$

(a)   (b)

Fig. 15.16: Computation of phase angles

Accordingly the sine of the phase angle (θ) between the two signals is equal to the ratio between the Y-axis intercept represented by y_1 and the maximum vertical deflection represented by y_2. So $\sin\theta = \dfrac{y_1}{y_2}$ and from the Figs. 15.16a and b it is clear that phase difference (θ) in Fig. 15.16a is 30° or 330° and in Fig. 15.16b, θ is equal to 150° or 210° respectively.

ILLUSTRATIVE EXAMPLES

Example 1

An alpha particle of mass 6.65×10^{-27} kg and charge twice that of an electron travels at right angles to a magnetic field with speed 6×10^5 m/sec. The flux density of the field is 0.2 W/m^2. [BE]

   i. Calculate the force acting on alpha particle.

   ii. Also calculate its acceleration.

Charge on an electron $e = 1.6 \times 10^{-19}$C

Solution

  i. Force $F = B \times q \times v$

$$= 0.2 \times (2 \times 1.6 \times 10^{-19}) \times 6 \times 10^5$$

$$= 3.84 \times 10^{-14} \text{ N}$$

  ii. Acceleration

$$a = \frac{\text{force}}{\text{mass}} = \frac{3.84 \times 10^{-14}}{6.65 \times 10^{-27}} = 5.77 \times 10^{12} \text{ m/sec}^2.$$

Example 2

An electron beam passes through a magnetic field of 2×10^{-3} W/m^2 and an electric field of 3.4×10^4 volts/metre, both acting simultaneously at the same point. If the path of electron remains undeviated, calculate the speed of electrons. If the electric field be removed, what will be the radius of electron path? (mass of electron = 9.0×10^{-31} kg and charge = 1.6×10^{-19} coulomb). [BE]

Solution

In equilibrium $Bev = Ee$

or $v = \dfrac{E}{B} = \dfrac{3.4 \times 10^4}{2 \times 10^{-3}} = 1.7 \times 10^7$ m/s

On removing the electric field

$$Bev = \frac{mv^2}{r}$$

or $r = \dfrac{mv}{Be}$

$$r = \frac{9.0 \times 10^{-31} \times 1.7 \times 10^7}{2 \times 10^{-3} \times 1.6 \times 10^{-19}}$$

$$= 4.78 \times 10^{-2} \text{ m.}$$

Example 3

An electron is at rest. Suddenly a uniform electric field of intensity 5000 N/C is switched on around the electron. The electron starts moving. Calculate its speed when it has travelled a distance of 2 cm.

Solution

$$a = \frac{eE}{m} \quad \text{and} \quad u = 0$$

$$v = \sqrt{u^2 + 2as} = \sqrt{2\frac{eE}{m}s}$$

$$= \sqrt{\frac{2 \times 1.6 \times 10^{-19} \times 5000 \times 0.02}{9.1 \times 10^{-31}}}$$

$$= 5.93 \times 10^6 \text{ m/s.}$$

Example 4

An electron travelling along +X-axis with velocity 5×10^6 m/s enters an electric field at the origin of the coordinate system. The electric field of strength 2000 N/C is in the +Y direction and it exists up to 6 cm. Calculate the vertical displacement of electron when it just leaves the electric field.

Solution

$$y_1 = \frac{-qEl^2}{2mv_x^2}$$

$$= \frac{-1.6 \times 10^{-19} \times 2000 \times (0.06)^2}{2 \times 9.1 \times 10^{-31} \times (5 \times 10^6)^2} = -2.53 \text{ cm}$$

Example 5

A positively charged particle ($q = 1.6 \times 10^{-19}$ C and $m = 1.67 \times 10^{-27}$ kg) with velocity 4×10^5 m/s enters an electric field of intensity 300 N/C, at the origin making an angle 35° with the +X direction. The electric field is in the −Y direction. Calculate the time required for the particle to reach the maximum height in the electric field.

Solution

$$v = 0,\ a = \frac{qE}{m} \text{ and } u = v_y = v\sin 35°$$

$$t = \frac{um}{qE} = \frac{4 \times 10^5 \times \sin 35° \times 1.67 \times 10^{-27}}{1.6 \times 10^{-19} \times 300}$$

$$= 7.98\ \mu s$$

Example 6

A proton is moving in a circular orbit of radius 30 cm in a magnetic field of 0.38 T. The magnetic field is perpendicular to the velocity of proton. Determine the orbital speed of the proton.

Solution

$$v = \frac{qBR}{m} = \frac{1.6 \times 10^{-19} \times 0.38 \times 0.3}{1.672 \times 10^{-27}}$$

$$= 11 \times 10^6 \text{ m/s}$$

Example 7

The velocity of a charged particle ($q = 1.6 \times 10^{-19}$ C and $m = 1.67 \times 10^{-27}$ kg) entering the magnetic field of strength 0.8 T is $[(4 \times 10^6)\ i + (3 \times 10^6)j]$ m/s. The particle describes a helix. Calculate the pitch of the helix and radius of the trajectory.

Solution

$$v_\parallel = 4 \times 10^6 \text{ m/s}$$

and

$$v_\perp = 3 \times 10^6 \text{ m/s}$$

$$p = v_\parallel\ \frac{2\pi m}{qB}$$

$$= \frac{4 \times 10^6 \times 2\pi \times 1.67 \times 10^{-27}}{1.6 \times 10^{-19} \times 0.8}$$

$$= 0.328 \text{ m}$$

$$R = \frac{mv_\perp}{qB}$$

$$= \frac{1.67 \times 10^{-27} \times 3 \times 10^6}{1.6 \times 10^{-19} \times 0.8}$$

$$= 0.03914 \text{ m}$$

REVIEW QUESTIONS

1. What is the effect of magnetic field on a moving charge particle? Describe any one application of this effect.

2. What is the effect of perpendicular electric field on the motion of charged particles? Obtain the appropriate relation for linear deflection.

3. Describe the construction of a CRO. Draw a labelled diagram. Describe their functions.

4. Describe the working of a CRO. How will you measure the frequency of an unknown signal using the principle of Lissajous figure.

5. Write short notes on:

   i. Cathode ray oscilloscope,

   ii. Determination of e/m of electron

   iii. Aston mass spectroscope.

PROBLEMS

1. Calculate the energy of an electron if it describes a circle of radius 0.303 m in magnetic field of 5×10^{-4} T.   [*Ans.* 2.02×10^3 eV]

2. Calculate the magnetic flux density required to accelerate protons in a cyclotron in which the potential difference applied between the dees has a frequency of 7×10^6 Hz.   [*Ans.* 0.46 T]

3. A cyclotron accelerates deuterons to 4 MeV. To what energy will the same cyclotron accelerate protons and α-particles? [*Ans.* 8 MeV for both]

4. An electron travelling at 2×10^7 m/s enters a magnetic field of flux density 2×10^{-3} T. Show that the radius of the path described by it is 5.688×10^{-2} m.

OBJECTIVE QUESTIONS

1. 1 eV = _____ J.   [1.602×10^{-19}]

2. A charge particle of charge e and mass m is moving in a transverse magnetic field B, the resonance frequency n = _____.   [$Be/2\pi m$]

3. The instrument which is based on the motion of charged particle in the crossed electric and magnetic field is _____.   [CRO]

4. _____ focussing is used in modern CRO.
[Electrostatic]

5. The voltage of the grid in CR tube is _____ with respect to the cathode.   [negative]

6. The function of trigger circuit in a CRO is to have _____ display on the screen.   [permanent]

7. Aston's mass spectrograph is useful for the investigation of _____.   [isotopes]

8. In Aston's mass spectrograph, all particles of the same e/m are brought to the same _____ irrespective of their _____.
[focus, velocities]

MULTIPLE CHOICE QUESTIONS

1. Relation between electron volt (eV) and joule (J) is
(a) 1 eV = 1.602×10^{-12} J

(b) 1 eV = 1.602×10^{-19} J
(c) 1 eV = 1.602×10^{-23} J
(d) 1 eV = 1.602×10^{7} J

2. The force acting on the electron of charge e due to transverse magnetic field (B) is
(a) Be (b) BeV^2
(c) BeV (d) BeV^3

3. The path of a charged particle in a transverse electric field is
(a) circular   (b) elliptical
(c) parabolic   (d) linear

4. In a transverse magnetic field (B) when an electron of mass m is moving in a circular path of radius r with constant V, then
(a) $e/m = V/RB$ (b) $e/m = BVR$
(c) $e/m = B/VR$ (d) $e/m = BR/V$

5. When an electron is moving with velocity v in crossed electric field (E) and magnetic field (B) and if force acting on the electron by electric and magnetic field simultaneously are same, then
(a) $v = E \times B$
(b) $v = E/B$
(c) $v = (e/m)RB$
(d) none of the above

Answers

| 1. (b) | 2. (c) | 3. (c) | 4. (a) | 5. (b) |

16 Crystallography and Crystal Imperfections

16.1 INTRODUCTION

A most remarkable feature of matter in the solid state is the tendency of the constituent atoms of a great many solids to arrange themselves in an ordered periodic pattern. A solid material with such a regular arrangement is said to be *crystalline*. A solid material without such a structure is called *non-crystalline or amorphous* (Fig. 16.1).

A crystalline material may be either in the form of *single crystal* or an aggregate of many crystals usually known as polycrystalline separated by well-defined boundaries called as *grain boundaries*. Polycrystalline material is stronger than ordinary one because crystals in polycrystalline materials have different orientations with respect to each other and grain boundaries obstruct the movement of dislocations.

Single crystals represent a material in its ideal condition and are produced artificially from their vapour or liquid state.

Whiskers are very thin filaments, hair-like single crystals of about 13 mm length and 10^{-4} cm diameter. These are produced as dislocations of free crystals and are without any structural defects. This is why these are far stronger than polycrystals of same material (Fig. 16.1).

Most solid matter is crystalline, i.e. nature favours the crystalline state of solids. The term crystal is limited to the description of any solid with an ordered atomic arrangement for a given structure. The energy of the ordered atomic arrangement is lower than that of an irregular packing of atoms.

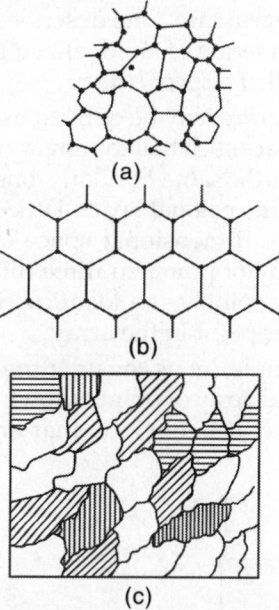

Fig. 16.1: (a) Amorphous or non-crystalline structure
(b) Crystalline or highly ordered structure
(c) Polycrystalline structure exhibiting grain and grain boundary

16.2 PERIODIC ARRAY OF ATOMS: SPACE LATTICE, BASIS UNIT CELL

A crystal is a regularly repeated structure on an atomic scale, i.e. a three dimensional periodic array of atoms. An ideal crystal is constructed by the infinite repetition in space of identical structural units. In other words, we can say that there exists some smallest grouping that repeats itself exactly in all directions in the crystal, so that the environment at one location is identical in all respects to the environment at a corresponding location somewhere else. In the simplest crystals such as silver, copper, aluminium, etc. the structural unit is a single atom. Often the structural unit is several atoms or molecules up to 10^2 in inorganic crystals and 10^4 in protein crystals. The structure of all crystals is described in terms of a lattice with a group of atoms attached to each lattice point. The group is termed *basis*. The *basis* is repeated in space to form the crystal structure.

A crystal is a three dimensional body. A regular and periodic three dimensional pattern of atoms or molecules in space is called the crystal structure. One describes the crystal structure in terms of an idealized geometrical concept called a *space lattice*.

A *space lattice* may be defined as an array of points in space such that the environment about each point is the same. Similarly, one may argue for three dimensional space lattice. One may define three dimensional space lattice as an infinite array of points in three dimensions in which every point has an identical environment as any other point in the array.

For example, let us consider the case of two dimensional array of points shown in Fig. 16.2. From figure, it is obvious that environment

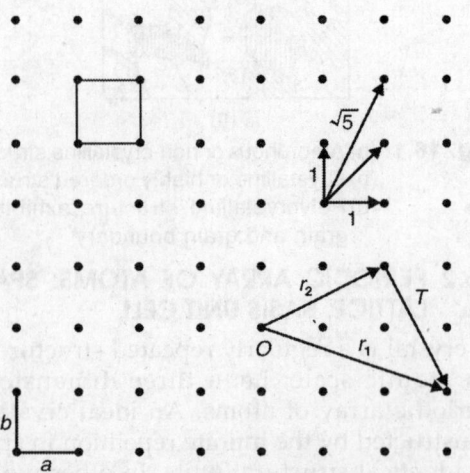

Fig. 16.2: Two dimensional array of points

about any two points is the same and hence it represents a space lattice. One can define the space lattice in a mathematical form as follows:

Choose any arbitrary point O as origin and consider the position vectors r_1 and r_2 of any two lattice points by joining them to point O (Fig. 16.2). The difference T of the two vectors r_1 and r_2 satisfy he following relation

$$T = n_1 a + n_2 b$$

where a and b are fundamental translational vectors characteristic of the array and n_1 and n_2 are integers. The array of these points is a two dimensional lattice.

We may note that a crystal lattice refers to the geometry of a set of points in space whereas the structure of the crystal refers to the actual ordering of its constituent ions, atoms, molecules in space.

16.3 CRYSTAL LATTICE AND LATTICE TRANSLATION VECTORS

An ideal crystal, i.e. one without structural defects, such as vacancies, impurities, or grain boundaries is composed of atoms arranged on a lattice defined by three fundamental translation vectors a, b, c, such that the atomic arrangement looks the same in every respect when viewed from any point r as when viewed from the point.

$$r' = r + u a + v b + w c \qquad (16.1)$$

Here u, v, w are arbitrary integers. The set of points r' specified by Eq. (16.1) for all values of integers u, v, w defines a lattice. A lattice is a regular periodic arrangement of points in space. A lattice is a mathematical abstraction and the crystal structure is formed only when a basis of atoms is attached identically to each lattice point. The simple logical relation between lattice, basis and crystal structure can be expressed as (Fig. 16.3a)

(Lattice + Basis = Crystal structure)

The lattice and the translation vectors a, b, c are said to be primitive if any two points r and r', from which the atomic arrangement looks the same (i.e. we are unable to determine that we are not still at r) always satisfy Eq. (16.1) with a suitable choice of integers u, v, w. This definition of the primitive translation vectors shows that there is no cell of smaller volume that could serve as a building block for the crystal structure. We define the *lattice translation operator* T in terms of three fundamental translation vectors a, b, c as follows:

$$T = ua + vb + wc \qquad (16.2)$$

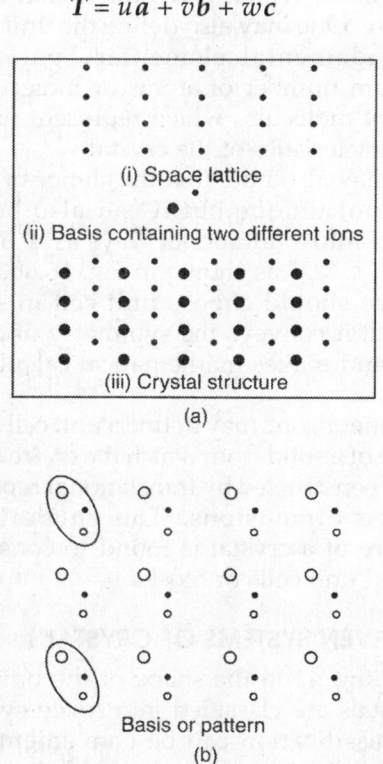

(i) Space lattice

(ii) Basis containing two different ions

(iii) Crystal structure

(a)

Basis or pattern

(b)

Fig. 16.3: Generation of crystal structure from lattice and basis

Obeying the translation operation means that when the operation T is applied to any point r (measured from some arbitrary origin) in the material, the resulting point r' given by Eq. (16.1) is exactly identical in all respects to the original point r. If r' is not identical to r for any arbitrary choice of u, v, w, the vectors a, b, c are not translation vectors. In order for an assembly of atoms to be classified as a crystal structure, it must be possible to find three translation vectors a, b, c which satisfy Eq. (16.1) such that r' is indistinguishable from r. The primitive translation vectors a, b, c are

often called *crystal axes* or *basis vectors*. It is not necessary that the vectors a, b and c be mutually $\perp^r$. In general, these vectors can have any angle between them. Figure 16.4 shows a part of a general oblique lattice in two dimensions.

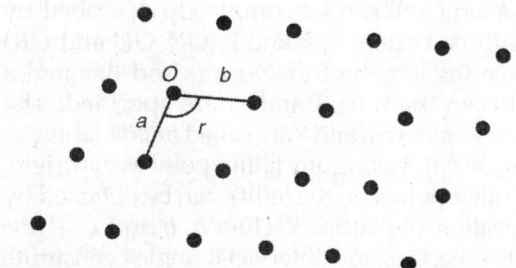

Fig. 16.4: Part of general oblique lattice in two dimensions

The parallelopipe formed by the translation vectors a, b, c is called a *primitive cell* (Fig. 16.5). With suitable crystal translation operations, one can define an elementary volume of the crystal, which can be termed unit cell. A primitive cell is also a unit cell having the minimum volume. Only one lattice point is associated with each primitive cell. From Fig. 16.5, it is evident that there are eight lattice points at the corners of the primitive cell, but these eight points are shared by eight primitive cells. Since each lattice point is the

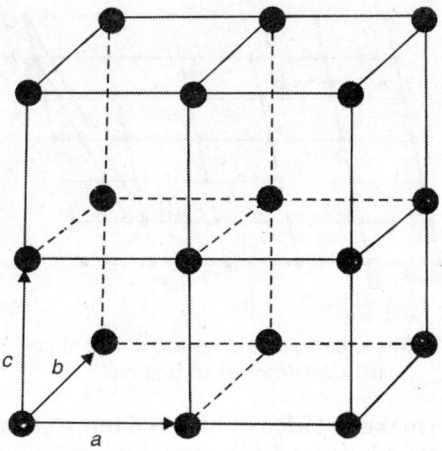

Fig. 16.5: Primitive cell

position of the basis of the crystal, the number of atoms in a primitive cell is equal to the number of atoms in the basis. In general, it is not essential that a unit cell be a primitive cell. Sometimes a non-primitive cell is more closely related with the symmetry of the crystal and it is chosen as a *unit cell*.

A unit cell can be completely described by the three vectors *a*, *b* and *c* (*OP*, *OQ* and *OR*) when the length of the vectors and the angles between them (α, β and γ) are specified. The three angles α, β and γ are called *interfacial angles* (Fig. 16.6a). Taking any lattice point as the origin, all other points on the lattice can be obtained by repeating the lattice vectors *a*, *b* and *c*. These lattice vectors and interfacial angles constitute the lattice parameters of a unit cell. Clearly, if the values of these intercepts and interfacial angles are known one can easily determine the form and actual size of the unit cell.

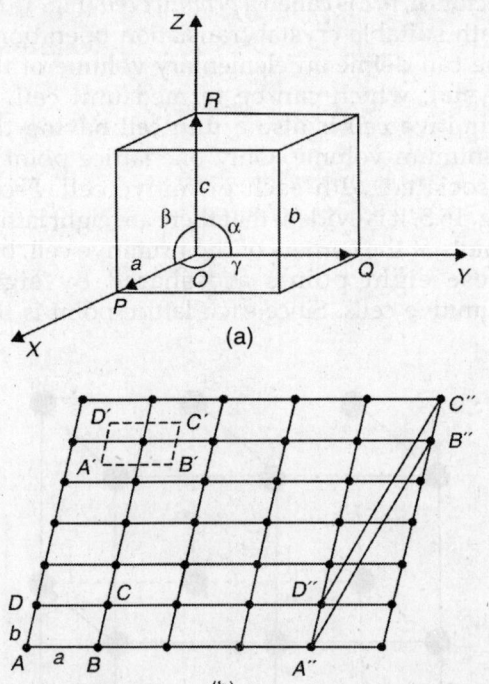

Fig. 16.6: (a) Lattice parameters of a unit cell
(b) Construction of unit cell

To make the idea of unit cell more clear, let us consider a two dimensional crystal in which the atoms are arranged as shown in Fig. 16.6b.

Let us consider a parallelogram such as *ABCD* with sides *a* (= *AB*) and *b* (= *AD*). Now, by rotating this parallelogram by any integral multiple of vectors *a* and *b*, one can obtain the entire crystal lattice. This fundamental unit *ABCD* is called a unit cell. Obviously, a unit cell is the smallest geometrical figure, the repetition of which gives the actual crystal structure. One may also define the unit cell as the fundamental elementary pattern of minimum number of atoms or molecules or group of molecules which represent fully all the characteristics of the crystal.

We may also note that the choice of a unit cell is not unique but it can also be constructed into a number of ways as *A' B' C' D'* or *A" B" C" D"* as shown in Fig. 16.6b. However, we should choose unit cell in such a way that it conveys the symmetry of crystal lattice and makes mathematical calculations easy.

In general, one may define a unit cell as that volume of a solid from which the entire crystal may be constructed by translational repetition in three dimensions. The entire lattice structure of a crystal is found to consists of identical unit cells or blocks.

16.4 SEVEN SYSTEMS OF CRYSTALS

Depending upon the shape of the unit cells, all crystals are classified into seven systems. This classification can be conveniently expressed in terms of the relations between the axes of the unit cells. The axes are designated as *a*, *b*, *c* and the angles between them as α, β, γ. As shown in Fig. 16.7, the angle between *b* and *c* is taken as α, between *c* and *a* as β and between *a* and *b* as γ. Table 16.1 gives the seven crystal systems together with the essential symmetric elements.

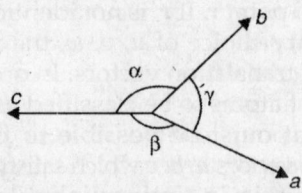

Fig. 16.7: Axes of the unit cell

Table 16.1: Seven systems of crystals

S. No.	Name of the crystal system	Relation of lengths of axes of unit cell	Relation of angle between axes	Number of lattice types
1.	Cubic	$a = b = c$	$\alpha = \beta = \gamma = 90°$	3(P, I, F)
2.	Trigonal	$a = b = c$	$\alpha = \beta = \gamma \neq 90°, <120°$	1 (P)
3.	Hexagonal	$a = b \neq c$	$\alpha = \beta = 90°, \gamma = 120°$	1 (P)
4.	Tetragonal	$a = b \neq c$	$\alpha = \beta = \gamma = 90°$	2 (P, I)
5.	Orthorhombic	$a \neq b \neq c$	$\alpha = \beta = \gamma = 90°$	4 (P, C, I, E)
6.	Monoclinic	$a \neq b \neq d$	$\alpha = \gamma = 90° \neq \beta$	2 (P, C)
7.	Triclinic	$a \neq b \neq c$	$\alpha \neq \beta \neq \gamma$	1 (P)

P → Primitive, C → Base centred
I → Body centred, F → Face centred.

From Table 16.1, it is evident that the cubic system is the simplest while the triclinic crystals are least symmetrical. Cubic crystals are the commonest and more than half of the naturally found crystals belong to this system. On the basis of nature of forces which binds the atoms in a molecule, crystals can be classified under the four categories: (a) ionic crystals (b) molecular crystals (c) covalent crystals (d) metallic crystals.

16.5 BRAVAIS LATTICES

There are various ways of positioning structureless points in space such that all points have identical surroundings. These are called Bravais lattices. There are five Bravais lattices in two dimensions and fourteen in three dimensions. For a cubic system we have the following three types of Bravais lattices.

   i. Primitive or simple cubic: There is one lattice point at each of the eight corners of the unit cell. This type of cell is called primitive or simple cubic cell (P) of the system (Fig. 16.8a).

   ii. Body centred cubic: There is one lattice point at each of the eight corners and one lattice point is at the centre of the cubic cell. This type of cell is called body centred cell (I) and it is shown in Fig. 16.8b.

   iii. Face centred cubic: There is one lattice point at each of the eight corners and one lattice point is at the centres of each of the six faces of the cubic cell. Such a cell is called face centred cubic cell (F) and it is shown in Fig. 16.8c.

$$a = b = c, \quad \alpha = \beta = \gamma = 90°$$

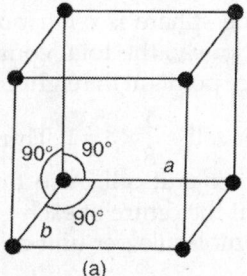

(a)

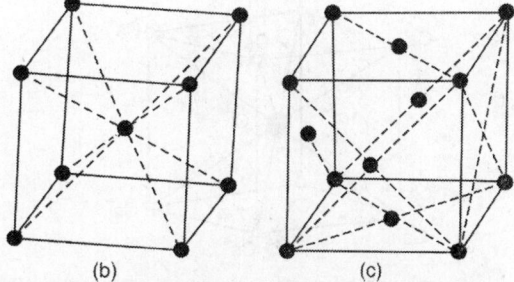

(b)   (c)

Fig. 16.8: Three types of Bravais lattices (a) Primitive (b) Body centred (c) Face centred

Cubic P, cubic I and cubic F lattices are also referred to simple cubic, body centred cubic (*bcc*) and face centred cubic (*fcc*) lattices respectively.

16.6 EXAMPLE OF CRYSTAL STRUCTURE: CALCULATION OF LATTICE POINTS PER UNIT CELL

   i. Simple cubic lattice: CsCl is an ionic crystal and it is a typical example of simple cubic (*sc*) or cubic P lattice. Lattice parameter of CsCl crystal is $a = b = c = 4.11$ Å. The structure of CsCl unit cell is shown in Fig. 16.9. The structure of CsCl may be considered as resulting from the combination of two simple cubic sublattices, one of Cs^+ ions and other of Cl^- ions situated at the body centre of each unit cell and *vice versa*. It is obvious that similar type of ions (Cs^+ in this case) form a simple cubic lattice having one atom at each of the eight corners of the cube. Since each cell is surrounded on all sides by other similar cells, the atoms at its corners are also shared by the adjoining cells. The corner atom of the cell is common to seven other similar unit cells adjacent to it, i.e. the sphere representing a corner atom is shared equally by eight unit cells. This means that only 1/8th of the sphere is contributed to each cell. Obviously, the total contribution of eight lattice points at the eight corners of a simple cubic cell is $\frac{1}{8} \times 8 = 1$. Thus, there is one Cs^+ ion per unit cell. Also there is one ion of Cl^- at the centre of each cell. Thus, there is one molecule per unit cell.

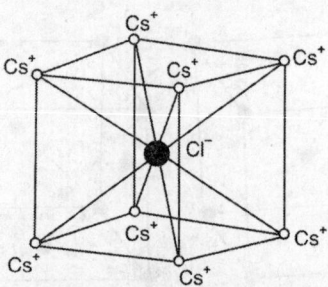

Fig. 16.9: CsCl crystal structure

ii. Face centred cubic (*fcc*) lattice: NaCl is the familiar and best example of *fcc* lattice. The lattice constant for NaCl crystal is $a = b = c = 5.63$ Å. NaCl is an ionic crystal and its unit cell is shown in Fig. 16.10. Each unit cell consists of eight corner ions, each being a member of eight cells. Also there is one ion at the centre of each of the six faces of the cell which is shared by the adjoining face of the neighbouring cell. Obviously, only half the sphere representing the ion is the share of each cell. Therefore,

the total number of atoms (of one kind) per unit cell is

$$8/8 + 6/2 = 4$$

Similarly, there are 4 ions of the other kind per unit cell. Thus, in *fcc* lattice there are four molecules per unit cell.

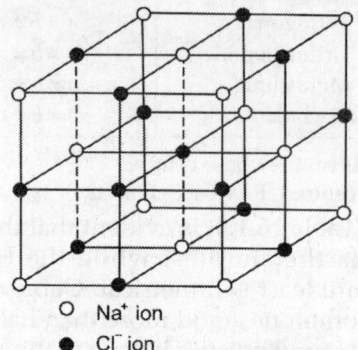

○ Na^+ ion
● Cl^- ion

Fig. 16.10: NaCl crystal structure

iii. Body centred cubic lattice (*bcc* lattice) or cubic *I* lattice: α-iron with $a = b = c = 2.86$ Å is the best example of the *bcc* lattice. In a unit cell of *bcc* lattice, there are eight atoms at the eight corners, each of them is a member of 8 cells. Also there is one atom at the centre of the body. Obviously, in *bcc* lattice, the total number of atoms per unit cell is $8/8 + 1 = 2$.

One can obtain the total number of atoms in a unit cell (N) by a generalized formula

$$N = N_B \frac{N_F}{2} + \frac{N_C}{8} \qquad (16.3)$$

Here $N_B \rightarrow$ the number of body centred or interior atoms, $N_F \rightarrow$ the number of face centred atoms and $N_C \rightarrow$ the number of corner atoms. Thus,

$N = 0 + 0 + 8/8 = 1$ in the *sc* lattice
$N = 0 + 6/2 + 8/8 = 4$ in the *fcc* lattice
$N = 1 + 0 + 8/8 = 2$ in the *bcc* system.

16.7 COORDINATION NUMBER (CN)

It is defined as the number of nearest neighbours to a given atom in a crystal lattice. Let us calculate CN for *sc*, *bcc* and *fcc* lattices:

i. *sc* lattice: It contains one atom at each of the eight corners of the unit cell (Fig. 16.8a).

Obviously, if we take the atom at one corner of the unit cell as the centre, it will be found to have 2 atoms along each coordinate axis, one on either side as its immediate neighbour. Therefore, the coordination number

$$CN = 2 \times 3 = 6.$$

Thus, we obtain the coordination number of a simple cubic or *P* cubic lattice as 6. The distance between the nearest neighbours is *a*, where *a* is the side of the unit cell.

In the case of CsCl simple cubic lattice (Fig. 16.9), we see that there are 8 nearest neighbouring ions of opposite kind for an ion of one kind. Obviously, the coordination number for this particular case is 8. The nearest distance between two neighbouring ions of the opposite kind is $a\sqrt{3}/2$.

ii. *fcc* **lattice:** It contains one ion at each of the eight corners of the unit cell and one atom at the centre of each of the six faces. If one takes the ion at the face centre as origin, then one finds that the face is common to two cubes and that there are 12 points surrounding it situated at a distance equal to half the face diagonal of the unit cube. Obviously, in *fcc* lattice, the coordination number is 12 and the distance between two nearest neighbours is $a/\sqrt{2}$. Alternatively, if one considers a corner atom of a unit cell as the origin, it has as its immediate neighbour one atom in the centre of its own upper face and 3 others in the upper faces of the adjoining cells, i.e. 4 atoms in the horizontal of the *XY* plane. The same is the case in the *XZ* and *YZ* planes. Obviously, the CN number is 3 × 4 = 12. Let us consider the case of NaCl (Fig. 16.10). If we consider the ions of only one kind, the number of nearest Na$^+$ ion to a Na$^+$ ion is 12 and the distance between the two nearest Na$^+$ ions is $a/\sqrt{2}$. NaCl structure shows that each Na$^+$ ion has six nearest neighbours of the opposite kind, i.e. Cl$^-$ ions and that the

separation is $a/2$. Therefore, the coordination number for NaCl is 6 and the distance between the nearest neighbouring atoms of the opposite kind is $a/2$.

iii. *bcc* **lattice:** It contains one atom at each of the eight corners of the unit cell and one atom at the body centre. If one takes the body centre point as origin, it is observed that the number of nearest neighbours is 8 and their distance from body centre is half the diagonal of the unit cube, i.e. equal to $a\sqrt{3}/2$. Obviously, the coordination number for *bcc* lattice is 8 and the distance between two nearest atoms is $a\sqrt{3}/2$.

16.8 LATTICE CONSTANT

One can calculate the lattice constant for cubic crystals as follows:

Let the lattice constant for cubic crystals be *a*. Obviously, each side of the cube will be *a*. The volume of unit cell = a^3. If ρ be the density of the crystal, then mass of each unit cell = $V\rho = a^3\rho$.

If *M* be the molecular weight and *N* the Avogadro number (i.e. the number of molecules per kg mole of the substance), then mass of each molecule = M/N.

Let *n* be the number of molecules (lattice points per unit cell), then the mass in each unit

$$\text{cell} = \frac{nM}{N}$$

$$\therefore \qquad a^3\rho = \frac{nM}{N}$$

or

$$a = \left(\frac{nM}{N\rho}\right)^{1/3} \qquad (16.4)$$

16.9 ATOMIC RADIUS

For the purpose of estimation of atomic radius, one can assume that the atoms are spheres in contact in a crystal. One can define atomic radius *r* as *half the distance between nearest neighbours in a crystal of pure element.* The atomic radius is expressed in terms of cube edge a. Now, we will calculate the atomic radius for the following cases:

i. Simple cubic lattice: Figure 16.11 shows the front view of a simple cubic lattice. There are eight atoms per unit cell situated at eight corners. If the side of the unit cell be a and r be the radius, then from Fig. 16.11, we have

$$a = 2r \quad \text{or} \quad r = a/2$$

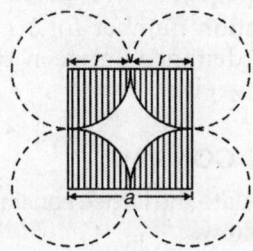

Fig. 16.11: Front view of simple cubic lattice

ii. Body centred cubic (bcc) lattice: There are eight atoms at the corners of the unit cell of bcc, each forming a number of eight cells and one atom per cell at the body centre. One can calculate r with the help of Fig. 16.12.

From Fig. 16.12, we have

$$(DF)^2 = (DA)^2 + (AF)^2$$

or
$$(4r)^2 = a^2 + (a\sqrt{2})^2$$

or
$$16r^2 = 3a^2$$

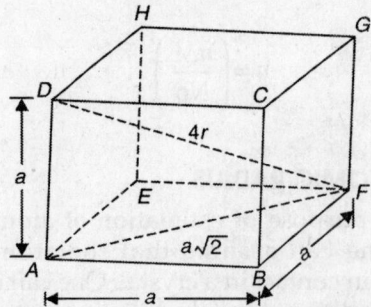

Fig. 16.12: Body centred cubic lattice

$$\therefore \quad r = \frac{\sqrt{3}}{4}a$$

or
$$a = \frac{4r}{\sqrt{3}}.$$

iii. Face centred cubic (fcc) lattice: In this lattice, there are eight corner atoms in the unit cell and one atom at the centre of each face. The front view of fcc structure unit cell is shown in Fig. 16.13a, whereas its cut view is shown in Fig. 16.13b.

From Fig. 16.13b, we have

$$(DB)^2 = (DC)^2 + (CB)^2$$

or
$$(4r)^2 = a^2 + a^2 = 2a^2$$

or
$$r = \frac{\sqrt{2}a}{4} \quad \text{or} \quad a = \frac{4r}{\sqrt{2}}$$

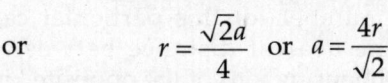
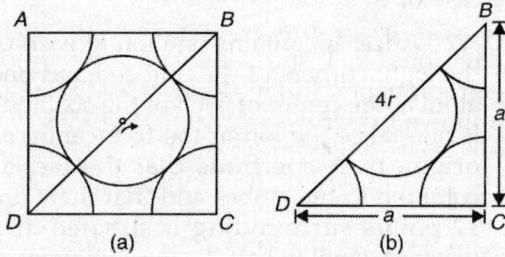

Fig. 16.13: (a) *fcc* structure unit cell (b) Cut view

16.10 CRYSTAL OR LATTICE PLANES AND MILLER INDICES

Because of the crystalline nature of solids, many physical properties of solids are anisotropic, i.e. depend on directions within the crystal and such properties may be characterized by tensors. For this reason, the specification of planes and direction in crystals is an important aspect of crystal structure analysis.

A crystal lattice may be considered as an aggregate of a set of parallel equidistant planes passing through the lattice points. These are called lattice, planes (Fig. 16.14). For a given crystal lattice, these sets of planes can be chosen in a number of different ways, the spacing between the successive planes accordingly varies, as also the density of crystal lattice points per unit area in each plane. For a particular lattice these sets of planes may be chosen in different ways as shown in Figs 16.4a, b, c, and d.

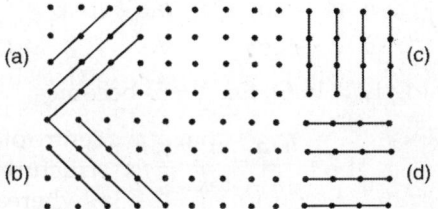

Fig. 16.14: A crystal lattice as a set of parallel equidistant planes passing through the lattice points

The basic necessity in crystal analysis is the ability with which one can describe relative orientations of lines and planes. One can describe it by a suitable frame of reference. In the case of two dimensions, for this purpose X and Y axes and in the case of three dimensions, X, Y and Z axes are selected on the graph paper. In Fig. 16.15, two directions are shown by arrows in two dimensions.

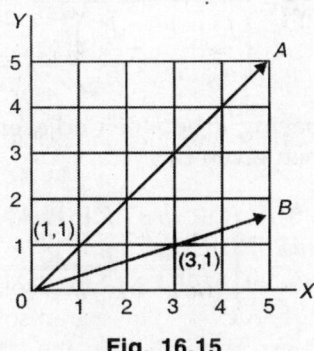

Fig. 16.15

We can see that in Fig. 16.15, these directions pass through origin and ends at A and B respectively. One can describe the directions by assigning the coordinates of the first whole numbered point (x, y) through which each passes. We can see that for direction OA, it is (1, 1) and OB it is (3, 1). Figure 16.16 shows the three dimensional picture. In this case also, one can describe the directions by the coordinates of first whole numbered point $[x, y, z]$. Generally, square brackets are used

to indicate a direction. Few directions are as follows:

$$OA\ [110],\ OB\ [100],\ OC\ [111],\ OD\ \left[0\ 1\ \frac{1}{2}\right]$$

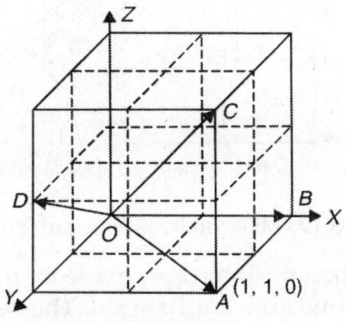

Fig. 16.16

We may note that the digits in the square brackets indicate the indices of that direction. The negative indices are represented by a bar over that digit. The planes are specified as follows:

i. Find the intercepts of the plane on the three crystal axes X, Y, Z. Let it be pa, qb and rc, where a, b and c are the primitives along X, Y and Z axes respectively and p, q, r may be either small integers or simple fractions.

ii. Take the reciprocals of p, q, r, i.e. p^{-1}, q^{-1} and r^{-1}.

iii. Find the smallest possible integers h, k, l such that
$$h : k : l = p^{-1} : q^{-1} : r^{-1}$$

This is done by multiplying the reciprocals by LCM of their denominators.

The numbers h, k, l are called the *Miller indices* of a given set of planes and the plane is specified as (hkl). It is worthwhile to mention that these indices refer not to a particular plane but to a set of parallel planes. To make the idea of Miller indices more clear, let us consider the Miller indices in a cubic crystal as shown in Fig. 16.17. We may remember that when the face is parallel to an axis, its intercept on the axis is infinite. Let us consider the face CGFB. This face cuts the X-axis at point F and we can see that it is parallel to Y and Z axes. If we take the sides of the cube as one unit its

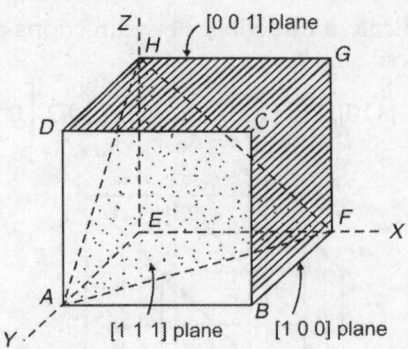

Fig. 16.17: Miller indices in a cubic crystal

length, then the intercepts made by this plane on the three axes are $1 : \infty : \infty$. The reciprocal of these cuts are $\dfrac{1}{1} : \dfrac{1}{\infty} : \dfrac{1}{\infty}$, i.e. $1 : 0 : 0$. Obviously, the Miller indices of this face are $(1\,0\,0)$. Now the face *HAF* cuts equal intercepts on the three axes, and therefore the indices are $1 : 1 : 1$. Clearly, the Miller indices of this plane are $(1\,1\,1)$. Similarly, we can see that the Miller indices of the plane *DHGC* are $(0\,0\,1)$.

16.11 SPACING OF PLANES IN CRYSTAL LATTICES

Let us consider the spacing between neighbouring planes of the $(h\,k\,l)$ system. The actual spacing d between adjacent planes can be calculated by taking any lattice point as origin, erecting axes in *a*, *b* and *c* directions, and finding the $\perp^r$ distance between this origin and the plane whose intercepts are $a/h, b/k, c/l$. From Fig. 16.18, it is evident that

$$d = OP = \frac{a}{h}\cos\alpha = \frac{b}{k}\cos\beta = \frac{c}{l}\cos\gamma \quad (16.5)$$

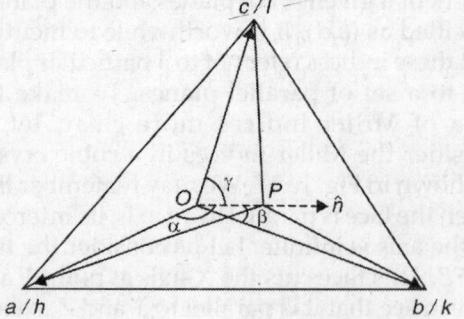

Fig. 16.18: Spacing of planes in crystal lattice

Here, α, β, γ are the angles between the normal to the plane and *a*, *b* and *c* axes respectively. If $\hat{n}$ is the unit normal vector to the plane, then

$$\hat{n}.a = a\cos\alpha,\ \hat{n}.b = b\cos\beta$$

and $\qquad \hat{n}.c = c\cos\gamma \qquad\qquad (16.6)$

Using Eq. (16.6), Eq. (16.5) becomes

$$d = \frac{\hat{n}\cdot a}{h} = \frac{\hat{n}\cdot b}{k} = \frac{\hat{n}\cdot c}{l} \qquad (16.7)$$

In an orthogonal lattice, taking the X-axis along *a*, the Y-axis along *b* and the Z-axis along *c*, the equation of the $(h\,k\,l)$ plane whose intercepts are $(a/h, b/k, c/l)$ is

$$f(x,y,z) = \left(\frac{hx}{a} + \frac{ky}{b} + \frac{lz}{c}\right) = 1 \qquad (16.8)$$

If $f(x, y, z) = $ constant in the equation of a surface, then ∇f is a vector normal to the surface, the unit normal $\hat{n}$ is given by

$$\hat{n} = \frac{\nabla f}{|\nabla f|} = \frac{\dfrac{h}{a}\hat{i} + \dfrac{k}{b}\hat{j} + \dfrac{l}{c}\hat{k}}{\left(\dfrac{h^2}{a^2} + \dfrac{k^2}{b^2} + \dfrac{l^2}{c^2}\right)^{1/2}} \qquad (16.9)$$

and the spacing d between adjacent $(h\,k\,l)$ planes is then given by

$$d = \frac{\hat{n}\cdot a}{n} = \frac{1}{n}\left(\frac{\dfrac{h}{a}\hat{i} + \dfrac{k}{b}\hat{j} + \dfrac{l}{c}\hat{k}}{\sqrt{h/a^2 + k^2/b^2 + l^2/c^2}}\right)\cdot a$$

or $\quad d = 1\Big/\sqrt{\dfrac{h^2}{a^2} + \dfrac{k^2}{b^2} + \dfrac{l^2}{c^2}} \qquad (16.10)$

Interplaner spacing d is also written as d_{hkl}.

Equation (16.10) is applicable to primitive lattices and for cubic, orthorhombic and tetragonal systems only, where the axes are mutually $\perp^r$ ($\alpha = \beta = \gamma = 90°$). In the case of cubic system, $a = b = c$ and hence for the cubic *P* lattice, Eq. (16.10) simplifies to

$$d = a/\sqrt{h^2 + k^2 + l^2} \qquad (16.11)$$

Obviously, planes with low index numbers have wide interplaner spacing compared to those having higher index numbers.

For a tetragonal crystal, $a = b$ so that Equation (16.10) becomes

$$d_{hkl} = \left[\frac{h^2 + k^2}{a^2} + \frac{l^2}{c} \right]^{-1/2} \quad (16.12)$$

The spacing between (1 0 0), (1 1 0) and (1 1 1) planes in the case of a simple cubic lattice from Eq. (16.11) are given by

$$d_{100} = a, \; d_{110} = a/\sqrt{2} \; \text{and} \; d_{111} = a/\sqrt{3}$$

Obviously, the ratio of spacing of the possible lattice plane in a simple cubic lattice are given by

$$d_{100} : d_{110} : d_{111} = a : a/\sqrt{2} : a/\sqrt{3}$$

$$= 1 : 1/\sqrt{2} : 1/\sqrt{3}$$

Spacing of Lattice Plane in the *bcc* Lattice

From Fig. 16.19, it is obvious that in *bcc* lattice, as compared with cubic *P* lattice, additional parallel planes (dotted) arise half way in the case of (1 0 0) and (1 1 1) planes. The spacing between (1 0 0), (1 1 0) and (1 1 1) planes in *bcc* lattice are given by

$$d_{100} = a/2, \; d_{110} = a/\sqrt{2} \; \text{and} \; d_{111} = a/\sqrt{3}$$

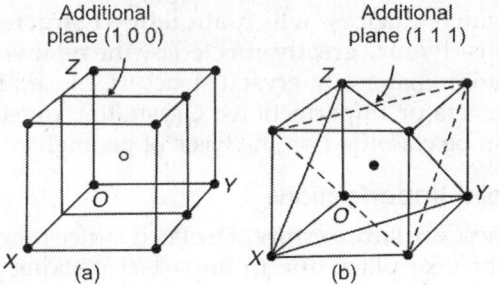

Fig. 16.19: Spacing of lattice plane in *bcc* lattice

Spacing of Lattice Plane in *fcc* Lattice

From Fig. 16.20, it is obvious that as compared with *P* cubic lattice, additional parallel planes (dotted) arise half way in the case of (1 0 0)

and (1 1 0) orientations. Therefore, the spacing between (1 0 0), (1 1 0) and (1 1 1) planes in *fcc* lattice will be

$$d_{100} = a/2, \; d_{110} = a/2\sqrt{2} \; \text{and} \; d_{111} = a/\sqrt{3}$$

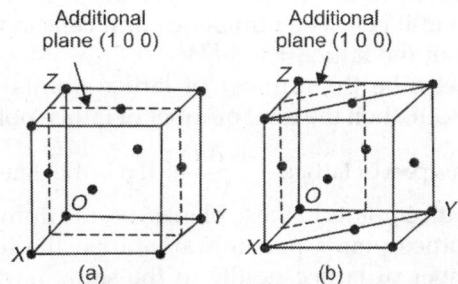

Fig. 16.20: Spacing of lattice plane in *fcc* lattice

16.12 DENSITY OF LATTICE POINTS OR PACKING FRACTION IN A MILLER PLANE

If the density of lattice points in a Miller plane is δ and the spacing between the planes is d which is given by Eq. (16.10), then the relation between δ and d can be expressed as

$$\frac{d}{\delta} = \text{constant} \quad (16.13)$$

If $(h\,k\,l)$ and $(h'\,k'\,l')$ are two alternative Miller planes, d and d' respectively are the interplaner spacings, δ and δ' are the densities of lattice points in these Miller planes, then

$$\frac{d}{d'} = \frac{\delta}{\delta'} \quad (16.14)$$

The proof is as follows:

Let unit volume be obtained in one case from m planes, each of area A and in the other case from m' planes, each of area A'. We have by equality of volumes,

$$mAd = m'A'd' \quad (16.15)$$

Also from the equality of the number of lattice points,

$$m\delta A = m'\delta' A' \quad (16.16)$$

From Eqs. (16.15) and (16.16), one obtains Eq. (16.14).

Density of Lattice Points in a Lattice Plane

Consider N successive parallel lattice planes in a lattice having spacing d and area of cross-section A.

The volume of the part of lattice under consideration = NAd. Let V be the volume of each unit cell. The number of unit cells in this part of the lattice = NAd/V.

Let n be the number of lattice points per unit cell, then the total number of lattice points in this part of lattice = $\dfrac{NAd}{V} n$. If ρ is the density of lattice points in these planes, i.e. the number of lattice points per unit area, then the total number of lattice points in the same part of lattice = $A\rho N$.

$$A\rho N = \frac{NAd}{V} n$$

or $$\rho = \frac{nd}{V} \qquad (16.17)$$

In the case of cubic, tetragonal and orthorhombic lattices $V = abc$,

$$\rho = \frac{nd}{abc} \qquad (16.18)$$

For primitive lattice in each of these systems, there is one lattice point pet unit cell, i.e. $n = 1$. Therefore, we have for these cases

$$\rho = \frac{d}{abc} \qquad (16.19)$$

16.13 DEFECTS OR IMPERFECTIONS IN CRYSTALS

So far, we have described perfectly regular crystal structures, called ideal crystals and obtained by combining a basis with an infinite space lattice. In ideal crystals, atoms were arranged in a regular way. However, the structure of real crystals differs from that of ideal ones. Real crystals always have certain defects or imperfections, and therefore, the arrangement of atoms in the volume of a crystal is far from being perfectly regular.

Natural crystals always contain defects, often in abundance, due to the uncontrolled conditions under which they were formed. The presence of defects which affect the colour can make these crystals valuable as gems, as in ruby (chromium replacing a small fraction of the aluminium in aluminium oxide Al_2O_3). Crystals prepared in laboratory will also always contain defects, although considerable control may be exercised over their type, concentration and distribution.

The importance of defects depends upon the material, type of defect, and properties which are being considered. Some properties, such as density and elastic constants, are proportional to the concentration of defects, and so a small defect concentration will have a very small effect on these. Other properties, e.g. the colour of an insulating crystal or the conductivity of a semiconductor crystal, may be much more sensitive to the presence of small number of defects. A defect free, i.e. ideal silicon crystal would be of little use in modern electronics. The use of silicon in electronic devices is dependent upon small concentrations of chemical impurities such as phosphorus and arsenic which give its desired properties.

There are some properties of materials such as stiffness, density and electrical conductivity which are termed structure-insensitive, are not affected by the presence of defects in crystals while there are many properties of greatest technical importance such as mechanical strength, ductility, crystal growth, magnetic hysteresis, dielectric strength, condition in semiconductors, which are termed structure sensitive, are greatly affected by the relatively minor changes in crystal structure caused by defects or imperfections. Crystalline defects can be classified on the basis of geometry.

Point Imperfections

These are lattice errors at isolated lattice points that take place due to imperfect packing of atoms during crystallization. The point imperfections also take place due to vibrations of atoms at high temperatures. These are completely local in effect.

  i. *The simplest point defect is vacancy*: This refers to an empty (unoccupied) site of a crystal lattice, i.e. a missing atom or vacant atomic site. Such defects may arise either from imperfect packing during original crystallisation or from thermal vibrations of the atoms at higher temperatures (Fig. 16.21a).

ii. *Interstitial imperfections*: In a closed packed structure of atoms in a crystal, if the atomic packing factor is low, an extra atom may be lodged within the crystal structure. This is known as interstitial position, i.e. voids. An extra atom can enter the interstitial space or void between the regularly positioned atoms only when it is substantially smaller than the present atoms, otherwise it will produce atomic distortion. The defect caused is known as interstitial defect. We may note that vacancy and interstitialcy are inverse phenomena (Fig. 16.21b).

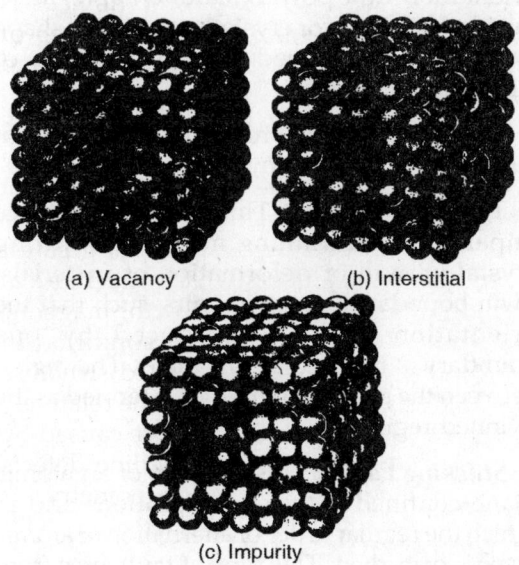

(a) Vacancy   (b) Interstitial

(c) Impurity

Fig. 16.21: Point defects (three types) in a crystal (a) Vacancy (b) Interstitial (c) Impurity. Point defect in a crystal disturbs the regularity of the crystal structure in its vicinity

iii. *Frenkel defect*: Whenever a missing atom, which is responsible for vacancy occupies an interstitial site (responsible

for interstitial defect), the defect caused is known as Frenkel defect (Fig. 16.22a).

iv. *Schottky defect*: These imperfections are similar to vacancies. This defect is caused, whenever a pair of positive and negative ions is missing from a crystal. This type of imperfection maintains a charge neutrality (Fig. 16.22b).

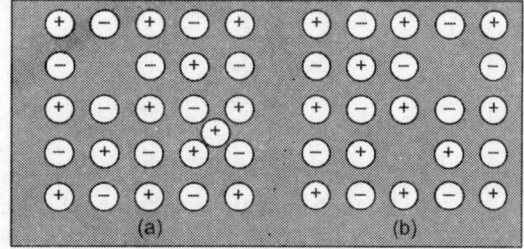

(a)   (b)

Fig. 16.22: Point defects in an ionic crystal: (a) Frenkel defect (b) Schottky defect. Both these defects preserve electrical neutrality

v. *Substitutional defect*: Whenever a foreign atom replaces the parent atom of the lattice and thus occupies the position of parent atom, the defect caused is called substitutional defect.

Phonon

When the temperature is raised, thermal vibrations takes place. This results in the defect of a symmetry and deviation in shape of atoms. This defect has much effect on the magnetic and electric properties.

Line Defects or Dislocations

Line imperfections are called dislocations. A linear disturbance, i.e. one dimensional imperfections in the geometrical sense of the atomic arrangement, which can very easily occur on the *slip plane* through the crystal is known as dislocation (Fig. 16.23). The most important kinds of linear defects are *edge* and *screw* dislocations (Figs 16.24 and 16.25). Both of these types are formed in the process of their deformation. Both these defects are the most striking imperfections and are responsible for the useful property of *ductility* in metals, ceramics and crystalline polymers.

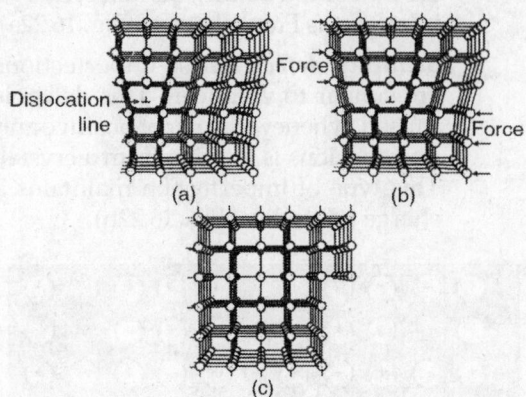

(a)   (b)

(c)

Fig. 16.23: Due to the motion of a dislocation through a crystal slip occur (a) Initial configuration of crystal (b) As the atoms in the layer under it successively shift their bonds with those of the upper layer one line at a time the dislocation moves to the right (c) Permanent deformation position of the crystal

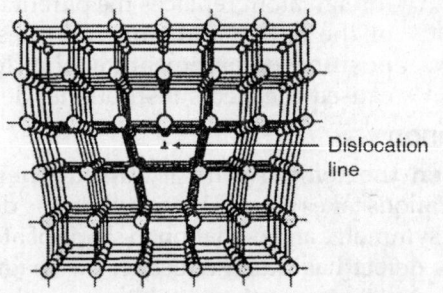

Fig. 16.24: An edge dislocation in a simple cubic crystal

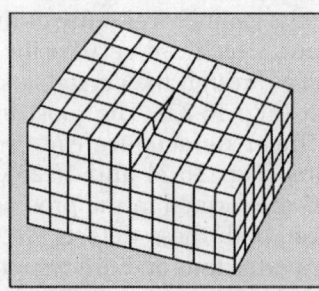

Fig. 16.25: Geometry of screw dislocation

The effect of dislocations is especially pronounced on the strength of materials.

Surface and Grain Boundary Defects

These imperfections are of a structural nature and arise from a change in the stacking of atomic planes on or across boundary and are two dimensional. This may be one of the orientations or of the stacking sequence of the planes. The most important kinds of surface defects are high-angle and low-angle boundaries, stacking faults and twin boundaries. Basically surface imperfections are of two types: *external and internal*.

Grain Boundaries: Engineering materials may be either polycrystalline or single crystal type. A crystalline alloy contains an enormous quantity of fine grains. Grain boundary imperfections are those surface imperfections which separate crystals or grains of different orientation in a polycrystalline aggregation during nucleation or crystallization. The shape is usually influenced by the presence of surrounding grains.

Tilt Boundaries: This is another type of surface defect called low-angle boundary.

Twin Boundaries: This is planer surface imperfection. Twinning may result during crystal growth or deformation of materials. Twin boundaries occur in pairs, such that the orientation change introduced by one boundary is restored by the other. The region between the pair of boundaries is termed as the twinned region.

Stacking Faults: This is a part of an atomic plane confined between dislocations and in which the regular order of alternation of atomic layer is disturbed. This type of fault arise from the stacking of one atomic plane out of sequence on another while the lattice on either side of the fault is perfect.

Volume Imperfection

Volume imperfections, e.g. cracks may arise when there is only small electrostatic dissimilarity between the stacking sequences of close packed planes in metals. Moreover, when clusters of atoms are missing, a large vacancy or void is created which is also a volume imperfection.

16.14 CRYSTAL STRUCTURES

Sodium Chloride (NaCl)

The crystal structure of NaCl crystal is shown in Fig. 16.26. We note that in this structure, sodium atom loses its outer electron and acquires an excess of positive charge while the chlorine atom accept one electron of sodium and so acquires an excess of negative charge. Due to electrostatic forces between their excessive charges, the two ions attract each other. Moreover, due to strong forces of repulsion as their outer electron shells come into close proximity, the two ions cannot approach each other within less than a certain distance. The equilibrium is attained when attraction and repulsion forces balance each other. Obviously, now the ions cannot approach each other any more. The geometrical structure of NaCl is simple cubic, with ions of sodium (Na) and chlorine (Cl) arranged alternatively at the corners of the cube (Fig. 16.26). In terms of the Bravais lattice, NaCl is a *face centred cube*. The basis consists of Na atom and one chlorine (Cl) atom separated by one-half the body diagonal of the cube. One finds that there are four molecules of NaCl in each unit cube, with ions occupying the following positions.

$$\text{Na}: 0, 0, 0 \qquad \frac{1}{2}, \frac{1}{2}, 0 \qquad \frac{1}{2}, 0, \frac{1}{2} \qquad 0, \frac{1}{2}, \frac{1}{2}$$

$$\text{Cl}: \frac{1}{2}, \frac{1}{2}, \frac{1}{2} \qquad 0, 0, \frac{1}{2} \qquad 0, \frac{1}{2}, 0 \qquad \frac{1}{2}, 0, 0$$

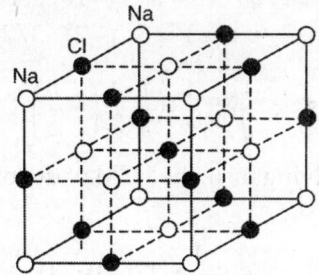

Fig. 16.26: Sodium chloride crystal structure

In NaCl crystal, each ion is surrounded by six nearest neighbours of the positive nature, i.e. coordination number is 6 and 12 next nearest neighbours of the same kind as the reference ion.

LiH, KBr, KCl, MgO, AgBr, BaO, etc. are other crystals which also have NaCl type structure.

Diamond Structure

Figure 16.27 shows the unit cell of diamond structure, which is a *fcc* structure with basis of two atoms, one located at $(0, 0, 0)$ and other at $(a/4, a/4, a/4)$ associated with each lattice point. We note that diamond structure is a combination of two interpenetrating face centred cubic sublattices. One sublattice has its origin at $(0, 0, 0)$ and the other at $(a/4, a/4, a\,4)$, i.e. one quarter way along the body diagonal. We find that each atom in the diamond structure has **four** nearest neighbours, occupying the corner point of a regular tetrahedron, to which it is bonded by strong covalent bonds. In diamond structure each atom has only 4 nearest neighbours and hence it is loosely packed structure. Other examples of having this type of structure are carbon, silicon, germanium, etc.

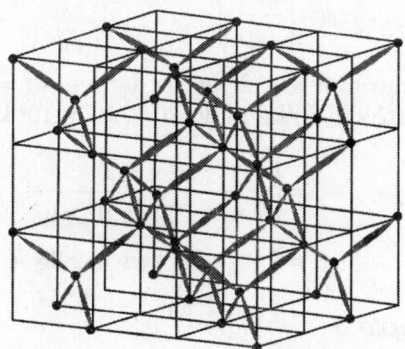

Fig. 16.27: Lattice structure of diamond

ILLUSTRATIVE EXAMPLES

Example 1

Calculate the lattice constant of rocksalt (NaCl) crystal. Given mol. wt. of NaCl = 58.45, $\rho = 2170 \text{ kg/m}^3$ and Avogadro number $N = 6 \times 10^{26}$ per kg mole.

Solution

NaCl belongs to *fcc* lattice and hence there are 4 molecules per unit cell. Mol. wt. of NaCl is 58.45. Hence, the mass in each unit cell is

$$= n\frac{M}{N} = 4 \times \frac{58.45}{6 \times 10^{26}} \text{ kg}$$

Let *a* be the lattice constant, then mass in each unit cell

$$= a^3 \times 2170 \text{ kg}$$

$$\therefore \quad a^3 \times 2170 = \frac{4 \times 58.45}{6 \times 10^{26}}$$

$$\therefore \quad a = \left(\frac{4 \times 58.43}{6 \times 10^{26} \times 2170}\right)^{1/3}$$

$$= 5.643 \times 10^{-10} \text{ m} = 5.643 \text{ Å}.$$

Example 2

The density of α-iron is 7.87×10^3 kg/m³ and its atomic weight is 55.8. If α-iron crystallizes in *bcc* space lattice, find the value of lattice constant *a* [$N = 6.02 \times 10^{26}$ per kg mole].

Solution

$$a = \left(\frac{nM}{N\rho}\right)^{1/3}$$

Given that $n = 2$ for *bcc* lattice, $M = 55.8$, $N = 6.02 \times 10^{26}$/kg mole, $\rho = 7.87 \times 10^3$ kg/m³

$$\therefore \quad a = \left(\frac{2 \times 55.8}{7.87 \times 10^3 \times 6.02 \times 10^{26}}\right)^{1/3}$$

$$= 2.86 \times 10^{-10} \text{ m} = 2.86 \text{ Å}.$$

Example 3

A substance with *fcc* lattice has density 6250 kg/m³ and molecular weight 60.2. Calculate the lattice constant *a*. Given Avogadro number is 6.02×10^{26}/kg mole.

Solution

$$a = \left(\frac{nM}{N\rho}\right)^{1/3}$$

Given that $n = 4$ for *fcc* lattice, $M = 60.2$, $N = 6.02 \times 10^{26}$/kg mole, $\rho = 6250$ kg/10³

$$\therefore \quad a = \left[\frac{4 \times 60.2}{6.02 \times 10^{26} \times 6250}\right]^{1/3}$$

$$= 4 \times 10^{-10} \text{ m} = 4 \text{ Å}.$$

Example 4

In a crystal, a lattice plane cuts intercepts of $1a$, $2b$ and $3c$ along the three axes where *a*, *b* and *c* are primitive vectors of the unit cell. Determine the Miller indices of the given plane.

Solution

Since the given plane cuts intercepts pa, qb, rc along the three axes, we have

$$pa : qb : rc = a : 2b : 3c$$

Here *a*, *b* and *c* are primitive vectors of the unit cell

$$\therefore \quad p : q : r = 1 : 2 : 3$$

$$\therefore \quad \frac{1}{p} : \frac{1}{q} : \frac{1}{r} = \frac{1}{1} : \frac{1}{2} : \frac{1}{3} = \frac{6 : 3 : 2}{6}$$

$$= 6 : 3 : 2$$

Obviously, the Miller indices of the plane are $(3 : 3 : 1)$.

Example 5

In an orthorhombic crystal, a lattice plane cuts intercepts of $3a$, $-2b$ and $3c/2$ along three axes. Deduce the Miller indices of the plane, where *a*, *b*, *c* are primitive vectors of the unit cell.

Solution

We have $\quad pa : qb : rc = 3a : (-2b) : 3c/2$

$$\therefore \quad p : q : r = 3 : (-2) : 3/2$$

or $\quad \dfrac{1}{p} : \dfrac{1}{q} : \dfrac{1}{r} = \dfrac{1}{3} : \left(-\dfrac{1}{2}\right) : \dfrac{2}{3}$

Multiplying by 6, the LCM of denominators, one obtains

$$\frac{1}{p} : \frac{1}{q} : \frac{1}{r} = 2 : (-3) : (4)$$

Obviously, the Miller indices of the plane are (2, 3, 4).

Example 6

In a simple cubic crystal , find the ratio of intercepts on the three axes by (1, 2, 3) plane.

Solution

Let the plane cut intercepts of lengths l_1, l_2, l_3 on the three crystal axes, then

$$l_1 : l_2 : l_3 = pa : qb : rc \qquad (1)$$

Here a, b and c are the primitive vectors of the unit cell and numbers p, q and r are related to the Miller indices (h, k, l) of the plane as

$$\frac{1}{p} : \frac{1}{q} : \frac{1}{r} = h : k : l$$

Since the given crystal is simple cubic, and hence

$$a = b = c$$

and $$h = 1, k = 2, l = 3$$

$$\therefore \quad p : q : r = \frac{1}{h} : \frac{1}{k} : \frac{1}{l} = \frac{1}{1} : \frac{1}{2} : \frac{1}{3}$$

Multiplying RHS by the least common multiple of denominators, one obtains

$$p : q : r = 6 : 3 : 2$$

∴ Equation (1) gives

$$l_1 : l_2 : l_3 = 6a : 3a : 2a = 6 : 3 : 2$$

$$(\because \; a = b = c)$$

Example 7

A plane of a crystal makes intercepts 1.49, 2.58 and 1.25 mm along the crystallographic axes. If the primitive a, b, c are 3.01, 5.12 and 4.8 Å, find the Miller indices of the plane.

Solution

$$p : q : r = \frac{1.49}{3.01} : \frac{2.58}{5.12} : \frac{1.25}{4.8}$$

$$= \frac{1}{2} : \frac{1}{2} : \frac{1}{3}$$

$$\therefore \; p^{-1} : q^{-1} : r^{-1} = 2 : 2 : 3$$

Obviously, the Miller indices are (2, 2, 3).

Example 8

In a simple cubic crystal, find the ratio of **(i)** intercepts on the three axes by (1 2 3) plane **(ii)** the spacings of (1 1 0) and (1 1 1) planes.

Solution

i. Let pa, qb, rc be the intercepts on the three axes respectively. For a simple cubic lattice $a = b = c$.

$$\therefore \qquad pa : qb : rc = pa : qa : ra = p : q : r$$

Given $$h = 1, k = 2, l = 1$$

But $$\frac{1}{p} : \frac{1}{q} : \frac{1}{r} = h : k : l = 1 : 2 : 3$$

$$\therefore \qquad p : q : r = 1 : 1/2 : 1/3 = 6 : 3 : 2$$

Obviously, the intercepts on the three axes by (1 2 3) planes are in the ratio of $6 : 3 : 2$

ii. The spacing between $(h\ k\ l)$ planes in a simple cubic lattice is given by

$$d_{hlk} = a/\sqrt{h^2 + k^2 + l^2}$$

$$d_{110} = a/\sqrt{1^2 + 1^2 + 0^2} = a/\sqrt{2}$$

$$d_{111} = a/\sqrt{1^2 + 1^2 + 1^2} = a/\sqrt{3}$$

$$\therefore \quad d_{110} : d_{111} = a/\sqrt{2} : a/\sqrt{3} = \sqrt{3} : \sqrt{2}.$$

Example 9

In a tetragonal lattice $a = b = 2.5$ Å, $c = 1.8$ Å. Deduce lattice spacing between (1 1 1) planes.

Solution

The lattice spacing

$$d_{hkl} = \left[\frac{h^2}{a^2} + \frac{k^2}{b^2} + \frac{l^2}{c^2} \right]^{-1/2}$$

Here $h = 1, k = 1, l = 1, a = b = 2.5$ Å, $c = 1.8$ Å

$$d_{111} = \left[\frac{1}{(2.5)^2} + \frac{1}{(2.5)^2} + \frac{1}{(1.8)^2} \right]^{-1/2}$$

$$= 1.26 \text{ Å}.$$

Example 10

Calculate the interplanar spacing for a (3 2 1) plane in a simple cubic lattice, where lattice constant is 4.2×10^{-10} m.

Solution

$$d_{hkl} = 1 \bigg/ \sqrt{\frac{h^2}{a^2} + \frac{k^2}{b^2} + \frac{l^2}{c^2}}$$

For a simple cubic lattice $a = b = c$

$\therefore \qquad d_{hkl} = 1 \big/ \sqrt{h^2 + k^2 + l^2}$

Given that $h = 3, k = 2, l = 1, a = 4.2 \times 10^{-10}$ m

$\therefore \qquad d_{3\,2\,1} = \dfrac{4.2 \times 10^{-10}}{\sqrt{3^2 + 2^2 + 1^2}} = \dfrac{4.2}{\sqrt{14}}$ Å

$\qquad\qquad = 1.123$ Å.

REVIEW QUESTIONS

1. Explain the concept of lattice, basis and crystal structure. How are they related?

2. What is a crystal lattice? With the help of an example distinguish between crystal lattice and crystal structure.

3. What is a space lattice? Describe briefly the seven systems of crystals. Mention and explain with examples the types of lattice in a cubic system. [BE]

4. Explain lattice plane. How can you describe the lattice planes in terms of Miller indices? With the help of the neat diagram in *sc* unit cell, show planes (2 1 1) and (1 1 1).

5. What is coordination number? Calculate the coordination number for simple cubic, *bcc* and *fcc* lattices.

6. What is Bravais lattice? What is the maximum number of Bravais lattices possible? How will you account for the existence of thousands of structures from these lattices? [AIME]

7. Draw neat sketches of unit cells of *s c*, *bcc* and *fcc* crystal structures. Calculate the number of atoms in each case.

8. Find the numbers of atoms per unit cell in *fcc* structure. [AMIE]

9. Explain packing factor or packing density. Show that the packing factor for simple lattice, body centred lattice and face centred lattice are $\pi/6$, $\sqrt{3}\pi/8$ and $\sqrt{2}\pi/6$ respectively. [BE]

10. Define atomic radius. Calculate the atomic radii in case of *s c*, *fcc* and *bcc* lattices. [Diploma]

11. What are Miller indices of a crystal plane? Show that the spacing between consecutive planes defined by Miller indices (*h k l*) is given by

$$d_{hkl} = \left[\frac{h^2}{a^2} + \frac{k^2}{b^2} + \frac{l^2}{c^2} \right]^{-1/2}$$

where *a*, *b* and *c* are primitive vectors along three mutually perpendicular axes.

12. Explain Bravais lattices and Miller indices in crystallographic notations. Show analytically that the (1 1 1) plane are perpendicular to [1 1 1] direction in a simple cubic crystal.

PROBLEMS

1. A potassium chloride crystal has a density of 1.98×10^3 kg/m^3. The molecular weight of KCl is 74.55. Show that the distance between the adjacent atoms is 3.14×10^{-10} m. [BE]

2. Draw (1 1 0) and (1 1 1) planes and (1 1 0) and (1 1 1) directions in a *sc* crystal.

3. Draw the planes (0 2 0), (1 2 0) and (2 2 0) in a *fcc* structure.

4. Determine the Miller indices of a set of a parallel planes which makes intercepts in the ratio $3a : 4b$ on X and Y axes and are parallel to Z-axis. a, b and c are primitive vectors of the lattice. [BE]

5. The lattice constant for a simple cubic lattice is 4.2×10^{-10} m. Calculate the interplaner spacing for (3 2 1) plane . [*Ans.* 1.123 Å]

6. In a crystal whose primitives are 1.2 Å, 1.8 Å and 2 Å, a plane (2 3 1) cuts an intercept 1.2 Å along X-axis. Show that the length of the intercepts along Y and Z axes are 1.2 Å and 4 Å respectively. [Diploma]

7. Lead is a face centred cubic with an atomic radius of 1.746 Å. Show that the spacing of (i) (2 0 0) planes (ii) (2 2 0) planes are 2.465 Å and 1.74 Å respectively. [AMIE]

8. In a tetragonal lattice $a = b = 2.5$ Å. Show that the lattice spacing between (1 1 1) planes is 1.26 Å.

SHORT ANSWER QUESTIONS

1. How are various crystal structures specified?

Ans. The various crystal structures are specified in terms of parallelopipe unit cells, which are characterized by geometry and atom positions within.

2. What is a space lattice?

Ans. A space lattice is an infinite array of points, all with identical surroundings.

3. How a crystal structure is obtained?

Ans. A crystal structure is obtained by combining a space lattice with a basis. The basis must give the number of atoms per lattice point, their types, mutual orientations and distance of orientation.

4. How may types of space lattices are there?

Ans. Fourteen, however, crystal structure run into thousands.

5. What are the three relatively simple crystal structures in which most common metals exist?

Ans. fcc, bcc and *hcp.*

6. Mention two features of crystal structure.

Ans. (i) Coordination number or number of nearest neighbour atoms and (ii) atomic packing (the fraction of solid sphere volume in the unit cell). Coordination number and atomic packing fraction are the same for *fcc* and *hcp* crystal structures, each of which may be generated by the stacking of closed packing planes of atoms.

7. What is difference between crystalline and non-crystalline solids?

Ans. Atoms in crystalline solids are positioned in an orderly and repeated pattern whereas in non-crystalline or amorphous materials there is random and disordered atomic distribution.

8. How crystal directions and crystal planes are denoted?

Ans. These are denoted by Miller indices. A family of crystal directions or planes includes all possible combinations of the indices, both positive and negative. For a given crystal structure, planes having identical atomic packing yet different Miller indices belong to the same family.

9. What are single crystals?

Ans. Single crystals are materials in which the atomic order extends uninterrupted over the entire specimen; under some circumstances, they may have flat faces and regular geometric shapes. We may note that vast majority of crystalline solids, however, are polycrystalline, being composed of many small crystals or grains having different crystallographic orientations.

10. How the space lattice and its dimensions can be determined from the powder method for X-ray diffraction?

Ans. From the position of lines in a powder pattern and from the extinction rules for different cubic crystals, the space lattice and its dimension can be determined.

11. What do you understand by crystal imperfections?

Ans. All solid materials contain large number of imperfections or deviations from crystalline perfection, and several types of them may be classified according to their geometry, and size.

12. What are point defects?

Ans. Points defects are those associated with one of two atomic positions, including vacancies (or vacant lattice sites), self-interstitials (host atoms that occupy interstitial sites), and impurity atoms. They are Frenkel and Schottky imperfections in ionic crystals.

13. What are the edge and screw dislocations?

Ans. These are the limiting types of line imperfections. One can resolve any general dislocation into edge and screw components.

OBJECTIVE QUESTIONS

1. The number of atoms present in the unit cell of *hcp* structure is _____. [6]

2. The minimum number of ions in the unit cell of an ionic crystal with *fcc* space lattice is _____. [8]

3. It *r* be the radius of the atom in a crystal crystallizing in the simple cubic structure, then the nearest neighbour distance is _____. [2*r*]

4. The nearest neighbour distance in *bcc* structure is _____. [$a\sqrt{3}/2$]

5. The unit cell with three lattice parameters is _____. [triclinic]

6. Aluminium crystallizes in _____ structure. [*fcc*]

7. The atomic diameter of an *fcc* crystal (lattice parameter *a*) is _____. [$a\sqrt{2}/2$]

8. The order of coordination number in *fcc, fcc* and *hcp* cells is _____. [8, 12, 12]

MULTIPLE CHOICE QUESTIONS

1. The number of lattice points in a primitive cell are
   (a) 3/2　　　　　(b) 1/2
   (c) 2　　　　　(d) 1

2. The number of lattice point in a rhombohedral unit cell is
   (a) 1　　　　　(b) 2
   (c) 4　　　　　(d) 16

3. The number of atoms present in the unit cell of *hcp* structure is
   (a) 2　　　　　(b) 4
   (c) 6　　　　　(d) 12

4. The minimum number of ions in the unit cell of an ionic crystal with *fcc* space lattice is
   (a) 16　　　　　(b) 12
   (c) 8　　　　　(d) 2

5. If *r* be the radius of the atom in a crystal crystallizing in the simple cubic structure, then the nearest neighbour distance is
   (a) 8*r*　　　　　(b) 4*r*
   (c) 3*r*　　　　　(d) *r*

6. The space lattice with two parameters does not belong to the crystal system
   (a) triclinic　　　　　(b) rhombohedral
   (c) hexagonal　　　　　(d) tetragonal

7. The number of diad axis of symmetry elements that are present in a cubic crystal are
   (a) 6　　　　　(b) 3
   (c) 2　　　　　(d) 1

8. The nearest neighbour distance in *bcc* structure is
   (a) $a/\sqrt{2}$　　　　　(b) $2a/\sqrt{3}$
   (c) $a\sqrt{3}/2$　　　　　(d) $2a/\sqrt{3}$

9. The unit cell with three lattice parameter is
   (a) monoclinic　　　　　(b) tetragonal
   (c) triclinic　　　　　(d) orthorhombic

10. Which one of the following metals crystallizes in *fcc* structure
    (a) zinc　　　　　(b) sodium
    (c) aluminium　　　　　(d) cesium chloride

11. The atomic diameter of an *fcc* crystal (lattice parameter *a*) is
    (a) $\dfrac{a}{2}$　　　　　(b) $a\sqrt{2}$
    (c) $\dfrac{a\sqrt{2}}{2}$　　　　　(d) $\dfrac{a}{2\sqrt{2}}$

12. The number of Bravais lattices with two lattice points are
    (a) 5　　　　　(b) 6
    (c) 3　　　　　(d) 1

13. The packing factor of diamond cubic crystal structure is
    (a) 90%　　　　　(b) 45%
    (c) 34%　　　　　(d) none of these

14. The unit cell has $a = 5\,\text{Å}, b = 8\,\text{Å}, c = 3\,\text{Å}, \alpha = 90°$, $\beta = 65°$ and $\gamma = 54°$. The space lattice for this unit cell is
    (a) monoclinic　　　　　(b) rhombohedral
    (c) triclinic　　　　　(d) orthorhombic

15. The faces in a tetragon are
    (a) 24　　　　　(b) 12
    (c) 6　　　　　(d) 2

16. Magnesium crystallizes in *hcp* structure. If the lattice constant is 0.32 nm, the nearest neighbour distance in magnesium is
    (a) 0.64 nm　　　　　(b) 0.32 nm
    (c) 0.16 nm　　　　　(d) 0.8 nm

17. Zinc crystallizes in *hcp* structure. If *r* be the radius of zinc atom, the height of the unit cell is
    (a) $r\sqrt{3/8}$　　　　　(b) $2r\sqrt{8/3}$
    (c) $r\sqrt{\dfrac{8}{3}}$　　　　　(d) *r*

18. The lattice constant of a *bcc* unit cell with atomic radius of 1.24 Å is
    (a) 1.432 Å　　　　　(b) 2.864 Å
    (c) 1.754 Å　　　　　(d) 1.432 Å

Answers

1. (d)	2. (a)	3. (c)	4. (c)	5. (c)
6. (a)	7. (a)	8. (c)	9. (c)	10. (c)
11. (c)	12. (a)	13. (d)	14. (c)	15. (b)
16. (b)	17. (b)	18. (b)		

17

Bonding in Solids (Crystal Bonding)

17.1 INTRODUCTION

A solid is defined as a piece of matter in which the atoms and the molecules comprising it are very closely packed. The materials in solid form exhibit quite a lot of interesting properties. It is clear that the reason for the close packing of the atoms and molecules in a solid is the existence of some types of the binding forces between them. The natural question then arises, what is the origin of these binding forces which bind the different particles in different ways, giving rise to different properties in different solids. Thus, the most convenient basis for classification of crystals is the character of the interatomic binding force(s) in various types of crystalline materials. In 1940, H Seitz classified solids into five types according to the bonding of atoms. The theme adopted by Seitz is generally accepted for classification. We shall also use this classification and list the solid types starting from the solids with largest binding energy to the most weakly bound.

The distinction in the said five categories is not sharp one, because some materials may belong to more than one class. Keeping this in mind, we discuss the various categories of crystals, in turn, in this chapter and illustrate how binding occurs in each category.

17.2 BONDING IN CRYSTALS

From the very existence of solids, we may draw two general conclusions: (i) there must act an attractive force between the atoms or molecules in a solid which keeps them together (ii) there must be repulsive forces acting between the atoms as well, since large

Table 17.1: Bonds in crystals

Solids	Type of bond	Formation	Binding energy (eV/atom)	Typical examples
Covalent	Covalent— homopolar or pair bonds	Electron shared between two atoms	2–6	Carbon (diamond), germanium, silicon, etc.
Ionic	Ionic	Electron transfer and coulomb interaction between cations and anions	0–2	Alkali halides
Metallic	Metallic	Freely moving, electrons in an array of positive ions	1–5	All metals and alloys
Molecular (Van der Waals)	Molecules	Weak attractive forces due to dipole interaction between pairs	0.002–0.1	Noble gases
Hydrogen bond	Hydrogen atom attracted between two other atoms	Electrostatic bond of H atom with an electronegative atom	0.5	Ice, organic compounds, biological materials

external pressures are required to compress a solid to appreciable extent. In order to understand the importance of these two types of forces, let us consider the simplest system, e.g. simple pair of atoms A and B which form a stable chemical compound. Let us assume that the potential energy of atom B due to the presence of atom A is given by

$$u(r) = -\frac{\alpha}{r^n} + \frac{\beta}{r^m} \qquad (17.1)$$

Here r is the distance between the nuclei of the two atoms, α, β, m and n are constant characteristics for the molecule AB. The zero of the energy is chosen such that for the finite separation, $u = 0$. In Eq. (17.1), the first term $\frac{\alpha}{r^n}$ which is negative, corresponds to the energy associated with the forces of attraction. The second term, $\frac{\beta}{r^m}$ which is positive, corresponds to the forces of repulsion. The force between the two atoms A and B as a function of r is given by

$$F(r) = \frac{du}{dr} = \frac{\alpha n}{r^{n+1}} - \frac{\beta m}{r^{m+1}} = 0 \qquad (17.2)$$

The energy (U) and the force (F) between two atoms A and B which form a chemical compound are plotted in Fig. 17.1.

The stable configuration of the system corresponds to the minimum in $u(r)$ curve which occurs for the particular separation $r = r_0$. The corresponding energy $u(r_0)$ is called the binding energy and is negative, i.e.

B.E. $= -u(r_0)$. This means that the energy in magnitude equals to $u(r_0)$ is required to dissociate the molecule, i.e. to separate the two atoms. As such this energy is called the **dissociation energy**. Dissociation occurs at high temperatures or as a result of other processes in which the molecules can absorb sufficient energy. The dissociation energies are of the order of 1 eV to few eV.

At the equilibrium separation $r = r_0$, the potential energy must be minimum or the first derivative of Eq. (17.1) must vanish, i.e.

$$\left.\frac{du}{dr}\right|_{r=r_0} = 0 = \frac{n\alpha}{r^{n+1}} - \frac{m\beta}{r^{m+1}}$$

From this it follows that

$$\beta = \alpha \frac{n}{m} [r_0]^{\{(m+1)-(n+1)\}}$$

or $[r_0]^{m-n} = \left(\frac{\beta m}{\alpha n}\right) \qquad (17.3)$

This means that at $r = r_0$ the attractive and repulsive forces balance each other, i.e. $F(r_0) = 0$. Thus, equilibrium state energy is

$$u(r_0) = \frac{\alpha}{r_0^n} + \frac{\beta}{r_0^m} = -\frac{\alpha}{r_0^n}\left[1 - \frac{n}{m}\right] \qquad (17.4)$$

It must be noted that though the attractive and repulsive forces are equal, the attractive and repulsive energies are not equal since $n \neq m$. If $m \gg n$, the total binding energy is essentially the energy of attraction and is determined by $-\frac{\alpha}{r^n}$.

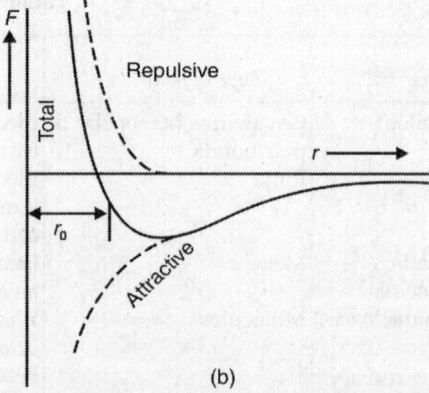

(a)   (b)

Fig. 17.1: Interatomic forces and potential energy for a system of two atoms as a function of their separation r. (a) Potential energy (b) Interatomic force

From Fig. 17.1, it is evident that a minimum in the energy curve is a must for the stability of the curve. The minimum in the energy curve is possible only if $m >> n$ and may be shown by employing the condition

$$\left.\frac{\partial^2 u}{\partial r^2}\right|_{r=r_0} > 0$$

This leads to

$$\frac{n(n+1)}{r_0^{n+2}}\alpha + \frac{m(m+1)}{r_0^{m+2}}\beta > 0$$

and substituting the value of r_0 from Eq. (17.3), we have $m > n$.

Thus, we conclude that the formation of a chemical bond requires that the repulsive forces be of short range rather than attractive ones.

Although the energy cannot, in general, be represented accurately by a relation of the type Eq. (17.1), but the above treatment provides some useful qualitative conclusions about the bonding of atoms in the solids.

Stability of the Solid

Figure 17.1 can be used to explain the stability of the solid. At large separations, the atoms do not interact with each other so that $u = 0$. As the atoms approach each other, they experience attractive forces due to each other, primarily because of the attraction of positive and negative charges in the atoms. As the interatomic separation decreases to the order of one or two atomic diameters, the attractive forces start to become appreciable, and also the repulsive forces come into play. The repulsive forces arise as a consequence of the Pauli's exclusion principle and has the effect of increasing the potential energy. Since this increase is not consistent with the stability of the system, the further reduction in the interatomic separation is checked. At still smaller distances the energy is dominantly repulsive. It is, thus, now quite clear that because of the repulsive forces the atoms cannot reduce their interatomic separation in an arbitrary fashion. This result explains the importance of the repulsive forces. In absence of the repulsive forces, the atoms would ultimately interpenetrate into each other because of the increasing attractive forces and would occupy the same site. The equilibrium is then established only when a compromise between the attractive and repulsive forces is reached so that F is equal to zero. It is clear from Fig. 17.1 that this occurs at $r = r_0$ and this situation corresponds to a minimum in the energy curve at this point.

17.3 IONIC BOND

Certain neutral atoms tend to loose their valence electron relatively easily because of the smallness of the ionisation potentials, and thus, acquire a net positive charge, then they are called **ions**. These atoms are called **electropositive** and their ions are known as **cations**. Most of these atoms form **metallic crystals**. For example, Na atom with the electronic configuration $1s^2 2s^2 2p^6 3s^1$ has the first ionisation potential of 5.13 eV. On removal of $3s^1$ electron, the Na^+ cation assumes the inert gas configuration $1s^2 2s^2 2p^6$ and ion becomes stable with essentially spherical distribution of electron charge (due to 10 electrons) around the sodium nucleus (containing 11 protons and 12 neutrons) with 11 units of positive charge. The net charge on the ion is then +1 in units of electronic charge.

Atoms having nonmetallic properties like O, N, F, Cl, etc. are known as electronegative atoms as they have a tendency to accept and bind additional electrons and thereby form negatively charged ions. These ions are known as anions. For example, Cl with $Z = 17$ has the electronic configuration $1s^2 2s^2 2p^6 3s^2 3p^5$ and can form an inert gas (argon structure) just by accepting one extra electron and thereby forming the ion Cl^- with a net negative charge of -1.

The positive (Na^+) and negative (Cl^-) ions then attract electrostatically and are bound together by an ionic bond. In addition to this attractive force, there is also a repulsive force, which becomes operative when the two negatively charged electron clouds of the ions try to overlap; when the two ions try to come near to each other, the molecule is at the equilibrium distance when the two opposing forces balance each other. The binding or cohesive energy is of the order of 5 to 10 eV. The charge distribution of each ion is spherically symmetric. Consequently, ionic solids crystallize in closed packed structures of which the NaCl and CsCl structures (Fig. 17.2) are the commonest having cubic structures. The melting points are quite high of the order of 800°C for NaCl.

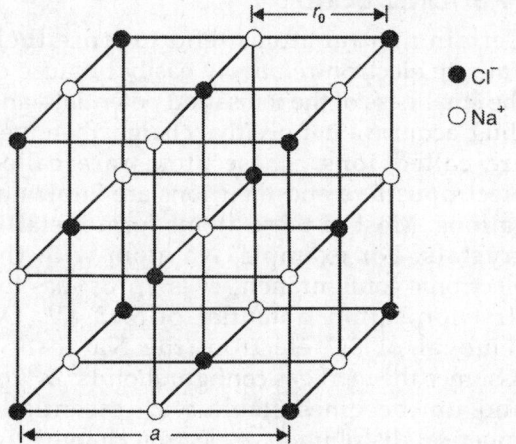

Fig. 17.2: NaCl crystal structure

The ionic bond is strong when compared with other bonds, a typical value for the binding energy of a pair of atoms being about 5 eV. This strength is attributed due to the strength of the coulomb force responsible for the bonding. Experimentally, this strength is characterised by the high melting temperatures associated with ionic crystals. Thus, the melting temperature for the ionic crystal NaCl is 801°C while the melting temperatures for Na and K metals are 97.8°C and 63°C respectively.

They are poor conductors of electricity at normal temperatures because the valence electrons are bound quite tightly to the ionic nuclei. At elevated temperatures, however, the ions themselves become mobile and ionic conductivity results.

The positive and negative ions have the same electrical structure as the nearest inert gas atoms, except that they are electrically charged. In case of NaCl, Na^+ have the neon electron structure and Cl^- have the argon electron structure. The charge on the ion is, therefore, spherically distributed (with some distortion near the region of contact with neighbouring atoms of course) and an ion is thus like a small charged sphere and the packing of the ions is closest one. Because ions having like charges repel each other, a stable packing is obtained in which ion is surrounded with ions having an opposite charge. A periodic array results in which the environment of all similar atoms is the same and the sum of all positive and negative charges adds up to zero, so that ionic solid is electrically neutral. They usually crystallize in the relatively close packed NaCl and CsCl structures (Fig. 17.2). The structure actually depends upon the radii of the two ions.

They are usually transparent to visible light, while exhibiting a single characteristic optical reflection peak in the far infrared region of the spectrum. These crystals are often quite soluble in ionising solvents such as water. In the solutions, these compounds usually dissociate into free ions.

An ideal ionic crystal is, thus, a salt which has a simple crystal structure (NaCl was proved by Bragg to have an interpenetrating face centred cubic lattice with Na^+ and Cl^- ions alternating in the lattice sites). The sizes of the Na^+ and Cl^- ions regarded as spheres are different. Na^+ ion has diameter of 1.9 Å and that of Cl^- ion is 3.62 Å; the diameter ratio being roughly 1 : 2 as obtained by Pauling. In the NaCl structure, each anion is surrounded by 6 cations and *vice versa* and the lattice

constant is 2.76 Å $= \left(\dfrac{1.9 + 3.62}{2} \right)$ which is the

distance between ions of opposite sign. The ionic radius is an emperical concept.

Typical ionic crystals are the halides, alkali metal oxides and sulphides, alkaline earth halides, metal oxides halides and sulphides, etc. Examples are NaCl, CaF_2, MgO, AgCl, ZnS, Al_2O_3, TiO_2, etc.

Potential Energy Diagram of the Ionic Molecule

Because of the considerable overlap between the electronic shells of the ionic crystal, the interatomic binding forces arise from two opposing contributions, first there is the coulomb attraction interatomic potential energy of the form

$$u_{att} = \frac{q^2}{4\pi\varepsilon_0 r} \qquad (17.5)$$

where q is the charge of an ion and r is the distance of separation between them. Second is the interaction due to the overlap of the electronic shells of neighbouring atoms. In general, the repulsive energy is of the form

$$u_{rep} = \frac{\beta}{r^m} \qquad (17.6)$$

More exact form of the repulsive energy in the ionic crystal is given by Born-Mayer as

$$u_{rep} = A \exp\left(-\frac{r}{b}\right) \qquad (17.7)$$

where A and b are empirical constants. The resulting interaction energy of two ions is

$$u = A \exp\left(-\frac{r}{b}\right) - \frac{q^2}{4\pi\varepsilon_0 r} \qquad (17.8)$$

This energy is shown in Fig. 17.3.

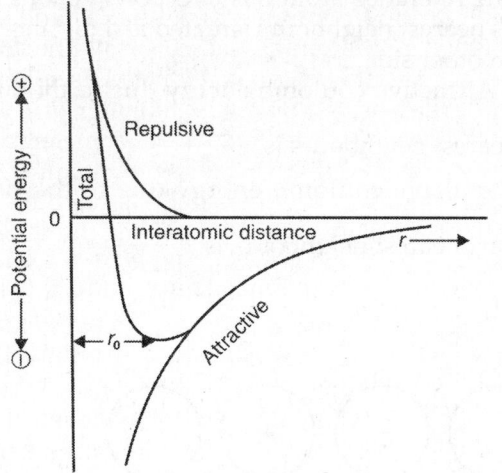

Fig. 17.3: Variation of potential energy with separation r.

17.4 BINDING ENERGY OF IONIC CRYSTAL

The most common crystal structure found for ionic crystal is NaCl. We shall consider it for the calculation of binding energy of ionic crystal. Let u_{ij} be the interaction energy between ith and jth ions of the crystal. We define a sum u_i which represents the total energy of interaction between ion i and all other ions of the crystal, i.e.

$$u_i = \sum_{g} u_{ij} \qquad (17.9)$$

where the summation includes all ions except $j = i$ because the binding between ions i and j itself is not possible. The interaction term u_{ij} in Eq. (17.9) may be written as the sum of two terms (i) coulomb interaction energy $\pm \dfrac{Ke^2}{r_{ij}}$ $\left(\text{here } K = \dfrac{1}{4\pi\varepsilon_0}\right)$ where r_{ij} is the distance between the ions i and j (ii) energy resulting from a repulsive force which becomes effective only at very small interionic separation. Thus, the total energy of interaction between the two ions i and j becomes

$$u_{ij} = A \exp\left(\frac{r_{ij}}{b}\right) \pm \frac{Ke^2}{r_{ij}} \qquad (17.10)$$

where + sign is taken for the like charges and − sign for unlike charges. The repulsive term $A \exp\left(-\dfrac{r_{ij}}{b}\right)$ describes the fact that each ion resists overlapping with the electron distribution of the neighbouring ions. We consider the strength A and range b of the repulsive term in Eq. (17.10) as constants to be determined from observed values of lattice constant and compressibility. As stated earlier, we have used the exponential form of the empirical repulsive potential rather than $\dfrac{a}{r_{ij}^n}$ form. This is done because the exponential form of repulsive potential may give a better representation of the repulsive interaction. For $r_{ij} = b$ the repulsive interaction reduces to e^{-1} of the value at $r = 0$.

Substituting Eq. (17.10) in Eq. (17.9), one obtains

$$u_i = \sum_j u_{ij} = \sum_j \left[A \exp\left(-\frac{r_{ij}}{b}\right) \pm \frac{Ke^2}{r_{ij}} \right] \quad (17.11)$$

In the NaCl structure, the value of u_i does not depend on whether the reference ion i is a positive or negative ion as long as it is not near the surface. Neglecting surface effects, one may write the total lattice energy of a crystal composed of N molecules or $2N$ ions as

$$U = Nu_i \quad (17.12)$$

Here N rather than $2N$ occurs in Eq. (17.12) as one must count each pair of interaction only once, i.e. each bond once. The total lattice energy U given by Eq. (17.12) is the energy required to separate the crystal into individual ions at an infinite distance apart.

For convenience, we introduce the dimensionless quantity P_{ij} such that

$$r_{ij} = P_{ij} R \quad (17.13)$$

Here R is the nearest neighbours separation in the crystal. Equation (17.11) then becomes

$$u_i = \sum_j \left[A \exp\left(-\frac{P_{ij}R}{b}\right) \pm \frac{Ke^2}{P_{ij}R} \right] \quad (17.14)$$

If we include the repulsive interaction only among nearest neighbours, we have

$$u_i = \begin{cases} A \exp\left(-\dfrac{P_{ij}R}{b}\right) - \dfrac{Ke^2}{R} & \text{(nearest neighbours)} \\[3mm] \pm \dfrac{1}{P_{ij}} \dfrac{Ke^2}{R} & \text{(otherwise)} \end{cases}$$

Thus,

$$U = \sum_j u_{ij} = Nu_i = N\left(ZAe^{\frac{-R}{b}} - \alpha \frac{Ke^2}{R} \right) \quad (17.15)$$

where Z is the number of nearest neighbours of any ion and

$$U = \sum_j \pm \frac{1}{P_{ij}} \quad (17.16)$$

is called **Madelung constant**. The sum should include the nearest neighbours contribution which is given by Z. In the theory of ionic crystals, the value of Madelung constant is of prime importance. Clearly the value of Madelung constant depends on the lattice structure. Madelung constant is really nothing but a correction factor that tells how much of an error will be there if only nearest neighbours are considered.

Madelung Constant for a Linear Chain Ionic Solid

We have said that the value of the Madelung constant depends on the crystal structure and for the sake of simplicity we will first calculate the constant for a linear ionic solid. In such a linear chain, the ions of alternate sign are arranged along an infinite line as shown in Fig. 17.4a. Let us consider a negative ion as reference ion and let R be interatomic spacing. The reference atom has two positive ions as its nearest neighbours situated at a distance R on other side.

Attractive coulomb energy due to the two nearest neighbours is $-\dfrac{2Ke^2}{R}$.

Repulsive coulomb energy due to the two next nearest neighbours is $+\dfrac{2Ke^2}{2R}$.

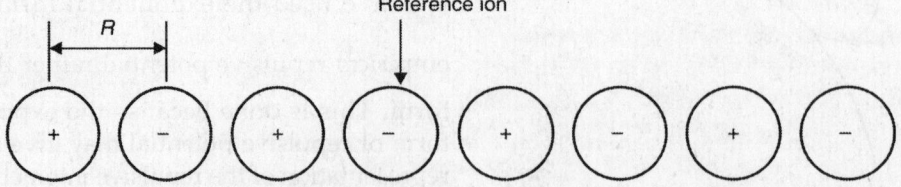

Fig. 17.4a: Madelung constant for a linear chain ionic solid

Attractive coulomb energy due to the two next nearest neighbours is $-\dfrac{2Ke^2}{3R}$.

Thus, the total coulomb energy due to all the ions in the linear chain ionic solid is

$$U = -\frac{2Ke^2}{R} + \frac{2Ke^2}{2R} - \frac{2Ke^2}{3R} + \dots$$

$$= -\frac{2Ke^2}{R}\left[1 - \frac{1}{2} + \frac{1}{3} - \frac{1}{4} + \dots\right] \qquad (17.17)$$

and comparison with Eq. (17.16) gives

$$\alpha = 2\left[1 - \frac{1}{2} + \frac{1}{3} - \frac{1}{4} + \dots\right] = 2\log_e 2 \qquad (17.18)$$

We see that the total coulomb energy of a linear chain ionic lattice is directly proportional to $\dfrac{Ke^2}{R}$. The constant of proportionality is the Madelung constant per molecule for the ionic solid. If n is the number of molecules per unit volume of the solid, the Madelung constant for one dimensional solid comes out to $2n \log 2$.

17.5 DETERMINATION OF THE REPULSIVE EXPONENT

As already discussed, the repulsive forces become noticeable when the electron shells of neighbouring ions begin to overlap, and they increase strongly in this region with decreasing values of r. These forces can best be studied on the basis of wave mechanics because they are nonclassical in nature. Born in his early work made the simple assumption that the repulsive energy between two ions as function of their separation could be expressed by a power law of the type $\dfrac{B'}{r^m}$,

where B' and m are constant characteristics of the ions in solid under consideration. Here m is called the repulsive exponent. Thus, we may write for the repulsive energy of an ion due to the presence of all other ions as

$$u_{\text{rep}} = \frac{B}{r^m} \qquad (17.19)$$

where B is related to B' by a numerical factor. In view of the fact that repulsive force depend so strongly on the distance between the particles, the repulsive energy given by Eq. (17.19) is mainly determined by the nearest neighbours of the central ion. The total energy of one ion due to the presence of all other ions is then expressed as

$$u(r) = -\frac{Ae^2}{r} + \frac{B}{r^m} \qquad (17.20)$$

Here r is interatomic distance and $A = K\alpha$.

The value of the repulsive exponent m can be obtained from the compressibility data. The compressibility (β) of a solid is given by the expression

$$\beta = -\frac{1}{V}\left(\frac{\partial V}{\partial P}\right) \qquad (17.21)$$

The first law of thermodynamics gives

$$du = \delta Q - \delta W$$

where δQ and δW are the heat flow and work done by the system respectively. If we consider experiments carried at near absolute zero, δQ can be neglected because $\delta Q = C_v dT$ and C_v is vanishingly small near absolute zero. Therefore,

$$du = -\delta W = -P\,dV$$

or $$P = -\frac{du}{dV}$$

and $$\frac{dP}{dV} = -\frac{d^2u}{dV^2} \qquad (17.22)$$

Substituting in Eq. (17.21), **one** obtains at absolute zero,

$$\frac{1}{\beta_0} = -V_0\left(\frac{\partial P}{\partial V}\right) = V_0\frac{d^2u}{dV^2} \qquad (17.23)$$

where V_0 is the volume of the crystal corresponding to an interionic distance r_0 (V corresponds to the variable r, the interatomic distance).

For NaCl structure, since r is the interionic distance and the lattice constant $a = 2r$ (Fig. 17.2), NaCl has face centred cubic structure. In a unit cell, a corner Na$^+$ ion is shared by 8 unit cells and there are 8 corners so that there is $\frac{1}{8} \times 8 = 1$ ion of Na^+. Each Na^+ ion, the face centred is shared by two unit cells and there are 6 such ions yielding $6 \times \frac{1}{2} = 3\ Na^+$ ions, thus, making a total of 4 Na^+ ions in the unit cell. There are 12 Cl^- ions in the middle of the edges and one at the body centre of the unit cell. Since each edge is shared by 4 unit cells, the number of Cl^- ions in the unit cell is $12 \times \frac{1}{4} + 1 = 4$. Hence, a unit cell of side a contains 4 molecules of NaCl. The volume per molecule is then $\frac{1}{4}a^3 = \frac{1}{4}(2r)^3 = 2r^3$. If we assume a mole of NaCl the number of molecules is N_0, i.e. equal to Avogadro number and thus,

$$V = 2N_0 r^3 \qquad (17.24)$$

$$\therefore \qquad \frac{dr}{dV} = \frac{1}{6N_0 r^2}$$

and $\left(\frac{dr}{dV}\right)^2_{r=r_0} = \frac{1}{36N_0^2 r_0^4} \qquad (17.25)$

We have

$$\frac{du}{dV} = \frac{du}{dr}\frac{dr}{dV}$$

and $\quad \frac{d^2u}{dV^2} = \frac{du}{dr}\frac{d^2r}{dV^2} + \frac{d^2u}{dr^2}\left(\frac{dr}{dV}\right)^2 \quad (17.26)$

But at equilibrium distance, we have $\frac{du}{dr} = 0$ and from Eq. (17.20), we have

$$\frac{du}{dr} = \frac{Ae^2}{r^2} - \frac{mB}{r^{m+1}} = 0 \text{ for } r = r_0$$

$$\therefore \qquad \frac{B}{r_0^m} = \frac{Ae^2}{mr_0}$$

Thus, equilibrium **cohesive** energy is

$$u(r_0) = -\frac{Ae^2}{r_0} + \frac{B}{r_0^m} = -\frac{Ae^2}{r_0} + \frac{Ae^2}{mr_0}$$

$$= \frac{Ae^2}{r_0}\left[1 - \frac{1}{m}\right] \qquad (17.27)$$

and equilibrium compressibility β_0 is given by

$$\frac{1}{\beta_0} = V_0 \frac{d^2u}{dV^2} = V_0 \left(\frac{d^2u}{dr^2}\right)_{r=r_0} \left(\frac{dr}{dV}\right)^2_{r=r_0} \quad (17.28)$$

But $\quad \frac{d^2u}{dr^2} = -\frac{2Ae^2}{r^3} + \frac{m(m+1)B}{r^{m+2}}$

$$\therefore \left(\frac{d^2u}{dr^2}\right)_{r=r_0} = -\frac{2Ae^2}{r_0^3} + \frac{m(m+1)B}{r_0^{m+2}}$$

$$= -\frac{2Ae^2}{r^3} + \frac{(m+1)Ae^2}{r_0^3} = \frac{Ae^2}{r_0^3}(m-1) \quad (17.29)$$

and substituting the values of $\left(\dfrac{d^2u}{dr^2}\right)_{r=r_0}$,

$\left(\dfrac{dr}{dV}\right)_{r=r_0}$ and V_0, we have

$$\frac{1}{\beta_0} = 2N_0 r_0^3 \frac{Ae^2}{r_0^3}(m-1)\frac{1}{36N_0^2 r_0^4}$$

$$= \frac{Ae^2(m-1)}{18N_0 r_0^4} \qquad (17.30)$$

From the experimental measurement of the bulk modulus as a function of temperature and by extrapolation to absolute zero, one obtains β_0 and hence from Eq. (17.30) the value of parameter m. For NaCl, $r_0 = 2.91$ Å and one obtains $m \approx 9$. It is found that for alkali halides m varies from 6 to 10. The equilibrium cohesive energy is then obtained using Eqs. (17.27) and (17.30).

Though there is marked variation in the value of m from one crystal to another, even then an appreciation error in m leads to a relatively small error in the lattice energy which is proportional to $\left(1 - \dfrac{1}{m}\right)$. If we change m by unity, $u(r_0)$ changes by only 1 or 2 percent. According to Eq. (17.27) and in view of the relatively large values of *m*, most of the lattice energy is due to the coulomb interaction and the repulsion contributes only a relatively small fraction. On the other hand, the repulsive and attractive forces balance each other at $r = r_0$ and also they are equal in magnitude.

17.6 COVALENT BONDING

The characteristic feature of the covalent bond is that electrons are shared rather than transferred from one atom to another. Covalent bonds are formed generally by the formation of shared electron pairs between the valence electrons in the incomplete outer shell of atoms.

The simple covalent bond occurs between atoms in the molecules H_2, CH_4, CCl_4. It is a strong bond and the bond between the two carbon atoms in the diamond has a cohesive (binding) energy of the order 7.3 eV with respect to separated neutral atoms. This is comparable with cohesive energy of ionic crystals. Because of this many authors choose not to make any sharp distinction between the ionic and covalent bonds.

Covalent crystals are quite hard but brittle, they have variable electrical resistance, i.e. quite sensitive to the presence of impurity atoms, the resistivity being low at high temperatures where conduction is almost as good as in some metals. The optical properties of the covalent crystal are characterised by high refractive index (μ) and high dielectric constant. The crystals being transparent to long wavelength radiations, but opaque to shorter wavelengths.

The classic examples of covalently bonded crystals are the group IV semiconductors, C, Si and Ge. The nature and origin of covalent bond have received considerable attention recently.

The most striking feature of the covalent bond is its strong directional properties. Thus, C, Si and Ge have the diamond lattice structure in which each carbon atom, say, is connected to four neighbouring carbon atoms at tetrahedral angle to one another even though this arrangement gives a low filling of space.

The other important characteristic property of the covalent bond is its saturability. Saturability means that each atom can form covalent bonds only with limited number of neighbours. For example, each hydrogen atom can form covalent bonds only with one of its neighbours. The electron pair constituting such a bond has antiparallel spins and occupies one quantum state. A third atom, in this case, instead of being attracted will be repelled.

The covalent bond is usually formed from two electrons, one from each atom participating in the bond as illustrated below.

$$:\dot{C}\dot{l}: + \cdot\dot{C}\dot{l}: \rightarrow :\dot{C}\dot{l}::\dot{C}\dot{l}:$$

Unpaired electrons

The electrons forming the covalent bond tend to be partly localised in the region between the atoms joined by the bond. In accordance with the Pauli principle, the spins of the two electrons in the covalent bond are antiparallel.

The binding of molecular hydrogen is a simple example of covalent bond. The strongest binding occurs when the spins of the two electrons are antiparallel. The binding depends on the relative spin orientation not because there are strong magnetic dipole force between the spins but because the Pauli principle modifies the distribution of charge according to the spin orientation. A spin dependent coulomb energy of the system is found. This is called the **exchange interaction**.

Covalent bonding also prevails in crystals of the from $A^N B^{8-N}$ composed of elements *A* and *B* with *N* and $8 - N$ valence electrons per atom of *A* and *B*, i.e. with 8 valence electrons per atom pair. The nature and origin of this bond have been excellently characterised by Philips.

17.7 METALLIC BONDING

Nearly 70% of the elements in the periodic table are metals. Metals are characterised by the availability of large number of valence electrons which can be easily removed from the parent atoms. A simple and rather crude picture of metallic bonding can be conceived as arising because of the fact that barring the repulsions which are also important, the energy of the valence electrons is lowered when they come close to more than one nucleus. These valence electrons called the electron gas are supposed to move about the crystal lattice freely more or less like gas molecules. The positive ion cores may be supposed to be held together by some sort of resultant electrostatic attraction between the fixed positive ions and the mobile negative electrons. The free electrons are mainly responsible for the high thermal and electrical conductivity, since they can absorb easily any energy from lattice vibrations or electro-magnetic radiations and thereby increase their energy and mobility. Metals are opaque since these free electrons absorb photons in the visible region and get excited to higher energies. High reflectivity of metals is due to the fact that these electrons re-emit this energy when they fall back to lower energy levels. The above theory due to Drude–Lorentz though successful in explaining the general nature of metals is found to be quite inadequate to account for several experimental results like temperature dependence of resistance, paramagnetism of free electrons, electronic specific heat, etc. The dependence of metallic character of solids on the number of valence electrons is another feature which the above theory fails to explain. Elements like sodium, potassium, etc. which have only one electron in the outermost orbit are marked by metallic character. When there are 2 or 3 electrons in the outermost orbit as in the case of magnesium, aluminium, etc., the elements are just metallic. When the number of valence electrons reach 4 as in the case of germanium, tin, lead, etc., the metallic character becomes much less. When the number of valence electrons become 6 or 7, the elements definitely loose their metallic character.

With the advent of quantum mechanics, a better picture of the metallic bonding was evolved. It is convenient to regard the metallic bonding as an unsaturated covalent bond with possible hybridization in some structure. Let us now consider the bonding between hydro-gen atoms and lithium atoms both belonging to the I group. Hydrogen is monovalent and has one electron in its outer orbit. When two hydrogen atoms come nearest, the valence electrons of the two hydrogen atoms share the same shell of the molecule, since it has two allowed quantum levels, one for spin up and other for spin down. The shell becomes complete. This electron sharing between the two atoms results in the lowering of electron energies, and hence a bonding of the two hydrogen atoms occurs with the consequent for motion of the hydrogen molecule. When the 3rd hydrogen atom comes nearer to the hydrogen molecule, it is no longer attracted since there is no room in the shell. It has to occupy only a higher energy level, which means that the force is repulsive, i.e. the bonding between that force is repulsive. We say that the bonding between the two hydrogen atoms is saturated. A hydrogen atom cannot bond with more than one hydrogen atom. It is a typical covalent bond, where the number of bonding is exactly equal to the number of valance electrons. Let us now consider the monovalent lithium atoms whose electron configuration is $1s^2 2s^1$. When another lithium atom comes nearer, we must expect it to form a covalent bond with the original atom since there is a vacant quantum state in the 2s subshell. We must also expect the bonding to become saturated since the 2s subshell becomes complete. But in the lithium atom, there are 6 more quantum states belonging to 2p subshell, whose energies are only slightly greater than the energies of 2s level. So when a third lithium atom comes near the molecule, the third electron occupies a different quantum state without violating Pauli's exclusion principle. This means that the third electron can also form an electron pair bond with the electron of the first atom. It turns out that quite a large number of atoms can surround a single lithium atom. Since the central atom has only unpaired electron, this electron must take turns to form electron pair bonds with its nearest neighbours. Since lithium crystallises to become a body centred cubic crystal in which each atom has 8 nearest neighbours, the central atom has to form covalent bonds with the 8 neighbours in turn. A covalent bond requires 2 electrons per bond. But lithium atom has to satisfy itself with only a quarter electron pair available per bond. The bonding is far from saturation. This situation is described as unsaturated covalent bond. This is the reason why as many atoms as possible form cluster around metallic atom in an effort to attain saturation. But geometric considerations limit the number of closest neighbours. In general, metallic atoms have the largest number of closest neighbours and metals are, in general, close packed structures. The unsaturated nature of the bonding is the cause of weakness in metals. The binding energies of Na and Al metals are 1.12 eV and 3.23 eV respectively as compared to 7.4 eV in the case of covalent bonded diamond. Because of the large number of vacancies in other orbits, metals do not exhibit preferences in forming the bonds. This helps easy rearrangement of atoms without loss of strengths, i.e. the metals can be easily deformed. Because of the large number of vacancies the electron can wander from atom to atom without being attached to any particular atom as in ionic crystal or covalent crystals. Thus the, valency electrons behave in a manner similar to the molecules in a gas.

From quantum mechanics point of view, we say that the bonding arises because of overlap of the orbitals which results in lowered energy states. In the case of metallic atoms, the wave functions of the valence electrons are spread out more in space. This means that the probability of occupation of the valance electrons, i.e. $|\psi^2|$ functions are also spread out. This results in the increase of the mean radius of the electron in the free atom. In addition, the kinetic energy of the valence electron is lowered by the extended ψ function. The spatial extension of the wave function is a consequence of the increase in de Broglie wavelength of the valence electron. From the de Broglie relation $\lambda = \dfrac{h}{p}$, we find that increased wavelength results in decreasing

momentum and hence decreasing kinetic energy. Further potential energy is also lowered because in the solid state, the valence electrons come closer to one or other nucleus than they are in a free atom. It is thus lowering of both the potential and the kinetic energies of the valence electrons that is responsible for the metallic bonding. In general, smaller the number of valence electrons an atom has, more loosely they are held and hence greater is the metallic bonding. As the number of valence electrons increases, the tightness with which they are held to the nucleus increases and hence they become more localised in space, increasing the covalent nature of the bonding. The transition metals like Fe, Ni, W, Ti, etc. have incomplete d subshells. This leads to the formation of **hybrid covalent bonds** due to the arrangement of electrons in the s and d levels. This accounts for their high bonding energies of the order of 4.5 eV.

How to Predict Predominance of Metallic or Covalent Bond

We have seen that lowering of the kinetic and the potential energies of the valence electrons results in the metallic bonding. The relative amount of lowering of the potential and kinetic energies varies from metal to metal. The potential energy effect is larger in monovalent metals, but the kinetic energy effect is larger in most other metals. We can say, in general, metallic bonding is favoured if the number of valence electrons per atom is small, since in that case only the lowest kinetic energy states need to be occupied, due to the smaller number of valence electrons. Lower kinetic energy configuration means large extension of spatial wave functions and which again means that the valence electrons can have more electron gas which is the special feature in metals. The elements on the right side of the periodic table are definitely non-metallic because the valency electrons became more lightly bound to the nucleus and the electron gas behaviour is completely absent. For elements in the middle of the periodic table, the competition between these two structures is very severe. Diamond exhibits almost pure covalent bonding whereas silicon and germanium are more metallic. Tin exhibits two forms: metallic tin (white) is stable above 13°C, covalent tin (grey) is stable below this temperature. Lead is mostly metallic.

17.8 MOLECULAR OR VAN DER WAALS BONDING

Though the inert gases do not have any valence electrons because of their completed shell structures to form any type of bonding, they do become liquids and solids at sufficiently low temperatures and high pressures. This is explained by the fact that an atom surrounded by revolving electrons constitute fluctuating dipoles or rotating dipoles. When two such atoms come near, there is dipolar interaction between these fluctuating dipoles leading to attractive forces. Higher order polar interactions also become possible but they are extremely feeble. The bonding which arises from the fluctuating dipole nature of the atom is called the van der Waals bond. Van der Waals forces arise out of induced dipoles, permanent dipoles and dispersion forces. For atoms or molecules which have no permanent dipoles, the interaction energy due to induced dipoles is given by

$$u_i = -\frac{2\mu^2\alpha}{r^6} \qquad (17.31)$$

where α is polarizability of the atom or the molecule, μ is the induced dipole moment and r is the interatomic distance.

When the atom or the molecule has permanent dipole moment, the energy due to the dipolar interaction is given by

$$u_p = -\frac{2\mu^4}{3r^6kT} \qquad (17.32)$$

where μ is the permanent dipole moment, k is Boltzmann's constant and T is temperature in Kelvin.

Dispersion forces are understood better from quantum mechanical considerations. This is the interaction between the fluctuating dipolar

atoms or molecules. The interaction energy u_d is given by

$$u_d = -\frac{3\alpha^2 h\nu_0}{4r^6} \qquad (17.33)$$

ν_0 is the oscillation frequency of the atom or the molecule before the interaction. So the total van der Waals attraction energy is given by

$$u = u_i + u_p + u_d$$

or $$u = -\frac{2\mu^2\alpha}{r^6} - \frac{2\mu^4}{3r^6kT} - \frac{3\alpha^2 h\nu_0}{4r^6} \qquad (17.34)$$

Equation (17.34) predicts values of right order of magnitude for many substances. Of course, van der Waals bonds are very weak as compared to covalent, metallic and ionic bonds. But they play an important role in determining the final arrangement of groups of atoms in solids specially in polymers.

Weak van der Waals forces occur in graphite. Grahpite consists of carbon atoms in tetragonal arrays with each atom bonded to three atoms. Figure 17.4b shows graphite structure.

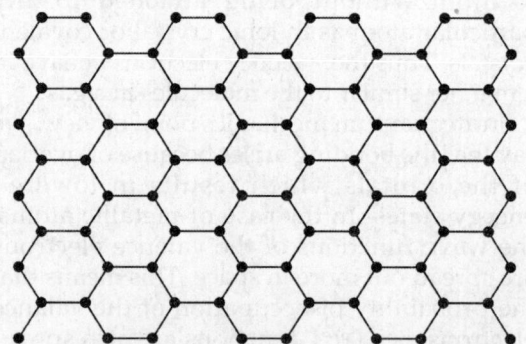

Fig. 17.4b: Graphite structure. It consists of carbon atoms in tetragonal arrays with each atom bonded to three others. The layers are bonded together by weak van der Waals forces.

17.9 HYDROGEN BONDING

The hydrogen bond is formed between an atom of hydrogen and an extremely electronegative atom, e.g an atom of oxygen, fluorine, nitrogen, chlorine. Such an atom attracts the bonding electrons and becomes negatively charged. The hydrogen atom after loosing the bonding electron assumes a positive charge. The hydrogen bond is a result of electrostatic attraction of these charges. Hydrogen bond in ice is illustrated in Fig. 17.5.

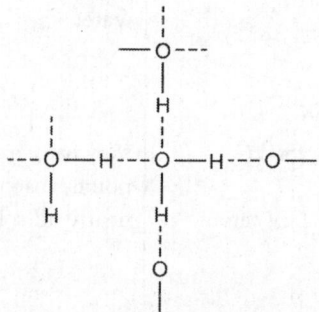

Fig. 17.5: Hydrogen bond

The hydrogen bond is essentially ionic in character and the cohesive energy may be discussed in terms of coulomb interactions between charges and Leonard–Jones potentials. The binding energy of ice or water is almost 0.5 eV.

The hydrogen bond is the cause of association of molecules of liquid (such as H_2O, acids, spirits, etc.) which results in greater viscosity, higher boiling point, abnormal thermal expansion, etc. Bonding in H_2O molecule is shown in Fig. 17.6. Should there be no hydrogen bond between the molecules of water, its boiling point at atmospheric pressure would not be +100°C but 80°C and its viscosity would be lower by almost an order of magnitude. When water is heated above 0°C, hydrogen bond is broken.

It should be pointed out that in real solids not one type of bonds discussed above ever exist purely by itself. Practically there is always a superposition of two or more types of bonds, one of them as a rule plays a dominant part in determining the structure and the properties of the solid.

The various crystal types are summarized in Table 17.2.

ILLUSTRATIVE EXAMPLES

Example 1

An atom A has an ionisation energy of 5 eV and another atom B has an electron affinity of 4 eV. What is the energy released or absorbed in placing the two atoms 5Å apart?

Solution

Since ionization energy of atom A is 5 eV that means the cost of removing one electron from atom A is 5 eV, i.e. when 5 eV energy is supplied the atom A becomes positive ion A^+.

The electron affinity of B is 4 eV, means when this atom absorbs an electron, it releases an energy of 4 eV and becomes electronegative.

These two will attract and when they are placed at a distance of 5Å, the coulombian potential energy then becomes

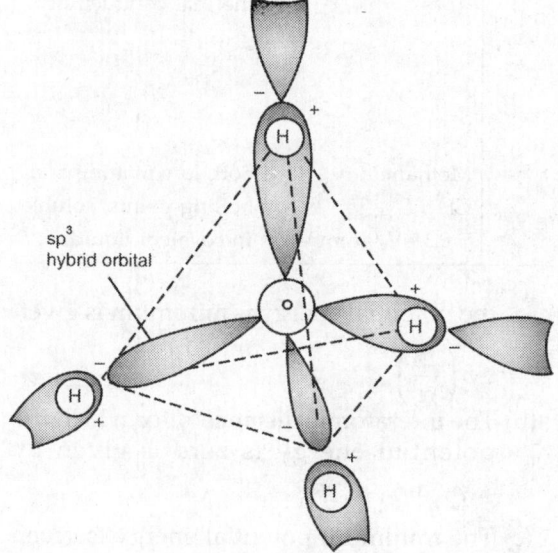

sp³ hybrid orbital

Fig. 17.6: Bonding in H_2O molecule. In an H_2O molecule, the four pairs of valence electrons around the oxygen atom (six contributed by the O atom and one each by the H atom) occupy four sp³ hybrid orbitals that form a tetrahedral pattern. We may note that each H_2O molecule can form H bonds with four other H_2O molecules.

$$u_c = -\frac{Kq_1q_1}{r} \qquad q_1 = q_2 = e = 1.6 \times 10^{-19} C$$

Table 17.2: Crystal types

Type		Bond	Example	Properties
Ionic	Negative ion / Positive ion	Electrostatic attraction	Sodium chloride NaCl, $E_{cohesive}$ = 3.28 eV/atom	Hard; high melting points; may be soluble in polar liquids such as water
Covalent	Shared eletrons	Shared electrons	Diamond C, $E_{cohesive}$ = 7.4 eV/atom	Very hard; high melting points; insoluble in nearly all solvents
Metallic	Metal ion / Electron gas	Electron gas	Sodium Na, $E_{cohesive}$ = 1.1 eV/atom	Ductile; metallic luster; high electrical and thermal conductivity
Molecular	Instantaneous charge separation in molecule	van der Waals forces	Methane CH_4, $E_{cohesive}$ = 0.1 eV/atom	Soft; low melting and boiling points; soluble in covalent liquids

$$= -\frac{9\times10^9 \times 1.6\times10^{-19} \times 1.6\times10^{-19}}{5\times10^{-10}} \text{ J}$$

$$= -\frac{9\times10^9 \times 1.6\times10^{-19}}{5\times10^{-10}} \text{ eV} = -2.88 \text{ eV}$$

Thus, total energy absorbed

$$= +5 - 4 + 2.88 = +3.88 \text{ eV}.$$

Example 2

If the potential energy function is expressed as

$$u(r) = -\frac{\alpha}{r^6} + \frac{\beta}{r^{12}}$$

show that:

(a) The intermolecular distance r_0 for which

the potential energy is minimum is given by $\left(\dfrac{2\beta}{\alpha}\right)^{1/6}$.

(b) The interatomic distance r_0' for which the potential energy is zero is given by $r_0 \, (2)^{-1/6}$.

(c) The minimum potential energy is given by $u_{min} = -\dfrac{\alpha^2}{4\beta}$.

Solution

(a) The condition for the minimum potential energy is

$$\left(\frac{du}{dr}\right) = 0$$

$$\therefore \quad \frac{\partial}{\partial r}\left(-\frac{\alpha}{r^6}+\frac{\beta}{r^{12}}\right)=\frac{6\alpha}{r^7}-\frac{12\beta}{r^{13}}=0$$

or $\quad r^6=\dfrac{2\beta}{\alpha}\quad$ or $\quad r=\left(\dfrac{2\beta}{\alpha}\right)^{1/6}$

The interatomic distance for minimum potential energy u_{min} is r_0

$$\therefore \quad r=r_0=\left(\frac{2\beta}{\alpha}\right)^{1/6} \tag{1}$$

(b) If $r=r_0'$ for which $u(r)=0$, then

$$-\frac{\alpha}{r_0^6}+\frac{\beta}{r_0'^{12}}=0$$

or $\quad r_0'^6=\dfrac{\beta}{\alpha}=\dfrac{r_0^6}{2}\;$ from Eq. (1)

or $\quad r_0'=r_0\left(\dfrac{1}{2}\right)^{1/6}=r_0\,(2)^{-1/6} \tag{2}$

(c) The minimum potential energy is

$$u_{min}=-\frac{\alpha}{r_0^6}+\frac{\beta}{r_0^{12}}=-\frac{\alpha}{r_0^6}\left[1-\frac{\beta}{\alpha}\left(\frac{1}{r_0}\right)^6\right]$$

$$=-\frac{\alpha}{\left(\dfrac{2\beta}{\alpha}\right)}\left[1-\frac{\beta}{\alpha}\cdot\frac{\alpha}{2\beta}\right]\left(\because r_0^6=\frac{2\beta}{\alpha}\right)$$

$$=-\frac{\alpha^2}{2\beta}\cdot\frac{1}{2}=-\frac{\alpha^2}{4\beta}\,.$$

Example 3

Sort out from the following substances (giving reasons) those having (i) ionic (ii) covalent (iii) metallic (iv) van der Waals bonds. What sort of structure (compact or loose) do you expect for them?

Cu, Zn, Pb, Ni, Mg, Ne, Ar, NaCl, NaI, CuNi and SiC.

Solution

   i. NaCl and NaI must have ionic bonds because Na atom have one single electron in its outermost shell and both Cl and I atoms have in their outermost

shell 7 electrons each, i.e. one short for completion of the shell. The former, therefore, easily transfer its lone outer shell electron to the outer shell of the latter, and we thus have an Na^+ ion and a Cl^- ion with the outer shells of both types of ions complete. The electrostatic attraction between the two oppositely charged ions then provides the ionic bond between them. Since the ionic bond is the strongest of all other bonds both NaCl and NaI must have a compact or close-packed structure.

  ii. SiC must have covalent bond, since both Si and C have four electrons in their outermost shells which form covalent bonds. These bonds being next only in strength to ionic bonds (on account of their directional nature), the structure of SiC too must be quite compact.

 iii. Metals Cu, Zn, Pb, Ni and Mg, as also the metallic alloy CuNi must naturally have metallic bonds. These bonds though weaker than ionic and covalent bonds are, however, non-directional. These metals and the alloy (CuNi) must also have a compact and close-packed structure.

 iv. Both Ne and argon, being inert gases must have van der Waals bonds, which are the weakest of all. Since they are non-directional, the structure of Ne and argon in the solid state must be expected to be faily compact.

Example 4

The ionisation energy of lithium is 5.4 eV and the electron affinity of bromine is 3.5 eV. (i) What is the net amount of energy required to form a Li^+ and Br^- ion pair? (ii) Considering them as a joint charge, how close together must a Li^+ and Br^- ion be for the total energy of the pair to be zero?

Solution

   i. Energy required to transfer a electron from a Li to a Br atom or to form Li^+ and Br^- pair

$$= 5.4\,\text{eV} - 3.5\,\text{eV} = 1.9\,\text{eV}$$

ii. Let Li^+ and Br^- be point charges with charge $(+e)$ and $(-e)$ and r be the distance between them. Then potential energy of charges is

$$-\frac{Kq_1q_2}{r} = \frac{Ke^2}{r} \text{ where } K = 9 \times 10^9 \frac{Nm^2}{C^2}$$

$\therefore$ Potential energy $= \dfrac{Ke}{r} \, eV$

For total energy of the pair to be zero, we have

$$\frac{Ke}{r} = 1.9$$

or $\qquad r = \dfrac{Ke}{1.9}$

$$r = \frac{9\times 10^9 \times 1.6 \times 10^{-10}}{1.9} = 7.58 \times 10^{-10} \text{ m}.$$

Example 5

The potential energy of a system of two atoms is given by the relation

$$U = \frac{A}{r^2} + \frac{B}{r^{10}}$$

A stable molecule is formed with the release of 8 eV energy when the interatomic distance is 2.8 Å. Find A and B and the force needed to dissociate this molecule into atoms and the interatomic distance at which the dissociation occurs.

Solution

We have the potential energy of the system

$$U = \frac{A}{r^2} + \frac{B}{r^{10}} \qquad (1)$$

$\therefore \qquad F = -\dfrac{dU}{dr} = -\dfrac{2A}{r^3} + \dfrac{10B}{r^{11}} \qquad (2)$

At equilibrium distance r_0, $F = 0$. Therefore, Eq. (2) gives

$$A = \frac{5B}{r_0^8} = \frac{5B}{(2.8\times 10^{-10})^8} = 1.32 \times 10^{17} B \qquad (3)$$

$\therefore$ Bond energy or the dissociation energy, U_0 is obtained from Eq. (1) by putting $r = r_0$,

$$U_0 = -\frac{A}{r_0^2} + \frac{B}{r_0^{10}} = -\frac{A}{r_0^2}\left(1 - \frac{B}{Ar_0^8}\right)$$

Using Eq. (3), one obtains

$$U_0 = -\left(\frac{A}{r_0^2}\right)\left(1 - \frac{B}{5}\right) = -\frac{4}{5}\frac{A}{r_0^2}$$

Negative sign signifies the release of energy.

$\therefore \quad A = \left(\dfrac{5}{4}\right)(2.8 \times 10^{-10})^2\,(8.0 \times 1.6 \times 10^{-19})$

$$= 1.256 \times 10^{-37} \, Jm^2$$

and $\qquad B = 9.52 \times 10^{-115} \, Jm^2$

The critical interatomic distance r_c is obtained from the condition

$$\left.\frac{dF}{dr}\right|_{r=r_c} = 0$$

From Eq. (3), $\dfrac{6A}{r_0^4} - 110\dfrac{B}{r_c^{12}} = 0$

It yields $r_c = \left(\dfrac{110B}{6A}\right)^{1/8}$ $\qquad (4)$

$$= \left(\frac{110\times 9.52\times 10^{-115}}{6\times 1.256\times 10^{-37}}\right)^{1/8} = 3.3 \times 10^{-10} \text{ m}$$

Putting $r = r_c$ in Eq. (2), the force required to dissociate the molecule is

$$F = -\left(-\frac{2}{r_c^3}\right)\left(A - \frac{5B}{r_c^8}\right)$$

Using Eq. (4), it becomes

$$F = \left(-\frac{2}{r_c^3}\right)\left[A - 5B\left(\frac{6A}{110B}\right)\right] = \left(\frac{2}{r_c^3}\right)\left(\frac{8A}{11}\right)$$

$$= \frac{2}{(3.3\times 10^{-10})^2}\left(\frac{8\times 1.256\times 10^{-37}}{11}\right)$$

$$= -5.08 \times 10^{-9} \text{ N}$$

The negative sign indicates that the force existing at $r = r_c$ is the force of attraction. The force needed to dissociate the molecule is 5.08×10^{-9} N.

Example 6

The lattice energy of KCl crystal containing N-molecules of KCl is given by

$$U = -N\left(\frac{Mq^2}{4\pi\varepsilon_0 R} - \frac{B}{R^n}\right).$$

Find the repulsive exponent n. Given nearest neighbour equilibrium distance, $R_0 = 3.14$ Å, compressibility of KCl, $K = 5.747 \times 10^{-11}$ m²/N, Madelung constant $M = 1.748$.

Solution

$$U = -N\left(\frac{Mq^2}{4\pi\varepsilon_0 R} - \frac{B}{R^n}\right).$$

Differentiating, we obtain

$$\frac{dU}{dR} = \frac{NMq^2}{4\pi\varepsilon_0 R^2} - \frac{nNB}{R^{n+1}} \qquad (1)$$

For $R = R_0$, $dU/dR = 0$, this yields

$$B = \frac{Mq^2}{4\pi\varepsilon_0 n} R_0^{n-1} \qquad (2)$$

The structure of KCl is identical to that of NaCl. The compressibility is reciprocal to the bulk modulus, and hence

$$\frac{1}{K} = \left(\frac{1}{18NR_0}\right)\frac{d^2U}{dR^2}\bigg|_{R=R_0} \qquad (3)$$

From Eq. (1), we have

$$\frac{d^2U}{dR^2}\bigg|_{R=R_0} = -\frac{2NMq^2}{4\pi\varepsilon_0 R_0^3} + \frac{n(n+1)NB}{R_0^{n+2}}$$

$$= -\frac{2NMq^2}{4\pi\varepsilon_0 R_0^3}(n-1) \text{ [using Eq. (2)]}$$

$$n = 1 + \frac{18R_0^4(4\pi\varepsilon_0)}{KMq^2}$$

$$= 1 + \frac{18\times(3.147\times10^{-10})^4}{5.747\times10^{-11}\times1.748\times(1.6\times10^{-19})^2\times9\times10^9}$$

$$n = 1 + 7.63 = 8.63$$

Example 7

The force of attraction between ions of Na and Cl is 3.02×10^{-9} N when the two ions just touch each other. Given ionic radius of Na$^+$ ion is 0.95 Å, $e = 1.6 \times 10^{-19}$ C, $\varepsilon_0 = 8.854 \times 10^{-12}$ C²/N m². Find the radius of Cl$^-$ ion.

Solution

We have $F_1 = \dfrac{Z_1 Z_2 e^2}{4\pi\varepsilon_0 r^2}$ (force of attraction)

$\therefore \qquad F = 3.02 \times 10^{-19}$

$$N = \frac{(+1)(-1)(1.6)^2 \times (10^{-19})^2}{4\pi \times 8.854 \times 10^{-12} r^2}$$

$$= \frac{0.23}{r^2} \times 10^{-26} \text{ N}$$

$\therefore \qquad r^2 = \dfrac{0.023 \times 10^{-26}}{3.02 \times 10^{-9}} = 0.0076 \times 10^{-17}$

$\therefore \qquad r = 0.276 \times 10^{-9}$ m $= r_{Na^+} + r_{Cl^-}$

$\therefore \qquad r_{Cl^-} = 2.76 \times 10^{-10} - 0.95 \times 10^{-10}$

$$= 1.81 \times 10^{-10} \text{ m}$$

$$= 1.81 \text{ Å}$$

$Z_1 = +1$ for Na$^+$

$Z_2 = -1$ for Cl$^-$

$e = 1.6 \times 10^{-19}$ C

$\varepsilon_0 = 8.854 \times 10^{-12}$.

Example 8

The ionic radii of Mg^{++} and S$^-$ respectively are 0.65 Å and 1.8 Å. Calculate the force of attraction between these ions. Given : $e = 1.6 \times 10^{-19}$ C, $\varepsilon_0 = 8.854 \times 10^{-12}$ C²/N m².

Solution

Force of attraction

$$F_1 = \frac{Z_1 Z_2 e^2}{4\pi\varepsilon_0 r^2}$$

$$= \frac{2 \times (-2) \times (1.60 \times 10^{-19})^2}{4\pi \times 8.854 \times 10^{-12} \times (2.49 \times 10^{-10})^2}$$

$$= \frac{4 \times 2.56 \times 10^{-38}}{111.27 \times 6.2 \times 10^{-32}} \text{N} = 0.015 \times 10^{-6} \text{N}$$

$$= 1.5 \times 10^{-8} \text{ m}$$

$$Z_1 = + 2 \text{ for Mg}^{++}$$

$$Z_2 = -2 \text{ for S}^-$$

$$r = \text{Radius of Mg}^{++} \text{ ion} + \text{radius of S}^-$$

$$= (0.65 + 1.8 \text{ Å})$$

$$= 2.49 \times 10^{-10} \text{ m}$$

Example 9

For NaCl, the Na atom has an ionization energy 5.14 eV and the Cl atom has an electron affinity 3.61 eV. The equilibrium separation between the ion-pair is 0.282 nm. Show that the energy required to transfer an electron from Na to Cl is about – 7.395 eV.

Solution

We have the coulomb energy of NaCl as

$$E = \frac{1.75e}{4\pi\varepsilon_0 r_0} \text{ eV}$$

$$= \frac{1.6 \times 10^{-19} \times 1.75}{4\pi \times 8.85 \times 10^{-12} \times 0.282 \times 10^{-9}} = 8.925 \text{ eV}$$

Now, the energy required to transfer an electron from Na to Cl

$$= (5.14 - 3.61 - 8.925) \text{ eV} = -7.395 \text{ eV}.$$

GLIMPSES

1. Atoms in molecules/solids are held together by chemical bonds. This holding together phenomenon is called as *bonding*.

2. In any solid, the mutual interatomic forces are basically *electrostatic* in nature, and the primary differences among different types of solids depend on the ways in which the valence electrons of constituent elements are distributed, i.e. it depends on the bonding.

3. On the basis of bonding type, there are five types of solids: (i) *ionic* solids (e.g. alkali halides, alkaline oxides) (ii) *covalent* solids (diamond, silicon, etc.) (iii) *van der Waals* bonded molecules (O_2, H_2, Kr, Xe, solid He) (iv) *hydrogen bonded* solids (i.e. some fluorides and compounds having water of crystallization) (v) *metallic* solids (various metals and alloys).

4. In a simple diatomic molecule, two atoms exert attractive forces between them which keep them together and repulsive forces between them since external energy is required to bind them together. At equilibrium, the two forces become equal. At this stage the system has minimum energy given by

$$V(r) = \frac{-a}{r^n} + \frac{b}{r^m}$$

where r is the distance between two nuclei and a, b, m, and n are constant characteristics of the molecule.

5. The energy at the equilibrium position $V(r_0)$ is called the *binding energy* or the *cohesive energy* of the molecule.

6. Based on bond strength, atomic bonds are classified into *primary* and *secondary*. Primary bonds have energies in the range 0.1 to 10 eV. Ionic, covalent and metallic bonds belong to this category. Secondary bonds have energies in the range 0.1 to 0.5 eV. Hydrogen and van der Waals bonds are examples of secondary bonds.

7. An *ionic bond* is formed when atoms that have low ionization energies, and hence can lose electrons readily, interact with other atoms that have high electron affinity. This bond is formed between an electropositive element (Group I or Group II elements like Na, K, Mg, etc.) and an electronegative element (Group VI or Group VII elements like Cl, Br, O, etc.).

8. A *covalent bond* is formed when one or more pairs of electrons are shared between the incomplete outer shells of two atoms, e.g. H_2 molecule. The spins of two electrons in a shared-pair are antiparallel. The covalent bond is also known as a *homopolar* or *electron pair bond*.

9. The *metallic bond* is formed between electropositive elements having low ionisation energy. The binding energy of metallic bonds are usually in the range of 1–3 eV. The free electrons are responsible for the high electrical and thermal conductivity of metals.

10. A hydrogen bond is formed when a hydrogen atom makes a bond with an extremely electronegative atom (O, F, N, Cl, etc.).

11. van der Waals bonds are usually observed in inert gases. The weak short range forces that are weaker than the atomic bonding forces, existing between atoms and molecules are called van der Waals forces. These bonds are formed due to dipole attraction.

REVIEW QUESTIONS

1. Describe the nature and origin of various forces existing between the atoms of a crystal. Explain the formation of a stable bond using the potential energy versus interatomic distance curve.

2. List the various types of bonds occurring in crystals. Discuss any one of them in brief.

3. Explain electron cloud. What is the role of electron cloud in metallic bond?

4. List the different types of bonds occuring in a crystal. Describe the characteristics of metallic bonds.

5. Describe the binding of atoms in metals.

6. Distinguish between ionic and metallic bonds in solids, illustrate with example.

7. Why are close packed structures experienced mostly in metals and not in ionic and covalent bonds?

8. Explain the origin of metallic bonding. How does it differ from ionic bonding? Explain with suitable examples.

9. How are atoms held together in a metallic bond? Explain diagramatically.

10. What is hydrogen bond? How is it different from a dipole bond? Explain the role of hydrogen bond during the formation of ice.

11. Derive an expression for the binding energy of an ionic crystal and obtain the expression for Madelung constant. Evaluate Madelung constant for a linear ionic crystal.

12. What are various types of bonding in materials? Explain the different types of bonding and illustrate them with suitable examples.

13. How do the melting points of the ionic solids vary?

14. Why do inert gases get liquified and solidified at very low temperatures?

15. Define cohesive energy and determine its value for crystal of inert gases.

16. Using the concept of wave mechanics, explain why a covalent bond tends to be the strongest in the directions when $|\psi^2|$ is a maximum.

17. Explain why carbon atoms in diamond bond covalently, while lead atoms bond metallically, even though carbon and lead have four valence electrons each.

18. Explain with the help of suitable sketches the various types of bonding in crystals.

19. Describe the ionic bonding and giving example show that the bond energy of ionic bonding decreases as the size of electrons increases. Explain this behaviour.

20. What is a secondary bond ? Explain the concept of dipole bonding. Why are some compounds non-polar even if they have dipoles?

21. Compare and contrast metallic, covalent and ionic bonds. Give examples.

22. What do you understand by percentage ionic factor of a covalent bonds?

23. Explain in which of the ionic and covalent bonds wave characteristics of electrons play important role.

24. Describe the essential features of the following with one example of each category: (a) metallic bond (b) ionic bond (c) covalent bond.

25. Explain briefly: (a) covalent bonding (b) ionic bonding.

26. It is observed that CCl_4 has no net dipole moment. What inference do you draw about C-Cl bond in this compound?

27. Describe compounds which show mixed ionic and covalent bonds.

28. The potential energy of a diatomic molecule is given in terms of the interatomic distance r by the expression: $U(r) = -\dfrac{a}{r^m} + \dfrac{b}{r^n}$, where symbols have usual meaning. Derive an expression for equal spacing of the atoms and hence obtain the dissociation energy.

29. Explain the origin of van der Waals force in molecular crystals. How does the binding due to van der Waals forces differ from the binding the valence crystals.

PROBLEMS

1. The potential energy of a system of two atoms is given by $U = \dfrac{-\alpha}{r^4} + \dfrac{\beta}{r^{12}}$. Calculate the amount of

energy released when the atoms form a stable bond. Also determine the bond length.

$$\left[Ans. \left(\frac{4\alpha^3}{27\beta}\right)^{1/2}, \left(\frac{3\beta}{\alpha}\right)^{1/8} \right]$$

2. Assume that the energy of two particles in the field of each other is given by the following function: $U = -\frac{a}{r} + \frac{b}{r^8}$, where a and b are constants and r is the distance between the centres of the particles. Show that if particles are pulled apart, the molecules will break as

soon as $r = \left(\frac{36b}{a}\right)^{1/7} = r_0 \, (4.5)^{1/7}$.

3. The interaction energy of a system of two atoms is given by $U = -\frac{A}{r^6} + \frac{B}{r^{12}}$. The atoms form a stable bond with bond length of 3Å and bond energy of 1.8 eV. Calculate A and B. Compute the forces required to break the molecule and the critical interatomic distance for which it occurs.

[*Ans.* 4.19×10^{-76} Jm⁶, 1.53×10^{-133} Jm¹², 2.56×10^{-9} N, 3.33×10^{-10} m]

4. Assume that the energy of two particles in the field of each other is given by

$$U = -\frac{2\mu^2\alpha}{r^6} - \frac{2\mu^4}{3r^6 kT} - \frac{3\alpha h v_0}{4r^6} \text{ where } \alpha, \beta \text{ are}$$

constants and r is the distance between centres of the particles. Show that the two particles form a

stable compound for $r = r_0 = \left(\frac{8\beta}{\alpha}\right)^{1/7}$ and that in the stable configuration, the energy attraction is 8 times the energy of repulsion.

5. Consider a line of alternate positive and negative ions each carrying a charge q. The repulsive potential energy between neighbours

is given by $\frac{A}{r^n}$. Show that for a total of $2N$ ions, the equilibrium potential energy of the system

is $U_0 = -\left(2Nq^2 \ln\frac{2}{r_0}\right)\left(1-\frac{1}{n}\right)$, where r_0 is the equilibrium inter-ionic distance.

6. The potential energy U of a system of two atoms varies as a function of their distance of

separation r as $U = -\frac{A}{r^n} + \frac{B}{r^m}$. Show that in equilibrium: (i) $r = r_0 = (mB/nA)^{1/(m-n)}$ (ii) the energy of attraction is m/n times the energy of repulsion (iii) the bond energy,

$$U_0 = \frac{A}{r_0^n}\left(\frac{m-n}{m}\right).$$

7. The distance between the nearest positive and negative ions in KCl is 3.14 Å. The structure is similar to NaCl. Find the ionic bond energy in the solid in kilocalories per mole.

SHORT ANSWER-QUESTIONS

1. How atomic bonding in solids are considered?
Ans. These are considered in terms of attractive and repulsive forces.

2. How many types of primary bonds are there?
Ans. Three: ionic, covalent and metallic.

3. Give names of secondary bonds.
Ans. van der Waals and hydrogen bonds. These are weak in comparison to primary bonds.

4. Mention the range of energies of primary and secondary bonds.
Ans. Primary bond energies are in the range of 100–1000 kJ/mol, whereas that of secondary bonds are in the range of 1–50 kJ/mol.

5. How non-directional ionic bond is developed?
Ans. For ionic bonds, electrically charged ions are formed by the transference of valence electrons from one atom to another. The forces are coulombic.

6. What is covalent bonding?
Ans. In a covalent bonding there is sharing of valence electrons between adjacent atoms.

7. What is a metal?
Ans. A metal is an array of positive ions which are held together in a cloud of free electrons.

8. What is metallic bonding?
Ans. With metallic bonding, the valence electrons form a 'sea of electrons' that is uniformly dispersed around the metal ion cores and acts as a form of glue for them. The metallic bond is directional and generally weaker than ionic and covalent bond.

9. What is the relation between melting and boiling points of materials and bond strength?
Ans. Melting and boiling points of materials increase with increasing bond strength. Strongs and directional bonds result in hard and brittle solids.

10. How secondary bonds are formed?

Ans. Secondary bonds are formed from attractive forces between electric dipoles, of which there are two types— induced and permanent.

11. What happens when hydrogen covalently bonds to a non-metallic element such as fluorine?

Ans. Highly polar molecule is formed.

MULTIPLE CHOICE QUESTIONS

1. Sharing of electrons between neighbouring atoms results in

(a) metallic bond   (b) ionic bond

(c) covalent bond   (d) none of above

2. Primary bonds have energy range in kJ/mol

(a) 10–100   (b) 10–1000

(c) 1–10   (d) 100–5000

3. The nature of binding for a crystal with alternate and evenly spaced positive and negative ions is

(a) dipole   (b) ionic

(c) metallic   (d) covalent

4. The atomic bond in NaCl is

(a) ionic   (b) metallic

(c) covalent   (d) van der Waal

5. Thermal expansion of materials arises from

(a) thermal vibrations

(b) weak bonds

(c) strong bonds

(d) asymmetry of potential energy curve

6. To break one H–Cl bond, 4.4 eV of energy is required. This energy is equal to

(a) 420 kJ/mol   (b) 420×10^3 kJ/mol

(c) 420 J/k-mol   (d) 42 J/k-mol

7. The electron affinity of helium in kJ/mol is

(a) 4   (b) 16

(c) – 16   (d) 0

8. The length of H–H bonds is

(a) 1 nm   (b) 0.1 m

(c) 0.037 nm   (d) 0.074 nm

9. The Fe–Fe bond length is 2.48 Å. The radius of iron atom is

(a) 0.62 Å   (b) 1.24 Å

(c) 2.48 Å   (d) 3.96 Å

10. Which of the following elements is a covalently bonded crystal?

(a) germanium   (b) aluminium

(c) lead   (d) sodium chloride

11. Hydrogen bonds are stronger than

(a) ionic bonds

(b) metallic bonds

(c) covalent bonds

(d) van der Waals bonds

12. Which of the following relation represent the partial energy of a diatomic molecule?

(a) $Ar^m + Br^n$ (b) $Ar^m - Br^n$

(c) $A/r^m - B/r^n$ (d) $-A/r^m + B/r^n$

13. The radius of an anion is r_a and that of cation is r_c, the bond length is

(a) $\sqrt{3} \; (r_c + r_a)$ (b) $r_a - r_c$

(c) $(r_c + r_a)$ (d) $(r_c + r_a)/2$

14. If 0.28 nm is the spacing between the nearest neighbour ions in NaCl crystal lattice, the unit cell parameter is

(a) 5.6 Å   (b) 1.6 Å

(c) 0.7 Å   (d) 1.4 Å

15. The bond energy of NaCl molecule is given by the relation :

(a) $V = e/4\pi \, \varepsilon_0 r_0^2$ (b) $V = -e/4\pi \, \varepsilon_0 r_0^2$

(c) $V = e^2/4\pi\varepsilon_0 r_0^2$ (d) $V = -e^2/4\pi\varepsilon_0 r_0$

16. Which of the following gives the lattice energy per k-mol of NaCl crystal? (symbols have their usual meanings)

(a) $-\dfrac{AN_A e^2}{4\pi\varepsilon_0 r_0}\left(\dfrac{n-1}{n}\right)$ (b) $\dfrac{AN_A r^2}{4\pi\varepsilon_0 r_0}\left(\dfrac{n}{n-1}\right)$

(c) $-\dfrac{AN_A e^2}{4\pi\varepsilon_0 r_0^2}\left(\dfrac{n-1}{n}\right)$ (d) $\dfrac{AN_A r^2}{4\pi\varepsilon_0 e}\left(\dfrac{n}{n-1}\right)$

17. Metallic bond is not characterized by

(a) opacity

(b) ductility

(c) high conductivity

(d) directionality

18. Mixed ionic-covalent bonds are found in

(a) high strength materials

(b) semiconductors

(c) heat insulators

(d) none of the above

19. The bond energy of C–C covalent bond in kJ/mol is
    (a) 370,800   (b) 640
    (c) 1640   (d) zero

20. The sublimation energy of sodium is
    (a) 1.08×10^3 kJ/kmol
    (b) 10.8×10^3 kJ/kmol
    (c) 108×10^3 kJ/kmol
    (d) 0.108×10^3 kJ/kmol

21. Which of the following has hydrogen bonding?
    (a) HF   (b) C
    (c) CH_4 (d) CsCl

22. A semiconductor has
    (a) van der Waals bonding
    (b) covalent bonding

(c) metallic bonding
(d) ionic bonding

23. The directional interatomic bond is
    (a) van der Waals
    (b) metallic
    (c) covalent
    (d) ionic

24. The cohesive energy in case of van der Waals bonding is
    (a) $0.1 - 0.8$ eV/atom
    (b) $0.002 - 0.1$ eV/atom
    (c) $5 - 8$ eV/atom
    (d) $8 - 12$ eV/atom

25. The solids which are always opaque to visible radiation:
    (a) covalent   (b) ionic
    (c) metallic   (d) van der Waals

Answers

1. (c)	2. (b)	3. (b)	4. (a)	5. (d)
6. (b)	7. (d)	8. (d)	9. (b)	10. (a)
11. (b)	12. (d)	13. (c)	14. (a)	15. (d)
16. (a)	17. (d)	18. (b)	19. (a)	20. (c)
21. (b)	22. (b)	23. (c)	24. (b)	25. (c)

18 X-Rays

18.1 INTRODUCTION

In 1985, X-rays were discovered by Rontgen while studying the phenomenon of discharge of electricity in rarified gases. An induction coil was connected to an evacuated tube in which at the ends of the tube there was a cathode and an anode. The tube was covered with thin black paper. During the experiment he observed that a paper screen washed with barium-platino-cyanide continued to fluorescent whether treated side or other he put towards the discharge tube. The fluorescence was observable 2 metres away from the apparatus. Rontgen soon convinced that the agency, causing fluorescence, origined in the discharge tube at the point where cathode-stream struck the tube. As the nature of these new rays was unknown, Rontgen called them X-rays.

With the discovery of X-rays, Rontgen proceeded to study these rays. These rays are not deflected by magnetic or electric fields. Hence, these do not possess any charge. These cause many other substances to fluoresce, e.g. calcium compounds, uranium glass, rock salt, etc. and also affect photographic plate. These travel in straight lines, causing shadow of the objects placed in their path. He produced a radiograph of a human hand showing clearly the bones against the flesh and asserted that X-rays would prove invaluable in diagnosis in medical practice.

18.2 PRODUCTION OF X-RAYS: COOLIDGE TUBE

The X-rays are produced by bombarding high energy electrons on a matter. The tube used for the purpose of producing X-rays is called an X-ray tube. Rontgen designed X-ray tube, but it has been improved soon. The basic principle of the production of X-rays demands the following parts in the tube:

   i. A source of electron
  ii. A target of high atomic weight
 iii. High tension to accelerate electrons.

Nowadays Coolidge tube (Fig. 18.1) is used for the production of X-rays. This tube was designed by an American physicist Coolidge in 1913. In this tube, the intensity and quality of X-rays can be controlled separately. It consists of a cathode and a tungsten filament F from which electrons are emitted. The filament is placed inside a metal cup C through which a narrow beam of electrons moves towards the target A, placed at an angle of 45° with the path of the beam, made of tungsten or molybdenum.

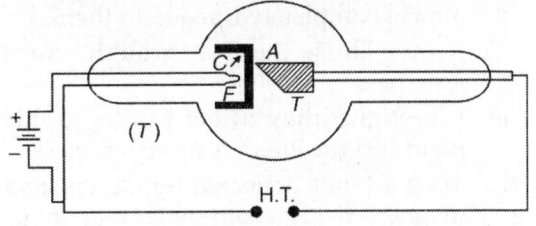

Fig. 18.1: Coolidge X-ray tube

The target *A* is fixed to a copper rod projecting outside and is kept cold by circulating cold water since enormous amount of heat is generated during the working of the tube. Filament and target both are placed inside a vacuum glass tube evacuated to a pressure of about 10^{-8} m of mercury. The filament is heated by a current 3–5 A which emits electrons. When a high voltage of the order of 60 kV to 1000 kV is applied to the target (also called anti-cathode), then this electron beam is accelerated towards the target, and hit it to produce *X*-rays. By adjusting the amount of current in the filament, the number of emitting electrons and hence intensity of emitted *X*-rays can be controlled. The penetrating power of the *X*-rays depends upon the energy of incident electrons. Hence by adjusting proper voltage of cathode and anti-cathode, the energy of electrons and hence the penetrating power of *X*-rays can be controlled. By using accelerated electrons from a betatron, by which electrons can be accelerated up to energies of the order of 10^8 eV, highly energetic and penetrating *X*-rays are produced. These *X*-rays have been used to investigate the structure of atomic nuclei.

18.3 PROPERTIES OF *X*-RAYS

The important properties of *X*-rays are as follows:

i. They are highly penetrating rays and can penetrate through most substances. However, their penetrability is different in different substances. *X*-rays are able to penetrate more in low density substances whereas less in high density substances, e.g. ordinary glass is quite transparent to *X*-rays, but lead glass is almost completely opaque to them.

ii. *X*-rays ionise the gas through which they pass.

iii. Like light, they affect photographic plate and the effect is more intense.

iv. They are not deflected by electric and magnetic fields which shows that unlike cathode rays or positive rays they are not beams of charged particles.

v. They cause fluorescence on several materials, e.g. zinc sulphide, a plate coated with barium platino-cyanide, etc. becomes luminous when exposed to *X*-rays.

vi. They exhibit the phenomenon of reflection, refraction, interference, diffraction and polarisation.

vii. They travel in straight lines with the speed of light.

viii. They liberate photoelectrons from the metals. When they strike metals, *secondary X-rays* are produced.

ix. *X*-rays have a harmful effect upon human body. A long exposure of any part of human body to *X*-rays kills the body tissues and causes incurable sores.

From the above characteristics, it is obvious that *X*-rays are electromagnetic radiations of wavelengths much shorter than those of visible or ultraviolet radiations. One can measure these wavelengths by diffraction experiments. Wavelength of *X*-rays are $\sim10^{-2}$ Å to 100 Å.

18.4 MEASUREMENT OF THE INTENSITY OF *X*-RAYS

One can measure the intensity of *X*-rays by photographic and ionization methods. The determination of the intensity by ionization method is more accurate. For such measurements, ionization chamber is usually employed (Fig. 18.2).

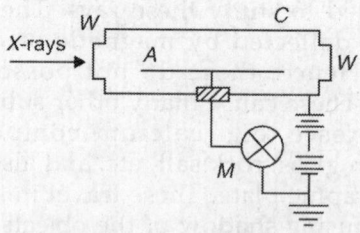

Fig. 18.2: Ionization chamber

C is a cylindrical chamber made of brass. There are two open ends which are closed by two thin sheets *W*, *W* of some low *Z* metal, e.g. aluminium. These are called windows of the chamber. A beam of X-rays can enter the chamber through one of the windows and come out through the other window. The length of the chamber is usually from 20 cm to 100 cm. The chamber is filled with some gas or vapour, e.g. methyl bromide, air, etc.

While passing through the chamber, the X-ray beam ionizes the gas within it. This causes the production of positive ions and negative electrons in the gas. Inside the chamber, a metal rod A insulated from the wall and parallel to the axis of the cylinder is placed. A potential difference of few hundred volts is maintained between A and C. Due to the movement of positive ions toward cathode and negative ions towards anode, ionization current is produced in the chamber. The metal rod A is connected to an electron meter M which can measure the very small ionization current produced in the chamber by the X-rays. It is found that the ionization current is linearly proportional to the intensity of X-rays passing through the chamber. Obviously, the measurement of the ionization current gives a measure of the X-ray beam.

18.5. VARIATION OF THE X-RAY INTENSITY WITH WAVELENGTH (CONTINUOUS AND CHARACTERISTIC X-RAYS)

It is reported that X-rays of different wavelengths are emitted from the X-ray tube. If one measures the intensity I of the X-rays as a function of λ (wavelength) and variation of I with λ is plotted graphically, then one obtains the graphs as shown in Figs. 18.3 and 18.4.

When the anode (i.e. the target) is made of some high Z metal, e.g. tungsten ($Z = 74$), it is found that I (intensity of the emitted X-rays) varies continuously with λ up to potential differences of the order of 50,000 volts between the anode and the cathode (Fig. 18.3). For a given potential difference, there is a minimum wavelength (λ_m) below which no X-rays are emitted. The minimum wavelength is given by

$$\nu_{max} = \frac{(dE)_{max}}{h}$$

or

$$\frac{c}{\lambda_m} = \frac{\text{KE of electron}}{h}$$

$$\therefore \quad \frac{1}{\lambda_m} = \frac{Ve}{ch} \text{ or } \lambda_m = \frac{hc}{eV} \quad (18.1)$$

where V is the tube potential and e, the charge on electron. Putting the values of h, c and e, one obtains

$$\lambda_m = \frac{1.24 \times 10^{-6} \text{ m}}{V \text{ (volts)}} = \frac{12413 \text{ Å}}{V \text{ (volts)}}$$

Obviously, λ_m is inversely proportional to the applied voltage. Relation (18.1) is known as *Duane–Hunt relation*.

With increasing wavelengths, the intensity at first increases and after attaining a maximum it begins to decrease again. From Fig. 18.3, it is evident that λ_m decreases with increasing potential difference (pd). For a given wavelength, the intensity is higher when the pd is higher. Following conclusions can be made from Fig. 18.3:

i. At small wavelength side there are no radiations emitted for given potential difference. Thus, each continuous spectrum shows a sharp limit on shorter wavelength side.

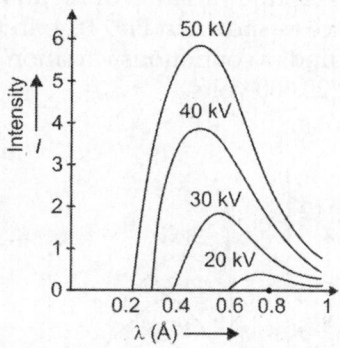

Fig. 18.3: Variation of X-ray intensity versus wavelength emitted from tungeston at different voltages. Continuous X-ray spectrum

ii. The intensity of the spectrum increases for all wavelengths, at all potentials of the tube. After attaining certain maxima, it starts decreasing for higher wavelengths.

iii. The total energy emitted increases with the applied voltage. Furthermore, the maxima of intensity shifts towards the shorter wavelength as applied voltage is increased.

iv. The total intensity of continuous X-rays for given tube potential V is proportional to atomic number Z of the target. For a given target, total intensity depends upon V^2 and electron beam current.

Further, it has been found that minimum wavelength limit of continuous X-rays does not depend upon the element of the target.

When an electron passes through the electric field of the atomic nucleus, it loses energy by radiation. This energy appears as a continuous X-ray spectrum and is called bremsstrahlung or breaking radiation. This means continuous radiations are actually produced by the 'braking' or slowing down of incident charged particles. The continuous distribution of wavelengths emitted is analogous to the continuous distribution of wavelengths in visible white light. Obviously, this type of X-radiation is also sometime called *white radiation*.

When the target is of low Z, e.g. molybdenum ($Z = 42$), the variation of I with λ has the appearance as shown in Fig. 18.4. In this case also, one finds a continuous variation of I with λ for $V < 20,000$ volts.

However, at higher potential difference ~25,000 V or even higher, several discrete peaks appear, superimposed upon the continuous background at certain definite wavelengths. Experimentally, it is observed that for targets with still lower Z, e.g. copper ($Z = 29$), such peaks begin to appear at sill low potential difference ($V \gtrsim 8000$ volts). We must note that the wavelengths at which the peaks appear for the copper target are different from those for the molybdenum target. This means the characteristic X-rays are the characteristics of the target materials and are different for different target elements.

The characteristic X-ray spectrum for molybdenum target (Fig. 18.4) exhibit two sharp peaks of high intensity occurring at wavelengths 0.71 Å and 0.63 Å. Barkla called these as K lines. Barkla also showed the presence of L-lines which are less penetrating (Fig. 18.5). Afterwards M-series has also been discovered. Mosely called these K-lines as K_α and K_β lines. These lines generally occur in the form of small groups. The group in the short wavelength region is termed K-series, the next group in the direction of increasing wavelength as L-series, then M-series and so on. Further, there is a definite short wavelength limit to the continuous X-ray spectrum.

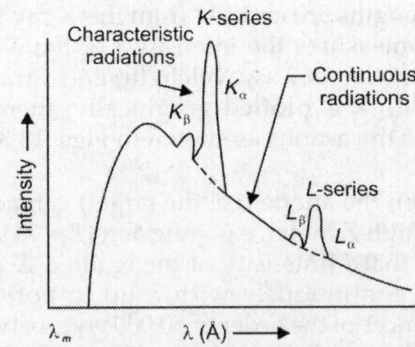

Fig. 18.5: The variation of intensity with λ at sufficiently high voltage

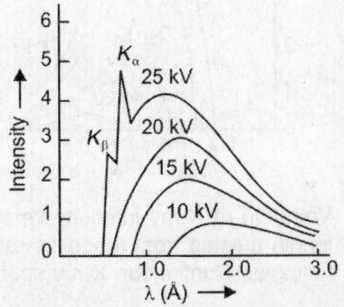

Fig. 18.4: Intensity versus wavelength curves for X-ray emitted from molybdenum target (Z=42) for different voltages. There apears peaks at higher voltages

18.6 ORIGIN OF CONTINUOUS SPECTRUM

According to electromagnetic theory of light, the accelerated or decelerated light charged particles, e.g. electrons or protons emit electromagnetic radiations of different frequencies. A part of the kinetic energies of these particles is transformed into the energy of the emitted radiations. When the electrons accelerated within the X-ray tube hit the anode (i.e. target), their motion becomes decelerated, i.e. their velocities decrease. During the process, the electrons experiences strong coulomb's force of attraction due to the nucleus of the target atom and is suddenly slowed down, at the same time suffering a deflection in its path. As a result, electrons emit electromagnetic radiations with a continuous distribution of wavelengths starting from a minimum.

Let an incident electron moving with velocity v be slowed down to a velocity v', then by the conservation of energy principle, the frequency of the emitted photon is given by

$$\frac{1}{2}mv^2 - \frac{1}{2}mv'^2 = h\nu = \frac{hc}{\lambda} \qquad (18.2)$$

where $h\nu$ is the energy of the photon of frequency ν (i.e. wavelength $\lambda \; \because \; \nu = c/\lambda$). If the electron is completely stopped during a collision, v (velocity) $= 0$, so that the maximum energy of the emitted photon will be equal to

$$E_m = h\nu_m = hc/\lambda_m = eV \qquad (18.3)$$

where V is potential difference between the cathode and the anode within the X-ray tube. Obviously, the electrons emitted from cathode acquire an energy eV when they hit the target (anode). Obviously, the wavelength of the emitted radiation depends on the fraction of the incident electron energy (eV) that is transformed into X-radiation. We must remember that many of the impinging electrons penetrate several atomic layers in the target before they are brought to rest. Thus not all the impinging electrons give rise to X-ray photons of the wavelength in accordance with Eq. (18.3). As a result, a continuous spectrum is obtained with the minimum wavelength limit set by Eq. (18.3). The increase in applied potential difference decreases the minimum wavelength (λ_{min}).

Using Eq. (18.3), one can estimate the maximum ν_m and λ_m as follows:

$$\lambda_m = \frac{hc}{eV} = \frac{12413}{V(\text{volts})} \; \overset{\circ}{A}$$

For $V = 10,000$ volts, one obtains $\lambda_m = 1.24$ Å and for $V = 50,000$ volts, $\lambda_m = 0.248$ Å.

It is worthwhile to mention that sometimes X-rays are classified according to their penetrating power. The most penetrating radiation is termed *hard radiation* whereas the radiation with relatively less penetrating power is called *medium radiation*. The radiation with very low penetrating power is termed *soft radiation*.

One can understand the nature of the spectral distribution of the continuous X-rays on the basis of the theory of bremsstrahlung based on electromagnetic theory.

Due to deceleration of the electron beam in the direction of its motion within the target inside the X-ray tube, we have to take the spectral energy distribution of the bremsstrahlung X-ray for electrons of different energies and assign different weights to these curves. These are based on the adsorption of X-rays within the X-ray tube, in air and in window of the inonization chamber, other variation of reflecting power of the crystal grating for different wavelengths and different ionizing powers of the beams of different wavelengths.

One finds that the spectral energy distribution of the bremsstrahlung radiation is independent of the frequency and can be expressed as

$$I_\lambda = I\nu(v^2/c^2) \propto 1/\lambda^2 \qquad (18.4)$$

Equation (18.4) explains the observed spectral distribution of the continuous X-rays.

18.7 ORIGIN OF THE CHARACTERISTIC X-RAYS SPECTRUM (X-RAYS PEAKS)

When an impinging electron has sufficiently large energy to enable it to penetrate into the interior of a target atom, it may knock off an electron in the K-shell of the target atom (Fig. 18.6a). Obviously, the atom is excited. The atom can remain in the excited state only for a very short interval of time ($\sim 10^{-8}$ s). An electron from the *L*-shell may jump into the vacant space in the *K*-shell as shown in Fig. 18.6b. The difference in energy between *K* and *L*-shells is emitted in the form of *X*-ray photon. The resulting line is the first line of the *K*-series. If an electron from the *M*-shell jumps into the vacant space in the *K*-shell,

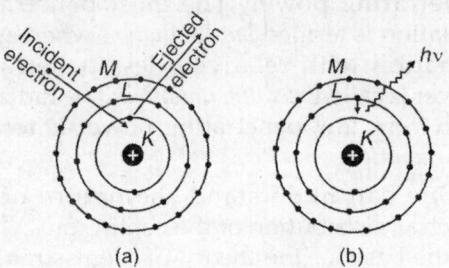

(a)   (b)

Fig. 18.6: Knocking off electron from *K*-shell

one obtains the second line of the *K*-series. Obviously, the jumps from the various outer shells of the atoms to the *K*-shell constitute the group of lines in the *K*-series. These radiations have the maximum energy and are the most penetrating. Similarly, the spectral lines in the *L*-series arise from the transitions of the electrons from the outer shells to the *L*-shell and so on. Clearly, the frequencies of *L*-series are smaller than those of the *K*-series. As the energies of various levels in target atom depends on the element of which the target is made, the wavelengths (or the frequencies) of the lines given out are different in the case of different elements. This is the reason that these spectral lines are called characteristic lines of the *X*-ray spectrum. One can better understand the production of *X*-ray spectral lines from the energy level diagram shown in Fig. 18.7. Here *K*, *L*, *M*, etc. show the various energy levels of the orbits. A jump from *L*-shell to *K*-shell gives rise to K_α line and so on. The *X*-ray characteristic lines originating from the transitions between the different *X*-ray levels are usually designated by the symbols α, β, γ, etc. For example,

K-level to *L*-level transition → K_α line

L-level to *M*-level transition → L_α line

L-level to *N*-level transition → L_β line

and so on. The frequencies of these lines are given by

$$v(K_\alpha) = \frac{E_K - E_L}{h}$$

$$v(L_\alpha) = \frac{E_L - E_M}{h} \tag{18.5}$$

and so on. One can also produce the characteristic radiation by allowing the continuous *X*-rays generated by the bremsstrahlung process in the *X*-ray tube to be scattered from different elements outside the tube. The bremsstrahlung process can be represented symbolically by the equation electron → electron + photon. These *X*-ray photons can eject electrons from the inner *K*, *L*, *M*, *N*, etc. orbits of the target atom of the scatterer if their energy hv is greater than the binding energies of the electrons in these orbits. Obviously, the vacancy thus created in an inner orbit is subsequently filled up by the transition of an electron from an outer orbit. The surplus energy due to such transtion in an atom is carried away by a characteristic *X*-ray photon. The characteristic radiation is also sometimes called *fluorescent radiation*. The K_α and K_β lines of molybdenum have wavelengths of about 0.07 and 0.06 nm respectively.

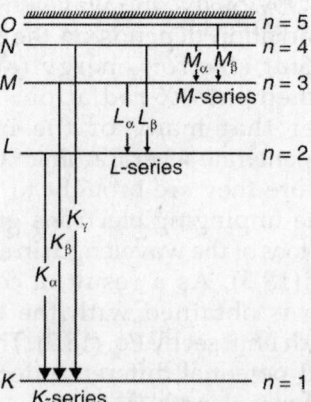

Fig. 18.7: Origin of characteristic spectrum K_α, K_β, etc. due to electron jump from higher level

18.8 FINE STRUCTURE OF *X*-RAY LINES

Careful measurement and detailed investigation has revealed that the characteristic X-ray lines are not single, but they have fine structures of their own. It is found that each of them is composed of a number of closely lying components. One can understand their origin by considering the three quantum numbers n, l and j associated with each electronic level. For a given principal quantum number n which determines the positions of the different electrons shells, e.g. $K(n = 1)$, $L(n = 2)$, $M(n = 3)$, $N(n = 4)$, etc., l (the azimuthal quantum number) can assume n different values: $l = 0, 1, 2, 3, ..., (n - 1)$. Again for a

given l, j (the total quantum number) can assume the two values: $j = l + 1/2$ or $j = l - 1/2$, whereas for $l = 0$, j can have only one value, viz. $j = 1/2$. This has led to the conclusion that K level is single, L level a triplet, M level a quintuplet and so on.

The two selection rules which govern the transition between the different X-ray levels are:

$$\Delta l = \pm 1 \text{ and } \Delta j = 0, \pm 1 \qquad (18.6)$$

Figure 18.8 shows the fine structure of the X-ray levels and the possible transitions between them. From the figure, it is evident that K_α line is composed of the two components K_{α_1} and K_{α_2}; K_β line is composed of

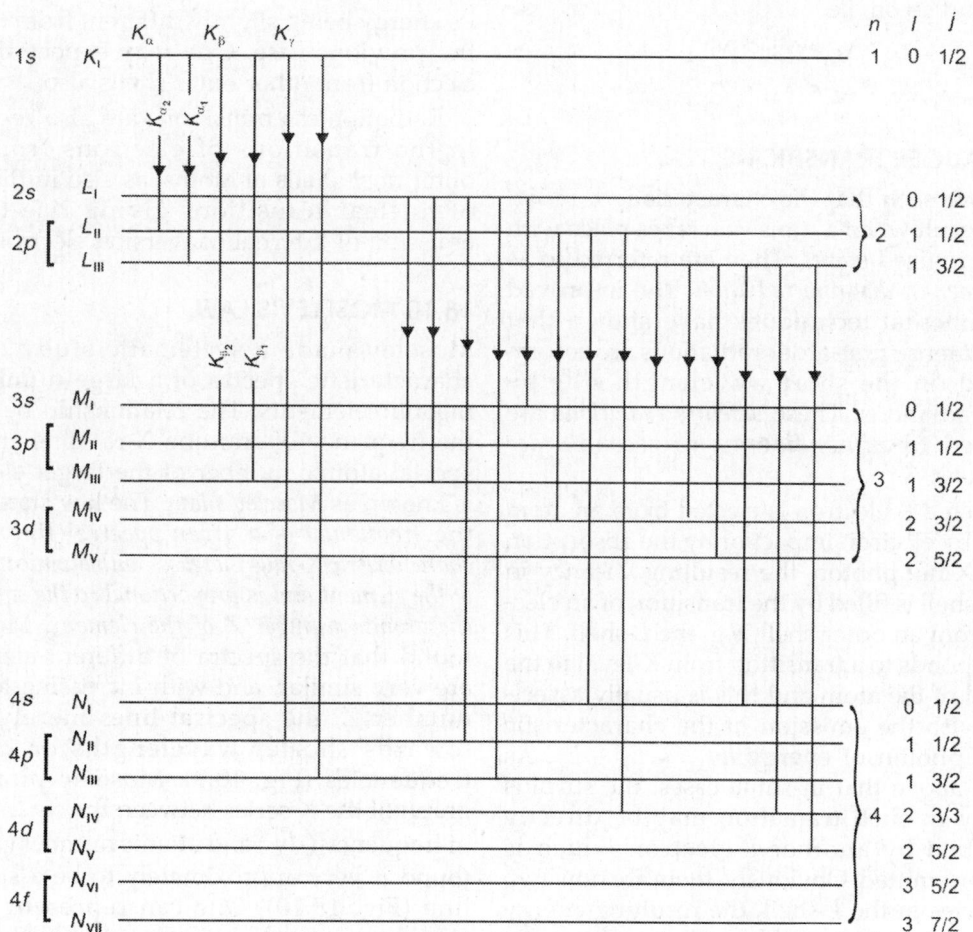

Fig. 18.8: Fine structure of *X*-ray lines and the possible transitions between them

two components K_{β_1} and K_{β_2}. We can see from Fig. 18.8 that K_{α_1} originated due to transitions from the K level to the L_{111} level for which $n = 2, l = 1$ and $j = 3/2$. We note that K_{α_2} component originates from the transition between K level to the L_{11} level for which $n = 2$, $l = 1$ and $j = 1/2$. One finds that the splitting of L, M, N, etc. lines are more complex.

We know that the binding energy of an electron in the K-shell is much higher than that of an electron in the L-shell and hence the energies and obviously, the frequencies of the K lines are much higher than those of the L lines so that the wavelengths of K lines are much shorter than those of the L lines. Similarly, one finds that the wavelengths of L lines are much shorter than those of the M lines and so on, i.e.

$$\nu_K > \nu_L > \nu_M > \nu_N \cdots$$

or $\qquad \lambda_K < \lambda_L < \lambda_M < \lambda_N \cdots$

18.9 AUGER TRANSITIONS

We have seen that the characteristic X-radiations result when a transition takes place with singly ionized states. They are referred to as *first-order* or *diagram radiation*. The improved experimental techniques have shown that these intense first-order radiations are accompanied on the short wavelength side by faint radiations. These *satellite radiations* are referred to as *nondiagram* or *second-order radiations*.

When a K-electron is ejected from an atom either by electron impact or by the absorption of an X-ray photon, the resulting vacancy in the K-shell is filled by the transition of an electron from an outer shell, e.g. the L-shell. This corresponds to a transition from K-level to the L-level of the atom and this is usually associated with the emission of the characteristic X-ray photon of energy $h\nu_{KL} = E_K - E_L$. As stated above that in some cases, the surplus energy in such transition may be directly absorbed by another L-electron which is thereby emitted. Obviously, there are now two vacancies in the L-shell, the resulting energy level of the atom will be different from the L level and can be referred to as LL level, its

energy being given by E_{LL}. One can easily find that the kinetic energy of the ejected second electron will then be

$$\frac{1}{2}mu^2 = E_K - E_{LL} \qquad (18.7)$$

Such a non-radiative transition resulting in the emission of electrons from the same atom first observed by Auger in 1925 are called *Auger transitions* and the effect is known as *Auger effect*.

We must note that the surplus energy in the transition of the first L-electron to the vacant K-shell may also be absorbed by an M-electron instead of a second L-electron as stated above. Obviously, in this case, the second electron would be an M-electron and its energy being slightly different from that in the previous case. One may expect similar ejection from other outer shells also.

Radiationless transitions are also reported in the transitions of electrons from the outermost shells of atoms, as also in the case of nuclear transitions giving rise to the emission of internal conversion electrons.

18.10 MOSELEY'S LAW

Moseley made a systematic study of the characteristic spectra of a large number of metallic elements. The relationship between the frequency of emitting X-ray line and the special atomic number of the target element is known as *Moseley's law*. The law states that the *frequency of a given spectral line of the characteristic X-rays increases with atomic number of the element and is proportional to the square of the atomic number Z of the element*. Moseley found that the spectra of different elements are very similar, and with increasing atomic number Z, the spectral lines merely shift towards shorter wavelengths or higher frequencies (Fig. 18.9). Moseley plotted a graph of the K-series between the square root of frequency ($\sqrt{\nu}$) and atomic number (Z) and found it very approximately to be a straight line (Fig. 18.10). One can represent these straight line graphs by an equation of the form (Moseley's law),

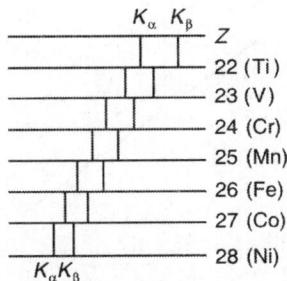

Fig. 18.9: Shifting of spectral lines towards shorter wavelengths

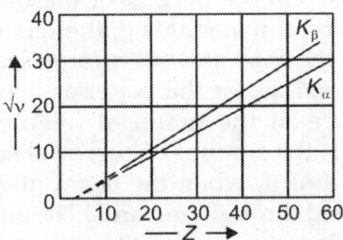

Fig. 18.10: Variation of square root of frequency ($\sqrt{\nu}$) with atomic number (Z)

$$\sqrt{\nu} = c_1(Z - b) \qquad (18.8)$$

where c_1 and b are constants for a given series. The constant b is called the screening factor. The X-ray spectrographs of all elements of periodic table prove the validity of Moseley's law. The law is applicable to other series also but with different values of c_1 and b. Rewriting Eq. (18.8)

$$\nu = C(Z - b)^2 \qquad (18.9)$$

where C is another constant. For K_α radiation, it is reported that $C = \dfrac{3}{4}Rc$, where R is Rydberg constant, c is the velocity of light in vacuum. For K_α radiation, $b = 1$ and hence one can write

$$\nu_{K_\alpha} = \frac{3}{4}Rc(Z - 1)^2 = Rc(Z - 1)^2 \left[\frac{1}{1^2} - \frac{1}{2^2}\right] \qquad (18.10)$$

One can see that Eq. (10) is similar to the expression for the first line of the Lyman series of hyrdrogen, the only difference being the presence of the factor $(Z - 1)^2$ in the present case in place of $Z^2 = 1$ for the case of hydrogen atom.

We have seen earlier that Mendeleev had arranged the elements in the periodic table in the order of increasing atomic weights. However, Mendeleev's table had a few anomalies, e.g. the properties of argon, cobalt and tellurium were not in agreement with the properties of the other elements of their group.

Moseley's work on X-ray spectra clearly indicated that an element is distinguished by its atomic number Z (number of positive charges on the nucleus of the atom). Rutherford in the same year also confirmed that the atomic number was the fundamental constant which decided the chemical properties of the atom. Moseley then pointed out that the elements in the periodic table of elements, must be arranged in the order of increasing atomic number (instead of atomic weight). In the light of this, Mendeleev changed the positions of certain elements in the periodic table. For example, Mendeleev found that the atomic number of cobalt ($^{58.9}_{27}\text{Co}$) is less than that of nickel ($^{58.7}_{28}\text{Ni}$) even though its atomic weight is greater. Keeping this in view, he arranged them in the order Co, Ni instead of Ni, Co. Similarly, Mendeleev placed argon ($^{40}_{18}\text{A}$) before potassium ($^{39}_{19}\text{K}$) and tellurium ($^{127.6}_{52}\text{Te}$) before iodine ($^{127}_{53}\text{I}$). When he made these changes, the anomalies of Mendeleev's periodic table disappeared.

Moseley had left certain gaps while arranging the elements in the order of increasing atomic number, e.g. at Z = 43 and 72. He predicted that the corresponding rare earth elements could exist but had not then beendiscovered. These elements were subsequently discovered and are named as technetium and hafnium respectively.

18.11 ABSORPTION OF X-RAYS

When a parallel beam of X-rays is incident over a slab of matter, a portion of it is being absorbed by the slab and the remaining part is transmitted. The absorbed portion of X-rays depends upon its intensity and thickness of the slab. Let a beam of X-rays of wavelength λ and intensity I be incident over the slab of the material of uniform thickness dx. If corresponding diminution in intensity is dI, we have

$$dI \propto -I dx$$

or $\qquad dI = -\mu I dx \qquad (18.11)$

where the constant of proportionality μ is called the linear coefficient of absorption of the material and is equal to the fractional decrease in the intensity of the X-rays per unit thickness of the absorber.

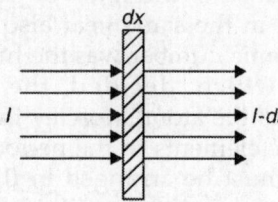

Fig. 18.11: Absorption of X-rays

Arranging Eq. (18.11) and integrating it, one obtains

$$\int \frac{dI}{I} = -\int \mu \, dx$$

or $\qquad \log_e I = -\mu x + C$

where C is integration constant, which can be determined from initial conditions. Thus, at $x = 0$, the intensity $I = I_0$. Substituting, we have

$$\log_e I_0 = C$$

If after traversing a distance x in the slab, the corresponding intensity is I, we have

$$\log_e I = -\mu x + \log_e I_0$$

or $\qquad \log_e \left(\dfrac{I}{I_0} \right) = -\mu x$

or $\qquad I = I_0 \exp(-\mu x) \qquad (18.12)$

Using Eq. (18.12) and knowing initial intensity and intensity after traversing a distance x in the material, μ can be determined. μ has the dimension of m^{-1}. The plot of $\log I$ against x is a straight line of slope $-\mu$ (Fig. 18.12).

The intensity of X-rays depends upon the energy transmitted through unit area placed perpendicular to the direction of propagation, whereas its absorption depend upon the number of atoms of the absorbing material. Hence, linear coefficient of absorption also

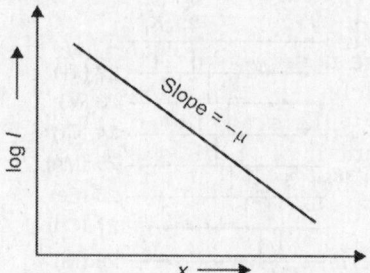

Fig. 18.12: The plot of log I versus x

depends upon the density of the material. If the density of material is ρ, the quantity μ/ρ is called the *mass absorption coefficient* (μ_m) of the material. μ_m of the material depends on the nature of the material (element) and represents the amount of energy absorbed by unit mass of it, when the beam of X-rays is transmitted through unit area. The absorption by specific atom is called the *atomic absorption coefficient* (μ_A) and is given by

$$\mu_A = \frac{\mu/\rho}{N/M} = \frac{\mu M}{\rho N} \qquad (18.13)$$

where M and N are atomic weight of the material and Avogadro number respectively.

18.12 SCATTERING OF X-RAYS

When a beam of X-rays is allowed to fall on a plate of some chosen element, then along with the absorption of these rays, which transformed into heat, etc. a portion is transmitted through the plate. The rays going out from the plate consist of a part of primary X-rays and part of secondary radiations. The secondary radiations contains four important types of radiations:

Scattered X-rays

These are similar to incident X-rays which have only changed their direction of propagation by passing through the absorbing material. Like primary rays, these are heterogeneous and independent of the nature of the absorbing material.

Characteristic X-rays

These are typical of the nature of the absorber material. These are homogeneous and do not depend on the quality of the primary rays.

Scattered β-rays

These are nothing but fast moving electrons (β-particles) produced by photoelectric process. Like scattered X-rays, these are independent of the nature of the absorber material but depend on the quality of the primary or incident rays.

Characteristic β-rays

These are similar to scattered β-rays with the exception of being characteristic of the absorber material.

Figure 18.13 depicts the phenomenon of interaction of X-rays with matter.

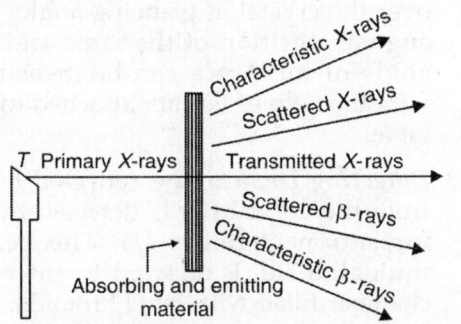

Fig. 18.13: Interaction of X-rays with matter

18.13 *X-RAY DIFFRACTION AND CRYSTAL AS THREE DIMENSIONAL GRATING*

Soon after the discovery of X-rays in 1895, the nature of X-rays became a matter of concern. Results of early experiments indicated that if X-rays consisted of waves, their wavelengths were of the order of 10^{-8} to 10^{-9} cm. Laue suggested that the ordered arrangement of atoms in a crystal must make it to act as a three-dimensional grating. Since the spacing between the layers of these atoms in a crystal is of the order of 10^{-8} cm (which is of the order of wavelength of X-rays), so a crystal would be suitable for the diffraction of X-rays.

Laue Experiment

The experimental arrangement of Laue and his co-workers is shown in Fig. 18.14. A thin pencil of X-rays after passing through aligned slit was allowed to pass through a thin plate of crystal of zinc blende. The transmitted beam

is received on a photographic plate. After the exposure of several hours, when the plate was developed, it has been found that in addition to the central spot, it consists of other fainter spots arranged regularly (Fig. 18.14). These spots are known as *Laue spots*.

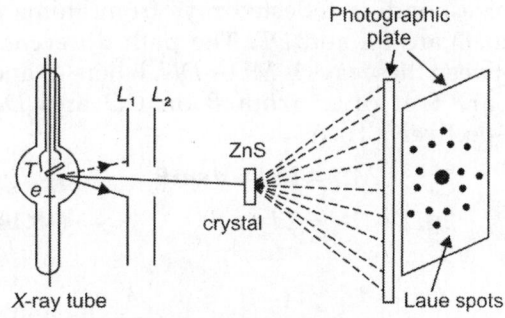

Fig. 18.14: Laue experiment

Each spot in the Laue pattern arises due to the constructive interference between the waves reflected from one of the various sets of parallel planes in the crystal. Figure 18.15 shows two such sets of planes by means of continuous and dotted lines. The corresponding planes will be $\perp^r$ to the plane of the paper. A simple interpretation was provided by W L Bragg.

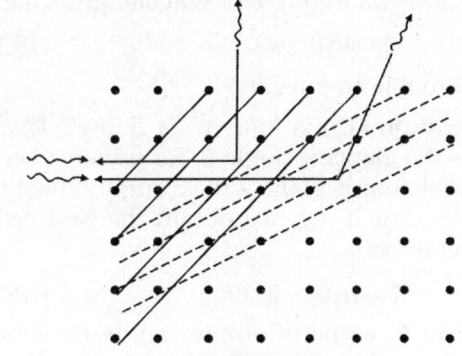

Fig. 18.15: Laue spots

Bragg's Law

Consider a set of parallel planes of the crystal separated by spacing d (Fig. 18.16). The crystal acts as a series of parallel reflecting planes. Let a beam of monochromatic X-rays strike the crystal. Since *X*-rays can penetrate the crystal, there would be partial reflection from every plane till it would be absorbed completely. The rays reflected from various planes will interfere constructively if they are in same phase. Consider two parallel rays *AB* and *CD* incident over the planes at glancing angle θ. Corresponding reflected rays from atoms *B* and *D* are *BE* and *DF*. The path difference between these rays is *MD* + *DN*, where *M* and *N* are feet of ⊥ʳ from *B* on *CD* and *DF* respectively. Then

$$MD = DN = d \sin\theta$$

$$\therefore \quad MD + DN = 2d \sin\theta \qquad (18.14)$$

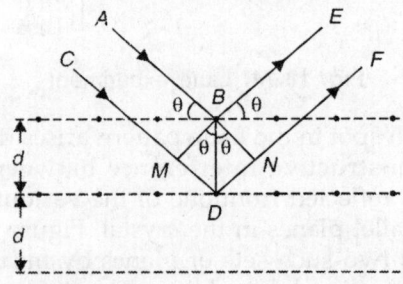

Fig. 18.16: Derivation of Bragg's law

Now the two sets of scattered waves reinforce each other when the path difference is an integral multiple of wavelength λ, i.e.

$$2d \sin\theta = n\lambda \qquad (18.15)$$

Here *n* is an integer.

Relation (15) is known as *Bragg's law*. It gives the glancing angle θ, for which a set of parallel atomic plane will strongly reflect the *X*-rays. For *n* = 1, we obtain the first order reflection as

$$2d \sin\theta_1 = \lambda \qquad (18.16)$$

Here θ₁ is the minimum angle for which reflection will occur. For *n* = 2, i.e. second order reflection, we have

$$2d \sin\theta_2 = 2\lambda \qquad (18.17)$$

Obviously, the Laue spots can easily be explained. Using Bragg's relation (18.15) and knowing lattice space *d*, the wavelength (λ) of *X*-rays can be determined.

Bragg's X-ray Spectrometer

In 1913, W H Bragg and his son W L Bragg developed the *X*-ray spectrometer for the measurement of *X*-ray wavelengths and spacing between the lattice planes of crystals. It consists mainly of following three parts:

1. **Source of X-rays (T):** It is an *X*-ray tube from which collimated fine pencil of *X*-rays is obtained.

2. **Turntable:** It is a rotating table similar to the ordinary spectrometer. Over this turntable is mounted the crystal (*C*). A fine pencil of *X*-rays is allowed to fall over the crystal at glancing angle. The angular position of the table and the angle of incidence can be measured with the help of vernier attached to the table.

3. **Detecting Device:** The reflected beam from the crystal (*C*) is detected by an ionisation chamber (*I*). The beam, through a slit, is allowed to enter the chamber filled with ethyl bromide. The ionisation current is measured by the electrometer (*E*). The arm carrying ionisation chamber can be rotated and be recorded with the help of vernier. For recording the intensity of reflected beam, in place of ionisation chamber, a photographic plate may also be used. Figure 18.17 shows the essential parts of the spectrometer. The ionisation current produced by the *X*-rays is a measure of the intensity of *X*-rays diffracted by an angle 2θ.

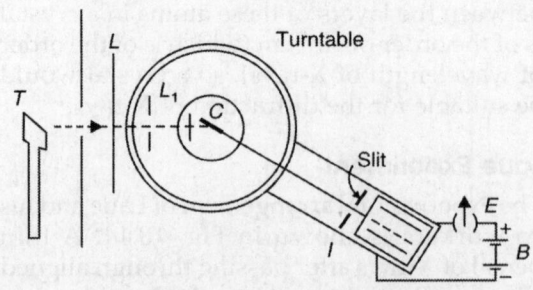

Fig. 18.17: Bragg's spectrometer

At each particular setting of the crystal, i.e. a certain value of θ, only the particular wavelength λ which satisfied the Bragg's relation

$$2d \sin\theta = n\lambda$$

is reflected to the photographic plate or ionisation chamber. Thus, if the crystal spacing d is known, then corresponding to the known value of θ, the wavelength of X-rays can be determined. Alternatively, if the X-ray wavelength is known, the crystal lattice spacing can be determined.

To obtain a spectrum of X-ray wavelength produced by a certain source, the ionisation current (i) is measured as a function of the glancing angle (θ). Figure 18.18 shows a typical plot obtained with a platinum target. The ionisation current is seen to increase suddenly at some sharply defined angles; the three peaks A_1, B_1 and C_1 represent first order X-ray spectrum lines. Second order of these lines appear at A_2, B_2 and C_2 at glancing angles whose sines are twice those of A_1, B_1, C_1. The relative intensities of the second order spectral lines are also in the same ratio as those of the corresponding lines in the first order. This line spectrum is superimposed on continuous spectrum.

The X-ray spectral lines are characteristic of the nature of the material used as target. The choice of the crystal to be used is governed by the range of wavelengths to be measured, the case with which a good crystal surface can be obtained and the reflecting power of the crystal. Commonly used crystals are calcite, rock salt and quartz.

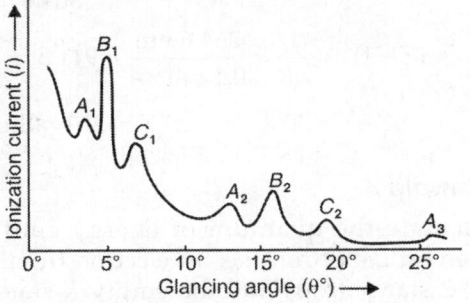

Fig. 18.18: Ionization current (i) as a function of glancing angle θ.

Determination of Wavelength

To determine wavelength of X-rays, the grating spacing d must be known accurately. As an example consider the case of NaCl crystal. X-ray study of this crystal shows that the sodium and chlorine atoms occupy alternate corners of the cube lattice. Hence, on the average the volume occupied by every atom is d^3 with mass ρd^3 where ρ is the density. Thus,

$$\rho d^3 = \frac{M}{2N}$$

Here M is molecular weight of sodium chloride = 58.454

∴ Mass of one mole of NaCl = 58.454 gm

N_0 is Avogadro number = 6.03×10^{23}

Since each mole of any substance consists of $N_0 = 6.03 \times 10^{23}$ molecules, the mass of each NaCl molecule

$$= \frac{58.454}{6.03 \times 10^{23}} = 9.71 \times 10^{23} \text{ gm/mole}$$

or mass of half a molecule or that of an elementary cube of NaCl

$$= \frac{9.71 \times 10^{-23}}{2} = 4.855 \times 10^{23} \text{ gm}$$

The density of NaCl = 2.16 gm/cm³

∴ Volume of the elementary cube of NaCl

$$= \frac{\text{Mass}}{\text{Density}}$$

$$= \frac{4.855 \times 10^{23}}{2.16} \text{ cm}^3$$

But volume of elementary cube = d^3

$$\therefore \quad d^3 = \frac{4.855 \times 10^{23}}{2.16}$$

$$\therefore \quad d = \left(\frac{4.855 \times 10^{23}}{2.16} \right)^{1/3}$$

$$= 2.82 \times 10^{-8} \text{ cm} = 2.82 \text{ Å}$$

ILLUSTRATIVE EXAMPLES

Example 1

Calculate the shortest wavelength of X-rays produced in a tube when the applied voltage is 12.4 kV.

Solution

Given that $h = 6.63 \times 10^{-34}$ Js
$c = 3 \times 10^8$ m/s
$e = 1.6 \times 10^{-19}$ C

We have $eV = h\nu_{max} = \dfrac{hc}{\lambda_{min}}$

$\therefore \qquad \lambda_{min} = \dfrac{hc}{eV}$

$$= \dfrac{6.63 \times 10^{-34} \times 3 \times 10^8}{1.6 \times 10^{-19} \times 12.400}$$

$$= 1 \times 10^{-10} \text{ m} = 1 \text{ Å}$$

Example 2

What should be the kinetic energy of an electron so that its de Broglie wavelength is same as that of X-rays produced in an X-ray tube operating at 30 kV?

Solution

We have $\lambda_{min} = \dfrac{hc}{eV}$ \hfill (1)

If p be the momentum and k be the kinetic energy of the electron of mass m, we have

$$k = \dfrac{p^2}{2m} \qquad\qquad (2)$$

But $p = \dfrac{h}{\lambda}$, where λ is the de Broglie wavelength of the electron

$\therefore \qquad k = \dfrac{h^2}{2m\lambda^2}$

Using (1), one obtains

$$k = \dfrac{e^2V^2}{2mc^2}$$

Here $h = 6.63 \times 10^{-34}$ Js, $m = 9.1 \times 10^{-34}$ kg, $e = 1.6 \times 10^{-19}$ C, $V = 30000$ volt

$$\therefore \quad k = \dfrac{(1.6 \times 10^{-19})^2 \times (30000)^2}{2 \times (9.1 \times 10^{-13}) \times (3 \times 10^8)^2} = 1.4 \times 10^{-16} \text{ J}$$

Example 3

The series limit of the Balmer series for hydrogen is 364.6 nm. Find the atomic number of the element which gives X-ray wavelengths down to 0.1 nm.

Solution

The Balmar series of hydrogen is given by

$$\dfrac{1}{\lambda} = R\left(\dfrac{1}{2^2} - \dfrac{1}{n^2}\right); \ n = 3, 4, 5, ..., \infty$$

Now, for the series limit ($n = \infty$), one gets

$$\dfrac{1}{\lambda_\infty} = \dfrac{R}{4}$$

or $\qquad R = \dfrac{4}{\lambda_\infty} = \dfrac{4}{364.6 \, \text{nm}}$

We have the X-ray wavelengths of the K-series as given by

$$\dfrac{1}{\lambda} = R(Z-1)^2\left(\dfrac{1}{1^2} - \dfrac{1}{n^2}\right)$$

The minmum wavelength of this series occurs when $n = \infty$ and thus

$$\dfrac{1}{\lambda} = R(Z-1)^2$$

(as $\lambda = 0.1$ nm and $R = \dfrac{4}{364.6}$ nm)

or $\quad (Z-1)^2 = \dfrac{1}{\lambda R} = \dfrac{364.6 \, \text{nm}}{0.1 \, \text{nm} \times 4} = 911.5$

or $\qquad Z - 1 = 30.2 \qquad \therefore \ Z = 31$ (gallium)

Example 4

Calculate the quantum of energy emitted when an L-electron (i.e. an electron from the $n = 2$ state) drops into the empty K-state in copper ($Z = 29$). Given $R = 109737 \times 10^2$ m^{-1}; $1\text{m}^{-1} = 1.239 \times 10^{-2}$ eV.

Solution

The electron drops from $n = 2$ to $n = 1$ state, we have

$$\frac{1}{\lambda} = R(Z-b)^2 \left[\frac{1}{1^2} - \frac{1}{2^2} \right]$$

For the K-line, $b = 1$. Thus,

$$\frac{1}{\lambda} = \frac{3}{4} R(Z-1)^2$$

$$= \frac{3}{4} \times (109737 \times 10^2 \, \text{m}^{-1})(29-1)^2$$

$$= 6.4525 \times 10^9 \, \text{m}^{-1} = 7994.6 \, \text{eV}.$$

Example 5

The wavelength of L_α X-ray lines of silver and platinum are 4.1538 Å and 1.3216 Å respectively. An unknown substance emits L_α X-ray with a wavelength of 0.966 Å. If the atomic number of silver and platinum be 47 and 78 respectively, determine atomic number of the unknown substance.

Solution

We have Moseley's law

$$\sqrt{\nu} = K(Z - b) = \sqrt{\frac{C}{\lambda}}$$

$$\therefore \quad \sqrt{\frac{3 \times 10^8}{4.1538 \times 10^{-10}}} = K(47 - b) \quad (1)$$

and $\sqrt{\dfrac{3 \times 10^8}{1.3216 \times 10^{-10}}} = K(78 - b) \quad (2)$

For unknown substance,

$$\sqrt{\frac{3 \times 10^8}{0.966 \times 10^{-8}}} = K(Z - b) \quad (3)$$

Dividing (1) by (2), one obtains

$$\sqrt{\frac{1.3216}{4.1538}} = \frac{47 - b}{78 - b}$$

$$\therefore \quad b \approx 7 \text{ (app.)}$$

Dividing (3) by (1), one obtains

$$\sqrt{\frac{1.3216}{0.966}} = \frac{Z - 7}{78 - 7}$$

or $1.142(78 - 7) = Z - 7$

$$\therefore \quad\quad\quad Z = 90.$$

Example 6

X-rays of wavelength 0.36 Å diffracted in a Bragg spectrometer at an angle of 4°48′. Find the effective value of atomic spacing.

Solution

Given $\lambda = 0.36 \times 10^{-8}$ cm; $\theta = 4°48'$

$\sin 4°48' = 0.0837$

We have from Bragg's law

$$\lambda = 2d \sin \theta$$

$$\therefore \quad\quad d = \frac{\lambda}{2 \sin \theta}$$

$$\therefore \quad d = \frac{0.36 \times 10^{-8}}{2 \times 0.0837} = 2.15 \times 10^{-8} \text{ cm} = 2.15 \text{ Å}.$$

Example 7

The glancing angle for the first order reflection from a nickel crystal is 9.5°. If the wavelength of X-rays is 0.58 Å, what is the spacing between the atomic planes of the crystal?

Solution

Given $n = 1$, $\lambda = 0.58 \times 10^{-10}$ m, $\theta = 9.5°$

$\sin 9.5° = 0.165$

$$2d \sin \theta = n\lambda$$

$$\therefore \quad\quad d = \frac{n\lambda}{2 \sin \theta}$$

$$= \frac{0.58 \times 10^{-10}}{2 \times 0.165} = 1.76 \times 10^{-10} \text{ m} = 1.76 \text{ Å}$$

REVIEW QUESTIONS

1. Give an account of the production and properties of X-rays. How can it be established that these are electromagnetic radiations ?

2. What are continuous and characteristic X-rays? Discuss their salient features.

3. How are continuous X-rays produced? What is the significance of the short wavelength limit and how it is related with the voltage applied across the X-ray tube?

4. What is characteristic X-ray spectrum? Explain its origin.

5. Discuss Moseley's law and explain it on the basis of Bohr atom model. What do you understand by bremsstrahlung?

6. Derive Moseley's law on the basis of Bohr's theory and discuss its importance.

7. The optical spectra of elements show periodic characteristic with increasing Z, while X-ray characteristic spectra show continuous order through all Z-values. Explain.

8. Mention the applications of Moseley's law.

9. How Moseley's law helped in rearranging various elements on a rational basis?

10. How can you account for the origin of characteristic lines in different K and L series of an X-ray spectra.

11. What factors govern intensity and penetrating power of X-rays? What changes will take place in the character of X-ray spectrum with the increase of tube potential?

12. Give a brief account of experiments to establish (i) the transverse character and (ii) the particle character of X-rays.

13. Discuss the diffraction of X-rays by a three dimensional lattice with special reference to Laue spots. Interpret them.

14. Explain on the basis of Bragg's law, the diffraction of X-rays produced by crystals.

PROBLEMS

1. An X-ray tube works at a potential difference of 100,000 V. Only 0.1% of the energy of cathode rays is converted into X-ray radiation and heat is generated in the target at the rate of 120 calorie per second. What current does the tube pass and what is the energy and velocity of an electron when it reaches the target?

[**Hint:** Operating potential = 10^5 V, electron energy = 10^5 eV = 1.6×10^{-14} J, velocity of the

electron $v = \left(\dfrac{2 \times 1.6 \times 10^{-14}}{9.1 \times 10^{-31}} \right)^{1/2} = 1.875 \times 10^8$ m/s.

Since the velocity is quite high, one has to apply relativistic correction. We have for KE,

$E = c^2(m - m_0) = c^2 m_0(\beta - 1)$ $\beta = 1 \Big/ \sqrt{1 - \dfrac{v^2}{c^2}}$

$\therefore 1 - \beta = E/m_0 c^2 = \dfrac{1.6 \times 10^{-14}}{9.1 \times 10^{-31} \times 9 \times 10^{16}} = \dfrac{1.6}{81.9}$

$= \dfrac{1}{5.12}$ $\therefore \beta = 1.1953$ and thus $1 - \dfrac{v^2}{c^2} = \dfrac{1}{\beta^2} \cong 0.7$

or $v = 1.643 \times 10^8$ m/s. If n electrons reach the target in one second, the amount of n electrons converted into heat = $n \times 159.84 \times 10^{-14}$J.

$\therefore n \times 159.84 \times 10^{-14} = 120 \times 4.2$. $\therefore n = 3.16 \times 10^{-14}$/s, current = $3.16 \times 10^{14} \times 1.6 \times 10^{-19}$ C/s = 50.6 μA]

2. What are the most energetic X-rays emitted in each case when 40 keV electrons bombard (i) Mo target ($Z = 42$) (ii) Cu target ($Z = 29$)?

[**Ans.** 40 keV, whichever be the target]

3. Electrons are accelerated in television tubes through potential differences of about 10 kV. Find the highest frequency of the electromagnetic waves emitted when these electrons strike the screen of the tube. What kind of waves are these ?

[**Ans.** 2.41×10^{18} Hz, X-rays]

4. An X-ray tube operating at 66 kV, is producing continuous X-rays. Calculate the shortest wavelength limit of X-rays.   [**Ans.** 0.19 Å]

5. The minimum wavelength recorded in the continuous X-ray spectrum of a 100 kV tube is 12.35×10^{-12} m. Find the value of Planck's constant.   [**Ans.** $h = 6.59 \times 10^{-34}$ Js]

6. The K_α line from molybdenum ($Z = 42$) has a wavelength of 0.7078 Å. Calculate the wavelength of K_α line of zinc ($Z = 30$).

[**Ans.** 1.4148 Å]

7. If the short series limit of the Balmer series for hydrogen is 3646 Å, calculate the atomic number of the element which gives X-ray wavelengths down to 1 Å. Identify the element.

[**Ans.** $Z = 31$, Gallium]

8. If the K, L, M energy levels of platinum involved in the emission of the K-series of X-rays of this element are 78000, 12000 and 30000 eV respectively, calculate the wavelengths of the K_α and K_β lines of platinum.

[**Ans.** $\lambda K_\alpha = 0.188$ Å, $\lambda K_\beta = 0.165$ Å]

9. A material whose K-absorption edge is 0.15 Å is irradiated with 0.13 Å X-rays. Calculate the maximum kinetic energy of photoelectrons that are emitted from the K-shell. [**Ans.** 12.68 keV]

10. Monochromatic *X*-rays of wavelength 0.124 Å are scattered from a block of carbon. Calculate the maximum energy of recoil electron.

[*Ans.* 28.13 keV]

11. The glancing angle of the first order spectrum in Bragg's spectrometer is found to be 8°. Calculate the wavelength of *X*-rays if the spacing between the planes be 2.83 × 10^{-10} m.

[*Ans.* 0.79 Å]

SHORT ANSWER QUESTIONS

1. What is the order of wavelength of *X*-rays?

Ans. 10^{-10} m.

2. *X*-rays photon collides with an electron and bounces off. Comment on its new frequency.

Ans. The new frequency of the photon which bounces off is either equal or less than the frequency of the incident photon.

3. The anode in the Coolidge tube has high atomic weight. Why?

Ans. (i) To produce hard *X*-rays and (ii) High melting point to withstand the high temperature.

4. Of which wavelength will the energy of *X*-rays be higher (*i*) 6 Å (*ii*) 36 Å.

Ans. The energy of 6 Å *X*-rays will be higher and it will be 6 times that of 36 Å.

5. On what factor the shortest wavelength of *X*-rays emitted from an *X*-ray tube depend?

Ans. The shortest wavelength of the *X*-rays emitted depends on the energy of the electrons incident on the target, which in turn depends on the potential they have fallen. We have $\lambda_{min} = hc/eV$ ($\because$ h, c, e are constants).

6. Why *X*-rays cannot be diffracted by means of an ordinary grating?

Ans. The shortest wavelength of the *X*-rays (compared to the grating constant of optical grating) makes it difficult to observe *X*-ray diffraction with ordinary gratings.

7. Does there exist a sharp limit on the short wavelength side for each continuous *X*-ray spectrum.

Ans. Yes, the sharp limit on the short wavelength side is dependent on the voltage applied to the incident electrons, and is given by , $\lambda_{min} = hc/eV$.

OBJECTIVE QUESTIONS

1. The target *T* in a Coolidge tube consists of a _____ block in which a piece of tungsten or molybdenum is fitted. [copper]

2. The *K*, *L*, *M*, etc. series constitute the _____ line spectra, which is the characteristic of a particular material. [*X*-ray]

3. *X*-rays are _____ waves of very _____ wavelength.

[electromagnetic, short]

4. Mathematically, Moseley equation is _____.

[$v = b (Z - a)^2$]

5. Intensity of *X*-ray beam decreases _____ with the thickness of absobing material.

[exponentially]

6. *X*-rays are used to _____ abnormal internal tissues. [destroy]

7. *X*-ray can be used to determine the _____ number and identification of various chemical _____. [atomic, elements]

MULTIPLE CHOICE QUESTIONS

1. Molybdenum is used as a target element for production of *X*-rays because it is

(a) a heavy element and can easily absorb high velocity electrons

(b) a heavy element with a high melting point

(c) an element having high thermal conductivity

(d) heavy and can easily detect electrons

2. The characteristic *X*-ray radiation is emitted when

(a) the electrons are accelerated to a fixed energy

(b) the source of electrons emits a mono-energetic beam

(c) the bombarding electrons knock out electrons from the inner shell of the target atoms and one of the outer electrons falls into this vacancy

(d) the valence electrons in the target atoms are removed as a result of the collision

3. Hydrogen atom does not emit *X*-rays because

(a) its energy levels are close to each other

(b) its energy levels are too far apart

(c) it is too small in size

(d) it has a single electron

4. Bragg's equation will have no solution, if

(a) $\lambda > 2d$ (b) $\lambda < 2d$

(c) $\lambda < d$ (d) $\lambda = d$

5. The energy of X-ray photon is 3.3×10^{-16} J. Its frequency per second would be
   (a) 5×10^{17}
   (b) 5×10^{-18}
   (c) 6.62×10^{18}
   (d) 2×10^{-18}

6. In producing X-rays, a beam of electrons accelerated by a potential difference V is made to strike a metal target. For what value of V, the X-rays have the lowest wavelength, $\lambda_m = 3094$ Å?
   (a) 10 kV
   (b) 20 kV
   (c) 30 kV
   (d) 40 kV

7. In radiotherapy, X-rays are used to
   (a) detect bone fracture
   (b) treat cancer by control
   (c) detect heart disease
   (d) detect fault in radio receiving circuits

8. Moseley's law relates
   (a) frequency and applied voltage
   (b) frequency and atomic number
   (c) wavelength and intensity of X-rays
   (d) wavelength and angle of scattering

9. The frequency of K_α line of a source of atomic number Z is proportional to
   (a) Z^2
   (b) $(Z-1)^2$
   (c) $1/Z$
   (d) Z

10. The wavelength of K_α line from an element of atomic number 41 is λ. Then the wavelength of K_α line of an element of atomic number 21 is
    (a) $\lambda/4$
    (b) 4λ
    (c) $3.08\,\lambda$
    (d) $0.26\,\lambda$

    [**Hint:** $\dfrac{v_{41}}{v_{21}} = \dfrac{41-1}{21-1} = 4 \quad \therefore \lambda_{41} : \lambda_{21} = 1 : 4$]

Answers

1. (b)	2. (c)	3. (d)	4. (a)	5. (a)
6. (d)	7. (b)	8. (b)	9. (b)	10. (a)

19

Quantum Mechanics

19.1 INTRODUCTION

The dynamical properties of the observable objects have been successfully explained by the classical mechanics as there was no restriction on the values of the dynamical property. But the concepts of classical mechanics failed to describe the actual behaviour of atomic and molecular dimensions as experimental observations for such systems show that only a well defined set of values are observable, e.g. an electron moving around the nucleus of an atom has only a discrete set of values for energy. This discreteness has been termed as quantization. The explanation of this quantization required the development of a new system of mechanics called quantum mechanics.

19.2 BLACKBODY RADIATIONS

Radiations are emitted by every substance as a result of the vibration of its particles and the character of radiation depends on the nature and temperature of the substance. The ability of a body to radiate is closely related to its ability to absorb radiation, as at any constant temperature the body will be in thermal equilibrium with its surroundings and must be absorbing radiation at the same rate as its limits.

Different solids emit radiations at different rates at the same temperature; the rate is maximum when the solid is perfectly black, i.e. behaves as a blackbody. A blackbody can

absorb all the radiations that fall on it. For practical purposes, an isothermal cavity with a small aperture through which radiations from outside may be admitted and absorbed completely due to repeated reflections inside the enclosure is considered a black body. Radiations emerging from the small hole of such a hollow enclosure is, therefore, called blackbody radiations. The intensity of radiations, i.e. total radiations emitted per unit surface area from a black body depends only on temperature (T) and is independent of the nature of the solid.

19.3 SPECTRAL DISTRIBUTION OF ENERGY IN THE BLACKBODY RADIATION

The radiations emitted by a black body is not confined to a single wavelength but spreads over a wide spectrum of wavelength. The experimentally observed dependence of the energy density on wavelength (λ) at different temperatures is shown in Fig. 19.1. A close investigation reveals the following important facts:

i. At a given temperature, the energy is not uniformly distributed in the radiation spectrum.

ii. At a given temperature the density of radiant energy increases with increase in wavelength and becomes maximum at a particular wavelength (λ_m). With further increase in wavelength, the density of radiant energy decreases.

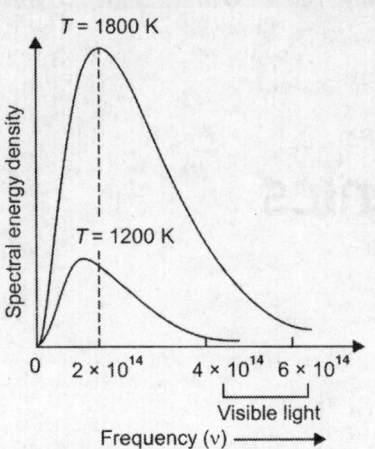

Fig. 19.1: Experimental dependence of the energy density on frequency and temperature

iii. An increase in temperature increases the energy of all spectral components. The short wavelengths cutoff advance towards the origin as the temperature increases and the peak of the curve shifts to shorter wavelength, i.e. λ_m (the wavelength for which energy emitted is maximum) decreases.

The shift of the peak of the curve was found to obey the following empirical relationship, commonly called *Wien displacement law*, i.e.

$$\lambda_m T = \text{constant} \tag{19.1}$$

With the help of classical thermodynamics, Wien has further shown that the amount of energy contained in the spectral region included within the wavelength λ and $\lambda + d\lambda$ emitted by a black body is given by

$$E_1 d_1 = \frac{A}{\lambda^5} e^{-B/\lambda T} \, d\lambda \tag{19.2}$$

where A and B are constants.

Wien law works well only for short wavelengths. There are considerable deviations from Wien's law at long wavelength and high temperatures. From Eq. (19.2), one obtains finite energy even for $T = \infty$ (infinite). Rayleigh argued that it is unlikely that energy (E) should be finite for $T = \infty$.

Later, Rayleigh and Jeans adopted a more rigorous method based on classical physics and came up with the following relation:

$$E_\lambda d\lambda = C\lambda^{-4} T e^{-A/\lambda T} \tag{19.2a}$$

where C is another constant.

This *Rayleigh-Jeans law* describes the experimental curve quite well in the long wavelength region but the energy density E_λ increases rapidly as λ decreases and approaches infinity for shorter wavelengths. This failure, for explaining black body radiation, was such a crushing blow to classical physics that it is historically referred as the *ultraviolet catastrophe*. It led Max Planck to discover that only if light emissions is a quantum phenomenon, then only the correct formula for the energy density of black body radiation be obtained.

The failure of Rayleigh-Jeans law to explain the spectral energy curves of black body radiation indicated that the fundamental assumptions of classical physics required suitable modifications.

19.4 PLANCK'S LAW OF RADIATION

Being well aware of the shortcomings of both Rayleigh-Jeans and Wien radiation laws, Planck in October 1900, proposed a remarkably well relation for the observed energy curves of the black body radiation. The formula could be written as

$$\rho(\nu)d\nu = \frac{a\nu^3}{\exp(b\nu) - 1} \tag{19.3}$$

where a and b are numerical constants. This formula had no theoretical background and was obtained empirically by the trial and error method to explain the observed results. However, a few weeks later, i.e. in December 1900, Planck announced the theoretical derivation of his radiation law [Eq.(19.3)]. For this, he proposed that a black body chamber was filled not only with radiation but also with molecules of a perfect gas, and dipole oscillators of molecular dimensions constituting the black body absorbed energy from the radiation and transferred it partly or wholly to the gas molecules when the latter collided with them. This was a round about way of explaining the emission and absorption of radiation by black body but there was no other way as the mechanism of emission and absorption of radiation by atoms and molecules by direct exchange was not known at that time. Planck calculated that the number of oscillators per unit volume was $8\pi v^2/c^3$, so that the radiation density was given by

$$\rho(v) = \frac{8\pi v^2}{c^3} E(v) \qquad (19.4)$$

where $E(v)$ is the mean energy of the oscillator emitting radiation of frequency v. Now, if $E(v)$ were to be determined by equipartition law, i.e. $E(v) = k_B T$, then Eq. (19.4) would lead to the Rayleigh–Jeans law. Planck gave up the hypothesis of continuous emission of radiation by oscillators and assumed that they emitted energy only when they acquired certain minimum energy E or an integral multiple of it, nE, that is to say an oscillator having an amount of energy nE could emit only after it had absorbed the amount $(n + 1)E$, and after emitting the amount of energy E it reverted back to the previous state of energy nE. Thus, the radiation of energy E could be emitted by only those oscillators which has energy $E, 2E, 3E,..,nE$. Further, E was directly proportional to v, i.e.

$$E = h v \qquad (19.5)$$

where the constant of proportionality, h was called the Planck's constant. The above idea can be expressed in the form of following revolutionary postulates which have become the foundation of quantum theory of radiation.

i. The amount of energy emitted or absorbed by an oscillator is proportional to its frequency. Calling the constant of proportionality h, one can write for the change in oscillator energy

$$\Delta\varepsilon = h v$$

ii. An oscillator cannot have an arbitrary energy but must occupy one of the discrete set of energy states given by

$$\varepsilon_n = nh v$$

where n is an integer or zero. It was assumed that the ground state corresponds to the zero energy state. The value of $h = 6.62619 \times 10^{-34}$ Joule sec. Planck's constant (h) is a universal constant which plays important role in all quantum phenomena.

The previous picture of a continuum of oscillator states is now replaced by a discrete set of *quantized* states. Furthermore, the amount of energy emitted or absorbed is also quantised, since each quantum must correspond to the energy difference between two states of a given oscillator. Each quantum of electromagnetic energy is called a photon. Using Planck's quantum hypothesis, we now derive the famous Planck's radiation formula.

Let N be the total number of Planck's oscillator and E_T is their total energy, then the average energy per oscillator $\bar{\varepsilon}$, is given by

$$\bar{\varepsilon} = \frac{E_T}{N} \qquad (19.6)$$

Let $N_0, N_1, N_2, ... N_n, ...,$ etc. be the number of oscillators having energies $0, \varepsilon, 2\varepsilon, 3\varepsilon, .., n\varepsilon,$ etc. respectively, then

$$N = N_0 + N_1 + N_2 +... + N_n +..$$

$$= \sum_{n=0}^{\infty} N_n \qquad (19.7)$$

and

$$E_T = 0 + \varepsilon N_1 + 2\varepsilon N_2 +... + n\varepsilon N_n +...$$

$$= \sum_{n=0}^{\infty} n\varepsilon N_n \qquad (19.8)$$

According to Maxwell–Boltzmann distribution function, the number of resonators or oscillators having energy nE will be

$$N_n = N_0 \exp\left(-\frac{n\varepsilon}{k_B T}\right) \qquad (19.9)$$

Obviously, the higher energy states are, thus, less likely to be populated, and putting

$$y = \exp\left(\frac{-\varepsilon}{k_B T}\right)$$

$$N = \sum_{n=0}^{\infty} N_n y^n$$

$$= N_0(1 + y + y^2 + \ldots) = N_0\left(\frac{1}{1-y}\right) \quad (19.10)$$

Similarly $\quad E_T = EN_0 \sum_{n=0}^{\infty} n y^n$

$$= EN_0(y + 2y^2 + 3y^3 + \ldots)$$

$$= EN_0 y(1 + 2y + 3y^2 + \ldots)$$

$$= \frac{EN_0 y}{(1-y)^2} \quad (19.11)$$

Hence $\quad \bar{\varepsilon} = \dfrac{yE}{1-y} = \dfrac{E\exp(-\varepsilon/k_B T)}{\exp(\varepsilon/k_B T) - 1} \quad (19.12)$

Using Eq. (19.12) for E, the energy density relation (19.5) becomes

$$\rho(v)dv = \frac{8\pi v^2}{c^3 \exp(\varepsilon/k_B T) - 1} dv$$

Since $\quad E = hv$

$$\rho(v)dv = \frac{8\pi h v^3}{c^3} \frac{1}{\exp(cv/k_B T)} dv \quad (19.13)$$

This is famous Planck's radiation law. This has the same form as Eq. (19.3) if a is put equal to $8\pi h/c^3$ and b equal to h/k_B. One can express Eq. (19.13) in terms of wavelength by using the relation $v = c/\lambda$ and hence

$$dv = -\frac{c}{\lambda^2} d\lambda$$

$$\rho(\lambda)d\lambda = \frac{8\pi hc}{\lambda^5} \frac{1}{\exp(ch/\lambda k_B T) - 1} d\lambda \quad (19.14)$$

For small temperatures and short wavelengths, λT is small and hence $\exp(ch/\lambda k_B T) \gg 1$. Obviously, 1 in the denominator of Eq. (19.14) can be neglected and hence it reduces to

$$\rho(\lambda)d\lambda = \frac{8\pi hc}{\lambda^5} \exp(-ch/\lambda k_B T) d\lambda \quad (19.15)$$

This is Wein's law with $a = 8\pi ch$ and $b = ch/\lambda k_B$. For higher temperatures and larger wavelengths,

$$\exp\left(\frac{ch}{\lambda k_B T}\right) = 1 + \frac{hc}{\lambda k_B T} \quad (19.16)$$

Using Eq. (19.16), Eq. (19.14) becomes ($\because T$ is large)

$$\rho(\lambda)d\lambda = \frac{8\pi k_B T}{\lambda^4} d\lambda \quad (19.17)$$

This is Rayleigh-Jeans law with $a = 8\pi k_B$.

Obviously, the Planck's theory of radiation incorporates all that is valid from the older theories as special cases. It, thus, serves as an excellent example of a conceptual advancement which opened exciting new frontiers while still preserving much of the older physics.

Planck's derivation introduced a new constant h, the value of which was obtained by fitting the formula (17) to the observed curves for blackbody radiation. The value of $h = 6.626 \times 10^{-34}$ Js.

The most important aspect of Planck's hypothesis was the new idea of discontinuous emission of radiation. According to classical physics, a system can absorb or emit any amount of energy whereas in Planck's hypothesis energy emitted by a blackbody is restricted by the relation $E = hv$. To this definite amount of energy hv, Planck gave the name *quantum*. Thus, while in classical physics energy is continuous variable, in quantum physics it is *quantized*.

Let us now reflect for a moment on the significance of Planck's postulate. It states that the energy of a harmonic oscillator expressed in terms of its frequency of vibration, may on an average occur in units no smaller than $E = nhv$ where $n = 0, 1, 2, \ldots$. It does not say what is the energy of the oscillator at any instant, nor does it say how the value of h is to be calculated.

From this we may conclude that:

i. Planck's postulate is empirical, i.e. it contains a parameter h whose value cannot be calculated except through the comparison of some expression in which it appears with the appropriate experimental results. There is no theory of Planck's constant. The only physical reason we can give for its existence is that it is invaluable for the elucidation of microscopic phenomena.

ii. The postulate is statistical in that it refers only to the time-averaged behaviour of the microscopic oscillator. Nothing is said about energy except that it varies directly with the frequency if evaluated over long period of time compared with v^{-1}.

iii. The postulate does not in itself suggest that a flaw exists in the Maxwell's theory of electromagnetism. It states only what energy an oscillator must have on an average if it is to remain in thermodynamic equilibrium with a radiation field. We should not conclude from this that the latter must necessarily exhibit discrete energies when interacting with matter. We have verified the quantum hypothesis only for the emission of radiation by a heated solid!

19.5 PHOTOELECTRIC EFFECT

The importance of the Planck's hypothesis of energy quantization was quickly realised by Albert Einstein. Einstein extended this idea to all kinds of electromagnetic radiations and explained the important phenomenon of photoelectric effect.

When ultraviolet light of X-rays strike a piece of metal, it frees electrons from its surface, and it is experimentally possible to count those electrons and to measure their energy. Quantitative investigation of the effect shows that the light intensity determines the number of electrons thus freed but has no influence on their energy. This is contrary to what we would expect from Maxwell's electromagnetic theory. Since the intensity of a light wave is proportional to the square of the amplitude of its electric vector (E), the energy of electrons is, surprisingly, determined only by the colour of the light, i.e. its frequency (v).

The following facts must be explained by any satisfactory theory of the photoelectric effect:

i. If the incident light has no frequency component above a certain threshold frequency, no photoelectrons will be emitted.

ii. The threshold frequency appears to depend on the properties of the metal.

iii. Provided that the threshold frequency is exceeded, photoelectrons will be emitted instantly, regardless of how low the intensity of light source is.

iv. The photoelectric current (number of electrons emitted per second) increases with the intensity of light but is independent of the frequency.

v. The maximum kinetic energy of the photoelectrons is independent of the intensity of light and depends only on its frequency.

Attempts to explain the above phenomena by means of classical radiation theory met with defeat. Classically, an electron would be ejected as soon as it could accumulate enough energy to overcome the force which binds it to the metal. This energy is called the *work function* of the metal. Since an electromagnetic wave has an energy density associated with it, one would expect that the metal should accumulate enough energy and ultimately, to eject an electron, regardless of the frequency or the intensity of the source. However, the classical picture was completely at a loss to explain all of the above experimental facts except number four. From the classical point of view, it was reasonable to expect the photoemission to increase when the energy density of the incident radiation is increased.

Einstein used Planck's idea of energy quantization to explain the above mentioned observations. He proposed that electromagnetic radiation (light) could be considered to have properties of both waves and particles. He called the particles of light as 'photons'. Each photon of light possesses an energy equal to hv (quantum of light). Before discussing the Einstein's photoelectric effect, we would like to discuss of the experimental study of the photoelectric effect.

When a radiation of high frequency, e.g. ultraviolet light or X-rays is incident on a clean metal surface, electrons are emitted from the surface of the metal. The emitted electrons are called photoelectrons and the phenomenon is called the photoelectric effect. The entire range of electromagnetic radiations from gamma rays and X-rays to the ultraviolet, the visible and the infrared produce this effect. The photoelectric effect is also observed in solids, liquids and gases. This effect was experimentally observed by Hertz in 1887.

Experimental Arrangement to Study the Photoelectric Effect

A simple experimental arrangement for the study of photoelectric effect is shown in Fig. 19.2.

The emitting surface, i.e. photosensitive plate A is mounted opposite to a metal plate B in a highly evacuated glass tube C. The two plates A and B form two electrodes, to which a variable potential difference can be applied. The evacuated glass tube is fitted with quartz window D, through which ultraviolet light from the source S is transmitted and allowed to fall on the plate A.

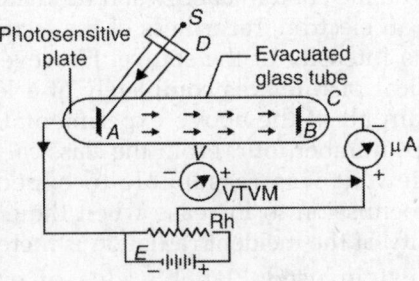

Fig. 19.2: Experimental arrangement to study the photoelectric effect

When the applied potential difference between the plates A and B is such that A is at negative potential with respect to B, the photoelectrons emitted from A are accelerated towards plate B.

The resulting photoelectric current I flowing in the circuit is measured by the μA (micro ammeter) and the accelerating potential difference V is measured by vacuum tube voltmeter. If the potential difference is reversed so that plate B is now at negative potential with respect to A, the photoelectrons are repelled towards A by the retarding potential. Consequently, the photoelectric current is reduced. It is found that the photoelectric current depends on the following factors: (i) the frequency of the incident radiation (ii) the intensity of incident radiation (iii) the potential difference between the electrodes and (iv) the nature of the emitting surface.

The important experimental observations are as follows:

Effect of Frequency on Photoelectric Current

The collector plate B is made sufficiently positive with respect to the emitter photosensitive plate A. The surface of plate A is illuminated with monochromatic light of different frequencies. It is observed that the photoelectric current is produced only when the frequency ν of the incident light is greater than a certain minimum value ν_0. This minimum value of the frequency is called the *threshold frequency* for the given surface.

Effect of the Intensity of Incident Light

To study the effect of the intensity of the incident light, the collector plate B is made sufficiently positive with respect to the emitter plate A. By keeping the frequency of the incident light and the potential difference V constant, the photoelectric current I is measured for the various intensities of the incident light. Figure 19.3 shows that the photoelectric current is proportional to the intensity of the incident light.

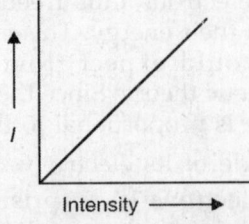

Fig. 19.3: Variation of the photoelectric current (I) with the intensity of the incident light

Effect of Potential Difference

Keeping the frequency and the intensity X_1 of the incident light constant, the potential difference between the electrodes is varied. Starting with a small positive potential, it is reduced to zero and then made negative with the help of a reversing key. The variation of photoelectric current with the potential difference is shown in Fig. 19.4. When the collector plate B is positive, the photoelectric current remains constant. When the collector plate is made more and more negative, the photoelectric current goes on decreasing until it is stopped entirely. The retarding potential that stops the photo current is called the stopping potential. If the intensity of the incident light is increased to X_2, then initial photo current for potential V_0 is found to be the same for the light of the same frequency.

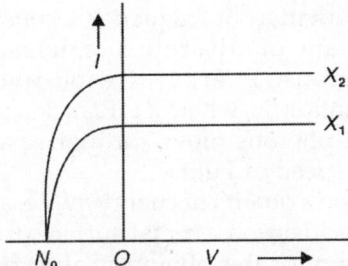

Fig. 19.4: Variation of the photoelectric current (I) with the potential difference

The above experimental observation is explained by assuming that when light on a certain frequency $v(v > v_0)$ is incident on a emitting surface, photoelectrons with kinetic energies ranging from zero to a certain maximum value are emitted from the surface. The photoelectrons which are emitted with the maximum kinetic energy are stopped by the stopping potential V_0 volts. Obviously, the work done by the retarding potential on the faster electrons must be equal to their kinetic energy, i.e.

$$eV_0 = \frac{1}{2} mv^2_{max} \qquad (19.18)$$

where m is the mass of the electron, e is the electronic charge and v_{max} is the maximum velocity of emission of the electrons.

Effect of Frequency on the Stopping Potential

To study the effect of the frequency of the incident light on the stopping potential, the intensities of light of different frequencies are adjusted to produce the same maximum (I_m) value of the photoelectric current when collector B is positive. The potential of collector B is then reduced in steps and made zero. It is then made more and more negative with the help of the reversing key. For the given photocathode, the graphs obtained are as shown in Fig. 19.5. From Fig. 19.5, it is evident that the stopping potential increases as the frequency of the incident light is increased. Thus, if the frequencies are in the increasing order, i.e. $v_1 > v_2 > v_3$, then the corresponding stopping potentials are also in the increasing order, i.e. $V_{01} > V_{02} > V_{03}$.

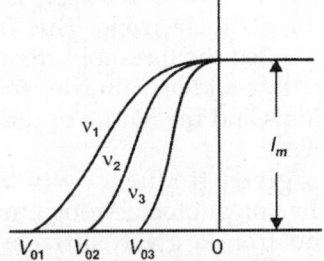

Fig. 19.5: Stopping potential versus intensity of incident light

Effect of Frequency on the Maximum Kinetic Energy

Magnitudes of the stopping potential at different frequencies are converted into the maximum kinetic energies. A plot of maximum kinetic energy against the frequency is shown in Fig. 19.6. A straight line is obtained. The straight line (a) meets the frequency axis at a certain frequency v_0.

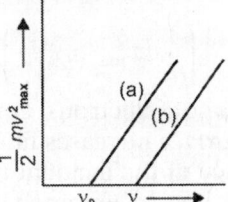

Fig. 19.6: Maximum K.E. versus frequency

This reveals that v_0 is the minimum frequency for which the maximum kinetic energy is zero. This minimum value of frequency is called the threshold frequency for the emitting surface.

Thus, the threshold frequency for a given emitting surface is defined as minimum frequency of incident light, below which no photoelectrons are emitted, however intense the incident light may be.

From Fig. 19.6, it is evident that for another emitting surface, a straight line (b) parallel to (a) is obtained.

On the basis of above experimental observations, we can summarize the following important characteristics of the photoelectric effect:

i. For a given photosensitive surface, there is a minimum frequency v_0 of incident radiation, below which there is no emission of photoelectrons. This frequency (v_0) is called the threshold frequency for the surface and its value depends on the materials and the nature of the emitting surface.

ii. For a given frequency $v(v > v_0)$, the number of photoelectrons emitted per second from a given surface, under a constant accelerating potential difference is directly proportional to the intensity of incident radiation.

iii. When radiation of a given frequency $v(v > v_0)$ is incident on a given surface, photoelectrons with kinetic energies ranging form zero to certain maximum value are emitted from the surface. The stopping potential for the photoelectrons and hence the maximum kinetic energy is independent of the intensity of the incident radiation.

iv. The maximum kinetic energy

$$\left(\frac{1}{2} m v_{max}^2 = e V_0 \right)$$

of the photoelectrons emitted from a given surface increases linearly with the frequency of the incident radiation. The straight line so obtained for the given surface meets the frequency axis at the threshold frequency v_0 for the surface. The slope of the straight line is the same for all emitting surfaces and is equal to Planck's constant h.

v. The emission of photoelectrons from photosensitive surface is an instantaneous process. As soon as radiation of frequency greater than the threshold frequency is incident on a given photosensitive surface, there is emission of photoelectrons.

vi. The rate of emission of the electrons is independent of temperature. Clearly, the photoelectrons phenomenon is entirely different to thermionic emission.

Einstein's Photoelectric Equation

Einstein's photoelectrons theory is based on the following:

i. A radiation of frequency v consists of a stream of discrete particles, called photons. Energy of each photon or quanta is hv, where h is Planck's constant. The photons move through space with the speed of light.

ii. When a quanta of energy hv, i.e. a photon is incident on a metal surface, the entire energy of the photon is absorbed by a single electron without any time lag. The probability of absorbing two or more photons at the same time by a metal surface is negligible.

Suppose that radiation of frequency v is incident on the surface of a metal in vacuum. The entire energy of a photon, i.e. hv is absorbed by an electron of the metal. We must note that the incident radiation penetrates several atomic diameters into the metal and electrons at different depths absorb this amount of energy from the incident photons. The energy hv absorbed by the electrons in the metal is used in the following two ways:

1. A part of the photon energy hv is used up by the electron to do a certain minimum amount of work to overcome attractive forces of positive ions of the metal. The minimum amount of energy is called the *photoelectric work function* and

usually denoted by W of the metal. The work function (W) depends on the nature of the emitting surface.

2. The remaining energy ($h\nu - W$) appears as the kinetic energy comes from the surface, it does not suffer any collision with another electron. In this situation, it will be emitted from the surface into vacuum with this energy as maximum kinetic energy $\frac{1}{2}mv_{max}^2$, where m is the mass of the electron and v_{max} its maximum speed for the given frequency. Thus, for such an electron, we have

$$\frac{1}{2}mv^2 = h\nu - W \qquad (19.19)$$

The above equation is called Einstein's photoelectric equation. Suppose the frequency of the incident radiation is reduced to a certain value ν_0 such that an electron is just emitted from the surface with zero kinetic energy.

Then we have, when $\nu = \nu_0$, $\frac{1}{2}mv_{max}^2 = 0$.

Then from Eq. (19.19), we have

$$0 = h\nu_0 - W$$

or $\qquad W = h\nu_0 \qquad (19.20)$

Substituting $W = h\nu_0$ from Eq. (19.20) in Eq. (19.19), we have

$$\frac{1}{2}mv_{max}^2 = h\nu - h\nu_0 = h(\nu - \nu_0) \qquad (19.21)$$

Eq. (19.21) is another form of Einstein's photoelectric equation. The frequency ν_0 is called the *threshold frequency*. On the basis of Einstein's theory, one can define the threshold frequency as follows:

It is the frequency of incident photon which has just sufficient energy to liberate an electron with zero kinetic energy from the emitting metal surface.

It is worthwhile to mention that the energy of light photon incident on the surface of the metal is not divided, but the whole energy of a photon is used up by the same electron in ejecting from the parent atom and imparting kinetic energy.

Explanation of the characteristics of the photoelectric effect from Einstein's equation

i. Eq. (21) shows that the graph of $\frac{1}{2}mv_{max}^2$ against ν is a straight line for an emitting surface. For all emitting surfaces, the straight lines will have the same slope equal to Planck's constant h.

ii. When the intensity of incident radiation is increased, then the increased number of photons in the incident radiation will cause the emission of more photo-electrons per second. Obviously, the photoelectric current is proportional to the intensity of the incident radiation. The increase the intensity does not affect the maximum kinetic energy.

iii. From Eqs. (19.20) and (19.21), we have:

(a) For $\nu < \nu_0$ photoelectrons cannot be emitted.

(b) For $\nu = \nu_0$ photoelectrons are just emitted with zero kinetic energy.

(c) For $\nu > \nu_0$ photoelectrons are emitted with kinetic energies ranging from zero to a maximum value.

(d) For $\nu > \nu_0$ and the intensity of incident light is increased, the number of photoelectrons emitted per second will increase.

Some recent important observations about the photoelectric effect are:

i. Only a small fraction (~5%) of the incident photon succeeds in ejecting photoelectrons from photometal while rest of the photons are absorbed by the system as a whole and generate thermal energy.

ii. Photoelectric effect is also observed for isolated atoms in the form of gas, e.g. Na, K vapour and this process is called photoionization.

iii. The energy required for ejecting the electrons may also be provided by heating the metal, which results in the thermionic emission of the electrons.

iv. The advent of lasers capable of emitting coherent radiation at high power levels, two or multiphoton photoemission has now been observed. Theory predicts that the double quantum photocurrent should be proportional to the square of the power of the incident radiation as opposed to the nearly linear relationship that holds for the single quantum photo-effect. In a multiphoton photoelectric effect, the kinetic energy of the photoelectron is given by

$$\frac{1}{2}mv^2 = Nh\nu - eV_s \qquad (19.22)$$

and the critical frequency is eV_s/Nh which is smaller than the corresponding frequency for single photon process by a factor of $1/N$.

19.6 COMPTON EFFECT

AH Compton in 1923, while studying the scattering of X-rays by a block of paraffin, observed that the wavelength of the scattered radiation was greater than the wavelength of incident radiation and the shift in wavelength was dependent on the angle of scattering. The wavelength of the scattered radiation is found maximum at right angles to the incident beam. This effect is called Compton effect. Essentially, it is a demonstration of the scattering of a photon by an electron as a particle-particle scattering.

Theory

The theory of Compton effect was given by Compton and also by Debye at about the same time. Compton considered the radiation to be quantised and assumed electron of the atom of scattering material to be loosely bound (almost free) and at rest before collision with the photon. Let λ and ν be the wavelength and frequency of the incident radiation. According to the quantum theory of radiation, the energy of each photon associated with this radiation is $h\nu$. When this photon of energy $h\nu$ strikes a loosely bound (almost free) electron at rest, then as a result of collision the electron gains momentum and recoils in a direction making an angle θ with the direction of incident radiation (Fig. 19.7). At the same time, the

photon is scattered at an angle ϕ with the direction of the incident radiation with lesser energy $h\nu'(\nu' > \nu)$. It is assumed that law of conservation of energy and law of conservation of momentum hold good.

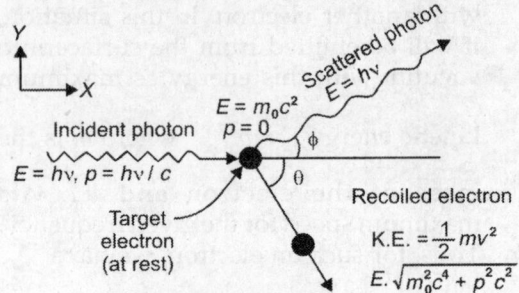

Fig. 19.7: The Compton effect

Let m_0 be the rest mass of the electron and m be the mass of the electron, when it is recoiled with a velocity v, then from the special theory of relativity, we have

$$m = \frac{m_0}{\sqrt{1 - (v^2/c^2)}} \qquad (19.23)$$

The momentum of the photon of energy $h\nu$ is given by

$$p = \frac{E}{c} = \frac{h\nu}{c} \qquad (19.24)$$

The energy of the electron (at rest) before collision will be m_0c^2 and after collision it will be mc^2. From the law of conservation of energy, we have

$$h\nu + m_0c^2 = h\nu' + mc^2 \qquad (19.25)$$

From the law of conservation of momentum, in the X-direction

$$\frac{h\nu}{c} = \frac{h\nu'}{c}\cos\phi + mv\cos\theta \qquad (19.26)$$

and in Y-direction

$$0 = \frac{h\nu'}{c}\sin\phi + mv\sin\theta \qquad (19.27)$$

From Eqs. (19.26) and (19.27), one obtains

$$mvc\cos\theta = h(\nu - \nu'\cos\phi) \qquad (19.28)$$

$$mvc\sin\theta = -h\nu'\sin\phi \qquad (19.29)$$

Squaring and adding Eqs. (19.28) and (19.29), one obtains

$$m^2 v^2 c^2 = h^2 \left[(v - v' \cos \phi)^2 + (v' \sin \phi)^2 \right] \quad (19.30)$$

Squaring Eq. (25), one obtains

$$m^2 c^4 = \left[(v - v') + m_0 c^2 \right]^2$$

$$= h^2 (v^2 - 2vv' + v'^2) + m_0^2 c^4 + 2h m_0 c^2 (v - v') \quad (19.31)$$

Subtracting Eq. (19.30) from Eq. (19.31), one obtains

$$m^2 c^4 \left(1 - \frac{v^2}{c^2} \right) = 2h^2 \, vv' (\cos \phi - 1)$$

$$+ m_0^2 c^4 + 2h m_0 c^2 (v - v') \quad (19.32)$$

But
$$mc^2 = \frac{m_0 c^2}{\sqrt{1 - v^2/c^2}}$$

or
$$m^2 c^2 \left(1 - \frac{v^2}{c^2} \right) = m_0^2 c^4 \quad (19.33)$$

Using Eq. (19.33), Eq. (19.32) becomes

$$2h^2 vv'(1 - \cos \phi) = 2h m_0 c^2 (v - v') \quad (19.34)$$

$$h(1 - \cos \phi) = m_0 c^2 \left(\frac{1}{v'} - \frac{1}{v} \right)$$

$$= m_0 c (\lambda' - \lambda) \quad (19.34a)$$

Obviously, the shift in wavelength of the scattered photon as compared to that of the incident photon is

$$\lambda' - \lambda = \Delta \lambda = \frac{h}{m_0 c} (1 - \cos \phi) \quad (19.35)$$

Also
$$\lambda' - \lambda = \Delta \lambda = \frac{h}{m_0 c} \sin^2 \frac{\phi}{2} \quad (19.35a)$$

This is Compton expression for the shift in the wavelength of the scattered X-rays. $h/m_0 c$ is denoted by λ_c and is called *Compton wavelength*. λ_c has the dimensions of length and it has value

$$\frac{h}{m_0 c} = \frac{6.6 \times 10^{-34}}{9.1 \times 10^{-31} \times 3 \times 10^8} = 0.0242 \times 10^{-10} \text{ m}$$

$$= 0.242 \text{ Å}$$

Thus
$$\Delta \lambda = \lambda_c (1 - \cos \phi) \quad (19.36)$$

From Eq. (19.35), it is evident that λ' is always greater than λ as $\cos \phi$ is always less than 1. Furthermore, $\Delta \lambda$ is also dependent on $\cos \phi$. Thus, maximum value of $(1 - \cos \phi)$ = $[1 - (-1)] = 2$ at $\phi = 180°$, hence the maximum value of the shift in wavelength can be

$$(\Delta \lambda)_{max} = 0.0484 \text{ Å}$$

Thus, to have an appreciable relative change in wavelength, i.e. $(\Delta \lambda / \lambda)$, the wavelength (λ) of the incident X-ray must be quite small. Obviously, the Compton effect is easily observable for smaller wavelengths.

Direction of Recoil Electron

Dividing Eq. (19.29) by Eq. (19.28), one obtains

$$\tan \theta = \frac{hv' \sin \phi}{h(v - v' \cos \phi)} = \frac{v' \sin \phi}{(v - v' \cos \phi)} \quad (19.37)$$

Using Eq. (19.34), one obtains

$$\frac{1}{v'} = \frac{1}{v} + \frac{h}{m_0 c^2} (1 - \cos \phi)$$

or
$$= \frac{1}{v} + \frac{h}{(m_0 c^2)} 2 \sin^2 \frac{\phi}{2}$$

or
$$\frac{1}{v'} = \frac{1 + \left(\dfrac{hv}{m_0 c^2} \right) 2 \sin^2 \dfrac{\phi}{2}}{v}$$

or
$$v' = \frac{v}{1 + \left(\dfrac{hv}{m_0 c^2} \right) 2 \sin^2 \dfrac{\phi}{2}} \quad (19.38)$$

Writing $\beta = hv/m_0 c^2$, Eq. (19.38) becomes

$$v' = \frac{v}{1 + 2\beta \sin^2 \dfrac{\phi}{2}} \quad (19.39)$$

Substituting the value of v' in Eq. (19.37), one obtains

$$\tan\theta = \frac{\dfrac{v\sin\phi}{\left[1+2\beta\sin^2\phi/2\right]}}{\dfrac{v-(v\cos\phi)}{\left(1+2\beta\sin^2\phi/2\right)}}$$

or $\quad \tan\theta = \dfrac{2\sin\dfrac{\phi}{2}\cos\dfrac{\phi}{2}}{2\sin^2\dfrac{\phi}{2}+2\beta\sin^2\dfrac{\phi}{2}}$

or $\quad \tan\theta = \dfrac{2\sin\dfrac{\phi}{2}\cos\dfrac{\phi}{2}}{2(1+\beta)\sin^2\dfrac{\phi}{2}} = \dfrac{\cot\dfrac{\phi}{2}}{1+\dfrac{hv}{m_0c^2}}$ (19.40)

Obviously, the recoil angle θ depends on the scattering angle if $\phi = 0° = 90°$ and if $\phi = 180°$; $\theta = 0°$. Obviously, as ϕ varies from $0°$ to $180°$, θ varies from $90°$ to $0°$. This shows that the electron can get recoiled only in the onward direction at angles less than $90°$, whereas a photon gets scattered in all directions. Since in most cases $hv = 10$ keV, and therefore $hv \ll m_0c^2$, one obtains a simple relation

$$\theta \approx \frac{1}{2}\ (\pi - \phi),$$

i.e. the direction in which the recoiled electron bisects the angle which is supplementary to the angle ϕ made by the scattered photon momentum with the initial photon momentum.

Kinetic Energy of Recoil Electron

The kinetic energy of the recoiled electron also depends on the scattering angle ϕ. One can easily show it as follows:

The kinetic energy of the recoil electron is the difference between the energies of the incident and scattered photons, i.e. KE of recoiled electron = $hv - hv'$.

Substituting the value of v' from Eq. (19.39), one obtains

$$KE = hv - h\left[\frac{v}{1+2\beta\sin^2\dfrac{\phi}{2}}\right]$$

$$= hv\left[\frac{2\beta\sin^2\dfrac{\phi}{2}}{1+2\beta\sin^2\dfrac{\phi}{2}}\right]$$ (19.41)

Experimental Verification of Compton Effect

The experimental arrangement to demonstrate the Compton effect is shown in Fig. 19.8. A beam of monochromatic X-rays of wavelength (λ) is made incident on a target. The distribution of intensity with wavelength of the monochromatic X-rays scattered at various angles is measured by X-ray spectrometer. The intensity distribution versus wavelengths curves for various scattering angles are shown in Fig. 19.9. Obviously, the curves have two peaks, one corresponding to modified X-rays

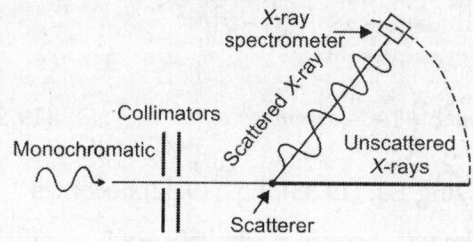

Fig. 19.8: Compton effect

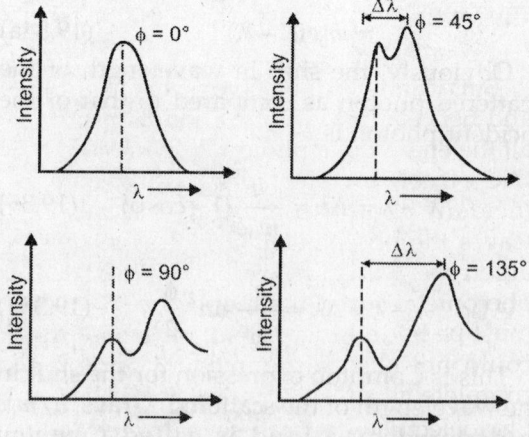

Fig. 19.9: Intensity distribution versus wavelength curves for various scattering angles

and the other corresponding to unmodified radiations. The difference between two peaks on wavelength axis gives the Compton effect or shift.

From the various curves, it is obvious that greater is the scattering angle, greater is the Compton shift in accordance with Eq. (19.35). At $\phi = 90°$, $\Delta\lambda = 0.024$ Å is in good agreement with the value obtained from Compton formula given by Eq. (19.35).

Comparison of Compton Effect and Photoelectric Effect

i. In Compton effect, a photon cannot transfer all of its energy to a free electron, whereas in photoelectric effect, an electron absorbs the incident photon completely before it leaves the photo-metal surface. To understand it clearly, let us consider a Compton collision between a photon and free electron (initially at rest). The incident photon has an energy $h\nu$ and a momentum $h\nu/c$. Let us suppose that this photon transfers its total energy (and momentum) to the electron who acquires a velocity v. From the law of conservation of momentum and energy, we have

$$h\nu = \frac{1}{2}mv^2 \text{ and } \frac{h\nu}{c} = mv$$

Eliminating $h\nu$ between the above two equations, one obtains

$$v = 2c$$

But v cannot exceed c (velocity of light) and hence a free electron cannot absorb all the energy of a photon. In photoelectric effect, an electron absorbs the incident photon completely before it leaves the photometal surface as photo electron. Thus, the electron in the photometal surface cannot be free, i.e. it must be bound to an atom, otherwise it could not absorb the incident photon completely.

ii. Photoelectric effect can be explained by mere quantum considerations, whereas one has to use relativistic considerations in applying laws of conservation of momentum and energy in case of explaining Compton effect with quantum theory. Obviously, we can say that Compton effect is *relativistic photoelectric effect*.

19.7 MATTER WAVES

The photoelectric effect and Compton effect provided conclusive evidence that electromagnetic radiation possesses corpuscular nature. At the same, the propagation experiments based on interference and diffraction also proved the existence of wave nature for these radiations. Obviously, radiation presents itself in two forms, particle or corpuscle and wave. In view of this, it is rather difficult to accept the conflicting ideas that radiation is a wave which is spread out cover space and also a particle which is localized at a point in space.

de Broglie Matter Waves

The theoretical physicist, de Broglie, in 1924 put forth a very bold suggestion that *wave–particle duality* is a general property, and like radiation matter must also exhibit dual (i.e. particle like and wave like) characteristic. His suggestion was partly based on the intuitive feeling that nature is symmetrical and if radiation can have dual characteristic of being wave and particle, then matter particles like electrons, photons, neutrons, atoms or molecules have an associated wave with them. This is called as matter wave or de Broglie wave.

Thus, the wavelength of the matter wave is

$$\lambda = \frac{h}{p} = \frac{h}{mv} \qquad (19.42)$$

where m is the mass of the material particle, v its velocity and p is momentum. In case, particle velocity is comparable with the velocity of light, then m is replaced by relativistic mass, i.e.

$$m = \frac{m_0}{\sqrt{1 - \dfrac{v^2}{c^2}}}$$

The de Broglie wavelength can be expressed in terms of energy

$$K = \frac{1}{2} mv^2 \quad \text{or} \quad v = \sqrt{mK}$$

Thus $\quad \lambda = \dfrac{h}{\sqrt{2mK}}$

If a charge particle carrying a charge q is accelerated through a potential difference of V volts, then

$$K = qV$$

or $\quad \lambda = \dfrac{h}{\sqrt{2mqV}}$

From special theory of relativity, the energy E and momentum p of a particle of rest mass m_0 are related as

$$E^2 = p^2c^2 + m_0^2 c^4 \qquad (19.43)$$

For a photon, $E = h\nu$ and rest mass $m_0 = 0$, hence

$$E^2 = p^2 c^2 \text{ or } E = pc \quad \therefore p = \frac{h\nu}{c} = \frac{h}{\lambda} \quad (19.44)$$

Obviously, Eq. (19.44) connects the corpuscular mode of description to the wave nature.

de Broglie Wavelength of Matter Waves

From Planck's theory of radiation, the energy of a photon is given by

$$E = h\nu = hc/\lambda \qquad (19.45)$$

$\therefore \qquad \lambda = hc/E \qquad (19.46)$

From special theory of relativity,

$$E = mc^2 \qquad (19.47)$$

Mass of the photon $m = h\nu/c^2$

Momentum of the photon

$$mc = p\frac{h\nu}{c} = \frac{h}{\lambda} \quad \because p = \frac{h}{\lambda}$$

or $\qquad \lambda = h/p \qquad (19.48)$

de Broglie postulated that what is true for an energy packet which we call as quantum or photon is also true for the material particle. Thus, for a particle of mass m moving with velocity v, we have $p = mv$ and the de Broglie wave associated with this particle is

$$\lambda = \frac{h}{mv} \qquad (19.49)$$

Here m stands for the relativistic mass. From Eq. (19.49), it is clear that:

i. If m or v is large, the de Broglie wave associated with the particle is small.

ii. The de Broglie wave associated with a material particle or photon is independent of any charge associated with it.

The expression for the de Broglie wavelength associated with an accelerated electron in the non-relativistic case will be

$$\lambda = \frac{h}{\sqrt{2m_0 eV}}$$

$$m_0 = 9.1 \times 10^{-31} \text{kg}$$

$$h = 6.6 \times 10^{-34} \text{ Js}, e = 1.6 \times 10^{-19} \text{ C}$$

$\therefore \qquad \lambda = \dfrac{6.6 \times 10^{-34}}{\sqrt{2 \times 9.1 \times 10^{-31} \times V}}$

$$\lambda = \frac{12.27 \times 10^{-10}}{\sqrt{V}} = \frac{12.27}{\sqrt{V}} \text{Å} = \sqrt{\frac{150}{V}} \text{ Å}$$

$$(19.50)$$

If electrons are accelerated by 100 volts, then the wavelengths of de Broglie waves associated with electrons will be 1.227 Å. The wavelength is of the same order of magnitude as the distance between the atomic planes in crystals. Crystals have been widely used for studying the diffraction of X-rays whose wavelengths are of the order of 1 Å. This immediately suggests that crystals can be used as diffraction grating to detect the existence of matter waves associated with electrons. Davisson and Germer in 1927 carried out a series of experiments and were first to demonstrate the existence of matter waves.

19.8 THE DAVISSON AND GERMER EXPERIMENT

The experimental arrangement used by Davisson and Germer is shown in Fig. 19.10. It consists of an electron gun which comprises a tungsten filament *F* heated by a low tension battery. The electron emitted by the filament are accelerated in an electric field of known potential difference from a high tension battery. The electrons are collimated to a fine beam by allowing them to pass through a system of suitable slits. This entire arrangement to produce a fine beam of electrons accelerated to a desired velocity is called electron gun.

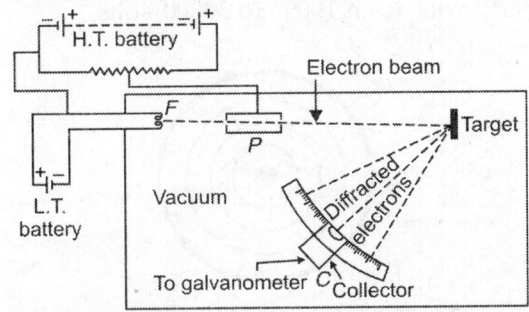

Fig. 19.10: Davisson and Germer experiment

The collimated beam of electrons is made to strike a nickel target capable of rotating about an axis perpendicular to the plane of the paper. The electrons are scattered in all directions by the atoms of nickel crystal. The intensity of the electron beam scattered in a given direction is measured by allowing it to enter in a Faraday cylinder called the collector, which can be rotated about the same axis as the target. The collector is connected to a galvanometer whose deflection is proportional to the intensity of the beam entering the collector. The whole apparatus is enclosed in an evacuated chamber

In one set of observations, the (111) face of the target nickel crystal was presented to the incident beam of electrons. The potential difference between the filament *F* and the electron gun was adjusted at 40 volts and then gradually increased in small steps. For each value of potential difference, polar graphs were drawn between angle θ (the angle between the incident beam and the beam entering the collector, known as *colatitude*) and the collector current. In these graphs, the length of the radius vector corresponding to a certain angle was drawn directly proportional to the collector current. For the potential difference of 40 volts, a smooth curve was obtained. As the potential difference was gradually increased, a little bump began to appear on the curve which increased to a maximum for *V* = 54 volts and θ = 50° (Fig. 19.11). Beyond 54 volts, the bump again decreased to become insignificant at about 68 volts. The pronounced maxima observed corresponding to accelerating potential of 54 volts at an angle of 50° with the electron beam provided a convincing proof that electron were associated with waves which after scattering from the regularly spaced atoms of the nickel crystal gave rise to a constructive interference.

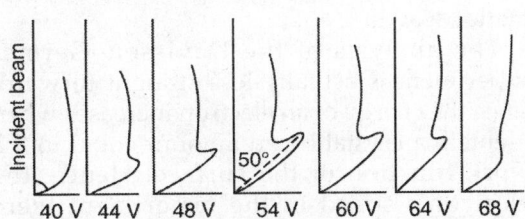

Fig. 19.11: Polar graphs between angle θ and intensity of scattered beam in Davisson–Germer experiment

In the above case, when the collector current is maximum at scattering (colatitude) angle of 50° corresponding to accelerating potential of 54 volts, the incident and the scattered beam evidently makes an angle of 65° with the family of Bragg's planes as shown in Fig. 19.12. X-ray diffraction studies reveal that the spacing of planes in this family is 0.91 Å. Taking *n* = 1, θ = 65°, one can calculate the wavelength of the incident wave by applying Bragg's law, $2d \sin\theta\, n\lambda$, we have

$$2 \times 0.91 \times \sin 65° = 1\,\lambda$$

or $\lambda = 2 \times 0.91 \times 0.9063 = 1.662$ Å

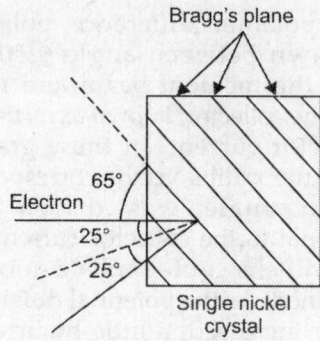

Fig. 19.12

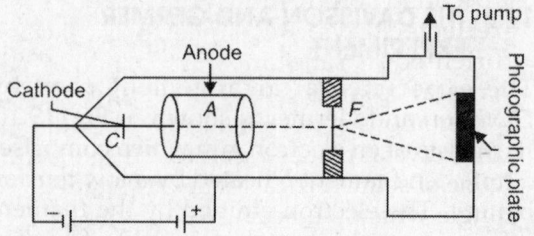

Fig. 19.13: Thomson's experiment for electron diffraction

Now applying de Broglie's relation for matter waves, we have

$$\lambda = \frac{h}{mv} = \frac{12.27}{\sqrt{V}}$$

Since $V = 54$ volts, $\lambda = \frac{12.27}{\sqrt{54}} = 1.673$ Å

Obviously, there is an excellent agreement between the two result. Thus, Davisson and Germer experiment may be considered to be a verification of the de Broglie hypothesis of matter waves.

The analysis of the Davisson–Germer experiment is actually less straight forward, since the energy of an electron increases when it enters a crystal by an amount equal to the work function of the surface. Hence, the electron's speed in the experiment were greater inside the crystal and the corresponding de Broglie wavelengths shorter than the corresponding values outside. An additional complication arises from interference between waves diffracted by different families of Bragg planes, which restricts the occurrence of maxima to certain combinations of electron energy and angle of incidence rather than merely to any combination they obey the Bragg equation.

19.9 GP THOMSON'S EXPERIMENT

GP Thomson, using X-ray powder diffraction method, obtained electron diffraction patterns. These results provided another experimental evidence for the existence of matter waves. A schematic diagram of Thomson's experiment is shown in Fig. 19.13.

The electrons emitted by the cathode C are accelerated and channelised by applying a potential difference between the cathode C and the cylindrical anode A. The potential difference is applied by a high tension battery and varies from 10000 to 50000 volts.

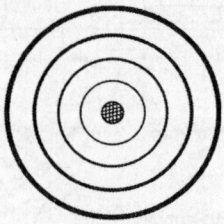

Fig. 19.14: Electron diffraction pattern

The accelerated fine beam of electrons is then made to bombard against a thin film of metal like gold, silver, aluminium, etc. After passing through the film, the electrons are received upon a photographic plate. The developed photographic plate shows a pattern consisting of a series of well defined concentric rings about a central spot (Fig. 19.14). It is worthwhile to mention that the diffraction pattern can be disturbed by a magnetic field, showing that the pattern was produced by electrons and not by any radiation.

In Thomson's experiment, one can imagine the metallic film to be composed of a very large number of tiny crystals oriented at random. Due to the random orientation of these millions of micro crystals, a given electron beam is intercepted by a particular micro crystal, this set of crystal planes is having suitable angle of orientation so as to satisfy Bragg's law ($2d \sin\theta = n\lambda$) and this gives rise to constructive interference between the reflected rays. But since the crystals are oriented at random, this particular set of crystal planes shall have all possible orientations about the incident beam. Due to this, crystal planes will lie on the surface of a cone of semi-vertical angle 2θ, where θ is the angle which the beam makes with this set of crystal planes as shown in Fig. 19.15. Other sets of crystal planes suitably oriented so as to satisfy Bragg's condition will reflect the beam along the surfaces of cones of different semi-vertical angles. The intercepts of these semi-vertical cones with the photographic film or plate give rise to diffraction pattern as shown in Fig. 19.15. The wavelength of electron waves can be calculated using the Bragg's relation and the value of lattice constant known from X-ray data.

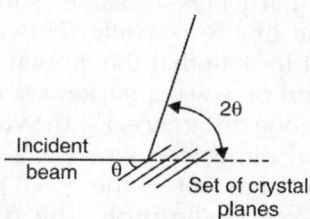

Fig. 19.15

Instead of following this method, Thomson determined the wavelength from de Broglie's relation and calculated the lattice constant with the help of Bragg's law. The value of lattice constant so obtained by Thomson was found to be in good agreement with the value determined from X-ray measurements confirming the existence of matter waves.

19.10 PROPERTIES OF MATTER WAVES

Following are the properties of matter waves:

i. Wavelength is inversely proportional to the mass and velocity of the particle.

ii. When $v = 0$ then $\lambda = \infty$, wave becomes indeterminate and $v = \infty$ then $\lambda = 0$. Thus, matter waves are generated by the motion of particles.

iii. $\lambda = h/mv$ is independent of charge, thus these are produced irrespective of particle being charged or not. This reveals that matter waves are not electromagnetic waves as electromagnetic waves are produced only through motion of charged particles.

iv. The velocity of matter waves is dependent on the velocity of the matter particle and contrary to electromagnetic waves these donot have constant velocity.

v. The velocity of matter wave is greater than the velocity of light.

A particle in motion associated with matter wave has two different velocities—one going mechanical motion v and another group or wave velocity.

Schrodinger postulated that a moving material particle is equivalent to wave packet rather than a single wave. A wave packet comprises a group of waves, each with slightly different velocity and wavelength, with phases and amplitudes so chosen that they interfere constructively over only a small region of space where the particle can be located, outside of which they produce destructive interference so that the amplitude reduces to zero rapidly. Obviously, a wave packet is formed as a result of the superposition of many plane waves of different frequencies grouped around some central frequency. Since the envelope of the packet must have a finite spatial extent, the sum of the amplitudes of all of its plane wave components must be zero everywhere except where the packet is localized. Form of a typical wave packet in one dimension is shown in Fig. 19.16.

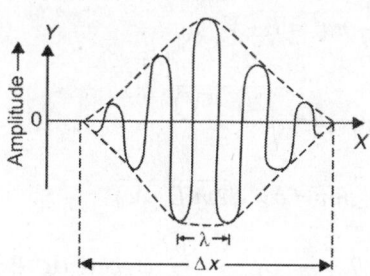

Fig. 19.16: A wave packet in one dimension

The velocity with which the wave packet moves is called the group velocity. The energy carried by the packet as its group velocity is, thus, analogous to the kinetic energy of a free particle of the same velocity. The individual waves forming the wave packet possess an average velocity, u called the *phase velocity*. One can easily show that the velocity of the material particle v is same as *group velocity* of wave packet.

Let us suppose that v is the velocity of the moving particle, g is the group velocity, u is the velocity of an individual wave train comprising the group or wave packet and λ is the wavelength of the individual wave. The group velocity g of a wave packet is found to be given by

$$g = u - \lambda \frac{du}{d\lambda} \qquad (19.51)$$

Writting Eq. (19.51) as

$$g = \lambda^2 \left(\frac{u}{\lambda^2} - \frac{1}{\lambda}\frac{du}{d\lambda} \right)$$

$$= -\lambda^2 \left(\frac{1}{\lambda}\frac{du}{d\lambda} - \frac{u}{\lambda^2} \right) = -\lambda^2 \frac{d}{d\lambda}\left(\frac{u}{\lambda} \right)$$

$$= -\lambda^2 \left(\frac{df}{d\lambda} \right) \left[\because f = \frac{u}{\lambda} \text{ (here } f \text{ is frequency)} \right]$$

$$\therefore \qquad \frac{1}{g} = -\frac{1}{\lambda^2}\left(\frac{d\lambda}{df} \right) = \frac{d}{df}\left(\frac{1}{\lambda} \right) \qquad (19.52)$$

Let E denotes the total energy and V the potential energy of the particle. If m is the mass of the particle, its kinetic energy is given by

$$\frac{1}{2}mv^2 = E - V$$

or $$v = \sqrt{\frac{2(E-V)}{m}} \qquad (19.53)$$

$$mv = p = \sqrt{2m(E-V)}$$

But $\lambda = \dfrac{h}{p}$ or $\dfrac{1}{\lambda} = \dfrac{p}{h}$ (from de Broglie's relation)

$$\therefore \qquad \frac{1}{\lambda} = \frac{\sqrt{2(E-V)}}{h} \qquad (19.54)$$

Substituting $1/\lambda$ from Eq. (19.54) in Eq. (19.52), one obtains

$$\frac{1}{g} = \frac{d}{df}\left(\frac{\sqrt{2m(E-V)}}{h} \right) = \frac{1}{h}\frac{d}{df}\sqrt{2m(hf-V)}$$

$$[\because E = hf]$$

$$= \frac{1}{h}\frac{1}{2}\left[2m(hf-V) \right]^{-1/2} 2mh = \sqrt{\frac{m}{2(hf-V)}} = \frac{1}{v}$$

$$[\text{from Eq. (19.53)}]$$

$$\therefore \qquad g = v \qquad (19.55)$$

Obviously, the group velocity g of de Broglie wave group or wave packet associated with a moving particle is the same as the classical velocity v of the particle. However, it is important to note that the motion of particle and motion of a wave packet are equivalent with only one difference, i.e. the wave packet has spatial dimension whereas a particle (a mass-point) has not. If the wave packet can be made small enough, the distinction between the two becomes insignificant.

19.11 THE UNCERTAINTY PRINCIPLE

The description of material particles in terms of waves and the setting up of a connection between energy and momentum, on one side, and frequency and wavelength, on the other, lead to an indefiniteness that had not been previous suspected in mechanics of particles. This indefiniteness has become famous as the *uncertainty principle*, or *indeterminacy principle* put forward by Heisenberg in 1927. This principle expresses a fundamental limit to the simultaneous determination of certain pairs of variables, such as position and momentum, energy and time. The principle can be stated for different pairs of variables as follows:

The product of uncertainly Δx (or possible error) in the x-coordinate of a particle, in motion, at some instant, and the uncertainly Δp_x in the x-component of the momentum, at the same instant, is of the order of or greater than $\hbar$, i.e.

$$\Delta p_x \Delta x \geq \hbar \left(\hbar = \frac{h}{2\pi} = 1.054 \times 10^{-34} \text{ Js} \right) \quad (19.56)$$

An expression of this type also holds good for other components of linear momentum and for angular momentum and angular position, i.e.

$$\Delta p_y \Delta y \geq h$$

$$\Delta p_z \Delta z \geq h \quad\quad\quad\quad\quad (19.56a)$$

It is important to mention here that there is no restriction on Δx or on Δp_x, but only on their product. Also, there is no restriction on product like $\Delta p_z \Delta y$.

Significance: One can draw the following conclusions from Eq. (19.56):

i. If the position coordinate x of a particle in motion is accurately determined at some instant, so that $\Delta x = 0$, then at the same instant the uncertainty Δp_x in the determination of the momentum of the particle becomes infinite.

ii. If the momentum p_x of a particle is accurately determined at some instant, so that $\Delta p_x = 0$, then at the same instant the uncertainty Δx in the determination of the position coordinate becomes infinite.

Obviously, if an experiment is designed to measure x or p_x accurately, the other quantity will become completely uncertain. One can measure both the quantities by means of an experiment, but only within certain limits of accuracy specified by the uncertainty relation.

iii. For a particle of mass m moving with velocity v, the product of uncertainty Δx and Δv is given by

$$\Delta x \, \Delta v \geq \frac{\hbar}{m}$$

For a heavy particle $\hbar/m \approx 0$ is very small, and therefore, the product Δx and Δv of the two uncertainties becomes very small. For such particles both x and v can be determined accurately.

For very heavy particles, if m is such that $\hbar/m \approx 0$, the uncertainties vanish and one can measure x and v with perfect accuracy. Obviously, this is the limiting case of classical mechanics. This means classical mechanics is true for heavy bodies and uncertainties are a characteristic of atomic and subnuclei particles.

The uncertainty relation for the simultaneous measurement of energy E and time t can be expressed as

$$\Delta E \Delta t \geq \hbar \quad\quad\quad\quad\quad (19.57)$$

Here again there are no restrictions on the accuracy with which E or t can be measured, but only on the product $\Delta E \Delta t$

Significance: The physical significance of the energy-time uncertainty relation is quite different from that of the position momentum uncertainty relation. If ΔE is the maximum uncertainty in the determination of the energy of a system in a particular state, then the minimum time interval for which the system remains in the state is given by

$$\Delta t = \frac{\hbar}{\Delta E}$$

And, if a system remains in a particular state for a maximum time instant Δt, then the minimum uncertainty in the energy of the system in the state is given by

$$\Delta E = \frac{\hbar}{\Delta t}$$

The uncertainty relation for angular momentum and angular displacement is often expressed as

$$\Delta J \Delta \theta \geq \hbar \quad\quad\quad\quad\quad (19.58)$$

19.12 EXPERIMENTAL ILLUSTRATIONS

The problem of uncertainty principle is a logical consequence of the dual behaviour of matter. A deeper insight into the nature of the uncertainty principle can be accomplished by the following experiments:

Localization Experiment (γ-ray Microscope)

Let us consider an experiment whereby a particle (say electron) might be observed in a microscope. Light of frequency v falls upon the particle and is scattered. If it is to be observed, the scattered photon must enter the objective of the microscope as shown in Fig. 19.17. There is, therefore, an uncertainty in its momentum given by $\Delta p = p \sin \alpha$ (from Fig. 19.17), $(\tan \alpha = \Delta p / p)$ when α is small, we can write $\tan \alpha = \sin \alpha = \Delta p / p$, since it can go anywhere within a cone of semi-vertical angle α. The particle p recoils with an equal and opposite amount of momentum p in accordance with the Compton effect. The position of the particle is defined by the resolving power of the microscope such that

$$\Delta x = \frac{\lambda}{\sin \alpha} \qquad (19.59)$$

Again we have $\sin \alpha = \dfrac{\lambda}{\Delta x} = \dfrac{\Delta p}{p}$ and therefore $\Delta p \Delta x = \hbar$.

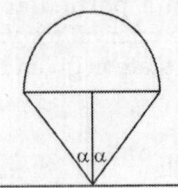

Fig. 19.17

To improve the resolving power, we can increase the aperture α of the microscope, but in doing so Δp is also increased. Decreasing the wavelength λ would also increase the resolving power, but a corresponding larger amount of momentum would be transferred to the particle from the photon due to a large Compton effect. Similar relationships exist for the momentum along each of the cartesian axes.

Determination of Position of a Particle Using a Single Slit

Let us suppose that we want to determine the cartesian coordinates of a particle. For simplicity, let us determine the x-coordinate only. Let us place a slit in the path of the particle with the width of the slit in the X-direction. If now the particle is observed on the other side of the slit, then the particle while crossing the slit must have x-coordinates lying between x_1 and x_2 positions (Fig. 19.18). As a particular example, let us suppose that we have a beam of electrons moving along the Z-direction and the width of the slit be d. With respect to the passage of an electron through the slit, the x-component of the position of the electron has uncertainty given by

$$\Delta x = x_2 - x_1 = d \qquad (19.60)$$

The wavelength associated with an electron is given by de Broglie relation $\lambda = h/p$ where p is the momentum of the electron.

As the electron emerges from the single slit, it is diffracted in an unpredictable direction and acquires certain velocity and momentum in the X-direction. Now, according to the phenomenon of diffraction from a single slit, most of the electrons in the beam will fall inside the central maximum, i.e. the diffracted electron has a large probability of hitting the screen in the region BC. From the theory of diffraction, we have

$$AB = \lambda L / d \qquad (19.61)$$

where L is the distance of the screen from the slit.

For an electron falling somewhere in the central maximum, the momentum in the X-direction can have any value between p_x and $-p_x$. Obviously, the uncertainty in the x-component of the momentum can be written as

$$\Delta p_x \approx p_z \qquad (19.62)$$

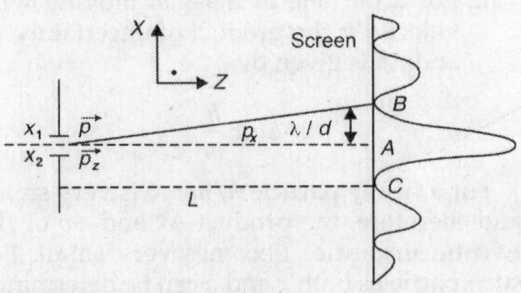

Fig. 19.18

It is worthwhile to note that the uncertainty Δp_x actually be larger than that given by Eq. (19.62) as electron can fall into other maxima also. From the geometry of the figure, we have

$$\frac{p_x}{p_z} = \frac{\lambda L/d}{L} = \frac{\lambda}{d} \qquad (19.63)$$

Obviously,

$$\Delta p_x \approx p_x \frac{\lambda}{d} p_z \qquad (19.64)$$

Here p_z represents the initial momentum of the electron and it is related to λ as

$$p_z = \frac{h}{\lambda} \qquad (19.65)$$

$$\therefore \qquad \Delta p_x \approx \frac{\lambda}{d} \frac{h}{\lambda} \approx \frac{h}{d} \qquad (19.66)$$

$$\therefore \qquad \Delta x \Delta p_x \approx d \frac{h}{d} \approx h$$

or $\qquad \Delta x \Delta p_x > \dfrac{h}{2x}$ (or $\hbar$) $\qquad (19.67)$

Obviously, the uncertainty in x can be made as small as desired by reducing the width of the slit but decrease in the width of the slit increases the width of the central maximum, i.e. it increases the uncertainty in p_x so that the product $\Delta x \, \Delta p_x$ is always greater than $\hbar$.

19.13 APPLICATIONS OF UNCERTAINTY PRINCIPLE

According to uncertainty principle, it is not possible to measure simultaneously the position and momentum of the particle with the desired accuracy. On the other hand, according to classical mechanics, it is possible to know the position and momentum of any particle with desired accuracy provided its initial position and momentum are known. Further, quantum mechanics introduced the laws of probability and therefore probability takes the place of certainty in classical mechanics. Moreover, the quantum mechanics introduces a new concept of particle according to which the term particle is something composed of a very deep and essential unison

avoidable uncertainty. The principle of uncertainty explains a large number of facts which could not be explained by classical mechanics. A few of them are discussed below:

i. *Non existence of free electrons in the nucleus*: We know that the maximum possible kinetic energy of an electron emitted by a radioactive nuclei is about 4 MeV.

ii. The rest mass of an electron is
$$m_0 = 9.11 \times 10^{-31} \text{ kg and}$$

iii. The diameter of the nucleus is 2×10^{-14} m.

If an electron exists inside the nucleus, it can be anywhere within the diameter of the nucleus. Obviously, the maximum uncertainty Δx in the position of the electron is the same as the diameter of the nucleus, i.e.

$$\Delta x = 2 \times 10^{-14} \qquad (19.68)$$

According to Heisenberg uncertainty relation, the product of the uncertainty Δx in the position of the electron and the uncertainty Δp_x in the x-component of its momentum is given by

$$\Delta x \, \Delta p_x \geq \hbar$$

The uncertainty in the momentum p_x is

$$\Delta p_x \geq \frac{\hbar}{\Delta x}$$

Minimum uncertainty in the momentum is given by

$$\Delta p_x \geq \frac{\hbar}{\Delta x} = \frac{6.63 \times 10^{-34}}{2\pi \times 2 \times 10^{-14}} = \frac{6.63 \times 10^{-20}}{4\pi}$$

$$= 5.28 \times 10^{-21} \text{ kg m/sec.} \quad (19.69)$$

This means that if the electron exists inside the nucleus, then its minimum momentum should be

$$E_{\min} = 5.28 \times 10^{-21} \text{ kg-m/sec}$$

According to theory of relativity, the total relativistic energy E of the particle is given by

$$E^2 = p^2 c^2 + m_0^2 c^4 \qquad (19.70)$$

For the electron having minimum momentum, the minimum energy is given by

$$E_{\min}^2 = p_{\min}^2 + m_0^2 c^4 \qquad (19.71)$$

$$= (5.28 \times 10^{-21} \times 3 \times 10^8)^2$$
$$+ (9.11 \times 10^{-31})^2 \times (3 \times 10^8)^4$$
$$= 2.5 \times 10^{-24} + 6.27 \times 10^{-27} \, \text{J}$$

Obviously, the second term is much smaller than the first and hence it can be neglected, then

$$E_{min} = \sqrt{2.5 \times 10^{-24}} = 1.58 \times 10^{-12} \, \text{J}$$

$$= \frac{1.58 \times 10^{-12}}{1.6 \times 10^{-19}} \, \text{eV} = 0.9875 \times 10^7 \, \text{eV}$$

$$= 9.875 \times 10^6 \, \text{eV} = 9.875 \, \text{MeV}$$

Obviously, if a free electron exists in the nucleus, it must have a minimum energy of about 9 MeV.

As stated in the beginning that the maximum kinetic energy which a β-particle (an electron), emitted from a radioactive nuclei can have is 4 MeV. This clearly indicates that free electrons cannot exist inside nuclei.

Ground State Energy of H Atom

Suppose the electron in the hydrogen atom moves around the nucleus, i.e. proton in a circular orbit of radius r. The maximum uncertainty in the determination of its position with respect to the proton can be taken equal to r, i.e. $\Delta r = r$.

From the uncertainty relation, the minimum uncertainty Δp in the simultaneous determination of its momentum p in the plane of the orbit is given by

$$p = \frac{\hbar}{\Delta r} = \frac{\hbar}{r}$$

Since the momentum p cannot be less than the uncertainty Δp and hence the minimum possible momentum is

$$p = \frac{\hbar}{r}$$

The kinetic energy K of the electron is given by

$$K = \frac{p^2}{2m} = \frac{\hbar^2}{2mr^2} \tag{19.72}$$

and the electrostatic potential energy is given by

$$U = \frac{e^2}{4\pi\varepsilon_0 r} \tag{19.73}$$

Therefore, the total energy of the electron in the hydrogen atom is

$$E = \frac{\hbar^2}{2mr^2} + \frac{e^2}{4\pi\varepsilon_0 r} \tag{19.74}$$

The ground state energy must be a minimum value of E. Suppose r_1 is the distance of the electron from the proton in the ground state.

For $E = E_{min}$, $\left(\dfrac{dE}{dr}\right)_{r=r_1} = 0$

$$\therefore \quad -\frac{\hbar^2}{mr_1^3} + \frac{e^2}{4\pi\varepsilon_0 r_1^2} = 0$$

or $\dfrac{e^2}{4\pi\varepsilon_0} = \dfrac{\hbar^2}{mr_1}$

$$\therefore \quad r_1 = \frac{4\pi\varepsilon_0\hbar^2}{me^2} = \frac{4\pi\varepsilon_0\hbar^2}{me^2 \times 4\pi^2} = \frac{\varepsilon_0 h^2}{\pi me^2} \tag{19.75}$$

Eq. (19.75) is the well-known expression for the radii of the Bohr orbit. Using the values:

$$\hbar = 6.63 \times 10^{-31} \, \text{Js}, \, m = 9.11 \times 10^{-31} \, \text{kg}$$

$$e = 1.6 \times 10^{-19} \, \text{C}, \, \varepsilon_0 = 8.85 \times 10^{-12} \, \text{C}^2/\text{Nm}^2$$

one obtains $\quad r_1 = 5.31 \times 10^{-11} \, \text{m}$

This is the radius of the hydrogen atom in the ground state.

The expression for the energy E_1 in the ground state is obtained by substituting

$$r = r_1 = 4\pi\varepsilon_0\hbar^2/me^2$$

From Eq. (19.74), we have

$$E_1 = \frac{\hbar}{2mr_1^2} - \frac{e^2}{4\pi\varepsilon_0 t_1}$$

$$= \frac{\hbar^2}{2m}\left(-\frac{me^2}{4\pi\varepsilon_0\hbar^2}\right)^2 - \frac{e^2}{4\pi\varepsilon_0}\left(-\frac{me^2}{16\pi\varepsilon_0\hbar^2}\right)$$

$$= \frac{me^4}{32\pi^2\varepsilon_0^2\hbar^2} - \frac{me^4}{16\pi^2\varepsilon_0^2\hbar^2} = -\frac{me^4}{32\pi^2\varepsilon_0^2\hbar^2}$$

$$= \frac{me^4 \times 4\pi^2}{32\pi^2\,\varepsilon_0^2\,h^2} = -\frac{me^4}{8\varepsilon_0^2 h^2} \tag{19.76}$$

Substituting the numerical values of various constants appearing in Eq. (19.76) and using 1 eV = 1.6 × 10⁻¹⁹ J, one obtains

$$E_1 = -13.6 \text{ eV}$$

This is the ground state energy of the hydrogen atom.

Ground State Energy of Harmonic Oscillator

The Hamiltonian of a linear harmonic oscillator is given by

$$\hat{H} = \hat{H}(\hat{x},\hat{p}) = \frac{\hat{p}^2}{2m} + \frac{1}{2}k\hat{x}^2 \tag{19.77}$$

where $\hat{x}$ and $\hat{p}$ are Hermitian operators corresponding to x and p respectively.

Let ψ represents eigenvector of $\hat{H}$ belonging to eigenvalue E. Also, let $\hat{A}$ and $\hat{B}$ be, respectively, the operators corresponding to position and momentum, one can write

$$\Delta x = \left\{\left\langle \bar{x}^2\right\rangle\right\}^{1/2} \text{ and } \Delta p = \left\{\left\langle p^2\right\rangle\right\}^{1/2} \tag{19.78}$$

where angular bracket $\langle\,\rangle$ represents the average. Since both x^2 and p^2 are positive for the oscillator, we have

$$\Delta x = 0; \Delta p > 0$$

so that $\Delta x\,\Delta p > 0$

According to Heisenberg uncertainty principle,

$$\Delta x\,\Delta p \geq \hbar/2, \tag{19.78a}$$

Now, $E = \left\langle\psi\left|\hat{H}\right|\psi\right\rangle = \left\langle\hat{H}\right\rangle$

$$= \frac{1}{2m}\left\langle\hat{p}^2\right\rangle + \frac{1}{2}k\left\langle\hat{x}^2\right\rangle$$

$$= \frac{1}{2m}(\Delta p)^2 + \frac{k}{2}(\Delta x)^2$$

$$\geq \frac{1}{2m}(\Delta p)^2 + \frac{\hbar^2 k}{8}\frac{1}{(\Delta p)^2} \tag{19.78b}$$

The minimum value of E is given by the minimum of the expression,

$$= \frac{1}{2m}(\Delta p)^2 + \frac{\hbar^2 k}{2}\frac{1}{(\Delta p)^2}$$

i.e. $E_{\min} = E_0 = \left[\frac{(\Delta p)^2}{2m} + \frac{\hbar^2 k}{8(\Delta p)^2}\right]_{\min}$

$$= \frac{(\Delta p_0)^2}{2m} + \frac{\hbar^2 k}{8(\Delta p_0)^2} \tag{19.79}$$

where $(\Delta p_0)^2 = \frac{1}{2}\sqrt{\hbar^2 km} = \frac{m}{2}\hbar\omega_c \tag{19.80}$

ω_c being the classical frequency of the oscillator given by

$$\omega_c = \sqrt{\frac{k}{m}} \tag{19.81}$$

Thus $E_0 = \frac{1}{2}\hbar\omega_c \tag{19.82}$

The above energy, i.e. $E_0 = \frac{1}{2}\hbar\omega_c$ is called the *lowest energy* or *zero point energy*.

Natural Width of Spectral Line

According to Heisenberg's uncertainty principle of energy and time,

$$\Delta E\Delta t \geq \hbar$$

Since the lifetime of electron in an excited orbit is finite ($\approx 10^{-8}$ sec), therefore the energy levels of the atom given by ($\Delta E \approx h/\Delta t$) must have a finite width. This means that the excited energy levels have a finite energy spread thereby indicating that the radiation emitted when an electron jumps truly monochromatic. In other words, the spectral lines can never be infinitely sharp but must have a natural finite width.

19.14 SCHRODINGER'S WAVE MECHANICS

The experiments described in preceding sections tell us that classical physics is inadequate for atomic scale phenomena and for interaction between electron and electromagnetic waves. The idea expressed in Planck's equation $E = h\nu$ and de Broglie's equation $\lambda = h/mv$ are basic to all these problems, but it is not possible to build a complete understanding of atomic phenomena directly upon these ideas.

We require a theory that enables us to predict and explain more complicated atomic phenomena. The theory that accomplishes this is the wave mechanics or quantum mechanics and the heart of the theory is Schrodinger's wave equation. Schrodinger in 1925, using de Broglie's ideas of matter waves set up a wave equation to describe the new mechanics of the particles and successfully applied it to various atomic problems. An equally successful matrix algebra formulation, different in form but basically equivalent to the Schrodinger treatment was developed by Heisenberg. A relativistic treatment of the quantum mechanics was pioneered by P. A. M. Dirac and led to a new and broader picture of nature including the concept of antimatter. In the present chapter, we shall restrict ourselves to the study of the Schrodinger's wave mechanics.

Wave Equation

In classical mechanics, wave equation is a second order differential equation in space and time. Solution of this wave equation represent wave disturbances in a medium. Obviously, a wave equation is the usual basis of mathematical theory of wave motion. As an example of wave equation, we can consider the equation of an electromagnetic wave travelling in the X-direction, i.e.

$$\frac{\partial^2 E_y}{\partial x^2} = \frac{1}{c^2} \frac{\partial^2 E_y}{\partial t^2} \qquad (19.83)$$

where E_y is the y-component of the electric intensity.

We must note that a differential equation for the wave associated with a particle in motion cannot be derived from first principles.

One can develop the equation by any one of the following procedures:

i. The classical equations of motion are transformed into a wave equation in accordance with wave properties of matter based on de Broglie's hypothesis.

ii. A complex variable quantity, called the wave function, is assumed to represent a plane simple harmonic wave associated with a free particle and the classical expression for the total energy is used.

iii. A particle at a given position and at a given time represented by a wave packet which is obtained by superposition of a group of plane waves of nearly the same wavelength, which interfere destructively everywhere except at the wave packet and the classical expression for the total energy is used.

iv. In the classical expression for the total energy of a particle, the dynamic quantities are replaced by their corresponding operators. These operators are allowed to operate upon the wave function.

Any one of the above procedures can be followed to develop the Schrodinger equation.

Wave Function

The variable quantity characterizing de Broglie waves associated with a material particle is called the *wave function* denoted by the symbol ψ. The value of the wave function associated with a moving body at the particular point (x, y, z) in space at time t is related to the likelihood of finding the body there at that time. In quantum mechanics, we are concerned with this wave function ψ of the body.

We assume that a wave function is a complex quantity and, therefore, it cannot have a direct physical meaning. It may be expressed in the form

$$\psi(x, y, z, t) = a + ib \qquad (19.84)$$

where a and b are real functions of the variables (x, y, z, t).

One can obtain the complex conjugate of ψ which is denoted by $\psi*$ by changing i to $-i$, i.e.

$$\psi*(x, y, z, t) = a - ib \qquad (19.85)$$

Multiplying Eq. (19.84) by Eq. (19.85), one obtains

$$\psi(x, y, z, t)\,\psi^*(x, y, z) = a^2 + b^2 \qquad (19.86)$$

On the LHS of Eq. (19.86) the product $\psi\psi^*$ is denoted by P, i.e.

$$P = \psi(x, y, z, t)\,\psi^*(x, y, z, t)$$
$$= |\psi(x, y, z, t)|^2 \qquad (19.87)$$

$$\therefore\ |\psi(x, y, z, t)|^2 = a^2 + b^2 \qquad (19.88)$$

Obviously, the product of ψ and ψ^* is real and positive if $\psi \neq 0$. Its positive square root is denoted by $|\psi(x, y, z, t)|^2$ and is called the **modulus** of ψ. The quantity $|\psi(x, y, z, t)|^2$ is called the **probability density.** Maxborn has used the result (19.88) and interpreted $\psi(x, y, z, t)$ as follows:

For the motion of a particle, the quantity

$$PdV = \psi(x, y, z, t)\,\psi^*(x, y, z, t)dV$$
$$= |\psi(x, y, z, t)|^2 dV \qquad (19.89)$$

is the *probability* that the particle will be found in a volume element dV surrounding the point at position (x, y, z) at time t. Thus, **the total probability** for finding the particle anywhere in space in question is proportional to the integral of $|\psi(x, y, z, t)|^2$ all over the space. For the motion of a particle in one dimension the quantity

$$Pdx = \psi(x, t) = |\psi(x, t)|^2 dx \qquad (19.90)$$

In this case $|\psi(x, t)|^2$ is called the probability per unit distance.

Thus, the total probability is

$$P_x = \int_{-\infty}^{\infty} |\psi(x, t)|^2\, dx \qquad (19.91)$$

for the one dimensional space along the X-axis. For three dimensional space, we have

$$P = \int_{-\infty}^{\infty} |\psi(x, y, z, t)|^2\, dV \qquad (19.92)$$

The above integrals must be finite in order to represent a real particle.

The above discussion clearly reveals that the quantum laws and results of their measurements have probabilitistic interpretation.

The wave function $\psi(r, t)$ is sometimes called a *probability density* for the particle as

$$P(x, t) = \frac{|\psi(x, t)|^2}{\int_{-\infty}^{\infty} |\psi(x, t)|^2\, dx} \qquad (19.93)$$

where

$$\int_{-\infty}^{\infty} P(x, t)dt = 1 \qquad (19.94)$$

For physical acceptable wave functions, it is always possible to introduce an appropriate factor in the wave function such that

$$\int_{-\infty}^{\infty} |\psi(x, t)|^2\, dx = \int_{-\infty}^{\infty} \psi^*(x, t)\psi(x, t)dx = 1$$
$$(19.95)$$

When Eq. (19.95) is true, wave function ψ is said to be normalised. A wave function which satisfies the condition (19.95) is said to be normalised (to unity).

The normalising condition for the wave function for the motion of a particle in three dimensions is

$$\int_{-\infty}^{\infty} |\psi(x, y, z)|^2\, dV = 1 \qquad (19.96)$$

where ψ is a function of the space coordinates (x, y, z).

If a wave function does not satisfy the normalization condition (19.96), then it must be multiplied by a constant factor called the normalisation factor. For example,

$$\int \psi^2 dx = N (N \neq 1)$$

The normalisation factor will be $1/\sqrt{N}$ and the normalised wave function will be $\dfrac{1}{\sqrt{N}}\psi$.

One can easily see that $\dfrac{1}{\sqrt{N}}\psi$ will be a good wave function as ψ itself.

From Eq. (19.93), it is obvious that when a wave function ψ is normalised, the probability density is simply the square of its amplitude.

To arrive at a result consistent with physical observation, several additional requirements are imposed on the wave function ψ:

i. ψ must be finite for all values of x, y, z.

ii. ψ must be well behaved, i.e. single valued and continuous everywhere, obviously, for each set of values of (x, y, z), ψ have one value only.

iii. ψ must be continuous in all regions except in those regions where the potential energy $V(x, y, z) = \infty$.

iv. If a particle can be present in the states described by the wave functions ψ_1, ψ_2, $\psi_3, .., \psi_n$, it can also be in all states described by the wave function constructed from ψ_1, ψ_2, $\psi_3, ..., \psi_n$ by the linear transformation

$$\lambda = a_1\psi_1 + a_2\psi_2 + ... + a_n\psi_n \quad (19.97)$$

where $a_1, a_2, a_3, ..., a_n$ are arbitrary complex numbers. This is known as *principle of superposition*.

v. The wave function ψ must approach zero as x tends to $\pm \infty$.

vi. A wave function satisfying the following condition is said to be an *orthogonal wave function*

$$\int_{-\infty}^{\infty} \psi_0^*(x)\psi_j(x)dx = 0 \quad (i \neq j) \quad (19.98)$$

The *orthogonality condition* guarantes the noninterference of the wave function representing different states.

vii. Degeneracy: There may, however, be situations when two wave functions ψ_j and ψ_k correspond to the same energy E. This is a case of degeneracy. The system in such an energy state is said to be in a degenerate state. In such a situation, the degenerate wave functions ψ_j and ψ_k will not necessarily be *orthogonal*.

viii. Orthonormal set: The normalisation and orthogonality conditions may be combined as follows:

$$\left.\begin{array}{l} f(\psi_i\psi_j)dv = 1 \text{ if } i = j \\ f(\psi_i\psi_j)dv = 0 \text{ if } i \neq j \end{array}\right\} \quad (19.99)$$

Wave functions satisfying Eq. (19.99) are said to form *orthonormal* set of wave functions.

Schrodinger Time Dependent Wave Equation

In quantum mechanics, the wave function ψ corresponds to the displacement y of the wave motion in a string, but ψ is not a measurable quantity and may, therefore, be complex. We shall, therefore, assume ψ in the X-direction to be specified by

$$\psi = Ae^{i\omega(t - x/v)} \quad (19.100)$$

If we replace ω by $2\pi v$ and $v = \lambda v$ in the above formula, we get

$$\psi = Ae^{-2\pi i (vt - x/\lambda)} \quad (19.101)$$

We know that v and λ can be written in terms of total energy E and momentum p of the particle. Since

$$E = hv \text{ and } \lambda = h/p \quad (19.102)$$

We have

$$\psi = Ae^{-\left(\frac{2\pi i}{h}\right)(Et - px)} \quad (19.103)$$

This expression for wave function is correct only for freely moving particles while we are interested in situations where the motion of particle is subjected to various restrictions. An important example is an electron bound to an atom by electric field of its nucleus. Let us now obtain a fundamental differential equation for ψ which can be solved under various specific restrictions. Differentiating the above equation twice with respect to x, we get

$$\frac{\partial^2 \psi}{\partial x^2} = -\frac{4\pi^2 p^2}{h^2}\psi \quad (19.104)$$

and once with respect to t, we get

$$\frac{\partial \psi}{\partial t} = -\frac{2\pi i E}{h}\psi \quad (19.105)$$

At speeds smaller than that of light, the total energy E of a particle is the sum of its kinetic energy $p^2/2m$ and its potential energy V, where V is in general a function of position and time

$$E = p^2/2m + V \quad (19.106)$$

Multiplying both sides of the equation by ψ, we obtain

$$E\psi = \frac{p^2\psi}{2m} + V\psi \qquad (19.107)$$

But from Eqs. (19.105) and (19.104), we get

$$E\psi = -\frac{h}{2\pi i}\frac{\partial\psi}{\partial t} \quad \text{and} \quad p^2\psi = -\frac{h^2}{4\pi^2}\frac{\partial^2\psi}{\partial x^2} \qquad (19.108)$$

Substituting these values in Eq. (19.107), one obtains

$$\frac{h}{2\pi i}\frac{\partial\psi}{\partial t} = \frac{h^2}{8\pi^2 m}\frac{\partial^2\psi}{\partial x^2} - V\psi \qquad (19.109)$$

This is the time dependent form of the Schrodinger's equation. In three dimensions, the equation is written as

$$\frac{h}{2\pi i}\frac{\partial\psi}{\partial t} = \frac{h^2}{8\pi^2 m}\left(\frac{\partial^2\psi}{\partial x^2} + \frac{\partial^2\psi}{\partial y^2} + \frac{\partial^2\psi}{\partial z^2}\right) - V\psi$$

or $\quad -i\hbar\dfrac{\partial\psi}{\partial t} = \dfrac{\hbar^2}{2m}\nabla^2\psi - V\psi \qquad (19.110)$

where $\hbar = \dfrac{h}{2\pi}$ and the Laplace operator

$$\nabla^2 = \left(\frac{\partial^2}{\partial x^2} + \frac{\partial^2}{\partial y^2} + \frac{\partial^2}{\partial z^2}\right)$$

Schrodinger Time Independent Wave Equation

In many problems, the force acting upon the particle and hence the potential energy does not depend explicitly upon time and vary with the position of the particle only. In that case, the Schrodinger equation may be further simplified by eliminating all dependence on time. We know that one dimensional wave function ψ of an unrestricted particle may be written as

$$\psi = Ae^{-\left(\frac{2\pi i}{h}\right)(Et - px)}$$

or $\quad = Ae^{-\left(\frac{2\pi iE}{h}\right)t}e^{\left(\frac{2\pi ip}{h}\right)x}$

$$= \phi e^{-\left(\frac{2\pi iE}{h}\right)t} \qquad (19.111)$$

This shows that ψ is the product of a time dependent function $e^{-(2\pi iE/h)t}$ and a position dependent function ϕ. Putting the value of ψ in time dependent Schrodinger equation, we get

$$-E\phi e^{-\left(\frac{2\pi iE}{h}\right)t} = \frac{h^2}{8\pi^2 m}e^{-\left(\frac{2\pi iE}{h}\right)t}\frac{\partial^2\phi}{\partial x^2}$$

$$-V\phi e^{-\left(\frac{2\pi iE}{h}\right)t} \qquad (19.112)$$

Dividing by the common factor $e^{-\left(\frac{2\pi iE}{h}\right)t}$, we get

$$\frac{\partial^2\phi}{\partial x^2} + \frac{8\pi^2 m}{h^2}(E - V)\phi = 0 \qquad (19.113)$$

This is the steady-state form of the Schrodinger equation. It can also be written in three dimensions as

$$\frac{\hbar^2}{2m}\nabla^2\phi + (E - V)\phi = 0 \qquad (114)$$

Justification of the Wave Equation

Let us consider a broad wave packet representing a rather sharply defined momentum and moving from a region of space in which the particle is free toward a region in which a force is present. For simplicity, we shall restrict to a one dimensional description. We also suppose that the forces are conservative, so that classically the particle has and maintains throughout its motion, a constant energy

$$E = \frac{p^2}{2m} + V \qquad (19.115)$$

Accordingly, the momentum is a function of position. Fourier decomposition of incident wave packet is

$$\psi(x,0) = \frac{1}{\sqrt{2x}}\int_0^\infty a(k)e^{ikx}dk \qquad (19.116)$$

By omitting the range of integration from $-\infty$ to 0, we have included the assumption that the wave packet moves toward positive x and contains no negative k components. Using the substitution

$$k = \sqrt{\frac{2mE}{h^2}} \qquad (19.117)$$

one can transform Eq. (19.117) into an integral over the energy of the free particle, i.e.

$$\psi(x,0) = \int_0^\infty f(E)e^{ik(Ex)}dx \qquad (19.118)$$

where $f(E)$ is assumed to be a smoothly varying positive function of E with a single peak. Obviously, Eq. (19.118) is a rather specialized wave packet. When this wave packet travels in force free space, it spreads in time as follows

$$\psi(x,t) = \int f(E)e^{\frac{i}{h}(px-Et)}dE \qquad (19.119)$$

Now the question arises that how this wave packet will move when it enters the region where a force is present.

Let us consider that the frequency remains unaltered when the wave packet approaches the region of changing potential energy. This is basic assumption and recommends itself by analogy with other types of wave motion. This is further supported by principle of conservation of energy, i.e. constant frequency according to the equation $E = h\nu = \hbar\omega$.

In the spirit of above remarks, we may compare the motion of a particle through a force field with the propagation of a wave through a strange medium whose detailed microscopic properties we do not propose to analyse here, but which can somehow be set into "sympathetic" vibrations of the same frequency and which thus interacts with the incident wave to modify its properties of propagation, i.e. its velocity of propagation and its wavelength. Thus, the form of wave packet moving through a region of force may be represented by

$$\psi(x,t) = \int f(E)e^{\frac{i}{h}[a(x,E)-Et]}dt \qquad (19.120)$$

This is a generalization of Eq. (19.119) in which $a(x, E)$ is a yet undetermined function. This function must be chosen so that Eq. (19.120) represents a wave packet classically. Since $f(E)$ is assumed to be a smooth positive function of E, the peak of the amplitude of Eq. (19.120) occurs if x and t are chosen so as to make the phase $f(E) - Et$ stationary when $f(E)$ is near its peak. This ensured that the Fourier components near the peak of $f(E)$ are as much in phase as possible and are, therefore, added constructively. Mathematically, the phase is stationary if

$$\frac{\partial}{\partial E}\big[a(x,E)-Et\big]=0 \qquad (19.121)$$

After differentiation, E in this equation is to be evaluated at the energy which corresponds to the peak of $f(E)$. Eq. (19.121) leads to the condition

$$t = \frac{\partial a}{\partial E} \qquad (19.122)$$

On the other hand, classically for a conservative motion

$$t = \int^x \frac{dx'}{V(x')} = \int^x \frac{\partial p(x',E)}{\partial E}dx' \quad (19.123)$$

where $p(x, E)$ is the momentum as defined by Eq. (19.115). Comparison of Eq. (19.122) and Eq. (19.123) gives the possible choice

$$a(x,E) = \int^x p(x',E)dx' \qquad (19.124)$$

Thus, we obtain

$$\psi(x,t) = \int^x f(E)e^{\frac{i}{h}\left[\int^x p(x',E)dx'-Et\right]}dx \quad (19.125)$$

If the particle is free and moves with constant momentum p corresponding to a wavelength $\lambda = h/p$, Eq. (19.125) is an exact equation instead of an approximation. It is reasonable to assume that Eq. (19.124) is better approximation, the more nearly $\lambda(x) = h / |p(x)|$ is independent of x.

$$\left|\frac{d\lambda}{dx}\right| \ll 1 \qquad (19.126)$$

If this is true, the wave is quasi-harmonic and $\lambda(x)$ can be considered as its slowly varying wavelength. Eq. (19.126) can also be written as

$$\lambda \left| \frac{dp}{dx} \right| << p^2 \qquad (19.127)$$

or

$$\lambda \left| \frac{dp}{dx} \right| << \frac{[p(x)]^2}{m} \qquad (19.128)$$

This condition implied that the potential energy changes very little over the distance of a wavelength. The particle can then be considered as nearly free over a distance of many wavelengths. For this region $\hbar \frac{dp}{dx}$ can be neglected compared to p^2 in expressions derived from Eq. (19.125).

One obstacle in the way of applying Eq. (19.125) extensively is that it may be quite inaccurate in regions where the kinetic energy varies rapidly with x, particularly near the classically turning points of the motion, where $p(x, E) = 0$ and where Eq. (127) can never be satisfied. Hence, Eq. (19.125) does not provide us with an adequate formula for the description of the particle motion, where it is least classical and most interesting.

It is at this point that we ask ourselves whether a more general satisfactory wave function could not be found if we abandoned the explicit construction of wave packets, which represents quite special semiclassical wave forms. A wave packet must always be given in terms of a particular initial wave $\psi(x, 0)$. Let us ask instead of some general relation which $\psi(x, t)$ must satisfy, no matter what the particular initial condition be. Our experience with classical physics suggests that such a region is most likely going to be differential equation for $\psi(x, t)$. In general, we expect that ψ waves are complex quantities and that $\psi(x, 0)$ alone, without knowledge of time derivative, determines the future of the wave. Hence, we shall attempt to obtain a first order differential equation in t. Following this idea, let us differentiate Eq. (19.125) with respect to time, we have

$$\frac{\partial \psi}{\partial t} \approx -\int f(E) \frac{i}{\hbar} E e^{\left[\frac{i}{\hbar} \left(\int p \, dx' - Et \right) \right]} dE \qquad (19.129)$$

We would like to eliminate in the initial condition $f(E)$ from RHS of Eq. (19.129). This can almost be accomplished, if we note

$$\frac{\partial \psi}{\partial t} \approx \frac{i}{\hbar} \int f(E) p(x, E) e^{\left[\frac{i}{\hbar} \left(\int^x p \, dx' - Et \right) \right]} dE$$

and

$$\frac{\partial^2 \psi}{\partial x^2} \approx -\frac{i}{\hbar^2} \int f(E) p^2 (x, E) e^{\left[\frac{i}{\hbar} \left(\int^x p \, dx' - Et \right) \right]} dE$$

$$+ \frac{i}{\hbar} \int f(E) \frac{\partial p}{\partial x} e^{\left[\frac{i}{\hbar} \left(\int^x p \, dx' - Et \right) \right]} dE \qquad (19.130)$$

But according to Eq. (19.126) the last term is negligible in comparison to the first. Using Eqs. (19.128), (19.129) and (19.130), one obtains at least as an approximation

$$i\hbar \frac{\partial \psi}{dt} = \frac{\hbar}{2m} \frac{\partial^2 \psi}{\partial x^2} + V(x)\psi \qquad (19.131)$$

Eq. (19.131) is valid even when Eq. (19.125) is not. In fact Eq. (19.131) is the quantum mechanical wave equation and thus justifies the Schrodinger's wave equation.

Probability Current Density

Immediately doubts arise over the consistency of the probabilistic interpretation of the wave function as discussed above. If probability of finding the particle in some bounded region of space decreases as time goes on, then the probability of finding it outside this region must increase by the same amount. The probability interpretation of the ψ waves can be made consistently only if this conservation of probability is guaranteed. This requirement is fulfilled owing to Gauss's integral theorem, if it is possible to define a probability current density J which together with the probability density $p = \psi^* \psi$ satisfies a continuity equation

$$\partial \rho / \partial t + \Delta \cdot J = 0 \qquad (19.132)$$

Exactly as in the case of the conservation of matter in hydrodynamics, or conservation of charge in electrodynamics, a relation of the form (19.132) can easily be deduced from the wave equation. The probability equation

$$p(r)dr = \int_V \psi^* dV \qquad (19.133)$$

gives the probability of finding the particle in a region of space of volume V bounded by the surface A. In order to determine the probability flow of the particle, we must know the changes in the probability with time within the region. Differentiating Eq. (133) with time, one obtains

$$\frac{dp}{dt} = \frac{d}{dt}\int_V \psi^*\psi dV = \int_V \left(\frac{d\psi^*}{dt}\psi + \psi^*\frac{d\psi}{dt}\right)dV$$

$$(134)$$

Time independent Schrodinger's equation and its complex conjugates are

$$\frac{\hbar}{2m}\nabla^2\psi + V\psi = i\hbar\frac{\partial\psi}{\partial t} \qquad (19.134a)$$

$$--\frac{\hbar^2}{2m}\nabla^2\psi^* + V\psi^* = i\hbar\frac{\partial\psi^*}{\partial t} \qquad (19.134b)$$

Multiplying Eq. (19.134a) by ψ^* and Eq. (19.134b) by ψ and then subtracting Eq. (19. 134a) from Eq. (19.134b), one obtains

$$-\frac{\hbar^2}{2m}\psi\nabla^2\psi^* + \frac{\hbar^2}{2m}\psi^*\nabla^2\psi = i\hbar\psi\frac{\partial\psi^*}{dt} - i\hbar\psi^*\frac{\partial\psi}{\partial t}$$

or $\psi\dfrac{\partial\psi^*}{dt} + \psi^*\dfrac{\partial\psi}{\partial t} = -\dfrac{\hbar^2}{2im}\left(\psi^*\nabla^2\psi - \psi\nabla^2\psi^*\right)$

$$(19.135)$$

From Eqs. (134) and (135), one obtains

$$\frac{dp}{dt} = -\frac{\hbar}{2im}\int_V \left(\psi^*\nabla^2\psi - \psi\nabla^2\psi^*\right)dV$$

$$(19.136)$$

The integral appearing in Eq. (19.136) can be transformed into integral over the surface V by Green's theorem, we have

$$\frac{dp}{dt} = \frac{\hbar}{2im}\int_A (\psi^*\nabla\psi - \psi\nabla\psi^*)dt \qquad (19.137)$$

or $\qquad \dfrac{dp}{dt} = \nabla\cdot J = 0 \qquad (19.138)$

where $J = \hbar/2im\left(\psi^*\nabla\psi - \psi\nabla\psi^*\right)$ is called the probability current density.

Clearly Eq. (19.138) is analogous to Eq. (19.132). Therefore, it is quite reasonable to interpret J as a probability current density. If Eq. (19.138) is multiplied by charge of the particle, then ep gives the *charge density* and ej gives the *current density* of charge in quantum mechanical description. One can express Eq. (19.138) as

$$\frac{dp}{dt} - \nabla\cdot J = -\int_A J\cdot dA \qquad (19.139)$$

This shows that the rate of change of probability of the particle inside a surface is equal to the negative of probability current density through the surface A. Eq. (19.139) is generally useful for the determination of flux of particle in the scattering problems. The idea of probability current density is generally useful for the description of unbounded systems.

We must note that the operator $-i\hbar\nabla$ represents the momentum, even when a force is present. Then the velocity will be represented by the operator

$$-\frac{i\hbar}{m}\nabla \quad \text{or} \quad \frac{\hbar}{im}\nabla$$

and it is apparent that

$$J = \text{real part of } \left(\psi^* \frac{\hbar}{im}\nabla\psi\right)$$

or real part of $\psi\dfrac{\hbar}{im}\nabla\psi^*$, while this interpretation of J is suggestive, it must be realised that it is not susceptible to direct measurement in the sense in which P is.

19.15 SIMPLE SOLUTIONS OF SCHRODINGER EQUATION

We have developed Schrodinger's wave equation for systems of fundamental particles or atoms in non-relativistic limit. When the total energy of such a system is independent of the time, the solution of Schrodinger equation reduces to the solution of the energy eigen equation ($H\psi = E\psi$). The total solution is then simply the energy eigen function ψE_n multiplied by the time factor, $\exp[-(i/\hbar)E_n t]$. The energy eigen states are also referred to as stationary state, since the probability density for such a state is constant in time. Furthermore, the expectation value of any operator that is itself not a function of time is constant for a stationary state. Now we will discuss it in detail.

Schrodinger Equation and Stationary State Solution

Schrodinger's time dependent one dimensional wave equation is

$$-\frac{\hbar^2}{2m}\frac{\partial^2\psi(x,t)}{\partial x^2} + V(x)\psi(x,t) = i\hbar\frac{\partial\psi(x,t)}{\partial t}$$
(19.140)

Wave function $\psi(x, t)$ can be expressed as the product of two functions, i.e. one involving time alone while the other coordinate alone, then one can write

$$\psi(x, t) = u(x) f(t)$$
(19.141)

Using Eq. (19.141), Eq. (19.140) becomes

$$-\frac{\hbar^2}{2m}\frac{d^2u(x)}{dx^2}f(t) + V(x)u(x)f(t)$$

$$= i\hbar\frac{df(t)}{dt}u(x)$$

Dividing by $u(x) f(t)$, one obtains

$$\frac{1}{u(x)}\left[-\frac{\hbar^2}{2m}\frac{d^2u(x)}{dx^2}+V(x)u(x)\right]$$

$$= i\hbar\frac{1}{f(t)}\frac{df(t)}{dt}$$
(19.142)

In the above equation, the is independent of time but a function of x whereas is independent of x but a function of time t only. This will be possible only when each side of Eq. (19.142) will be equal to a same constant,

say E. This constant has the dimensions of energy. Thus, we have

$$\frac{1}{u(x)}\left[-\frac{\hbar^2}{2m}\frac{d^2u(x)}{dx^2}+V(x)u(x)\right]=E$$
(19.143)

and $\dfrac{i\hbar}{f(t)}\dfrac{df(t)}{dt}=E$
(19.144)

Equation (19.144) can be rewritten as

$$\frac{df(t)}{f(t)} = -\frac{i}{\hbar}Edt$$

which on integration yields

$$\ln f(t) = \frac{iE}{\hbar}t + \ln C$$

or $f(t)=Ce^{-iEt/\hbar}$
(19.145)

Eq. (19.143) can be put as follows:

$$\frac{d^2u(x)}{dx^2} + \frac{2m}{\hbar^2}[E-V(x)]u(x)=0$$
(19.146)

Eq. (19.146) represents Schrodinger's time independent or steady state equation, whose solution is given by

$$u(x) = \sum_n a_n u_n(x)$$
(19.147)

On substituting Eqs. (19.145) and (19.147) in Eq. (19.141), one obtains

$$\psi(x,t) = \sum_n a_n\psi_n(x,t)$$

$$= \sum_n a_n\psi_n(x)\exp\left[-\frac{i}{\hbar}E_n t\right]$$
(19.148)

Eq. (19.148) represents the general solution of the one dimensional time dependent Schrodinger's equation corresponding to various values of the constant energy E of a particle.

Stationary State Solutions

If in a particular state, the probability distribution function $\psi\psi^*$ does not depend upon time, then the state of the system is said to be stationary state.

Suppose we now consider the probability distribution function $\psi\psi^*$ for a system in the state represented by the following wave function:

$$\psi(x,y,z,t) = \sum_{n=1}^{\infty} a_n \psi_n(x,y,z)\exp\left(-\frac{i}{\hbar}E_n t\right)$$

(19.149)

The complex conjugate of Eq. (19.149) is

$$\psi^*(x,y,z,t) = \sum_{n=1}^{\infty} a_n \psi_n^*(x,y,z)\exp\left(-\frac{i}{\hbar}E_n t\right)$$

(19.150)

On multiplying Eqs. (149) and (150), one obtains the probability distribution function $\psi\psi^*$, i.e.

$$\psi\psi^* = \left\{\sum_{n=1}^{\infty} a_n \psi_n(x,y,z)\exp\left[\frac{-i}{\hbar}E_n t\right]\right\}$$

$$\times \left\{\sum_{m=1}^{\infty} a_m^* \psi_m^*(x,y,z)\exp\left[\frac{-i}{\hbar}E_m t\right]\right\}$$

or $\psi\psi^* = \sum a_n a_n^* \psi_n(x,y,z) \psi_n^*(x,y,z)$

$$+ \sum_m \sum_n{}' a_n a_m^* \psi_n(x,y,z)$$

$$\psi_m^*(x,y,z)\exp\left\{\frac{i}{\hbar}(E_m - E_n)t\right\} \quad (19.151)$$

In the first term in Eq. (19.151), the terms with $m = n$ have been included while in the second term of the same equation, the terms with $m = n$ have been excluded, i.e. the prime on the double summation indicates that the terms with $m = n$ are excluded. As time t enters the probability function $\psi\psi^*$ in Eq. (19.151), it means that ψ expressed by Eq. (19.149) is not a stationary state solution. If a_n's are zero for all values except for one value of E_n, then $\psi\psi^*$ will be independent of time t. In such a case, the state represented by wave function ψ would be stationary state. For this reason, the wave function will contain only a single term and will be represented by

$$\psi_n(x,y,z,t) = \psi_n(x,y,z)\exp\left(-\frac{i}{\hbar}E_n t\right)(19.152)$$

The solution represented by Eq. (19.152) is stationary states solution, i.e. when the particle is in a particular state of energy, the probability density is independent of time. The probability density for a stationary state or an eigen state is constant with respect to time and the expectation value for the energy of an eigen state is precisely the energy of the state for all time. One can easily prove that the expectation value of any operator which does not explicitly depend upon time is also independent of time for a stationary state.

Boundary and Continuity Conditions on the Wave Functions

Schrodinger steady state equation in three dimensions is

$$\frac{\partial^2\psi}{\partial x^2} + \frac{\partial^2\psi}{\partial y^2} + \frac{\partial^2\psi}{\partial z^2} + \frac{2m}{\hbar^2}(E-V)\psi = 0 \quad (19.153)$$

Schrodinger equation (steady state) in one dimension is

$$\frac{\partial^2\psi(x)}{\partial x^2} + \frac{2m}{\hbar^2}(E-V)\psi(x) = 0 \quad (19.154)$$

The time independent or steady state Schrodinger wave equation is a second order linear differential equation in x [in general in $r(x, y, z)$]. In general, Schrodinger steady state equation can be solved only for certain values of the energy E. What is meant by this statement has nothing to do with any mathematical difficulties that may be present, but is something much more fundamental. To "solve" Schrodinger's equation for a given system means to obtain a wave function ψ that not only obeys the equation and whatever boundary conditions there are but also fulfils the requirements for an acceptable wave function—namely that it and its derivatives, i.e. ψ and $\partial\psi/\partial x$ be continuous, finite and single-valued. If there is no such wave function, the system cannot exist in a steady state. Hence, while obtaining the solutions of steady state Schrodinger equation, it will be required that the wave function and its gradient are continuous, finite and single-valued at every point in space.

Thus, energy quantization appears in wave mechanics as a natural element of the theory and energy quantization in the physical world is revealed as universal phenomenon characteristic of all stable systems.

Free Particle System

A free particle is one which is not subjected to any force or potential barrier and is free to move in limitless space. For a linear motion of the particle, the Hamiltonian H is given as

$$H = -\frac{\hbar^2}{2m}\frac{d^2}{dx^2} + V \qquad (19.155)$$

and the wave function ψ (which is also eigen function of H here) is $\psi(x)$.

Since the particle is not subjected to any external force, its potential energy V is constant, which for convenience may be assumed to be zero. Thus, the Schrodinger steady state equation for the free particle may be written explicitly as

$$\frac{d^2\psi}{dx^2} + \frac{2m}{\hbar^2}E\psi = 0 \qquad (19.156)$$

The general solution of Eq. (19.156) is

$$\psi = Ae^{ikx} + Be^{-ikx} \qquad (19.157)$$

where $\qquad k = \sqrt{\dfrac{2m}{\hbar^2}E} \qquad (19.158)$

and A and B are constants.

Eq. (19.157) gives the steady state or time independent part of the wave function. The complete wave function for particle is given by

$$\psi(x, t) = \psi(x)e^{-iEt/\hbar}$$

$$= \left(Ae^{ikx} + Be^{-ikx}\right)e^{-iEt/\hbar} \qquad (19.159)$$

$$= Ae^{ikx}e^{-i\omega t} + Be^{ikt}e^{-i\omega t} \qquad (\because \omega = E/\hbar)$$

$$= Ae^{-i(\omega t - kx)} + Be^{-i(\omega t + kx)} \qquad (19.160)$$

Eq. (19.160) represents the sum of two stationary solutions in principle of superposition, i.e. a continuous plane simple harmonic wave.

The first term on the RHS of Eq. (19.160) represents the wave travelling in the positive X-direction. The wavelength of each travelling plane wave is $\lambda = 2\pi/k$. The phase velocity of these waves is

$$v = v\lambda = \frac{E}{h}\frac{2\pi}{k} = \frac{E}{\hbar k}$$

According to de Broglie's hypothesis, these waves correspond to a particle of momentum

$$p = \frac{h}{\lambda} = \frac{hk}{2\pi} = \hbar k$$

and energy

$$E\left(= \frac{p^2}{2m} = \frac{\hbar^2 k^2}{2m} = \frac{\hbar^2 2mE}{2m\hbar^2} = E\right)$$

Therefore, for the motion of the particle in the positive X-direction, we may have

$$\psi(x) = Ae^{ikx} \qquad (19.161)$$

or $\qquad \psi(x, t) = Ae^{-i(\omega t - kx)} \qquad (19.162)$

The momentum operator, $\dfrac{\hbar}{i}\dfrac{\partial}{\partial x}$ operating on the wave function $\psi(x, t)$ gives

$$\frac{\hbar}{i}\frac{\partial\psi}{\partial x} = \frac{\hbar}{i}\left[A(ik)e^{-i(\omega t - kx)}\right]$$

$$= \hbar k Ae^{-i(\omega t - kx)}$$

$$= \hbar k\psi = p_x\psi \qquad (19.163)$$

This equation shows that the wave function $\psi(x, t)$ for the particle is an eigen function of the linear momentum operator and the momentum p_x is the eigenvalue of the operator. Hence, the momentum remains sharp with the value p_x.

Now, as the constant k^2 and hence energy E are completely arbitrary, i.e. no constraint has been applied on these, the particle can have any energy between 0 and $+\infty$, i.e. the particle has a continuous energy spectrum. The two sets of waves represented by Eq. (160) are travelling in opposite directions in $+X$ and $-X$ directions. Thus, corresponding to the same eigenvalue E, there are two eigen functions as with the energy operator $i\hbar\partial/\partial t$, one obtains

$$i\hbar\frac{\partial}{\partial t}\left(Ae^{ikx}e^{-iEt/\hbar}\right) = E\left(Ae^{-ikx}e^{-iEt/\hbar}\right) \quad (19.164)$$

and $i\hbar\dfrac{\partial}{\partial t}\left(Ae^{-ikx}e^{-iEt/\hbar}\right) = E\left(Ae^{-ikx}e^{-iEt/\hbar}\right)$

$$(19.165)$$

i.e. in the two cases, we have

$$i\hbar\frac{\partial\psi}{\partial t} = E\psi \quad (19.166)$$

Obviously, ψ has two different values but the eigenvalue E remains the same. Such eigen functions are termed *degenerate functions* and in the present case of a free particle, the degeneracy is two fold.

The probability p of finding the particle between x and $x + dx$ is given by

$$p\,dx = \psi(x, t)\psi^*(x, t)dx = A^2 dx \quad (19.167)$$

Therefore, the probability density P for the position of the particle with the definite value of momentum is constant over the X-axis, i.e. all positions of the particle are equally probable. This conclusion is also obtained from the principle of uncertainty.

According to the interpretation of the wave function, the probability of finding the particle somewhere in space must be equal to 1, i.e.

$$\int_{-\infty}^{\infty} \psi(x,t)\psi^*(x,t)dx = 1 \quad (19.168)$$

In this case, $\psi(x, t)\,\psi^*(x, t)dx = \text{constant} = A^2$. Hence, the wave function for the particle cannot be normalized and A must remain arbitrary. The difficulty arises because we are dealing with an arbitrary case. In practice, we cannot have an absolutely free particle. This particle will always be confined within an enclosure in the laboratory, and hence its position can be determined with absolute accuracy.

Particle in Bound States

When the motion of a particle is confined to a limited region such that the particle moves back and forth in the region, the particle is said to be in a bound state. One dimensional motion of a particle assumed to take place with zero potential energy over a fixed distance and the potential energy is assumed to become infinite are the simplest examples of all motions in bound states. But in practice such a motion is not possible.

Particle in One Dimensional Box

(Particle in an infinitely deep one dimensional potential well)

Let us consider the one dimensional motion along X-axis of a particle of mass m in a hollow rectangular box having perfectly rigid walls. Let the origin be at one corner of the box and the X-axis be perpendicular to the parallel opposite walls as shown in Fig. 19.19. Let a be the distance between the walls so that the motion along the X-axis is confined between $x = 0$ and $x = a$. Suppose inside the box, i.e. in the region $0 < x < a$, there is no force acting on the particle so that in this region, i.e. inside the box, the potential energy $V(x)$ is zero. When the particle collides with the perfectly rigid walls there is no loss of energy so that the total energy E of the particle remains constant. In order to leave the region, the particle will have to do an infinite amount of work. Since this is not possible, it cannot exist outside the box. Hence for $x \geq 0$ and $x \geq a$, i.e. outside the box, the wave function $\psi(x) = 0$.

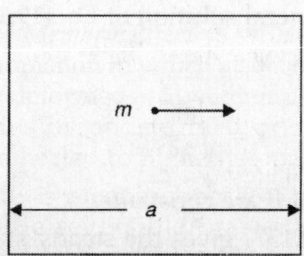

Fig. 19.19: One dimensional motion of a particle in a hollow rectangular box

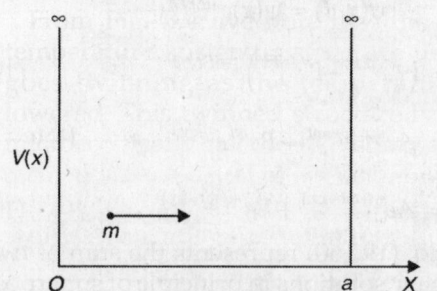

Fig. 19.20: Square well potential of infinite depth

The potential energy curve for such a particle is shown in Fig. 19.20. Because of its appearance it is called a *square well potential of infinite depth*.

Wave Equation for a Particle

For the motion of the particle along the X-axis inside the box where $V(x) = 0$, the time independent Schrodinger wave equation is

$$\frac{d^2\psi}{dx^2} + \frac{2m}{\hbar^2}E\psi = 0 \quad [\because V(x)=0] \quad (19.169)$$

where E is the total energy of the particle. Eq. (19.169) can be expressed as

$$\frac{d^2\psi}{dx^2} + K^2\psi = 0 \quad (19.170)$$

where

$$K^2 = \frac{2mE}{\hbar^2} \quad (19.171)$$

For a particular value of the energy E, K is a constant.

Solution of the Equation

The general solution of Eq. (19.170) is

$$\psi(x) = A\sin Kx + B\cos Kx \quad (19.172)$$

where A and B are constants of integration. We have the boundary conditions as:

   i. At $x = 0$, $\psi(x) = 0$
   ii. At $x = a$, $\psi(x) = 0$

From the first condition, one obtains $B = 0$. Therefore, from Eq. (19.172), one obtains

$$\psi(x) = A\sin Kx = 0 \quad (19.173)$$

Using second boundary condition, one obtains

$$\psi(a) = A\sin Ka = 0 \quad (19.174)$$

We cannot take $A = 0$ because there will then be no solution. Hence, Eq. (19.174) is satisfied only when $Ka = n\pi$.

$$K = \frac{n\pi}{a} \quad (19.175)$$

where $n = 1, 2, 3, \ldots$, i.e. n is an integer. We cannot take $n = 0$ because for $n = 0$, $K = 0$, $E = 0$ and hence $\psi(x) = 0$ everywhere in the box. This means that a particle with zero energy cannot be present inside the box, i.e. a particle in the box cannot have zero energy. Hence, the wave functions for the motion of particle in the region $0 < x < a$ are given by

$$\psi_n(x) = A\sin\frac{n\pi x}{a} \quad \text{[where } n \text{ is even] (19.176)}$$

Eigenvalues of Energy

Substituting the value of K from Eq. (19.171) in Eq. (19.175), one obtains

$$\sqrt{\frac{2mE}{\hbar^2}} = \frac{n\pi}{a}$$

or

$$E = \frac{n^2\pi^2\hbar^2}{2ma^2}$$

$$E_n = \frac{n^2\pi^2\hbar^2}{2ma^2} \quad (19.177)$$

Eq. (19.177) represents the eigenvalues of the energy. These values are called the energy levels of the particle. Using $\hbar = h/2\pi$, one can rewrite Eq. (19.177) as

$$E_n = \frac{n^2h^2}{8ma^2}$$

From Eq. (19.177), one can draw the following conclusions:

   i. The lowest energy of the particle is obtained by putting $n = 1$ in Eq. (19.177). One obtains for $n = 1$ as

$$E_1 = \frac{\pi^2\hbar^2}{2ma^2} \quad (19.178)$$

It is called the ground state energy level of the particle. The value of E_n in terms of E_1 are given by

$$E_n = n^2E_1 \quad (19.179)$$

In the above discussion, the n numbers are called quantum numbers while the E values are called energy levels. The quantum states with higher values of $n (= 2, 3, 4, \ldots, \text{etc.})$ are termed excited states.

According to quantum mechanics (but not in accordance to classical mechanics), the kinetic energy is not equal to zero in ground state.

ii. The possible values of energy of the particle in the potential box are discrete corresponding to $n = 1, 2, 3,....$ The classical mechanics does not need any such requirement of discrete energy levels for bound systems. The quantization of energy conclusion is universal, i.e. whenever a particle is confined to move in a limited space, its energy is quantized. For example, the energies of electrons bound to atoms and molecules are quantized.

iii. *The energy levels:* The quantization of energy is shown graphically in Fig. 19.21. The allowed energy levels are shown as horizontal lines. That these lines are straight emphasizes the fact that energy (E) is the same irrespective of the position (x) of the particle. Since for a particular value of n, the energy (E) does not change with x, the latter is often termed *constant of motion*. Further, since H or E is independent of time (t), the system under consideration is a good and simple example of conservative system.

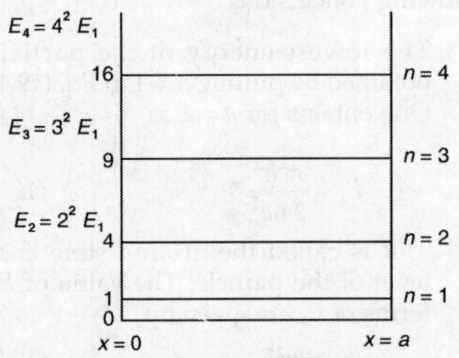

Fig. 19.21: The energy levels

The lowest value of E is not zero but $\pi^2 h^2 / 2ma^2$ (for $n = 1$). Evidently, the particle in a box will continue to move to and fro in the region 0 to L even at the temperature of absolute zero; for this region it is called the *zero point energy*. This conclusion is universal and valid for any system undergoing to and fro motion.

The dependence of E on n^2 results in the increased spacing between energy levels as n increases.

$$\Delta E = E_{n+1} - E_n = \left[(n+1)^2 - n^2 \right] E_1$$

$$= (2n+1)E_1 \tag{19.180}$$

Since the energy E_1 is proportional to $1/a^2$, smaller the box greater are the energy spacing (ΔE), i.e. the quantization is more pronounced. Alternatively, as the box is made wider, ΔE decreases and in the limited case of $L \to \infty$ (free particle) the quantization vanishes.

The wave functions ψ_n corresponding to E_n are called eigen functions.

Normalization of the Wave Functions

The wave functions for the motion of the particle are

$$\psi_n(x) = A \sin \frac{n\pi x}{a} \tag{19.181}$$

in the region $0 < x < a$ and $\psi_n = 0$ in the region $x < 0, x > a$. This wave function is still incomplete as much as the constant A is not determined. A can be determined by making use of the fact that the total probability of the particle somewhere in the box must be unity, i.e.

$$\int_0^a \left| \psi_n(x)^2 \right| dx = 1 \tag{19.182}$$

or $$\int_0^{a^2} A^2 \sin^2 \frac{n\pi x}{a} dx = 1$$

or $$A^2 \int_0^{a^2} \frac{1}{2} \left(1 - \cos \frac{2n\pi x}{a} \right) dx = 1$$

or $$\frac{A^2}{2} \left[x - \frac{a}{2\pi n} \sin \frac{2n\pi x}{a} \right]_0^a = 1$$

The second term of the integrated expression becomes zero at both the limits. Therefore,

$$\frac{A^2 a}{2} = 1 \quad \text{or} \quad A = \sqrt{\frac{2}{a}} \qquad (19.183)$$

The normalized wave function is thus

$$\psi_n = \sqrt{\frac{2}{a}} \sin \frac{n\pi x}{a} \qquad (19.184)$$

Note: Actually $\psi = \pm\sqrt{2/a}\sin nx/a$. However, since probability is given by $|\psi|^2$ and not by ψ, it is immaterial whether positive or negative A is chosen. Customarily, we select the positive series of normalized ψ functions.

Form of the Wave Functions

The expression for the real form of the wave function can be used conveniently to draw its graphical representations. The wave functions for the first three values of n are shown in Fig. 19.22. It is evident that the wave function ψ_1 has two nodes at $x = 0$, $x = a$. The wave function ψ_2 has three nodes at $x = 0$, $x = a/2$ and $x = a$. The wave function ψ_3 has four nodes $x = 0$, $x = a/3$, $x = 2a/3$ and $x = a$. Thus, the wave function ψ_n will have $(n + 1)$ nodes. Obviously, the increasing number of nodes is consistent with increasing energy of the system.

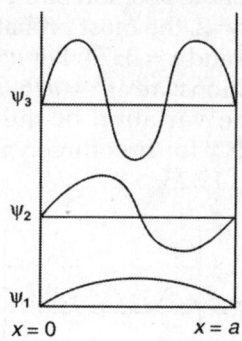

$$\psi_3$$

$$\psi_2$$

$$\psi_1$$

$$x = 0 \qquad\qquad x = a$$

Fig. 19.22: Form of wave functions for three values of n

Orthogonality of the Wave Function

Let $\psi_n(x)$ and $\psi_m(x)$ be the normalized wave functions of the particle in the interval $(0, a)$ corresponding to the different energy levels E_n and E_m respectively.

These wave functions are

$$\psi_n(x) = \sqrt{\frac{2}{a}} \sin \frac{n\pi x}{a}$$

$$\psi_m(x) = \sqrt{\frac{2}{a}} \sin \frac{m\pi x}{a}$$

where m and n are integers. In the present case, the wave functions are real, therefore,

$$\psi_n^*(x) = \psi_n(x) \quad \text{and} \quad \psi_m^*(x) = \psi_m(x)$$

Now we have, when $m \neq n$

$$\int_0^a \psi_m^* \psi_n \, dx = \frac{2}{a} \int_0^a \sin \frac{n\pi x}{a} \, dx$$

$$= \frac{1}{a} \int_0^a \left[\cos \frac{(m-n)\pi x}{a} - \cos \frac{(m+n)\pi x}{a} \right] dx$$

$$= \frac{1}{a} \left[\frac{a}{\pi(m-n)} \sin \frac{(m-n)\pi x}{a} \right.$$

$$\left. - \frac{a}{\pi(m+n)} \sin \frac{(m+n)\pi x}{a} \right]_0^a = 0$$

Obviously, the functions are mutually orthogonal in the interval $(0, a)$. These functions are also normalized in this interval. The wave functions which are normalized and mutually orthogonal in an interval, are said to form an orthogonal set in this interval. Since the wave functions are zero outside the interval $(0, a)$, they are also orthogonal in the whole range of x, i.e. in the interval $(-\infty, \infty)$.

Eigen Functions of Momentum

From Eq. (19.177), we have the possible values of energy as

$$E_n = \frac{n^2 \pi^2 \hbar^2}{2ma^2} \qquad (19.185)$$

and the wave function of a particle in a box whose energy is E is

$$\psi = A \sin \frac{\pi x}{a} \qquad (19.186)$$

But $\qquad \sqrt{\dfrac{2mE}{\hbar^2}} = \dfrac{x}{a}$ $\qquad$ (19.187)

Substituting the value of a from Eq. (19.187) in Eq. (19.186), we have

$\therefore \qquad \psi = A\sin\sqrt{\dfrac{2mE}{\hbar^2}}\, x$ $\qquad$ (19.188)

Substituting E_n for E yields

$\qquad \psi_n = A\sin\dfrac{n\pi x}{a}$ $\qquad$ (19.189)

for the eigen function corresponding to the energy eigenvalues E_n. Substituting

$$A = \sqrt{2/a},\ \text{we have}$$

$\qquad \psi_n = \sqrt{\dfrac{2}{a}}\sin\dfrac{n\pi x}{a}$ $\qquad$ (19.190)

We can easily see that these eigen functions meet all the requirements for each quantum number n, ψ_n is a single-valued function of x and ψ_n and $\partial\psi_n/\partial x$ are continuous. Furthermore, the integral of $|\psi_n|^2$ over all space is finite, as one can see by integrating $|\psi_n|^2 dx$ from $x = 0$ to $x = a$ (since the particle by hypothesis is confined within these limits).

$$\int_{-\infty}^{\infty}|\psi_n|^2 dx = \int_0^L |\psi_n|^2 dx$$

$$= A^2 \int_0^L \sin^2\left(\dfrac{n\pi x}{a}\right)dx$$

or $\qquad 1 = A^2\dfrac{a}{2}$

$\therefore \qquad A = \sqrt{\dfrac{2}{a}}$

Eigenvalues of the Momentum

The eigenvalues of the momentum P_n along the X-axis are given by

$$P_n^2 = 2mE_n = 2m\dfrac{n^2\pi^2\hbar^2}{2ma^2} = \dfrac{n^2\pi^2\hbar^2}{a^2}$$

or $\qquad P_n = \dfrac{n\pi\hbar}{a} = \dfrac{nh}{2a}$ $\qquad$ (19.191)

Equation (19.191) shows that the eigenvalues of the momentum of the particle are discrete and the difference between the momenta corresponding to two consecutive energy levels is constant and equal to $h/2a$.

Probability of Location of the Particle Over a Small Range between x and x+dx

According to quantum mechanics, the probability $P(x)\,dx$ that the particle be found over a small distance dx at position x is given by

$$P(x)dx = \int |\psi_n|^2 dx = \dfrac{2}{a}\sin^2\left(\dfrac{n\pi x}{a}\right)dx \quad (19.192)$$

therefore, the probability density for the one dimensional motion is

$$P(x) = \dfrac{2}{a}\sin^2\left(\dfrac{n\pi x}{a}\right) \qquad (19.193)$$

The probability density is maximum when

$$\dfrac{n\pi x}{a} = \dfrac{\pi}{2}, \dfrac{3\pi}{2}, \dfrac{5\pi}{2}$$

or $\qquad x = \dfrac{a}{2n}, \dfrac{3a}{2n}, \dfrac{5a}{2n}$ $\qquad$ (19.194)

Obviously, for the state defined by $n = 1$, the most probable position of the particle is at $x = a/2$. For $n = 2$, the most probable positions are at $x = a/4$ and $x = 3a/4$. For $n = 3$, the most probable positions are at $x = a/6$, $x = 3a/6$ and $x = 5a/6$. The variation of the probability densities with x for first three values of n are shown in Fig. 19.23.

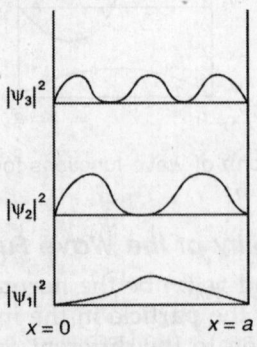

Fig. 19.23: Probability density versus x

Thus, the quantum mechanical result is quite different from the classical one. According to classical mechanics, a particle in such a potential box would travel with a uniform velocity from one wall to the other and at the walls it would be perfectly reflected. Therefore, the probability of finding the particle within a small distance dx anywhere in the one dimensional box is the same and is equal to dx/a. The probability $1/a$ is represented by a straight line at height $1/a$ above the X-axis.

The problem of particle in a one dimensional box is not merely an exercise in mathematics, but it can serve as a good model for the calculation of approximate energies in atomic and molecular systems, besides accounting for the effect of quantization.

Particle in a Rectangular Three Dimensional Box

We have seen that a particle moving freely in a one dimensional box (PE = 0) serves as a very convenient model for several types of atomic and molecular systems. Calculations, though approximate, agree fairly well with observed results. Electronic motions in atoms and molecules are, however, three dimensional and a three dimensional box model should be more appropriate. Though electronic motions in atoms and molecules are complicated due to some other factors, let us see how far the results of quantum mechanical treatment of a single particle moving in a three dimensional rectangular box are of interest.

Let a particle of mass m be in motion in a rectangular potential box as shown in Fig. 19.24. With sides of lengths a, b, c parallel

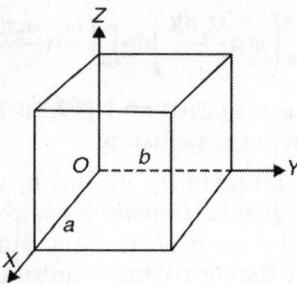

Fig. 19.24: Particle in a three dimensional rectangular box

to the X, Y and Z axes respectively, suppose there is no force acting on the particle inside the box, so that in the region $0 < x < a, 0 < y < b$ and $0 < z < c$, the potential energy $V(x, y, z) = 0$ and outside the box $V(x, y, z) = \infty$.

Wave Equation for the Particle

For the motion of the particle inside the box, the Schrodinger time independent equation reduces to

$$\nabla^2 \psi + \frac{2m}{\hbar^2} E\psi = 0 \qquad (19.195)$$

$$\frac{\partial^2 \psi}{\partial x^2} + \frac{\partial^2 \psi}{\partial y^2} + \frac{\partial^2 \psi}{\partial z^2} + \frac{2m}{\hbar^2} E\psi = 0 \qquad (19.196)$$

This is a partial differential equation where the wave function ψ is a function of x, y and z. The three dimensional Eq. (19.196) can be reduced to ordinary one dimensional equation by the method of separation of variables. We assume that the wave function $\psi(x, y, z)$ is equal to the product of three functions x, y and z each of which is a function of one variable only. Thus, we have

$$\psi(x, y, z) = X(x)\, Y(y)\, Z(z) \qquad (19.197)$$

Substituting Eq. (19.197) in Eq. (19.196), one obtains

$$YZ\frac{d^2 X}{dx^2} + ZX\frac{d^2 Y}{dy^2} + XY\frac{d^2 Z}{dz^2}$$

$$+ \frac{2mE}{\hbar^2} XYZ = 0 \qquad (19.198)$$

We have used ordinary derivatives instead of partial derivatives because each of the functions x, y and z is a function of one variable only. Dividing Eq. (19.198) by XYZ, one obtains

$$\frac{1}{X}\frac{d^2 X}{dx^2} + \frac{1}{Y}\frac{d^2 Y}{dy^2} + \frac{1}{Z}\frac{d^2 Z}{dz^2} + \frac{2m}{\hbar^2} E = 0 \qquad (19.199)$$

In Eq. (19.199), $2mE / \hbar^2$ is constant for a particular value of kinetic energy. Since the velocity of the particle, being a vector quantity, can be resolved into three components along the coordinate axes, the kinetic energy E can be expressed as the sum of the corresponding terms E_x, E_y and E_z. Hence

$$E = E_x + E_y + E_z \qquad (19.200)$$

From Eqs. (19.199) and (19.200), we have following three ordinary differential equations, each comprising a single variable.

$$\frac{1}{X}\frac{d^2X}{dx^2} + \frac{2mE_x}{\hbar^2} = 0 \qquad (19.201)$$

$$\frac{1}{Y}\frac{d^2Y}{dx^2} + \frac{2mE_y}{\hbar^2} = 0 \qquad (19.202)$$

$$\frac{1}{Z}\frac{d^2Z}{dx^2} + \frac{2mE_z}{\hbar^2} = 0 \qquad (19.203)$$

Each of these equations has a form similar to the Schrodinger equation $d^2\psi/dx^2 + K^2\psi = 0$ for a particle in one dimensional box with similar boundary conditions. In other words, the problem of a particle in three dimensional box is reduced to one of three particles in three separate one dimensional boxes of lengths a, b and c.

The **boundary** condition applicable to the solution is $X(0) = X(a) = 0$. So the eigenvalues of E_x are given by

$$E_x = \frac{\pi^2\hbar^2}{2ma^2}n_x^2 \qquad (19.204)$$

where $n = 1, 2, 3, \ldots$ and the corresponding normalized eigen functions are given by

$$X(x) = \sqrt{\frac{2}{a}}\sin\frac{n_x\pi x}{a} \qquad (19.205)$$

The solutions for y and z are of the same form. One obtains

$$Y(y) = \sqrt{\frac{2}{b}}\sin\frac{n_y\pi y}{b} \qquad (19.206)$$

$$E_y = \frac{\pi^2\hbar^2}{2mb^2}n_y^2 \qquad (19.207)$$

and $\quad Z(z) = \sqrt{\frac{2}{c}}\sin\frac{n_z\pi z}{c} \qquad (19.208)$

$$E_z = \frac{\pi^2\hbar^2}{2mc^2}n_z^2 \qquad (19.209)$$

where n_x, n_y and n_z are now the quantum numbers for the three separate boxes.

Eigenvalues of Energy

Substituting the expressions for E_x, E_y and E_z in Eq. (19.200), one obtains

$$E_{n_x,n_y,n_z} = \frac{\pi^2\hbar^2}{2m}\left[\frac{n_x^2}{a^2} + \frac{n_y^2}{b^2} + \frac{n_z^2}{c^2}\right] \qquad (19.210)$$

where $n_x = 1, 2, 3, \ldots$, $n_y = 1, 2, 3, \ldots$ and $n_z = 1, 2, 3, \ldots$

Eq. (19.210) gives the eigenvalues of the energy of the particle. These values are called the energy levels of the particle. The lowest (ground state) energy E_1 corresponds to $n_x = n_y = n_z = 1$.

Wave Functions

The total normalized wave functions inside the box for the stationary states are given by

$$\psi_{n_x, n_y, n_z}(x, y, z) = X(x)\,Y(y)\,Z(z)$$

$$= \sqrt{\frac{8}{abc}}\sin\frac{n_x\pi x}{a}\sin\frac{n_y\pi y}{b}\sin\frac{n_z\pi z}{c} \qquad (19.211)$$

where n_x, n_y and n_z are integers. The wave functions are zero outside the box. It may easily be proved that the wave function is normalized because

$$= \frac{8}{abc}\int_0^a\left(\sin\frac{n_x\pi x}{a}\right)^2 dx$$

$$\int_0^b\left(\sin\frac{n_y\pi y}{b}\right)dy\int_0^c\left(\sin\frac{n_z\pi z}{c}\right)^2 dz = 1$$

From Eqs. (19.210) and (19.211), one can draw following conclusions:

i. Three integers n_x, n_y and n_z which are called *quantum numbers* are required to describe each stationary state. If we change the sign of the quantum numbers, there is no change in the energy and in the wave function except that the minus sign appears on RHS of Eq. (19.211). Obviously, all the stationary states are given by the positive integral values of n_x, n_y and n_z. No quantum number can be zero, because if any one of them is taken zero, then $\psi(x, y, z) = 0$ which would mean that the particle does not exist in the box.

ii. The lowest possible energy, i.e. the ground state occurs when $n_x = n_y = n_z = 1$ and it depends on the values of a, b and c.

iii. If the particle is confined in a cubical box in which $a = b = c = 1$ (say), the eigenvalues of energy are given by

$$E_{n_x, n_y, n_z} = \frac{\pi^2 \hbar^2}{2mL^2}\left(n_x^2 + n_y^2 + n_z^2\right) \quad (19.212)$$

In this case, the energy of the particle in the ground state is given by

$$E_{111} = \frac{3\pi^2 \hbar^2}{2mL^2} \quad (19.213)$$

No other state will have this energy and this state has only one wave function. Therefore, the ground state and the energy level is said to be non-degenerate. We note that E_{111} is three times that for a particle in one dimensional box of length L.

The corresponding wave function is

$$\psi_{111} = \sqrt{\frac{8}{L^2}} \sin\frac{\pi x}{L} \sin\frac{\pi y}{L} \sin\frac{\pi z}{L} \quad (19.214)$$

(the subscripts in E and ψ denote the quantum numbers)

iv. *Degeneracy:* An interesting situation arises in a cubical box. In cubical box, the energy depends on the sum of the squares of the quantum numbers. Consequently, the particle having the same energy in an excited state will have several different stationary states or different wave functions. Such states and energy levels are said to be degenerate.

For example, we consider the first excited state when only one of the

quantum numbers is 2 while the others are 1 each, i.e.

n_x	n_y	n_z
2	1	1
1	2	1
1	1	2

Obviously, there are three independent states corresponding to the following three combinations of n_x, n_y and n_z: (2, 1, 1), (1, 2, 1) and (1, 1, 2). Each state has the same energy as

$$E_{211} = E_{121} = E_{112} = \frac{6\pi^2 \hbar^2}{2mL^2} \quad (19.215)$$

but a different wave function

$$\psi_{211} = \sqrt{\frac{8}{L^3}} \sin\frac{2\pi x}{L} \sin\frac{\pi y}{L} \sin\frac{\pi z}{L}$$

$$\psi_{121} = \sqrt{\frac{8}{L^3}} \sin\frac{\pi x}{L} \sin\frac{2\pi y}{L} \sin\frac{\pi z}{L} \quad (19.216)$$

$$\psi_{112} = \sqrt{\frac{8}{L^3}} \sin\frac{\pi x}{L} \sin\frac{\pi y}{L} \sin\frac{2\pi z}{L}$$

Obviously, the first excited state is triply degenerate or threefold degenerate. The first three energy levels of a particle in a cubical box are shown in Fig. 19.25. The degeneracy arises whenever there is an integral relation among a, b and c. But if $n_x = n_y = n_z$, the state will be non-degenerate. A slight distortion of the system may lead to the breakdown of degeneracy.

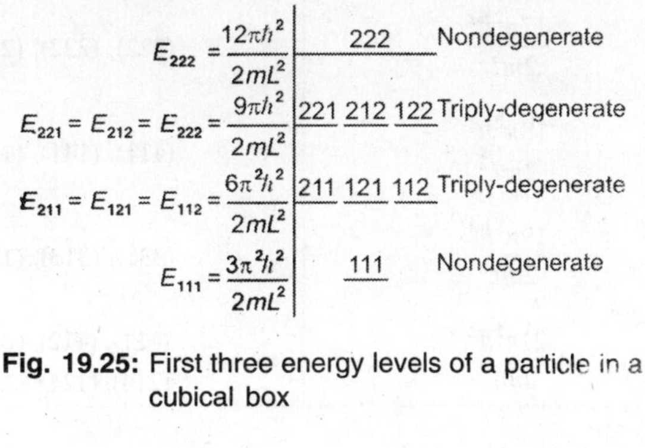

$$E_{222} = \frac{12\pi\hbar^2}{2mL^2} \quad \underline{\quad 222 \quad} \text{ Nondegenerate}$$

$$E_{221} = E_{212} = E_{222} = \frac{9\pi\hbar^2}{2mL^2} \quad \underline{221\ 212\ 122} \text{ Triply-degenerate}$$

$$E_{211} = E_{121} = E_{112} = \frac{6\pi^2\hbar^2}{2mL^2} \quad \underline{211\ 121\ 112} \text{ Triply-degenerate}$$

$$E_{111} = \frac{3\pi^2\hbar^2}{2mL^2} \quad \underline{\quad 111 \quad} \text{ Nondegenerate}$$

Fig. 19.25: First three energy levels of a particle in a cubical box

Degree of degeneracy: The number of independent wave functions for the stationary states of an energy level is called the degeneracy of the level.

For a particle in a cubical box, number of energy levels with the corresponding quantum numbers and the degree of degeneracy are given in Table 19.1.

Physical Representation of the Wave Function

The function ψ involves three variables x, y and z and therefore, it cannot be represented graphically. A qualitative understanding of the wave function can be achieved, however, from the nature of the expression (211), the

value of the function in the ground state ($n_x = n_y = n_z = 1$) is maximum when $x = a/2$, $y = b/2$ and $z = c/2$ (i.e. at the box centre) and is positive everywhere. In the excited states, the function changes sign and exhibits modes in some particular regions. For example, for the state ψ_{211}, the function vanishes at $x = a/2$ irrespective of the values of y and z, i.e. along the YZ plane which is, therefore, the nodal plane. Similarly, the XY and XZ planes passing through $b/2$ and $c/2$ are nodal planes for the states ψ_{121} and ψ_{112} respectively.

One can arrive at a better picture by considering $|\psi|^2$ function, i.e. probability. For the state ψ_{111} in a cubical box, the probability of finding the particle is maximum at the

Table 19.1: Energy levels, quantum numbers and degree of degeneracy for a particle in a cubical box

Energy levels	Quantum numbers	Degree of degeneracy
$\dfrac{3\pi^2\hbar^2}{2mL^2}$	(111)	Nondegenerate
$\dfrac{6\pi^2\hbar^2}{2mL^2}$	(211), (121), (112)	Threefold degenerate
$\dfrac{9\pi^2\hbar^2}{2mL^2}$	(221), (212), (122)	Threefold degenerate
$\dfrac{11\pi^2\hbar^2}{2mL^2}$	(311), (131), (113)	Threefold degenerate
$\dfrac{12\pi^2\hbar^2}{2mL^2}$	(222)	Nondegenerate
$\dfrac{14\pi^2\hbar^2}{2mL^2}$	(123), (132), (213) (231), (312), (321)	Sixfold degenerate
$\dfrac{17\pi^2\hbar^2}{2mL^2}$	(322), (232), (223)	Threefold degenerate
$\dfrac{18\pi^2\hbar^2}{2mL^2}$	(411), (141), (114)	Threefold degenerate
$\dfrac{19\pi^2\hbar^2}{2mL^2}$	(331), (313), (133)	Threefold degenerate
$\dfrac{21\pi^2\hbar^2}{2mL^2}$	(421), (412), (241) (214), (124), (142)	Sixfold degenerate

centre of the box and decreases uniformly towards each side. The probability distribution may be shown pictorially as in Fig. 19.26. In the next higher state, the probability distribution exhibits two equal halves and the probability is zero along the vertical planes passing through the box centre (nodal planes) as described above.

The state (2, 2, 2) has three vertical nodal planes along each side, i.e. $yz = xz$ and xy. It is apparent that the energy level increases the probability distribution and spread out more and more inside the box and in the limiting case tends to corresponds to the classical prediction.

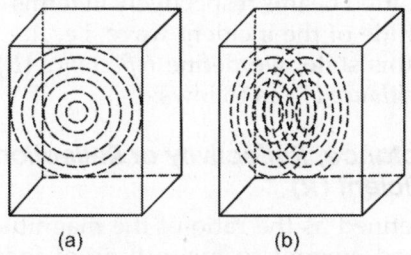

(a)   (b)

Fig. 19.26: Probability distribution

Potential Step (Single Step) Barrier

Let us consider potentials that are constant in time and that are also constant throughout prescribed regions of space. The simplest of these is the step potential shown in Fig. 19.27 in which the one dimensional space is divided into two regions such that $V = 0$ for $x < 0$ and for $V = V_0$ for $x > 0$. When a force field acting on a particle is zero or near zero everywhere except in a limited region, it is known as step

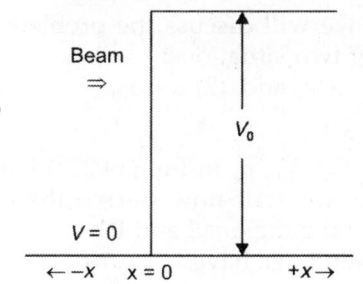

Fig. 19.27: Step potential

potential. The potential function of a potential step may be represented as

$$V(x) = \begin{cases} 0 & \text{for } x < 0 \\ V_0 & \text{for } x > 0 \end{cases}$$

Suppose the beam of electrons of energy E move from left to right, i.e. along the positive direction of X-axis. Two situations arises, namely $E < V_0$ and $E > V_0$, separately as these lead to different physical situations. We will restrict to one dimension only. We will apply quantum mechanics to this problem according to which electrons behave like a wave when moving from left to right and face a potential hill at $V(x) = V_0$ at $x = 0$.

This problem is analogous to light striking a sheet of glass where the wave is partly reflected and partly transmitted. Here also the electrons at the discontinuity will be partly reflected and partly transmitted. Let us associate a wave function $\psi(x, t)$ with the particle. The wave function has to be obtained by solving the Schrodinger wave equation.

The Schrodinger wave equation for region I, where $V(x) = 0$, is

$$\frac{\partial^2 \psi}{\partial x^2} - \frac{2m}{\hbar^2} E \psi = 0 \qquad (19.217)$$

The Schrodinger wave equation for the region II is

$$\frac{\partial^2 \psi}{\partial x^2} - \frac{2m}{\hbar^2}(E - V_0)\psi = 0 \qquad (19.218)$$

The general solution of Eqs. (19.217) and (19.218) will be

$$\psi_1 = A e^{ip_1 x / \hbar} + B e^{-ip_1 x / \hbar} \qquad (19.219)$$

$$\psi_2 = C e^{ip_2 x / \hbar} + D e^{-ip_2 x / \hbar} \qquad (19.220)$$

where ψ_1 and ψ_2 are the wave functions for the regions I and II respectively, A, B, C and D are constants of integration to be determined by the boundary conditions; p_1 and p_2 are the momenta in the region I and II respectively and are given by

$$p_1 = \sqrt{2mE}$$

$$p_2 = \sqrt{2m(E - V_0)} \qquad (19.221)$$

In Eq. (19.219), the first term represents a wave advancing in the positive direction of X-axis in the first region, i.e. incident wave and the second term represents a wave moving in the negative direction of X, i.e. reflected wave.

In Eq. (19.220), the first term represents a wave advancing in the positive direction of X-axis in the region II, i.e. transmitted wave and the second term represents a wave moving in the negative X-direction in the region II, i.e. reflected wave. Since discontinuity occurs only at $x = 0$ in the region II and after which there occurs no discontinuity in this region. This means that the reflection will not take place in this region, i.e. $D = 0$. Hence, Eq. (19.220) becomes as follows:

$$\psi_2 = Ce^{ip_2 x/\hbar} \qquad (19.222)$$

Boundary Conditions

In order for ψ to be finite, E and V must be finite because infinite energies do not occur in nature. Then from Schrodinger equation, it follows that $\partial^2 \psi / \partial x^2$ is not necessarily continuous but is everywhere finite. In order for $\partial^2 \psi / \partial x^2$ to be finite, $\partial \psi / \partial x$ should be continuous everywhere. By the continuity of $\partial \psi / \partial x$, we mean

$$\frac{\partial \psi_1}{\partial x} = \frac{\partial \psi_2}{\partial x} \text{ at } x = 0, \text{ i.e.}$$

$$\left(\frac{\partial \psi_1}{\partial x}\right)_{x=0} = \left(\frac{\partial \psi_2}{\partial x}\right)_{x=0} \qquad (19.222a)$$

This is the first boundary condition. In order for $\partial \psi / \partial x$ to be continuous, ψ must be continuous. The continuity of ψ implies that $\psi_1 = \psi_2 = 0$ at $x = 0$, i.e.

$$(\psi_1)_{x=0} = (\psi_2)_{x=0} \qquad (19.222b)$$

This is the second boundary condition. On applying the boundary condition (222b) to Eq. (219) and Eq. (220), one obtains

$$A + B = C \qquad (19.223)$$

On differentiating Eqs. (19.219) and (19.222), one obtains

$$\frac{\partial \psi_1}{\partial x} = \frac{ip_1}{\hbar}\left[Ae^{ip_2 x/\hbar} - Be^{-ip_2 x/\hbar}\right] \quad (19.224)$$

$$\frac{\partial \psi_1}{\partial x} = \frac{ip_2}{\hbar}Ce^{ip_2 x/\hbar} \qquad (19.225)$$

On applying the boundary condition (19.222a)

$$p_2 C = p_1[A + B] \qquad (19.226)$$

Solving Eqs. (19.223) and (19.226) for B and C, one obtains

$$C = \frac{2p_1 A}{p_1 + p_2} \qquad (19.227)$$

and

$$B = \frac{p_1 - p_2}{p_1 + p_2} = A \qquad (19.228)$$

Obviously, in Eqs. (19.227) and (19.228), B anc C denote the amplitude of reflected and transmitted beams respectively in terms of the amplitude of the incident wave, i.e. A.

At this state, we define *reflectance* (R) and *transmittance* (T) as follows:

Reflectance, Reflectivity or Reflection Coefficient (R)

It is defined as the ratio of the magnitude of reflected current to magnitude of incident current, i.e.

$$R = \frac{\text{Magnitude of reflected current}}{\text{Magnitude of incident current}} \quad (19.229)$$

Transmittance (T)

At the potential discontinuity, transmittance is defined as the ratio of the magnitude of transmitted current to the magnitude of incident current, i.e.

$$T = \frac{\text{Magnitude of reflected current}}{\text{Magnitude of incident current}} \quad (19.230)$$

Now, we will discuss the problem under following two situations:
(1) $E > V_0$ and (2) $E < V_0$.

Case 1. $E > V_0$

When $E > V_0$, p_2 in Eq. (19.222) is real. In this case, we will now derive the current density in the regions I and II.

In region I, we have

$$\psi_1 = Ae^{ip_1 x/\hbar} + Be^{-ip_1 x/\hbar} \qquad (19.231)$$

Complex conjugate of Eq. (19.231) will be

$$\psi_1^* = A^* e^{ip_1 x / \hbar} + B^* e^{-ip_1 x / \hbar} \qquad (19.232)$$

On differentiating Eqs. (19.231) and (19.232) with respect to x, one obtains

$$\frac{\partial \psi_1}{\partial x} = \frac{ip_1}{\hbar} \left[A e^{ip_1 x / \hbar} - B e^{-ip_1 x / \hbar} \right] \qquad (19.233)$$

$$\frac{\partial \psi_1^*}{\partial x} = -\frac{ip_1}{\hbar} \left[A^* e^{ip_1 x / \hbar} - B^* e^{-ip_1 x / \hbar} \right] \quad (19.234)$$

The probability current density is defined as

$$J = \frac{\hbar}{2im} \left[\psi^* \nabla \psi - \psi \nabla \psi^* \right]$$

Applying the above relation to region I, one obtains

$$(J_x)_{\text{I}} = \frac{\hbar}{2im} \left[\psi_1^* \frac{\partial \psi_1}{\partial x} - \psi_1 \frac{\partial \psi_1^*}{\partial x} \right] \qquad (19.235)$$

On substituting Eqs. (19.231), (19.232), (19.233) and (19.234), one obtains

$$(J_x)_{\text{I}} = \frac{\hbar}{2im} \left\{ \left(A e^{-ip_1 x / \hbar} + B^* e^{ip_1 x / \hbar} \right) \right.$$

$$\times \left(\frac{ip_1}{\hbar} \right) \left(A e^{ip_1 x / \hbar} - E e^{ip_1 x / \hbar} \right) \right\}$$

$$- \left\{ \left(A e^{ip_1 x / \hbar} + B e^{-ip_1 x / \hbar} \right) \times \left(\frac{-ip_1}{\hbar} \right) \right\}$$

$$\times \left\{ \left(A^* e^{-ip_1 x / \hbar} - B^* e^{ip_1 x / \hbar} \right) \right\} \right]$$

$$= p_1 \frac{(AA^* - B^* B)}{m} = \frac{p_1}{m} \left[|A|^2 - |B|^2 \right] (19.236)$$

From Eq. (19.236), it is evident that the current in the region I is equal to the difference between two terms, of which the first one proportional to $P_1 A^2$ represents the incident beam travelling from left to right, whereas the second one which is proportional to $p_1 |B^2|$ represents the reflected beam travelling from right to left.

$\therefore$ The probability current of the incident beam

$$= |A|^2 \frac{p_1}{m} \qquad (19.237)$$

and the probability current of the reflected beam

$$= |B|^2 \frac{p_1}{m} \qquad (19.238)$$

In region II, we have

$$\psi_2 = C e^{ip_2 x / \hbar} \qquad (19.239a)$$

Its complex conjugate is given by

$$\psi_2^* = C e^{-ip_2 x / \hbar} \qquad (19.239b)$$

On differentiating Eqs. (19.239a) and (19.239b) with respect to x, one obtains

$$\frac{\partial \psi_2}{\partial x} = \frac{ip_2}{\hbar} C e^{-ip_2 x / \hbar} \qquad (19.239c)$$

and $$\frac{\partial \psi_2^*}{\partial x} = \frac{-ip_2}{\hbar} C^* e^{-ip_2 x / \hbar} \qquad (19.239d)$$

The probability in region II is

$$(J_x)_{\text{II}} = \frac{\hbar}{2im} \left[\psi_2^* \frac{\partial \psi_2}{\partial x} - \psi_2 \frac{\partial \psi_2^*}{\partial x} \right] \qquad (19.239e)$$

On substituting Eqs. (19.239a) to (19.239d) in Eq. (15.239e), one obtains

$$(J_x)_{\text{II}} = \frac{\hbar}{2im} - \left[\left\{ C^* e^{-ip_2 x / \hbar} \left(\frac{ip_2}{\hbar} \right) C e^{ip_2 x / \hbar} \right\} \right.$$

$$\left. - \left\{ C e^{-ip_1 x / \hbar} \left(\frac{-ip_2}{\hbar} \right) C e^{ip_2 x / \hbar} \right\} \right]$$

$$= \frac{p_2}{2m} \left[CC^* + CC^* \right] = \frac{p_2}{m} \left(CC^* \right) = \frac{|C|^2 p_2}{m}$$

$$(19.240)$$

It is obvious from Eq. (19.240) that there is only transmitted wave in region II, and hence Eq. (19.240) represents the transmitted current.

Now, we find the expressions for reflectance and transmittance for the case when $E > V_0$ or p_1 is real.

From Eq. (19.229), we have

Reflectance (R)

$$= \frac{\text{Magnitude of reflected current}}{\text{Magnitude of incident current}}$$

On substituting Eqs. (19.237) and (19.238) in the above relation, one obtains

$$R = \frac{|B|^2}{|A|^2} \frac{p_1/m}{p_1/m} \qquad (19.240a)$$

From Eq. (19.227), one obtains

$$B = \frac{p_1 - p_2}{p_1 + p_2} A$$

or $\dfrac{B}{A} = \dfrac{p_1 - p_2}{p_1 + p_2}$

On substituting the above in Eq. (19.240a), one obtains

$$R = \frac{(p_1 - p_2)^2}{(p_1 + p_2)^2} \qquad (19.241)$$

From Eq. (19.230), one obtains

Transmittance

$$(T) = \frac{\text{Magnitude of transmitted current}}{\text{Magnitude of incident current}}$$

Substituting Eqs. (19.237) and (19.240) in the above, one obtains

$$T = \frac{|C|^2}{|A|^2} \frac{p_2/m}{p_1/m}$$

Using Eq. (19.227), one obtains

$$T = \left(\frac{2p_1}{p_1 + p_2}\right)^2 \frac{p_2}{p_1} = \frac{4p_1 p_2}{(p_1 + p_2)^2} \qquad (19.242)$$

From the definitions of reflectance and transmittance, it follows that the sum of reflectance and transmittance must be equal to unity. Using Eqs. (19.241) and (19.242), one can easily verify it as follows:

$$R + T = \frac{(p_1 - p_2)^2}{(p_1 + p_2)^2} + \frac{4p_1 p_2}{(p_1 + p_2)^2}$$

$$= \frac{(p_1 + p_2)^2}{(p_1 + p_2)^2} = 1 \qquad (19.243)$$

From Eq. (19.241), it is obvious that reflectance (R) approaches zero as p_2 approaches p_1 and unity as p_2 approaches zero.

But $\quad p_2 = \sqrt{2m(E - V_0)}$

and $\quad p_1 = \sqrt{2mE}$

From the above expressions, it follows that p_2 would approach zero when V_0 is zero, i.e. R would be zero when V_0 becomes equal to zero. This means that some reflection must take place even if $E \geq V_0$. The reflectance would be large only when V_0 is almost comparable in size with E. We must note that it is the wave nature of matter which gives rise to the property of reflection from a sudden change in potential. However, the classical theory does not give rise to this property, when $E > V_0$. Obviously, the property of reflection from a sudden change in potential is purely a quantum mechanical effect.

Case 2: $E < V_0$

When $E < V_0$ is imaginary, i.e.

$$p_2 = i\sqrt{2m(V_0 - E)} \qquad (19.244)$$

Its complex conjugate would be

$$p_2^* = i\sqrt{2m(V_0 - E)} = -p_2 \qquad (19.245)$$

Since p_2 is imaginary, it is possible to calculate the probability current associated with wave function ψ_2 in the following way:

$$\psi_2 = Ce^{ip_2 x/\hbar} \qquad (19.246)$$

and $\psi_2^* = C^* e^{-ip_2 x/\hbar}$ ($\because p_2$ is imaginary) (19.247)

On differentiating Eqs. (19.246) and (19.247), one obtains

$$\frac{\partial \psi_2}{\partial x} = \frac{ip_2}{\hbar} Ce^{ip_2 x/\hbar} \qquad (19.248)$$

$$\frac{\partial \psi_2^*}{\partial x} = \frac{ip_2^*}{\hbar} Ce^{ip_2^* x/\hbar} \qquad (19.249)$$

In this case, the probability current is given by

$$J_2 = \frac{\hbar}{2im}\left[\psi_2^* \frac{\partial \psi_2}{\partial x} - \psi_2 \frac{\partial \psi_2^*}{\partial x}\right] \qquad (19.250)$$

or $J_2 = \dfrac{\hbar}{2im}\Big[C^* e^{-ip_2^* x/\hbar}\left(ip_2/\hbar\right) C e^{ip_2 x}$

$$-Ce^{ip_2 x/\hbar}\left(-ip_2^*/\hbar\right)C^* e^{-ip_2^* x/\hbar}\Big] \qquad (19.251)$$

But $p_2^* = p_2$

$$\therefore \quad J_x = \frac{\hbar}{2im}\Big[C^* e^{ip_2 x/\hbar}\left(\frac{ip_2}{\hbar}\right) C e^{ip_2 x/\hbar}$$

$$CC^*\left(ip_2/\hbar\right)e^{ip_2 x/\hbar} e^{ip_2 x/\hbar}\Big] = 0 \quad (19.252)$$

This means that the transmittance current is zero. Thus,

Transmittance

$$(T) = \frac{\text{Magnitude of transmitted current}}{\text{Magnitude of incident current}}$$

$$= \frac{0}{\text{Magnitude of incident current}} = 0$$

$$\therefore \quad T = 0 \qquad (19.253)$$

Now,

Reflectance

$$(R) = \frac{\text{Magnitude of reflected current}}{\text{Magnitude of incident current}}$$

$$= \frac{(BB^*)p_1/m}{(AA^*)p_1/m} \qquad \text{[using Eq. (19.236)]}$$

$$= \frac{\left(\dfrac{p_1-p_2}{p_1+p_2}\right) A \left(\dfrac{p_1-p_2}{p_1+p_2}\right)^* A^*}{AA^*}$$

$$= \frac{(p_1-p_2)(p_1-p_2)}{(p_1+p_2)(p_1+p_2)}$$

[by using Eq. (19.228) and its complex conjugate]

$$R = \left(\frac{p_1-p_2}{p_1+p_2}\right)\left(\frac{p_1-p_2^*}{p_1+p_2^*}\right)$$

$$= \frac{(p_1-p_2)(p_1+p_2)}{(p_1+p_2)(p_1+p_2)} \qquad \text{[since } p_2^* = -p_2\text{]}$$

$$R = 1 \qquad (19.254)$$

From Eqs. (19.253) and (19.254), it is evident that the entire wave gets reflected when $E < V_0$, i.e. no electrons are transmitted but all are reflected.

The Rectangular Potential Barrier: Penetration of a Barrier: The Tunnel Effect

A rectangular one dimensional potential barrier of height V_0 and width a for a particle is shown in Fig. 19.28. It extends over the region II from $x = 0$ to $x = a$ in which the potential energy $V(x)$ of the particle is constant and is equal to V_0. On both the sides of the barrier, i.e. in region I and region III, $V(x) = 0$. Obviously, particle in regions I and III experiences no force on it, i.e. it behaves as a free particle. Potential field can be represented as

$V(x) = 0$	$x < 0$	Region I
$V(x) = 0$	$0 < x < a$	Region II
$V(x) = 0$	$x > a$	Region III

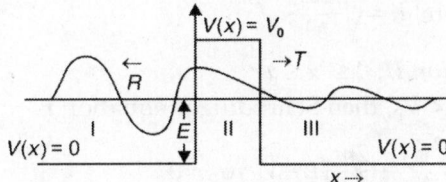

Fig. 19.28: The rectangular potential barrier

Suppose a beam of particles (each of mass m) travelling parallel to the X-axis from left to right is incident on the potential barrier. In the region I and III, the energy E of a particle is wholly kinetic and in the region II it is partly kinetic and partly potential. If the kinetic energy of the particle $E > V_0$, then according to classical mechanics the probability of any particle reaching the region III after crossing the region II is zero. However, according to quantum mechanics, the transmission probability has a small but definite value. This behaviour is called tunnel effect. The phenomenon is known as tunnelling.

The situation shown in Fig. 19.28 represents the encounter between a free particle and a potential barrier. The incident particle from region I can be turned back by the barrier or it can penetrate through it. Obviously, in the region I the desired wave function has to represent a particle moving to the right (incident particle) and a particle moving to the left (reflected particle). In region III, the wave function has to represent only a particle moving to the right, i.e. transmitted particle.

Let $\psi_1(x)$, $\psi_2(x)$ and $\psi_3(x)$ be the wave functions for the motion of particle in regions I, II and III respectively. In these regions, the time independent Schrodinger wave equations are as follows:

Region I: $-\infty < x < 0$

In this region, particle is free, i.e. $V = 0$. Schrodinger equation can be written as

$$\frac{d^2\psi_1}{dx^2} + \frac{2mE}{\hbar^2}\psi_1 = 0$$

or $\quad \dfrac{d^2\psi_1}{dx^2} + K^2\psi_1 = 0 \qquad (19.255)$

where $K = \sqrt{\dfrac{2mE}{\hbar^2}}$.

Region II: $0 \leq x \leq a$

If $E < V_0$, then Schrodinger equation is

$$\frac{d^2\psi_2}{dx^2} + \frac{2m}{\hbar^2}(E - V_0)\psi_2 = 0$$

or $\quad \dfrac{d^2\psi_2}{dx^2} - \dfrac{2m}{\hbar^2}(V_0 - E)\psi_2 = 0 \qquad (19.256)$

where $\beta = \sqrt{\dfrac{2m(V_0 - E)}{\hbar^2}}$

Region III: $a \leq x \leq \infty$

$$\frac{d^2\psi_3}{dx^2} + \frac{2mE}{\hbar^2}\psi_3 = 0$$

or $\quad \dfrac{d^2\psi_3}{dx^2} + K^2\psi_3 = 0 \qquad (19.257)$

where $K = \sqrt{\dfrac{2mE}{\hbar^2}}$.

The general solutions of Eqs. (19.255), (19.256) and (19.257) are

$\psi_1 = Ae^{ikx} + Be^{-ikx} \qquad (19.258)$

$\psi_2 = Ce^{\beta x} + De^{-\beta x} \qquad (19.259)$

$\psi_3 = Ge^{ikx} + He^{-ikx} \qquad (19.260)$

Since there is no particle coming from left in the region III, we must have $H = 0$. Therefore,

$\psi_3 = Ge^{ikx} \qquad (19.260a)$

We can interpret the terms in the above equations as follows:

In Eq. (19.258), the term $A\exp(ikx)$ is a wave of amplitude A travelling in the positive X-direction and the term $B\exp(-ikx)$ is the wave of amplitude B reflected in the negative X-direction from the potential barrier, when the incident wave falls on the barrier. In Eq. (19.259), the term $D\exp(-\beta x)$ is an exponentially decreasing wave function representing a non-oscillatory disturbance which moves through the barrier in the positive X-direction, and the term $C\exp(\beta x)$ is the reflected disturbance within the barrier, it is an exponentially decreasing wave function. Eq. (19.260) represents the transmitted wave in the region III. Obviously, this wave travels in the positive X-direction.

Expressions for the Coefficients A and B in Terms of the Coefficient G

The boundary conditions require the continuity of $\psi(x)$ and $\partial\psi/\partial x$ so that

   i. At $x = 0$, we have

$$\psi_1(0) = \psi_2(0)$$

$\therefore$ From Eqs. (19.258) and (19.259), we get

$$A + B = C + D \qquad (19.261)$$

We also have

$$\left(\frac{\partial \psi_1}{\partial x}\right)_{x=0} = \left(\frac{\partial \psi_2}{\partial x}\right)_{x=0}$$

$$\therefore \quad Aik - Bik = C\beta - D\beta$$

or $(A-B) = \dfrac{\beta}{ik}(C-D)$ \hfill (19.262)

From Eqs. (19.261) and (19.262), one obtains

$$A = \frac{1}{2}\left[\left(1+\frac{\beta}{ik}\right)C + \left(1-\frac{\beta}{ik}\right)D\right] \quad (19.263)$$

and $B = \dfrac{1}{2}\left[\left(1-\dfrac{\beta}{ik}\right)C + \left(1+\dfrac{\beta}{ik}\right)D\right]$

$$\hfill (19.264)$$

ii. At $x = a$, we have

$\psi_2(a) = \psi_3(a)$

∴ From Eqs. (19.259) and (19.260), one obtains

$$Ce^{\beta a} + De^{-\beta a} = Ge^{ika} \quad (19.265)$$

We also have

$$\left(\frac{\partial \psi_2}{\partial x}\right)_{x=a} = \left(\frac{\partial \psi_3}{\partial x}\right)_{x=a}$$

$$\therefore \quad C\beta e^{\beta a} - D\beta e^{-\beta a} = Gike^{ika}$$

or $\quad Ce^{\beta a} - De^{-\beta a} = \dfrac{ik}{\beta}Ge^{ika}$ \hfill (19.266)

From Eqs. (19.265) and (19.266), one obtains

$$C = \frac{1}{2}\left(1+\frac{ik}{\beta}\right)e^{-\beta a}Ge^{ika} \quad (19.267)$$

$$D = \frac{1}{2}\left(1+\frac{ik}{\beta}\right)e^{\beta a}Ge^{ika} \quad (19.268)$$

Substituting the values of C and D in Eq. (19.263), one obtains

$$A = \frac{1}{2}\left[\left(1+\frac{\beta}{ik}\right)\frac{1}{2}\left(1+\frac{ik}{\beta}\right)e^{-\beta a}\right.$$

$$\left. + \left(1+\frac{\beta}{ik}\right)\frac{1}{2}\left(1-\frac{ik}{\beta}\right)e^{\beta a}\right]Ge^{ika}$$

On simplification, one obtains

$$A = \left[\left(\frac{e^{\beta a}+e^{-\beta a}}{2}\right)\right.$$

$$\left. -\frac{1}{2}\left(\frac{\beta}{ik}+\frac{ik}{\beta}\right)\left(\frac{e^{\beta a}-e^{-\beta a}}{2}\right)\right]Ge^{ika}$$

Using the following trigonometric relations:

$$\frac{e^{\beta a}+e^{-\beta a}}{2} = \cosh\beta a$$

and $\quad \dfrac{e^{\beta a}-e^{-\beta a}}{2} = \sinh\beta a,$

one obtains

$$A = \left[\cosh\beta a - \frac{1}{2}\left(\frac{\beta}{ik}+\frac{ik}{\beta}\right)\sinh\beta a\right]Ge^{ika}$$

or $A = \left[\cosh\beta a + \dfrac{i}{2}\left(\dfrac{\beta}{k}-\dfrac{k}{\beta}\right)\sinh\beta a\right]Ge^{ika}$

$$\hfill (19.269)$$

Similarly, substituting the values of C and D in Eq. (19.264), one obtains

$$B = \frac{1}{2}\left[\left(1-\frac{\beta}{ik}\right)\frac{1}{2}\left(1+\frac{ik}{\beta}\right)e^{-\beta a}\right.$$

$$\left. + \left(1+\frac{\beta}{ik}\right)\frac{1}{2}\left(1-\frac{ik}{\beta}\right)e^{\beta a}\right]Ge^{ika}$$

$$= \frac{1}{2}\left[-\frac{1}{2}\left(\frac{\beta}{ik}-\frac{ik}{\beta}\right)e^{-\beta a} + \frac{1}{2}\left(\frac{\beta}{ik}-\frac{ik}{\beta}\right)e^{\beta a}\right]Ge^{ika}$$

$$= \frac{1}{2}\left[\left(\frac{\beta}{ik}-\frac{ik}{\beta}\right)\left\{\frac{e^{\beta a}-e^{-\beta a}}{2}\right\}\right]Ge^{ika}$$

$$= -\frac{i}{2}\left(\frac{\beta}{k}+\frac{k}{\beta}\right)\sinh\beta a\,Ge^{ika} \quad (19.270)$$

Transmission Probability

The flux of particles in the incident beam is given by

S_1 = Probability density of particles in the incident beam × Particle velocity

$$= \psi_1(x)\, \psi_1(x)^*\, v_1$$

$$= \left(Ae^{ikx}\right)\left(Ae^{ikx}\right)^* v_1 = Ae^{ikx} A^* e^{-ikx} v_1$$

$$= AA^* v_1$$

where A^* is the complex conjugate of A and v_1 is the particle velocity in the incident beam. Similarly, the flux of particles in the transmitted beam is given by

$$S_2 = \left(Ge^{ikx}\right)\left(Ge^{ikx}\right)^* v_1 = GG^* v_1$$

The transmission probability T is defined as

$$T = \frac{GG^* v_1}{AA^* v_1} = \left(\frac{G}{A}\right)\left(\frac{G}{A}\right)^* = \left|\frac{G}{A}\right|^2 \qquad (19.271)$$

Here $\left(\dfrac{G}{A}\right)^*$ is the complex conjugate of $\dfrac{G}{A}$.

The quantity T is also referred to as the transmission coefficient. From Eq. (19.269), one obtains

$$\frac{A}{G} = \left[\cosh \beta a + \frac{i}{2}\left(\frac{\beta}{k} - \frac{k}{\beta}\right)\sinh \beta a\right]e^{ika}$$

and

$$\left(\frac{A}{G}\right)^* = \left[\cosh \beta a - \frac{i}{2}\left(\frac{\beta}{k} - \frac{k}{\beta}\right)\sinh \beta a\right]e^{-ika}$$

Therefore,

$$\left(\frac{A}{G}\right)\left(\frac{A}{G}\right)^* = \cosh^2 \beta a + \frac{1}{4}\left(\frac{\beta}{k} - \frac{k}{\beta}\right)^2 \sinh^2 \beta a$$

or $\quad \left|\dfrac{A}{G}\right|^* = \cosh^2 \beta a + \dfrac{1}{4}\left(\dfrac{\beta}{k} - \dfrac{k}{\beta}\right)^2 \sinh^2 \beta a$

Using $\cosh^2 \beta a = 1 + \sinh^2 \beta a$, one obtains

$$\left|\frac{A}{G}\right|^2 = 1 + \left[\frac{1}{4}\left(\frac{\beta}{k} - \frac{k}{\beta}\right)^2 + 1\right]\sinh^2 \beta a$$

$$= 1 + \frac{1}{4}\left[\left(\frac{\beta}{k} - \frac{k}{\beta}\right)^2 + 4\right]\sinh^2 \beta a$$

or $\quad \dfrac{1}{T} = 1 + \dfrac{1}{4}\left(\dfrac{\beta}{k} - \dfrac{k}{\beta}\right)^2 \sinh^2 \beta a \qquad (19.272)$

Using the values of $k = \sqrt{\dfrac{2mE}{\hbar^2}}$ and

$$\beta = \sqrt{\frac{2m(V_0 - E)}{\hbar^2}}, \text{ one obtains}$$

$$\frac{\beta}{k} = \sqrt{\frac{V_0 - E}{E}} \text{ and } \frac{k}{\beta} = \sqrt{\frac{E}{V_0 - E}}$$

Substituting these values in Eq. (19.272), one obtains on simplification,

$$\frac{1}{T} = 1 + \frac{V_0^2}{4E(V_0 - E)}\sinh^2 \beta a \qquad (19.273)$$

When $E < V_0$, the transmission probability is given by

$$T = \frac{1}{1 + \dfrac{V_0^2}{4E(V_0 - E)}\sinh^2 \beta a} \qquad (19.274)$$

Equation (19.274) is the expression for transmission probability or the transmission coefficient in the case when $E < V_0$. This expression leads to the following conclusions:

Case 1: $E < V_0$

i. Classically, a particle with energy $E < V_0$ cannot penetrate the potential barrier and it must rebound from the barrier. Eq. (19.274) shows that quantum mechanically, there is a finite probability for a particle to tunnel through a potential barrier of height V_0 even when the initial kinetic energy E of the particle is less than V_0.

ii. When $\beta a \gg 1$, then $e^{\beta a}$ is large but $e^{-\beta a} = 1/e^{\beta a}$ is very small compared to 1. Obviously, one can neglect $e^{-\beta a}$ compared to $e^{\beta a}$.

$$\therefore \ \sinh^2 \beta a = \left(\frac{e^{\beta a} - e^{-\beta a}}{2}\right) \approx \frac{1}{4}\exp(2\beta a)$$

Then $\dfrac{1}{T} \approx 1 + \dfrac{V_0^2}{4E(V_0 - E)} \dfrac{1}{4}(2\beta a)$

$\approx \dfrac{V_0^2 \exp(2\beta a)}{16E(V_0 - E)}$

$\therefore \quad T \approx \dfrac{16E}{V_0^2}(V_0 - E) \dfrac{1}{4} \exp(2\beta a)$

$\approx \dfrac{16E}{V_0}\left(1 - \dfrac{E}{V_0}\right) \exp(-2\beta a)$

(19.275)

Eq. (19.275) is true only when $E < V_0$. Obviously, when $\beta a \gg 1$, T becomes very small and it decreases exponentially with increase in the thickness of the barrier. Eq. (19.275) also shows that if a is constant and if E/V_0 decreases, then T decreases exponentially. This variation is shown by the curve in Fig. 19.29.

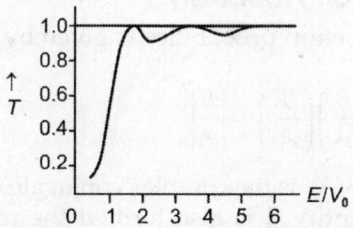

Fig. 19.29: Transmissivity (T) versus E/V_0 curve

iii. $E \approx V_0$, i.e. when the particle energy E approaches the potential energy V_0 of the top of the barrier, then

$\beta a = a\sqrt{\dfrac{2m(V_0 - E)}{\hbar^2}} \ll 1$

$\sinh^2 \beta a \approx a\sqrt{\dfrac{2m(V_0 - E)}{\hbar^2}}$

and $\sinh^2 \beta a \approx \dfrac{2m(V_0 - E)}{\hbar^2}a^2$

Hence $\dfrac{1}{T} \approx 1 + \dfrac{V_0^2}{4E(V_0 - E)} \dfrac{2m(V_0 - E)}{\hbar^2}a^2$

$\approx 1 + \dfrac{2mV_0^2 a^2}{4E\hbar^2}$

Since $E \to V_0$, one obtains

$\dfrac{1}{T} \approx 1 + \dfrac{mV_0 a^2}{2\hbar^2}$

or $T = \dfrac{1}{\dfrac{mV_0 a^2}{2\hbar^2}}$

(19.276)

Obviously, $T < 1$. Thus, when E approaches V_0, T increases and when $E \approx V_0$, the transmission probability is given by Eq. (19.276) and it is less than 1, i.e. $T < 1$.

Case 2: $E > V_0$

When $E > V_0$,

then $\beta = \sqrt{\dfrac{2m(V_0 - E)}{\hbar^2}} = i\sqrt{\dfrac{2m(E - V_0)}{\hbar^2}} = i\alpha$

where $\alpha = \sqrt{\dfrac{2m(E - V_0)}{\hbar^2}}$

$\sinh \beta a = \dfrac{e^{\beta a} - e^{-\beta a}}{2} = \dfrac{e^{i\alpha a} - e^{-i\alpha a}}{2}$

$\beta = i\left(\dfrac{e^{i\alpha a} - e^{-i\alpha a}}{2i}\right) = i \sin \alpha a.$

Squaring both the sides, one obtains
$\sinh^2 a = -\sin^2 \alpha a.$

Substituting this value in Eq. (19.273), one obtains

$\dfrac{1}{T} = 1 - \dfrac{V^2}{4E(V_0 - E)}\sin^2 \alpha a$

$= 1 + \dfrac{V_0^2}{4E(E - V_0)}\sin^2 \alpha a$

(19.277)

$\therefore \quad T = \dfrac{1}{1 + \dfrac{V^2}{4E(E - V_0)}\sin^2 \alpha a}$

(19.278)

The above expression for the transmission probability or the transmission coefficient in the case when $E > V_0$ leads to the following conclusions:

i. When E approaches V_0, then

$$\alpha a = a\sqrt{\frac{2m(E-V_0)}{\hbar^2}} \text{ will be} \ll 1$$

$$\therefore \quad \sin \alpha a \approx a\sqrt{\frac{2m(E-V_0)}{\hbar^2}}$$

and $\quad \sin^2 \alpha a = \dfrac{2m(E-V_0)a^2}{\hbar^2}$

Hence $\quad \dfrac{1}{T} \approx 1 + \dfrac{V_0^2}{4E(E-V_0)} \dfrac{2m(E-V_0)a^2}{\hbar^2}$

$$\approx 1 + \frac{2mV_0^2 a^2}{4E\hbar^2}$$

Since $E \rightarrow V_0$, one obtains

$$\frac{1}{T} \approx 1 + \frac{mV_0^2 a^2}{2\hbar^2}$$

or $\quad T \approx \dfrac{1}{1 + \dfrac{mV_0^2 a^2}{2\hbar^2}}$

We note that the above expression is the same as that obtained in the previous case for $E < V_0$.

ii. Eq. (19.278) shows that for $E < V_0$, $T = 1$ when

$\alpha a = n\pi$, where $n = 1, 2, 3, \ldots$

or $\quad a = \dfrac{n\pi}{\alpha} = \dfrac{n\pi}{\sqrt{2m(E-V_0)/\hbar^2}}$

$$= \frac{n\pi\hbar}{\sqrt{2m(E-V_0)}} \qquad (19.279)$$

or $\quad a = \dfrac{nh}{2\sqrt{2m(E-V_0)}}$

But $h/2\sqrt{2m(E-V_0)}$ is the de Broglie wavelength λ of the particle in the region of kinetic energy ($E - V_0$). Hence $T = 1$, when

$$a = n(\lambda/2) \qquad (19.280)$$

This oscillation is the analog of such phenomenon as the optical transmission through a nonreflecting thin film or the use of matching stubs in wave guides or the Ramsauer-Townsend effect in the scattering of low energy electrons from noble gas atoms.

This shows that if the width of the barrier is an integral multiple of half wavelength, there is perfect transmission of the incident beam of the particles.

Eq. (19.278) also shows that as E increases, T oscillates between the maximum value 1 and a value less than 1 as shown in Fig. 19.29.

Reflection Probability

The reflection probability is given by

$$R = \left(\frac{B}{A}\right)\left(\frac{B}{A}\right)^* = \left|\frac{B}{A}\right|^2 \qquad (19.281)$$

where $(B/A)^*$ is the complex conjugate of (B/A). The quantity R is also termed the reflection coefficient. Dividing Eq. (19.270) by Eq. (19.269), one obtains

$$\frac{B}{A} = \frac{-\dfrac{i}{2}\left(\dfrac{\beta}{k}+\dfrac{k}{\beta}\right)\sinh\beta a}{\cosh\beta a + \dfrac{i}{2}\left(\dfrac{\beta}{k}+\dfrac{k}{\beta}\right)\sinh\beta a}$$

and $\quad \left(\dfrac{B}{A}\right)^* = \dfrac{-\dfrac{i}{2}\left(\dfrac{\beta}{k}+\dfrac{k}{\beta}\right)\sinh\beta a}{\cosh\beta a - \dfrac{i}{2}\left(\dfrac{\beta}{k}-\dfrac{k}{\beta}\right)\sinh\beta a}$

$$\therefore \quad R = \left(\frac{B}{A}\right)\left(\frac{B}{A}\right)^*$$

$$= \frac{-\frac{1}{4}\left(\frac{\beta}{k}+\frac{k}{\beta}\right)^2 \sinh^2 \beta a}{\cosh^2 \beta a + \frac{1}{4}\left(\frac{\beta}{k}-\frac{k}{\beta}\right)^2 \sinh^2 \beta a}$$

$$= \frac{-\frac{1}{4}\left(\frac{\beta}{k}+\frac{k}{\beta}\right)^2 \sinh^2 \beta a}{1+\frac{1}{4}\left(\frac{\beta}{k}-\frac{k}{\beta}\right)^2 \sinh^2 \beta a} \qquad (19.282)$$

The above equation is the expression for the reflection coefficient. From Eq. (19.272), we have the expression for the transmission coefficient

$$T = \frac{1}{1+\frac{1}{4}\left(\frac{\beta}{k}-\frac{k}{\beta}\right)^2 \sinh^2 \beta a} \qquad (19.283)$$

Adding Eqs. (19.282) and (19.283), one obtains

$$R + T = 1 \qquad (19.284)$$

The quantum mechanical tunnel effect or the transmission coefficient depends, as is clear from Eq. (19.276), on the width a and height V_0 of the potential barrier in addition to the mass of the particle. The tunnel effect has a great significance in physics and chemistry. It provides explanations for the following phenomena:

i. the emission of α particles from a radioactive nucleus, e.g.
$$Ra \to Rn + \alpha$$
ii. the field emission of electrons from a cold metallic surface
iii. the electrical breakdown of insulators
iv. the reverse breakdown of semiconductor diodes
v. the switching action of a tunnel diode.

ILLUSTRATIVE EXAMPLES

Example 1

Calculate the value of the retarding potential needed to stop the photoelectrons ejected from a metal surface of work function 1.2 eV with light of frequency 5.5×10^{14} sec^{-1}.

Solution

From Einstein photoelectric equation, we have
$$eV_0 = h\nu - W$$

Here $e = 1.6 \times 10^{-19}$C, $\nu = 5.5 \times 10^{14}$ sec^{-1}

$$W = 1.2 \text{ eV} = 1.2 \times 1.6 \times 10^{-19}\text{J} = 1.92 \times 10^{-19}\text{J}$$

$$h = 6.6 \times 10^{-34}\text{Js}$$

$\therefore 1.6 \times 10^{-19} V_0 = 6.6 \times 10^{-34} \times 5.5 \times 10^{14}$

$$- 1.92 \times 10^{-19}$$

$$1.6 \times 10^{-19} V_0 = 1.71 \times 10^{-19}$$

$$V_0 = \frac{1.71 \times 10^{-19}}{1.6 \times 10^{-19}} = 1.06 \text{ volt}$$

Example 2

In an experiment, tungsten cathode which has a threshold 2300 Å is irradiated by ultraviolet light of wavelength 1800 Å. Calculate (i) the maximum energy of emitted photoelectrons and (ii) the work function for tungsten. Give your results in electron volts. Given 1 eV = 1.6×10^{-19}J and velocity of light $c = 3 \times 10^8$ m/s.

Solution

We have from Einstein's photoelectric equation

$$\frac{1}{2}mv_{max}^2 = h\nu - hc = h\left(\frac{c}{\lambda} - \frac{c}{\lambda_0}\right)$$

Here λ_0 is the threshold wavelength

$$\therefore \frac{1}{2}mv_{max}^2 = hc\left(\frac{\lambda_0 - \lambda}{\lambda \lambda_0}\right)$$

$$\therefore \frac{1}{2}mv_{max}^2 = 6.6 \times 10^{-34} \times 3 \times 10^8$$

$$\times \left[\frac{2300 \times 10^{-10} - 1800 \times 10^{-10}}{2300 \times 10^{-10} \times 1800 \times 10^{-10}}\right]$$

$$= \frac{55 \times 10^{-19}}{23 \times 1.6 \times 10^{-19}} = 1.485 \text{ eV}$$

∴ Work function

$$W = h\nu_0 = \frac{hc}{\lambda_0} = \frac{6.6 \times 10^{-34} \times 3 \times 10^8}{2300 \times 10^{-10}}$$

or $W = 8.608 \times 10^{19} = \dfrac{8.608 \times 10^{-19}}{1.6 \times 10^{-19}}$ eV

$$= 5.38 \text{ eV}$$

Example 3

Find the maximum wavelength of light that can liberate electrons from potassium. Work function of potassium is 2.24 eV.

Solution

$$\lambda_0 = \frac{hc}{W}$$

where $W = 2.24$ eV $= 2.24 \times 1.6 \times 10^{-19}$ J

$$\therefore \quad \lambda_0 = \frac{6.6 \times 10^{-34} \times 3 \times 10^8}{2.24 \times 6 \times 10^{-19}} \text{ m}$$

$$= 5524 \times 10^{-10} \text{ m} = 5524 \text{ Å}.$$

Example 4

A metallic surface, when illuminated with light of wavelength λ_1, emits electrons with energies upto a maximum value E_1, and when illuminated with light of wavelength λ_2, where $\lambda_2 < \lambda_1$, it emits electrons with energies up to a maximum value E_2. Prove that Planck's constant h and the work function ϕ of the metal are given by

$$h = \frac{(E_2 - E_1)}{C(\lambda_1 - \lambda_2)}$$

and $\quad \phi = \dfrac{E_2 \lambda_2 - E_1 \lambda_1}{\lambda_1 - \lambda_2}$

Solution

The maximum kinetic energy E_1 is given by

$$E_1 = h\nu_1 - \phi$$

$$E_1 = \frac{hc}{\lambda_1} - \phi \tag{1}$$

Similarly, the maximum kinetic energy E_2 is given by

$$E_2 = \frac{hc}{\lambda_2} - \phi \tag{2}$$

Subtracting Eq. (1) froms Eq. (2), one obtains

$$E_2 - E_1 = hc\left(\frac{1}{\lambda_2} - \frac{1}{\lambda_1}\right) = hc\left(\frac{\lambda_1 - \lambda_2}{\lambda_1 \lambda_2}\right)$$

$$\therefore \quad h = \frac{(E_2 - E_1)\lambda_1 \lambda_2}{c(\lambda_1 - \lambda_2)} \tag{3}$$

To obtain the expression for work function ϕ, one obtains from Eqs. (1) and (2)

$$E_1 \lambda_1 = hc - \phi \lambda_1 \tag{4}$$

$$E_2 \lambda_2 = hc - \phi \lambda_2 \tag{5}$$

Subtracting Eq. (4) from (5), one obtains

$$E_2 \lambda_2 = E_1 \lambda_1 = \phi(\lambda_1 - \lambda_2)$$

$$\therefore \quad \phi = \frac{E_2 \lambda_2 - E_1 \lambda_1}{\lambda_1 - \lambda_2}$$

Example 5

A photon of wavelength 3 Å suffers Compton scattering by a free electron originally at rest. If the angle of scattering is 90°, what is the kinetic energy of the recoil electron ?

Solution

$$\Delta \lambda = \frac{\lambda}{m_0 c}(1 - \cos\phi)$$

$$= 0.0242(1 - 0) = 0.0242 \text{ Å}$$

$\therefore$ Compton wavelength $\lambda_0 = \dfrac{\lambda}{m_0 c} = 0.0242 \text{Å}$

By the law of conservation of energy

$$h\nu + m_0 c^2 = h\nu + mc^2$$

But $\quad mc^2 = m_0 c^2 + \dfrac{1}{2} m_0 v^2 \quad$ (for $v = c$)

KE of recoil electron $= h(\nu - \nu') = \dfrac{1}{2} m_0 v^2$

$$= hc\left(\frac{\lambda' - \lambda}{\lambda \lambda'}\right) \approx \frac{hc \Delta \lambda}{\lambda(\lambda + \Delta \lambda)}$$

Substituting the proper values, one obtains

$$KE = \frac{6.6 \times 10^{34} \times 3 \times 10^{8} \times 0.0242 \times 10^{-10}}{3 \times 10^{-10} \times 3.0242 \times 10^{-19}} \, J$$

$$= \frac{6.6 \times 3 \times 0.0242 \times 10^{-16}}{3 \times 3.0242 \times 1.6 \times 10^{-19}} = 33.28 \text{ eV}.$$

Example 6

In an experiment of Compton scattering, the incident radiation has wavelength 2 Å, while wavelength of radiation scattered through 180° is 2.048 Å. Calculate (i) the wavelength of scattered radiation if they are viewed at an angle of 60° to the direction of incidence (ii) the energy of the recoil electron which scatter radiation through 60°.

Solution

We have $\quad \lambda' - \lambda = \dfrac{h}{m_0 c}(1 - \cos\phi)$

When $\qquad \phi = 180°$

$\therefore \qquad \cos\phi = \cos 180° = -1$

$\qquad \lambda = 2 \text{ Å} = 2 \times 10^{-10} \text{m}$

$\qquad \lambda' = 2.048 \text{ Å} = 2.048 \times 10^{-10} \text{m}$

$\therefore \quad (2.048 - 2) \times 10^{-10} = \dfrac{h}{m_0 c}(1 + 1)$

or $\qquad \dfrac{h}{m_0 c} = \dfrac{0.048 \times 10^{-10}}{2}$

$\qquad\qquad = 0.024 \times 10^{-10} \text{ m} = 0.024 \text{ Å}$

i. $\qquad \phi = 60°, \cos\phi = \cos 60° = 1/2$

$\therefore \quad \lambda' = \lambda + \dfrac{h}{m_0 c}(1 - \cos 60°)$

$\qquad = 2 \times 10^{-10} + 0.024 \times 10^{-10} \times \left(1 - \dfrac{1}{2}\right)$

$\qquad = 2 \times 10^{-10} + 0.012 \times 10^{-10} \text{ m}$

$\qquad = 2.012 \text{ Å}$

ii. Energy of recoil electron which scatter photon through 60° is

$$(KE)_{recoil} = (m - m_0)c^2$$

$$= h(\nu - \nu') = hc\left(\frac{1}{\lambda} - \frac{1}{\lambda'}\right)$$

$$= \frac{hc(\lambda' - \lambda)}{\lambda\lambda'}$$

$$= \frac{6.6 \times 10^{-34} \times 3 \times 10^{8} \times (2.012 - 2) \times 10^{-10}}{2 \times 10^{-10} \times 2.012 \times 10^{-10}}$$

$$= 5.91 \times 10^{-18} J.$$

Example 7

Calculate the maximum percentage change in wavelength due to Compton scattering for incident photons of wavelengths 1 Å and 10 Å. What inference do you draw from the result?

Solution

The wavelength shift by Compton scattering is given by

$$\Delta\lambda = \frac{h}{m_0 c}(1 - \cos\phi)$$

For maximum value of $\Delta\lambda$, $\phi = \pi$

$$\therefore \quad (\Delta\lambda)_{max} = \frac{h}{m_0 c}(1 + 1)$$

$$= 2 \times 0.242 = 0.0484 \text{ Å}$$

Maximum percentage change in wavelength

$$= \frac{(\Delta\lambda)_{max}}{\lambda} \times 100 = \frac{0.0484}{1} \times 100 \quad \text{(for } \lambda = 1 \text{ Å)}$$

$$= 4.8\%$$

and for $\lambda = 10 \text{ Å}$, maximum change in wavelength

$$= \frac{0.0484}{10} \times 100 = 0.48\%$$

Obviously, $(\Delta\lambda)_{max}/\lambda$ is very small for larger wavelengths ~10 Å and hence it is very difficult to detect this change in wavelength, i.e. Compton effect can be observed only for photons whose wavelength is sufficiently small.

Example 8

Show that:

(a) the maximum recoil energy of a free electron of rest mass m_0, when struck by a photon of frequency ν is given by

$$K_{max} = \frac{(h\nu)^2}{h\nu + (1/2)m_0c^2}$$

(b) If λ is the wavelength of the photon and λ_c is the Compton wavelength of the electron, then show that

$$K_{max} = \frac{2m_0c^2\lambda^2}{\lambda^2 + 2\lambda_c\lambda}$$

Solution

(a) The kinetic energy of the recoil electron

$$K = h\nu \frac{\dfrac{h\nu}{m_0c^2}(1-\cos\phi)}{1+\dfrac{h\nu}{m_0c^2}(1-\cos\phi)}$$

$$= \frac{(h\nu^2)\,2\sin^2\phi/2}{m_0c^2 + h\nu\,2\sin^2\phi/2}$$

$$= \frac{2(h\nu^2)}{2h\nu + m_0c^2\cosec^2\phi/2}$$

K will be maximum, when $\cosec^2\phi/2$ is minimum. The minimum value of $\cosec^2\phi/2$ is 1 and this value occurs when $\phi/2 = 90°$, i.e. when $\phi = 180°$.

∴ The maximum value of K is given by

$$K_{max} = \frac{2(h\nu)^2}{2h\nu + m_0c^2} = \frac{(h\nu)^2}{h\nu + \dfrac{1}{2}m_0c^2} \qquad (1)$$

(b) From Eq. (1), one obtains

$$K_{max} = \frac{2(m_0c)^2(h/m_0c)^2\nu^2}{2(m_0c)(h/m_0c)\,\nu + m_0c^2}$$

Substituting $\dfrac{h}{m_0c} = \lambda_0$ and $\nu = \dfrac{c}{\lambda}$ and

simplifying, one obtains

$$K_{max} = \frac{2(m_0c)^2\lambda_c^2\dfrac{c^2}{\lambda^2}}{2\lambda_c\dfrac{c}{\lambda+c}}$$

$$= \frac{2m_0\lambda_c^2c^2}{2\lambda\lambda_c + \lambda^2} = \frac{2m_0c^2\lambda^2}{\lambda^2 + 2\lambda_0\lambda}.$$

Example 9

Show that the de Broglie wavelength associated with an electron of energy V electron volts is approximately $\dfrac{12.27}{\sqrt{V}}$ Å.

Solution

We have $\lambda = \dfrac{h}{\sqrt{2meV}}$

Substituting the values of h, m and e, one obtains

$$\lambda = \frac{6.6\times10^{-34}}{\sqrt{2\times9.1\times10^{-34}\times1.6\times10^{-19}V}}$$

$$= \frac{12.27\times10^{-10}}{\sqrt{V}} = \frac{12.27}{\sqrt{V}}\text{Å}.$$

Example 10

A particle of rest mass m_0 has a kinetic energy K. Show that its de Broglie wavelength is given by

$$\lambda = \frac{hc}{\sqrt{K\left(K+2m_0c^2\right)}}$$

Hence, calculate the wavelength of an electron of kinetic energy 1 MeV. What will be the value of λ if $K \ll m_0c^2$?

Solution

The de Broglie wavelength associated with a particle moving with velocity v is given by

$$\lambda = \frac{h}{mv} \qquad (1)$$

where m is relativistic mass of the particle given by

$$m = \frac{m_0}{\sqrt{1 - v^2/c^2}} \qquad (2)$$

$$\therefore \quad 1 - \frac{v^2}{c^2} = \frac{m_0^2}{m^2}$$

or

$$\frac{v^2}{c^2} = 1 - \frac{m_0^2}{m^2} = \frac{(m^2 - m_0^2)}{m^2}$$

or

$$mv = c\sqrt{m^2 - m_0^2} \qquad (3)$$

From Eqs. (1) and (3), we obtain

$$\lambda = \frac{h}{c\sqrt{m^2 - m_0^2}} = \frac{hc}{c^2\sqrt{m^2 - m_0^2}} \qquad (4)$$

But $\quad c^2\sqrt{m^2 - m_0^2} = \sqrt{c^4(m - m_0)(m + m_0)}$

$$= \sqrt{\{(m - m_0)c^2\}\{c^4(m - m_0)c^2 + 2m_0c^2)\}}$$

$$= \sqrt{K(K + 2m_0c^2)}$$

$$\therefore \quad K = (m - m_0)c^2$$

$$\therefore \quad \lambda = \frac{hc}{\sqrt{K(K + 2m_0c^2)}}$$

Given $m_0c^2 = 0.5$ MeV, so that when $K = 1$ MeV, we have

$$\therefore \quad \lambda = \frac{6.62 \times 10^{-34} \times 3 \times 10^8}{1.414 \times 1.6 \times 10^{-13}}$$

$$= 8.78 \times 10^{-13} \text{ m} = 8.78 \times 10^{-3} \text{ Å}$$

If $K \ll m_0c^2$, one can replace $K + m_0c^2$ by m_0c^2 and hence one obtains

$$\lambda = \frac{hc}{\sqrt{2m_0Kc^2}} = \frac{h}{\sqrt{2m_0K}}.$$

Example 11

Show that the de Broglie wavelength for a material particle of rest mass m_0 and charge q accelerated from rest through a potential difference of V volts relativistically is given by

$$\lambda = \frac{h}{\sqrt{\left[2m_0qV\left(1 + \frac{qV}{2m_0c^2}\right)\right]}}$$

Solution

In the given case, the kinetic energy $T = qV$ as in nonrelativistic case, but $T \neq \frac{1}{2}mv^2$. Obviously, one cannot find momentum directly from T. Instead, one has to use the relativistic formula.

$$E^2 = p^2c^2 + m_0^2c^4$$

$$E = T + m_0c^2 = qV + m_0c^2$$

where m_0 = rest mass of electron and T = kinetic energy

$$\therefore \quad p^2c^2 = E^2 - m_0^2c^4 = \left(qV + m_0c^2\right)^2 - m_0^2c^4$$

$$= q^2V^2 + 2m_0c^2qV$$

$$\therefore \quad p^2 = 2m_0qV\left(1 + \frac{qV}{2m_0c^2}\right)$$

or

$$p = \sqrt{\left[2m_0qV\left(1 + \frac{qV}{2m_0c^2}\right)\right]}$$

$\therefore$ de Broglie wavelength will be

$$\lambda = \frac{h}{p} = \frac{h}{\sqrt{\left[2m_0qV\left(1 + \frac{qV}{2m_0c^2}\right)\right]}}.$$

Example 12

Assume that the uncertainty in position of a hydrogen molecule of mass about 2×10^{-27} kg is of the order of the diameter about 10^{-10} m. Determine the uncertainty in momentum.

Solution

The uncertainty in momentum is given by

$$\Delta p_x \geq \frac{\hbar}{\Delta x} \geq \frac{1.054 \times 10^{-34}}{10^{-10}} \geq 1.054 \times 10^{-24} \, \text{kg m/sec.}$$

Example 13

The average period that elapes between the excitation of an atom and the time it emits radiation is 10^{-8} second. Find the uncertainty in the energy emitted and the uncertainty in the frequency of light emitted. Given

$$\hbar = 1.054 \times 10^{-34} \, \text{Js and } h = 6.63 \times 10^{-34} \, \text{Js.}$$

Solution

The uncertainty in the energy is given by

$$\Delta E = \frac{\hbar}{\Delta t} = \frac{1.054 \times 10^{-34}}{10^{-8}} = 1.054 \times 10^{-26} \, \text{J}$$

The uncertainty in the frequency of light emitted is given by

$$\Delta v = \frac{\Delta E}{h} = \frac{1.054 \times 10^{-26}}{6.63 \times 10^{-34}} = 1.59 \times 10^{7} \, \text{Hz.}$$

Example 14

A nucleon is confined to a nucleus of radius 5×10^{-15} m. Calculate the minimum uncertainty in the momentum of the nucleon. Also calculate the minimum kinetic energy of the nucleon. Given, mass of nucleon = 5×10^{-15} m and Planck's constant $h = 6.6 \times 10^{-34}$ Js.

Solution

We have $(\Delta p)_{min} (\Delta x)_{max} \approx h$

Given $(\Delta x)_{max} = 2 \times 5 \times 10^{-15} \, \text{m} = 10^{-14} \, \text{m}$

$$(\Delta p)_{min} = \frac{h}{(\Delta x)_{max}} = \frac{6.6 \times 10^{-34}}{10^{-14}}$$

$$= 6.6 \times 10^{-20} \, \text{kg m/sec}$$

Since p cannot be less than $(\Delta p)_{min}$, so we have

$$p_{min} = (\Delta p)_{min}$$

$$E_{min} = \frac{p_{min}^2}{2m} = \frac{\left(6.6 \times 10^{-20}\right)^2}{2 \times 1.67 \times 10^{-27}} = 1.32 \times 10^{-12} \, \text{J}$$

$$= 8.25 \times 10^6 \, \text{eV} = 8.25 \, \text{MeV.}$$

Example 15

Find the lowest energy of an electron confined to move in one dimensional potential box of length 1 Å.

Solution

Given $\quad E_n = \frac{n^2 \pi^2 \hbar^2}{2 m a^2}$

$$\therefore \quad E_1 = \frac{1^2 \pi^2 \hbar^2}{2 m a^2}$$

For $\quad E_1 = \frac{\pi^2 \times (1.054 \times 10^{-34})^2}{2 \times 9.1 \times 10^{-31} \times (10^{-10})^2}$

$$E_1 = \frac{(1.054)^2 \times (3.14)^2 \times 10^{-17}}{2 \times 9.1} = 6 \times 10^{-18} \, \text{J}$$

$$E_1 = \frac{6 \times 10^{-18}}{1.6 \times 10^{-19}} \, \text{eV} = 35.5 \, \text{eV.}$$

Example 16

The size of the nucleus is 10^{-14} m. Treating it as a one dimensional box show why electron does not exist inside the nucleus.

Solution

Zero point energy $E_1 = \frac{\pi^2 \hbar^2}{2 m a^2}$

$$= 5.44 \times 10^{-10} \, \text{J}$$

$$\left(\hbar = 1.054 \times 10^{-34} \, \text{Js}, m = 9.1 \times 10^{-31} \, \text{kg,} \right.$$

$$\left. a = 10^{-14} \, \text{m} \right)$$

The coulombic attraction energy, on the other hand, is

$$V = \frac{e^2}{r} = -2.3 \times 10^{-14} \, \text{J}$$

The coulombic attraction energy is too weak for the nucleus to hold an electron.

Example 17

Calculate the length of a one dimensional box for which the difference between the lowest energy levels of a molecule becomes comparable to its average kinetic energy at a given temperature.

Solution

$$E_{n+1} - E_n = (2n+1)\frac{\pi^2\hbar^2}{2ma^2} = \frac{1}{2}k_BT$$

where k_B is Boltzmann's constant

$$\therefore \quad a = \left[\frac{(2n+1)\pi^2\hbar^2}{mk_{B/T}}\right].$$

Example 18

For a particle in the states $n = 1, 2$ and 3 of a one dimensional box of length L, find the probability that the particle is in the region $0 < x < a/4$.

Solution

Probability $P = \dfrac{2}{a}\displaystyle\int_0^{1/4} \sin\frac{2n\pi x}{a}\,dx$

$$\text{(using } \sin^2 x = \frac{1-\cos 2x}{2}\text{)}$$

$$= \frac{1}{4} - \frac{\sin n\pi/2}{2n\pi}$$

For $n = 1$, $P = \dfrac{1}{4} - \dfrac{1}{2}\pi = 0.0908$

For $n = 2$, $P = 0.25$ and

For $n = 3$, $P = 0.3031$

Example 19

A particle of mass $m = 10^{-29}$ kg is confined to move in a box of length 2 Å whose ends are at $x = 0$ and $x = 2$ Å. What is the probability of finding the particle between 1.6000 Å and 1.6001 Å for $n = 1$ and $n = 2$.

Solution

$$P = \frac{2}{a}\int_{1.6}^{1.6001} \sin\frac{2n\pi x}{a}\,dx$$

$$= \frac{1}{a}\left[x - \left\{a\left(\sin\frac{2n\pi x}{a}\right)\right\}\middle/2n\pi\right]_{1.6000}^{1.6001}$$

For $n = 1$,

$$P = \frac{1}{2.0}\left[0.0001 - \frac{\begin{array}{c}2\times\sin(2\pi\times1.6001/2)\\-\sin(2\pi\times1.6001/2)\end{array}}{2\pi}\right]$$

$$= 3.45 \times 10^{-5}$$

Similarly, for $n = 2$, one obtains $P = 9.05 \times 10^{-5}$.

Example 20

(a) Calculate the lowest energy of an electron confined in a cubical box of each side 1 Å.

(b) Find the temperature at which the average energy of the molecules of a perfect gas would be equal to the lowest energy of the electron, $k_B = 1.38 \times 10^{-23}$ J/K.

Solution

(a) The possible energies of a particle in a cubical box of each side a are given by

$$E = \frac{3\pi^2\hbar^2}{2ma^2}\left(n_x^2 + n_y^2 + n_z^2\right)$$

For the lowest energy

$$n_x = n_y = n_z = 1$$

$$\therefore \ E_1 = \frac{3\pi^2\hbar^2}{2ma^2} = \frac{3\pi^2\times(10^{-10})^2}{2\times9.11\times10^{-31}\times(10^{-10})^2}$$

$$= \frac{3\times(1.054)^2\times(3.14)^2\times10^{-17}}{2\times9.11}$$

$$= 1.803\times10^{-17}\ \text{J}$$

$$= 108\ \text{eV}$$

(b) Let T K be the required temperature, then

$$\frac{3}{2}k_BT = \frac{3\pi^2\hbar^2}{2ma^2} = 18.03\times10^{-18}\ \text{J}$$

$$\therefore \quad T = \frac{2}{3} \times \frac{18.03 \times 10^{-18}}{1.38 \times 10^{-23}} = 8.71 \times 10^5 \text{ K}.$$

Note: The above model is not acceptable for H atom as the experimental value of ground state energy E_{111} is –13.6 eV.

Example 21

A beam of mono-energetic electrons strikes the surface of a metal at normal incidence. Calculate the reflection probability of these electrons if $E = 0.1$ eV and $V_0 = 8$ eV.

Solution

Since the electrons with energy 0.1 eV encounters a potential drop –8 eV at the metal surface, they will all enter the metal according to classical mechanics and because of the law of conservation of energy they will acquire a final momentum $p_1 = \sqrt{2m(E - V_0)}/\hbar$ after doing so. From potential step expressions, we have

$$R = \frac{(p_2 - p_1)^2}{(p_2 + p_1)^2} \quad \text{and} \quad T = \frac{4p_1 p_2}{(p_2 + p_1)^2}$$

Thus for $E = 0.1$ eV, $V_0 = 8$ eV

$$R = \left[\frac{\sqrt{E} - \sqrt{(E - V_0)}}{\sqrt{E} + \sqrt{(E - V_0)}}\right]^2 = \left[\frac{1 - \sqrt{(1 - V_0/E)}}{1 + \sqrt{(1 - + V_0/E)}}\right]^2$$

$$= \left(\frac{8}{10}\right)^2 = 0.64.$$

Example 22

Find the width of the potential barrier for an α-particle emitted with kinetic energy 5.5 MeV from $^{222}_{86}$Rn.

Solution

The width of the potential barrier due to the nucleus is given by $a = r_1 - r_0$ where r_0 is the effective nuclear radius and r_1 is the radius from the centre of the nucleus, at which the potential energy of the α-particle is equal to the kinetic energy of the emitted α-particle.

Here $r_0 = 1.5 \times 10^{-15} A^{1/3} = 1.4 \times 10^{-15} \times (222)^{1/3}$

$$= 9.08 \times 10^{-15} \text{ m}.$$

r_1 is obtained from

$$E = \frac{2(Z - 2)e^2}{4\pi\varepsilon_0 r_1} \quad \therefore \quad r_1 = \frac{2(Z - 2)e^2}{4\pi\varepsilon_0 E}$$

$$= \frac{2 \times (86 - 2) \times (1.6 \times 10^{-19})^2}{4\pi \times 8.85 \times 10^{-12} \times 5.5 \times 1.6 \times 10^{-13}}$$

$$= 34.97 \times 10^{-15} \text{ m}$$

Now $a = r_1 - r_0 = (43.97 \times 10^{-15} - 9.08 \times 10^{-15})$ m

$$= 43.98 \times 10^{-15} \text{ m}.$$

Example 23

A stream of electrons, each of energy $E = 3$ eV is incident on a potential barrier of height $V_0 = 4$ eV. The width of the barrier is 20 Å. Calculate the percentage transmission of the beam through this barrier.

Solution

The probability of transmission through a potential barrier is given by

$$T = \frac{16E}{V_0}\left(1 - \frac{E}{V_0}\right)e^{-2\beta a}$$

where

$$2\beta a = 2a\sqrt{\frac{2m(V_0 - E)}{\hbar^2}} = \frac{2a}{\hbar^2}\sqrt{2m(V_0 - E)}$$

Given $E = 3$ eV $= 3 \times 1.6 \times 10^{-19}$J

$$V_0 = 4 \text{ eV} = 4 \times 1.6 \times 10^{-19}\text{J}$$

$$a = 20 \text{ Å} = 20 \times 10^{-10} = 2 \times 10^{-9} \text{ m}$$

$$\therefore \quad 2\beta a = \frac{2a}{\hbar}\sqrt{2m(V_0 - E)} = 20.49$$

$$\therefore \quad T = \frac{16E}{V_0}\left(1 - \frac{E}{V_0}\right)e^{-2\beta a}$$

$$= \frac{16 \times 3}{4}\left(1 - \frac{3}{4}\right)e^{20.49}$$

$$= 12 \times \frac{1}{4} \times \frac{1}{(2.718)^{20.49}} = 3.797 \times 10^{-9}$$

$\therefore$ Percentage transmission
$$= 3.797 \times 10^{-7}\%.$$

Example 24

An α-particle having energy 10 MeV approaches a potential barrier of height equal to 30 MeV. Determine the width of the potential barrier if the transmission coefficient is 2×10^{-3}. Given mass of a α-particle $= 4 \times 1.67 \times 10^{-27}$ kg and $\hbar = 1.054 \times 10^{-34}$ J s.

Solution

$$T = \frac{16E}{V_0}\left(1 - \frac{E}{V_0}\right)e^{-2\beta a}$$

and $\quad e^{2\beta a} = \frac{16E}{TV_0}\left(1 - \frac{E}{V_0}\right)$

$\therefore \quad 2\beta a = 2.303 \log_{10}\dfrac{16E}{TV_0}\left(1 - \dfrac{E}{V_0}\right)$

or $\quad a = \dfrac{1}{2\beta} \times 2.303 \log_{10}\dfrac{16E}{TV_0}\left(1 - \dfrac{E}{V_0}\right)$

$$\beta = \frac{1}{h}\sqrt{2(E - V_0)}$$

$$= \frac{1}{1.054 \times 10^{-34}} \times \sqrt{2 \times 4 \times 1.67 \times 10^{-27}}$$

$$\times \sqrt{(30 - 10) \times 10^6 \times 1.6 \times 10^{-19}}$$

Hence

$$\beta = \frac{10^{14}}{1.054 \times 10^{-34}} \times \sqrt{2 \times 4 \times 1.64 \times 20 \times 1.6}$$

$$= \frac{20.68 \times 10^{14}}{1.054} = 19.62 \times 10^{14}$$

$$a = \frac{10^{-14}}{2 \times 19.62}$$

$$\times 2.3031 \log_{10}\frac{16 \times 10}{2 \times 10^{-3} \times 30}\left(1 - \frac{10}{30}\right)$$

$$= \frac{10^{-14}}{2 \times 19.62} \times 2.3031 \log_{10}\frac{16}{9} \times 10^3$$

$$= \frac{10^{-14} \times 2.303 \times 3.25}{39.24} = 1.91 \times 10^{15} \text{ m}$$

PROBLEMS

1. Explain the distribution of energy in blackbody spectrum. How far the classical theories explain the distribution? Write a note on Planck's law of distribution.

2. State Wien's displacement law and Rayleigh-Jean's law of radiation and indicate how far the two laws could explain the curves.

3. Discuss Planck's hypothesis for a blackbody radiation. Show that the average energy of Planck's oscillator of frequency ν in thermal equilibrium with a heat reservoir of temperature T is given by

$$\bar{\varepsilon} = \frac{h\nu}{[\exp(h\nu/k_BT) - 1]}$$

4. Discuss the distribution of energy in blackbody radiation and explain the success and limitations of the classical theory in explaining it. Derive Planck's radiation formula, explaining how the shortcomings of classical theory were overcome.

5. What is photoelectric effect? Give an account of the photoelectric emission of the electrons. Give Einstein's interpretation for the same.

6. How Einstein's photoelectric equation explains the phenomenon of photoelectric effect? What information is derived for the nature of light?

7. What are the laws of photoelectric effect? How has photoelectric effect been explained by Einstein?

8. What is Compton effect? Derive an expression for the change in wavelength of a photon when it is scattered by an electron. Calculate the maximum change in wavelength in a Compton scattering experiment.

9. Describe Compton effect. Derive an expression for the frequency of the scattered photon in terms of the frequency of incident radiation and scattering angle.

10. (a) Explain why Compton effect is not observed experimentally for visible rays?

(b) Both photoelectric effect and Compton effect arise due to action of photons on electrons but the two effects are not the same. Explain.

(c) Explain the presence of unmodified radiation in Compton scattering.

11. Explain the difference between photoelectric effect and Compton effect.

12. Explain Compton scattering. Derive an expression for Compton scattering.

13. What is de Broglie hypothesis? Show that the wavelength λ associated with a particle of mass m and kinetic energy E is given by $\lambda = h / \sqrt{2meE}$, where h is Planck's constant.

14. Describe Davisson and Germer's experiment and indicate the important conclusions.

15. (a) Draw the schematic diagram of Davisson and Germer apparatus to study the electron diffraction. Explain how the experiment was carried out.

(b) Show that the results of this experiment are closely in agreement with the electron wavelength calculated by the de Broglie relation.

16. State Heisenberg's uncertainty principle. Is this principle the outcome of the wave description of a particle? Give atleast one experiment to prove its validity.

17. (a) What is correct statement of Heisenberg's uncertainty principle? Discuss its significance and importance.

(b) From the uncertainty principle, show that the electrons cannot reside in the nucleus of the atom. Given that nuclear radius = 10^{14} m, and the maximum possible kinetic energy of an electron in an atom = 4 MeV.

18. Show that the photon in a 1240 nm infrared light beam have energies of 1 eV.

$$\left[\text{Hint: } h\nu = \frac{hc}{\lambda} = \frac{6.63 \times 10^{-34} \times 3 \times 10^8}{1240 \times 10^{-9}} \right.$$

$$= 1.6 \times 10^{-9} \text{ J} = 1 \text{ eV} \Big]$$

19. Calculate the energy of a photon of blue light of wavelength 450 nm.   [*Ans.* 2.76 eV]

$$\left[\text{Hint: } E = \frac{hc}{\lambda} = \frac{6.63 \times 10^{-34} \times 3 \times 10^8}{450 \times 10^{-9}} \right.$$

$$= 4.42 \times 10^{-9} \text{ J} = 2.76 \text{ eV} \Big]$$

20. In order to break a chemical bond in the molecules of human skin, causing sunburn, a photon of energy of about 3.5 eV is required. To what wavelength does this correspond?

21. Calculate the average energy of an oscillator with frequency 10 Hz (s^{-1}) at a temperature 1000 K.   [*Ans.* 5.5 × 10^{-20} J]

22. Planck's law of distribution of energy in the blackbody radiation is given by the equation

$$E(\nu)d\nu = \frac{8\pi h\nu^3}{c^3} \frac{d\nu}{\exp(h\nu/k_B T) - 1}$$

where $E(\nu)$ is the energy associated with the radiation of frequency v. Show that (a) the total energy of all radiations from the blackbody is proportional to the fourth power of the temperature of the body and (b) the wavelength corresponding to the maximum energy density is inversely proportional to the temperature.

23. The work function of sodium metal is 2.3 eV. What is the longest wavelength of light that can cause photoelectron emission from sodium?   [*Ans.* λ = 540 nm]

24. What potential difference must be applied to stop the fastest photoelectrons emitted by a nickel surface under the action of ultraviolet light of wavelength 2000 Å? The work function of nickel is 5.01 eV.

[**Hint:** Energy of photon

$$= \frac{hc}{\lambda} = \frac{6.6 \times 10^{-34} \times 3 \times 10^8}{2000 \times 10^{-10}}$$

$$= 9.95 \times 10^{-19} \text{ J} = 6.21 \text{ eV}$$

Then from the photoelectric equation, the energy of the fastest emitted electron is 6.21 eV – 5.01 eV = 1.20 eV. Obviously, a negative retarding potential of 1.2 eV is required]

25. Will photoelectrons be emitted by a copper surface, of work function 4.4 eV, when illuminated by visible light?   [*Ans.* No]
[**Hint:** Threshold

$$\lambda = \frac{hc}{W_{min}} = \frac{6.63 \times 10^{-34} \times 3 \times 10^8}{4.4 \times 1.6 \times 10^{-19}} = 282 \text{ nm} .$$

Obviously, visible light (400 nm to 700 nm) cannot eject photoelectrons from copper]

26. Suppose that a 3.64 nm photon going in the +X-direction collides head on with a 2 × 10^5 m/s electron moving in the −X-direction. If the collision is perfectly elastic, find the conditions after collision.

[Hint: From the law of conservation of momentum,

Momentum before collision = Momentum after collision

$$\frac{h}{\lambda_0} - mv_0 = \frac{h}{\lambda} - mv$$

But $h/\lambda_0 = mv_0$ in this case and hence $h/\lambda = mv$. Also for a perfectly elastic collision,

$$(KE)_{before} = (KE)_{after}$$

$$\frac{hc}{\lambda_0} + \frac{1}{2}mv_0^2 = \frac{hc}{\lambda} + \frac{1}{2}mv^2$$

Using $h/\lambda_0 = mv = mv_0$, the photon also rebounds with its original wavelength].

27. A photon ($\lambda = 0.400$ nm) strikes an electron at rest and rebounds at an angle of 150° to its original direction. Find the speed and wavelength of the photon after the collision. **[*Ans.* 0.4045 nm]**

28. An electron falls from rest through a potential difference of 100 V. What is its de Broglie wavelength? **[*Ans.* 12.3 nm]**

29. A photon of frequency v is scattered by an electron of mass m_0. The scattered photon of frequency v' travels in a direction inclined at 90° with the direction. Prove that the de Broglie wavelength of the recoil electron is given by

$$\lambda_r = \frac{C}{\sqrt{v^2 + v'^2}}$$

30. Show that the wavelength λ associated with an electron of mass m_0 and kinetic energy k is given by

$$\lambda = \frac{h}{\sqrt{2m_0K}}$$

31. What potential difference is required in an electron microscope to give electrons a wavelength of 0.5 Å?
[Hint: KE of electron

$$= \frac{1}{2}mv^2 = \frac{1}{2}m\left(\frac{h}{m\lambda}\right)^2 = \frac{h^2}{2m\lambda^2}$$

$$= \frac{\left(6.63 \times 10^{-34}\right)^2}{2 \times 9.1 \times 10^{-31} \times \left(0.5 \times 10^{-10}\right)^2}$$

$$= 9.66 \times 10^{-17} \text{ J}$$

But KE = qV

$$\therefore \quad V = \frac{KE}{q} = \frac{9.66 \times 10^{-17} \text{ J}}{1.6 \times 10^{-19} \text{ C}} = 600 \text{ V} \Big]$$

32. A beam of X-rays is scattered by loosely bound electrons at 45° from the diffraction of the beam. The wavelength of the scattered X-rays is 0.22 Å. What is the wavelength of X-rays in the direct beam? **[*Ans.* 0.2129 Å]**

$$\Big[\textbf{Hint:} \; \lambda = \lambda' - \frac{h}{m_0 c}(1 - \cos\phi)$$

$$= 0.22 - 0.02426\left(1 - 1/\sqrt{2}\right)\Big]$$

33. X-rays with $\lambda = 2.00$ Å are scattered from a carbon block and the scattered radiation is viewed at 90° to the incident beam. Calculate the Compton shift. **[*Ans.* 0.02426 Å]**

34. An X-ray photon of frequency 1.5×10^{19} Hz is scattered by a free electron and the frequency of the scattered photon is 1.2×10^{19} Hz. Find the kinetic energy imparted to the electron. **[Hint:** $K = hv - hv'$] **[*Ans.* 1.989×10^{-15} J]**

35. Assume that an electron is inside a nucleus of radius 10^{-15} m. From the uncertainty principle, estimate the kinetic energy of the electron.
[*Ans.* 0.95×10^{10} eV]

$$\Big[\textbf{Hint:} \; \Delta x \sim 10^{-15}m\Delta x \Delta p \sim \frac{h}{4\pi\Delta x}$$

$$= \frac{6.6 \times 10^{-34} \text{ kg m}}{4\pi \times 10^{-15} \text{ sec}}$$

Since the magnitude of the momentum cannot be less than that of uncertainty, the momentum of the electron will at least be equal to Δp. Thus, the minimum KE of the electron will be

$$\frac{(\Delta p)^2}{2m} = \frac{\left(6.63 \times 10^{-19}\right)^2}{2 \times \left(9.1 \times 10^{-31}\right)(4\pi)^2} \text{ J} \sim 0.95 \times 10^{10} \text{ eV}\Big]$$

36. Deduce the time independent Schrodinger equation. Explain the physical significance of wave function.

37. (a) Derive the time independent Schrodinger wave equation for a particle.

(b) Give the physical meaning of the wave function. What do you understand by normalised wave function?

38. (a) Obtain the time independent Schrodinger wave equation for a particle.

(b) Separate the wave function into time dependent and time-independent parts and obtain the steady state Schrodinger equation.

39. Show that one-dimensional Schrodinger equation has a solution of the form $\phi(x)\,\psi(t)$ if the potential energy of the particle associated with the wave function is independent of time. Show that the most general solution of Schrodinger equation for such a particle can be written as

$$\phi = \sum_{n=1}^{\infty} a_n \psi_n(x) e^{-iEt/\hbar}$$

40. (a) Why should any wave function be finite, continuous and differentiable at every point?

(b) When and why do we call a solution of Schrodinger equation a stationary one? Show that they occur only if the potential energy does not depend explicitly on the time.

41. Obtain Schrodinger time dependent wave equation and separate it into space and time dependent parts. Given the probability interpretation of the wave function, show that the probability density ρ and the probability current density J satisfy the continuity equation. Explain the physical significance of this equation.

$$\frac{d\rho}{dt} + \nabla \cdot J = 0$$

42. Explain why is it possible to choose probability current-density in quantum mechanics by the following expression.

$$J = \frac{e\hbar}{2m}\left(\psi^* \nabla \psi - \psi \nabla \psi^*\right)$$

The symbols have their usual meanings.

43. Distinguish between phase velocity and group velocity. Show that for a nonrelativistic free particle the phase velocity is half of the group velocity.

44. Write down Schrodinger equation for a particle in a one dimensional box. Solve it to obtain eigen functions and show that the eigenvalues are discrete.

45. Write one dimensional Schrodinger time independent equation for a particle moving in a potential $V(x)$. Explain its significance. What are stationary state solutions?

46. Write down Schrodinger's equation for a particle in a three dimensional box. Solve it to obtain eigen functions and show that the eigenvalues are discrete. Explain clearly the meaning of degeneracy of levels.

47. Explain clearly what do you understand by degeneracy of levels. Explain the statement "for a degenerate energy level there can be more than one different wave functions". Illustrate it by taking the example of particle in a box.

48. Formulate the problem of square well potential by writing the appropriate potential and Schrodinger equations. Obtain the solutions and an expression for the transmissibility (transmission coefficient). Discuss the expression with special emphasis on resonance.

49. A particle travelling with energy E along X-axis has in its path a rectangular potential barrier of height $V > E$ and width a. Calculate the transmission coefficient of the particle and discuss briefly its application to the observed phenomenon of α-decay in nuclei.

50. Determine the transmission coefficient for a particle of energy $E < V_0$ for a rectangular one dimensional potential barrier given by

$$V = 0 \text{ for } x < -a \text{ and } x > a$$
$$V = V_0 \text{ for } -a < x < a$$

Discuss briefly its application to the observed phenomenon of α-decay.

51. Explain the problem of leaking of a particle through a rectangular potential barrier of finite width and explain the theory of α-decay.

52. (a) What do you mean by tunnelling through a barrier? A particle travelling with energy E along X-axis has a potential barrier defined as

$$V(x) = \begin{cases} 0 \text{ for } x < 0 \\ V_0 \text{ for } 0 < x < a \\ 0 \text{ for } x > a \end{cases}$$

Derive an expression for the equation and transmission coefficients of the particle.

(b) Show that coefficient of reflection of particles incident on barrier is same whether they approach from right or left.

53. A beam of particles with energy E is incident on a potential barrier with potential function

$$V(x) = \begin{cases} 0 & \text{for } x < 0 \\ V_0 & \text{for } 0 < x < a \\ 0 & \text{for } x > a \end{cases}$$

Show that there is a finite probability of transmission even if $E < V_0$.

54. Describe the tunnelling phenomenon in quantum mechanics. Obtain the transmission coefficient for a particle of energy E crossing the barrier defined by $V(x) = 0$ for $x < 0$, $V(x) = V_0$ for $0 < x < a$ and $V(x) = 0$ for $x > a$. Comment on the solution when the thickness is a multiple of the de Broglie wavelength.

55. Describe the tunnelling phenomenon in quantum mechanics. Show that tunnelling increases if the height and width of the barrier are reduced.

56. The width and depth of a finite square well potential are a and V_0 respectively. Derive an expression for the transmission coefficient of a particle having positive energy. What are transmission resonances?

57. Show that the wave functions for a particle in one dimensional box, $\psi = A \sin n\pi x / L$, are not eigen functions of the linear momentum operator p_x, but they can be written as combinations of the eigen functions of p_x.
   [**Hint:** $\sin n\pi x / L = 1/2i \, (\exp)(in\pi x/\ell) - \exp(-in\pi x/\ell)$
   Both exponential terms on the right hand side can be easily shown to be eigen functions of p_x with eigenvalues $\pm 2\pi n\hbar/2L$).]

58. Verify that $\psi = A \sin kx + B \cos kx$ is a general solution for a particle in a one dimensional box having infinite potential walls.

59. Find the energies of the six lowest energy levels of a particle in a cubical box. Which of the levels are degenerate?
   [**Ans.** 3, 6, 9, 11, 12 and $14 \times \pi^2 \hbar^2 / 2mL^2$, 1st and 5th are nondegenerate; 2nd, 3rd and 4th are threefold degenerate and the 6th is sixfold degenerate.]

60. Determine the degree of degeneracy of the energy level $\dfrac{38\pi^2 \hbar^2}{2mL^2}$ of a particle in a cubical potential box.
   [**Ans.** The quantum numbers (611), (161), (116), (532), (523), (352), (325), (253), and (235) give the same energy levels, therefore, the level is ninefold degenerate.]

61. An electron is confined to one dimensional box of side 1 Å. Obtain the first four eigenvalues in eV of the electron.
   [**Ans.** 38 eV, 152 eV, 342 eV, 608 eV]

62. An electron is confined to move between two rigid walls separated by 10^{-9} m. Find the de Broglie wavelengths representing the first three allowed energy states of the electron and the corresponding energies.

63. Calculate the de Broglie wavelength associated with an electron associated by 5000 volts. Given mass of the electron = 9.1×10^{-28} gm, Planck's constant = 6.6×10^{-34} J s. [Raj. BE 90]

64. (a) What is Compton scattering? Obtain an expression for shift in wavelength of the scattered photon by Compton scattering? [Raj. BE 94, 97, 98]

   (b) Calculate maximum percentage change in wavelength due to Compton scattering for incident photons of wavelengths 1 Å, 10 Å and 100 Å. What inference do you draw from this calculation? [BE]

65. (a) Explain Heisenberg's uncertainty principle for (i) momentum and position (ii) energy and time. Illustrate its validity with the help of one example of each.

   (b) Derive Schrodinger time independent wave equation. What is the physical significance of wave function ψ used in this equation? [Raj. BE 94]

   (c) Explain why Compton effect is not observed experimentally for visible rays. Explain Compton scattering.

   (d) State Heisenberg uncertainty principle. Explain the validity by any thought experiment.

   (e) What is tunnel effect? [Raj. BE 96]

66. Light described at a place by the equation:

   $$E = (100 \text{ volt/m}) \left[\sin\left(5 \times 10^{15}\right)t + \sin\left(8 \times 10^{15}\right)t \right]$$

   falls on a metal surface having work function 2.0 eV. Calculate the maximum kinetic energy of the photo electrons and the stopping potential (Planck's constant $h = 6.63 \times 10^{-34}$ J s). [Raj. BE 97]

67. Describe Schrodinger time independent wave equation and eigen physical interpretation of wave function ψ. [Raj. BE 97]

68. Write down Schrodinger equation and solve it to determine the eigenvalues and eigen functions for a particle in a box. [Raj. BE 99]

69. Explain the meaning of the terms degeneracy and tunnelling. [Raj. BE 98]

70. Write short notes on:

    (a) Heisenberg uncertainty principle. [Raj. BE 98]

    (b) Photoelectric effect. [Raj. BE 99]

71. (a) Derive Schrodinger time independent wave equation. Give physical interpretation and essential requirements of wave function ψ used in this equation.

    (b) Consider an electron whose total energy 5 eV approaching a barrier whose height is 6 eV and width is 7 Å. Find out de Broglie wavelength of incident electron and probability of transmission through the barrier. (mass of electron = 9.1×10^{-31} kg, Planck's constant = 6.6×10^{-34} J). [Raj. BE 2000]

72. A measurement establishes the position of a proton with an accuracy $\pm 10^{-11}$ m. Find the uncertainty in the proton's position one second later. Assume proton's speed $v \ll c$. (mass of proton = 1.67×10^{-27} kg) [Raj. BE 2000]

73. Write short note on Compton effect and its significance. [Raj. BE 2000]

74. (a) What is Compton effect? Derive an expression for wavelength shift in Compton effect. [Raj. BE 2002]

    (b) Write short note on tunnel effect. [Raj. BE 2002]

75. (a) What is Compton scattering? Obtain an expression for shift in wavelength of the scattered photon by compton scattering.

    (b) In Compton scattering, the energy of an incident X-ray photon is 150 keV and that of scattered photon is 130 keV. Determine the angle of scattering.

    (c) State Heisenberg's uncertainty principle. [Raj. BE 2002]

76. Derive the Schrodinger time dependent equation. Explain the following:

    i. Hamiltonian of a free particle

    ii. Physical significance of wave function ψ.

    iii. Normalized and orthogonalized wave functions. [Raj. BE 2002]

77. (a) Describe Heisenberg's uncertainty principle and apply it to explain nonexistence of electron in nucleus?

    (b) Derive Schrodinger time independent wave equation. What is the physical significance of wave function ψ used in this equation?

(c) Find the lowest energy of an electron confined to move in a one dimensional potential box of length 1 Å. [Raj. BE 2000, 2001]

78. (a) Obtain an expression for the density of states for free electron gas in metal and hence find expression for the Fermi energy.

    (b) Calculate Fermi energy in copper assuming that each copper atom contributes one free electron to the electron gas. Given density of copper 8.94×10^3 kg/m^3 and atomic mass is 63.5 amu.

    (c) Write down Schrodinger equation for a particle of mass m trapped in one dimensional box of size a. Solve it for energy eigen values and eigen functions. How does solutions modify if particle were in three dimensional cubical box of side a? Find lowest energy of the following states: (i) nondegenerate, (ii) doubly degenerate, (iii) triply degenerate and (iv) sixfold degenerate for 3D cubical box. [Raj. BE 2001]

79. (a) What is tunnel effect? Write down Schrodinger equation for potential barrier problem and steps to find out the transmission coefficient of a particle having less energy than the height of potential barrier.

    (b) Show that the value of energy which a photon must have so that it may transfer half of its energy to an electron at rest is about 256 keV in a Compton scattering experiment. [Raj. BE 2003]

80. A measurement establishes the position of a proton with an accuracy of $\pm 10^{-11}$ meter. Find the uncertainty in the proton's position one second later. Assume proton's speed $v < c (m_p = 1.67 \times 10^{-27}$ kg). [Raj. BE 2000]

SHORT ANSWER QUESTIONS

1. What will be the de Broglie wavelength of an electron with a velocity of 10^7 m/s?

Ans. $\lambda = \dfrac{h}{mv} = \dfrac{6.63 \times 10^{-34} \text{ Js}}{(9.1 \times 10^{-31} \text{ kg}) \times (10^7 \text{ m/s})}$

$= 7.3 \times 10^{-11}$ m

2. What is a wave function?

Ans. The quantity whose variations make up matter waves is called the wave function (ψ). The value of the wave function associated with a moving body at the particular point (x, y, z) in space at the time t is related to the likelihood of finding the body at the time.

3. Why the wave function ψ itself has no direct physical significance?

Ans. ψ cannot be interpreted in terms of an experiment. The probability that something be in a certain place at a given time must lie between 0 (the object is definitely not there) and 1 (the object is definitely there). An intermediate probability, say 0.2 means that there is a 20% chance of finding the object. But the amplitude of a wave can be negative as well as positive and negative probability is meaningless. Hence, ψ by itself cannot be an observable quantity.

4. What $|\psi|^2$ signifies?

Ans. The probability of experimentally finding the body described by the wave function ψ at the point (x, y, z) at the time t is proportional to the value of $|\psi|^2$ there at t. A large value of $|\psi|^2$ means the strong possibility of the body's presence, while a small value of $|\psi|^2$ means the slight possibility of its presence. As long as $|\psi|^2$ is not actually zero (0) somewhere, however, there is a definite chance however small, of detecting it there.

5. A 10 g marble is in a box 10 cm across. What will be its permitted energies?

Ans. $m = 10 \text{ g} = 1 \times 10^{-2} \text{ kg}, L = 10 \text{ cm} = 1.0 \times 10^{-1} \text{ m}$

$$E_n = \frac{(n^2)(6.63 \times 10^{-34} \text{Js})^2}{(8)(1.0 \times 10^{-2} \text{kg})(1.0 \times 10^{-1} \text{m})^2}$$

$$= 5.5 \times 10^{-64} n^2 \text{J}$$

The minimum energy the marble can have is corresponding to $n = 1$. A marble with this kinetic energy has a speed of only 3.3×10^{-31} m/s and, therefore, one cannot distinguish it experimentally from a stationary marble.

6. Justify the statement: Classical mechanics is an approximation of quantum mechanics.

Ans. The fundamental difference between classical mechanics and quantum mechanics lies in what they describe. In classical mechanics, the future history of a particle is completely determined by its initial position and momentum together with the forces that act upon it. In the everyday world, one can determine these well enough for the predictions of Newtonian mechanics to agree with what we find.

Quantum mechanics also arrives at relationship between observable quantities, but the uncertainty principle suggests that the nature of an observable quantity is different in the atomic realm. Cause and effect are still related in quantum mechanics, but what they concern needs careful interpretation. In quantum mechanics, the kind of certainty about the future characteristic of classical mechanics is impossible because the initial state of a particle cannot be established with sufficient accuracy.

We may note that the quantities whose relationships quantum mechanics explore are probabilities. Classical mechanics turn out just an approximate version of quantum mechanics.

7. Write the condition for a normalized wave function ψ.

Ans. $\int |\psi|^2 \, dv = 1$

8. Can Schrodinger equation be derived from other basic principles of physics?

Ans. No, Schrodinger equation is a basic principle in itself.

9. Is Schrodinger equation linear in the wave function ψ? What is the meaning of it?

Ans. Yes, Schrodinger's equation is linear in the wave function ψ. By this we mean that the equation has terms that contain ψ and its derivatives but no terms independent of ψ or that involves higher power of ψ or its derivatives.

10. Is Schrodinger equation linear in probabilities?

Ans. No.

11. What an operator in quantum mechanics signify?

Ans. An operator tells us what operation to carry out on the quantity that follows it.

12. What is an important property of Schrodinger steady state equation?

Ans. If Schrodinger steady state equation

$$\frac{\partial^2 \psi}{\partial x^2} + \frac{2m}{\hbar^2}(E - V)\psi = 0$$

has one or more solutions for a given system, each of these wave functions corresponds to a specific value of energy E. Thus energy quantization appears in wave mechanics as a natural element of theory, and energy quantization in the physical world is revealed as a universal phenomenon characteristic of all stable systems.

OBJECTIVE QUESTIONS

1. The average energy of Planck's oscillator of frequency v is _____. $[hv/(e^{hv/kT} - 1)]$

2. For the study of Compton effect, the best suited rays are _____. [X-rays]

3. Bohr's quantum condition is _____. $[mvr = nh/2\pi]$

4. Quantum mechanical operator of momentum p is _____. $\left[\dfrac{\hbar^2}{i}\nabla\right]$

5. Quantum mechanical operator of kinetic energy T is _____. $\left[-\dfrac{\hbar^2}{2m}\nabla^2\right]$

6. If A is an operator operating on a state function ψ, then the possible values of any physical quantity of a system (e.g. energy, angular momentum, etc.) in operator equation $A\psi = a\psi$ are given by _____. [eigenvalue a]

7. If $A = 3x^2$ and $B = \dfrac{d}{dx}$, then _____.

$[AB \neq BA]$

8. The energy of a particle of mass m in a one dimensional box of length $2L$ with ends at $\pm 2L$, i.e. the origin at the middle point is _____.

$\left[E_n = \dfrac{n^2 h^2}{16\mu L^2}\right]$

9. The average value of p_x^2 for a particle in a box is _____. [*Ans.* $2\ mE$]

10. For a particle in a box, the fractional difference in the energy between adjacent eigenvalues is _____. $\left[\dfrac{\Delta E}{E_n} = \dfrac{2n+1}{n^2}\right]$

MULTIPLE CHOICE QUESTIONS

1. Wien's displacement law is
   (a) $\lambda_m T$ = constant (b) λ_m/T = constant
   (c) λ_m/T^2 = constant (d) $\lambda_m T^2$ = constant

2. The average energy of Planck's oscillator of frequency v is
   (*a*) $hv\ [e^{hv}/kT - 1]$ (*b*) $hv/[e^{hv/kT} - 1]$
   (*c*) hv (*d*) hv/kT

3. Planck's radiation formula in terms of wavelength is

   (a) $\rho(\lambda)\ d\lambda = \dfrac{8\pi hc}{\lambda^5}d\lambda$

   (b) $\rho(\lambda)\ d\lambda = \exp\left[-ch/\lambda kT\right]d\lambda$

   (c) $\rho(\lambda)\ d\lambda = \dfrac{8\pi hc}{\lambda^5}\exp[-ch/\lambda cT]d\lambda$

   (d) $\rho(\lambda)\ d\lambda = \left[\dfrac{8\pi hc}{\lambda^5} - 1\right]d\lambda$

4. Einstein's photoelectric equation is

   (a) $hv = \left(\dfrac{1}{2}mv^2\right)_{max}$ (b) $hv = hv_0 + \left(\dfrac{1}{2}mv^2\right)_{max}$

   (c) $hv = 2hv_0$ (d) $hv = hv_0/2$

5. The shift in the wavelength in Compton effect depends
   (a) only on the energy of the incident photon
   (b) only on the scattering angle
   (c) on scattering angle as well as energy of the incident photon
   (d) none of the above

6. The change in wavelength of an X-ray photon when it is scattered through an angle of $90°$ by a free electron is of the order of
   (a) 10^{-8} m (b) 10^{-10} m
   (c) 10^{-12} m (d) 10^{-15} m

   [**Hint:** $\Delta\lambda = \dfrac{h}{m_0 c}\ (1 - \cos\phi)$

   $= \dfrac{6.6\times10^{-34}}{9.1\times10^{-31}\times3\times10^8}\ (1 - \cos 90°) = 2.4\times10^{-12}\,\text{m}]$

7. For the study of Compton effect, the best suited rays are:
   (a) gamma rays (b) visible rays
   (c) X-rays (d) ultraviolet rays

8. In Compton scattering for a given photon of energy E, the maximum kinetic energy gained by scattered electron is
   (a) $2E^2/(2E + m_0c^2)$ (b) $2E^2/m_0c^2$
   (c) $E^2/2m_0c^2$ (d) $4E^2/(E + 2m_0c^2)$

9. The de Broglie wave group associated with a body in motion travels with the velocity
   (a) with which the body is moving
   (b) half of the de Broglie wave group velocity
   (c) one fourth of the velocity with which the body is moving
   (d) none of the above

10. Bohr's quantum condition is
    (a) $mvr = h/2\pi$ (b) $mvr = (n + 1)\,h/2\pi$
    (c) $mvr = nh/2\pi$ (d) $mvr = n^2h/2\pi$

11. The uncertainty in the location of a particle is equal to its de Broglie wavelength. The uncertainty in its velocity is
    (a) $v/4\pi$ (b) $v/2$
    (c) $v/8\pi$ (d) $v/16\pi$

    [**Hint:** $\Delta x \Delta p = h/4\pi$, we have $\Delta x = \dfrac{h}{p}$

    $\therefore \dfrac{h}{p}\Delta p = \dfrac{h}{4\pi}$ or $\Delta p = \dfrac{p}{4\pi}$ or $\Delta(mv) = \dfrac{mv}{4\pi}$

    $\therefore \Delta v = \dfrac{v}{4\pi}$]

12. Assume that we have two small opaque bodies at large distance from one another supported by fine threads in a large evacuated enclosure whose walls are opaque and kept at a constant temperature. In such a case the bodies and walls can exchange energy only by means of radiation. Let e represents the rate of emission of radiation energy by a body and let a represents the rate of absorption of radiation energy by a body. At equilibrium,
    (a) $\dfrac{e_1}{a_1} = \dfrac{e_2}{a_2} = 0$ (b) $\dfrac{e_1}{a_1} = \dfrac{e_2}{a_2} = 1$
    (c) $\dfrac{e_1}{a_1^2} = \dfrac{e_2}{a_2^2} = 1$ (d) $\dfrac{e_1^2}{a_1} = \dfrac{e_2^2}{e_2} = 1$

13. An atom can radiate at any time after it is excited. It is found that in a typical case the average excited atom has a lifetime of about 10^{-8}s, i.e. during this period it emits a photon and is de-excited. The minimum uncertainty Δv in the frequency of the photon is
    (a) $8 \times 10^6\ \text{s}^{-1}$ (b) $10^4\ \text{s}^{-1}$
    (c) $5 \times 10^8\ \text{s}^{-1}$ (d) $8 \times 10^9\ \text{s}^{-1}$

14. The Laplacian operator is
    (a) ∇ (b) ∇^2
    (c) ∇^3 (d) ∇^4

15. Eigenvalues of a Hermitian operator are
    (a) imaginary
    (b) real
    (c) can be real or imaginary
    (d) zero

16. Eigen functions of a Hermitian operator corresponding to different eigenvalues are
    (a) orthogonal
    (b) orthonormal
    (c) orthogonal as well as orthonormal
    (d) none of the above

17. Quantum mechanical operator of momentum p is
    (a) $\dfrac{\hbar}{i}\nabla$ (b) $-\dfrac{\hbar^2}{2m}\nabla^2$
    (c) $\dfrac{\hbar}{i}\nabla^2$ (d) $\nabla^2/8\pi$

18. Quantum mechanical operator of kinetic energy T is
    (a) $\dfrac{\hbar}{i}\nabla$ (b) $-\dfrac{\hbar^2}{2m}\nabla^2$
    (c) $\hbar\nabla$ (d) $\hbar\nabla^2/8\pi$

19. Quantum mechanical operator for the x component of angular momentum L_x is
    (a) $\dfrac{h}{2\pi i}\left(x\dfrac{\partial}{\partial y} - y\dfrac{\partial}{\partial x}\right)$ (b) $\dfrac{h}{2\pi i}\left(z\dfrac{\partial}{\partial x} - x\dfrac{\partial}{\partial z}\right)$
    (c) $\dfrac{h}{2\pi i}\left(y\dfrac{\partial}{\partial z} - z\dfrac{\partial}{\partial y}\right)$ (d) none of the above

20. If A is the operator operating on a state function ψ, then the possible values of any physical quantity of a system (e.g. energy, angular momentum, etc.) in the operator equation $A\psi = a\psi$ are given by
    (a) eigen function ψ
    (b) eigenvalue a
    (c) eigenvalue a as well as eigen functions ψ
    (d) none of the above is corrcnt

21. $\psi\,(x) = A\exp\,(\pm 2\pi i\,p_x\,x/h)$ is a well behaved function with A as constant and p_x is the x-

component of momentum. The eigen function of momentum operator is $(k = 2\pi p_x/h)$

(a) $\psi = A \exp(\pm ikx)$ (b) $\psi = A \exp(ikx)$

(c) $\psi = A \exp(-ikx)$ (d) none of the above

22. The eigenvalues of p_x in previous question are

(a) $\dfrac{kh}{2\pi}$

(b) $\dfrac{k}{2\pi}$

(c) $\dfrac{k^2}{h^2}$

(d) zero

23. If $A = 3x^2$ and $B = \dfrac{d}{dx}$, then

(a) $AB = BA$ (b) $AB \neq BA$

(c) $AB = BA/2$ (d) $AB = 2BA$

24. If two operators A and B are Hermitian, then their product (AB) is also Hermitian if and only if A and B

(a) commute   (b) do not commute

(c) are non-zero   (d) none of the above

Answers

1. (a)	2. (b)	3. (c)	4. (b)	5. (c)
6. (c)	7. (c)	8. (a)	9. (a)	10. (c)
11. (a)	12. (b)	13. (a)	14. (a)	15. (b)
16. (c)	17. (b)	18. (a)	19. (a)	20. (b)
21. (a)	22. (b)	23. (b)	24. (d)	

20

Free Electron Theory of Metals

20.1 INTRODUCTION

We know that metals are one of the important class of solids and more than two-thirds of elements are metals. Obviously, nature favours metallic state of elements. Various attempts have been made to model a metal from the beginning of 20th century, which may satisfactorily explain the striking physical properties of metals. In the most simplified model, it is assumed that the outermost valence electrons in a metal behave like free electrons, e.g. we have the electronic configuration of copper and magnesium as $1s^2 2s^2 2p^6 3s^2 3p^6 3d^{10} 4s^1$ and $1s^2 2s^2 2p^6 3s^2$ respectively. The outermost $4s^1$ and $3s^2$ electrons in Cu and magnesium respectively are assumed to behave as free electrons. The movement of free electrons, in this simple model of free electron gas, is considered to be independent of the details of nuclei and all the other 'core' electrons in the metal. The free electrons in metals are called *conduction electrons*, since they are responsible for the electrical conduction in metals. We may note that the collective body of 'conduction electrons contributed by individual metal atoms is interpreted as the free electron gas' that occupies most of the volume of the metal. This concept of free electron gas when applied to metals, explains forces of cohesion and repulsion, binding the energy levels and the behaviour of conductors, insulators and magnetic materials.

The first version of the free electron model was introduced by P Drude in the early 1900s, with improvement soon after by H A Lorentz. This is now known as *Drude–Lorentz free electron theory of metals*. The other theories of metals are *Sommerfeld's free electron quantum theory* and *zone theory*.

20.2 DRUDE–LORENTZ THEORY

Drude's theory is essentially based on the classical kinetic theory of gases. According to Drude, the metal must have two types of particles as against only one type in the simplest gases. The discovery of electrons bearing negative charge made it mandatory to accept the presence of positively charge entities (or particles) to fulfil the condition of charge neutrality. In this model, it is assumed that when metal atoms come together from a metal, the valence electrons get liberated and move freely within the metal. The remainder of the atom is a positive ion carrying the major portion of the atomic mass. Drude considered these particles as heavy and immobile. When the metal is subjected to an external electric field, the free electrons referred to as conduction electrons in metals move in the background of immobile positive metal ions. The movement of electrons obeys the laws of the classical kinetic theory of gases. Lorentz in 1909, applied Maxwell–Boltzmann statistics to the electron gas with the following two assumptions:

1. The mutual repulsion between the negatively charged electrons is negligible.
2. The potential field due to positive ions within the crystal can be assumed to be constant everywhere.

The combined ideas of Drude and Lorentz constitute the Drude–Lorentz theory. As this is based on the classical Maxwell–Boltzmann (MB) statistics and therefore it is also called as the classical theory.

Since the electrons move freely inside the metals irrespective of the crystal structure, the ratio of the electrical conductivity σ to the thermal conductivity, K should be constant for all metals at a constant temperature, i.e.

$$\frac{\sigma}{K} = \text{constant} \qquad (20.1)$$

This is called the *Wiedemann–Franz law* and has been realised in practice.

This theory explained a number of properties of a metal, e.g. electrical conductivity, thermal conductivity, lustre and opacity of metals. The opacity is due to absorption of all the incident electromagnetic radiations by free electrons which are then set into forced oscillations. The electrons return to their normal states by emitting the same amount of energy in all directions, thus producing metallic lustre.

The main drawbacks of this theory are:

i. The theory correctly predicted the room temperature resistivity of various metals but the temperature dependence of resistivity could not be established accurately. The theory predicted that resistivity varies as $\sqrt{T}$ whereas actually it is found to vary linearly with temperature T.

ii. The theory yielded incorrect magnitudes of the specific heat and paramagnetic susceptibility of metals. These properties of metals are based on interactions of free electrons with the external sources of energy which may be thermal or magnetic in nature. The application of MB statistics to this theory allows all the free electrons to gain energy which results

in much higher values of heat capacity and also paramagnetic susceptibility. The classical theory is further unable to account for the occurrence of mean free paths ($\sim 10^8$ to 10^9 interatomic spacings or more than one centimetre) at low temperatures.

The above shortcomings of Drude–Lorentz theory were removed by Sommerfeld in 1928. He applied Fermi–Dirac (FD) statistics instead of MB statistics, and treated the problem quantum mechanically. The possible electronic energy states in the potential energy box and the distribution of electrons in these states are then determined using quantum statistics.

20.3 FERMI–DIRAC DISTRIBUTION FUNCTION

We have stated that Sommerfeld theory differs from the classical theory in the replacement of MB statistics. Fermi–Dirac distribution function is commonly expressed as a function of electron energy as

$$f(E) = \frac{1}{\exp\left[(E-\mu)/k_B T\right]+1} \qquad (20.2)$$

where $f(E)$ is known as the FD distribution function. It denotes the probability that an orbital of energy E be populated in an ideal electron gas in thermal equilibrium, μ is the chemical potential. The highest energy level populated at absolute zero is designated as E_F and is called as *Fermi level*.

At $\underset{T \to 0}{\text{Lt}}\ \mu = E_F$, Eq. (20.2) takes the form

$$f(E) = \frac{1}{\exp\left(\dfrac{E-E_F}{k_B T}\right)+1} \qquad (20.3)$$

where E_F is *Fermi energy*. For different temperatures, the plot of $f(E)$ versus E is shown in Fig. 20.1.

At absolute zero, Eq. (20.3) gives

$$f(E) = 1 \text{ for } E \leq E_{FO}$$
$$= 0 \text{ for } E > E_{FO} \qquad (20.4)$$

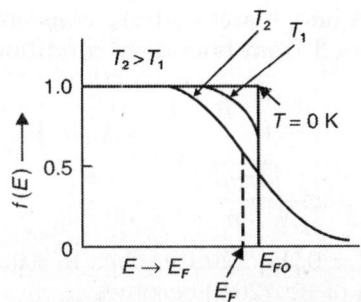

Fig. 20.1: A schematic plot of Fermi distribution function for three different temperatures. In the plot, the variation of Fermi energy with temperature is also shown

Obviously, all the energy states below E_{FO} are occupied and all the states above it are empty. This shows that Fermi distribution function is a *step function* and at 0 K, the Fermi level E_{FO} represents the highest filled energy level.

For temperature greater than 0 K but less than the melting point of the metal such that $k_B T \ll E_F$, the Fermi distribution function loses its character. We may note that the probability of occupation, $f(E)$ decreases gradually from 1 to 0 near E_F. This indicates that some of the states below E_F are empty while some others above it are filled. This is because some of the electrons from the energy status below E_F gain thermal energy and get excited to the states above E_F. From Eq. (20.3), at $E = E_F$, one obtains

$$f(E_F) = \frac{1}{2}$$

Clearly, for temperatures greater than 0 K, the Fermi level may be defined as the level where the probability of occupation is $\frac{1}{2}$. We may note that unlike E_{FO}, it is not the topmost filled level, instead it lies between the filled levels and empty levels. Moreover, the position of the Fermi level is not fixed but changes with temperature. The following approximate relationship exist between E_F and E_{FO}:

$$E_F = E_{FO} \left[1 - \frac{\pi^2}{12} \left(\frac{k_B T}{E_{FO}} \right)^2 \right] \qquad (20.5)$$

Since the spacing between the levels in case of metals is quite small ($\sim 10^{-19}$ eV), the highest filled energy level is usually taken as the Fermi level.

The Fermi energy E_F is often expressed in terms of temperature by the relation

$$E_F = k_B T_F \qquad (20.6)$$

where T_F called the *Fermi temperature*, turns out to be of the order of tens and thousands of Kelvin because the Fermi energy per electron is several volts. We have $\dfrac{k_B T}{E_F} \cong 0.01$ at room temperature for $E_F = 3$ eV.

20.4 SOMMERFELD FREE ELECTRON MODEL

One cannot totally ignore the electron–ion interaction even in metals because the conduction electrons remain confined within the volume of the metal. Considering the free nature of valence electrons as assumed in the classical theory, Sommerfeld treated the problem quantum mechanically using the FD statistics in place of classical MB statistics. The basic assumptions of this model are mentioned below:

   i. The valence electrons in a metal are free.

  ii. Valence electrons in a crystal are confined to move within the boundaries of the crystal. Clearly, the electrons within the crystal have a lower potential energy than outside. However, the potential energy of an electron is uniform or constant within the crystal (Drude theory).

 iii. The electrons are free to move within the crystal, but are prevented from leaving the crystal boundaries by very high potential energy barriers on its surface.

 iv. The allowed energy levels of an electron bound to single atom are quantized.

  v. The electronic specific heat of metals is very low.

Using this model, we describe the one dimensional and three dimensional cases separately.

Free Electron Gas in One Dimensional Box

Let us consider an electron of mass m bound to move in a one dimensional crystal of length L as shown in Fig. 20.2. The electron is prevented from leaving the crystal by the presence of a large potential energy at its surface. The potential energy everywhere within the crystal is assumed to be constant and equal to zero.

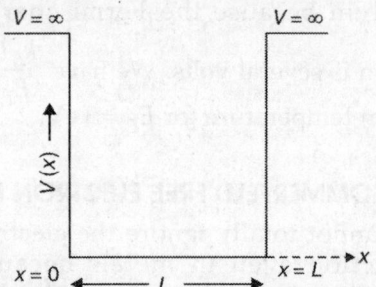

Fig. 20.2: One dimensional potential well bounded by infinite potential energy barriers

We have

$$V(x) = 0 \text{ for } 0 < x < L$$

$$V(x) = \infty \text{ for } x \leq 0 \text{ and } x \geq L \quad (20.7)$$

Sommerfeld assumed that the potential of an electron in a metal is uniform. He applied the one dimensional Schrodinger equation, i.e.

$$\frac{d^2\psi_n}{dx^2} + \frac{8\pi^2 m}{h^2}(E_n - V)\psi_n = 0 \quad (20.8)$$

to calculate the total energy E_n, where ψ_n is the wave function of the electron occupying the nth state. E_n represents the kinetic energy of the electron in the nth state and V is its potential energy. We may note that the potential energy everywhere within the crystal is assumed to be constant and equal to zero. Thus, the Schrodinger equation (20.8) takes the form

$$\frac{d^2\psi_n}{dx^2} + \frac{8\pi^2 m}{h^2}E_n\psi_n = 0 \quad (20.9)$$

The general solution of Eq. (20.9) is

$$\psi_n(x) = A\sin kx + B\cos kx \quad (20.10)$$

where A and B are arbitrary constants to be determined from boundary conditions. One obtains $A = \sqrt{\dfrac{2}{L}}$, $B = 0$

and $\quad k = \sqrt{\dfrac{4\pi m E_n}{n}} \quad (20.11)$

For $x = 0$, Eq. (20.10) gives $B = 0$ and the solution of Eq. (20.9) becomes

$$\psi_n(x) = A\sin kx \quad (20.11a)$$

Also, since $\psi_n(L) = 0$, Eq. (20.11a) yields

$$\sin kL = 0 \quad (20.11b)$$

or $\quad k = \dfrac{n\pi}{L}$

where $n = 1, 2, 3,\dots$. Now the Eq. (20.11a) for the allowed wave function takes the form

$$\psi_n(x) = A\sin\left(\frac{n\pi}{L}\right)x \quad (20.11c)$$

The allowed discrete energy values are given by

$$E_n = \frac{h^2}{8mL^2}n^2 \quad (20.12)$$

where $\quad n = 1, 2, 3,\dots$

Thus, $\quad E_n \propto n^2 \quad (20.13)$

The number n is quantum number. The energy spectrum consists of discrete energy levels where the spacing between the levels can be determined by the values of n and L. It decreases with increasing L. If L is of the order of a few centimetres, the energy levels form almost a continuum. But if L has atomic dimensions, the spacing between the levels becomes appreciable. Figure 20.3 shows E_n versus n plot. The energy levels and wave functions corresponding to $n = 1, 2, 3$ and 4 are shown in Fig. 20.4. Thus, one finds that if the total number of electrons to be accommodated is seven, the energy levels with $n < 4$ would be occupied while the level with $n > 4$ would be empty. The topmost filled energy level at 0 K temperature is known as *Fermi level* and the energy corresponding to this level is called as the *Fermi energy* E_F.

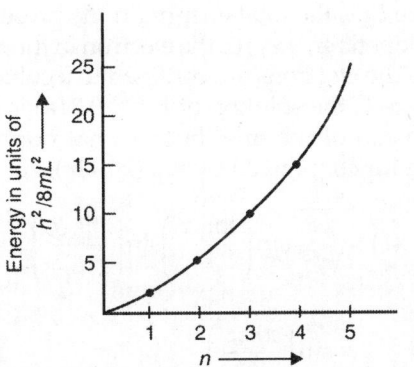

Fig. 20.3: For a one dimensional crystal, E_n versus n curve

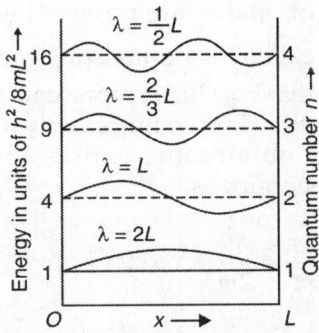

Fig. 20.4: In a one dimensional crystal first four wave functions (solid lines) and the corresponding energy levels (broken lines) of an electron

If N be the total number of electrons to be accommodated on the line, then for even n, we have

$$2n_F = N \qquad (20.14)$$

where n_F represents the principal quantum number of the Fermi level. Now, for $N = n_F$, one obtains from Eq. (20.12).

$$E_F = \frac{\hbar^2}{2m}\left(\frac{\pi n_F}{L}\right)^2 = \frac{\hbar^2}{2m}\left(\frac{N\pi}{2L}\right)^2 \qquad (20.15)$$

For $N/L = 0.5$ electrons/Å

$$= 5 \times 10^9 \text{ electrons/m},$$

Equation (20.15) gives

$$E_F = 3.7 \times 10^{-19} \text{ J} = 2.4 \text{ eV}$$

Clearly, if 5×10^9 electrons are accommodated on one meter length of the line, the energy of the topmost electron would be 2.4 eV.

Total Energy

The total energy, E_0, of all the N electrons in the ground state is determined by summing up the energies of the individual electrons. For N electrons, the number of filled energy level is $N/2$ and one obtains

$$E_0 = 2\sum_{n=1}^{N/2} E_n = \frac{1}{3}\frac{\hbar^2}{2m}\left(\frac{N\pi}{2L}\right)^2 N$$

$$= \frac{1}{3}N E_n \qquad (20.16)$$

Thus, for one dimensional crystal, the average kinetic energy in the ground state is one-third of the Fermi energy.

Density of States

One can define the density of states as the number of electronic states present in a unit energy range. If $D(E)$ is density of states, then

$$D(E) = \frac{dn}{dE} \qquad (20.17)$$

where dn represents the number of electronic quantum states present in the energy interval E and $E + dE$. For a free electron gas, since each energy level contains two electronic states, one with spin up and other with spin down, we have the actual density of states twice the value given by Eq. (20.17). Thus, we have

$$D(E) = 2\frac{dn}{dE} \qquad (20.18)$$

From Eq. (20.12), one obtains

$$\frac{dE}{dn} = \frac{\hbar^2}{2m}\left(\frac{\pi}{L}\right)^2 2n = \frac{h^2 n}{4mL^2} \qquad (20.19)$$

Using Eq. (20.19), Eq. (20.18) takes the form

$$D(E) = 2\left(\frac{4mL^2}{nh^2}\right) = \frac{8mL^2}{h^2}\cdot\frac{1}{n}$$

Again, from Eq. (20.12), one obtains

$$\frac{1}{n} = \left(\frac{h^2}{8\pi L^2 E}\right)^{1/2} \qquad (20.20)$$

$$\therefore \ D(E) = \left(\frac{8mL^2}{h^2 E}\right)^{1/2} = \frac{4L}{h}\left(\frac{m}{2E}\right)^{1/2}$$

Figure 20.5 shows the plot of $D(E)$ versus E. From figure, we can see that all the levels present below the Fermi level are filled and all those present above it are empty. However, this type of situation, in fact, exists at absolute zero. This means at 0 K, the Fermi level divides the filled and unfilled levels in the metallic crystal.

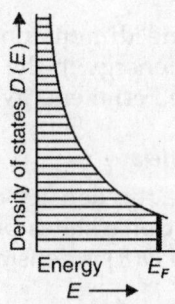

Fig. 20.5: Variation of electronic states density $D(E)$ with energy for one dimensional metallic

Free Electron Gas in Three Dimensions

Equation (20.14) is the expression for energies in one dimension only. The corresponding three dimensional case would be one in which an electron moves in all directions so that quantum numbers n_x, n_y and n_z are required corresponding to the three coordinate axes. For simplicity, a three dimensional crystal may be regarded as a cubical box having length of the edge equal to L. One can write the free particle Schrödinger equation in three dimensions as

$$\frac{-\hbar^2}{2m}\left(\frac{\partial^2}{\partial x^2} + \frac{\partial^2}{\partial y^2} + \frac{\partial^2}{\partial z^2}\right)\psi_k(r) = E_k\psi_k(r)$$

$$(20.21)$$

or $\quad \nabla^2\psi_k(r) + \dfrac{2m}{\hbar^2}E_k\psi_k(r) = 0 \qquad (20.21a)$

where E_k is the total energy (in the present case E_k is kinetic energy) of the electron in the K-state. Since the electrons are confined to a cubical box of edge L, the solution of Eq. (20.21a) is just an extension of the one dimensional normalized wave function given by Eq. (20.11c), i.e.

$$\psi_k(r) = \sqrt{\frac{8}{L^3}}\sin\left(\frac{\pi n_x x}{L}\right)\sin\left(\frac{\pi n_y y}{L}\right)$$

$$\times \sin\left(\frac{\pi n_z z}{L}\right) \qquad (20.22)$$

or $\quad \psi_k(r) = Ae^{ik\cdot r} = Ae^{i(k_x x + k_y y + k_z z)} \qquad (20.22a)$

where n_x, n_y and n_z are positive integers and $\sqrt{8/L^3}$ is called the normalizing constant. We may note that Eq. (20.22) represents a standing wave solution. Proceeding exactly in the same way, one obtains the expression for the allowed energies as

$$E_k = \frac{\hbar^2 k^2}{2m} = \frac{\hbar^2}{2m}\left(k_x^2 + k_y^2 + k_z^2\right) = n^2\frac{1}{2m}\left(\frac{\hbar}{2L}\right)^2$$

with $n^2 = n_x^2 + n_y^2 + n_z^2 \qquad (20.23)$

where the magnitude of the wave vector k is related to the wavelength λ as

$$k = \frac{2\pi}{\lambda} \qquad (20.24)$$

where $\quad \lambda = \dfrac{h}{mv} = \dfrac{2L}{n} \qquad (20.24a)$

Applying normalization condition

$$\int_0^v \psi^*(r)\psi(r)dV = 1$$

or $\quad \displaystyle\int_0^L\int_0^L\int_0^L A^2 e^{-ik\cdot r}e^{ik\cdot r}dx\,dy\,dz = 1$

or $\quad A = \left(1/L^3\right)^{1/2} = (1/V)^{1/2}$

Thus, the normalized wave function is

$$\psi(r) = \left(\frac{1}{V}\right)^{1/2}e^{ik\cdot r} \qquad (20.25)$$

For an electron moving in the free space, we use $E_k = \dfrac{\hbar^2 k^2}{2m}$ to express free energy. But when a free electron moves in a potential box, the electron energy is meaningfully given by

$$E_k = n^2 \frac{1}{2m}\left(\frac{\hbar}{2L}\right)^2.$$

The components (k_x, k_y, k_z) or (n_x, n_y, n_z) together with the spin quantum number m are the quantum numbers of the problem. We may note that the allowed energy values in the three dimensional space produce constant spherical energy surfaces.

$$E_k = \frac{\hbar^2 k^2}{2m} = \text{constant} \qquad (20.26)$$

One can easily see that the density of states in this case is given by

$$D(E) = \frac{1}{8}\frac{4\pi}{3}\left(\frac{2mEL^2}{\pi^2\hbar^2}\right)^{3/2}$$

$$= \frac{V}{6\pi^2}\left(\frac{2m}{\hbar^2}\right)^{1/3} E^{3/2} \qquad (20.27)$$

where $V = L^3$ is the volume of the cube. If $g(E)dE$ is the density of the allowed states between E and $E + dE$ and $D(E)$ is the total density of states, then

$$D(E) = \int_0^E g(E)dE$$

or $\quad g(E) = \dfrac{d}{dE}D(E)$

$$= \frac{V}{4\pi^2}\left(\frac{2m}{\hbar^2}\right)^{3/2} E^{1/2} \qquad (20.28)$$

This is the density of states without including the factor of two for spin up and down degeneracy. Including this, the density of states takes the form

$$g(E) = \frac{V}{2\pi^2}\left(\frac{2m}{\hbar^2}\right)^{3/2} E^{1/2} \qquad (20.29)$$

Fermi Level

Let us consider the occupancy of the allowed energy states at absolute zero. Let us add electrons to the allowed energy states. First we will fill the ground state and then the higher excited states will be filled, until the last electron is added at a level, above which all other states are unoccupied. This level is defined as Fermi level as shown in Fig. 20.6. Clearly, Fermi energy E_F is defined as the energy of the topmost filled level. One can calculate it from the total number of free electrons N as

$$N = \int_0^{E_F} g(E)dE = \frac{V}{3\pi^2}\left(\frac{2m}{3\pi^2}\right)E_F^{3/2}$$

or $\quad E_F = \dfrac{\hbar^2}{2m}\left(3\pi^2\dfrac{N}{V}\right)^{2/3} \qquad (20.30)$

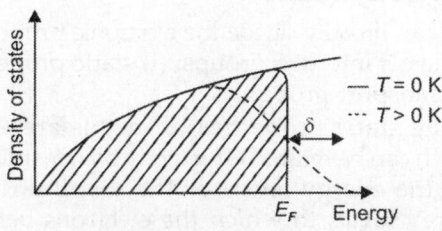

Fig. 20.6: Density of states as a function of energy at absolute zero (solid curve) and at $T > 0$ K (dotted curve)

Equation (20.30) will lead for Fermi surface, the Fermi vector as

$$E_F = \frac{\hbar^2}{2m}K_F^2$$

or $\quad K_F = \left(\dfrac{3\pi^2 N}{V}\right)^{1/3} \qquad (20.31)$

The electron velocity v_F at the Fermi surface is

$$v_F = \frac{\hbar K_F}{m} = \frac{\hbar}{m}\left(\frac{3\pi^2 N}{V}\right)^{1/3} \qquad (20.32)$$

Now, the Fermi temperature is defined as

$$T_F = \frac{E_F}{k_B} = \frac{\hbar}{2mk_B}\left(\frac{3\pi^2 N}{V}\right)^{2/3} \quad (20.33)$$

As stated above at absolute zero, all the allowed states of energy up to Fermi level will be occupied. As the temperature is raised, a variation in electron energy is provided by the energy of the crystal.

We note that this variation is quite small as compared to the electron energy itself such that the energy distribution can only change slightly at a temperature higher than 0 K. From Fig. 20.6, it is clear that only electrons with energies near E_F will change their state on raising the temperature because thermal energy will be insufficient to excite electrons with lower energies to unoccupied states above E_F.

20.5 APPLICATIONS OF THE FREE ELECTRON GAS MODEL

One can broadly divide the electronic properties of metals into two groups: (i) static properties (ii) transport properties.

The static properties of metals are those which can be treated effectively by considering just the energy levels or the distribution of energy levels to which the electrons belong. More specifically, one can say that one needs to consider the overall change in potential of the electrons without investigating deeply the process producing the transitions in detail. Various electron emission properties, magnetic properties and properties like heat, capacity and contact potential falls under this category. These properties arise as a result of the excitations of electrons of metals by thermal energy, light energy, subjecting metals to strong electric fields, etc.

The transport properties of metals are those which one can treat by considering the detailed response of electrons to an external field, i.e. by considering the acceleration properties of electrons. Electrical and thermal conductivities, Hall effect, thermoelectricity, etc. fall under this category.

Now, we consider the electronic specific heat of metals.

Electronic Specific Heat of Metals

In accordance with the classical statistical mechanics, the average kinetic energy of a free electron is given by

$$\bar{E}_0 = \frac{3}{2}k_B T \quad (20.34)$$

If the metal contains N free electrons, then the total kinetic energy takes the form

$$E = N\bar{E}_0 = \frac{3}{2}Nk_B T .$$

Therefore, the electronic specific heat is given by

$$C_V = \frac{3}{2}Nk_B \quad (20.35)$$

From the measurement of optical reflection coefficient of metals, we can infer that the number of electrons per atom in metals is of the order of unity. The value of electronic specific heat is obtained as $3R/2$ or 3 cal/g/K when we consider one gram atom of the metal. On the other hand, the specific heat associated with the lattice vibrations at high temperatures is obtained as about $3R/2$ or 6 cal/g/K. Thus, we can easily conclude that the specific heat of metals should be about 50% greater than the specific heat of insulators. Obviously, this is quite contradictory to the experimental observations which indicate that the electronic contribution to the specific heat is very small. In accordance with Eq. (20.35), the electronic contribution at room temperature is not more than 0.01 of the value given by this equation. Moreover, as T approaches zero, this contribution decreases linearly to zero. This reveals that all the electrons might not contribute to the specific heat, instead only small fraction of the total number of electrons might contribute.

In accordance with quantum statistics, only those electrons contribute to the specific heat which lie within an electron range $k_B T$ below the Fermi level. This is due to the fact that when the electrons gets heated up to a temperature T, these electrons acquire energy of the order of $k_B T$ and jump to the empty higher energy states. However, the deep living electrons cannot do so because the unfilled energy states are not available to these electrons for excitation. Clearly, these electrons as such do not contribute to the specific heat. One finds that the number of electrons which contribute to the specific heat is of the order of $N\left(\dfrac{k_B T}{E_F}\right)$

or $N\left(\dfrac{T}{T_F}\right)$, where T_F denotes the Fermi temperature. E_F is defined by the equation

$$E_F = k_B T_F$$

Using the effective value of N, Eq. (20.35) takes the form

$$C_V = \frac{3}{2} N K_B \left(\frac{T}{T_F}\right) \qquad (20.36)$$

This reveals that C_V is proportional to T and at low temperatures (below the Debye temperature and Fermi temperature), lattice specific heat follows T^3 law. For $T = 300$ K and $T_F = 3000$ K, one obtains from the above equation,

$$C_V = (0.001)\left(\frac{T}{T_F}\right) N k_B$$

The specific heat calculated from the above relation agrees with the experimental observation. Obviously, application of quantum statistics modifies the thermal behaviour of free electrons in a simple and satisfying manner.

At low temperatures below the Debye temperature, and the Fermi temperature, the heat capacity of a metal is given by

$$C_V = \gamma T + \lambda T^3 \qquad (20.37)$$

where the first term represents the electron contribution and the second term is the usual Debye term contributed by photons. We may note that the photon or ionic contribution dominates at high temperature but drops spectacularly at low temperatures where it becomes even smaller than the electronic contribution.

ILLUSTRATIVE EXAMPLES

Example 1

Show that the energy difference between the $n_x = n_y = n_z = 1$ level and reset higher energy level for free electrons in a solid cube of side 0.01 m is 1.1×10^{-14} eV.

Solution

We have volume $= (0.01)^3 = 10^{-6}$ m^3

$$E_1 = \frac{h^2}{8mL^2}\left(n_x^2 + n_y^2 + n_z^2\right) = \frac{3}{8}\frac{h^2}{mL^2}$$

$$E_2 = \frac{h^2}{8mL^2}\left(n_x^2 + n_y^2 + n_z^2\right) = \frac{6}{8}\frac{h^2}{mL^2}$$

$$\therefore\ E_2 - E_1 = \frac{3h^2}{8mL^2} = \frac{3\times\left(6.6\times10^{-34}\right)^2}{8\times9.1\times10^{-31}\times(0.01)^2}$$

$$= 1.8 \times 10^{-33}\text{ J} = 1.1 \times 10^{-14}\text{ eV}.$$

Example 2

Show that the number of energy states available for the electrons in a cubical box of 10^{-2} m side lying below an energy of 4 eV is about 36×10^{21}.

Solution

We have the number of energy states below 4 eV as

$$\int_0^4 g(E)\,dE = \frac{V}{2\pi^2}\left(\frac{m}{\hbar^2}\right)^{3/2}\int_0^4 E^{1/2}\,dE$$

$$= \frac{V}{2\pi^2}\left(\frac{2m}{\hbar^2}\right)^{3/2}\frac{3}{2}E^{3/2}\Bigg|_0^4$$

$$= \frac{(0.01)^3}{2\pi^2}\left(\frac{2\times9.1\times10^{-31}}{1.05\times1.05\times10^{-64}}\right)^{1/2}$$

$$\times\frac{2}{3}\times 8\left(1.6\times10^{-19}\right)^{3/2}$$

$$= 36 \times 10^{21}$$

Example 3

The Fermi velocity of electrons in a metal is 0.86×10^6 m/s. Show that the Fermi energy and Fermi temperature of the metal are 2.1 eV and 2.4×10^4 K respectively.

Solution

We have

$$E_F = \frac{1}{2}mv_s^2 z$$

$$= \frac{\frac{1}{2} \times 9.1 \times 10^{-31} \times (0.86)^2 \times 10^{12}}{1.6 \times 10^{-19}} \text{ eV} = 2.1 \text{ eV}$$

$$T_F = \frac{E_F}{k_B} = \frac{2.1 \times 1.6 \times 10^{-19}}{1.38 \times 10^{-21}} = 2.4 \times 10^4 \text{ K}.$$

REVIEW QUESTIONS

1. Describe the main features of the free electron gas model. How does it help to explain the lattice heat capacity of metals?

2. What are the shortcomings of Drude model of a metal?

3. Using free electron model, calculate density of energy states for free electron in the range E to $E + \Delta E$.

4. What is density of energy states in metals? Derive an expression for the density of energy states and hence obtain Fermi energy of metals.
   [BE]

5. In a long chain molecule of length 5 Å, electrons may be treated as free to move along the length. Calculate the zero point energy, the energy gap between the first two energy states of the electron and also the wavelength arising from this transition.

6. Write short notes on:
   i. Sommerfeld's theory of metals
   ii. Fermi energy
   iii. Density of states.

7. What is Fermi level? What does it signify? How does it change with temperature and concentration of electrons?

8. Derive an expressions for the Fermi energy and density of states for a free electron gas in three dimensions.

SHORT ANSWER QUESTIONS

1. What is free electron quantum theory of metals?

Ans. According to free electron model of metals, the valence electrons of atoms constituting metal move freely within the limits of metal. Electrons

in a metal may be treated as a gas whose constituent particles are non-interacting spin 1/2 fermions in a three dimensional box, i.e. a Fermi gas.

2. Name three kinds of statistics.

Ans.   i. Maxwell–Boltzmann (MB) or classical statistics.

    ii. Fermi–Dirac (FD) or quantum statistics.

    iii. Bose–Einstein (BE) or quantum statistics.

3. What are fermions?

Ans. The particles which have half odd integral values of spin, $\frac{1}{2}\hbar$, $\frac{3}{2}\hbar$, $\frac{5}{2}\hbar$, ..., etc. are called fermions.

4. What is the nature of fermions?

Ans. These are individualists in nature and obey Pauli exclusion principle. If a quantum state is already occupied by a fermion, no other fermion will settle in that state.

5. What are bosons?

Ans. These are the particles having integral values of spin 0, $\hbar$, $2\hbar$, $3\hbar$,.... Photons, phonons (quantum of acoustical vibrations) and helium atoms are bosons.

6. What is the nature of bosons?

Ans. Bosons are collectivists in nature and do not obey Pauli exclusion principle and hence can occupy the same quantum state in any number.

7. What is the difference between classical and quantum statistics?

Ans. Classical statistics is applied to classical objects and under appropriate conditions it may be applied to quantum object too. Quantum statistics apply only to quantum objects.

8. Write the expression for electrical conductivity of metals on the basis of Drude's model.

Ans. $\sigma = \dfrac{ne^2 \wedge}{(3mk_BT)^{1/2}}$, where $\wedge$ is the electron mean free path.

9. How experimentally the DC electrical conductivity of metal varies?

Ans. $\sigma \propto T^{-1}$

10. The Drude model correctly describes the Widemann–Franz law, but gives a wrong magnitude of the Lorentz number L. Comment on it.

Ans. The value of L obtained in the Drude theory is given by

$$L = \frac{k}{\sigma T} = \frac{3}{2}\left(\frac{k_B}{e}\right)^2$$

where k is the thermal conductivity. This value is about half the measured value.

11. What is the relation between the DC electrical conductivities in the Drude model and Lorentz model?

Ans. $\sigma_D = \left(\frac{3\pi}{8}\right)^{1/2} \sigma_L \quad \begin{array}{l} D \to \text{Drude} \\ L \to \text{Lorentz} \end{array}$

OBJECTIVE QUESTIONS

1. In the classical free-electron model, in a metal containing N ion cores per unit volume, each having effective scattering radius R, the mean free path of an electron would be $\lambda = \underline{\hspace{1cm}}$.

$$\left[\frac{1}{\pi}R^2 N\right]$$

2. DC electrical conductivity (σ) of a metal is observed to vary with temperature as $\sigma \propto \underline{\hspace{1cm}}$. This reveals that all the free electrons of a metal do not take part in $\underline{\hspace{1cm}}$, rather only the electrons near the Fermi level are involved in $\underline{\hspace{1cm}}$ and $\underline{\hspace{1cm}}$ conduction.

[$\sigma \propto T^{-1}$, conduction, thermal, electrical]

3. The Boltzmann transport equation is concerned with the determination of the $\underline{\hspace{1cm}}$ distribution function of the system under influence of $\underline{\hspace{1cm}}$ fields.

[steady-state, external]

4. Sommerfeld gave the theory of electrical conduction in metals using $\underline{\hspace{1cm}}$ statistics.

[Fermi–Dirac (FD)]

5. Sommerfeld's theory of electrical conduction gave results in confirmation with $\underline{\hspace{1cm}}$ law.

[Wiedemann–Franz]

MULTIPLE CHOICE QUESTIONS

1. The density of allowed states between E and $E + dE$ is proportional to
   - (a) E
   - (b) E^2
   - (c) $E^{3/2}$
   - (d) $E^{1/2}$

2. The classical expression for electrical resistivity exhibiting its temperature dependence is
   - (a) $\rho \propto 1/T^2$
   - (b) $\rho \propto T^{1/2}$
   - (c) $\rho \propto T^2$
   - (d) $\rho \propto 1/T$

3. Mobility of the electron is
   - (a) reciprocal of conductivity
   - (b) flow of electron per unit electric field
   - (c) average electron drift velocity per unit electric field
   - (d) none of the above

4. At absolute zero temperature, all the allowed states of energy up to Fermi level will be
   - (a) occupied
   - (b) partly occupied and partly empty
   - (c) empty
   - (d) half empty and half occupied

5. Wiedemann–Franz law is
   - (a) $\sigma/\sigma_T = LT$
   - (b) $\sigma/\sigma_T = T/L$
   - (c) $\sigma_T/\sigma = L/T$
   - (d) $\sigma_T/\sigma = LT$

6. At temperature above zero, for $E \ll E_F$, the Fermi–Dirac function approaches
   - (a) unity
   - (b) zero
   - (c) $e^{-E/k_B T}$
   - (d) infinity

7. The Fermi energy of a given metal is 1.4 eV. The Fermi temperature of the metal is
   - (a) 1.6×10^5 K
   - (b) 1.6×10^6 K
   - (c) 1.6×10^3 K
   - (d) 1.6×10^4 K

8. Most widely used conducting materials are
   - (a) gold and silver
   - (b) copper and aluminium
   - (c) germanium and aluminium
   - (d) tungsten and platinum

Answers

1. (d)	**2.** (b)	**3.** (c)	**4.** (a)	**5.** (d)
6. (a)	**7.** (d)	**8.** (b)		

21

Semiconductors and Applications

21.1 INTRODUCTION

One can classify the solid materials into three main groups from the consideration of their current carrying capabilities: conductors, semiconductors and insulators. Conductors have an abundance of free electrons that act as charge carriers, which means that they have high conductivity. On the other hand, an insulator has hardly any free electrons and offers very low level of conductivity. However, a semiconductor is a material that has a conductivity level somewhere between extremes of a conductor and an insulator, we often use the term resistivity (inverse of conductivity). Typical resistivity values for the said three broad categories of materials are given in Table 21.1.

It is also observed that at 20°C, semiconductors have resistivity lying between that of metals and insulators. However, at a very low temperature (nearly absolute zero) a semiconductor behaves as an insulator. We may note that semiconductors in very pure

form have *negative temperature coefficient* of temperature, i.e. the number of charge carriers increases rapidly with the rise in temperature. Semiconductors have the special feature of having two types of charge carriers: (i) *electrons* (ii) *holes*. Holes in semiconductors are absence of electrons in covalent bonds and act equivalently as positive charge carriers. Moreover, semiconductors have the unique property, such that their conductivity increases by several orders of magnitude even by minute amount of doping with certain impurities. Truly speaking, this property of semiconductors have been exploited to develop various electronic devices. In the development of semiconductor devices, silicon (Si), germanium (Ge) and gallium arsenide (GaAs) have received broadest range of interest. Moreover, one can alter the characteristics of semiconductors significantly through the application of heat or light. It has open a vast area of application of semi-conducting materials as heat and light

Table 21.1: Typical electrical resistivity values of few conductors, semiconductors and insulators
(at 20°C in Ω cm)

Conductors	Semiconductors	Insulators
Copper (Cu) ~1.7×10^{-6}	Silicon (Si) ~50×10^3	Mica ~7×10^{12}
Silver (Ag) ~1.6×10^{-6}	Germanium (Ge) ~50	Glass ~10^8–10^9
Aluminium (Al) ~2.8×10^{-6}	Indium antimonide (InSb) ~200	Diamond ~10^{12}
Iron (Fe) ~10×10^{-6}		
Nichrome (NiCr) ~100×10^{-6}		

sensitive devices. Semiconductors have the following characteristic properties.

Semiconductors have some unique electrical characteristics that render them especially useful. At room temperature, the electrical conductivity of a metal is about 10^8 siemens/m while that of an insulator like diamond is about 10^{-10} siemens/m. The electrical conductivity of a semiconductor lies in the range of 10^5 to 10^{-4} siemens/m. The electrical properties of semiconductors are extremely sensitive to the presence of even minute concentrations of impurities. Semiconductors have the following additional characteristic properties:

i. A pure (intrinsic) semiconductor has a negative temperature coefficient of resistance, i.e. the resistance of semiconductors decreases with increase in temperature and *vice versa*. This behaviour of semiconductors is contrary to the positive temperature coefficient of resistance exhibited by a metal.

ii. Upon irradiation by light, a semiconductor exhibit a photovoltaic effect or a change in resistance.

iii. With respect to a metal, semiconductors have high thermoelectric power of signs both positive and negative.

iv. In general, there are two types of semiconductors. Those in which electrons and holes are produced by *thermal activation* are called *intrinsic* or pure semiconductors, e.g. silicon and germanium. In other type, the current carriers, holes or free electrons are produced by the addition of very small quantities of elements of group III or V of the periodic table and are known as *extrinsic* semiconductors. Extrinsic semiconductors may be either N- or P-type depending on whether electrons or holes respectively, are the predominant charge carriers. Donor impurities introduce excess electrons, whereas acceptor impurities introduce excess holes. The junction between a P-type and an N-type semiconductor possesses rectification properties.

v. The electrical conductivity of semiconducting materials is particularly sensitive to impurity type and content as well as to temperature.

The band gap of a semiconductor is smaller than that of an insulator. At 0 K, semiconductors have a completely filled valence band, separated from an empty conduction band by a relatively narrow forbidden gap, generally less than 2 eV. The band gap in semiconductors lies in the range 0.2 to 2.5 eV, whereas the band gap of an insulator like diamond is 6 V. Due to this smaller band gap, electrons are thermally excited from the valence band to conduction band at ordinary temperatures in a semiconductor. This accounts for a larger electrical conductivity of an intrinsic (pure) semiconductor than an insulator. The two elemental semiconductors are Si and Ge having band gap energies of approximately 1.1 eV and 0.7 eV respectively. Both are found in group IVA of the periodic table and are covalently bonded. Ge was the key semiconducting material for the majority of the early solid state devices. Subsequently, it has been replaced by Si in many applications. Si has a better thermal stability, is readily available and also has an advanced technology. In addition to Ge and Si, a host of semiconducting materials like gallium arsenide (GaAs), indium phosphide (InP), indium antimonide (InSb), indium arsenide (InAs), lead sulphide (PbS), cadmium sulphide (CdS), lead telluride (PbTe), zinc telluride (ZnTe), mercury (II) indium telluride ($HgIn_2Te_4$), zinc selenide (ZnSe), cadmium selenide (CdSe), mercury (II) selenide (HgSe), magnesium antimonide (Mg_2Sb_2), magnesium iodide (MgI_2), etc. also exhibit semiconducting properties. These semiconductor materials have also been used for several electronic devices, e.g. GaAs is used in transistors, lasers, microwave and millimetre wave guides, PbS and PbTe are used in infrared detectors, CdS in light meters, CdTe in the detection of nuclear radiation, and so on.

An important subject of scientific and technological interest is *amorphous semiconductors*. In an amorphous substance, the atomic arrangement has some short range but no long range order. The representative amorphous semiconductors are selenium, germanium and silicon in their amorphous states, and arsenic and germanium chalcogenides, including such ternary system as Ge-As-Te. Some amorphous semiconductors can be prepared by a suitable quenching procedure from the melt. Amorphous films can be obtained by vapour deposition.

Alloys of semiconducting compounds also find wide range of important applications. For example, gallium arsenide phosphide $GaAs_{1-x}P_x$ is used in light-emitting diodes (LED), indium gallium arsenide ($In_xGa_{1-x}As$) in microwave and optoelectronic devices, cadmium mercury telluride ($Cd_xHg_{1-x}Te$) in infrared detectors, etc.

The nanosized powders of silicon, silicon nitride (SiN), silicon carbide (SiC) and their thin films have been considered for applications in opto-electronic devices and quantum optic devices. SiC and SiN are also used as advanced ceramics with controlled microstructures because their strength and toughness increase when the grain size diminishes.

Today no society can be called modern or developed unless it has sizeable electronics industry. And there can be no electronics industry without the semiconductors and related technologies. Semiconductors form the backbone of electronics. Semiconductors affect all walks of life whether it is communications, computers, biomedical, power, aviation, defence, enter-tainment, etc. The transistors, integrated circuits (ICs), lasers and detectors, sensors and other semiconductor devices through the items of daily use touch our life everyday. A brief account of the structure of semiconductors and some simple and popular semiconductor devices are presented in this chapter.

21.2 INTRINSIC SEMICONDUCTORS

If the electrical conductivity of a semiconductor is entirely due to motion of charge carriers (electrons and holes) which are generated by thermal excitations from the valence band to the conduction band, i.e. as a result of thermal disruption of covalent bonds, the semiconductor is referred to as *intrinsic* (or *pure*) semiconductors. In intrinsic semiconductors, electrons and holes are always equal in number. A semiconductor is not truly intrinsic unless its impurity content is less than 1 part impurity in 100 million parts of semiconductor.

Ge and Si are two important elemental semiconductors. Each atom in a Ge crystal has four valence electrons. The inner ionic core of the Ge atom has a positive charge of +4 units of electronic charge. The four valence electrons are held by covalent bonds with the valence electrons of four nearest neighbour Ge atoms. A simplified two dimensional representation of a Ge crystal at absolute zero temperature is shown in Fig. 21.1. Since the valence electrons bind one atom to another, they are not available for electrical conduction in the absence of any thermal disruption of chemical bonds at 0 K. Obviously, at 0 K a pure or intrinsic semiconductor behaves like an insulator. In the band picture, we can describe the situation by saying that at 0 K the valence band is completely filled and the conduction band is completely empty.

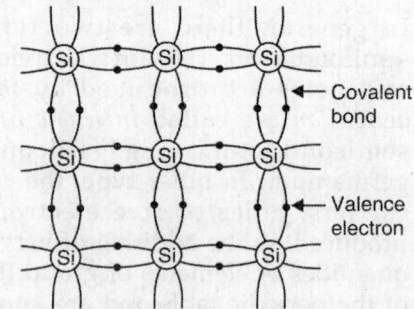

Fig. 21.1: A simplified representation of the crystalline structure of a semiconductor silicon (Si) crystal at absolute zero temperature

At room temperature, some of the valence electrons have enough thermal energy to break their valence or covalent bonds. These electrons, thus, become *free* to move at random throughout the crystal. The energy (E_g) needed to break a covalent bond at room temperature is about 1.1 eV in silicon and 0.72 in Ge. Each electron that leaves the covalent bond creates a *vacant site* or *hole* at its original position. A vacancy in the covalent bond caused by a free electron is represented by a tiny circle (Fig. 21.2a). This vacancy or deficiency of electron is called a *hole*. In the energy band picture, the electrons are thermally excited from the valence band into the conduction band where they free (Fig. 21.2b). The vacancies thus formed in the valence band are holes. A hole is equivalent to a net positive charge equal to that of electron. Whenever a free electron is generated, a hole is created simultaneously at its original position, i.e. free electrons and holes in a semiconductor are always generated in pairs. Obviously, the concentration of free electrons and holes will always be equal in an intrinsic semiconductor. This type of generation of free electron-hole pairs in semiconductors is termed as *thermal generation*. Since a hole has a positive charge, it moves in a direction opposite to that in which an electron moves when an external field is applied (Fig. 21.3).

energy. When an external field is applied to the semiconductor, a drift velocity is superimposed on the random thermal motion of the charge carriers, i.e. electrons and holes. The drift of the electrons in the conduction band and that of holes in the valence band produce an electric current. The electrons move towards the positive electrode, whereas the holes towards the negative electrode (Fig. 21.3). The current produced by the movement of the electrons and holes in opposite directions and since the electron carries a negative charge and the hole a positive charge, thus, the conventional current flows within the semiconductor from the positive electrode to the negative electrode. The energy of a hole is measured downward from the top of the valence band.

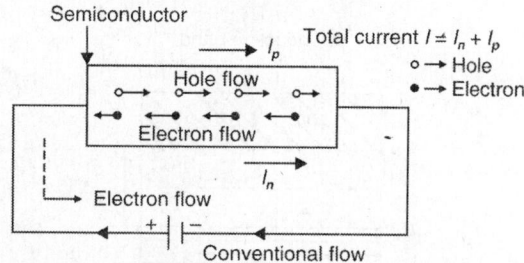

Fig. 21.3: Conduction in intrinsic semiconductor

The motion of the electrons in the valence band may be considered to be equivalent to the motion of holes in the opposite direction. Obviously, the holes also contribute to the conductivity. When an electron-hole pair is thermally created, a valence electron in a neighbouring atom can have sufficient thermal energy to jump into the position of the hole and reconstruct the covalent bond. In doing so, the electron leaves a hole in its initial position. Effectively, the hole moves from one position to the other position. Thus the holes move in the direction opposite to that of valence electrons. That is why a hole behaves like a free positive charge equal in magnitude to the electronic charge. The electrons in the conduction band move more easily than the electrons which cause the motion of holes in the valence band. Hence, the contribution to

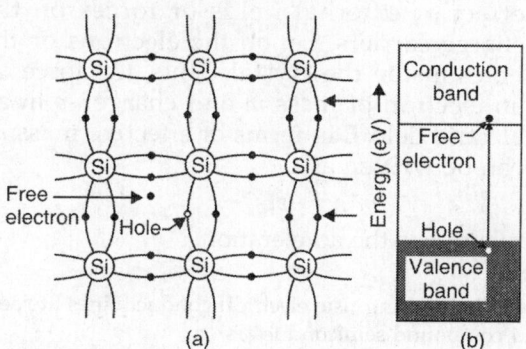

(a) (b)

Fig. 21.2: Creation of an electron–hole pair by the rupture of a covalent bond at room temperature (a) Two dimensional lattice of Si crystal at room temperature (b) Energy band picture of creation of electron–hole pair in Ge crystal at room temperature

The electrons in the conduction band and holes in the valence band moves in a random fashion within the crystal due to their thermal

the electric current by the electrons in the conduction band is more than that by the holes in the valence band. The salient features of an intrinsic semiconductor can be summarised as follows:

i. The number of electrons in the conduction band is equal to the number of holes in the valence band. In equilibrium, the electron concentration n and hole concentration p are equal, i.e. $n = p = n_i$, where n_i is termed intrinsic concentration.

ii. The Fermi level lies in the energy gap exactly between the valence and conduction bands ($E_F = E_g/2$) (Fig. 21.4). The band gaps for some of the elemental and compound semiconductors are given in Table 21.2.

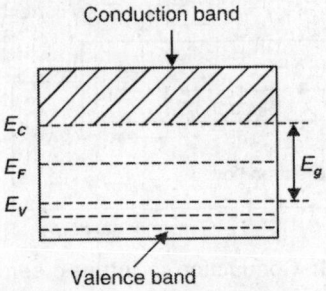

Fig. 21.4: Fermi level for an intrinsic semiconductor

iii. The contribution of the electrons to the electric current is more than that due to the holes.

iv. About 1 atom out of 10^3 atoms of an intrinsic semiconductor contributes to the conduction.

v. An electron and hole can behave as a pair bound to each other. Such a bound pair is usually referred as *exciton*. An exciton is electrically neutral and so does not take part in electrical conduction.

Effective Mass

When an external field is applied to a semiconductor, the charge carriers, i.e. the electrons and the holes, experience forces due to the external applied field and also due to the internal periodic field produced by the crystal. If the external applied field is much weaker than the internal field, the effect of the latter is to modify the mass of the carriers in such a way that the carriers respond to the applied field with this modified mass obeying the laws of classical mechanics. This modified mass of the carriers is termed the *effective mass* of the carriers and usually denoted by m^*. m^* is usually different from the electronic mass (m) in vacuum. The *effective mass* approximation avoids the quantum nature of the problem and allows us to use classical mechanics to study the effect of external fields or forces on the charge carriers, i.e. on the electrons or the holes inside the crystal. Thus, the force on an electron of mass m and charge $-e$ in an electric field E in terms of effective mass m^* can be written as

$$m^* a = -eE$$

where a is the acceleration.

Table 21.2: Band gap energies, electron and hole mobilities and intrinsic electrical conductivities at room temperature (300 K) for few elemental and compound semiconductors

Material	Band gap (eV)	Electrical conductivity [$(\Omega\,m)^{-1}$]	Electron mobility (m^2/Vs)	Hole mobility (m^2/Vs)
Si	1.11	4×10^{-2}	0.14	0.05
Ge	0.67	2.2	0.38	0.18
GaP	2.25	–	0.03	0.015
GaAs	1.42	10^{-6}	0.85	0.04
InSb	0.17	2×10^4	7.7	0.07
CdS	2.40	–	0.03	–
ZnTe	2.26	–	0.03	0.01

Recombination of Electrons and Holes

This is the process in which the free electrons in the conduction band jump into the valence band to combine with holes. In this process of recombination, the electron-hole (*e-h*) pair is destroyed. The rate of recombination is approximately proportional to the product of electron concentration and hole concentration. In this process, the minimum energy released in the form of electromagnetic radiation is equal to the band gap (E_g). We have

$$E_g = h\nu$$

where h is Planck's constant and ν is the frequency of the radiation. The wavelength of the radiation is given by

$$\lambda = \frac{c}{v} = \frac{ch}{E_g}$$

where c (= 3×10^8 m/s) is the velocity of light in free space. The generation and recombination of electrons and holes are illustrated in Fig. 21.5.

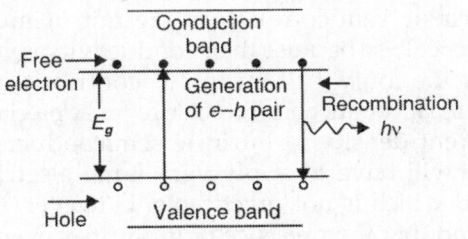

Fig. 21.5: The generation and recombination of electrons and holes in a semiconductor

While some electron–hole pairs are lost due to recombination, new *e–h* pairs are generated due to thermal excitation. For an intrinsic semiconductor at a constant temperature, the rate of combination and the rate of generation of *e–h* pairs are equal, and hence the electron and hole concentrations remain constant at their thermal equilibrium value. With the rise in temperature, the thermal equilibrium value of the electron and the hole concentrations also increases.

Intrinsic Conductivity

Since there are two types of charge carriers (free electrons and holes) in the intrinsic semiconductor, its *specific conductance* is the sum of the conductivities σ_n due to free electrons and σ_p due to holes. Thus, the electric conductivity of an intrinsic semiconductor is

$$\sigma_i = \sigma_n + \sigma_p = ne\mu_n + pe\mu_h \qquad (21.1)$$

where μ_n and μ_h are electron and hole mobilities respectively. The mobility μ is defined as the drift velocity per unit electric field. The magnitude of μ_h is always less than μ_n for semiconductors. The mobilities of electrons and holes are μ_n = 1250 cm²/Vs and μ_h = 480 cm²/Vs. n and p are concentrations (number of charge carriers per cubic meter) of electrons and holes respectively in a semiconductor. Since for an intrinsic semiconductor

$$n = p = n_i \qquad (21.2)$$

one obtains

$$\sigma_i = n_i e \,(\mu_n + \mu_h) \qquad (21.3)$$

The mobilities depend on temperature as a modest power law, i.e. μ goes as $1/T^{3/2}$

or $\qquad \mu_n = \alpha T^{-3/2}$ and $\mu_h = \beta T^{-3/2}$

Thus $\qquad \mu_n + \mu_h = (\alpha + \beta)T^{-3/2} = \gamma T^{-3/2}$

Equation (3) becomes

$$\sigma_i = \gamma n_i T^{-3/2} = \gamma (en_i)T^{-3/2}$$

Now $\qquad n_i = CT^{-3/2} \exp(E_g/2k_BT)$

with $\qquad C = 2 \left(\dfrac{2\pi mk_B}{h^2} \right)^{3/2}$

$$= \left[\frac{2\pi \times 9.1 \times 10^{-31} \times 1.38 \times 10^{-23}}{(6.626)^2 \times 10^{-68}} \right]^{3/2}$$

$$= 4.83 \times 10^{21}$$

Hence $n_i = 4.83 \times 10^{21} T^{-3/2} \exp(-E_g/2k_BT)$

Using this value of n_i, one obtains

$$\sigma_i = \gamma Ce \exp(-E_g/2\,k_BT)$$

or $\qquad \sigma_i = B \exp(-E_g/2\,k_BT) \qquad (21.4)$

or $\rho_i = \dfrac{\exp\left(E_g/2k_BT\right)}{B} = A\exp\left(E_g/2k_BT\right)$

$$(21.5)$$

where ρ_i is the *intrinsic resistivity*. Taking logarithms on both the sides of Eq. (21.5), one obtains

$$\log \rho_i = \frac{E_g}{2k_BT} + \log A \qquad (21.6)$$

Equation (21.6) suggests us a method of determining the energy gap (E_g) of an intrinsic semiconductor. Eq. (21.6) explains why the conductivity of intrinsic semiconductor varies exponentially with increase in temperature.

A measurement of intrinsic resistivity of the material at various temperatures helps us to plot the results on a semilogarithmic paper. The slope of the curve is ($E_g/2k_BT$) and hence the energy gap can be determined (Figs 21.6a and b).

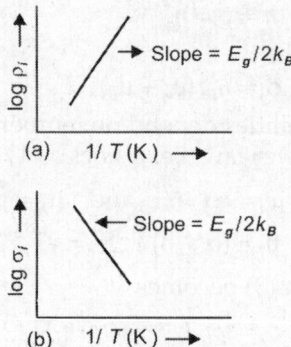

(a)   1/ T (K) ⟶

(b)   1/ T (K) ⟶

Fig. 21.6: (a) log ρ_i versus 1/T (b) log σ_i versus 1/T

Figure 21.7 shows the plot of ρ_i versus $1/T$ for some intrinsic semiconductors. E_g for Ge and Si turns out to 0.72 eV and 1.2 eV respectively. The ratio of the conductivity of copper to that of Ge is 3×10^7 while the corresponding ratio of conduction electron density is 2×10^9. We must remember that the mobility of electrons in Ge(= 0.4 m²/Vs) is about 100 times that in copper. The mobility of holes in Ge is -0.2 m²/Vs, so that electrons carry about 2/3rd of the current while holes carry the remainder.

Thus, the conductivity (or resistivity) of a semiconducting material, in addition to being dependent on electron/and or hole concentrations, is also a function of charge carriers mobilities, i.e. the ease with which electrons and holes are transported through the crystal. Furthermore, magnitudes of electron and hole mobilities are influenced by the presence of those some *crystalline* defects that are responsible for the scattering of electrons in metals-thermal vibrations (i.e. temperature) and impurity atoms (if any).

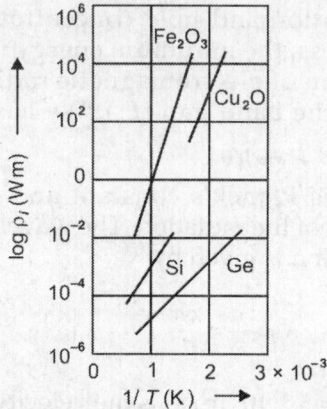

Fig. 21.7: Plot of log ρ_i versus 1/T for some semi-conductors in the intrinsic range

21.3 EXTRINSIC SEMICONDUCTORS

Intrinsic semiconductors are not of much practical use because their conductivity is very low, i.e. only 1 atom in 10^9 contributes to electrical conduction. To achieve an appreciable current density in intrinsic semiconductors, one will have to apply very large electrical field which is not practicable. However, it is found that the presence of impurities even to the extent of 1 in 10^9 alters significantly the conduction of Ge and Si. The impurities that are used have known properties, and can be deliberately introduced in carefully controlled proportions. As a result, one obtains a semiconductor of any predetermined conductivity. For instance, the conductivity of Si is increased thousand times by the addition of 10 parts per million of boron. The conduction that occurs then is termed *impurity conduction* and is of paramount importance in the operation of a semi-conductor device. This process of adding an impurity to an intrinsic semiconductor is called *doping*. The impurity that is added is called *dopant*. When these impurities or foreign atoms are added into the semiconducting structure, the available quantum states are altered; one or more new energy levels may appear in the band structure of the semiconductor. This introduces significant changes in the properties of semiconductors. The resulting material is called as a *doped, impure* or *extrinsic* semiconductor.

Extrinsic semiconductors are of two kinds, one in which the impurity contributes additional electrons in the conduction band and the other in which it contributes additional holes in the valence band. In Ge or Si, addition of group V elements like phosphorous (P), antimony (Sb) and arsenic (As) produces excess free electrons, whereas group III impurity elements like indium (In), boron (B), aluminium (Al) and gallium (Ga) creates excess holes. In a compound semiconductor, e.g. InSb, a group VI element tellurium (Te) generates excess free electrons and group II element zinc (Zn) produces excess holes. We may note that when small amounts (1 part in 10^7 approximately) of tetravalent or pentavalent impurity is added during crystal formation of a semiconductor, the impurity atoms lock into the crystal lattice since they are not greatly different in size from Ge or Si atom and the crystal is not unduly distorted. Depending upon the impurities, an extrinsic semiconductor can be of two types: *N-type* and *P-type*.

N-type Semiconductors

Germanium and silicon are tetravalent. The impurity atoms may be either pentavalent or tetravalent, i.e. from group V and III of the periodic table. If a small quantity of a pentavalent impurity (having 5 electrons in the outermost orbit) like arsenic (As), antimony (Sb) or phosphorous (P) is introduced in germanium, it replaces equal number of germanium atoms without changing the physical state of the crystal. Each of the four out of five valence electrons of impurity, say arsenic, enters into covalent bonds with germanium, while the fifth valence electron is set free to move from one atom to the other

as shown in Fig. 21.8. The impurity is called *donor* impurity as it donates electron and the crystal is called N-type semiconductor. A small amount of arsenic (impurity) injects billions of free electrons into germanium, thus increasing its conductivity enormously. In N-type semiconductors, the *majority* carriers of charge are the electrons and holes are *minority* carriers. This is because when donor atoms are added to a semiconductor, the extra free electrons give the semiconductor a greater number of free electrons than it would normally have. And, unlike, the electrons that are freed because of thermal agitation, donor electrons do not produce holes. As a result, the current carriers in a semiconductor doped with pentavalent impurities are primarily negative electrons.

The impurity atom has five valence electrons. After donating one electron, it is left with +1 excess charge. It then becomes a positively charged immobile ion. It is immobile because it is held tightly in the crystal by the four covalent bonds as shown in Fig. 21.8.

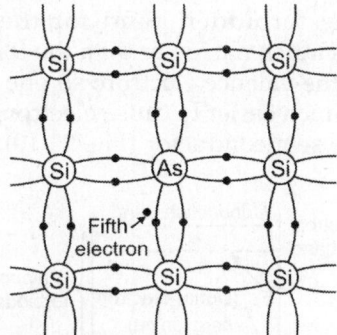

Fig. 21.8: N-type semiconductor

It is important to understand that in N-type semiconductors, although electrons (negatively charged) are the majority carriers, yet the semiconductor doped with impurity remained electrically neutral. Free electrons and holes are generated in pairs due to thermal energy and negative charge of electrons donated by impurity atoms is exactly balanced by positive charge of the immobile ions. Representation of an N-type semiconductor is shown in Fig. 21.9.

We have not shown silicon or germanium atoms in this figure. One should assume them as a continuous structure over the whole background. The fixed or immobile ions are regularly distributed in the crystal structure. The electrons and the holes, being free to move, are shown randomly distributed at any moment.

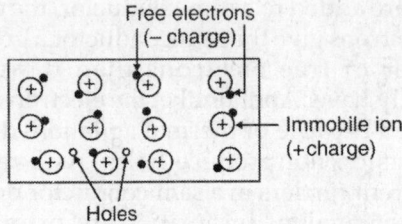

Fig. 21.9: Representation of an N-type semiconductor

Since N-type semiconductors have extra free electrons, and pure semiconductors do not, the energy band diagram for a doped semiconductor is slightly different from that of a pure semiconductor. In effect, another energy level exists; a level for the donor electron, which is closer to the conduction band. The forbidden band for the donor electron is much narrower than the forbidden band for the valence electron; so one can see that it is much easier to cause electron flow in an N-type semiconductor (Fig. 21.10).

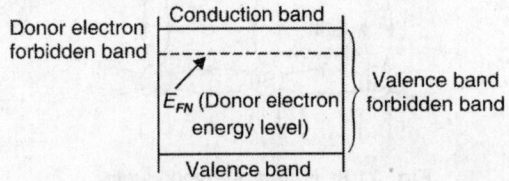

Fig. 21.10: Excess free electrons in N-type semi-conductors produce a donor energy level close to the conduction band

P-type Semiconductors

When a trivalent impurity (having 3 electrons in outermost orbit) like indium (In), boron (B) or gallium (Ga) is added in a germanium intrinsic semiconductor, the impurity atoms will displace some of the germanium atoms in the crystal during its formation as shown in Fig. 21.11. In this case, only three out of the four possible covalent bonds are filled while the fourth bond is vacant and the vacancy acts as a hole. Hence, a hole moves relative to the electron in a direction opposite to the direction of electron, when an electron moves from one bond to the other. This trivalent impurity known as the acceptor or P-type impurity injects into the crystal billions of holes and the majority carriers of the charge are the holes responsible for the conductivity of the crystal. For this reason such crystals are called P-type semiconductors or P-type crystals. P-type semiconductor can be represented as shown in Fig. 21.12.

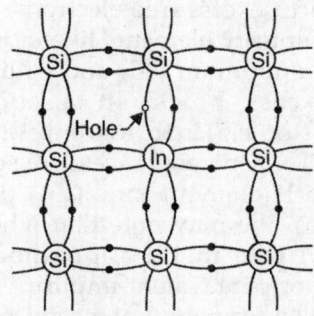

Fig. 21.11: P-type semiconductor

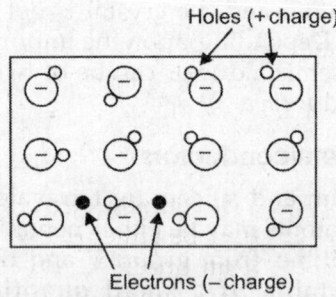

Fig. 21.12: Representation of P-type semiconductor

The energy band diagram of P-type semiconductor also differs from that of the pure semiconductor. Since there is an extra number of holes, which tend to attract electrons, they aid in starting current flow. As a result, the acceptor energy level is also somewhat higher than that of the valence band. However, it is not as high as the donor level (Fig. 21.13).

P-type semiconductor will conduct easily than pure semiconductors, but not quite as easy as N-type semiconductors.

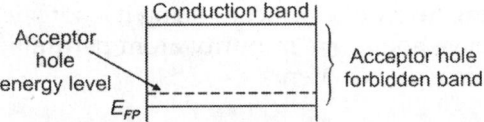

Fig. 21.13: Excess holes in P-type semiconductors introduce an acceptor energy level close to valence level

Effect of Temperature

When the temperature of N-type semiconductor is raised, the number of electron-hole pairs due to thermal excitation from the valence band to the conduction band will increase. The number of electrons coming from the donor level will remain constant as the donor atoms are already ionised. Obviously, at very high temperature the concentration of thermally generated free electrons from the valence band will be much larger than the concentration of free electrons contributed by the donors. At this situation, the hole and the electron concentrations will be nearly equal and the semiconductor will behave like an intrinsic one. On the basis of the same argument, one can say that a P-type semiconductor will also behave as intrinsic semiconductor at very high temperature. Generalizing, we can say that as the temperature of an extrinsic semiconductor increases the semiconductor passes from an extrinsic to an intrinsic one.

Effect of Doping on Electrical Conductivity

The electrical conductivity of a semiconductor increases significantly on doping. Let us consider Ge at room temperature (~300 K).

The intrinsic carrier concentration (n_i) for Ge is $7.2 \times 10^{19}/m^3$. The conductivity of Ge in terms of electron mobility (μ_n) and hole mobility (μ_h) is

$$\sigma = n_i e (\mu_n + \mu_h)$$
$$= 7.2 \times 10^{19} \times 1.6 \times 10^{-19} (0.38 + 0.18)$$
$$= 6.68 \ \Omega^{-1} m^{-1}$$

The values of electron and hole mobilities for Ge are taken from Table 21.3.

On doping Ge with a pentavalent impurity of concentration of 1 ppm, we have

$$N_d = 4.42 \times 10^{28} \times 10^{-6}$$
$$= 4.42 \times 10^{22}$$

Obviously, $N_d \sim 600 n_i$

Now, we calculate the hole and electron concentration in the doped Ge at 300 K, using

$$p_0 = n_i^2 / n_0 \text{ and } n_0 \approx N_d$$

So, $n_0 \approx 4.4 \times 10^{22}/m^3$ and $p_0 \approx 1.2 \times 10^{17}/m^3$ where N_d is the donor impurity density.

Ignoring the hole contribution to the conductivity of doped Ge, one obtains

$$\sigma = n_0 e \mu_n = 4.4 \times 10^{22} \times 1.6 \times 10^{-19} \times 0.38$$
$$= 2675.2 \ \Omega^{-1} m^{-1}$$

This shows that on doping, the electrical conductivity of Ge increases significantly, i.e. by a factor of ~400.

Charge Carrier Densities in Extrinsic Semiconductors

We have seen that the introduction of impurities in pure semiconductors increases the density of one type or another type of charge carriers. The product of holes and electrons in a semiconductor is constant depending on the width of energy gap and temperature and hence the introduction of the impurities results in an increase in the density of one type of carrier and a decrease in the density of the other type of carrier. In an extrinsic semiconductor, the carriers introduced by the impurities are called *majority carriers* and the other type are called *minority carriers*. It is important to note that the low value for minority carrier density is due to added recombination.

Let N_d be the donor impurity density, N_a the acceptor impurity density, p the density of holes and n the density of electrons in an extrinsic semiconducting material. From the condition of charge neutrality, we have

$$N_d + p = N_a + n$$

or

$$n = (N_d - N_a) + p$$

$$= (N_d - N_a) + \frac{n_i^2}{n} \quad (\because np = n_i^2)$$

or $\quad n^2 - (N_d - N_a)n - n_i^2 = 0 \quad\quad (21.7)$

Solving the quadratic equation in n, one obtains

$$n = \frac{(N_d - N_a) \pm \sqrt{(N_d - N_a)^2 + 4n_i^2}}{2} \quad (21.8)$$

or $\quad n \approx (N_d - N_a)$ when $(N_d - N_a) >> n_i$

Obviously, the electron density (n) in the N-type semiconductor equals the difference in the donor and acceptor impurity densities when they are large compared to the intrinsic density, n_i. Similarly, the hole density (p) in a P-type semiconductor is given by

$$p = (N_a - N_d) \text{ when } (N_a - N_d) >> n_i$$

Donor and Acceptor States

When an impurity atom from group V of the periodic table, say phosphorous is added to a pure Ge or Si crystal as a pentavalent impurity, these impurity atoms enter the lattice by substitution for normal atoms, and not in interstitial positions. These impurity atoms contribute five electrons per atom to the valence band, i.e. we have an extra electron per impurity atom. These additional electrons (which cannot be accommodated in the valence band of the original lattice) occupy some discrete energy levels just below the conduction band; the separation may be a few tenths of an electron volt. These excess electrons are released by the impurity atoms and excited into the conduction band. The excited electrons then contribute to the electrical conductivity of the semiconductor. Conversely, the impurity may consist of atoms having *fewer* electrons than of a semiconductor (Fig. 21.14a). For the cases in which Si and Ge are the host substances, the impurity atoms could be boron or aluminium, each of which contributes only three electrons. In this situation, the impurity introduces vacant discrete energy levels, very close to the top of the valence band. Therefore it is easy to excite some of the more energetic electrons in the valence band into the impurity levels.

This process produces vacant states, or holes, in the valence band. These holes then act as positive electrons (Fig. 21.14b). We must note that to produce significant changes in the conductivity of a semiconductor, it is sufficient to have about one impurity atom per million semiconductor atoms.

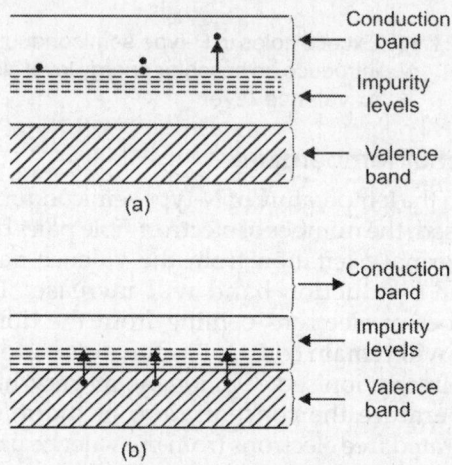

Fig. 21.14: Impurities in a semiconductor (a) Donors or N-type (b) Acceptors or P-type

We have already seen that the crystal as a whole remains neutral because the electron remains in the crystal. We have further seen that the band diagram of N-type or P-type semiconductor differs from that of the pure semiconductor. The band diagram for N-type or P-type semiconductor explains clearly why the conduction becomes possible in impure semiconductor at comparatively low temperatures.

Fermi Level in Extrinsic Semiconductor

We have read that Fermi level is situated in the middle of the band gap in an intrinsic semiconductor as the electron and hole densities are equal. When the intrinsic semiconductor is doped, the carrier densities change, consequently the position of the Fermi level also changes. The shift in the position of the Fermi level can easily be related to the majority carrier density in an extrinsic semiconductor if it is assumed that the addition of impurities do not affect the densities of energy states in the conduction and valence bands.

Let N_c and N_v denote the density of states in the conduction band and valence band respectively. We have for an intrinsic semiconductor

$$
\left.
\begin{aligned}
n &= N_c \exp\left[\frac{E_{fi}-E_c}{kT}\right] \\
\text{and} \quad p &= N_v \exp\left[\frac{E_c-E_{fi}}{kT}\right]
\end{aligned}
\right\} \quad (21.9)
$$

Here E_{fi} is the energy associated with the Fermi level in an intrinsic semiconductor. For an intrinsic semiconductor, we have $n = p$, and therefore from Eq. (21.9), we have

$$
\frac{N_c}{N_v} = \exp\left[\frac{E_c - E_v - 2E_{fi}}{kT}\right]
$$

Let E_{fi} be the energy associated with the Fermi level in an N-type semiconductor having an electron density n, then we have

$$
n = N_c = \exp\left[\frac{E_{fn}-E_c}{kT}\right]
$$

and

$$
p = N_c = \exp\left[\frac{E_v-E_{fi}}{kT}\right]
$$

$$
\therefore \quad \frac{n}{p} = \frac{n^2}{n_i^2} = \frac{N_c}{N_v}
$$

$$
= \exp\left[\frac{2E_{fn}-E_c-E_v}{kT}\right] = \exp\left[\frac{2\left(E_{fn}-E_{fi}\right)}{kT}\right]
$$

or

$$
n = n_i \exp\left[\frac{E_{fn}-E_{fi}}{kT}\right] \quad (21.10)
$$

Similarly, one obtains for a P-type semiconductor

$$
p = n_i \exp\left[\frac{E_{fi}-E_{fp}}{kT}\right] \quad (21.11)
$$

Thus, the shift in the Fermi level in the N-and P-type semiconductor can be expressed as

$$
\left.
\begin{aligned}
E_{fn}-E_{fi} &= kT\ln\frac{n}{n_i} \\
E_{fi}-E_{fp} &= kT\ln\frac{p}{p_i}
\end{aligned}
\right\} \quad (21.12)
$$

Figure 21.15 represents the shift in the Fermi level in the N-and P-types semiconductors.

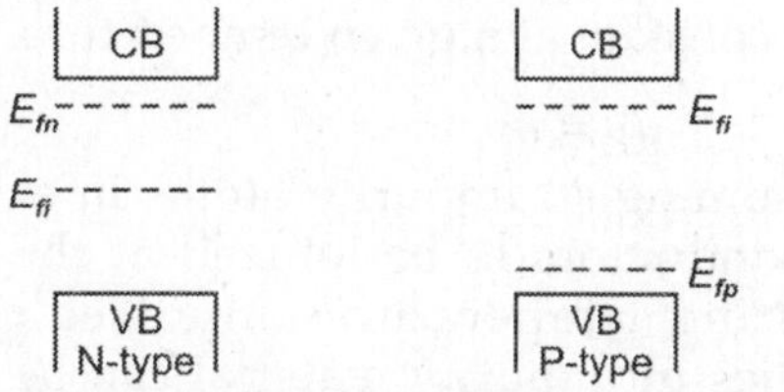

Fig. 21.15: Shift in the Fermi level in the N-and P-types semiconductors

Thermal Ionization of Extrinsic Semiconductors

When the temperature of an extrinsic semiconductor is raised above 0 K, the impurity atoms get ionized. Due to ionization the donor impurity atoms give rise to electrons in the conduction band and the acceptor impurity atoms give rise to holes in the valence band. These electrons or holes along with those generated by intrinsic action, then serve as the current carriers at temperature. Two cases of interest are mentioned below:

1. Conduction electron concentration is equal to

$$
n = n_0 \exp[(E_F - E_g)/kT] \approx N_d \quad (21.13)
$$

This shows that under the present conditions the concentration of conduction electrons is approximately equal to the concentration of donors. This means that all the donors are ionized. Equation (21.13) suggests that at room temperature, the impurity concentration for Si and Ge up to 10^{14} to 10^{16} donors per cc suits this range, i.e. if we dope an intrinsic semiconductor crystal with this concentration of donors, one can certainly predict that one will have $\sim 10^{16}$ electrons/cc.

2. At higher temperature, the carrier concentration is proportional to $\sqrt{N_d}$.

Charge Densities in Extrinsic Semiconductors

The density of impurity atoms in N- and P-type materials is so low compared to the density of semiconductor atoms that the rate of thermal pair generation is not affected appreciably by the presence of impurity atoms. In the case of intrinsic semiconductors, we have seen that the concentration product was a constant at a given temperature.

$$np = n_i^2 \qquad (21.14)$$

Assuming all impurity atoms in extrinsic semiconductors to be ionized at the usual operating temperatures, the free charge densities in impurity material can be based upon N_D, the donor atom density in N-material, or N_A, the acceptor atom density in P-material.

The electrical neutrality of the material demands

$$p = N_D = n + N_A \qquad (21.15)$$

The LHS of Eq. (21.15) gives the total positive charge as the sum of holes in the valence bonds and the positive charge associated with the donor atoms that have given up electrons and become positive ions. The RHS of Eq. (21.15) sums the negative charge of the electron density n and the negative charge due to electrons held by the ionized acceptor atoms.

Using Eq. (21.14), one can write

$$n = \frac{n_i^2}{p} \quad \text{and} \quad p = \frac{n_i^2}{n} \qquad (21.16)$$

Only donor impurities are introduced in N-type material, so $N_A = 0$. The donor density will be made much larger than the density of intrinsic holes, i.e.

$$N_D >> p$$

and in N-type material the electron density is written from Eq. (21.15) as

$$n \cong N_D \qquad (21.17)$$

On the similar reasoning for P-type material, we have $N_D = 0$ and $N_A >> n$. One obtains density relations in P-type material as

$$n \cong \frac{n_i^2}{N_A} \qquad (21.18)$$

$$p \cong N_A \qquad (21.19)$$

From the above results, we can conclude that the density of majority carrier approximates the impurity atom density at usual ambient temperatures and the density of minority carriers reduces below the intrinsic level. This means that the increased electron density in N-type material raises the probability that an electron will meet and recombine with a hole and so the hole density decreases to maintain n^2 constant.

21.4 SEMICONDUCTOR DEVICES

A semiconductor device can be defined as a unit which consists, partly or wholly, of semiconducting materials and which can perform useful functions in electronic apparatus and solid state research. Examples of semiconductor devices are semiconductor diodes (P-N junction), transistors, integrated circuits (ICs), etc. Si, Ge and GaAs are most commonly used materials for the fabrication of semiconductor devices. For convenience, the properties of these semiconductors are summarized in Table 21.3.

Almost all semiconductor devices are comprised of a single crystal semiconductor incorporating two or more semiconducting regions of different impurity density. The difference in the electric fields and carrier densities associated with differently doped regions, called junctions, permit a wide range of essentially nonlinear conductivity effects in devices incorporating two, three or more distinct regions. Most semiconductor devices can be understood through the simplest of such junctions, called the P-N junction, which is a system of two semiconductors in physical contact, one with excess of electrons (N-type) and other with excess of holes (P-type) (Table 21.3).

P-N Junction

When a P-type semiconductor is brought into contact with N-type semiconductor as the process of crystallisation is taking place, the

Table 21.3: Properties of Si and Ge semiconductors

Property	Symbol	Unit	Value Germanium (Ge)	Silicon (Si)	GaAs
Atomic number			32	14	
Atomic weight			72.6	28.08	144.63
Density		kg/m^3	5.32×10^3	2.33×10^3	5.32×10^3
Atom concentration		atoms/m^3	4.4×10^{28}	5×10^{28}	2.21×10^{28}
Relative dielectric constant	ε_r		16	11.8	10.9
Band gap at 0 K	E_{g0}	eV	0.785	1.21	1.43
Band gap at 300 K	E_g	eV	0.72	1.1	1.32
Intrinsic carrier (generation) at 300 K	n_i	$\dfrac{Carriers}{m^3}$	2.5×10^{19}	1.5×10^{16}	9.0×10^{12}
Crystal structure			Diamond	Diamond	Zinc blende
Lattice constant	a	Å	5.65748	5.43086	5.6534
Melting point		°C	936	1420	1250
Minority carrier lifetime		second (s)	$\sim 10^{-3}$	$\sim 2.5 \times 10^{-3}$	$\sim 10^{-8}$
Breakdown field		V/m	$\sim 10^7$	$\sim 3 \times 10^7$	$\sim 4 \times 10^5$
Diffusion constant	D_n (electron)	m^2/s	0.009842	0.003367	0.001036
	D_n (holes)		0.004662	0.001295	0.000906
Effective density of states in the conduction band	N_c	m^{-3}	1.04×10^{25}	2.8×10^{25}	4.7×10^{22}
Effective density of states in the valence band	N_v	m^{-3}	6.4×10^{24}	1.02×10^{25}	7.0×10^{22}
Intrinsic conductivity	σ_i	Sm^{-1}	2.2428	0.4325×10^{-3}	1.2832×10^{-6}
Mobility (drift) at 300 K	μ_n (electron)	$\dfrac{m^2}{Vs}$	0.38	0.13	0.85
	μ_h (hole)	$\dfrac{m^2}{Vs}$	0.18	0.05	0.04
Work function	W	Volt (V)	4.4	4.8	4.7
Raman phonon energy		eV	0.037	0.063	0.035

resulting combination is called a P-N junction. This junction has important properties and is, in effect, the basis of modern semiconductor theory and practice. Most semiconductor devices contain one or more P-N junctions. The most important characteristic of a P-N junction is its ability to conduct current in one direction only. In the reverse direction, it offers very high resistance.

P-N Junction with No External Voltage

Figure 21.16 shows a P-N junction just immediately after it is formed. There is no external voltage connected to the P-N junction. Since N-type material has a high concentration of free electrons while P-type material has a high concentration of holes, the following processes are initiated:

i. At the junction, holes from the P-region diffuse into the N-region and free electrons from the N-region diffuse into the P-region. This process is called *diffusion*. Holes combine with the free electrons in the N-region whereas electrons combine with the holes in the P-region.

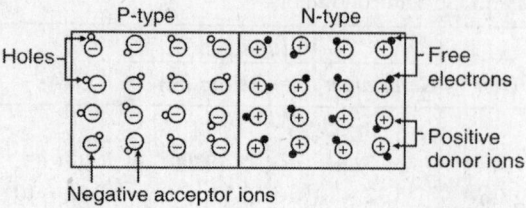

Fig. 21.16: A P-N junction when just formed

ii. The diffusion of holes from P-region to N region and electrons from N-region to P-region across the junction takes place because they move haphazardly due to thermal energy and also because there is a difference in their concentrations (the P-region has more holes whereas N-region has more free electrons) in the two regions.

iii. As the free electrons move across the junction from N-type to P-type, positive donor ions are uncovered, i.e. they are robbed off free electrons. Hence, a positive charge is built on the N-side of the junction. At the same time, the free electrons cross the junction and uncover the negative acceptor ions by filling in the holes. Therefore, a net negative charge is established on P-side of the junction. When a sufficient number of donor and acceptor ions are uncovered, further diffusion is stopped. It is because now, a barrier is set up against further movement of charge carriers. This is called *potential barrier or junction barrier*. The potential barrier is of the order of 0.1 to 0.3 volts. Figure 21.17 shows the electrostatic potential difference across the P-N junction. How this potential barrier is developed? It is because now positive charge (ions) on N-side repels holes to cross from P-type to N-type and negative charge (ions) on N-side repels free electron to enter from N type to P-type. Because of this difference in potential established between the two sections, which inhibits further electron-hole combinations at the junction, and

the Fermi level of the two sides is in the same level as shown in Fig. 21.17c.

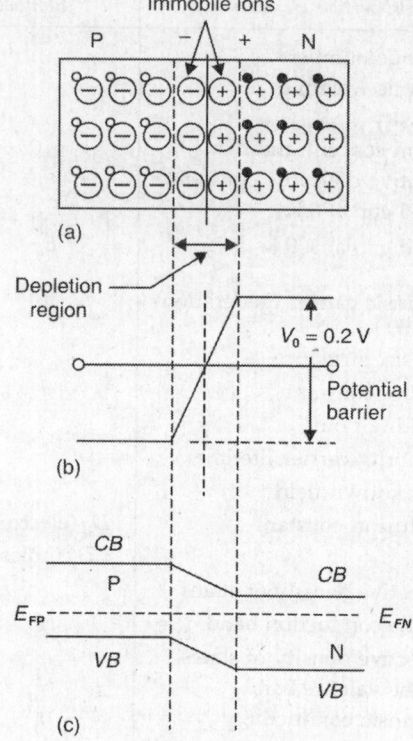

Fig. 21.17: Potential barrier across the P-N junction

iv. The region across the P-N junction in which the potential changes from positive to negative is called the *depletion region*. The width of this region is of the order of 6×10^{-8} m. Since this region has immobile (fixed) ions which are electrically charged, it is also called as the *space-charge region*. Outside this region on each of side of the junction, material is still neutral.

v. The potential barrier for a silicon P-N junction is about 0.7 V, whereas for a germanium P-N junction it is approximately 0.3 V.

The potential barrier discourages the diffusion of majority carriers across the junction. However, the potential barrier helps minority carriers (few free electrons in the

P-region and few holes in the N-region) to drift across the junction. The minority carriers are constantly generated due to thermal energy. But electric current cannot flow since no circuit has been connected to the P-N junction.

Forward and Reverse Biasing

Forward Biasing

We have seen that the natural tendency of the majority carriers (free electrons in the N-section and holes in the P-section) was to combine at the junction. This is how the *depletion region* and *potential barrier* were formed. Actually, the combination of electrons and holes at the junction allows electrons to move in the same direction in both the P-and N-sections. In the N-section, free electrons move toward the junction; in the P-section, for the holes to move toward the junction, valence electrons move away from the junction. Therefore, electron flow in both the sections is in the same direction. This, of course, would be the basis of current flow.

With the P-N junction alone, the action stops because there is no external circuit and because of the potential barrier that builds up. So, for current to flow, a battery can be connected to the diode to overcome the potential barrier. And the polarity of the battery should be such that the majority carriers in both sections are driven toward the junction. When the battery is connected in this way, it provides forward bias, causing *forward or high current* to flow, because it allows the majority carriers to provide the current flow.

To apply forward bias, positive terminal of the battery is connected to P-type and negative terminal to N-type as shown in Fig. 21.18a. The applied forward potential establishes an electric field which acts against the potential barrier field. Obviously, the resultant field is weakened and the barrier height is reduced at the P-N junction (Figs 21.18b and c). Since potential barrier height is very small (~0.2 V), hence a small forward voltage is sufficient to completely eliminate the barrier. Obviously, at some forward voltage, the potential barrier at the P-N junction can be eliminated altogether. Then the junction resistance will

become almost zero and a low resistance path is established for the entire circuit. Thus, a large current is generated in the circuit even for the small potential difference applied. Such a circuit is called *forward bias circuit* and the current is called forward current. The salient features of the forward bias circuit are as follows:

i. At some forward voltage, the potential barrier is eliminated altogether.

ii. The P-N junction offers low forward resistance (r_f) to current flow.

iii. The magnitude of current in the circuit due to the establishment of low resistance path depends upon the applied forward voltage and it reduces as the voltage is increased thereby forward current increases. It is given by

$$I = I_0\left(e^{V/\eta V_T} - 1\right) \qquad (21.20)$$

where V = applied voltage

$V_T = \dfrac{T}{11600}$ here T = temperature in Kelvin

η = 1 for Ge and 2 for Si

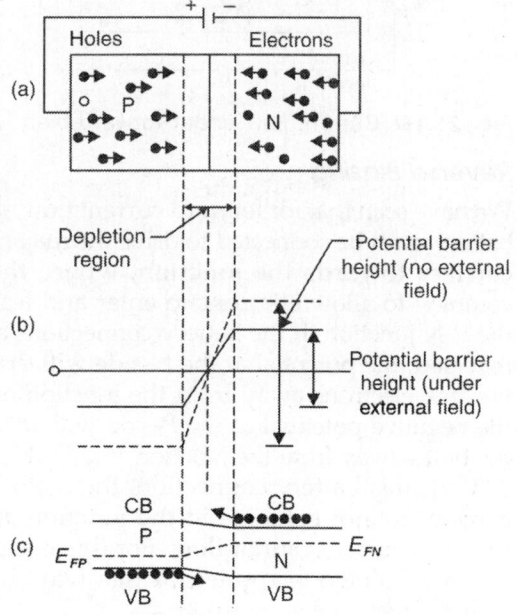

Fig. 21.18: P-N junction showing forward biasing

The mechanism of current flow in a forward biased P-N junction is as follows:

i. The free electrons from the negative battery terminal continue to arrive into the N-region while the free electrons in the N-region move towards the P-N junction.

ii. The electrons travel through the N-region as free electrons. Obviously, current in N-region of the P-N junction is by free electrons.

iii. When these free electrons reach the P-N junction, they combine with holes and become valence electrons. Since a hole is in the covalent bond, and hence when a free electron combines with a hole, it becomes a valence electron.

iv. Current in the P-region is by holes. The electrons travel through P-region as valence electrons.

v. These valence electrons after leaving the crystal flow into the positive terminal of the battery.

The current flow in a forward biased P-N junction is illustrated in Fig. 21.19.

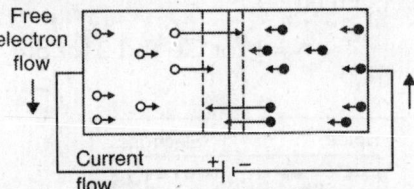

Fig. 21.19: Current flow under forward bias

Reverse Biasing

We have seen that for forward current flow, the battery must be connected to drive the majority carriers towards the junction, where they combine to allow electrons to enter and leave the P-N junction. If the battery connections are reversed, the potential at the N-side will draw the free electrons away from the junction and the negative potential at the P-side will attract the holes away from the junction.

With this battery connection, the majority carriers cannot combine at the junction and majority current cannot flow. For this reason, when a voltage is applied in this way, it is called reverse bias.

However, reverse bias can cause a *reverse current* to flow because minority carriers are present in the semiconductor sections. Remember that although the P-section is doped to have excess holes, yet some electrons are freed because of thermal agitation. Also, although the N-section is doped to have excess free electrons, some electrons are freed to produce holes in the N-section. The free electrons in the P-section and the holes in the N-section are the minority carriers. Now, with reverse bias, one can see that the battery potentials repel the minority carriers toward the junction. As a result, these minority carriers cross the P-N junction in exactly the same way that the majority carriers did with forward bias. However, since there are much fewer minority carriers than there are majority carriers, this minority current, or reverse current as it is usually called, is much less and is of the order of µA. Reverse bias P-N junction is shown in Fig. 21.20 The salient features of the reverse biased P-N junction are as follows:

i. The height of the potential barrier is increased and width of depletion region also increases (Figs 21.20b and c).

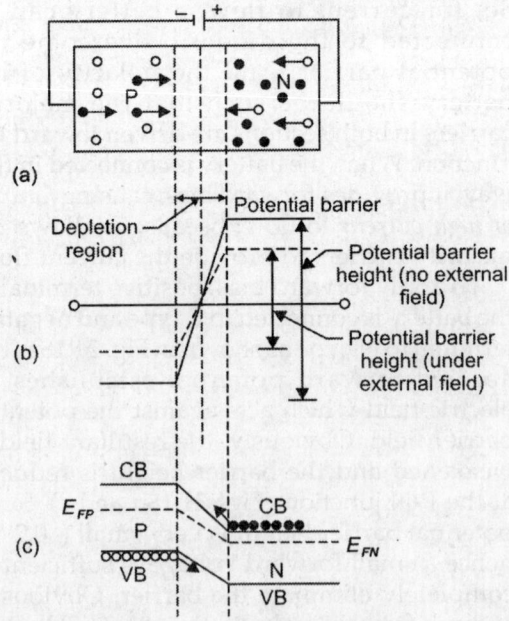

Fig. 21.20: P-N junction showing reverse biasing

ii. The reverse bias P-N junction offers very high resistance to current flow. This resistance is called reverse resistance (R_r).

iii. Due to the establishment of high resistance path, very small current flows in the circuit. This current is usually called as the *leakage current or reverse saturation current*.

The current in the above situation is given by

$$I = I_0\left(e^{V/\eta V_T} - 1\right) \qquad (21.21)$$

Since applied voltage V is negative, hence as it increases, the first term in RHS reduces very fast and current from Eq. (21.21) reduces to $I = (-)I_0$ which is quite low and constant.

Volt–Ampere Characteristic of P-N Junction

The P-N junction can be represented by a symbol of arrow and dash as shown in Fig. 21.21. The arrow head represents the P-section of the crystal and shows the direction of flow of holes or conventional current. Since the P-N junction diode's resistance changes according to the direction of current flow and hence it is called a *nonlinear* device. Basically, its nonlinearity is dependent on the polarity of the applied voltage.

P O———▶|———O N

Fig. 21.21: Symbol of the P-N junction diode

A graph between the potential difference across the P-N junction and the current through the junction is called the *V–I* characteristics of the P-N junction direction diode and is shown in Fig. 21.22.

In the forward direction, though considerably more current flows and the current for the most part increases linearly as the bias voltage is increased, in the forward direction, then, the P-N junction can be considered a linear device over a large portion of its operating curve. The small portion of the curve that is just above zero bias is nonlinear. This results because both majority and minority currents actually comprise the overall current. Since the minority carriers are low energy carriers, majority current starts

first and then as the voltage is raised, minority current joins in, causing a nonlinear rise in current. But as the voltage is increased further, minority current becomes saturated since there are only few minority carriers. The curve then follows the majority current increase which is linear.

Because of the nonlinearity of the curve, if

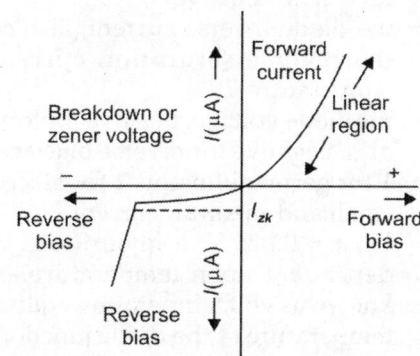

Fig. 21.22: *V–I* characteristics of a P-N junction diode

a very small signal voltage is applied to the diode so that it only operates around the knee, the signal will be distorted. The signals must be large enough so that they operate mostly over the linear part of the curve.

When reverse bias is applied, a slight reverse current flows. This reverse current increases only negligibly as the bias voltage is increased a lot (20 to 25 volts). At this stage, current suddenly rises in the reverse direction due to the breakdown of the crystal, i.e. covalent bonds of the crystal are broken in very large number. This breakdown reverse bias is called the *breakdown* or *zener voltage*. Diodes are also designed to produce a useful wide range of zener breakdown region and are used in special voltage-regulating circuits.

Effect of Temperature on P-N Junction Diodes

We know that temperature rise boosts the generation of electron–hole pairs in semiconductors and increases their conductivity. From the consideration of the energies of the carriers crossing the depletion region in a P-N junction diode, an involved calculation yields the following relation between diode current, voltage and temperature.

$$I = I_0 \left[\exp\left(\frac{qV}{\eta kT}\right) - 1 \right] = I_0 \left[\exp\left(\frac{V}{\eta V_T}\right) - 1 \right]$$

$$(21.22)$$

where

$I =$ the diode current (forward if positive, reverse if negative)

$I_0 =$ the diode reverse current, also called the reverse saturation current, at temperature T

$V =$ the diode voltage, positive for forward bias, negative for reverse bias; in volts

$\eta =$ 1 for germanium and 2 for silicon for small and medium current

$V_T = kT/q = 0.025$ V, a quantity in volts, dependent upon temperature and is known as volt temperature equivalent

$T =$ temperature of the diode junction (K).

On increasing the temperature for forward characteristic (Fig. 21.22), it shifts to left showing increase in current for same voltage and it shifts to right when temperature is decreased showing biased current.

We have studied that a P-N junction conducts easily when forward biased and practically does not conduct when reverse biased (very small current flows). This characteristic of P-N junction (semiconductor diode) is very similar to that of vacuum diode and we may also say that it acts like a switch which conducts current in one direction in the ON position and does not conduct in the OFF position.

For an ideal semiconductor diode, the (V, I) characteristics may be viewed as made of two straight lines as shown in Fig. 21.23a. Such a volt–ampere characteristics consisting of straight line segments is said to be piecewise linear.

However, for simplification we assume the (E, I) characteristic of solid state diode (so also vacuum diode) as shown in Fig. 21.23c.

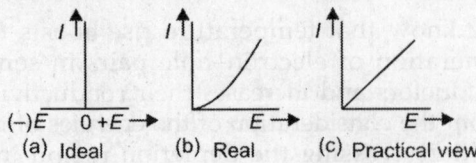

(a) Ideal   (b) Real   (c) Practical view

Fig. 21.23: Diode characteristics

Rectifiers

A rectifier may be defined as a device which converts AC voltage/current into DC voltage/current.

Half Wave Rectifier

It makes use of a unidirectional conduction device like a diode. Whenever an AC signal $e = E_m \sin \omega t$ is applied across a circuit consisting of a P-N diode junction with some resistance R_L called load (Fig. 21.24) during the positive half cycle of input AC signal, the diode conducts and there is current through load, but during negative half cycle we find no current through it. Thus, the output voltage across load appears during positive half of input cycle. We call that the half wave is rectified in the output.

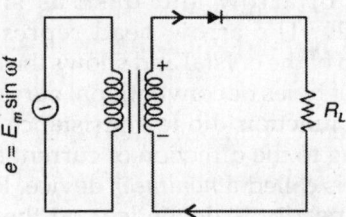

Fig. 21.24: Half wave rectifier

The shape of the current through R_L consists of half sinusoidal wave as shown in Fig. 21.25. So also the shape of output voltage (e_0) will be similar to it.

The current flowing through diode during positive half cycle

$$i_b = \frac{\text{Applied voltage}}{\text{Total resistance of the circuit}}$$

$$i = i_b \frac{E_m \sin \omega t}{R_S + r_f + R_L} \quad \text{when } 0 < \omega t < 2\pi \quad (21.23)$$

and $\quad i_b = 0 \qquad$ when $\pi < \omega t < 2\pi \quad (21.24)$

Here R_S is the resistance of secondary coil, r_f the forward resistance of diode and R_L the load.

If R_S and r_f are very small in comparison with the load R_L, then Eq. (21.23) can be written as

$$i = i_b \frac{E_m}{R_L} \sin \omega t \quad \text{when } 0 < \omega t < \pi \quad (21.24a)$$

or $\quad i = i_b = I_m \sin \omega t$

where $I_m = \dfrac{E_m}{R_L}$ is called the peak value of current.

The output voltage across R_L

$e_0 = i_b \times R_L$ when $0 < \omega t < \pi$

or $\quad e_0 = \left[\dfrac{E_m}{R_S + r_f + R_L} \sin \omega t \right] R_L \quad (21.25)$

$\qquad = (I_m R_L) \sin \omega t \qquad\qquad (21.25a)$

A Fourier analysis of the half sinusoidal voltage pulse appearing across the load yield

$e = \dfrac{E_m}{\pi} + \dfrac{E_m}{2} \sin \omega t - \dfrac{0.2 E_m}{3\pi} \cos 2\omega t \ (21.26)$

Thus, a diode is a frequency converter in which an input frequency ω is changed to a large number of output frequencies. In rectification, we desire the DC or zero frequency component and it is $\dfrac{E_m}{\pi}$.

The DC or average current I_{DC} is given by

$I_{DC} = \dfrac{1}{2\pi} \left[\int_0^\pi i \, d(\omega t) + \int_\pi^{2\pi} i \, d(\omega t) \right] \quad (21.27)$

The second term in the bracket is zero. There is no conduction during π to 2π interval of input. So

$I_{DC} = \dfrac{1}{2\pi} \int_0^\pi i \, d(\omega t) \qquad\qquad (21.28)$

Substituting for i from Eq. (21.23), one obtains

or $\quad I_{DC} = \dfrac{1}{2\pi} \int_0^\pi \dfrac{E_m}{R_L} \sin \omega t \, d(\omega t)$

$\qquad = \dfrac{(-)E_m}{2\pi R_L} \big[\cos \omega t \big]_0^\pi$

or $\quad I_{DC} = \dfrac{E_m}{\pi R_L} = \dfrac{I_m}{\pi} \qquad\qquad (21.29)$

where R_S and r_f have been neglected. This current has been shown as a dashed straight line in Fig. 21.25c.

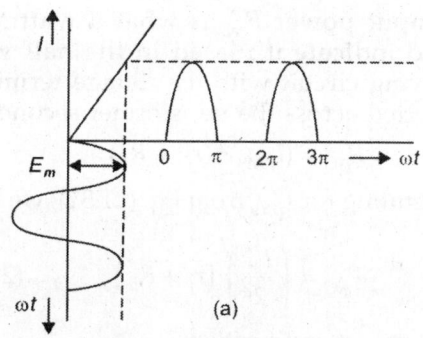

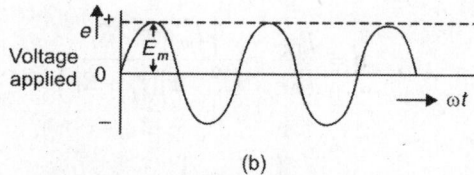

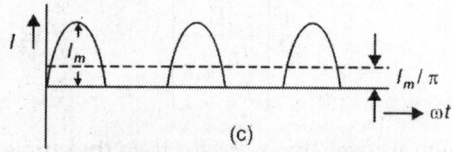

Fig. 21.25: Half wave rectification

The effective value of the current in the output is the RMS value of current. It is given by

$I_{RMS} = \sqrt{\dfrac{1}{2\pi} \int_0^{2\pi} i^2 d(\omega t)} \qquad (21.30)$

or $\quad I_{RMS} = \sqrt{\dfrac{1}{2\pi} \int_0^\pi i^2 d(\omega t) + \int_\pi^{2\pi} i^2 d(\omega t)}$

Substituting for i and noting that there is no conduction for the interval π to 2π, i.e. second term is zero, we have

$I_{RMS} = \dfrac{I_m}{2} \qquad\qquad (21.31)$

Efficiency of Rectifier

The efficiency of rectifier is defined as the ratio of the DC output power to the AC input power.

The useful DC output power is obtained across the load R_L and so

$P_{DC} = I_{DC}^2 R_L = \left(I_m / \pi \right)^2 R_L \qquad (21.32)$

AC input power P_{AC} is what a wattmeter would indicate if placed in the half wave rectifying circuit with its voltage terminals connected across the transformer secondary

$$P_{AC} = (I_{RMS})^2 \, (r_f + R_L)$$

Substituting for I_{rms} from Eq. (21.31), we have

$$P_{AC} = \left(\frac{I_m}{2}\right)^2 (r_f + R_L) \qquad (21.33)$$

Efficiency of rectification

$$\eta = \frac{P_{DC}}{P_{AC}} = \frac{\left[I_m^2 / \pi^2\right] R_L}{\left(I_m^2 / 4\right)\left(r_f + R_L\right)}$$

$$= \frac{4}{\pi^2} \left[\frac{1}{\dfrac{r_f}{R_L} + 1}\right] \qquad (21.34)$$

If we assume that $r_f << R_L$, then the efficiency

$$\eta = \frac{4}{\pi^2} = 0.406 \qquad (21.35)$$

Theoretically, the maximum value of efficiency of a half wave rectifier is 40.6% (under no diode loss), i.e. under the best conditions only 40.6% of AC input power is converted into DC power in the load. The rest exists as AC power in the load.

Ripple Factor

The objective of a rectifier is to convert the AC into DC. A measure of the purity of DC output is the ripple factor r which is defined as the ratio of two current (or voltage) components.

$$r = \frac{\text{Effective (or RMS)}}{\text{value of AC component}}{\text{DC component}}$$

or

$$r = \frac{(I_r)_{RMS}}{I_{DC}} = \frac{(V_r)_{RMS}}{V_{DC}} \qquad (21.36)$$

Let I_{RMS} be the RMS value of i, the total half wave rectified output current. Similarly, let

$(I_r)_{RMS}$ be the RMS value of the ripple current i_r consisting of the alternating components.

Now

$$I_{RMS}^2 = I_{DC}^2 + (I_r)_{RMS}^2$$

or

$$(I_r)_{RMS} = \sqrt{I_{RMS}^2 - I_{DC}^2}$$

Hence

$$r = \frac{(I_r)_{RMS}}{I_{DC}} = \sqrt{I_{RMS}^2 - I_{DC}^2}\Big/ I_{DC}$$

or

$$r = \sqrt{\left(\frac{I_{RMS}}{I_{DC}}\right)^2 - 1} \qquad (21.37)$$

From Eqs. (21.31) and (21.29), one obtains

$$I_{RMS} = \frac{I_m}{2}$$

and

$$I_{DC} = I_m / \pi$$

Substituting these values in Eq. (21.37) we have ripple factor

$$r = \sqrt{\left(\frac{I_m / 2}{I_m / \pi}\right)^2 - 1}$$

or

$$r = \sqrt{\frac{\pi^2}{4} - 1} = 1.21 \qquad (21.38)$$

Expressed as percentage, this indicates that the amount of AC present in the output is 121% of the DC voltage. Hence, half wave rectifier is not very successful in converting AC to DC.

Voltage Regulation

The degree to which a power supply varies in output voltage under conditions of load variations is measured by the voltage regulation which is usually expressed as percentage.

$$\% VR = \left[\frac{V_{No\,load} - V_{Full\,load}}{V_{Full\,load}}\right] \times 100 \qquad (21.39)$$

Peak Inverse Voltage

The peak voltage appearing across the diode during (reverse) negative half cycle is called peak inverse voltage. This value is $(-)E_m$ during interval π to 2π.

Full Wave Rectifier

While discussing half wave rectifier, we have noted that it suffers from many disadvantages, mainly as follows:

 i. There is excessive ripple   ($\approx 121\%$)

 ii. Low efficiency   ($\approx 40.6\%$)

 iii. DC saturation of transformer secondary coil.

Many of the above defects may be overcome by the addition of another diode, i.e. by employing two half wave rectifiers; the output current can be made to flow during the entire cycle of the input voltage. We call such an arrangement as full wave rectifier (Fig. 21.26).

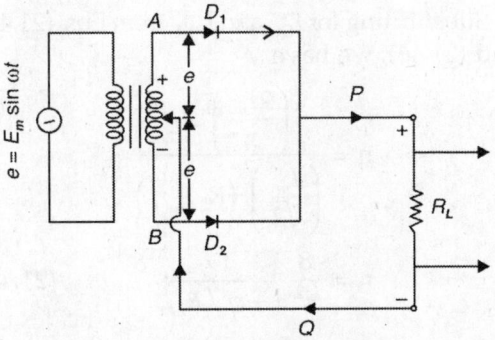

Fig. 21.26: Full wave rectifier

Here the secondary of the transformer is centre-tapped and as such it has twice the voltage from line to line when an AC voltage is applied across the primary coil of the transformer. The one half appears across diode D_1 and R_L which is in series with it. During positive half of input voltage the upper terminal A is positive and the diode D_1 conducts and current flows through R_L. The upper end P of load R_L is positive. Path of current is AD_1PQB. The diode D_2 does not conduct since the lower terminal B is negative.

The shape of current is half sinusoidal, so is the output voltage across R_L as shown in Fig. 21.27.

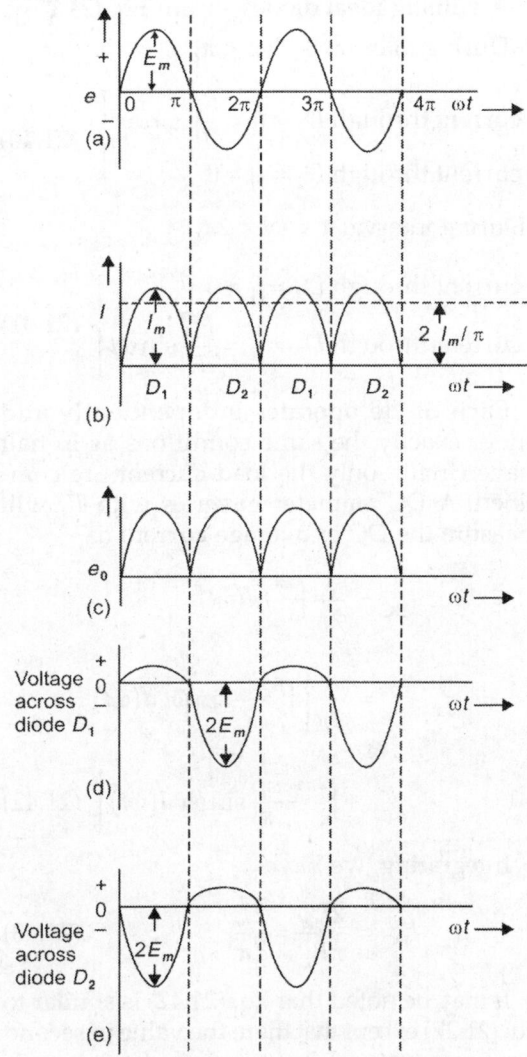

Fig. 21.27: Full wave rectification

During negative half of input voltage, the lower end B is positive, so diode D_2 conducts and current flows through R_L. Here again the upper terminal P of R_L is positive and path of current is BD_2PQC.

The shape of the current and output voltage is similar to positive half cycle. Thus, during both positive half and negative half of the input voltage, the current through R_L flows such that upper end P is always positive.

Thus, output voltage is unidirectional (Fig. 21.27c) and we have load current of half sine pulses.

Assuming ideal diodes—from Eq. (21.23):

During interval $0 < \omega t < \pi$,

$$\left.\begin{array}{l} \text{current through } D_1 = i_1 = \dfrac{E_m}{R}\sin\omega t \\[2mm] \text{current through } D_2 = i_1 = 0 \end{array}\right\} \quad (21.40)$$

During interval $\pi < \omega t < 2\pi$,

$$\left.\begin{array}{l} \text{current through } D_2 = i_1 = 0 \\[2mm] \text{current through } D_2 = i_2 = \dfrac{E_m}{R}\sin\omega t \end{array}\right\} \quad (21.41)$$

Each diode operates independently and under exactly the same conditions as in half wave circuit, only the load current are combined. A DC ammeter in series with R_L will measure the DC or average current as

$$I_{DC} = \frac{1}{2\pi}\int_0^{2\pi} i\,d(\omega t)$$

$$= \frac{1}{2\pi}\left[\int_0^{\pi}\frac{E_m}{R}\sin\omega t\,d(\omega t)\right.$$

$$\left. +\int_{\pi}^{2\pi} -\frac{E_m}{R}\sin\omega t\,d(\omega t)\right] \quad (21.42)$$

Integrating, we have

$$I_{DC} = \frac{2E_m}{\pi R_L} = \frac{2I_m}{\pi} \quad (21.43)$$

It may be noted that Eq. (21.42) is similar to Eq. (21.27) except that there the value of second term was zero, also the value of I_{DC} here is twice that of I_{DC} for a half wave rectifier. So the output DC voltage is

$$E_{DC} = I_{DC}\times R_L = \frac{2E_m}{\pi} \quad (21.44)$$

A Fourier analysis of the voltage wave form applied across the load may be analysed and it yields

$$e = \frac{2E_m}{\pi} - \frac{4E_m}{3\pi}\cos 2\omega t - \frac{4E_m}{15\pi}\cos 4\omega t \quad (21.45)$$

From Eq. (21.45), we note that the lowest frequency present in the output is $2\omega t/2\pi$ unlike $\omega/2\pi$ in case of half wave rectifier, which makes removal of AC harmonics easier with filter circuits.

The effective value of the current in the output is the RMS value of current and is given by

$$I_{RMS} = \sqrt{\frac{1}{2\pi}\int_0^{2\pi} i\,d(\omega t)}$$

Substituting for $i = \dfrac{E_m}{R_L}\sin\omega t = I_m\sin\omega t$, we have

$$I_{RMS} = \frac{I_m}{\sqrt{2}} \quad (21.46)$$

which is different from half wave rectifier as given by Eq. (21.31).

Efficiency of Rectification

It is defined the same way as in the case of half wave rectifier, so

$$\eta = \frac{P_{DC}}{P_{AC}} = \frac{I_{DC}^2 R_L}{(I_{RMS})^2\cdot(R_L + r_f)}$$

Substituting for I_{DC} and I_{RMS} from Eqs. (21.43) and (21.46), we have

$$\eta = \frac{\left(\dfrac{2I_m}{\pi}\right)^2 R_L}{\left(\dfrac{I_m}{\sqrt{2}}\right)^2 (R_L + r_f)}$$

or

$$\eta = \frac{8}{\pi^2}\frac{1}{1 + r_f/R_L} \quad (21.47)$$

Assuming $r_f \ll R_L$, we have

$$\eta = \frac{8}{\pi^2} = 0.812$$

which is double of the half wave rectifier.

Ripple Factor

As given in Eq. (21.35), here also

$$r = \frac{(I_r)_{RMS}}{I_{DC}}$$

and

$$I_{RMS}^2 = I_{DC}^2 + (I_r)_{RMS}^2$$

Hence

$$r = \sqrt{\left(\frac{I_{RMS}}{I_{DC}}\right)^2 - 1}$$

Substituting for I_{RMS} and I_{DC} from Eqs. (21.46) and (21.43), we have

or
$$r = \sqrt{\frac{\pi^2}{8} - 1} = 0.48 \qquad (21.48)$$

which when expressed as percentage indicates that the amount of AC present in the output is 48% of the DC voltage. Thus, a full wave rectifier is more successful in the removal of AC component from output or converting AC into DC.

Peak Inverse Voltage

It is the voltage across the diode during its non-conduction period.

Thus, both the diodes D_1 and D_2 have their own peak inverse voltage as shown in Figs 21.27d and e.

The PIV for D_1 is $(-)2E_m$ during interval π to 2π.

The PIV for D_2 is $(-)2E_m$ during interval 0 to π.

It is for this reason that diodes employed in full wave rectifier are more robust than employed in half wave rectifier.

21.5 ZENER DIODES

We have seen that in the breakdown region, large changes in diode current produce only small changes in diode voltage. So a semiconductor P-N diode designed to operate in the breakdown region may be employed as a constant voltage device. The diodes used in such a manner are called *avalanche breakdown* of *zener diodes*.

They are used as a voltage regulator. The voltage source V and resistance R are so selected that the diode operates in the breakdown region. The diode voltage in this region which is also the voltage across the load R_L is called zener voltage (V_Z) and the diode current is called zener current (I_Z). As the load current (I_L) or the supply voltage changes, the diode accommodates itself to these changes and maintains nearly constant load voltage (V_Z).

The diode will continue to regulate the voltage until the diode current falls to knee current I_{zk} (Fig. 21.28). Depending upon the nature of the semiconductor and its doping, the breakdown voltage in diode ranges from about 3 volts to several hundred volts. The breakdown phenomenon is reversible and harmless so long as the safe operating temperature is maintained.

The mechanism of diode breakdown at reverse voltage is explained below:

i. **Avalanche breakdown:** In this mechanism, the minority charge carriers (electrons in P-type and holes in N-type) acquire sufficient energy from the applied reverse voltage to produce new charge carriers by removing valence electrons from the covalent bonds. The new carriers, in turn, produce additional charge carriers and the process multiplies to give large reverse current. The diode is then said to be in the region of *avalanche breakdown*, usually, a junction with a broad depletion layer (therefore a low field intensity) breakdown by this mechanism. With the increase in temperature, the vibrations of the atoms in the crystal increase which increases the possibility of collisions of the charge carriers with the lattice atoms and reduces the possibility for the

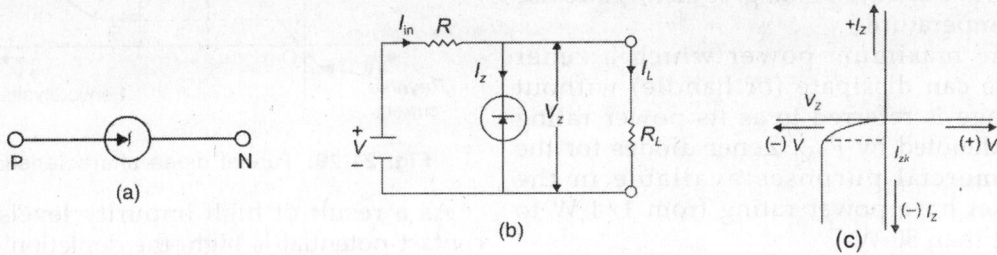

Fig. 21.28: (a) Symbol of zener diode (b) Circuit of a zener diode (c) A typical *I–V* characteristics of a zener diode

carriers to gain sufficient energy to start avalanche process. Thus, at high voltages, the avalanche process is prominent and does take place to cause diode breakdown. The operating voltages in such diodes range from several volts to a few hundred volts.

ii. **Zener breakdown:** In this mechanism, the breakdown is initiated through a direct rupture of covalent bonds rather than avalanche process due to the existence of strong electric field across depletion layer. A junction having a narrow depletion layer (and hence high field intensity $E = \dfrac{V_r}{d}$) will breakdown by this mechanism. An increase in temperature increases the energy of the valence electrons and makes it easier for these electrons to escape from these covalent bonds. Smaller applied voltage is, therefore, required to pull these electrons from the crystal lattice. The *zener effect plays an important role only in diodes with breakdown voltage below about 6 volts.* The zener diode is always used in reverse biased condition.

The range of voltages about the breakdown voltage in which a zener diode conducts in reverse direction is called *tolerance.*

The breakdown voltage of zener diode depends upon operating temperature. It is found to decrease with increase in junction temperature. This is due to the increased reverse current, (i.e. increase in minority carriers) that flows with increasing temperature. The decrease in breakdown voltage is about 2 mV per degree centigrade rise in temperature.

The maximum power which a zener diode can dissipate (or handle) without damage is referred to as its power rating and denoted by P_{ZM}. Zener diodes for the commercial purposes available in the market have power rating from 1/4 W to more than 50 W.

The opposition offered to the current flowing through the zener diode in the operating region is known as *zener resistance* (R_z) or zener impedance (Z_z).

Zener diodes find wide commercial and industrial applications, e.g. voltage stabilizer, meter protection, wave shaping, etc.

Constant Current Diodes

These are the diodes that work exactly opposite to zener diodes. These diodes keep the *current constantly flowing through them when* the voltage changes, i.e. instead of holding the voltage constant, these diodes hold the current constant. The range of voltage over which a diode can keep the current constant is known as *voltage compliance.* These diodes are optimised for a particular voltage compliance.

21.6 TUNNEL DIODE

It is a device just like a P-N junction, which offers negative resistance under certain bias conditions. The negative resistance of the diode is due to tunnelling and it is called a tunnel diode or *Esaki diode.* It is made very much like an ordinary P-N junction diode, except that both the P and N-regions are heavily doped $(1 : 10^4)$, thousand times more than ordinary diode. It is used as an active device in electronic circuits in the frequency range of few megahertz. The semiconductors with very high impurity concentration are referred to as degenerate semiconductors. Typical tunnel diode characteristic is shown in Fig. 21.29.

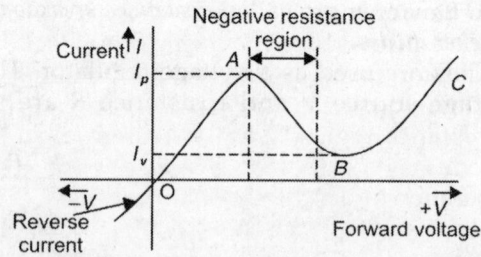

Fig. 21.29: Tunnel diode characteristic

As a result of high impurity levels, the contact potential is high, the depletion layer is very narrow and the Fermi levels lie in the conduction band from N-side and in the valence band of the P-side. Under these circumstances, a very abrupt transition from P to N-type material is achieved within the crystal. Since depletion region is very narrow, this gives rise to extremely large electric fields.

Tunnel Effect

When a P-N semiconductor is heavily doped (it has many majority carriers and ions) and under forward biased, the hole and valence electron random drift is heavy. As a result, it is not uncommon for a large number of electrons to fill holes and release energy to only a few other valence electrons. These few electrons, then, have their energy levels raised considerably so that they can cross from the N to the P-section and current increases as shown by *OA* (Fig. 21.29) even with little or no applied voltage. This action which seems to allow a valence electron to cross a potential barrier without enough applied external energy is called the *tunnel effect* because it seems as though the valence electron 'tunnels' through the forbidden band. On further increasing the voltage, the barrier height decreases and Fermi level E_{FN} is much raised so the current decreases as shown by *AB* (Fig. 21.29). This region is called *negative resistance region*. On further increasing the voltage, the conduction band of N-type is in level with CB of P-type and free electrons of N-type easily cross to P-type thereby current again increases like ordinary diode as shown by *BC* (Fig. 21.29).

The negative resistance region *AB* (Fig. 21.29) allows the diode to be used as an oscillator. It can also be used as an electronic switch since it has a good response in negative resistance region. By its nature, the tunnel diode has a rather high reverse current but operation under this condition is not generally used.

Our main interest in the tunnel diode is its application as a very high speed switch. Since tunnelling takes place at the speed of light, the transient response is limited only by total shunt capacitance (junction plus stray wiring capacitance) and peak driving current. Switching times of the order of a nanosecond are reasonable and times as low as 50 picoseconds have been obtained.

The advantages of a tunnel diode are (i) low cost (ii) low noise (iii) high speed (iv) environmental immunity (v) low power.

The disadvantages of the diode are its low output voltage swing and the fact that it is a two terminal device unlike ordinary diode in which current flows only when forward biased. Because of this latter feature, there is no isolation between the input and output and it leads to serious circuit design problems.

However, we may have a special type of tunnel diode whose peak value current is of the order of valley current. Such a diode is called *backward diode*.

21.7 PHOTODIODE

A photodiode is a P-N junction diode packed into a transparent plastic working under reverse biased condition.

Principle

A very small current flows through a P-N junction diode when it is reverse biased. It is because minority charge carriers take part in conduction. The number of minority carriers depend upon the working temperature. However, when photons (light) of suitable frequency are incident on the junction, the number of charge carriers increases. But this happens only when light radiations touch the junction which is not the case in ordinary diode.

Working

In a photodiode, visible light is focussed on the junction through a lens. Light on being incident produces free electrons and holes. Thus, the number of charge carriers increases and hence current increases. Generally, a photodiode is optimized for its sensitivity to light. As the light intensity increases, more and more charge carriers are generated and reverse bias current increases.

In this respect, a photodiode acts as a photo detector, a device which converts incoming light signal into electrical signal.

Applications

Some of the important applications of photodiodes are (i) photo detection (both visible and invisible) (ii) demonstration (iii) logic circuits (iv) switching (v) optical communication system.

21.8 SOLAR CELLS

Solar cells, which convert the sunlight directly into electricity, at present furnish the most important long duration power supply for satellites and space vehicles. These cells have also been successfully employed in small-scale terrestrial applications. Today, the solar cell is considered a major device for obtaining energy from the sun. It can convert sunlight directly into electricity with high conversion efficiency, provide nearly permanent power at low operating cost and is virtually free of pollution. Recently, research and pollution development of low-cost, flat panel solar cells, thin film devices, concentrator systems and many innovative concepts have increased. We can expect that in near future, the costs of small solar power modular units and solar power plants will be economically feasible for large-scale use of solar energy.

The first solar cell was developed by Chapin *et al.* in 1954 using a diffused silicon P-N junction. Subsequently, the cadmium sulphide solar cell was developed by Raynolds *et al.* in 1954. Till date, solar cells have been made of many other semiconductors, using various device configurations and employing single-crystal, polycrystal and amorphous thin film structures.

Photovoltaic Effect and Solar Cells

The photovoltaic effect can be observed in nature in a variety of materials, but the materials that have shown the best performance in sunlight are the semiconductors. In photovoltaic conversion, the solar radiation falls on devices called solar cells which convert the sunlight directly into electricity. The principal advantages associated with solar cells are that they have no moving parts, require little maintenance and work quite satisfactorily with beam or diffuse radiation.

21.9 LIGHT-EMITTING DIODE (LED)

LED is a solid state (P-N junction diode) light source which has replaced incandescent lamps in many day to day applications.

LED is just not an ordinary P-N junction diode where silicon is used. Here we use compounds having elements like gallium, arsenic and phosphorous which are semi-transparent unlike silicon which is opaque. (Gallium arsenide gives infrared radiations and gallium arsenide phosphide gives visible light either red or yellow).

The advantages of an LED are:

i. Low voltage of operation
ii. Long life (more than 20 years)
iii. Fast ON-OFF switching (10^{-9} sec).

The other uses of LED are:

1. For indicating power ON/OFF (power level indicators)
2. Optical switching applications
3. Solid state video displays
4. Optical communication—Energy coupling circuits.

All natural colours are composed from three primary colours namely red, green and blue. LED in red and green have been available but the blue one has been missing. Blue light emitters have several potential applications. They can be used for full colour displays of large area. They can be used in traffic lights as replacements for ordinary bulbs resulting in huge power savings as well as cost. Use of blue lasers can also result in higher density storage of information in optical CD-ROMs. Blue light emitters have been fabricated using several materials including ZnO, ZnSe and SiC. These attempts have been reasonably successful. However, maximum success in last few years has been achieved with GaN. This material has proved to be very useful as it is also capable of operating at high power density, high temperature and unfriendly environment. GaN having a band gap of 3.4 eV can give continuously varying band gap by combining with AIN to get up to 6.2 eV and with InN to get a band gap down to 1.9 eV. The high thermal conductivity and superior stability of this material makes it ideal for several applications over other competing materials.

21.10 LIQUID CRYSTAL DISPLAY (LCD)

Just like LED, we have another type of display that uses seven segment called liquid crystal Display also known as *electroluminescent display*. It consists of a thin layer of normally transparent liquid crystal material between two electrodes.

When an electric field is applied, the liquid crystal material between the two electrodes becomes turbulent, reflecting and scattering ambient light. It provides excellent brightness under high ambient light conditions and requires only 50 μW of power per segment, thereby total power for one complete display of 7 segments is 350 μW. This power is much less than that of an LED display, but the life expectancy is not as high as that for LED which is minimum 10,000 hours.

These displays are used in watches, pocket calculators, pocket televisions and portable instrument displays.

21.11 THERMISTORS AND BARRETTERS

We have read that the electrical conductivity of a semiconductor changes significantly with temperature and has a negative temperature coefficient of resistivity. This property is utilised in a device called *thermistor*, whose resistance is temperature sensitive, usually decreasing with temperature. Conventional wire-wound metallic resistors have positive temperature coefficient of resistivity. Commercial thermistors are usually made of sintered mixtures of Mn_2O_3, NiO_2 and Co_2O_3. A thermistor consists of a semiconductor bead of approximately 0.04 cm in diameter. Two thin wires are attached to the bead to provide for the two terminals. Diameter of the wire is approximately 0.25 μm.

Thermistors find use in control systems operated by temperature changes, in the measurement of microwave power, in thermometry and as a thermal relay. In electronic circuits, thermistors have been used to compensate for the change in resistance with temperature of ordinary components where variation of component values cannot be tolerated.

The *I-V* characteristic of a thermistor has negative slope. Devices that exhibit, in some region, a negative slope in their *I-V* characteristics are useful for making oscillators, amplifiers and switching circuits. However, thermistors are not suitable in these applications because their response characteristics are too slow. There are certain bulk semiconducting compounds which have negative resistance characteristics over a limited range of operating parameters utilizing mechanism unrelated to the temperature sensitivity of the resistivity. These materials have been used to obtain devices based on *Gunn effect*.

A heavily doped semiconductor shows metallic properties. It has a positive temperature coefficient of resistivity owing to the decrease in carrier mobility with increasing temperature. Such a device is called *sensistor*. Thermistors also find extensive use as sensing elements in microwave power measuring equipments and as temperature sensors of electronic thermometers. Thermistors are capable of yielding power information over the power range of 10^{-5} mW to 20 mW with a typical burn out level of 400 mW.

A *barretter* has a positive temperature coefficient of resistance, consists of an approximately mounted piece of *Wallaston* platinum wire having diameter of approximately 1.25 μm. Barretters are capable of yielding power information over the range of 10^{-5} mW to 20 mW. A typical burnout level is 20 mW. For low level *rf* power application below 10^3 MHz, 0.001 A, *littlefuse* may be used as a barretter.

21.12 GUNN EFFECT AND GUNN DIODE

Ridley–Watkins and Hilsum independently predicted that semiconductor materials under certain conditions can offer differential negative resistance. This differential negative resistance is a bulk effect and due to transfer of electrons from one valley to another in the conduction band. Gunn while experimenting on a sample of N-type GaAs and some other III-V compounds, found that the current through the sample increased linearly with voltage till a certain threshold voltage. Beyond the threshold voltage, a number of current pulses appeared with a time interval proportional to the length of the sample. This threshold field is high (~400 V/mm). The oscillations lie in the microwave range and set in the negative differential conductance (NDC) region where the current decreases with increase in electric field E (Fig. 21.30). This behaviour is a consequence of the band structure of these materials. In the region of velocity field curve, where v decreases with increasing E, the differential mobility (dv/dE) of the electrons becomes negative. The reason is following.

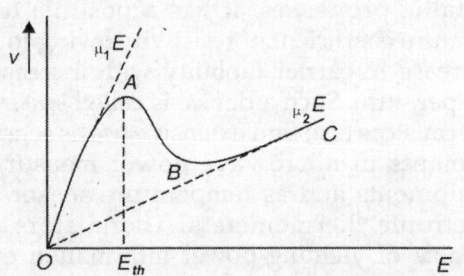

Fig. 21.30: A possible characteristic of electron drift velocity versus field for a semiconductor exhibiting the transferred electron mechanism

nGaAs has a direct band gap. The band gap is 1.4 eV. The free electrons in nGaAs normally occupy the lowest energy states in the conduction band. In this situation, effective mass (m_e^*) and mobility (μ_n) of electrons enhances. So as a result of the transfer of the electrons, current begins to decrease with increase in field because $T = n_1 e \mu_1 E + n_2 e \mu_2 E$, where n_1 and n_2 are the concentration of electrons having the mobility μ_1 and μ_2

respectively. The average drift velocity of an electron is

$$v = \left(\frac{n_1 \mu_1 + n_2 \mu_2}{n_1 + n_2} \right) \quad (21.49)$$

$(n_1 + n_2)$ being the total electron concentration.

If the transition of the electrons from the high-mobility state to the low-mobility state occurs rapidly over a range of field E, one observes that v diminishes with field E beyond a certain threshold field E_{th}, shown by region AB in Fig. 21.30. When all the electrons move to the low-mobility state, the drift velocity is $v = \mu_2 E$. One finds that v continues to increase slowly with E. This builds up the successive oscillations.

Owing to the occurrence of the NDM in the velocity–field curve, Gunn diodes are used as sources of microwaves especially where high power is not a requirement. Gunn diodes are also used as local oscillators for mixers in microwave receivers over the frequency range 1 to 100 GHz. A Gunn device can carry one of two possible currents just before the threshold voltage, depending on the presence or absence of a domain. This can, therefore, be used as a high-speed binary logic.

Gunn diode is a versatile semiconductor device. These are commercially available for pulsed operation yielding a power output of 5 W in the frequency range 5.0 to 12.0 GHz.

21.13 IMPATT, TRAPATT AND QWITT DIODES

IMPATT (Impact Ionization Avalanche Transit Time) diodes employ impact ionization and transit time properties of semiconductor structures to produce negative resistance at microwave frequencies. TRAPATT (Trapped Plasma Avalanche Triggered Transit) diode is an IMPATT related device. These are quite new devices and used as oscillators in the microwave region. These devices work well in the breakdown region. Although IMPATT operation can be obtained in simpler structures, the Read diode is best suited for illustration of basic principles.

QWITT diode is a low-noise injection mechanism with superior high frequency characteristics. The length of the transit time as well as the shape of the current pulse can be optimized to obtain the best power-frequency performance from the QWITT diode. This diode should extend the normal frequency limit associated with transit-time devices, while providing higher output power than simple quantum-well RTD oscillations.

21.14 PIN DIODES

This is a P-N junction with greatly improved times. Obviously, this is a P-N junction with a doping profile tailored so that an intrinsic layer, i.e. the I-region is sandwitched between a P-layer and an N-layer as shown in Fig. 21.31.

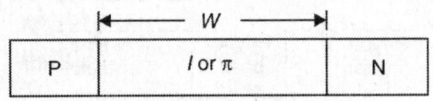

Fig. 21.31: PIN diodes

In practice, however, the idealized I (intrinsic)-region is approximated by either a high resistivity P-layer (referred to as π-layer) or a high resistivity N-layer (referred to as v layer). The resistivity of I-layer is typically 10^3 Ωm.

The PIN diode has found wide applications in microwave circuits. It can be used as a microwave switch with essentially constant depletion layer capacitance and high power handling capability. When PIN diodes are used as microwave switches and when they are biased in the OFF condition, the bias is usually beyond the swept-out voltage (usually 2 V). PIN diodes can also be used as a *variolosser* (variables attenuator) by controlling the device resistance which varies approximately linearly with the forward current. PIN diodes can also modulate signals up to the GHz range. We must note that the forward characteristics of a thyristor in its ON state closely resemble those of PIN diode.

21.15 TRANSISTOR

Transistors are three terminal solid-state devices and are extremely important semiconductor devices in today's microelectronic circuitry.

In a way, they have revolutionized the entire field of electronics. They are capable of two primary types of functions. First, transistors can perform the same operation as their vacuum tube precursor, the triode valve, i.e. they can amplify an electrical signal (i.e. as amplifier). In addition, they serve as switching devices in computers for the processing and storage of information. The major types of transistors are: the junction (or bimodal) transistor and the *metal oxide semiconductor field-effect transistor* (MOSFET).

Junction Transistor

This is composed of two P-N junctions arranged back to back in either the N-P-N or P-N-P configuration. The charge carriers in a junction transistor have positive as well as negative polarity and so the nomenclature *bipolar junction transistor* is used.

Bipolar Junction Transistor

A semiconductor diode described earlier has one P-N junction. If we grow another junction either on N or P-side, then we get a P-N-P or an N-P-N device having two junctions.

In actual practice, a junction transistor consists of a silicon (or germanium) crystal, in which a layer of N-type silicon (germanium) is sandwitched between two layers of P-type silicon (germanium) and thereby we get a P-N-P transistor. Alternatively, it may consist of a layer of P-type between two layers of N-type material and we get a N-P-N type transistor. The semiconductor sandwitched is extremely small and is hermatically sealed against moisture inside a metal case.

The three terminals taken from each section of a semiconductor are called (i) emitter (ii) base (iii) collector (Fig. 21.32a). The middle section (base) is very thin in comparison with the other two.

Emitter as the name goes emits or supplies charge carriers (electrons or holes as the case may be). It is always forward biased with respect to base so that it may supply a large number of carriers. As a P-N-P transistor we have holes as majority carriers and as a N-P-N we have electrons as majority carriers.

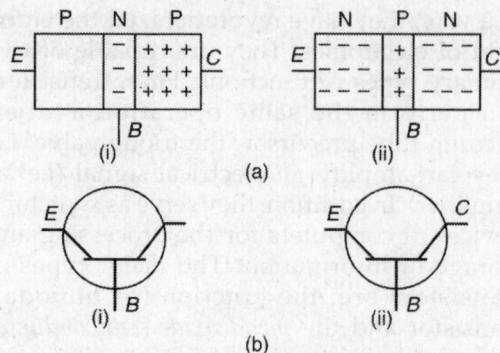

Fig. 21.32: (a) (i) Transistor (ii) Block diagram
(b) (i) P-N-P (ii) N-P-N

Base is the middle region which forms the two junctions between emitter and collector. The base–emitter junction is forward biased while base–collector junction is reverse biased (Fig. 21.33).

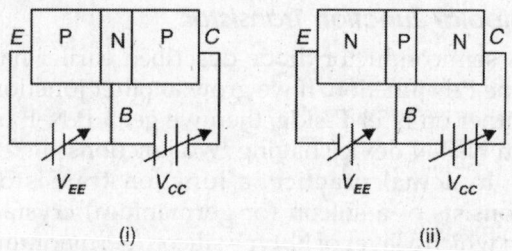

Fig. 21.33: Forward and reverse biasing

Forward biasing allows a low resistance for emitter circuit and reverse biasing provides high resistance. In this respect, a transistor transfers a signal from low resistance region to high resistance region so that the name [transfer + resistor] owes its origin.

Doping in this region is also minimum with respect to emitter or collector.

Collector, the region on the other side of base, has the maximum area of the three and collects the majority carriers injected from the emitter so the name collector goes. It is reverse biased with respect to base.

Although the area of collector region is maximum of the three, yet in practice the area is shown symmetrical to the emitter region (Fig. 21.32a).

The doping is also maximum in this region.

Transistor Action

As soon as the two junctions are formed, a potential difference exists across the two junction as shown in Fig. 21.34a. This is under zero bias (no external battery) condition. Such diagrams are called potential hill diagrams.

On applying proper biasing, i.e. emitter forward biased and collector being reverse biased, the level of the potential hill changes as shown by dashed curve (Fig. 21.34b).

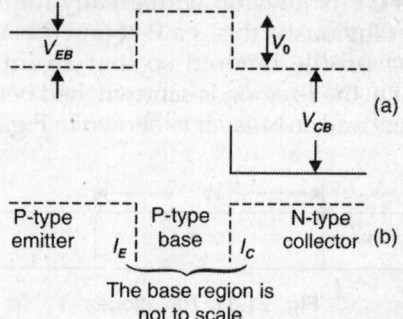

The base region is not to scale

Fig. 21.34: Barrier height of transistor (a) Zero biased dash curve (b) Under proper bias solid curve

The forward biasing of emitter junction lowers the emitter–base potential barrier height whereas the reverse biasing of collector junction increases the collector–base potential barrier height. The lowering of emitter–base potential barrier height causes the majority carriers to flow from emitter towards base (holes in case of P-N-P and electrons in case N-P-N transistors).

It is to be noted that a few electrons flow from the base to emitter in the former case and holes in latter.

Flow of majority carriers across emitter–base junction constitutes the emitter current I_E (Fig. 21.35a). Here we are considering common emitter configuration. So as the holes enter the base region they have a tendency to combine with the electrons present there. However, only a very small percentage (less than 5%) of them combine, since the base region is very thin and doping is also very low. These electrons which combine with incoming holes are replaced by the battery V_{EE} and it constitutes base current I_B which is very small ($\approx \mu A$).

The remaining holes (95% or more) reach the collector region where they are swept away by accelerating field of reverse bias V_{CC} and constitute the collector current I_C. In this way almost all the emitter current (composed of holes, i.e. majority carriers) flows towards the collector circuit. It is clear that emitter current is the sum of base current and collector current

$$I_E = I_B + I_C \qquad (21.50)$$

It may be noted that if emitter junction is not biased ($I_E = 0$), then current flowing across collector junction is very very small since it is constituted of minority carriers and so current flowing in the collector circuit is very small ($\approx \mu A$).

So far we have considered a P-N-P transistor where majority current is composed of holes; the same treatment applies for N-P-N transistor where the current is composed of electrons as the majority carriers.

We can now take into account the flow of various charge carriers and the resulting current in the E, B and C regions of the P-N-P device.

i. At emitter–(a) holes injected by the emitter into base (I_{pE})

    (b) electrons injected by the base into emitter (I_{nE}) due to forward bias of the emitter–base junction, so $I_E = I_{pE} + I_{nE}$.

ii. At base   Majority of the holes injected into base from emitter diffuse towards collector, the remaining few recombine with free electrons present in base ($I_{pE} - I_pc_1$).

iii. At collector–(a) If the emitter junction is not forward biased while collector junction is reverse biased, then collector current is composed only of minority carriers and is equal to reverse saturation current I_{CO} as in case of diode.

$$I_{CO} = I_{nCO} + I_{pCO}$$

(b) When emitter junction is forward biased a large number of holes I_{pC_1} reach collector.

So $I_C = I_{CO} - I_{pC_1}$

Here the minus sign has been chosen to show that two components have opposite direction, also I_{pC_1} is some fraction of emitter current I_E and we may write

$$I_C = I_{CO} - \alpha I_E \qquad (21.51)$$

where α is defined as the fraction of total emitter current which represents holes that have travelled from emitter to collector so

$$\alpha = (I_{pC_1} / I_E) \qquad (21.52)$$

For a P-N-P transistor, I_E is positive and both I_C and I_{CO} are negative. Considering common base configuration, all the above components have been shown in Fig. 21.35b.

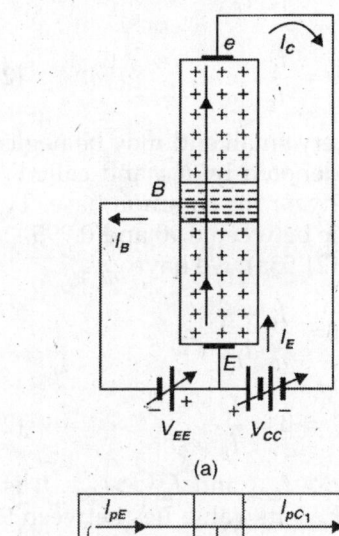

(a)

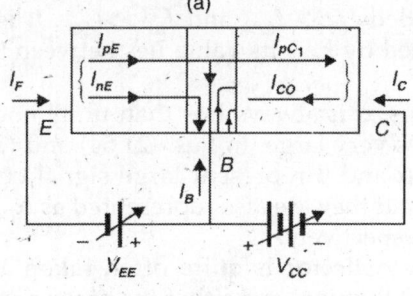

(b)

Fig. 21.35: (a) Common emitter configuration of main current components in a transistor (P-N-P) (b) Various components of current, common base configuration

In Eq. (21.51), if we consider magnitude and do not consider the direction, we may write

$$I_C = \alpha I_E + I_{CO} \qquad (21.52a)$$

Using Eq. (21.50), we have

$$I_C = \alpha(I_C + I_B)I_{CO}$$

or $\quad (1 - \alpha)I_C = \alpha I_B + I_{CO}$

or $\qquad I_C = \dfrac{\alpha}{1-\alpha} I_B + \dfrac{I_{CO}}{1-\alpha} \qquad (21.53)$

The quantity $\dfrac{\alpha}{1-\alpha}$ is defined as a new constant β and so

$$I_C = \beta I_B (\beta + 1)I_{CO} \qquad (21.53a)$$

Also from Eq. (21.52a),

$$\alpha = \frac{I_C - I_{CO}}{I_E}$$

or $\qquad \alpha = \dfrac{I_C}{I_E} \qquad (21.54)$

since I_{CO} is very small and may be neglected.

It is also denoted by h_{FB} and called *large-signal current gain* for common base. Typical values of it lie between 0.90 and 0.995.

From Eq. (21.53a), we have

$$\beta = \frac{I_C - I_{CO}}{I_E}$$

or $\qquad \beta_{DC} = \beta = \dfrac{I_C}{I_B} \qquad (21.55)$

provided $I_C \gg I_{CO}$ and $I_B \gg I_{CO}$. It is also denoted by h_{FE}. Its value lies between 9 and 199.

Thus, α is always less than unity and β is always very large. In Eqs. (21.54) and (21.55) both α and β represent large signal current gain and they are also represented as α_{DC} and β_{DC} respectively.

The collector is quite often taken as the output terminal and either the base or emitter is taken as input terminal to get a voltage gain from the device by way of *resistance transfer* action between the reverse biased and forward biased junctions.

Action of Transistor as Amplifier
Voltage Amplifier

The reverse biased collector junction is equivalent to a high resistance (r_c) due to small current and the forward biased emitter junction is equivalent to a low input resistance (r_i) because of large emitter current. If for instance, we take emitter E as the input terminal (Fig. 21.36) and apply a signal V_S so that in addition to the biasing current I_E an additional current ΔI_E flows, then we shall get at the collector an extra current ΔI_C in addition to DC current I_C.

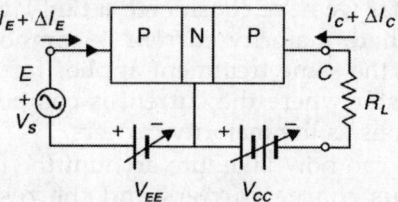

Fig. 21.36: Transistor amplifier, common base configuration

Hence, from Eq. (21.54)

$$\Delta I_C = i_C = \alpha \Delta I_E$$

Although ΔI_C is less than ΔI_E according to definition of α, yet the situation is different as we consider the voltage output at collector, where a load R_L is also connected

$$V_o = R_L \Delta I_C = [\alpha \Delta I_E]R_L$$

The signal voltage at input is

$$V_i = r_i \Delta I_E$$

so voltage gain

$$A_v = \frac{V_o}{V_i} = \frac{(\alpha \Delta I_E)R_L}{\Delta I_E r_i} = \frac{\alpha R_L}{r_i}$$

Now, although α is less than 1, but generally $R_L \gg r_i$

So $\qquad V_o \gg V_i$

i.e. output voltage is quite greater than input voltage.

Thus, it gives rise to voltage amplification. The product of signal voltage and current at the output terminal and input terminal indicates that the power at output is also larger than at input. Hence, the transistor also gives power gain. But here it is to be noted that the current gain is less than unity.

Current Amplifier

If we feed a signal in common emitter configuration, a current ΔI_B results at base terminal in addition to the bias base current I_B. This, in turn, will cause an AC component of collector current ΔI_C to flow in collector region in addition to collector current I_C. This doping of junction and width of base are so chosen that a small base current change can handle a large collector current change. Hence, a transistor has current gain in common emitter configuration. It should be noted that in the present case it is the base terminal which has been taken as input terminal whereas in the former case it was the emitter terminal which was taken as input terminal.

Three Transistor Configurations

As discussed earlier, we may take any of the three terminals as input terminal and make use of transistor as amplifier. So a transistor can be connected in three configurations known as:

   i. Common base configuration (CB)
   ii. Common emitter configuration (CE)
   iii. Common collector configuration (CC)

Common Base Configuration

Here the base terminal is common to both input as well as output terminals as shown in Fig. 21.33 or Fig. 21.37a. The emitter terminal is taken as input and collector is taken as output terminal.

Common Emitter Configuration

Here the emitter terminal is common to both input as well as output terminals as shown in Fig. 21.37b. The base terminal is taken as input and collector terminal as output.

Common Collector Configuration

Here the collector terminal is common to both input as well as output terminals as shown in Fig. 21.37c. The base terminal is input terminal and emitter terminal is output terminal.

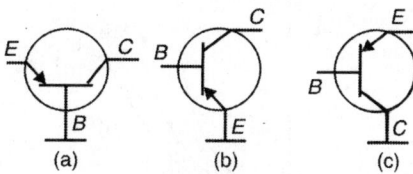

Fig. 21.37: (a) CB configuration (b) CE configuration (c) CC configuration

This configuration is popularly known as *emitter follower.*

Characteristic Curves of a Transistor
Common Base Configuration

The circuit to draw the characteristics of the transistor in CB configuration is shown in Fig. 21.38. Here the emitter junction is forward biased by the battery V_{EE} and the corresponding input voltage and current is read by the voltmeter V_{EB} and ammeter I_E.

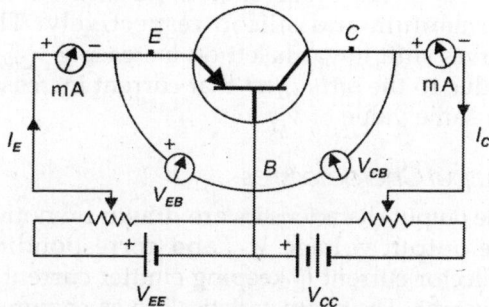

Fig. 21.38: Circuit for drawing characteristics of transistor in CB configuration

The collector junction is reverse biased by battery V_{CC} and the corresponding output voltage and current is read by the voltmeter V_{CB} and ammeter I_C.

Input Characteristics

The input characteristics are drawn by noting input voltage V_{EB} and corresponding current I_E keeping V_{CB} constant. The characteristics are shown in Fig. 21.39.

They are similar to the forward biased diode characteristics and represented by

$$(V_{EB}\ I_E)_{V_{CB} = \text{constant}}$$

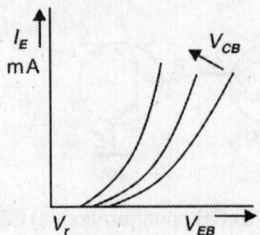

Fig. 21.39: Input characteristics CB configuration

The inverse of the slope of characteristics

$$\frac{1}{\text{slope}} = \frac{\Delta V_{EB}}{\Delta I_E}\bigg|_{V_{CB} = \text{constant}}$$

gives input resistance which is denoted as h_i.

The shape of the curve can be easily explained by the fact that here the emitter junction behaves like forward biased diode so on slightly increasing the voltage V_{EB}, the current I_E increases a lot. Also there exists a cut in voltage V_r below which emitter current is very small or zero. It is 0.2 and 0.7 for germanium and silicon respectively. The curves shift towards left on increasing V_{CB}, it is due to the *early effect* that current increases for same value of V_{EB}.

Output Characteristics

The output characteristics are drawn by noting the output voltage V_{CB} and corresponding collector current I_C keeping emitter current I_E constant. The characteristics are as shown in Fig. 21.40 and represented by

$$(V_{CB} I_C)_{I_E = \text{constant}}$$

So we get different curves for different values of I_E.

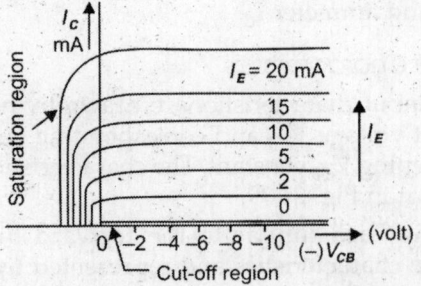

Fig. 21.40: Output characteristics CB configuration

The family of characteristics is divided into three regions.

1. ***Active region***: It is customary to plot V_{CB} along (+)ve X-axis although the values of V_{CB} are negative.

   For the first curve when $I_E = 0$, i.e. when input terminals are open, the collector current $I_C = I_{CO}$ the reverse saturation current (which is μA for germanium and nanoamperes for silicon) of the diode.

   Suppose now that input circuit is on and I_E flows then a fraction of it αI_E would flow towards the collector region and collector current in this case is $(\alpha I_E + I_{CO})$ and it is independent of collector voltage since I_{CO} is very small and mainly depends on I_E, so the curves appear to be practically flat. They have very small slope. Its value gives h_0 or (I/r_0).

2. ***Saturation region***: The region to the left of ordinate $V_{CB} = 0$ and above $I_E = 0$ characteristics is known as saturation region, where both emitter and collector junctions are forward biased. Here V_{CB} is positive for P-N-P transistor so the forward biasing of collector is responsible for large change in collector current for a small change in V_{CB} and I_C increases exponentially with voltage.

3. ***Cut-off region***: The characteristics for $I_E = 0$ passes through origin but is otherwise similar to the other characteristics. This characteristic is not coincident with (X-axis) voltage axis although in diagram it appears so, since the gap is too small to be shown on the scale chosen. The region below $I_E = 0$ characteristic for which both emitter and collector junctions are reverse biased is called cut-off region.

Common Emitter Configuration

The circuit to draw the characteristics of the transistor in CE configuration is shown in Fig. 21.41. Here the emitter junction is forward biased by the battery V_{BB} and the corresponding input voltage and current is read by voltmeter V_{BE} and ammeter I_B.

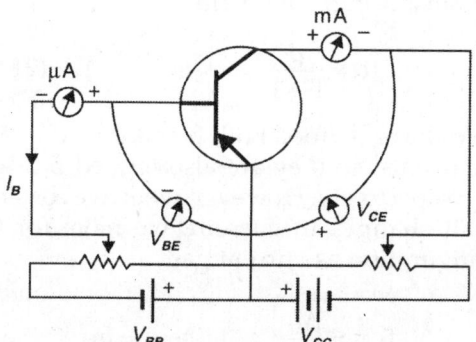

Fig. 21.41: Circuit for drawing characteristics of transistor in CE configuration

The collector junction is reverse biased by the battery V_{CC} and the corresponding output voltage and current is read by voltmeter V_{CE} and ammeter I_C.

Input Characteristics

The input characteristics are drawn by noting the input voltage V_{BE} and the corres-ponding base current I_B keeping output voltage V_{CE} constant (Fig. 21.42). The input characteristics are essentially that of a forward biased diode. They shift to left on increasing V_{CE}, i.e. making collector junction less negative. The inverse of their slope gives input resistance r_i (or h_i)

$$= \frac{\Delta V_{BE}}{\Delta I_B}\bigg|_{V_{CE} = \text{constant}}$$

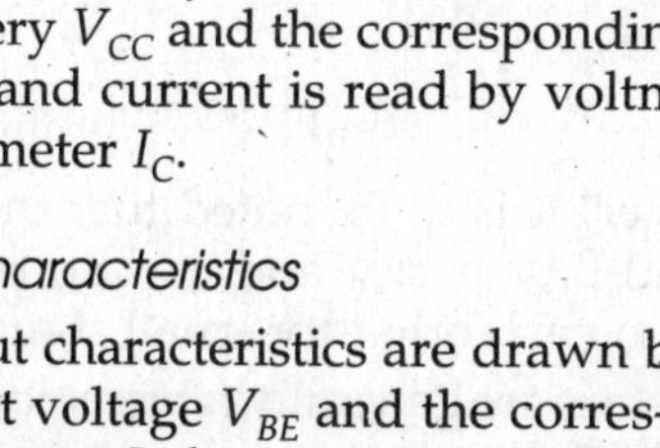

Fig. 21.42: Input characteristics CE configuration

Output Characteristics

The output characteristics are drawn by noting output voltage V_{CE} and the collector current I_C keeping base current I_B constant. The characteristics are as shown in Fig. 21.43 and

represented by $(V_{CE}, I_C)_{I_B = \text{constant}}$. So we get different curves for different values of I_B.

Just like common configuration, the family of curves may be divided into three regions.

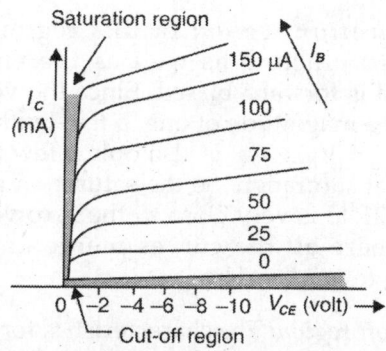

Fig. 21.43: Output characteristics CE configuration

1. **Active region:** Here collector junction is reverse biased and emitter junction is forward biased. In Fig 21.43, the active region is the area to the right of ordinate V_{CE} = a few tenths of volt and above $I_B = 0$. For given value of I_B, the emitter injects certain number of charge carriers into base region. When collector–emitter voltage V_{CE} is small, the collector is unable to collect all charge carriers. So, on increasing V_{CE}, collector current first increases and then rate of increase is quite small and I_C is nearly constant and it would change only on changing I_B.

From Eq. (21.53),

$$I_C = \frac{I_{CO}}{1-\alpha} + \frac{\alpha}{1-\alpha} I_B$$

and from Eq. (21.53a)

$$I_C = (1+\beta)I_{CO} + \beta I_B \text{ since } \beta = \frac{\alpha}{1-\alpha}$$

If α was truly a constant, then according to Eq. (21.53), I_C would be independent of V_{CE} and curves of Fig. 21.43 would be horizontal. But it is not so and there is a slope in the characteristics. Thus, it represents that a slight change in α causes a large change in β and the variation in α or β is responsible for **the** slope of the characteristic with V_{CE}. Their slope gives h_0 or $1/r_0$

$$= \frac{\Delta I_C}{\Delta V_{CE}}\bigg|_{I_B = \text{constant}}$$

2. **Saturation region:** In this region, the collector junction as well as the emitter junction is forward biased. Since the voltage V_{BE} has a magnitude of only a few tenths of a volt, $V_{CE} = V_{BE} - V_{BC}$ is also only a few tenths of volt at saturation. So the saturation region in Fig. 21.43 is very close to the zero voltage axis where all the curves merge and fall rapidly towards origin.

3. **Cut-off region:** The characteristics for $I_B = 0$ passes through origin and we find that appreciable collector current exists under the conditions.

From Eqs. (21.50) and (21.53),

$$I_C = I_E = \frac{I_{CO}}{1 - \alpha} = I_{CEO}.$$

and I_{CEO} is quite large in comparison with I_{CO} because the value of α is very close to unity. So there is appreciable gap between the X-axis and the curve for $I_B = 0$ unlike the case in CB configuration. We may attribute I_{CEO} as reverse biased saturation current.

Relation between α and β: We know that

$$I_B = I_E - I_C$$

and by definitions

$$\alpha = I_C / I_E \text{ and } \beta = I_C / I_B$$

Now expression for β can be written as

$$\beta = \frac{I_C}{I_B} = \frac{I_C}{I_E - I_C}$$

Dividing both numerator and denominator by I_E, we have

$$\beta = \frac{I_C / I_E}{\dfrac{I_E}{I_E} - \dfrac{I_C}{I_E}} = \frac{\alpha}{1 - \alpha} \qquad (21.56)$$

Transposing, we obtain

$$\alpha = \frac{\beta}{\beta + 1} \qquad (21.57)$$

We have defined both β and α as the ratio of currents, so they are also termed β_{DC} and α_{DC} respectively. However, when we consider small changes in current, we have for CE configuration as current gain

$$\beta_{AC} = \frac{\Delta I_C}{\Delta I_B}\bigg|_{V_{CE} = \text{constant}} = h_{fe}$$

and for CB configuration as current gain

$$\alpha_{AC} = \frac{\Delta I_C}{\Delta I_E}\bigg|_{V_{CE} = \text{constant}} = h_{fb}$$

However, it is to be noted that the values of β_{AC} and β_{DC} or α_{AC} and α_{DC} do not differ much from each other for small signals.

Transistors are generally fabricated with the help of the following four basic techniques: (i) grown technique (ii) alloy or fused technique (iii) diffusion technique and (iv) epitaxial technique.

In many circuits such as those required in high speed counters and computer, a transistor is made to act as a *switch*. It is ON when it is in saturation, and OFF when it is in cut-off condition. By applying pulse voltages, the transistor is caused to switch over from one state to the other.

Field Effect Transistor (FET)

This is a semiconductor device with the output current controlled by an electric field. Since the current in FET is carried predominantly by one type of carriers, it is known as *unipolar transistor*. Obviously, FET is different from the bipolar transistor (BJT) which involves two types of carriers, i.e. both electrons and holes.

FET has several forms, e.g. junction field effect transistor (JFET), metal oxide semiconductor field effect transistor (MOSFET), etc.

The different classes of FETs are characterised by a high input impedance. These devices are used in controlled switching between conducting and non-conducting states in digital circuits. FETs are also thermally stable. The main disadvantage of the FET is its relatively small gain-bandwidth product in comparison with conventional transistor.

A schematic diagram of a FET is shown in Fig. 21.44. The JFET consists of a uniformly doped semiconductor bar N-type (or P-type) of Si or GaAs. The bar has ohmic contacts at the two ends and semiconductor junctions in its two sides. If the semiconductor bar is N-type, the JFET is called an N-channel JFET and if the bar is P-type, the device is termed P-channel JFET. Two sides of the semiconductor bar are heavily doped with impurities opposite to that of bar, i.e. P-type impurities for a N-type bar and *vice versa*. For a N-type JFET, we can denote the three regions by P^+, N and P^+, where P^+ denotes a very heavily doped P-region. By applying a voltage V_{DD} between the two ends of a semiconductor bar, a current is allowed to flow along the length of the bar. The central N-region through which the majority carriers (electrons) flow from the source (S) to the collector or drain (D) is called the channel. The P^+ regions form the *gate*. At each of the P-N junctions, there is a depletion layer extending into the N-region. These layers penetrate more and more as V_{GG} (reverse bias) is increased, thus controlling the flow of electrons in the channel, i.e. the drain current I_D. The characteristics curves ($I_D - V_{DD}$ curves) are shown in Fig. 21.45. When I_D is increased from zero at constant V_{GS}, I_D increases linearly at first confirming to the Ohm's law. With

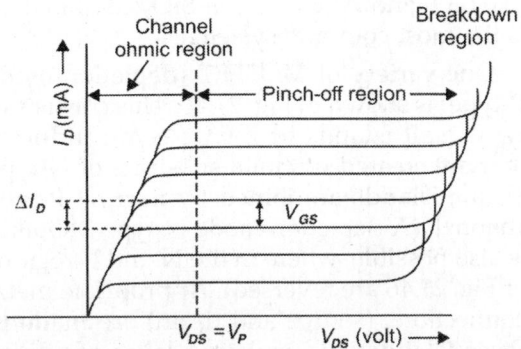

Fig. 21.45: I_D-V_{DS} static characteristics of an *N*-channel JFET

further increase in V_{DS}, the current attains saturation and this is called the *pinch-off* or the *saturation region*. Beyond a certain value of V_{DS}, I_D increases very rapidly indicating the breakdown region. The name *field* effect is used for the device because the transverse field produced by the gate gives the *effect* of controlling the drain current.

JFETs are less temperature sensitive and are not affected much by radiations. The input resistance of JFET is very high. One can achieve much lower switching time in switching devices.

Metal Oxide Semiconductor Field Effect Transistor (MOSFET)

Like JFET, the MOSFET is also field effect transistor, whose drain current (I_d) is controlled by the voltage on the gate. MOSFET and JFET differ physically as well as in operation. The MOSFET is also referred to as the IGFET (insulated gate field effect transistor) and MISFET (metal insulator semiconductor field effect transistor). The MOSFET is commercially more important than the JFET since MOS devices are suitable for large scale integration. MOSFET are of two types: (i) depletion type (ii) enhancement type. The MOSFET is an important power device.

MOSFETs can be N-channel or P-channel types. MOSFETs have been constructed with various semiconductors, e.g. silicon and gallium arsenide, and with different insulators

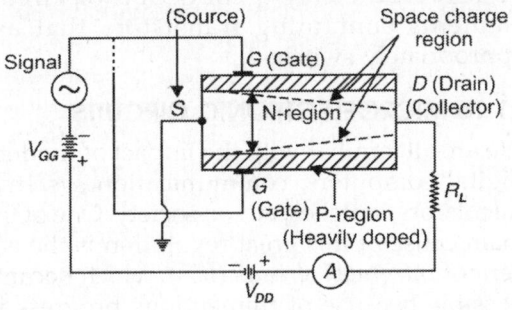

Fig. 21.44: Basic structure of an *N*-channel JFET

like SiO_2 and Al_2O_3. The Si-SiO_2 combination is the most common system.

One variety of MOSFET (depletion-mode P-type) is shown in Fig. 21.46. This consists of two small islands of P-type semiconductor that are created within a substrate of N-type Si, the islands are joined by narrow P-type channel. (A depletion mode N-type MOSFET is also possible, where in the N-and P-regions of Fig. 21.46 are reversed). Appropriate metal connections (source and drain) are made to these islands; an insulating layer of silicon dioxide is formed by the surface oxidation of the silicon. A final connector (gate) is then fashioned into the surface of this insulating layer.

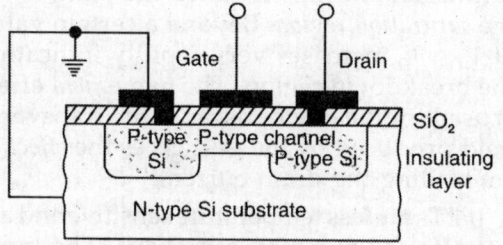

Fig. 21.46: Schematic cross-section view of a depletion-mode P-type MOSFET transistor

The conductivity of the channel is varied by the presence of an electric field imposed on the gate. For example, imposition of a positive electric field on the gate will drive charge carriers (in this case holes) out of the channel, thereby reducing the electrical conductivity. Thus, a small alteration in the electric field at the gate will produce a relatively large variation in current between the source and the drain. In some respects, then, the operation of a MOSFET is very similar to that described for the junction transistor. The primary difference is that the gate current is exceedingly small in comparison to the base current of a junction transistor. Obviously, MOSFETs are, therefore, used where the signal sources to be amplified cannot sustain in appreciable current.

Another important difference between MOSFETs and junction transistors is that,

although majority carriers dominate in the function of MOSFETs (i.e. holes for the depletion-mode-P-type MOSFET (Fig. 21.46)), minority carriers do play a role with junction transistors (i.e. injected holes in the N-type base region).

The MOSFETs can be used in any of the circuits wherever JFET is used with the bulk (substrate) connected to source. We may note that handling of MOSFET is not so easy. It requires special precautions because of very fine layer of SiO_2 between gate and channel. It is very susceptible to high voltages and can get punctured. Even static electric charge can puncture it. This is why MOSFETs are protected by a shorting ring that is wrapped around all four terminals which must remain until the device is mounted and soldered properly.

Semiconductors in Computers

In addition to their ability to amplify an imposed electrical signal, semiconductor diodes and transistors may also act as *switching* devices, a feature utilized for arithmetic and logical operations and also for storage of information in computers. Computer numbers and functions are expressed in terms of binary code, i.e. number written to the *base* 2. Within this scheme or framework, numbers are represented by a series of two states (sometimes designated 0 and 1). Now, diodes and transistors within a digital circuit operate as switches that also have two state—ON and OFF, or conducting and non-conducting; OFF corresponds to one binary state, and ON to the other state. Obviously, a single number may be represented by a collection of circuit elements containing transistors that are appropriately switched.

21.16 MICROELECTRONIC CIRCUITS

We are all familiar with the impact of modern digital computers, communication systems, calculators, watches, etc. on society. One of the main cause of this great revolution is the advent of *integrated circuits* (ICs), which became possible because of tremendous progress in semiconductor technology in recent years. The operation of these systems and many others is based on the principle of digital techniques and these systems are called as *digital systems*. Development in ICs technology have made it possible to fabricate complex digital circuits, such as *microprocessors*, memory units, etc. on tiny chips of silicon. The wondership—the microprocessor has been the most fantastic development of recent years. Inexpensive microelectronic circuits are mass produced by using some very igneous fabrication techniques. The process begins with the growth of relatively large cylindrical single crystals of high purity silicon from which thin circular wafers are cut. Many microelectronic or ICs, sometimes referred to as *chips*, are prepared on a single wafer. A chip is rectangular, typically of the order of 6 mm (1/4 inch) on a side and contains thousands of circuit elements; diodes, transistors, resistors and capacitors. Presently, microprocessor chips containing 500 million transistors are being produced, and this number doubles about every 18 months.

Microelectronic circuits consist of many layers that lie within or are stacked on top of the silicon wafer in a precisely detailed pattern. Using a photolithographic technique, for each layer, very small elements are masked in accordance with a microscopic pattern. Circuit elements are constructed by the selective introduction of specific materials (by diffusion or ion implantation) into unmasked regions to create localized N-type, P-type, high resistivity or conductive areas. This procedure is repeated, layer by layer, until the total integrated circuit has been fabricated, as illustrated in the MOSFET scheme.

21.17 MICROELECTROMECHANICAL SYSTEMS (MEMSs)

Almost all applications in industry, defence, medicine and other fields require sensing and control of various parameters. In recent years, taking advantage of the silicon IC technology, a new field has emerged which attempts to combine *sensor*, *actuator* and the *control* circuit

on as one integrated unit. In this sense, it emulates a biological system (Fig. 21.47). These are known as *smart sensors, microsystems technology* (MST) or *microelectromechanical system* (MEMS). It has been possible to fashion miniature mechanical devices such as gears, motors, springs, etc. Their combination with sensing and actuating functions has given engineers and scientists the tools to build microsystems that could not be imagined earlier. Some examples of the *smart structures* and systems are given in Fig. 21.48.

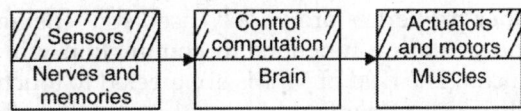

Fig. 21.47: Integrated sensor–actuator systems with controller are analogous to biological systems

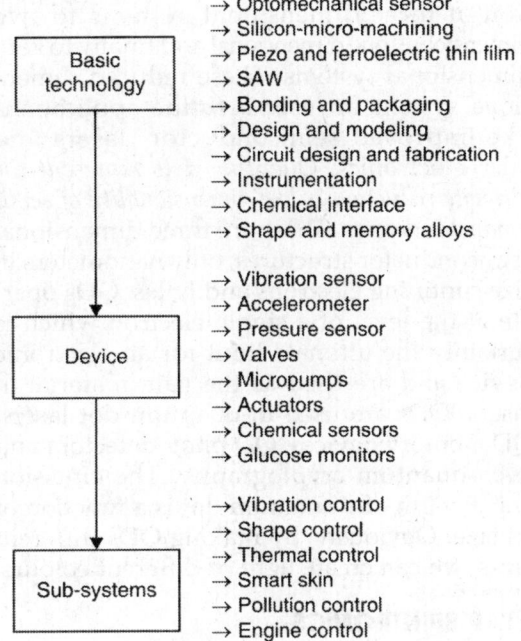

Fig. 21.48: Few examples of smart structures and systems and their applications

MEMS devices are small in size, lightweight, low cost, reliable, with large batch fabrication technology. They generally consist of sensors that gather environmental information such as pressure, temperature, acceleration, etc. integrated electronics to process the data collected and actuators to influence and control the environment in the desired manner.

The MEMS technology involves a large number of materials. Silicon forms the backbone of these systems due to its excellent mechanical properties as well as mature microfabrication technology including lithography, etching and bonding. Other materials having piezoelectric, piezoresistive, ferroelectric and other properties are widely used for sensing and actuating functions in conjunction with silicon. The field of MEMs is expected to touch all aspects of our lives during this decade with revolutions in aviation, automobiles, pollution control and industrial processes.

21.18 QUANTUM DOTS (QDs)

Rapid progress in the fabrication of semiconductor structures has resulted into the reduction of three dimensional systems to two dimensional, one dimensional and finally to zero dimensional systems. These reduced dimensional systems are used in future applications like improved semiconductor lasers and microelectronics. *Quantum dots represent the ultimate reduction in the dimensionality of semiconductor devices*. These are three dimensional semiconductor structures, only nanometers in size confining electrons and holes. QDs operate at the level of a single electron which is certainly the ultimate limit for an electronic device and are used as the gain material in lasers. QDs are used in quantum dot lasers, QD memory devices, QD photodetectors and even quantum cryptography. The emission wavelength of a quantum dot is a function of its size. Obviously, by making QDs different sizes, we can create light of different colours.

21.19 SPINTRONICS

A revolutionary new class of semiconductor electronics based on the spin of freedom could be created. The performance of conventional devices is limited in speed and dissipation whereas spintronics devices are capable of much higher speed at very low power. Spintronics transistors may work at a faster speed, are also smaller in size and will consume less power.

The study of electron spin in materials is called spintronics. Spintronics is based on the direction of spin and spin-coupling. Spins can be oriented in one direction or the other, called spin-up or spin-down. When electron spins are aligned in one direction, these create a net magnetic moment as seen in magnetic materials like iron and cobalt. Magnetism is an intrinsic physical property associated with the spin of electrons in a material. The electron spin may exist not only in the up or down state but also in infinitely many intermediate states because of its quantum nature depending on the energy of the system. This property may lead to highly parallel computation which could make a *quantum computer* work much faster for certain types of calculations. In quantum mechanics, an electron can be in both spin-up and spin-down states, at the same time. The mixed state could form the base of a computer built around not binary bits but the *quantum bits* or *qubit*. It is any combination of a 1 or a 0. The simplest device using spin-dependent effect is a sandwich with two magnetic layers surrounding a non-magnetic metal or insulator. If the two magnetic layers are different, then the magnetization direction of one can be rotated with respect to the other. This leads to the utilization of these structures as sensor elements and for memory elements. Scientists and engineers are now trying to use the property of the electron-like spin rather than charge to develop new generation of microelectronic devices which may be more versatile and robust than silicon chips and circuit elements. Spins appear to be remarkably robust and move relatively easily between semiconductors. In case of electron transport from one material to another, the spins do not lose its orientation or scatter from impurities or structural defects.

ILLUSTRATIVE EXAMPLES

Example 1

A half wave rectifier, having a diode of resistance(r_f) = 10^3 Ω and a load of 10^4 Ω rectifies an alternating voltage of 300 volts peak value. Calculate:

i. Peak, average and RMS value of current
ii. DC power input
iii. AC input power
iv. Efficiency of rectifier.

Solution

i. Peak value of current flowing

$$I_m = \frac{E_m}{r_f + R_L} = \frac{300}{1000 + 10000}$$

or I_m = 27.31 mA

ii. Average of DC current flowing

$$I_{DC} = \frac{i_m}{\pi} = \frac{27.31}{3.14} \text{ mA}$$

or I_{DC} = 8.68 mA

iii. RMS value of current flowing

$$I_{RMS} = \frac{I_m}{2} = \frac{27.31}{2}$$

$$= 13.65 \text{ mA}$$

iv. DC power output $I_{DC} = I_{DC}^2 \times R_L$

$$= \left(8.68 \times 10^{-3}\right)^2 \times 10000$$

$$= 0.754 \text{ watt}$$

v. AC input power $P_{AC} = (I_{RMS})^2 \times (R_L + r_f)$

$$= \left(13.65 \times 10^{-3}\right)^2 \times (11000)$$

$$= 2.05 \text{ watt}$$

vi. Efficiency of rectification

$$\eta = \frac{P_{DC}}{P_{AC}} = \frac{0.754}{2.05} = 36.81\%$$

Example 2

A half wave rectifier consists of a diode having dynamic resistance of 1 Ω at its operating current and a transformer whose open circuit secondary voltage is 12.6 V, 60 Hz. It has secondary of resistance 3 ohms.

i. What is the no load DC output voltage?
ii. What is the DC output voltage when the load draws a current of 100 mA?
iii. What is the % voltage regulation of the rectifier?

Solution

i. No load DC voltage = $\dfrac{E_m}{\pi} = \dfrac{(E_{RMS})\sqrt{2}}{\pi}$

$$= \frac{12.6 \times 1.4}{3.14}$$

ii. Full load $V_{DC} = V_{No\,load} - I_{DC}(R_S + r_f)$

$$= 5.66 - 100 \times 10^{-3}(3 + 1)$$

$$= 5.26 \text{ V}$$

iii. % voltage regulation

$$= \frac{V_{No\,load} - V_{Full\,load}}{V_{Full\,load}} \times 100$$

$$= \frac{5.66 - 5.26}{5.26} \times 100$$

$$= \frac{0.4}{5.26} \times 100 = 7.6$$

Example 3

A DC power supply is known to have a ripple factor of 10%. If the DC output voltage is 10 volt. What is the RMS value of ripple voltage?

Solution

Riple factor $\gamma = \dfrac{10}{100} = \dfrac{(V_r)_{RMS}}{V_{DC}}$

RMS value of ripple voltage

$$(V_r)_{RMS} = \frac{10}{100} \times V_{DC}$$

$$= \frac{10}{100} \times 10 = 1V.$$

Example 4

A 120 V, 60 Hz voltage is applied to the primary of a 1 : 5 step-up transformer used in a full wave rectifier having a load of 1 kΩ. Neglecting the voltage drop across the diodes calculate:

   i. DC voltage across the load

   ii. DC current through the load

   iii. DC power delivered to the load

   iv. AC ripple voltage and its frequency

   v. The PIV across each diode.

Solution

Since the transformer is step up, the voltage across the two ends of secondary

$$= 120 \times 5 = 600 \text{ V}$$

Voltage from centre tapping to one end

$$= \frac{600}{2} = 300 \text{ V}$$

i. $V_{DC} = \dfrac{2E_m}{\pi}$ since there is no drop across

   diode voltage across load

$$= \frac{2 \times 300 \times \sqrt{2}}{\pi} = 269.3 \text{ V}$$

ii. DC current flowing through load

$$I_{DC} = \frac{V_{DC}}{R_L} = \frac{269.3}{1000} = 269.3 \times 10^{-3} \text{ A}$$

$$= 269.3 \text{ mA}$$

iii. DC power delivered to load

$$P_{DC} = 72 \text{ W}$$

iv. AC ripple voltage

$$(V_r)_{RMS} = \sqrt{(V)_{RMS}^2 - (V_{DC})^2}$$

$$= \sqrt{(300)^2 - (269.3)^2} = 130 \text{ V}$$

Frequency of ripple voltage

$$= 2 \times 60 = 120 \text{ Hz}$$

v. Peak inverse voltage across each diode

$$= 2E_m$$

$$= 2 \times 300 \times 1.4 = 840 \text{ V}.$$

Example 5

Find the conductivity and resistivity of a pure silicon crystal at temperature 300 K. The density of electron hole pair per cc at 300 K for a pure silicon crystal is 1.072×10^{10} and the mobility of electron $\mu_n = 1350 \text{ cm}^2/\text{volt sec}$ and hole mobility $\mu_h = 480 \text{ cm}^2/\text{volt sec}$.

Solution

Conductivity of pure silicon crystal is given by

$$\sigma = n_i e\,(\mu_n + \mu_h), \; n_i = 1.072 \times 10^{10}$$

$$\sigma_i = 1.072 \times 10^{10} \times 1.6 \times 10^{-19}(1350 + 480)$$

$$= 3.14 \times 10^{-6} \text{ mho/cm}$$

$$\mu_n = 1350 \text{ cm}^2/\text{Vs}$$

$$\mu_h = 480 \text{ cm}^2/\text{Vs}$$

$$e = 1.6 \times 10^{-19} \text{ C}$$

Resistivity of silicon crystal is given by

$$\rho_i = \frac{1}{\sigma_i} = \frac{1}{3.14 \times 10^{-6}} = 3.18 \times 10^{-5} \,\Omega\text{m}$$

$$= 3.18 \times 10^{-3} \,\Omega\text{m}$$

Example 6

A silicon wafer is doped with phosphorus of concentration 10^{13} atoms/cm^3. If all the donor atoms are active, what is its resistivity at room temperature? The electron mobility is 1200 cm^2/volt sec, charge on the electron is 1.6×10^{-19} coulomb.

Solution

$$\sigma = \mu e n \qquad\qquad \left| \mu = 1200 \text{ cm}^2/\text{volt sec} \right.$$

$$\sigma = 1200 \times 1.6 \times 10^{-19} \times 10^{13} \; \left| e = 1.6 \times 10^{-19} \text{ C} \right.$$

$$= 19.2 \times 110^{-14} \text{ mho/cm} \quad \left| n = 10^{13} = N_P \right.$$

Resistivity $\rho = \dfrac{1}{\sigma} = \dfrac{1}{19.2 \times 10^{-4}} = 5.2 \times 10^2 \,\Omega\text{m}.$

Example 7

Find the resistance of an intrinsic germanium rod 1 cm long, 1 mm wide and 1 mm thick at temperature of 300 K. For germanium

$$n_i = 2.5 \times 10^{13},$$

$$\mu_n = 3900 \text{ cm}^2/\text{volt sec at 300 K.}$$

Solution

$$\sigma = n_i e \, (\mu_e + \mu_h)$$

$$= 2.5 \times 10^{13} \times 1.6 \times 10^{-19} \, (0.39 + 0.19)$$

$$= 2.32 \text{ mho/cm}$$

or $\quad \rho = \dfrac{1}{232} \, \Omega m$

Now, Resistance

$$R = \rho_i = \left[\frac{\text{Length}}{\text{Area of cross-section}} \right]$$

$$= 2.32 \times \frac{10^{-6}}{10^{-2}}$$

$$\therefore \quad R = \frac{1}{2.32 \times 10^{-4}} \text{ ohm} = 4.31 \text{ k}\Omega$$

Example 8

A sample of germanium is made of P-material by adding acceptor atoms at a rate of one atom per 4×10^8 germanium atoms. The acceptor density is assumed to be zero and $n_i = 2.5 \times 10^{19}$ per m^3 at 300 K. There are 4.4×10^{28} germanium atoms/m^3. The acceptor density is found to be 1.1×10^{20} atoms/m^3.

Solution

$$n_i^2 = n_p = n_p \, N_a$$

$$\therefore \quad n_p \approx \frac{n_i^2}{N_a} = \frac{6.25 \times 10^{38}}{1.1 \times 10^{20}} = 5.6 \times 18^{18}$$

$$\therefore \quad \frac{n_p}{n_i} = \frac{5.6 \times 10^{18}}{2.5 \times 10^{19}} = 0.22$$

REVIEW QUESTIONS

1. What are semiconductors? How they differ from metals and insulators? What are their characteristics properties? [BE]

2. Why an increase in temperature of semiconductor increases its conductivity? [BE]

3. Explain the terms: (i) intrinsic semiconductors (ii) extrinsic semiconductors. [AMIE]

4. What is impurity conduction in semiconductors? Explain how the presence of a small impurity in a semiconductor modifies its conduction properties.

5. Define the terms: (i) electron–hole pair (ii) donor (iii) acceptor. Give examples of each with suita-ble materials. [AMIE]

6. Describe N-type and P-type semiconductors. [AMIE]

7. Describe a P-N junction and transistor with neat sketches. [BE]

8. Explain the process of recombination of electrons and holes in a semiconductor. If the electron–hole pair recombine, how can their concentration remain constant at a particular temperature?

9. What are extrinsic semiconductors? How can they be formed?

10. At high temperature an extrinsic semiconductor behaves like an intrinsic one, why?

11. Is it possible for compound semiconductors to exhibit intrinsic behaviour? Explain.

12. For each of the following pairs of semiconductors decide which will have the smaller band gap energy E_g and then give reason: (a) carbon (C) and Ge (b) AlP and InSb (c) GaAs and ZnSe (d) ZnSe and CdTe (e) CdS and NaCl.

13. (a) Explain why no hole is generated by the electron excitation involving a donor impurity atom.

    (b) Explain why no free electron is generated by the electron excitation involving an acceptor impurity atom.

14. Will each of the following elements act as a donor or acceptor when added to the indicated semiconducting material? Assume that the impurity elements are substitutional.

Impurity	Semiconductor
N	Si
B	Ge
Zn	GaAs
S	InSb
In	CdS
As	ZnTe

PROBLEMS

1. For intrinsic gallium arsenide, the room temperature electrical conductivity is 10^{-6} $(\Omega m)^{-1}$, the electron and hole mobilities are respectively 0.85 and 0.04 m^2/Vs. Compute the intrinsic concentration n_i at room temperature.

   [*Ans.* 7.0×10^{12} m^{-3}]

2. Calculate the electrical conductivity of intrinsic silicon at 150°C (423 K). [*Ans.* 0.52 $(\Omega m)^{-1}$]

3. The high-purity silicon is added to $10^{23}/m^3$ arsenic atoms: (a) Is this material N-type or P-type? (b) Calculate the room temperature electrical conductivity of this material (c) Compute the conductivity at 100° C (373 K).

[*Ans.* (a) N-type (b) 1120 $(\Omega m)^{-1}$ (c) 640 $(\Omega m)^{-1}$]

4. The following electrical characteristics have been found for both intrinsic and P-type extrinsic indium phosphide (InP) at room temperature:

	$\sigma(\ (\Omega m)^{-1}$	$N(m^{-3})$	$P(m^{-3})$
Intrinsic	2.5×10^{-6}	3.0×10^{13}	3.0×10^{13}
Extrinsic (N-type)	3.6×10^{-5}	4.5×10^{14}	2.0×10^{12}

Calculate electron and hole mobilities.

[*Ans.* $\mu_e = 0.50\ m^2/Vs$, $\mu_h = 0.02\ m^2/Vs$]

5. Mobilities of electrons and holes in a sample of intrinsic germanium at room temperature are 3600 cm²/Vs and 1700 cm²/Vs respectively. If the electron and hole densities are each equal to 2.5×10^{13} per cc, calculate its conductivity.

[*Ans.* 2.12 mho/m]

SHORT ANSWER QUESTIONS

1. Why semiconductive materials are of limited value in their intrinsic state?

Ans. Due to the limited number of free electrons in the condition band and holes in the valence band.

2. How many types of extrinsic (impure) semiconductive materials are there?

Ans. There are two types of extrinsic semiconducting materials, N-type and P-type. These are the key building blocks for most types of electronic devices.

3. What is doping?

Ans. The conductivity of silicon and germanium can be drastically increased by the controlled addition of impurities to the intrinsic (pure) semiconductive material. This process is called *doping*. This increases the number of current carriers (electrons or holes). The two categories of impurities are N-type and P-type.

4. What is reverse bias?

Ans. Reverse bias is the condition that essentially prevents current through the diode.

5. What are the values of forward biased barrier potential for silicon and germanium diodes? How these are affected by forward current?

Ans. A forward biased barrier potential is typically 0.7 V for a silicon diode and 0.3 V for a germanium diode. These values increase slightly with forward current.

6. What is a solar cell?

Ans. A solar cell is a photodiode operated in its photovoltaic mode.

7. What are optocouplers?

Ans. Optocouplers are device that use light to optically couple a signal between two electrically isolated points. Optocouplers are also called optoisolators.

8. What is a phototransistor?

Ans. A phototransistor is a light detecting device that is also called a photosensor. It is a transistor whose base current is supplied by the carriers generated due to the incident or striking light.

9. What is photodarlington?

Ans. A photodarlington is a phototransistor packed with another transistor connected in the Darlington configuration.

OBJECTIVE QUESTIONS

1. Normal conductors have resistivities that increase with increasing temperature, but in semiconductors the resistivity _____ with increasing temperature. [decreases]

2. In a semiconductor there exists a small band gap, _____ between the valence band and the conduction band. [1 eV]

3. Some semiconductors are N-type, meaning that the majority carriers are _____ electrons. [negative]

4. Some semiconductors are P-type, meaning that the majority carriers are positive _____. [holes]

5. Materials of P-type and N-type can be placed end to end to form a _____ diode. [P-N junction]

MULTIPLE CHOICE QUESTIONS

1. N-type germanium is obtained on doping intrinsic germanium by
   (a) phosphorous   (b) aluminium
   (c) boron   (d) gold

2. Depletion region is a zone which contains
   (a) holes only
   (b) electrons only
   (c) both electrons and holes
   (d) neither electrons nor holes

3. In a semicoductor diode, arrow represents
   (a) N-type material
   (b) P-type material
   (c) both P and N-type materials
   (d) none of these

4. Zener diode is used for
   (a) rectification
   (b) amplification
   (c) stabilization
   (d) none of the above

5. For a tunnel diode a decrease in current causes
   (a) voltage constancy
   (b) decrease in voltage
   (c) increase in voltage
   (d) none of the above

6. P-N junction is formed when P-type semicon-ductor and N-type semiconductor are joined
   (a) together
   (b) physically
   (c) to get homogeneous material chemically
   (d) in such a manner that electrons and holes diffuse to give depletion layer

7. The depletion region of a junction diode is formed
   (a) when forward bias is applied to it
   (b) when the temperature of the junction is reduced
   (c) under reverse bias
   (d) during the manufacturing process

8. The width of the depletion layer of a junction
   (a) is independent of applied voltage
   (b) is increased under reverse bias
   (c) decreases with light doping
   (d) increase with heavy doping

9. The LED or the light-emitting diode
   (a) is made from one of the two basic semicon-ducting materials, silicon or germanium
   (b) is made from the semiconducting com-pound gallium arsenide phosphide
   (c) emits light when forward biased
   (d) emits light when reverse biased

10. The P-side of a junction diode is earthed and the N-side is given a potential of 2 V. The diode will
    (a) not conduct
    (b) conduct partially
    (c) breakdown
    (d) conduct fully

11. For detecting light intensity, we use a/an
    (a) photodiode in reverse bias
    (b) photodiode in forward bias
    (c) LED in reverse bias
    (d) LED in forward bias

12. When a P-N junction diode is forward biased, the flow of current across the junction in mainly due to
    (a) diffusion of charges
    (b) drift of charges
    (c) depends on the nature of the material
    (d) both drift and diffusion of charges

13. A P-N junction diode cannot be used
    (a) as a rectifier
    (b) for increasing the amplitude of an ac signal
    (c) for getting light radiation
    (d) for converting light energy into electrical energy

14. A strong electric field across a P-N junction that causes covalent bonds to break apart is called
    (a) reverse breakdown
    (b) avalanche breakdown
    (c) lever breakdown
    (d) low voltage breakdown

15. A light-emitting diode produces light when
    (a) forward biased
    (b) reverse biased
    (c) unbiased
    (d) none of the above

16. A solar cell is an example of
    (a) photo emissive cell
    (b) photo radiation cell
    (c) photo radiation cell
    (d) photo conductive cell

17. When holes leave the P-material to fill electrons in the N-material the process is called
    (a) diffusion
    (b) depletion
    (c) avalanche breakdown
    (d) zener breakdown

18. A diode which has zero breakdown voltage is known as
    (a) tunnel diode   (b) zener diode
    (c) Schottky diode   (d) backward diode

19. A zener diode is used as
    (a) a coupler
    (b) a rectifier
    (c) an amplifier
    (d) a voltage regulator

20. An empirical formula relating the resistance of many semiconductors to temperature is

    (a) $\log R + \dfrac{K}{\log R} = A + \dfrac{B}{T}$

    (b) $\log R + K \log R = \dfrac{B}{T}$

    (c) $\log R + \dfrac{K}{\log R} = A$

    (d) $\log R = A + \dfrac{B}{T}$

Answers

1. (a)	2. (d)	3. (b)	4. (c)	5. (c)
6. (d)	7. (d)	8. (b)	9. (b, c)	10. (d)
11. (a)	12. (a)	13. (b)	14. (c)	15. (b)
16. (c)	17. (a)	18. (a)	19. (d)	20. (a)

22 Superconductivity

22.1 INTRODUCTION

When cooled to sufficiently low temperatures, a large number of metals and alloys can conduct electric current without resistance. Obviously, these specific materials undergo a phase transition to a new superconducting state characterized by the complete loss of DC resistance below a well defined critical temperature, T_C. Thus, zero resistivity ($\rho = 0$), i.e. infinite conductivity is observed in a superconductor at all temperatures below critical temperature ($\rho = 0$ for all $T < T_C$). However, if we pass a current higher than the critical current density J_C, superconductivity disappears. This limits the maximum current which the material can sustain and is an important problem for applications of superconducting material. For elements, the transition temperature, T_C lies generally below 10 K. Figure 22.1 shows resistance versus temperature for a low temperature superconductor. At the transition temperature, T_C the resistance drops abruptly to an unmeasurably small value. The transition from normal to the superconducting phase is often sharp and occurs between 10^{-2} K and 10^{-4} K. This behaviour is remarkably different from the steadily decreasing resistance of non-superconducting metals (Fig. 22.2) and suggests the existence of a physically different superconducting state. In pure metals, the zero resistance state can be reached within a temperature range of 1 mK. In the case of impure metals, the transition to the superconducting

state may be considerably broadened. A transition width ≈ 0.05 K was observed for impure tin. The resistivity of a superconductor to direct current is zero as far as it can be measured. The estimates of the resistivity in the superconducting phase place it at less than 4×10^{-25} Ω m, which is essentially zero for all practical purposes. A striking way to

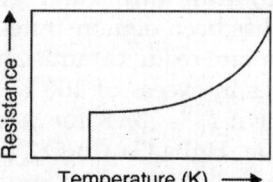

Fig. 22.1: Resistance versus temperature for a low temperature superconductor

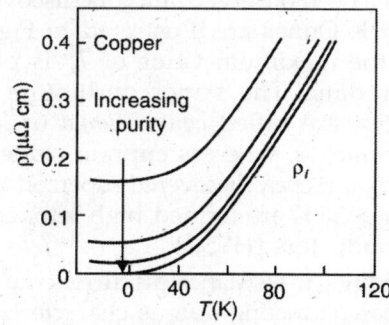

Fig. 22.2: Resistivity of copper for residual ratio (RRR) values of 10, 30 and 100. At very low temperatures, the intrinsic resistivity $\rho(T)$ due to electron–phonon interaction approaches zero. RRR is defined as RRR = ρ(273 K)/ρ(4 K)

demonstrate zero resistivity is to induce a current around a close ring of a superconducting metal. Experiments have been performed in which a 'persistent current' has run for over two and a half years without any measurable decay. The time dependence of the current I in the loop is given by $I(t) = I_0 e^{-t/\tau}$, where I_0 is the initial value of the current and t is the time which has elapsed since the supercurrent has been induced. The ratio of resistance R and self inductance L of the superconducting loop determines the time constant τ for the decay of the current. Above the critical temperature T_C, the metal is in the normal state and resistance is proportional to T^5. In many metals, the exponent is between 2 and 6, considerably different from the value 5 predicted by Bloch theory.

The critical temperature, T_C varies from superconductor to superconductor but lies between less than 1 K and approximately 20 K for metals and metal alloys. Until 1986 the maximum T_C was observed in an alloy of niobium, aluminium and germanium. Recently it has been demonstrated that some complex cuprate oxide ceramics have critical temperatures in excess of 100 K. Today, the highest known T_C is 133 K for mercury based cuprate oxide, $HgBa_2Ca_2Cu_3O_{8+\delta}$. When this compound is subjected to high pressure ~30 GPa, the onset of T_C increases to ~164 K. The dramatic evolution of critical temperatures that have been observed since its discovery in 1911 by K Onnes are illustrated in Fig. 22.3 where the maximum value of T_C is plotted against date. The superconductors with $T_C < 25$ K are called *conventional* or *low T_C superconductors*, whereas cuprate oxides and some other recently discovered superconductors with $T_C > 25$ K are termed high temperature superconductors (HTSC).

In addition to resistanceless current transport, the superconducting state is characterized by perfect diamagnetism, i.e. $B = 0$ inside the superconductor. The magnetic inductance becomes zero inside the superconductor when it is cooled below T_C in a weak external magnetic field; the magnetic flux is expelled from the interior of the superconductor (Fig. 22.4). This

effect is called the *Meissner–Oschsenfeld effect* after its discoverers and it is the ultimate practical test in any new material. We must note that there always exists some critical field H_C above which superconductivity disappears. Supercondctivity disappears and the material returns to the normal state if one applies an external magnetic field of strength greater than H_C.

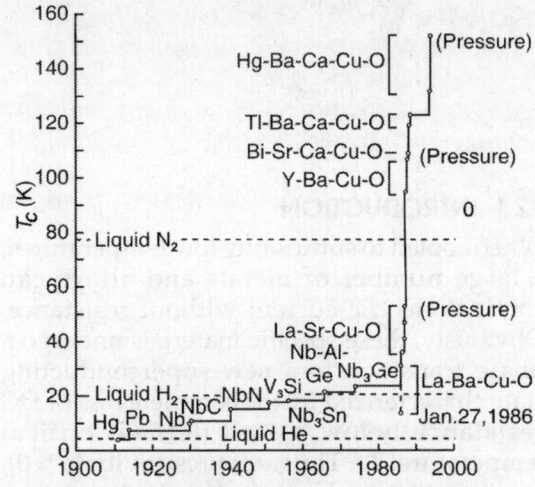

Fig. 22.3: The evolution of superconductivity critical temperature since its discovery

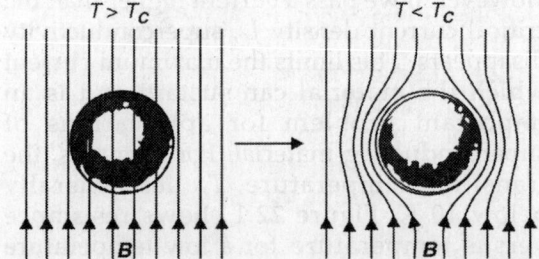

Fig. 22.4: Expulsion of a weak, external magnetic field from the interior of the superconducting material

On the basis of magnetic response, superconducting materials may be divided into two classes designated as type-I and type-II. Type-I materials, while in the superconducting state, are completely diamagnetic, i.e. in an applied magnetic field, field will be excluded from the body of material (Miessner effect). Several metallic elements including aluminium, lead, tin and mercury belong to type-I group.

Type-II superconductors are completely diamagnetic at low applied fields and field exclusion is total. However, the transition from the superconducting state to the normal state is gradual and occurs between lower critical and upper critical fields, designated as H_{C_1} and H_{C_2} respectively (Fig. 22.5). The magnetic flux lines begin to penetrate into the body of the material at H_{C_1} and with increasing applied magnetic field, this penetration continues; at H_{C_2} field penetration is complete. For fields between H_{C_1} and H_{C_2}, the material exists in what is termed a mixed state—both normal and superconducting regions are present.

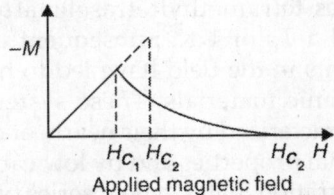

Fig. 22.5: Variation of magnetization as a function of the magnetic field for (a) Type-I super-conductor (dotted line) (b) Type-II ideal superconductor (solid line)

Type-II superconductors are preferred over type-I for most practical applications by virtue of their higher critical temperatures and critical magnetic fields.

The three material parameters, T_C, H_C and J_C are of extreme importance in the practical applications of superconductivity. Figure 22.6 shows schematically the boundary in temperature, magnetic field and current density space separating normal and superconducting

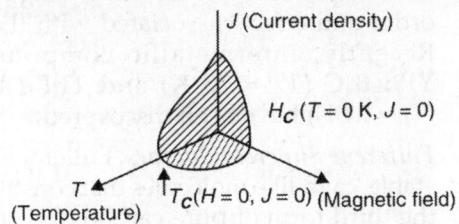

Fig. 22.6: Critical temperature (T_C), current density (J_C) and magnetic field boundary separating superconducting and normal conducting states

states. The position of this boundary will, of course, depend on the material. For temperature, magnetic field and current density values lying between the origin and this boundary, the material will be superconductive; outside the boundary, conduction is normal.

22.2 SUPERCONDUCTING MATERIALS

Superconducting Elements

Metallic elements are mostly superconductors. Their T_C are typically of the order of a few Kelvin. Among metals, niobium exhibits the highest critical temperature of the pure elements, $T_C = 9.2$ K. Noble metals, copper, silver and gold are excellent conductors of electricity at ambient temperatures, but are not superconductors due to very low temperatures at all. Magnetic metallic elements do not exhibit superconductivity.

The best known semiconductors, Si and Ge, become superconductors under a pressure ~2 kbar with $T_C = 7$ K and 5.3 K respectively. Other elements that become superconductors under pressure include P, As, Se, Y, Sb, Te, Ba, Bi, Ce and U.

Binary Alloys and Compounds

In most alloys and compounds, the critical temperature is usually somewhat higher. Nb compounds, like Nb_3Sn, Nb_3Ge and in particular, Nb-Ti, are of technological interest. While the maximum current density that can pass through the standard water-cooled copper wire at 300 K is about 2000 A /cm^2, one can pass very high current densities of up to 10^4 A/cm^2 in high magnetic fields of 10 tesla at 4.2 K through a wire made of Nb-Ti alloy without destroying superconductivity. This enables the construction of powerful super-magnets which provide a basis for a range of large scale applications like energy storage or levitation of trains, etc.

Transition metals combined with other elements often produce binary alloys or compounds with T_C higher than those of starting elements. The intermetallic compounds and ordinary compounds usually exhibit high T_C.

Intermetallic Compounds

Among the intermetallic superconductors, the most favourable group is the one based on the A_3B compound. In the cubic A-15 structure, six binary compounds have T_C over 17 K. The highest known T_C prior to 1986, close to $T_C = 23$ K is obtained in Nb_3Ge stabilized by traces of oxygen or aluminium; it exhibits the upper critical field of 38 T. The A-15 structure exists in about 70 binary compounds. H_{c_2} (T) values for A-15 superconducting compounds are very high. H_{c_2}(T), the upper critical field values as the temperature approaches 0 K for A-15 compounds are also very high, e.g. 44 T for a composition Nb_{79} $(Al_{73}Ge_{27})_{21}$, 32 T for Nb_3Al, 39 T for Nb_3Ge, 23 T for Nb_3Sn, 21 T for V_3Ga and 25 T for V_3Si.

i. *Chevrel Phases:* In 1971, Chevrel *et al.*, discovered a series of ternary molybdenum chalcogenides having the general formula $M_xMO_6X_8$ where M represents any of a large number (nearly 40) of metallic elements and rare earths (RE) throughout the periodic table, x has value between 1 and 4, depending on the element M, and X is a chalcogen; sulphur (S), selenium (Se) or **tellurium** (Te). The highest $T_C = (15$ K) in the series is obtained in $PbMO_6S_8$ with unusual high H_{c_2} value $= 60$ T $= 600$ kilogauss. This was the highest value of H_{c_2} prior to the discovery of HTSC cuprates. The large values of H_{c_2} as compared with Nb_3Sn and NbTi make this material interesting for making superconducting wires. Critical current density (J_C) as high as ~3×10^{-5} A/cm^2 have been reported at 4.2 K and this provides an impetus for making wires out of these very brittle materials. It has been reported that antiferromagnetism of the rare earth can coexist with superconductivity like in Gd, Tb, Dy, Tr compounds where T_C is 1.4, 1.65, 2.1 and 1.85 K and T_N (Neel temperature) is 0.84, 0.9, 0.4 and 0.15 K respectively. The compound $HoMo_6S_8$ exhibits coexistence of superconductivity and ferromagnetism. The material is superconducting only between two critical temperatures, $T_{C_1} = 2$ K and T_{C_2} below 0.65 K. Below 0.65 K, the material is ferromagnetic.

ii. *Tetragonal Rare Earth Rhodium Borides:* This series has general formula $RERh_4B_4$. The compounds for RE = Y, Er, Tm and Lu exhibit superconductivity. The compound $ErRh_4B_4$ is a typical re-entrant ferromagnetic superconductor ($T_{C_1} = 8.67$ K, $T_{C_2} = 0.775$ K warming and 0.710 K cooling and Curie temperature 1.2 K).

iii. *Organic Superconductors:* These are a novel group of materials. The first organic superconductor $[TMTSF]_2PF_6$, where TMTSF denotes tetramethyltetraselenafulvalene, had a T_C of 1 K. Subsequent developments in the field have led to higher T_C organic materials. These systems were characterized by their nearly one dimensional properties and by low carrier concentration. Later, a new series of organic materials with two dimensional characteristics was discovered: $(BEDT-TTF)_2X$, where BEDT-TTF denotes bisethylenedithio-tetrathiafulvalene. The κ-modification of the X = $Cu(NCS)_2$ has $T_C = 10.4$ K and κ-$(BEDT-TTF)_2Cu[NCN]_2Br$ has $T_C = 11.2$ K, which is probably the highest known T_C so far observed in these materials.

iv. *Rare Earth Transition Metal Borocarbides:* Superconductivity was reported recently in a series of compounds with the formula RNi_2B_2C with a maximum T_C of 16.5 K for R = Lu. These materials display both superconductivity and magnetic order and effects associated with them. Recently, intermetallic compounds YPd_5B_3C ($T_C = 23$ K) and $ThPd_3B_3C$ ($T_C = 21$ K) have been discovered.

v. *Fullerene Superconductors:* Fullerene are stable, cage-like molecules that constitute the third form of pure carbon; the other two are diamond and graphite. The archetype fullerene is C_{60}. When C_{60} is doped with alkaline metals, superconductivity is observed with $T_C = 18$ K, 30 K, 33 K and

40 K for K_3C_{60}, Rb_3C_{60}, $RbCs_2C_{60}$ and Cs_3C_{60} respectively. Hole doped C_{60} (for $C_{60}/CHBr_3$ with 3 to 3.5 holes per C_{60} molecule) was recently reported to be superconducting with a T_C as high as 117 K, although the nature of experiment meant that the supercurrents were confined to the surface of the C_{60} crystal, rather than probing the bulk.

vi. *Heavy Fermion Superconductors:* A small number of compounds which include one Ce compound $CeCu_2Si_2$ and few uranium compounds— UBe_{13}, UPt_3, URu_2Si_2, UNi_2Al_3, UPd_2Al_3 exhibit superconductivity with $T_C < 1$ K. These compounds are characterized by enormous volumes of linear coefficient of the electronic specific heat ($C_e = \gamma T$) which can be as high as ~1 J mol^{-1} K^{-2} and a corresponding large electron mass ($m \sim 10^2 - 10^3$ m_e, where m_e is free electron mass). Most of these systems exhibit the coexistence of superconductivity and antiferromagnetism.

viii. *Itinerant Electron Ferromagnetic Superconductors:* The intermetallic compound Y_9Co_7 has been shown to exhibit an interesting interplay between very weak ferromagnetism and some form of superconductivity. The system shows ferromagnetism below 6–8 K and at a lower temperature (~3 K) superconductivity sets in. Recently, superconductivity in UGe_2 is reported below 1 K on the border of weak ferromagnetism.

viii. *Quantum Spin Ladder Materials:* These materials consist of ladders made of AFM chains of $s = 1/2$ spins coupled by interchain AFM bonds. $SrCu_2O_3$ and $LaCuO_{2.5}$ 2-leg ladder materials, whereas $Sr_2Cu_2O_5$ is a 3-leg ladder material. Superconductivity has apparently been discovered in the ladder material $Sr_{0.4}Ca_{13.6}Cu_{24}O_{41.84}$ under pressure with $T_C \approx 12$ K at 3 GPa.

ix. *Magnesium Diboride (MgB₂):* Very recently, superconductivity is reported at 39 K in the simple binary ceramic compound MgB_2. This is probably highest T_C yet determined for a non copper oxide bulk superconductor.

x. *Sr₂RuO₄:* This superconducting compound has the same structure as the $La_{2-x}M_xCuO_4$ (M = Ba, Sr, Ca, Na) HTSC cuprates. While the T_C of Sr_2RuO_4 is only ~1K, this compound is of considerable interest because it is the only layered perovskite superconductor without copper.

xi. *Iron-pnictide Superconductors:* Recently several series of new layered iron-pnictide superconductors with T_C above 40 – 57.3 K has attracted a lot of scientific interest. The discovery of T_C as high as 57.3 K in La-doped iron based compound $Sm_{0.95}O_{0.85}F_{0.15}$ Fe As is quite surprising since iron ions in many compounds have magnetic moments and they normally form an ordered magnetic moment.
All iron-pnictide superconductors include a two dimensional (2D) Fe Pn (Pn: Pnictogen atom) layer with a tetragonal structure at room temperature. Therefore, at a glance their physical properties are considered to be highly two dimensional, similar to those of cuprates.

xii. *High Temperature Cuprate Superconductors:* Until 1986, most of the superconducting compounds were metals and alloys. However, some oxide superconductors were known for decades, but their transition temperature were rather low. This was mainly due to a low number of carriers in the metallic state. Two known exceptions were $LiTi_2O_4$ and $BaPbBiO_3$ with T_C ~13 K. This was quite unusual as their densities of carriers were also very small. The major breakthrough came in 1986 when J G Bednorz and K A Muller discovered superconductivity with a T_C (onset) of 35 K a new record in a mixed phase copper oxide ceramic containing $La_{2-x}Ba_xCuO_4$. The end of 1986 and the beginning of 1987 was marked by synthesis of rare earth metal oxides with the discovery of yttrium barium copper oxide, $YBa_2Cu_3O_7$, which has a critical temperature of about 92 K. This was a significant breakthrough as it meant that for the first time the world has witnessed the existence of a superconductor with a T_C above that of liquid nitrogen (boiling point 77 K). Nitrogen is much more abundant than helium, much less expensive and liquid nitrogen cryogenic systems using helium refrigeration. One application which could benefit from nitrogen cooling is the development of hybrid microelectronic technology (semiconductor superconductor devices) both gallium arsenide and silicon can be tailored to perform better at liquid nitrogen temperatures.

In early 1988, bismuth (Bi) and thallium (Tl) based cuprates $Bi_2Sr_2Ca_2Cu_3O_{10}$ (T_C = 110 K) and $Tl_2Ba_2Ca_2Cu_3O_{10}$ (T_C = 125 K) were discovered. These new HTSC cuprates may have some advantages over ceramic superconductors containing rare earths. Since the critical current density increases as T/T_C decreases, a T_C far above the opening temperature of liquid nitrogen (77 K) is advantageous. Moreover, the new materials are more stable than the rare earth cuprate superconductors; they do not lose oxygen or react with water.

The maximum value of T_C has now increased to 133 K for mercury based cuprate, $HgBa_2Ca_2O_{8+\delta}$. When this compound is subjected to high pressure ~30 GPa, the onset of T_C increases to ~164 K (more than half way to room temperature). While $HgBa_2Ca_2O_{8+\delta}$ cannot be used in applications at such high pressures, this striking result suggests that values of T_C in the neighbourhood of 160 K, or even higher, are attainable in cuprate oxides. More than 100 different cuprate materials, many of which are superconducting, have been discovered since 1986. Several of the more important are listed in Table 22.1, along with the maximum values of T_C observed in each class of materials.

Table 22.1

i. Some important classes of HTSC cuprates along with the maximum value of T_C observed in each class.
ii. Examples of the abbreviated names (nicknames) used to denote superconducting cuprate materials.

i.	Material	Maximum T_C (K)
	$La_{2-x}MCuO_4$; M = Ba, Sr, Ca, Na	~40
	$Ln_{2-x}MCuO_4$; Ln = Pr, Nd, Sm, Eu; M = Ce, Th	~25
	$YBa_2Cu_3O_{7-\delta}$	92
	$LnBa_2Cu_3O_{7-\delta}$	~95
	$RBa_2Cu_4O_8$	~80
	$Bi_2Sr_2Ca_{n-1}Cu_nO_{2n+4}$ (n = 1, 2, 3, 4)	(n = 3) 110
	$TlBa_2Ca_{n-1}Cu_nO_{2n+3}$ (n = 1, 2, 3, 4)	(n = 4) 122
	$Tl_2Ba_2Ca_{n-1}Cu_nO_{2n+4}$ (n = 1, 2, 3, 4)	(n = 3) 122
	$HgBa_2Ca_{n-1}Cu_nO_{2n+2}$ (n = 1, 2, 3, 4)	(n = 3) 133

ii.	Material	Nickname	Maximum T_C (K)
	$YBa_2Cu_3O_{7-\delta}$	YBCO; YBCO-123; Y-123	92
	$Bi_2Sr_2Ca_2Cu_3O_{10}$	BSCCO; BSCCO-2223; Bi-2223	110
	$Tl_2Ba_2Ca_2Cu_3O_{10}$	TBCCO; TBCCO-2223; Tl-2223	122
	$HgBa_2Ca_2Cu_3O_8$	HBCCO; HBCCO-1223; Hg-1223	133

22.3 HTSC CUPRATE MATERIALS CHARACTERISTICS

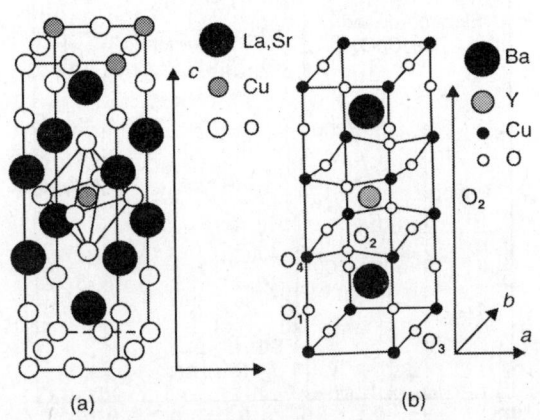

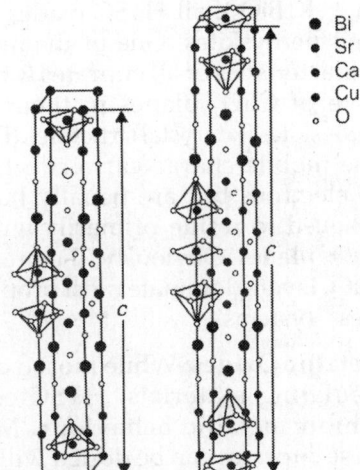

(a) Lattice structure of $La_{2-x}Sr_xCuO_4$

(b) Orthorhombic unit cell of $YBa_2Cu_3O_{7-\delta}$. A special feature of the Y-123 structure are CuO chains in *b* direction

(c) Pseudo-tetragonal unit cells for the Bi-2212 and Bi-2223 structures

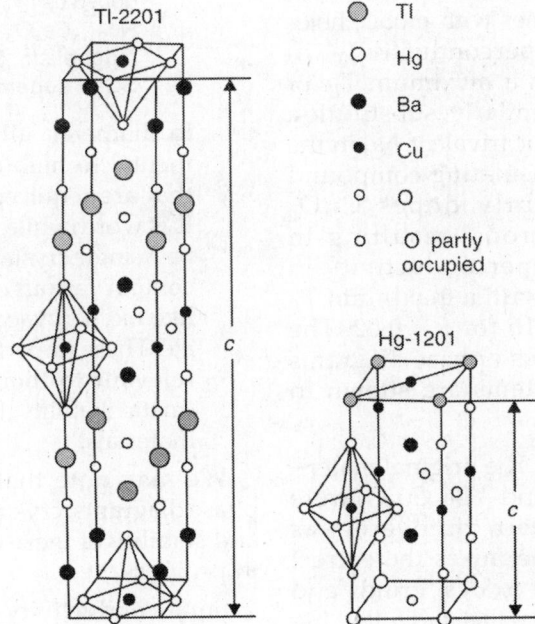

(d) Tetragonal unit cells of $Tl_2Ba_2CuO_{6+\delta}$ and $HgBa_2CuO_{4+\delta}$. The distance between neighbouring CuO_2 planes is much smaller in the Hg-1201 than in the Tl-2201 compound

Fig. 22.7: Crystal structures of HTSC cuprate superconductors (a) Lattice structure of $La_{2-x}Sr_xCuO_4$ (b) Orthorhombic unit cell of $YBa_2Cu_3O_{7-\delta}$ (c) Pseudo-tetragonal unit cells for Bi-2212 and Bi-2223 structures (d) Tetragonal unit cells of $Tl_2Ba_2CuO_{6+\delta}$ and $HgBa_2CuO_{4+\delta}$.

i. **Highly Anisotropic, Layered Structures:** Except for one material (isotropic, cubic $Ba_{1-x}K_xBiO_3$), all HTSC oxides are layered perovskites. One of the important characteristics of all cuprates is the presence of CuO_2 planes in their layered perovskite-like crystal structures (Fig. 22.7). The mobile charge carriers which can be electrons but are usually holes, are believed to reside primarily within the CuO_2 planes. Obviously, the presence of CuO_2 layers dominate most properties of these systems.

ii. **Metallic Oxides:** While most oxides are insulating materials, HTSC cuprates exhibit metallic behaviour. Many of these cuprates can be doped with charge carriers and rendered superconducting by substitution of appropriate elements into an insulating parent compound. For example, substitution of divalent Sr for trivalent La in the antiferromagnetic insulator La_2CuO_4 dopes the CuO_2 planes with mobile holes and produces superconductivity in $La_{2-x}Sr_xCuO_4$ with a maximum T_C of ~40 K at $x \approx 0.17$. Similarly, substitution of tetravalent Ce for trivalent Nd in the antiferromagnetic insulating compound Nd_2CuO_4 apparently dopes CuO_2 planes with electrons, resulting in electron doped superconductivity in $Nd_{2-x}Ce_xCuO_{4-y}$ with a maximum T_C of ~25 K at $x \approx 0.15$ for $y \approx 0.02$. The temperature T versus x phase diagrams for both these systems are shown in Fig. 22.8.

iii. **Ceramic Materials:** The original materials $La_{2-x}Sr_xCuO_4$ and $YBa_2Cu_3O_7$ were synthesized by their discoverers as ceramic pellets. One mixes the correct ratio of constituent oxides, grinds and sinters them, make a pallet and following a calcining procedure (at ~950°C) cools it down in oxygen. Typical ceramics, HTSC oxides also contain grains, grain boundaries, twins, voids and other imperfections. Even some of the best thin films may consist of grains a few microns

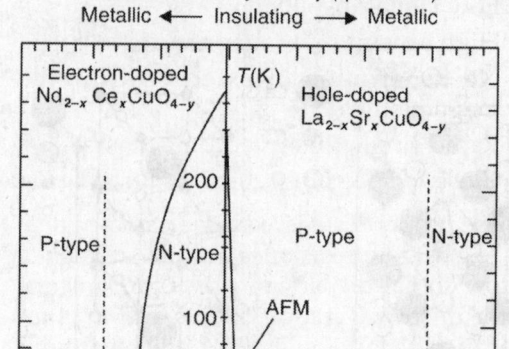

Fig. 22.8: Temperature-dopant concentration $(T - x)$ phase diagram delineating the regions of superconductivity and antiferromagnetic-ordering of the Cu^{2+} ions for the hole-doped $La_{2-x}Sr_xCuO_4$ and electron-doped $Nd_{2-x}Ce_xO_{4-y}$ systems. AFM→ Antiferromagnetism, SG→ spin glass phase, and SC→ superconducting phase

in diameter, all these are mostly detrimental to high critical current densities that are required for applications.

It is worthwhile to mention that even the best single crystals of HTSC cuprates often contain various defects and imperfections like oxygen vacancies, twins, impurities, etc. These imperfections are not only very relevant to their basic thermodynamic (meta) stability, but also intrinsic to these materials.

We may note that the understanding of phase diagrams, crystal chemistry, preparation and stability of these oxides is still very much in progress.

Superconductivity is not only of theoretical interest, but also of technological importance. The special properties of superconductor materials are:

• Zero resistance to direct current
• High current density
• Low resistance at high frequency

- Low signal dispersion
- High sensitivity to magnetic field
- No penetration of externally applied magnetic field
- Rapid single flux quantum transfer
- Close to speed of light signal transfer

These properties enable important applications, e.g. (i) low resistance at high frequency and low signal dispersion play a fundamental role for microwave components and in communication technologies (ii) high sensitivity to magnetic fields makes it possible to produce superconductive sensors with a sensing capability 1000 times higher than conventional devices (iii) the capability of driving off external magnetic fields holds the potential for applications in magnetic levitation systems for transportation (iv) important applications in digital electronics and high speed computing derived from other peculiar superconductor properties such as the Josephson effect and the sharp transition to superconducting status.

Superconducting materials are the basis of extremely sensitive magnetic field and electromagnetic radiation detectors and, as such, have found widespread use in many scientific research fields and in medicine. Magnetic levitation is one of the most facinating demonstrations of superconductivity. With the recent discovery of high temperature superconductors, the number of technological applications of superconductivity is expected to increase enormously.

22.4 CHARACTERISTIC PROPERTIES OF SUPERCONDUCTORS

Zero Resistivity, i.e. Infinite Conductivity ($\rho = 0$ for all $T < T_C$)

The DC (zero frequency) electrical resistance of a superconductor at all temperatures below a critical temperature T_C is practically zero (Fig. 22.9). In the first approximation, the transition is not accompanied by any change in structure of property of the crystal lattice and has been interpreted as an electronic phase transition.

The transition from the normal to the superconducting state occurs sharply in pure metals, i.e. low temperature superconductors (Fig. 22.1) but not so in some impure, deformed and HTSC oxides. Bi, Tl and Hg-based cuprate superconductors are chemically complex materials, in which there may exist several superconducting phases in one specimen (Fig. 22.9a). A two-step transition reflects the presence of at least two superconducting phases. Figure 22.9a shows resistance versus temperature for a single and multiphase high temperature superconductors. The transition width ΔT_C for single-phase high temperature superconductors is typically ≈ 1 K. In epitaxial $YBa_2Cu_3O_7$ films ΔT_C values as small 0.3 K has been achieved.

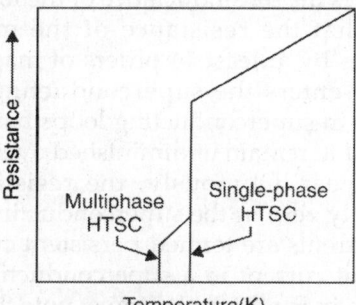

Fig. 22.9a: Resistance versus temperature curves of a single and a multiphase high temperature superconductor

If we assume the usual Ohm's law ($V = RI$) describing the superconductor state

$$E = \rho J \qquad (22.1)$$

where E represents the electrical field, ρ the resistivity and J the electric current density in the sample, then zero resistivity implies zero electric field.

So, if we take the Maxwell equation

$$\text{curl } E = \frac{\partial B}{\partial t}$$

We have $\quad \dfrac{\partial B}{\partial t} = 0 \qquad (22.2)$

for $\rho = 0, E = 0$, i.e. for a superconducting state.

Persistent Current

One way to test for zero resistivity is to measure the current in a loop of superconducting wire in the absence of an electric field. If the resistivity is truly zero, then the current does not diminish with time. Current can be induced by directing a static magnetic field through a loop when it is in the normal state. The material is then cooled below the critical temperature T_C, so that it becomes superconducting. Removal of the magnet alters the flux through the loop, thereby inducing a current. If the ring had a finite resistance R, the current would decrease according to

$$I(t) = I_0 \exp(-Rt/L)$$

where L is the self-inductance of the loop. It is found that the resistance of the material decreases by at least 14 orders of magnitude when it enters the superconducting state, currents in superconducting loops have been observed to remain undiminished over longer than a year. Obviously, the resistivity is practically zero in the superconducting state. Such currents are termed persistent currents. Persistent current in a superconducting ring is shown in Fig. 23.9b. We must note that zero resistivity and persistent current are one and the same property of the superconductors.

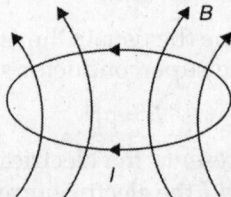

Fig. 22.9b: Persisent current in a superconducting ring

Meissner–Ochsenfeld Effect (*B* = 0 Inside the Superconductor)

The magnetic inductance becomes zero inside the superconductor when it is cooled in weak external field (Fig. 22.4). The effect is called the Meissner–Ochsenfeld effect. From Eq. (22.2), we see that the magnetic induction in the interior of the sample has to be constant as a function of time. The final state of the sample would have been different if it were cooled under an applied external field or if the field were applied after the sample has been cooled below T_C. In the former case, the field would have remained within the sample, while in the latter it would have been zero. For the specimen to be in the same thermodynamic state, independent of the precise sequence that one uses in cooling or in applying the field, the superconducting metal always expels the fields from its interior and has

$$B = 0$$

The superconducting state of a metal exists only in a particular range of temperature and field strength. The condition for the superconducting state to exist in the metal is that some combination of temperature and field strength should be less than a critical value. Superconductivity of the metal will disappear if the temperature of the specimen is raised above its T_C or if a sufficiently strong magnetic field is employed. There always exists some critical field H_C, above which superconductivity disappears. This field is temperature dependent (Fig. 22.10) and the empirical relation which describes well this dependence is

$$H_C(T) = H_0[1 - (T/T_C)^2] \qquad (22.3)$$

where H_0 is the value of H_C at absolute zero.

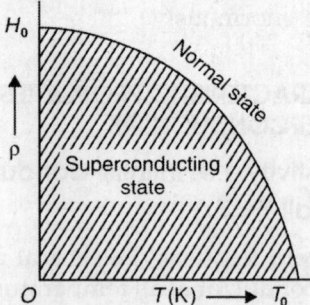

Fig. 22.10: The critical magnetic field at which superconductivity of a metal disappears

Measurements of the H_C required to destroy the superconductivity near $T = 0$ K shows that the condensation energy (energy difference between the normal and the superconducting states) is of the order $\frac{1}{2}\,\mu H_C^2\,(0)$. We have

$$\frac{H_C^2}{8\pi} = \frac{1}{2}\,N\,(0)\,\Delta^2 \qquad (22.3a)$$

where Δ is called the superconducting energy gap and is determined from the following relation

$$\Delta = 2\,\hbar\omega_D\,\exp\left(-\frac{1}{N(0)|g|}\right) \qquad (22.3b)$$

where ω_D is the Debye phonon frequency, $N(0)$ is the electron density of states at the Fermi level and $|g|$ is the electron–phonon interaction strength.

We have already mentioned in Section 22.1 that the superconducting materials that completely expel magnetic flux until they become completely normal are called *type-I superconductors*. For a *type-II superconductor*, there are two critical fields: the lower H_{C_1} and the upper H_{C_2} (Fig. 22.5). So in applied field smaller than H_{C_1}, the type-II superconductor behaves just like a type-I superconductor below H_C. Above H_{C_1}, the flux partially penetrates into the material until the upper critical field, H_{C_2} is reached. Above H_{C_2}, the material returns to the normal state. Between H_{C_1} and H_{C_2} the superconductor is said to be in *mixed state*. For all applied fields $H_{C_1} < H < H_{C_2}$, magnetic flux partially penetrates the superconducting specimen in the form of tiny microscope filaments called *vortices*. The diameter of vortex in conventional superconductor is typically 100 nm.

Type-II superconducting materials can withstand strong applied magnetic fields without returning to the normal state due to partial flux penetration. Superconductivity can and does persist in the mixed state upto the upper critical field, H_{C_2}, which is sometimes as high as 60 tesla (Chevrel phases) or even ~150 tesla in HTSC cuprates. At fields higher than H_{C_2}, the superconductor returns to the normal state.

Nb compounds, chevrel phases and HTCS cuprates are all type-II superconductors. These superconductors are important for technological applications. The reason for this is that the creation of vortices keeps the magnetic energy smaller than the condensation energy, so the overall free energy of the mixed superconducting state remains more favourable than the normal state even up to high magnetic fields. Since the supercurrent can flow in the mixed state through the superconducting regions between vortices, type-II superconductors allow to construct wires needed for high field magnets.

Thermal Properties of Superconductors
Entropy

Entropy of all superconductors decreases considerably upon cooling below T_C. The entropy is a measure of the degree of disorder of a given system and hence this decrease in a superconductor signifies that the superconducting state is more ordered than the normal state. The fraction of electrons that is thermally excited in the normal state becomes ordered in the superconducting state. The entropy variation is relatively weak, of the order of $10^{-4}\,k_B/$atom ($k_B \rightarrow$ Boltzmann constant).

Heat Capacity or Specific Heat

The specific heat C_n of a normal metal consists of two contributions, C_n^e from the electrons in the conduction band and C_n^l from the lattice. Thus,

$$C_n = C_n^e + C_n^l = \gamma T + \beta T^3 \qquad (22.4)$$

The first term in Eq. (22.4) is linearly proportional to T while the second term is proportional to T^3. The specific heat of the superconductor (C_{es}), changes at T_C in a characteristic way (Fig. 22.11). In zero magnetic field, there appears a discontinuity at T_C. At temperatures well below T_C, the heat capacity fits an exponential form

$$C_{es}(T < T_C) = A\exp(-\Delta/k_B T) \qquad (22.5)$$

where $\Delta = \dfrac{E_g}{2}$, E_g is called the *energy gap*. Such an exponential temperature dependence is a hallmark of a system with a gap Δ in the spectrum of allowed energy states, separating the excited states from the ground states by energy Δ. Although there are also materials with gapless superconductivity, most materials of interest do have an energy gap. The relation between Δ and T_C is given by Bardeen–Cooper–Schrieffer (BCS) theory of superconductivity.

$$2\Delta = 3.5 k_B T_C \qquad (22.6)$$

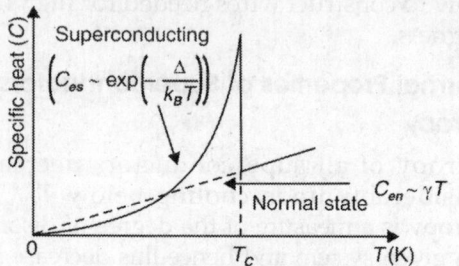

Fig. 22.11: Schematic representation of the specific heat of a metal in the normal and the superconducting phase. There is a characteristic jump of the transition point

Heat capacity measurements provided the first indications of such a gap in superconductors and one of the key features of BCS theory is its prediction of such a gap in superconductors.

Estimates show that the gap in conventional superconductors with $T_C < 20$ K is of the order of 1 meV, while in HTSC oxides with $T_C \sim 100$ K, $\Delta \sim$ several (1–10) meV. We may note that in superconductor 2Δ corresponds to the energy needed to break a Cooper pair, whereas in semiconductors the gap in the energy spectrum corresponds to the energy difference between the valence and the conduction band (~1 eV).

Thermal Conductivity

The thermal conductivity of superconductors undergoes a continuous change between the two phases and usually lower in a superconducting phase and at very low temperatures approaches zero. This suggests that the electronic contribution drops, the superconducting electrons possibly plays no part in heat transfer. The thermal conductivity

of tin ($T_C = 3.73$ K) at 2 K is 16 W cm^{-1}K^{-1} for the superconducting phase and 34 Wcm^{-1}K^{-1} for the normal phase.

Thermoelectric Properties

A combined thermal and electric effect of interest and practical importance is the Peltier effect, which is the basis of operation of thermocouples used for temperature measurement. If the two junction regions of a loop made of two different metals are maintained at different temperatures, an electrical current is driven around the loop. A greatly weakened form of this current occurs in the superconducting state because some 'normal' electrons remain at any non-zero temperature. However, the presence of superconductivity shorts out the usual thermoelectric voltage.

Acoustic Attenuation

When sound wave propagates through a metal, the microscopic electric fields due to the displacements of the ions can impart energy to the electrons near the Fermi level thereby removing energy from the wave. One can express this by the attenuation coefficient α of acoustic waves. The ratio of α for superconducting and normal state is given by

$$\frac{\alpha_s}{\alpha_n} = \frac{2}{1 + \exp(\Delta / k_B T)} \qquad (22.7)$$

At low temperatures, one finds

$$\frac{\alpha_s}{\alpha_n} = 2\left[\exp(\Delta / k_B T)\right] \qquad (22.8)$$

High Frequency Electromagnetic Properties

For all frequencies much higher than the frequency corresponding to the energy gap, $E_g = h\nu$ where ν is the frequency in Hz, the electromagnetic response in the superconducting state is identical to the response of the normal state. Remembering that 1 eV ~ 10^{14} Hz, one can easily understand that the change in the frequency response occurs and at $\nu \sim 10^{11}$ Hz and ~10^{12} Hz in the conventional and HTSC oxides respectively.

High frequency electromagnetic properties of superconductors differ from zero frequency (or low frequency) behaviour. In the radio frequency ($<10^8$ Hz) and microwave frequency range ($\sim10^8$–10^{11} Hz) the resistance of the superconductor to current flow is not zero. However, the resistance and the accompanying energy loss are still rather small.

In the optical region of the spectrum ($\sim10^{15}$ Hz), one finds no difference in electromagnetic response between the normal and the superconducting state so one does not see any change in the appearance of the sample as it undergoes the transition. In the range $\sim10^{11}$–10^{12} Hz (depending on the material), there is a sharp increase in the absorption of electromagnetic radiation by the superconductor. This is due to the existence of energy gap (E_g) in the electronic energy spectrum.

We may note that the sharp rise in the absorption occurs at the energy ($E = h\nu$) of a single photon which is just sufficient to produce an excitation (by depairing weakly coupled Cooper pairs).

HTSC oxides exhibit light excitations on the T_C values. These materials are ceramic in nature, black in colour and P-type (hole-type) in carriers. Though these materials are black in colour, they do not behave like blackbody. Their optical absorption spectra are continuous but consist of discrete bonds corresponding to different energy levels. These are responsible for additional *photoinduced* charge carriers and giving rise to change in T_C. One may call it as photodoping and superconductors as photoinduced superconductors.

The Energy Gap

We have seen that the heat capacity in the superconducting state well below T_C varies with temperature in an exponential manner in accordance with relation (22.5) with $\Delta = bk_B T_C$, where b is a constant. Near the transition temperature, the half width ($\Delta = E_g/2$) of the energy gap is approximately

$$\Delta(T) = 3.2k_B T_C (1 - T/T_C)^2 \qquad (22.9)$$

Such a temperature dependence is characteristic of a system that has an energy gap

in its spectrum of allowed energy states. One can determine the energy gap of a superconductor by knowing the absorption of electromagnetic waves. One can detect this gap by photo absorption, quantum tunnelling and other experiments. A plot of the temperature dependence of energy gap parameter $\Delta(T)$ is shown in Fig. 22.12. We note that the energy gap is zero at $T \to T_C$ and reaches a maximum value $\Delta_0(T)$ as the temperature lowered towards 0 K.

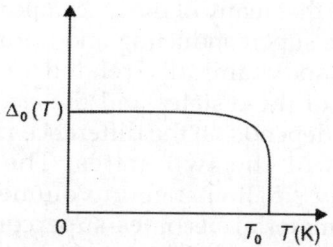

Fig. 22.12: The energy gap $\Delta(T)$ versus temperature. $\Delta(T)$ vanishes with infinite slope as $T \to T_C$ leading to the second order phase transition

Isotope Effect

It has been observed that the critical temperature (T_C) of superconductors varies with the ionic or isotopic mass. The relation valid for some simple metals is given by

$$T_C \propto M^{-\alpha}$$

$$\text{or} \qquad M^\alpha T_C = \text{constant} \qquad (22.10)$$

where M is the atomic mass of the isotope and α is roughly 0.5. Obviously, the existence of the isotope effect indicated that, although superconductivity is an electronic phenomenon, it nevertheless depends in an important way on the vibrations of the crystal lattice in which the electrons move. Fortunately, not until after the development of the BCS theory it was discovered that the situation is more complicated than it had appeared. For some superconductors the exponent of M, i.e. α is not $-1/2$, but near zero, e.g. Ru and Zr and for at least one it is positive.

The isotope effect, as studied by substitution of ^{18}O for ^{16}O in HTSC cuprates, is very weak. The substitution of ^{18}O for ^{16}O in Eq. (22.10) gives $\alpha \approx 0.02$ in YBCO and $\alpha \approx 0.15$ in LBCO. This has prompted the exploration of non-phonon electronic coupling mechanisms responsible for superconductivity in these cuprate systems. However, a weak isotope effect is not conclusive.

Mechanical Effects

When a superconducting material is mechanically stressed, it is found experimentally that both T_C and H_C are slightly altered. One can easily see that many of the mechanical properties of the superconducting and normal states are thermodynamically related to the free energies of these states and the critical field strength depends on the difference in the free energies of the two states. There is an extremely small change in volume when a normal material becomes superconducting and the thermal expansion coefficient and bulk modulus of elasticity must also be slightly different in the superconducting and normal states. However, the effects are extremely small.

Absence of Effects

We have seen that most of the electronic properties of a superconductor are profoundly affected by the transition to the superconducting state, while many other properties are changed very little if at all. These include the mechanical and elastic properties, tensile strength, sound velocity and density among others.

Characteristic Phenomenological Parameters

Penetration Depth (λ)

While studying Meissner effect, we mentioned that the superconductor expels a (weak) magnetic field B from its interior, i.e. $B = 0$ in the interior of a superconductor. The finer experiments reveal that the field B penetrates into the superconductor within a very thin surface layer. Consider the boundary of a semi-infinite slab. When the external field is applied parallel to the boundary, the applied field does not suddenly drop to zero at the surface of the

superconductor but decays exponentially according to the relation

$$H(x) = H(0)\exp(-x/\lambda) \qquad (22.11)$$

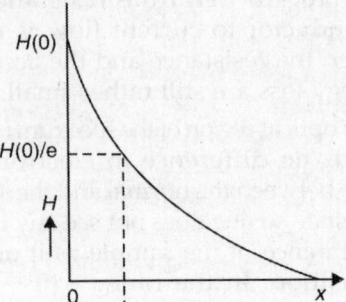

Fig. 22.13: Decay of the magnetic field in the interior of a superconductor

where $H(0)$ is the value of the magnetic field at the surface and λ is a characteristic length known as the penetration depth; λ is the distance for H to fall from $H(0)$ to $H(0)/e$. In most of the superconductors, λ is of the order of 500 Å. λ depends on the material and temperature, the latter variation being given approximately by

$$\lambda = \lambda_0 \left[1 - \left(\frac{T}{T_C} \right)^4 \right]^{-1} \qquad (22.12)$$

where λ_0 is the penetration depth at zero temperature for the particular material and is typically of order 500 Å.

If a superconducting film or filament is thinner than λ, its properties are significantly different from those of the bulk material. In particular, the value of H_C increases as thickness decreases and the special property of type-II superconductor arise from this.

The Coherence Length (ξ)

It is a measure of the distance over which the gap parameter (Δ) can vary, for instance in a spatially varying magnetic field or near a superconductor-normal metal boundary. It is also referred to as the distance between two electrons of the Cooper pair within the highly coherent superconducting state. The intrinsic or BCS coherence length ξ_0 is defined as

$$\xi_0 = \frac{\hbar v_F}{\pi \Delta} \qquad (22.13)$$

where v_F is the Fermi velocity and Δ is the energy gap.

Using order of magnitude values for v_F and Δ, one obtains $\xi = 1600$ Å in pure Al, $\xi = 380$ Å in pure Nb, but only about 10 Å in the new HTSC oxides. Table 22.2 provides values of λ and ξ for few selected superconductors.

When a small impurity is added to a metal, λ increases very rapidly while ξ decreases.

Ginzburg---Landau Parameter (κ)

The ratio of the characteristic length λ and ξ is called the Ginzburg–Landau ratio κ

$$\kappa = \frac{\lambda}{\xi} \qquad (22.14)$$

Close to T_C, κ is independent of temperature and it allows one to distinguish between type-I and type-II superconductors.

If κ < 0.7, material is type-I superconductor and if κ > 0.7, the material is type-II superconductor. The exact critical value of κ which separates type-I from type-II behaviour is $1/\sqrt{2} \approx 0.7$.

In the latter case, the magnetic flux does penetrate the sample in the form of the cylindrical tubes called vortices. The vortices have a radius λ and destroy superconductivity locally within a cylinder of radius ξ. It is energetically favourable for type-II super-conductors to let the flux penetrate partially in the form of vortices.

Flux Quantization

This refers to the fact that the magnetic flux threading a superconductor loop cannot have an arbitrary value; it has to be a multiple of $\phi_0 = h/2e$. The flux quantum is defined as

$$\phi = \frac{h}{2e} = 2.0678 \times 10^{-15} \text{ weber} \qquad (22.15)$$

Superconducting devices can measure this tiny variation of magnetic flux which is exceedingly important in metrology and advanced instrumentation.

22.5 CRITICAL TEMPERATURE (T_C)

The temperature at which the DC electrical resistance vanishes is called the critical temperature of the superconductor and denoted by T_C. About 23 metals and several alloys and compounds have been found to exhibit superconductivity. Metals which exhibit superconductivity are not particularly good conductors in the normal (above T_C) and the value of T_C vary from 0.4 K to 11.2 K and examples are Pb (7.2 K), Hg (4.2 K), Sn (3.7 K) and Al (1.2 K). Some of the compounds which exhibit superconductivity are composed of elements which themselves are not super-conducting, e.g. CuS. The highest transition temperature known so far is 133 K for layered mercury cuprate oxide compound. Table 22.2 gives values of T_C for various materials chosen to illustrate a wide range of properties.

Table 22.2: T_C, H_{C_2}, λ and ξ for few selected superconductors

Material	T_C (K)	E_{C_2}(Tesla)	λ (Å)	ξ(Å)
Al	1.1	0.02 (also exhibit type-I superconductivity)	500	16000
Nb	9.2	0.2	400	380
Nb-Tl	9.5	14	600	450
Nb_3Sn	18.3	24	800	35
Rb_3C_{60}	29.3	~50	1600	~20
$La_{1-x}Sr_xCuO_4$	38	~65	2500	~15
$YBa_2Cu_3O_7$	92	~120	4000	~10

We shall learn from BCS theory that T_C is given by the relation

$$k_B T_C = \hbar \omega_D \exp\left[1 - \frac{1}{|g| N(0)}\right]$$

where ω_D is the Debye phonon frequency, $N(0)$ is the electron density of states at the Fermi level and $|g|$ is the electron–phonon interaction strength. The search for a room temperature superconductor is centred on how to control the parameter $|g|$ in the fabrication of a superconducting alloy. Recent discovery of high temperature super-conductivity in layered cuprates has raised the question, what is the possible mechanism of superconductivity in these systems?

22.6 JOSEPHSON EFFECTS

In 1962, BD Josephson predicted theoretically that Cooper pair tunneling through a very thin insulating layer (~2 nanometer) is possible. His theory predicted in addition to the Giaever current, the existence of supercurrent, arising from tunneling of the bound electron pairs. Such a junction is called a *weak link*. The effects of pair tunneling include:

i. **DC Josepheson Effect:** A DC current flows across the junction resulting Cooper pair tunneling in the absence of any electric or magnetic field, i.e. at zero voltage across the insulating layer. One obtains the following expression for current through the contact for two identical superconductors as

$$I = I_C \sin \delta$$

where I_C is the critical current for the Josephson junction and δ is the phase difference of the two Cooper wave functions. Theoretical results for voltage across the junction yield

$$V_t = \frac{\pi \Delta}{2e} \tag{22.16}$$

For metals most commonly used, as electrodes, e.g. lead and niobium, V_t is of the order of ~2.5 mV.

ii. **AC Josephson Effect:** Josephson also predicted that if a constant non-zero voltage V were maintained across the tunnel barrier, an alternating supercurrent could flow through the barrier in addition to the DC current produced by the tunneling of unpaired electrons. This effect has been utilized in a precision determination of $\hbar/e$. Further, an r_f voltage applied with the DC voltage can then cause a DC current across the junction. The frequency of the AC Josephson current is given by the Josephson relation

$$\omega = \frac{2eV}{\hbar} \tag{22.17}$$

where V is the DC voltage applied across the junction.

Josephson Junctions

These are sophisticated sandwich structures of superconducting films (usually of Nb) separated by extremely thin (~10 Å) insulating oxide layers. In a suitable submicroscopic circuit, they act as the fastest switching elements available today.

Josephson junction have also the lowest power consumption of their operating temperature, so they might provide the basis for the architecture of some of the fastest computers of the future. They are already used in the fastest commercially available oscilloscopes that operates at 10 GHz.

22.7 PROPERTIES OF HTSC OXIDES

All HTSC oxides possess the main properties of elemental conventional superconductors, including zero resistivity, Meissner effect, flux quantization, Josephson effect, etc. In addition, HTSC oxides are characterized by (i) highly anisotropic layered structures (ii) short zero temperature coherence length (ξ = 10 Å) (iii) transport properties in the normal state significantly different from those of normal metals, with strongly anisotropic resistivity, anomalous magnetoresistance, pseudogap, etc. This is why HTSC oxides are termed *nonconventional superconductors*.

T_C, λ, ξ, H_{C2} and H_{C1} for few selected HTSC cuprates are given in Table 22.3.

22.8 THERMODYNAMICS OF A SUPERCONDUCTOR

The transition between the normal and superconducting state is thermodynamically reversible. This change occurs because the Gibbs free energy in the superconducting state, G_S, is lower than its value G_N in the normal state. Gibbs free energy (G) is defined as

$$G(T, p) = U - TS + PV \qquad (22.18)$$

where U, T, S, P and V are internal energy, temperature, entropy, pressure and volume of the system respectively.

Let us consider that a superconductor is placed in a magnetic field, H_a. In analogy with Eq. (22.18), the Gibbs free energy in this case can be expressed by adding the magnetic contribution to the energy density of a magnetized specimen. If susceptibility, χ is independent of H, the magnetic energy is

$$U_M = -\int_0^{H0} \mu_0 HDM$$

$$= -\int_0^{H0} \mu_0 H\chi dH = -\frac{1}{2}\mu_0 x H_a^2 \qquad (22.19)$$

Now, neglecting the term PV in Eq. (22.18) for a magnetic solid and adding the magnetic energy term of Eq. (22.19), one obtains

$$G(T, H_a) = U - TS - \frac{1}{2}\mu_0\chi H_a^2 \qquad (22.20)$$

In the superconducting state $\chi = -1$ and hence

$$G_S(T, H_a) = G_S(T, 0) + \frac{1}{2}\mu_0 H_a^2 \qquad (22.21)$$

In the normal state, χ is very small and we may assume $\chi = 0$. Thus,

$$G_N(T, H_a) = G_N(T, 0) \qquad (22.22)$$

Normal and superconducting phases are at equilibrium at $H_a = H_C$. This means

$$G_S(T, H_C) = G_N(T, H_C) = G_N(T, 0)$$

$$= G_S(T, 0) + \frac{1}{2}\mu_0 H_C^2 (0) \quad (22.23)$$

or $\quad G_N(T, 0) - G_S(T, 0) = \frac{1}{2}\mu_0 H_C^2 \qquad (22.24)$

From Eqs. (22.21), (22.22) and (22.23), it is obvious that for $H_a > H_C$, $> G_S > G_N$ while for $H_a < H_C$, $G_S < G_N$, and we note that the superconducting state is stable for $H_a > H_C$. Figure 22.14 shows free energy versus applied magnetic fields. We can see that the transition from the superconducting state to the normal state for $H_a > H_C$ occurs because G_S exceeds G_N at this magnetic field.

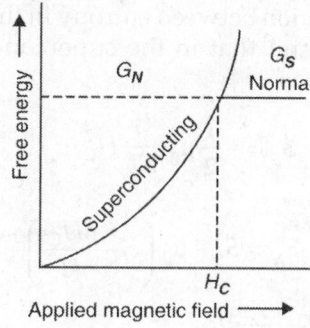

Fig. 22.14: Free energy versus applied field

Table 22.3: Critical temperature (T_C), penetration depth (λ_i), coherence length (ξ_i) and the critical fields H_{C1} and H_{C2} of few HTSC cuprates ($i = ab$ or c)

HTSC cuprates	T_C (K)	λ_{ab} (Å)	λ_C (Å)	ξ_{ab} (Å)	H_{C2}/T	H_{C1}/T
LBCO	38	800	4500	35	80	15
YBCO	94	1500	7000	15	150	40
BSCCO	110	2500	9000	13	260	32

From Fig. 22.14, we note that the normal and superconducting phases are in equilibrium along the critical field. For all the points along this curve, Eq. (22.23) is satisfied and hence for these points

$$dG_N(T, H_C) = dG_S(T, H_C)$$

or $\left(\dfrac{\partial G_N}{\partial T}\right)_{H_C} dT + \left(\dfrac{\partial G_N}{\partial H_C}\right)_T dH_C$

$$= \left(\dfrac{\partial G_N}{\partial T}\right)_{H_C} dT + \left(\dfrac{\partial G_S}{\partial H_C}\right)_T dH_C \quad (22.25)$$

We have $G = U - TS + \dfrac{1}{2}\mu_0 H_a^2$ when H_a is constant.

$$dG = dU - TdS - SdT \text{ and } dU = TdS - C_v dT.$$

Thus $\quad S = -\left(\dfrac{\partial G}{\partial T}\right)_{H_C} \quad (22.26)$

and $\quad C_v^e = T\left(\dfrac{\partial G}{\partial T}\right)_{H_C} \quad (22.27)$

The relation between entropy in the normal state, S_N, and that in the superconductivity state, S_S is as

$$S_N - S_S = -\dfrac{1}{2}\mu_0 \dfrac{d}{dT} H_C^2$$

or $\quad S_N = S_S - \mu_0\left(H_C \dfrac{dH_C}{dT}\right) \quad (22.28)$

From Eq. (22.3),

$$\dfrac{dH_C(T)}{dT} = -2H_C(0)\dfrac{T}{T_C^2} < 0 \quad (22.29)$$

Since $H_C (dH_C/dT)$ is always negative and hence S_S is lower than S_N. Obviously, the superconducting state is more ordered than the normal state.

Using Eqs. (22.27) and (22.28), one obtains the difference in specific heats C_N and C_S in the normal and superconducting states respectively as

$$C_N - C_S = -T\mu_0\left[\left(\dfrac{dH_C}{dT}\right)^2 + H_C \dfrac{d^2 H_C}{dT^2}\right]$$

$$= \dfrac{2H_C^2(0)\mu_0}{T_C}\left[\dfrac{T}{T_C} - 3\left(\dfrac{T}{T_C}\right)^3\right] \quad (22.30)$$

Specific heat in the superconducting state has no linear term, as such we may identify

$$C_N = \gamma T = \dfrac{2H_C^2(0)}{T_C^2}\mu_0 T \quad (22.31)$$

with $\quad \gamma = \dfrac{2H_C^2(0)}{T_C^2}\mu_0 \quad (22.32)$

Thus,

$$C_S = \dfrac{6H_C^2(0)\mu_0}{T_C}\left(\dfrac{T}{T_C}\right)^3 = 3\gamma T_C\left(\dfrac{T}{T_C}\right)^3 \quad (22.33)$$

Relation (22.33) explains the experimental result on the specific heat of the conventional superconductors.

22.9 THEORY OF SUPERCONDUCTIVITY

One of the simplest idea which was put forward in 1934 for describing superconductivity was the *two-fluid model*. Some properties can be understood with the simple assumption that some electrons behave in the normal way as nearly free electrons, while other exhibit anomalous behaviour. Developing this idea, F and H London were able to describe the electrodynamics of what is now called type-II superconductors.

One cannot describe the unique characterization of the superconducting state, i.e. $B = 0$ in the applied magnetic field by considering the superconductor as a medium of zero resistivity. This means that one will have to modify the Ohm's law drastically, in order to describe the zero resistivity and Meissener effect in the superconducting state, without modifying the Maxwell's equations. We have Ohm's law

$$J = \sigma E \quad \left(\because \sigma = \dfrac{1}{\rho} = \text{conductivity,}\right.$$
$$\left. J = \text{current density}\right) \quad (22.34)$$

In the superconductivity state, electron behaves differently, so it was assumed that J (current density) is directly proportional to the vector potential A of local magnetic field, where $B = \nabla \times A$ (B magnetic induction, and $\nabla = \hat{i}\dfrac{\partial}{\partial x} + \hat{j}\dfrac{\partial}{\partial y} + \hat{k}\dfrac{\partial}{\partial z}$, i.e. $B = \text{curl}A$ and electric field intensity $E = -\dfrac{\partial A}{\partial t}$, we have

$$J \propto A \qquad (22.35)$$

where $\quad B = \nabla \times A = \text{curl}A \qquad (22.36)$

The constant of proportionality in Eq. (22.35), thus has taken in SI unit as $-1/\mu_0\lambda_L^2$, where λ_L is a constant with dimensions of length and of value $\lambda_L^2 = m/\mu_0 ne^2$. Hence, Eq. (22.35) takes the form

$$J = -\frac{A}{\mu_0\lambda_L^2} = -\frac{ne^2}{m}A \qquad (22.37)$$

instead of $J = \sigma E$ as in Ohm's law [Eq. (22.34)], where n is the free electron density, e the electronic charge, m the electronic mass. Equation (22.37) is called the *London equation*.

One can also derive London equation in a superconducting state in the following manner. The resistivity $\rho = 0$ in a superconducting state and so the electrons can be freely accelerated. The acceleration is given by

$$\frac{dv}{dt} = \frac{eE}{m} \qquad (22.38)$$

where v is the velocity of the free electrons. Multiplying both sides of Eq. (22.38) by ne, one obtains

$$\frac{d}{dt}(nev) = \frac{ne^2}{m}E = \frac{ne^2}{m}\frac{(-\partial A)}{\partial t}$$

We have $nev = J$, current density. Hence, by integrating the above equation with respect to t, one obtains

$$J = -\frac{ne^2}{m}A = \frac{1}{\mu_0\lambda_L^2}A \qquad (22.39a)$$

which is same as Eq. (22.37), the London equation, where constant of integration is taken to be zero.

One can explain Meissner effect in superconducting state by London equation as follows:

Taking curl on both sides of Eq. (22.37), we have

$$\nabla \times J = -\frac{ne^2}{m}\nabla \times A = -\frac{\nabla \times A}{\mu_0\lambda_L^2} = -\frac{B}{\mu_0\lambda_L^2} \qquad (22.39b)$$

Now considering curl of the Maxwell's equation

$$\nabla \times B = \mu_0 J, \text{ we have}$$

$$\nabla \times (\nabla \times B) = \mu_0(\nabla \times J)$$

or $\nabla \times (\nabla \times B) = \nabla(\nabla \cdot B) - \nabla^2 B = \mu_0(\nabla \times J)$

However, $\nabla \cdot B = 0$ from Maxwell's second equation.

Thus, one finds

$$\nabla \times (\nabla \times B) = -\nabla^2 B = \mu_0(\nabla \times J) \qquad (22.39c)$$

Comparing Eq. (22.39b) and Eq. (22.39c), one obtains

$$\nabla^2 B = \frac{B}{\lambda_L^2} = \text{constant} \qquad (22.39d)$$

Equation (22.39d) accounts for the *Meissner effect*, as it *does not allow a solution uniform in space*, which clearly reveals that a uniform magnetic field cannot exist in a superconductor. No doubt, uniform magnetic field B exists outside the superconductor, it gradually decreases exponentially to zero, according to the solution of Eq. (22.39c) as

$$B(x) = B(o)\exp(-x/\lambda_L) \qquad (22.39e)$$

where λ_L is the *depth of penetration* of the magnetic field and called as *London penetration depth* having dimension of length and of magnitude as

$$\lambda_L^2 = \frac{m}{\mu_0 ne^2} \qquad (22.39f)$$

We may note that an applied external magnetic field H will penetrate a thin film, if the thickness is much less than λ_L, i.e. in a *thin film*, Meissner effect is incomplete. For bulk specimen of superconductors, Meissner effect is complete. Obviously, the shape of the specimen has very important effect, which are simple only when the specimen has the shape of a long cylinder whose axis is parallel to the applied magnetic field.

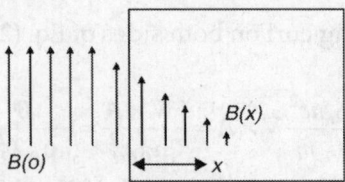

Fig. 22.15: Penetration depth

The London theory is not applicable to situations in which the number of superconducting electrons, n_S varies; it does not link n_S with the applied voltage or current.

Ginzburg and Landau (G–L) in 1950, using the Landu's theory of second order phase transitions, proposed a highly innovative phenomenological theory of superconductivity — *the Ginzburg–Landau theory*. They first derive two equations which can be used to calculate both the distribution of fields and the variation of the number of superconducting electrons. This theory describes a superconductor close to T_C. The order parameter $\psi(r) = |\psi| \exp i\phi(r)$ is a complex scalar and $\phi(r)$ is small. They introduced two unknown parameters α and β which can be estimated from experimental measurements. The characteristic quantities of the materials λ, ξ or H_C, α and β can be eliminated and replaced by two of the measured characteristic quantities (λ, ξ or H_C). While both the characteristic lengths λ and ξ diverge at T_C, their ratio $\kappa = \lambda/\xi$, which is called the G–L parameter does not depend on temperature when close to T_C. If $\kappa < 1/\sqrt{2}$, one is dealing with type-I superconductors and if $\kappa > 1/\sqrt{2}$ one has a type-II superconductors. As flux expulsion is incomplete in type-II superconductor, the mixed state exists between H_{C_1} and H_{C_2}. One of the greatest success of G–L theory was prediction of the existence of type-II superconductors. An interesting relation can be obtained for the $H_{C_1} H_{C_2}$ as

$$H_{C_1} H_{C_2} = H_C^2 \log \kappa \qquad (22.40)$$

G–L theory, although originally phenomenological, proved to be exact and very powerful.

Microscopic BCS Theory

The microscopic theory of superconductivity was formulated by Bardeen, Cooper and Schrieffer (BCS theory) in 1957, almost five decades after the discovery of the phenomenon in 1911 by Kammerlingh Onnes. It is an elegant but mathematically complex theory. We, therefore, present a brief discussion of the results of this theory.

The first clue to the basic interaction between electrons which gives rise to superconductivity was provided by the isotope effect. Two different isotopes of the same metal exhibit different T_C's [Eq. (22.10)]. Why is the mass of an atom involved in a purely electronic property? Obviously, the motion of ions has something to do with superconductivity.

The second clue was found by Leon Cooper. He showed that a normal metal (with standard metallic properties) could not be formed if there was a small attraction between electrons. In such a case, two electrons would form pairs, however, small the attractive interaction. And, if electrons did form pairs, completely different properties for the whole ensemble of electrons would be observed.

Thus, the central idea is that motion of ions can lead to attractive interaction between electrons. How can this happen? When an electron moves among the positive ions of lattice in a solid, it interacts with the lattice vibrations (called *phonons*) and attracts them. However, the motion of ions is slow, so the electron advances a great deal while the ions somwhat converge towards each other. They build a region of positive charge, which before relaxing attracts another electron. So, due to this slow response of ions, there appears an effective attractive interaction between electrons, called the *electron–phonon interaction* (Fig. 22.16). Of course, the repulsive coulomb interaction is also present and there is a delicate 'competition' between the coulomb interaction and electron–phonon interaction which makes some metals superconducting and other non-superconducting.

Cooper in 1956 showed that two electrons with an attractive interaction can bind together to form a bound pair (often called a *Cooper pair*) if they are in the presence of a high density

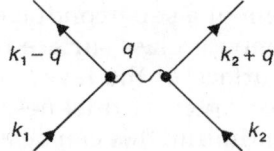

Fig. 22.16: Electron–phonon interaction. An electron of wave vector k_1 emits a virtual phonon q which is absorbed by electron k_2, k_1 is thus scattered as $k_1 - q$ and k_2 as $k_2 + q$. If this phonon energy exceeds electronic energy, the electron–phonon interaction is attractive

fluid of other electrons, no matter how weak the interaction is. The two partners of Cooper pair have opposite momenta (K) and spin angular momenta, i.e. Cooper pair composed of particle pair $K\uparrow$ and $-K\downarrow$. This pairing is known as s-wave pairing. There are other forms of particle pairing possible, e.g. d wave, $d + s$ wave, p wave, etc. with the BCS theory. It is now established that superconducting gap in HTSC oxides is of $d_{x^2-y^2}$ symmetry. These electrons due to their charge also repel each other with a coulomb interaction, V_C but BCS assume that in the superconducting phase electron–phonon mediated interaction V_{ph} dominates over V_C and net interaction V is positive, i.e.

$$V = -V_{ph} + V_C < 0 \qquad (22.41)$$

The main assumptions of BCS theory are:

i. The superconducting ground state can be expressed in terms of Cooper pairs so that the states ($K\uparrow$, $-K\downarrow$) are simultaneously occupied or empty.

ii. Various interactions in the normal and superconducting states are identical and only the effective screened coulomb interaction has to be considered.

iii. The effective interaction is zero, except when two electrons of wave vector K and K' have energies close to the Fermi energy. Then the attractive interaction is taken as a constant $-V$. If ξ_k is the energy measured from the Fermi energy, or more precisely from the chemical potential μ, then:

$$\xi_k = \varepsilon_k - \mu \qquad (22.42)$$

In order for the electrons to attract each other, the energies of both the electrons have to satisfy the criterion

$$|\xi_k| = \hbar\omega_D \ll E_F < k_B\theta_D \qquad (22.43)$$

where θ_D is the Debye temperature.

BCS in 1957 showed how to construct a wave function in which all of the electrons (at least, all of important ones) are paired. Once this wave function is adjusted to minimize the free energy, it can be used as a basis for a complete microscopic theory of superconductivity.

At zero temperature, electrons in a superconductor are paired. In the superconducting state, excitations are obtained by breaking up pairs which costs a minimum energy $2\Delta(T)$. The excitation spectrum is given by

$$K_k = \sqrt{\xi_k^2 + \Delta^2} \qquad (22.44)$$

where Δ is a fundamental quantity introduced by BCS is called the gap or the superconducting order parameter. Δ depends on the temperature and obeys the relation, known as the *self-consistent BCS equation.*

$$\frac{1}{VN(E_F)} = \int_0^{\hbar\omega D} d\xi \left(\xi^2 + \Delta^2\right)^{-1/2}$$

$$\tan h\left[\frac{1}{2k_BT}\left(\xi^2 + \Delta^2\right)^{1/2}\right] \quad (22.45)$$

where $\hbar\omega_D$ is the characteristic energy of the attractive potential and $N(E_F)$ is the density of states as Fermi level.

When $\Delta(T) = 0$, i.e. for $T = T_C$ the energy of the normal and the superconducting states are equal. An important BCS relation which relates $\Delta(0)$ and T_C is

$$2\Delta(0) = 3.52 k_B T_C \qquad (22.46)$$

Close to T_C we have an approximate relation

$$\frac{\Delta(T)}{\Delta(0)} = 1.74 \left(1 - \frac{T}{T_C}\right)^{1/2} \qquad (22.47)$$

The critical temperature (T_C) of simple conventional superconductor is given by the well-known BCS formula

$$k_B T_C = 1.13 \ \hbar\omega_D \exp\left[\frac{-1}{\lambda_{ep}}\right] \qquad (22.48)$$

λ_{ep} is the dimensionless electron–phonon coupling parameter. Its value for conventional superconductors is very close to 0.3. ω_D is Debye or characteristic frequency, varies from one metal to another but only over a small range of value. Instead of ω_D, one can use the Debye temperature θ_D ($\because k_B \theta_D \approx \hbar \omega_D$), θ_D ranges from 100 K to 500 K. Such a range of θ_D (and $\lambda_{ep} \sim 0.3$) implies a maximum BCS value of $T_C \sim 25$ K.

The extension of BCS weak coupling theory to strong coupling was presented by Eliashberg in 1960. MacMillan in 1968 obtained an approximate solution of Eliashberg's equations, good for $\lambda \leq 2$. BCS theory alone cannot explain the anomalous behaviour of HTSC oxides in normal and superconducting states. Several theoretical proposals have been advanced, e.g. Anderson's Resonance Valence Bond (RVB) theory; *t-J* model, interlayer tunneling model, Spin–Fermion model, Varma's Marginal Fermi Liquid Model, SO(2) and SO(5) symmetry models, Boson Fermion model. None of these models provide satisfactory explanation of the anomalous features of HTSC oxides. However, the efforts to understand the HTSC oxides pairing mechanism, with enhancing future prospects for new HTSC materials and novel applications are in progress.

22.10 QUANTUM TUNNELING

This is a quantum mechanical process which permits electrons to penetrate from one side to the other through an extremely thin potential barrier to electron flow. Tunneling had been considered to be a possible electron transport mechanism between metal electrodes separated by either a narrow vacuum or a thin insulating film usually made of oxides. In 1960, Giaever demonstrates for the first time that if one or both of the metals were in a superconducting state, the current-voltage curve in such metal tunnel junctions revealed many details of that state. Giaever's technique was sensitive enough to measure the most important feature of the BCS theory—the energy gap which forms when the electrons condense into correlated bound pairs called as Cooper pairs.

A thin film of a superconductor is deposited on a smooth glass surface by evaporation. The surface of the layer is coated by an insulator layer of thickness 10 nm. In case of aluminium, this can be achieved by oxidizing the layer first deposited. Let us suppose that a second metal coating (normal metal) is made on the top of the insulating layer. Electrodes are connected to the metal states on either side of the insulating layer and then a gradually increasing potential is applied. This will raise the Fermi level on the right side. Until a certain critical voltage V_C is attained, there will be no current in circuit. The current begins to grow afterwards as shown in Fig. 22.18. One can explain this on the basis of quantum tunneling. We know that quantum mechanically, an electron on one side of the potential barrier has finite probability of tunneling through it, if there is an allowed state of equal energy is available on the other side of the potential barrier. The density of states function in the energy space for a sandwich consisting of a superconductor, an insulator and a normal metal (all at absolute zero) is shown in Fig. 22.17. We know that in a normal conductor, the electrons fill up to the Fermi level.

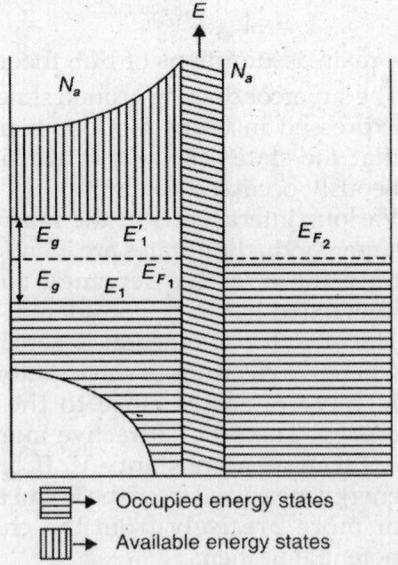

→ Occupied energy states

→ Available energy states

Fig. 22.17: Superconductor oxide normal junction at 0 K

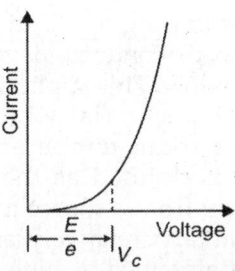

Fig. 22.18: Current (I)– voltage (V) characteristic curve for Giaever tunneling

The Fermi level in the superconductor is in the middle of the band gap. When the sandwich is formed, the Fermi levels are aligned (Fig. 22.17). Let us suppose that a voltage is applied across the barrier so as to raise the Fermi level on the right side of the junction. Until the Fermi level E_{F_2} on the right side of the junction raised up to the level E'_1 electrons cannot move from right to the left as there are no available energy levels on the left side; since in that portion the band gap exist. Obviously, once the Fermi level E_{F_2} rises above the level of E'_1 electrons can easily tunnel through the barrier from the right to the left since there are energy states available for the electrons on the left band side. Thus, the current increases. However, there will be no current until the voltage becomes available equal to the band gap ($E'_1 - E_2$). This has been confirmed experimentally.

The tunneling phenomenon has been exploited in many fields. For example, small area tunnel-junctions are used for mixing and synthesis of frequencies ranging from DC to the infrared region of the spectrum. This leads to absolute measurement of frequency in the infrared and provided the most accurate determination of the speed of light.

To study the non equilibrium superconducting properties, two tunnel junctions, one on top of the other sharing the middle electrode are used. Tunnel diodes are also used as a spectroscopic tool to study the phonon and plasmon spectra of the metals and the vibrational spectra of complex organic molecules introduced inside the insulating barriers.

22.11 APPLICATIONS OF SUPERCONDUCTIVITY

Magnets

Superconducting materials are used in a variety of ways where both stable magnetic fields are required or when tiny magnetic fields are to be detected. Superconductors are used to make electromagnets which can produce a high intensity magnetic field. A conventional electromagnet is made by passing a current through a metallic wire wound in the shape of a solenoid. The magnetic field due to a solenoid is given by

$$B = \mu_0 \, nI$$

where n is the number of turns per unit length and I is the current. The magnetic field can be increased by having a ferrite core. For a wire of given diameter, the magnetic field produced by a current carrying conductor is determined by the maximum current density. The maximum current density in a copper wire is 4×10^6 amps/m². With a superconducting wire the current density can be increased to 10^{11} amps/m². In addition, the use of iron core within the solenoid can be dispensed with. These factors considerably reduce the weight and size of the electromagnet. This also yields a highly homogeneous magnetic field (one part in 10^8) over the desired volume. Since there is no resistance, there is no need to maintain the current. The disadvantages of the superconducting electromagnet are the cost and size of the refrigeration system needed to cool the superconducting wire below the transition temperature. Type II superconductors are used for making magnets on account of their high critical density. Care also has to be taken to see that in the vortex state the vortex lines are pinned. The movement of vortex lines leads to dissipation of energy. Due care must be taken to prevent vibrations which lead to movement of vortex lines. Superconducting magnets are used in the following applications:

Magnetic Separation and Purification of Iron Ores

In this process, ferromagnetic or paramagnetic particles are extracted from a stream of non magnetic material by the action of the magnetic field. The force acting on the particle is given by $F = \mu H \nabla H$, where H is the magnetic field, ∇H is the field gradient and μ is the magnetic permeability.

Electromagnetic Clamps

Permanent magnets and electromagnets are used to attract and hold ferromagnetic objects, to clamp objects and assemblies in many manufacturing processes. Wherever permanent electromagnets are used, superconducting magnets can be replaced on account of the higher magnetic field generated by them. In aerospace industry, electromagnetic dent pullers are used for surface finishing. In a dent puller, a high magnetic field gradient is applied across the metallic sheet. Normally pulsed fields are employed. By varying the magnetic field, its gradient across the surface and the pulse rate, it is possible to remove the dents.

Particle Accelerators

In nuclear physics and particle physics, it is necessary to accelerate the particles to very high speeds and this is done by the application of electric fields and magnetic fields. Superconducting magnets are the preferred choice on account of their strength.

Plasma Confinement in Nuclear Fusion Reactors

Nuclear fusion reaction has to be carried out at high temperatures and pressures and in such conditions the nuclei have to be confined in a small volume using magnetic field. Superconducting magnets are the obvious choice for such experiments.

Electrical Equipment

Current Leads

On account of their high current density, superconductors are ideally suited for applications where high currents are needed.

Power Transmission

Low loss transmission of electric power using superconducting cables has been viewed as one of the most promising applications. The DC resistance of a superconductor is zero while the AC resistance is low. This application requires the finding of a material which is superconducting at room temperature. Power transmission is defined as the transfer of electrical energy from a source to a load over conductors that carry relatively large currents, while being maintained at high voltage, the power transmitted in the product of voltage and current. A power system consists of a generator located in a remote area, a transformer to raise the voltage and lower the current output of the generator, a transmission line or cable to transfer the power to a developed area and to receive the transmitted power and transform it from a high voltage to a lower distribution voltage. Finally, the power is sent over many distribution lines to loads (customers) in the area. At each customer point a final transformer will lower the voltage again to the lower level. There is a loss in power transmission due to Joule heating. The advantage of using superconductor transmission lines is that they can carry large current with lower current losses. The only cost involved is the cooling of superconductor below its transition temperature.

Fault Current Limiters

Any electrical circuit is at a risk of short circuit condition and some sort of protection is regularly employed. The simple fuse box and modern circuit breakers are examples of this. Fault currents are transient currents that flow through an electrical power system when a short circuit occurs. Current breakers can open in about 50 ms. Fault currents exceed normal currents by more than an order of magnitude. Hence, even within 50 ms, there is a good possibility that the system might be damaged in short circuit condition. FCL is meant to prevent this damage. It has variable impedance installed in series with a circuit breaker (Fig. 22.19). The impedance increases instantaneously when there is a short circuit. Hence, FCL prevents the fault currents from reaching a maximum value. Typically current exceeds only by a factor of 2 above the normal.

The function of FCL is to protect the circuit just long enough for the circuit breaker to activate. With a superconductor FCL under normal conditions the impedance is zero. In the event of a fault current, the critical current is exceeded and the superconductor reverts to normal conducting state. The switching time is less than a nanosecond. It then becomes a series resistor. The energy dissipated in the superconductor during a fault current can be reduced by employing shunt resistance.

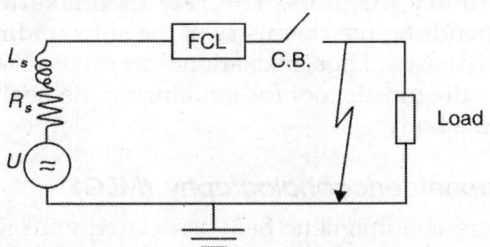

Fig. 22.19: Fault current limiter

Motors and Generators

Electric motors form a major portion of many electrical appliances. A motor converts electricity into power by rotating shaft which is turned to obtain mechanical work. A generator starts with a rotating shaft and produces electricity. In conventional motors, steel is used to increase the magnetic field produced by the motor coils. Iron saturates magnetically at 2.2 T. Hence, the maximum field strength in a conventional device is 2 T. In general, the power output of any rotating machine can be expressed as $P = C \times V \times B \times J$ where V is the volume, B is the peak air gap residual magnetic flux density, J is the linear current density at the core gap and C is a constant. The main advantage of using superconductor is that they can create an air gap magnetic field without any losses. This reduces power consumption and volume.

Transformers

They consist of inductively coupled coils of usually copper or other materials wound over high permeability steel cores. Transformers are used to change voltage/current ratio. The step up transformer steps up the voltage to reduce long distance transmission losses. In step down transformer, the voltage is stepped down for distribution and end use. On account of the high current density of superconductor coils, better flux linkage between the primary and secondary coils is possible. The superconductor transformer is smaller in size and lower in weight. The main disadvantage is the need to cool the primary and secondary coils of the transformers.

Superconducting Magnetic Energy Storage (SMES)

The power distribution is not usually reliable. There could be voltage fluctuations or intermittent power. Hence, it is always desirable to store energy so that it can be tapped when needed. Superconducting magnetic energy storage is one possible answer to improve the power quality problem. Figure 22.20 shows the conceptual diagram of the source and load interface with SMES. Through an AC/DC persistent current flowing through the superconductor coil, SMES stores energy in the magnetic field of the coil. When a superconducting coil is used there are no resistive losses and a DC current persists as long as the coil is below the transition temperature.

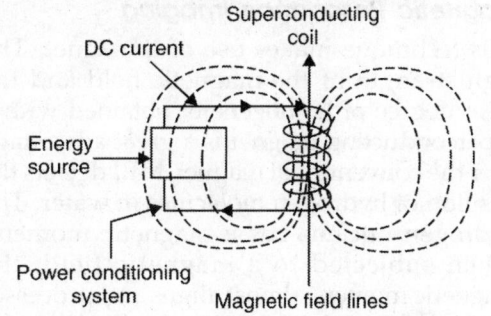

Fig. 22.20: Superconducting magnetic energy storage

Electronics
Microwave Resonators and Filters

Monolithic microwave integrated circuits (MMIC) are used in many microwave devices such as resonators and filters. However, MMICs are quite different from conventional ICs. The conventional ICs contain very high packing densities whereas as in these packing densities are small. An MMIC whose elements are formed on an insulating substrate such as glass or ceramics is known as film integrated circuit. The low surface resistance of the superconductor makes them ideal for fabrication of MMICs. Resonators and filters have become the most active application for high temperature superconductor microwave circuits. According to structures, resonators are classified into three categories: one dimensional, planar type and three dimensional type or cavity type. Filters can also be classified in the same way. MMICs are used in cellular telephone and communication systems.

Logic Circuits

The presence or the absence of flux quanta in a superconducting loop can be used to represent the logic states 0 and 1. They are known as rapid single flux quantum logic (RSFQ) circuits. They offer a unique combination of high speed, low power and the complexity of a LSI type circuit with Josephson junctions. On account of their high switching speeds (<10 ps), they are ideally suited for ultra high speed digital signal processing.

Medical Diagnostics

Magnetic Resonance Imaging

This technique makes use of a magnet. The high strength of the magnetic field and the great degree of homogeneity obtained with a superconducting magnet is a great advantage over the conventional magnet. MRI detects the position of hydrogen molecules in water. The hydrogen nucleus has a magnetic moment. When subjected to a magnetic field, the magnetic moment almost aligns and processes about the direction of the magnetic field at a characteristic frequency called the Larmor frequency. The Larmor frequency is proportional to the magnetic field and is in the RF frequency range. A second magnetic field which is small and at the RF frequency causes the hydrogen to flip from parallel state to antiparallel state. Removal of the RF field causes hydrogen to relax emitting a RF signal that can be detected. Mathematical processing

of the signal yields information about the protons. If instead of using a constant DC field, a small gradient is superposed on the top of the constant field, the Larmor frequency will vary with the position. By analyzing the emitted RF signal with respect to frequency, good spatial resolution can be obtained. The time variation of 1 in 10^9 and spatial variation of 1 part in 10^5 in magnetic field has been achieved using superconducting magnets which is far superior to what is obtained with ordinary magnets. The rate of relaxation depends on the chemistry of the surrounding body tissue. Thus, relaxation time can be used as a diagnostic tool for monitoring the health of the body.

Magnetoencephalography (MEG)

There is a magnetic field associated with the flow of current. In the body neurons and muscle fibres both generate current when they are activated. The terms neuron firing refers to the sudden passage of a pulse of current along a neuron. That sets up a magnetic field because a finite cluster of charges would have moved a finite distance. Typically a neuron is 2 mm long and 2 µm in diameter. The current flow when a single neuron is fired cannot be detected since the field generated is very small. However, in the brain neurons are clustered together, somewhat aligned in certain patterns and act synchronously in groups. When a group of 10000 parallel neurons fire simultaneously, the net current is sufficiently intense to generate a magnetic field. This can be detected outside the skull using SQUIDS. The neurons parallel to the surface of the skull generate the detectable field; the neurons perpendicular to that surface have their magnetic fields concealed within the cranium. Fortunately, the majority of the brain neurons is of the former variety and give a signal when they are fired. The important factor is that many of the neurons in a particular region are strongly interconnected and fire simultaneously. Typically a volume of 0.1 mm^3 can generate a distinguishable magnetic field. MEG uses SQUIDS to detect magnetic fields arising from the currents within the brain and produces a map of brain's magnetic activity. Typically the fields are 80×10^{-15} T up. MEG scans are helpful to detect tumors in the brain. Mapping of the magnetic activity is also helpful to treat epileptic patients.

Nondestructive Testing

Eddy Current Testing (ECT)

Eddy current testing is one of the techniques employed in nondestructive testing. In this technique, an alternating current frequency 1–10 kHz is made to flow in a coil which in turn produces an alternating magnetic field around it. This coil when brought close to the electrically conducting surface of a metallic material to be inspected induces an eddy current flow in the material due to electromagnetic induction. These eddy currents are generally parallel to the coil winding. The presence of any defect or discontinuity in the material disturbs the eddy current flow. The depth of penetration (δ) is given by

$$\delta = \frac{10^6}{2\pi} \sqrt{\frac{10\rho}{\mu f}}$$

where δ is the penetration depth in mm, μ is the relative permeability, f is the frequency and ρ is the resistivity in SI units. The detector coil senses anomalies in the magnetic field of the current which are caused by their spatial diversion when they encounter flows such as corrosion and flaws. SQUIDs can be employed to measure the minute changes in the magnetic fields.

Particle and Photon Detectors

Energetic particles and photon can be detected using superconductors. When a photon or a charged particle is absorbed in a superconducting film operated below the transition temperature, the energy of the particle is sufficient to break the Cooper pairs. The excess number of electrons thus generated can be measured using a superconducting tunnel junction. If the energy of the particle is sufficient, then the superconductor reverts to normal conducting state. This change can be monitored to detect the particle or the photon.

In the metal-superconductor junction, the absorption of a charged particle or photon can raise the temperature which in turn changes the electron distribution. This again is reflected in the magnitude of the tunnel current.

Magnetic Levitation

Maglev Vehicle

Diamagnetic materials are repelled by the magnetic field. Superconducting materials are perfect magnetic materials and hence experience large repulsive forces. The Meissner effect provides a force to lift the superconductor above a magnet.

The height h at which a spherical shaped superconductor levitates above a magnet of thickness a is given by

$$h = \left[\frac{B^2(a) a^2}{4\pi\rho g}\right]^{1/3}$$

where ρ is the density of the superconductor and g is the acceleration due to gravity. The flux density B is assumed to vary linearly and inversely with the height above the magnet.

The repulsive forces can also be generated when a magnet is in motion over a conducting surface. A magnet in motion generates eddy currents on the surface. There is a force of repulsion between the magnetic field due to the eddy currents and the magnet due to Lenz's law. But whenever the magnet is in motion there is also a drag force. The force on the moving magnet is complicated due to the presence of eddy currents. *Thus, a permanent magnet can be levitated above a rotating aluminum disc.* The force on a magnet moving over a nonmagnetic conducting plane can be conveniently resolved into two components. One component is the lift force perpendicular to the plane and the other is a drag force opposite to the direction of motion. At low velocity, the drag force is proportional to velocity v and considerably greater than the lift force which is proportional to v^2. As the velocity increases the drag force reaches a maximum (referred to as drag peak) and then decreases as $v^{-1/2}$. The lift force, on the other hand, approaches an asymptotic value.

Two different kinds of Maglev vehicles have been designed (Fig. 22.21). The difference is in the way the vehicle attaches itself to the guide way. In electromagnetic suspension systems (EMS), attractive levitation forces are made use of while in the electrodynamic systems (EDS), repulsive forces between the magnets are used. In EMS, attractive forces between the electromagnets and the ferro-magnetic (steel) guide way provides the lift. Since the force of attraction increases with decreasing distance, such systems are inherently unstable and the magnet currents must be carefully controlled to maintain the desired suspension height. The magnet to guide way spacing is ~cm. The advantage is that it is possible to maintain magnetic suspension even when the vehicle is standing still. In EDS, the repulsive forces between moving magnets and the eddy currents they induce in conducting (aluminum) guide way. The repulsive levitation force is inherently stable with distance and comparatively large levitation heights ~20–30 cm are attainable by using superconducting magnets.

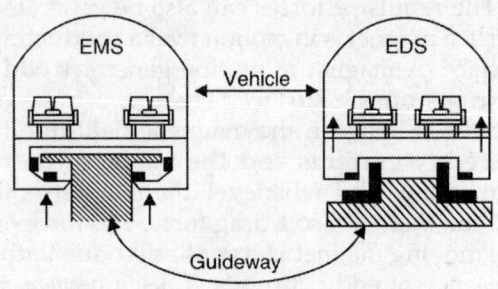

Fig. 22.21: Maglev vehicle

Magnetic Bearing

In a magnetic bearing, the repulsive force between two similar magnetic poles is used, to get rid of the contact friction forces arising in the conventional bearing systems. An ideal bearing should have stability when the two surfaces are in motion. The repulsive force between the magnets in the vertical direction is invariably associated with instabilities in the lateral direction. The problem can be over-come by having a superconducting cylinder

enclosing the magnetic bearing as shown in Fig. 22.22. Flywheels are devices for storing mechanical energy. Flywheel mounted using a magnetic bearing could work as an efficient storage device (Fig. 22.22).

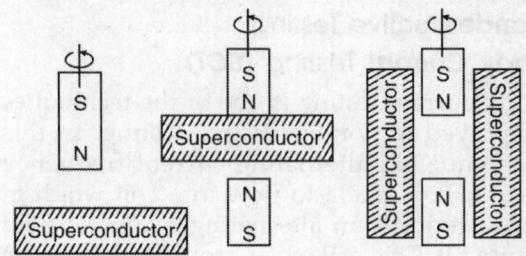

Fig. 22.22: Magnetic bearing

Vibration Isolators

Magnetic levitation can also be used for designing vibration isolators. Vibration isolators isolate the mechanical system which is a source of noise due to moving parts. The noise is both air borne as well as structure borne. The structure borne noise can be prevented by mounting the mechanic system on platforms supported by strong springs. Magnetic levitation provides a mechanism to lift the system above the ground, and thus prevent the noise being propagated along the ground.

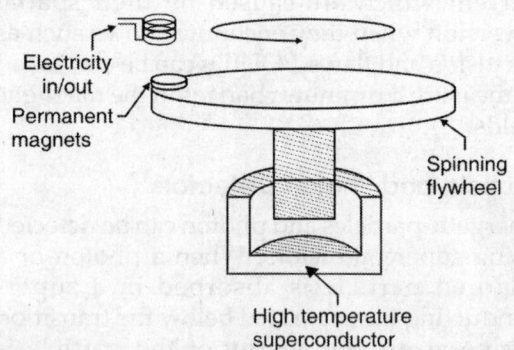

Fig. 23.23: Flywheel mounted on a magnetic bearing

Thin film applications in electronics, together with superconducting magnets, are usually considered the most important area of superconductivity-based technology. Most of these applications are based on Josephson effect that enables the construction of the fastest nanoscopic switches, Josephson junctions and related device structures, SQUIDs (superconducting quantum interference devices). SQUIDs explore subtle quantum interference effects— an analysis of superconducting loop shows that the magnetic flux that can thread the loop is quantized in quantum units of flux which have a value of 2×10^{-15} weber. SQUIDs are suitably processed superconducting loops which detect minute changes in magnetic flux, i.e. they are high sensitivity magnetic flux detectors that can be used in the finest precision instruments at the forefront of meteorology.

The HTSC oxides provide us with superconductors which can even be employed at the temperature of liquid nitrogen ($T_B = 77$ K) or even higher temperature. While the widespread use of HTSC cuprate superconductors in technology has not yet been realized, steady and significant progress has been made towards this objective during the past 15 years. Recent developments indicate that HTSC cuprates will begin to have a significant impact on technology during the next 5 to 10 years. The applications in superconducting electronics that are likely to be realized on this time scale include SQUIDs, NMR (nuclear magnetic resonance) coils, wireless communications subsystems, MRI coils and NMR microscopes and digital instruments. In the area of superconducting wires and tapes, applications that appear to be feasible within this same time period include power transmission lines, motors and generators, transformers, current limiters, magnetic energy storage, magnetic separation, research magnetic systems and current leads.

An example of recent progress in the area of superconducting wires and tapes is the development of flexible superconducting ribbons consisting of deposits of YBCO on textured substrates which have critical current densities $J_C \sim 10^6$ A/cm^2 in fields up to 8T at 64 K, a temperature that can be achieved by pumping on liquid nitrogen. The performance of these prototype conductors in storing magnetic fields already surpasses that of NbTi and Nb$_3$Sn, which are currently used in commercial superconducting wires at liquid helium temperatures in a comparable field range.

It is certain that HTSC oxides have opened new opportunities for technological applications of superconductivity. It is also certain, however, that a great deal of development will be necessary before we have in our hands suitable conductors for use in, say, magnet construction or for microcircuitry. HTSC oxides will have some impact on electronics related applications— initially in passive microwave devices and SQUIDs. The mixtures of HTSC oxides with silver may offer somewhat improved properties and might provide the basis for future wires and cables. It is economic question that will play a crucial role here.

The requirements for critical current density (J_C) in a given magnetic field for few selected applications are presented in Table 22.4.

Table 22.4: Requirements for J_C in a given magnetic field for several applications of superconductors

Application	H (Tesla)	J_C (A/cm^2)
Interconnects	0.1	5×10^6
AC transmission lines	0.2	10^5
DC transmission lines	0.2	2×10^4
SQUIDs	0.1	2×10^2
Motors and generators	~4	~10^4
Fault current limiters	>5	>10^5
Power generation	5–7	2×10^4
Power storage	5–10	10^3–10^4
MRI scanners	2	10^4
Mineral separation	2–5	10^4

A brief account of potential applications of super conductivity in vital areas is given in Table 22.5.

We may note that most high current applications of superconductors require $J_C > 10^5$ A/cm^2 and $T_{use} \sim T_C/2$ and for weak current and most applications in electronics

$$T_{use} \sim \frac{2}{3} T_C$$

T_C, upper critical field (H_{C_2}), λ and ξ of some technologically important superconductors are given in Table 22.6.

Table 22.5: Potential applications of superconductivity in vital areas

Field	Applications
• Magnets	• High field magnets • Magnetic levitation • Magnetic shielding • NMR (Magnetic diagnostics and spectro scopy) • Large machines (colliders, magnetic fusion confinement, r.f. cavities • Ore refining
• Energy or power-related	• Electric power storage and transmission • Magnetic energy storage • Power production by magnetic fusion and magnetohydrodynamics (MHD)
• Transportation	• High speed maglev trains • Ship-drive system
• Electronic and other small devices	• SQUIDs • Bolometer • Electromagnetic shielding • Josephson devices (square-law detector, parametric amplifier, mixer)
• Computer and information processing	• Optoelectronics • Voltage standard • Active superconducting elements (FETs) • Malchel filters • Semiconductor–superconductor hybrids (A-D converters)

Table 22.6: Important parameters of few selected superconductors

Material	T_C(K)	Penetration depth λ (nm)	Coherence length ξ (nm)	Upper critical field H_{C_2} (T)
Nb	9.2	40	38	0.2
Nb-Ti	9.2	60	40	14
NbN	16	250	4	16
Nb$_3$Sn	18	80	3	24
YBa$_2$Cu$_3$O$_7$	92	150/600	1.5/0.4	150/40
Bi$_2$Sr$_2$Ca$_2$Cu$_3$O$_{10}$	110	200/1000	1.4/0.2	250/30

ILLUSTRATIVE EXAMPLES

Example 1

The critical temperature of mercury is 4.2 K.

(a) Calculate the energy gap in electron volts at $T = 0$.

(b) Calculate the wavelength of a photon whose energy is just sufficient to break up Cooper pairs in mercury at $T = 0$. In what region of the electromagnetic spectrum are such photons found?

Solution

(a) The Cooper pair binding energy, or gap energy, is

$$E_g = 3kT_C$$

$$= 3 \times 1.4 \times 10^{-23} \text{ J/K} \times 4.2 \text{ K}$$

$$= 1.8 \times 10^{-22} \text{ J} \simeq 1.1 \times 10^{-3} \text{ eV}$$

(b) $E_g = h\nu = hc/\lambda$

$$\therefore \quad \lambda = \frac{hc}{E_g} = \frac{6.6 \times 10^{-34} \text{Js} \times 3 \times 10^8 \text{ m/s}}{1.8 \times 10^{-22} \text{ J}}$$

$$= 1.1 \times 10^{-3} \text{ m}$$

Obviously, these photons are in the very short wavelength part of the microwave region.

Example 2

Does the metal look like a superconductor to electromagnetic waves having wavelengths shorter than that found in Example 1(b)?

Solution

No, since the energy content of shorter wavelength photons is sufficiently high to break up the Cooper pairs, or excite the conduction electrons through the energy gap into the non-superconducting states above the gap.

Example 3

How much current can a lead wire, 1 mm in diameter, carry in its superconducting state at 4.2 K? (Given T_C for Pb is 7.2 K and $B_C(0) = 0.0803$ Wb/m²)

Solution

We have $B_C(T) = B_C(0)\left[1 - \left(\dfrac{T}{T_C}\right)^2\right]$

Given $B_C(0) = 0.0803$ Wb/m²

$$T = 4.2 \text{ K}, \ T_C = 7.2 \text{ K}, \ 2r = 10^{-3} \text{ m}$$

$$\therefore \quad B_C(4.2) = 0.0803\left[1 - \left(\frac{4.2}{7.2}\right)^2\right] = 0.0548T$$

Now $I_C = \dfrac{2\pi r B_C(T)}{\mu_0} = \dfrac{\pi(2r) B_C(T)}{\mu_0}$

$$= \frac{\pi(10^{-3})0.0548}{4\pi \times 10^{-7}} = \frac{548}{4} = 137 \text{ A}.$$

Example 4

For a superconductor, $T_C = 3$ K and $n_S = 10^{28}/\text{m}^3$. Find the penetration depth at 0 K and 1 K.

Solution

$$\lambda_0 = \left(\frac{m}{\mu_0 n_S e^2}\right)^{1/2} = \left(\frac{m}{\mu_0 n_S}\right)^{1/2} \cdot \frac{1}{e}$$

We have $m = 9.1 \times 10^{-31}$ kg

$$\mu_0 = 12.56 \times 10^{-7} \text{ SI units}$$

$$n_S = 10^{28} \text{ m}^{-3}$$

and $e = 1.6 \times 10^{-19}$ C

$$\therefore \quad \lambda_0 = \left(\frac{9.1 \times 10^{-31}}{12.56 \times 10^{-7} \times 10^{28}}\right)^{1/2} \times \frac{1}{1.6 \times 10^{-19}}$$

$$= \frac{0.85 \times 10^{-26}}{1.6 \times 10^{-19}} = 0.53 \times 10^{-7} \text{ m} = 530 \text{ Å}$$

$$\lambda_{T=3K} = \frac{\lambda_0}{\left[1 - \left(\dfrac{I}{T_C}\right)^2\right]^{1/2}} = \frac{530 \times 10^{-10}}{\left[1 - \left(\dfrac{1}{3}\right)^4\right]^{1/2}}$$

$$= \frac{530 \times 10^{-8} \times 9}{8.95}$$

$$= 533 \times 10^{-10} \text{ m}$$

$$= 533 \text{ Å}.$$

Example 5

A lead wire has a critical magnetic field of 6.5×10^3 A m⁻¹ at 0 K. The critical temperature is 7.18 K. At what temperature the critical field would drop to 4.5×10^3 A m⁻¹? The diameter of the wire is 2 mm. What is the critical current density at that temperature?　　　[BTech]

Solution

$$\frac{[H_C - H_0]}{H_0} = -\left(\frac{T}{T_C}\right)^2$$

$$\frac{[H_0 - H_C]}{H_0} = \frac{T^2}{T_C^2}$$

$$\frac{[H_0 - H_C]}{H_0} T_C^2 = T^2$$

i.e. $T^2 = \dfrac{(6.5 \times 10^3 - 4.5 \times 10^3)}{6.5 \times 10^3} \, 7.18 \times 7.18$

$= 15.86$ K

i.e. $T = 3.98$ K

The critical current density

$$J_C = \frac{H_C \times 2\pi r}{(\pi r^2)} = \frac{4.5 \times 10^3 \times 2}{1 \times 10^{-3}}$$

$= 9.0 \times 10^6 \, \text{A m}^{-2}$ at 3.98 K.

Example 6

The transition temperature of Nb is 9.5 K. The critical magnetic field induction density is 0.2 T at 0 K. What is its stabilization energy? What is the magnetic field induction at 3.5 K?

[BTech]

Solution

$$\Delta U = \frac{(B_C)^2}{2\mu_0} = \frac{(0.2)^2 \times 7}{2 \times 4 \times 22 \times 10^{-7}}$$

$= 159 \times 10^2$ J

$$B = B_0 \left[1 - \left(\frac{T}{T_C} \right)^2 \right]$$

$$= 0.2 \left[\frac{(9.5)^2 - (3.5)^2}{(9.5)^2} \right] = 0.173 \text{ T}$$

REVIEW QUESTIONS

1. What is superconductivity? Mention some important property changes that occur in materials when they undergo phase change from normal to the superconducting state.

2. What are type-I and type-II superconducting materials? Give three examples of each. Why type-II materials are preferred for applications of superconductivity? [BE]

3. Define critical temperature, critical field and critical current density for a superconductor.

[AMIE]

4. Explain the behaviour of a superconductor in a magnetic field. What is Meissner effect? How does the critical magnetic field varies with temperature?

5. What is isotope effect in superconductors? How it provided a clue to the microscopic BCS theory of superconductivity? [BE]

6. Describe the effects of (i) magnetic field (ii) frequency (iii) isotopes on superconductors.

7. What are the characteristic lengths of a super-conductor? How penetration depth varies with temperature? [BE]

8. Give a brief account of thermodynamics of a superconductor.

9. What is BCS theory of superconductivity? Write its salient features.

10. Write a short note on potential applications of superconductors. [BE]

11. What are DC and AC Josephson effects? Explain their importance.

12. What do you understand by quantum tunneling? Explain its significance. [BE]

13. What are HTSC oxides? How are they different from conventional superconductors? What are their future prospects from applications of superconductivity point of view?

PROBLEMS

1. (a) Show from Maxwell's equations that resistivity $\rho = 0$ (a perfect conductor) implies that $B = 0$ inside the material (b) Show, from Maxwell's equations, that $B = 0$ inside a material (a superconductor) implies that the resistivity (ρ) of the material is zero, i.e. $\rho = 0$.

2. Show that Meissner effect implies perfect conductivity, but *vice versa* is not true.

3. The resistivities of Cu, Ag and Au at room temperature in units of 10^{-8} Ω m are 1.6, 1.5 and 2.4 while those of Ti, Zr and Hf are 89, 45 and 32 respectively. Explain why the former do not show superconductivity, whereas the latter becomes superconductors?

4. The critical temperature (T_C) for Hg with isotopic mass 199.5 is 4.185 K. Calculate its critical temperature when its isotope mass changes to 203.4. [*Ans.* 4.139 K]

5. Show that the frequency of the electromagnetic waves radiated by a Josephson junction having a voltage of 650 μV across its terminals is 3.15×10^{11} Hz.

SHORT ANSWER QUESTIONS

1. What is the relaxation time for a superconductor?

Ans. Infinite

2. What is the susceptibility of a superconductor?

Ans. Negative and unity.

3. What is superconducting electron density at absolute zero?

Ans. Finite.

4. What is critical field for a superconductor?

Ans. The magnetic field above which superconductivity disappears is called critical field B_C, given by

$$B_C(T) = B_C(0)\left[1 - \frac{T^2}{T_C^2}\right]$$

For type-II superconductor, there are two critical fields, e.g. B_{C_1} (lower critical field and B_{C_2} (upper critical field).

5. For a material to be superconductor, what should be its two distinctive properties?

Ans. Zero resistivity, i.e. infinite conductivity and $B = 0$ inside the superconductor.

6. What should be the Ginzburg–Landau parameter, $\kappa = \lambda/\xi$ for a type-II superconductor?

Ans. $\kappa > 1$

7. What is the BCS relation for critical temperature (T_C) for simple conventional weak coupling superconductors?

Ans. $k_B T_C = 1.13\hbar\omega_D\exp[-1/\lambda N(0)]$

8. What is the BCS relation between zero temperature gap $\Delta(0)$ and critical temperature, T_C?

Ans. $2\Delta(0) = 3.552\, k_B T_C$ and close to T_C.

$$\Delta(T) = 1.74\,\Delta(0)\left[1 - \frac{T}{T_C}\right]^{1/2}$$

OBJECTIVE QUESTIONS

1. A superconductor has no _____ for all $T < T_C$. [resistivity]

2. Meissner–Ochsenfeld effect shows that _____ inside the superconductor. [$B = 0$]

3. BCS theory assumes a spherical _____ and isotropic energy gap. [Fermi surface]

4. Critical magnetic fields of a superconductor _____ if temperature decreases. [increases]

5. Mercury is a type _____ superconductor. [I]

6. Superconducting electron density is _____ at absolute zero. [finite]

7. Nb_3Sn is a type _____ superconductor. [II]

8. The surface energy for a type-I is always _____. [positive]

9. Flux quantum in a superconductor is _____. [$h/2e$]

10. The total charge on a Cooper pair in a superconductor is _____. [$-2e$]

11. A Cooper pair is a system of two electrons bound by exchange of _____ between them. [phonon]

MULTIPLE CHOICE QUESTIONS

1. The temperature at which a metal becomes superconductor is called

(a) Curie temperature

(b) Debye temperature

(c) Critical temperature

(d) Neel temperature

2. The transition temperature of most superconducting elements falls within the range

(a) 0–143 K (b) 0–10 K

(c) 0–23 K (d) 0–50 K

3. The relation between T_C and H_C for a superconductor is

(a) $H_C = H_0\left[1 - \left(\frac{T}{T_C}\right)^2\right]$

(b) $H_C = H_0\left[1 + T_C^2\right]$

(c) $H_C = H_0\left[(T - T_C)^2\right]$

(d) $H_C = \dfrac{H_0}{\left(1 - \dfrac{T}{T_C}\right)^2}$

4. A type-I superconducting material when placed in a magnetic field will

(a) expel all the magnetic lines of forces passing through it

(b) attract the magnetic field towards its centre

(c) not influence the magnetic field

(d) none of the above

5. A superconducting material on being subjected to critical magnetic field (H_C)
   (a) changes to normal state
   (b) remain unaffected
   (c) exhibit both normal as well as superconducting state properties
   (d) none of the above

6. The width of the energy gap of a superconducting material is maximum at
   (a) T_C K   (b) 0 K
   (c) $\left(\dfrac{T_C}{2}\right)$ K   (d) $\left(\dfrac{T_C}{3}\right)$ K

7. In superconductivity state
   (a) entropy decreases and thermal conductivity increases
   (b) entropy increases and thermal conductivity decreases
   (c) entropy and thermal conductivity both decrease
   (d) entropy and thermal conductivity both increase

8. A superconductor has
   (a) a negative susceptibility
   (b) a positive susceptibility
   (c) sometimes negative and sometimes positive susceptibility
   (d) none of the above

9. Critical magnetic field of a superconductor
   (a) increases with rise in temperature
   (b) decreases with rise in temperature
   (c) does not depend on superconducting transition temperature
   (d) increases with decrease in temperature

10. Electron density of a superconductor
    (a) is infinite at absolute zero
    (b) is zero at absolute zero
    (c) is finite at absolute zero
    (d) increases with the rise in temperature from absolute zero to transition temperature

11. The isotope effect coefficient for a superconductor
    (a) is zero
    (b) is generally greater than zero
    (c) is generally in the range from 0.2 to 0.6
    (d) is generally much greater than 1

12. Which of the following is a multiple connected superconductor?
    (a) a ring   (b) a solid sphere
    (c) a solid cylinder   (d) a solid rod

13. A Cooper pair is a system of two electrons bound by exchange
    (a) of a photon between them
    (b) of a phonon between them
    (c) of a proton between them
    (d) of a neutron between them

Answers

1. (c)	2. (b)	3. (a)	4. (a)	5. (a)
6. (b)	7. (b)	8. (a)	9. (d)	10. (c)
11. (c)	12. (a)	13. (b)		

23 Dielectrics

23.1 INTRODUCTION

These are the materials or insulators which have the unique characteristic of being able to store electric charge. The electrons in these substances are localized in the process of bonding the atoms together. Obviously, covalent or ionic bonds, a mixture of both, or van der Waals bonding between closed shell atoms give rise to solids or gases which exhibit dielectric or insulating properties. Dielectric materials may be gases, liquids or solids with the exception of air which is the insulating material between the bare conductors of the overhead electric grid system. Liquid dielectrics are used mainly as impregnants for high voltage paper, insulated cables and capacitors as filling and cooling media for transformers and circuit breakers. Most common properties of dielectric materials are (i) dielectric constant (ii) dielectric strength (iii) insulation resistance (iv) surface resistivity (v) loss factor (vi) tangent of loss factor in terms of a capacitor or phase difference (vii) polar and non-polar materials.

Materials, which are capable of separating electrical conductors circuit breakers, e.g. silicon, oils, liquid dielectrics have high dielectric constant, high resistance, high dielectric strength when moisture and impurities are removed from them. They have high temperature dissipation capacity and least dielectric losses. Their dielectric constant is greater than one. Following materials are important from engineering point of view:

Mica: It is the widely used insulating material in switch gears armature windings, electrical heating devices like iron, hot plates, etc. It is also used in capacitors for high frequency application. Mica is an inorganic compound of silicates of aluminium, soda potash and magnesia. It is crystalline in nature and can be easily split into very thin falt sheets. The two important types of mica are: (i) *muscovite* (ii) *phlogopite*. Mica has good dielectric strength and mechanical strength. Its dielectric constant varies between 5 and 7.5, loss tangent between 0.0003 and 0.015 and dielectric strength between 700 and 1000 kV/mm.

Asbestos: It is also used as an insulator in the form of paper, tape, cloth and board. Asbestos is widely used in panel boards, insulating tubes and cylinders in the construction of air cooled transformers. Asbestos is an inorganic material, which is used to designate a group of naturally occurring fibre materials. Asbestos has good dielectric and mechanical properties.

Ceramics: These are generally non-metallic inorganic compounds, e.g. silicates, aluminates, oxides, carbides, borides, nitrides and hydroxides. Ceramics used as dielectrics may be broadly described as alumina, porcelains, ceramics, titanates, etc. These have excellent dielectric and mechanical properties. The dielectric constant of most commonly used ceramics varies between 4 and 10. These are used in switches in plug holders, thermocouples, cathode heaters, vacuum type ceramic metal seals, etc. Ceramic capacitors may be operated at high temperatures and can be moulded into any shape and size.

Electric grade ceramics are used for the manufacturing of insulators, terminal blocks, plates, frames, coils, etc. They must have low losses, good insulating properties and high strength.

Insulators for operation at low frequencies are made of electric-grade porcelain which possesses fairly good electrical properties. A drawback is that it has high losses which increase sharply on heating above 200 °C, and a low strength.

Insulating parts for operation at high frequencies are mostly made of steatite, a talc bace material. Steatites contain no harmful impurities and retain their properties at temperatures up to 100 °C. They can be pressed easily, shrink on burning by only 1–2%, and are suitable for making parts in which dense but porous structure and accurate dimensions are essential. In contrast to other kinds of ceramics, steatite can be cut quite easily (after burning). Among their drawbacks are cracking under the action of rapidly varying temperatures and some difficulties involved in the burning.

Glass: It is an inorganic insulating material, which comprises complex system of oxides. Silica (SiO_2) is the most essential constituent of many commercially used glasses. It is fused with alkali (like potash, soda, etc.) and some base (like lime, lead oxide, etc.). The silica glass (having 100% SiO_2) is the best insulating material. The dielectric constant of glass varies between 3.7 and 10 and loss tangent between 0.0003 and 0.01 and dielectric strength between 2.5 and 50 kV/mm. Glass is used in electric bulbs, X-ray tubes, mercury switches, electronic valves as insulating material. It is also used in capacitors as dielectric material.

Resins: These are organic polymers and may be natural or synthetic. The synthetic resins are produced artificially. The commonly used synthetic resins are polyethylene, polystyrene, polyvinyl chloride, acrylic resins, teflon, nylon, etc. These have good dielectric and mechanical properties. The dielectric constant of resins varies between 2 and 4.5, the loss tangent between 0.0002 and 0.04 and dielectric strength is quite high. These are used in transformers, high frequency capacitors. These are also used as a dielectric material in DC capacitors.

Rubber: These are organic polymers and may be natural or synthetic. The natural rubber obtained from rubber tree has limited applications due to its resistance to low and high temperatures. The synthetic rubbers are produced artificially by *copolymerisation* of isobutylene and isoprene. These have good electrical and thermal properties. The dielectric constant of rubber varies between 2.5 and 5 and loss tangent between 0.01 and 0.03. Rubbers are used as an insulating material for electric wires, tapes, cables, coatings, motor windings, transformers, etc.

Gaseous dielectrics, e.g. nitrogen, hydrogen, etc. have dielectric constant 1.0.

The best dielectrics are polystyrene, polyethylene, polymide, mineral oil, pure alumina and pure silica.

23.2 DIELECTRIC CONSTANT

In an electric field, the charge density D is directly proportional to the applied field, i.e.

$$D = \varepsilon E \qquad (23.1)$$

where ε is the dielectric constant or permittivity of the material placed between the electrodes. For vacuum,

$$D = \varepsilon_0 E \qquad (23.1a)$$

where $\varepsilon_0 = 8.854 \times 10^{-12}$ farad/m is called the absolute permittivity.

One can define the relative permittivity (ε_r) of a dielectric as:

i. The ratio of the electric field density produced in the medium to that produced in vacuum by the same electric field strength.

ii. The ratio of capacitance of a condenser containing a given dielectric to the same condenser with vacuum as dielectric. Thus, relative permittivity can be expressed as

$$\varepsilon_r = \frac{\varepsilon}{\varepsilon_0} \qquad (23.2)$$

Now, charge density

$$= \varepsilon_0 \varepsilon_r E = \varepsilon_0 E + \varepsilon_0 (\varepsilon_r - 1)E$$

where $\varepsilon_r - 1 = \chi$ is called *electric susceptibility*.

χ = Bound charge density/free charge density.

Permittivity is influenced by permeability (μ), stress distribution, temperature and frequency.

The basic phenomenon that takes place in dielectric materials can be easily understood if we consider a solid dielectric located in the region between the two plates of a *parallel plate condenser (capacitor)*.

Let us consider a parallel plate capacitor connected to a DC voltage source, V_0 as shown in Fig. 23.1a. The DC source charges the plates of capacitor till the potential difference across the plates acquire the voltage V_0. Let the charges on the two plates are $+Q_0$ and $-Q_0$, when the potential difference between two plates is V_0.

The *capacitance* C_0 related to the quantity of charge stored on either plate of the capacitor is given by

$$C_0 = \frac{Q_0}{V_0} \qquad (23.3)$$

Next, we disconnect the plates of the capacitor from the voltage source V_0. Obviously, the circuit is open and therefore the magnitude of the charge Q_0 on either plate must remain constant. Let us suppose that we now introduce a dielectric *ABCD* in the form of a rectangular parallelepiped between the two plates (Fig. 23.1b) and dielectric fills the space between the plates completely. Now, the potential difference between the plates falls to V. When we pull out the dielectric slab, the potential difference returns to the original value V_0. This experimental observation clearly reveals that the original charges on the plates have not been affected by the introduction of the dielectric between the plates. One can therefore conclude that the value of the capacitance increased from C_0 to C in the presence of the dielectric between the plates, i.e.

$$C = \frac{Q_0}{V} \qquad (23.4)$$

Let us write $V = \dfrac{V_0}{\varepsilon_r}$ $\qquad (23.5)$

$$\therefore \qquad C = \frac{Q_0}{V_0/\varepsilon_r} = \varepsilon_r C_0 \qquad (23.6)$$

or $\qquad \varepsilon_r = \dfrac{C}{C_0}$ $\qquad (23.7)$

ε_r is called *relative permeability* or dielectric constant and in the case of dielectrics its value is greater than unity. ε_r is a dimensionless quantity. Moreover, ε_r is independent of the size and shape of the dielectric. From Eq. (23.6), it is obvious that the charge storing capacity of a capacitor is enhanced by the dielectric which permits a capacitor to store ε_r times more charge for the same potential difference.

The dielectric constant of a dielectric is usually measured by measuring and comparing the capacitance of a capacitor containing the

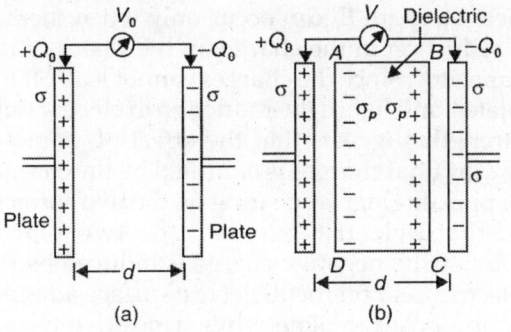

(a)　　　　(b)

Fig. 23.1: Parallel plate capacitor (a) When there is no dielectric present, i.e. there is vacuum between the plates and plates are charged to charge Q_0 and a DC potential difference V_0 is established (b) When dielectric (*ABCD*) is introduced, Q_0 remains constant whereas V_0 falls to V

dielectric medium to the capacitance of the same capacitor with vacuum (or air) as the medium. For a parallel plate capacitor having area of each plate A m^2, the distance between the two plates be d m and relative permeability of the dielectric medium ε_r, the capacitance is given by

$$C_0 = \frac{\varepsilon_0 A}{d} \qquad (23.8)$$

(when the medium between the two plates is vacuum or air).

where $\varepsilon_0 = 8.85 \times 10^{-12}$ F/m is the permittivity of free space. When the capacitor is completely filled with a dielectric, the capacitance is

$$C = \frac{\varepsilon_0 \varepsilon_r A}{d} \qquad (23.9)$$

Energy Stored in a Capacitor

We know that work is done when two equal and opposite charges are separated and the work done is stored in the form of electric potential energy. We can recover this energy if the charges are allowed to assume their original configuration. A capacitor gets charged by a voltage source (say by a battery) at the expense of energy stored in the battery, i.e. chemical energy. The work done by the voltage source is given by

$$W = \frac{1}{2}\frac{Q^2}{C} = \frac{1}{2}C_0 V_0^2 = U \qquad (23.10)$$

The work W is stored in the form of potential energy (U) in the electric field between the plates of the capacitor. If the capacitor is discharged, the energy can be recovered from the electric field, e.g. if we connect an electric motor to a charged capacitor, current will flow through the motor, the shaft of the motor will turn and mechanical work will be done. The total amount of mechanical work equals to $\frac{1}{2}C_0 V_0^2$, when the capacitor is completely discharged.

Since the electric field $E_0 (= V_0/d)$ is constant between the plates of the capacitor, the *energy density* U_0, which is stored energy per unit volume, will be constant. Thus,

$$U_0 = \frac{U}{Ad} = \frac{\frac{1}{2}C_0 V_0^2}{Ad} = \frac{1}{2}\varepsilon_0 \left(\frac{V_0}{d}\right)^2 \qquad (23.11)$$

$$= \frac{1}{2}\varepsilon_0 E_0^2 \qquad (23.12)$$

When a dielectric medium is present between the plates of the capacitor, the energy density is given by

$$U_0 = \frac{1}{2}\varepsilon_0 \varepsilon_r E_0^2 \qquad (23.13)$$

Induced Charges

According to Gauss's theorem, the flux of the electric field E_0 through a closed surface is equal to the algebraic sum of charges enclosed by this surface divided by ε_0. We have

$$E_0 = \frac{\sigma}{\varepsilon_0} \qquad (23.14)$$

where σ is the magnitude of *surface charge density* on either plate of the capacitor.

When a dielectric is introduced between the plates of a capacitor, there occurs a reduction in potential difference across the plates, i.e. it implies a decrease in the electric field strength E_0 in the region. The decrease in the electric field strength E_0 can occur only when there is a reduction in the charge on the plates of the capacitor. Since the charges cannot leak off the plates, and hence the reduction in electric field strength suggests that the effect of some of the original charges is annulled by the charges of opposite sign appearing on the two surfaces of the dielectric between the two plates. Obviously, negative charge is induced by the electric field on the dielectric surface adjacent to the positive plate while a positive charge of equal magnitude is induced on the dielectric surface adjacent to the negative plate. Clearly, the electric field is inducing charges on the surfaces of dielectric. When charges of opposite polarity are induced on the surface of dielectric, we say that dielectric is *polarized*. Due to induced surface charges on the dielectric, an induced electric field E_i is developed, which opposes the external applied field E_0 (Fig. 23.2). Thus, the net electric field E_0 in the dielectric has a magnitude given by

$$E = E_0 - E_i \qquad (23.15)$$

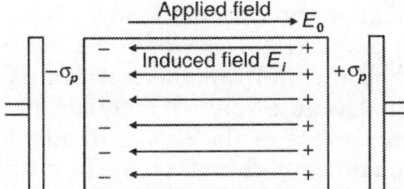

Fig. 23.2: Polarization of a dielectric. Induced charges appear on the opposite side of the dielectric. An induced (internal) field E_i directed opposite to the applied field E_0 appears within the dielectric

Let σ_p be the magnitude of the induced surface charge density (Fig. 23.1b). Obviously, this charge density nullifies the effect of part of the free charge density σ present on the plates of the capacitor.

The relation between the induced electric field E_i and induced charge σ_p is

$$E_i = \frac{\sigma_p}{\varepsilon_0} \qquad (23.16)$$

Now, the net surface charge density that contributes to the electric field E within the dielectric is $(\sigma - \sigma_p)$. On applying Gauss's theorem, one obtains

$$E = \frac{\sigma - \sigma_p}{\varepsilon_0} \qquad (23.17)$$

Using $C/C_0 = \sigma_p$ and $\dfrac{E_0}{E} = \varepsilon_r$, one finds

$$E = \frac{E_0}{\varepsilon_r} = \frac{\sigma}{\varepsilon_0\,\varepsilon_r} = \frac{\sigma}{\varepsilon} \qquad (23.18)$$

where $\qquad \varepsilon = \varepsilon_0 \varepsilon_r \qquad (23.19)$

Now, equating Eqs. (23.17) and (23.18), one finds

$$\frac{\sigma - \sigma_p}{\varepsilon_0} = \frac{\sigma}{\varepsilon_0 \varepsilon_r}$$

or $\qquad \sigma_p = \sigma\left(1 - \dfrac{1}{\varepsilon_r}\right) \qquad (22.20)$

Since $\varepsilon_r > 1$, we have $\sigma_p < \sigma$. We may note that for moderately large values of ε_r, σ_p is nearly equal to σ. The presence of polarization charge diminishes the field E within the dielectric causing $E \gg E_0$.

23.3 POLARISATION

We consider an electrically neutral slab of an isotropic dielectric inserted between the plates of a parallel plate capacitor. Let us imagine that the dielectric is divided into a large number of identical cells of volume dV. Now, under the action of applied external field, charges are induced in each cell and each cell acquires a charge dq. We define the intensity of polarization as the total dipole moment per unit volume of the material. Thus, one can write

$$P = \sum \frac{d\mu}{dV} = \frac{\Sigma d\mu}{V} \qquad (23.21)$$

The magnitude of polarization is directly proportional to the intensity of the electric field, E. Thus, we have

$$P = \chi \varepsilon_0 E \qquad (23.22)$$

where χ is the proportionality constant and is called *dielectric susceptibility* of the material. χ characterizes the ease with which a dielectric material can be influenced by an external field.

In the dielectric within each cell, the induced charge appear on the front and rear faces of each cell. These are called *bound charges*. We may note that the bound charges are not free to migrate through the dielectric. The common wall between two adjacent cells carries a charge $+q$ from one cell and $-q$ from the other cell. Obviously, in the interior of the dielectric these charges cancel. However, the charges on the opposite faces of the dielectric do not cancel. Figure 23.3 depicts this situation and shows that the layer of charge nearest to the boundary of the slab remains and constitutes the polarization charge of surface density σ_p. We can say that the polarized dielectric is equivalent to a big *dipole* consisting of polarization charges separated by a distance d,

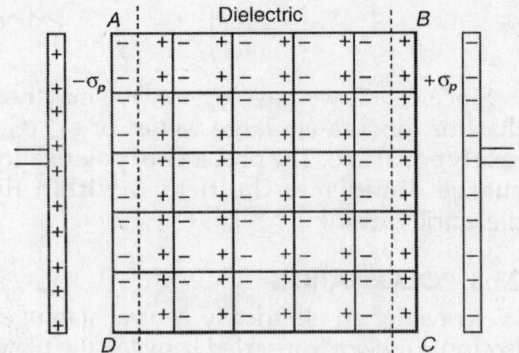

Fig. 23.3: A dielectric slab may be considered to be made up of a large number of identical and infinitely small cells with a dipole moment μ. The bound charges on the walls of the small cells cancel throughout the cell and leaving surface charges only. The magnitude of bound charges left on the sides of the slab is σ_p

which is also the thickness of the slab. The dipole moment of entire slab *ABCD* is

$$\mu = (A\sigma_p)d = \sigma_p V \qquad (23.23)$$

where *A* is the surface area of the slab and *V* its volume. We have from Eq. (23.21),

$$P = \frac{\mu}{V}$$

$$\therefore \qquad P = \sigma_p \qquad (23.24)$$

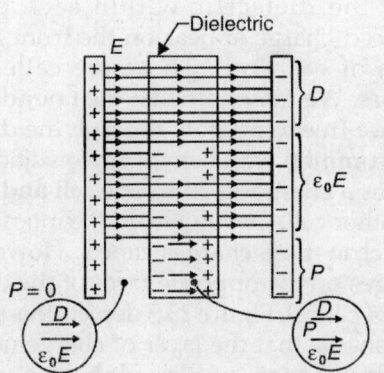

Fig. 23.4: The electric field vector *E* is connected to the free charges, the polarization vector *P* is connected to polarization charges and the displacement vector *D* to all charges

From Eq. (23.24), it is evident that polarization is equal to the surface density of the induced charges in the dielectric. We may note that polarization is vector quantity and its direction is from negative charges to positive charges. We may see that (Fig. 23.4) direction of *P* is the same as the electric field. Due to polarization, an internal field E_i is produced in the dielectric, which opposes the external field and therefore weakens it. We may note that the absolute value of *P* equals the sum of the projections of the dipole moments of all the elementary volumes on the field direction.

23.4 RELATION BETWEEN *E*, *P* AND *D*

If E_i is the field due to polarization of a dielectric, which is opposite in its direction to the external field, then the effective field *E* in the dielectric is given by

$$E = E_0 - E_i = \frac{\sigma}{\varepsilon_0} - \frac{\sigma_p}{\varepsilon_0}$$

Using Eq. (23.24), we obtain

or $$\varepsilon_0 E = \sigma - \sigma_p \qquad (23.25)$$

Let us introduce an auxiliary vector *D*, called the *displacement vector*. Its magnitude is equal to the surface density of charges, i.e.

$$D = \sigma \qquad (23.26)$$

$$\therefore \qquad \varepsilon_0 E = D - P$$

$$\therefore \qquad D = \varepsilon_0 E + P \qquad (23.27)$$

Relation between ε_r and χ:

We have $P = \sigma - \varepsilon_0 E = \varepsilon_0 \varepsilon_r E - \varepsilon_0 E$

or $$P = \varepsilon_0 (\varepsilon_r - 1) E \qquad (23.28)$$

Comparing Eqs. (23.22) and (23.28), we find

$$\chi = \varepsilon_r - 1$$

or $$\varepsilon_r = 1 + \chi \qquad (23.29)$$

Thus, like susceptibility, the dielectric constant is also a measure of polarization of the material. Larger the polarization per unit resultant field, greater will be the dielectric constant of the dielectric.

23.5 INDUCED DIPOLES

When a dielectric is placed in an electric field, the question arises, what will be its effect on an atom?

The nucleus of an atom is about 10^{-15} m in diameter and one can regard it as a point. The electron cloud is about 10^{-10} m in diameter and we may assume that its negative charge is concentrated at its centre. Obviously, the centres of gravity of positive and negative charges in an atom coincide (Fig. 23.5a). This means that such an atom does not produce any electric field of its own.

Let the atom be placed in an electric field of field strength E, then the electron cloud will be displaced in the direction of E. Let this distance be d with respect to the nucleus (Fig. 23.5b). Obviously, the centres of gravity of positive and negative charges in the atom no more coincide (Fig. 23.5b). Clearly, the atom is now equivalent to a system of two charges, $q = Ze$, of equal magnitude but opposite in sign separated by a distance d (Fig. 23.5b). Such a system is usually called an electric dipole or simply a dipole.

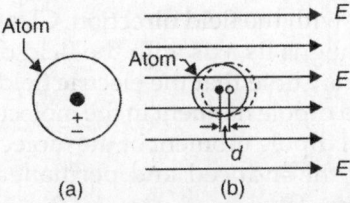

(a)   (b)

Fig. 23.5: (a) The centres of positive and negative charges in an atom coincide and atom does not possess electric dipole moment. (b) Atom placed in an electric field of strength E. Electron cloud is displaced with respect to the nucleus. The centres of negative and positive charges no longer coincide and are separated. An electric dipole is induced

We may note that dipole induced in the atom due to the action of external field is electrically neutral as a whole. However, the induced dipole sets up its own electric field which is opposite in direction to the electric field (Fig. 23.6). The product of the magnitude of the charge (q) and separation distance between any two charges (d) is called the dipole moment (μ), i.e.

$$\mu = qd \qquad (23.30)$$

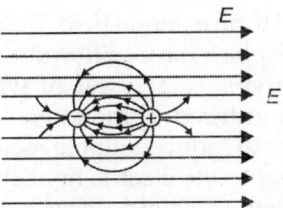

Fig. 23.6: The electric dipole as a whole is electrically neutral but it sets up its own electric field in space which is opposite in direction to the external field

The dipole moment (μ) is a vector directed along the axis of the dipole from the negative charge to the positive charge (Fig. 23.7).

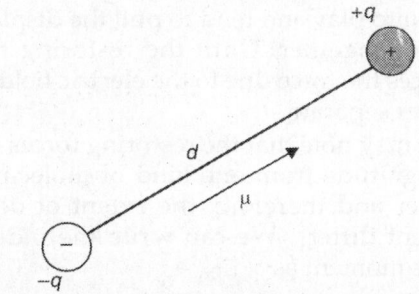

Fig. 23.7: An electric dipole. This consists of two equal and opposite charges separated by a distance d. The dipole moment μ is directed from negative to positive charge

Generalizing the above, we can say that any neutral system of N point charges $Q_1, Q_2,...,$ Q_N occupying a volume having linear dimensions d acts as a dipole. However, the sum of the charges ΣQ_i in the volume should be equal to zero so that the system remains neutral. Now, the dipole moment of such a neutral system of point charges can be represented as

$$\mu = \sum_i Q_i \cdot r_i \qquad (23.31)$$

where r_i is a vector drawn from the origin of coordinate system to the position of charge Q_i. We may note that when the system is really neutral, then μ will be independent of the choice of the origin of the coordinate system.

We have already said that dielectric materials are made up of atoms and molecules which are neutral systems. Whenever an atom or molecule is subjected to an external field, the electric field tends to displace the equilibrium positions of bound charges, as a result of which dipole moment is induced in the atom or molecule. The amount of induced dipole moment will be proportional to the electric field strength, and hence larger the field greater will be the displacement of charges and hence larger the induced dipole moment.

Since the charges are displaced along the the field direction and hence the dipole moment is induced in the same direction. The molecule is said to be *polarized* by the electric field. When a molecule gets polarized, restoring forces due to coulomb attraction come into play and tend to pull the displaced charges together. Until the restoring force balances the force due to the electric field, the charges separate.

We may note that the restoring forces vary in magnitude from one kind of molecule to another and therefore, the extent of dipole moment differs. We can write the induced dipole moment as

$$\mu = \alpha E \qquad (23.32)$$

where α is the proportionality constant and called as the *polarizability* of the molecule. The polarizability (α) characterizes the capacity of the electric charges in the molecule to suffer displacement in external electric field. 'α' has the dimensions of volume. As soon as the electric field is switched off, the induced dipole moment vanishes.

23.6 PERMANENT DIPOLES

There are some molecules known as polar molecules. In these molecules, the centres of gravity of the charges of opposite sign are separated even in the absence of an external electric field. Such molecules are said to have intrinsic dipole moment and carry permanent dipoles.

When a molecule having intrinsic dipole moment is placed in a uniform electric field E as shown in Fig. 23.8, the electric field E exerts a force $+qE$ on the charge $+q$ and $-qE$ on $-q$. Since the two forces acting on the dipole are equal and opposite to each other, and hence

the net force on it is zero. Thus, there is no translational force on the dipole in a uniform field. However, the two forces are antiparallel and constitute a couple which tends to rotate the dipole. The torque on the dipole is

$$\tau = qEd \sin\theta = \mu E \sin\theta$$

or $$\tau = \mu \times E \qquad (23.33)$$

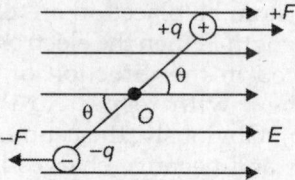

Fig. 23.8: An electric dipole in uniform electric field experiences a torque equal to $\mu E \sin\theta$, where θ is the angle between μ and E

Summarizing, we can say that a dipole in a uniform electric field does not undergo translational motion but rotates in an attempt to align with the field direction. Clearly, a free dipole aligns its axis with the field direction.

We may note that the electric field can also induce a dipole moment in the molecule. Thus, the total dipole moment of the molecule is the sum of the induced and permanent dipole moments. Thus,

$$\mu_{total} = \mu_{ind} + \mu_{per}$$

In case of polar molecules, $\mu_{ind} << \mu_{per}$ and

$$\mu_{total} \approx \mu_{per}$$

23.7 PHASE DIFFERENCE AND ELECTRIC LOSS

A dipole tends to align itself along the direction of applied electric field and for AC fields tends to follow the field and be in a phase with it. However, the interaction of this dipole with other dipoles in the medium prevents this and this leads to dielectric loss. This loss appears as heat. This dielectric loss is connected with ε_r'', the imaginary part of the dielectric constant. The rate of loss of energy in unit volume of the material is obtained as

$$W = \frac{1}{2}\varepsilon_0 \varepsilon_r'' \omega E_0^2 \qquad (23.34)$$

Obviously, the energy loss is proportional to ε_r''. Usually the dielectric loss is expressed in terms of a quantity called *the loss tangent*, $\tan \delta$. This is defined as $\tan \delta = \varepsilon_r''/\varepsilon_r'$. The angle δ is the complement of angle ϕ, i.e. $90° - \phi = \delta$, where ϕ is the angle between the applied field and the resultant current vector (Fig. 23.9)

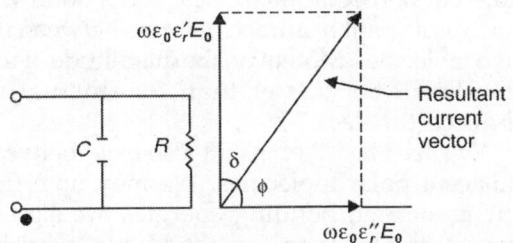

Fig. 23.9: Vector relationship between the field vector E_0, the current vector $\omega \varepsilon_0 \varepsilon_r' E_0$ which leads the field by 90° and the current vector $\omega \varepsilon_0 \varepsilon_r'' E_0$ which is in phase with the field. The angle δ is shown

From alternating current studies, we have

$$\text{Power} = V_R I_R \cos \phi$$

where V_R and I_R are RMS values of voltage and current, and for a circuit containing capacitance C and resistance R, we have

$$\tan \delta = \frac{1}{\omega C R} \tag{23.35}$$

Using Eq. (23.2), one obtains for a dielectric

$$W = \frac{1}{2} \varepsilon_0 \varepsilon' \tan \delta \, \omega E_0^2 = \frac{1}{2} \frac{E_0^2}{\rho} \tag{23.36}$$

where ρ is the resistivity. Thus,

$$\tan \delta = \frac{1}{\omega \varepsilon_0 \varepsilon' \rho} \tag{23.37}$$

The energy losses in a dielectric material are due to: (i) *ionization* (ii) *leakage current* (iii) *polarization* (iv) *structural inhomogeneity*.

Ionization losses occur in gases and solids having pores with entrapped gases. With the rise in field strength applied to a gas, a stage is reached when the gas molecules gets ionized due to collisions. This leads to enhanced conduction leading to dielectric losses. At low or ordinary electric fields, the conductivity is low

and hence the loss is also small [Eq. (23.36)]. For example, for $\rho \approx 10^{16}$ Ωm, $\varepsilon_r' \approx 1$ and at 100 Hz, $\tan \delta < 2 \times 10^{-8}$. In case of solids with gas inclusions, the loss tangent increases with voltage from the voltage required for the ionization of gas molecules to the value of voltage at which ionization is complete. When such inclusions are high, there are chances of the failure of the dielectric.

Due to the leakage current, the losses in highly conducting liquids and solids are high [Eq. (23.35)]. In case of neutral liquids, the losses will be small as their conductivity is low. Purity of oil is quite important as otherwise dielectric losses increase due to conduction. Pure transformer oil has $\tan \delta \approx 0.0005$.

Dipole losses in the radiofrequency region are usually due to dipole rotation or ions jumping from one equilibrium position to another. The dielectric loses associated with ions, the frequencies of which fall in the infrared region, are usually referred to as *infrared absorption*. Similarly, the losses in the optical region, associated with the electrons, are referred to as *optical absorption*.

23.8 DIELECTRIC STRENGTH

As the voltage applied across a dielectric is increased, dielectric loses its insulation property at a certain voltage called the *breakdown voltage* (V_{br}). The corresponding field strength (E_{br}) is given by

$$E_{br} = \frac{V_{br}}{t} \tag{23.38}$$

Here t is the thickness of the material. This voltage, V_{br} per unit thickness of the material is called the *dielectric strength* and usually expressed in volts/millimetre or kV/millimetre or mega volt/m. The breakdown of a dielectric is due to the collision of accelerated **electrons** or ions with molecules. At relatively **high fields**, the electrons in the dielectric gain enough energy to knock other charged particles and make them available for conduction.

Dielectric failure involves deterioration and electric breakdown or cascading. Resistance to electronic breakdown in a dielectric is *intrinsic dielectric strength* of the material.

Intrinsic breakdown in a dielectric is caused by imperfection. When the applied potential across the dielectric becomes high enough, a few electrons are broken loose at points where their bonds are strained by the presence of imperfection. Upon being freed, these electrons are accelerated rapidly through the material. The dielectric material loses completely its insulating capacity as a result of this process.

Obviously, dielectric strength is structure sensitive property of the material. We may note that the actual thickness of the material affects the breakdown potential (V_{br}) per unit thickness, i.e. dielectric strength. Usually thicker materials possess lower dielectric strength than thinner materials.

The fundamental breakdown mechanisms in a solid are: (i) *intrinsic breakdown* (ii) *thermal breakdown* (iii) *discharge breakdown* (iv) *electrochemical breakdown*.

23.9 POWER FACTOR

The ratio of power loss in a material to the product of applied voltage and current is called power factor. The power factor and loss angle depend upon the nature of the material, applied voltage, humidity, temperature and frequency.

23.10 POLAR AND NON-POLAR MATERIALS

Permanent dipole moment exist in some molecules by virtue of an asymmetrical arrangement of positively and negatively charged regions. Such molecules are termed polar molecules. For example, in the HCl molecule (Fig. 23.10), the electron of hydrogen atom spends more time moving around the Cl atom than around the H atom. Therefore, the centre of negative charges does not coincide with that of positive charges, and the molecule has a dipole moment directed from the Cl atom to the H atom, that is, we may write H^+Cl^-. The electric dipole of HCl molecule is $p = 3.43 \times 10^{-30}$ C m. In the CO molecule (Fig. 23.10b), the charge distribution is only slightly asymmetric and the electric dipole moment is relatively small, about 0.4×10^{-30} C m, with the carbon atom corresponding to the positive and the oxygen atom to the negative end of the molecule.

Polar molecules can also induce dipoles in adjacent non-polar molecules, and a bond will form as a result of attractive forces between the two molecules. Moreover, the magnitude of this bond will be greater than for fluctuating induced dipoles.

Van der Waals forces will also exist between adjacent polar molecules. We may note that the associated bonding energies are significantly greater than for bonds involving induced dipoles.

The strongest secondary bonding type, the hydrogen bond, is a special case of polar molecule bonding. We find that it occurs between molecules in which hydrogen is covalently bonded to fluorine (as in HF molecule), oxygen (as in H_2O molecule) and nitrogen (as in NH_3 molecule). For each H—F, H—O or H—N bond, the single hydrogen electron is shared with the other atom. Obviously, the hydrogen end of the bond is essentially a positively charged bare proton that is unscreened by any electrons. This highly positively charged end of the molecule is capable of a strong attractive force with the negative end of an adjacent molecule (Fig. 23.11). Clearly, this single proton forms a bridge between two negatively charged atoms. We may note that the magnitude of the hydrogen bond is generally greater than that of the other types of secondary bonds and may be as high as 0.52 eV/molecule (= 51 kJ/mol). We find that the melting and boiling temperatures for hydrogen fluoride and H_2O are abnormally high in light of their low molecular weights, as a consequence of hydrogen bonding.

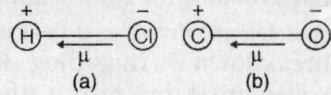

(a)   (b)

Fig. 23.10: Polar diatomic molecules

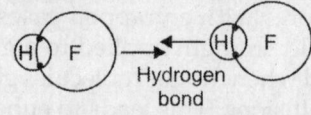

Fig. 23.11: Representation of hydrogen bond in HF molecule

Non-polar Materials

Most of the hydrocarbons are non-polar. In non-polar materials, the molecules which are usually diatomic and composed of two atoms of the same type may be represented as positive nuclei of charge q surrounded by a symmetrically distributed negative electron cloud of charge $-q$. In the absence of an applied electric field, the centres of gravity of the positive and negative charge distributions coincide. When the molecules are subjected to an external electric field the positive and negative charges experience electric force tending to move them apart in the direction of external electric field. The distance moved is very small ($\sim 10^{-10}$ m) since the displacement is limited by restoring forces which increases with increasing displacement. The centres of positive and negative charges no longer coincide and the molecules are said to be polarized.

Polarization

When a material is placed in an electric field (Fig. 23.12), e.g. between the plates of a condenser, the field strength of charged particles within the material interact with the electric field. If the material is a conductor, some of the free electrons simply move to the side nearest the positive electrode until they totally counteract the applied electric field. Obviously, no field is left within the material. The displacement of charged particles occurs simultaneously, bringing about equilibrium.

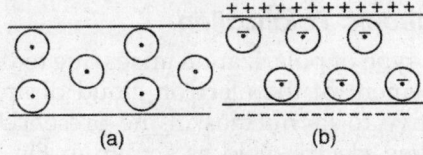

(a)   (b)

Fig. 23.12: Electronic polarization (a) In the absence of field (b) In the presence of electric field, the formation of induced dipole

If the material is non-conducting or insulator (dielectric), electrons can only be displaced locally as they are bound to the individual atoms. This local displacement of electrons, however, is sufficient to polarize the material. The negative electron cloud is displaced in each atom relative to the positive nucleus, thereby creating a small induced dipole whose negative pole is toward the positive side of the electric field. All dielectric materials are subjected to such *electronic polarization* (Fig. 23.12).

Induced polarization also occurs in ionic materials. In an ionic crystal, for example the negative ions are attracted towards the positive side and *vice versa* as illustrated in Fig. 23.13. NaCl is a good example of ionic polarization. Its interionic separation is 0.1 nm and dipole moment is of the order of 10^{-29} Cm.

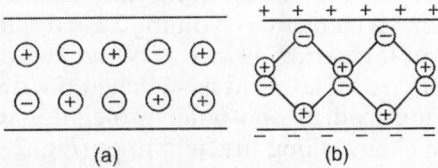

(a)   (b)

Fig. 23.13: Ionic polarization (a) In the absence of external field (b) Ions displacement towards opposite charged plates in an external field

There are many molecular structures which have permanent electronic dipoles due to asymmetric nature of their atomic structures. Few examples of such molecules are H_2O, HCl, polyvinylchloride, etc. Some molecules also contain permanent charges which produce dipole moments in the individual molecules. There are many polymers of this type (polymers) and glasses contain dipoles. The permanent molecular dipoles in such materials can rotate about their axis of symmetry to align with an applied field which exert a torque in them. This is known as *orientation polarization*. This type of polarization is rarely perfect as thermal fluctuations do not permit all the dipoles to align with the applied field. With the rise in temperature the degree of polarization decreases. In the process of orientation, polarization carriers collect at the boundaries between the constituent materials resulting in local distortion of the applied field. This is called as *space–charge polarization*.

The dipole moment of a single pair of charges is given by

$$\mu = qd \qquad (23.39)$$

where q is one of the charges and d is the distance between charges.

Atomic View of Polarization

Let us consider a slab of a dielectric located between the two plates of a parallel plate capacitor. When there is no external field, each volume of the dielectric has no dipole moment. When the dielectric is a non-polar material, the constituent molecules of the dielectric do not possess intrinsic dipole moment and if the material is a polar material, the individual molecular dipoles are randomly oriented so that in an elementary volume $\Sigma\mu = 0$. Clearly, the polarization is zero. When we apply electric field, i.e. field is switched on, dipoles are induced in non-polar molecules which form chains along the field lines (Fig. 23.14).

We have the polarization

$$P = N\mu = N \propto E \qquad (23.40)$$

$$\text{(for non-polar molecules)}$$

where N is the number of molecules per unit volume.

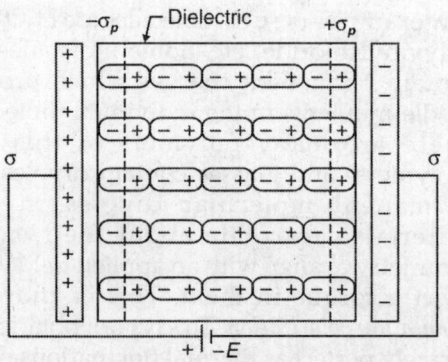

Fig. 23.14: Atomic view point of polarization of a dielectric. The induced charge density (σ_p) on the dieletric is less than the free charge density (σ) on the plates

In case of polar dielectrics, the molecular dipoles experience a torque that tends to align them with a electric field direction. Total alignment is not achieved due to the disordering effects of the thermal agitation. In the direction of the field, and average alignment $<\mu>$ is achieved. The polarization is thus given by

$$P = N <\mu> \text{ (for polar molecules) (23.41)}$$

Obviously, the action of electric field brings the dipoles into a certain ordered arrangement in space. One finds that the ends of adjacent dipoles carrying opposite charges neutralise each other. However, the charges of dipole ends terminating on the opposite face of the slab remain uncompensated. This is why the application of an electric field to the dielectric produces a displacement of charges within the dielectric material through a progressive orientation of intrinsic or induced dipoles. This is termed dielectric polarization.

Sources of Polarization

As stated earlier, the net polarizability of a dielectric material results mainly from the following three types of contributions:

   i. Electronic polarization

   ii. Ionic polarization

   iii. Dipolar or orientational polarization

The extent to which a particular polarizability contributes depends on the nature of dielectric and the frequency of the applied electric field. One or two of the said polarizations are always present and at a particular temperature or frequency of the applied field. However, one or another type of source of polarization may contribute in a large measure to the total polarization.

Electronic Polarization

This type of polarization arises due to the displacement of the electron cloud of an atom relative to its nucleus in the presence of an applied electric field as shown in Fig. 23.15. Electronic polarization occurs in all dielectrics and sets in over a very short period of time of the order of 10^{-14} to 10^{-15} s and independent of temperature. This type of polarization occurs in all dielectrics for any state of aggregation. The electronic polarization is given by

$$P_e = N\alpha_e E \qquad (23.42)$$

where α_e is the electronic polarizability. One can obtain the contribution of P_e to the dielectric constant as follows:

$$\varepsilon_r = 1$$

or

$$\varepsilon_r = 1 + \frac{N\alpha_e}{\varepsilon_0} \qquad (23.43)$$

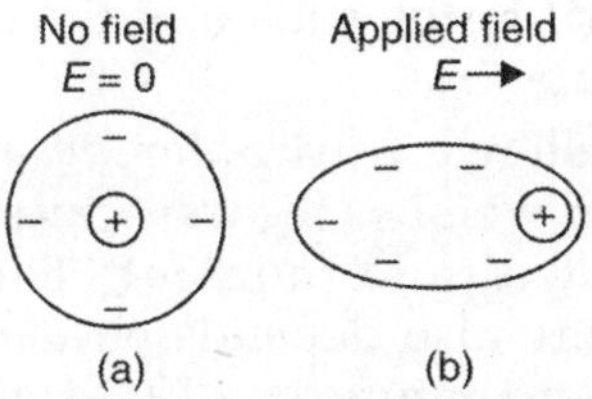

No field   Applied field
$E = 0$ $E \rightarrow$

(a)   (b)

Fig. 23.15: Electronic polarization in the presence of an external electric field

The expression (23.43) indicates the dielectric constant due to electronic polarization alone, and thus gives the dielectric of a non-polar gaseous dielectric. This reveals that the dielectric constant of a non-polar gas depends on the polarizability of a molecule and the number of molecules in unit volume of the dielectric. For a monoatomic gas, the electronic polarization is given by

$$\alpha_e = 4\pi\varepsilon_0 R^3 \qquad (23.44)$$

$$\therefore \qquad \varepsilon_r = 1 + 4\pi N R^3 \qquad (23.45)$$

where R denotes the radius of the atom.

Ionic Polarization

The ionic polarizability arises due to displacement of a charged ion relative to other ions in a solid or ionic crystal. In other words, we can say that ionic polarization occurs by the elastic displacement of positive and negative ions from their equilibrium positions, e.g. NaCl. NaCl molecule consists of Na^+ ions bound to Cl^- ions through ionic bond. If the interatomic distance is d, the molecule exhibits an intrinsic dipole moment equal to ed. When an external electric field is applied to NaCl molecule, the Na^+ and Cl^- are displaced in opposite direction until ionic bonding forces stop the process.

The dipole moment of the molecule increases consequently. When we reverse the field direction, the ions move closer and again the dipole moment undergoes a change.

Clearly, dipoles are induced. Assuming the forces near equilibrium as simple harmonic, the displacement Δx in the presence of electric field E is given by

$$\beta\Delta x \cong eE \qquad (23.46)$$

where β is force constant. The induced dipole moment is proportional to the applied field, i.e.

$$\mu_i = \alpha_i E \qquad (23.47)$$

where α_i is known as *ionic polarizability*. We may note that the ions also experience electronic polarization. For most materials, ionic polarization is less than the electronic polarization. Typically,

$$\alpha_i = \alpha_e / 10$$

We know that ionic polarization is given by

$$P_i = N\alpha_i E \qquad (23.48)$$

For $\beta \cong 20\ \mathrm{N\,m^{-1}}$, $\alpha_i = 10^{-39}\ \mathrm{F\,m^2}$. The ionic contribution is important at low frequencies. NaCl has $\varepsilon_r \cong 5.6$ at low frequencies, whereas the value reduces to about 2.25 at optical frequencies. Ionic polarizability takes about 10^{-11} to 10^{-14} s to build up and almost independent by temperature variations. Figure 23.16 illustrates the phenomenon of ionic polarization.

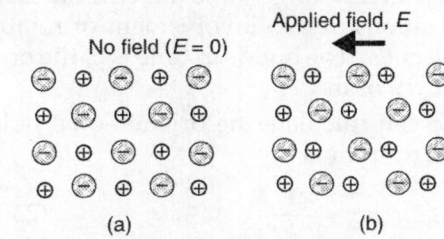

Fig. 23.16: Ionic polarization: (a) $E = 0$, unpolarized dielectric (b) In an applied field, E, ionic polarization takes place due to relative motion of electrically charged positive (+) and negative (–) ions

Orientation or Dipolar Polarization

This is the characteristic of polar dielectrics which consist of molecules having permanent dipole moment, e.g. water molecule H_2O. In the absence of an external electric field, the dipoles have random orientations and there is no net polarization (Fig. 23.17a). However, when the electric field is applied, the dipoles orient themselves and the field produce orientational or dipolar polariza-tion. In the case of the electronic and ionic polarizations, external field force is balanced by a restoring force due to coulomb attraction, but for orientation polarization restoring forces do not exist. However, the dipolar alignment is counteracted by thermal agitation and an equilibrium state is reached wherein the different dipoles make all possible angles varying from 0 to π radians with the field direction (Fig. 23.17b). Even in the case of liquids or gases, where all molecules are free to rotate, a complete alignment cannot be achieved due to the randomizing effect of the temperature. However, it is estimated that it is enough if one molecular dipole in 10^5 completely aligns with the field to produce orientation polarization of the order of electronic polarization.

Clearly, orientation polarization is strongly temperature dependent. The rotation of polar molecules in the case of solids may be highly restricted by the lattice forces and leads to a great reduction in their contribution to orientation polarization. Due to this reason, the dielectric constant of water is about 80, whereas that of ice is about 10 only.

The process of orientation polarization takes relatively longer time than the other two polarizations due to involvement of rotation of molecules. The build-up time is of the order of $\sim 10^{-10}$ s or more.

One can calculate the orientational polarization α_o from

$$\alpha_o = \frac{\mu^2}{3kT} \qquad (23.49)$$

Orientational polarization is given by Debye's law as

$$P_0 = \frac{N\mu^2 E}{3kT} \qquad (23.50)$$

Obviously, the orientational polarization is inversely proportional to the temperature and proportional to the square of the permanent dipole moment.

The existence of dipolar polarizability depends on whether the molecules possess a permanent dipole moment. The dipolar polarizability also depends on temperature. At room temperature, $\alpha_d \sim 10^{-39}$ F m^2 which is comparable to electronic polarizability.

Total Polarization

The total polarizability of a dielectric is given by

$$\alpha = \alpha_e + \alpha_i + \alpha_o \qquad (23.51)$$

or

$$\alpha = \alpha_e + \alpha_i + \frac{\mu^2}{3kT} \qquad (23.52)$$

The total polarization P may be expressed as

$$P = P_e + P_i + P_o$$
$$= N\left[\alpha_e + \alpha_i + \frac{\mu^2}{3kT}\right]E \qquad (23.53)$$

One can express the dielectric constant of a polar gaseous dielectric as

$$\varepsilon_0(\varepsilon_r - 1) = N\left(\alpha_e + \alpha_e\alpha_i + \frac{\mu^2}{3kT}\right) \qquad (23.54)$$

A plot of ε_r or P with $1/T$ is a straight line (Fig. 23.18)

As stated earlier, it is possible for one or more of the contributions to the polarization to be either negligible in magnitude relative to the others or absent. For instance, the orientation polarization does not exist in non-polar dielectrics and ionic polarization in covalently bonded materials. In polar dielectrics, the electronic polarization is negligible as compared to dipolar polarization.

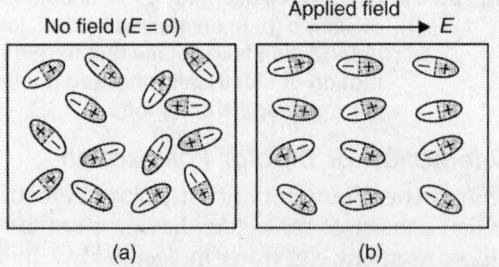

No field ($E = 0$)   Applied field $\longrightarrow E$

(a)   (b)

Fig. 23.17: Dipolar polarization (a) No field ($E = 0$), molecular dipoles are randomly oriented (b) Dipoles are partially aligned in the applied field

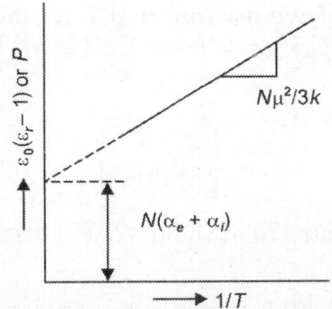

Fig. 23.18: Variation of P or ε_r with $1/T$. The permanent dipole moment can be determined from the slope of the curve

Internal Field in Solids

The internal field (E_i) defined as the electric field acting at the location of given atom, is equal to the sum of the electric fields created by the neighbouring atoms and the applied field. We know that the atoms in gases are in constant random motion and are separated by sufficiently large distances. Therefore, one can neglect the interaction between atoms. Obviously, when an external field E is applied, the intensity of the electric field experienced by a given atom in the gas equals to the applied field E. In solids and liquids, the atoms are so close that they touch each other leading to a strong interaction between them. Since the atoms are surrounded on all sides by other polarized atoms, the internal intensity of the field at a given point of the material is

$$E_i = E + E' \qquad (23.55)$$

One can evaluate the value of E' by the summation of all the effects of the surrounding atoms. Let us consider a one dimensional solid consisting of a string of equidistant identical atoms, each of polarizability α_e (Fig. 23.19)

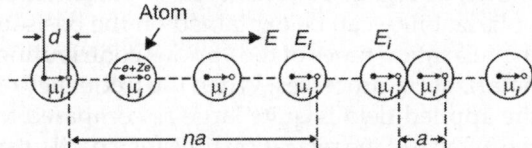

Fig. 23.19: A one dimensional solid. The internal field E_i is different from the external field, $E\ (=V/l)$.

Let the external field E is applied in a direction parallel to the string. Let us determine the internal field E_i experienced by any one of the atom (say A). Obviously, the field experienced by all other atoms will also be same, and from the consideration of symmetry, E_i will be parallel to E. The induced dipole moment in each of the atoms of the string is as

$$\mu_{\text{ind}} = \alpha_e E_i \qquad (23.56)$$

Now, the field at A due to the dipole induced in an atom located at distance na from it is given by

$$E_n = \frac{Ze}{4\pi\varepsilon_0}\left[\frac{1}{(na)^2} - \frac{1}{(na+d)^2}\right]$$

$$= \frac{Ze}{4\pi\varepsilon_0}\left[\frac{(na+d)^2-(na)^2}{(na+d)^2(na)^2}\right]$$

$$= \frac{2Zed}{4\pi\varepsilon_0(na)^3} \qquad (\because d \ll na)$$

$$= \frac{\mu_i}{2\pi\varepsilon_0(na)^2} \qquad (\because \mu_i = Zed)$$

$\therefore$ The total field E_i at A is obtained as

$$E_i = E + \frac{\mu_i}{2\pi\varepsilon_0}\left[2\sum_{n=1}^{\infty}\frac{1}{(na)^3}\right]$$

In order to take into account the atoms to the left and right of A, we have multiplied by a factor of 2. Now,

$$E_i = E + \frac{\mu_i}{\pi\varepsilon_0\alpha^3}\sum_{n=1}^{\infty}\frac{1}{n^3}$$

$$= E + \frac{1.2\mu_i}{\pi\varepsilon_0\alpha^3} \qquad (23.57)$$

Clearly, the combined effect of induced dipoles of neighbouring atoms is to produce a net field at the location of a given atom which is larger than the applied field. From Eq. (23.57), we can see that greater the polarizability a_e or smaller the intermolecular spacing a, larger is the internal field.

One can determine the local field in three dimensional solid with the help of the structure of the solid. An accurate calculation of the internal field in solids and liquids is, in general, very complicated. One finds the general expression for E_i very much similar to the expression (23.57) where the number N density of atoms replaces $1/a^3$. As $P = N\mu_i$, and hence we can write the general expression as

$$E_i = E + \frac{\gamma P}{\varepsilon_0} \qquad (23.58)$$

where γ is the proportionality constant and termed the *internal field constant*.

We may note that the value of γ is independent of the internal arrangement of atoms in dielectric. In general, $\gamma \approx 1$. In the case of an infinite chain of molecules, $\gamma = 1.2/\pi$. One can evaluate γ in other simple cases also. In case of crystals having cubic symmetry, $\gamma = 1/3$ and the internal field is given by

$$E_i = E + \frac{P}{3\varepsilon_0} \qquad (23.59)$$

This is called as *Lorentz field*.

Clausius–Mossotti Relation

We now consider the simple case of an elemental solid dielectric which exhibit only electronic polarizability. Silicon, germanium and diamond are such solids which are made up of single type of atoms. The electronic polarizability per atom, α_e is related to the bulk polarization P as

$$\alpha_e = \frac{P}{NE_i} \qquad (23.60)$$

where N is the number of atoms/m^3 and E_i is the local field. Using Eq. (23.58), Eq. (23.60) becomes

$$\alpha_e = \frac{P}{N\left(E + \dfrac{\gamma P}{\varepsilon_0}\right)} \qquad (23.61)$$

But

$$E = \frac{P}{\varepsilon_0(\varepsilon_r - 1)} \qquad (23.61a)$$

Now, if we assume that internal field is Lorentz field, $\gamma = 1/3$ and Eq. (23.61) becomes

$$\alpha_e = \frac{P}{N\left(E + \dfrac{P}{3\varepsilon_0}\right)} \qquad (23.62)$$

Using Eq. (23.61a), Eq. (23.62) becomes

$$\alpha_e = \frac{P}{N\left[\dfrac{P}{\varepsilon_0(\varepsilon_r - 1)} + \dfrac{P}{3\varepsilon_0}\right]}$$

or

$$\frac{N\alpha_e}{E_0} = \frac{3(\varepsilon_r - 1)}{(\varepsilon_r + 2)}$$

or

$$\frac{\varepsilon_r - 1}{\varepsilon_r + 1} = \frac{N\alpha_e}{3\varepsilon_0} \qquad (23.63)$$

This is known as the *Clausius–Mossotti relation*. It relates the dielectric constant to the atomic polarizability for non-polar solids provided the condition of cubic symmetry holds. In a more general form, it is expressed as

$$\frac{\varepsilon_r - 1}{\varepsilon_r + 1} = \frac{1}{3\varepsilon_0}\sum_j N_j \alpha_j \qquad (23.64)$$

The measured values of ε_r for diamond, silicon and germanium are 5.68, 12 and 16 respectively.

Frequency Dependence of Total Polarizability

The total polarizability of a dielectric is given by

$$\alpha = \alpha_e + \alpha_i + \alpha_o \qquad (23.65)$$

It decreases with increase in frequency as shown in Fig. 23.20. This type of variation of polarizability can be explained on the basis of the relaxation times of the various contributing polarization processes. When the frequency of the applied field is quite large as compared to the inverse of the relaxation time for a particular polarization process, the contribution of that process to polarizability is negligible. As relaxation time is maximum for the dipolar process and minimum for the electronic process, the dipolar contribution disappears first followed by ionic electronic contributions.

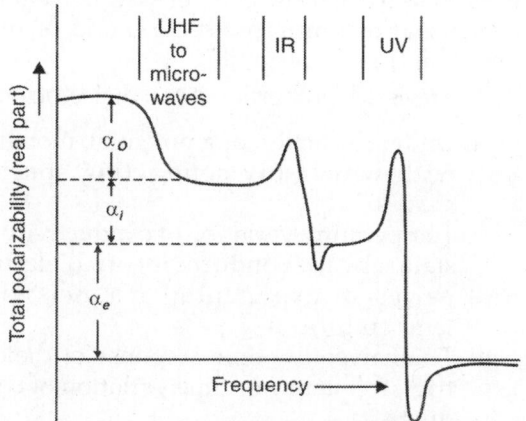

Fig. 23.20: Frequency dependence of total polarizability. The curve also shows the various contributions

Frequency Dependence of the Dielectric Constant

In several practical situations, a dielectric is subjected to an alternating field. As an AC changes its direction with time and with each direction reversal, the polarization components are required to follow the field reversals in order to contribute to the total polarization of the dielectric. Obviously, the total polarization depends on the ability of dipoles to orient themselves in the direction of field during each alteration of the field. The relative permeability which is a measure of the polarization shows marked differences in the behaviour at different frequencies of the electric field for a polar dielectric as shown in Fig. 23.21. We can see that in audio frequency region, all the three types of polarization are possible and the dielectric is characterized by a polarizability $\alpha = \alpha_e + \alpha_i + \alpha_o$ and the polarization $P = P_e + P_i + P_o$. At low frequencies, one finds that the dipoles will get sufficient time to orient themselves completely along the instantaneous direction of the field. This orientation of the dipoles occurs first in one direction and then in the other, following the

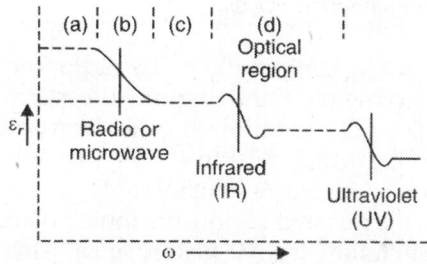

Fig. 23.21: Variation of relative permeability with angular frequency of the alternating field applied across a dielectric. In the optical region (d) only electronic polarization contributes to ε_r, in the microwave region (c) both electronic and ionic polarization contribute to ε_r and in radio frequency region (a) orientation polarization also contributes to ε_r

changes in the direction of the applied electric field E (Fig. 23.22). The average time taken by the dipoles to orient in the electric field direction is known as *relaxation time* (τ). The reciprocal of τ, i.e. τ^{-1} is called the *relaxation frequency* (τ^{-1}). When the frequency of the applied electric field is much higher than the relaxation frequency of the dipoles, the dipoles cannot reverse fast enough. However, when the dipole relaxation time (τ) is less than half the period of electric field T, i.e. $T/2$, the dipole can follow alterations in AC electric field and contribute to orientation polarization. Consequently, the orientation polarization, which is effective at low frequencies, gets damped out for higher AC field frequencies

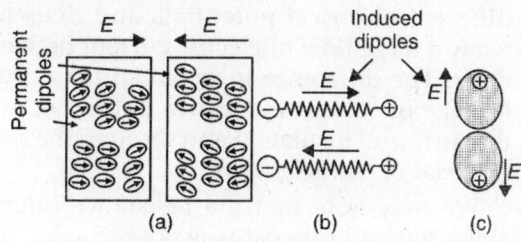

Fig. 23.22: The behaviour of (a) Aermanent (b) and (c) Induced dipoles in an applied AC electric field. The orientation of dipoles first takes place in one direction and then in other in accordance with the changes in the direction of the electric field

($f_{field} > f_{relax}$). Normally in the radio frequency or microwave band region, the permanent dipoles fails to follow the field reversal and the polarization falls to $P = P_i + P_e$ only. As a result, ε_r decreases considerably.

In the infrared region, the ionic polarization fails to follow the AC field reversals due to the inertia of the system and hence the contribution of ionic polarizability becomes almost negligible and total polarization $P = P_e$ only. In the optical region, the electron cloud follows the AC electric field variations and the dielectric material exhibits an electronic polarizability α_e and the relative permeability,

$$(\varepsilon_r)_{opt.reg.} = n^2 \qquad (23.66)$$

i.e. $(\varepsilon_r)_{opt. reg.}$ is equal to the square of the refractive index n of the dielectric. In the ultraviolet region, the electron cloud too fails to follow the electric field alterations and electronic field contribution to the polarization ceases and total polarization, $P = 0$. Above the ultraviolet range, ε_r approaches unity, i.e.

$$(\varepsilon_r)_{X\text{-ray}} = 1 \qquad (23.67)$$

We take the example of water. The dielectric constant of water at low frequency is generally referred to as *static dielectric constant*. At room temperature, its value is about 80. In the optical region, it falls to about 1.8.

23.11 INSULATION RESISTANCE

This separates a number of conductors at different electrical potentials and does not allow a large flow of electric current between them. The difference in potential may cause leakage of current along two paths, i.e. over the surface of insulation and through the solid material of the insulators.

We may note that the resistance offered along the said two paths is not the same due to the nature of the material. The former is due to the surface resistivity of the material, whereas the latter is the result of the volume resistivity of the material. The combined effect of two is called the insulation resistance of the material.

We may define the surface resistivity as the resistance between the two opposite edges of a square of unit area of insulation surface. The volume resistivity is defined as the resistance presented to the flow of an electric current by a material of unit cross-section and of unit length at 0°C.

The resistive properties of real dielectrics are:

i. Under the action of a potential, mobility with which any conducting species moves.

ii. Temperature variation of conductivity to state why the conductivity of a dielectric varies in exponential manner with temperature.

iii. To analyse the time response of dielectrics, frequency or time variation of conductivity.

iv. While designing an equipment, the knowledge of breakdown or the highest voltage a dielectric will withstand is of interest. Usually insulators breakdown at 10^6 V/m to 10^9 V/m to DC at 20° C.

23.12 CLASSIFICATION OF DIELECTRICS

One can classify dielectrics into three groups as follows:

Class	Effect of electric field
Simple dielectrics	Creates dipoles
Paraelectrics	Orients dipoles
Ferroelectrics	Orients domains of aligned permanent dipoles

Ferroelectrics

The group of dielectric materials called *ferroelectrics* exhibit spontaneous polarization i.e. polarization in the absence of an electric field. In a sense, ferroelectrics are the electric analog of the ferromagnets, which may display permanent magnetic behaviour. In ferroelectrics, the polarization can be changed and even reversed by an external electric field. The reversibility of the spontaneous polarization is due to the fact that the structure of a ferroelectric crystal can be derived from a non-polarized structure by small displacement of ions. In most ferroelectric crystals, this non-polarized structure becomes stable if the crystal is heated above a critical temperature, the *ferroelectric Curie temperature* (T_c), i.e. the crystal undergoes a phase transition from the polarized phase (ferroelectric phase) into an unpolarized phase (paraelectric phase). The change of the spontaneous polarization at T_c can be continuous or discontinuous. The T_c of different types of ferroelectric crystals range from a few degrees absolute to a few hundred degrees absolute.

From a practical standpoint, ferroelectrics can be divided into two classes. In this first class of ferroelectrics, polarization can occur along only one crystal axis, e.g. Rochelle salt, KH_2PO_4, $(NH_4)_2SO_4$, guanidine aluminium sulphate hexahydrate, glycine sulphate, colemanite and thiourea. In this second class of ferroelectrics, spontaneous polarization can occur along several axes that are equivalent in the paraelectric phase, e.g. $BaTiO_3$-type (or perovskite type) ferroelectrics, $Cd_2Nb_2O_7$, PbN_2O_6, certain alums, such as methyl ammonium alum and $(NH_4)_2 Cd_3(SO_4)_3$.

One can also place ferroelectrics as *proper ferroelectrics and improper ferroelectrics*. In proper ferroelectrics, for example $BaTiO_3$, KH_3PO_4 and Rochelle salt, the spontaneous polarization is the order parameter. In improper ferroelectrics, the spontaneous polarization

can be considered as a by-product of another structural phase transition. The properties of some representative ferroelectrics are given in Table 23.1.

The spontaneous polarization in ferroelectrics can occur in at least two equivalent crystal directions, thus a ferroelectric crystal consists in general of regions of homogeneous domains. Ferroelectrics of the first class consist of domains with parallel and antiparallel polarization (Fig. 23.23a), whereas ferroelectrics of the second class can assume much more complicated domain configurations (Fig. 23.23b). The region between two adjacent domains is called a *domain wall*. Within this wall, the spontaneous polarization changes its direction.

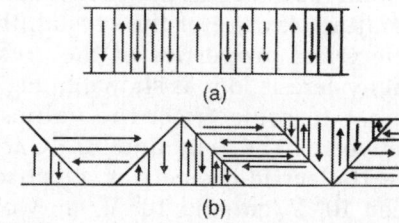

Fig. 23.23: Domain configurations (simplified) encountered in ferroelectric (a) First class (b) Second class

Table 23.1: Properties of some representative ferroelectric crystals

Group	Crystal	Curie temperature T_c (K)	P_s $C/m^2 \times 10^{-2}$	At T (K)
Ilmenites and	GeTe	670	–	–
Perovskites	$LiNbO_3$	1480	71	296
	$KNbO_3$	710	30	600
	$BaTiO_3$	393	26	300
	$SrTiO_3$	32	3	4.2
KDP type	KH_2PO_4 (KDP)	123	4.7	100
	KD_2PO_4	213	5.5	100
	$Rb_2H_2PO_4$	147	5.6	90
	KH_2AsO_4	97	5.0	78
TGS type	$(NH_2CH_2COOH)_3$ H_2SO_4 (Triglycine sulphate)	322	2.8	275
Rochelle salt type	$NaKC_4H_4O_6$ $4H_2O$ (Rochelle salt)	296 (upper) 255 (lower)	0.25	275

When an electric field is applied to specimen of a ferroelectric crystal, the polarization first rises rapidly with applied field to a value above which the dependence is linear. Linear extrapolation to zero field gives the *saturation of spontaneous polarization*. On subsequently reducing the field to zero, remanent (residual) polarization remains. The negative field to reduce the polarization to zero is called the coercive field. Obviously, a ferroelectric crystal can be switched by the application of an electric field and a *hysteresis loop* is associated with the switching. The existence of a dielectric hysteresis loop in a dielectric material implies that the substance possesses a spontaneous polarization, P_s and the value of spontaneous polarization (depending upon the shape of hysteresis loop) depends upon a number of factors such as the dimension of the specimen, temperature, texture of the crystal, thermal and electrical properties of the crystal. A typical hysteresis loop is shown in Fig. 23.24. For most ferroelectrics, the values of P_s are between 10^{-7} C/cm^2 and 10^{-4} C/cm^2. In non-ferroelectric dielectrics, electric field between 10^5 V/cm and 10^8 V/cm would be necessary in order to achieve such large polarizations.

χ_E in the paraelectric phase (above Curie temperature, T_c) is related to the temperature by the Curie–Weiss law,

$$\chi_E = \frac{C}{T - T_c} \tag{23.68}$$

where C is Curie constant and T_c is Curie–Weiss temperature.

As a rule, the dielectric constant ε measured along a ferroelectric axis increases in the paraelectric phase. When the Curie temperature, T_c is approached in many ferroelectrics, this increase can be approximated by Curie–Weiss law,

$$\varepsilon = \frac{T}{T - T_0} \tag{23.68a}$$

where T_0 is equal or some what smaller than T_c. For BaTiO$_3$, this law holds unaltered up to frequencies of 24×10^{10} Hz. Ferroelectric crystals will have an extremely large dielectric constant.

One of the most common ferroelectric is barium titanate (BaTiO$_3$).

The spontaneous polarization of this crystal is a consequence of the positioning of the Ba^{2+}, Ti^{4+} and O^{2-} ions within the unit cell (Fig. 23.25). The Ba^{2+} ions are located at the corners of the unit cell, which is of tetragonal symmetry (a cube that has been elongated slightly in one direction). The dipole moment

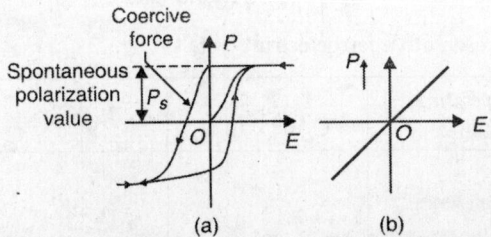

(a)   (b)

Fig. 23.24: Net polarization *P* of a ferroelectric crystal versus externally applied electric field (a) Hysteresis (b) *P–E* relation above Curie temperature

The hysteresis loop of a ferroelectric material changes its shape as the temperature is increased. The height and width decrease with increase in temperature. At a certain temperature known as *ferroelectric Curie temperature* (T_c), the loop merges to a straight line and the ferroelectric behaviour of the material disappears. The electric susceptibility

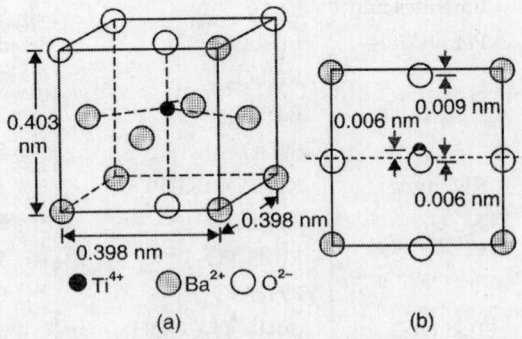

(a)   (b)

Fig. 23.25: BaTiO$_3$ unit cell (a) In an isomeric projection (b) Looking at one face, exhibiting the displacement of Ti^{4+} and O^{2-} ions from the center of face

results from the relative displacements of the O^{2-} and Ti^{4+} ions from their symmetrical positions as shown in the side view of the unit cell. We can see that O^{2-} ions are located near but slightly below the centres of each of the six faces, whereas the Ti^{4+} ion is displaced upward from the unit cell corner. Obviously, a permanent ionic dipole moment is associated with each unit cell. However, when $BaTiO_3$ is heated above its *ferroelectric* T_c (=120 °C), the unit cell becomes cubic, and all ions assume symmetric positions within the cubic unit cell; the material now has a perovskite crystal structure and ferroelectric behaviour ceases. Ceramic materials and especially ferroelectrics are the most important ones among dielectrics. Ceramics may have very diversified electric properties and are almost insusceptible to ageing and heating.

Spontaneous polarization of this group of materials results as a consequence of interactions between adjacent permanent dipoles wherein they mutually align, all in the same direction. For example with $BaTiO_3$, the relative displacements of O^{2-} and Ti^{4+} ions are in the same direction for all the unit cells within some volume region of the specimen. At room temperature, ε for $BaTiO_3$ may be as high as 5000. Consequently, capacitors made from these ferroelectric materials can be significantly smaller than capacitors made from other dielectric materials. This is why these materials are widely used in the manufacture of miniatured capacitors.

23.13 ELECTROSTRICTION

This is mechanical deformation which always accompanies polarization in dielectric. An electric field polarizes any material by inducing dipole moments. This displacement of charges from their equilibrium positions alter the mechanical dimensions of a solid; it causes electrostriction. However, mechanical stress applied to a neutral material cannot induce dipole moment, i.e. *electrostriction has no inverse*. If a mechanical distortion creates a voltage, the effect must be caused by permanent dipole moment anchored in the structure without a centre of symmetry.

23.14 PIEZOELECTRICITY OR PRESSURE ELECTRICITY

Electricity, or electric polarity, resulting from the application of mechanical pressure on a dielectric crystal is called piezoelectricity or pressure electricity. The application of a mechanical stress produces in certain dielectric (electrically non-conducting) crystal an electric polarization (electric dipole moment per cubic meter) which is proportional to this stress. If the crystal is isolated, this polarization manifests itself as a voltage across the crystal, and if the crystal is short-circuited, a flow of charge can be observed during loading. Conversely, application of a voltage between certain faces of the crystal produces a mechanical distortion of the material. This reciprocal relationship is referred to as the piezoelectric effect. The phenomenon of generation of voltage under mechanical stress is referred to as the direct piezoelectric effect, and the mechanical strain produced in the crystal under electric stress is called the *converse piezoelectric effect*. The piezoelectric effect is demonstrated in Fig. 23.26

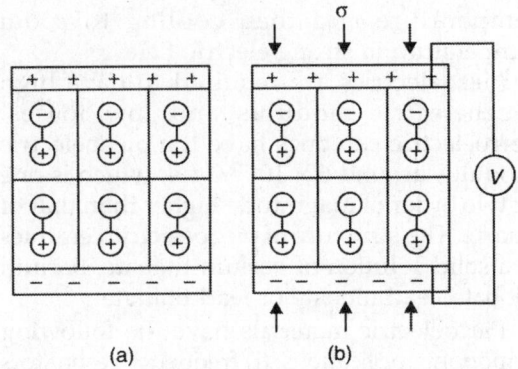

(a)   (b)

Fig. 23.26: Piezoelectric effect (a) Dipoles within a piezoelectric material (b) A voltage is generated when the material is subjected to a compressive stress

Piezoelectricity occurs only in insulating materials. Only few ceramic materials exhibit this property. Piezoelectric materials are used extensively in *transducers* for converting a mechanical strain into an electrical signal.

Such devices include microphones, phonograph pick-ups, vibration-sensing element, and the like. The converse effect, in which a mechanical output is derived from an electrical signal input, is also widely used in such devices as sonic and ultrasonic transducers, headphones, loudspeakers, and cutting heads for disk recording. Both the direct and converse effects are employed in device in which the mechanical resonance frequency of the crystal is of importance. Such devices include electric wave filters and frequency control elements in electronic oscillator circuits.

The necessary condition for the piezoelectric effect is the absence of symmetry in the crystal structure. Of the 32 crystal classes, 21 lack a centre of symmetry, and with the exception of one class, all of these are piezoelectric. Piezoelectric materials include titanates of barium and lead, zirconate ($PbZrO_3$), ammonium dihydrogen phosphate ($NH_4H_2PO_4$), natural quartz. As stated earlier, this property is a characteristic of materials having complicated crystal structure with a low degree of symmetry. One may improve the piezoelectric behaviour of a polycrystalline specimen by heating above its Curie temperature and then cooling to room temperature in strong electric field.

Piezoelectrics are required to have high piezoelectric modulus and low losses. Ferroelectric ceramics have the piezoelectric modulus around 6×10^{-10} C/N, which is one or two order of magnitude higher than that of quartz. The structure of piezoelectric ceramics is a solid solution of barium titanate, barium niobate, lead niobate or lead titanate.

Piezoelectric materials have the following important applications: (i) frequency resonators (ii) gramophone pick-ups (iii) filters (iv) ultrasonic flaw detectors (v) underwater sonar transducers (vi) air transducers (earphones, microphones, hearing aids, etc.).

23.15 USES OF DIELECTRICS

We have seen that dielectric materials are electrically insulative, yet susceptible to polarization in the presence of an electric field.

This polarization phenomenon accounts for the ability of the dielectrics to increase the charge storing capacity of capacitors. Now, we can summarize the main uses of dielectrics as follows: (i) piezoelectric and electro-optic devices (ii) in capacitors, resistors and strain gauges (iii) thermionic valves, radiation detectors, electric devices, dielectric amplifier (iv) dielectrics are usually used as ordinary insulators in power cables, signal, electric motors, etc. (v) dielectics are used in transformers and various forms of switchgear and generators where the dissipation problem of heat is active, and a common way of getting rid of it is to insulate with a transformer oil, i.e. mineral oil.

ILLUSTRATIVE EXAMPLES

Example 1

Show that the relative dielectric constant of a barium titanate crystal, which when inserted in a parallel plate condenser of area 10 mm × 10 mm and distance of separation of 2 mm gives a capacitance of 10^{-9} F is 2259.

Solution

We have

$C = 10^{-9}$ F, $d = 2 \times 10^{-3}$ m, $\varepsilon_0 = 8.854 \times 10^{-12}$
$A = 10 \times 10^{-3} \times 10 \times 10^{-3} = 10^{-4}\, m^2$

$$C = \frac{\varepsilon_r \varepsilon_0 A}{d}$$

or

$$\varepsilon_r = \frac{Cd}{\varepsilon_0 A}$$

$$= \frac{10^{-19} \times 0^2 \times 10^{-3}}{8.854 \times 10^{-12} \times 10^{-4}} = 2259$$

Example 2

Using the given data for $BaTiO_3$ crystal, calculate the polarization. The shift of the titanium ion from the body centre is 0.06 Å. The oxygen anions of the side faces shift by 0.06 Å, while the oxygen anions of the top and bottom faces shift by 0.08 Å, all in a direction opposite to that of the titanium ion.

Solution

Let us calculate the dipole moments due to the effective number of each type of ion in the unit cell of $BaTiO_3$ as:

i. Dipole moment due to two O^{2-} ions on the four side faces,

$$\mu_1 = Qd = 2 \times 2 \times 1.6 \times 10^{-19} \times 0.06 \times 10^{-10} \, C \, m$$

ii. Dipole moment due to one O^{2-} on top and bottom

$$\mu_2 = Qd = 1 \times 2 \times 1.6 \times 10^{-19} \times 0.08 \times 10^{-10} \, C \, m$$

iii. Dipole moment due to one Ti^{4+} ion at body centre

$$\mu_3 = Qd = 1 \times 4 \times 1.6 \times 10^{-19} \times 0.06 \times 10^{-10} \, C \, m$$

∴ Total dipole moment $= 1.02 \times 10^{-29} \, C \, m$

Now, the polarization is the total dipole moment per unit volume. Now, ignoring the effect due to barium ions, one obtains

$$P = \frac{\mu}{A} = \frac{1.02 \times 10^{-29}}{4.03 \times (3.98)^2 \times 10^{-30}}$$

$$= 0.16 \, C \, m^2$$

Example 3

The atomic weight and density of sulphur are 32 and 2.08 gm/cm^3 respectively. The electronic polarizability of the atom is $3.28 \times 10^{-40} \, F \, m^2$. If solid sulphur has cubical symmetry, what will be its relative dielectric constant?

Solution

$$\frac{\varepsilon_r - 1}{\varepsilon_r + 2} = \frac{N\alpha_e}{3\varepsilon_0}$$

∴

$$N = \frac{N_A P}{M}, \frac{\varepsilon_r - 1}{\varepsilon_r + 2} = \frac{N_A \alpha_e}{3M\varepsilon_0}$$

∴

$$\frac{\varepsilon_r - 1}{\varepsilon_r + 2} = \frac{\left(6.023 \times 10^{26}\right)\left(2.08 \times 10^3 \, kg/m^3\right) \times \left(3.28 \times 10^{-40} \, F/m^2\right)}{3 \times 32 \left(8.854 \times 10^{-12} \, F/m\right)}$$

$$= 0.483$$

$$\varepsilon_r = \frac{1.966}{0.517} = 3.8$$

Example 4

Calculate the electronic polarizability of argon atom. Given $\varepsilon_r = 1.0024$ at NTP and $N = 2.7 \times 10^{25}$ atoms/m^3. [BTech, BE]

Solution

$$P = \varepsilon_0(\varepsilon_r - 1)E$$

$$P = N\alpha_e E$$

∴

$$\alpha_e = \frac{\varepsilon_0(\varepsilon_r - 1)}{N}$$

$$= \frac{\left(8.85 \times 10^{-12} \, F/m\right)(1.0024 - 1)}{2.7 \times 10^{25}/m^3}$$

$$= 7.9 \times 10^{-40} \, F \, m^2$$

Example 5

For ice, relaxation time is given as 18×10^{-6} s at $22\,°C$. Calculate the frequency when the real and imaginary parts of the complex dielectric constant will become equal. What will be the phase difference between the current and voltage at this frequency? Which of the two will be leading in phase?

Solution

The real and imaginary parts of dielectric constant will be equal when $\omega\tau = 1$.

$$\omega = \frac{1}{\tau}$$

or

$$f = \frac{1}{2\pi\tau} = \frac{1}{2 \times 3.142 \times 18 \times 10^{-6} \, s}$$

∴

$$f = 8.8 \, kHz$$

The phase difference between the current and voltage at 8.8 kHz is given by $\phi = (90° - \delta)$

$$\delta = \tan^{-1}(\varepsilon_r''/\varepsilon_r') = \tan^{-1} 1 = 45°$$

$$\phi = 45°$$

Example 6

The number of atoms in volume of one cubic metre of hydrogen gas is 9.8×10^{26}. The radius of the hydrogen atom is 0.53 Å. Calculate the polarizability and relative permittivity. [BE]

Solution

$$\alpha_e = 4\pi\varepsilon_0 R^3$$
$$= 4(3.142) \times (8.85 \times 10^{-12}\ \text{F/m})$$
$$\times (0.53 \times 10^{-10}\ \text{m})^3$$
$$= 1.66 \times 10^{-41}\ \text{F}\,\text{m}^2$$
$$\varepsilon_r = 1 + 4\pi N R^3$$
$$= 1 + 4(3.142) \times (9.8 \times 10^{26}/\text{m}^3)$$
$$\times (0.53 \times 10^{-10}\ \text{m})^3$$
$$= 1.0018.$$

Example 7

The value of ε_r for glass is 6.75 at frequency 10^9 Hz. What mechanisms are contributing towards dielectric constant? What percentage may be attributed to ionic polarizability? Given refractive index of glass is 1.5.
[BTech, BE]

Solution

We have the polarization at 10^9 Hz,

$$P = P_i + P_e = \varepsilon_0(\varepsilon_r - 1)E$$

At optical frequencies, we have

$$\varepsilon_r = n^2,\ P_e = \varepsilon_0(n^2 - 1)E$$

and
$$P_i = \varepsilon_0(\varepsilon_r - 1)E - \varepsilon_0(n^2 - 1)E$$
$$= \varepsilon_0(\varepsilon_r - n^2)E$$

% of ionic polarizability

$$= \frac{P_i}{P_e + P_i} \times 100 = \frac{\varepsilon_0\left(\varepsilon_r - n^2\right)E}{\varepsilon_0\left(\varepsilon_r - 1\right)E} \times 100$$

$$= \frac{\varepsilon_r - n^2}{\varepsilon_r - 1} \times 100 = \frac{6.75 \times 2.25}{6.75 - 1}$$

$$= 78.3\%$$

Example 8

A parallel plate capacitor has an area of 7.45×10^{-4} m² and the plates are separated by a distance of 2.45×10^{-3} m, across which a potential of 10 V is applied. If a material with

dielectric constant 6 is introduced between the plates, determine the capacitance, the charge stored on each plate, the dielectric displacement D and polarization. [BE]

Solution

$$C = \frac{\varepsilon_r \varepsilon_0 A}{d}$$
$$A = 7.45 \times 10^{-4}\ \text{m}^2$$
$$d = 2.45 \times 10^{-3}\ \text{m}$$
$$V = 10\ \text{volt}$$
$$\varepsilon_r = 6,\ \varepsilon_0 = 8.85 \times 10^{-12}$$
$$\therefore \quad C = \frac{6 \times 8.85 \times 10^{-12} \times 7.45 \times 10^{-4}}{2.45 \times 10^{-3}}$$
$$= 1.61 \times 10^{-11}\ \text{F}$$
$$Q = CV$$
$$= 1.61 \times 10^{-11} \times 10 = 1.61 \times 10^{-10}\ \text{C}$$

Polarization

$$P = \varepsilon_0\left(\varepsilon_r - 1\right)E = \varepsilon_0\left(\varepsilon_r - 1\right)\frac{V}{d}$$

$$= 8.85 \times 10^{-12}(6-1) \times \frac{10}{2.45 \times 10^3}$$

$$= 1.8 \times 10^{-7}\ \text{C/m}^2$$

Example 9

A parallel plate capacitor consists of two plates each of area 5×10^{-4} m². They are separated by a distance 1.5×10^{-3} m and filled with dielectric of relative permeability 6. Calculate the charge on the capacitor if it is connected to a 100 V DC supply. [BE]

Solution

$$Q = CV \tag{1}$$

and
$$C = \frac{K\varepsilon_0 A}{d} \tag{2}$$

From Eqs. (1) and (2),

$$Q = \frac{K\varepsilon_0 A V}{d}$$
$$\varepsilon_0 = 8.85 \times 10^{-12}$$
$$K = 6$$
$$A = 5 \times 10^{-4}\ \text{m}^2$$

$$d = 1.5 \times 10^{-3} \text{ m}$$
$$V = 100 \text{ volt}$$

$$\therefore \quad Q = \frac{8.85 \times 10^{-12} \times 6 \times 5 \times 10^{-4} \times 100}{1.5 \times 10^{-3}}$$

$$= 1.77 \times 10^{-9} \text{ C.}$$

REVIEW QUESTIONS

1. What are dielectrics? Derive an expression for internal field for one dimensional dielectric solid. [Visvesvaraya Tech Univ 2004]

2. Define dielectric polarization. Derive an expression for internal field in case of solid and liquid dielectric by considering one dimensional array of atoms.
   [Visvesvaraya Tech Univ 2003]

3. (a) Discuss different polarization mechanisms.
   (b) Deduce an expression for internal field in case of one dimensional array of atoms in solids. [Visvesvaraya 2003]

4. Explain the behaviour of dielectrics under static electric fields. Derive the relation between the polarization P and the external field E. [BE]

5. Explain the significance of D, E and P in $D = \varepsilon_0 E + P$ and derive the relation. [BE]

6. Define dielectric susceptibility and polarizability of a dielectric. Derive the relation connecting the two. [BTech]

7. Explain electronic and ionic polarizability and show that the electronic polarizability for a monoatomic gas increases as the size of the atoms becomes larger. [BE]

8. What is ionic polarizability? Explain why it is found to be rather insensitive to temperature. [BE, BTech]

9. Explain why certain molecules possess permanent electric dipole moment even in the absence of an applied electric field. [BE]

10. What is meant by a permanent dipole moment for a polyatomic gas? Show how the potential energy varies with its orientation in the electric field. Write an expression for the orientation polarizability. [BE]

11. Write the expression for total polarization P for a polyatomic gas. How can you determine permanent dipole moment of molecules if P is known at different temperatures? [BTech]

12. Define dipole moment and classify dielectric materials on its basis. Give suitable examples in support of your answer. [BE]

13. What is dipolar polarizability? Obtain an expression for dipolar polarizability of a dielectric at moderate temperature. [BE]

14. What are different mechanisms of polarization in a dielectric? Explain in brief with examples. [BE]

15. What are polar and non-polar molecules? Explain electrical breakdown in solid dielectrics.

16. What is dielectric loss? Show that dielectric loss is given by $\tan \delta = \varepsilon_r'' / \varepsilon_r'$. What do you understand by loss tangent as referred to polar dielectrics? [BE, BTech]

17. What are the various mechanisms that contribute to dielectric loss? Sketch the loss spectrum for a polar material. [BE]

18. Explain the behaviour of a dielectric in an alternating field and write down the expression for energy absorbed per second in dielectric material when it is subjected to an alternating electric field of amplitude E. [BE]

19. What are the main causes for the dielectric losses occurring in RF, IR and visible region of the electromagnetic spectrum? [BE]

20. Describe the characteristic properties of ferroelectric materials. What do you understand by polarization catastrophe? [BE]

21. Explain the meaning and origin of piezoelectricity. Justify the statement— *All ferroelectric crystals are piezoelectric, but all piezoelectric crystals are not necessarily ferroelectric.*

22. What are ferroelectric materials? How do dielectric constant and polarization change with temperature in a ferroelectric material? [BE]

23. In what respects ferroelectrics differ from ordinary dielectrics? Explain the phenomenon of spontaneous polarization in ferroelectrics. [BTech]

24. What are ferroelectric materials? Sketch the variation of spontaneous polarization with temperature in ferroelectrics. Do ferroelectrics exhibit domains? [BE]

25. What are direct and inverse piezoelectric effects? Explain their importance and practical applications. [BE]

26. Explain: (a) How is dielectric loss utilized in cooking food in microwave oven. (b) How does a gas lighter and quartz watch utilize piezoelectric phenomenon in their functioning. [BE]

27. Write short notes on:

    i. Dielectric materials

    ii. Polar and non-polar dielectrics

    iii. Ferroelectric materials

    iv. Characteristics of ferroelectric $BaTiO_3$

PROBLEMS

1. A parallel plate capacitor of area 650 mm^2 and a plate separation of 4 mm has a charge of 2×10^{-10} C. Show that the resultant voltage across the capacitor when a material of dielectric constant 3.5 is introduced between the plates is 39.7 volts.

$$\left[\textbf{Hint:} \ C = \frac{\varepsilon_0 \varepsilon_r A}{d} \ \text{and} \ C = Q/V \ , \ V = \frac{Qd}{\varepsilon_0 \varepsilon_r A} \right.$$

$$= \frac{2 \times 10^{-10} \times 4 \times 10^{-3}}{8.85 \times 10^{-12} \times 3.5 \times 650 \times 10^{-6}} = 39.7 \text{ V} \ \left. \right]$$

2. A parallel plate capacitor has circular plates of radius 8 cm and plate separation 1 mm. If the potential difference between the plates is 100 volts, show that the charges on the plates are $+1.78 \times 10^{-8}$ C and -1.78×10^{-8} C respectively.

$$\left[\textbf{Hint:} \ C = \frac{\varepsilon_0 A}{d} = \frac{8.85 \times 10^{-12} \times \pi \times (0.08)^2}{10^{-3}} \right.$$

$$Q = CV = \frac{8.85 \times 10^{-12} \times \pi \times (0.08)^2}{10^{-3}} \times 100$$

$$= 1.78 \times 10^{-8} \text{ C} \ ]$$

Thus, the charges on the plates are $+1.78 \times 10^{-8}$ C and -1.78×10^{-8} C respectively.

3. A parallel plate capacitor consists of two plates each of area 5×10^{-4} m^2 separated by a distance of 1.5×10^{-3} m and filled with a dielectric of relative permeability 6. The capacitor is connected to a 100 V DC source. Show that the charge on the capacitor is 1.77×10^{-9} C.

4. A parallel plate capacitor of area 2 sq m with a medium of relative permeability 7 is charged to a potential of 100 volts. Calculate the capacitance and the energy stored in the condenser. Given the distance between the plates is 10^{-4} m.

$$\left[\textbf{Hint:} \ C = \frac{\varepsilon_0 \varepsilon_r A}{d} = \frac{8.85 \times 10^{-12} \times 7 \times 2}{10^{-4}} \right.$$

$$= 1.24 \times 10^{-6} \text{ F} = 1.24 \text{ μF}, \ E = \frac{1}{2} CV^2$$

$$= \frac{1}{2} \times 123.9 \times 10^{-8} \times (100)^2 = 61.95 \times 10^{-4} \text{ J} \ ]$$

SHORT ANSWER QUESTIONS

1. What are dielectric materials?

Ans. These are insulators which possess high electrical resistivity.

2. What is the significance of relative permittivity, ε_r?

Ans. ε_r is an important characteristic of a dielectric and it is a measure of polarization of dielectric subjected to electric field.

3. What does polarization of dielectrics signifies?

Ans. Polarization results in a dielectric due to inducement of charges and accounts for the ability of dielectrics to increase the charge storing capability of capacitors.

4. What is dielectric susceptibility?

Ans. The ease with which a dielectric is polarized in the presence of an electric field is called dielectric susceptibility.

5. How will you define polarizability?

Ans. The polarizability P is defined as the dipole moment per unit volume of the material and is related to the electric susceptibility, χ_e, as

$$P = \varepsilon_0 \chi_e E$$

6. What do you understand by electric dipole and dipole moment?

Ans. An electric dipole is said to exist when there is a net spatial separation of positively and negatively charged entities. In a molecule, an electric dipole is said to exist due to inherent asymmetry of molecule or may be induced in it due to the action of an external field.

7. What is dipole moment?

Ans. A dipole moment is the product of charge q on the dipole and the distance of separation d between the charges, i.e.

$$\mu = qd.$$

8. What is the effect of induced dipoles?

Ans. Bulk polarization of the dielectric.

9. What is atomic polarizability?

Ans. The atomic polarizability (α) is the dipole moment (P) per unit electric field at an atom and is given by the relation

$$P = \alpha E_{\text{loc}}$$

10. What does polarizability and dielectric constant of the material signify? How these two parameters are related with each other?

Ans. The polarizability is an atomic property whereas the dielectric constant is a macroscopic property of the material. The two parameters are related to each other by the Clausius–Mossotti relation given as

$$\frac{\varepsilon_r - 1}{\varepsilon_r + 2} = \frac{N\alpha}{3\varepsilon_0}$$

11. What are the main contributions to the net polarizability of dielectric material?

Ans. i. *Electronic polarization*: This arises due to shift of electron cloud with respect to the nucleus in the atom subjected to an electric field:

$$P_e = N\mu_{ind} = N\alpha_e E$$

ii. *Ionic polarization*: This occurs due to shift of positive and negative ions from their equilibrium positions

$$P_i = N\alpha_i E$$

iii. *Dipolar or orientation polarization*: This consists in partial alignment of molecular dipoles with the applied field E

$$P_0 = \frac{N\mu^2}{3kT}E$$

At low frequencies of the applied electric field, all the three contributions are present. As the frequency increases, the dipolar contribution disappears first followed by ionic and electronic contributions at higher frequencies. At optical frequencies, the polarization is mainly due to electronic contribution.

12. Write the local field within a dielectric.

Ans.
$$E_i = E + \frac{\gamma P}{\varepsilon_0}$$

In case of cubic symmetry,

$$E_i = E + \frac{P}{3\varepsilon_0} \quad \text{(Lorentz field)}$$

13. What is dielectric loss?

Ans. Electric power is absorbed by the dielectric from the source whenever a particular polarization mechanism tends to become ineffective, i.e. disappears. The maximum power absorption will occur at the relaxation frequency and known as *dielectric loss*. Power dissipation per unit volume of dielectric is given by

$$P_L = \omega\varepsilon_0\varepsilon_r E^2$$

and dielectric loss in terms of loss angle δ as

$$\tan \delta = \frac{\varepsilon_r''}{\varepsilon_r'}$$

14. What is dielectric strength?

Ans. This is defined as the maximum electric field which can be applied across a dielectric without causing an electrical conduction in it.

15. What is electrostriction?

Ans. This refers to the change in dimensions of a dielectric subjected to an electric field.

16. What are ferroelectric materials?

Ans. These are the materials which exhibit spontaneous polarization, i.e. polarization in the absence of an applied field. These are analogous to ferromagnetic materials. These are basically *non-linear* dielectrics, in which polarization does not vary linearly with the applied field. These substances exhibit Curie temperature, domains and hysteresis. The examples are barium titanate, the Rochelle salt, etc.

17. What is piezoelectric effect?

Ans. The polarization of a dielectric in response to mechanical deformation is called direct piezoelectric effect. Mechanical deformation of a dielectric caused by an external voltage is called the inverse piezoelectric effect. The absence of the centre of inversion is the prerequisite for the occurrence of piezoelectricity. All ferroelectric crystals are piezoelectric, but all piezoelectric crystals are not necessarily ferroelectric. Example of piezoelectric crystal is quartz.

OBJECTIVE QUESTIONS

1. A forbidden energy gap, E_g is very _____ in dielectrics and excitation of electrons from the normally full valence band to empty conduction band is _____ under ordinary conditions.
[large, not possible]

2. The relative permeability, ε_r is a _____ quantity which is always greater than _____ in case of dielectrics. [dimensionless, unity]

3. Based on dipole moment, dielectrics can be broadly classified into two major groups _____ and _____ dielectrics.
[polar, non-polar]

4. Ionic polarization in ionic crystals is brought about by the elastic displacement of _____ and negative ions from their equilibrium positions. [positive]

5. The orientation polarization is characteristic of _____ dielectrics which consists of molecules having permanent _____.
[polar, dipole moment]

6. The average time taken by the dipoles to reorient in the field direction is known as the _____. [relaxation time]

7. The mechanical deformation of a dielectric induced by an electric field is called _____. [electrostriction]

8. Ferroelectrics are anisotropic crystals which exhibit _____ polarization. [spontaneous]

MULTIPLE CHOICE QUESTIONS

1. Choose the correct relationship for a dielectric
   (a) $\vec{D} = \epsilon_0 \vec{P}$
   (b) $\vec{D} = \epsilon_0 \vec{E} + \vec{P}$
   (c) $\vec{D} = \epsilon_0 \vec{E}$
   (d) $\vec{D} = \vec{P} \times \vec{E}$

2. Dielectrics are
   (a) superconductors
   (b) electric insulators
   (c) materials that work under low voltages
   (d) none of the above

3. If ϵ_r is the relative permeability of an isotropic dielectric medium, then the electric susceptibility is
   (a) $\epsilon_r - 1$
   (b) ϵ_r
   (c) $\epsilon_r \epsilon_0$
   (d) ϵ_r / ϵ_0

4. A dielectric
   (a) does not contain free charges or electrons
   (b) does not contain molecules
   (c) contains free charges or electrons
   (d) when placed between the plates of a capacitor, electric field increases at every point in the medium

5. The internal field in a dielectric that has a polarization $\vec{P}$ under the action of an external field $\vec{E}$ is
   (a) $\vec{P}/3\epsilon_0$
   (b) $3\epsilon_0 \vec{E}/\vec{P}$
   (c) $\vec{P}/\vec{E}$
   (d) $\vec{P}/3\epsilon_0$

6. Which one of the following is a non-polar molecule?
   (a) $CHCl_3$
   (b) CO_2
   (c) polyethylene
   (d) petroleum oils

7. If μ_0 and ϵ_0 represent the permeability and the permittivity of free space respectively, then the velocity of electromagnetic waves in free space is
   (a) μ_0/ϵ_0
   (b) $\sqrt{\mu_0/\epsilon_0}$
   (c) $\sqrt{\mu_0 \epsilon_0}$
   (d) $\dfrac{1}{\sqrt{\mu_0 \epsilon_0}}$

8. When a dielectric medium of relative permittivity ϵ_r is under the action of an external field $\vec{E}$, the total effective field intensity $\vec{E}'$ is
   (a) $\left(\dfrac{\epsilon_r - 1}{3}\right)\vec{E}$
   (b) $\dfrac{3(\epsilon_r - 2)}{(\epsilon_r + 1)}\vec{E}$
   (c) $\left(\dfrac{\epsilon_r - 2}{3}\right)\vec{E}$
   (d) $\dfrac{(\epsilon_r - 1)}{(\epsilon_r + 2)}\vec{E}$

9. Amongst the following, Clausius–Mosotti relation is
   (a) $\dfrac{\epsilon_r - 1}{\epsilon_r + 2} = \dfrac{Nd}{3\epsilon_0}$
   (b) $\epsilon_r - 1 = \dfrac{Nd}{3\epsilon_0}$
   (c) $\epsilon_r + 2 = \dfrac{Nd}{3\epsilon_0}$
   (d) $(\epsilon_r - 1)\epsilon_0 \vec{P} = \vec{E}$

10. The polarization that is affected by temperature is
    (a) electronic polarization
    (b) dipolar polarization
    (c) ionic polarization
    (d) none of the above

11. If η represents the refractive index of a medium, α_{el}, the electronic polarizability and N, the total number of atoms, in the medium, then
    (a) $\eta^2 = \dfrac{N\alpha_{el}}{3\epsilon_0}$
    (b) $\eta^2 + 1 = N\alpha_{el}$
    (c) $\eta^2 + 2 = \dfrac{N\alpha_{el}}{3\epsilon_0}(\eta^2 - 1)$
    (d) $\dfrac{\eta^2 - 1}{\eta^2 + 2} = \dfrac{N\alpha_{el}}{3\epsilon_0}$

Answers

1. (b)	2. (b)	3. (a)	4. (a)	5. (a)	6. (b)
7. (d)	8. (c)	9. (a)	10. (b)	11. (d)	

24 Digital Electronics

24.1 INTRODUCTION

A typical NPN or PNP transistor operating as an amplifier produces a voltage gain of around 200. Thus, an output of around 3.0 V can be obtained from an input voltage of around 15 mV. While this can be considered good for some applications, much smaller voltages need to be amplified in many other situations. An amplifier with much higher gain is necessary for these applications. A common approach towards increasing the gain is to use more than one transistor, and link them up in such a way that the output from one serves as input for the next, and so on. Such *multistage amplifiers*, as they are called, are rather expensive to put together, especially when very large gains are desired.

As a switch, a single transistor can be very effective to control a simple operation, e.g. turning on a lamp. Where many similar switching operations are to be undertaken randomly, many transistor switches are required. Again, putting together a large bank of single transistors is rather expensive and time consuming.

Modern technology has addressed the above problems with the manufacture of the *integrated circuit*, simply referred to as IC. As silicon is more common than germanium in solid state devices, the IC is also commonly referred to as the *silicon chip*. Even in one of its simplest forms, the integrated circuit can behave in a manner equivalent to a large number of transistors and diodes put together. Some ICs produce amplification factors of the order of 3000 or more. These are known as *operational amplifiers*. Other ICs used for switching do also perform the switching action of a large number of transistors and diodes put together. In many cases, ICs are manufactured to perform these functions as desired. As technology has developed, today's electronic component market has an increasing range of these devices, designed to play the many switching and amplifying roles in circuits. We find some of these control circuits in the household washing machine, hi–fi system automated cameras and many other devices.

The silicon chip is normally in the shape of a rectangular block with metal connecting legs (usually called pins) along each of the longer sides. These legs provide connections to the various elements of the circuitry inside the chip. The number of legs of ICs are usually 8, 14, 16, 20 and more. Figure 24.1 is a top view of some of the common chips.

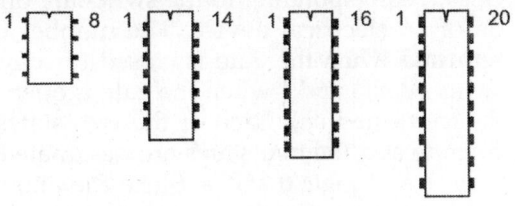

Fig. 24.1: Some integrated circuits

Each integrated circuit comes with a set of notes indicating the kind of chip it is, the pin identifications, showing the connections of the input, the bias and output voltages. The booklet on the device would also show the recommended operating conditions—temperature, voltage limits, gain, etc.

24.2 DIGITAL ELECTRONICS

Integrated circuits fall into a number of categories. These include *operational amplifiers, logic gates, timers* and *microprocessors*. Operational amplifiers employ a large batch of miniature multistage transistor amplifiers connected such that overall gain in the order of thousands are obtained. With these devices, millivolt inputs can be amplified to produce outputs of the order of volts.

Whilst operational amplifiers give amplified outputs, logic gates produce the same output—typically 5 volts—when triggered. The electronic gate is either open or closed, switching a device on or off. The electronics industry today has a wide range of these gates, some having one, two or more inputs and outputs. Some of these chips are used as timers, switching a device in some state for a prescribed period (or at specific times). This latter application underpins the behaviour of many automated gadgets we take for granted today.

A logic gate is an electronic circuit which makes logic decisions. Today, *logic gates* are at the heart of the modern branch of electronics known as *digital electronics*. The electronics of valves and amplifiers give out *analog* outputs, i.e. outputs of volts or millivolts and amps or milliamps. Logic gates are either open or closed, corresponding to the switching off or on of an electrical device. The number 0 is returned when the gate is closed (or device switched *off*) and 1 when the gate is open (or device turned *on*). Each of the two states is discrete and the two states are designated as logic 1 and logic 0 states. Since the allowed states are only two, they are known as binary states. In logic, a statement is characterised as *true or false*.

The input-output relationship of the binary variables for each logic gate can be represented in a tabular form in a truth table. Obviously, a *truth table is a compact way of representing the statements that define the values of dependent variables*.

Since digital devices and digital circuits operate in the binary number system (0 and 1), and hence they make it possible to use boolean algebra as a mathematical tool for analysis and design of digital circuits and system.

A boolean variable is a quantity that may be equal to either 0 or 1. In the digital logic field, several other terms are used synonymously with 0 and 1. Some of the more common ones are listed in Table 24.1.

Table 24.1: Various names of two states in digital circuits

Logic 0	Loigc 1
False	True
Off	On
Low	High
No	Yes
Open switch	Closed switch
Down	Up
Cold	Hot
South	North

There are four systems of arithmetic, which are often used in digital systems. These are: (i) decimal (ii) binary (iii) hexadecimal (iv) octal.

In any number system, there is an ordered set of symbols known as *digits*. Collection of these digits makes a number which in general has two parts, integer and fractional and are set apart by a radix point (·). Hence, one can represent a number system as

$$(N_b) = \frac{a_{n-1} a_{n-2} a_{n-3} \ldots a_1 a_0}{\text{Integer portion}} \cdot \frac{a_{-1} a_{-2} a_{-3} \ldots a_{-m}}{\text{Fractional portion}}$$

Radix point

where $N \rightarrow$ a number
$b \rightarrow$ radix or base of the number system

$n \rightarrow$ number of digits in integer portion

$m \rightarrow$ number of digits in fractional portion

$a_{n-1} \rightarrow$ most significant digit (MSD)

$a_{-m} \rightarrow$ least significant digit (LSD)

and $\qquad 0 \leq (a_i \text{ or } a_{-f}) \leq b - 1$

One can define the *base* or *radix* of a number as the number of different digits which can occur in each position in the number system.

A number system which utilizes ten distinct digits, i.e. 0, 1, 2, 3, 4, 5, 6, 7, 8, and 9 is known as *decimal number system*. It represents numbers in terms of a group of ten, as shown in Fig. 24.2

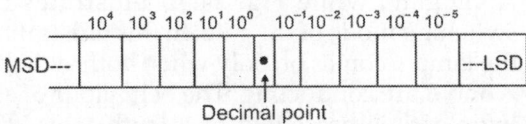

Fig. 24.2: Decimal position values as powers of 10

The decimal number system has a base or radix of 10. For example

$$359_{10} = 3 \times 100 + 5 \times 10 + 9 \times 1$$
$$= 3 \times 10^2 + 5 \times 10^1 + 9 \times 10^0$$

We can see that in the above example, 9 is the *least significant digit* (LSD) and 3 is the most significant digit (MSD). We now consider another example:

$$1536.469_{10} = 1 \times 10^3 + 5 \times 10^2 + 3 \times 10^1$$
$$+ 6 \times 10^0 + 4 \times 10^{-1} + 6 \times 10^{-2}$$
$$+ 9 \times 10^{-3}$$

We can see that powers are numbered to the left of the decimal point starting with 0 and to the decimal point starting with –1.

One can write the general rule for representing numbers in the decimal system by using positional notation as

$$a_n a_{n-1} \ldots a_2 a_1 a_0 = a_n 10^n + a_{n-1} 10^{n-1} + \ldots$$
$$+ a_2 10^2 + a_1 10^1 + a_0 10^0$$

where n is the number of digits to the left of the decimal point.

Binary Number System

A number system that uses only two digits, 0 and 1 is called binary number system. This system is also called a *base two system*. The two symbols 0 and 1 are known as bits (binary digits).

The binary system groups numbers by two and by powers of two (Fig. 24.3).

One can express the *weight* or place value of each position in terms of 2 and as $2^0, 2^1, 2^2, 2^3, \ldots$, etc. The least significant digit has a weight of $2^0 (= 1)$. The second position to the left of the least significant digit is multiplied by $2^1 (= 2)$. The third position has weight equal to $2^2 (= 4)$. Clearly, the weights are in the ascending powers of 2 or 1, 2, 4, 8, 16, 32, 64, …, etc.

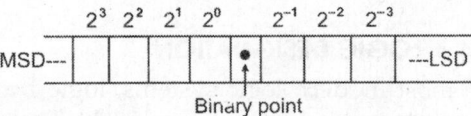

Fig. 24.3: Binary position values as a power of 2

The numeral 10_{two} (one, zero, base two) stands for two, the base of the system.

To make the binary system more clear, we consider following examples:

$$1071_2 = 1 \times 2^3 + 0 \times 2^2 + 7 \times 2^1 + 1 \times 2^0$$
$$= 8 + 0 + 14 + 1 = 23_{10}$$

$$10701.011_2 = 1 \times 2^4 + 0 \times 2^3 + 7 \times 2^2 + 0 \times 2^1$$
$$+ 1 \times 2^0 + 0 \times 2^{-1} + 1 \times 2^{-2} + 1 \times 2^{-3}$$
$$= 16 + 0 + 28 + 0 + 1 + 0.25 + 0.125$$
$$= 45.375_{10}$$

We can see that in each binary, the value increases in powers of two starting with 0 to the left of the binary point and decreases to the right of the binary point starting with powers –1.

Interestingly, binary system is used in digital computers because all electrical and electronic circuits can be made to respond to the two states concept. For example, a switch can be either opened or closed; only two possible states exist; a transistor can be made to operate in *cut-off* or *saturation*; a signal either *high* or *low*; a magnetic tape can be either magnetised or non-magnetised; a punched tape can have a hole or no hole. Obviously, in these examples each device is operated in any one of the two possible states and the intermediate condition does not exist. Clearly, 0 can represent one of the states and 1 can represent the other. Clearly, binary numbers are most convenient to use in analysing or designing digital circuits.

One can convert a binary number into decimal number by multiplying 1 or 0 by the weight corresponding to its position and adding all the values. For example

$$110121_2 = 1 \times 2^5 + 1 \times 2^4 + 0 \times 2^3 + 1 \times 2^2$$
$$+ 2 \times 2^1 + 1 \times 2^0$$
$$= 32 + 16 + 0 + 4 + 1 = 57_{10}$$

Similarly, one can convert a number from decimal to binary system.

24.3 LOGIC DESIGNATION

In most modern logic systems, logic 1 and 0 are represented by voltage levels. There are generally accepted rules for the definition of these logic levels in digital systems. *Positive logic* (or active high levels) means that the most positive logic voltage level (also called as the high level) is defined to be the logical 1 state and the most negative logic voltage level (also called as the low level) is defined to be the logical 0 state. *Negative logic* (or active low level) is just the opposite, the most positive (high) level is a 0, and the most negative (low) level is a 1. For example, if the voltage levels are –0.2 V and –5 V, then in a positive logic system, the –5 V level represents a zero and the –0.2 V level represents a one. Conversely, if the voltage levels are 0.1 V and 5 V, then in a negative logic system, the 5 V level represents a zero and the 0.2 V represents a one.

The effect of changing from one logic designation to the other is that all logic functions are complemented, e.g. a +ve AND becomes a –ve OR, a +ve NOR becomes a –ve NAND, etc. The simplest approach for converting the logic designation (positive or negative) is to replace all 0's with 1's and all 1's with 0's in the truth table for the device, and then determine the resulting logic function.

We may note that the choice of positive or negative logic is made by the individual logic designer, largely as a matter of personal preference. There is, basically, no real advantage to either designation. Usually, most logic designers and writers use positive logic. The terms "positive" and "negative" logic have no place in pure logic theory.

24.4 LOGIC GATES
AND Gate

This is also called *all* or *nothing* gate. The gate accepts two or more inputs. An output of zero is returned if any of the inputs is zero. An output of 1 is returned when all of the inputs are 1. The simplest AND gate has two inputs. Figure 24.4a shows the circuit symbol for the AND gate, while Fig. 24.4b illustrates the switching mode of the gate. In this illustration, the lamp *L* comes on only when both switches *A* and *B* are conducting. The behaviour of any logic gate is presented in a truth table. The truth table for the AND gate is also given in the figure.

The three-input and the four-input AND gates work on the same all-or-nothing principle, giving output only when there is a voltage on all of the inputs.

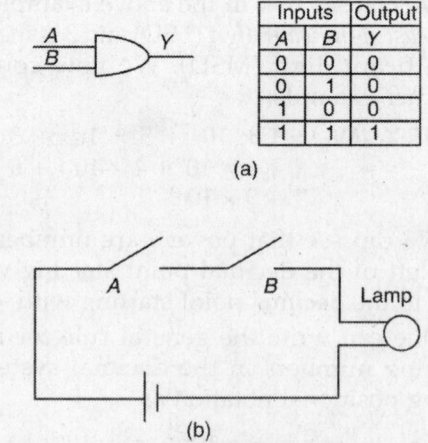

Inputs		Output
A	B	Y
0	0	0
0	1	0
1	0	0
1	1	1

(a)

(b)

Fig. 24.4: The two-input AND gate (a) Symbol and truth table (b) Circuit using switches

OR Gate

The OR gate is also called the *any* or *all* gate. The gate accepts two or more inputs. An output of zero is returned only if there is zero input on all gates. An output of 1 is returned if any of the inputs is 1. Figure 24.5a shows the circuit symbol for the OR gate, whilst Fig. 24.5b illustrates the switching mode of the gate. The lamp comes on when any or both switches A and B are conducting. The behaviour of this gate is shown in the adjoining truth table.

The three-input and the four-input OR gates work on the same any-or-all principle, giving output when any input is on.

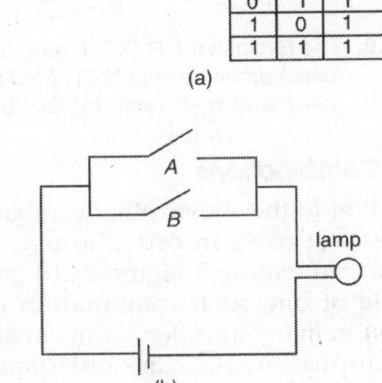

Inputs		Output
A	B	Y
0	0	0
0	1	1
1	0	1
1	1	1

(a)

(b)

Fig. 24.5: The two-input OR gate (a) Symbol and truth table (b) Circuit using switches

NOT Gate

This gate is also called the inverter or inverting gate. A NOT gate has only one input, and one output. An output of 1 is returned for zero input and *vice versa*. Figure 24.6a shows the circuit symbol for the NOT gate, and its truth table. Two NOT gates connected in series would produce an output of 1 for an input of 1. This is shown in Fig. 24.6b. The behaviour is shown in the truth table.

In the analysis of the performance of the NOT gate, a common term used is **complement**. In normal circuitry, the opposite of a device

being "on" is being "off". In digital electronics, "0" corresponds to the "off" state and "1" corresponds to the "on" state. The complement of "0" is hence "1" and the complement of A is written as $\bar{A}$.

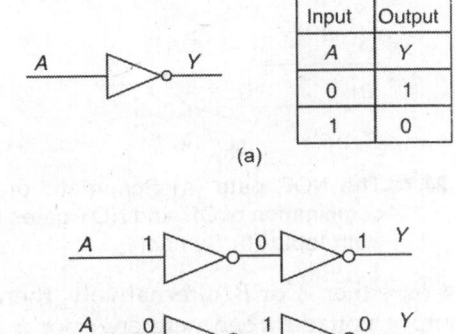

Input	Output
A	Y
0	1
1	0

(a)

(b)

Fig. 24.6: The NOT gate (a) Symbol and truth table (b) Two inverters in series

Combining Logic Gates

Figure 24.6b gives a simple case of combining logic gates. Logic gates of any types can be combined to give any desired output when the inputs vary. This has provided modern electronics with an almost endless set of combinations for the control of complicated circuits. These circuits are used in many devices—from household gadgets to laboratory instruments. Special useful gates have resulted from combining two or more basic gates. These are the NOR gate and the NAND gate.

NOR Gate

The NOR gate performs the inverse of the OR gate. A simple equation representing the NOR gate is

$$NOR = OR \cdot NOT$$

and Fig. 24.7a is a schematic of the combination that gives the NOR function, whilst Fig. 24.7b is the circuit symbol for the NOR gate.

From its truth table, it is seen that this combination conducts only when there is no

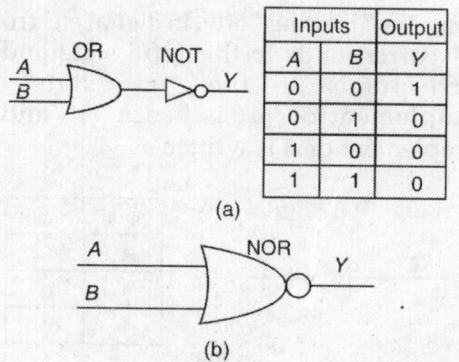

Inputs		Output
A	B	Y
0	0	1
0	1	0
1	0	0
1	1	0

(a)

(b)

Fig. 24.7: The NOR gate (a) Schematic of the combination of OR and NOT gates and truth table (b) Symbol

input on either A or B; alternatively, there is an output voltage when neither A nor B has an input. In practice, many electronic logic devices perform the NOR function easily, thereby making it possible to achieve the NOR function without the need for two gates, as this is often more expensive.

NAND Gate

NAND is short form for NOT AND. The expression for NAND is

$$\text{NAND} = \text{AND} \cdot \text{NOT}$$

This gate is formed by putting an AND gate and a NOT gate together. Figure 24.8 shows the combination, the truth table, and the usual circuit symbol. It is clear from the truth table that this combination does not conduct when both A and B have inputs.

Inputs		Output
A	B	Y
0	0	1
0	1	1
1	0	1
1	1	0

(a)

(b)

Fig. 24.8: The NAND gate (a) Schematic of the combination of AND and NOT gates and truth table (b) Symbol

Exclusive-OR (or XOR) Gate

The Exclusive-OR (abbreviated as XOR) gate is referred to as the *any but not all* gate.

There is an output for any input, but no output for all "0" or all "1" inputs. Figure 24.9 shows the XOR combination, the truth table, and usual circuit symbol.

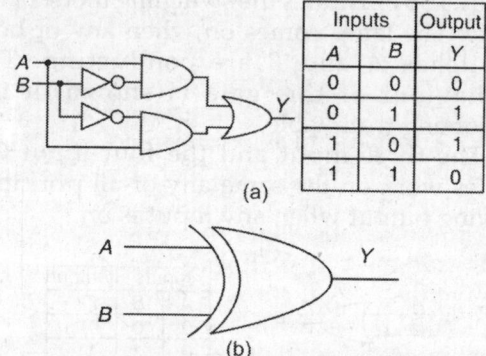

Inputs		Output
A	B	Y
0	0	0
0	1	1
1	0	1
1	1	0

(a)

(b)

Fig. 24.9: The Exclusive-OR (XOR) gate (a) Schematic combination of NOT, AND and OR gates and truth table (b) Symbol

Other Combinations

In addition to the above, other combinations of gates are used in order to obtain the performance required. Figure 24.10 gives an example of one such combination that is common in many intruder alarm circuits.

The inputs A, B, ... are obtained from tamper circuits which come on when their respective switches are opened. There is always an output when any or all of the tamper switches are opened. In operation, this output is used to trigger a buzzer or siren.

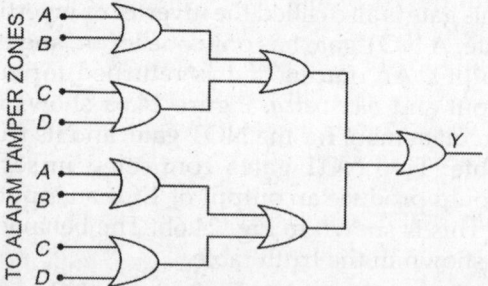

Fig. 24.10: An example of gate combination in an intruder alarm circuit

In addition to operating switches, logic gates can be combined in circuits to perform arithmetic operations such as addition, subtraction, multiplication, division, powers, logarithmic, exponential and trigonometric functions. These advancements has given rise to the pocket calculators, advanced teller machines used at supermarket outlets and in complicated scientific and industrial applications. At the very advanced end of digital electronics is the *microprocessor*. This is a single chip fabricated in such a manner that it does the functions of a large number of chips put together. This single chip can sometimes have as many as 40 connecting points or more. The microprocessor can be programmed to perform many complicated switching operations, perform arithmetic operations such as addition, subtraction, multiplication and division. It also has the ability to store the results of these operations in certain memory locations for use when required. The microprocessor allows the operator to *program* the operations desired. The microprocessor then carries out the instructions at the *input* and gives an *output* information.

The microprocessor is today at the heart of many complicated electronic devices. Examples are in toys, remote control units for certain household appliances, washing machines, household intruder alarm circuits, dishwashers, microwave ovens, and many more. The most intricate use is in the microcomputer.

In the technology today, we take the operations of switches and buttons so much for granted. It is nice to note that what brings about the desired results have been possible with the revolutionary advances in the field of electronics. Bearing in mind that the beginning of it all was the diode, we come to appreciate the huge developments in the field of electronics—a technology that has had immense impact on our lives today.

24.5 BOOLEAN ALGEBRA

This is a system of mathematical logic. This has a set of rules, laws and theorems by which logical operations can be expressed symbolically in equation form and can be manipulated mathematically. In boolean algebra, every number is either a 0 or 1. There are no negative or fractional numbers. There are basically three boolean operators namely plus (+), times (×) and overbar (–) which may be thought of as codes for the basic gates representing the three basic logic operations AND, OR and INVERT. Following set of axioms are used to build a number of useful theorems (proofs of these axioms are beyond the scope of this book).

Axiom 1	$0.0 = 0$	Axiom 7	$1 + 0 = 1$
Axiom 2	$0.1 = 0$	Axiom 8	$1 + 0 = 1$
Axiom 3	$1.0 = 0$	Axiom 9	$\overline{1} = 0$
Axiom 4	$1.1 = 1$	Axiom 10	$\overline{0} = 1$
Axiom 5	$0 + 0 = 0$		
Axiom 6	$0 + 1 = 1$		

Axiom 8 is known as *idempotent relation*.

Basic Laws of Boolean Algebra

One will have to follow certain well developed laws and rules to apply boolean algebra properly. These laws are:

i. **The commutative law of addition**
This law for two variables is written algebraically as:

$$A + B = B + A$$

$$A \cdot B = B \cdot A$$

One can extend the commutative properties to any number of variables, i.e.

$$A + B + C = B + A + C$$

and $ABC = BAC = CAB$ and so on.

ii. **The associative Laws**
The associative law of addition for three variables is

$$A + (B + C) = (A + B) + C$$
$$A \cdot (BC) = (AB) \cdot C$$

One can extend the associative properties to any number of variables

$$(A + B + C) + D = (A + B) + (C + D)$$
$$= (A + B + C) + D$$
$$= A + B + C + D$$
$$A \cdot (BCD) = (AB) \cdot (CD)$$
$$= (ABC) \cdot D = ABCD$$

iii. The distributive laws

For three variables, the distributive laws can be written as

$$A \cdot (B + C) = A \cdot B + A \cdot C$$

$$A + BC = (A + B) \cdot (A + C)$$

$$A + (\bar{A} \cdot B) = A + B$$

Rules for Boolean Algebra

i. Duality property

- Changing each OR sign to an AND sign, i.e. $A + 0 = A$
- Changing each AND sign to an OR sign, i.e. $A - 1 = A$
- Complementing any 0 or 1 appearing in the boolean expression, for example

$$A + (B + C) = AB + AC$$

One can obtain the dual reduction by changing OR and AND operation.

ii. Complement property

$$A \bar{A} = 0$$

$$A + \bar{A} = 1$$

iii. Absorption property

$$A + AB = A$$

$$A(A + B) = A$$

iv. Idempotency property

Idempotency means the property of sameness, i.e.

$$A + A = A$$

$$A A = A$$

We may note that the laws $A + 0 = A$, $A + A = A$, $A + \bar{A} = 1$, $1 \cdot 0 = 0$, etc. are the duals of the laws $A \cdot 1 = A$, $A \cdot A = A$, $A \cdot \bar{A} = 0$, $0 + 1 = 1$, etc. We define duality as the changing of all AND signs to OR signs, all OR signs to AND signs, all 1's to 0's, and all 0's to 1's. Duality is one of the governing rules of boolean algebra, and it can be stated that *if an equation is true, then its dual is also true.*

We may note that the rule of duality is an extremely valuable tool for logic design, especially for complementing logic (boolean) functions and expressions as they contain both explicit and implicit parentheses. When applying the rule of duality, this grouping must be maintained. For example,

$$Y = \bar{A}(B + C\bar{D}) + C\bar{B}D$$

Now, with explicit and implicit parentheses, the above equation must be written as

$$\bar{Y} = \left\{ \bar{A} \left[B + \left(C\bar{D} \right) \right] \right\} + \bar{B}D$$

Application of the rule of duality gives

$$\bar{Y} = \left[A + \bar{B} \left(\bar{C} + D \right) \right] \left(B + \bar{D} \right)$$

ILLUSTRATIVE EXAMPLES

Example 1

Convert the binary number 110111 to decimal.

Solution

$$110111_2 = 1 \times 2^5 + 1 \times 2^4 + 0 \times 2^3 + 1 \times 2^2$$
$$+ 1 \times 2^1 + 1 \times 2^0$$
$$= 1 \times 32 + 1 \times 16 + 0 \times 8 + 1 \times 4$$
$$+ 1 \times 2 + 1 \times 1$$
$$= 32 + 16 + 0 + 4 + 2 + 1 = 55_{10}$$

Example 2

Convert 199_{10} into its binary equivalent.

Solution

There are good number of methods for converting a decimal number to a binary number. The first method is simply to subtract values of powers of 2 which can be subtracted from the decimal number until nothing remains. The value of highest power of 2 is subtracted first, then the second highest and so on. First the value of highest power of 2 which can be subtracted from 29 is $2^4 = 16$. Thus, $29 - 16 = 13$. Now, the value of highest power of 2 which can be subtracted from 13 is 2^3, then $13 - 2^3 = 13 - 8 = 5$. The value of highest power of 2 which can be subtracted from 5 is 2^2, then $5 - 2^2 = 5 - 4 = 1$. The remainder after subtraction is 1^0 or 2^0. Therefore, the binary representation for 29 is

$$29_{10} = 2^4 + 2^3 + 2^2 + 2^0$$

$$= 16 + 8 + 4 + 0 \times 2 + 1$$

$$= 1 \quad 1 \quad 1 \quad 0 \quad 1$$

$$\therefore \quad [29]_{10} = [11101]_2$$

The most popular method is *double-dabble method* which is also called as *divide-by-two* method. In this method, the decimal number is repeatedly *divided* by 2, and the remainder after each division is used to indicate the coefficient of the binary number to be formed. We may note that the binary number derived is written from the bottom up. For example, convert 199_{10} into its binary equivalent

$199 \div 2$	99 + remainder 1	**(L S B)**
$99 \div 2$	49 + remainder 1	
$49 \div 2$	24 + remainder 1	
$24 \div 2$	12 + remainder 0	
$12 \div 2$	6 + remainder 0	
$6 \div 2$	3 + remainder 0	
$3 \div 2$	1 + remainder 1	
$1 \div 2$	0 + remainder 1	**(M S B)**

The binary representation of 199 is 11000111. One can check the result

$$[11000111]_2 = 1 \times 2^7 + 1 \times 2^6 + 0 \times 2^5 + 0 \times 2^4$$
$$+ 0 \times 2^3 + 1 \times 2^2 + 1 \times 2^1 + 1 \times 2^0$$
$$= 128 + 64 + 0 + 0 + 0 + 4 + 2 + 1$$
$$= [199]_{10}$$

We may note that the first remainder is the LSB and last remainder is the MSB. However, this method will not work for *mixed numbers*.

Example 3

Convert 14.625_{10} to binary.

Solution

First the integral part 14 is converted into binary and then we convert the fractional part 0.625 into binary as below:

Integral part

$$14 \div 2 = 7 + 0$$
$$7 \div 2 = 3 + 1$$
$$3 \div 2 = 1 + 1$$
$$1 \div 2 = 0 + 1$$

Fractional part

$$0.625 \times 2 = 1.250 \text{ with a carry of 1}$$
$$0.250 \times 2 = 0.500 \text{ with a carry of 0}$$
$$0.5000 \times 2 = 1.000 \text{ with a carry of 1}$$

The binary equivalent is $[1110.101]_2$.

Example 4

Prove $A + BC = (A + B)(A + C)$

Solution

We have

$$A + BC = A \cdot 1 + BC \qquad (\because A \cdot 1 = A)$$
$$= A(1 + B) + BC \qquad (\because A + 1 = 1)$$
$$= A \cdot 1 + AB + BC \quad [\because A(B + C) = AB + BC]$$
$$= A \cdot (1 + C) + AB + BC \qquad [\because 1 + A = 1)$$
$$= A \cdot 1 + AC + AB + BC$$
$$= A \cdot A + AC + AB + BC \qquad [\because A \cdot A = A)$$
$$= A(A + C) + B(A + C)$$
$$= (A + C)(A + B)$$

Example 5

Simplify $A + A\bar{B} + \bar{A}B$.

Solution

$$A + A \cdot \bar{B} + \bar{A} \cdot B = (A + A \cdot \bar{B}) + \bar{A}B$$

$$= A(1 + \bar{B}) + \bar{A}B$$

$$= A + \bar{A} \cdot B$$

REVIEW QUESTIONS

1. Using examples where necessary, explain the following terms used in electronics:
   i. Logic gates
   ii. Truth tables
   iii. Integrated circuits
   iv. Microprocessor.

2. (a) List the three basic gates used in digital electronics, describing the function of each (b) Give two examples of how basic gates can be combined.

3. A two-input AND gate has two trains of pulses on its inputs A and B. The levels of pulses at different times intervals are shown, the duration of each time interval being 100 ms.

   Complete the following table by filling in the output level for each time interval.

4. Complete Table 24.2 for a two-input OR gate, with the same inputs.

Table 24.2

Time (ms)	Level of inputs		Output level
	A	B	Y
100	0	1	
200	1	1	
300	1	0	
400	1	1	
500	1	0	
600	0	0	
700	0	1	
800	0	0	

5. (a) Draw a sketch of the four-input OR gate.
(b) Draw the truth table for the four-input OR gate.

6. Prepare the truth table for each of the logic combinations depicted in Fig. 24.11

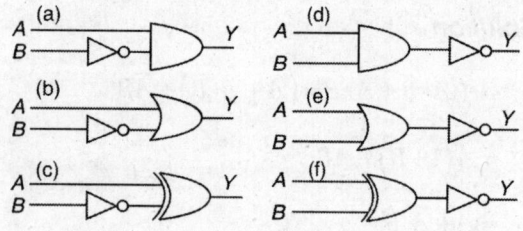

Fig. 24.11

7. In the circuit segment shown in Fig. 24.12, the output of each logic combination is connected to a lamp. For each combination state the input conditions that would switch the lamp on. Verify using a truth table. Assume all gates are correctly biased.

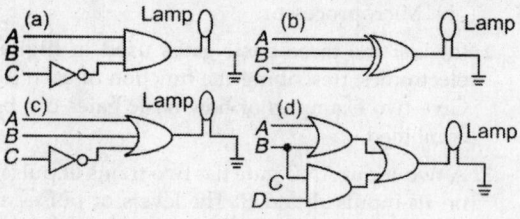

Fig. 24.12

8. Prepare the truth table for each of the logic combinations depicted in Fig. 24.13.

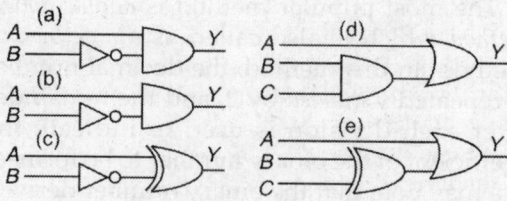

Fig. 24.13

SHORT ANSWER QUESTIONS

1. What are the two states of digital circuits?

Ans. The two states may be ON or OFF, true or false, high or low.

2. On which system, digital circuits operate?

Ans. Binary system

3. What is a binary system?

Ans. This system has a base 2 and uses only two bits 0 and 1.

4. Are binary arithmetic operations comparable to decimal operation?

Ans. Yes

5. What are the different ways in which negative numbers are expressed in the binary system?

Ans. There are three different ways in the binary systems in which negative numbers are expressed. These are: sign magnitude notation, one's complement notation and two's complement notation.

6. Does digital circuits can perform logical functions apart from binary arithmetic operations?

Ans. Yes

7. What is a logic gate?

Ans. This is an electronic circuit, which has two or more inputs and only one output.

8. What are the basic logic gates?

Ans. AND, OR and NOT gates.

9. What do you understand by one's complement addition and subtraction?

Ans. One's complement addition uses an end-around carry. One's complement subtraction requires inverting and then adding.

10. What do you understand by 2's complement addition and subtraction?

Ans. 2's complement addition ignores end-around carry bit and 2's complement subtraction requires complementing and then adding.

11. What is boolean algebra?

Ans. The binary logic applied in the design and analysis of digital circuits is called boolean algebra.

12. Which theorems are generally used for simplifying logic expressions?

Ans. De-Morgan's theorems.

13. Which number system is used in computers?

Ans. Computer uses binary number system to perform arithmetic and other operations dealing with numbers.

14. What is an AND gate?

Ans. An AND gate is a device whose output is 1 if and only if all its inputs are 1.

15. What is an OR gate?

Ans. An OR gate is device whose output is 1 if at least one of its inputs is 1.

16. What is a NOT gate?

Ans. The NOT gate or Inverter is the simplest among all the gates. This is a gate whose output is 1 when its input is 0 and whose output is 0 when its input is 1.

Nuclear Physics (Atomic Nucleus and Related Phenomena)

25.1 ATOMIC NUCLEUS (THE NUCLEUS AS A PART OF THE ATOM

The atomic nucleus contains only two types of particles namely proton and neutron. It is true further that particles may be ejected from the nuclear assembly but they do not exist independently within the nucleus. Protons and neutrons are designated by the common name of the nucleons. The other particles of the modern physics are never found as independent entities within the nucleus. The atomic nucleus is not only the seat of the mass but also the origin of the energy of the atom. The atomic nuclei is occupying a small region at the centre of the atom with a diameter of the order of 10^{-14} m. The nuclear radius of the various nuclei can be estimated from the following relation.

$$R = r_0 A^{1/3} \qquad (25.1a)$$

where r_0 is a constant ($= 1.2 \times 10^{-15}$ m), same for all nuclei. Various measurements for r_0 range between 1.0 m and 1.5 × 10^{-15} m. Looking at the smallness of nucleus, we use the *femtometer*, abbreviated fm, with 1 fm = 10^{-15} m. An alternative term for 10^{-15} m, Fermi. The term Fermi, has the same abbreviation, fm is more commonly used than femtometer. We may note that a nucleus has no clearly defined boundary. Obviously, the nuclear radius has only arbitrary meaning. Volume of an atomic nuclear radius, $V = 1.2 \times 10^{-45} A/m^3$. The mass of an atomic nucleus of mass number A is approximately

$$M = 1.66 \times 10^{-27} A \text{ kg}$$

Making use of these values, one obtains the average density of nuclear matter as

$$\rho = \frac{M}{V} = 1.49 \times 10^{18} \text{ kg/m}^3$$

which is independent of mass number A. This density of the nucleus is about 10^{15} times greater than the density of matter in bulk, and gives an idea of the degree of compactness of the nucleons in a nucleus. It also shows that matter in bulk is essentially empty, since most of the mass is concentrated in the nucleus.

The primary constituents of nuclei are the proton and neutron and the nuclear mass is roughly the sum of its constituent proton and neutron masses (the slight difference being the binding energy of nucleus). The nuclear charge is $+e$ times the number (Z) of protons.

The atomic (and nuclear) mass number A is the total integral number of protons and neutrons in a nucleus. We will designate an atomic nucleus by the symbol $^A_Z X_N$, where Z is atomic number (number of protons), A is mass number ($= Z + N$), N is neutron number (number of neutrons) and X is symbol of chemical element. Each nuclear species with a given Z and A is called a **nuclide**. When we write the nuclide symbol for the better known elements, Z is often omitted, e.g. ^{16}O, ^{14}N, etc.

We may note that the chemical properties of an atom are determined by its electron configuration. Since the number of electrons and protons are equal in a neutral atom and hence the chemical properties are essentially determined by Z. Obviously, the dependence of the chemical properties on N is negligible.

Atomic masses are measured in mass units which are denoted by symbol u. Atomic mass units are defined in terms of the mass of the isotope ^{12}C, whose atomic is defined to be exactly 12 u.

$1u = 1.6605 \times 10^{-27}$ kg $= 931.49$ MeV/c^2. The masses of proton, neutron and electron are given in Table 25.1

25.2 ATOMIC NUMBER, MASS NUMBER AND ISOTOPES

A particular nucleus is identified by the number of protons it has or its atomic number Z and mass number A. Thus, the number of neutrons is $N = A - Z$. A nuclide is designated by the symbol of the chemical element to which it belong, according to the value of Z, with superscript to the left indicating the value of the mass number A, such as ^{12}C, ^{16}O, ^{235}U. Sometimes it is convenient to write the atomic number Z as a subscript to the left, such as $^{12}_{6}C$, $^{16}_{8}O$, $^{235}_{92}U$. Due to great variety of nuclides, they are classified in three categories as mentioned below:

1. **Isotopes:** These are the nuclides having the same atomic number Z but different neutron number N and different mass number A, for e.g. $^{1}_{1}H$ and $^{2}_{1}H$ are isotopes of hydrogen, $^{238}_{92}UI$ and $^{234}_{92}UII$ are natural radioisotopes of $_{92}U$. Similarly $^{230}_{90}$Ionium, $^{232}_{90}Th$, $^{234}_{90}UX_1$ are natural isotopes. Let us now consider the radioactive disintegration.

$$^{230}_{90}\text{Ionium} \xrightarrow{\alpha} {}^{226}_{88}\text{Ra}$$

Obviously, $^{226}_{88}$Ra is the product of disintegration of ionium. It is found that the chemical properties of $^{230}_{90}$Ionium turned out to be similar to those of naturally occurring radioactive element thorium $\left(^{232}_{90}Th \right)$, i.e. when these two elements mixed could not be separated chemically even with extremely most sensitive methods. Also, ionium and thorium have been found to be spectroscopically identical. Clearly, $^{230}_{90}$Ionium and $^{232}_{90}Th$ are isotopes. Thus, one can say that isotopes mean the same position in the periodic table. In general, we can say that two species $^{A}_{Z}X$ and $^{A'}_{Z}Y$ which have different atomic masses (A and A') inspite of being different elements and have same atomic number (Z) and moreover identical chemical properties, then these different species are called *isotopes*.

2. **Isobars:** These are the nuclides having the same total number of nucleons (or A) but differ in Z and N. Examples of isobars are

$$^{14}_{6}C \text{ and } ^{14}_{7}N, \quad ^{234}_{90}UX_1, \quad ^{234}_{91}UX_2, \quad ^{234}_{92}UII$$

3. **Isotones:** These are the nuclides having the same number of neutrons N but different Z and A. Examples of isotones are $^{13}_{6}C$ and $^{14}_{7}N$. We may note that isotones having a given value of N obviously do not all correspond to the same chemical element. The analysis of the properties of **isotopes, isotones** and **isobars** helps us to disclose several

Table 25.1

Particle	Rest energy (MeV)	Charge	Atomic mass unit (u)	Spin, Magnetic moment
Proton (P)	938.272	$+e$	1.0072765	½, 2.79 μ_N
Neutron (n)	939.566	0	1.0086649	½, -1.91 μ_N
Electron (e)	0.51100	$-e$	5.4859×10^{-4}	½, -1.00116 μ_B

features of nuclear structure. Such analysis helps us, e.g. to predict what will happen to the stability of a nucleus when an extra neutron (n) or proton (p) is added to it.

Size and Shapes of Nuclei

The nuclear radius may be approximated from a spherical charge distribution to be

$$R = r_0 A^{1/3}$$

where $r_0 = 1.2 \times 10^{-15}$ m. Various measurements for r_0 range between 1.0 and 1.5×10^{-15} m. Since nucleus is so small, we use **femtometer** (fm), with 1 fm = 10^{-15} m, Fermi (fm). The term Fermi, which has the same abbreviation, fm, is more commonly used than femtometer.

Hofstadter and coworkers at Stanford University in the 1950s performed the first precision electron scattering measurement of the nuclear charge distribution using electron energies from 100 MeV to 500 MeV. In order to figure out the actual shape of most nuclei, one needs a particle having a short wavelength. The de-Broglie wavelength of a 500 MeV electron is about 2.5 fm, and by now, measurements have been made with much shorter wavelengths using higher energy electrons. These measurements are approximately described for all but the lightest nuclei by the Fermi distribution for the nuclear charge density $\rho(r)$ as

$$\rho(r) = \frac{\rho_0}{1 + \exp[(r-R)/a]} \quad (25.1b)$$

Figure 25.1 shows the shape of this distribution. Here ρ_0 is the central nuclear density, R is the distance at which the nuclear density has dropped to 50% of its central value.

Intrinsic Spin

The neutron and proton are fermions and spin quantum number of each is $f = 1/2$. The spin quantization rules for neutron and proton are same as for the electron.

Intrinsic Magnetic Moment

Since the proton's charge is positive and hence the proton's intrinsic magnetic moment points

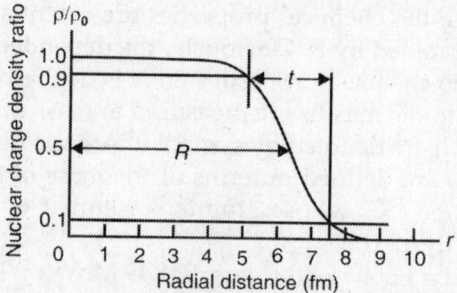

Fig. 25.1: The shape of the Fermi distribution for the nuclear charge density given by Eq. (25.1b) where $\rho_0(r)$ is the central charge density, R is the distance at which the nuclear density has dropped to 50% of its central value

in the same direction as its intrinsic spin angular momentum. We may note that this is in contrast with the electron, where the spin the magnetic moment point in the opposite direction. The unit of nuclear magnetic moment is nuclear magneton (μ_N). μ_N is defined by analogy to the Bohr magneton for electrons as follows:

$$\mu_N = \frac{e\hbar}{2m_p}$$

We may note that divisor in calculating μ_N is the proton mass m_p, which makes the nuclear magneton some 1800 times smaller than the Bohr magneton.

We have μ_p (proton magnetic moment) = 2.79 μ_N. We can see that this contrasts strongly with the magnetic moment of the electron $\mu_e = -1.00116\mu_s$. We may note that neutron, which is electrically neutral, also has a magnetic moment, $\mu_n = -1.91\mu_N$. Negative sign clearly indicates that the magnetic moment, points opposites to the neutron spin. It is interesting to note the large deviation from unity of the proton's magnetic moment and the fact that the neutron even has a magnetic moment clearly indicates that nucleons are more complicated structurally than electrons. The non-zero neutron magnetic moment clearly implies that the neutron has negative and positive internal charge components at different radii, and thus have a complex internal charge distribution.

25.3 NUCLEAR BINDING ENERGY

Mass spectroscopic studies of masses of the isotopes reveal that the mass number A of isotopes are all nearly whole numbers, the slight variation being only in the second place of decimal. This is called as **whole number rule.** This rule leads to the general acceptance that atomic nuclei are made up of hydrogen nuclei (protons) and neutrons having almost the same mass. Although atomic masses are close to whole numbers but they differ from integers by small amounts, i.e. they depart from the whole number rule systematically. The light nuclei have masses slightly higher than the nearest integer, e.g. $^1_1H = 1.100813$, $^4_2He = 4.00408$, $^6_3Li = 6.01614$. Large number of medium nuclei have masses somewhat lower, e.g. $^{20}Ne = 19.9988$, $^{35}_{17}Cl = 34.933$, $^{98}Mo = 97.944$. Very heavy nuclei have slightly higher mass again, e.g. $^{200}Hg = 200.028$, $^{205}Tl = 205.036$.

The difference between atomic mass and mass number of an atom is called **mass defect** (Δm), i.e.

$$\Delta m = M - A \qquad (25.1c)$$

where M is the measured mass and A is mass number. Obviously, the mass defect represents the deviation of atomic mass from the whole number A. The deviations of the masses of nuclides from whole numbers was an important problem. Aston expressed these deviations in the form of a quantity called **packing fraction** (f) defined by

$$f = \frac{\text{Actual atomic mass} - \text{Mass number}}{\text{Mass number}}$$

$$= \frac{M_{Z,A} - A}{A} = \frac{\Delta}{A} \qquad (25.2)$$

The packing fraction is defined as the mass defect per nucleon in its nucleus. f is a very small quantity. Aston multiplied the values of f by 10^4 to obtain figures which are more convenient to record. Thus, the usual quoted values of packing fractions are expressed as

$$f = \frac{\Delta}{A} \times 10^4 \qquad (25.3)$$

Concluding, one may say that packing fraction (f) is the mean gain or loss per particle in atomic mass, in order to pack the nuclear elementary particle to form a stable nucleus. The variation of packing fraction (f) with mass number for a large number of nuclides is shown in Fig. 25.2. The packing fraction, with the exception of these for ^4_2He, $^{16}_8O$ and $^{12}_6C$ fall on or near the smooth curve. The values are high for elements of low mass number, apart from the nuclides mentioned. From the graph, it is evident that the values of packing fractions at first fall rapidly with increasing mass number, pass through a relatively flat minimum and then rise slowly but steadily, for the heavy elements beyond $A = 180$. The knowledge of packing fraction was very useful in the study of isotopic masses, but they do not have a precise theoretical significance and precise physical meaning.

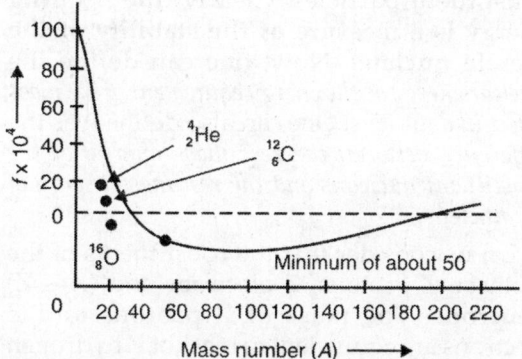

Fig. 25.2: Variation of atomic packing fraction $(f \times 10^4)$ with mass number (A)

25.4 BINDING ENERGY

When a nuclide is formed from its constituent particles, it is found that sum of the mass of the nuclide is less than the sum of the masses of its constituent particles in the free state. For example, deuterium atom $^2_1H(D)$ has a mass of 2.014102 u, while the rest mass of a free neutron and free electron together is equivalent to 2.016490 u, which is 0.002388 u greater. Obviously, when protons and neutrons combine to form a nucleus, a loss in mass results. This loss in mass is released in the form of equivalent energy when the nucleus is formed. If Δm is the decrease in mass when number of protons, neutrons and electrons combine to form an atom, then the energy released ΔE will be in accordance with the Einstein's relation

$$\Delta E = (\Delta m)c^2 \tag{25.4}$$

where c is the velocity of light. We may note that the same amount of energy would be needed to break the atom into its constituent particles. In other words, one can say that if Δm is the increase in mass due to the combination of protons, neutrons and electrons to form the atom, then the energy $E = \Delta mc^2$ will be absorbed in this process and when the atom breaks, this amount of energy will be released. The loss in mass (Δm) is called the *mass defect* and its energy equivalent (ΔE) is called the *binding energy* of the nucleus. Obviously, ΔE is also the energy that would be needed to break the nucleus into its constituent particles. Clearly, **the binding energy is a measure of the stability of the atomic nucleus.** Now, one can define the *nuclear energy as the energy equivalent of the mass defect in a nucleus.* One can also define it as the *difference between the rest mass energy of the constituent nucleons and the rest mass energy of the nucleus.*

Let us consider that the constituents of the atom be Z protons, Z electrons and $(A - Z)$ neutrons. The mass of Z protons and Z electrons are equivalent to mass of Z hydrogen atoms. Now, one can write the mass defect as

$$\Delta m = ZM_H + (A - Z) M_N - M_{Z,A}$$

where M_H is the mass of the hydrogen atom in u, M_N is the mass of neutron in amu and $M_{Z,A}$ is the atomic mass, the actual mass of the atom in amu.

We know that the mass of an electron is very negligible, and therefore the main contribution to the mass of the atom is due to the nucleus, i.e. due to the mass of nucleons (i.e. mass of protons and mass of neutrons). Thus, one can write the mass defect as

$$\Delta m = ZM_P + (A - Z) M_N - M_{Z,A}$$

where M_P is the mass of the proton in amu. Now the binding energy (B.E.) of the nucleus is

$$\text{B.E.} = \Delta mc^2 [ZM_P + (A - Z) M_N - M_{Z,A}]c^2$$

Since 1 amu = 931.48 MeV, and therefore B.E. of the nucleus is

$$\text{B.E.} = 931.48 [ZM_P - (A - Z) M_N - M_{Z,A}] \tag{25.5}$$

Now the binding energy per nucleon is obtained as

$$\frac{\text{B.E.}}{A} = 931.48 \left[\frac{Z}{A} M_P - \left(1 - \frac{Z}{A} \right) M_N - \frac{M_{Z,A}}{A} \right]$$

A graph of B.E./A (binding energy per nucleon) as a function of the mass number is shown in Fig. 25.3. From the graph, we note that with the exception of ^4_2He, $^{16}_8\text{O}$, and $^{12}_6\text{C}$, the values of B.E./A lie on or close to a single curve. We may note that B.E. is a measure of the stability of the nucleus. From the graph, one can easily conclude the following:

i. B.E./nucleon for very light nuclides, e.g. H, He, etc. is very small.

ii. B.E./nucleon rises sharply at first and then gradually until it reaches a maximum value of 8.79 MeV at $A = 56$ corresponding to $^{56}_{26}\text{Fe}$.

iii. The maximum of B.E./nucleon curve is quite flat and B.E./nucleon is still 8.4 MeV at $A = 140$.

iv. For higher mass numbers, the value of B.E./A drops slowly to about 7.6 MeV at the highest mass numbers, e.g. uranium at $A = 238$. Looking at the pacing fraction curve (Fig. 25.2), we note that its two ends show that packing fraction f is positive for light and heavy nuclides due to which B.E./A is also less for light and heavy nuclides. This fact immediately suggests that energy will be released if light nuclei can somehow be fused to form heavier ones.

This process of releasing nuclear energy by fusing lighter nuclei to form a medium size nucleus is known as *fusion* and is the source of energy in stars. Energy will also

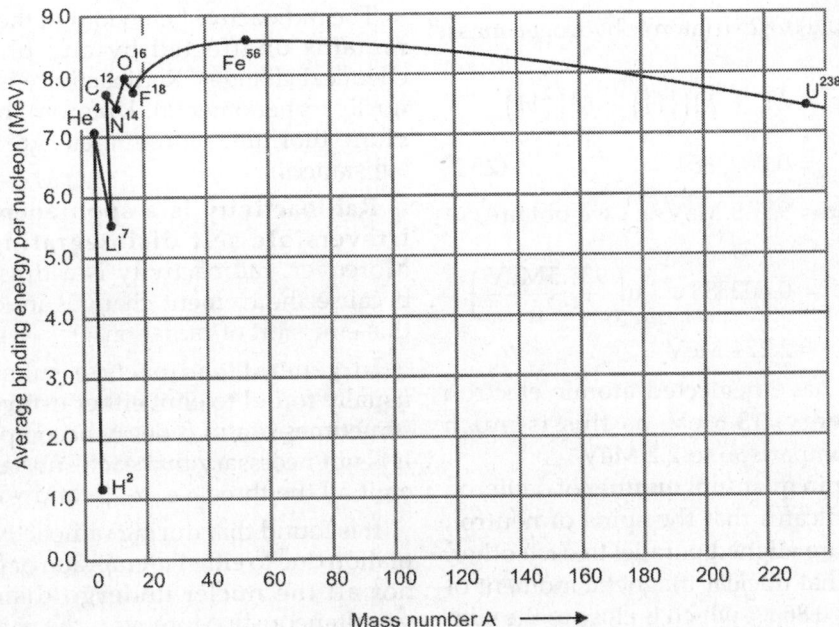

Fig. 25.3: Variation of binding energy per nucleon (B.E./A) with mass number A

be released if heavy nuclei can somehow be split up into lighter ones. This method of releasing nuclear energy by breaking up of a heavy nucleus is known as *fission* and is the basis for the operation of nuclear power reactors.

v. The B.E./A has an average value of about 8 MeV over a considerable range of mass numbers, where packing fraction f is negative.

vi. The positive value of B.E./A for all values of A reveals the stability of the nuclei and at the same time also shows that there will be attractive nuclear forces between the nucleons. We know that nuclei does not collapse, and therefore these forces must become repulsive at very close distance between nucleons.

vii. We note that B.E./A vary in peculiar manner from nucleus to nucleus for very lighter nuclides. Further, there is a rapid increase in B.E./A for light nuclei with a notable peak at A = 4n, e.g. for C-12 and O-16. Obviously, this reflects the peculiar stability of the α $\left(^4_2\mathrm{He}\right)$ particle structure.

viii. The B.E./A is remarkably constant (up to 10%) above A > 20. This reveals an important property of the nuclear forces, i.e. saturation characteristics of these forces. This shows that B.E. is nearly proportional to A, i.e. the nuclear constituents interact with only a limited number of their neighbouring nucleons. Obviously, nuclear forces are saturated.

ix. When Z or N equals to the magic numbers 2, 8, 20, 50, 82, 126 where B.E. is very high, discontinuities in the B.E. curve have been reported.

25.5 THE DEUTERON ($^2_1$H)

A deuteron consists of one proton and one neutron and after proton, this is the simplest nucleus. The deuteron mass is 2.013553 u and the mass of a deuterium atom is 2.014102 u. The difference B_d of the deuteron can be calculated from

$$M\left(^2_1\mathrm{H}\right) = M_n + M\left(^1_1\mathrm{H}\right) - B_d/c^2 \quad (25.6)$$

Here $M\left(^2_1\mathrm{H}\right) = 2.014102\,\mathrm{u}$ (atomic deuteron mass), $M_n = 1.008665\,\mathrm{u}$ (neutron mass)

$M\left(_{1}^{1}H\right) = 1.007825 \, u$ (atomic hydrogen mass)

$$B_d/c^2 = M_n + M\left(_1^1H\right) - M\left(_1^2H\right)$$

$$= 0.0023884 \qquad (25.7)$$

Taking $u = 931.5 \, MeV/c^2$, we obtain

$$B_d/c^2 = 0.002388 \; c^2. \; u\left(\frac{931.5 \, MeV}{c^2 u}\right)$$

$$= 2.224 \, MeV$$

Here, we have neglected atomic electron binding energy, 13.6 eV, as this is much smaller in comparison to 2.2 MeV.

Nuclear spin quantum number of deuteron is 1. This indicates that the spins of neutron and photon are aligned parallel to each other. We can see that nuclear magnetic moment of a deuteron is $0.86\mu_N$, which is close to the sum of the values for the free proton and neutron: $2.79\,\mu_N - 1.91\,\mu_N = 0.88\,\mu_N$. Obviously, this supports our hypothesis of parallel spins.

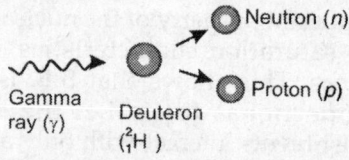

Fig. 25.4: A gamma ray photon of energy greater than 2.22 MeV is able to dissociate a deuteron into a neutron and a proton. This experiment of photodisintegration confirms that the binding energy of deuteron is 2.22 MeV

25.6 RADIOACTIVITY

Natural Radioactivity

Henry Becquerel in 1896 discovered the phenomenon of natural radioactivity. The atomic nuclei of certain substances, e.g. uranium and thorium minerals, radium, plutonium emit radiations which ionize gases and affect photographic plates. Such substances are said to be *radioactive* substances and the phenomenon is known as *natural radioactivity*. All the naturally occurring radioactive elements lie in the range of atomic number $Z = 81$ to 92.

The radioactive behaviour of the substances remains unaffected by any physical and chemical changes. Radioactivity is basically a nuclear phenomenon. Uranium, radon, polonium, thorium, actinium, etc. are radioactive substances.

Radioactivity is a spontaneous and an irreversible self disintegrating activity. Moreover, radioactivity is a drastic process because the element changes in kind due to this emission of radiation.

In general, radioactive substances are usually found to emit either α-rays or β-rays, sometimes α and β decay accompany γ-rays. It is not necessary that each substance should emit all the three, i.e. α-, β- and γ-rays.

It is found that during radioactive transformation due to either *d*-emission or β-emission, not all the nuclei undergo disintegration simultaneously Moreover, the rate at which the decay takes place depends on the number of parent atoms available at a given nuclei.

α-Rays

It is doubly ionized positively charged helium $\left(_2^4He\right)$ atom with its both the electrons removed.

β-Rays

This consists of electrons $\left(_{-1}^{0}e\right)$ with negative charge and almost zero mass. They move with very high velocities. We may note that β+-ray denotes positron $\left(_{+1}^{0}e\right)$ with positive charge, same in magnitude as an electron but with opposite charge and almost zero mass as an electron.

γ-Rays

They are electromagnetic waves (or photons) of very short wavelength ($\approx 0.4 - 0.004$ Å). Their frequency is higher than that of X-rays. They do not carry any charge.

Radioactive Decay or Disintegration

Although nuclear forces are very strong but heavy elements like uranium, thorium, plutonium, radium, etc. are unstable, i.e. these elements are constantly breaking up or disintegrating into fresh, radioactive atoms with the emission of α-, β- or γ-rays from their nuclei. In other words, one can say that an unstable nucleus disintegrates to achieve a new configuration which is stable or one which leads to a stable one. We may note that in the process of disintegration, the parent (original) atom disappears and gives rise to daughter (new) atom. These new atoms are also, in general, radioactive and hence spontaneously disintegrates, in turn, thereby leading to a chain of different radioactive elements forming a radioactive series, until an inactive element (usually lead) is reached. There are three well-known naturally occurring radioactive disintegration series: **(i) uranium radium series (ii) thorium series (iii) actinium series.** This spontaneous breaking up of the radioactive nucleus is termed *radioactive* disintegration.

We can represent α and β disintegrations as

$$_Z^A X \xrightarrow{\alpha\left(_2^4 He\right)} {}_{Z-2}^{A-4}Y + {}_2^4 H$$

where X is an element of mass number A and atomic number Z. This disintegrates into another element Y having mass number less than four, i.e. $A - 4$ and atomic number less then 2, i.e. $Z - 2$. Obviously, in this process α-particle $\left(_2^4 He \, nucleus\right)$ is emitted. An example of above disintegration is decay of $_{92}^{238} U$ nuclei.

$$_{92}^{238} U \xrightarrow[_2^4 He]{\alpha} {}_{90}^{234}Th + {}_2^4 He$$

Similarly, when a radioactive element disintegrates by the emission of β-rays, it is turned into another new element whose mass number does not change but atomic number (number of proton in the nucleus) increases by 1. In general, β−-disintegration can be expressed by the following nuclear reaction.

$$_Z^A X \xrightarrow[_{-1}^{0}e]{\beta^-} {}_{Z-1}^{A}Y + {}_{-1}^{0}e + \gamma\text{-rays}$$

An example of β disintegration is

$$_{82}^{214} RaB \xrightarrow[_{-1}^{0}e]{\beta^-} {}_{83}^{214}RaC + {}_{-1}^{0}e + \gamma\text{-rays}$$

Laws of Radioactive Disintegration

On the basis of experimental observations, Rutherford and Soddy in 1902 proposed the theory of radioactive disintegration. The theory accounts for the radioactive disintegration of uranium, radium and actinium. The important features of this theory are:

i. The radioactive disintegration is a spontaneous phenomenon producing new radioactive elements. The process continues till stable element is reached. Further, the disintegration can neither be accelerated nor be retarded. The phenomenon is a *statistical one* and entirely dependent upon the *law of change*.

ii. The atoms of radioactive element constantly break into new atoms with the emission of α-rays, β-rays, which are usually but not necessarily accompanied by γ-radiation.

iii. The radioactive atom has definite probability of disintegration. The number of atoms disintegrating in a certain time is directly proportional to the number of parent atoms remaining at that instant.

Let N be the number of atoms present in the radioactive substance at any instant t. Let dN be the number that disintegrates in a short interval dt. Then the rate of disintegration is $-dN/dt$, and is proportional to N. One can write

$$-\frac{dN}{dt} \propto N$$

or

$$\frac{dN}{dt} = -\lambda N = \frac{A}{C} \qquad (25.8)$$

where the proportionality constant λ is a constant for a given substance and is known as *transformation constant* or *decay constant* or *radioactive constant* or *disintegration constant*.

The constant λ is independent of all external conditions like temperature, pressure. The negative sign in Eq. (25.8) indicates that there is a decrease in the original number of atoms in present element. In Eq. (25.8), A is called the activity and C is the detection coefficient. We may note that N and A are always interchangeable by the relation $\dfrac{A}{C} = -\lambda N$. Equation (25.8) can be rewritten as

$$\frac{dN}{N} = -\lambda dt \qquad (25.9)$$

If at $t = 0$, $N = N_0$ and at $t = t$, $N = N$, then by integrating Eq. (25.9) within the limits 0 to t, one obtains

$$\int_{N_0}^{N} \frac{dN}{N} = -\lambda \int_{t_0}^{t} dt$$

or $\left[\log_e N\right]_{N_0}^{N} = -[\lambda t]_0^t$

or $\dfrac{\log_e N}{\log_e N_0} = -\lambda t$

or $\dfrac{N}{N_0} = e^{-\lambda t}$

or $N = N_0 e^{-\lambda t} \qquad (25.10)$

Equation (25.10) shows that the number of atoms of a given radioactive substance decreases exponentially with time, provided no new atoms are added or introduced, i.e. more rapidly at first and slowly afterwards (Fig. 25.5). From the graph, it is obvious that N is theoretically never zero but $N \to 0$ as $t \to \infty$. Equation (25.10) is experimentally verified for many radioactive elements. Figure 25.6 shows the decay of a radioactive element A and growth of other element B. Here the relative activity N/N_0 is plotted against decay time. The quantity dN/dt, i.e. rate of decay is known

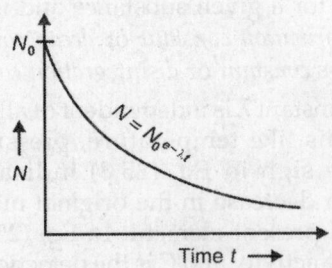

Fig. 25.5: Radioactive decay

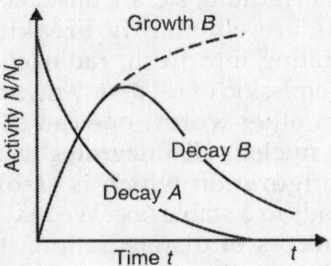

Fig. 25.6: Radioactive decay and growth

as activity or the strength of the radioactive substance. Using Eq. (25.10) and putting $t = \dfrac{1}{\lambda}$, one obtains

$$N = N_0 e^{-1/\lambda}\,\lambda \ \text{or} \ \frac{N}{N_0} = \frac{1}{e} \quad (25.11)$$

Hence, the radioactive constant is the reciprocal of the time during which the original number of atoms of a radioactive substance reduce to $1/e$ of its value.

Half-life

The half-life period T of a radioactive substance is the time in which the amount of radioactive substance is reduced to half of its original value, i.e.

$$\frac{N}{N_0} = \frac{1}{2}$$

In general, $\dfrac{N}{N_0} = e^{-\lambda t}$

When $t = T$, then $\dfrac{N}{N_0} = e^{-\lambda T} = \dfrac{1}{2}$

or $\lambda T = \log 2 = \log_{10} 2 \times 2.3026$

$\qquad = 0.03010 \times 2.306 = 0.6931$

or $T = \dfrac{0.6931}{\lambda} \qquad (25.12)$

This is the relation between half-life and decay constant. The half-life for radium is about 1600 years and that for uranium is 5000 million years.

Average Life (τ) or Mean Life of a Radioactive Atom

The average life of a radioactive atom is equal to the sum of the lifetime of all the atoms divided by the total number of atoms.

Mathematically, average life of radioactive substance is

$$\tau = \frac{1}{\lambda} = \frac{T}{0.693} \qquad (25.13)$$

i.e. mean life τ of a radioactive atom is equal to the reciprocal of its disintegration constant and it is greater than its half-life period. The reason is that the last few atoms of a radioactive substance may last for a very long period to time.

Units of Radioactivity

The unit of radioactivity is *Curie*, which is defined as the activity of one gm of radium in which 3.7×10^{10} atoms disintegrate per second, i.e.

$$1 \text{ Curie} = 3.7 \times 10^{10} \text{ disintegrate/sec}$$

Radioactivity is also measured in **Rutherford** which is defined as the activity of 1 gm of radium in which 10^6 atoms disintegrate per second. The smaller units of radioactivity are

1 mC (milli Curie) = 10^{-3} Curie = 3.7×10^7 dis/sec

1μC (micro Curie) = 10^{-6} Curie = 3.7×10^4 dis/sec

1 mRd (milli Rutherford) = 10^3 Rutherford (Rd)
$$= 10^3 \text{ dis/sec}$$

1μRd (micro Rutherford) = 10^{-6} Rutherford (Rd)
$$= 1 \text{ dis/sec.}$$

25.7 ALPHA, BETA AND GAMMA DECAY

Radioactive decay may occur for a nucleus when some other combination of nucleons has a lower mass. Consider the radioactive nucleus $^A_Z X$ called the parent and have the mass $M\left(^A_Z X\right)$. In the decay, two or more products can be produced. In the case of two products, let the mass of the lighter one be M_y and the mass of the heavier one (usually called the daughter) be M_D. According to principle of conservation of energy, we have

$$M\left(^A_Z X\right) = M_D + M_y + \frac{Q}{c^2} \qquad (25.14)$$

where Q is the energy released and is equal to the total kinetic energy of the reaction products. In terms of mass, Q is given by

$$Q = \left[M\left(^A_Z X\right) - M_D - M_y \right] c^2 \quad (25.15)$$

We may note that the disintegration energy Q is the negative of the binding energy (B.E.). Interestingly, the binding energy normally refers to stable nuclei, whereas Q is normally used with unstable nuclei. In case, if B.E. > 0, a nuclide is bound and stable, if $Q > 0$, a nuclide is unbound, unstable and decay. If we examine the naturally abundant radioactive nuclei, one finds that decays emitting nucleons do not occur, as the masses are such that $Q < 0$.

Alpha Decay (α-decay)

Alpha $\left(^4_2 He\right)$ nuclei is particularly stable. The binding energy of α nuclei is 28.3 MeV. In a α nuclei, the combination of two neutrons and two protons is particularly strong because of the pairing effects. If the last two protons and two neutrons in a nucleus are bound by less than 28.3 MeV, then an α-decay is energetically possible. We have

$$^A_Z X \xrightarrow{\alpha} {}^{A-4}_{Z-2} Y + \alpha$$

$$Q = \left[M\left(^A_Z X\right) - M\left(^{A-4}_{Z-2} Y\right) - M\left(^4_2 He\right) \right] c^2 \quad (25.16)$$

If $Q > 0$, then α-decay is possible. The familiar α-decay reaction is

$$^{230}_{92} U \longrightarrow \alpha + {}^{226}_{90} Th$$

$$M\left(^{230}_{92} U\right) = 230.033924\,u$$

$$M\left(^4_2 He\right) = 4.0026034\,u$$

$$M\left(^{226}_{90} Th\right) = 226.0248914\,u$$

$$\therefore \quad Q = [M(^{230}\text{U}) - M(^{226}\text{Th}) - M(^{4}\text{He})]c^2$$

$$= [230.033927\,\text{u} - 226.02489\,\text{u}$$

$$- 4.002603\,\text{u}]c^2 \left(\frac{931.5\,\text{MeV}}{c^2\text{u}}\right)$$

$$= 6.0\ \text{MeV}.$$

Obviously, $Q > 0$ and hence α-decay is allowed. The mass of the products is less than the mass of the decaying nuclide. Many of the nuclei above $A = 150$ are found susceptible to α-decay. These heavy nuclei have increasingly stronger coulomb repulsion as protons are added. The expulsion of two protons along with two neutrons in the form of α-particle may decrease this coulomb energy and make the resulting nucleus more stable.

One may ask, why any nuclei with $A > 150$ exists? We may note that nuclei are not necessarily made up of a collection of α-particles. For α-decay to occur, two neutrons and two protons group together within the nucleus prior to decay. Further, the α-particle, even when formed, has great difficulty in overcoming the nuclear attraction from the remaining nucleons to escape. Figure 25.7 shows the potential energy diagram.

Normally, the barrier height V_B for α-particle is greater than 20 MeV. The kinetic energy of α-particles emitted from nuclei ranges from 4 MeV to 10 MeV. Classically, it is impossible for the α-particles to escape as the P.E. barrier is greater than the kinetic energy of α-particles. If we project 5 MeV α-particles onto a heavy nucleus, we find that α-particle is repelled by the coulomb force as shown in Fig. 25.7 and does not get close enough to feel the attraction of the short range nuclear force. Classically, it is virtually impossible for an α-particle to reach the nucleus. Now the question arises, how can the α-particle ever surmount the barrier if it is trapped inside the potential barrier. However, quantum mechanically, the α-particles are able to tunnel through the barrier. Quantum mechanically, there is small but finite change for the α-particle to appear on the other side of the barrier. We may note that the probability depends critically on the

barrier height and width. Figure 25.8 shows that a higher energy α-particle, E_2 has a much higher probability than does a lower energy α-particle, E_1 to tunnel through the barrier. We may note that the higher tunnelling probability corresponds to a shorter lifetime for the radioactive nuclide. A comparison of lifetimes of various α-emitters with the kinetic energies of the α-particles is shown in Fig. 25.9.

We can easily see that there is a strong correlation between lower energies and greater difficulty of escaping.

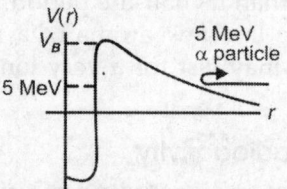

Fig. 25.7: The potential energy barrier for an α-particle. One can see that the coulomb barrier V_B is much greater than typical α-particle energies produced by radioactive sources. We may note that classically a 5 MeV particle inside the nucleus or scattered from outside cannot penetrate the potential barrier.

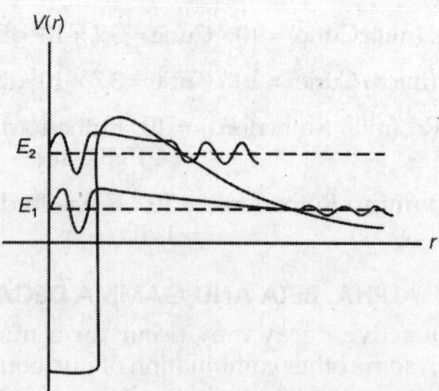

Fig. 25.8: Quantum mechanically, α-particles can tunnel through the barrier. A higher energy α-particle E_2 has a much higher probability (shorter lifetime) than a lower energy α-particle E_1. We may note that the waves shown in figure are schematics only

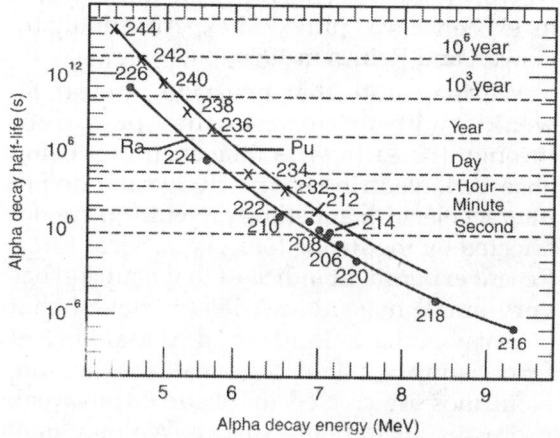

Fig. 25.9: A plot of half-lives for several radioactive α-emitters of radium and plutonium isotopes versus their α-energy. The two curves reveal that the higher energy α-particles result from nuclei having a much shorter lifetime, these α-particles have a higher probability of tunneling through the coulomb barrier

We may note that when a radioactive parent at rest α-decays to an α-particle and a heavy daughter (*D*) (Fig. 25.10), conservation of mass–energy and momentum still must occur. We have

$$Q = K_\alpha + K_D \qquad (25.17)$$

$$p_\alpha = p_D$$

$$K_\alpha = Q - K_D = Q - \frac{p_D^2}{2M_D}$$

$$= Q - \frac{p_\alpha^2}{2M_D}$$

$$K_\alpha = Q - \frac{2M_\alpha K_\alpha}{2M_D} = Q - \frac{M_\alpha}{M_D} K_\alpha$$

or $K_\alpha\left(1 + \dfrac{M_\alpha}{M_D}\right) = Q$

or $K_\alpha = \dfrac{M_D}{M_D + M_\alpha} Q = \left(\dfrac{A-4}{A}\right) Q \qquad (25.18)$

Since the parent mass *A* is normally over 150, α-particle takes most of the energy.

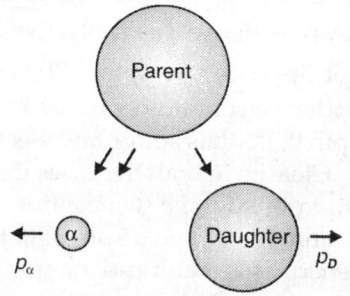

Fig. 25.10: α-decay of radioactive parent to an α-particle and a heavy daughter. Conservation of mass–energy and momentum still obeys. The momentum of α-particle and daughter are equal and opposite

Beta Decay (β-decay)

β⁻-decay

Some nuclides are not stable, and therefore radioactive decay occurs. We have seen that in α-decay the parent nucleus reverts to daughter nucleus that is down 2 units in atomic number *Z* and 4 units in mass number *A*. In many cases, it is observed that the α-decay leaves the daughter nucleus farther from the line of stability than the parent. Unstable nuclei may move closer to the line of stability by undergoing β⁻-decay, for example, the β-decay of neutron

$$^1_0 n \rightarrow {}^1_1 p + \beta^- \left({}^{\;0}_{-1} e\right) \qquad (25.19)$$

We know that electrons cannot exist within the nucleus, so when β⁻-decay occurs for nuclide, β⁻-particle is created at the time of decay. The β⁻-decay of ^{14}C to ^{14}N, a stable nucleus is

$$^{14}_{\;6} C \rightarrow {}^{14}_{\;7} N + \beta^- \qquad (25.20)$$

Like α-decay, this decay produces two products and one expect to measure a mono-energetic electron spectrum. However, the electron energy spectrum from the β⁻-decay of ^{14}C (Fig. 25.11) shows a continuous energy spectrum up to a maximum energy. The experimental result remains a major puzzle for several years. In addition to strange energy spectrum, there was no spin conservation.

We may note that in $_0^1n$ decay, the spin half $_0^1n$ cannot decay to two spin 1/2 particles, a proton and an electron. In case of ^{14}C decay, ^{14}C has spin 0, ^{14}N has spin 1 and electron has spin 1/2 · Clearly, in both the cases there is no conservation of spin, i.e. we cannot combine spin 1/2 and 1 to obtain a spin of 0. Both the electron energy spectrum and the spin angular momentum posed major challenge before scientists to understand β⁻-decay.

W Pauli in 1930 proposed correct explanation, which suggested that a third particle later named as 'neutrons' (v) must be produced in β⁻-decay. v has spin quantum number 1/2 , charge 0 and carries away the additional energy missing in Fig. 25.11. In Fig. 25.11, an occasional electron is detected with K.E., K_{max} required to conserve energy, but in the great majority of cases, the electron's K.E. is less than K_{max}. This reveals that the neutrino has little or no mass, and its energy may be all kinetic, like the photon. We may note that the photon cannot be the massing particle as it has spin 1. Pauli's suggestion seemed to explain the difficulties faced in β⁻-decay spectrum, and all circumstantial evidence supported the v hypothesis.

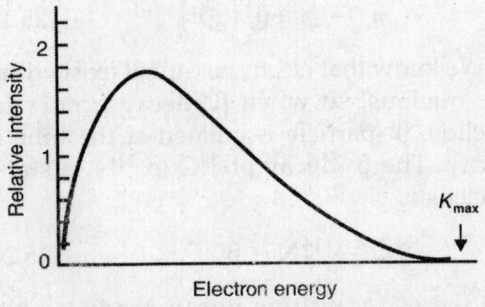

Fig. 25.11: β⁻-decay of ^{14}C, the relative intensity of electrons as a function of kinetic energy. If there were only two products in β⁻-decay, the electron energy would be monoenergetic and the energy equals to K_{max}

However, the detection of v was difficult and its existence was provided experimentally by Cowan and Reines in 1956.

We may note that neutrino interact so weakly with matter that they pass right through the earth with little chance of being absorbed. Neutrino has no charge and do not interact electromagnetically. Neutrino are not affected by the strong force, i.e. nuclear force. Recent experiment indicates that neutrino has very small mass. Now, it is believed that β⁻-decay is the reflection of a special kind of force, simply called the weak interaction. Neutrinos are created (or absorbed) in weak processes, including β⁻-decay. We may note that the electromagnetic and weak forces are two manifestations of the **electroweak** force.

Now, we also know about antineutrinos ($\bar{v}$). The β⁻-decay of a free neutron and ^{14}C is how correctly written as

$$_0^1n \longrightarrow {}_1^1p + \beta^- + \bar{v} \ (\beta^-\text{-decay})$$

β⁺-decay

The nucleus ^{14}O is unstable and having an excess of protons. It decays by emitting a positron (β⁺) to become a stable ^{14}N. The reaction is

$$^{14}O \longrightarrow {}^{14}N + \beta^+ + v \ (\beta^+\text{-decay})$$

Nuclei near ^{14}N with an excess neutron (^{14}C) or proton (^{14}O) will decay to ^{14}N by the appropriate beta decay.

In general, β⁺-decay reaction can be expressed as

$$_Z^A X \longrightarrow {}_{z-1}^A Y + \beta^+ + v \ (\beta^+\text{-decay}) \quad (25.21)$$

The disintegration energy Q for β⁺-decay is

$$Q = \left[M\left(_Z^A X\right) - M\left(_{z-1}^A Y\right) - 2m_e \right] c^2$$

$$(\beta^+\text{-decay}) \quad (25.22)$$

We may note that in this case the electron masses do not cancel when atomic masses are used. The mass of β⁺ is equal to that of β⁻.

Electron Capture

For β-decay, there is one other important possibility. Since inner K-shell and L-shell electrons are tightly bound and their (classical) orbits are highly elliptical, these electrons spend a reasonable amount of time passing through the nucleus thereby enhancing the possibility of atomic electron capture (EC). A proton in the nucleus absorbs the e^-, producing a neutron and a neutrino. One can write the reaction for a proton as

$$p = e^- \rightarrow n + v$$

The general reaction can be expressed as

$$^A_Z X + e^- \longrightarrow ^{A}_{Z-1} Y + v \text{ (electron capture)}$$
$$(25.23)$$

When one of the inner atomic electrons is captured, another electron will take its place, and thus produce a series of characteristic atomic X-ray spectra. This is a signature of electron capture as these X-rays are produced in the absence of any other kind of radiation. We may note that X-rays can also be produced in other kinds of nuclear decay because the decay product may knock out electrons (e.g. an α-particle).

Interestingly, electron capture has the same effect as positron decay, a proton is converted to a neutron. We way note that electron capture occurs more frequently for higher Z nuclides as the inner atomic electron shells are more tightly bound and there is greater probability of an electron being absorbed. One can write the disintegration energy Q for electron capture as

$$Q = \left[M\left(^A_Z X\right) - M\left(^{A}_{Z-1} Y\right) \right] c^2 \text{ (electron capture)}$$
$$(25.24)$$

Since $Q > 0$ for β⁺-decay or electron capture (EC) to occur, here will be some cases where EC is possible, but not β⁺-decay due to the difference in Eqs. (25.21) and (25.24).

Gamma Decay (γ-decay)

A nucleus in an excited state decays by emitting a photon, just like an atom. But the much higher energy associated with nuclear processes puts such photons in the gamma ray region of the spectrum. Since the γ-ray photon is neutral and massless, it does not change the type of nucleus, and therefore one may write

$$^A_Z X^* \rightarrow ^A_Z X + \gamma \text{ (gamma decay)} \quad (25.25)$$

The γ-ray energy hv is given by the difference of the higher energy state $E >$ (or E_2) and the lower energy $E <$ (or E_1)

$$hv = E > - E < = E_2 - E_1 \quad (25.26)$$

In order to conserve momentum, the nucleus normally must absorb some of this energy difference. However, for a nucleus initially at rest, Eq. (25.26) is a very good approximation.

25.8 NUCLEAR FORCE

What hold nucleons in a nucleus together? The protons all carry a positive charge of electricity and they must repel one another. Yet the nucleons, both protons and neutrons are tightly bound in the nucleus. Obviously, there must also be a very strong force of attraction. This force cannot be electrical in nature since such a force would lead to repulsion. Furthermore, an electrical force could not involve the neutrons since they have no electrical charge. There is evidently an entirely different type of force operating in a nucleus, and it must be so strong that it can then overcome the electrical repulsion of the closely packed protons. This special nuclear force is an aspect of what is called the strong interaction. It is the strongest type of force known. The nature of the nuclear force, holding nucleons in nuclei, has not yet fully been clarified. At the same time, much data have been obtained on the physical properties of the nuclei, as well as on the interaction of free nucleons in collisions over a wide range of kinetic energies from 10^{-4} to 10^{-10} eV. The analysis of the observed phenomena allows the following conclusions to be drawn about the force between nucleons. We may note that the energy of the nuclear forces and of coulomb interaction between the protons in the nucleus is equal to the *binding energy*.

   i. *Nuclear forces are charge independent forces:* The nuclear forces acting between two protons (p, p) or between two neutrons (n, n) are the same. This reveals that nuclear forces are of a non electrical nature.

ii. Nuclear forces are forces of attraction: Nuclear forces are extremely strong attractive forces over a distance of about nuclear radius (10^{-15} to 10^{-14} m); beyond this distance the force drops sharply almost to zero. The magnitude of nuclear forces is one million times greater than the force needed to separate an extra nuclear electron from the atom.

iii. *Nuclear forces are short range action forces:* The nuclear forces only operate at distances between the nucleons that are comparable (in order of magnitude) with the size of the nucleons. These distances are called range R ($\simeq 1.5 \times 10^{-15}$ m) of nuclear forces. Although the exact dependence on distance is not well established, there is good evidence that the force vanishes for all practical purposes at distances greater than a few times 10^{-15} m. According to Yukawa, the nuclear force between two nucleons is due to the interchange of particles called *mesons*. Yukawa represented the nuclear force field by a potential $V_p(r)$ given by

$$V_p(r) = -V_0 r_0 \frac{e^{-r/r_0}}{r} \qquad (25.27)$$

where V_0 and r_0 are empirical constants. Constant r_0 is the range of the nuclear force and V_0 gives the strength of the interaction. However, there are some doubts about that the nuclear interaction can be described in terms of potential energy functions in the same way that we have been able to explain the gravitational and electromagnetic interactions, which falls off rapidly at $r = 1/\alpha$.

According to this theory, the force between the two nucleons is the result of constant exchange of mesons between them.

iv. *Nuclear forces are spin dependent:* Nuclear forces depend on the mutual orientation of spins of various nucleons and are different in parallel and antiparallel spins.

v. *Nuclear forces are strongest in nature:* The force between the nucleons are the strongest forces known in nature. The gravitational and the electromagnetic-interacting forces were known to man long before the knowledge of nuclear forces. The gravitational forces are the weakest forces ~10^{-40} of the strong interacting nuclear forces.

vi. *Nuclear forces get readily saturated by surrounding nucleons:* Nuclear forces are the only ones known in nature that show saturation effect. The ability of nuclear forces to act upon other particles attain a point of saturation when a nucleon gets completely surrounded by other nucleons. However, the nucleons those are located outside the surrounding nucleons do not 'feel' the interaction of the surrounded nucleons.

Summarising, we can say that nuclear forces between nucleons are attractive in nature when they are $0.5 - 2.5$ fm (1 fm $= 10^{-15}$ m) apart. These forces are short range forces having maximum value at about ~2×10^{-15} m and falls off sharply with distance. Obviously, they become negligible beyond this range. Nuclear forces are charge independent, and therefore nuclear force between p-p, n-n and p-n are almost the same. Nuclear forces have the property of saturation, i.e. a particular nucleon interacts with a limited number of nucleons around it and other surrounding ones remain unaffected. This is why nuclear forces become saturated over short distance. Nuclear forces are *spin dependent*, i.e. depend on the mutual orientation of spins of various nucleons and are different in parallel and anti-parallel spins.

In addition to *strong nuclear* short range *force* which is far stronger than coulomb repulsion force, there is a short range force far more weaker than the nuclear force. It may be as small as 10^{-14} of strong nuclear force. This is termed *weak interaction short range nuclear force* as indicated by *beta decay*. We may note that it is not of gravitational or coulomb type. We may also note that weaker the force, larger must be the system in order that it might be of importance, e.g. strong interacting nuclear force holds the nucleons, the electromagnetic force holds the larger systems of atoms and molecules, while the gravitational force becomes important in astral bodies and systems. Now, we can summarise the chief forces of nature as *strong nuclear* force, *weak interacting* force, *electromagnetic* force and *gravitational* force.

We may note that in accordance with Yukawa's theory of nuclear forces, protons and neutrons do not exist independently within a nucleus but constantly exchange charges by emission and absorption of π-*mesons* or *pions* (within the nucleus), in very short intervals ~10^{-23} to 10^{-24} s. As the exchange occurs in a very short time and hence no visible change in nucleonic mass would be observed in accordance with uncertainty principle. This gives rise to rapid exchange of meson or mesonic field between protons and neutrons in which meson acts as a quantum of nuclear force. We may also note that this process is analogous to exchange of photons between charge particles in electromagnetic interactions. Many observable properties of the nuclear force have been successfully explained by a model based on the exchange of pions.

25.9 STABLE NUCLEI OR NUCLEAR STABILITY

Not every combination of protons and neutrons will stick together indefinitely. Too many protons and electrical repulsion wins out, sooner or later the nucleus decays by emitting a chunk of nuclear material. In larger nuclei, most protons are far apart and therefore experience electrical repulsion more strongly than nuclear attraction (Fig. 25.12).

To hold these nuclei together, therefore, requires more neutrons, which contribute attractive nuclear force but not electrical repulsion. So larger nuclei tend to have a higher ratio of neutrons to protons. Even this effect has its limits, and the result is that there are no stable nuclei for $Z > 83$.

A nucleus containing A nucleons is said to be **stable** if its mass is smaller than that of any other possible combination of A nucleons. The binding energy of a nucleus $^A_Z X$ against dissociation into any other possible combination of nucleons, for e.g. nuclei R and S, is given by

$$\text{B.E.} = \left[M(R) + M(S) - M\left(^A_Z X\right) \right] c^2 \quad (25.28)$$

In particular, the energy required to remove one proton (or neutron) from a nuclide is called the proton (or neutron) **separation energy,** and Eq. (25.28) is quite useful for finding this energy. We may note that even if B.E. is negative for a particular dissociation, there may be other reasons why the nucleus is **stable.**

Figure 25.12 shows known stable nuclei as well as many known unstable nuclei that are long-lived enough to be observed. We may note that Fig. 25.12 also includes nuclides that decay in a millisecond or less. The line representing the stable nuclides is called the **line of stability.** From the line of stability, one can extract the following facts:

 i. It appears that for $A \le 40$, nature prefers the number of protons and neutrons in the nucleus to be about the same, $Z = N$. However, for $A \ge 40$, there is a decided preference for $N > Z$. One can understand this difference as follows. The strength of the nuclear force is independent of whether the particles are *n-n*, *n-p* or *p-p*. Equal number of neutrons and protons may give the most attractive average internucleon nuclear force, but we must consider coulomb force as well. We may note that as the number of protons increases, the coulomb force between all the protons becomes stronger and stronger until it eventually affects the binding significantly.

 ii. One can calculate the **electrostatic energy** required to contain a charge Ze evenly spread throughout a sphere of radius R

by determining the work required to bring the charge inside the sphere from infinity and comes out to be

$$\Delta E_{coul} = \frac{3}{5}\left(\frac{(Ze)^2}{4\pi\varepsilon_0 R}\right) \quad (25.29)$$

In case of a single proton, Eq. (25.29) gives for self-energy,

$$\Delta E_{coul} = \frac{3}{5}\left(\frac{e^2}{4\pi\varepsilon_0 R}\right) \quad (\because Z=1) \quad (25.30)$$

Obviously, Eq. (25.30) represents the work done to assemble the proton itself and we do not want to include it in the electrostatic self-energy of a nucleus composed of Z protons. This means that we must subtract Z such terms from the total given in Eq. (25.29) to give us the total coulomb repulsion energy in a nucleus

$$\Delta E_{coul} = \frac{3}{5}\left(\frac{Z(Z-1)e^2}{4\pi\varepsilon_0 R}\right) \quad (25.31)$$

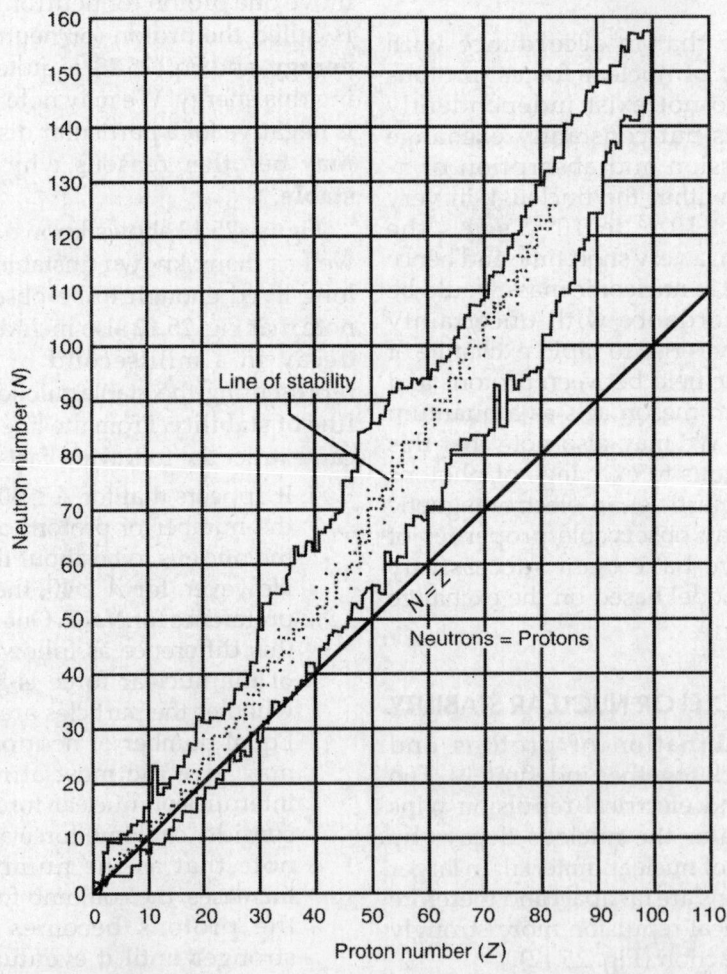

Fig. 25.12: A plot of the known nuclides with neutron number N versus proton number Z. The solid points represent stable nuclides and the shaded area represents unstable nuclei. A smooth line through the solid points would represent the line of stability

The delicate balance between neutrons and protons results in about 400 known stable nuclei, collectively called nuclides. Figure 25.12 reveals that there are no stable nuclei with $Z > 83$ because of the increasingly larger coulomb force. The heaviest known stable nucleus is $^{209}_{83}$Bi. All nuclei with $Z > 83$ and $A > 209$ will envetually decay spontaneously into some combination of smaller masses. Adding one proton to a heavy nucleus adds a constant amount of nuclear B.E., but the repulsive coulomb energy increases as $\Delta E_{coul} \approx (Z + 1)^2 - Z^2 \approx 2Z$. Since the coulomb force is long range, the proton interacts electromagnetically with the protons within the nucleus already there. And because this repulsive energy increases with Z, nuclei with higher Z eventually become unstable. However, the neutrons dilute the coulomb repulsion slightly because they intersperse among the protons, causing the protons to be slightly farther apart.

Figure 25.12 reveals that most stable nuclides have both even Z and even N (called even–even nuclides), e.g. $^2_1$H, $^6_3$Li, $^{10}_5$B and $^{14}_7$N. All the other stable nuclides are odd–even or even–odd, i.e. with either an odd number of Z or N. Obviously, nature apparently prefers nuclei with even numbers of protons and neutrons.

Pauli exclusion principle helps to understand the above empirical observation. Neutrons and protons are distinguishable fermions, hence they separately obey the exclusion principle. Only two neutrons (or protons) may coexist in each spatial (quantum state), one with spin up and the other with spin down. Obviously, each nuclear energy level is then able to hold two particles whose spins are paired to 0. We may note that this configuration of opposite spins is particularly stable as placing the same number of particles in any other arrangement will produce a (less stable) state of higher energy. Thus, there lies the preference for even N and Z.

25.10 NUCLEAR MODELS

We have seen how the right ratio of neutrons to protons is essential for stable nuclei and why that ratio increases for larger nuclei. Figure 25.12 summarizes this information. But a closer look at Fig. 25.12 reveals that there are more stable nuclei for even values of Z, and some those with the so-called magic numbers 2, 8, 20, 28, 50, 82 and 126 protons and neutrons have more stable nuclei, why?

Answering this question and explaining the decay mechanisms and lifetimes of unstable nuclei require a theory of nuclear structure. There is still no complete nuclear theory analogous to atomic theory that explains all aspects of all nuclei. Our still imperfect knowledge of the nuclear force and the tight packing of nucleons render a useless simple two-particle model like the one for hydrogen. Instead, nuclear physicists resort to several models to explain different aspects of nuclear structure. Together, these models provide a good understanding of nucleus and accurately predict nuclear properties, although not with the precision available in atomic physics.

Liquid Drop Model

Bohr and coworkers in 1930 were able to explain many nuclear phenomena by treating the nucleus as a collection of interacting particles in a liquid drop. This model of the nucleus is known as the liquid drop model. It provides a reasonable approximation for heavier nuclei, whose many nucleons behave somewhat like the molecules in a drop of liquid. Weizsacker in 1935 proposed the **semi-empirical mass formula** based on the liquid drop model. It is written in terms of the total binding energy as

$$B\left(^A_Z X\right) = a_v A - a_A A^{2/3} - \frac{3}{5}\frac{Z(Z-1)e^2}{4\pi\varepsilon_0 r}$$

$$- a_s \frac{(N-Z)^2}{A} + \delta \qquad (25.32)$$

This is actually the binding energy. The first term on RHS of Eq. (25.32) indicates that the binding energy is approximately the sum of all the interactions between nucleons. Since the nuclear force is short range and each nucleon interacts only with its nearest neighbours, this interaction is proportional to A, the total number of nucleons.

The second term, called the **surface effect,** is simply a correction to the first term (similar to surface tension) because the nucleons on the nuclear surface are not completely surrounded by other nucleons. The surface nucleons do not have saturated interactions, and one will have to apply a correction proportional to the liquid drop surface, $4\pi R^2$. Since $R \sim A^{1/3}$, the correction is proportional to $A^{2/3}$. The third term is coulomb energy.

The fourth term is due to **symmetry energy.** In the absence of coulomb forces, nucleus prefers to have $N = Z$. This term has a quantum-mechanical origin, depending on exclusion principle. We may note that the sign of this term is independent of the sign of $(N - Z)$.

The last term is due to the pairing energy and reflects the fact that the nucleus is more stable for even-even nuclides. One can determine this term empirically. One set of values given by Fermi for the parameters of Eq. (25.32) is

$$a_v = 14 \text{ MeV volume}$$
$$a_A = 13 \text{ MeV surface}$$
$$a_s = 19 \text{ MeV symmetry}$$

pairing

$$\delta = \begin{cases} +\Delta \text{ for even-even nuclei} \\ 0 \text{ for odd } A(\text{even-odd,} \\ \text{odd -even}) \text{ nuclei} \\ -\Delta \text{ for odd- odd nuclei} \end{cases}$$

where $\Delta = 33 \text{ MeV. } A^{3/2}$

A liquid drop nucleus can rotate, vibrate, and change shape as long as its volume does not change and the resulting quantized energy levels predict nuclear gamma ray spectra that are in good agreement with the observation. The liquid drop model also helps in explaining nuclear fission. But it cannot account for the dramatic effects of small changes in nucleon number, particularly the role of magic numbers.

Nuclear Shell Model

Mayer and Jensen in 1940 advanced the idea that nucleus has a shell structure similar to that of atoms. The shells occur because neutrons and protons obey the exclusion principle, and the magic numbers correspond to closed shell configurations analogous to the electronic structure of inert gases. Closed shell nucleons are tightly bound, making a magic nucleus particularly stable. Additional nucleons beyond a closed shell stay largely on the outskirts of the nucleus, where they are more readily excited to higher energy levels. Neutrons and protons behave independently in the shell model, and each has its own set of quantum numbers. Closed shell structure therefore occurs with magic numbers of either protons or neutrons. Some nuclei like $^{40}_{20}\text{Ca}$ ($Z = 20, N = 20$), are *doubly magic* and show exceptional stability.

Collective Model

This model, advanced by Aage Bohr, combines aspects of the liquid drop and shell models, emphasizing the collective quantum-mechanical behaviour of the nucleons. One remarkable prediction of this model is that large, non-magical nuclei may be more stable if they take non-spherical shapes.

Active area of nuclear structure research involves the creation and exploration of exceptionally heavy or neutron rich nuclei. The creation of elements 115 and 116 in the early 2000 s suggests that physicists are approaching a region of longer-lived nuclei dubbed the *islands of stability*, which may be associated with a new magic number of 184. Creation of silicon-42 in 2005, whose relative stability implies that its atomic number $Z = 14$ becomes magic in this neutron-bloated ($N = 28$) species. Until we develop a complete nuclear theory, experiments like these will continue to challenge physicists with nuclear surprises.

25.11 NUCLEAR REACTIONS AND Q VALUE

A change in the structure of nuclei can be brought about by bombarding them with fast moving particles, e.g. neutron, proton, alpha particle, etc. In this process, the characteristic or identity of incident nuclear particle undergoes a change which is called nuclear reaction. One can analyse a nuclear reaction quantitatively in terms of masses and energies of the nuclei and the particles involved. Consider a nuclear reaction represented by

$$X + x \rightarrow Y + y \qquad (25.33)$$

where X is the target nucleus, x is the bombarding particle, Y is the product nucleus and y is the product particle. Assuming that the target nucleus X is at rest, the law of conservation of energy demands that

$$M_X c^2 + (E_x + m_x c^2) = (E_Y + M_Y c^2)$$
$$+ (E_y + m_y c^2) \qquad (25.34)$$

where M_X, m_x, M_Y and m_y represent the masses of the target nucleus, incident particle, product nucleus and product particle respectively. The E's represent the kinetic energies. On rearranging Eq. (25.34), one obtains

$$Q = E_Y + E_y - E_x = [M_X + m_x - M_Y - m_y]c^2 \quad (25.35)$$

where $Q = E_Y + E_y - E_x$ is called the *energy balance* of the reaction or Q value of the reaction and can be determined from the energy difference or the mass difference. We may note that here excited state of product nucleus and the emission of more than one particle is not taken into account.

If the Q value for a reaction is positive, kinetic energy of the products is greater than that of the reactants, the reaction is then said to be *exothermic* (or *exoergic*) nuclear reaction. In other words, the total mass of the reactants is greater than that of the products. If the Q value for a reaction is negative, it is called *endothermic* (or *endoergic*) reaction. Equation (25.35) further suggests that whatever change occurs in total kinetic energy of the system of particles, it must be balanced by equal change in rest mass energy.

25.12 ARTIFICIAL TRANSMUTATION OR ARTIFICIAL DISINTEGRATION

The general term *transmutation* is used for the conversion of one element into another. When by some artificial means, one type of atom is converted into another type of atom the process is known as *artificial transmutation*. We may note that this process is entirely different from the spontaneous transmutations in radioactive elements. The artificial transmutation is also called as artificial disintegration because the transmuted element disintegrates into a new element. Obviously, *the process of providing new stable nuclei from other stable nucleus is known as artificial transmutation or artificial disintegration of elements.*

Rutherford's Experiment for Artificial Disintegration

Rutherford in 1919 performed the earliest experiments on artificial nuclear disintegration. The disintegration was caused by α-particles from the RaC and resulted in the emission of a highly energetic proton from ordinary nitrogen nucleus.

When N_2 nucleus captures α-particles, it is changed into unstable nucleus which immediately disintegrates into a stable oxygen nucleus by ejecting a proton. The dinsintegration is represented by the reaction

$$^{14}_{7}N + {}^{4}_{2}He \rightarrow \left[{}^{18}_{9}F \right] \rightarrow {}^{17}_{8}O + {}^{1}_{1}H$$

Thus, an atom of nitrogen is transformed into an atom of oxygen. This process is called transmutation of elements.

In the abbreviated notation, the above reaction can be expressed as $^{14}N\,(\alpha, p)\,^{17}O$. This is read as: a nitrogen nucleus (^{14}N) (target nucleus) interacts with alpha particle (α) (incident or projectile), a proton (p) is ejected and an oxygen nucleus (^{17}O) (recoil nucleus) is left behind.

25.13 ARTIFICIAL RADIOACTIVITY

In 1934, Curie–Jiliot bombarded aluminium and boron with α-particles. They found that these materials continued to emit radiation after the source of α-particles removed. This shows that the product of disintegration must be radioactive and as usual it undergoes nuclear transformation. Moreover, the intensity of this induced activity was found to decay exponentially. This phenomenon is calld *artificial* or *induced radioactivity*. The determination of the deflections of the particles emitted in magnetic and electric field showed that these are positively charged particles called *positrons* (or *positive electrons*).

The reaction for aluminium can be represented as follows:

$$\,_{13}^{27}\text{Al} + \,_{2}^{4}\text{He} \rightarrow \left[\,_{15}^{30}\text{P}\right] + \,_{0}^{1}n$$

Obviously, due to α-particle bombardment on aluminium, neutrons are emitted and an unstable isotope of phosphorous (half-life 2.5 minutes) is formed. This isotope of phosphorus then decays spontaneously giving stable silicon and positron $\left(\,_{+1}^{0}e\right)$ represented by the following reaction:

$$\,_{15}^{30}\text{P} \rightarrow \,_{14}^{30}\text{Si} + \,_{+1}^{0}e + \gamma$$

Some other artificial radioactive reactions are:

i. $\,_{5}^{10}\text{B} + \,_{2}^{4}\text{He} \rightarrow \left[\,_{7}^{14}\text{N}\right] \rightarrow \,_{7}^{13}\text{N}^{*} + \,_{0}^{1}n$

[*unstable (half-life 10 min)]

$$\,_{7}^{13}\text{N}^{*} \rightarrow \,_{6}^{13}\text{C} + \,_{+1}^{0}e + \gamma$$

ii. $\,_{11}^{33}\text{Na} + \,_{2}^{4}\text{He} \rightarrow \,_{13}^{26}\text{Al}^{*} + \,_{0}^{1}n$ [*unstable]

$$\,_{13}^{26}\text{Al} \rightarrow \,_{12}^{26}\text{Mg} + \,_{+1}^{0}e + \gamma$$

iii. $\,_{7}^{14}\text{N} + \,_{2}^{4}\text{He} \rightarrow \,_{9}^{17}\text{F}^{*} + \,_{0}^{1}n$ [*unstable]

$$\,_{9}^{17}\text{F}^{*} \rightarrow \,_{8}^{17}\text{O} + \,_{+1}^{0}e + \gamma$$

The elements with *mark in the above reactions are the *artificial radioactive isotopes* of the respective elements.

Practically by bombarding almost all the elements, with different particles, can be made radioactive, the different particles used as projectiles are α-particles, neutrons, protons, deuterons and γ-radiations. The α-particles have a small probability of colliding with the nuclei, and hence they are very little employed for artificial radioactivity. Mose useful projectiles are neutrons, obtained from nuclear reactor.

The study of artificial radioactivity helps to clarify the concepts of constitution and stability of nuclei. The artificial radioactive nuclides decay either by electron emission, or positron emission or orbital electron capture depends on the availability of disintegration energy. We may note that the nuclear reactions induced by protons, deuterons, neutrons and photons can also result in radioactive products. Moreover, there seems to be some correlation between the radioactivity of an artificial nuclide and the process and means of transformation by which it is produced. Electron emission is common for activities produced by (n, p), (n, α), (n, γ) and (d, p) nuclear reactions. These nuclear reactions decrease the positive charge to mass ratio of the nucleus, e.g.

$$\,_{49}^{115}\text{In}(n, \gamma)\,_{49}^{116}\text{In}^{*}$$

$$\,_{49}^{116}\text{In}^{*} \rightarrow \,_{50}^{116}\text{Sn} + \,_{-1}^{0}e$$

and $\,_{7}^{14}\text{N}(n, p)\,_{6}^{14}\text{C}^{*}, \,_{6}^{14}\text{C}^{*} \rightarrow \,_{7}^{14}\text{N} + \,_{-1}^{0}e$

Positron emission is common for the activity produced by (p, n), (d, n) (α, n) and (γ, v) nuclear reactions, which increase the positive charge to mass ratio of the nucleus, e.g.

$$\,_{6}^{13}\text{C}(p, \gamma)\,_{7}^{13}\text{N}^{*}, \,_{7}^{13}\text{N}^{*} \rightarrow \,_{6}^{14}\text{C} + \,_{+1}^{0}e$$

and $\,_{28}^{58}\text{Ni}(p, n)\,_{29}^{58}\text{Cu}^{*}, \,_{29}^{58}\text{Cu}^{*} \rightarrow \,_{28}^{58}\text{Ni} + \,_{+1}^{0}e$

We may note that usually (α, p), (p, α) and (d, α) reactions lead to stable products.

This discovery of artificial or induced radioactivity opened a new excited front of research and now hundreds of radioactive nuclides have been made by various methods, e.g. iodine has only one stable species $\,_{53}^{127}\text{I}$ containing 53 p and 74 n. However, there are 17 artificially radioactive isotopes of iodine which are possible, containing from 68 to 86 neutrons. We may note that the discovery of radiation from the natural and artificial radioactive nuclides has shown that analogous to atomic energy levels, **nucleus also has energy levels.** One can study these energy levels within the nucleus with the help of *nuclear spectroscopy.* Nuclear spectroscopy is an important source of information about the study of structure of the nucleus. Presently, more than 1200 radio nuclides have been identified.

25.14 RADIOISOTOPES

We have seen that in addition to the naturally occurring radioisotopes such as radium, thousands of others have been made artificially.

Radioisotopes of Transuranium Elements

All elements with $Z > 92$ are called transuranium elements because they all lie beyond $^{238}_{92}U$ in the periodic table. The first of these is neptunium. Neptunium is detected when a beam of slow neutrons is incident on $^{238}_{92}U$, then the common radioisotope ^{239}U is formed which gives β^--emission and the final product is neptunium $\left(^{239}_{93}Np\right)$ having $Z = 93$.

Neptunium changes into plutonium $\left(^{239}_{94}Pu\right)$ with spontaneous emission of β^--particles. Plutonium is 94th element. In this way, members of periodic table are increasing and we have presently 109 elements.

An interesting feature of the transuranic elements is that they form a series beginning with actinium, referred to as the actinide series, which is analogous to the rare earth series. The absence of the transuranic elements in nature, with the possible exceptions of neptunium and plutonium, may be attributed to their extremely short half-life compared with the age of the earth.

Use of Radioisotopes

Radioisotopes have numerous applications in medicine, agriculture, industry and pure research. Many applications employ a special technique known as *tracer technique*. The important applications of radioactive isotopes mainly depends on the fact that the chemical properties of the radioisotopes of a given element are essentially identical. One can detect a radioisotope easily by its radioactivity and stable isotope by a mass spectrometer.

Tracer Technique

A small amount of a radioisotope is introduced into the material to be studied and with the help of GM counter, its path is traced. For example, a leakage in an underground water-pipe can be easily detected by this method. A small quantity of γ-emitter radiosodium ^{24}Na is introduced into the pipe at its inlet. After the liquid has gone through the pipe, the ground around the leak will have larger γ-ray activity which can be detected by moving a GM counter on the ground. Similarly, in order to locate a blockage in an underground sewage pipe, one will have to introduce a rubber ball having ^{24}Na into the pipe. A GM counter above ground will give the position of the ball when it has come to rest.

One can also use radioisotopes in transporting different oils through underground pipe to distant places. When the type of the oil flowing through the pipe is changed, a small quantity of radioisotope is mixed exactly at the position where the change takes place. A Geiger counter is placed near the other end of the pipe line, which gives a signal when the radioisotope passes.

Tracer technique is also employed in the field of medicine in a number of ways. For example, a doctor can find out any obstruction in the circulation of the blood in the human body. Doctor injects radiophosphorous (^{32}P) into the blood of patient and examines the movement of the blood by detecting radiations emitted by ^{32}P with the help of GM counter. Doctor can thus locate clots of blood present in the body. In a similar way, one can also study the passage of a particular element in the body and the rate at which it accumulates in different organs. For example, phosphorous accumulates in bones and iodine in thyroid gland. When the thyroid gland suffers from some disease, its rate of accumulation of iodine changes. In order to investigate it, radio iodine (^{131}I) is given orally to the patient and the radiation emitted by his thyroid gland is measured externally by a GM counter at suitable intervals over the following 48 hours or so. Thus, one can diagnose an overactive or underactive (ailing) thyroid gland.

Tracer technique finds use in agriculture to study the rate and direction of movement of an element in a plant. For this, a radioisotope of that element is injected in the ground near the plant. After a few days the plant is laid on a photographic paper to produce an autoradiograph. The positions reached by the element are represented by the dark areas in the radiograph. This technique provided valuable information regarding the optimum season for fertilizing crops and for poisoning weeds.

Tracer technique is used for testing the uniformity of mixtures in industry. For testing a chocolate mixture, a small quantity of short-lived radioisotope, such as ^{24}Na or ^{36}Mn is added to the primary ingredients. Several different samples of the final product are then tested for radioactivity with the help of a GM counter. If each sample yields the same counting rate, then the mixing has been uniform. One can use this method in mixing processes occurring in the manufacture of chocolate, soap, cement, paint, fertilizers, cattle food and medical tablets.

The tracer technique is found to be extremely sensitive in testing the sealing process in making envelopes for radio valves. A sample valve is filled with radiokrypton (^{85}Kr) and a GM counter is held outside the valve. The GM counter detects even an extremely poor leakage.

This technique also finds use in research to study the exchange of atoms between various molecules, and to investigate the solubility and vaporization of materials.

Besides the uses of radioisotopes employing the tracer technique, there are hundreds of other uses in various fields, as mentioned below:

Biological Uses

i. A plant, when grows, absorbs phosphorus both from the soil and from fertilizer. To study the effect of fertilizer on the plant, one will have to know what properties of phosphorus comes from each other. If we are using radioisotope in fertilizer, we can easily find the exact proportion easily.

ii. It is also helpful for blood transfusion, since one can detect whether the given blood is suitable for the patient or not.

Medical Uses

Radioisotopes have been widely used in locating and detecting the presence of *tumours* and particularly *brain tumours* which are quite difficult to detect.

To study the cases of *restricted circulation of blood*, radiosodium has been used. Using radio-iodine or radiosodium, pumping action of the heart has been studied.

Radioisotopes have been used to study virtually every organ and tissue in the body. ^{3}H, ^{14}C and ^{32}P, all β^--emitters are widely used in medical research. More than 1100 radioisotopes are available for clinical use. By far the most widely used is ^{99m}Tc. An isomer of technetium with a half-life of 6 hours, produces a 140 keV γ-ray when it decays to the ground state of ^{99}Tc. A patient is given intravenous injection of ^{99m}Tc, e.g. TcO$_4$. The ^{99m}Tc is trapped by cancerous cells, e.g. in the thyroid or salivary glands, choroid plexus of the brain or the gastric mucosa. Few minutes after the injection, patient is scanned by an array of NaI γ-ray detectors. This helps to pinpoint the ^{99m}Tc activity. This is only one out of the several applications of ^{99m}Tc.

The radioactive $^{67}_{31}$Ga is useful as a tumour–localizing agent for Hodgkin's disease.

^{153}Gd radioisotope is used for detection of bone mineral loss or osteoporosis in elderly people. ^{192}Ir is used for a large number of cancer treatments and also for industrial radiography for welds in steel, oil well rigs and pipelines. ^{125}I reeds is implanted into the prostate to inhibit cancer. In addition to ^{125}I, which has a half-life of 60 days, ^{103}Pd (half-life 17 days) is also used. Both these nuclides emit useful low-energy γ-rays.

Radioisotopes are also used in tomography. This is a technique for displaying images or practically any part of the body to look for abnormal physical shapes or for functional characteristic of organs. Making use of the detectors together with computers, 3 dimensional images can be obtained. Tomography now includes single photon emission computed tomography (SPECT), positron emission tomography (PET), magnetic resonance imaging (MRI), etc.

Industrial Uses

There are many different uses to which radioisotopes are put in industry.

i. In industry, alloys are frequently subjected to different treatments such as annealing, quenching, etc. and radioisotopes have been used to study the effect of such treatment.

ii. The movement of atoms in metal within the crystal lattice, i.e. *self diffusion* in metal has been studied using radioisotopes. During blending of different materials, uniformity of mixing has been achieved by labeling one of the constituents with radioactive tracer.

Agricultural Uses

i. Radiations from certain radioisotopes are used for killing insects which damage the food grains. Certain seeds and canned food can be stored for longer periods by gently exposing them to radiations.

ii. Better yield of milk from cows, and more eggs from hens have been obtained on the basis of information gained by mixing radioisotopes with their diet.

iii. Radioisotopes are also used for determining the function of fertilizer in different plants. This increases the agricultural yield.

iv. Certain seeds, when exposed to feeble radiations, develop into different varieties of plants, e.g. new and exciting colours have been given to some of the flowering plants.

Radiocarbon Dating

It our atmosphere, radioactive ^{14}C is produced by the bombardment of ^{14}N by neutrons produced by cosmic rays.

$$^{1}_{0}n + ^{14}_{7}N \rightarrow ^{14}_{6}C + ^{1}_{1}p$$

In the atmosphere, a natural equilibrium of ^{14}C to ^{12}C exists. As shown in Fig. 25.13, all living organisms use or breathe CO_2 from the atmosphere.

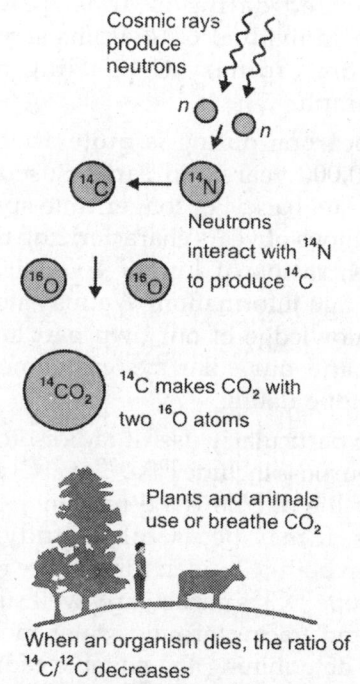

Fig. 25.13: Neutrons produced from cosmic rays react with ^{14}N in the atmosphere to produce ^{14}C. ^{14}C nuclei enter into living organisms in the form of CO_2. After the death of living organism, the ratio of $^{14}C/^{12}C$ in it decreases according to the decay rate of ^{14}C

However, after the death of living organisms, their intake of ^{14}C ceases and the ratio $^{14}C/^{12}C$ (= R) decreases due to decay of ^{14}C. Measuring the ratio of ^{14}C to a stable ^{12}C in a sample of once living matter and comparing with the ratio found in living material then provides the age. Archaeologists, art historians, geologists and others use radioactive decay to date ancient objects. For ages up to a few tens of thousands of years, the 5730-year-isotope C-14 is especially useful.

The cosmic ray flux at the earth varies with solar activity and so, therefore, does the $^{14}C/^{12}C$ ratio. Scientists correct for this effect with data from growth rings in ancient trees, which provide an independent measure of age. Measuring the actual radioactivity takes a fairly large sample, so today the most sophisticated dating is done instead by counting individual C-14 atoms separating them from ordinary C-12 using a mass spectrograph.

Radiocarbon dating is quite accurate to about 20,000 years and can be used about 50,000 years back. For longer time spans, up to the billions of years characterizing the ages of rocks, ratios of longer lived isotopes provide age information. We may note that much knowledge of our own part and own planets and our solar system comes from radioisotope dating.

Other particularly useful radioisotopes for dating purpose include ^{10}Be, ^{26}Al, ^{36}Cl and ^{130}I. The half-life of ^{10}Be is 1.5 million years and perhaps, it may be useful in studying the evolution both of humans and of the ice ages. The isotope ^{36}Cl is particularly well suited for dating and tracing ground water movement and for determining the suitable radioactive waste depositories. The dating of ^{10}Be and ^{26}Al in marine sediments has confirmed their extraterrestrial origin, possibly from comets.

25.15 NUCLEAR REACTIONS

One can bring a change in the structure of the nuclei by bombarding them with fast moving particles. In this process, the characteristics or identity of incident nuclear particles undergo a change which is called a nuclear reaction. Usually, the following nuclear particles are used: $\alpha\left(^{4}_{2}\text{He}\right)$, proton $p\left(^{1}_{1}\text{H}\right)$, deuteron $\left(^{2}_{1}\text{D}\right)$, neutron $\left(^{1}_{0}n\right)$, high energy photons, i.e. γ-rays, etc. On the basis of the nature of projectile particle and outgoing particle, one can classify the nuclear reactions as follows:

Reactions Induced by α-particles

i. (α, p) reaction: One can write the general (α, p) reaction as

$$^{A}_{Z}X + ^{4}_{2}\text{He}(\alpha) \rightarrow \left[^{A+4}_{Z+2}\text{Cn}\right] \rightarrow ^{A+3}_{Z+1}Y + ^{1}_{1}\text{H}$$

(Cn → Compound nucleus)
Example:

$$^{27}_{13}\text{Al} + ^{4}_{2}\text{He} \rightarrow \left[^{31}_{15}\text{P}\right] \rightarrow ^{30}_{14}\text{Si} + ^{1}_{1}\text{H}$$

ii. (α, n) reaction: General reaction

$$^{A}_{Z}X + ^{4}_{2}\text{He} \rightarrow \left[^{A+4}_{Z+2}\text{Cn}\right] \rightarrow ^{A+3}_{Z+2}Y + ^{1}_{0}n$$

Examples:

$$^{9}_{4}\text{Be} + ^{4}_{2}\text{He} \rightarrow \left[^{13}_{6}\text{C}\right] \rightarrow ^{12}_{6}\text{C} + ^{1}_{0}n$$

$$^{14}_{7}\text{N} + ^{4}_{2}\text{He} \rightarrow \left[^{18}_{9}\text{F}\right] \rightarrow ^{17}_{9}\text{F} + ^{1}_{0}n$$

We may note that (α, n) reaction is a well-known reaction used for the detection of neutron by Chadwick in 1932.

Reactions Induced by Protons

i. (p, α) reaction: General reaction

$$^{A}_{Z}X + ^{1}_{1}\text{H} \rightarrow \left[^{A+1}_{Z+1}\text{Cn}\right] \rightarrow ^{A-3}_{Z-1}Y + ^{4}_{2}\text{He}$$

Examples:

$$^{7}_{3}\text{Li} + ^{1}_{1}\text{H} \rightarrow \left[^{8}_{4}\text{Be}\right] \rightarrow ^{4}_{2}\text{He} + ^{4}_{2}\text{He}$$

$$^{27}_{13}\text{Al} + ^{1}_{1}\text{H} \rightarrow \left[^{28}_{14}\text{Si}\right] \rightarrow ^{24}_{12}\text{Mg} + ^{4}_{2}\text{He}$$

(p, α) reaction provided one of the earliest quantitative proof of the famous Einstein's relation, $E = mc^2$.

ii. (p, n) reaction: General reaction

$$^{A}_{Z}X + ^{1}_{1}\text{H} \rightarrow \left[^{A+1}_{Z+1}\text{Cn}\right] \rightarrow ^{A}_{Z+1}Y + ^{1}_{0}n$$

Examples:

$$^{11}_{5}\text{B} + ^{1}_{1}\text{H} \rightarrow \left[^{12}_{6}\text{C}\right] \rightarrow ^{11}_{6}\text{C} + ^{1}_{0}n$$

$$^{18}_{8}\text{O} + ^{1}_{1}\text{H} \rightarrow \left[^{19}_{9}\text{F}\right] \rightarrow ^{18}_{9}\text{F} + ^{1}_{0}n$$

We may note that in (p, n) reactions, the mass change is usually negative, so the reactions are *endothermic*.

iii. (p, γ) reaction: General reaction

$$^{A}_{Z}X + ^{1}_{1}\text{H} \rightarrow \left[^{A+1}_{Z+1}\text{Cn}\right] \rightarrow ^{A+1}_{Z+1}Y + \gamma$$

Example: $^{27}_{13}\text{Al} + ^{1}_{1}\text{H} \rightarrow ^{28}_{14}\text{Si} \rightarrow ^{28}_{14}\text{Si} + \gamma$

Transmutation by Deuterons

i. (d, α) reaction: General reaction

$$_Z^A X + {}_1^2 H \rightarrow \left[{}_{Z+1}^{A+2} Cn \right] \rightarrow {}_{Z-1}^{A-2} Y + {}_2^4 He$$

Example: ${}_8^{16}O + {}_1^2 H \rightarrow {}_9^{18}F \rightarrow {}_7^{14}N + {}_2^4 He$

These reactions are, in general, *exothermic* as the mass change is usually positive.

ii. (d, p) reaction: General reaction

$$_Z^A X + {}_1^2 H \rightarrow \left[{}_{Z+1}^{A+2} Cn \right] \rightarrow {}_{Z+1}^{A+1} X + {}_1^1 H$$

Examples:

${}_6^{12}C + {}_1^2 H \rightarrow {}_7^{14}N \rightarrow {}_6^{13}C + {}_1^1 H$

${}_3^7 Li + {}_1^2 H \rightarrow \left[{}_4^9 Be \right] \rightarrow {}_3^8 Li + {}_1^1 H$

We may note that here in (d, p) reactions, isotopes will form. The Q value of these reactions is usually positive, and obviously the reactions are in general *exothermic*.

iii. (d, n) reaction: General reaction

$$_Z^A X + {}_1^2 H \rightarrow \left[{}_{Z+1}^{A+2} Cn \right] \rightarrow {}_{Z+1}^{A+1} Y + {}_0^1 n$$

Examples:

${}_3^7 Li + {}_1^2 H \rightarrow \left[{}_4^9 Be \right] \rightarrow {}_4^8 Be + {}_0^1 n$

${}_6^{12}C + {}_1^2 H \rightarrow \left[{}_7^{14}N \right] \rightarrow {}_7^{13}N + {}_0^1 n$

Transmutation by Neutrons

i. (n, α) reaction: General reaction

$$_Z^A X + {}_0^1 n \rightarrow \left[{}_Z^{A+1} Cn \right] \rightarrow {}_{Z-2}^{A-3} Y + {}_2^4 He$$

Example: ${}_3^6 Li + {}_0^1 n + \left[{}_3^7 Li \right] \rightarrow {}_1^3 H + {}_2^4 He$

ii. (n, p) reaction: General reaction

$$_Z^A X + {}_0^1 n \rightarrow \left[{}_Z^{A+1} Cn \right] \rightarrow {}_{Z-1}^A Y + {}_1^1 H$$

Examples:

${}_7^{14}N + {}_0^1 n \rightarrow \left[{}_7^{15}N \right] \rightarrow {}_6^{14}C + {}_1^1 H$

${}_{12}^{24}Mg + {}_0^1 n \rightarrow \left[{}_{12}^{25}Mg \right] \rightarrow {}_{11}^{24}Na + {}_1^1 H$

iii. $(n, 2n)$ reaction: General reaction

$$_Z^A X + {}_0^1 n \rightarrow \left[{}_Z^{A+1} Cn \right] \rightarrow {}_Z^{A-1} X + {}_0^1 n + {}_0^1 n$$

Example:

${}_{13}^{27}Al + {}_0^1 n \rightarrow \left[{}_{13}^{28}Al \right] \rightarrow \left[{}_{13}^{26}Al \right] + 2\,{}_0^1 n$

The reaction is, in general, *endothermic* as the mass change in these reactions is always negative.

iv. (n, γ) reaction: General reaction

$$_Z^A X + {}_0^1 n \rightarrow \left[{}_Z^{A+1} Cn \right] \rightarrow {}_Z^{A+1} X + \gamma$$

Examples:

${}_{13}^{27}Al + {}_0^1 n \rightarrow \left[{}_{13}^{28}Al \right] \rightarrow {}_{13}^{28}Al + \gamma$

${}_{92}^{238}U + {}_0^1 n \rightarrow \left[{}_{92}^{239}U \right] \rightarrow {}_{92}^{239}U + \gamma$

${}_1^2 H + {}_0^1 n \rightarrow \left[{}_1^3 H \right] \rightarrow {}_1^3 H + \gamma$

The Q value of these reactions is always positive, and hence these reactions are *exothermic* with excess energy being carried away by γ-rays.

The slow neutrons, almost in all cases, induce (n, γ) reactions (radiative capture), and the product nuclei are often radioactive. These reactions are important source of artificial radioisotopes.

Transmutation by Photon (Photo Disintegration)

i. (γ, n) reaction: General reaction

$$_Z^A X + \gamma \rightarrow \left[{}_Z^A Cn \right] \rightarrow {}_Z^{A-1} X + {}_0^1 n$$

Examples:

${}_1^2 H + \gamma \rightarrow \left[{}_1^2 H \right] \rightarrow {}_1^1 H + {}_0^1 n$

${}_4^9 Be + \gamma \rightarrow \left[{}_4^9 Be \right] \rightarrow {}_4^8 Be + {}_0^1 n$

The reactions are *endothermic* as the Q value is negative.

ii. (γ, p) reaction: General reaction

$$_Z^A X + \gamma \rightarrow \left[{}_Z^A Cn \right] \rightarrow {}_{Z-1}^{A-1} X + {}_1^1 H$$

Example:

$$^{25}_{13}\text{Mg} + \gamma \rightarrow \left[^{25}_{13}\text{Mg}\right] \rightarrow ^{24}_{11}\text{Na} + ^{1}_{1}\text{H}$$

We may note that same nuclides can be transmuted in a number of different ways, depending upon the particle used to produce transmutation and upon the particle given off during the nuclear reaction.

25.16 NUCLEAR FISSION

Nuclear fission is a process in which a heavy nucleus after capturing a neutron splits up into two lighter nuclei of comparable masses. The product nuclei are called *fission fragments*. The process is accompanied by the release of a few fast neutrons and a huge amount of energy in the form of the kinetic energy of the fission fragments, and also as γ-rays. One of the typical fission reaction is

$$^{235}_{92}\text{U} + ^{1}_{0}n \rightarrow \left[^{236}_{92}\text{U}\right] \rightarrow ^{144}_{56}\text{Ba} + ^{89}_{36}\text{Kr} + 3\,^{1}_{0}n + Q$$

Of course, many different fission reactions are possible, with many different fission products.

Both ^{144}Ba and ^{89}Kr are highly unstable and they reach stability by a series of β- and γ-emissions. The value of Q is found about to be 200 MeV per fission of ^{235}U atom. Recalling that 1 MeV = 1.6×10^{-13} J, the energy released per fission of ^{235}U is 3.2×10^{-13} J.

We may note that in a nucleus, there is a competition between the nuclear force, which holds the nucleus together, and the electrostatic repulsion of the nuclear protons, which tries to tear the nuclear apart. For most nuclei, the nuclear force dominates this competition, but for heavier nuclei there is a delicate balance between the nuclear forces, a balance that is easily upset. We may further note that the energy necessary to cause fusion is typically about 6 MeV.

Later researches have shown that besides uranium other nuclei are also fissionable. $^{232}_{90}\text{Th}$ and $^{231}_{90}\text{Pa}$ can be fissioned by fast neutrons, whereas transuranic element $^{239}_{94}\text{Pu}$ and an artificial isotope of $^{233}_{92}\text{U}$ can be fissioned by thermal neutrons.

Besides neutrons, accelerated protons, deuterons and α-particles can also induce fission in the nuclei of thorium, uranium and transuranic elements. Even γ-rays can cause nuclear fission which is known as *photofission*. Some lighter elements can also be fissioned by very high energy photons and deuterons.

Spontaneous fission (in which no bombarding particle is required) of thorium, uranium and transuranic elements has also been detected. It is a process (like natural radioactivity) in which a nucleus splits into two fragments of its own accord. However, the probability of this type of fission is very very small.

There are three features of nuclear fission that make it useful as a means to generate electrical energy.

1. **Energy dissipation:** Most of the energy released in a nuclear fission process is released as kinetic energy of the fission fragments. These relatively heavy fragments do not travel very far through the reactor fuel element before they dissipate most of their kinetic energy in collisions with the atoms of the fuel element. One can extract the energy as heat and used to boil water; the resulting steam can then be used in conventional way to drive a turbine to generate electricity.

2. **Neutron multiplicity:** The second important feature of the fission reaction that makes it useful is that the average number of neutrons produced is greater than one, making possible the *chain reaction*. How much greater than one it must be, in order to achieve a chain reaction, depends on the construction of the nuclear reactor.

3. **Prompt and delayed neutrons:** The third advantage of nuclear fission process is the one that enables an operator or the mechanical system to control the reaction and keep it from proceeding too rapidly.

The two neutrons emitted in the fission process

$$^{233}_{92}U + n \rightarrow ^{93}_{37}Rb + ^{141}_{55}Cs + 2n$$

are called *prompt neutrons* and constitute 99% of the total neutrons ejected in the process. About 1% of the neutrons in the fission process are *delayed neutrons* emitted following the decays of the heavy fragments while reaching the stable state. The best known example of this is

$$^{87}_{35}Br \xrightarrow{\ \beta^- \ } ^{87}_{36}Kr \xrightarrow{\ n \ } ^{87}_{36}Kr$$

the half-life of which is 55.6 s. The delayed neutrons which increase the mean lifetime of all fission neutrons are of importance in controlling the rate of reactions in a nuclear reactor.

Theory of Nuclear Fission

Bohr and Wheeler proposed that a heavy nucleus behaves like a liquid drop and that the fission may be explained as a result of oscillations brought about by impinging neutrons. Some of the properties of nuclear forces (saturation, short range) are analogous to the properties of the nuclear forces which hold a liquid drop together. Attractive forces between nucleons, like a the attractive forces between molecules, give rise to surface tension

and the spherical state. Any disturbance of this state will require an external force, which distorts the sphere into an ellipsoid. If the force is large enough, the ellipsoid narrows into a dumb-bell shape and finally breaks at the neck into two major portions with some additional small drops. Figure 25.14 explains the supposed mechanism of nuclear fission.

When an atomic nucleus undergoes fission, the incident neutron combines with the nucleus to form a compound nucleus which is highly energetic. The extra energy of the nucleus is partly the kinetic energy of the incident neutron but largely the added binding energy of the incident neutron. This excitation energy appears to initiate a series of rapid oscillations in the drop. When the excitation energy is low, the oscillations about the spherical shape are small, so that at its maximum deformation the nucleus adopts the ellipsoidal shape (Fig. 25.14b). The restoring force of the nucleus arises from the short-range internucleon forces. If the oscillations become so violent, the stage D (Fig. 25.14b) is reached, and as such half is now positively charged, the final fission into stage E is inevitable. Clearly, there is a threshold energy or a critical energy required to produce stage D after which the nucleus cannot return to state A, because of coulomb repulsion between the two positive parts. During the fission a large

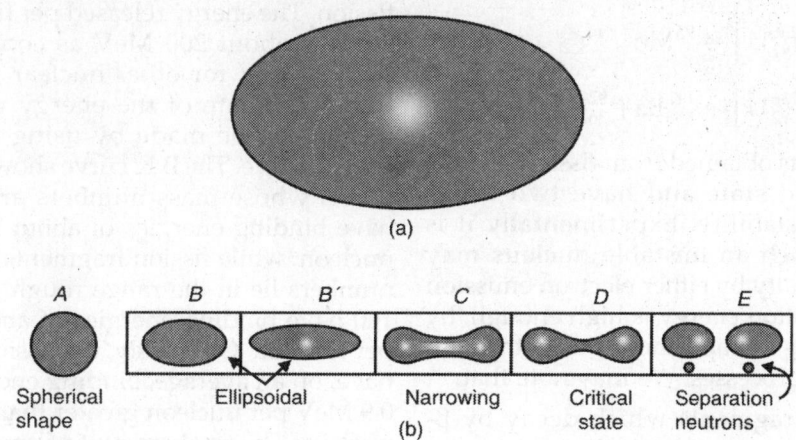

(a)

A	B	B'	C	D	E
Spherical shape	Ellipsoidal		Narrowing	Critical state	Separation neutrons

(b)

Fig. 25.14: (a) The elongated shape of heavy nucleus (b) Sequence of nuclear shapes in nuclear fission in liquid drop model of nucleus

amount of energy is released in fission process. The product nuclides must be in highly excited states at the instant of fission. Part of this energy is released almost immediately and carried away by the prompt neutrons. The remaining excess energy is released by neutrons or α-particles, or β-decays or γ-ray emission.

The value of critical deformation energy E_{crit} was first calculated by Bohr and Wheeler on liquid drop model and that they obtained the condition for spontaneous process as $\dfrac{Z^2}{A} \geq 45$. This condition shows that U^{238} requires more total excitation energy than U^{235} to initiate fission.

Fission Products

In a given nucleus, the fission may occur in a number of different ways. In general, the fissionable nucleus gives two fission fragments which thereafter decay by electron emission into stable end products. What particular fragments are produced by the given nucleus is a matter of chance. The uranium nucleus is capable of splitting in about 40 different ways. About 97% of the U^{235} nuclei undergoing nuclear fission yield products, which falls into two groups: *light group* (A = 85 to 104) and a *heavy group* (A = 130 to 149). The following reactions make it clear:

$$^{235}_{92}U + {}^{1}_{0}n \rightarrow \left[{}^{236}_{92}U \right] \rightarrow {}^{95}Mo + {}^{139}La + 2\,{}^{1}_{0}n + Q$$

$$^{235}_{92}U + {}^{1}_{0}n \rightarrow \left[{}^{236}_{92}U \right] \rightarrow {}^{141}_{56}Ba + {}^{92}_{36}Kr + 3\,{}^{1}_{0}n + Q$$

The products obtained from fission are in a highly excited state and have two many neutrons for stability. Experimentally it is found that such an unstable nucleus may approach stability by either electron emission or if the excitation energy is high enough, by ejection of one or more neutron or through both of these processes. We may note that:

i. Fission fragments which decay by β-emission start a short radioactive series involving the successive emission of electrons until a static stable product is

reached. Such type of series is called *fission decay series*. We may note that in above equations ^{95}Mo and ^{139}La are the stable products. Following is an example of fission series of one fission product.

$$^{140}_{54}Xe \xrightarrow{\beta^-} {}^{140}_{55}Cs \xrightarrow{\beta^-} {}^{140}_{56}Ba$$
$$\downarrow \beta^-$$
$$^{140}_{58}Ce \xleftarrow{\beta^-} {}^{140}_{57}La$$
$$\text{(Stable)}$$

ii. We have seen that neutrons are emitted in fission. In the second of above equation, it is thought that a compound nucleus $\left({}^{236}_{92}U \right)$ is first formed, which split into two fragments. Each of these fragments has too many neutrons for stability, and also excess energy (~6 MeV or even greater) needed to expel a neutron. After its formation, the excited, unstable nucleus consequently ejects one or more neutrons with γ-rays within a very short time. This explanation seems to be consistent with experimental result.

Energy Released in Nuclear Fission

We have seen that when a heavy nucleus undergoes a fission process, a large amount of energy is released together with the emission of neutrons which further produced fission. This is the most striking properties of fission. The energy released per fission of ^{235}U atom is about 200 MeV as compared with several MeV for other nuclear reactions. A rough estimate of the energy released per fission can be made by using the binding energy curve. The B.E. curve shows that heavy nuclei whose mass numbers are above 240 have binding energies of about 7.6 MeV per nucleon, while fission fragments whose mass numbers lie in the range roughly from 70 to 160 have binding energies of about 8.5 MeV per nucleon. Obviously, the fission fragments have, on an average, binding energy of about 0.9 MeV per nucleon greater than the (heavy) nucleus. The total amount of energy released per fission of a compound nucleus ^{236}U, which has 236 nucleons, should be roughly equal to the product of number of nucleons (236) multiplied by the excess B.E./particle (0.9 MeV), i.e. $236 \times 0.9 = 212$ MeV or roughly 200 MeV $= 3.2 \times 10^{-11}$ J. We can also calculate the total energy released per fission from the nuclear masses of ^{236}U and pair of fission products for the following reaction:

$$^{235}\text{U} + {}^1_0n \rightarrow \left[{}^{236}\text{U}\right] \rightarrow {}^{95}\text{Mo} + {}^{139}\text{La} + 2\,{}^1_0n$$

Now, Q value of the reaction

$$Q = (\text{Mass of the reactants} - \text{Mass of products})c^2$$

$$= (236.133 - 235.919) \times 931 \text{ MeV}$$

The reaction is *exothermic* as the Q value is positive and the amount of energy released per fission is obtained as

$$E = 0.124 \times 981 = 198 \text{ MeV}$$

$$\simeq 200 \text{ MeV} = 3.2 \times 10^{-11} \text{ J}$$

Although this energy seems quite small but we must remember that this is the energy released due to fission per nucleus of ^{235}U. There are N atoms in 1 gram atom of ^{235}U, where $N = 6.02 \times 10^{26}$ (Avogadro number).

Thus, 1 kg of ^{235}U has $\dfrac{6.02 \times 10^{26}}{235} = 2.56 \times 10^{24}$ atoms. If all the nuclei of uranium atoms in 1kg are fissioned, the released energy would be $2.56 \times 10^{24} \times 3.2 \times 10^{-11} = 8.2 \times 10^{10}$ J. It would produce about 2.28×10^4 kWH of electrical energy. We may note that there are about 30 different ways in which the compound nucleus (^{236}U) can divide, but mass excess is approximately the same, for all these processes and in each case 200 MeV is the average amount of energy released per fission and in good agreement with experimental observations.

Table 25.2

Atomic mass of ^{235}U	= 235.124 u
Atomic mass of ${}^1_0n$	= 1.009 u
Total mass of the reactants	= 236.133 u
Atomic mass of ^{139}La	= 138.955 u
Atomic mass of ^{95}Mo	= 94.946 u
Atomic mass of 2 ${}^1_0n$	= 2.018 u
Total mass of the products	= 235.919 u

Nuclear Fission as a Source of Energy

We have seen that large amount of energy is released in fission reaction and also with this more than one neutron is produced. This has made it possible to use the fission process as a source of energy. During each fission process, on the average 2.5 neutrons are emitted which attack other atoms causing fission. The number of neutrons goes on multiplying rapidly during fission process till entire fissionable material is disintegrated. The reaction gets self sustained and is known as a *chain reaction*. The chain reaction may be of two types:

Uncontrolled Chain Reaction

When more than one of the neutrons emitted in a particular fission process cause further fission, then amount of energy is liberated in each fission process and within a very short time the released energy takes a tremendous magnitude and is released as a violent explosion. This process takes place in a *nuclear bomb*. Such type of chain reaction is called an *uncontrolled chain reaction*.

Controlled Chain Reaction

If by some means, one can control the fission reaction in such a way that only one of the neutrons emitted in a fission causes another fission, then fission rate remains constant and the energy is released steadily. Such a chain reaction is called a *controlled chain reaction* and used in *nuclear reactor*, which can be used as a source of neutrons or of nuclear power.

Energy Release

We now calculate the energy released by the complete fission of 1 kg of uranium. We have energy released/fission = 200 MeV.

$$1 \text{ MeV} = 1.6 \times 10^{-13} \text{ J} = 1.6 \times 10^{-13} \text{ Ws}$$

$\therefore$ Energy released per fission

$$= 200 \times 1.6 \times 10^{-13} \text{ J} = 3.2 \times 10^{-11} \text{ J}$$

$$= 3.2 \times 10^{11} \text{ Ws}$$

Avogadro's number = Number of atoms/kg atoms

$$= 6.02 \times 10^{26}/\text{kg-atoms}$$

Now, the energy released by the fission of 1 kg atoms of ^{235}U or by 235 kg of ^{235}U is

$$= 3.2 \times 10^{-11} \times 6.02 \times 10^{26} = 1.93 \times 10^{16} \, J$$

$\therefore$ Energy released per kg of ^{235}U

$$= 1.93 \times 10^{16}/235 = 8.2 \times 10^{13} J = 8.2 \times 10^{13} Ws$$

Power output $(P)/s = 8.2 \times 10^{13}$ W

Power output/day, $P = \dfrac{8.2 \times 10^{13}}{2.4 \times 60 \times 60}$

$$= 9.49 \times 10^8 \, W = 10^3 \, MW$$

Obviously, the power output from 1 kg of ^{235}U is

$$= 10^3 \, MW$$

Clearly, 1 kg ^{235}U on complete fission will supply 1000 MW of power/day. We may note that this is equivalent to the output of a thermal power plant which consumes about 300 tonnes of coal/day.

Fission rate of ^{235}U for producing 1 W of electric power = $1/3.2 \times 10^{-11} = 3.1 \times 10^{10}$ fission/s.

Time required for one fission

$$= 1/3.1 \times 10^{10}$$

$$= 3.2 \times 10^{-9} \, s$$

This clearly reveals that fission rate is very high and large amount of fission energy is released in an extremely short interval of time, i.e. 10^{-9}s . Obviously, this much of production of energy leads to a very powerful explosion unless its rate of production is controlled.

We have already mentioned that controlled chain reactions are essential for building a nuclear reactor. In order to achieve a chain reaction with uranium, we will have to take care of following two competing processes:

i. Fission of uranium nuclei, with the emission of more neutrons than are captured. On an average, 2.5 neutrons per fission of uranium nuclei are released.

ii. Non fission capture of uranium and other materials and escape of neutrons without being captured.

If the loss of neutron by (ii) process is less than or equal to the surplus neutrons produced

by (i) process, a nuclear chain reaction occurs, otherwise a chain reaction is not possible.

The *size of the sample* is one of the most crucial factors in making favourable condition for neutron balance in a reaction. We may note that if the size of the sample is less, then neutron will go out of the sample without producing fission, whereas if the size is too big then production rate of neutron will be more than escape, so explosion will occur. Obviously, there is a *critical size* of the sample, which is essential for the production of neutrons in a controlled chain reaction. Critical size calculation of the sample is one of the main problems in designing a nuclear reactor.

25.17 NUCLEAR REACTOR OR NUCLEAR PILE

A *nuclear reactor*, or a *nuclear pile*, is a device in which a self-sustaining controlled chain reaction is produced in a fissionable material. Obviously, nuclear reactor is a source of controlled energy.

The fissionable material, known as *fuel*, plays the key role in the operation of a nuclear reactor. Uranium, preferably enriched with the isotope ^{235}U is used as fuel. Natural uranium contains 98.28% abundance of ^{238}U and 0.72% abundance of ^{235}U. It will release about 2.5 fast neutrons with average energy 2 MeV, if some spontaneous fission occurs. For these fast neutrons, the fission cross, section (probability) of two isotopes of uranium ^{238}U and ^{235}U are almost same. However, the abundance of ^{238}U is very large, and hence the fission neutrons cause further fissions which will mainly occur in ^{238}U. We may note that these neutrons can suffer (i) fission, (ii) radioactive capture and (iii) elastic or inelastic scattering. However, the last two processes are more probable, and therefore chain reaction is not possible in natural uranium. It is reported that $^{239}_{94}Pu$ will be produced in a non fission capture of neutrons by ^{238}U (Fig. 25.15).

One can make possible the chain reaction in natural uranium by using **moderator.** A moderator slows down the neutrons to thermal energies by elastic collisions between

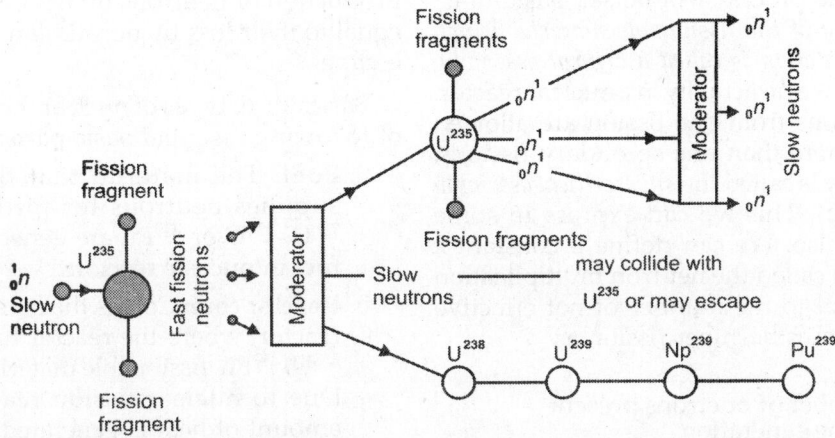

Fig. 25.15: A chain reaction exhibiting different steps occuring in fission process

its nuclei and the fission neutrons. The thermal neutrons have a very high probability of producing fission in ^{235}U nuclei. Graphite and heavy water (D_2O) are most suitable moderators. Beryllium oxide, hydrides of metals and organic liquids are other materials used as moderators. We may note that the nuclei of these substances absorb neutrons only to a slight extent. When heavy water, D_2O is used as moderator, uranium can be dissolved in it. This type of nuclear reactor is known as *homogeneous nuclear reactor*.

Uranium is in the form of rods, when graphite is used as moderator and U-rods are distributed in a regular way throughout the graphite thereby forming a lattice. This type of reactor is called *heterogeneous reactor* (Fig. 25.16).

Critical Size for Nuclear Reactor

On average, if a chain reaction is to be sustained in a lump of uranium, at least one of the 2.5 neutrons born per fission must be preserved for further fission, i.e. if all the neutrons in excess of one per fission are

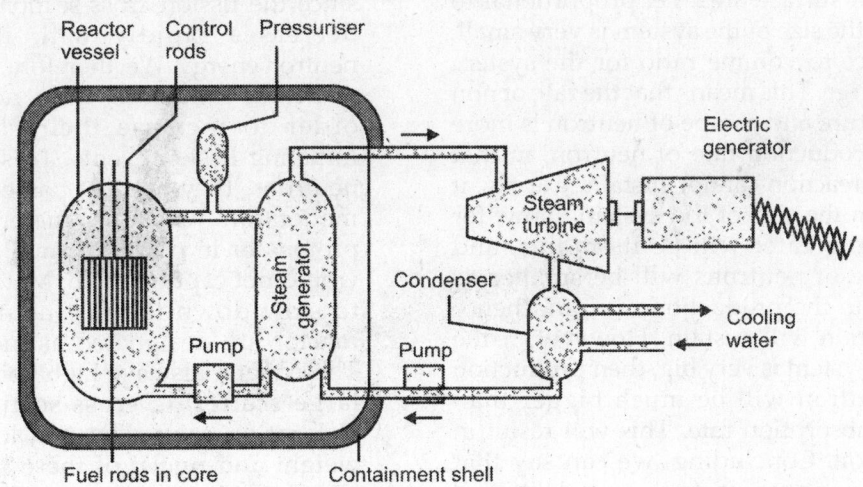

Fig. 25.16: A heterogeneous nuclear reactor

removed, the process will be self-sustaining. *The assembly of the fissionable material which meets this criterion is called a critical assembly.* This is what we do actually in a nuclear reactor. If the neutrons from one fission are allowed to induce more than one secondary fissions, the assembly is called the *supercritical assembly* (atom bomb). This we can express in some other way also. We can define a constant *k* (sometimes) called the neutron multiplication factor), equal to the number of net effective fission neutrons born per fission as

$$k = \frac{\text{Number of nuetrons present in one generation}}{\text{Number of neutrons present in the previous generation}}$$

If $k < 1$, the chain reaction will **slow down** and stop. If $k = 1$, the reaction will proceed at a **steady rate.** If $k > 1$, however the reaction will grow to an **explosive rate.** These situations are called as *subcritical*, *critical* and *supercritical* respectively.

The need for a favourable neutron balance sets certain conditions on the size of the reactor. If uranium fuel is distributed in a regular way throughout the assembly, the production of neutrons depends on the volume of the system, i.e. proportional to r^3, whereas the probability of escape of neutron depends on surface area, i.e. proportional to r^2. Now, if the size of the system is very small, then surface to volume ratio for the system ($= 1/r$) is large. This means that the rate of non fission capture and escape of neutron is more than the production rate of neutron, in that case chain reaction cannot sustain. Clearly, it depends on the size of the system. Now, for larger size, greater will be its radius, and escape rate of neutrons will be smaller as compared to the production rate, and hence chain reaction will sustain. However, if the size of the system is very big, then production rate of neutron will be much bigger than escape or absorption rate. This will result in an explosion. Concluding, we can say that for effective nuclear reactor we need a *critical size* of the system for which $k = 1$, i.e. the

production of neutrons by fission will be just equal to their loss by non-fission capture and escape.

Basically all types of nuclear reactors consist of following essential basic parts:

  i. **Fuel:** The material that fissions and supplies neutrons for further fissions ^{235}U, ^{238}U or ^{239}Pu are generally used as fuel in nuclear reactors.

 ii. **Reactor core:** Core is the main part of the reactor, where the reactor fuel like ^{235}U, ^{238}U, ^{239}Pu, fissionable materials are used. Due to nuclear fission reaction, huge amount of heat is generated in the core. Usually the reactor core has a right circular cylinder with about a diameter of few meters.

Fuel elements are generally made up of uranium rods. Reactor core, in general, has fuel elements, moderator, control rods, and cooling material. All these are housed in a pressure vessel as shown in Fig. 25.16.

iii. **Moderators:** A major difficulty in trying to produce a chain reaction is the energy of the neutrons emitted in the fission process. Typically, the kinetic energies of these neutrons are a few MeV, such energetic neutrons have a relatively low probability of inducing new fissions, since the fission cross-section generally decreases rapidly with increasing neutron energy. We, therefore, must slow down, or moderate, these neutrons in order to increase their chances of initiating fission events. To slow down neutrons, they are just passed through moderators which are material rich in protons or in nuclei of small mass that would not capture them. Materials used to slow down the fast neutrons in a reactor are designated as moderators. These materials have high boiling point, large scattering cross-section, small absorption coefficient and low atomic weight and nuclei of these substances absorb neutrons only to a slight extent. Commonly used materials as moderators are heavy water (D_2O), graphite, beryllium oxide, etc.

iv. **Reflectors:** Reflectors surrounding the reactor core, containing the fissionable material and moderator are of the same substances as used for moderator. The efficiency of the reflector increases rapidly with its thickness and reaches its limiting value when this thickness is several times the mean free path of the neutrons in the substance. The reactor system including reflector, must be enclosed in a shield, usually of concrete. This reduces the intensity of the reduction of neutrons and γ-rays leaving the reactor.

v. **Cooling system:** This is intended for removing the heat energy released in the reactor core. The heat is evolved from the kinetic energy of the fission fragments when they are slowed down in the fuel and moderator. The coolant or heat transfer agent (water, steam, He, CO_2, air and certain molten metals and alloys) is pumped through the reactor core. Then, through a heat exchanger the coolant transfers heat to the secondary thermal system of their reactor. The energy generated in a reactor is to be converted into electric power. For this purpose, energy generated is transferred from the coolant to a working liquid, to produce steam or hot gas. The regulating vapour or gas is then used to generate electricity by means of turbines. Ordinary water and heavy water, both are found to be good coolants, as they can serve both as moderator and coolant. A coolant should have high specific heat, high boiling point, cheap, easy to pump and chemically stable.

vi. **Controlled and safety systems:** The control and safety systems enable the chain reaction to be controlled, prevent if from spontaneously running away and also protect the space surrounding the reactor against intensive neutron flux and gamma rays existing in the reactor core.

The first aim is achieved by pushing control rods of boron or cadmium into the reactor core. These rods have a large neutron absorption cross-section. The second aim is achieved by surrounding the reactor with massive layers of substances that strongly absorb neutrons and gamma rays and also by providing completely closed coolant circuits without any leakage. One can explain the control of the thermal reactor by a quantity K_e known as **effective multiplication factor,** which controls the condition of reaction at steady state. How the neutron concentration would increase from generation to generation provided no neutrons leaked out from the system can be known from the multiplicative factor K_∞. To obtain a controlled chain reaction in a reactor, one will have to combine K_∞, in some way with the quantity which tells something about neutron leakage, and that is called *effective multiplication factor* (K_e) defined as

$$K_e = \frac{P}{A+L}$$

where P is production rate of neutrons, A is absorption rate of neutrons, L is leakage rate of neutrons, $K_e = 1$ denotes steady state, $K_e = 0$ critical condition, $K_e > 1$ is supercritical and $K_e < 1$ is subcritical condition.

We have mentioned earlier that production rate is proportional to r^3, the volume of the reacting system, whereas the escape or leakage rate is proportional to r^2, the surface area of the reacting system. Therefore,

$$K_e \propto \frac{r^3}{r^2} \propto r$$

where r is the radius of the reacting system. Clearly, the effective neutron multiplication factor (K_e) depends on r, i.e. the size of the nuclear reacting system. For maintaining controlled chain reaction in a nuclear reactor, i.e. to make effective multiplication factor $K_e = 1$, one needs a critical size of the system for which the production of neutrons by fission process will be just balanced by the non fission capture and escape. Now

P(production rate of neutron) $= NF$

where N is average number of neutron emission per fission and F is the rate of fission process.

$$K_e = \frac{NF}{A+L} = \frac{N\left(\dfrac{F}{A}\right)^*}{1+\left(\dfrac{L}{A}\right)}$$

We may note that the ratio F/A depends upon the amount of fissionable and non fissionable materials and on their cross-sections of fission and neutron capture.

One can control the fission rate in a nuclear reactor by means of controlling rods. Controlling rods are generally made up of cadmium or boron steel. By adjusting the position of control rods in the core, one can adjust the rate of absorption of excess neutrons, i.e. F/A can be controlled. Thus, one can maintain the value of $K_e = 1$, i.e. obtain controlled fission. If the value $K_e < 1$, then one will have to add fresh fissionable material.

Thermal Power Nuclear Reactor

Primarily, the power reactors are intended for the production of electricity, i.e. usable power. These reactors transform the heat generated from the fission into electricity. The heat so generated is allowed to pass to a fluid passing through the core of the reactor. The hot fluid is then allowed to pass through a heat exchanger where energy is transferred and utilized by proper means.

Figure 25.16 shows the basic principle of one type of power reactor. A definite quantity of pure enriched uranium forms the centre of heat energy source. By keeping the value of $K_e = 1$, the fission rate and hence the rate of production of heat energy is controlled, which is achieved with the help of controlling cadmium rods. One can decrease the temperature if desired, by pushing little further the cadmium rods so that they may absorb more neutrons thereby decreasing the

fission rate (F/A). Cadmium rods are pulled out a little further for raising the temperature. When the reactor or pile becomes subcritical, one will have to add fresh fissionable material.

As a primary coolant, liquid sodium is used. Liquid sodium is not only a dense material with good thermal conductivity but it also has very low absorption cross-section for thermal neutrons. To lower the melting point, sodium is usually alloyed with potassium. This primary coolant absorbs energy, which is transferred to water in a heat exchanger. The steam that is generated drives a turbine. The turbine is connected to an electric generator. The electric power so generated can be used for running factories, lighting cities or for other purposes. There are a good number of power reactors operating in India. Besides producing energy, a nuclear reactor is also a powerful source of neutrons and of intense induced radioactivity.

Fast Breeder Reactors

The interest in fast breeder reactors for commercial power production lies in the prospect that they can serve as breeders. Such reactors could not only produce useful power but also regenerate more fossil material than is consumed. Ultimately use is to be made of the relatively large amounts of ^{232}Th and ^{238}U, that are available in nature. Obviously, breeder reactors must assume a prominent role in the nuclear power program.

In a breeder reactor, nuclei of one fissionable material are transformed as a result of nuclear reactions into nuclei of another fissionable material. Conversion of ^{232}Th and ^{238}U into ^{233}U and ^{239}Pu respectively as a result of neutron absorption is as follows:

$$^{232}\text{Th} + {}_{0}^{1}n \rightarrow {}^{233}\text{Th} + \gamma$$

$$\text{(fast) 22 min} \downarrow \beta^{-}$$

$$^{233}\text{Pa}$$

$$\downarrow 27\,d$$

$$^{233}\text{U} + \beta^{-}$$

and $^{238}U + {}_0^1 n \rightarrow {}^{239}U + \gamma$

(fast) 24 min $\downarrow \beta^-$

^{239}Np

2.3 $d \downarrow$

$^{239}Pu + \beta^-$

Clearly, more fissionable material is produced than is consumed. It is inevitable that some neutrons should be lost by leakage or by absorption, but so long as more than two neutrons are produced in fission there is a possibility that more nuclear fuel may be produced than is consumed. This is the basic principle of the breeder reactor and the process is called *breeding*. If one could made it work, a *breeder reactor* could make new fuel for itself and in addition, a stockpile of fissionable material could be built up for use as a fuel in other new reactors.

Breeder reactors promises low fuel costs and efficient utilization of fuel resources through breeding. Use of sodium as a coolant permits achieving high temperatures at normal pressure; also sodium is a very efficient heat transfer medium. The handling of sodium introduces some design and operating complications. Experimental breeder reactor (EBR) has been built to study the feasibility of power production and breeding in such a reactor.

. Research in the field of fast breeder reactors are in progress. The value of n for ^{233}U and neutrons is 2.28 ± 0.02, which makes the breeding of ^{233}U from ^{232}Th in thermal reactor system feasible. Research and developments are in progress in such systems. One design of such system involves the use of $^{233}U\,O_2SO_4$ in D_2O, whereas another design involves the use of a solution of ^{233}U Bi in liquid bismuth with graphite as a moderator. It is reported that ^{233}U - ^{232}Th system could also be used in a fast breeder reactor, but the greater value of v for ^{239}Pu favours the use of latter material in fast breeder reactors.

If we could achieve the conversion of ^{238}U and ^{232}Th into fissionable materials, the resulting increase in available nuclear fuels would be a step in the direction of fast breeder economic nuclear power reactor. We may note that much information is needed concerning the cross-sections for nuclear reactions of fast and intermediate energy neutrons with fissionable and fertile materials. Obviously, a lot of research work is to be done in the field of breeder reactors. Presently, several breeder reactors are operating. They have proved to be extremely expensive and have had operating problems.

Nuclear Reactor Problems

Today, nuclear power supplies about 18% of the world's total electrical energy. Nuclear power plant construction continues around that world. However, this is affected after recent Tsunami's destruction in Japan. Problems certainly exist with nuclear power plants. Nevertheless, ongoing uncertainties about the list of catastrophic nuclear accidents, long term waste storage and its disposal, terrorism, environmental problems and weapon proliferation continue to haunt the nuclear power industry. Proliferation especially is a very real concern. Although nuclear power and weapons development are different enterprises, they share infrastructure and educated technological elite that can be put to either purpose. If nuclear power is to advance, it will need to do so under strict international guarantees against division of materials and expertise to weapons production. For large expansion of nuclear power, scientists will have to overcome four critical problems— lower costs, improved safety, better nuclear waste management, and lower proliferation risk.

25.18 NUCLEAR FUSION AND THERMONUCLEAR REACTIONS

When two or more very light nuclei moving at very high speeds are fused together to form a single nucleus then the process is called as *nuclear fusion*. The mass of the product nucleus is less than the sum of the masses of the nuclei which are fused. The lost mass is converted into energy which is released in the process according to the relation $E = mc^2$. This property of the light nuclei is shown by the binding energy curve (Fig. 25.3), in which the average binding energy per nucleon rises rapidly with increase in the mass number in the range of low mass number nuclei. Fusion reactions take place at very high temperature ($\sim 10^8$ K), so it is also called thermonuclear reaction. *Hydrogen bomb* is based on this principle and also called as **nuclear fusion bomb**.

For example, two deuterons (heavy hydrogen nuclei) can be fused to form a triton (tritium nucleus) according to the following fusion reaction:

$$^2_1H + {}^2_1H \rightarrow {}^3_1H + {}^1_1H + 4\,\text{MeV}$$

The tritron so formed can further combine with a third deuteron to form an α-particle ($^4_2$He nucleus) according to following reaction

$$^3_1H + {}^2_1H \rightarrow {}^4_2He + {}^1_0n + 17.6\,\text{MeV}$$

The net result of the two reactions is the burning of three deuterons and the formation of an α-particle, a ^1_0n and $^1_1$H (proton). The total energy released is 21.6 MeV. Obviously, the energy released per deuteron burnt is 7.2 MeV. Most of the energy liberated in the fusion reaction appears as kinetic energies of neutron and proton. We may note that for fusion reaction to take place, the component nuclei must be brought to 10^{-14} m. In order to approach so closely, nuclei participating in reaction should be imparted high energies so that two positively charged deuterons may overcome the repulsive force and fused together. The only practical way of fusing the two nuclei of deuterons together is to raise their temperature to the order of $10^8\,°$C, typical of sun's interior. At the required temperature, a fusion reactor's fuel will be in the form of *plasma*, which is fully ionized gas. Plasma is the *fourth state of matter*.

The energy output in fusion reaction is much less than the energy released in fission reaction, which is about 200 MeV. However, this does note mean that fusion is a weaker source of energy in comparison to fission. In fact, the energy output per unit mass of the material consumed is much greater in the case of the fusion of the light nuclei then in the case of fission of heavy nuclei.

A successful fusion reactor has three basic requirements to fulfil as mentioned below:

1. The plasma temperature must be high so that an adequate number of the ions have the speed essential to come close enough together to react despite their mutual repulsion. Taking into account that many ions have speed well above the average and that tunnelling through the potential barrier reduces the ion energy needed, the minimum temperature for igniting a D-T plasma is about 100 million.

2. The plasma density (n ions/m^3) must be high to ensure that collisions between nuclei are frequent.

3. The plasma of reacting nuclei must remain together for a sufficiently long time τ, how long depends on the product $n\tau$, the confinement quality parameter. In the case of D-T plasma with $kT \sim 10\,\text{keV}$, $n\tau$ must be greater than roughly $10^{20}\,\text{s/m}^3$ for break even more than that for ignition.

The approach of magnetic confinement of plasma (tokamak, etc.) to the controlled release of fusion energy has thus far shown the most promising use of strong magnetic field to confine the reactive plasma. Laser beams have also received the most attention for inertial confinement of plasma but electron and proton beams have promise as well. However, magnetic confinement seems closer to the good of achieving a working fusion reactor.

A possible fusion reactor is shown in Fig. 25.17.

The Source of Stellar Energy: Stellar Thermonuclear Reaction

Bethe in 1939 suggested that the production of stellar energy is by thermonuclear reactions in which protons are continuously transformed into helium nuclei. For comparatively low stellar temperatures, he proposed the proton-proton cycle represented by following chain reactions:

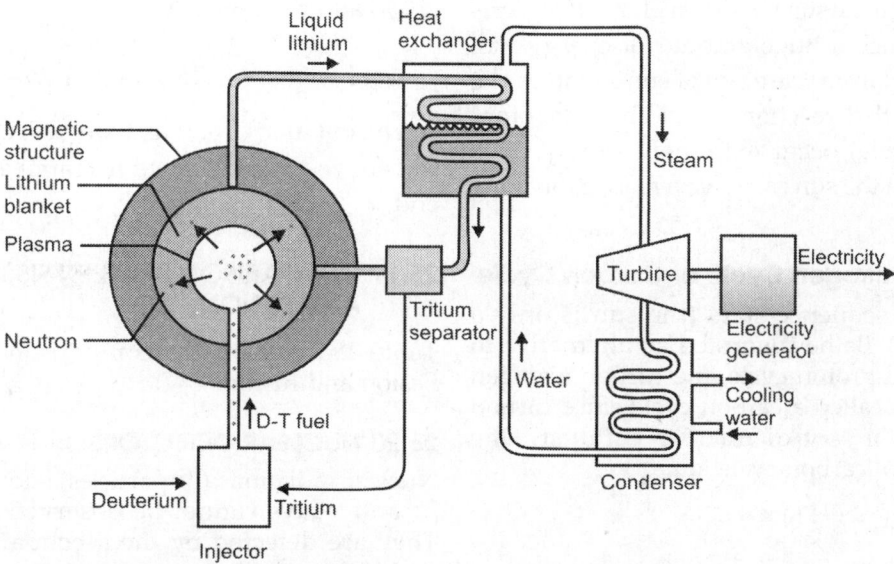

Fig. 25.17: Configuration of a possible fusion reactor using magnetic confinement. The tritium is obtained from neutron reaction with lithium and tritium must be separated prior to going back to fusion reactor as fuel in fission reactors, the heat produced by fusion is used to run a turbine, which in turn generates electric power.

$$_1^1H + _1^1H \rightarrow _1^2D + _{+1}^0e\,(\beta^+) + v + 0.42\ \text{MeV} \quad \text{(i)}$$
$$\text{(Neutrino)}$$

$$_1^2D + _1^1H \rightarrow _2^3He + \gamma + 5.5\ \text{MeV} \quad \text{(ii)}$$

$$_2^3He + _1^1H \rightarrow _2^4He + _{+1}^0e\,(\beta^+) + v \quad \text{(iii)}$$
$$\text{(Neutrino)}$$

$$_2^3He + _2^3He \rightarrow _2^4He + 2\,_1^1H + 12.8\ \text{MeV} \quad \text{(iv)}$$

In this cycle of reactions, fusion reactions (i) and (ii) must occur twice to yield two $_2^3He$ nuclei in (iv). The net result of (i), (ii) and (iv) reactions is that four protons are fused to produce an α-particle, two positrons, two neutrinos and two γ-photons.

$$4\,_1^1H \rightarrow _2^4He + 2\,_1^0e + 2v + 2\gamma$$

The energy released in the cycle is 24.6 MeV for the mass difference is 4 × 1.0078 ~ 4.0026 or 0.0286 μ. Since the neutrinos escape, the energy left is about 24.3 MeV after the fusion of $4\,_1^1H$ into $_2^4He$ nucleus.

If we add (i), (ii) and (iii) equations, we obtain

$$4\,_1^1H \rightarrow _2^4He + 2\,_1^0e\,(\beta^+) + 2v + 27\ \text{MeV} \quad \text{(v)}$$

This is considered to be an important source of energy in the sun. It predominates in states of comparatively low temperatures. The amount of energy radiated by sun is ≈ 10^{26} J/s. The age of the sun is estimated to be ≈ 5 × 10^9 years. Hydrogen and helium together form 90% by weight of sun's matter. The temperature of the core (centre of sun) is about 3 × 10^7 °C. Obviously, fusion reactions are continuously taking place in sun which is the source of the stellar energy.

The reaction (v) is exothermic and amount of energy released in this reaction is 26.7 MeV = 42.7 × 10^{-13} J.

One gram of sun's matter contains about 2 × 10^{23} protons. Obviously, the available energy supply from the sun would be 21.3×10^{10} J/gm. Clearly, such a huge amount of energy will be released from one gram of sun's matter. We may note that reactions cited above (proton-proton cycle) occur extremely slowly. This means that the sun can have a reasonably long life.

Carbon–Nitrogen Cycle or Carbon Cycle

For main sequence stars (the sun is only a small star), Bethe suggested an alternative to the proton-proton cycle, the carbon-nitrogen cycle. It is called a carbon cycle since carbon serves as a sort of nuclear catalyst. The reactions of carbon cycle are:

$$^{12}_{6}C + ^{1}_{1}H \rightarrow ^{13}_{7}\overset{*}{N} + \gamma$$

$$^{13}_{7}\overset{*}{N} \rightarrow ^{13}_{6}C + ^{0}_{+1}e(\beta^{+}) + v$$

$$^{13}_{6}C + ^{1}_{1}H \rightarrow ^{14}_{7}N + \gamma$$

$$^{14}_{7}N + ^{1}_{1}H \rightarrow ^{15}_{8}\overset{*}{O} + \gamma$$

$$^{15}_{7}N + ^{1}_{1}H \rightarrow ^{12}_{6}C + ^{4}_{2}He$$

On addition, we get

$$4\,^{1}_{1}H \rightarrow ^{4}_{2}He + 2\,^{0}_{+1}e(\beta^{+}) + 3\gamma + 2v + 26\,\text{MeV}$$

The initial $^{12}_{6}C$ acts as a kind of catalyst for the above process, since it reappears at the end.

25.19 COMPARISON BETWEEN FISSION AND FUSION

Table 25.3 depicts the comparison between fission and fusion.

25.20 NUCLEAR DETECTORS

Nuclear radiations (or particles) such as α, β^{-}, β^{+} and γ-rays cannot be observed directly. They are detected by the secondary effects which they produce in the materials through which they pass. There are such effects, e.g. ionization, photographic and fluorescence. Nuclear detecting instruments are based upon these effects. The instruments which are used for the detection of nuclear radiations are: (i) ionization chamber (ii) proportional counter (iii) Geiger–Muller counter (iv) cloud chamber,

Table 25.3: Comparison between fission and fusion

Fission	*Fusion*
1. Energy released due to fission of one nucleus of ^{235}U is about 200 MeV.	1. Energy released in a D-T fusion reaction is 17.59 MeV.
2. A heavy nucleus is split up into two nuclei, e.g. $^{235}_{92}$U into ^{95}Mo and ^{139}La.	2. Two lighter nuclei fused together at very high temperature, e.g. $^{2}_{1}$H, $^{3}_{1}$H.
3. Fission process is possible even at room temperature.	3. Fusion is possible only at very high temperature (~ millions K).
4. The links of fission prices are neutrons.	4. The links of this process are protons.
5. The energy per nucleon in fission process is 0.85 MeV or 200 MeV per nuclide.	5. The energy per nucleon is 6.75 MeV.
6. Several isotopes are formed as product particles in this process.	6. In general, isotopes are not formed in this process.
7. Availability of atoms used, e.g. ^{235}U, ^{238}U is limited and also costly.	7. Availability of atoms used, e.g. $^{2}_{1}$H, $^{3}_{1}$H is unlimited and at the same time cheap also.
8. Disposal of radioactive waste is a challenging problem.	8. In comparison to fission, radioactive waste material produced is very small.
9. Nuclear fission power reactor is already in operation.	9. Fusion reactor development research is in progress.

etc. Out of these four each of the first three detectors, used to detect radioactive radiation, is based on the production of ionization in a gas and the separation and collection of ions by means of an electrostatic field. One can explain the differences in these three systems with the help of Fig. 25.18. Figure 25.18 shows a cylindrical conducting chamber and insulated from it. The conducting chamber is filled with a gas at a pressure of about one atmosphere or even less than that. Between the wall and the central electrode through the resistance R shunted by capacitor C, a voltage V is applied as shown in Fig. 25.18. Relative to that of the central wall, the central electrode is at a positive potential.

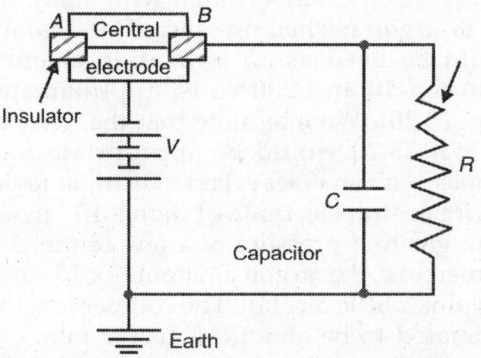

Fig. 25.18: Nuclear particle detector for detecting α- and β-particles

When a charged particle enters the ionization chamber, ionization of gas takes place resulting in an ion pair formation. Due to the presence of voltage V, positive ions move to the chamber walls and negative ions (electrons) move to the central electrode. Now, for a given initial ionization, how many ion pairs are collected or how many electrons reach the central electrode with the variation of the applied voltage, is shown in Fig. 25.19, where we have plotted the curves of total ion collection as functions of applied voltage. For convenience, we have taken the logarithm to the base 10 of the number (log n) of ion pairs as the ordinate. When there is no voltage across the electrodes, the ions will recombine and obviously, no charge will appear on the capacitor. With the increase in voltage, say, to

a few volts, there is competition between the loss of ion pairs by recombination and the removal of ions by collection on the electrodes. Some electrons will also reach the central electrode. Now, at an applied voltage V_1 (~10 V), it is found that the loss of ions by recombination becomes negligible and all electrons reach central electrode, log n attains the value unity. Now, on further increasing the voltage from V_1, n stays constant until a voltage V_2 is reached which may be some tens or hundreds of volts depending on the experimental conditions. The number of ion pairs collected is independent of the applied voltage between V_1 and V_2 region and curve obtained is horizontal. This region is called the *ionization chamber region*.

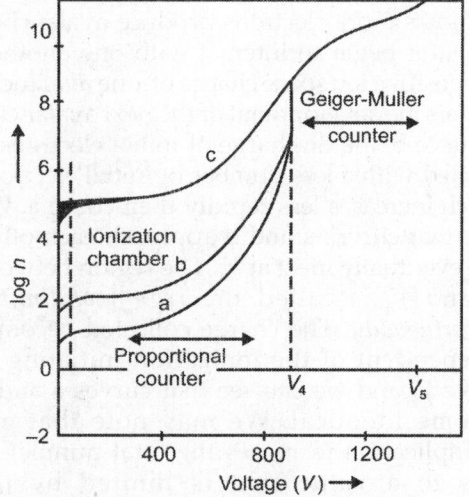

Fig. 25.19: Variation of total ion pair (log n) with applied voltage (V)

Now, the voltage is further increased above V_2. Because of the phenomenon of gas amplification or gas multiplication, n increases above 10. Now, the electrons released in the primary ionization acquire enough energy to produce additional ionization, when they collide with gas molecules and increases roughly exponentially with applied voltage V. Close to the central electrode, each initial electron produces a small *avalanche* of electrons with most of these secondary electrons. Curve b results, if number of ion pairs are formed initially is more. Curve b is parallel to curve a in the ionization chamber region and we can see that it is separated from it by one unit on the log n scale. We may note that above applied voltage V_2, the behaviour of the two curves a and b is quite interesting. However, for some range of voltages, up to V_3, each electron acts independently and gives its own avalanche, now they are being affected by the presence of other electrons. On the log n scale, the two curves a and b continue to be parallel with a difference between them of one unit. The number of ion pairs collected between V_2 and V_3 is then proportional to the initial ionization. This region is known as *proportional counter operation region*.

The gas multiplication effect continues to increase very rapidly above the applied voltage V_3 and as more electrons produce avalanches, the latter begin to interact with one another. The positive ion space charge of one avalanche inhibits the development of the next avalanche. In curve b, the discharge of initial electrons is affected with a less number of initial electrons which increases less rapidly than curve a. We note that curves a and b approach each other and eventually meet at V_4. The region between V_3 and V_4 is called the *region of limited proportionality*. The charge collected becomes independent of the ionization initiating at above V_4 and we can see that curves a and b become identical. We may note that gas multiplication increases the total number of ions to a value that is limited by the characteristics of the chamber and the external circuit. The region above the applied voltage V_4 is called the region of Geiger–Muller (GM) counter operation. The region of GM counter operation ends at a voltage V_5 where the discharge tends to propagate itself indefinitely. Obviously, the applied voltage V_5 marks the end of the useful voltage scale and the region above V_5 being that of continuous discharge.

Curve c is obtained if the initial ionization is very large. Curve c is similar to curves a and b, and parallel to them between V_2 and V_3, the proportional counter region ends sooner instead of at V_3. Obviously, the extent of that region depends on the initial ionization.

On the basis of the behaviour of the ions of the gas in the electrostatic field particles detection instruments have been developed as follows:

Ionization Chamber

This is mainly used for detecting the α- and β-rays and to compare the activities of sources emitting these rays. This operates at voltages in the range V_1 and V_2, and moreover it is characterized by complete collection, without gas amplification, of all the electrons initially liberated by the passage of the particle. Subject to certain conditions, it will give a pulse proportional to the number of these particles, i.e. electrons. Basically, it consists of a chamber fitted with a gas, like air or argon normal pressure. This chamber could be used as an ionization chamber between 10 and 200 volts, approximately (Fig. 25.20). We may note that the numbers in Fig. 25.20 would be appropriate for a counter with an outer cylinder 0.01m in radius, a wire central electrode of radius 10^{-4} m and with gas to a pressure of a few centimeters of mercury, e.g. argon at about 6×10^{-2} m of Hg plus a little alcohol. The counter length is estimated to be about 0.1 m. We may note that the electrodes of an ionization chamber

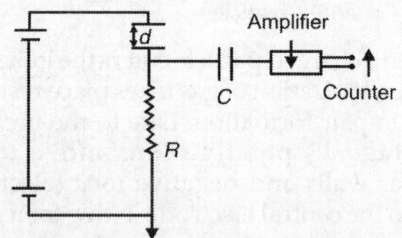

Fig. 25.20: Ionization chamber

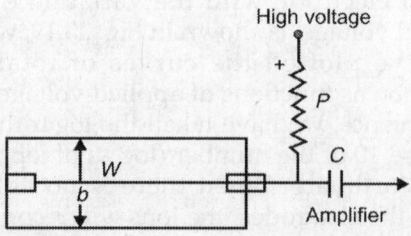

Fig. 25.21: The proportional counter

can also be parallel plates. Ionization chambers are usually of two types: (i) non-integating (ii) integrating. First type of ionization chamber consists of a cylindrical conducting chamber with metallic electrode *AB* located on the axis of the chamber and insulated from it. The second type consists of a cylindrical chamber with two electrodes *A* and *B* placed at some distances from each other.

An ionization chamber is much less sensitive to β-particles (in comparison to α-particles) because β-particles produce fewer pairs of ions in their passage through the chamber. It is too intensive for γ-rays as these rays produce very little ionisation in the gas.

For detecting γ-rays, an ionization chamber of thick wall high atomic number (Pt, Bi) is employed. The γ-rays impinging on the walls of the chamber eject high speed electrons which produce ionisation in the gas.

We may note that the ionisation chamber does not give a count of individual particles, but an average effect of a large number of particles.

Proportional Counter

This is used to detect and measure the energies of particles (α and β) and also of photons (γ- and X-rays). This operates in the region V_2 to V_3 and is characterised by a gas multiplication independent of the number of initial electrons. This counter functions satisfactorily at voltages 500 to 800 volts or above at pressure of about one atmosphere. Although gas multiplication is utilized in this counter, the pulse is always proportional to the initial ionization. The use of proportional counter permits both the counting and energy determination of particles which do not produce enough ions to yield a detectable pulse in the region of voltages V_1 to V_2. These pulses are displayed on an oscilloscope screen. The pulse height gives a measure of the energy of particle (or photon) entering the counter. Obviously, the proportional counter is thus able to distinguish directly (by virtue of the height of the pulse produced) between α- and β-particles and γ-rays since they differ so much in their ionizing powers. Clearly this counter offers advantages for pulse they measure.

Geiger–Muller (GM) Counter or GM Tube

This is the most versatile, accurate and useful of the detecting instruments used for detecting and measuring the energies of α, β, γ and X-rays. It is sufficiently sensitive to detect individual α- and β-particles. It operates in the voltage region V_3 to V_5 and is characterized by the speed of the discharge throughout the entire length of the counter, resulting in a pulse independent of the initial ionization. GM counter is widely used in industry and medicine to locate radioisotopes and in the laboratory for comparing the activity of radioactive specimen.

Usually a GM counter consists of a fine wire (usually tungsten) placed along the axis of a hollow metallic cylinder electrode (cathode) enclosed in a glass tube (Fig. 25.22) The tube contains a suitable mixture of 90% argon at 10 cm pressure and 10% ethyl alcohol vapour at 1 cm pressure. We may note that different mixtures of gases at different pressures are used in different designs.

A potential difference between 800 V and 2000 V, is usually applied between cylinder and wire which acts as anode. However, the value of this applied voltage is adjusted to be somewhat below the breakdown potential of the gaseous mixture.

When a charged particle passes through the GM counter tube, it ionizes the gas molecules, thereby releasing a number of ion pairs. The central wire attracts the electrons (–ve ions) while the cylindrical electrode attracts the positive ions and the resulting pulse of a current through the tube produces a voltage pulse ($\approx 10\,\mu V$) across resistance *R*. The pulse is amplified by an electronic pulse amplifier to about 5 to 10 V. This amplified pulse is finally passed on either to a scalar or to a rate-meter. The scalar records the arrival of each individual pulse separately and so provides the exact number of particles entering the GM tube in a given time interval. The ratemeter records the number of pulses in a given time, i.e. it gives the average rate at which the particles enter the GM tube. Scalar is usually

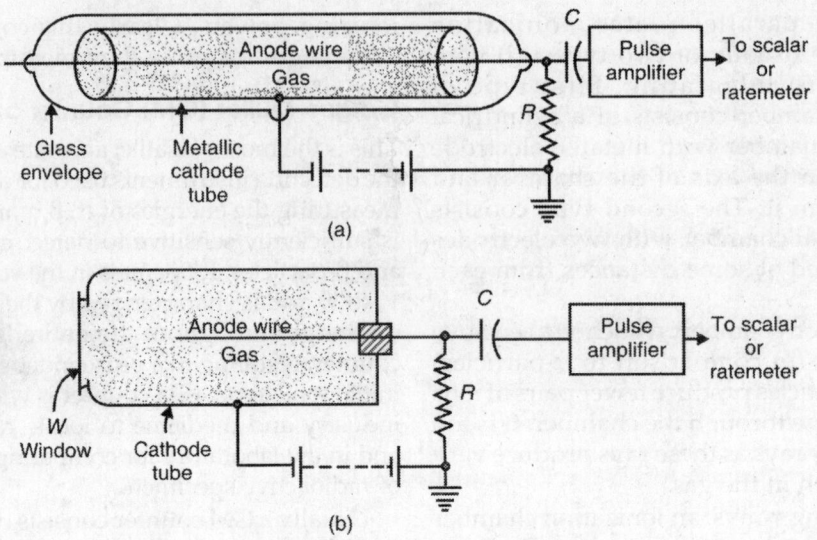

(a)

(b)

Fig. 25.22: Geiger–Muller counter

used in conjuction with the GM tube for very high pulse rates. The combination of GM tube with the recorder is called GM counter.

The magnitude of the pulse produced dos not depend upon the nature or energy of the entering particle, and hence a GM tube does not distinguish between the arrival of an α-particle and a β-particle. We may note that a proportional counter does it. However, the particles can be distinguished by finding what absorber placed in their path will prevent them from entering the GM tube.

GM counter discussed so far is suitable for β- and γ-rays measurement and it is not suitable for α-particles, because the walls of glass tube are too thick and therefore α-particles cannot penetrate through the walls. For detection of α-particles, the *Geiger-point counter* shown in Fig. 25.22b is generally used. It mainly consists of an axially mounted fine wire inside a tube which is maintained at a positive potential with respect to the GM tube. This GM tube contains dry air at atmospheric pressure and has an extremely thin foil window at one end. The α-particles to be detected enter the tube through the window (*W*).

The successful operation of GM counter depends upon the proper voltage to the electrodes. Figure 25.23 represents the counts per minute as a function of voltage. From Fig. 25.23, it is obvious that if the voltage is less than 1000 V, there is no discharge, i.e. no secondary ionization. With the further increase in voltage, secondary ionization takes place. Now, the number of impulses increases almost linearly with applied voltage. This region is most suitable for **proportional counters**. On further increase of voltage to about 1200 V, the number of impulses remains constant over a certain region known as *working plateau region*. In this region, the magnitude of impulses becomes independent of the amount of original ionization and is a

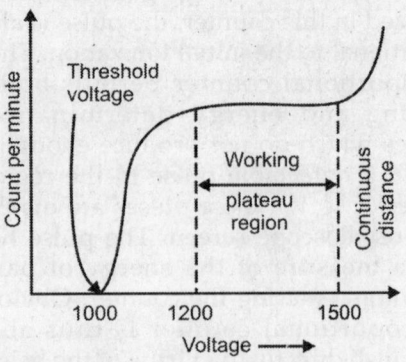

Fig. 25.23: Counts per minute of GM counter as a function of voltage

function of potential, nature of gas, resistance R and geometrical conditions of apparatus. This is the most suitable region for GM counter. Beyond this voltage, i.e. above this region, a continuous discharge will take place, which is undesirable and hence avoided.

25.21 PARTICLE ACCELERATORS

Cyclotron

Cyclotron is first orbital type of accelerator in which charged atomic particles move under the action of a strong magnetic field in circular paths. This was invented by Lawrence in 1932. The important feature of the cyclotron is that when the particles are to be accelerated, the phase of the accelerating field is always such as to speed them up. The machine is capable of giving energetic ions of larger intensity than that of linear accelerator. The principle of this accelerator is based upon resonance of electric and magnetic fields, and therefore it is called as **magnetic resonance accelerator**. However, the name cyclotron is much more popular as the acceleration of particles involves the cyclic motion.

To understand the basic **principle** of a cyclotron, let us consider a charged particle of charge e and mass m, moving with uniform velocity v, in a transverse magnetic field B. This transverse magnetic field will exert a force on the particle perpendicular to the plane of v and B both. The magnitude of this force is evB. Under the influence of this force, the particle will move along a circular path, and will be subjected to centrifugal force given by $\dfrac{mv^2}{r}$, where r is radius of curvature of the path. Thus,

$$\frac{mv^2}{r} = evB \text{ or } \frac{v}{r} = \omega = \frac{e}{m} \cdot B \qquad (25.36)$$

where ω is angular velocity of rotation of particle.

The time T required by the particle in completing one revolution will be given by

$$T = \frac{2\pi r}{v} = \frac{2\pi m}{eB} \qquad (25.37)$$

Equation (25.37) shows that irrespective of the velocity of the particle and for given value of B and e/m, the time of revolution remains constant.

The cyclotron consists of two short, hollow, semicircular cylinders D_1 and D_2 mounted rigidly along their diameters, separated from each other by a gap as shown in Fig. 25.24a. Because of their shapes, these semicircular cylinders are called *dees*. The dees are connected to a high frequency oscillator, which is really a short wave radio transmitter supplying energy to the dees. The entire arrangements of dees is placed inside a vacuum chamber within the pole pieces of a strong magnet. The complete cross-sectional diagram is shown in Fig. 25.24b. To obtain charged particles, say protons, at the centre of dees is placed a hot filament from which, by passing electric current, thermions (electrons) can be obtained. Now, when a trace of hydrogen gas is admitted to the evacuated chamber, these electrons ionize some of the hydrogen atoms by collision, producing protons.

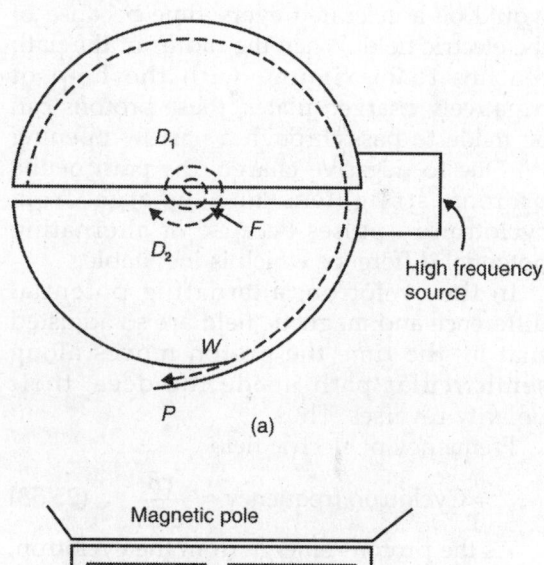

Fig. 25.24: (a) Position of dees of a cyclotron (b) Cross-sectional diagram of a cyclotron

Consider protons in the neighbourhood of F, and the dees are connected to high frequency oscillator. At the instant when D_1 is negative and D_2 is positive, the protons will be accelerated towards D_1. As soon as the protons enter the space inside D_1, the electric field would not interact with its motion and under the influence of strong magnetic field their path would be circular. If after making half revolution, the potential difference between the dees D_1 and D_2 is reversed so that D_1 becomes positive and D_2 negative and the protons reach in the gap between dees D_1 and D_2, these would be accelerated towards D_2, (as D_2 is negative) and repelled by D_1 causing the protons to increase in their speed. Hence is inside the space of D_2 the particles will move in a circular path of larger radius. Now, again after completing half revolution in D_2 as the protons reach in the gap, again the polarity of dees reverses and the protons would be accelerated towards D_1. In this way, every time when the protons reach in the gap, these would be accelerated every time because of the electric field. When the radius of the path reaches to maximum, with the help of negatively charged plates, these protons can be made to pass through a narrow opening W. Due to negative charge, the path of the protons straightens and emerge from cyclotron in pulses because of alternating potential difference which is inevitable.

In the cyclotron, alternating potential difference and magnetic field are so adjusted that by the time the proton moves along semicircular path inside the dees, their polarity reverses. Thus,

Frequency of electric field

$$= \text{Cyclotron frequency} = \frac{eB}{2\pi m} \quad (25.38)$$

As the protons emerge from the cyclotron, their velocity and hence energy would be maximum. At the same time, the radius of curvature of the path equals that of radius R of the dees. Thus,

$$Bev_{max} = \frac{mv_{max}^2}{R} \text{ or } v_{max} = \frac{BeR}{m} \quad (25.39)$$

Hence, maximum energy, E_{max} is given by

$$E_{max} = \frac{1}{2}mv_{max}^2 = \frac{1}{2}\frac{B^2e^2R^2}{m} \quad (25.40)$$

Thus, we see that in a cyclotron the energy can be increased to any amount by increasing cyclotron radius. However, it is not possible. Because, as the velocity of the particles increases, the mass of the particles also increases according to the relativistic relation

$$m = \frac{m_0}{\sqrt{1 - \dfrac{v^2}{c^2}}} \quad (25.41)$$

This relativistic mass increases with velocity limits the maximum energy obtainable from cyclotron. It is because of this reason electrons cannot be accelerated by cyclotron.

Lawrence's first cyclotron with magnetic pole diameter 2.5 inches provided proton of energy 80 keV. At Berkeley, California, USA, so many cyclotrons of different diameters of pole pieces have been built. A 60 inch pole piece cyclotron is capable of giving deuterons at 24 MeV and helium atom to about 40 MeV. The magnetic field applied in 60 inch cyclotron was 1.6 W/m^2 or 16000 gauss.

Usefulness and Limitations

The cyclotron can energise particles like protons, deuterons, α-particles to several million electron volts ($\sim 10^{-50}$ MeV). Main advantage of cyclotron is that it does not demand any excessively high **voltage** source of producing high energy particles. Moreover, no accelerating tube capable of withstanding high potential gradient is necessary.

The most serious limitations imposed on the action of a cyclotron is, however, due to the relativistic increase in mass of the accelerated particle in accordance with the relation (25.41).

The frequency is given by

$$f = \left(\frac{q}{m}\right)\frac{B}{2\pi} = \left(\frac{q}{m}\right)\frac{Bc^2}{2\pi c^2}$$

$$= \frac{Bqc^2}{2\pi\left(m_0c^2 + W_k\right)} \quad (25.42)$$

where W_k is kinetic energy of the accelerated particle at speed v.

From Eq. (25.42), it is obvious that as the speed of the particle increases, f decreases and W_k increases. Thus, in a fixed frequency cyclotron, the particle as it spirals out from the source, will lag in phase behind the accelerating dees voltage. Obviously, it would thus cease to get accelerated. This means that one must suitably modify the frequency of oscillation to ensure resonance acceleration in the gaps. One cannot achieve this without upsetting the condition for low energy particles.

We may note that with electrons, the relativistic change in mass is more pronounced than with positive ions, e.g. proton, deuterons even for low energies.

Other limitations of the cyclotron that limit the maximum energy gain by the accelerating particle is set by the engineering difficulties and expense. The maximum kinetic energy of the particles when they leave the cyclotron is

$$W_m = \frac{1}{2}mv_{\max}^2 = \frac{1}{2}m\left(\frac{r_m qB}{m}\right)^2 = \frac{1}{2}\frac{q^2B^2}{m}r_m^2$$

Obviously, $W_m \propto r_m^2$. This means that for a given particle the size of a cyclotron increases more rapidly than the corresponding increase in energy.

The compensation for relativistic mass gain has been achieved in *synchrocyclotron*.

Betatron

It has been stated above that due to large relativistic increase in mass of the electrons at low energy, a cyclotron cannot be used to accelerate them. Kerst in 1941 invented another accelerator to accelerate β-particles or fast moving electrons, known as betatron. The fundamental difference between a betatron

and cyclotron is that in a betatron a rapidly changing magnetic field is used to accelerate the electrons and their path is an orbit of constant radius. The action of betatron depends upon the same principle as that of transformer, in which an alternating current applied to a primary coil induces a current in the secondary coil. In the primary coil, oscillating magnetic field is produced due to alternating current which in turn produces oscillating potential in the secondary coil. In this way, betatron is a transformer in which electron located inside an annular, doughnut-shaped vacuum chamber plays the role of secondary coil of one turn, and placed within the poles of laminated steel of electromagnet. The electrons gain energy by induction, because of the change of flux ϕ, with time, linking with the orbit.

Let us consider an electron moving in a circular orbit of radius r, where the total magnetic flux through the orbit is ϕ and the flux density at the orbit is B. This magnetic field B is perpendicular to orbital plane. The emf induced in the orbit is given by

$$\varepsilon = -\frac{d\phi}{dt} \quad (25.43)$$

The work done on the electron of charge e, in one revolution is given by εe which equals $e\dfrac{d\phi}{dt}$. If F is tangential force on the orbit, the work can also be given by $F \cdot 2\pi r$. Hence,

$$F = \frac{e}{2\pi r} \cdot \frac{d\phi}{dt} \quad (25.44)$$

Due to influence of this applied force, the energy of the electron will increase and hence it would try to move in an orbit of larger radius. If an orbit of constant radius is to be maintained, this increase in radius of curvature of the orbit be resisted. A radial force due to magnetic field and acting perpendicular to the direction of motion of the electron keeps it moving in the orbit of constant radius. This inward radial force is Bev, and must be equal to the centrifugal force mv^2/r.

Thus,

$$Bev = \frac{mv^2}{r} \qquad (25.45)$$

The tangential force on the electron is for the orbit of constant radius, is given by

$$F = \frac{d}{dt}(mv) = \frac{d}{dt}(Ber) = er\frac{dB}{dt} \qquad (25.46)$$

Hence, from Eqs. (25.44) and (25.46)

$$\frac{e}{2\pi r} \cdot \frac{d\phi}{dt} = er\frac{dB}{dt} \text{ or } d\phi = 2\pi r^2 \, dB \qquad (25.47)$$

Integrating $\int_0^\phi d\phi = 2\pi r^2 \int_0^B dB$

or $\qquad \phi = 2\pi r^2 B \qquad (25.48)$

This is known as **betatron condition**. The result shows that the total flux ϕ within the radius of the orbit r, must be twice the value which would be obtained if flux density B were uniform over whole area of the orbit and also this should be proportional to B. This condition holds for relativistic energies as well as non relativistic energies. Such a flux distribution is obtained by specially shaped countered pole piece faces, where the flux density at the centre of the orbit is greater than it is at the circumference of the orbit.

The construction of betatron is illustrated in Figs 25.25 and 25.26a. The electrons are produced from a hot filament and are given primary acceleration by the application of electric field of the order of a few thousands volts. These are then injected in the highly evacuated doughnut tube DD made of glass or ceramic at G. In the coils of electromagnet and alternating current from mains supply of normal frequency, say 50 c/sec, is passed. Thus, increasing flux in a given direction is obtained for quarter cycle only. This is shown in Fig. 25.26b. At any instant when magnetic field is just rising from zero in first quarter cycle, induces a potential in the doughnut tube, increasing the energy of the electrons. When the strength of the field passes its maximum

and start decreasing in the next quarter cycle, the direction of induced emf reverses and the electrons slow down. To avoid this effect the electrons are removed from the stable orbit. This is done by discharging a capacitor through the primary coils or through auxiliary coils around primary, at the instant when electron receives the desired amount of energy. Due to this extra current momentarily the total flux ϕ increases rapidly, but flux density B increases less rapidly and causes orbital radius

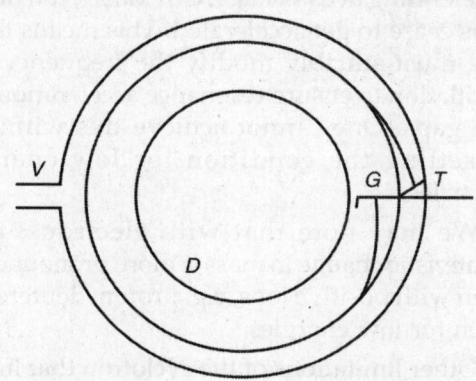

Fig. 25.25: Doughnut shaped annular chamber in a betatron

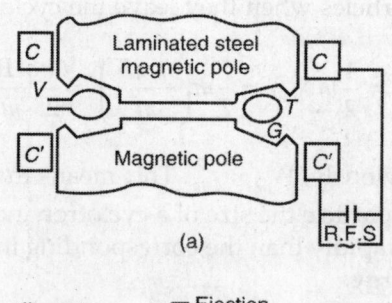

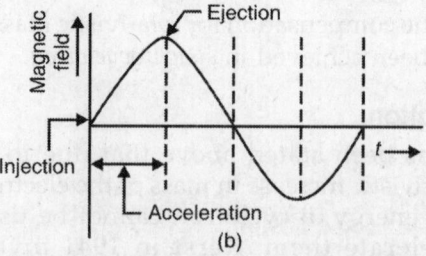

Fig. 25.26: (a) Position of D.T. in the pole pieces of electromagnet (b) Acceleration cycle in a betatron

r to increase. Thus, when the radius of final orbit increases the beam can be made to hit the target T, producing hard X-rays or γ-rays depending upon their energies in pulses.

The energy of the electron beam can be calculated from the average induced emf and the total number of revolutions made by the electron. Let the flux variation be given by

$$\phi = \phi_0 \sin \omega t$$

where ϕ_0 is amplitude and ω is angular frequency of changing flux. Hence, the energy per revolution is given by

$$e\frac{d\phi}{dt} = \omega e \phi_0 \cos \omega t \qquad (25.49)$$

Hence, the average energy per revolution

$$= \omega e \phi_0 \frac{1}{T/4} \int_0^{T/4} \cos \omega t \, dt$$

$$= \frac{2\omega e \phi_0}{\pi} \qquad (25.50)$$

This represents energy per revolution. To calculate the number of revolutions, we can approximately consider particle to travel with the velocity of light c, because electron of energy 1 MeV moves with velocity $0.94c$. Furthermore, these accelerate only during quarter of alternating cycle, i.e. for time

$$\frac{1}{4}T, \text{ viz. } \frac{1}{4}\frac{2\pi}{\omega} = \frac{\pi}{2\omega}$$

Distance traversed in this time $= \dfrac{c\pi}{2\omega}$

Distance traversed in one revolution $= 2\pi r$

$\therefore$ Number of revolutions

$$= \frac{c\pi}{2\omega} \cdot \frac{1}{2\pi r} = \frac{c}{4\omega r} \qquad (25.51)$$

Energy Acquired by the Electron

The momentum of electron, $mv = \dfrac{E}{c}$ (since the electron moves at relativistic velocity)

Also $\qquad mv = BeR$

$\therefore \qquad \dfrac{E}{c} = BeR$

or $\qquad E = BeRc \qquad (25.52)$

With betatrons, electrons at 300 MeV have been obtained. The 100 MeV betatron of General Electric Research Laboratories has 66 inches orbit of electron and 76 inches pole piece face diameter. This is operated with 60 c/sec alternating current. At the University of Illinois, electrons with energies more than 300 MeV have been obtained.

Any accelerated charge particle loses energy by radiation. Hence, an electron also loses energy by radiation and the rate of loss of radiating energy depends upon fourth power of the energy it is acquiring. This is turn sets limitations on the energies to which electrons can be accelerated in a betatron. The maximum energy is reached when the energy lost per turn by radiation equals the maximum energy acquired by the electron per turn under practical conditions. Maximum electron energy can be reached to about 1000 MeV.

Uses and Limitations

Betatrons can energise electrons to the value of about 1000 MeV and are quite successful accelerators for electrons. Betatron is relatively small size accelerator and therefore quite suitable for laboratory work. Apart from its efficient use as an electron accelerator, it is quite useful and powerful source for X-ray generator. This is why betatron is used in hospitals for *therapeutic X-rays*. We may note that betatron can produce X-rays so short in wavelength as to surpass the γ-rays from radioactive sources.

Betatron suffers from the problem of *orbital stability*. One can achieve the stability by properly designing the machine. The main drawbacks of this accelerator are: (i) *huge magnet* (diameter exceeding ~2 m) is required for supplying the variable flux to accelerate the electrons (ii) an accelerated electron loses energy by radiation and the rate of loss of energy increases with the fourth power of energy so that a limit is set for the high energy electrons by *radiation loss* and (iii) the maximum energy that a betatron under practical condition prevailing in the machine is about 1 BeV (= 1000 MeV).

The disadvantages a betatron suffers from accelerating electrons have been eliminated to a great extent in *electron synchrotron*. In fact, in electron synchrotron, one combines the accelerating system of the cyclotron with the ring-shaped pulsating guide field of a betatron.

A comparison between cyclotron and betatron is presented in Table 25.4.

Table 25.4: A comparison between cyclotron and betatron

Cyclotron	Betatron
1. In a fixed frequency cyclotron, alternating magnetic field is not employed, i.e. magnetic field is constant.	1. An oscillating field is used in betatron. This field is also called *induction field*.
2. In a cyclotron, particles are accelerated in helical orbits and the orbital radius goes on increasing and equals the radius of the dees in the limit.	2. In a betatron, the acceleration of the particles takes place in a circular orbit of almost constant radius.
3. For accelerating electrons, a fixed-frequency cyclotron is not suitable.	3. Betatron is specially designed and meant for accelerating electrons.
4. A cyclotron suffers from orbital stability problem.	4. Betatron also suffers from the problem of orbital stability.
5. Cyclotron cannot be used as an X-ray generator.	5. Betatron is highly efficient generator for X-rays.

ILLUSTRATIVE EXAMPLES

Example 1

The half-life of radioactive cobalt 60 is 5.26 years. Calculate the activity of one gram sample of cobalt. Also express the activity in Curie and Rutherford.

Solution

Activity $\quad A = \lambda N$

$$\lambda = \frac{0.693}{T_{1/2}}$$

$$= \frac{0.693}{5.26 \times 365 \times 24 \times 60 \times 60} \sec^{-1}$$

$$= 4.17 \times 10^{-9} \sec^{-1}$$

For cobalt,

$$N = \frac{6.02 \times 10^{23}}{60} = 1.002 \times 10^{22} \text{ dis/s}$$

$A = \lambda N = 4.17 \times 10^{-9} \times 1.002 \times 10^{22} \text{ dis/s}$

$$= 4.2 \times 10^{13} \text{ dis/s}$$

Activity (A) in Curie

$$= \frac{4.2 \times 10^{13}}{3.7 \times 10^{10}} = 1130 \text{ Curie}$$

Activity (A) in Rutherford

$$= \frac{4.2 \times 10^{13}}{10^6} = 4.2 \times 10^7 \text{ Rutherford.}$$

Example 2

A wooden piece of great antiquity weighs 50 gm and shows ^{14}C activity of 320 dis/minute. Estimate the length of the time which has elapsed since this wood was part of a living tree. Assume that living tree shows a ^{14}C activity of 12 dis/minute/gm. Given half-life of ^{14}C is 5730 years.

Solution

Let us suppose that living tree had N_0 radioactive atoms just before it died. Its activity was

$$A_0 = \lambda N_0$$

When the tree died, the activity of the tree decreases exponentially. Thus, we have

$$A = -\frac{dN}{dt} = \lambda N = \lambda N_0 \, e^{-\lambda t} \qquad (1)$$

$$\therefore \qquad \frac{A}{A_0} = e^{-\lambda t}$$

Given $\quad A_0 = \text{dis/min/gm.}$

$$A = \frac{320}{50} \text{ dis/min/gm}$$

and $\quad \lambda = \dfrac{0.693}{5730} /\text{year}$

From Eq. (1), we have

$$\log_e\left(\frac{A}{A_0}\right) = -\lambda t$$

or $\quad t = \dfrac{1}{\lambda} \times 2.303 \log_{10}\left(\dfrac{A_0}{A}\right)$

$$= \frac{5730}{0.693} \times 2.303 \log_{10}\left(\frac{50 \times 12}{320}\right) \text{ years}$$

$$= 5170 \text{ years.}$$

Example 3

Calculate the Q value of the reaction $^{14}_{7}\text{Na}$ (α, p) $^{17}_{8}\text{O}$. Given masses are as follows:

Mass of $^{14}_{7}\text{Na}$ = 14.00753 u

Mass of $^{4}_{2}\text{He}$ = 4.00386 u

Mass of $^{17}_{8}\text{O}$ = 17.00450 u

Mass of $^{1}_{1}\text{H}$ = 1.00813 u

Solution

The given reaction is

$$^{14}_{7}\text{Na} + {}^{4}_{2}\text{He} \rightarrow {}^{17}_{8}\text{O} + {}^{1}_{1}\text{H} + Q$$

Now, substituting the given values of masses, one obtains

$$14.00753 + 4.00386 = 17.00450 + 1.00813 + Q$$

or $Q = 14.00753 + 4.00386 - 17.00450 - 1.00813$

$$= -0.00124 \text{ u} = -0.00124 \times 931 \text{ MeV}$$

$$= -1.154 \text{ MeV.}$$

Obviously, the reaction is endoergic.

Example 4

Assuming that energy released by the fission of a single uranium atom is 200 MeV, calculate the number of fissions per sec required to produce 1 W of power.

Solution

Given energy released per fission

$$= 200 \text{ MeV} = 2 \times 10^8 \times 1.6 \times 10^{-19} \text{ J}$$

$$= 3.2 \times 10^{-11} \text{ J} = 3.2 \times 10^{-11} \text{ W}$$

Number of fissions required to produce 1 W

of power $= \dfrac{1}{3.2 \times 10^{-11}} = 3.125 \times 10^{10}$ fissions.

Example 5

Calculate the energy released in the fission of the following reaction:

$$^{235}_{92}\text{U} + {}^{1}_{0}n \rightarrow {}^{95}_{42}\text{Mo} + {}^{139}_{57}\text{La} + 7\left({}^{0}_{-1}e\right) + 2\,{}^{1}_{0}n + Q$$

Given:

Mass of ^{235}U = 235.0439 u

Mass of neutron $\left({}^{1}_{0}n\right)$ = 1.0087 u

Mass of ^{139}La = 138.9061 u

Solution

Total initial mass = 236.0526 u

Total final mass = 235.8332 u

Mass converted into the energy

$\Delta m = 236.0526 - 235.8332 = 0.219$ u

∴ Energy equivalent

$$= 0.219 \times 931 = 204.3 \text{ MeV}$$

∴ $\quad Q = 204.3 \text{ MeV.}$

Example 6

A power reactor is developing energy at the rate of 30000 kW. How many atoms of ^{235}U undergo fission per second? How many kg of ^{235}U would be consumed in 1000 hours of operation. Assuming that on an average 200 MeV energy is released per fission?

Solution

Given that the rate of development of energy by power reactor = 30000 kW = 3×10^7 W

$$= 3 \times 10^7 \text{ J/s}$$

Energy released per fission = 200 MeV

$$= 200 \times 10^6 \times 1.6 \times 10^{-19} \text{ J} = 3.2 \times 10^{-11} \text{ J}$$

Number of atoms undergoing fission/s

$$= \frac{3 \times 10^7}{3.2 \times 10^{-11}} = 9.6 \times 10^{17}$$

Number of atoms whose fission takes place in 1000 hours

$$= 9.6 \times 10^{17} \times 1000 \times 60 \times 60$$

$$= 3.384 \times 10^{23}$$

∴ 6.02×10^{23} atoms of ^{235}U weigh

235×10^{-3} kg

∴ 3.384×10^{23} atoms of ^{235}U will weigh

$$= \frac{235 \times 10^{-3}}{6.02 \times 10^{23}} \times 3.384 \times 10^{23}$$

$$= 132.1 \times 10^{-3} \text{ kg} = 0.1321 \text{ kg.}$$

Example 7

Calculate the energy liberated when a single helium nuclei is formed by the fusion of two deuterium nuclei.

Solution

The desired equation is

$$^2_1H + {}^2_1H \rightarrow {}^4_2He + Q$$
$$\text{(D)} \quad \text{(D)} \quad (\alpha)$$

Mass of 2D atoms = $2 \times 2.01478 = 4.02956$ u

Mass of $^4_2$He atom = 4.00388 u

Mass defect = (4.02956 − 4.00388) u

$$= 0.02568 \text{ u}$$

Energy released, $Q = 0.02568 \times 931.8$ MeV

$$= 24 \text{ MeV (approx.).}$$

Example 8

Calculate the power output of a nuclear reactor which consumes 10 kg of ^{235}U per day. Given that the average energy released per ^{235}U fission is 200 MeV.

Solution

Number of atoms in 1 kg of ^{235}U

= 6.02×10^{23}, i.e. Avogadro number

$$= \frac{6.02 \times 10^{23}}{235}$$

or Number of atoms in 10 kg

$$= \frac{6.02 \times 10^{23}}{235} \times 10 = 2.56 \times 10^{22}$$

Since energy released per fission = 200 MeV, therefore fission energy produced by these atoms = $2.56 \times 10^{22} \times 200$ MeV

$$= 2.56 \times 10^{22} \times 200 \times 1.6 \times 10^{-13} \text{ J}$$

$$= 819.2 \times 10^9 \text{ J}$$

Time taken to consume 10 kg uranium 235

$$= 1 \text{ day} = 24 \times 3600 \text{ s}$$

∴ Power produced

$$= \frac{819.2 \times 10^9}{24 \times 3600} = 9.48 \times 10^6 \text{ W.}$$

Example 9

The magnetic flux density in a cyclotron is 2T. What should be the frequency in order to accelerate protons or charge 1.6×10^{-19} C and mass 1.79×10^{-27} kg?

Solution

$$n = \frac{Bq}{2\pi m} = \frac{2 \times 1.6 \times 10^{-19}}{2 \times 3.14 \times 1.79 \times 10^{-27}}$$

$$= 2.845 \times 10^7 \text{ Hz} = 28.45 \text{ MHz.}$$

Example 10

Prior to emerging from a cyclotron, deuterons describe a circle of radius 0.32 m. The frequency of the applied voltage is 10 MHz. Calculate the speed of the deuterons emerging from the cyclotron. You may neglect relativistic effect.

Solution

$$f = \frac{Bq}{2\pi m} \qquad (1)$$

$$R = \frac{mv}{BQ} \qquad (2)$$

Multiplying Eqs. (1) and (2), we have

$$Rf = \frac{v}{2\pi}$$

∴ $$v = 2\pi Rf = 2\pi \times 0.32 \times 10 \times 10^6$$

$$= 2.01 \times 10^7 \text{ m/s.}$$

Example 11

A betatron has a magnetic current supply of frequency 60 Hz and the peak magnetic flux density at the orbit is 0.5 Wb/m^2. If the radius of the electron orbit is 0.75 m, determine: (i) the final energy of the electrons (ii) the total time of flight of electrons.

Solution

We have the momentum of electron at the orbit

$$p = Ber$$

We have to treat the electron relativistically. The final energy of the electron is

$$E = mc^2 = pc = Berc$$
$$= 0.5 \times 1.6 \times 10^{-19} \times 0.75 \times 3 \times 10^8 \, J$$

$$= \frac{0.5 \times 1.6 \times 10^{-19} \times 0.75 \times 3 \times 10^8}{1.6 \times 10^{-19}} \, eV$$

$$= 1.25 \times 10^8 \, eV = 112.5 \, MeV$$

The total number of revolutions n by the electron during acceleration time (one quarter cycle of magnetic flux) is

$$n = \frac{c}{4\omega r_0} = \frac{c}{8\pi v r_0}$$

$$= \frac{3 \times 10^8}{8 \times 3.14 \times 60 \times 75} = 2.65 \times 10^5$$

Total time of flight of electron

$$= \frac{\text{Total distance covered}}{\text{Velocity}}$$

$$= \frac{2\pi r_0 n}{c}$$

$$= \frac{2.65 \times 10^5 \times 3.14 \times 0.75}{3 \times 10^8}$$

$$= 4.16 \times 10^{-3} \, s.$$

Example 12

The maximum magnetic field at the orbit in a certain betatron was 0.4 T and operating at 50 c/s with a stable orbit of diameter 1.5 m. Calculate the average energy gained per revolution and the final energy of the electrons. You may assume that the electron moves nearly with the speed of light.

Solution

The distance traversed by electrons during acceleration time in one quarter cycle is

$$d = \frac{1}{4}cT = \frac{c}{4f} \text{ where } f \text{ is the frequency of AC.}$$

Let r be the radius of the stable orbit, then the number of revolutions made by electrons in one quarter cycle is given by

$$n = \frac{d}{2\pi r} = \frac{c}{2\pi r f}$$

$$= \frac{3 \times 10^8}{8 \times 3.14 \times 0.75 \times 50}$$

$$= 3.18 \times 10^5$$

Now, if E be the final energy acquired, the momentum p is give by,

$$p = \frac{E}{c} \text{ (relativistic)}$$

If B be the magnetic field at the orbit and v the final velocity acquired, we have

$$Bev = \frac{mv^2}{r}$$

or

$$p = mv = Ber = \frac{E}{c}$$

$\therefore$ Final energy $E = Berc$

$$= 0.4 \times 1.6 \times 10^{-19} \times 0.75 \times 37 \times 10^8$$

$$= 1.44 \times 10^{-11} \, J$$

$$= \frac{1.44 \times 10^{-11}}{1.6 \times 10^{-13}} \, MeV = 90 \, MeV$$

$\therefore$ Average energy/revolution

$$= \frac{90 \times 10^6}{3.18 \times 10^5} \, eV = 283 \, eV.$$

Example 13

Find the number of ion pairs by 10 MeV proton. If in the proportionality region, the amplification is 10^3, current pulse time is 10µs and resistance between electrodes is 10^4, find voltage pulse height. The amount of energy required to produce one in pair is 34 eV.

Solution

Number of ion pairs by 10 MeV protons

$$n = \frac{10 \times 10^6}{34} = 2.94 \times 10^5$$

Total ion pairs in proportionality region
$$= 2.94 \times 10^5 \times 10^3 = 2.94 \times 10^8$$

Hence, charge on electrode due to these pairs,

$$q = 2.94 \times 10^8 \times 1.6 \times 10^{-19} \text{ C}$$

or $\qquad q = 4.704 \times 10^{-11} \text{ C}$

$$\therefore \quad \text{Current} = \frac{q}{t} = \frac{4.704 \times 10^{-11}}{10 \times 10^{-6}}$$

$$= 4.704 \times 10^6 \text{ A}$$

$\therefore \quad$ Voltage pulse height $= i \times R$

$$= 4.704 \times 10^6 \times 10^4 = 4.704 \times 10^{-2}$$

$$= 0.47 \text{ V}.$$

Example 14

A halogen quenched GM counter works at 1 kV. Anode wire has radius 0.2 mm and radius of cathode is 20 mm. The guarantee period for the counter is 10^9 constant. Find maximum radial field. If the counter is operated 30 hrs a week and count 3000 counts/min, find the life of the counter (take 1 year = 50 weeks).

Solution

The field at an axial distance r is given by

$$E_r = \frac{V_0}{e \log_e \left(\dfrac{b}{a} \right)}$$

This would be maximum near the surface of wire

$$\therefore \quad (E_r)_{max} = \frac{1 \times 10^3}{0.2 \times \log_e 100}$$

Here $\qquad b = 20 \text{ mm}, a = r = 0.2 \text{ mm}$

or $\qquad (E_r)_{max} = \dfrac{1 \times 10^3}{0.2 \times 2.3026 \times 2}$

$$= 1.087 \times 10^3 \text{ V/mm}$$

If its life is x, we have
$$10^9 = x \times 50 \times 30 \times 60 \times 3000$$

$$x = \frac{10^3}{5 \times 6 \times 3 \times 3} = \frac{1000}{270} \text{ years}$$

$$= 3.7 \text{ years}.$$

GLIMPSES

1. The primary constituents of nuclei are the photon and the neutron and the nuclear mass is roughly the sum of its constituent proton and neutron masses (the slight different being the binding energy of the nucleus).

2. Each nuclear species with a given Z and A is called a **nuclide**.

3. A nuclear species can be represented by $^A_Z X$, where A is mass number, i.e. the total number of protons. The total number of neutrons in a nucleus is the neutron number N, where $A = N + Z$.

4. Nuclides with the same Z but different A and N values are called **isotopes**, e.g. $^2_1 H$, $^3_1 H$ are isotopes of $^1_1 H$.

5. Nuclides with the same neutron number are called **isotones**, e.g. $^{14}_6 C$, $^{15}_7 N$, $^{16}_8 O$ and $^{17}_9 F$.

6. Nuclides with the same value of A are called **isobars**, e.g. $^{16}_6 C$, $^{17}_7 N$ and $^{16}_8 O$.

7. Masses are measured in terms of atomic mass unit u.

8. Assuming that a nucleus is spherical, its radius is
   $r = r_0 A^{1/3}$, where $r_0 = 1.2$ fm

9. Electron scattering is useful to measure the size and shape of nuclei.

10. The properties of nucleons are:

Property	Neutron	Proton
Mass (u)	1.008665	1.007276
Charge (e)	0	+1
Spin ($\hbar$)	½	½
Magnetic moment ($e\hbar/2m_p$)	– 1.91	+ 2.79

11. Nuclei are stable because of the **nuclear force** between nucleons. This short range (~3 fm) force dominates the coulomb repulsive force at distances of less than about 2 fm and is independent of charge and the study of the deuteron and nucleon–nucleon scattering indicates that the nuclear force is attractive.

12. **Stable isotopes** require a delicate balance between protons and neutrons, with near equal numbers for lighter stable nuclei and more neutrons for heavier nuclei.

13. **Unstable** isotopes are radioactive and decay by shedding particles.

14. The curve of **binding energy** shows that energy can be released by either **fusion** of ligher nuclei or **fission** of heavier nuclei. The total binding energy for a nuclide is

$$B\left(^A_Z X\right)=\left[Nm_n + Zm\left(^1\text{H}\right) - M\left(^A_Z X\right)\right]c^2$$

15. The von Weizsocker semi-empirical mass formula is useful in predicting the nuclear binding energy. There are no stable nuclei with $Z > 83$ or $A > 209$. Nuclei tend to be more stable with an even number of protons and/or neutrons. Nuclei near ^{56}Fe have the highest binding energy per nucleon and the average binding energy per nucleon for most nuclei is about 8 MeV.

16. Radioactive processes include α-decay, β-decay and γ-decay. **Alpha decay** emits a ^{4}He nucleus:

$$^A_Z X \rightarrow ^{A-4}_{Z-2}Y + ^4_2\text{He}$$

Beta decay emits an electron or a positron and antineutrino or a neutrino

$$^A_Z X \rightarrow ^A_{Z+1}Y + ^{\ 0}_{-1}e\left(\beta^-\right) + \bar{v}$$

17. In the **electron capture** process, the nucleus of an atom absorbs one of its own electrons (usually from the K shell) and emits a neutrino.

18. **Gamma ray** emits a high energy photon (gamma ray) when an excited nucleus drops to a lower energy state.

$$X^* \rightarrow X + \gamma$$

19. If a radioactive material contains N_0 radioactive nuclei at $t = 0$, the number N of nuclei containing at time t is

$$N = N_0\, e^{-\lambda t}$$

where λ is **decay constant**. The **decay rate** or **activity**, of a radioactive substrate is given by

$$R = \left|\frac{dN}{dt}\right| = N_0\,\lambda e^{-\lambda t} = R_0 e^{-\lambda t}$$

where $R_0 = N_0\lambda$ is the activity at $t = 0$.

20. The half-life $T_{1/2}$ is defined as the time interval required for half of a given number of radioactive nuclei to decay, where

$$T_{1/2} = \frac{\ln 2}{\lambda} = \frac{0.693}{\lambda}$$

21. Radioactivity is measured in **becquerels**, with 1 Bq equals to one decay per second. Sieverts (S_v) measure the biological effects of radiation.

22. There are only four radioactive series. For example, two of them begin with uranium isotopes, ^{235}U and ^{238}U.

23. Radioisotopes are useful to date objects like the age of the earth and ancient objects. Radiocarbon ^{14}C is one of the most useful.

24. Nuclear reactions can occur when the target nucleus X is bombarded by a particle 'a', resulting in nucleus 'y' and an outgoing particle 'b':

$$a + x \rightarrow y + b \quad \text{or} \quad X(a, b)\, y$$

The rest energy transformed to kinetic energy in such a reaction, called the **reaction energy Q**, is

$$Q = [M_a + M_x - M_y - M_b]c^2$$

A reaction for which Q is positive is called **exothermic**. A reaction for which Q is negative is called **endothermic**. The minimum kinetic energy of the incoming particle necessary for such a reaction to occur is called the **threshold energy**.

25. Different nuclear reaction mechanisms include compound nucleus, coulomb excitation, and direct reactions, among others. Nuclei have excited states, which may appear as responses in a compound nucleus reaction. The lifetimes T of nuclear states are related to their width t by the uncertainty principle $T_t \approx \dfrac{\hbar}{2}$.

26. Heavy nuclei, e.g. $^{235}_{92}$U and $^{239}_{94}$Pu may fission into two nearly equal fission fragments when struck by low energy neutrons

$$^1_0 n + ^{235}_{92}\text{U} \rightarrow X + Y + b^1_0 n$$

The extra neutrons produced in fission can sustain a **chain reaction** provided there is a **critical mass** of fissile material. Exponentially growing chain reactions power fission weapons, while controlled fission occurs in **nuclear reactors** used for power generation.

27. One may build nuclear reactions in different ways for special purposes. A breeder reactor produces more fissionable fuel than it consumes.

28. Nuclear fusion powers the sun and starts but has proved elusive on earth except in thermonuclear weapons. Isotopes of hydrogen, ^2_1H and ^3_1H, appear to be most useful. Although fusion is the source of sun's energy, it has yet been controlled on Earth. Inertial confinement or magnetic confinement fusion may one day provide us with nearly limitless energy.

29. Nuclear science have plentiful applications and includes medicine, art, archaeology, crime detection, mining, oil, mineral studies and production and small power systems.

30. The search for new elements continues in the quest for the island of stability for super heavy elements.

REVIEW QUESTIONS

1. What is radioactivity? What types of radiations are emitted in radioactive disintegration?

2. What is radioactive decay? Explain the laws of radioactive disintegration.

3. What is half-life? Obtain an expression for mean lifetime of a radioactive element in terms of half-life.

4. What do you understand by activity of a radioactive substance? Show how it varies.

5. Explain the terms: decay constant, half-life and mean life.

6. Define the Curie and Becquerel.

7. Explain, how is the rate of disintegration affected when a solution of radioactive substance is heated.

8. Write notes on (i) activity of radioactive substances (ii) natural radioactivity (iii) radioactive dating (age of the earth).

9. What are radioactive isopotes? Give some of their uses.

10. Write notes on (i) radioactive tracer technique and its uses (ii) uses of radioactive isotopes in medicine, agriculture and industry (iii) C-14 dating.

11. What are transuranic elements?

12. What are nuclear reactions? Give some examples of nuclear reactions produces by $\alpha, p, \,^2_1H$, n and photon.

13. What do you understand by the Q value of a nuclear reaction? Calculate Q value for the following reaction:

$$^4_2He + \,^{14}_7N \rightarrow \,^{17}_8O + \,^1_1H + Q$$

Given: Mass of ^4_2He = 4.00388 u, $^{14}_7N$ = 14.00755 u, $^{17}_8O$ = 17.00453 u, ^1_1H = 1.00815 u

[*Ans.* – 1.16 MeV]

14. What are exothermic (exoergic) and endothermic nuclear reactions?

15. Explain the terms: mass defect, packing fraction, binding energy and binding energy per particle. What is the significance of binding energy?

16. Explain the meaning of 1 u. What is the energy equivalent of 1 u in MeV. [*Ans.* 931 MeV]

17. Draw a curve showing the variation of binding energy per nucleon against the mass number. Use this curve to explain the release of energy in fusion of light nuclei and fission of heavy nucleus.

18. The principle of releasing nuclear energy is the same in a nuclear reaction and in an atom bomb. Why then the control rods are used only in a nuclear reactor?

19. What is nuclear fission? Explain it on the basis of liquid drop model of the nucleus. Why is it possible to produce the fission of ^{235}U with slow neutrons whereas it is necessary to use fast neutrons to produce the fission of ^{238}U?

20. What is a chain reaction? Explain uncontrolled and controlled chain reactions. Mention the condition for a self sustained chain reaction and how they are achieved.

21. Explain the function of a moderator to sustain a chain reaction.

22. Draw a labelled diagram of a nuclear reactor and explain its working and mention its important uses.

23. What is a nuclear fusion process? Give few examples.

24. What is the difference between nuclear fission and fusion?

25. Explain the origin of solar energy.

26. Explain the principle of breeder reactor.

27. Write a short note on GM counter.

28. Explain construction and working of Geiger–Muller counter.

29. Write a short note on scintillation counter.

30. Write a short note on photomultiplier tube.

31. (a) What do you mean by dead time in GM counter? Mention some of its applications. Draw a neat diagram of the GM counter.

(b) Describe the construction and working of a scintillation counter.

32. Through a schematic diagram, explain operation of a scintillation counter. How shall you select phosphor for protons, β-particles, slow neutrons and fast neutrons. Obtain an expression for output voltage pulse when radiation of energy E falls on the counting system.

33. Describe construction and working of Geiger–Muller counter. Explain the terms: dead time and quenching.

34. Draw curves to show qualitative variation of ionization current (assuming incident particle energy E in one case and $2E$ in another) versus voltage applied to an ionization chamber. Indicate operation regions of proportional and Geiger–Muller counters and give physical explanation to curves. What factors govern the counting rate of a GM counter?

35. Explain the principle of cyclotron. Give its construction, working and theory. Derive the relation for the energy of the particles emerging from a cyclotron.

36. Show that the kinetic energy of the particle in a cyclotron is independent of the voltage applied.

37. Write notes on:

    i. Nuclear reactor   ii. Betatron

    iii. Cyclotron   iv. Nuclear stability

    v. Nuclear fission   vi. Nuclear fusion

    vii. Fast breeder reactor

    viii. Nuclear binding energy curve

    ix. Geiger–Muller counter

    x. Nuclear models

PROBLEMS

1. ^{40}K has a half-life of 4×10^8 years. Find its decay constant. [*Ans.* 1.75×10^{-9}/years]

2. The half-life of radium is 1600 years. How long will it take for 15/16 of a sample of radiation to decay? [*Ans.* 1600 years]

3. The half-life of radon is 3.8 days. Find when its 1/2th would remain. [*Ans.* 14.45 days]

4. Find decay constant of the radioisotope whose half-life is 5 hours. [*Ans.* 3.85×10^{-5}/s]

5. The half-life of tritium is 12.5 years. Find the fraction of the amount of a pure sample after 25 years. [*Ans.* 1/4th]

6. Calculate the B.E./nucleon for (a) ^{36}S (b) ^{42}Ca.

7. $^3_1$H and $^4_2$He have masses 3.016050 u and 3.016030 u respectively. Find their binding

energies in MeV. What is the cause of the difference between their values?
[*Ans.* 8.48 MeV, 8.65 MeV]

8. The mass of $^{35}_{17}$Cl is 34.9800 u. Calculate its binding energy and B.E. per nucleon. Given that mass of ^1_0n = 1.008665 u and $^1_1$H = 1.007825 u.
[*Ans.* 288 MeV, 8.22 MeV]

9. The binding energy of $^{35}_{17}$Cl is 298 MeV, calculate its mass in u. [*Ans.* 34.97 u]

10. Calculate the Q value of the following nuclear reaction:

$$^7_3\text{Li} + ^1_1\text{H} \rightarrow ^4_2\text{He} + ^4_2\text{He} + Q$$

Mass of $^7_3$Li = 7.016005 u, $^1_1$H = 1.007825 u and $^4_2$He = 4.00260 u. [*Ans.* 17.34 MeV]

11. Calculate the Q value of the reaction:

$$^9_4\text{Be} + ^4_2\text{He} \rightarrow ^{12}_6\text{C} + ^1_0n + Q$$

Mass of $^9_4$Be = 9.015060 u and $^4_2$He = 4.006674 u, ^1_0n = 1.008986 u and $^{12}_6$C = 12.003316 u.
[*Ans.* 10.83 MeV]

12. The fission of $^{235}_{92}$U releases approximately 200 MeV. What percentage of original mass $^{235}_{92}$U + ^1_0n disappears? [*Ans.* 0.1%]

13. A reactor is developing nuclear energy at the rate of 30,000 kW. How many kg of ^{235}U would be used in 1000 hours of operating that on an average energy of 200 MeV is released per fission. [*Ans.* 0.132 kg]

14. What is the energy released when 1 kg of nuclear fuel is consumed if the following fusion reaction is possible.

$$^2_1\text{H} + ^2_1\text{H} \rightarrow ^4_2\text{He} \qquad [\textit{Ans.}\ 5.74 \times 10^{14}\ \text{J}]$$

SHORT ANSWER QUESTIONS

1. What hold the nucleons together inside a nucleus?

Ans. Nuclear force.

2. What is the relation between the radius r of a nucleus and its nucleons A.

Ans. $r = r_0 A^{1/3}$.

3. What is artificial transmutation of elements?

Ans. An artificial disintegration of nuclei is a process in which a nucleus is transformed into a different species by its reaction with an energetic particle or photon.

4. What are transuranic elements?

Ans. Artificially produced elements beyond $Z = 92$ produced in the laboratory by the bombardment of certain heavier nuclei with appropriate particles are called transuranic elements.

5. Do the nuclear forces obey inverse square law?

Ans. No.

6. Write five examples of isotopes.

Ans. Isotopes of hydrogen: ^1_1H, ^2_1H and ^3_1H

Isotopes of oxygen: $^{16}_8O$, $^{17}_8O$ and $^{18}_8O$

Isotopes of chlorine: $^{35}_{17}Cl$ and $^{37}_{17}Cl$

Isotopes of potassium: $^{39}_{19}K$ and $^{40}_{19}K$

Isotopes of uranium: $^{235}_{92}U$ and $^{238}_{92}U$

7. Write four examples of isobars.

Ans. $^{13}_6C$, $^{13}_7N$; $^{40}_{19}K$, $^{40}_{20}Ca$; $^{40}_{18}A$, $^{40}_{20}Ca$; $^{24}_{12}M$, $^{24}_{11}Na$

8. Give one example of isotones.

Ans. $^{31}_{14}Si$ and $^{32}_{15}P$

9. What do you understand by background radiations?

Ans. Background radiations are those radioactive radiations that everyone is exposed to due to the presence of natural radioactive substances on the earth as well as the cosmic radiations, and the radioactive substances present in the body.

Internal background radiations are from the radioactive substances K-40 and C-14 present in the entire body. External background radiations are from cosmic rays and the radiations from the radioactive rocks on the earth.

10. What is a nuclear reactor?

Ans. It is a device in which a self sustaining controlled chain reaction is produced in a fissionable material. Obviously, a nuclear reactor or a nuclear pile is a source of controlled release of nuclear energy which is utilized for several useful purposes.

11. What are the essential units of a nuclear power plant?

Ans. (i) Nuclear reactor (ii) Heat exchanger (iii) Steam turbine (iv) Electric generator or dynamo.

12. What is nuclear fusion?

Ans. A nuclear reaction in which lighter atoms combine together to form heavier nuclei resulting in the release of tremendous amount of energy.

13. The binding energy of ^7_3Li and ^4_2He are 39.2 MeV and 28.24 MeV respectively. Which one of the two nuclei is more stable?

Ans. B.E. per particle of ^7_3Li = $\dfrac{39.2}{7}$ = 5.6 MeV

B.E. per particle of ^4_2He = $\dfrac{28.24}{4}$ = 7.06 MeV

Helium is more stable because its B.E. per particle is more.

14. What is hydrogen bomb?

Ans. Like atom bomb, hydrogen bomb is a weapon of mass destruction, employing thermonuclear reactions, which can release enormous energy within few minutes. The extremely high temperature required for the uncontrolled fusion reactions in a hydrogen bomb is produced by a fusion bomb containing ^{235}U or ^{239}Pu. Obviously, hydrogen bomb is a fission–fission bomb.

15. What is the order of the magnitude of temperature enabling a fusion reaction to take place?

Ans. About 10^6 K.

16. What are the fundamental parts of a nuclear reactor used for generation of electricity?

Ans. (i) Nuclear fuel (ii) Moderator (iii) Coolant (iv) Shield (v) Controlling rods.

17. What is the function of a moderator in a nuclear reactor?

Ans. Moderator is used to slow down the neutrons. Heavy water, graphite or beryllium oxide is used as moderator.

18. What is the function of controlling rods in a nuclear reactor?

Ans. Cadmium rods are used as controlling rods to control the rate of fission in the nuclear reactor. These rods absorb extra neutrons produced in the fission reaction.

19. What is a chain reaction?

Ans. A chain reaction is a series of nuclear fissions whereby the neutrons produced in each fission cause additional fissions releasing enormous amount of energy.

20. What do you understand by a nuclear reaction?

Ans. Any reaction that tends to change the configuration of the nucleus of an atom can be termed a nuclear reaction.

21. How many types of nuclear reactions are there?

Ans. (i) Nuclear fission reaction, (ii) Nuclear fusion reaction.

22. What are fission reactions?

Ans. Nuclear fission is a process in which the heavy nucleus of a radioactive substance like uranium is split into lighter nuclei by the bombardment of a low energy (slow moving) neutron, the reaction being accompanied by the release of energy and a two or three or more neutrons.

23. How U-235 isotope is used to determine the age of earth or rocks?

Ans. By determining the present ratio of U-235 to stable lead isotope.

24. Give few examples of uses of radioisotopes.

Ans.   i. $^{60}_{27}$Co is used to kill cancer cells in the blood of a human body.

   ii. $^{24}_{11}$Na is used to detect a clot in the blood of a human body.

   iii. $^{131}_{53}$I is used to test the functioning of the thyroid gland.

   iv. $^{60}_{27}$Co like radioisotopes is used to detect the internal flow in cast materials.

   v. Radioisotopes are used in agriculture in the selection of fertilizers.

   vi. Radioisotopes are used by irradiating seeds to prevent unnecessary mutations of plants.

   vii. Perishable food items can be kept fresh for longer durations by exposing them to mild γ-radiations from a radioisotope.

   viii. With the help of suitable detectors radioactive gases are injected into the atmosphere and study is made of the course of monsoon winds and the change they undergo.

25. What is carbondating?

Ans. The process of estimating the age by measuring the disintegration of C-14 is known as carbondating.

26. Why are γ-rays emitted in nuclear processes and not in orbital electron transitions?

Ans. The energy of γ-rays is of the order of MeV. Energies of this magnitude occur in nuclear processes and not in electron orbital transitions.

27. Why is the energy distribution of β-decay continuous?

Ans. The phenomenon of β-decay arises due to the conversion of a neutron in the nucleus into a proton, electron and antineutrino. Because the energy available in beta is shared by the electron and antineutrino in all possible ratios as they come out of the nucleus. This is why the β-ray spectrum is continuous.

$$^{1}_{0}n \rightarrow ^{1}_{1}H + ^{0}_{-1}e + \bar{\nu}$$

OBJECTIVE QUESTIONS

1. Nuclear forces are non _____ and non _____.   [electric, gravitational]

2. $1\,u =$ _____ MeV.   [931]

3. $^{35}_{17}Cl +$ _____ $= ^{32}_{16}S + ^{4}_{2}He$. $\left[^{1}_{0}n\right]$

4. The total amount of energy released per fission of ^{235}U nucleus is roughly _____MeV. [200]

5. The chain reaction is a process in which the nuclear _____ of an atom induces _____ in another atom which again induces fission in another atom and so on.

            [fission, nuclear fission]

6. Nuclear fusion is a process in which two _____ nuclei are _____ together to form a heavier and stable nucleus.   [lighter, fused]

7. Source of stellar energy is _____.

            [nuclear fusion]

8. In proton–proton cycle, protons are continuously transformed into _____ nuclei.   [helium]

9. In carbon–nitrogen cycle, the synthesis of _____ nuclei with carbon (acting as a nuclear catalyst) takes place.   [hydrogen]

10. A nuclear reactor is a device in which _____ is produced under a self sustaining controlled _____.   [nuclear fission, chain reaction]

11. The penetrating power of α-particle is _____ than that of β-particle.   [less]

12. Nuclei of heavy atoms contain more _____ than _____.   [protons, neutrons].

13. The β-particle is emitted when a _____ inside the nucleus changes to _____. [neutron, proton].

14. Atoms of different elements whose nuclei have the same atomic weight but different atomic number are called _____.   [isobars]

15. The radioactive isotope used in the treatment of cancer is a good source of _____ radiations.

            [γ]

16. Radioactive isotope of carbon is _____. [^{14}C]

17. $^{3}_{1}H +$ _____ $\rightarrow ^{4}_{2}He + ^{1}_{0}n + Q$. $\left[^{2}_{1}H\right]$

18. _____ is responsible for keeping the nucleons together in a nucleus.   [nuclear force]

19. The common product formed in the artificial transmutation by a proton is _____. [helium]

20. In plants and animals the ratio of ^{14}C to ^{12}C is _____.   [constant, i.e. 10^{-12}]

21. When an α-particle is ejected, the atomic number of the atom decreases by _____. [2]

MULTIPLE CHOICE QUESTIONS

1. The nuclear radius may be approximated from a spherical charge distribution to be
(a) $R = r_0 A^{2/3}$ (b) $R = r_0 A^{1/3}$
(c) $R = r_0 A^{-2/3}$ (d) $R = r_0 A^{-1/3}$

2. Nuclear magneton is:
(a) $\mu_N = \dfrac{e\hbar}{2m_p}$ (b) $\mu_N = \dfrac{e\hbar}{2m_e}$

(c) $\mu_N = \dfrac{e\hbar}{2}$ (d) $\mu_N = \dfrac{e\hbar}{2(m_p + M_e)}$

3. A deuteron nucleus consists of
(a) one proton and one electron
(b) one neutron and one electron
(c) one proton and one neutron
(d) none of the above

4. In scattering neutrons from protons, a deuteron is sometimes formed in the nuclear reaction
(a) $n + p \rightarrow d$ (b) $n + p \rightarrow d + \gamma$
(c) $n + p \rightarrow d + {}_{-1}^{0}e$ (d) $n + p \rightarrow d + {}_{+1}^{0}e$

5. The radioactive decay law is
(a) $N(t) = N_0 e^{\lambda t}$ (b) $N(t) = N_0 e^{-\lambda t}$
(c) $N(t) = \lambda N_0 t$ (d) $N(t) = N_0 e^{-\lambda t_{1/2}}$

6. The average (or mean) lifetime τ is calculated to be

(a) $\tau = \dfrac{1}{\lambda}$ (b) $\tau = \lambda$

(c) $\tau = e^{-\lambda}$ (d) $\tau = e^{\lambda}$ [1]

7. Which one of the following is correct?
(a) $n \rightarrow p + \beta^- + \bar{v}$ (b) $n \rightarrow p + \beta^-$
(c) $n \rightarrow p + \beta^- + v$ (d) $n \rightarrow p + v + \bar{v}$

8. Which one is a electron capture reaction?
(a) $p \rightarrow n + {}_{+1}^{0}e$ (b) $p + e^- \rightarrow n + v$
(c) $p \rightarrow n + {}_{-1}^{0}e$ (d) $^{14}O \rightarrow {}^{14}N + \beta^+ + v$

9. Radioactive ^{14}C is produced in our atmosphere by the bombardment of ^{14}N by neutrons produced by cosmic rays. This can be expressed as
(a) $n + {}^{14}N \rightarrow {}^{14}C + p$
(b) $n + {}^{14}N \rightarrow {}^{14}C$
(c) $n + {}^{14}N \rightarrow {}^{14}C + {}_{+1}^{0}e$
(d) $n + {}^{14}N \rightarrow {}^{14}C + {}_{-1}^{0}e$

10. The total self energy of a sphere of radius R containing a charge Ze evenly spread throughout the sphere is given by
(a) $\Delta E_{coul} = \dfrac{3}{5}\dfrac{(Ze)^2}{4\pi\varepsilon_0 R}$ (b) $\Delta E_{coul} = \dfrac{1}{2}\dfrac{(Ze)^2}{4\pi\varepsilon_0 R}$

(c) $\Delta E_{coul} = \dfrac{Ze}{4\pi\varepsilon_0 R^2}$ (d) $\Delta E_{coul} = \dfrac{1}{4}\dfrac{(Ze)^2}{4\pi\varepsilon_0 R}$

11. An example of induced fission is
(a) ${}_1^2H + {}_1^2H \rightarrow {}_0^1n + {}_1^3He + 3.3\,MeV$
(b) $n + {}_{92}^{235}U \rightarrow {}_{92}^{236}U^* \rightarrow {}_{40}^{90}Zr + {}_{52}^{134}Te + 3n$
(c) $^2H + {}^2H \rightarrow n + 4He + 17.6\,MeV$
(d) $^3He + {}^3He \rightarrow {}^4He + {}^1H + {}^1H + 12.86\,MeV$

12. Which of the following radiations has the highest penetrating power?
(a) α (b) β
(c) γ (d) X-rays

13. Which one of the following is an example of isobars?
(a) ${}_{13}^{14}Si, {}_{15}^{32}P$ (b) ${}_7^{14}N, {}_7^{15}N$
(c) ${}_6^{13}C, {}_7^{13}N$ (d) none of these

14. Higher value of decay constant indicates
(a) slower decay   (b) faster decay
(c) half-life period   (d) none of above

15. Which of the following is/are subatomic particles?
(a) positron   (b) proton
(c) neutrino   (d) all of the above

16. To convert ${}_{90}^{232}Th$ into ${}_{90}^{228}Th$, the least number of α-and β-particles that have to be emitted are
(a) 2, 1   (b) 1, 1
(c) 1, 2   (d) 2, 2

17. $^A_Z X$ and $^A_{Y+1} Y$ are two radioactive elements. X emits one α-particle and Y emits one β-particle. The daughter nuclei formed are

(a) isotones

(b) isotopes

(c) isobars

(d) none of the above

18. $^1_1H + ^7_3Li \rightarrow$ _____ $+ ^4_2He$

(a) ^1_1H　　　　　　　　(b) ^4_2He

(c) ^9_4Be　　　　　　　　(d) ^8_4Be

19. When $^{238}_{92}U$ disintegrates to form $^{206}_{82}Pb$, number of α- and β-particles emitted are

(a) 8α, 6β　　　　　　　(b) 6α, 8β

(c) 8α, 8β　　　　　　　(d) 6α, 6β

Answers

1. (b)	2. (a)	3. (c)	4. (b)	5. (b)
6. (a)	7. (a)	8. (b)	9. (a)	10. (a)
11. (b)	12. (c)	13. (c)	14. (b)	15. (d)
16. (c)	17. (a)	18. (b)	19. (a)	

26 Nanomaterials

26.1 INTRODUCTION

Nanoscience and nanotechnology primarily study the synthesis, characterization, exploration and exploitation of nanomaterials. Nanomaterials are characterized by at least one dimension in the nanometer (1 nm = 10^{-9} m) range. Nanostructures constitute a bridge between molecules and infinite bulk systems. Individual nanostructures include clusters, nanocrystals, nanowires, quantum dots and nanotubes, whereas collection of nanostructures involves arrays, assemblies and super lattices of individual nanostructures. The physical and chemical properties of nanostructured materials can differ significantly from the atomic-molecular or bulk materials having same composition. The uniqueness of the structural characteristics, energetics, response, dynamics and chemistry of nanostructured materials constitutes the basis of nanoscience. Suitable control of the properties and response of these materials can lead to new devices and technologies, i.e. nanotechnology.

Nanotechnology can be considered as a group of emerging technologies in which structure of matter is controlled at the nanometer scale, to produce new, interesting nanostructured materials and nano devices that have useful and unique properties. Nanotechnology imposes only limited control of structure at the nanometre scale producing useful products.

The impact of nanotechnology is already being felt in the form of new computer memories that provide rapid access to stored data, that can hold more of this data than the minidrives used in iPODs and do not need any external power source to retain the data. It is being felt in the form of prototypes for photovoltaic cells that can literally be sprayed onto buildings or computers to provide cheap power sources. And it is being felt in the form of nanoengineered gels that spread the recovery of damaged nerve cells.

Nanoscience and nanotechnology have grown explosively in the past two decades due to increasing availability of methods of synthesis of these materials as well as tools of characterization and manipulation. Several innovative methods of synthesizing nanoparticles and nanotubes and their assemblies are now available. The size-dependent electrical, optical and magnetic properties of individual nanostructures, metals and other materials are being understood in a better manner. These size dependent properties of these materials can be utilized to tailor them as per the requirements of a specific application.

There is great vitality in the area of nanoscience and nanotechnology and immense opportunities. Truly speaking, nanoscience is a interdisciplinary area covering physics, chemistry, biology, materials science, and engineering and technology.

Nanoscience and nanotechnology has tremendous scope for applications over a wide range. This is likely to benefit various industrial sectors including chemical and electronic industries as well as manufacturing. Health care, medical practice, space research and environmental protection will definitely benefit from nanoscience.

Studies are being conducted on the potential use of nanomaterials in diverse applications including hydrogen storage, ion sensing and gas sensing, surface-modified nanoparticles for enhanced oil recovery adsorption of chemical and biological agent on to nanoparticles, active electrode materials for lithium-ion batteries, light emitting devices, dental composition, etc. Stronger and lighter metals could be used in space as well as implants in the human body. Drugs for immediate effect in a body can be produced and could be delivered to specific location in a very short time. Electronic devices of smaller size, e.g. transistors can be fabricated, helping computers operate faster. Nanotechnology also finds applications in producing quantum dot lasers or tiny lasers, high density storage devices, nanoelectromechanical systems (NEMS), etc. Nanoscience and nanotechnology has many applications in biological sciences too. The list of applications is increasing day by day with the progress of research in this field. One of the major and difficult problems facing the design of nanostructure based systems in understanding how to interconnect and address them. Obviously, the success of this field will depend on the development of new devices and manufacturing technologies.

One can also see nanotechnology at work in nature. There are some species living in the sea that produce shells surrounding them by making use of nanoscale calcium carbonate particles. These shells are quite strong and do not yield to pressure easily. Moreover, cracks produced on the surface of these shells do not propagate inside and thus the animals remain safe.

The field of nanoscience and nanotechnology has matured so rapidly that it is probably hard to find a segment of any technical subject where the implications of nanomaterials have not been explored at least to a preliminary extent. There is every reason to believe that the field will

make must progress in the coming decade. Figure 26.1 describes different fields of nanotechnology that are emerging.

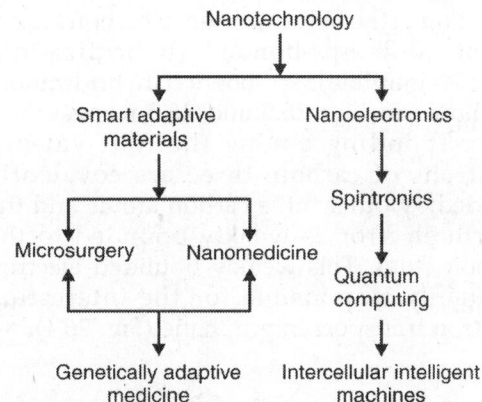

Fig. 26.1: Emerging fields of nanotechnology

26.2 NANOSCALE MIRACLES

The key element of nanoscience and nanotechnology is carbon, single sixth chemical element denoted by the symbol 'C' occupying a unique position in the periodic table of elements. Sometimes nanotechnology is also termed *carbon nanotechnology*. Carbon is a fundamental element in the living world, i.e. it is the basis of existence of life on the earth. We know that this element, as compound forms is present in the human body, food items, atmosphere, plants, fuels, etc. Carbon has four electrons in its outermost orbit, i.e. its valency is 4. Thus, carbon can bind itself or other atoms and form different compounds. Carbon atoms can form complicated network of molecules, which is the basis of organic chemistry. Carbon is the only element in the periodic table that has allotropes from zero dimension to three dimensions. *Diamond, graphite and amorphous carbon* are considered to be three main allotropic forms of carbon. Out of these three allotropic forms, two well known solid forms of carbon, i.e. diamond and graphite exhibit entirely different properties. Each of them is quite useful in its own way. Diamond is a hard material and known as a precious gem, largely used in jewelery. Diamond owes its sparkle partly to its high refractive index. Diamond is a poor conductor of electricity. On the other hand, graphite is remarkably soft and fair electrical conductor used in pencils. The difference in their properties is due to variation in the hybridization of carbon atoms. Diamond is sp^3-bonded (hybridization) whereas graphite is sp^2-bonded (hybridization) as shown in Figs 26.2 and 26.3 respectively. In sp^2-bonding among the four valence electrons of carbon, three are covalently bonded by other three carbon atoms and the fourth electron is weakly bounded to the carbon atom. This weakly bounded electron is mainly responsible for the interesting electron transport in graphene (Fig. 26.4).

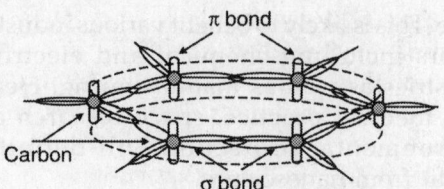

Fig. 26.4: Schematic of basic hexagonal bonding structure in graphene sheet where carbon atoms are denoted with circles, the σ-bonds are connecting carbon nuclei in plane and π-bonds are out of the plane

The electronic structure or the manner in which electrons are placed according to their energy are different in both diamond and graphite materials. This is why their physical properties are different. We may remember that the electronic structure of a solid depends on its atomic number and crystal structure or ordering of atoms in the lattice.

There are other three phases or allotropic forms of carbon that exist only in nano form. Graphene, refers to a monolayer of 2d lattice of sp^2 bonded carbon atoms with bond length approximately 0.14 nm arranged in a honeycomb lattice (with bonding similar to graphite). Graphene could be produced only recently in 2004 by micromechanical cleavage of graphite. We know that graphite is a layered structure with strong 2d bonds and very weak (van der Wall type) interlayer coupling. This property of graphite is used to cleave graphite crystal and 2d graphene is produced. Graphene exhibits ballistic transport and quantum hall effect observed only in two dimensions. We may not that these effects are not observed in 3D bulk solids. Obviously, graphene is truly a 2D crystal and is a gapless semiconductor. Graphene is found to be a very promising candidate to make ballistic field effect transistor. Moreover, the electronic structure of graphene is quite peculiar and one can make its use for the development of carbon transistor. We may remember that almost all semiconductor devices are based on silicon and germanium.

Today, there exist a whole family of other forms of carbon, e.g. the hollow and *cagelike*

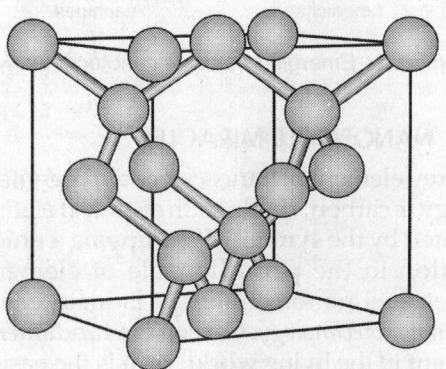

Fig. 26.2: Diamond structure

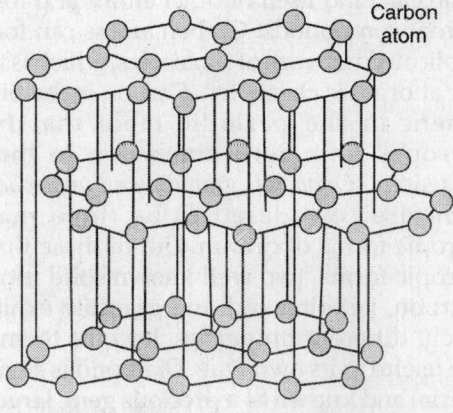

Fig. 26.3: The structure of graphite

Buckminsterfullerene molecule also called as buckyball or C_{60} fullerene. This is a zero dimensional carbon allotropic form having sp^2-like graphite bonding. The balls resemble geodesic dome, designed by Buckminster Fuller, and the buckballs or fullerenes are named in his honour. C_{60} is the smallest fullerene molecule in which two pentagons share an edge. Now, there are at least thirty or more forms of fullerenes and also an extended family of carbon nanotubes (CNTs). Amongst these, C_{60} is the first spherical carbon molecule with carbon atoms arranged in a soccer ball shape, i.e. the structure of C_{60} is truncated icosahedrons which resembles a soccer ball. Figure 26.5 is a schematic of C_{60} fullerene.

great diversity of synthetic protocols combined with the high number of chemical reactions that have been mainly applied to C_{60} has led to the formation of a wide variety of functionalized fullerenes. Minute quantities of fullerenes in the form of C_{60}, C_{70}, C_{76} and C_{84} molecules are produced in nature. They have been observed in carbon root and also found to be formed by lightning discharges in the atmosphere.

Figure 26.6 shows a slightly elongated and the spherical carbon molecule, has seventy carbon atoms and is known as C_{70}. The structure of C_{70} has extra six-membered carbon ring. Similarly, there are also a good number of other potential structures containing the same number of carbon atoms.

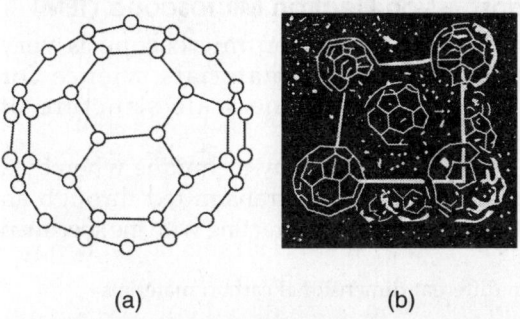

(a)   (b)

Fig. 26.5: (a) The structure of C_{60} molecule (b) Face centered cubic crystal of C_{60} fullerite

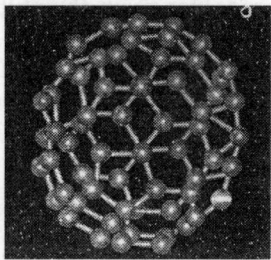

Fig. 26.6: Structure of roughy ball shaped C_{70} fullerene

The structure of C_{60} molecule have 20 hexagons and 12 pentagons with a carbon atom at the vertices of each polygon and a bond along each polygon edge. The structure contains 60 carbon atoms in which five-membered ring is isolated by six-membered ring. Although the bonding in C_{60} is basically sp^2, but due to the curved surface of the icosahedron, some sp^3 hybridization is also present. C_{60} has a unique property of being able to carry something inside it. When doped with alkali metal atom, C_{60} fullerene molecule exhibits superconductivity. We may note that C_{60} molecule shows wave–particle duality, and hence it is a truly quantum mechanical system.

The main advantage gained upon functionalization of fullerenes is a substantial increase in their solubility. The existence of a

However, their peculiar shape depend on whether five membered rings are isolated or not, or whether seven membered rings are present or not. So far no significant applications of fullerenes have been developed, yet scientists are very excited about the one new allotrope of carbon which is **carbon nanotube** (CNT). The current interest of researchers on CNTs is a direct consequence of the synthesis of **Buckminster fullerene,** C_{60} and other fullerenes. They are molecules composed entirely of carbon in the form of a tube (Fig. 26.7).

Fig. 26.7: Carbon nanotube structure

We have seen that carbon is the most amazing element in the periodic table. In the past 30 years, we have experienced the renaissance in three materials families: **graphite intercalation compounds** (GICs), **fullerene solids** and **nanotubes**. GICs provided a wealth of detailed physical and chemical information, while immediate applications were frustrated by cost, lack of air stability and, most importantly, the lack of new or greatly improved applications and properties. Though the foundations of physics and chemistry remain unshaken, the Li ion battery industry has certainly benefited greatly from this knowledge. Fullerene solids fared better, but no large volume application has yet emerged. However, the elucidation of electronic properties in this highly correlated system has made invaluable contributions to fundamental science. Nanotubes may offer the best prospects yet. Much of the current research and commercialization of nano-technology rely on tubes, balls and wires prepared from carbon atoms/or other materials. The carbon based tubes can be CNTs, bucky tubes and very long tubes, often referred to as **nanowires**. The combination of enhanced properties encourages the drive to multifunctional materials. Extension to other tubular and nano wire-like materials provide ample scope for new discoveries.

The various parameters of carbon atoms in different dimensional carbon materials are given in Table 26.1.

Nanomaterials have already shown a high potential that some of their applications in biology, medicine and drug delivery mechanisms have raised several ethical issues. We will discuss them in subsequent sections.

26.3 STUDY OF NANOSCALE MATERIALS
Transmission Electron Microscope (TEM)

Transmission electron microscope is very important tool in materials science for investigating the fine scale structure of materials.

TEM is a microscopy technique whereby a beam of electrons is transmitted through an ultrathin specimen, interacting with the specimen

Table 26.1: Various parameters of carbon atom in different dimensional carbon materials

Parameters	Fullerene	Carbon nanotubes	Graphite	Diamond
Dimension	Zero	One	Two	Three
Hybridization	sp^2	sp^2	sp^3	sp^3
Crystal structure	*fcc*	*fcc*	Hexagonal	Cubic
Lattice constant (Å)	14.15	—	$a = 2.456$ Å	3.513 Å
			$c = 6.696$ Å	—
Colour	Black solid	Black	Grey	Colourless
Density (g/cm^3)	1.72	1.2 – 2.0	2.26	3.51
Specific gravity	1.7 – 1.9	2	2.2	3.52
Bond length (Å)	1.4	1.44	1.42	1.54
Bond energy (eV/mole)	> 25	> 25	25	15
Melting point (°C)	> 800 (sublimes)	3652	3652	3550
Thermal conductivity (Wm^{-1}K^{-1})	0.4	—	80 – 230	900 – 2320
Young's modulus (TPa)	—	1 – 1.2	—	—
Tensile strength (GPa)	—	> 60	—	—
Band gap (eV)	—	0.5 eV ($d = 1.5$ mm)	—	5.47
		15 meV ($d \approx 1.5$ nm)	—	
Electrical conductivity (A/cm^2)	—	10^9	—	3.513

magnified and focussed by an objective lens and appears on an imaging screen, a fluorescent screen in most TEMs plus a monitor, or on a layer of photographic film, or to be detected by a sensor such as a CCD camera. The first practical TEM was build by Arebus and Hillier at the University of Toronto in 1938.

Figure 26.8 shows schematically a TEM illustrating the basic principle of its operation. An electron beam is produced by a heated tungsten filament at the top of an evacuated column and is accelerated down the column by a high voltage (usually from 75 to 120 kV). Electromagnetic coils are used to condense the electron beam and then it is passed through a very thin (i.e. about 100 nm thick or less) section of a thin metal specimen placed on the specimen stage (Fig. 26.8). The specimen area examined must be very thin so that some of the electrons entering are able to pass through it. As the electrons pass through the specimen, some are absorbed and some of them are scattered so that they change direction. Differences in crystal atomic arrangement will cause electron scattering. After the electron beam has passed through the specimen, it is focussed with the objective coil and then enlarged and projected on a fluorescent screen (Fig. 26.8). A region in a metal specimen that tends to scatter electrons to a high degree will appear dark on the viewing screen. Thus, dislocations that have an irregular linear atomic arrangement will appear as dark lines on the electron microscope screen.

A TEM works much like a slide projector. A projector shines a beam of light through (transmits) the slide, as the light passes through, it is affected by the structures and objects on the slide. These effects result in only certain parts of the light beam being transmitted through certain parts of the slide. This transmitted beam is then projected into the viewing screen forming an enlarged image of the slide.

TEMs work the same way except that they shine a beam of electrons (like the light) through the specimen (like the slide). Whatever part is transmitted is projected into a phosphor screen for the user to see. The working of a typical TEM is as follows:

The source at the top represents the electron gun producing a stream of monochromatic electrons.

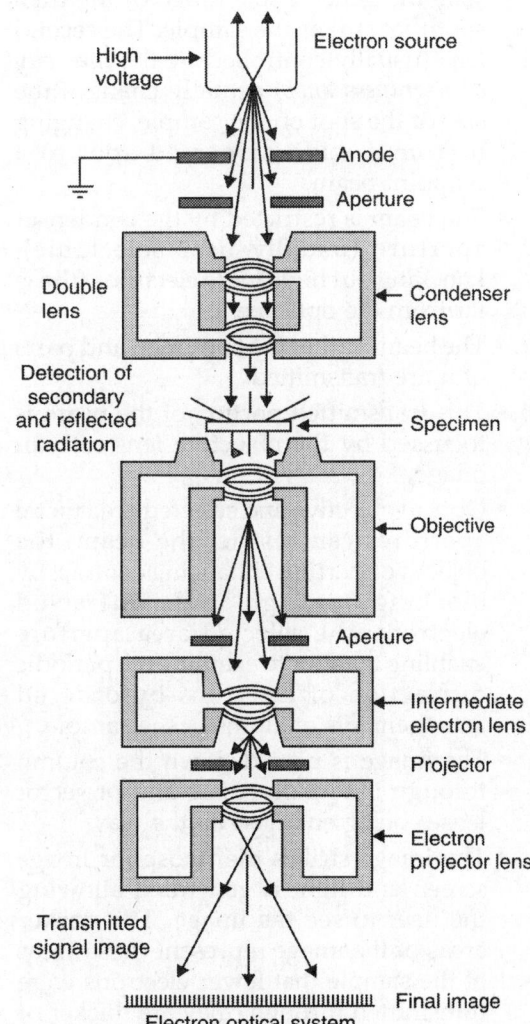

Fig. 26.8: Schematic arrangement of electron lens system in a transmission electron microscope (TEM). All the lenses are enclosed in a column that is evacuated during operation. The path of the electron beam from the electron source to the final projected transmitted image is indicated by arrows. A specimen thin enough to allow an electron beam to be transmitted through it is placed between the condenser and objective lenses as indicated

- This stream of monochromatic electrons is focussed to a small, thin, coherent beam by the use of condenser lenses. The first lens largely determines the *spot size*, the general size range of the final spot that strikes the sample. The second lens (usually controlled by the *intensity of brightness knob*) actually changes the size of the spot on the sample, changing it from a wide dispersed spot to a pinpoint beam.
- The beam is restricted by the condenser aperture (usually user selectable), knocking out high angle electrons (those far from the optic axis).
- The beam strikes the specimen and parts of it are transmitted.
- This transmitted portion of the beam is focussed by the objective lens into an image.
- Optical objective and selected area metal apertures can restrict the beam, the objective aperture enhancing contrast by blocking out high-angle diffracted electrons, the selected area aperture enabling the user to examine the periodic diffraction of electrons by ordered arrangements of atoms in the sample.
- The image is passed down the column through the intermediate and projector lenses being enlarged all the way.
- The image strikes the phosphor image screen and light is generated allowing the user to see the image. The darker areas of the image represent those areas of the sample that fewer electrons were transmitted through (they are thicker or denser). The lighter areas of the image represent those area of the sample that more electrons were transmitted through (they are thinner or less dense).

Applications of TEM

The TEM is used heavily in both materials science/metallurgy and the biological sciences. In both cases, the specimen must be very thin and able to withstand the high vacuum present inside the instrument.

For biological specimen, the maximum specimen thickness is roughly 1 micrometre. To withstand the instrument vacuum, biological specimen are typically held at liquid nitrogen temperatures after embedding in vitreous ice or fixated using a negative staining material such as uranyl acetate or by plastic embedding. Typical biological applications include tomographic reconstructions of small cells or thin sections of larger cells and 3D reconstructions of individual molecules via single particle reconstruction.

Limitations of TEM

There are a number of drawbacks to the TEM technique. Many materials require extensive sample preparation to produce a sample thin enough to be electron transparent which makes TEM analysis a relatively time consuming process with a low output of samples. The structure of the sample may also be changed during the preparation process. Also the field of view is relatively small, raising the possibility that the region analysed may not be characteristic of the whole sample. There is a possibility that the sample may be damaged by the electron beam, particularly in the case of biological materials.

Scanning Electron Microscope (SEM)

Figure 26.9 shows the schematic arrangement of SEM. Unlike the TEM, where electrons of the high voltage beam form the image of the specimen, the SEM produces images by detecting low-energy (<50 eV) secondary electrons that are emitted from the surface of the specimen due to excitation by the primary electron beam. In SEM, the electron beam is allowed to move across the substance with detectors generating an image by mapping the detected signals with beam position.

Usually, the TEM resolution is about an order of magnitude higher than the SEM resolution; however, because the SEM image relies on surface processes rather than transmission, it is possible to image bulk samples and also has a much greater depth of view. Thus, SEM can produce images that are better representation of the 3D structure of the sample.

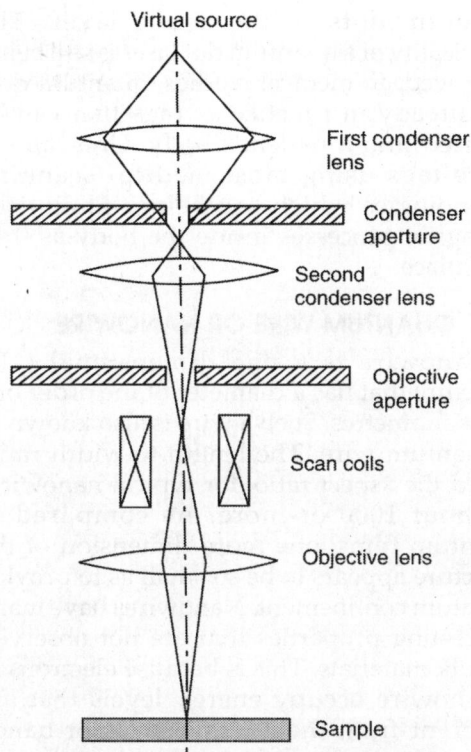

Virtual source

First condenser lens

Condenser aperture

Second condenser lens

Objective aperture

Scan coils

Objective lens

Sample

Fig. 26.9: Schematic diagram of SEM

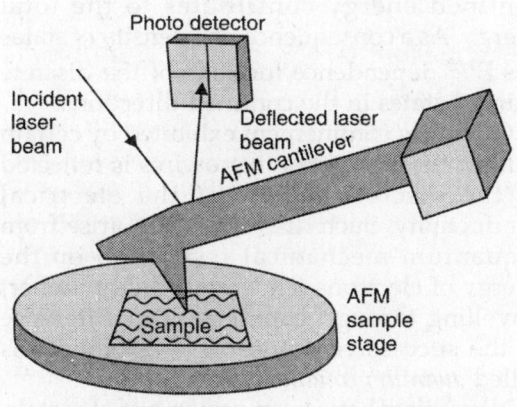

Photo detector

Incident laser beam

Deflected laser beam

AFM cantilever

Sample

AFM sample stage

Fig. 26.10: Schematic diagram of AFM

Scanning probe microscopes are of two types: (i) **atomic force microscope** (AFM) (ii) **scanning tunneling microscope** (STM). In the AFM technique, a tiny tip moves up and down in response to the electromagnetic forces between the atoms of the surface and those on the tip.

The motion of this tip is recorded and used to generate an image of the atomic surface using a computer. Figure 26.10 shows a schematic arrangement of AFM.

26.4 QUANTUM DOTS

These are zero dimensional (0D) structures, 3–60 nm in size in which the carriers are confined in all the three directions. The size of quantum dots depends upon the application for which it is synthesized. Common shapes generally include pyramids, cylinders and spheres. Different synthesis route produce different kinds of quantum dots. The energy states of quantum dots are quantized in all directions and the density of states is represented by series of discrete, sharp peaks resembling that of an atom. On the basis of this com-parison, one can label quantum dots as **artificial atoms.** Usually quantum dots are formed by a definite number of atoms. Quantum dots are typically represented by atomic clusters or **nanocrystallites.** On the basis of coupling of quantum dots with external electric circuits, one can put them under two classes: (i) **open dots** (ii) almost isolated or **closed quantum dots**. In an open quantum dot, the coupling is strong and the movement of electrons across dot-lead junctions is allowed classically. When point contacts connecting a dot with an external circuit are pinched off, effective barriers are formed and conductance takes place by tunneling. These are almost isolated or closed quantum dots.

An ensemble of quantum dots spatially arranged with a period (the distance between the nearest dots) comparable or shorter than the electron de Broglie wavelength is called a **quantum dot solid**. The basic properties of such dense quantum dot solid are expected to reproduce features inherent in solids, i.e. formation of energy bands in perfect lattice and existence of localized and delocalized electron states in disordered quantum dot structures. The quantum dot solid can be considered as a special form of **colloidal crystal**. Electronic and optical properties of quantum dot solids can be determined by electron confinement in each dot and collective effects arising from periodic special organization of the dots.

Quantum dots can be fabricated with either a top-down technique or bottom-up technique. The top-down technique produces quantum dots with limited uniformity in diameters. Uniformity in diameters is crucial for the production of large arrays of dots that have consistent properties. Unfortunately, top-down approaches such as lithography are limited by the diffraction limit and cannot create dense arrays of quantum dots.

The most common way to produce a quantum dot is through the bottom-up approach. Bottom-up approach is a proper way to produce quantum dots in dense arrays that self-assemble in an orderly manner. The sites required for the nucleation of quantum dots seem to be provided by the high strain developed on the substrates. Such strain is developed at the interface between the substrate and the deposited layer. It is easier to produce quantum dots having consistent properties using this approach.

A quantum dot is at least 10 times larger in size than an atom. A quantum dot emits only specific wavelength of electromagnetic radiation depending on its size. Quantum theory predicts that with decreasing size of quantum dots there will be a corresponding increase in energy of emitted light.

The ability to control the emission property of a quantum dot has tremendous technological applications as well as of great academic interest to researchers. Unlike any other quantum structure, quantum dots have excellent confinement properties. This makes quantum dots extremely efficient in emitting light. Quantum dots have been the source of some of the world's most powerful lasers produced till date. **Quantum dot lasers** combine the length scales defining the confinement of photons (100 nm) and the length scales defining the confinement of electrons and holes (10 nm). Moreover, quantum dot lasers exhibit reduced temperature resistivity. For fabrication, it is necessary to form high quality, uniform quantum dots in the active layer. The practicality of a quantum dot laser is still being improved. In medical studies, quantum dots are already in practice as **tags** that can be inserted into a patient's body. One can see these tags using most medical scanning technologies which can help pinpoint the biological processes inside the body as they take place.

26.5 QUANTUM WIRE OR NANOWIRE

A nanowire is a one dimensional (1D) structure that has a diameter of the order of a few nanometres. Such a wire is also known as a quantum wire. The length-to-width ratio, called the aspect ratio, for typical nanowires is about 1000 or more. As compared to quantum films, one more dimension of the structure appears to be so small as to provide quantum confinement. Nanowires have many interesting properties that are not observed in bulk materials. This is because electrons in a nanowire occupy energy levels that are different from the energy levels or bands observed in bulk materials. Charge carriers are free to move only along the wire. Thus, only one kinetic component along with the confined energy contributes to the total energy. As a consequence, the density of states has $E^{1/2}$ dependence for each of the discrete pairs of states in the confined directions.

Quantum confinement exhibited by certain nanowires, e.g. carbon nanowires is reflected in the discrete values of the electrical conductivity. Such discrete values arise from a quantum mechanical restriction on the energy of electrons (and hence their number) travelling through nanowire. The difference in the successive values of conductivity is called *quantum conductivity*.

Nanowires have been grown out of metals, traditional semiconductors such as silicon and gallium, and a variety of polymers.

Conductivity of Nanowires

The conductivity of a nanowire is observed to be much less in comparison to the corresponding conductivity of the bulk material. This is due to a variety of reasons. One reason is the scattering of electrons from the boundaries of the wire. When the width of the wire is smaller than the mean free path of the free electrons in the bulk material, the scattering of electrons from the boundaries increases. In copper, for example, the mean free path (at 300 K) is 40 nm. Copper nanowires less than 40 nm width will shorten the mean free path to the wire width, resulting in reduced conductivity. Silicon nanowires coated with nickel have shown improved conductivity.

Nanowire conductivity is found to be strongly influenced by the edge effects. The edge effects are due to atoms that are at the nanowire surface and are not fully bonded to neighbouring atoms as the atoms within the bulk of the nanowire. The unbounded atoms are often a source of defects within the nanowire and seems to cause the nanowire to conduct electricity more poorly in comparison to the bulk material. As a nanowire reduces in size, the surface atoms get more numerous as compared to the atoms within the nanowire, and edge effects become more prominent.

The conductivity of quantum wire can undergo quantization. This can be explained on the basis of the quantization of energy of the electrons travelling through a nanowire. The conductivity of nanowire is thus the effect of the sum of the transport by separate *channels* of different quantized energy levels. The thinner the nanowire is, the smaller the number of channels available to the transport of electrons. One can verify this by measuring the conductivity of a nanowire while stretching it. Its conductivity decreases in a stepwise manner while its diameter shrinks.

Uses of Nanowires

Applications of nanowires include sensors, interconnects, memory devices (potentially), logic devices and processors.

It is possible that semiconductor nanowire crossings may be important for the future of digital computing. No doubt, there are many other uses of nanowires, the only ones that actually take advantage of physics in the nanometre regime seems to be the applications in fabrication of electronic devices.

Conducting nanowires offer the possibility of connecting molecular-scale devices in a **molecular computer**. Dispersions of conducting nanowires in different polymers are being investigated for use as transparent electrodes for flexible flat-screen displays.

Due to their high value of Young's modulus, application of nanowires in mechanically enhancing composites is also being investigated. Since nanowires appear in bundles, they may be used as tribological additives to improve friction characteristics and reliability of electronic transducers and actuators.

26.6 MODELLING OF QUANTUM SIZE EFFECT

The experimentally observed size effect can be explained with the help of the following theoretical models:

i. **Weak confinement region ($R \gg a_B$):** The region wherein size of the cluster, R is greater than the Bohr radius, a_B of the exciton in the bulk material is termed weak confinement region. In this region, due to size quantization of exciton, (exciton is a bound pair of electron and hole formed due to excitation of electron at an elevated energy level due to interaction with electromagnetic interaction) there is small increase, ΔE in the exciton energy. This is given by $\Delta E = \hbar^2 \pi^2 / 2MR^2$, where M is the mass of the exciton and is given by $M = m_e^* + m_h^*$, where m_e^* and m_h^* are effective masses of electron and hole respectively and R is the radius of the cluster.

ii. **Medium confinement region ($R \approx a_B$):** The region wherein size of the cluster is comparable to the Bohr radius of the exciton in the bulk material is termed medium confinement region. This is an usual situation observed in certain semiconductor compounds. In this region, due to size quantization of exciton there is increase in the exciton energy. This increase in energy is given by

$$\Delta E = \frac{\hbar^2 \pi^2}{2m_e^* R^2} \qquad (26.1)$$

iii. **Strong confinement region ($R \ll a_B$):** The region wherein size of the cluster is smaller than the Bohr radius of the exciton in the bulk material is termed strong confinement region. There is an increase in the exciton energy due to size quantization of exciton. This increase in energy is given by

$$\Delta E = \frac{\hbar^2 \pi^2}{2\mu R^2} \qquad (26.2)$$

where μ is the reduced effective mass of the electron.

The **EMA model** was further improved by Brus in 1984 and called as **Brus model** or **single-band effective-mass model**. In 1986, Kayanuma further improved this model and found to be in agreement with experimental results.

The **empirical tight binding model** (ETBM) proposed by Wang and Herron in 1990 was successful in explaining many phenomena observed in optical spectra.

Empirical pseudo-potential model (EPM) developed by Ram Krishna and Friesner in 1991 is well established and found to be a reliable approach. **Effective bond order model** (EBOM) developed by Einevoll in 1992 helped in calculating energy levels for particular phase of certain semiconductor compounds. However, it may be noted that all the models are not applicable for cluster sizes smaller than or equal to 1 nm.

26.7 SURFACE AND INTERFACE EFFECTS

Significant changes are reported in the properties of nanomaterials due to the spatial confinement of charge carriers in three dimensions. The surface properties of nanomaterials also play an important role in determining the properties of these materials. The larger surface-to-volume ratio observed at the nanoscale is one of the reasons for the typical properties of these materials. The number of surface atoms in a nanomaterial is appreciably larger than those in the bulk. This causes the defects to be located at the surface.

These surface atoms and defects seem to play a major role in deciding the properties of these materials.

In a small cluster, a larger fraction of atoms occupy surface positions, e.g. a 50 Å CdS cluster has about 15% of its atoms lying on its surface. The effect of relatively large proportion of atoms on or near the surface of clusters becomes prominent in the study of surface photons, absorbed species and enhancement in the reactivity by using catalyst. The existence of this vast interface between the cluster and the surrounding medium seems to have a profound effect on the properties of the cluster. Whatever is the method of preparation, in most of the cases, the clusters have an imperfect surface. This leaves defects on the surface. These defects act as trap for the movement of the charge carriers after the photoexcitation. The presence of the trapped electrons or holes can modify the optical properties of these clusters and can given rise to *nonlinear optical effects*. This can also alter the nature of the photochemical reactions which are of considerable interest in the field of photocatalysis. Controlling the cluster surface and, in turn, the cluster properties is a major challenge before researchers and provides a unique opportunity in this exciting and promising field.

The surface-to-volume ratio is important in the process of catalysis. The surface area of a catalyst is generally expressed in the units of square metre per gram and called as the **specific surface area** (S).

$$S = \frac{\text{Area}}{\text{Volume} \times \text{Density}} \qquad (26.3)$$

The specific surface area per gram for geometrical shapes generally observed in nanomaterials are as follows:

$S = 6 \times 10^3/pd$ for a sphere with diameter d

$S = 6 \times 10^3/pa$ for a cube with side a

$S = 2 \times 10^3/pl$ for a thin disc whose length l is very small compared to its diameter d

$S = 4 \times 10^3/pd$ for a long cylinder with its diameter d very small compared to its length l.

26.8 PROPERTIES OF NANOMATERIALS

Research on nanomaterials has been stimulated by their technological applications. These materials exhibit entirely new and special properties. Technologically useful properties of these materials are not limited to their structural, chemical or mechanical behaviours. For example, multilayers represent examples of materials in which one can modify or tune a property for specific application by sensitively controlling the individual layer thickness.

Very different physical laws dominate the nanoscale materials. Properties of materials change. Some of these materials are better conductors of electricity or heat, some are mechanically stronger, some have different magnetic properties, and some reflect light better or change colours as their size changes. Nanoscale materials have a large surface area available for interactions compared to bulk materials of equal volume. We know that nanomaterials are formed by grouping a small number of molecules together. These are called **nanoclusters** or simply **clusters**. The smaller number of molecules in a cluster also gives rise to some properties that are observed only in the nanomaterials.

Optical Properties

Confining electrons to small geometries give rise to **particle in a box** energy levels. This quantum confinement creates new energy states and results in the optoelectronic properties of semiconductors. Because of their role in quantum dots, nanoparticles made of elements which are normal constituents of semiconductors have been the subject of much study with particular emphasis on their electronic properties.

Optical properties such as the colour of a material depends on its electronic structure. As the size of the material is reduced, the number of molecules that form the material gets smaller. This results in smaller size of the clusters that form the nanomaterial. The energy difference between two energy levels is inversely proportional to the size of the box.

The size of a cluster (atomic clusters) are large molecules containing typically from 10 to 2000 atoms. They have specific properties (mainly due to their large surface-to-volume ratio), which are size dependent and different from both those of atoms and the bulk materials in this case. The energy levels of the electrons in the constituent molecules are no longer continuous as against the case of a solid. The energy levels get spread. This separation affects the energies required for the transitions of electrons to the excited states and the subsequent emission when the electron falls back to the ground state. This energy difference determines the colour of the light emitted. Clusters of different sizes, therefore, show different colours. The energy levels for electrons in clusters with different sizes are shown in Fig. 26.11.

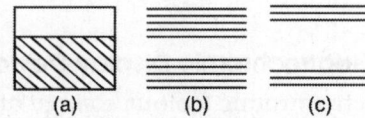

(a) (b) (c)

Fig. 26.11: Change in separation of energy level with size of cluster (a) Bulk metal (b) Large metal cluster (c) Small metal cluster

For example, a piece of gold cut in smaller size is observed to change its colour in the nanosize regime. Depending on the size, it can turn red, blue, yellow or some other colour. Materials having different size reflect and absorb light differently. Semiconductor cadmium selenide, in nanoscale size, can turn to any colour from red to blue depending on its size.

Similar changes in colours have been reported in the nature also. For example, leaves of a tree or plant appear green because the atomic arrangement in them is such that it absorbs all colors except green which is reflected. When leaves die, chlorophyll, the major constituent responsible for their green colour, breaks down changing the atomic arrangement and they appear brown. Colours observed on a peacock feather are also due to the nanosize particles on it.

Electrical Properties

Nanomaterials can store considerably more energy than the conventional materials due to their large surface area. This property of nanomaterials finds very interesting and useful applications, e.g. high-energy-density batteries.

We know that conventional and rechargeable batteries are used in almost all applications that require electric power. The energy density (storage capacity) of these batteries is quite low and they require frequent recharging. Nanocrystalline materials are found to be good as separator plates in batteries as they can hold considerably more energy than the separator plates made from conventional bulk materials, e.g. nickel-metal hydride batteries made of nanocrystalline nickel and metal hydrides are likely to require far less frequent recharging and last much longer.

Large Electrochromic Display Devices

The electrochromic (colour change of matter caused by an electric current) device consists of materials whose optical properties can be changed by passing electric current through them or by applying an electric field. These devices are similar to *liquid crystal displays* (LCD) commonly used in calculators and watches and are primarily used in public hoardings and ticker boards to display information. The resolution, brightness and contrast of the display of these devices depend on the grain size of the material used in them. Nanomaterials, such as tungstic oxide gel (tungsten oxide), are being explored for this purpose.

Chemical Properties

The electronic structure of nanomaterials depends on their size and therefore their ability to react chemically depends on their size. The higher surface-to-volume ratio is also responsible for their higher chemical reactivity. For example, gold is fairly inert when in bulk form, but at the nanoscale it is highly reactive.

Applications of Chemical Properties of Nanomaterials

One of the major factors for the chemical applications of nanomaterials is the increase in their surface area. The chemical activity of the nanomaterials is found to increase with the increase in the surface area. Nanosize materials can be used as catalysts due to their enhanced chemical activity. Nanomaterials find applications in automobile catalytic converters and power generation equipment. Nanomaterials react with harmful and toxic gases, e.g. carbon monoxide and nitrogen oxide to prevent environmental pollution arising from burning gasoline and coal. Fuel cell technology is another important application of the noble metal nanoparticles where they are used as catalysts. The present day fuel cell catalysts are based on the platinum (Pt) group metals. Pt and Pt-Ru alloys are some of the most frequently used catalysts from this group. Obviously, the use of these metals is one major factor for higher costs of these cells. One possibility to produce economical catalysts is the use of bimetallic nanoparticles.

Gold nanoparticles are widely studied as they have several potential applications. For example, they have been used for the identification of the DNA sequences associated with diseases, as well as viruses. New devices for use in computers can also be fabricated by using these materials. Obviously, these materials would therefore be excellent components in nanoelectronics, optics and biology.

Mechanical Properties

Scientific challenges in nanoscience and nanotechnology include the development of nanomaterials with novel mechanical properties. The need for scratch, mar, and/or abrasion resistance is well established in various markets, including fingernail polishes, flooring, plastic glazing, headlamp covers and other automotive parts, transportation windows, and optical lenses, where clear scratch-resistant coatings are used. Many of the mechanical properties of materials are modified at the nanoscale. Hardness and elastic modulus, fracture toughness, scratch resistance and fatigue strength are some properties that are modified. Energy dissipation, mechanical coupling within arrays of components and mechanical non-linearities are influenced by structuring components at the nanometre scale.

A number of important questions regarding the mechanical behaviour of nanocrystalline metals are yet to be answered; fundamentally the underlying mechanisms of deformation at both low and high temperatures are unknown. The major research challenge in this area currently lies in the areas of synthesis and processing. Until nanocrystalline materials can be produced which are dense, free of flaws, of high purity and of sufficient quantity to carry out a variety of mechanical tests, measured values will be influenced by extrinsic rather than intrinsic properties. For high temperature applications, a great amount of research work needs to be performed in controlling grain sizes so that the correct balance of ductility, formability and strength are achieved. Inspite of these challenging problems, nanocrystalline materials offer a wide variety of opportunities for fabricating new materials with precisely engineered microstructures where strength and ductility can be finely tuned. Moreover, the ability to structure materials on the nanoscale level makes it possible to test fundamental ideas about grain boundaries, dislocations and deformations.

Mechanical Applications of Nanomaterials

Nanocrystalline materials possess unusual properties and their deformation behaviour has attracted intense activity in recent years.

Tougher and harder cutting tools can be developed using nanomaterials. Nanomaterials required for these purpose are made from tungsten carbide, tantalum carbide and titanium carbide. Such tools are found to be harder, much more wear-resistant, erosion-resistant and also last longer than their conventional (large-grained) counterparts. Moreover, for the miniaturisation of micro-electronic circuits, the industry needs

microdrills (drill bits with diameter less than the thickness of an average human hair or 100 μm) with enhanced edge retention and far better wear resistance. Nanocrystalline carbides are quite stronger, harder and wear-resistant and therefore they are currently being used in these microdrills.

Nanomaterials can be used to manufacture automobiles with greater fuel efficiency. Nanomaterials are quite stronger, harder, much more wear-resistant and hence they could be coated on spark plugs and other engine parts. Moreover, automobiles waste significant amounts of energy by losing the thermal energy generated by the engine. Nanocrystalline ceramics, e.g. zirconia and alumina retain heat much more efficiently. Engine cylinders coated with such materials are under consideration. This would result in complete and efficient combustion of the fuel.

One of the key properties of aircraft components is their fatigue strength. Fatigue strength decreases with the age of the component. Moreover, the fatigue strength increases with reduction in the grain size of the material. Nanomaterials provide such a significant reduction in the grain size over conventional materials that the fatigue life is increased by an average of 200–300%. In spacecraft, elevated-temperature strength of the material is crucial as the components (such as rocket engines, thrusters, and vectoring nozzles) operate at much higher temperatures than aircrafts and higher speeds. Nanomaterials are useful to enhance performance characteristics of aerospace components as well.

Ceramics are quite hard, brittle and at the same time difficult to machine even at high temperatures. However, with a reduction in grain size, the properties of ceramics change drastically. Ceramics made of nanocrystals are found to be ductile in nature. They can be pressed and sintered into various shapes at significantly lower temperatures. Zirconia in bulk form, for example, is a hard, brittle ceramic. At nanoscale, it turns into a super plastic material, i.e. it can be deformed to a great extent (up to 300% of its original length). These ceramics have nanocrystalline grains.

Ceramics based on silicon nitride (Si_3N_4) and silicon carbide (SiC) have been used in automotive applications as high-strength springs, ball bearings and valve lifters. They have also been used as components in high-temperature furnances as they possess good formability and machinability combined with excellent physical, chemical and mechanical properties.

Nanocrystalline materials are found to provide better thermal insulation. Aerogels are porous and extremely lightweight nanocrystalline materials. They can withstand 100 times their weight. They are currently being used for insulation in offices and homes. They are also being used as materials for *smart* windows, which darken when the sun is too bright and lighten otherwise.

Magnetic Properties

The magnetic properties of nanometer sized particles differ in several respects from the properties of their bulk counterparts. The large fraction of atoms located on surfaces or interfaces, for example, whose local environments differ greatly from those of the interior atoms leads to a blurring of the distinction between intrinsic and extrinsic properties. Since nanophase particles can be as much as 50% surface material, new magnetic properties, characteristics of surfaces and interfaces become important from applications point of view.

Very interesting magnetic properties are observed in small **clusters**. The magnetic moment of each atom in cluster interacts with the moments of other atoms and forces all the other moments to orient in one direction with respect to some symmetry axis of the cluster. The cluster thus has a net magnetic moment and it is magnetised. As the cluster size decreases, it becomes easier for the atoms in a cluster to orient their magnetic moments in one direction and display ferromagnetic behaviour. In some cases, non magnetic materials show a prominent increase in their magnetic moment when they contain less than 20 atoms.

The strength of a magnet is measured in terms of **coercivity** and **saturation magnetization**.

It is observed that these values increase with a decrease in the grain size and increase in the specific surface area (surface area per unit volume) of the grains. Obviously, nano-materials present magnetic properties that are not observed so far.

Magnetic Applications of Nanomaterials

Today information technologies ranging from personal computers to main frames use magnetic materials to store information on tapes, floppy diskettes and hard disks. Our seemingly insatiable appetite for more computer memory will probably be met by a variety of magnetic recording technologies based on nanocrystalline thin-film media and magneto-optic materials. Personal computers and several of our consumers and industrial electronics components are now powered largely by lightweight switch mode supplies using new magnetic materials technology that was unavailable two decades ago. Magnetic materials touch many other aspects of our lives.

Magnetic properties of nanometer sized particles of both ferrimagnetic and ferromagnetic materials have attracted considerable attention in recent years. Such interest has been generated for practical as well as theoretical reasons. Some of the application areas being explored are in high-density magnetic recording media, as ferrofluids, giant magnetoresistive systems. Some notable applications of nanostructured magnetic materials are storage applications, giant magneto resistance (GMR) recording head, and high power magnets.

Magnets made of nanocrystalline yttrium-samarium-cobalt grains possess very unusual magnetic properties due to their extremely large surface area. Such magnets are observed to possess high magnetic retention. Typical applications for these high-power rare-earth magnets include quieter submarines, automobile alternators, land-based power generators and motors for ships, ultrasensitive analytical instruments and magnetic resonance imaging (MRI) in medical diagnostics.

Melting Temperature

The large surface-to-volume ratio of nano-materials exhibit a very unusual effect on their thermal properties. They show substantially lower melting temperatures as compared to the solids. Negative heat capacities are also reported in some nanomaterials. The theoretical analysis and calorimetric investigations show that in the temperature range $10\,K \leq \theta_D$ (Debye temperature), the heat capacity of nanopowders is from 1.2 to 2 times higher than that of coarse grain bulk materials. One can easily explain these observations. When a large system, e.g. a solid is heated, the energy added is completely converted into potential energy of the atoms/molecules. The solid is continuously converted to liquid form due to this added energy. The kinetic energy of the atoms/molecules and therefore the temperature of the solid remain constant. A small system like a nanomaterial, on the other hand, tries to avoid partially molten states and converts some part of its kinetic energy into potential energy. Obviously, the kinetic energy of the nanomaterial decreases and it becomes cooler. Studies of the variation in the melting temperatures of gold reveal such behaviour. Variation in the melting temperature of gold with reduction in the particle size is shown in Fig. 26.12.

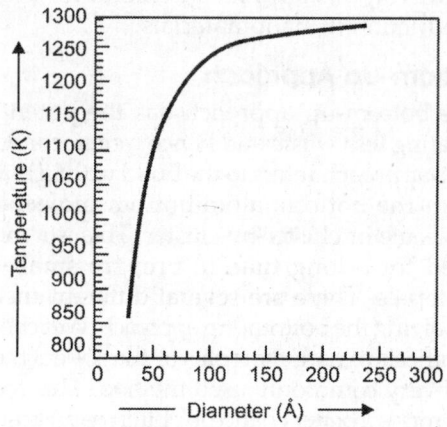

Fig. 26.12: Variation in the melting temperature of gold with reduction in the particle size

Structural Properties

One can determine the size of a nanocluster by the number of molecules that form it. Naturally, the arrangement of these molecules is quite different when the number of molecules forming the cluster changes.

The size dependence of the structural properties leads to changes in many other properties of nanomaterials. Structural properties of metallic nanocluster are studied using the XRD analysis.

Aluminium is face centered cubic in the bulk form. It is observed to be in the face centered cubic hexagonal close packed and the icosahedral structure at nanoscale. Similarly, indium is observed to undergo a transition from its tetragonal structure in the bulk form to the face centered. The changes in the structure of aluminium is shown in Fig. 26.13.

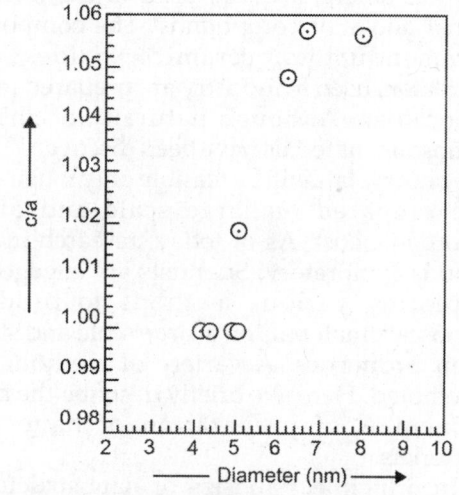

Fig. 26.13: Changes in the structure of aluminium

The lattice structure of nanometer sized crystallites in nanocrystalline samples synthesized by various methods showed that the lattice structure of the nanometer sized crystallites evidently deviates from the equilibrium state. One may classify the deviation as: (i) distorted lattice structures and stoichiometric line compounds (ii) formation of metastable phase below a critical crystalline size.

Structural changes in the nanomaterials also cause changes in the electronic structure of the clusters. The band levels and the valence energy change when the number of molecules in the cluster changes. The energy levels are in the form of a continuum when the material is in the bulk form. As the number of molecules is reduced, the energy levels become discrete and the material behaves in a different way. This change in the energy levels is shown in Fig. 26.13.

26.9 PREPARATION OF NANOSCALE MATERIALS

Prior to the invention of transistor, metals were treated as the most important solids. Metals are obtained from earth crust and then purified. Usually, metals are used in the form of alloys. Similarly, elemental semiconductors such as silicon are also used in the form of either alloys or compounds. The compound semiconductors, ceramics, oxides and insulators used in industry are prepared in the laboratory. Although naturally occurring nanoscale materials have been discovered, use of nanocrystals will be feasible only when they are prepared on large scale and at an affordable cost. As of today, research in this field is exploratory. Scientists are engaged in exploring various methods to produce nanoscale materials on a large scale and study their properties. A variety of methods are attempted. Here, we briefly describe the basic principles of preparation of nanoscale materials.

In principle, nanosize or nanostructured materials can be prepared by either arranging the atoms one by one, known as the *bottom-up approach*, or using the techniques such as atom force microcopies or one can cut down the bulk or scrape the macroscopic solid to obtain nanosize solid known at the *top-down approach*.

Top–Down Approach

This approach involves the breaking down of the bulk material into nanosize structures or particles. Truly speaking, these techniques are an extension of those that have been used for producing micron-sized particles. An example of such a technique is high energy wet ball milling. The material to be prepared is passed through a rolling mill consisting of two hard rollers. These rollers are generally made from hardened steel or are coated with tungsten carbide. The complete process is carried out in a sealed container and the container is vigorously shaken continuously. This method produces the nanoparticles in powder form.

There is another common method in which the bulk material is cut to the required size and shape using an electron beam. The process requires a very high vacuum and involves sophisticated equipments for control. Other radiations such as lasers, X-rays or UV rays are also used in this technique. The size of the nanomaterials to be fabricated is governed by the wavelength of the radiation used. Obviously, this puts some restrictions on the radiations that can be used for nanofabrication. Electron beams are generally used to fabricate nanoscale devices as the wavelength of the beam can be controlled by changing its energy.

The drawback of the top-down approach is the imperfection of surface structure and significant crystallographic damage to the processed patterns. Further, it can also introduce internal stress in the material produced, which can cause changes in the properties of the material. This causes difficulties in the device design and fabrication. However, this approach is useful for the bulk production of nanomaterials.

Bottom–Up Approach

The bottom-up approach has the potential of creating less waste and hence more economical. This approach refers to the build up of a material from the bottom: atom-bottom molecule-by-molecule or cluster-by-cluster. This is a method used for a long time to prepare small-sized materials. There are several different methods involving the bottom–up approach. Wet chemical synthesis of nanomaterials or the sol-gel method is a very commonly used method. This method promises a better chance to obtain nanostructures with less defects, more homogeneous chemical composition.

Chemical Methods

Chemical methods of preparing nanomaterials are quite simple in principle, versatile, and inexpensive. Large number of materials such as metals, semiconductors, insulators, ceramics and other complicated solids are grown in nanostructured form by employing these methods. The chemical reactions can be carried at low or at high temperatures. Low temperature methods use reactions of aqueous solutions while high temperature methods use different solvents. A salt containing cation and the one containing anion is mixed under controlled conditions of concentration, rate of reaction, temperature and pH of the solvents in presence of some organic molecules that work as binders. The large organic molecules bind to cations and/or anions and control the size of the precipitate. Depending on the concentration of cations, anions and organic molecules, nanocrystals of the required size can be formed in the solution. For example, one of the reactions for formation of zinc sulphide (ZnS) nanocrystals is: zinc acetate + sodium sulphide + thiophenol → zinc sulphide + sodium acetate + zinc thiophenol

$$Zn(CH_3COO)_2 + Na_2S + C_6H_5SH \rightarrow ZnS$$
$$+ 2Na(CH_3COO) + ZnSC_6H_5$$

Zinc sulphide is insoluble in organic solvents like methanol and hence precipitates out. However, sodium acetate is highly soluble in normal solvents and remains in the solution. Nanocrystal powder thus formed is separated out by using centrifugal machine and is washed several times so that the reaction by-products and unreacted solvents are removed. Different chemical reactions produce nanocrystals with different physical properties.

Nanocrystals can also be obtained by heating the powder of the material with glass, melting the mixture and re-solidifying the same. Many old churches in Rome, old palaces in India have decorative coloured glasses made from nanocrystals grown in this fashion. We can say that knowledge of growing nanocrystals is very old while they are looked upon with new focus for the last three decades.

The sol-gel process involves the formation of inorganic networks through the formation of a colloidal suspension (*sol*) and gelation of the sol to form a network in a continuous liquid phase (*gel*). The precursors for synthesizing these colloids consist usually of a metal or metalloid element surrounded by various reactive ligands. The starting material is processed to form a dispersible oxide and forms a sol in contact with water or dilute acid. Removal of the liquid from the sol yields the gel, and the sol/gel transition controls the particle size and shape. Calcination of the gel produces the oxide. Various steps involved in the sol-gel process can be summarized as follows:

i. Formation of different stable solutions of the alkoxide or solvated metal precursor (the sol).

ii. Gelation is the formation of an oxide or alcohol-bridged network (the gel) by a polycondensation or polyesterification reaction that results in a dramatic increase in the viscosity of the solution.

iii. Ageing of the gel (syneresis), during which the polycondensation reactions continue until the gel transforms into a solid mass, accompanied by contraction of the gel network and expulsion of solvent from gel pores. Sometimes the smaller particles merge with the larger particles and form coarse particles. This is called as **Ostwald ripening** or **coarsening.**

iv. **Drying of the gel:** In this step, water and other volatile liquids are removed from the gel network. However, this process is complicated due to fundamental changes in the structure of the gel.

v. **Dehydration:** In this step, the surface-bound M-OH groups are removed and the gel is stabilised against dehydration.

vi. **Densification and decomposition** of the gels at high temperature ($T > 800°C$). In this step, the pores of the gel network are collapsed and remaining organic species volatilised.

This synthesis method offers the possibility of synthesizing non metallic inorganic materials, e.g. glasses, glass ceramics or ceramic materials at very low temperatures compared to the high temperature process required by melting glass or firing ceramics.

It is very difficult to control the growth of particles and then stop the newly formed particles from mixing together. Moreover, we must ensure that the reactions are complete so that no undesired reactant is left on the product and completely removing any growth aids that may have been used in the process. However, the production rates of nanopowders are very low by this process. The main advantage of any bottom-up approach is that one can obtain same-sized particles.

Physical Methods

Most of the methods of preparing nanoscale materials discussed so far are physical methods of preparing nanoparticles. In addition to these methods, radio frequency (RF) plasma and molecular beam epitaxy (MBE) are the techniques also used to fabricate nanomaterials.

RF Plasma

Radio frequency plasma technique is used to synthesize metal nanoparticles. The metal is placed in a specially designed container and kept in a vacuum chamber. By using RF plasma, it is evaporated with the help of RF coils surrounding the container. By passing helium gas, high-temperature plasma is generated inside the RF coils. The metal vapours condense on to the helium gas atoms and diffuse towards a collector rod, i.e. at a lower temperature and form the nanoparticles. These are then passivated by allowing to pass some gas like oxygen through the chamber.

Molecular Beam Epitaxy

This is an advanced technique used to form nanoparticles. In this technique, a substance is evaporated under ultra-high vacuum (UVH $\approx 10^{-9}$ torr). The required material is evaporated at a very slow rate ($1\,\mu m/h$) and a thin film is deposited on a suitable substrate

(supporting material) heated to the required temperature. Semiconductor heterojunctions can be grown in a controlled manner by this technique. The substrate material and the material to be deposited must be extremely pure so that any contamination in the device is avoided. This technique offers a very good control over the thickness of the material, it is reasonably reproducible and the material prepared is also in its purest form. The UHV helps to make it possible to analyse and characterize the material deposited while it is being prepared. This method is highly advanced as well as expensive. However, nanodevices are fabricated by using this method.

26.10 BIOLOGICAL NANOMATERIALS BIOMIMICKING

Biomaterials have received a considerable amount of attention over the last three decades as a means of treating diseases and easing suffering. Biomaterials have found applications in about 8000 different kinds of medical devices, which have been used in repairing skeletal systems, returning cardiovascular functionality, replacing organs, and repairing and returning senses. Even though biomaterials have had a pronounced impact in medical treatment, a need still exists to be able to design and develop better polymer, ceramic and metal systems.

Nanotechnology is playing an important role in improving biomaterial performance.

Most research focussing on biological microelectromechanical systems (Bio MEMSs) is for their use in diagnosing devices and for the detection of DNA, viruses, proteins and other biologically derived molecules. Nanoscale Bio MEMS could allow for the realtime detection and analysis of signaling pathways, which would further our knowledge and understanding of the basic mechanisms and functions of the cell.

It is now believed that significant evidence exists that highlights the promise nanotechnology has for biological applications.

Many biological materials can be classified as biological nanomaterials, e.g. proteins, viruses and bacteria are some of the familiar examples. The building blocks of proteins are the amino acids. The size of each amino acid is about 0.6 nm. Out of more than 100 odd amino acids that occur naturally, only 20 are involved in the formation of proteins. A protein chain is formed by tying hundreds or sometimes thousands of amino acid molecules together. Such chains resemble nanowires. The basic building blocks for the DNA are four nucleotide molecules that bind together in the form of a double helix. Thus, the DNA in the form of a double helix is also an example of double nanowire.

Microelectromechanical Systems (MEMSs)

Microelectromechanical systems (MEMSs) are small integrated devices or systems that combine electrical and mechanical components. To make MEMSs devices, lithographic techniques are combined with metal deposition processes. MEMSs involve a mechanical response to an applied electrical signal, or an electrical response resulting from a mechanical deformation. The size of MEMSs devices or systems vary in the sub micrometre (or sub micron) level to the millimetre level. A system can have any number of such devices, from a few to millions. The same technology used for fabrication of integrated circuits is extended to fabricate the MEMSs. The MEMSs has mechanical elements such as beams, gears, diaphragms, and springs to devices along with the electrical components on it.

The main advantages of MEMSs devices are miniaturization, multiplicity and the ability to directly integrate these devices into microelectronics. Multiplicity refers to the large number of devices and designs that can be manufactured rapidly, e.g. miniaturization has enabled the development of micrometer-sized accelerometers for activating airbags in cars.

MEMSs devices have applications which include inkjet-printer cartridges, accelerometers, miniature robots, microengines, locks, inertial sensors, microtransmissions, micromirrors, microactuators, optical scanners, fluid pumps, transducers and chemical pressure and flow sensors. New applications of MEMSs devices

and systems are emerging as the existing technology is applied to the miniaturization and integration of conventional devices.

MEMSs devices can sense, control and activate mechanical processes on the microscale and function individually or in arrays to generate effects on the macroscale. The microfabrication technology enables fabrication of large arrays of devices, which individually perform simple tasks, but in combination can accomplish complicated functions.

Future applications of MEMSs will be driven by processes enabling greater functionality through higher levels of electronic-mechanical integration and also greater number of mechanical components working alone or together to enable a complex action. Future MEMSs products will like to have higher levels of electrical–mechanical integration and also more intimate interaction with the physical world. Advancing from their success as MEMSs sensors, MEMSs products will be embedded in larger non MEMSs devices such as printers, automobiles and biomedical diagnostic equipment and will also enable new and improved systems. MEMSs devices are manufactured using batch fabrication techniques similar to those used for integrated circuits, unprecedented levels of functionality, reliability and sophistication can be placed on a small silicon chip at a relatively low cost.

There are several devices and machines that have micrometer-sized elements.

Nanoelectromechanical Systems (NEMSs)

NEMSs are similar to MEMSs but smaller in size and are related to nanotechnology and nanomechanics. NEMSs machines and devices are in the early states of development and many are still in conceptual stages. Numerous computer simulations of possibilities and ideas have been advanced. Because of their nanoscale size, NEMSs could be useful to measure small displacements and forces at a molecule scale. There are two standard methods of fabrication of NEMSs machines and devices: *the top–down and the bottom–up approach*. The top-down approach is accepted as the most standard approach to NEMSs. This approach may be summarised as a set of tools designed to build a smaller set of tools, e.g. a millimetre-sized factory builds micrometre-sized factory, which in turn can build nanometre-sized devices and machines. The other approach can be thought of as putting together single atoms or molecules until a desired level of complexity and functionality is achieved in a device. Such an approach may utilize molecular self assembly or mimic molecular biology systems.

The first very large scale integration NEMSs device was demonstrated in 2000 by scientists at IBM. Its premise was an array of AFM tips can heat/sense a deformable substrate in order to function as a memory device. A combination of these approaches may also be used, in which nanoscale molecules are integrated into a top-down framework, e.g. the carbon nanotube nanomotor.

More recently, a technique named **nano-imprint lithography** has been developed that may provide a low-cost, high-production rate manufacturing technology. This can produce patterns on a surface having 10 nm resolution at low cost and high rates because it does not require the use of a sophisticated radiation beam generating patterns for the production of each surface.

By exploiting nanoscale effects, NEMSs offer a number of unique properties which pave the way to applications such as force sensors, chemical sensors and ultrahigh frequency resonators. The most interesting properties of NEMSs arise from the behaviour of active parts, which is typically in the form of cantilevers or doubly damped beams with nanoscale dimensions. In NEMSs, the charge controls the mechanical motion, and *vice versa*. The presence of mechanical motion decisions that the moving element to posses high-quality mechanical properties, including high Young's modulus and low density. While limitations in strength and flexibility compromise the performance of Si-based NEMSs actuators, carbon nanotubes (CNTs) are better candidate to realize the full potential of NEMSs, in part due to their one dimensional structure, high aspect ratio, perfect terminated surfaces and exceptional electronic and mechanical properties. These properties, now complemented by significant advances in growth and manipulation techniques, make CNTs the most promising building blocks for next generation NEMSs. The predicted behaviour of CNT based NEMSs switches-basis of NEMSs devices is favourable and electronic properties have been shown to be reversible with mechanical deformation by a local probe.

26.11 CARBON NANOTUBES (CNTs)

Carbon nanotubes were discovered in 1991 by Sumio Iijima at the NEC laboratories. Although other nanotubes based on boron nitride and molybdenum have also been reported, currently CNTs are by far the most important group. CNTs have occupied a special identity in the field of nanotechnology. Since their discovery, theoretical and experimental studies in different fields, e.g. mechanics, optics, electronics, etc. have focussed on both the fundamental physical properties and on the potential applications of nanotubes. In all fields there has been substantial progress over the last two decades. Basically, CNTs are naturally self organized nanostructure in the form of a tube composed of carbon atoms with completed bonds. The structure of CNTs consists of a single sheet of graphite, rolled into a tube, both ends of which are capped with C_{60} fullerene hemispheres. The tubes can be either open at their ends or capped at one or both ends with half a spheroidal fullerene. Figure 26.14 shows a schematic representation of a carbon nanutube. A nanotube is completely specified by what is referred to as a roll-up vector, which identifies *its helical nature* and diameter. Depending on its roll-up vector, a nanotube can have either metallic (Fig. 26.14a) or semi-conducting (Figs 26.14b and c) properties.

There are vast number of ways in which a graphite sheet can be rolled up to form a seamless cylinder and hence a wide variety of nanotubes exist. Nanotubes are characterized by their *chiral* (or wrapping) vector, c such that $c = na_1 + ma_2$, where a_1 and a_2 are the basis vectors of the graphite (more correctly graphene)

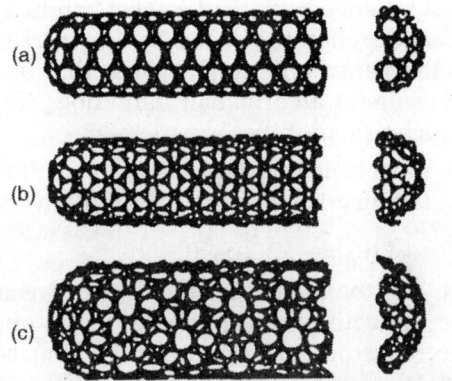

Fig. 26.14: The structure of carbon nanotubes
(a) Metallic (b) and (c) Semiconducting

lattice and *m, n* are integers (Fig. 26.15). The chiral vector spans the circumference of the tube formed by joining the dotted lines shown in Fig. 26.15. Figure 26.16 illustrates the established nomenclature of three different types of nanotubes: (a) the armchair (Fig. 26.16a) (b) the zigzag (Fig. 26.16b) (c) the chiral (Fig. 26.16c) nanotube.

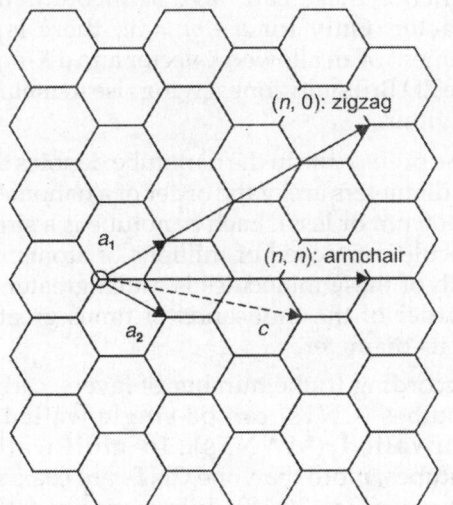

Fig. 26.15: The relationship between the graphite lattice basis vectors a_1, a_2 and the chiral vector $c = na_1 + ma_2$ is used to characterize carbon nanotubes. Two limiting cases are shown, $(n, 0)$ indices are associated with zigzag tubes whereas (n, n) indices are associated with armchair tubes. All other tubes are chiral

Closely following the synthesis of nanotubes came the remarkable theoretical prediction that their electronic properties

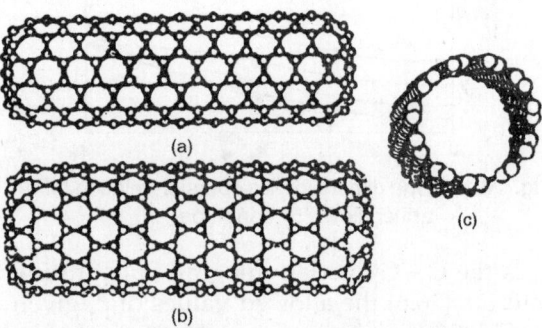

Fig. 26.16: Diagrams of three types of nanotubes
(a) Armchair (b) Zigzag (c) Chiral

could be changed between metallic and semiconducting simply by varying the tube diameter or its helicity, i.e. by changing the values of *n* and *m*. A key theoretical result is that armchair nanotubes are metallic, whereas for all other SWCNTs, when $n - m = 3l$ (l is an integer) the tubes are metallic; otherwise they are semiconducting.

These striking electronic properties can be understood within a tight-binding framework. In the direction along c (i.e. along circumference of the tube), periodic boundary conditions constrain the values of electron wave vector to those given by

$$c \cdot k = 2\pi q \ (q = 1,..., n) \qquad (26.4)$$

Taking the particular example of an armchair nanotube, this can be reduced to

$$k_x n 4\sqrt{3} \, a_0 = 2\pi q \qquad (26.5)$$

where $a_0 = (\sqrt{3}) \, 0.142$ nm is the graphite lattice constant and defines the chiral vector for the armchair tube [in the case $c = n \, (a_1 + a_2)$].

The well-known tight binding dispersion relation for a 2D graphite sheet is

$$E_{2D}(k_x, k_y) = +\gamma_0 \left\{ 1 + 4\cos\left(\frac{\sqrt{3}k_x a_0}{2}\right) \cos\left(\frac{k_y a_0}{2}\right) \right.$$

$$\left. + 4\cos^2\left(\frac{k_y a_0}{2}\right) \right\}^{1/2} \qquad (26.6)$$

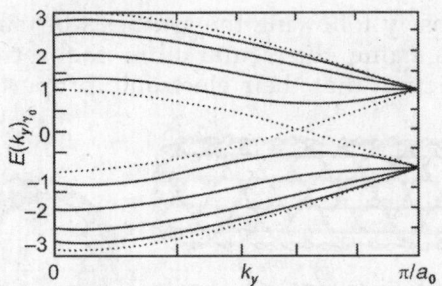

Fig. 26.17: One dimensional dispersion relation for an armchair (5, 5) nanotube

γ_0 is the C – C overlap integral. Substituting into Eq. (26.6) the allowed values of k_x given by Eq. (26.5) yields

$$E_{1D}^A \ (k_y) = \pm \ \gamma_0 \left\{ 1 \pm 4 \cos\left(\frac{q\pi}{n}\right) \cos\left(\frac{k_y a_0}{2}\right) \right.$$
$$\left. + \ 4 \cos^2\left(\frac{k_y a_0}{2}\right) \right\}^{1/2} \ (-\pi < k_y \, a_0 < \pi)$$

(26.7)

where the superscript A denotes that Eq. (26.7) is valid only for armchair tubes. It is assumed that the value of the overlap integral for graphite and nanotube are identical. The 1D dispersion relation given by Eq. (26.7) is shown graphically in Fig. 26.17 for the case of (5, 5) armchair SWNT. There are six dispersion curves for both the valence and conduction bands. In each case, four bands are doubly degenerate (bold continuous lines) and two bands are non degenerate (thin, dashed lines). Thus, ten electronic levels comprise the valence bands and the conduction bands as expected from the fact that there are ten hexagons around the circumference of a (5, 5) nanotube. There are two, particularly noteworthy, features of the dispersion relations shown in Fig. 26.17. First there is a very high degree of degeneracy at the Brillouin zone boundary. One can readily understand this from the fact that at $k_y a_0 = \pi$, Eq. (26.7) reduces to

$$E_{1D}^A = \pm \gamma_0$$

The second and more important feature of Figs 26.16a, b and c in terms of electronic properties of armchair nanotubes is that the highest valence band and lowest conduction band are degenerate at $k_y = \pm \, 2\pi/(3a_0)$ and each cross the Fermi energy $(E = 0)$ at that point. This is true for all armchair nanotubes. Thus, all armchair nanotubes are metallic.

We must note that Eqs. (26.5) to (26.7) represent taking a 1D slice in a direction given by

$$E(\theta) = - \, KV \cos^2 \theta$$

where K is the magnetic anisotropy constant, V the volume of the particle, θ the angle between the magnetization vector and an easy direction of magnetization and $E(\theta)$ is the magnetic anisotropy energy for a uniaxial particle, through the 2D graphite band structure. The crossing at $E = 0$ in Fig. 26.17 occurs because the corresponding 2D graphite band cross at the K points (i.e. the corners) of the 2D Brillouin zone. For armchair tubes, there is always an allowed K-vector that goes through K point of the 2D lattice. However, this alignment of the tube wave vector and K-point of the graphite Brillouin zone does not automatically occur for zigzag or chiral tubes and hence these can have semiconducting character. Only for $n - m = 3l$, there is an alignment of an allowed k-vector and a K-point of the 2D Brillouin zone, giving rise to metallic behaviour.

The prefix *nano* in the nanotube denotes that tube diameters are of the order of a nanometer (i.e. 100 nm or less). Each nanotube is a single molecule composed of millions of atoms; the length of these molecules is much greater (of the order of the thousands of times greater) than its diameter.

According to the number of layers, carbon nanotubes (CNTs) can be **single walled or multiwalled** (MWNTs). In multiwalled nanotubes, more than one CNTs are coaxially arranged (Fig. 26.18). Iijima realized that graphite could be bent to form MWCNTs with different helicities or chiralities which refer to the way hexagonal rings are arranged with respect to the tube axis (Fig. 26.19).

It is found theoretically that the electronic properties of carbon nanotubes depend on the diameter and helicity. In particular, all of the so called armchair type nanotubes are conductors (Fig. 26.19) and most zigzag nanotubes are semiconductors (Fig. 26.19).

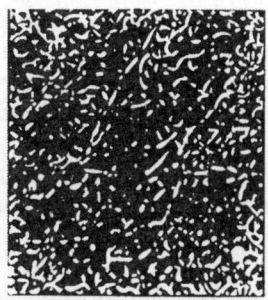

Fig. 26.18: SEM image of CNTs

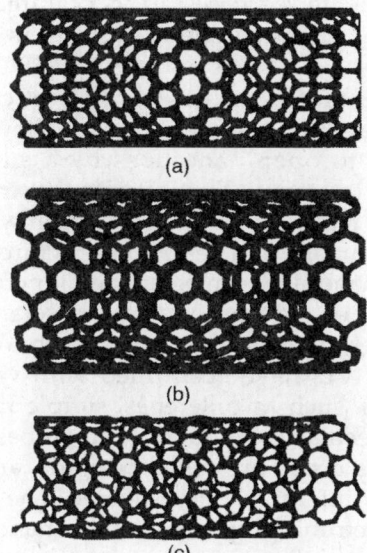

Fig. 26.19: Structures of carbon nanotube (a) Zigzag nanotube (b) Armchair nanotube (c) Chiral or helical nanotube

These nanotubes are extremely strong and stiff known fibres and relatively ductile. According to indirect measurements, it has been found that carbon nanotubes are 100 times stronger than steel and six times lighter. For single walled nanotubes, tensile strength ranges between 50 GPa and 200 GPa (approximately of order of magnitude greater than for carbon fibres); this is the strongest known material. Elastic modulus values are of the order of one tetrapascal [TPa (1 TPa = 103 GPa)], with fracture strain between 5% and 20%. Furthermore, nanotubes have relatively low densities. The outstanding properties of the CNTs include extremely high tensile strength, high modulus, flexibility, lightweight and high thermal conductivity. On the basis of unique properties, the carbon nanotube has been termed the *ultimate fibre* and is extremely promising as a reinforcement in composite materials. The high stiffness, aspect ratio and nanoscale tip make CNTs excellent probes for sensing, testing, manipulation and fabrication on the nanoscale.

Carbon nanotubes also have unique structure sensitive electrical characteristics. Depending on the orientation of the hexagonal units in the graphite plane, i.e. tube well, with the tube axis, the nanotube may behave electrically as either a metal or a semiconductor. Semiconducting and metallic nanotubes recently are put into use in field effect transistors (FET) and single electron transistors. The flat-panel and full-colour displays, i.e. TV and computer monitors, have been fabricated using carbon nanotubes as field emitters. These displays should be cheaper to produce and will have lower power requirements than CRT and liquid crystal displays. The mechanical strength of CNT is amazingly 600 times tougher than steel, which finds applications in microelectromechanical systems (MEMSs) and in aerospace. Progress has been made in using CNTs as field emission devices, e.g. diodes, transistors, etc. for high resolution display systems. Construction of nanoprobes and electrodes in biological and other applications are also being aimed.

Various Forms of Carbon Nanotubes

As already mentioned, the carbon nanotube is a generic term that includes both single and multiwalled nanotubes (SWNTs and MWNTs). The smallest nanotube reported till date has a diameter of only 4 Å. Since increasing the curvature of the graphite wall increases the strain energy, this is also predicted to be the smallest possible diameter a nanotube can have. Typical diameters for SWNTs are between 1 and 2 nm, corresponding to n and m numbers of around 10. In MWNTs, the distance between the adjacent layers is about 0.34 nm, which is much more than the atomic separation (1.42 Å) within the individual layers. While typical values for the outer diameter are 10–20 nm, the inner diameter is typically 3–5 nm. Thus, an ordinary MWNT has 10–30 layers, each of which are SWNTs of different diameters. The shell structure of MWNTs has not been studied thoroughly yet, but from geometrical considerations it is concluded that adjacent layers will generally be incommensurate. Also, larger the diameter of a MWNT is, more likely the outer layers are disordered and non circular.

The ends of nanotubes are generally closed in the growth process. While the wall of the carbon nanotube is made up of an exclusively hexagonal pattern, pentagons are needed to cap the tubes. More generally, pentagonal and heptagonal (5- and 6-cornered polygons respectively) defects enable the graphite sheet to take up more complicated structures than a simple tube, as we will discuss below.

Unless special setups are used for the growth of SWNTs, they are usually assembled into ropes by their mutual van der Waals attraction. For example, a rope of a typical diameter of about 10 nm contains ~100 SWNTs. Since multiwalled nanotubes have a much higher bending stiffness, they do not arrange similarly.

Nanotubes grown under suitable conditions have a very low concentration of defects over μm distances, that is, over hundreds or even thousands of interatomic spacings. However, if one were able to control the occurrence of defects, very useful and interesting nanotube structures would emerge. We will next show what kind of modified SWNTs and MWNTs have already been observed due to the occurrence of defects within the nanotubes. A single, stable defect structure in the hexagonal lattice of a graphite sheet is made up of a pentagon-heptagon pair, that is, 5- and 7-sided polygons. One such defect can cause a sharp bend in a SWNT. Defective nanotubes are especially interesting for electronic applications where the defect site may act as a tunnelling barrier. In the case of MWNTs, highly defective tubes may be obtained with certain synthesis conditions, for example with relatively low growth temperatures. While high-quality MWNTs are very straight and stiff, very defective ones have a continuous and smooth curvature. On the other hand, the curvature can be highly regular and so result in helices. A SWNT, or a single shell of a MWNT, can have a second nanotube branching out. The beginning point of the branch has similarly to the end sections pentagons dispersed among the hexagons to accommodate the necessary curvature. Such branching has been observed in a few cases till date.

As mentioned earlier, the ends of carbon nanotubes are usually closed. It is, however, possible to open nanotubes by a suitable chemical treatment. This is due to the stronger chemical reactivity of the end section with its larger curvature and pentagon structures. The opened nanotubes can be utilized for creating yet another side branch in the science of nanotubes, namely filled tubes. Both SWNTs and MWNTs have been filled with various materials, such as fullerenes, simple metals, and molecular compounds. Nanotubes with fullerenes inside are called peapods, and are presently intensively investigated. The study of the electronic properties of filled nanotubes is still in its infancy.

Chemistry of Carbon Nanotubes

The chemistry of carbon nanotubes, which encompasses the aspects of purification and functionalization, is a very important field for tailoring their structural and electronic properties and thus for the development of future applications. Purification aims at removing the non tubular material in raw nanotube samples. Several techniques, more or less complex, have been proposed, including solvent treatment followed by ultrafiltration, flocculation using aqueous surfactants, oxidation and acid washing coupled with centrifugation, resuspension in surfactant solution and cross flow filtration steps. These methods, however, lead to a mixture of nanotubes of differing lengths and still containing potential contamination with non tubular material. Polymer suspensions and chromatographic purification are useful to remove non tubular material. Field flow fractionation on purified, shortened nanotubes and size exclusion chromatography on raw nanotubes have demonstrated some success at producing fractions separated by length. Capillary electrophoresis was reported to allow separation based on tube length.

Several methods for attachment of oxygen containing groups to the nanotube surfaces have been developed. Carboxylic, carbonyl and hydroxyl groups wave found covering the nanotube walls. The concentration of acid groups on the surfaces of HNO_3 treated nanotubes was higher than that found for graphite treated under the same conditions. Sonication of nanotubes prior to the functionalization in acids increases the concentration of acid groups. Oxidation of nanotubes with a H_2SO_4–HNO_3 mixture also leads to a higher concentration of functional groups on the surface. Long-term acid treatment of nanotubes also induces modifications. Functional groups can be removed from the surface of nanotubes by heating. It should be noted that SWNTs obtained by different methods exhibit different behaviour in the course of acid treatment and subsequent oxidation. The sensitivity of SWNTs to chemical processing was investigated by TEM.

To enhance the possibilities of atomic force microscopes with nanotube probe tips, the tip ends were functionalized using both solutions and gases. Modification of tips using a discharge in N_2 results in the formation of nitrogen containing heterocycles at the tip ends. High degree of saturation of nanotube tips with functional groups is achieved by treatment with a low-pressure ammonia plasma followed by oxidation with $NaClO_3$ solution.

The reactions with silicon of individual nanotubes or nanotube bundles at 970 °C in ultrahigh vacuum resulted in the formation of Si-β-SiC (nanorod) nanotube heterojunctions.

Oriented or Localized Growth

The term *oriented* is used here to describe the formation of large quantities of CNTs aligned perpendicularly to the substrate surface, usually MWNTs of controlled dimensions (diameter 20–400 nm, length 0.1–60 μm). The key steps are substrate patterning, catalyst deposition and the appropriate use of different CVD routes.

The term *localized* describes the self assembled nanointerconnections made by CNTs between islands on prepatterned substrate. The directed growth of SWNTs parallel to the plane of a silicon substrate was reported. The SWNTs are suspended bridges grown from catalyst material placed on top of regularly patterned silicon tower structures. The key synthetic step in this approach involves developing a series of liquid-phase catalyst precursor materials, allowing for uniform film formation and large-scale catalyst patterning. Contact printing techniques were used to selectively deposit catalyst precursors on top of tac silicon tower arrays. Calcination led to the formation of catalyst particles confined on the tower tops and subsequent CVD (CH_3, 900°C, 20 min) growth yielded SWNTs emanating from the towers. Directed free-standing SWNT networks were formed by nanotubes growing to adjacent towers and becoming suspended above the surface. It is noted that a rational design of tower arrangements could lead to a variety directed SWNT architectures. The liquid phase catalyst precursors consist of three general components— inorganic chloride precursors, a removable triblock copolymer serving as the structure-directing agent for the chlorides, and an appropriate alcohol for dissolution of the inorganic and polymer compounds. The increase in yield was achieved through a so-called conditioning step that uses bulk amounts of catalysts placed near (upstream) the catalyst-patterned substrate in the CVD growth environment.

Electromechanical Properties

Along with their electronic properties, carbon nanotubes have been praised for their mechanical properties. The praise is based on high strength, flexibility and large aspect ration. A carbon nanotube is a freestanding conductor and its mechanical properties are highly relevant for the electronic properties and the potential technological applications of the tube. For example, nanotube based mechanical oscillators may have applications in high-frequency electronic circuits and carbon nanotubes are already used in scanning probe microscopy both as STM and AFM tips. Thin, strong and durable nanotubes come extremely close to an ideal scanning tip.

Mechanical properties of carbon nanotubes are closely related to the properties of a graphite sheet, but the tubular anisotropic form affects the mechanical behaviour. The basis is the graphite sp^2 bond, which is the strongest of chemical bonds. In nanotubes, the overall density of defects can be extremely low depending on the synthesizing method and prevailing synthesizing parameters. This has led to predictions of a very high axial strength.

Basic Mechanical Properties

Theoretical calculation of Young's modulus for individual SWNTs center around 1 TPa or slightly higher, but values as high as 5.5 TPa have been presented. The spread is due to different interaction models and also to differing values of nanotube wall thickness (if an isolated nanotube is not a well-defined quantity). Most of the theoretical attention has been on SWNTs because modelling the interlayer interaction in MWNTs is a complicated matter. Young's modulus values for MWNTs as well as SWNTs range from 0.97 TPa to 1.11 TPa with the value increasing slightly with the number of layers.

The small size of carbon nanotubes presents challenges also for experimental characterization. Nevertheless, measurements have been performed. The first Young's modulus measurement related thermal vibration amplitudes of MWNTs to their Young's modulus and obtained an average value of 1.8 TPa with a large spread. After that, with AFM techniques, values such as 1.28 ± 0.59 TPa for arc-discharge-produced tubes have been obtained. The latest measuring technique is due to Poncharal group, who induced vibrations on MWNTs by alternating potential and have measured the vibration frequencies. Young's modulus values between 0.7 TPa and 1.3 TPa for tubes with a diameter less than 12 nm and between 0.1 TPa and 0.3 TPa for thicker tubes were reported. The drop is explained by an onset of a wave-like bending mode of the nanotube wall at the inner arc of the bending. Measuring SWNTs is more complicated than measuring MWNTs due to their small diameters and the tendency to bundle. Measurement of individual SWNTs using the thermal vibrations method provide an average value of 1.25 TPa. Although the current measurements suffer from inaccuracies due to vibration amplitude measurement and assumptions made on AFM tip characteristics, the current agreement is that defect free nanotubes, both SWNTs and MWNTs have a Young's modulus value around slightly above 1 TPa, which is extremely high and sets nanotubes as the strongest known material albeit challenged by other nanotubular structures such as boron nitride tubes.

Theoretically, nanotube tensile strength is high and this is supported by calculations in which SWNTs support as high as 30% of axial strain before brittle failure and by more recent kinetic activation based calculations that give a yield strain of 17% with a chirality and temperature-dependent defect formation activation energy barriers. Tensile strength values ranging from 11 GPa to 63 GPa are reported. The role of the inner layers is not, however, clear and the failure process may be more complicated. For individual SWNTs, the experimental value of tensile strength is still an open question, but for bundles of SWNTs tensile strength values ranging between a few GPa and several tens of GPa depending on the bundle of measurement characteristic.

As a result of the high tensile strength, nanotubes can bear large strain before plastic deformation or brittle failure. The tubes buckle, flatten, form ripples and generally deform under strain but plastic deformation and relaxation occur only at elevated strain levels and temperatures. An example of simulated nanotube behaviour under strain is presented in Fig. 26.20.

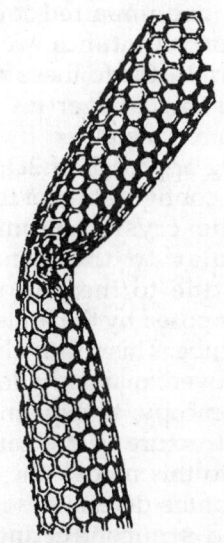

Fig. 26.20: A molecular dynamics simulations based example of nanotube behaviour under strain. The image shows portion of a (14, 0) nanotube of 2632 atoms (tube length is 20 nm) bent 36 degrees with constant bending rate at 300 K temperature and then, to enhance structural reconstruction, annealed in an elevated 3300 K temperature for a sequence of three 130 ps annealing and cooling simulation runs. In the bending process, the tube buckled but remained hexagonally bonded. The defects visible in the image relax the strain and are a result of the third annealing

Defects in Carbon Nanotubes

As in any material, defects play an important role in nanotube properties. Structurally, defects make the tube less strong and thus, in general, defects are not desirable from the purely mechanical point of view. However, they alter the electronic proper-

ties locally, which can be utilized in the creation of single tube devices. Defects are generated in the synthesizing process, and they can also be caused by mechanical manipulation, or for example, by ion or electron beam bombardment of the tube. The most typical structural defects are fivefold (pentagon) and sevenfold (heptagon) rings in the sixfold (hexagonal) lattice. Other types of typical defects are vacancies and miscellaneous bonding configurations such as amorphous diamond. Noncarbon based defects include substitutional atoms or atom groups. In addition to these, MWNTs exhibit diverse defects based on discontinuous inner layers. Defects may alter the tube from a straight tube to a bulging, kinked, spiral, or even more miscellaneous form. Defects and deformations induce scattering and localization, which result in conductivity drops, symmetry breaking based gap opening, and conductivity barriers. The scattering effects of isolated point defects are averaged over the circumference, and therefore their influence decreases with an increasing diameter and is small for all but the very thinnest tubes. However, if the defects or deformations accumulate locally, a tunnelling barrier may be formed. AFM manipulation has been used to make sharp bends in SWNTs, which has resulted in the formation of tunnelling barriers. The created changes are not, however, necessarily permanent. Nanotube defect manipulation offers interesting prospects both for devices with permanent characteristics and for devices with tunable properties, but a breakthrough requires more accurate and flexible mass manipulation of tube properties than is currently available.

Carbon Nanotubes as Mechanical Oscillators

Due to their small size, combined with a low mass density and high Young's modulus, nanotubes are predicted to have characteristic vibration frequencies in the gigahertz regime, which is extremely high for a mechanical vibrator. Such frequencies are not attainable with the current silicon-based micro-mechanical oscillator production techniques with which eigen frequencies in the 10 – 100 MHz regime can be attained. A first approximation for the fundamental modes of a nanotube vibrator can be obtained from elasticity theory and is for a thin cylindrical transversally vibrating rod that is clamped from one end.

$$f_1 = 0.28 \, \frac{d_1}{L^2} \sqrt{\frac{Y}{\rho}} \qquad (26.8)$$

The values for tube diameter $d_1 = 1$ nm, Young's modulus $Y = 1$ TPa and density $\rho = 750$ kg/m^3 are used. The mass density of a SWNT is not a well-defined quantity; the value presented here is a rough estimate based on the nanotube atomic spacial density and the volume encompassed by a tubular cylinder depicting the tube of 1 nm radius, to compare with graphite mass density 1900–2300 kg/m^3, a frequency of 10 MHz is obtained for a tube length of 1 μm and a frequency of 1 GHz for a length of 100 nm. For a rod clamped from both ends, the constant in front of the equation is altered according to the boundary conditions, but the overall dependency on d_1, L, Y and ρ remains. Based on the discussion above, nanotubes and especially SWNTs are capable of significantly smaller dimensions and higher frequencies than conventional silicon-based vibrators.

Currently, only a few studies of excited mechanical vibrations in nanotubes exist. Pancharal group applied alternating potentials to MWNTs and measured Young's modulus and Roulet group have performed a basic study of acoustoelectric coupling in ropes of SWNTs, and they reported dependence on the excitation power, either electron heating or phase coherence breaking is observed.

There are many things in the mechanical properties that have not been measured, and the measurements that exist suffer from a degree of inaccuracy. Most theory concentrates on SWNTs and although the SWNTs, often show more clearly the nanotube characteristic and desirable electronic properties, the theoretical aspects due to multiple layers still have many questions to answer, as do defectous-structure related matters. However, in nanoscale science and technology, electronic and mechanical behaviour will become more intimately connected and obviously, the electromechanical properties of carbon nanotubes will play an important role.

Electronic Structure and Properties

Tight Binding Model: The carbon atoms in nanotubes are sp^2 hybridized. Thus, each atom contributes one unpaired π-electron to the tube. Carbon nanotubes are electronically active materials due to these π-electrons. The basic electronic properties of the carbon nanotubes are calculated by extending the tight binding approach of 2D graphite to the geometrical configuration of the 1D nanotube system. The crystal momentum vector perpendicular to the nanotube axis is quantized due to the periodic boundary condition imposed by the finite circumference of the nanotube. This relatively simple model has been proven, mainly by low temperature STM spectroscopy, to explain well the basic electronic structure of the carbon nanotube. According to this model, the chirality of the carbon nanotube determines its division into metallic and semiconducting tubes. In the model, the curvature of graphitic wall of the nanotube is ignored. This has lately been shown to bring about important modifications to the model. Still the model survives well enough to continue to serve as a basis for classification of the nanotubes into metallic and semiconducting ones.

A brief presentation of the basic tight binding model of the single shell carbon nanotube is as follows. Due to the tubular 1D structure, the nanotube has N separate 1D energy bands. In a general form, these are derived from the graphite energy levels by the general formula

$$E_\mu(k) = E_{g2D} \left(k \frac{K_2}{|K_2|} + \mu K_1 \right)$$

$$\left(\mu = 0, ..., N-1, -\frac{\pi}{T} < k < \frac{\pi}{T} \right) \quad (26.9)$$

where E_{g2D} denotes the energy eigenvalues of a 2D graphite sheet, K_1 and K_2 are reciprocal lattice vectors corresponding to C and T respectively. Figure 26.21 shows how the carbon nanotube K-vectors are projected to the graphite momentum space lattice. In graphite, the conduction and valence bands meet at the so called K points in the momentum space, which is responsible for its semimetallic behaviour. In the tight binding picture, the metallicity of a carbon nanotube is determined by whether the 1D lines of allowed K-vectors run through these points.

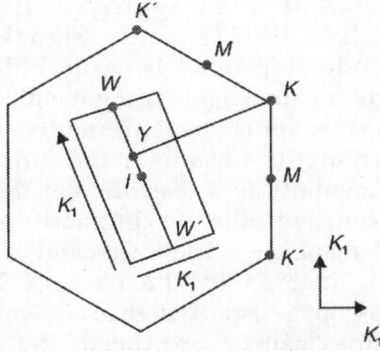

Fig. 26.21: Allowed K-vector or carbon nanotubes are formed by the vectors K_1 and K_2 in the reciprocal lattice space of graphite. The nanotube is metallic if the ratio of YK to K_1 is an integer

In an energy diagram, whether the energy bands will cross the Fermi energy or leave an energy gap depends on (n, m) that is the chirality. The armchair nanotubes, with indices (n, n), have two bands that cross each other and the Fermi energy at the points $k = \pm 2\pi/3a$, they are thus always metallic, while only some of the zigzag tubes are so. The general condition for the metallicity of a nanotube is that the factor $2n + m$ is a multiple of 3, a condition that indeed is fulfilled for the armchair tubes. The theory also predicts that one-third of all nanotubes are metallic while two-thirds are semiconducting. These rates of occurrence have not been confirmed by the presently performed experiments, which however is not surprising considering doping effects and the fact that almost nothing is known about the

possible preference in the growth processes for certain nanotube symmetries over others.

As a general result, in all metallic 1D systems with half-filled bands, the so called Peierls distortion will, by an atomic displacement, change the unit cell so that the band becomes filled and a gap opens up at the Fermi energy. The phenomenon is well known among conducting polymers, charge transfer salts, etc. In carbon nanotubes, however, it has been shown theoretically that the effect very quickly diminishes as a function of nanotube radius.

STM Imaging and Spectroscopy

The electronic structure of the SWNT that is predicted by the tight binding calculations described above was put to test at first primarily by scanning tunneling spectroscopy, done with low temperature STM. In such experiments, it is possible to image the nanotube with atomic resolution along the length of the tube and with a width of a few unit cells. The width of the imaging is, of course, limited by the curvature of the nanotube. This imaging reveals (in the best cases) the structure and diameter of the nanotube, enabling one to determine the (n, m) indices of the nanotube with some confidence. Furthermore, the STM tip be positioned at any point along the nanotube, separated by a tunnelling distance from the nanotube surface, and spectroscopic measurements be performed by taking the current-voltage characteristics. The I–V characteristics reveal the metallic or semiconducting character of the tube since the measured current is proportional to its density of states (DOS). This experimental technique, thus, addresses the fundamental question of the relation between the structure and the basic electronic properties of the individual nanotube. As seen by STM spectroscopy, in the zero bias region the metallic tube has a finite conductance while the semiconducting tube has zero, corresponding to the basic Fermi surface properties. At higher voltages, asymmetric peaks in the conductance appear. These are the Van Hove peaks due to the 1D DOS. The low-temperature STM experiments were carried out with SWNTs, and thus they cover only a very limited range of diameters between 1 and 2 nm. For this range, the measured E_g for semiconducting nanotubes followed roughly the predicted $1/d_1$ dependence, being in the range 0.3–0.7 eV.

In recent scanning tunnelling spectroscopy measurements on SWNTs, a curvature-induced *extra gap* was found in zigzag type metallic nanotubes, but not in tubes of armchair type. This gap is also inversely dependent on the nanotube radius. The measured results were theoretically anticipated in refined calculations. The curvature-induced gap has a size of a few 10s of MeV for subnanometer tube diameters. From this energy scale, we see that for all but the very smallest of possible tubes, the effect is washed out at room temperature.

Other scanning tunnelling spectroscopy experiments have been performed that explore other topics such as individual quantum states in short SWNTs and MWNTs, doped nanotubes, crossing nanotubes, defects and junctions in SWNTs and peapods (SWNTs filled with C_{60}).

26.12 EXPERIMENTAL STUDIES: RAMAN SPECTROSCOPY

Another important tool for nanotube characterization is Raman spectroscopy. Particularly, the strong, so called radial breathing mode is derived from the tubular structure of the nanotube and is absent from all other 2D graphitic forms of carbon. It is widely utilized to give a measure of the nanotube diameter. Due to the singularities in the DOS of the 1D nanotubes and the electron-phonon coupling, very strong resonant Raman effects are observable in studies on carbon nanotubes. This enables the use of Raman spectroscopy to investigate the electronic structure of carbon nanotubes, particularly the occurrence of semiconducting and metallic properties. A possibly very important recent development is the ability to measure the Raman signal from individual nanotubes with the so called micro-Raman technique. It was claimed to be possible to determine the (n, m) indices of the isolated nanotube with this technique. This, of course, opens up the possibility of transport measurements on a nanotube of known symmetry.

Optical Properties

The optical physics of nanotubes has not moved forward in pace with the other developments in the field. This is due to inability to make samples consisting of nanotubes of a single symmetry. Thus, all optical signals are a mixture of signals from a very large variety of nanotubes. Furthermore, since nanotubes tend to aggregate together, especially SWNTs, any signal from semiconducting tubes is easily hidden by interaction with neighbouring metallic tubes. Unlike transport measurements, optical measurements on single nanotubes are experimentally less feasible. On the other hand, however, the 1D singularities in the optical response of nanotubes give strong signals, just as in the case of Raman spectroscopy. Also, if aligned assemblies of nanotubes can be produced, the optical anisotropy of such a system is potentially huge. In a recent development, SWNTs were processed by sonication and surface-active chemicals to produce solutions of truly individually solvated tubes, coated with the surface-active hydrocarbon molecules. These SWNTs were rather short (due to the strong sonication), with an average length of only 130 nm. Perhaps for the first time, photoexcited luminescence from nanotube solutions were observed in this case. The luminescence spectrum was closely associated with the measured absorption spectrum, with only a minor red-shift. This work is intensively being pursued and is likely significantly widen research in carbon nanotube systems. Photoconductivity measurements on single nanotubes have so far not been successful.

The research on the optical properties of carbon nanotube may in the near future bring forward a number of potential applications within optoelectroates, thus adding to the list that already exists within other fields of nanotube based nanotechnology.

Electronic Transport Properties

Within the realm of electronic transport physics, the carbon nanotube offers exciting possibilities both as a testing ground in basic research as well as in nanotechnology. In the former case, the near perfection in structure for length exceeding 1μm makes the nanotube the best concrete example of a 1D quantum wire. On the other hand, in nanotechnology the prospect of devices made from a single nanotube, that is, intrananotube devices, offer the possibility of truly molecular level electronics. Even if molecular electronics is eventually pursued with smaller molecules, carbon nanotubes may be used as connecting wires between metal structures of ~100 nm size and molecules of ~1 nm size.

Although the carbon nanotube is mainly marketed as a perfect 1D object, this is not always the case. In fact, depending on the circumstances one may have to treat the nanotube as anything from a 0D to a 3D object. Obviously, if a SWNT is short enough, it becomes a 0D object or a quantum dot. It only the outer shell of a MWNT is considered, and the diameter of the tube is large enough with respect to some critical length scale of the particular situation, this outer shell becomes a 2D object. Finally, if we again consider a larger MWNT (in diameter), and the charge carrier coupling between the layers is strong enough, then we may have at hand an anisotropic 3D object. We can thus see that even without considering nanotubes of different chiralities and more elaborate topologies than the simple tubular one, the subject of electron transport in carbon nanotubes can be rather complicated.

As has been discussed, the energy gap in semiconducting SWNTs is ~0.5 eV, which is high enough with a good margin for the nanotube to function at room temperature as the active component, e.g. in field effect transistors. Other effects, such as the energy level quantization in short nanotubes with length from 10 nm to 100 nm require cryogenic tempera tures for the effects to be observable. Roughly, it can be said that in metallic nanotubes the interesting transport phenomena occur at low temperatures, while in semiconducting tubes much of the work is carried out at room temperature.

Bulk Transport

In this section, we review the work on the transport physics of single SWNTs, SWNT ropes, and MWNTs. This is clearly more significant than the subject of transport in macroscopic amounts of carbon nanotubes. However, the carbon nanotube may offer interesting applications as the conductive component in composites, when mixed together with an insulating host material. In order to have a composite conducting, the volume fraction of the conductive component has to exceed some critical value. Typically, tile conductive material consists of μm-sized particles of a more or less rounded shape. For conduction to occur, the particles have to touch each other frequently enough so that the conductive channels are formed over macroscopic distances. As prescribed by percolation theory, this occurs at a volume fraction of 16% whereby the conductivity of the composite rises very sharply (as a function of filling percentage) with several orders of magnitude. However, it has already been demonstrated with conductive material consisting of 1D objects, for example fibrils of polyaniline a conducting polymer, that there is no apparent sharp threshold for the onset of conduction, and more importantly, the composite becomes conducting at a much lower volume fraction of the conductive component. Obviously, this should also be achievable with nanotubes as the conductive components. In fact, nanotube-based composite systems are already being considered for commercial applications.

Carbon Nanocones

Theoretical studies predicted the formation of graphitic cones. Subsequently, isolated graphitic cones were produced by carbon condensation on a graphite substrate and by pyrolysis of heavy oils. More recently, single walled aggregates of conical graphitic structures have been prepared by laser ablation of graphite targets. In addition, conical structures consisting of other layered materials such as BN have also been prepared by reacting boron oxide vapours with multiwalled CNTs.

Recently, it is reported that pyrolysis of palladium precursors always produce conical nanofibres. An important feature of these new nanostructures is that they are held together by van der Waals forces since the fibre is composed of an arrangement of stacked cones which can be opened (lamp-shade), or closed. It is believed that nanocones may be good electron field emitters. Calculations of the electronic properties of nanocones reveal that there is a charge accumulation towards the tip and that there are localized states near the Fermi level. These features make them suitable as field emitters.

26.13 PRODUCTION METHODS FOR CNTs

Carbon nanotbes can be obtained by using four different methods: (i) arc discharge (ii) pyrolysis (iii) laser vaporization (iv) electrolysis.

The electric arc discharge setup is shown in Fig. 26.22. This method is very similar to the one used for obtaining fullerenes with two main differences— first, the pressure is higher, around 500 torr (for fullerenes the pressure is around 100 torr) and second, the nanotubes are found deposited in the cathode and not in the soot. This method produces well-ordered

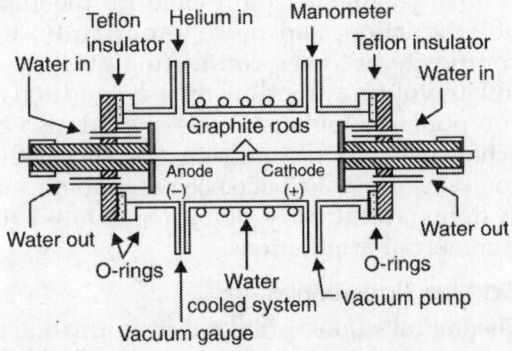

Fig. 26.22: Arc discharge setup

MWCNTs with diameters from 2 nm to 30 nm. The separation between graphite layers is around 3.4 Å, which is very close to that of graphite (3.35 Å) and corresponds to turbostratic graphite. The length of these

nanotubes can be up to 30 μm. The optimum conditions for generating nanotubes with this method consist of using currents of 150 amperes, voltages of 25 volts and graphite electrodes between 6 mm and 8 mm in diameter with a separation between them of 1 mm under helium atmosphere. The electric arc reaction is too violent, so it is very difficult to control the formation reaction.

The pyrolysis method (Fig. 26.23) consists of heating a hydrocarbon or a precursor which contains carbon under the presence of a catalyst such as nickel, cobalt or iron. Prior to the discovery of C_{60}, it was well known that carbon fibres could be obtained by pyrolytic methods using hydrocarbons. We must note that carbon fibres are not nanotubes since the fibres are bigger in diameter and exhibit a great amount of defects and impurities. Nevertheless, the application of carbon fibres is widespread, showing a strength comparable to steel.

In 1994, Ajayan group found that carbon nanotubes could be aligned by embedding

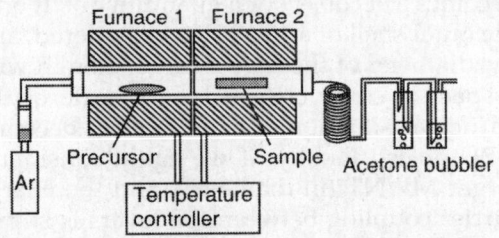

Fig. 26.23: Pyrolysis setup

them in a polymer, thus forming a composite material. When the polymer was cut, they observed an aligned arrangement of nanotubes. In this context, a better alternative is to use laser ablation of cobalt thin films and then pyrolyse them with an organic precursor such as 2-amino 4, 6-dichloro-5-tryazine. The nanotubes obtained in this way are aligned in bundles which can grow up to 40 μm and diameters between 30 nm and 50 nm.

The laser vaporization method involves firing a high power laser towards a graphite target inside a furnace at 1200°C. The condensation of material generated by the laser is responsible for the nanotube formation. If we add nickel or cobalt to the graphite target, we obtain single walled carbon nanotubes (SWCNTs). The SWCNTs show diameters around 14 Å forming a two dimensional crystal with a lattice constant of 17 Å. Arrangements of SWCNTs can also be obtained by electric arc discharge with mixtures of carbon, nickel and yttrium.

The electrolysis method is based on using graphite electrodes inside a molten salt such as lithium chloride (700°C) under an argon atmosphere. Depending on conditions, 20% to 30% consists of nanotubes. The depth of the cathode and the current (3–10 amps) play an important role in the formation of nanotubes.

26.14 KEY ISSUES IN NANOMANUFACTURING

Similar to large scale fabrication, nanomanufacturing research issues revolve around precursor materials; fabrication processes and characterization techniques; instrumentation and equipment; theoretical modelling and control; and design and integration of structures into devices and systems.

26.15 NANOSTRUCTURES IN MOTION

Micromechanics, micromachines and moving nanostructures represent a new *microrevolution*. However, the concepts related to micromechanics and micromachining of silicon are not new. A number of commercial products including accelerometers, pressure transducers, thermal print heads and ink jet nozzle arrays have been fabricated for internal industrial use and for the open market.

Since 1982, the field of micromechanics has expanded considerably, adding micrometers movable linkages, gears, microturbines, electrical drives and microweezers to an ever growing suite of micromechanical devices.

26.16 NANOMATERIAL ADVANTAGE

Nanocrystalline materials are three dimensional solids composed of nanometer-sized grains or crystallites. Because of their unique structure, which is characterized by ultrafine grains and a rather high density of crystal lattice defect, these materials have extraordinary fundamental properties that could be exploited to make next-generation superstrong materials.

Strength and ductility are the central mechanical properties of any material. They are determined by the physical nature of plastic deformation, which in conventional, coarse-grained metals is mainly carried by dislocations-line defects of regular crystal lattice within individual grains. Studies suggest, however, that the mechanism of plastic deformation in nanocrystalline materials may be different and so may lead to novel mechanical properties.

High strength, or hardness has already been observed in many nanomaterials. But in most cases, such nanomaterials have very low ductility; they fail when their shape is changed. Some even become brittle when a force or deforming load is applied. Strength and ductility are usually opposing characteristics; higher the strength, lower the ductility and *vice versa*. This correlation is associated with the nature of plasticity; the more difficult it is for dislocations to appear and to move, the stronger but brittle and less ductile is any crystalline material.

Very recently, Wang and coworkers reported their success in engineering— an unusual nanocrystalline material that combines two useful properties that are often mutually exclusive strength and ductility. They created a copper nanostructure by rolling the metal at temperature below 77 K and then heating it to around 450 K. The result was a *bimodal* structure of micrometer-sized grains (at a volume fraction of around 25%) embedded in a matrix of nanocrystalline grains. The material showed extraordinary high ductility, but also retained its high strength. The reason for this behaviour seems to be that while the nanocrystalline grains provide strength, the embedded larger grains stabilize the tensile deformation of the material.

Processing nanomaterials to achieve both strength and ductility can enhance the material's resistance to fatigue. More than anything else, it is fatigue that at present limits the lifetime, and hence the range of applications of many advanced materials. Ethics is the study of understanding what is good and what is bad. As nanotechnology steps into center stage and the bounty, its promises begin to become reality; it will raise several issues of ethics, public policy, law and social responsibility. When we try to understand a new technology like the nanotechnology, it is very natural to think about the good and the bad effects of such technology. Any technology can be used for constructive as well as destructive use. This is also true for the nanotechnology. Nanotechnology as of now is still in its infancy. A variety of applications are being proposed and explored. A majority of these are in the areas related to computer technology, defence, electronics, medicine and consumer products like cosmetics. These applications of nanomaterials can have different impacts on mankind and the world around. It is quite natural to think about the good and the bad effects of nanotechnology. When we think about such effects, we will be in a position to decide whether to take nanotechnology in one direction or the other. We shall discuss the good and the bad about nanotechnology and the ethical issues it can raise.

26.17 APPLICATIONS OF NANOTECHNOLOGY

We would not like to focus exclusively on potential dangers of nanotechnology. Let us begin by considering some of the good things that the nanotechnology can provide. For example, nanotechnology helps to clean up the environment. Paints based on nanomaterials could convert sunlight into usable energy and remove CO_2 from the atmosphere. There are other materials which could remove other pollutants from the air. Lighter but stronger materials could be developed from designer molecules. Aircraft made up of lighter materials with the strength of diamonds would be more fuel-efficient and safer. Clothing made up of stronger materials would last longer.

In future, nanobots (nanoscale robots) would travel through blood vessels clearing away plaque and entering cancerous cells to destroy them. Nanotechnology is expected to manufacture food and clean water

economically. Computer chips may be made at lower cost from chemical synthesis avoiding toxic by-products. Obviously, nanotechnology is quite promising. Scientists have started understanding nanoscience and would like to exploit it to the benefit of mankind. Human life can be made healthier and safer by the use of nanoscience and nanotechnology. *The potential benefits of nanoscience and nanotechnology are immense but the potential hazards are equally serious.*

Today, nanotech is making its way into our lives, producing spill- and wrinkle-resistant clothing, lighter and stronger sports equipment, self cleaning windows that break down dirt, etc. We will continue to see improvements in everyday products and advances in industries such as electronics. For instance, nanotechnology is helping to develop organic light-emitting diodes that can emit their own light, such as a glow worm. These would be brighter, faster, lighter, more energy efficient and would enable ultrathin displays viewable from any angle without loss of quality. Nanotechnology will be useful in many other ways in the near and distant future. The following are some of the examples:

- *Transportation*: New materials will make vehicles lighter and stronger. These vehicles will use less fuel and can withstand more damage. Space travel will become inexpensive and no longer limited to the elite few.

- *Computers*: As processing chips become smaller, faster and more powerful, computers will also shrink in size, allowing them to be truly ubiquitous and embedded in clothing and even in the human body.

- *Military*: Soldiers will wear an exoskeleton that can change its flexibility and become instant armor. Clothes will be able to store energy to deliver super-human strength when needed, such as for jumping over a 20 ft wall. Weapons can be miniaturised, and smart bullets can be programmed to hit specific targets.

- *Energy*: More powerful and smaller batteries will enable devices to become smaller and operate longer. Efficient molecule-sized solar cells can be mixed into road asphalt to continually harness the sun's energy. Fossil fuels and coal will be replaced with renewable energy sources, even turning garbage into fuel.

- *Environment*: One may be able to rebuild our thinning ozone layer and clean up the environment with nanobots that eat oil spills and other contaminants. More efficient manufacturing processes mean less pollution. We would not need to cut trees to produce paper anymore.

- *Medicine*: Surgeries will be performed with tools more than 1000 times precise than the sharpest scalpel today. Cosmetic surgery would not even require surgery—eyes can perhaps change colour and noses can change shape without cutting. One will be able to use medical monitoring devices to detect diseases. New drugs can target specific cancer cells and viruses to destroy them. They can also fix defective genes or otherwise deliver precise treatment. To the extent that getting old is like any other physical disorder, repairing cells slow, halt or reverse the ageing process.

- *Distant future*: Nanotechnology enables researchers to build materials atom by atom. Scientists, therefore, predict that they will be able to manufacture anything. Among other things, they can make food from dirt and water, eliminating famine worldwide.

26.18 LIMITATIONS OF NANOTECHNOLOGY

There is no doubt that there are many positive benefits from nanotechnology, but what about the bad effects of nanotechnology? Is nanotechnology ultimately worth the cost, or is the price too high. Let us examine some of the negative consequences predicted for nanotechnology.

- *Medical and health*: Nanoparticles have been shown to be absorbed in the livers of research animals and even cause brain damage in fish exposed to them within 2 days. If nanoparticles find their way into living cells, they can enter the food chain through bacteria and pose a health threat, e.g. mercury in fish, pesticides in vegetables, or hormones in meat. Carbon nanotubes look very much like asbestos fibres, but what happens if they get released into air? Being carbon-based, they would not be detected and obviously they are repelled by the usual defence mechanism of our bodies.

- *Environment*: Since nanomaterials are as strong as diamonds, question arises, how decomposable will they be? Will they litter our environment further or pose other disposal problems like nuclear waste and nuclear hazards or space litter? In the distant future, will self replicating nanobots be necessary to create the trillions of nanoassemblers needed to build any kind of product-run amok, spreading as quickly as virus?

- *Privacy*: With the shrink in size of products, eavesdropping devices too can become invisible to the naked eye and more mobile. This makes it easier to invade our privacy. If nano devices are planted into our bodies, mind controlling nanodevices may be able to affect our thoughts by manipulating brain processes.

- *Terrorism*: Capabilities of terrorists also advances with advances in capabilities of military. Thus, weapons become more powerful and portable. Obviously, these devices can also be turned against us. Nanotech may create new, unimaginable forms of torture causing disintegration of a person at the molecular level. Radical groups or terrorists could let loose nanodevices targeted to kill a specific person.

- *Society*: With all these limitations, several experts advocate a strong system to regular and monitor nanotechnology developments. Nanotechnology is a multi-disciplinary science. Nanotechnology could be used to restore environment to spread wealth, and to cure more illness. This depends on our action, working within the limits set by the real world.

Nanotechnology has become a very active and vital area of research, which is fast developing in industrial sectors and spreading to almost every field of science and engineering. Control of matter on a molecular scale is the main aim of all researchers working in this field. Fundamental physical, mechanical, magnetic and biological properties of materials are improved remarkably as the size of their constituent grains decreases to a nanometer scale. Nanostructured materials and associated processing technologies have opened up exciting new possibilities for future applications in aerospace, automotives, electronics, coatings, non volatile memories, sensors, acuators, optoelectronics, drug delivery, etc. One of the difficult problems facing the design of nanostructures-based stems is understanding how to interconnect and address them. No doubt, the success of nanoscience will depend on the development of new device and manufacturing technologies.

GLIMPSES

1. *Nanoscience* is, at its simplest, the study of fundamental principles of molecules and structures with at least one dimension roughly between 1 and 100 nanometers.

   These structures are known as nanostructures.

2. *Nanotechnology* is the application of nanostructures into useful nanodevices. Nanotechnology has applications in electronic devices including information storage retrieval, as well as life sciences.

3. *Molecular nanotechnology* is the self assembly of the molecules into an ordered and functional structure.

4. *Carbon nanotubes* (CNTs) are seamless graphene tubule structures with nanometer-sized diameters and high aspect ratios. This new classes of one-dimensional materials is shown to have exceptional chemical, thermal and novel electronic properties. Nanotubes are categorized as single walled nanotubes (SWNTs) and multiwalled nanotubes (MWNTs). SWNTs has just the single shell of hexagons, whereas in MWNTs, multiple layers are nested like the rings in a tree trunk.

5. *Nanotubes* have very high electrical and thermal conductivities, along with stability up to very high temperatures.

6. *The electronic* properties of nanotubes have been used to fabricate nanoscale transistors.

7. Another exciting nanotube application is in flat panel displays for televisions and computer monitors.

8. *Quantum dots* are nanometer-sized semiconductor crystals or electrostatically confined electrons. They are fabricated in semiconductor materials and have typical dimensions between nanometers to a few micron (10^{-6} m). These can be easily connected to electrodes and are therefore excellent tools to study atomic like properties.

9. *Nanowires* are structures that have a thickness or diameter constrained to tens of nanometers or less and an unconstrained length. Many different types of nanowires exit, including metallic (e.g. Ni, Pt, Au), semiconducting (e.g. Si, InP, GaN, etc.), and insulating (e.g. SiO_2, TiO_2). Molecular nanowires are composed of repeating molecular units either organic (e.g. DNA) or inorganic (e.g. $Mo_6S_{9x}I_x$).

REVIEW QUESTIONS

1. What is nanoscience and nanotechnology? What is the difference between these two?

2. Explain the working of SEM and TEM. Compare the two.

3. Discuss the impact of nanoscience in various fields. Give examples of some natural phenomena where nanoscience plays an important role.

4. What are quantum dots? How does the frequency of the light emitted from a quantum dot depends on the size of quantum dot?

5. What are nanowires? In what respects the electrical properties of nanowires different from conventional wires?

6. Discuss the mechanical properties of nanomaterials. How can these be useful in various applications?

7. Describe the sol-gel method of fabrication of nanomaterials.

8. How nanotechnology will affect the life of a common man?

9. What are carbon nanotubes? Explain their electrical properties.

10. Mention important properties of nanoparticles.

11. How would one make and stabilize quantum dots?

12. What are the different types of quantum dots investigated? What makes quantum dots luminescence attractive?

13. What are the important concerns of materials scientists in the nanoscience area?

14. How many types of nanotubes are there? Mention their characteristics.

15. What are the main allotropic forms of carbon? Compare their important properties.

16. What are the various techniques used to synthesize nanostructured materials/CNTs? Discuss one of them briefly.

17. From cars to medicine, nanotubes may be a miracle material . They are stronger than steel and as flexible as plastic, conduct energy better than almost any material ever discovered and can be made from unexotic raw materials such as methane gas". Comment on the above statement.

SHORT ANSWER QUESTIONS

1. What are nanomaterials?

Ans. These are the materials containing nanocrystals, i.e. their grain size is in the range of 1 to 100 nm. Nanomaterials may be metals, alloys, intermetallics or ceramics.

2. What do you understand by nanotechnology?

Ans. Nanotechnology deals with various structures of matter having dimensions of a billionth of a meter.

3. What is nanoscience and nanotechnology?

Ans. Nanoscience and nanotechnology primarily deal with the synthesis, characterization, exploration and exploitation of nanostructured materials characterized by at least one dimension in the nanometer (1 nm = 10^{-9} m) range.

4. How is nanoworld uniquely different?

Ans. The physical and chemical properties of nanostructures are distinctly different from those of a single atom (molecule) and bulk matter with the same chemical composition. These differences between nanomaterials and the molecular and condensed phase materials pertain to the spatial structures and shape, phase changes, energetic, electronic structure, chemical reactivity and catalytic properties of large, finite systems and their assemblies.

5. What are the most significant applications of nanomaterials?

Ans. Perhaps in nanodevices and nanoelectronics.

6. How is nanoscience interdisciplinary?

Ans. Nanoscience is truly interdisciplinary area covering physics, chemistry, biology, materials and engineering.

7. What is a carbon nanotube?

Ans. Carbon nanotubes are long cylinders of 3-coordinated carbon, slightly pyramidalized by curvature from the pure sp^2 hybridization of graphene, toward the diamond like sp^3. Infinitely long in principle, a perfect tube is capped at both ends by hemifullerenes, leaving no dangling bonds. A single walled carbon nanotube (SWNT) is one such cylinder, while multiwalled tube (MWNT) consists of many nested cylinders whose successive radii differ by roughly the interlayer spacing of graphite.

8. What are the advantages of MWNTs over SWNTs?

Ans. Multiwalled tubes have two advantages over their SWNTs cousins. The multishell structure is stiffer than the single wall one, especially in compression. Large scale syntheses by enhanced chemical vapour deposition (CVD) processes are many, while for SWNTs, only the Rice's HiPco process appears to be scalable.

9. What are the important concerns of materials scientists in the nanoscience area?

Ans. i. Nanoparticles or nanocrystals of metals and semiconductors, nanotubes, nanowires and nanobiological systems.

ii. Assemblies of nanostructures (e.g. nano-crystals and nanowires) and use of biological systems such as DNA as molecular nanowires and templates for metallic or semiconducting nanostructures.

iii. Theoretical and computational investigations that provide the conceptual framework for structure, dynamics, response and transport in nanostructures.

iv. Applications of nanomaterials in biology, medicine, electronics, chemical processes, high strength material, etc.

10. What are the present goals of the science and technology of nanomaterials?

Ans. To master the synthesis of nanostructures (nano-building units) and their assemblies of desired properties; to explore and establish nanodevice concepts, to generate new classes of high-performance nanomaterials, including biology inspired systems; and to improve techniques for the investigation of nanostructures.

11. What are nanocomposites?

Ans. Nanocomposites are the heterogeneous materials composed of at least one type of nanometer-sized particles.

12. What are van der Waals forces?

Ans. The van der Waals forces represent the weak (secondary) interaction between adjacent molecules or atoms and arise from their electric dipoles. The van der Waals forces are long-range (even for distances larger than tens of nanometer), and may be attractive or repulsive depending on the distance between interacting atoms or molecules. In particles, the van der Waals forces between nanotubes become attractive when the distance is larger than the equilibrium distance or repulsive when the distance is smaller than the equilibrium distance. The van der Waals forces between nearest nanotubes play a crucial role in physical properties of multiwalled carbon nanotubes and single wall carbon nanotubes ropes.

13. How many kinds of CNTs other than SWCNT and MWCNT have been developed so far?

Ans. i. Carbon nanocoil having unique 3D structures.

ii. Carbon nanofibres (CNFs): These are cylindrical or conical structures having length in the order of a few microns and diameter varying from 10 to 200 nm. The internal structure of CNFs varies and comprises different arrangements of modified graphene sheets.

iii. Carbon nanocone (CNCON): This has a perfect conical structure with symmetry different from other known carbon materials, including nanotubes, nanocoil and bucky balls.

iv. Carbon nanohorn (CNH): This is made of the same graphite structure as CNT. CNH is a single wall carbon cone having structures similar to those of nanotube caps.

v. Carbon nanoribbon (CNR): This is strip of a graphene sheet with morphology different from CNTs. It is highly graphitized and width of ribbon is in the range of 1 to 20 nm and length is in microscopic scale.

14. What is nanoelectronics?

Ans. This is the field of science and engineering dealing with the fabrication, study and application of nanosize electronic devices for information technologies. Generally, quantum effects determine the operational principles of these devices.

15. What is nano-fish memory device?

Ans. This is similar in structure to a MOSFET (MOS field effect transistor) except that it is a three terminal device with two gate electrodes, one on the top of the other. The top electrode forms the control gate, below which a floating gate is capacitively coupled to the control gate. The memory cell operation involves putting charge on the floating gate or removing it, corresponding to two logic levels. These devices utilize single or multiple nanoparticles as the charge storage element.

16. What do you understand by nanophotonics?

Ans. Nanophotonics is the field of science and engineering dealing with the fabrication, study and application of optical phenomena at the nanoscale. It deals with structures and process which are spatially localized in domains smaller than the wavelength of visible radiation. The field includes nanoscale confinement of radiation, nanoscale confinement of matter, nanoscale physical or photochemical transformations.

17. What are quantum wires?

Ans. These are 1D structures. As compared to quantum films, one more dimension of the structure appears to be so small as provide quantum confinement. Charge carriers are free to move only along the wire. Thus, only one kinetic component along with the confined energy contributes to the total energy. As a consequence, the density of states has $E^{1/2}$ dependence for each of the discrete pairs of states in the confined directions.

18. What are quantum dots?

Ans. These are 0D structures in which the carriers are confined in all three directions. Their energy state is quantized in all directions and the density of states is represented by series of discrete, sharp picks resembling that of an atom. Quantum dots are also called as artificial atoms. These are usually formed by a definite number of atoms. They are typically represented by atomic clusters or nanocrystallites. Considering the coupling of quantum dots with external electric circuits, they can be classified as open dots and almost isolated or closed quantum dots. In an open dot, the coupling is strong and the movement of electrons across dot-lead functions is classically allowed. When point contacts, connecting a dot with an external circuit, are pinched off, effective barriers are formed and conductance occurs only by tunnelling. These are almost isolated or closed quantum dots.

19. What do you understand by bucky ball, C_{60}?

Ans. This is a hollow cage carbon molecule consisting of 60 carbon atoms and has resemblance of its molecular structure to geodesic domes.

20. What is density matrix?

Ans. A representation of mixed states generated by right-multiplying the state vector by its adjoint.

21. What is lithography?

Ans. Literally, lithography means *stone writing*. This refers generally to any technique for forming a patterned structure on a solid surface.

22. What is photolithography?

Ans. This is a photochemical technique for etching specific patterns using projected light. It is most widely used technique today.

23. What is a nanocomputer?

Ans. A computer whose characteristic length scale is between $10^{-7.5}$ m and $10^{-10.5}$ m (i.e. between ~32 and ~ 0.32 nm, i.e. closer to 1 nm than to 1 micrometer or 1 picometer on logarithmic scale).

24. What are smart materials?

Ans. These are usually defined as those that can sense external stimuli and then respond to those stimuli by some means in real (or near real) time. Such materials are therefore a product that lies somewhere between a material and more complex nanodevice – a sensor, in fact. Some smart materials respond to light, others to pressure and so on. Some are self repairing. These materials find widespread applications today. These materials should be recognized along with the familiar metals, plastics, ceramics, composites, powder metals and special type and multifunctional materials, shape metal alloys (SMAs), michromechanical systems (MMSs), functionally graded materials (FGMs) and nanomaterials). By using these materials instead of adding mass, engineers can endow structures with built-in responses to a myriad of contingencies. In their various forms, these materials can perform as actuators, which can adapt to their environments by changing characteristics such as shape and stiffness, as sensors which provide the actuators with information about structural and environmental changes, smart materials are beginning to play an important role in civil engineering drawings for dams, bridges, highways and buildings. There are many more applications of these materials in process.

OBJECTIVE QUESTIONS

1. The prefix nano in the word nanotechnology means a _____ (1×10^{-9}). [billionth]

2. Nanostructures constitute a bridge between molecules and infinite _____ systems. [bulk]

3. The structure of nanoparticles of CdS, CdSe and such materials is affected by _____. [size]

4. The synthesis of nanomaterials includes control of size, shape and _____. [structures]

5. The most significant applications of nano-materials may be in nanodevices and _____. [nanoelectronics]

6. The skeleton of C_{60} consists of 20 hexagonal and _____ pentagonal rings fused all together. [12]

7. Composite materials reinforced with carbon or graphite are often used in _____ goods. [sporting]

8. The discovery and rapid evolution of carbon nanotubes has played a major role in triggering the explosive growth of _____ in nano-technology. [research and development]

9. Graphene refers to a nanolayer of sp^2 bonded _____ atoms. [carbon]

10. Carbon nanotubes are hollow cylinders of _____ sheets. [graphite]

11. A carbon nanotube made of single graphite layer rolled up into a hollow cylinder is called is _____. [SWNT]

12. A tube comprising several concentrically arranged cylinders is referred to as _____. [MWNT]

13. Multiwalled nanotubes have similar lengths to single walled tubes, but much larger _____. [diameters]

14. Nanotubes can behave like metals or _____. [semiconductors]

15. Graphene is a two dimensional honeycomb net, made up of _____ bounded carbon atoms. [sp^3]

16. Armchair tubes always show _____ properties. [metallic]

17. The fundamental band gap in semiconducting nanotubes ranges from _____ eV to _____ eV being dependent on small variations of the diameter and bonding angle. [0.4, 0.7]

MULTIPLE CHOICE QUESTIONS

1. Grain size of nanocrystals are in the range
   (a) 1 to 100 nm
   (b) 100 to 1000 nm
   (c) 10000 to 100000 nm
   (d) 1000 to 10000 nm

2. Nanomaterials are said to have
   (a) low strength
   (b) low toughness
   (c) less brittle
   (d) high strength, hardness, formability, toughness and more brittle

3. Nanomaterials exhibit
   (a) super plasticity
   (b) superconductivity
   (c) semiconducting properties
   (d) all the above

4. At nanoscale, the surface-to-volume ratio is
   (a) very high   (b) very low
   (c) 1 : 1   (d) 1 : 5

5. Carbon nanotubes are about 100 times stronger than steel but just
   (a) quite heavier than a steel piece of same size
   (b) about 1/6th of the weight of steel
   (c) about 30 times heavier than that of steel
   (d) equal in weight

6. Armchair, zigzag and chiral are the names of the different types of
   (a) single walled carbon nanotubes
   (b) multiwalled carbon nanotubes
   (c) buckyballs
   (d) nonw of the above

7. C_{60}, a turncated icosahedron consists of
   (a) 6 pentagons and 12 hexagons
   (b) 20 pentagons and 12 hexagons
   (c) 12 pentagons and 6 hexagons
   (d) 12 pentagons and 20 hexagons

8. In an armchair carbon nanotube, the relation between n and m is
   (a) $n \neq m$ (b) $n = m$
   (c) $n > m$ (d) $n < m$

9. The commercial method for the synthesis of carbon nanotubes is
   (a) high pressure codeposition
   (b) chemical vapour deposition
   (c) laser ablation technique
   (d) arc discharge method

10. For a zigzag carbon nanotube,
    (a) $m = 5$ (b) $m = 0$
    (c) $n = 0$ (d) $n = m$

11. Chemical vapour deposition technique in nanotechnology is used for
    (a) identification of nanoparticles
    (b) synthesis of carbon nanotubes
    (c) characterization of nanoparticles
    (d) determination of the size of nanoparticles

12. Properties of materials change remarkably as their size approaches to
    (a) macroscale   (b) microscale
    (c) nanoscale   (d) none of the above

Answers

1. (a)	2. (d)	3. (d)	4. (a)	5. (b)
6. (a)	7. (d)	8. (b)	9. (b)	10. (b)
11. (b)	12. (c)			

Appendix I

Part A

Composites

A.1 INTRODUCTION

Many of modern technologies require materials with unusual combinations of properties that cannot be met by the metals, conventional metallic alloys, ceramics and polymeric materials, e.g. materials needed for aerospace, underwater and transportation applications. For example, engineers working in aircraft industry are looking and searching for structural materials that have low densities, are strong, stiff, and abrasion and impact resistant, and are not easily corroded. Obviously, this is a rather formidable combination of characteristics. Usually strong materials are relatively dense; also increasing the strength or stiffness generally results in a decrease in impact strength. Therefore, in composites the whole is greater than the sum of the parts (synergy). Wood, celery, bamboo, and corn are all examples of nature's composites, where two materials combine to reinforce and bind together. We have already discussed in previous chapters, composites of sort which include multiphase metal alloys, ceramics and polymers. A composite is considered to be any multiphase material that exhibits a significant proportion of the properties of both constituent phases such that a better combination of properties is realized. This is termed the *principle of combined action*. According to this principle, better combinations are fashioned by the judicious combination of two or more distinct materials.

All composites generally have one thing in common: a *matrix* or *binder* combined with a reinforcing material. Obviously, a composite consists of a matrix material, dispered within which is a dispersion of one or more phases of another material. If the fibres are directionally oriented and continuous, the material is termed an *advanced composite*. Reinforced concrete is a good example of composite material. When concrete is reinforced with steel rebar, it becomes the matrix, which surrounds the reinforcing fibre, the rebar. Another example is reinforced fibre glass products such as fishing rods. Glass fibres are set in a thermosetting resin matrix. This produces a strong, lightweight, flexible fishing rod. Other fibres are produced from aramid (Kevlar and Nomex), boron, carbon, graphite and ultrahigh molecular weight polyethylene (spectra). The matrix for these materials is typically a thermosetting epoxy resin. These materials provide some exceptional increase in mechanical properties, sometimes three to six times greater than steel. Another example of a composite material is pearlitic steels. The microstructure of this material consists of alternating layers of α-ferrite and cementite. The ferrite phase is soft and ductile, whereas cementite is hard and very brittle. The combined mechanical characteristics of this composite, i.e. pearlite (reasonably high ductility and strength) are superior to those of either of the constituent phases. Other common varieties of composites include combinations— fibre-resin, fibre-ceramic, carbon-metal, metal concrete, metal resin and wood plastic. Most of the contemporary advanced composites use glass, kevlar (an aramid), or one of the various types of graphite fibres. *Fibre composites* form an important subset of this class of engineering materials. Obviously, a composite is a multiphase material that is artificially made, as to one that occurs naturally. In addition, the constituent phases in a composite must be chemically dissimilar and separated by a distinct interface. This is why most metallic alloys and many ceramics do not fit this definition because their multiple phases are formed as a consequence of natural phenomena.

There are many reasons for making composites— the incorporation of fibres into brittle ceramics produces a composite of enhanced toughness. Fillers, such as the presence of aggregate in concrete, reduce the overall cost of the product, and additionally improve the compressive strength. The second phase may, furthermore, be a gas, as in the manufacture of foamed products of low density.

On the basis of strength and stiffness alone fibre reinforced composite materials may not be superior to metals of comparable strength, but when the specific modulus (i.e. modulus per unit weight) and specific strength are considered, then their use implies that the weight of components can be reduced. This is an important factor in all forms of transport where reductions of weight result in greater energy savings.

In order to produce a new generation of extraordinary materials, scientists and engineers while designing composite materials have ingeniously combined various metals, ceramics and polymers.

One can classify composite materials as per simple scheme shown in Fig. A.1. There are three main divisions— particle-reinforced, fibre-reinforced and structural composites. We note that there exist at least two subdivisions for each main division.

The dispersed phase for particle reinforced composites is equiaxed (i.e. particle dimensions are approximately the same in all directions) and the dispersed phase for fibre reinforced composites has the geometry of a fibre (i.e. a large length-to-diameter ratio). Structural composites are combinations of composites and homogeneous materials.

Examples of these three groups include concrete, a mixture of cement and aggregate, which is a particulate composite; fibre glass, a mixture of glass fibres imbedded in a resin matrix, which is a fibre composite; and plywood, alternating layers of laminate veneers, which is a laminate composite. These are discussed in detail in section A.3.

A.2 GENERAL CHARACTERISTICS

i. Composite materials are superior to all other known structural materials in specific strength and stiffness, high temperature strength, fatigue strength

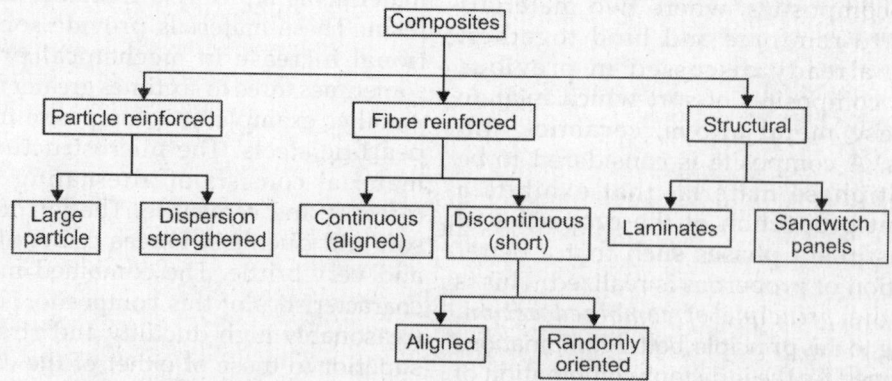

Fig. A.1: A simple classification scheme for the various composite types

and other properties. The desired combination of properties can be tailored in advance and realized in the manufacture of a particular material. Moreover, the material can be shaped in this process as close as possible to the form of final products or even structural units.

ii. Composite materials are complex materials whose components differ strongly from each other in the properties, are mutually insoluble or only slightly soluble and divided by distinct by distinct boundaries.

iii. The principle of manufacture of composites has been borrowed from nature. Trunks and stems of plants and bones of man and animals are examples of natural composites. In wood, cellulose fibres are bonded by plastic lignin; in bones, thin and strong fibres of phosphates are bonded by plastic collagen.

iv. The properties of composites mainly depend on the physico-mechanical properties of their components and the strength of bonds between them. A characteristic feature of composite materials is that the merits of their components are fully utilized. Composite materials may acquire certain valuable properties not found in the components. For obtaining the optimal properties in composites, their components are chosen so as to have sharply different, but complementary properties.

v. The base or matrix, of composites may consist of metals or alloys (metallic composites), polymers, carbon and ceramic materials (non metallic composites).

vi. The matrix is essentially the binding and shaping component in composites. Its properties determine to a large extent the process conditions for the manufacture of composite materials and the important operating characteristics, e.g. working temperature, fatigue strength, resistance to environmental effects, density and specific strength. Some composites have a combined matrix which consists of alternating layers (two or more) of different composition.

vii. Composites with combined matrix may be called *multi-matrix* or *multi-layer* composites (Fig. A.2a). Multimatrix composites can be characterized by a wider spectrum of useful properties. For example, use of titanium as an addition to aluminium may increase the strength of a composite material in directions transverse to fibres. Aluminium layers in a matrix diminish the density of composite material.

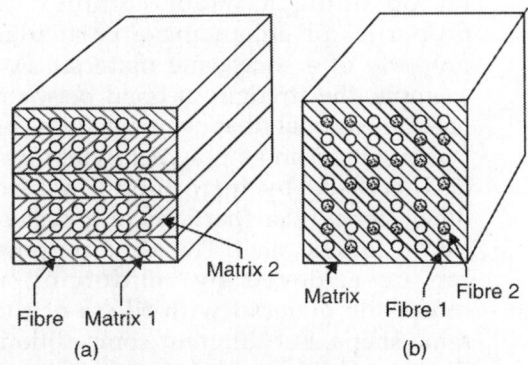

Fig. A.2: Schemes of (a) Multi-matrix (b) Polyfibre composite materials

viii. Fillers, i.e. other components are uniformly distributed in a matrix. These play the major part in strengthening of composites, and thus they are called *strengtheners*. Fillers should possess high values of strength, hardness and elastic modulus. These characteristics should be substantially higher than those of the matrix. With an increase of the elastic modulus and ultimate strength of a filler, the corresponding properties of a composite material also increase, but do not reach the value of the filler. Fillers are alternatively called reinforcing components. This is a broader term than strengthener; it does not specify the particular strengthening role of filler which may be used for improving other properties of a composite.

The properties of a composite material can also depend on the shape (geometry), dimensions, concentration and distribution of filler (reinforcement pattern).

With regards to their shape, fillers are divided into three main groups: (i) zero dimensional (ii) one dimensional (iii) two dimensional.

Based on the reinforcement pattern, composite materials are divided into three groups: with uniaxial, biaxial and triaxial reinforcement.

ix. Fillers of different shapes may be used for obtaining a wider complex of properties or enhancing a particular property of a composite material. For example, the strength of bond between one dimensional filler elements (glass or carbon fibres) and a polymer matrix can be increased by introducing a zero dimensional filler (particles of asbestos, silicon carbide, etc.). The same purpose can be achieved by reinforcing a composite material with fillers of the same shape, but different composition. For example, the modulus of elasticity of composite materials with a polymer matrix reinforced by glass fibres can be increased by additional reinforcement with boron fibres.

Composite materials containing two or more different fillers are termed *complex-reinforced composites* (Fig. A.2b).

A.3 PARTICLE-REINFORCED COMPOSITES

These can be further classified under two subgroups: (i) *large-particle* (ii) *dispersion-strengthened composites*. The distinction between these is based upon reinforcement or strengthening mechanism. The term *large* indicates that particle-matrix interactions cannot be treated on the atomic or molecular level rather continuum mechanics is used. The particulate phase for most of these composites is harder and stiffer than the matrix. In the vicinity of each particle, these reinforcing particles tend to restrain movement of the matrix phase. Obviously, the matrix transfers some of the applied stress to the particles, which bear a fraction of the load. We may note that the degree of reinforcement or improvement of mechanical behaviour depends on strong bonding at the matrix particle interface.

Particles for dispersion-strengthened composites are normally much smaller (diameter between 0.01 and 0.1 μm). Particle-matrix interactions occur on the atomic or molecular level and lead to strengthening. We may not that the mechanism of strengthening is similar to that for precipitation. The matrix bears the major portion of an applied load, whereas the small dispersed particle hinder or impede the motion of dislocations. Obviously, plastic deformation is restricted such that yield and tensile strengths as well as hardness improve.

Large-particle Composites

Fillers added to some polymeric materials produce large-particle composites. The fillers modify or improve the properties of the materials and/or replace some of the polymer volume with less expensive filler material.

Concrete is another familiar example of large-particle composite. Concrete is composed of cement (the matrix), and sand and gravel (the particulates).

We may note that particles can have quite a variety of geometries, but they should have approximately the same dimension in all directions (equiaxed). Particules should be small and evenly distributed throughout the matrix for effective reinforcement. Moreover, the volume fraction of the two phases influences the behaviour; mechanical properties are enhanced with increasing particulate content. The *rule of mixtures* equations reveal that the elastic modulus should fall between an upper bound given by

$$E_c(u) = E_m V_m + E_p V_p \tag{A.1}$$

and a lower bound or limit,

$$E_c(t) = \frac{E_m E_p}{V_m E_p + V_p E_m} \tag{A.2}$$

where E and V denote the elastic modulus and volume fraction, respectively and the subscript c, m and p represent composite, matrix and particulate phases respectively.

Large-particle composites are utilized with all three material types, i.e. metals, polymers and ceramics. Examples of ceramic-metal composite are *cermets*. Cemented carbide, which is composed of extremely hard particles of a refractory carbide ceramic such as tungsten carbide (WC) or titanium carbide (TiC), embedded in a matrix of a metal such as cobalt or nickel is the most common cermet. These composites are widely used as cutting tools for hardened steels.

Interestingly, both elastomers and plastics are frequently reinforced with various particulate materials. The use of many of the modern rubbers would have been severely restricted without reinforcing particulate materials, e.g. as carbon black. Carbon black consists of very small and essentially spherical particles of carbon. Carbon black is produced by the combustion of natural or oil in an atmosphere that has only a limited air supply. When carbon black is added to vulcanized rubber, this material which is very cheap, enhances tensile strength, toughness and tear and abrasion resistance. Automobile tire contains of the order of 15 to 30 volume percentage of carbon black.

In order that carbon black may provide significant reinforcement, the particle size must be evenly small, with diameters between 20 nm and 50 nm. Moreover, the particle must be evenly distributed throughout the rubber and must form a strong adhesive bond with the rubber matrix. Particle reinforcement using other materials, e.g. silica, is much less effective as this special interaction between the rubber molecules and particle surfaces does not exist.

Concrete

This is a common large-particle composite in which both matrix and disperse phases are ceramic materials. Broadly speaking, concrete implies a composite material consisting of an aggregate of the particles that are bound together in a solid body by some type of binding medium, i.e. a cement. The two most familiar concretes are those made with portland and asphaltic cements, where the aggregate is gravel and sand. Asphaltic cement is used primarily as a paving material on a wider scale, whereas portland cement concrete is used extensively as a structural building material.

Portland Cement Concrete

The ingredients for portland cement concrete are portland cement, a fine aggregate (sand), a coarse aggregate (gravel), and water. The aggregate particles act as a filler material to reduce the overall cost of concrete product as they are cheap, whereas cement is relatively costly. The ingredients have to be added in correct proportions so that one may achieve the optimum strength and workability of concrete mixture. One can achieve dense packing of the aggregate and good interfacial contact by having particles of two different sizes. The fine particles of sand should fill the void spaces between the gravel particles. Normally, these aggregates comprise between 60% and 80% of the total volume. However, the amount of cement-water paste should be sufficient to coat all the sand and gravel particles, otherwise the cementitious bonds will be incomplete.

This is a major material for construction. One can pour portland cement concrete in place and it hardens at room temperature, and even when submerged in water. However, as a structural material, it is found to be relatively weak and extremely brittle; the tensile strength is approximately 10 to 15 times smaller than its compressive strength. Moreover, large concrete structure can experience considerable thermal expansion and also contraction with temperature fluctuations. Water also penetrates into external pores, which can cause severe cracking in cold weather (as a consequence of freeze-thaw cycles).

Reinforced Concrete

One can increase the strength of portland cement concrete by additional reinforcement.

One can achieve this by means of steel rods, wires, bars (rebar), or mesh, which are embedded into the fresh and uncured concrete. Obviously, the reinforcement renders the hardened structure capable of supporting greater tensile, compressive and shear stresses. Considerable reinforcement is maintained even when cracks develop in the concrete.

The coefficient of thermal expansion for steel is nearly the same as that of concrete, and thus it serves as a suitable reinforcement. Moreover, steel is not rapidly corroded in the cement environment and also a relatively strong adhesive bond is formed between it and the curved concrete. One can enhance this adhesion by the incorporation of contours into the surface of the steel member. This permits a greater degree of interlocking.

One can also reinforce portland cement concrete by mixing into the fresh concrete fibres of a high modulus material, e.g. glass, nylon, steel and polyethylene. However, when exposed to cement environment some fibre materials experience rapid deterioration, and therefore care must be exercised in using this type of reinforcement.

There is another reinforcement technique of strengthening concrete. This involves the introduction of residual compressive stresses into the structural member. The resulting material is called *prestressed concrete*. The characteristic of brittle ceramic, i.e. they are stronger in compression than in tension is utilized in this method. Obviously, to fracture a prestressed concrete member, the magnitude of the precompressive stress must be exceeded by an applied tensile stress.

In one prestressed technique, inside the empty moulds high-strength steel wires are positioned and stretched with a high tensile force, which is maintained constant. After placing the concrete and allowing them to harden, the tension is released. As the wires contract, they put the structure in a state of compression because the stress is transmitted to the concrete through the concrete-wire bond that is formed.

There is also another technique known as *post tensioning*. In this technique, stresses are applied after the concrete hardens. Sheet metal or rubber tubes are situated inside and pass through the concrete forms, around which the concrete is cast. Steel wires are fed through the resulting holes after the cement has hardened and tension is applied to the wires through jacks attached and abutted to the faces of the structure. Also, a compressive stress is imposed on the concrete piece, this time by the jacks. To protect the wire from corrosion, the empty spaces inside the tubing are filled finally with a grout.

We must note that the concrete that is prestressed should be of a high quality and there must be a low shrinkage and low creep rate. Usually, prestressed concretes are prefabricated and used mostly for railway bridges and highways.

Dispersion-strengthened Composite

This type of composite contains small particulates or dispersions, which increase the strength of the composite by blocking the movement of dislocations. The *dispersoid* is typically a stable oxide of the original material. A common example is sintered aluminium powder (SAP). SAP has an aluminium matrix, which contains up to 14% aluminium oxide (Al_2O_3). This composite is produced with the *powder metallurgy process*, where the powders are mixed, compacted at high pressures, and sintered together. Sintering involves heating a material until the particles of the material fuse together. Only the edges of the particles are generally bonded together; the whole particle does not melt. In terms of sintered ceramics, the product is a strong, rigid, brittle product that exhibit good compressive strength, high melting points, and good heat resistance. Examples of dispersion-strengthened composites include Ag-CdO, used as an electrical contact material, Pb-PbO, used in battery plates and Be-BeO, used in nuclear reactor and aerospace components. The high temperature strength of nickel alloy may be enhanced significantly by the addition of about 3 volume percentage of thoria (ThO_2) as finely dispersed particles. This material is known as thoria-dispersed (or TD) nickel.

A.4 FIBRE-REINFORCED COMPOSITES

These are strong fibres embedded in a softer matrix to produce products with high strength-to-weight ratios. The matrix material transmits the load to fibres, which absorb the stress. Under an applied stress, fibre-matrix bond ceases at the fibre ends, yielding a matrix deformation pattern (Fig. A.3).

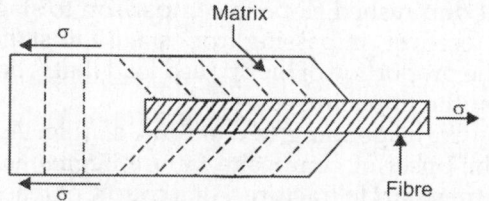

Fig. A.3: The deformation pattern in the matrix surrounding a fibre, subjected to an applied tensile load

In order to have effective strengthening and stiffening of the composite material, some critical fibre length is essential. This critical length l_c is dependent on the fibre diameter d and its ultimate (or tensile) strength σ_f and on the fibre-matrix bond strength (or the shear yield strength of the matrix, whichever is smaller) τ_c according to the following relation:

$$l_c = \frac{\sigma_f^* \, d}{2 \, \tau_c} \qquad (A.3)$$

The critical length, l_c is of the order of 1 mm for a number of glass and carbon fibre-matrix combinations and ranges between 20 and 150 times the fibre diameter.

Obviously, the strength of these composites comes from the bonding between the reinforcement fibres and the matrix. The length-to-diameter, or *aspect ratio* of the fibres used as reinforcement influences the properties of the composite. The higher the aspect ratio, the stronger the composite. Therefore, long, continuous fibres are better than short ones for composite construction. However, continuous fibres are more difficult to produce and place in the matrix. Shorter fibres are easier to place in the matrix but offer poor reinforcement. Some trade-off is made

when shorter, discontinuous fibres are used with aspect ratios greater than a specified minimum value. The greater the number of fibres, the stronger the composite. This holds true up to about 80% of the volume of the composite, where the matrix can no longer completely surround the fibres.

Fibres for which $l \gg l_c$ (normally $l > 15 \, l_c$) are called *continuous*, whereas fibres which have lengths shorter than this are termed *discontinuous*. The matrix deforms around the discontinuous fibre having length less than l_c such that there is virtually no stress transference and little reinforcement by the fibre. These are essentially the particulate composites as discussed earlier. In order to affect a significant improvement in strength of the composite, the fibres must be continuous.

The arrangement of *orientation* of the fibres, relative to each other, the fibre *concentration*, and *distribution* all have a significant influence on the strength and other properties of fibre-reinforced composites. There are two possible extremes with respect to orientation: (i) a parallel alignment of the longitudinal axis of the fibres in the single direction (ii) a totally random alignment. Continuous fibres are normally aligned as shown in Fig. A.4a; discontinuous fibres may be aligned as shown in Fig. A.4b and randomly oriented, or partially oriented as shown in Fig. A.4c. One can realize overall better composite properties when fibre distribution is uniform.

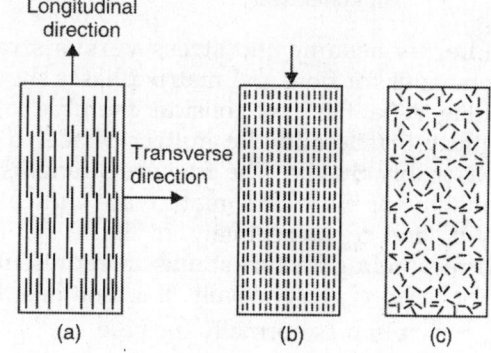

Fig. A.4: Representation of (a) Continuous and aligned (b) Discontinuous and aligned (c) Discontinuous and randomly oriented fibre-reinforced composites

Continuous and Aligned Fibre Composites

Mechanical behaviour of this type of composite depends on: (i) stress–strain behaviour of fibre and matrix phases (ii) the phase volume fractions (iii) the direction in which the stress or load is applied. Moreover, the properties of a composite having its fibres aligned are highly anisotropic, i.e. dependent on the direction in which they are measured. Let us first consider the situation shown in Fig. A.5a, wherein the stress is applied along the direction of alignment, i.e. the *longitudinal direction*.

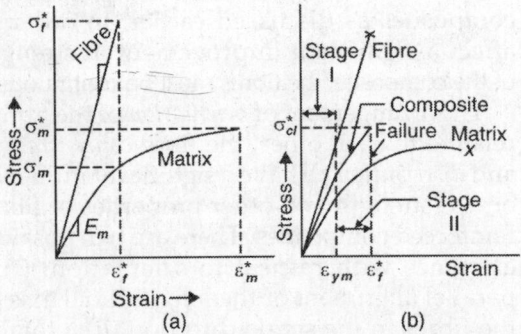

Fig. A.5: (a) Stress–strain curves for brittle fibre and ductile matrix materials. For both the materials, stresses and strains are noted (b) Stress–strain curve for an aligned fibre-reinforced composite, which is exposed to a uniaxial stress applied in the direction of alignment. We may note that curve for fibre and matrix materials of part (a) are also superimposed

Let us assume the stress versus strain behaviour for fibre and matrix phases shown in Fig. A.5a. Here we consider the fibre to be totally brittle and the matrix phase to be reasonably ductile. The fracture strengths in tension for fibre and matrix are indicated by σ_f^* and σ_m^* respectively, in Fig. A.5a. The corresponding fracture strains are represented by ε_f^* and ε_m^* respectively. It is assumed that $\varepsilon_m^* > \varepsilon_f^*$, which is normally the case.

A fibre-reinforced composite consisting of these fibre and matrix materials will exhibit the stress–strain behaviour shown in Fig. A.5b. In the initial stage (region I), both fibres and

matrix deform elastically. Normally, this portion of the curve is linear. Typically, for a composite of this type, the matrix yields and deform plastically as shown in Fig. A.5b at ε_{ym} while the fibres continue to stretch elastically, in as much as the tensile strength of the fibres significantly higher than the yield strength of the matrix. This constitutes the stage II. This stage is ordinarily very linear but of diminished slope in comparison to stage I. Moreover, in passing from stage I to stage II, the proportion of the applied load that is borne by the fibres increases.

Corresponding to a strain $\sim \varepsilon_f^*$ (Fig. A.5b), the onset of composite failure begins as the fibres start to fracture. For a couple of reasons, composite failure is not catastrophic— all the fibres do not fracture at the same time, since there will always be considerable variations in the fracture strength of brittle fibre materials. Moreover, even after fibre failure, the matrix is still intact in as much as $\varepsilon_f^* < \varepsilon_m^*$ (Fig. A.5a). Obviously, these fractured fibres, which are shorter than the original ones, are still embedded within the intact matrix, and consequently are capable of sustaining a diminished load as the matrix continues to plastically deform.

Elastic Behaviour—Longitudinal Loading

Let us study the elastic behaviour of a continuous and oriented fibrous composite that is loaded in the direction of fibre alignment direction (i.e. *longitudinal loading*). Assuming that the fibre-matrix interfacial bond is very good and the situation is that of *isostrain*, i.e. deformation of both matrix and fibres is same. Thus, the total load sustained by the composite, F_c is given by

$$F_c = F_m + F_f \qquad (A.4)$$

where F_m and F_f are loads carried by the matrix phase and the fibre phase respectively. Now, stress $F = \sigma A$, thus the expressions for F_c, F_m and F_f in terms of their respective stresses, i.e. σ_c, σ_m and σ_f and cross-sectional areas, i.e. A_c, A_m and A_f are possible. Equation (A.4) yields

$$\sigma_c A_c = \sigma_m A_m + \sigma_f A_f \qquad (A.5)$$

or $$\sigma_c = \sigma_m \frac{A_m}{A_c} + \sigma_f \frac{A_f}{A_c} \qquad \text{(A.6)}$$

Now, $A_m/A_c = V_m$ and $A_f/A_c = V_f$, where V_m and V_f are volume fraction of the matrix and fibres respectively. Here we have assumed that composite, fibre and phase lengths are all equal. Truly speaking A_m/A_c and A_f/A_c are the area fractions of the matrix and fibre phases. Since phase lengths for both matrix and fibre are taken equal, and hence we have taken A_m/A_c and A_f/A_c equivalent to V_m and V_f respectively.

Thus, Eq. (A.6) becomes

$$\sigma_c = \sigma_m V_m + \sigma_f V_f \qquad \text{(A.7)}$$

In the light of assumption of isostrain state, we have

$$\varepsilon_c = \varepsilon_m = \varepsilon_f \qquad \text{(A.8)}$$

Dividing each term in Eq. (A.7) by its respective strain, we obtain

$$\frac{\sigma_c}{\varepsilon_c} = \frac{\sigma_m}{\varepsilon_m} V_m + \frac{\sigma_f}{\varepsilon_f} V_f \qquad \text{(A.9)}$$

If composite matrix and fibre deformations are all elastic, then $\sigma_c/\varepsilon_c = E_c$, $\sigma_m/\varepsilon_m = E_m$ and $\sigma_f/\varepsilon_f = E_f$, where E's being the moduli of elasticity for the respective phases. Substituting in Eq. (A.9), we obtain an expression for the modulus of elasticity of a continuous and aligned fibrous composite in the direction of alignment, E_{cl} as

$$E_{cl} = E_m V_m + E_f V_f \qquad \text{(A.10a)}$$

or $$E_{cl} = E_m (1 - V_f) + E_f V_f \qquad \text{(A.10b)}$$

Since the composite consists of only matrix and fibre phases, i.e. $V_m + V_f = 1$.

Obviously, E_{cl} is equal to the volume-fraction weighted average of the moduli of the elasticity of the fibre and matrix phases. We may note that other properties, including density, also have this dependence on volume fractions.

We can also show that for longitudinal loading, the ratio of the load carried by the fibres to that carried by the matrix is

$$\frac{F_f}{F_m} = \frac{E_f V_f}{E_m V_m} \qquad \text{(A.11)}$$

Elastic Behaviour—Transverse Loading

A continuous and oriented fibre composite may be loaded in transverse direction (i.e. *transverse loading*). Obviously, load is applied at 90° to the direction of fibre alignment. For this situation, the stress σ to which the composite as well as both phases are exposed is the same, i.e.

$$\sigma_c = \sigma_m = \sigma_f = \sigma \qquad \text{(A.12)}$$

This state is called as *isostress state*. Now, the strain or deformation of entire composite ε_c is

$$\varepsilon_c = \varepsilon_m V_m + \varepsilon_f V_f \qquad \text{(A.13)}$$

or $$\frac{\sigma}{E_{ct}} = \frac{\sigma}{E_m} V_m + \frac{\sigma}{E_f} V_f \; (\because \varepsilon = \frac{\sigma}{E}) \quad \text{(A.14)}$$

where E_{ct} is the modulus of elasticity in the transverse direction. Dividing Eq. (A.14) by σ, yields

$$\frac{1}{E_{ct}} = \frac{V_m}{E_m} + \frac{V_f}{E_f} \qquad \text{(A.15)}$$

or $$E_{ct} = \frac{E_m E_f}{V_m E_f + V_f E_m}$$

$$= \frac{E_m E_f}{(1 - V_f) E_f + V_f E_m} \qquad \text{(A.16)}$$

Longitudinal Tensile Strength

Let us study the strength characteristics of continuous and aligned fibre-reinforced composites that are loaded in the *longitudinal direction*. In this case, strength is usually taken as the maximum stress on the stress–strain curve (Fig. A.5b). Usually, this point corresponds to fibre fracture, and indicates the onset of composite failure. We may note that the failure of this type of composite material is relatively complex process and there are several different failure modes possible. Again, the mode that operates for a specific composite will depend on fibre and matrix properties, and the nature and strength of the fibre–matrix interfacial bond.

For $\varepsilon_f^* < \varepsilon_m^*$ (Fig. A.5a), we note that fibres will fail before the matrix. Once the fibres have fractured, the majority of the load that was borne by fibres now gets transferred to the matrix. One obtains the expression for the longitudinal strength of the composite, σ_{cl}^* as

$$\sigma_{cl}^* = \sigma_m^* (1 - V_f) + \sigma_f^* V_f \qquad (A.17)$$

where σ_m^* is the stress in the matrix at fibre failure (Fig. A.5a) and σ_f^* is the fibre tensile strength.

Transverse Tensile Strength

The strength of continuous and unidirectional fibrous composites are highly anisotropic. Normally, such composites are designed to be loaded along the high strength, longitudinal direction. However, transverse tensile loads may also be present during in service applications. Under these prevailing circumstances, premature failure may result in as much as transverse strength is usually extremely low. Sometimes, it lies below the tensile strength of the matrix. Obviously, the reinforcing effect of the fibres is negative one.

Properties of both the fibre and matrix, the fibre-matrix bond strength, and the presence of voids have significant influence on the transverse strength. In order to improve the transverse strength of these composites, one will have to modify the properties of matrix.

Discontinuous and Aligned Fibre Composites

Although reinforcement efficiency is lower for discontinuous than for continuous fibres, discontinuous and aligned fibre composites (Fig. A.4b) are more in demand. Chopped glass fibres are used mostly, however, carbon and aramid discontinuous fibres are also used. One can produce these short fibre composites with moduli of elasticity and tensile strength approaching 90% and 50%, respectively, of their continuous fibre counterparts.

The longitudinal strength (σ_{cd}^*) for a discontinuous and aligned fibre composite having a uniform distribution of fibres and in which $l > l_c$, is given by

$$\sigma_{cd}^* = \sigma_f^* V_f \left(1 - \frac{l_c}{2l} \right) + \sigma_m' (1 - V_f) \qquad (A.18)$$

Here σ_f^* and σ_m' represent, respectively, the fracture strength of the fibre and the stress in the matrix when the composite fails (Fig. A.5a).

For $l < l_c$, i.e. fibre length is less than critical, then the longitudinal strength $(\sigma_{cd'}^*)$ is given by

$$\sigma_{cd'}^* = \frac{l\tau_c}{d} V_f + \sigma_m' (1 - V_f) \qquad (A.19)$$

where d is the diameter of the fibre and τ_c is smaller of the either, the fibre matrix bond strength or the matrix shear yield strength.

Discontinuous and Randomly Oriented Fibre Composites

Normally, fibres are used when the fibre orientation is random, short and discontinuous. Reinforcement of this type is shown in Fig. A.4c. One can use a *rule-of-mixture* expression for the elastic modulus similar to Eq. (A.10), as follows:

$$E_{cd} = K E_f V_f + E_m V_m \qquad (A.20)$$

Here K is a fibre efficiency parameter, which depends on V_f and the ratio E_f / E_m. However, its magnitude will be less than unity in the range 0.1 to 0.6. Thus, the modulus increases in some proportion to the volume fraction of fibre for random fibre reinforcement (as with oriented).

We have seen that aligned fibrous components are inherently anisotropic in that the maximum strength and reinforcement are achieved along the alignment (longitudinal) direction. Fibre reinforcement is virtually non existent in the transverse direction and fractures usually occurs at relatively low tensile stresses. The composite strength lies between these extremes for other stress orientations.

Laminar Composites

When multidirectional stresses are imposed within a single plane, aligned layers that are fastened together one on top of another at different orientations are frequently utilized. These are called *laminar composites*. These are generally designed to provide high strength and low cost at a lighter weight. A familiar laminar composite is plywood, where the veneers are joined by adhesives, typically phenolic or amine resins. The individual odd number of piles are staked so that the grain in each layer runs perpendicular to that of the layers above and below it (Fig. A.6). This technique offers plywood that is strong and yet cheaper. Some safety glass is a laminated structure, where an adhesive such as polyvinyl butyral is used between two outer layers of glass to keep the glass from flying when broken. Formica is another common laminate used for countertops. Laminates require two or more layers be bonded together. Laminations may also be constructed using fabric material such as cotton, paper, or woven glass fibres embedded in a plastic matrix. Obviously, a laminar composite has relatively high strength in a number of directions in the two dimensional plane; however, the strength in any given direction is, of course, lower than it would be if all the fibres were oriented in that direction. Modern ski is one example of a relatively complex *laminated structure*.

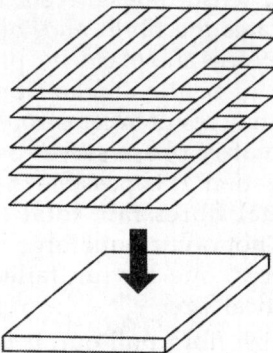

Fig. A.6: Laminar composite. The stacking of successive oriented, fibre-reinforced layers

Applications involving totally multidirectional applied stresses generally use discontinuous fibres, which are randomly oriented in the matrix material. We may note that the reinforcement efficiency is found only 1/5th that of an aligned composite in the longitudinal direction; however, the mechanical characteristics are isotropic.

For a particular composite, consideration of orientation and fibre length will depend on the level and nature of the applied stress as well as fabrication cost. Short-fibre composites (both aligned and randomly oriented) production rates are rapid, and intricate shapes can be formed that are not possible with continuous fibre reinforcement. Moreover, fabrication costs are considerably lower than for continuous and aligned.

Sandwich Structures

These have thin layers of facing materials over a low density material, or combcore, such as a polymer foam or expanded metal structure. A familiar sandwich-structured composite is corrugated cardboard. The corrugated paper core is covered by two faces of thin paper. In structures of this type, the facing material serves to fix the inner core in place. The core provides the strength. Typical face materials include aluminium alloys, fibre-reinforced plastics, titanium, steel and plywood.

Structurally, the core serves two purposes: (i) it separates the faces and resists deformations perpendicular to face plane (ii) it provides a certain degree of sheer rigidity along planes that are perpendicular to the faces. Foamed polymers, synthetic rubbers, inorganic cements, balsa wood, etc. materials and structures are used for cores. Core has lower stiffness and lower strength.

Another popular core consists of a *honeycomb* structure, which finds wide use in industries such as the aircraft industry, where higher strength and lower weight are important factors. The honeycomb structure consists of thin foils that have been formed into interlocking hexagonal cells, with axis oriented perpendicular to the face panels. The material used may be similar to the face material. These structures are lightweight, stiff and strong and can be filled to provide sound and vibration damping. The honeycomb structure is shown in Fig. A.7.

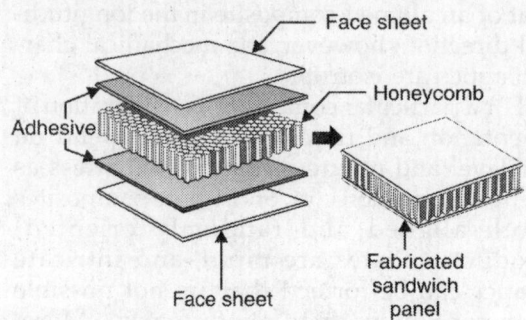

Fig. A.7: Honeycomb structure

The Fibre Phase

An important characteristics of most materials, especially of brittle ones, is that a small diameter fibre is much stronger than the bulk material. We have read that the probability of the presence of a critical surface flaw that can lead to fracture diminishes with decreasing specimen volume. This feature of materials is used to advantage in the fibre-reinforced composites. Moreover, the materials used for reinforcing fibres have high tensile strengths.

Fibres can be grouped into three different classes based on their *diameter* and *character*: *whiskers*, *fibres* and *wires*. Whiskers are not utilized extensively as a reinforcement medium because they are extremely expensive, and moreover, it is difficult and impractical to incorporate whiskers into a matrix. Graphite, silicon carbide, silicon nitride and aluminium oxide are few whisker materials.

Materials, which fall under the head *fibres* are either polycrystalline or amorphous and have small diameters. Fibrous materials are usually either polymers or ceramics, (e.g. the polymer aramids, glass, carbon, boron, aluminium oxide and silicon carbide). Fibre-reinforced composites have been used for centuries, e.g. straw has been used in the making of mud bricks since the Pharaohs. Steel rebar has been used for many years in reinforced concrete structure. Fibre glass cloth

and resin have been used to repair many automobiles fenders. These applications are among the more familiar.

Fine wires have relatively large diameters. Typical materials under this group include steel, molybdenum and tungsten. Wires are widely used as radical steel reinforcement in automobile tyres, in filament-wound rocket castings, and in wire-wound high pressure houses.

The Matrix Phase

The matrix phase of a fibrous composite may be a polymer, or ceramic and a metal. Metals and polymers are generally used as matrix materials because some ductility is desirable. To improve fracture toughness for ceramic-matrix composites, the reinforcing component is added.

The matrix phase serves several functions for fibre-reinforced composites: (i) it binds the fibres together and acts as a medium by which an externally applied stress is transmitted and distributed to the fibres. However, only a very small proportion of an applied load is sustained by the matrix phase. Also, the matrix material should be ductile. The elastic modulus of the fibre should be much higher than that of the matrix. (ii) matrix has to protect the individual fibres from surface damage as a result of mechanical abrasion or chemical reactions with the environment. These interactions may cause surface flaws capable of forming cracks, which may lead to failure at low tensile stress levels. Lastly, the matrix separates the fibres and, by virtue of its relative softness and plasticity, prevents the propagation of brittle cracks from fibre to fibre, which may result in catastrophic failure. Obviously, matrix phase serves as a barrier crack propagation. It is possible that some of the individual fibres fail, total composite fracture will not occur until large number of adjacent fibres, one having failed, form a cluster of critical size.

To minimize fibre pull-out, it is essential that adhesive bonding forces between fibre and matrix be high. In reality, in the choice of the matrix-fibre combination bonding strength is an important factor. The magnitude of this bond plays vital role in the ultimate strength of the composite. To maximize the stress transmittance from the weak matrix to the strong fibres, adequate bonding is essential.

Polymer-Matrix Composites (PMCs)

These materials consist of a polymer resin (here resin denotes a high-molecular weight reinforcing plastic) as the matrix, with fibres as the reinforcement medium. In light of their room-temperature properties, ease of fabrication, and cost, these PMCs are used in the greatest diversity of composite applications, as well as in huge quantities. In accordance with reinforcement type (i.e. glass, carbon and aramid), various classifications of PMCs, along with their applications and the various polymer resins that are used, are as follows:

Glass Fibre-Reinforced Polymer (GFRP) Composites

Fibre glass is a composite consisting of glass fibres, which may be continuous or discontinuous and contained within a polymer matrix. These are commonly produced in E glass (E is for electrical), because it draws well and has good strength and stiffness. A typical composition (Wt%) would be $52SiO_2$, $17CaO$, $14Al_2O_3$, $10Ba_2O_3$ with some oxides of Mg, Na and K, and molten glass is gravity fed into a series of platinum bushings each of which has several hundred holes in its base. Fine glass filaments are drawn mechanically as the glass exudes from the holes, then wound on to drums at speeds of several thousand metres per minute.

The strength of the glass fibres is dependent upon the surface damage arising when they rub against each other during processing. The application of a size coating early at an early stage during manufacture minimises this degradation in properties, by reducing the propensity for forming these 'Griffith' cracks. The size consists of an emulsified polymer in water, and also has the effect of binding the fibres together for ease of further processing.

We are familiar with many fibre glass applications, e.g. automotive and marine bodies, plastic pipes, industrial floorings and containers. In order to decrease vehicle weight and boost fuel efficiencies, the transport industries are utilizing to maximum extent of glass fibre-reinforced plastics. Many new applications are under investigation by the automotive industry.

Carbon Fibre-Reinforced Polymer (CFRP) Composites

These consist of small crystallites of graphite. The atoms in the basal planes are held together by very strong covalent bonds, and there are weak van der Waals forces between the layers. To obtain high modulus and high strength, the layer planes of the graphite have to the aligned parallel to the axis of the fibre, and the modulus of carbon fibres depends on the degree of perfection of alignment of atom planes. This varies considerably with the particular manufacturing route adopted, of which there are three main possibilities:

1. Starting with the polymer PAN (polyacrylonitride), which closely resembles polyethylene in molecular confirmation, it is converted into a fibre and then stretched to produce alignment of the molecular chains along the fibre axis. While still under tension, it is heated in oxygen to form cross-links between ladder-molecules and finally chemically reduced to give (at high temperatures) a graphic structure. The final graphitization temperature determines whether the fibres have maximum stiffness but a relatively low strength (Type-I fibres), or whether they develop maximum strength (Type-II fibres).

2. Alternatively, fibres may be produced by melt-spinning molten pitch. During the spinning process, the orifice causes the planar molecules to become aligned. It is then treated, whilst held under tension in order to maintain its preferred orientation, at temperatures up to 2000 °C to form the requisite grains of graphite.

3. It is also possible to stretch either of the fibre types described above during the graphitization stage, giving further orientation of the layers parallel to the fibre axis.

Carbon fibres have the highest specific modulus and specific strength of all reinforcing fibre materials. Carbon fibres retain their high tensile modulus and high strength at elevated temperatures. However, high temperature oxidation may be a problem. Carbon fibres are not affected by moisture or a wide variety of solvents, acids and bases at room temperature. Carbon fibres exhibit a divesity of physical and mechanical characteristics, allowing composites incorporating these fibres to have specific engineering properties.

On the basis of tensile modulus, carbon fibres can have four classes: *standard*, *intermediate*, *high* and *ultrahigh* moduli. Fibre diameters generally range between 4 µm and 10 µm. Both *continuous* and *chopped* forms of these fibres are available. Moreover, carbon fibres are usually coated with a protective epoxy size that also improves adhesion with the polymer matrix.

Carbon-reinforced polymer composites are extensively used in sports and recreational equipment (fishing rods, golf clubs), filament-wound rocket motor cases, pressure vessel and aircraft structural components—both military and commercial, fixed wing and helicopters.

Aramid Fibre-Reinforced Polymer Composites

Aramid fibres are high-strength, high modulus materials, especially desirable for their outstanding strength-to-weight ratios, which are superior to metals. Chemically, this group of materials is called as polyparaphenylene terephthalamide. There are a number of aramid materials. Trade names of two of these most common materials are kevlar and nomex. Kevlar has several grades (viz. kevlar 29, 49 and 149) that have different mechanical behaviours. Mer chemistry and mode of chain alignment for aramid (kevlar) are shown in Fig. A.8. These fibres have longitudinal tensile strengths and tensile moduli and they are higher than for other polymeric fibre materials, but they are relatively weak in compression. This material is also known for its toughness, impact resistance, and resistance to creep and fatigue failure. Although aramids are thermoplastics, they are nevertheless, resistant to combustion and stable to relatively high temperatures. The temperature range over which aramids retain their high mechanical properties is between −200°C and 200°C. Aramids are susceptible to degradation by strong acids and bases, but they relatively inert in other solvents and chemicals.

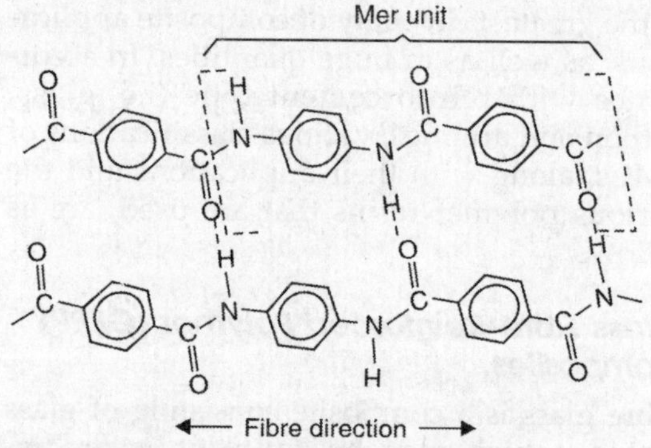

Fig. A.8: Mer and chain structures for aramid (kevlar) fibres

Usually, aramid fibres are used in composites having polymer matrices. Common matrix materials are the epoxies and polyesters. Aramid fibres are relatively flexible and somewhat ductile, they may be processed by most common textile operations. Aramid composites are used in ballistic products (bullet-proof vests), sporting goods, tyres, ropes, missile cases, pressure levels, and as a replacement for asbestos in automotive brake and clutch linings, and gaskets.

Other Fibre Reinforcement Materials

Other fibre materials are boron, silicon carbide and aluminium oxide. However, these fibre materials are used to much lesser degrees. Boron fibre-reinforced polymer composites find use in military aircraft components, helicopter robot blades, and some sporting goods. Silicon carbide and alumina fibres are used in tennis rackets, rocket nose cones and circuit boards.

Polymer-matrix Materials

Polyesters and vinyl esters are the most widely used and least expensive polymer resins. These matrix materials are basically used for fibre glass reinforced composites. Mutations of a large number of resin provide a wide range of properties of these materials. The epoxies are more expensive, and in addition to wide ranging commercial applications, also find extensive us in PMCs for aerospace applications. In comparison to polyesters and vinyl resinis, epoxies have better mechanical properties and resistance to moisture. Polymide resins are used for high temperature applications (~230 °C). High-temperature thermoplastic resins may find wider future aerospace application. These include polyetheretherketon (PEEK), polyphenylene sulphide (PPS) and polyetherimide (PEI).

Metal-matrix Composites (MMCs)

The matrix in these composites is a ductile metal. These composites can be used at high service temperatures than their base metal counterparts. The reinforcement in these materials may improve specific stiffness, specific strength, abrasion resistance, creep resistance, thermal conductivity and dimensional stability. In comparison to polymer-matrix composites, these materials have higher operating temperatures, non flammability and greater resistance to degradation by organic fluids. In comparison to PMCs, MMCs are much more expensive and therefore, their use is somewhat restricted.

Alloys of aluminium, magnesium, titanium and copper, and superalloys are used as matrix materials. The reinforcement may be in the form of particulates, both continuous and discontinuous fibres and whiskers (concentrations range from 10 vol% to 60 vol%) carbon, silicon carbide, boron, alumina and refractory metals are continuous fibre materials, whereas silicon carbide whiskers, chopped fibres of alumina and carbon, and particulates of silicon carbide and alumina are discontinuous reinforcements sense, cermets fall within MMCs.

At elevated temperatures, some matrix-reinforcement combinations are highly reactive. This problem can be resolved either by using a protective surface coating to the reinforcement or by modifying the matrix alloy composition.

Recently, automobile industries have begun to use MMCs in their products, e.g. some engine components have been introduced consisting of an aluminium-alloy matrix that is reinforced with alumina and carbon fibres. This MMCs is light in weight and can resist wear and thermal distortion. MMCs are also used in driveshafts (that have higher rotational speeds and reduced vibrational noise levels), forged suspension and transmission components and extruded stabilizer. MMCs are used by aerospace industry. Aluminium alloy metal-matrix composites, boron fibres are used as the reinforcement for the space shuttle orbiter, and continuous graphite fibres for Hubble telescope.

Using refractory metals such as tungsten, the high temperature creep and rupture properties of some of the superalloys (Ni and Co based alloys) may be enhanced by fibre reinforcement.

Ceramic-matrix Composites (CMCs)

The fracture toughnesses of ceramics have been improved significantly by the development of CMCs particulates, fibres or whiskers of one ceramic material that have been embedded into a matrix of another ceramic. The fracture toughnesses of CMCs materials lie between 6 and 20 MPa $\sqrt{m}$. Increasing fibre content improves strength and fracture toughness. There is a considerable reduction in the scatter of fracture strengths for whisker-reinforced ceramics relative to their unreinforced counterparts. These CMCs exhibit improved high temperature creep behaviour and resistance to thermal shock.

SiC whisker-reinforced alumina are being utilized as cutting tool inserts for machining hard metal alloys. We may note that tool lives for these materials are greater than for cemented carbides.

Carbon–Carbon Composites

Carbon–carbon composite, i.e. carbon fibre-reinforced carbon–matrix composite is one of the most advanced and promising engineering material. Obviously, in the material, both reinforcement and matrix are carbon. These materials have high-tensile moduli and tensile strengths that are retained to temperatures in excess of 2000 °C, resistance to creep and relatively large fracture toughness values. These materials have low coefficients of thermal expansion and relatively high thermal conductivities. These characteristics of these materials, coupled with high strengths, give rise to a relatively low susceptibility to thermal shock. The major drawback of these materials is a propensity to high-temperature oxidation. These materials are relatively new and expensive, and therefore, are not in wide use.

These materials are employed in rocket motors, as friction, materials in aircraft and high-performance automobiles, for hot-pressing moulds, in components for advanced turbine engines, and as ablative shields for re-entry vehicles.

Hybrid Composites

The composites obtained by using two or more different kinds of fibres in a single matrix are termed *hybrid*. Hybrid composites have a better all-round combination of properties than composites containing only a single fibre type. Although a variety of fibre combinations and matrix materials are used, but in the most common system, both carbon and glass fibres are incorporated into a polymeric resin. The carbon fibres are expensive, but they are strong and relatively stiff and provide a low density reinforcement. Glass fibres lack the stiffness of carbon, but they are inexpensive. The glass-carbon hybrid may be produced at a lower cost than either of the comparable all-carbon or all-glass reinforced plastics and also it is stronger and tougher.

The properties of hybrid composites are anisotropic. When these are stressed in tension, failure is usually non catastrophic, i.e. does not occur suddenly. The carbon fibres are the first to fail, at which time the load is transferred to the glass fibres. Upon failures of the glass fibres, the matrix phase will have to sustain the applied load. We may note that eventual composite failure concurs with that of the matrix phase.

Hybrid composites find applications in lightweight land, water and air transport structural components, lightweight orthopedic components and sporting goods.

REVIEW QUESTIONS

1. Define the term composite and explain what are advanced composites. Give few examples.

2. What factors influence the final properties of composites?

3. What is the difference between matrix and dispersed phases in a composite material?

4. Contrast the mechanical characteristics of matrix and dispersed phase for fibre reinforced composite materials.

5. What is the difference between cement and concrete?

6. Mention three important limitations that restrict the use of concrete as a structural material.

7. What are the functions of the matrix phase for a polymer-matrix fibre-reinforced composite?

8. For a polymer-matrix fibre-reinforced composite, compare the desired mechanical characteristics of matrix and fibre phases.

9. Why glass fibres are most commonly used for reinforcement?

10. What is the difference between carbon and graphite?

11. Why fibre glass-reinforced composites are utilized extensively?

12. What is a hybrid composite? Mention two important advantages of hybrid composites over normal fibre composites.

SHORT ANSWER QUESTIONS

1. What are composites?

Ans. These are artificially produced multiphase materials having a desirable combination of the best properties of the constituent phases. Usually, one phase (the matrix) is continuous and completely surrounds the other (the disperse phase).

2. What is concrete?

Ans. This is a type of large particle, composite consists of an aggregate of particles bonded together with cement.

3. What is portland cement concrete?

Ans. The aggregate consists of sand and gravel. The cementitious bond develops as a result of chemical reaction between the portland cement and water.

4. Out of the several composite types, for which the potential for reinforcement efficiency is greatest?

Ans. Fibre reinforced.

5. What types of fibres will be for short and discontinuous fibrous composites?

Ans. The fibres may be either aligned or randomly oriented.

6. Can polymer-matrix be reinforced with glass, carbon and aramid fibres?

Ans. Yes.

7. Which are the more advanced composites?

Ans. (i) Carbon–carbon (ii) hybrids.

8. Write two kinds of structural composites.

Ans. (i) Laminar composites (ii) Sandwich panels.

MULTIPLE CHOICE QUESTIONS

1. The cermets are examples of
   (a) ceramic-metal composites
   (b) carbon–carbon composites
   (c) metal-matrix composites
   (d) none of the above

2. The critical fibre length (l_c) is related to the fibre diameter d, ultimate (or tensile) strength σ_f^* and fibre-matrix bond strength τ_c as

(a) $l_c = \dfrac{\sigma_f^* d}{\tau_c^2}$

(b) $l_c = \dfrac{\sigma_f^* d}{2\tau_c}$

(c) $l_c = \dfrac{\sqrt{\sigma_f^* d}}{\tau_c}$

(d) $l_c = \sqrt{\sigma_f^* d\tau_c}$

3. Whiskers are
   (a) very thin steel wires
   (b) very thin copper wires
   (c) very thin single crystals
   (d) none of the above

4. The matrix phase of fibrous composites may be
   (a) a metal only
   (b) a polymer only
   (c) a ceramic only
   (d) a metal, polymer or ceramic

5. For metal-matrix composites, the matrix is a
   (a) ductile metal   (b) polymer
   (c) ceramic   (d) none of the above

6. A laminar composite is composed of
   (a) one dimensional sheets or panels
   (b) two dimensional sheets or panels
   (c) three dimensional sheets or panels
   (d) none of the above

7. Most advanced and promising composite is
   (a) carbon–carbon   (b) metal-matrix
   (c) polymer-matrix   (d) none of the above

8. Trade names for two most common aramid materials are
   (a) silicon carbide, silicon nitride
   (b) e-glass, aluminium oxide
   (c) kevlar, nomex
   (d) high strength steel, aluminium oxide

Answers

1. (a)	**2.** (b)	**3.** (c)	**4.** (d)	**5.** (a)
6. (b)	**7.** (a)	**8.** (c)		

Shape Memory Alloys and Nonlinear Materials

B.1 SHAPE MEMORY ALLOYS (SMAs)

Smart or Intelligent Materials

Smart or intelligent materials form a group of new and state-of-art materials now being developed that will have a significant influence on many of our technologies. The adjective 'smart' implies that these materials are able to sense changes in their environments and then respond to these changes in predetermined manners—traits that are also found in living organisms. In addition, the concept of smart materials is being extended to rather sophisticated systems that consists of both smart and traditional materials.

The field of smart materials attempts to combine the sensor (that detects an input signal), actuator (that performs a responsive and adaptive function) and the control circuit on as one integrated unit. Actuators may be called upon the change shape, position, natural frequency, or mechanical characteristics in response to changes in temperature, electric fields, and/or magnetic fields.

Usually, four types of materials are commonly used for actuators: shape memory alloys, piezoelectric ceramics, magnetostrictive materials, and electrorheological/magnetorheological fluids. Shape memory alloys are metals that, after having been deformed, revert back to their original shapes when temperature is changed. Piezoelectric ceramics expand and contract in response to an applied electric field (or voltage); conversely, these materials also generate an electric field when their dimensions are altered. The behaviour of magnetostrictive materials is analogous to that of piezoelectric ceramic materials, except that they are responsive to magnetic field. Also, electrorheological and magnetorheological fluids are liquids that experience dramatic changes in viscosity upon the application of electric and magnetic fields, respectively.

The combined system of sensor, actuator and control circuit on as one IC unit, emulates a biological system (Fig. B.1).

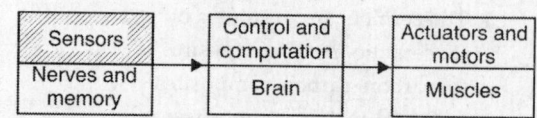

Fig. B.1: Integrated sensor-actuator systems with controller are analogous to biological systems

These are known as smart sensors, micro-system technology (MST) or microelectro-mechanical systems (MEMSs). Materials/devices employed as sensors include optical fibres, piezoelectric materials (including some polymers) and MEMS.

For example, one type of smart system is used in helicopters to reduce aerodynamic cockpit noise that is created by the rotating rotor blades. Piezoelectric sensors inserted into the blades, monitor blade stresses and deformations, feedback signals from these sensors are fed into a computer controlled adaptive device, which generates noise cancelling antidose.

MEMS devices are small in size, light-weight, low cost, reliable with large batch fabrication technology. They generally consists of sensors that gather environmental information such as pressure, temperature, acceleration, etc. integrated electronics to process the data collected and actuators to influence and control the environment in the desired manner.

The MEMS technology involves a large number of materials. Silicon forms the backbone of these systems also due to its excellent mechanical properties as well as mature micro-fabrication technology including lithography, etching and bonding. Other materials having piezoelectric, piezoresistive, ferroelectric and other properties are widely used for sensing and actuating functions in conjunction with silicon. The field of MEMS is expected to touch all aspects of our lives during this decade with revolution in aviation, pollution control and industrial processes.

Smart materials can respond to external stimuli, e.g. pressure, temperature, optical, electric field, magnetic field, moisture, etc.

Smart materials offer numerous possible applications in medical engineering, automobile engineering, civil engineering, aerospace and consumer electronics.

Among several smart materials, shape memory alloys, fibre optic sensors and piezoelectric materials are currently receiving wide attention.

Shape Memory Alloys (SMAs)

These are metals which exhibit two unique properties: (i) shape memory effect (SME) (ii) pseudo elasticity or super elasticity (SE).

Sharp memory alloys are a unique class of alloys or materials, which remember their shape even after severe deformation, i.e. when a sharp memory alloy is once deformed in the cold shape (martensite) these materials will stay deformed until heated, whereupon heating they will spontaneously return to their original predetermined hot shape (austenite).

It is reported that the structural changes at the atomic level contribute to their unique properties of the materials.

Types of Shape Memory Alloys

Shape memory alloys can be placed under two categories:

1. *One Way Shape Memory Alloys*: The materials which exhibit shape memory effect, (i.e. taking their own shape) only upon heating are termed one way shape memory alloys.

2. *Two Way Shape Memory Alloys*: There are some materials which exhibit shape memory effect both during heating and cooling. Such materials are termed two way shape memory alloys.

Crystal Structure of Shape Memory Alloys

These materials have two distinct types of crystal structure. At any particular instant the crystal structure or the phase of the SMA is determined by the temperature and internal stresses.

At high temperature, the phase of SMA is called *austenite* phase (microscopically they possess small *platelet structure*) and at low temperature, the phase is called the *martensite* (microscopically they possess needle-like structure). In both the cases, the material remains in solid form.

If possesses a highly symmetric structure such as the one in cubic system in austenite phase. However, in the martensite phase the structure is of low symmetry such as monoclinic. The basis for shape memory effect is that these materials can very easily transform to and from martensite.

One can broadly classify the shape memory alloys into three types: *thermal induced shape memory, stress induced shape memory* and *ferromagnetic shape memory* materials.

Temperature Induced Transformation

The phase transformation in smart material alloys do not take place at a particular temperature but over a range of temperature (Fig. B.2). This transformation is named as the *temperature induced transformation*.

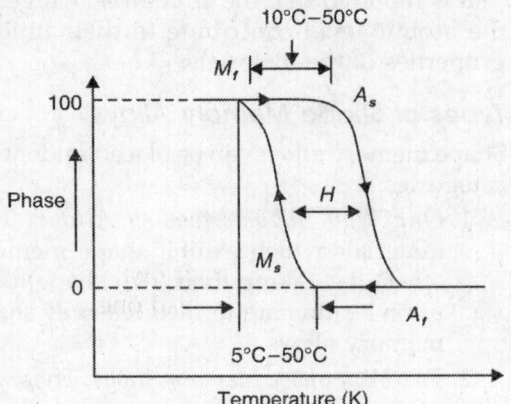

Fig. B.2: Temperature induced transformation in SMAs. Temperature M_s and M_f during cooling and A_s and A_f during heating

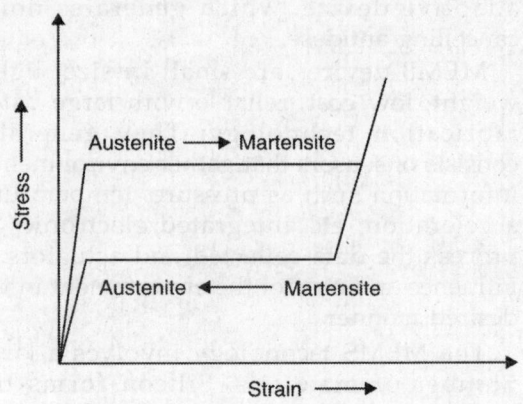

Fig. B.3: Stress induced transformation in SMA

The temperature induced transformation in SMAs is characterised by four temperatures, M_s and M_f during cooling and A_s and A_f during heating (Fig. B.2). M_s and M_f in Fig. B.2 indicate the temperatures at which the transformation from the parent phase austenite into martensite starts and ends respectively. Similarly, A_s and A_f in Fig. B.2 indicate the temperatures at which the reverse transformation from martensite to austenite starts and ends upon heating. The overall transformation has a temperature range of 10–15 °C depending on the chemical composition of the alloy. It is observed that this overall transformation exhibit hysteresis in which the transformations on heating and cooling do not develop. However, the hysteresis H depends on the composition of the alloy system.

Stress Induced Transformation

When stress is applied over the shape memory alloys, there is also a possibility for phase transformation.

The stress induced transformation in shape memory alloys takes place at a constant temperature which is above the A_f temperature, the martensite phase can be induced by applying stress over the austenite phase (Fig. B.3)

The austenite phase undergoes a large elastic deformation when stressed above a certain value. Stressed beyond the elastic limit causes permanent plastic strains. When the stress is removed, the material almost completely recovers to the parent austenite phase at a much lower value of stress (Fig. B.3)

Functional Properties of SMAs

There are two important characteristic properties of SMAs. These are:

i. *Shape Memory Effect (SME):* This is the phenomenon in which a specimen apparently deformed at lower temperature and regains its undeformed original shape when heated to higher temperature.

SME is a consequence of a crystallographic reversible martensitic phase transformation taking place in the solid state. Figure B.4 shows schematically the crystallographic formation of martensite and reversion to austenite on heating.

From Fig. B.4, we note that the high temperature austenitic structure undergoes twinning as the temperature is lowered. This twinned structure having microscopically needle-like structure is termed *martensite*. This phase of SMA is relatively soft and easily deformed phase which exists at low temperature.

Under deformation, i.e. on applying external stress, this phase takes a

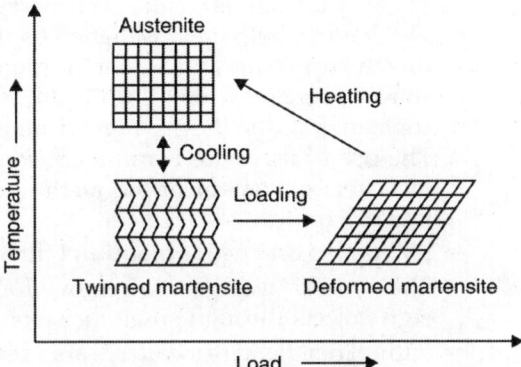

Fig. B.4: Shape memory effect in SMAs

particular shape termed deformed (twinned) martensite and undergoes a large elastic strain during this process. When heated in this condition, the deformed martensite returns to the stable austenite structure and in the process recovers the elastic strain. Clearly, the shape memory phenomenon is observed.

This shape memory effect is being implemented in making coffee pots, thermostats, vascular stents and hydraulic fittings for aeroplanes.

ii. *Superelasticity (SE) or Pseudoelasticity:* This refers to the ability of shape memory alloys to return to its original shape upon unloading after a substantial deformation. This superelastic effect is based on stress induced martensitic (SIM) transformation. This superelasticity in shape memory alloys occurs at a constant temperature when the alloy is completely composed of austenite phase (temperature is greater than A_f temperature).

The stress, i.e. load on shape memory alloy is increased till the austenite phase gets transformed into martensite phase simply due to loading. However, the martensite begins to transform back to austenite as soon as the load is decreased, since the temperature is still above A_f. As a result, it regains its original shape.

This effect, which causes material to be extremely elastic is called superelastic effect or pseudoelasticity. This effect is

nonlinear and also temperature and strain dependent. Superelasticity is a mechanical type of behavious of SMAs.

Figure B.5 shows the stress–strain behaviour of different phases at constant temperature for nickel-titanium alloy when tested below, within and above its transformation range.

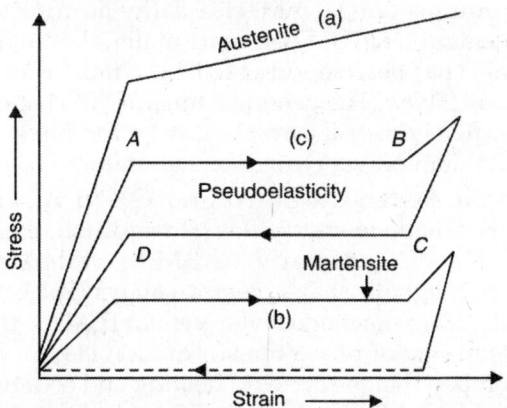

Fig. B.5: Stress–strain behaviour of different shapes of shape memory alloys at constant temperature

The austenite phase at a temperature above A_f is found to have much higher yield and flow stresses (Fig. B.5a). However, the martensite is easily deformed to several percent strain at quite a low stress (Fig. B.5b).

After removing the stress, this martensite phase returns back to the austenite phase upon heating (shown in Fig. B.5 by dashed line). However, no such shape recovering is found in the austenite phase upon straining and heating, because no phase change takes place.

The above mentioned two behaviours of SMAs are called as thermomechanical behaviour. An interesting stress–strain behaviour is depicted by Fig. B.5c where the material is completely composed of austenite phase ($t > A_f$). The martensite phase at this temperature can be stressed induced (curve *AB* in Fig. B.5). Again, on unloading, the material returns to its original shape (shown by curve *CD* in Fig. B.5) austenite without applying heat. This is a pure mechanical behaviour and is termed *superelasticity (SE)* or *pseudoelasticity*.

SE behaviour of SMAs is applied in making eye glass frames, cellular phone antennae, medical tools and orthodontic arches.

Important Smart Material Alloys Used Commercially

There are wide variety of alloys known to exhibit SME, but there are only few alloys which are used as commercial alloys. The most common smart material alloy is nickel-titanium (Ni-Ti). This smart material alloy of Ni-Ti has been reported to be the most useful of all SMAs. The generic name for Ni-Ti alloy family is *nitinol*, which stands for Nickel-Titanium Naval Ordnance Laboratory.

Cu-As-Ni, Cu-Zn-Al and Co-Ni-Al are other shape memory alloys. In addition, there is also a subfamily of SMAs called as ferromagnetic shape memory alloys (FSMAs) which are magnetically driven. In FSMAs, the frequency of phase transformation is high as compared to the cycles of heating and cooling. Only Ni-Mn-Ga ternary alloy has shown satisfactory performance in this category.

Mn-Cu, Ni-Ti-Hf, Ti-Pd-Ni and In-Tl, etc. are other examples of such SMAs.

Applications of SMAs

The crystal properties of SMAs make them suitable for use in a variety of fields. The important applications of SMAs are particularly in the following fields:

i. *Aircraft and Space Industry:* SMAs can be used in this industry as fine tuned helicopter blades and as lock ring electrical connectors. In addition, they also find applications in antenna opening high damping parts, triggering devices and hubble telescope.

ii. *Automobile Industry:* SMAs are used in this industry in making spring actuators, clutch systems, thermostats, oil pressure control unit and high pressure sealing plugs.

iii. *Medical Field:* SMAs find largest commercial applications in the field of bioengineering and medical science. The potential applications in these fields are:

- SMAs are used as dental arch wires. These wires help the misaligned teeth to return gradually to their original shape exerting a small and almost constant force on the misaligned teeth. The use of this wire is more efficient in faster movement of the teeth and less discomfort.
- SMAs also find use as blood clot filter.
- SMAs make tweezers to remove foreign objects through small incisions.
- Nitinol needle wire localizers are used to locate and mark breast tumours. This helps subsequent more exact and less invasive.
- SMAs find use in designing microsurgical instruments and microgrippers, etc.
- SMAs find use as guide wires for catheters through blood vessels.

iv. *Consumer Products:*

- These are used in making frames for eye glass, which offers improved comfort and flexibility. They also find use in phone antenna.
- Ni-Ti springs in coffee pots can be trained to open a valve and helps to release hot water at proper temperature.
- Nitinol finds use in robotic actuators and micromanipulators to stimulate human muscle motion.
- Using SMAs shape memory based toys and ornamental goods have been fabricated.
- SMAs can be used as couplers and fasteners.
- For industrial facilities, SMAs based fixed safety valves have been developed.
- Temperature sensitive SMA valves are widely used to instantly restrict water flow in sinks or showers.
- SMAs can be used to design safety valves. These valves provide emergency shutdown of process control lines that handle flammable and toxic fluid and gases.

Merits and Demerits of SMAs

SMAs show advantageous biocompatibility and find applications in diverse fields. They have good mechanical properties.

The major disadvantage of SMAs is that they are highly expensive to manufacture and machine it.

B.2 NONLINEAR MATERIALS

Active Materials

Today, the smallest feature size in production systems is about 250 nanometers—the smallest feature size in computer chips. Since atoms are an angstrom or so across and carbon nanotubes have a diameter as small as 0.7 nanometers, atomically precise molecular machines can be smaller than current MEMS devices by two to three orders of magnitude in each dimension, or six to nine orders of magnitude smaller in volume (and mass). For example, the size of the kinesin motor, which transports material in cells, is 12 nm. Han computationally demonstrated that molecular gears fashioned from single walled carbon nanotubes with benzyne teeth should operate well at 50–100 gigahertz. These gears are about two nanometers across. Han computationally demonstrated cooling the gears with an inert atmosphere. Srivastava simulated powering the gears using alternating electric fields generated by a single simulated laser. In this case, charges were added to opposite sides of the tube to form a dipole.

To make active materials, a material might be filled with nanoscale sensors, computers and actuators so the material can probe its environment, compute a response and act. Although this document is concerned with relatively simple artificial systems, living tissue may be thought of as an active material. Living tissue is filled with protein machines which gives living tissue properties (adaptability, growth, self-repair, etc.) unimaginable in conventional materials.

Nonlinear Optical Materials (NLO)

Photons can be used to generate and transmit data. Though it is beyond the scope of this text, there are also devices for storing data in an optical fashion. This emerging area using photons to replace what was previously performed by electrons is called *photonics*. The integration of photonic and electronic devices is called *optoelectronics*. In this section, we describe how the optical properties of materials can be used in yet another manner to create devices that can act as switches, multiplexers and mirrors and even change the wavelength of light. The materials that exhibit these effects and that are used in these photonic and optoelectronic devices are generally referred to as *nonlinear optical* (NLO) *materials*, because their outputs are not a linear response to some input. We are most familiar with linear responses of materials; for example, the intensity of photons emitted from a tungsten filament in a light bulb increases (more or less) linearly with the power we put into it. But even in this example, nonlinear behaviour can occur in the form of *saturation*, where further increases in power do not result in more output, or in the form of *breakdown*, when the filament burns out. The diodes described in the preceding sections are nonlinear electronic components—their output currents remain negligibly low until the driving voltages reach a threshold value whereupon the output current increases sharply.

The best way to describe a nonlinear optical response is first through some examples, then through some mathematics. Shortly after the invention of lasers, it was observed that a ruby laser beam passing through a quartz crystal produced a faint beam at the laser's second harmonic, that is, at twice the fundamental frequency of the ruby laser. A much simpler-to-perform example of nonlinear optical behaviour is found when the polarization of light passing through a crystal is modified upon application of an electric field to the crystal. The original of both of these NLO effects is related to the change in refractive indices that result by an applied electric field and the modulation of light beams by these field-dependent indices. The relationship between refractive index and polarizability is well established.

An example of an electro-optic switch based on NLO materials is shown in Fig. B.6. The switch comprises two parallel waveguides made of NLO materials. The waveguide channels have a different refractive index from the surrounding material. The light can be switched back and forth between the channels by applying and removing a voltage across the bottleneck. In the absence of an electric field, the light travelling through the lower waveguide interacts with the upper waveguide in a nonlinear manner at the bottleneck, causing the light to switch channels. Switching does not occur when an electric field is applied. The electric field polarizes the NLO material and alters the refractive indices of the two channels, such that the nonlinear interaction at the bottleneck is modified and the light stays in the lower waveguide.

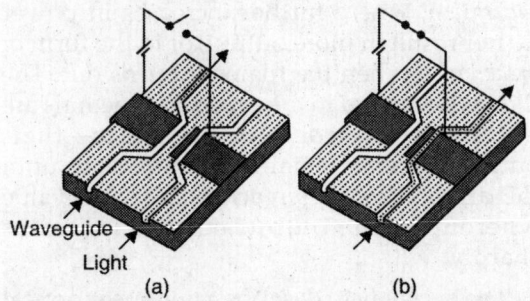

Fig. B.6: Schematic illustration of an electro-optic switch (a) An open circuit causes light to switch clannels (b) A closed switch keeps light in the same channel

Many current NLO devices are based on crystalline materials (such as lithium niobate for the electro-optic switch) and nonlinear optical glasses, but there is intense interest in utility polymers due to their inherently low densities and potential ease of processing. Some of the polymers that have been used for photonic devices are based on architectures that have highly polarizable electronic structures. Several polyimides fall into this category (Fig. B.7a), as do polyacrylates and polyurethanes with optically responsive side groups (Fig. B.7b).

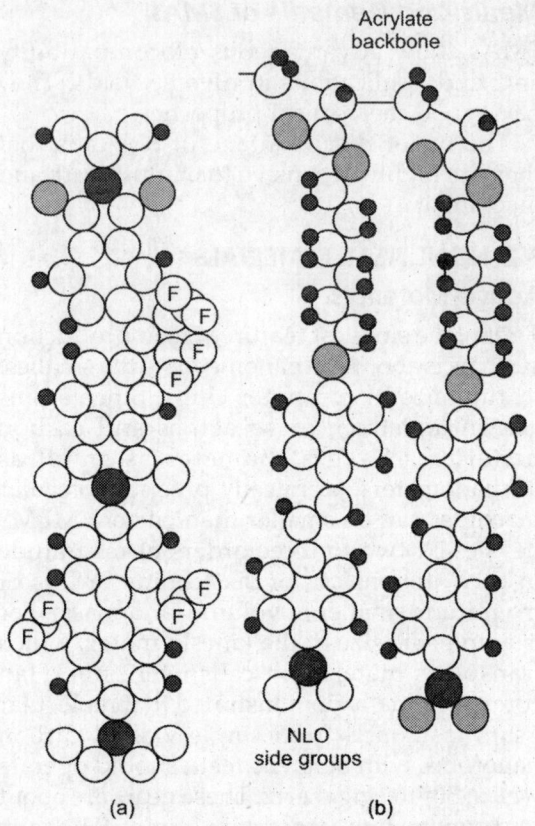

Fig. B.7: Examples of nonlinear optical polymers (a) Fluorinated polyimide (b) Polyacrylate with NLO side groups

Nonlinear optical (NLO) materials capable of generating the second harmonic frequency play an important role in the domain of optoelectronics and photonics. Nonlinear optical (NLO) crystals with high conversion efficiencies for second harmonic generation (SHG) and transparent in visible and ultraviolet ranges are required for numerous device applications. Within the last decade much progress has been made in the development of these NLO organic materials having large nonlinear optical coefficients. However, most of the organic NLO crystals are constituted by weak van der Waals and hydrogen bonds with conjugated π electrons. So they are soft in nature and difficult to polish and these materials also have intense absorption in UV region. In view of these problems, new types of hybrid NLO materials have been explored from organic-inorganic complexes with stronger ionic bond such as L-arginine phosphate (LAP) monohydrate and others have been synthesized and single crystals have been grown. LAP crystals are used for second, third and fourth harmonic generation from the fundamental radiation 1.06 μm and also sum and difference frequencies generation in wide spectral range from UV to IR. LAP has three times the nonlinearity, two or three times high damage threshold of KDP, chemical stability and also lower cost. A problem related with LAP growth is the presence of micro-organisms arising during growth, which leads to deterioration in L-histidine tetrafluoroborate (L-HFB), L-arginine tetrafluoroborate (L-AFB) and others.

GLIMPSES

- Shape memory alloys (SMAs) are metals which exhibit shape memory effect (SME) and pseudoelasticity or superelasticity (SE).
- Shape memory alloys remember their shape even after severe deformation, i.e. when a SMA is once deformed in the cold shape (martensite) these materials will stay deformed until heated, whereupon heating they will spontaneously return to their original predetermined hot shape (austenite).
- The structural changes at the atomic level contribute to unique properties of smart materials.
- The materials which exhibit shape memory effect (i.e. taking their own shape) only upon heating are called *one way shape memory.*
- Some materials exhibit shape memory effect during heating and cooling. Such materials are said to be *two way shape memory.*
- The shape memory materials could be broadly classified into thermal induced shape memory, stress induced shape memory and ferromagnetic shape memory materials.

- Superelasticity (SE) or pseudoelasticity refers to the ability of SMA to return to its original shape upon unloading after a substantial deformation.
- Materials with exceptional nonlinear optical properties are critical to the continuing development of photonic and electro-optical devices, such as those used in networking, communications and storage equipment. Most current technologies are based on inorganic materials with the appropriate optical properties. However, using inorganic materials has some major drawbacks. Primarily, inorganic materials used in optical systems are difficult and expensive to process. In addition, inorganic materials have a high dielectric constant, requiring larger poling voltages and often suffering from changes in the material's refractive index. Moreover, inorganic materials have a low electro-optic coefficient, which is less suitable for electro-optic modulation.
- Organic materials exhibit exceptional nonlinear optical properties. These materials are easily made, with active crystalline films forming spontaneously. The resulting organic material has a low dielectric constant, eliminating the need for poling while maintaining the refractive index.

REVIEW QUESTIONS

1. What are shape memory alloys? Explain the temperature induced and stress induced transformations.
2. Explain the crystal structure of shape memory alloys.
3. Mention important applications of shape memory alloys in different fields.
4. Classify shape memory alloys.
5. What are shape memory alloys? What causes the shape memory alloys to remember their shape?
6. How does phase transformation takes place in shape memory alloys? Explain superelasticity or pseudoelasticity in SMAs.
7. What are ferromagnetic shape memory alloys?
8. What are the advantages and disadvantages of shape memory alloys?

9. What are nonlinear materials? Mention their properties and applications.

10. What do you understand by passive and active nonlinear materials?

SHORT ANSWER QUESTIONS

1. What are shape memory alloys (SMAs)?

Ans. These are metals which exhibit two very unique properties: shape memory effect (SME) and superelasticity (SE). Interestingly, SMAs are unique class of materials, which remember their shape even after severe deformation.

2. How many types of shape memory alloys are there?

Ans. Two types: (i) one way shape memory alloys (ii) two way shape memory alloys.

The materials which exhibit shape memory effect only upon heating are termed one way shape memory alloys. However, those materials which exhibit the property both during heating and cooling are termed two way shape memory alloys.

3. What do you understand by smart (intelligent) materials?

Ans. These materials are a group of new and state-of-art materials now being developed that will have a significant influence on many of our technologies. The adjective *smart* implies that these materials are able to sense changes in their environments and then respond to these changes in predetermined manners—traits that are also found in living organism. In addition, this *smart* concept is being extended to rather sophisticated system that consists of both smart and traditional materials.

Components of a smart material (or system) include some types of sensor (that detects an input signal), and an actuator that performs a responsive and adaptive function. Actuators may be called upon to change shape, position, natural frequency or mechanical characteristic in response to changes in temperature, electric fields, and/or magnetic fields.

Four types of materials are commonly used for actuators: shape memory alloys, piezoelectric ceramic, magnetostrictive materials, and electrorheological/magnetorheological fluids.

Materials/devices employed as sensors include optical fibres piezoelectric materials (including some polymers) and micro electromechanical devices (MEMS).

4. What are different classes of SMAs?

Ans. Broadly speaking there are three classes: (i) thermal induced SMAs (ii) stress induced SMAs (iii) ferromagnetic SMAs.

5. Mention the phases or crystal structure exhibited by SMAs at low and high temperatures.

Ans. At low temperature is martensite phase and at high temperature is austenite phase.

6. Why SMAs remember their shape?

Ans. At the atomic level, the structural changes causes SMAs to remember their shape, i.e. the easy to and fro martensite transformation is found to be responsible for this.

7. How active materials are prepared?

Ans. A material might be filled with nanoscale sensors, computers and actuators so the material can probe its environment, compute a response and act.

8. What are nonlinear optical materials?

Ans. The integration of photonic and electronic devices is called optoelectronics. The materials that exhibit nonlinear optical behaviour and that are used in photonic and optoelectronic devices are generally referred to as nonlinear optical (NLO) materials.

MULTIPLE CHOICE QUESTION

1. Shape memory effect is exhibited by
   (a) shape memory alloys
   (b) normal metals
   (c) plasma
   (d) none of these

ANSWER

1. (a)

Appendix II

Quantum Information and Quantum Computing

CLASSICAL INFORMATION

Classical information is performed using **bits**, which are just two-state systems, with the two states 0 and 1. We may note that **bit is the fundamental unit of classical information. By grouping bits together one can represent arbitrary pieces of information and by manipulating these bits one can perform arbitrary computation.** Suppose we allow up to 64 unique symbols, to include all the upper case and lower case letters and the punctuation characters. Then we could replace each symbol with a unique string of six bits, since there are $2^6 = 64$ possible permutations. When we want to encode numbers into a series of bits, rather than translating each familiar symbol 0, 1,..., 8, 9 into a series of bits, we instead use a rule system such as binary that assigns a unique number to every possible series of bits. Such a scheme has perfect efficiency, using no more bits than are absolutely necessary to represent a given range of numbers.

In case of **image,** we can adequately encode by breaking the image up into a fine grid of points or pixels, and assigning a pure colour to each pixel. This process of approximating continuous media (images, sounds, movies) as a series of symbols is called **digitization,** and the stored entity is said to be **digital.** Thus, one can store any of these forms of knowledge as a stream of bits.

INFORMATION INSIDE A CLASSICAL COMPUTER

Modern computers store bits in different forms at different times. The representation depends on whether the information is being stored (short or long term), transmitted over some distance or processed. Take the example of long-term storage. This is usually done using a **magnetic disk,** which is essentially a surface that has been coated by a thin layer of magnetic particles. The surface is divided up into little areas, less than a micron in size, and each such area stores a single bit through the collective orientation of the many magnetic particles within. The value of the bit is set when a magnetic head comes and imposes an orientation on the particles by applying a strong field. Similarly, the value of the bit is read by detecting the weak magnetic field generated by the aligned particles.

Because many particles are being used to store a single bit, there are actually a large number of states of the physical system that are called "0" and similarly a large number that are called "1". There are also many states that do not correspond to a clear majority of particles being aligned in either direction. These states have *no meaning* in terms of representing classical information; if the computer is operating successfully, then such states will only occur transiently, as the bit value of the collective is switched from one valid state to another.

QUANTUM BIT OR QUBIT

We have read that real universe is governed by quantum mechanics, and when we use that theory to describe a system with two reliably distinguishable states we discover something remarkable. Any such system actually has an infinity of possible states. Anything that can store a bit is, in fact, much richer than that. The richer entity is called the **quantum bit** or **qubit** for short and this is the real unit of information. **Qubit is simply a quantum system with two orthonormal basis states,** which we shall call $|0\rangle$ and $|1\rangle$.

Let us see what it might mean to speak of a machine that can process **quantum information.** Whereas ordinary machines store and process bits, the new machine must be able to store and process these richer entities called qubits. Critically, a device does not become a **quantum computer** (a quantum computer is a device capable of processing quantum information) merely by being composed of sufficiently small components. If a device built at the atomic scale were incapable of successfully processing qubits, then it would remain a classical computer. Indeed, processes in living systems (such as DNA replication) are seen as an information processing but it is a classical molecular scale processing rather than quantum computing.

The theory of **quantum information processing** (QIP) is the *general theory* of information processing with real physical systems. Classical information processing is what happens when the system only uses a limited portion of what QIP allows. Of course, in practice this design choice has really been a necessity. Scientists are now trying very hard to create a machine that is not limited to classical information processing, although it is very difficult.

There are many possible physical implementations of a qubit, such as spin states of electrons or atomic nuclei, charge states of **quantum dots** (as the scale of a material is reduced to OD, discrete energy levels arise due to quantum confinement effect. Such structures are known as artificial atoms or quantum dots (QDs). Many mesoscopic or nanoscale systems such as semiconductor nanocrystals, metal nanoparticles, lithographically defined small islands on semiconductor heterostructures, and single walled carbon nanotubes (SWCNTs) with tunnel barriers are examples of QDs), atomic energy levels, vibrational states of groups of atoms, polarization states of photons, or paths in an interferometer. The use of **quantum mechanical superposition states** and **entanglement (quantum entanglement** is a quantum mechanical phenomenon in which the quantum states of two or more objects (particles) have to be described with reference to each other, even though the individual objects may be spatially separated) in a computer can theoretically solve important mathematical and physical problems much faster than classical computers. Electron spin states were identified early as an attractive realization of a qubit in a quantum computer, because they are relatively robust against decoherence (uncontrolled interaction with the environment). Advances in the field of semiconductor quantum dots have made this system a valuable host for the electron spin. The quantum state of the electron spin can be coherently controlled by applying short bursts of an oscillating magnetic field giving rise to oscillations of the spin state, which can be detected by measuring the quantum dot.

At this stage the physical implementation of qubit is not important— the idea of a qubit is to abstract the discussion away from physical details. Taking the standard approach of quantum information theory, one can begin by not worrying too much about the properties of these states, or even what their energies are; one can simply assume that they are **eigenstates** of the system's Hamiltonian with known **eigenvalues** (i.e. known energies). This approach allows one to concentrate on the fundamental properties of the system without considering all the tedious details.

One can in principle perform classical information processing on our quantum system by using the two states $|0\rangle$ and $|1\rangle$ as our logical states 0 to 1 and proceeding in the usual way, giving rise to the field of **reversible computing**. A **qubit** is not confined to these two states, but can be found in arbitrary superposition states. Although it is not immediately obvious what a state like

$$| \psi \rangle = \alpha | 0 \rangle + \beta | 1 \rangle$$

where α and β are complex numbers, actually means information processing terms, it is clear that quantum bits are in some sense more powerful than their classical equivalents. **Quantum information processing is, of course, the art of exploiting these superposition states to perform information processing tasks which are impossible for classical systems.** Just as the real power of classical information processing requires groups of bits, the real advantages of quantum information processing only become clear in systems with two or more qubits.

Obviously, the important features of QIP are different from those of classical computing and can be broken down into three categories: (i) linear superposition (ii) entanglement (iii) quantum parallelism.

i. **Linear superposition:** Contrary to classical bit, a quantum bit or **qubit** can take not only two discrete values 0 and 1, but also all possible **linear combination** of them. This is a consequence of a fundamental property of quantum states. It is possible to construct a **linear superposition** of quantum state $| 0 \rangle$ and quantum state $| 1 \rangle$.

ii. **Entanglement:** At a quantum level it appears that two quantum objects can form a single entity, even when they are well separated from each other. Any attempt to consider this entity as a combination of two independent quantum objects given by the tensor product of quantum states fails, unless the possibility of signal propagation at superluminal speed is allowed. These quantum objects that cannot be decomposed into a tensor product of individual independent quantum objects are called **entangled quantum objects**. In brief, we can say that quantum entanglement is a quantum mechanical phenomenon in which the quantum states of two or more objects (particles) have to be described with reference to each other even though the individual objects may be spatially separated. This leads to correlation between observable physical properties of the systems. For example, it is possible to have two particles in a single quantum state such that when one is observed to be spin-up, the other one will always be observed to be spin-down and *vice versa*, this despite the fact that it is possible to predict, according to quantum mechanics, which set of measurements will be observed. As a result, measurements performed on one system seem to be instantaneously influencing other system entangled with it.

iii. **Quantum parallelism:** This makes it possible to perform a large number of operations in parallel, which represents a key difference from classical computing. Namely, in classical computing it is possible to know the internal status of the computer. On the other hand, because of the **no-cloning theorem**, it is not possible to know the current state of a quantum computer. This property has led to the development of **Shor factorization algorithm.** Some other important quantum algorithms include the **Grover search algorithm**, which is used to perform a search for an entry in an unstructured data base; the **quantum Fourier transform**, which is the basis for a number of different algorithms; and **Simon's algorithm**. A quantum computer is able to encode all input strings of length N simultaneously into a single computational step. In other words, quantum computer is able simultaneously to pursue 2^N classical paths, indicating that quantum computer is significantly more powerful than a classical one.

Although QIP has opened up some fascinating perspectives as indicated above, there are certain limitations that need to be overcome before QIP becomes a commercial reality. The first is related to the **number of existing algorithms**, whose number is significantly lower than that of classical algorithms. The second problem is related to physical implementation issues. There are many potential technologies, e.g. NMR, ion traps, cavity quantum electrodynamics, photonics, quantum dots, and superconducting technologies, etc. Nevertheless, it is not clear which technology will prevail. On the other hand, for quantum computing applications there are many potential technologies that compete with each other. Moreover, presently the number of qubits that can be manipulated is of the order of tens, well below that needed for meaningful quantum computation, which is of the order of thousands. Another problem, which can be considered as the major difficulty, is related to **decoherence.** Decoherence is related to the interaction of qubits with environment that blur the fragile superposition states. It also introduces errors.

QUANTUM COMPUTATION

Quantum computation in its modern form is still a relatively young discipline. Basics of quantum computation are as follows:

Quantum state: A possible state in which a quantum mechanical system can exist is termed a **quantum state.** A quantum state can be described by either, a state vector, a wave function or a complete set of quantum numbers for a specific system. In terms of **bra-ket** notation introduced by Dirac, a quantum state $|\psi\rangle$ can be expressed as sum of **basis*** state $|\phi_i\rangle$ as

$$|\psi\rangle = \sum_i c_i |\phi_i\rangle \qquad (1)$$

where c_i represents the probability amplitude. In other words, every possible wave function ψ can be expanded in terms of an orthogonal set of states ϕ_i. This is also called as **principle of superposition.** The normalization condition demands that the sum of probability should be 1, i.e. $\sum_i |\phi_i\rangle|^2 = 1$.

The expectation value of a measurement A on a quantum state is given as

$$\langle A \rangle = \langle \psi | A | \psi \rangle$$
$$= \sum_i a_i \langle \psi | \alpha_i \rangle \langle \alpha_i | \psi \rangle$$
$$= \sum_i a_i |\langle \alpha_i | \psi \rangle|^2 \qquad (2)$$

Here $\langle \alpha_i \rangle$ are basis states for the operator.

Pure and Mixed Quantum States

A quantum state presented above is a pure state. A **mixed** quantum state is a statistical distribution of pure states. One can represent the mixed state (statistical distribution of pure states) by a **density operator** as

$$\rho = \sum_s p_s |\psi_s\rangle \langle \psi_s| \text{ with } p_s > 0$$

and $\sum_s p_s = 1$ $\qquad (3)$

where p_s is the fraction of each ensemble in pure state $|\psi_s\rangle$. We may note that the pure states in this expression need not be orthogonal, but if these are orthogonal then Eq. (3) represents simply the spectral decomposition of density operator. If we made measurement A on a system in a mixed state, the average outcome is

$$\langle A \rangle = \Sigma p_s \langle \psi_s | A | \psi_s \rangle$$
$$= \sum_s \sum_i p_s a_i |\langle \alpha_i | \psi_s \rangle|^2$$
$$= \text{Tr } (\rho A) \qquad (4)$$

where Tr denotes the trace of a matrix of an operator. (The trace of a matrix or operator Q satisfies $\text{Tr } Q = \sum_k q_{kk} = \sum_k \langle \phi_k, Q\phi_k \rangle$, where ϕ_k is any orthonormal basis and q_{kk} denotes the diagonal elements in a fixed matrix representation of A). We can see that in Eq. (4) two averaging are done, one is the average over basis states of pure state and the other is

---

*A basis is a set of vectors, the linear combination of whose can represent every vector in a given vector space.

the statistical average over the ensemble of pure states.

Entangled States

Let us consider two non interacting systems A and B with corresponding state vectors $|\psi\rangle_A$ and $|\phi\rangle_B$ in the respective **Hilbert space** (a **Hilbert space** is a finite or infinite complete vector space on the basic field of complex numbers. In this space, a scale product is defined such that it assigns a complex number to each pair of functions $\psi(x)$ and $\phi(x)$ out of a set of linear functions). The state vectors of a quantum mechanical system constitute a Hilbert space H_A and H_B. One can define each of these state vectors in terms of corresponding basis states, $|k_i\rangle_A$ and $|k_i\rangle_B$, by equations similar to Eq. (1). One can write the state of composite system as

$$|\psi\rangle_{AB} = |\psi\rangle_A \, |\phi\rangle_B$$

$$= \sum_{i,j} c_{ij} |k_i\rangle_A \otimes |k_j\rangle_B$$

The two states $|\psi\rangle_A$ and $|\phi\rangle_B$, are said to be separable, provided $c_{ij} = c_i^A \, c_j^B$, such that

$$|\psi\rangle_A = \sum_i c_i^A |k_i\rangle_A$$

and $$|\phi\rangle_B = \sum_i c_i^B |k_j\rangle_B$$

otherwise, these are non separable or **entangled states**. We may note that a pure state is said to be entangled state if it cannot be expressed as a product in any basis. For example, one can define a composite state as

$$|\psi\rangle_{AB} = \frac{1}{\sqrt{2}} \left(|0\rangle_A \otimes |1\rangle_B - |1\rangle_A \otimes |0\rangle_B \right)$$

This is an entangled state, where ($|0\rangle_A$, $|1\rangle_A$) are the respective basis states of H_A and H_B. If any measurement is done to observe the state of the composite system, say, by an observed associated to system A, then the possible result can be either 0 or 1 and the state vector will reduce, i.e. collapse to ($|0\rangle_A |1\rangle_B$) or ($|1\rangle_A |0\rangle_B$). The result of any subsequent measurement by

observer related to system B will, respectively, yield either 0 or 1. Obviously, the system B has been altered by the measurement performed by observer related to system A, although, we have assumed the two systems to be non interacting or spatially separated. Similar observation was reported while we discuss **double-slit** experiment with incident electrons as well as the **measurement theory.**

Bit and Qubit

As stated earlier, classically, the information is coded using **binary system**, i.e. in terms of **strings** 0 and 1 and basic unit of information is called a **bit**. As stated earlier, the basic unit of information in quantum computation is **qubit** and can be expressed in terms of state (basis) vectors $|0\rangle$ and $|1\rangle$. Physically, such states can be realized in case of spin 1/2 particles, i.e. **spin up** ($\uparrow$) and **spin down** ($\downarrow$) states. There are many other examples of such states, e.g. polarizations of photons as **vertical** and **horizontal polarization.** Mathematically, the state of such a two level quantum system is equivalent to a **two dimensional vector space** over the complex numbers, i.e.

$$|0\rangle = \begin{pmatrix} 1 \\ 0 \end{pmatrix} \quad \text{and} \quad |1\rangle = \begin{pmatrix} 0 \\ 1 \end{pmatrix} \tag{5a}$$

A classical string say (1001) can be represented as a product

$$|1001\rangle = \begin{pmatrix} 0 \\ 1 \end{pmatrix} \otimes \begin{pmatrix} 1 \\ 0 \end{pmatrix} \otimes \begin{pmatrix} 1 \\ 0 \end{pmatrix} \otimes \begin{pmatrix} 0 \\ 1 \end{pmatrix} \tag{5b}$$

We can think a pure qubit as a linear superposition of two basis states [Eq. (5a)] as, $|\psi\rangle = \alpha|0\rangle \pm \beta|1\rangle$, such that $\alpha^2 + \beta^2 = 1$. This means that the probability of qubit measured in state $|0\rangle$ is α^2 and in state $|1\rangle$ is β^2. Moreover, multiple qubits exhibit **quantum entanglement,** i.e. due to entanglement, a set of qubits exhibit superposition of different binary strings (e.g. say 0 1 0 1 0 and 1 1 1 1 1) simultaneously. We may note that this **quantum parallelism** is an important feature of quantum computation. Interestingly, each independent state of the quantum particle used in computer can follow its own independent computation path to conclusion while its other states are observed and changed.

Quantum Parallelism

Computers of present day generation are based on sequential computing. In such a computer, one finds a physical device on which only one operation can be performed at a time. Information processing then requires a long sequence of operations. One can enhance the speed of by **parallel computing** where the operations can be performed simultaneously on n physical devices. This means that the length of sequence of operations may be decreased by employing more physical devices. Further, the use of n processors yields n outputs, which can be extracted and analyzed, i.e. can be combined further as per requirement.

In **quantum computer,** there is only one physical device with a state that can be described by a superposition or entangled states. The logical operations or gates in a QC are implemented by unitary operators that can act on state vector and not on the individual pieces in the superposition. The result is single vector from which one can extract only information equivalent to that in n **classical bits.** Obviously, **the effective parallelism is inherent to the quantum computations.** However, it is difficult for one to extract the useful information after the operation because the accessible information is limited by principles of quantum measurement as stated earlier.

Computational Basis

One can define the state of a quantum information processor by a state vector $|\psi\rangle$ (pure state) in the appropriate **Hilbert space.** Customarily, the 0 and 1 are represented as a pair of orthonormal vectors [Eq. (5a)] The tensor product [Eq. (5b)] then yields an orthonormal basis for the chosen vector space,

$[(C)^{2^n}]$, and is referred to as **computational basis**. One can always write an arbitrary state vector as a superposition of elements of the computational basis. One can also define a **linear operator** specifying its action on a set of basis vectors, such as those in the computational basis. One can then extend it to arbitrary vectors in chosen vector space by linearity. The implementation of a gate requires that one finds a physical operation that has the desired effect. We may note that in any successful implementation the state of the system can be described by a state vector in chosen vector space and the implementation of gates affect the state of system and not its basis vectors.

We may note that an important step-in quantum computing is defining state vector and realizing it physically. Further, there are two common ways of implementing **qubit states:** (i) based on the use of spin $1/2$ particles, e.g. electrons or protons (ii) based on use of polarization state of single photon. Once the states of qubits are formed, the next step is to control these, i.e. implement the gates. This is the area of frontal research and also a challenging one.

GLIMPSES

1. Classical information processing is performed using **bits,** which are just two-state systems with the two states called 0 and 1. The corresponding basic element used in quantum information is the quantum bit or qubit. Qubit is simply a quantum system with two orthonormal basis states, $|0\rangle$ and $|1\rangle$.

2. The theory of quantum information processing (QIP) is the general theory of information processing with real physical systems.

3. Contrary to the classical bit, a qubit can take not only two discrete values 0 and 1, but also all possible linear combinations of them. This is a consequence of a fundamental property of states. It is possible to construct a **linear superposition** of quantum state $|0\rangle$ and quantum state $|1\rangle$.

4. **Quantum entanglement** is a quantum mechanical phenomenon in which the quantum states of two or more objects (particles) have to be described with reference to each other, even though the individual objects may be spatially separated.

5. The use of quantum mechanical superposition states and entanglement in a computer can theoretically solve important mathematical and physical problems much faster than classical computers.

6. A pure state is said to be **entangled** if it cannot be written as a product in any basis.

7. Physically, a **qubit** can be realized using spin components of spin 1/2 particles, i.e. two states can correspond to **spin up** and **spin down** states or alternatively, similar states can be realized using polarization of photons as **vertical** and **horizontal** polarization.

8. Quantum parallelism is inherent to the quantum computations. This makes it possible to perform a large number of operations in parallel, which represents a key difference from classical computing.

9. A **quantum computer** is simultaneously able to pressure 2^N classical paths indicating that a quantum computer is significantly more powerful than a classical one.

10. A single qubit is a state ket (vector)

$$|\psi\rangle = a|0\rangle + b|1\rangle$$

parameterized by two complex numbers a and b satisfying the normalization condition $|a|^2 + |b|^2 = 1$.

REVIEW QUESTIONS

1. What is bit and qubit. Explain the difference between them.

2. Explain classical information processing and quantum information processing. Mention fundamental features of QIP.

3. Explain the terms: (i) linear superposition (ii) entanglement (iii) quantum parallelism.

4. What do you understand by pure and mixed quantum states?

5. What is a quantum computer?

Suggested Readings

- Kakani, SL, *Mechanics*, 3rd edn, 2014, Viva Book, New Delhi.
- Beiser, A, *Modern Physics*, 6th edn, 2003, McGraw Hill.
- Alonso and Finn, *University Physics*, 2nd edn.
- Kakani, SL and Kakani Amit, *Material Science*, 3rd edn, 2014, New Age Int., New Delhi.
- Kakani, SL and Bhandari KC, *Electronics Theory and Applications*, 4th edn, 2013, New Age.
- Kakani, SL and Chandalia HM, *Quantum Mechanics*, 5th edn, 2015, Sultan Chand, New Delhi.
- Kakani, SL and Hemrajani C, *Solid State Physics*, 5th edn, 2015, Sultan Chand.
- Kakani, SL and Bhandari KC, *Optics*, 2nd edn, 2003, Sultan Chand.
- Kakani, SL and Kakani Shubhra, *Modern Physics*, 2nd edn, 2013, Viva Books.
- Kakani, SL and Hemrajani C, *Electromagnetics*, 2nd edn, 2015, CBS Pubs, New Delhi.
- Kakani, SL and Hemrajani C, *Waves and Oscillations*, 2002, CBS Pubs.
- Krane, KS, *Modern Physics*, 2nd edn, 1998, John Wiley.
- Beiser, A, *Perspectives of Modern Physics*, 5th edn, 1969, McGraw Hill.
- Kakani, SL and Hemrajani C, *Mathematical Physics*, 2nd edn, 2015, CBS Pubs, New Delhi.
- Bernstein *et al*, *Modern Physics*, 2000, Pearson Education.
- Kakani, SL and Kakani Shubhra, *Superconductivity*, 2nd edn, 2012, New Age Int.
- Kakani, SL, *Thermodynamics*, 2nd edn, 2007, Sultan Chand.
- Kakani, SL and Kakani Shubhra, *Nuclear Physics*, 2nd edn, 2013, Viva Books.
- Kakani, SL and Bhandari KC, *Electronic Devices and Circuits*, 2012, Viva Books.
- Kakani, SL and Kakani Shubhra, *Applied Physics*, 2014, Viva Books.
- Kakani, SL and Kakani Amit, *Engineering Materials*, 2013, New Age Int., New Delhi.
- Kakani, SL and Kakani Shubhra, *Carbon Nanotubes*, 2015, New Age Int., New Delhi.
- Kakani, SL and Hemrajani C, *Simplified Physics*, Vol. I, II 2014, 2015, Himalaya Pubs House, Mumbai.

Index

Reader's Note

Reader's Note